PROCEEDINGS OF THE TWENTY-EIGHTH ANNUAL SYMPOSIUM ON

COMPUTATIONAL GEOMETRY

(SCG'12)

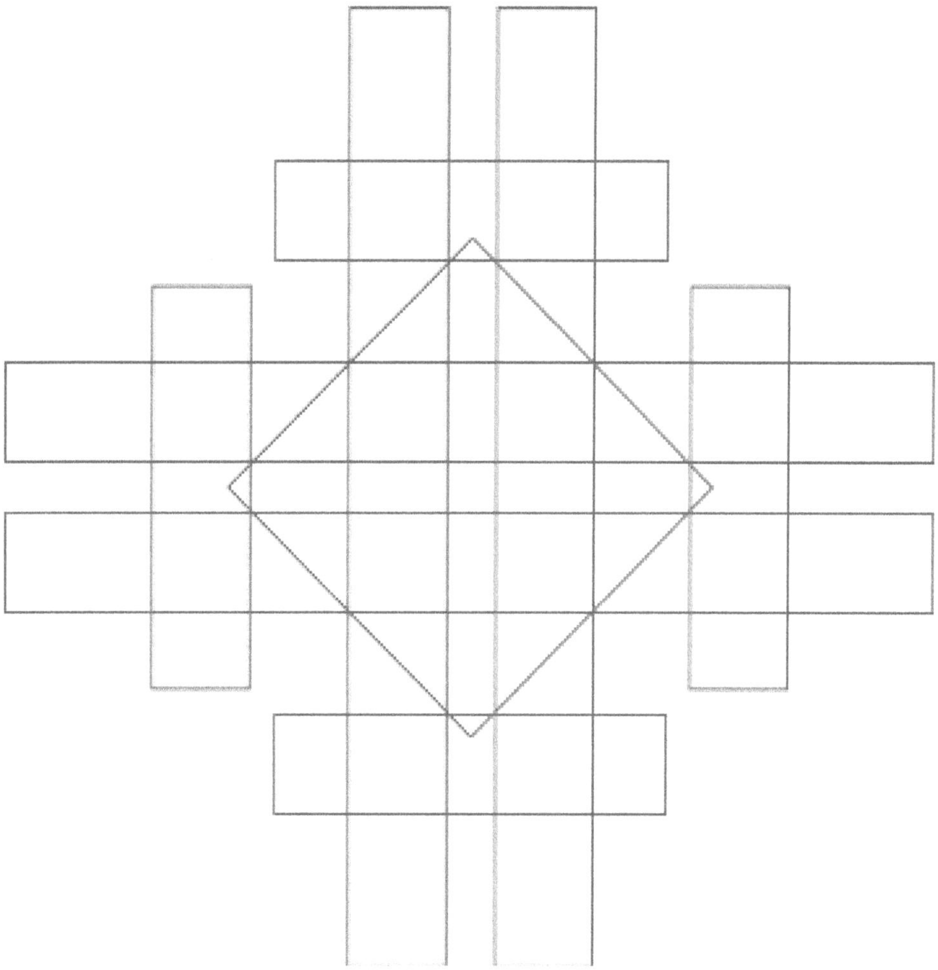

June 17–20, 2012
Chapel Hill, North Carolina, USA

Sponsored by ACM SIGACT & ACM SIGGRAPH

Association for
Computing Machinery

Advancing Computing as a Science & Profession

The Association for Computing Machinery
2 Penn Plaza, Suite 701
New York, New York 10121-0701

ISBN: 978-1-4503-1299-8

Additional copies may be ordered prepaid from:

ACM Order Department
PO Box 30777
New York, NY 10087-0777, USA

Phone: 1-800-342-6626 (USA and Canada)
+1-212-626-0500 (Global)
Fax: +1-212-944-1318
E-mail: acmhelp@acm.org
Hours of Operation: 8:30 am – 4:30 pm ET

ACM Order Number: 429120

Printed in the USA

Foreword

This volume contains the abstracts of the 44 research papers, 4 video presentations, and 3 invited talks scheduled for presentation at the 28th Annual Symposium on Computational Geometry, to be held at the University of North Carolina, Chapel Hill, June 17-20, 2012.

There were 126 paper submissions (with 2 withdrawn) among which the program committee selected 44 papers for presentation at the conference. The selection process mainly consisted of two phases. In the review phase, the program committee produced at least 3 reviews for each submitted paper. In many cases, program committee members called upon external subreviewers for assistance. In the deliberation phase, which lasted from January 17 to February 16, the program committee discussed the submitted papers and their reviews extensively, obtaining additional input from experts in some cases.

The on-line submission process and the review and deliberation phases were conducted using the electronic tools provided by EasyChair. We acknowledge that the entire complex process, from paper submission to final notification of decisions, was aided significantly by this system.

In addition to the paper submissions, six videos were submitted in response to the Call for Videos and Multimedia. All six were reviewed, and four of these were accepted for presentation. The videos use animation and graphics to communicate the behavior of specific algorithms. These proceedings contain a short, 2-page description for each of the videos accepted. The final videos for all accepted contributions will be archived at http://www.computational-geometry.org.

In many cases, the papers in these proceedings do not contain full details of all results claimed, due to the 10-page limit for regular submissions. We hope that most of these papers will later appear with full details in refereed journals. Each of the three invited talks is represented in this volume by a 1-page description of the topic, the speaker, and some background references. We hope that this information will provide the reader with the means to explore further.

This year, we are taking the initiative to confer three awards: two Best Paper awards and one Best Student Presentation award. The two Best Papers will be selected from the 44 accepted papers by a sub-committee of the program committee, based on the nominations of the program committee and the merits of the research results. The Best Student Presentation award will be determined at the symposium based on input from the attendees at the talks given by students. Student speakers must be authors or co-authors of the papers they present. We hope that all these awards will be seen as encouraging tokens for the authors.

We note that this year, as an experiment, the symposium is being held in the mornings over a 4-day period, along with an afternoon series of geometry related events collectively called CG:APT (Computational Geometry: Applications, Practice, and Theory). CG:APT has its own program committee. The aim of holding the symposium together with CG:APT is to welcome a broader group of participants, including graduate students and those working in areas where mutual interaction with computational geometry could prove fruitful.

Finally, we cannot overstate the significance of the role played by the program and video committees and the external reviewers in the selection process. Our community is truly privileged to have their expertise, insights, and diligence in selecting the papers and the videos. We wish to thank the authors of all submissions, the reviewers and subreviewers, the invited speakers, the attendees, the steering committee, the organizing committee, the volunteers, the people at ACM and Sheridan Printing Company who have assisted us with the proceedings, and the sponsors and supporters. Together, they make this conference possible. We also thank the organizers of CG:APT for the opportunity to interact with them on the design and coordination of these two events. We wish all the participants a stimulating geometry week in Chapel Hill.

Tamal Dey and Sue Whitesides
Program Committee Co-chairs

Christian Knauer
Video and Multimedia Committee Chair

Table of Contents

Parallel Session IVa: Monday morning, June 18, after the break

Parallel Session IVb: Monday morning, June 18, after the break

Invited Talk: Monday afternoon, June 18, Christopher Bishop

Session V: Tuesday morning, June 19, before the break

Parallel Session VIa: Tuesday morning, June 19, after the break

Parallel Session VIb: Tuesday morning, June 19, after the break

Invited Talk: Tuesday afternoon, June 19, Rien van de Weygaert

Parallel Session VIIa: Wednesday, June 20, before the break

Parallel Session VIIb: Wednesday, June 20, before the break

Session VIII: Wednesday, June 20, after the break

Video Session: Wednesday afternoon, June 20

SCG 2012 Symposium Organization

Local Arrangements: Jack Snoeyink *(University of North Carolina at Chapel Hill, USA) (chair)*

Steering Committee:
Mark de Berg *(TU Eindhoven, the Netherlands) (secretary/chancellor)*
Joseph S. B. Mitchell *(SUNY Stony Brook, USA)*
Günter Rote *(Freie Universität Berlin, Germany)*
Jack Snoeyink *(University of North Carolina at Chapel Hill, USA) (chair)*
Monique Teillaud *(INRIA Sophia Antipolis, France)*

Program Committee:
Pankaj Agarwal *(Duke University, USA)*
Dominique Attali *(CNRS Grenoble, France)*
Gill Barequet *(Technion-Israel Institute of Technology)*
Mark de Berg *(TU Eindhoven, the Netherlands)*
Danny Chen *(University of Notre Dame, USA)*
Tamal Dey *(The Ohio State University, USA) (co-chair)*
Vida Dujmović *(Carleton University, Canada)*
David Eppstein *(University of California Irvine, USA)*
Leonidas Guibas *(Stanford University, USA)*
Sylvain Lazard *(INRIA Nancy Grand Est, France)*
Dinesh Manocha *(The University of North Carolina at Chapel Hill, USA)*
Steve Oudot *(INRIA Saclay, France)*
Konrad Polthier *(Freie Universität Berlin, Germany)*
Edgar Ramos *(Universidad Nacional de Colombia Medellín, Colombia)*
Jian Sun *(Tsinghua University, China)*
Takeshi Tokuyama *(Tohoku University, Japan)*
Yusu Wang *(The Ohio State University, USA)*
Max Wardetzky *(University of Göttingen, Germany)*
Sue Whitesides *(University of Victoria, Canada) (co-chair)*

Video and Multimedia Committee:
Helmut Alt *(Freie Universität Berlin, Germany)*
Sándor Fekete *(TU Braunschweig, Germany)*
Panos Giannopoulos *(Universität Bayreuth, Germany)*
Xavier Goaoc *(INRIA Nancy Grand Est, France)*
Rolf Klein *(Universität Bonn, Germany)*
Christian Knauer *(Universität Bayreuth, Germany) (chair)*
Joseph S. B. Mitchell *(SUNY Stony Brook, USA)*
Bettina Speckmann *(TU Eindhoven, the Netherlands)*
Fabian Stehn *(Universität Bayreuth, Germany)*
Alexander Wolff *(Universität Würzburg, Germany)*

Additional Reviewers:

Eyal Ackerman
Marcel R Ackermann
Henry Adams
Peyman Afshani
Hee-Kap Ahn
Oswin Aichholzer
Luca Castelli Aleardi
Greg Aloupis
Nina Amenta
Boris Aronov
Esther Arkin
Sunil Arya
Sang Won Bae
Ulrich Bauer
Mikhail Belkin
Paul Bendich
Sergey Bereg
Silvia Biasotti
Janis Bode
Jean-Daniel Boissonnat
Alexandre Bos
Prosenjit Bose
Yacine Bouzidi
David Bremner
Mickaël Buchet
Kevin Buchin
Boris Bukh
Sergio Cabello
Paz Carmi
Frédéric Cazals
Erin Chambers
Timothy Chan
Frédéric Chazal
Bernard Chazelle
Chao Chen
Daniel Chen
Siu-Wing Cheng
Otfried Cheong
David Cohen-Steiner
Éric Colin de Verdière
Ovidiu Daescu
Erik D. Demaine
Olivier Devillers
Luc Devroye
Stefan von Deylen
Adrian Dumitrescu
Ramsay Dyer
Ioannis Emiris
Jeff Erickson
Will Evans
Sándor Fekete
Luis Felipe Barba Flores
Efi Fogel
Guilherme D da Fonseca
Stefan Funke
Joachim Giesen
Marc Glisse
Xavier Goaoc
Joachim Gudmundsson
Dan Halperin
Sariel Har-Peled
Jan-Henrik Haunert
John Hershberger
Michael Hoffmann
R Inkulu
Minghui Jiang
Jiongxin Jin
Matthew Kahle
Haim Kaplan
Matthew Katz
Akitoshi Kawamura
Mark Keil
Matias Korman
R Sharath Kumar
Carsten Lange
Stefan Langerman
Francis Lazarus
Audrey Lee-St. John
William Lenhart
Michael Lesnick
Joshua Levine
Bruno Lévy
Jian Li
Kuiyu Li
Jyh-Ming Lien
André Lieutier
Frank Lutz
Anil Maheshwari
Jiří Matoušek
Gopi Meenakshisundaram
Quentin Mérigot
Margaret Meyerhofer
Gary Miller
Dave Millman
Joe Mitchell

Additional Reviewers (continued):

Thomas Moelhave
Pat Morin
Guillaume Moroz
Dmitriy Morozov
David Mount
Wolfgang Mulzer
Vijay Natarajan
Frank Nielsen
Marc Noy
Yoshio Okamoto
René van Oostrum
Yota Otachi
Evanthia Papadopoulou
Sylvain Petitjean
Vincent Pilaud
Konstantin Poelke
Sheung-Hung Poon
Marc Pouget
Pedro Ramos
Ulrich Reitebuch
Jürgen Richter-Gebert
Günter Rote
Bardia Sadri
David Salinas
Swami Sankararmanan
Francisco Santos
Ketan Savla
Scott Schaefer
Stefan Schirra
André Schulz
Henrik Schumacher
Farhad Shahrokhi
Donald Sheehy
Bruce Shepherd
Jessica Sherette
Jonathan Shewchuk
Akiyoshi Shioura
Vin de Silva
Michiel Smid
Shakhar Smorodinsky
Frank Sottile
Bettina Speckmann
Svetlana Stolpner
Kokichi Sugihara
Andrew Suk
Pierluigi Taddei
Tiow-Seng Tan
Xuehou Tan
Shin-Ichi Tanigawa
Andrew Tausz
Monique Teillaud
Boris Thibert
Isabella Thiesen
Csaba Tóth
Ryuhei Uehara
Kasturi Varadarajan
Santosh Vempala
Antoine Vigneron
Uli Wagner
Charlie Wang
Haitao Wang
Anna Wawrzinek
Christophe Weibel
Rephael Wenger
Carola Wenk
Stephen Wismath
David R Wood
Ge Xia
Yuan Yao
Ke Yi
Mariette Yvinec
Rade Živaljević

SCG 2012 Sponsors and Supporters

Sponsors

SIGACT

ACMSIGGRAPH

Supporters

THE UNIVERSITY
of NORTH CAROLINA
at CHAPEL HILL

UNC
DEPARTMENT OF
COMPUTER SCIENCE

On Topological Changes in The Delaunay Triangulation of Moving Points*

Natan Rubin
Institut für Informatik
Freie Universität Berlin
Takustr. 9, 14195, Berlin, Germany

ABSTRACT

Let P be a collection of n points moving along pseudo-algebraic trajectories in the plane.[1] One of the hardest open problems in combinatorial and computational geometry is to obtain a nearly quadratic upper bound, or at least a subcubic bound, on the maximum number of discrete changes that the Delaunay triangulation $\mathrm{DT}(P)$ of P experiences during the motion of the points of P.

In this paper we obtain an upper bound of $O(n^{2+\varepsilon})$, for any $\varepsilon > 0$, under the assumptions that (i) any four points can be co-circular at most twice, and (ii) either no ordered triple of points can be collinear more than once, or no triple of points can be collinear more than twice.

Categories and Subject Descriptors

F.2.2 [**Analysis of algorithms and problem complexity**]: Non-numerical algorithms and problems—*Geometrical problems and computations*; G.2.1 [**Discrete mathematics**]: Combinatorics—*Counting problems*

General Terms

Algorithms, Theory

Keywords

Delaunay triangulation, Voronoi diagram, moving points, discrete changes, kinetic algorithms

1. INTRODUCTION

Delaunay triangulations. Let P be a finite set of points in the plane. Let $\mathrm{VD}(P)$ and $\mathrm{DT}(P)$ denote the Voronoi diagram and Delaunay triangulation of P, respectively. For a point $p \in P$, let $\mathrm{Vor}(p)$ denote the Voronoi cell of p. The Delaunay triangulation $\mathrm{DT} = \mathrm{DT}(P)$ consists of all triangles spanned by P whose circumcircles do not contain points of P in their interior. A pair of points $p, q \in P$ are connected by a Delaunay edge if and only if there is a circle passing through p and q that does not contain any point of P in its interior. Delaunay triangulations and their duals, Voronoi diagrams, are fundamental to much of computational geometry and its applications. See [5, 10, 11] for a survey and a textbook on these structures.

In many applications of Delaunay/Voronoi methods (e.g., mesh generation and kinetic collision detection) the points of the input set P are moving continuously. Even though the motion of the points is continuous, the combinatorial structure of the Voronoi and Delaunay diagrams changes only at discrete times when certain critical events occur.

For the purpose of kinetic maintenance, Delaunay triangulations are nice structures, because, as mentioned above, they admit local certifications associated with individual triangles. This makes it simple to maintain $\mathrm{DT}(P)$ under point motion: an update is necessary only when one of these empty circumcircle conditions fails—this corresponds to co-circularities of certain subsets of four points.[2] Whenever such an event happens, a single edge flip easily restores Delaunayhood.

Let n be the number of moving points in P. We assume that the points move with so-called pseudo-algebraic motions of constant description complexity, meaning (in particular) that any four points are co-circular at most s times, for some constant $s > 0$. By using lower-envelope techniques in certain parametric planes, Guibas et al. [12] showed a roughly cubic upper bound of $O(n^2\lambda_{s+2}(n))$ on the number of discrete (also known as *topological*) changes in $\mathrm{DT}(P)$, where $\lambda_s(n)$ is the maximum length of an (n, s)-Davenport-Schinzel sequence [19]. If each point of P is moving along a straight line and with the same speed, a slightly better upper bound of $O(n^3)$ can be established for the number of discrete changes experienced by $\mathrm{DT}(P)$ (see, e.g., [16]). A substantial gap exists between these upper bounds and the best known quadratic lower bound [19]. Closing this gap has been in the computational geometry lore for many years, and is considered as one of the major (and very difficult) problems in the field; see [9, 11].

The instances of the general problem for which the number of discrete changes in $\mathrm{DT}(P)$ is provably sub-cubic, are strikingly few [13, 16]. It is especially worth mentioning the result of Koltun [16] which deals with sets of points moving along straight lines with equal speeds such that all points start their motion from a fixed line. In this particular case one can show that any four points are co-circular at most *twice* and any three points are co-linear at most

---

*Email: rubinnat@tau.ac.il.

Work on this paper was partly supported by grant 338/09 from the Israel Science Fund and by the Minerval Fellowship Program.

[1]So, in particular, assuming general position, there are constants $s, c > 0$ such that any four points are co-circular at most s times, and any three points are collinear at most c times.

[2]We assume that the motion of the points is generic, so that no more than four points can become co-circular at a given time.

SCG'12, June 17–20, 2012, Chapel Hill, North Carolina, USA.

once. (In a slightly more general setting, which is not covered by [16], the points might start their motion from a circle. In that case, any three points can be collinear at most twice. The result of [16] is not topological and relies on the equations of point trajectories.)

Several relaxed formulations of the problem were studied. For example, Chew [6] proved that $\mathrm{VD}(P)$ undergoes a near-quadratic number of discrete changes if it is defined with respect to a "polygonal" distance function. More recent studies [4, 15] show how to maintain a triangulation of P in a near-quadratic time. Agarwal et al. [3] show how to efficiently maintain a so called *α-stable* subgraph $\mathrm{DT}(P)$, whose edges are robust with respect to small changes in the underlying norm. Acar et al. [1] show how to efficiently maintain the full Delaunay triangulation of a well-spaced *superset* of the n moving points.

Our result. We study the case in which (i) any four points of P are co-circular at most *twice* during the motion, and (ii) either every ordered triple of points can be collinear at most once[3] or every unordered triple can be collinear at most twice, and derive a nearly tight upper bound of $O(n^{2+\varepsilon})$, for any $\varepsilon > 0$, on the number of discrete changes experienced by $\mathrm{DT}(P)$ during the motion in either of these cases. We believe that our results constitute a substantial progress towards establishing nearly quadratic (or just sub-cubic) bounds for more general instances of the problem, such as the simple and natural instance of points moving along straight lines with equal speeds. In this case any four points admit at most *three* co-circularities, and any triple of points can be collinear at most twice. We believe that the tools developed in this paper can be extended to tackle this instance, and possibly also the general case, and we are currently exploring this research direction.

Proof overview. The majority of the discrete changes in $\mathrm{DT}(P)$ occur at moments t_0 when some four points $p, q, a, b \in P$ are co-circular, and the corresponding circumdisc contains no other points of P. We refer to these events as Delaunay co-circularities. Suppose that p, a, q, b appear along their common circumcircle in this order, so ab and pq form the chords of the quadrilateral spanned by these points. Right before t_0, one of the chords, say pq, is Delaunay and thus admits a P-empty disc whose boundary contains p and q. Right after time t_0, the edge pq is replaced in $\mathrm{DT}(P)$ by ab. Informally, this happens because the Delaunayhood of pq is violated by a and b: Any disc whose boundary contains p and q contains at least one of the points a, b. If pq does not re-enter $\mathrm{DT}(P)$ after time t_0, we can charge the event at time t_0 to the edge pq. We thus assume that pq is again Delaunay at some moment $t_1 > t_0$. In particular, at that moment the Delaunayhood of pq is no longer violated by a and b. Before this happens, either at least one of a or b must hit pq, or an additional co-circularity of a, b, p, q must occur during (t_0, t_1). Using our assumption that a, b, p, q induce at most two co-circularity events, we can guarantee (up to a reversal of the time axis) that the co-circularity at time t_0 is the last co-circularity of these 4 points. Thus, one of a, b, let it be a, must cross pq during (t_0, t_1).

Our goal is to derive a recurrence formula for the maximum number $N(n)$ of such Delaunay co-circularities induced by any set P of n points (whose motion satisfies the above conditions). To do so, we study, in Section 2, the set of all co-circularities that involve the disappearing Delaunay edge pq and some other pair of points of $P \setminus \{p, q\}$ and occur during the period (t_0, t_1) when pq is absent from $\mathrm{DT}(P)$. If we find at least $\Omega(k^2)$ "shallow" co-circularities[4], involving p, q, and another pair of points, whose respective circumdiscs contain at most k points of P, we charge them for the disappearance of pq. We use the standard probabilistic technique of Clarkson and Shor [7] to show that the number of Delaunay co-circularities, for which our simple charging works, is $O\left(k^2 N(n/k)\right)$. Informally, such Delaunay co-circularities contribute a nearly quadratic term to the overall recurrence formula (see, e.g., [2] and [17]). Similarly, if we find a "shallow" collinearity of p, q and another point (one halfplane bounded by the line of collinearity contains at most k points) we charge the disappearance of pq to this collinearity. A combination of the Clarkson-Shor technique with the known near-quadratic bound on the number of topological changes in the convex hull of P (see [19, Section 8.6.1]) yields a near-quadratic term to the bound in this case.

It thus remains to bound the number of Delaunay co-circularities for which p and q participate in fewer "shallow" co-circularities and in no "shallow" collinearity during (t_0, t_1). In this case, we show that one can restore the Delaunayhood of pq throughout (t_0, t_1) by removal of some subset A of $O(k)$ points of P. In particular, the point a, which crosses pq, must belong to A. In the smaller Delaunay triangulation $\mathrm{DT}\left((P \setminus A) \cup \{a\}\right)$, the edge pq undergoes a complex process referred to as a *Delaunay crossing by a*.

In Section 3, we derive a recurrence formula for the number of these Delaunay crossings. This is achieved by establishing several structural properties of these crossings, which (as we believe) are of independent interest. Combined with the analysis of Section 2, this yields the desired "near-quadratic" recurrence for the number of Delaunay co-circularities. To make our analysis more accessible for the reader, we first assume, in Section 3, that any three points define at most one ordered collinearity. In Section 4 we extend this analysis for the case in which any three points define at most two collinearities (of any kind).

2. THE OVERALL FRAMEWORK

Delaunay co-circularities. Let P be a collection of n points moving along pseudo-algebraic trajectories in the plane, so that any four points are co-circular at most *twice* during the motion. (We also constrain the number of times, and the patterns, in which a triple of points can become collinear; see the introduction and below for precise details.) In addition, we assume, without loss of generality, that the trajectories of the points of P satisfy all the standard general position assumptions; see the full version [18] for more details.

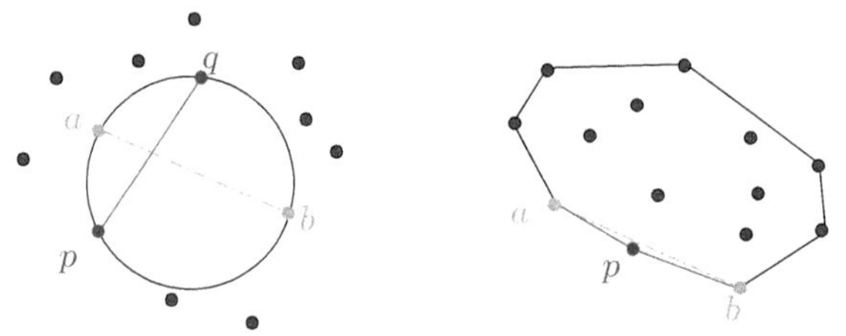

Figure 1: Left: A Delaunay co-circularity of a, b, p, q. An old Delaunay edge pq is replaced by the new edge ab. Right: A collinearity of a, p, b right before p ceases being a vertex on the boundary of the convex hull.

The Delaunay triangulation $\mathrm{DT}(P)$ changes at discrete time moments t_0 when one of the following two types of events occurs.

(i) Some four points a, b, p, q of P become co-circular, so that the cicrumdisc of p, q, a, b is *empty*, i.e., does not contain any point

[3]That is, there can be only one collinearity of an ordered triple (p, q, r) so that the points appear in this order along the common line.

[4]Each of these co-circularities would become a Delaunay co-circularity after removal of at most k points of P.

of P in its interior. We refer to such events as *Delaunay co-circularities*, to distinguish them from non-Delaunay co-circularities, for which the circumdisc of a, b, p, q is nonempty, that is, contains one or more points of P in its interior. See Figure 1 (left).

(ii) Some three points a, b, p of P become collinear on the boundary of the convex hull of P. Assume that p lies between a and b. In this case, if p moves into the interior of the hull then the triangle abp becomes a new Delaunay triangle, and if p moves outside and becomes a new vertex, the old Delaunay triangle abp shrinks to a segment and disappears. See Figure 1 (right). The number of such collinearities on the convex hull boundary is known to be at most nearly quadratic; see, e.g., [19, Section 8.6.1] and below.

We say that a co-circularity event at time t_0 involving a, b, p, q has *index* 1 (resp., 2) if this is the first (resp., second) co-circularity involving a, b, p, q.

We say that a co-circularity event has *level* k if its corresponding circumdisc contains exactly k points of P in its interior. In particular, the Delaunay co-circularities have level 0. The co-circularities having level at most k are called *k-shallow*.

Similar notations apply to collinearities of triples of points p, q, r. A collinearity of p, q, r is called *k-shallow* if the number of points of P to the left, or to the right, of the line through p, q, r is at most k. The standard probabilistic argument of Clarkson and Shor [7] implies that the number of such events, for $k \geq 1$, is $O(k^3 L(n/k))$, where $L(m)$ denote the maximum number of discrete changes on the convex hull of an m-point subset of P. As shown, e.g., in [19, Section 8.6.1], $L(m) = O(m^2\beta(m))$, where $\beta(\cdot)$ is an extremely slowly growing function.[5] We thus get that the number of k-shallow collinearities is $O(kn^2\beta(n/k)) = O(kn^2\beta(n))$.

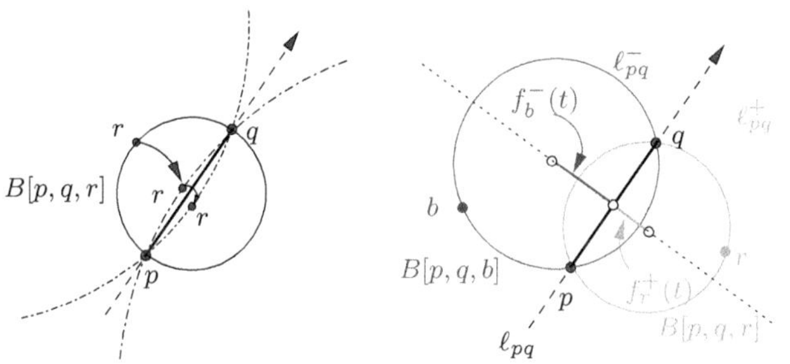

Figure 2: Left: The circumdisc $B[p, q, r]$ of p, q and r moves continuously as long as these three points are not collinear, and then flips over to the other side of the line of collinearity after the collinearity. Right: A snapshot at moment t. In the depicted configuration we have $f^-_b(t) < 0 < f^+_r(t)$.

The red-blue arrangement. For every ordered pair (p, q) of points of P, denote by ℓ_{pq} the line passing through p and q and oriented from p to q. Define ℓ^-_{pq} (resp., ℓ^+_{pq}) to be the halfplane to the left (resp., right) of ℓ_{pq}. Notice that ℓ_{pq} moves continuously with p and q (since, by assumption, p and q never coincide during the motion). Note also that ℓ_{pq} and ℓ_{qp} are oppositely oriented and that $\ell^+_{pq} = \ell^-_{qp}$ and $\ell^-_{pq} = \ell^+_{qp}$. Accordingly, we orient the edge pq connecting p and q from p to q, so that the edges pq and qp have opposite orientations.

Any three points p, q, r span a circumdisc $B[p, q, r]$ which moves continuously with p, q, r as long as p, q, r are not collinear. See Figure 2 (left). When p, q, r become collinear, say, when r crosses pq from ℓ^-_{pq} to ℓ^+_{pq}, the circumdisc $B[p, q, r]$ changes instantly from being all of ℓ^+_{pq} to all of ℓ^-_{pq}. Similarly, when r crosses ℓ_{pq} from ℓ^-_{pq} to ℓ^+_{pq} *outside* pq, the circumdisc changes instantly from ℓ^-_{pq} to ℓ^+_{pq}. Symmetric changes occur when r crosses ℓ_{pq} from ℓ^+_{pq} to ℓ^-_{pq}.

For a fixed ordered pair $p, q \in P$, we call a point a of $P \setminus \{p, q\}$ *red* (with respect to the oriented edge pq) if $a \in \ell^+_{pq}$; otherwise it is *blue*.

As in [12], we define, for each $r \in P \setminus \{p, q\}$, a pair of partial functions f^+_r, f^-_r over the time axis as follows. If $r \in \ell^+_{pq}$ at time t then $f^-_r(t)$ is undefined, and $f^+_r(t)$ is the signed distance of the center c of $B[p, q, r]$ from ℓ_{pq}; it is positive (resp., negative) if c lies in ℓ^+_{pq} (resp., in ℓ^-_{pq}). A symmetric definition applies when $r \in \ell^-_{pq}$. Here too $f^-_r(t)$ is positive (resp., negative) if the center of $B[p, q, r]$ lies in ℓ^+_{pq} (resp., in ℓ^-_{pq}). We refer to f^+_r as the *red function* of r (with respect to pq) and to f^-_r as the *blue function* of r. Note that at all times when p, q, r are not collinear, exactly one of f^+_r, f^-_r is defined. See Figure 2 (right). The common points of discontinuity of f^+_r, f^-_r occur at moments when r crosses ℓ_{pq}. Specifically, f^+_r tends to $+\infty$ before r crosses ℓ_{pq} from ℓ^+_{pq} to ℓ^-_{pq} outside the segment pq, and it tends to $-\infty$ when r does so within pq; the behavior of f^-_r is fully symmetric.

Let E^+ denote the lower envelope of the red functions, and let E^- denote the upper envelope of the blue functions. The edge pq is a Delaunay edge at time t if and only if $E^-(t) < E^+(t)$. Any disc whose bounding circle passes through p and q which is centered anywhere in the interval $(E^-(t), E^+(t))$ along the perpendicular bisector of pq is empty at time t, and thus serves as a witness to pq being Delaunay. If pq is not Delaunay at time t, there is a pair of a red function $f^+_r(t)$ and a blue function $f^-_b(t)$ such that $f^+_r(t) < f^-_b(t)$. For example, we can take f^+_r (resp., f^-_b) to be the function attaining E^+ (resp., E^-) at time t; see Figure 3 (left). In such a case, we say that the Delaunayhood of pq is *violated* by the pair of points $r, b \in P$ that define f^+_r, f^-_b. Note that in general there can be many pairs (r, b) that violate pq (quadratically many in the worst case).

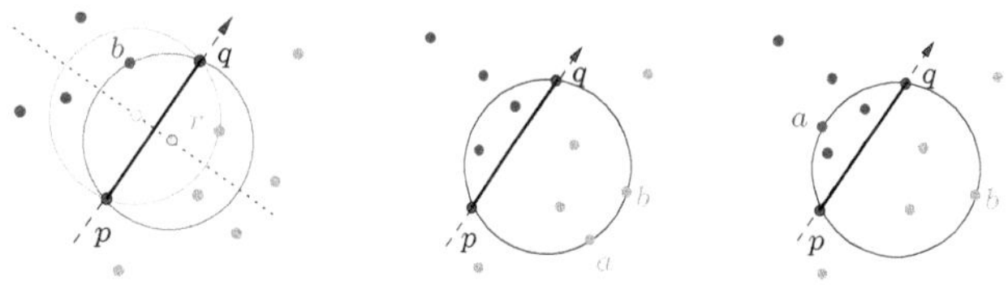

Figure 3: Left: A snapshot at fixed time t. The red and blue envelopes E^+, E^- coincide with the functions f^+_r, f^-_b, respectively. The edge pq is not a Delaunay edge because $E^+(t)$ (the hollow center) is smaller than $E^-(t)$ (the shaded center). Center and right: Red-red and red-blue co-circularities.

Hence, at any time when the edge pq joins or leaves $\mathrm{DT}(P)$, via a Delaunay co-circularity involving p, q, and two other points of P, we have $E^-(t) = E^+(t)$. In this case the two other points, a, b, are such that one of them, say a, lies in ℓ^+_{pq} and b lies in ℓ^-_{pq}, and $E^+(t) = f^+_a(t), E^-(t) = f^-_b(t)$.

Let $\mathcal{A} = \mathcal{A}_{pq}$ denote the arrangement of the $2n-2$ functions $f^+_r(t), f^-_r(t)$, for $r \in P \setminus \{p, q\}$, drawn in the parametric (t, ρ)-plane, where t is the time and ρ measures signed distance to the midpoint of pq along the perpendicular bisector of pq. We label each vertex of $\mathcal{A}$ as red-red, blue-blue, or red-blue, according to the colors of the two functions meeting at the vertex. Note that our general position assumptions imply that $\mathcal{A}$ is also in general position, so that no three function pass through a common vertex, and no pair of functions are tangent to each other. Note also that the

[5]Specifically, $\beta(n) = \frac{\lambda_{s+2}(n)}{n}$, where s is the maximum number of collinearities of any fixed triple of points. The pseudo-algebraicity of the motion implies that s is a constant, but we will restrict s further; see below.

functions forming $\mathcal{A}$ have in general discontinuities, at the corresponding collinearities. At each such collinearity, a red function f_r^+ tends to ∞ or $-\infty$ on one side of the critical time, and is replaced on the other side by the corresponding blue function f_r^- which tends to $-\infty$ or ∞, respectively.

An intersection between two red functions f_a^+, f_b^+ corresponds to a co-circularity event which involves p, q, a and b, occurring when both a and b lie in ℓ_{pq}^+. Similarly, an intersection of two blue functions f_a^-, f_b^- corresponds to a co-circularity event involving p, q, a, b where both a and b lie in ℓ_{pq}^-. Also, an intersection of a red fuction f_a^+ and a blue function f_b^- represents a co-circularity of p, q, a, b, where $a \in \ell_{pq}^+$ and $b \in \ell_{pq}^-$. We label these co-circularities, as we labeled the vertices of $\mathcal{A}$, as red-red, blue-blue, and red-blue (all with respect to pq), depending on the respective colors of a and b. See Figure 3 (center and right).

It is instructive to note that in any co-circularity of four points of P there are exactly two pairs (the opposite pairs in the co-circularity) with respect to which the co-circularity is red-blue, and four pairs (the adjacent pairs) with respect to which the co-circularity is "monochromatic". In the case of a Delaunay co-circularity, the two pairs for which the co-circularity is red-blue are those that enter or leave the Delaunay triangulation $\mathrm{DT}(P)$ (one pair enters and one leaves). The Delaunayhood of pairs for which the co-circularity is monochromatic is not affected by the co-circularity, which appears in the corresponding arrangement as a *breakpoint* of either $E^+(t)$ or of $E^-(t)$.

Overview. Consider a Delaunay co-circularity event at time t_0 at which an edge pq of $\mathrm{DT}(P)$ is replaced by another edge ab, because of a red-blue co-circularity (with respect to pq, and, for that matter, also with respect to ab) of level 0. Assume first that the co-circularity of p, q, a, b has index 2; the case of index 1 is handled fully symmetrically, by reversing the direction of the time axis.

There are at most $O(n^2)$ such events for which the vanishing edge pq never reappears in $\mathrm{DT}(P)$, so we focus on the Delaunay co-circularities (of index 2) whose corresponding edge pq rejoins $\mathrm{DT}(P)$ at some future moment $t_1 > t_0$. Note that in this case, if the co-circularity involved two other points a, b, then at least one of a, b must cross ℓ_{pq} between t_0 and t_1 otherwise p, q, a and b would have to become co-circular again, in order to "free" pq from non-Delaunayhood, which is impossible since our co-circularity has index 2. Hence a major part of the general strategy is to charge the reappearance of pq to crossings (that is, collinearities) involving pq, which occur between t_0 and t_1.

Nevertheless, before we get to charging collinearities, we first follow a more traditional charging scheme. We fix some sufficiently large constant parameter $k > 0$, and try to charge the disappearance of pq at time t_0 to at least $\Omega(k^2)$ k-shallow red-red, red-blue, and/or blue-blue co-circularities (all with respect to pq) that occur during the period (t_0, t_1) of the absence of pq from $\mathrm{DT}(P)$. The charging is almost unique (i.e., each co-circularity is charged at most a constant number of times in this manner). This in itself, combined with the Clarkson-Shor analysis technique, would lead to a recurrence with a nearly quadratic solution (see below for details, and see, e.g., [2] for similar situations). Unfortunately, this scheme does not always work because there might exist Delaunay co-circularities for which the respective red-blue arrangement (of the disappearing edge pq) contains relatively few shallow vertices of all three kinds during the interval (t_0, t_1). In these cases we try to charge the Delaunay co-circularity to a k-shallow collinearity event in which the edge pq or the line ℓ_{pq} is crossed by a point of P. This would also lead to a recurrence with a nearly quadratic solution. This however may also fail, and then we resort to a third kind of charging. Informally, we charge the co-circularity to a collinearity of p, q, and one of the other two points, a, b, that participated in the Delaunay co-circularity event at time t_0 (together with p and q). This collinearity is not shallow in the sense defined above, but is shallow in a different sense, in which, after removing at most k points of P it becomes a *Delaunay crossing*, a novel concept that we introduce for our analysis and which is discussed next.

Delaunay crossings. A *Delaunay crossing* is a triple $(pq, r, I = (t_0, t_1))$, where $p, q, r \in P$ and I is a time interval, such that (1) pq leaves $\mathrm{DT}(P)$ at time t_0, and returns at time t_1 (and pq does not belong to $\mathrm{DT}(P)$ during (t_0, t_1)), (2) r crosses the segment pq *at least* once during I, and (3) pq is an edge of $\mathrm{DT}(P \setminus \{r\})$ during I (i.e., removing r restores the Delaunayhood of pq during the entire time interval I).

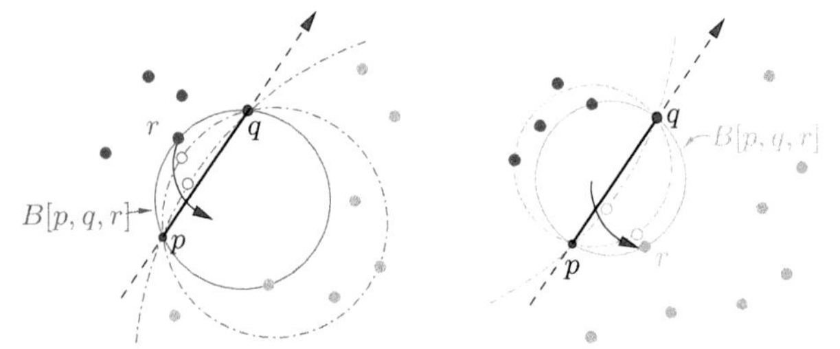

Figure 4: A Delaunay crossing of pq by r from ℓ_{pq}^- to ℓ_{pq}^+. Several snapshots of the continuous motion of $B[p, q, r]$ before and after r crosses pq are depicted (in the left and right figures, respectively). Hollow points specify the positions of r when $pq \notin \mathrm{DT}(P)$. The solid circle in the left (resp., right) figure is the Delaunay co-circularity that starts (resp., ends) I.

It is easy to see that the third condition is equivalent to the following condition, expressed in terms of the red-blue arrangement $\mathcal{A}_{pq}$ associated with pq: The point r participates only in red-blue co-circularites during the interval I, and these are the only red-blue co-circularities that occur during I. More specifically, note that r is red during some portion of I and is blue during the complementary portion (both portions are nonempty). During the former portion the graph of f_r^+ coincides with the red lower envelope E^+ (otherwise $E^+(t) < E^-(t)$ would hold sometime during I even after removal of r), so it can only meet the graphs of blue functions. Similarly, during the latter portion f_r^- coincides with the blue upper envelope E^-, so it can only meet the graphs of red functions. See Figure 4 for a schematic illustration of this behavior.

Notice that no points, other than r, cross pq during I (any such crossing would clearly contradict the third condition at the very moment when it occurs). Moreover, r does not cross ℓ_{pq} outside pq during I; otherwise pq would belong to $\mathrm{DT}(P)$ when r belongs to $\ell_{pq} \setminus pq$.

THEOREM 2.1. *Let P be a collection of n points moving as described above. Suppose that an edge pq belongs to $\mathrm{DT}(P)$ at moments t_0 and t_1, for $t_0 < t_1$. Let $k > 12$ be some sufficiently large constant.*[6] *Then one of the following conditions holds:*

(i) There is a k-shallow collinearity which takes place during (t_0, t_1), and involves p, q and another point r.

(ii) There are $\Omega(k^2)$ k-shallow red-red, red-blue, or blue-blue co-circularities (with respect to pq) which occur during (t_0, t_1).

(iii) There is a subset $A \subset P$ of fewer than $3k$ points whose removal guarantees that pq belongs to $\mathrm{DT}(P \setminus A)$ throughout (t_0, t_1).

[6]The constants is the $O(\cdot)$ and $\Omega(\cdot)$ notations do not depend on k.

Notice that we do not assume that pq leaves $\mathrm{DT}(P)$ at any moment during (t_0, t_1). A typical application of the theorem in our charging scheme is as follows.[7] Suppose that pq disappears from $\mathrm{DT}(P)$ at time t_0 and reappears at time $t_1 > t_0$. If Case (i) or Case (ii) arises then we can charge the disappearance of pq either to the k-shallow collinearity in Case (i), or to the $\Omega(k^2)$ k-shallow co-circularities in Case (ii). Clearly, any co-circularity or collinearity is charged at most a constant number of times. (Namely, it can be charged only for the disappearances of edges pq whose two vertices p, q participate in the event, and only for the disappearance immediately preceding the event, without any in-between reappearance.) Both charges are good in the sense that they generate a recurrence with a near-quadratic solution. The third case is more complicated to handle, and the entire Section 3 will be devoted to it.

Proof of Theorem 2.1 — sketch. We only give an outline of the complete proof, which we delegate to the full version [18]. Consider the portion of the red-blue arrangement associated with pq within the time interval (t_0, t_1). As above, refer to the parametric plane in which this arrangement is represented as the $t\rho$-plane, where t is the time axis and ρ measures signed distances from ℓ_{pq}. We define the *red* (resp., *blue*) *level* of a point $x = (t, \rho)$ in this parametric $\mathbb{R}^2$ as the number of red (resp., blue) functions that lie below (resp., above) x (in the ρ-direction). See Figure 5. It is easily checked that the level of a co-circularity event at time t, with circumcenter at distance ρ from ℓ_{pq}, is the sum of the red and the blue levels of (t, ρ).

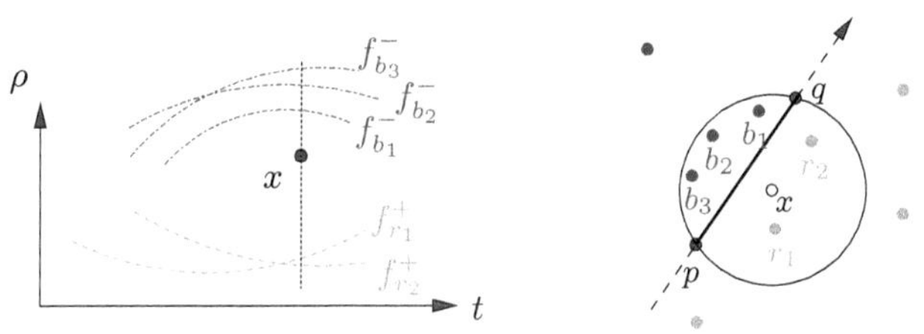

Figure 5: Left: The point $x = (t, \rho)$ lies below three blue functions and above two red functions, so its blue and red levels are 3 and 2, respectively. Right: The circumdisc centered at (signed) distance ρ from ℓ_{pq} and touching p and q at time t contains the three corresponding blue points and two red points.

First, suppose that at some moment $t_0 < t^* < t_1$ there is a disc D^* that touches p and q, and contains at least $k/3$ red points and at least $k/3$ blue points. Notice that, for every red point inside D^*, the corresponding disc $B[p, q, r]$ contains at least $k/3$ blue points. See Figure 6 (left). We argue that, for some constant fraction of such red points r, the boundary of $B[p, q, r]$ is crossed by $\Omega(k)$ red and blue points so that the resulting co-circularities are k-shallow (or else the point r is involved in a k-shallow collinearity with p and q). Informally, this happens because pq was Delaunay at time t_0, so $B[p, q, r]$ contained no blue points at time t_0, but it contains $k/3$ such points at time t^* (or another possibility is that the point r was blue at time t_0 and changes its color during (t_0, t_1)). In other words, the blue level of f_r^+ sharply increases during (t_0, t^*); see Figure 6 (right). Repeating this argument for $\Omega(k)$ red points inside D^*, we encounter either $\Omega(k^2)$ k-shallow co-circularities, or a k-shallow collinearity, all involving p and q.

Now assume that the above situation does not happen. In particular, there are always fewer than $k/3$ red points r whose respective discs $B[p, q, r]$ contain at least $k/3$ blue points. Moreover, if the disc $B[p, q, r]$ contains at least $k/3$ blue points then f_r^+ is among

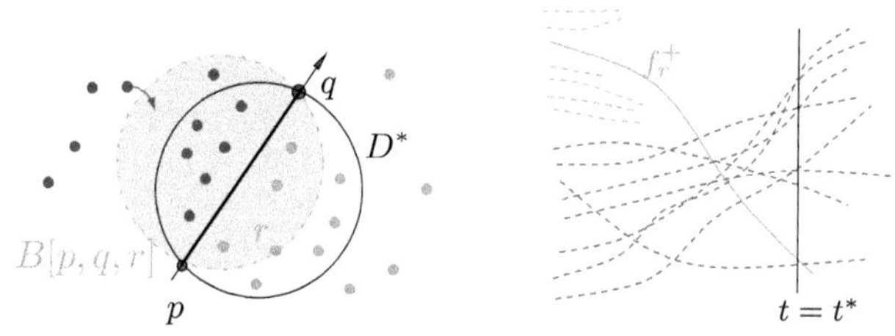

Figure 6: Proof of Theorem 2.1: The first case. For each red point r inside D^* the disc $B[p, q, r]$ contains at least $k/3$ blue points (left). The blue level of f_r^+ sharply increases from 0 to (at least) $k/3$ (right).

the $k/3$ "shallowest" red-functions (i.e., the $k/3$ red functions that attain the smallest values) for, otherwise, the disc $B[p, q, r]$ would contain at least $k/3$ red points in addition to $k/3$ blue ones; see Figure 7 (left). In other words, f_r^+ has level most $k/3$ in the arrangement of all red functions. Symmetric claims hold for the set of blue points (and their functions f_b^-). We show that the Delaunayhood of pq can be enforced throughout (t_0, t_1) by removing the set A_R of all red points r whose respective functions f_r^+ appear at level at most $k/3$ in the "red" arrangement (at some time during (t_0, t_1)), and the symmetric set A_B of blue points.[8] See Figure 7 (right). Finally, we argue that if the number of removed points in $A_R \cup A_B$ exceeds $3k$ then the "surplus" points participate in $\Omega(k^2)$ k-shallow co-circularities. Indeed, in this case we have $\Omega(k)$, say, red points r such that the red level of f_r^+ is at least k at time t_0, and it decreases to $k/3$ (or r changes its color during (t_0, t_1)). (A very similar argument was used in [2] to address a totally different problem.)

Figure 7: Proof of Theorem 2.1. The second case. Left: The red points r whose disc $B[p, q, r]$ contains at least $k/3$ points induce $k/3$-shallow functions f_r^+. Right: Removing "shallow" red and blue points restores the Delaunayhood of pq.

The number of Delaunay co-circularities. Let $N(n)$ be the maximum possible number of Delaunay co-circularities induced by any set P of n points whose motion satisfies the above assumptions. Consider a Delaunay co-circularity of index 2, at time t_0, where the Delaunayhood of $pq \in \mathrm{DT}(P)$ is violated by $a \in \ell_{pq}^-$ and $b \in \ell_{pq}^+$. If pq never rejoins $\mathrm{DT}(P)$, we (uniquely) charge this co-circularity to the edge pq. Hence we may assume that pq reappears in $\mathrm{DT}(P)$ at some future moment $t_1 > t_0$.

We are now ready to apply Theorem 2.1. Fix a sufficiently large constant $k > 12$. Assume first that one of the conditions (i) or (ii) holds, so we can charge the co-circularity of p, q, a, and b either to $\Omega(k^2)$ k-shallow co-circularities (each involving p, q, and some two other points of P), or to a k-shallow collinearity (involving p, q, and some third point of P). To bound the overall number of k-shallow co-circularities charged in this manner, we apply the probabilistic argument of Clarkson and Shor [7]. Specifically, we consider a random sample R of $\Theta(n/k)$ points of P and observe

[7]More sophisticated applications of the theorem arise in Section 3.

[8]A point of P can change its color during (t_0, t_1), and thus belong to both deleted sets.

that any k-shallow co-circularity (with respect to P) becomes a Delaunay co-circularity (with respect to R) with probability $\Theta(1/k^4)$. Hence, the overall number of k-shallow co-circularities is $O(k^4 N(n/k))$. Each k-shallow co-circularity is charged by only $O(1)$ Delaunay co-circularities in this manner, and it has to "pay" only $O(1/k^2)$ units every time it is charged. Similarly, as already argued, the number of k-shallow collinearities is $O(kn^2\beta(n))$, and each such collinearity is charged by at most $O(1)$ Delaunay co-circularities. Hence, there are at most $O(k^2N(n/k) + kn^2\beta(n))$ Delaunay co-circularities for which one of the conditions (i) or (ii) holds.

Assume then that condition (iii) holds for our co-circularity. We need the following lemma, whose easy proof is sketched in Figure 8.

LEMMA 2.2. *Assume that some four points p, q, a, b define a co-circularity of index 2 at time t_0, so that $a \in \ell_{pq}^-$ and $b \in \ell_{pq}^+$, after which the Delaunayhood of pq is violated*[9] *by a and b. Furthermore, suppose that pq enters $\mathrm{DT}(P)$ at some future time $t_1 > t_0$. Then either a crosses pq from ℓ_{pq}^- to ℓ_{pq}^+, or b crosses pq from ℓ_{pq}^+ to ℓ_{pq}^-, during (t_0, t_1).*

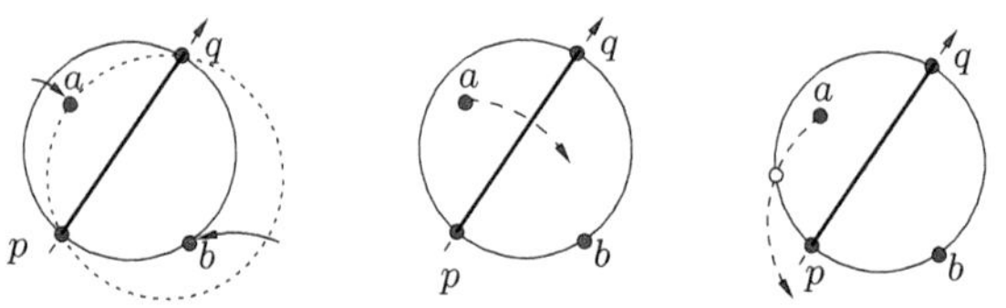

Figure 8: Proof of Lemma 2.2. The setup right after time t_0 when the *last* co-circularity of p, q, a, b occurred (left). The Delaunayhood of pq remains violated by a and b after time t_0 as long as none of a, b hits ℓ_{pq}. The first crossing of ℓ_{pq} by a or b must occur within pq (as depicted in the central figure) for, otherwise, we would first encounter a third (and forbidden) co-circularity of p, q, a, b (right).

By assumption, there is a set A of at most $3k$ points (necessarily including at least one of a or b) whose removal ensures the Delaunayhood of pq during (t_0, t_1). By Lemma 2.2, one the two points a, b, let it be a, crosses pq during (t_0, t_1). We sample at random (and without replacement) a subset $R \subset P$ of $O(n/k)$ points, and notice that the following two events occur simultaneously with probability at least $\Omega(1/k^3)$: (1) the points p, q, a belong to R, and (2) none of the points of $A \setminus \{a\}$ belong to R. Since a crosses pq during (t_0, t_1) (and pq is Delaunay at time t_0 and t_1), the sample R induces a Delaunay crossing (pq, a, I), for some time interval $I \subset (t_0, t_1)$. (If a crosses pq more than once, there may be several such crossings which occur at disjoint sub-intervals of (t_0, t_1).) We charge the disappearance of pq from $\mathrm{DT}(P)$ to this crossing (or to the first such crossing if there are several) and note that the charging is unique (i.e., every Delaunay crossing (pq, a, I) in $\mathrm{DT}(R)$ is charged by at most one disappearance of the respective edge pq from $\mathrm{DT}(P)$). Hence, the number of Delaunay co-circularities of this kind is bounded by $O(k^3C(n/k))$, where $C(n)$ denotes the maximum number of Delaunay crossings induced by any collection P of n points whose motion satisfies the above assumptions.

If the Delaunay co-circularity of p, q, a, b has index 1, we reverse the direction of the time axis and argue as above for the edge ab instead of pq. We thus obtain the following recurrence for the number of Delaunay co-circularities:

[9]In particular, the co-circularity at time t_0 is red-blue with respect to pq; see Figure 8.

$$N(n) \le c\left(k^2N(n/k) + k^3C(n/k) + kn^2\beta(n)\right), \qquad (1)$$

for some constant $c > 0$ that depends on the motion of the points but is independent of k.

Informally, (1) implies that the maximum number of Delaunay co-circularities is asymptotically dominated by the maximum number of Delaunay crossings (assuming it is at least quadratic).

3. THE NUMBER OF DELAUNAY CROSSINGS

In this section we derive a recurrence relation for the maximum number $C(n)$ of Delaunay crossings induced by a set P of n moving points as above, which expresses $C(n)$ in terms of the maximum number of Delaunay co-circularities of smaller-size sets, plus a nearly quadratic additive term. Plugging this relation into (1) yields a nearly quadratic upper bound on $N(n)$.

Let $(pq, r, I = (t_0, t_1))$ be a Delaunay crossing, as defined in the previous section. Specifically, pq disappears from $\mathrm{DT}(P)$ at time t_0, rejoins $\mathrm{DT}(P)$ at time t_1, and remains Delaunay throughout I in $\mathrm{DT}(P \setminus \{r\})$. To distinguish between the notion of a Delaunay crossing (pq, r, I), which lasts for the full time interval I, and the instance where r actually lies on the segment pq, we refer to the latter event by saying that r *hits* pq.

To simplify our description, we make the following assumption for the time being: Any point r crosses any edge $pq \in P \times P$ at most once. (Note that p, q, r might still define additional collinearity events, in which r is not the middle point of the collinearity.) In Section 4, we show how to replace this assumption by a more "natural" one[10]: Any three points are collinear at most twice (where one point may be the middle one in both collinearities). The following simple lemma lies at the heart of our analysis.

LEMMA 3.1. *If $(pq, r, I = (t_0, t_1))$ is a Delaunay crossing then, for any $s \in P \setminus \{p, q, r\}$ the points p, q, r, s define a co-circularity which takes place during I and is red-blue with respect to pq.*

PROOF. By definition, r crosses pq at some (unique) time $t_0 < t^* < t_1$, say from ℓ_{pq}^- to ℓ_{pq}^+. The disc $B[p, q, r]$ is P-empty at t_0 and at t_1 and moves continuously throughout $[t_0, t^*)$ and $(t^*, t_1]$. Just before t^*, $B[p, q, r]$ is the entire ℓ_{pq}^+, so every point $s \in P \cap \ell_{pq}^+$ at time t^* must have entered $B[p, q, r]$ during $[t_0, t^*)$, forming a co-circularity with p, q, r at the time it enters the disc. See Figure 9 (left). (As mentioned in Section 2, this co-circularity of p, q, r, s is red-blue with respect to pq, that is, the point s enters $B[p, q, r]$ through $\partial B[p, q, r] \cap \ell_{pq}^+$.) A symmetric argument (in which we reverse the direction of the time axis) shows that the same holds for all the points $s \in P$ that lie in ℓ_{pq}^- at time t^*; see Figure 9 (center). □

LEMMA 3.2. *The number of triples of points $p, q, r \in P$ for which there exist two time intervals I_1, I_2 such that both (pq, r, I_1) and (qr, p, I_2) are Delaunay crossings, is at most $O(n^2)$.*

PROOF. Assume that the Delaunay crossing of p and qr ends after that of r and pq; that is I_2 ends after I_1 (note that I_1 and I_2 need not be disjoint). Let s be a point of $P \setminus \{p, q, r\}$. By Lemma 3.1, the four points p, q, r, s define a co-circularity event during I_1. Similarly, the same four points p, q, r, s define a co-circularity

[10]For example, as is well known, it holds for points moving with constant speeds along straight lines.

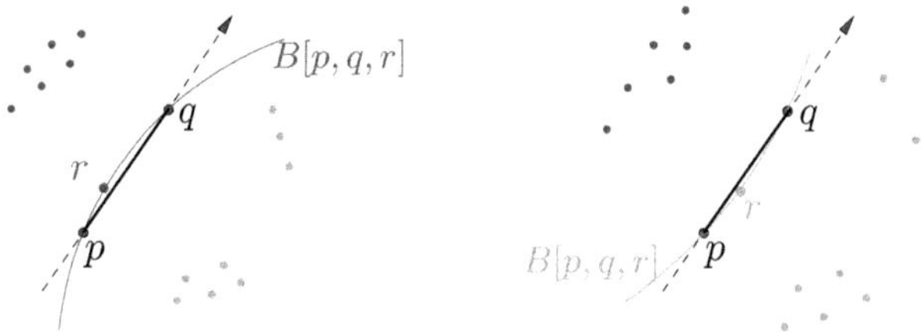

Figure 9: Left: Right before r crosses pq, the circumdisc $B = B[p,q,r]$ contains all points in $P \cap \ell_{pq}^{+}$. Right: After r crosses pq, B contains all points in $P \cap \ell_{pq}^{-}$.

event during I_2. We claim that the above two co-circularities are distinct. Indeed, the former co-circularity of p, q, r, s is red-blue with respect to the edge pq (which is crossed by r during I_1), so p and q are not adjacent in the co-circularity. On the other hand, the latter co-circularity is red-blue with respect to qr (which is crossed by p during I_2), so q and r are not adjacent in the co-circularity. However, both non-adjacencies cannot occur simultaneously in the same co-circularity, so these two co-circularities of p, q, r, s must be distinct.

Hence, the points p, q, r, s induce at least (by our assumption, exactly) two common co-circularity events before qr re-enters $\mathrm{DT}(P)$.

Thus, we cannot have a Delaunay crossing of qr by s after qr re-enters $\mathrm{DT}(P)$, for otherwise this would lead, according to Lemma 3.1 to a third co-circularity event involving p, q, r and s. Since this holds for every point $s \in P \setminus \{p, q, r\}$, the crossing of qr by p is the last Delaunay crossing of qr, so it can be charged uniquely to this edge. (Clearly, any two Delaunay crossings of the same edge qr take place at disjoint time intervals.) □

The following useful lemma is established in the full version [18]. Informally, it allows to distinguish Delaunay crossings from other collinearities induced by the points of P.

LEMMA 3.3. *If $(pq, r, I = (t_0, t_1))$ is a Delaunay crossing then each of the edges pr, rq belongs to $\mathrm{DT}(P)$ throughout I.*

Remark: To establish the Delaunayhood of pr and rq at times t_0 and t_1 it is sufficient to observe that the first and the last red-blue co-circularities (with respect to pq) during I, which involve r, are Delaunay co-circularities.

Our analysis is culminated by the following theorem, whose rather tedious proof incorporates all the machinery introduced in Sections 2 and 3.

THEOREM 3.4. *Let $12 < k_1 < k_2$ be a pair of constants. The maximum number $C(n)$ of Delaunay crossings satisfies the recurrence*[11]

$$C(n) \le c_1 \left(k_1^2 N(n/k_1) + k_1^4 k_2^2 N\left(\frac{n}{k_1 k_2}\right) + k_1^2 k_2 n^2 \beta(n) \right), \tag{2}$$

for some constant $c_1 > 0$ that depends on the motion of the points but is independent of k_1, k_2.

Notice that a single point can hit $\Omega(n^2)$ edges connecting various pairs of points of P.

Proof of Theorem 3.4 (sketch). Our goal is to show that (on average) a point r of P is involved in relatively few Delaunay crossings of edges pq, pa that share the same vertex p. (Below we refer to such edges as *p-edges*.) The excess is charged to shallow co-circularities and collinearities, and (finally) to the special triples of points which can be counted using Theorem 3.2.

[11]In the full version [18] we establish a somewhat stronger recurrence. Nevertheless, both recurrences lead to the same asymptotic bound on the number of Delaunay co-circularities.

Without loss of generality, we assume in what follows that, for any Delaunay crossing $(pq, r, I = (t_0, t_1))$, the point r crosses pq from ℓ_{pq}^{-} to ℓ_{pq}^{+} during I. (If r crosses pq in the opposite direction, we regard this crossing as $(qp, r, I = (t_0, t_1))$.) The following lemma defines a natural order on the Delaunay crossings of p-edges by the same point r.

LEMMA 3.5. *Let r be a point which is involved in Delaunay crossings of two p-edges, say pq and pa, and suppose that r hits pq before it hits pa. Then the Delaunay crossing of pq by r begins (resp., ends) before the beginning (resp., end) of the crossing of pa by r. Clearly, the converse statements hold too.*

PROOF. We only prove the part of the lemma concerning the ending times of the crossings, because the proof about the starting times is fully symmetric (by reversing the direction of the time axis). Note that r enters ℓ_{pq}^{+} only once during the Delaunay crossing of pq by r, namely, right after r hits pq. Indeed, by assumption, r cannot exit ℓ_{pq}^{+} by crossing pq again, and it cannot cross $\ell_{pq} \setminus pq$ because at that time pq, which is Delaunay in $\mathrm{DT}(P \setminus \{r\})$, would be Delaunay also in the presence of r, contrary to the definition of a Delaunay crossing. Hence, we may assume that r still lies in ℓ_{pq}^{+} when it hits pa during the Delaunay crossing of that edge. Indeed, otherwise the crossing of pq would by then be over, so the claim would hold trivially. In particular, $\vec{pa}$ lies clockwise to $\vec{pq}$ at that time.

It suffices to prove that the co-circularity of p, q, r, a, which (by Lemma 3.1) occurs during the Delaunay crossing of pa by r, takes place when the crossing of pq by r is already finished (and, in particular, after the co-circularity of p, q, r, a that occurs during the crossing of pq).

Before the Delaunayhood of pa is restored, we have a co-circularity p, q, r, a in which q leaves $B[p, a, r] \cap \ell_{pa}^{-}$ (see the proof of Lemma 3.1: right after the crossing, q, which lies then in ℓ_{pa}^{-}, is contained in $B[p, a, r]$ and has to leave that disc before it becomes empty). Notice that this is a red-blue co-circularity with respect to pa, and a red-red co-circularity with respect to pq; see Figure 10. Since no red-red or blue-blue co-circularities occur during a Delaunay crossing of an edge, the crossing of pq is already over. □

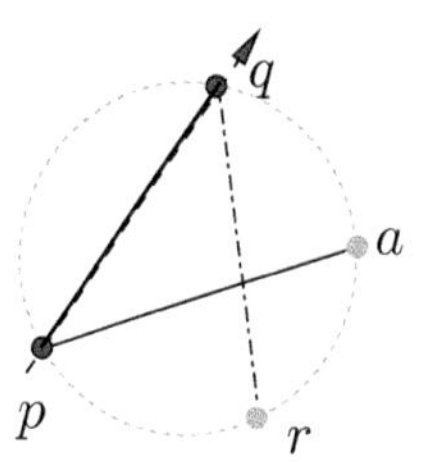

Figure 10: Proof of Lemma 3.5: The second co-circularity of p, q, r, a which occurs when r still lies in ℓ_{pq}^{+}. This is a red-red co-circularity with respect to pq, so the crossing of pq is already over.

The proof of Theorem 3.4 proceeds by bounding the maximum possible number $Q(n)$ of *special quadruples* p, q, a, r where r is involved in two *consecutive* Delaunay crossings $(pq, r, I = (t_0, t_1))$, $(pa, r, J = (t_2, t_3))$. That is, we require that no Delaunay crossing begins during (t_0, t_2) or ends during (t_1, t_3). Any Delaunay crossing is a part of a special quadruple, unless it is the last such

Delaunay crossing of a p-edge by r. Hence, we get the following easy bound on the number of Delaunay crossings in terms of special quadruples:

$$C(n) \le Q(n) + O(n^2) \qquad (3)$$

Overview. As in Section 2, we first try to charge a special quadruple p, q, a, r to sufficiently many shallow co circularities (or to a shallow collinearity) in the red-blue arrangements of the edges pr, rq, ra. (According to Lemma 3.3, each of these edges is Delaunay during some subinterval of $I \cup J$.) This is done by skillful application of Theorem 2.1 to the respective red-blue arrangements of these edges, with the constant parameter k equal to k_1 or k_2. If none of these chargings is successful, we obtain a "small" subset A of P so that some triple among $\{p, q, r\}, \{p, a, r\}, \{q, a, r\}$ induces *two* Delaunay crossings within $\mathrm{DT}(P \setminus A)$. We then use Lemma 3.2 to directly bound the number of such triples (and their respective quadruples).

Charging along pr. Consider the red-blue arrangement associated with the edge pr, during the time interval (t_1, t_3). Note that pr is Delaunay both at time t_1 and at time t_3 (recall Lemma 3.3), which allows us to use Theorem 2.1 (with k_1 instead of k). In cases (i) and (ii) of that theorem we charge the quadruple p, q, a, r to $\Omega(k_1^2)$ k_1-shallow co-circularities, or to a k_1-shallow collinearity (within the arrangement of pr).

In both cases, each shallow co-circularity or collinearity is charged at most a constant number of times. Indeed, consider the moment t^* when the charged event occurs, and notice that it involves p and r (together with one or two additional points of P). Then the moment t_1 (when the crossing of pq by r ends) is the last time before t^* when a Delaunay crossing of a p-edge by r is completed. Hence, having guessed p and r (in $O(1)$ ways), q is uniquely determined. (Clearly, a is uniquely determined by the ordered triple p, q, r.) Therefore, using the upper bounds on the number of k_1-shallow collinearities and co-circularities established in Section 2, and the fact that each co-circularity pays only $\Theta(1/k_1^2)$ units when it is charged in this manner, we get that the number of such special quadruples p, q, a, r for which the red-blue arrangement of pr (during (t_1, t_3)) satisfies condition (i) or (ii) of Theorem 2.1 is $O(k_1^2 N(n/k_1) + k_1 n^2 \beta(n))$.

Assume then that the red-blue arrangement of pr (during (t_1, t_3)) satisfies condition (iii) of Theorem 2.1. Then one can restore the Delaunayhood of pr during (t_1, t_3) by removing a set A_0 of at most $3k_1$ points of P (possibly including q or a). The standard probabilistic argument of Clarkson and Shor [7] allows to bound the number of special quadruples in terms of the maximum number $Q_1(n)$ of the (even more) special quadruples for which pr belongs to $\mathrm{DT}(P \setminus \{q, a\})$ throughout (t_1, t_3). Altogether, we obtain the following upper bound on $Q(n)$:

$$Q(n) \le c_2(k_1^4 Q_1(n/k_1) + k_1^2 N(n/k_1) + k_1 n^2 \beta(n)) \qquad (4)$$

It therefore remains to provide a suitable upper bound on the maximum possible number $Q_1(n)$ of the more "refined" quadruples p, q, a, r, which satisfy the above condition on the Delaunayhood of the edge pr. If q hits pr during (t_1, t_3), this additional condition implies that the triple p, q, r defines a second Delaunay crossing in $\mathrm{DT}(P \setminus \{a\})$; see Figure 11 (left). Lemma 3.2 suggests that this scenario happens for $O(n^2)$ quadruples p, q, a, r. (Notice that a is uniquely determined by the choice of p, q and r.) Therefore, we may from here on assume that q does not hit pr during (t_1, t_3).

Let p, q, a, r be any remaining special quadruple (which has not been "eliminated" by the previous chargings.) We claim that no

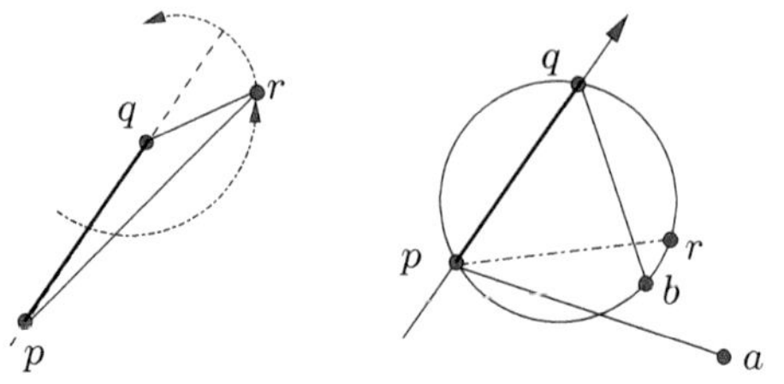

Figure 11: Left: The two Delaunay crossings of the triple p, q, r. Right: The second co-circularity of p, q, r, b, which occurs at time t_b during the crossing of bq. pr is hit by b before re-entering $\mathrm{DT}(P)$.

Delaunay crossing of a q-edge bq by r can end during the time interval (t_1, t_3) and before r crosses ℓ_{pq} again (if at all).[12] This is done by showing that, for any such Delaunay crossing, the corresponding point b would hit pr during (t_1, t_3), thereby contradicting the Delaunayhood of pr in $\mathrm{DT}(P \setminus \{q, a\})$ during (t_1, t_3).

Indeed, notice that the four points p, q, r, b define (at least) one co-circularity during the crossing of pq by r (recall Lemma 3.1), and another such co-circularity during the crossing of bq by r. To be specific, applying a symmetric version of Lemma 3.5, involving crossings of the q-edges qp, qb by r from right to left, and using the fact that the crossing of qb ends after that of qp, we conclude that r hits bq after it hits pq and that, by assumption, r lies in ℓ_{pq}^+ between these two events. Consequently, p lies in ℓ_{bq}^- when r hits qb. Hence, the latter co-circularity of p, q, r, b is red-blue with respect to bq, and it occurs at time t_b when r already lies in ℓ_{bq}^+, and p leaves $B[q, b, r] \cap \ell_{bq}^-$. Right after time t_b the edge pr cannot be present in $\mathrm{DT}(P)$, because its Delaunayhood is violated by q and b; see Figure 11 (right). By Lemma 3.3, the moment t_b occurs after the interval I, and before J, implying that these intervals are disjoint in this case. By Lemma 2.2, pr is crossed by at least one of q, b at some moment between t_b and $t_2 < t_3$, the beginning of J when pa disappears from $\mathrm{DT}(P)$ (note that pr is Delaunay at time t_2). Since the crossing of pr by q is excluded, it must be the case that b hits pr during the time interval (t_1, t_3), contrary to the previous assumption.

Charging along rq. If no Delaunay crossing of a q-edges bq by r occurs after time t_1, (pq, r, I) is the *last* such Delaunay crossing of a q-edge bq by r. Clearly, this happens for at most $O(n^2)$ quadruples p, q, a, r. Otherwise, let t_4 be the ending time of the "next" Delaunay crossing of some q-edge bq by r. (That is, the crossings of pq and bq induce a special "q-quadruple" q, p, b, r.) Notice that, according to Lemma 3.3, the edge rq belongs to $\mathrm{DT}(P)$ at time t_4.

We are now in position to apply Theorem 2.1 for the edge rq over the interval (t_1, t_4), with k_2 in the role of k. If one of the first two conditions of Theorem 2.1 is satisfied, we can charge the quadruple p, q, r, a to $\Omega(k_2^2)$ k_2-shallow co-circularities or to a k_2-shallow collinearity (all with respect to rq). Just as in the previous cases (of charging along pr), the crucial observation is that every charged collinearity or co-circularity involves q and r, and pq is the q-edge whose Delaunay crossing by r ends up latest before the charged event. Hence, the number of special quadruples that can be charged in this manner is at most $O(k_2^2 N(n/k_2) + k_2 n^2 \beta(n))$.

Now assume that condition (iii) of Theorem 2.1 holds. Then there is a subset A of at most $3k_2$ points whose removal restores the Delaunayhood of rq throughout (t_1, t_4). If p hits rq during (t_1, t_4),

[12]Recall that, in accordance with the convention introduced earlier in this section, we orient each such edge bq from b to q, since we are interested only in crossings by r from the left side ℓ_{bq}^- to the right side ℓ_{bq}^+.

it follows that the triple p, q, r induces two Delaunay crossings in the smaller Delaunay triangulation $\mathrm{DT}((P \setminus A) \cup \{p\})$. A combination of Lemma 3.2 with the Clarkson-Shor technique [7] provides an upper bound of $O(k_2 n^2)$ on the number of such triples. Hence, this scenario happens for at most $O(k_2 n^2)$ quadruples. It therefore suffices to handle the quadruples for which p does not hit rq during (t_1, t_4). The preceding analysis then implies that $t_4 > t_3$ (for, otherwise, there would be a crossing of a q-edge bq that ends during (t_1, t_4) and while r is still in ℓ^+_{pq}). Hence, we may assume that r remains in ℓ^+_{pq} during (t_1, t_3).

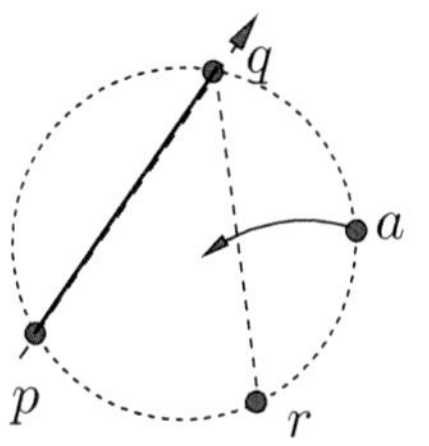

Figure 12: The edge rq is crossed by a after time t_a and before time t_4 (when rq is again Delaunay).

Notice that the four points p, q, r, a define (at least) one co-circularity during the crossing of pq by r (recall Lemma 3.1), and another such co-circularity during the crossing of pa by r. To be specific, applying Lemma 3.5 and using the fact that the crossing of pa ends after that of pq, we conclude that q lies in ℓ^-_{pa} when r hits pa. Hence, the second co-circularity of p, q, r, a occurs at time t_a when r already lies in ℓ^+_{pa}, and q leaves $B[p, a, r] \cap \ell^-_{pa}$; see Figure 10. Right after time t_a the edge rq cannot be present in $\mathrm{DT}(P)$, because its Delaunayhood is violated by p and a (so, according to Lemma 3.3, we have $t_a > t_1$). By Lemma 2.2, rq is crossed by one of p or a during (t_a, t_4). Since, by the previous assumption, p does not hit rq during (t_1, t_4), we are left with the case where a crosses rq during (t_a, t_4). Thus, $\mathrm{DT}((P \setminus A) \cup \{a\})$ contains a Delaunay crossing of rq by a, which takes place during some sub-interval of (t_a, t_4); see Figure 13. This case requires the following special care. Informally, we want to charge the quadruples of this kind using Theorem 2.1 and Lemma 3.2, in much the same way as above. However, to make this charging effective, we need to analyze the scenario in the reverse direction of the time axis.

Specifically, we reverse the direction of the time axis, and note that, in the *new* setting, we still encounter Delaunay crossings of pq and pa by r, but now the crossing of pa begins at time t_3 and ends at time t_2, and the crossing of pq begins at time t_1 and ends at time t_0. In this new setting, the roles of pq and pa are switched, so the Delaunay crossing of pa ends before that of pq (because, in the original time frame, I started before J). In particular, according to Lemma 3.5 no crossings of p-edges end during (t_2, t_0) (in the reverse time direction). Moreover, the additional assumption on the Delaunayhood of pr in $\mathrm{DT}(P \setminus \{q, a\})$ is still satisfied during the interval (t_2, t_0), which is contained in $I \cup (t_3, t_1)$.

We follow the same case analysis as in the proof so far with the same pair of parameters $0 < k_1 < k_2$, but with the differences just noted. For example, if a hits pr during (t_2, t_0) then the triple p, a, r defines two Delaunay crossings in $\mathrm{DT}(P \setminus \{q\})$, which allows to bound the number of such quadruples by $O(n^2)$ (using Lemma 3.2). Similarly, we use the Delaunayhood of pr in $\mathrm{DT}(P \setminus \{q, a\})$ to show, for the remaining quadruples, that no Delaunay crossing of a-edge ba by r can end in the interval (t_2, t_0) while r lies in ℓ^-_{pa}.

Charging along ra. Proceeding in the same spirit, we apply Theorem 2.1 to the edge ra over a suitable interval (t_2, t_5), where t_5 denotes the starting time of the "next" Delaunay crossing of an a-edge (after pa) by r. (If t_5 does not exist then we can charge our quadruple to the pair ra.) Arguing exactly as in the previous cases, we can show that, with the exception of some $O(k_2^2 N(n/k_2) + k_2 n^2 \beta(n))$ special quadruples (each charged to a k_2-shallow collinearity or $\Omega(k_2^2)$ k_2-shallow co-circularities), there is a subset B of at most $3k_2$ points whose removal guarantees the Delaunayhood of ra throughout (t_2, t_5). If p hits ra during (t_2, t_5) then Lemma 3.2 (combined with the standard Clarkson-Shor technique [7]) provides an upper bound of $O(k_2 n^2)$ on the number of such quadruples. Otherwise, r still lies in ℓ^+_{pa} at the moment of the second (previously, first) co-circularity of p, q, a, r, right after which the Delaunayhood of ra is violated by p and q; see Figure 13 (left). Notice that ra is again Delaunay by time t_5. Hence, according to Lemma 2.2, it is crossed by q during (t_2, t_5) (as the crossing by p has been outruled). See Figure 13 (right). Thus, there is a Delaunay crossing of ra by q in $\mathrm{DT}((P \setminus B) \cup \{q\})$, which takes place during (t_2, t_5), and this is the symmetric situation that we still need to handle.

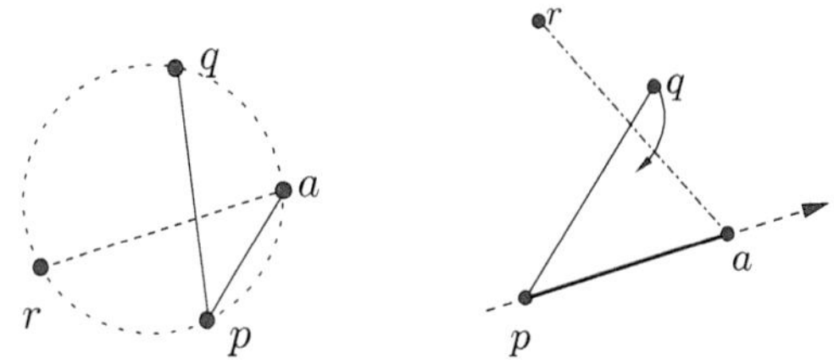

Figure 13: Left: The second (previously, first) co-circularity of p, q, r, a, after which the Delaunayhood of ra is violated (in the reversed direction of the time axis). Right: In the reversed direction of the time axis, ra is crossed by q before returning to $\mathrm{DT}(P)$.

Fortunately, after all these preparations, this final case is easy to handle. Indeed, from the prior analysis, we get that the diagram $\mathrm{DT}([P \setminus (A \cup B)] \cup \{a, q\})$ contains a Delaunay crossing of qr by a, and a Delaunay crossing of ra by q. Combining Lemma 3.2 with the probabilistic argument of Clarkson and Shor [7], as above, we get that the number of such triples q, r, a is at most $O(k_2 n^2)$. We charge each of the remaining quadruples to the triple q, r, a and observe that the choice of q, r, a uniquely determines p (and, therefore, the entire charging quadruple). Indeed, (pq, r, I) is the last Delaunay crossing of a q-edge (of the chosen orientation) by r to end before a hits rq.

Altogether, we have the following upper bound on $Q_1(n)$:

$$Q_1(n) \le c_3(k_2^2 N(n/k_2) + k_2^2 n^2 \beta(n)), \tag{5}$$

where c_3 is a positive constant which does not depend on k_2.

The desired Inequality (2) now follows (for large enough c_1) by combining the Inequalities (3), (4) and (5). □

We now substitute the inequality (2) into (1), and obtain the following recurrence for $N(n)$, in which we choose $12 < k \ll k_1 \ll k_2$:

$$N(n) \le c_4 \left(k^2 N\left(\frac{n}{k}\right) + k^3 k_1^2 N\left(\frac{n}{k_1 k}\right) \right.$$
$$\left. + k^3 k_1^4 k_2^2 N\left(\frac{n}{k_2 k_1 k}\right) + k k_1^2 k_2 n^2 \beta(n) \right),$$

where c_2 is a constant factor which does not depend on the choice of k, k_1, k_2.

Arguing as in earlier solutions of similar charging-based recurrences (see, e.g., [19, Section 7.3.2]), the recurrence solves to $N(n) = O(n^{2+\varepsilon})$, for any $\varepsilon > 0$.

In conclusion, we have the following theorem:

THEOREM 3.6. *Let P be a collection of n points moving along pseudo-algebraic trajectories in the plane, so that any four points of P are co-circular at most twice, and, for any ordered triple p, q, r, the point r can cross the segment pq at most once. Then the Delaunay triangulation $\mathrm{DT}(P)$ of P experiences at most $O(n^{2+\varepsilon})$ discrete changes throughout the motion, for any $\varepsilon > 0$.*

4. REPEATED ORDERED COLLINEARITIES

In the full version [18], we extend the analysis of Section 3 to the case where any three points of P can be collinear at most twice during the motion. In particular, and in contrast to the assumptions in Section 3, a point $r \in P$ can hit an edge $pq \in P \times P$ twice during the motion; moreover, this may occur during a single Delaunay crossing of pq by r.

As far as *single* Delaunay crossings (pq, r, I) are concerned, where r hits pq only once[13] during I, the vast majority of the analysis in Section 3 still holds. The only exception is the case analysis in the proof of Theorem 3.4 (appropriately restricted to such single crossings), which must be modified to handle certain new scenarious (or, more precisely, new variations of the old ones).

In addition, we study *double* Delaunay crossings (pq, r, I), where r hits pq exactly twice during I. This is done by a simplified version of the analysis that was used in the case of single Delaunay crossings. That is, we prove that, on average, any point r is involved in "few" such double crossings of edges that share the same vertex. The excess is charged to shallow co-circularities and collinearities, and (finally) to the special triples of points which can be counted using Theorem 3.2. We proceed by establishing a recursive bound on the number of special quadruples, each formed by a pair of "consecutive" double Delaunay crossings of adjacent edges pq and pa by the same point r. As in Section 3, the crucial step is applying Theorem 2.1 to the edges rq, ra over suitable time intervals. This leads to a simplified variant of the recursion in (2). Altogether, we obtain:

THEOREM 4.1. *Let P be a collection of n points moving along pseudo-algebraic trajectories in the plane, so that any four points of P are co-circular at most twice, and the points of any triple p, q, r can be collinear at most twice. Then the Delaunay triangulation $\mathrm{DT}(P)$ of P experiences at most $O(n^{2+\varepsilon})$ discrete changes throughout the motion, for any $\varepsilon > 0$.*

Discussion. In our analysis we have reduced the problem of bounding the number of Delaunay co-circularities to the more specific problem of bounding the number of Delaunay crossings. The aforementioned reduction easily extends to the case in which any four points of P can be co-circular at most *three times* during the motion. For these reasons, the author believes that the techniques introduced in this paper can be used to establish sub-cubic upper bounds for more general instances of the problem, such as the instance where the points are moving along straight lines with equal speeds.

Acknowledgements. I would like to thank my former Ph.D. advisor Micha Sharir whose help made this work possible. In particular, I would like to thank him for the insightful discussions, and, especially, for his invaluable help in the preparation of this paper.

[13]Recall that, in Section 3, this was assumed to be the only possible situation.

5. REFERENCES

[1] U. A. Acar, B. Hudson and D. Türkoglu, Kinetic mesh refinement in 2D, *Proc. 27th Annu. Sympos. Comput. Geom.*, 2011, 341–350.

[2] P. K. Agarwal, O. Cheong and M. Sharir, The overlay of lower envelopes in 3-space and its applications, *Discrete Comput. Geom.* 15 (1996), 1–13.

[3] P. K. Agarwal, J. Gao, L. Guibas, H. Kaplan, V. Koltun, N. Rubin and M. Sharir, Kinetic stable Delaunay graphs, *Proc. 26th Annu. Symp. on Comput. Geom.* (2010), 127–136.

[4] P. K. Agarwal, Y. Wang and H. Yu, A 2D kinetic triangulation with near-quadratic topological changes, *Discrete Comput. Geom.* 36 (2006), 573–592.

[5] F. Aurenhammer and R. Klein, Voronoi diagrams, in *Handbook of Computational Geometry*, J.-R. Sack and J. Urrutia, Eds., Elsevier, Amsterdam, 2000, pages 201–290.

[6] L. P. Chew, Near-quadratic bounds for the L_1 Voronoi diagram of moving points, *Comput. Geom. Theory Appl.* 7 (1997), 73–80.

[7] K. Clarkson and P. Shor, Applications of random sampling in computational geometry, II, *Discrete Comput. Geom.* 4 (1989), 387–421.

[8] B. Delaunay, Sur la sphère vide. A la memoire de Georges Voronoi, *Izv. Akad. Nauk SSSR, Otdelenie Matematicheskih i Estestvennyh Nauk* 7 (1934), 793–800.

[9] E. D. Demaine, J. S. B. Mitchell, and J. O'Rourke, The Open Problems Project, `http://www.cs.smith.edu/~orourke/TOPP/`.

[10] H. Edelsbrunner, *Geometry and Topology for Mesh Generation*, Cambridge University Press, Cambride, 2001.

[11] S. Fortune, Voronoi diagrams and Delaunay triangulations, In J. E. Goodman and J. O'Rourke, editors, *Handbook of Discrete and Computational Geometry*, CRC Press, Inc., Boca Raton, FL, USA, second edition, 2004, pages 513–528.

[12] L. J. Guibas, J. S. B. Mitchell and T. Roos, Voronoi diagrams of moving points in the plane, *Proc. 17th Internat. Workshop Graph-Theoret. Concepts Comput. Sci.*, volume 570 of *Lecture Notes Comput. Sci.*, pages 113–125. Springer-Verlag, 1992.

[13] D. P. Huttenlocher, K. Kedem and J. M. Kleinberg, Voronoi Diagrams of Rigidly Moving Sets of Points, *Inf. Process. Lett.* 43 (4) (1992), 217–223.

[14] C. Icking, R. Klein, N.-M. Lê and L. Ma, Convex distance functions in 3-space are different, *Fundam. Inform.* 22 (4) (1995), 331–352.

[15] H. Kaplan, N. Rubin and M. Sharir, A kinetic triangulation scheme for moving points in the plane, *Comput. Geom. Theory Appl.* 44 (2011), 191–205.

[16] V. Koltun, Ready, Set, Go! The Voronoi diagram of moving points that start from a line, *Inf. Process. Lett.* 89 (5) (2004), 233–235.

[17] V. Koltun and M. Sharir, 3-dimensional Euclidean Voronoi diagrams of lines with a fixed number of orientations, *SIAM J. Comput.* 32 (3) (2003), 616–642.

[18] N. Rubin, On topological changes in the Delaunay triangulation of moving points, `http://page.mi.fu-berlin.de/rubinnat/pubs/KDel2Full.pdf`.

[19] M. Sharir and P. K. Agarwal, *Davenport-Schinzel Sequences and Their Geometric Applications*, Cambridge University Press, New York, 1995.

Geometric Packing under Non-uniform Constraints*

Alina Ene[†]
Dept. Computer Science
University of Illinois,
Urbana-Champaign
201 N. Goodwin Avenue
Urbana, IL, 61801, USA
ene1@uiuc.edu

Sariel Har-Peled[‡]
Dept. Computer Science
University of Illinois,
Urbana-Champaign
201 N. Goodwin Avenue
Urbana, IL, 61801, USA
sariel@uiuc.edu

Benjamin Raichel[§]
Dept. Computer Science
University of Illinois,
Urbana-Champaign
201 N. Goodwin Avenue
Urbana, IL, 61801, USA
raichel2@uiuc.edu

ABSTRACT

We study the problem of discrete geometric packing. Here, given weighted regions (say in the plane) and points (with capacities), one has to pick a maximum weight subset of the regions such that no point is covered more than its capacity. We provide a general framework and an algorithm for approximating the optimal solution for packing in hypergraphs arising out of such geometric settings.

Using this framework we get a flotilla of results on this problem (and also on its dual, where one wants to pick a maximum weight subset of the points when the regions have capacities). For example, for the case of fat triangles of similar size, we show an $O(1)$-approximation and prove that no PTAS is possible.

See [14] for the full version of the paper.

Categories and Subject Descriptors

F.2.2 [**Analysis of Algorithms and problem complexity**]: Nonnumerical Algorithms and Problems—*Geometrical problems and computations*

General Terms

Algorithms, Theory

Keywords

Rounding scheme, Optimization, Independent set

*The full version of the paper is available from the arXiv [14].

[†]Work on this paper was partially supported by NSF grants CCF-0728782 and CCF-1016684.

[‡]Work on this paper was partially supported by a NSF AF award CCF-0915984.

[§]Work on this paper was partially supported by a NSF AF award CCF-0915984.

1. INTRODUCTION

Motivation and examples.

Consider the problem of ***obnoxious facility location*** [25, 4]; that is, you have to place several facilities, but these facilities are undesired (i.e., obnoxious). Facilities of this type include nuclear reactors, wind farms, airports, power plants, factories, prisons, etc. Facilities can also be semi-desirable – a customer might want to have supermarkets close to their home, but they do not want to have too many of them close by as they increase traffic, noise, etc. One natural way to model this geometrically is to associate each obnoxious facility with its region of undesirability. We also have customers (modeled as points), and each customer has a threshold of how many obnoxious facilities it is willing to accept covering it. Different customers may have different thresholds, for example because more affluent people have stronger political power and it is harder to place obnoxious facilities near their homes.

Naturally, if you allow only a single region to cover each customer, then this is a classical packing problem, and much work has been done on packing disks/balls [24]. However, there are many cases where allowing limited interaction between the packed regions is allowed (after all, these facilities are required for modern existence). As a concrete example of this type of problem, consider the placement of radio stations/cellphone towers. While airports allow only very limited levels of interference[1], higher levels of such interference is acceptable in residential neighborhoods. However, at a certain point there is going to be resistance to placing more wireless towers in residential areas, as these towers are viewed as causing cancer (this fear might be baseless, but it does not change the political reality of the difficulty of placing such towers). On the other hand, there is little resistance to placing such towers along highways in sparsely populated areas.

In this paper, we are interested in the modeling of such problems and in the computation of an efficient approximation to the optimal solution of such problems.

Modeling.

As hinted by the above, perhaps the most natural way to model this problem is as a generalization of the well known independent set problem.

[1]See `http://tinyurl.com/7td67v3` for a story of an airport closing down because of radio interference.

Independent set.

is a fundamental discrete optimization problem. Unfortunately, it is not only computationally hard, but it is even hard to approximate to within a factor of $n^{1-\varepsilon}$, for any constant ε [20] (under the assumption that $\mathsf{NP} \neq \mathsf{P}$). Surprisingly, the problem is considerably easier in some geometric settings. For example, there is a PTAS[2] [6, 15] for the following problem: Given a set of unit disks in the plane, find a maximum cardinality subset of the disks whose interiors are disjoint. Furthermore, a simple local search algorithm yields the desired approximation: For any $\varepsilon > 0$, the local search algorithm that tries to swap subsets of size $O(1/\varepsilon^2)$ yields a $(1-\varepsilon)$-approximation in $n^{O(1/\varepsilon^2)}$ time [7, 8].

The discrete independent set problem.

In this paper, we consider packing problems in geometric settings that are natural extensions of the geometric independent set problem described above. As a starting point, motivated by practical applications, we consider the discrete version of the geometric independent set problem in which, in addition to a set of weighted regions, we are given a set of points, and the goal is to select a maximum weight subset of the regions so that each point is contained in at most one of the selected regions. We refer to this problem as the ***discrete independent set*** problem. Chan and Har-Peled [8] studied this discrete variant and proved that one can get a good approximation if the union complexity of the regions is small.

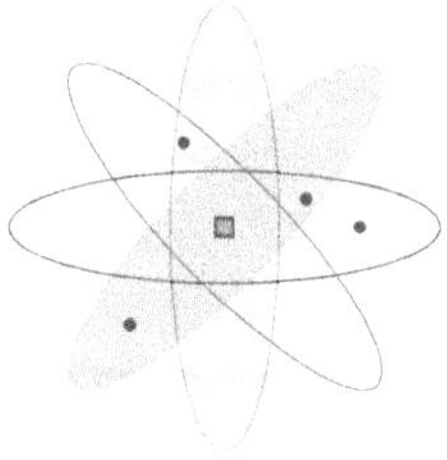

Note that the discrete independent set problem captures the continuous version of the independent set problem, since we can place a point in each face of the induced arrangement of the given regions. In fact, the discrete version is considerably harder (in some cases) than the continuous variant. The difficulty lies in that several regions forming a valid solution to an instance of a discrete independent set problem may contain a common point that is not part of the set of points given as input; the figure on the right shows an example in which the middle point, marked as a square, is covered twice by the given valid solution.

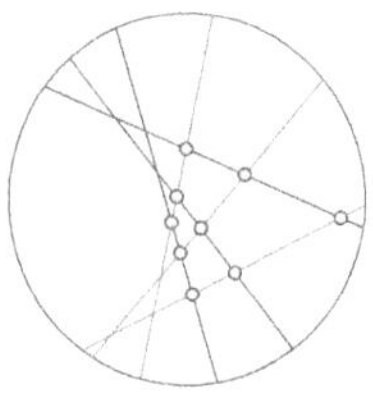

To illustrate the difference in difficulty, consider the case when the input consists of a set S of segments (in general position) with their endpoints on a circle, such that for every pair of segments, its members intersect. Clearly, in the continuous version, the maximum independent set of segments is a single segment. However, in this case, the discrete version captures the graph independent set problem. More precisely, we can encode any instance of independent set (i.e., a graph $G = (V, E)$) as an instance of this problem as follows. Every vertex $v \in V$ is mapped to a segment s_v of S, and every edge $uv \in E$, is mapped to the point $\mathsf{s}_u \cap \mathsf{s}_v$ (which is added to a set of points P). Clearly, an independent set of segments of S (in relation to the point set P) corresponds to an independent set in G. That is, the geometric discrete version is sometimes as hard as the graph independent set problem. For example, the figure on the right depicts the resulting instance encoding independent set for $K_{3,3}$.

[2]Polynomial time approximation scheme.

The packing problem.

In this paper, we are interested in the natural extension of the discrete independent set problem to the case where every point has a capacity and might be covered several times (but not exceeding its capacity). The resulting problem has a flavor of a packing problem, and is defined formally as follows.

Problem 1.1 (PACKREGIONS.) *Given a set $\mathcal{D}$ of regions and a set P of points such that each region r has a weight $w(r)$ and each point p has a capacity $\#(\mathsf{p})$, find a maximum weight subset $X \subseteq \mathcal{D}$ of the regions such that, for each point p, the number of regions in X that contain p is at most its capacity $\#(\mathsf{p})$.*

We emphasize that different points might have different capacities, which makes the problem considerably more challenging to solve than the unit capacities case (i.e., the discrete independent set problem). We also consider the following dual problem in which the points have weights and the regions have capacities.

Problem 1.2 (PACKPOINTS.) *Given a set $\mathcal{D}$ of regions and a set P of points such that each region r has a capacity $\#(r)$ and each point p has a weight $w(\mathsf{p})$, find a maximum weight subset $X \subseteq \mathsf{P}$ of the points such that each region r contains at most $\#(r)$ points of X.*

Hypergraph framework. These two problems can be stated in a unified way in the language of hypergraphs[3]. Given an instance of PACKREGIONS, we construct a hypergraph as follows: Each weighted region is a vertex, and all the regions containing a given point of capacity k become a hyperedge (consisting of these regions) of capacity k. A similar reduction works for PACKPOINTS, where the given weighted points are the vertices, and each region of capacity k becomes a hyperedge of capacity k consisting of all of the points contained in this region. Therefore the previous two problems are special cases of the following problem.

Problem 1.3 (HGRAPHPACKING.) *Given a hypergraph $\mathsf{G} = (\mathsf{V}, \mathsf{E})$ with a weight function $w(\cdot)$ on the vertices and a capacity function $\#(\cdot)$ on the hyperedges, find a maximum weight subset $X \subseteq \mathsf{V}$, such that $\forall f \in \mathsf{E}$ we have $|X \cap f| \leq \#(f)$.*

We will be interested primarily in hypergraphs with certain hereditary properties. A hypergraph property is *hereditary* if the sub-hypergraph induced by any subset of the vertices has the property; an example of a hereditary property of hypergraphs is having bounded VC dimension. Roughly, we are interested in hypergraphs having the ***bounded growth property***: For any induced sub-hypergraph on t vertices the number of its hyperedges that contain exactly k vertices is near linear in t and its dependency on k is bounded by $2^{O(k)}$, see Definition 2.1. Such hypergraphs arise naturally when considering points and "nice" regions in the plane.

[3]A ***hypergraph*** G is a pair (V, E), where V is a set of vertices and E is a collection of subsets of V which are called ***hyperedges***.

Our results.

- Main result. Our main result is an algorithm that provides a good approximation for HGraphPacking as a function of the growth of the hypergraph, see Theorem 3.10. Our result can be viewed as an extension of the work of Chan and Har-Peled [8] to these considerably more general and intricate settings. For simplicity, we focus on linear weight functions; in the full version of the paper [14] we show that our main result extends to the case in which the weight function is a non-negative submodular function.
- Regions with low union complexity. In Section 4, we apply our main result to regions that have low union complexity, and we get the following results:
  (A) If the union complexity of n regions is $O(nu(n))$ then we get an $O\left(u(n)^{1/\nu}\right)$-approximation for PackRegions, where ν is the minimum capacity of any point in the given instance. (That is, the problem becomes easier as the minimum capacity increases.) For the case where all the capacities are one, this is the discrete independent set problem, and our algorithm specializes to the algorithm of Chan and Har-Peled [8], which gives an $O(u(n))$-approximation.
  (B) As a special case, we get a constant factor approximation for PackRegions if the union complexity of the regions is linear. This holds for (i) fat-triangles of similar size, (ii) unit axis-parallel cubes in 3d, and (iii) pseudo-disks. See Corollary 4.3.
  (C) Similarly, since the union complexity of fat triangles in the plane is $O(n \log^* n)$ [16, 2], we get an $O\left((\log^* n)^{1/\nu}\right)$ approximation for such instances of PackRegions.
- Bi-criteria approximation. Our main result also implies a bi-criteria approximation algorithm. That is, we can improve the quality of the solution, at the cost of potentially violating low capacity regions. Formally, if the input instance $\mathsf{G} = (\mathsf{V}, \mathsf{E})$ of HGraphPacking has at most $F_k(t) = 2^{O(k)} F(t)$ edges of size k when restricted to any subset of t vertices, then for any integer $\phi \geq 1$, our algorithm yields an $\left(O\left((F(n)/n)^{1/\phi}\right), \phi\right)$-approximation to the given instance G of HGraphPacking. Specifically, the value of the generated solution X is at least $\Omega\left(\mathrm{opt}/(F(n)/n)^{1/\phi}\right)$, where opt is the value of the optimal solution, and for every hyperedge $f \in \mathsf{E}$, we have $|f \cap X| \leq \max(\phi, \#(f))$.
  As an example, for any set of n regions in the plane such that the boundaries of any pair of them intersect $O(1)$ times, the above implies that one can get an $\left(O\left(n^{1/\phi}\right), \phi\right)$-approximation for PackRegions.
- Axis-parallel boxes. The union complexity of axis-parallel rectangles can be as high as quadratic, and therefore we cannot immediately apply our main result to get a good approximation. Instead, we decompose the union of axis-parallel rectangles into regions of low union complexity, and this decomposition together with our main result gives us an $O(\log n)$ approximation for instances of PackRegions in which the regions are axis-parallel rectangles in the plane (see the full version of the paper [14]).
  A more involved analysis also applies to the three dimensional case, where we get an $O(\log^3 n)$ approximation for PackRegions for axis parallel boxes (see the full version of the paper [14]).
- Dual problem. We show in Section 4.2 that, by standard lifting techniques, we can apply our result for PackRegions, where the regions are disks, to the dual problem of PackPointsInDisks. However, for other regions, the dual problem PackPoints seems to be more challenging. Specifically, this is true for the case of axis-parallel rectangles. For this case, we first provide a constant factor approximation for *skyline* instances of the problem; a skyline instance is a set of rectangles that lie on the x-axis. Interestingly, if the set of rectangles is defined in relation to a set of points (and each rectangle contains only a few points), then one can define a near-linear (in the number of points) sized set of rectangles such that each original rectangle is the union of two new rectangles. Combining this with the skyline result and a sparsifying technique, we get an $(O(\log n), 2)$-approximation; that is, every rectangle b contains at most $\max(2, \#(\mathsf{b}))$ points of the solution constructed, and the total weight of the solution is $\Omega(\mathrm{opt}/\log n)$ (see the full version of the paper [14]). (Note that, by applying our general framework directly to this setting, we only get an $\left(O\left(n^{1/\phi}\right), \phi\right)$-approximation, for any integer $\phi > 0$.)
- Packing points into fat triangles. We provide a polylog bi-criteria approximation for the problem of packing points into fat triangles. This requires proving that one can compute, for a given point set, a small number of canonical subsets such that the point set covered by any fat-triangle (if the set is sufficiently small) is the union of a constant number of these canonical subsets. Proving this requires non-trivial modifications of the result of Aronov *et al.* [3]. In addition, we show that a measure defined over a fat triangle can be covered by a few fat triangles, each one of them containing only a constant fraction of the original measure. We believe these two results are of independent interest. Plugging this into the machinery previously developed for axis parallel rectangles yields the new approximation algorithm. See the full version of the paper [14] for details.
- PTAS for disks and planes. We adapt the techniques of Mustafa and Ray [22] in order to get a PTAS for instances consisting of unweighted disks and unit-capacity points: we lift the problem to 3d, we construct an approximate conflict graph (as done by Mustafa and Ray), and we use a local search algorithm. This result also implies a PTAS for PackPoints for unweighted points and uniform capacity halfspaces in $\mathbb{R}^3$. See the full version of the paper [14] for the details.
- Hardness. We show some hardness results for our problems. In particular, we show that PackPoints for fat triangles in the plane is as hard as independent set in general graphs (see Lemma 5.2). We also show that PackRegions is APX-hard (and thus there is no PTAS) for similarly sized fat triangles in the plane (thus "matching" the result of Corollary 4.3).

Main technical contribution.

In addition to the results mentioned above, our work further develops and extends the techniques for rounding linear programming relaxations for geometric packing problems.

Our algorithms use the randomized rounding with alteration technique to round a fractional solution rising out of a natural LP relaxation; this technique has been used in much more general settings [23]. The rounding uses the fractional solution to construct a random sample of the regions. The sampled regions might not form a feasible solution and therefore we need to pick a feasible subset of the sample. This step typically involves selecting an ordering in which to consider the sampled regions and greedily picking a feasible subset based on the ordering [7, 9, 26]. The main technical difficulty lies in finding the ordering of the regions. The main idea behind previous approaches is to build a conflict graph and argue that there exists a vertex of low degree; this vertex gives us a region that is a good candidate for the last region in the ordering, and we can recursively consider the remaining regions. We use a similar approach to construct an ordering, but the conflicts that are relevant in our settings are more complicated. For example, in the geometric independent set problem, a conflict is a pair of regions that overlap. However, in our setting, a conflict involves a larger number of regions and therefore we need to consider a conflict hypergraph. We show that there exists a vertex of low degree in this hypergraph and therefore we are able to extend some of the previous approaches to this more general setting. Unsurprisingly, this extension involves an analysis that is considerably more involved.

Previous work.

The work that is closest to ours is the paper of Chan and Har-Peled [7] which addresses an easier special case of the problems we consider. Fox and Pach [17] presented an n^{ε} approximation for independent set for segments in the plane. The usage of LP relaxations for approximating such problems is becoming more popular. In particular, Chalermsook and Chuzhoy [5] use a natural LP relaxation to get an $O(\log\log n)$-approximation for independent set of axis parallel rectangles in the plane. The geometric set cover problem and the more general problem, the geometric set multi-cover problem, have approximation algorithms that use ε-nets to round the natural LP relaxation; see [9] and references therein. Chan and Har-Peled [7] used local search to get a PTAS for independent set of pseudo-disks. Independently, Mustafa and Ray [22] used similar ideas to get a PTAS for hitting set of pseudo-disks in the plane. There is not much work on the hardness of optimization problems in the geometric settings we are interested in. [10] shows that the problem of independent set of axis-parallel boxes in three dimensions is APX-hard (the problem is known to be NP-Hard in the plane). See also [18, 19] and references therein for some recent hardness results. Naturally, in non-geometric settings, there is a vast literature on the problems and techniques we use, see [27]. As we already mentioned, our algorithms use the randomized rounding with alteration technique to round a fractional solution. This technique was used in [23] to find an approximate solution to packing integer programs (PIPs) of the form $\{\max wx : Ax \le b, x \in \mathbb{Z}_+^n\}$, where A is a matrix whose entries are either 0 or 1. The approximation guarantee given in [23] is $O(n^{1/B})$, where $B = \min_i b_i$.

Organization.

In Section 2 we define the problem and the associated LP relaxation, and describe some basic tools used throughout the paper. In Section 3 we present the approximation algorithm for the hypergraph case. In Section 4 we present various applications of our main result. In Section 5 we present some hardness results. Due to space limitations, some sections of the paper are contained only in the full version of the paper [14]. In particular, in [14] we extend our main result to the case in which the weight function is a submodular function. We also present algorithms for the case of packing rectangles and packing points into both rectanlges and fat triangles. Also we present a PTAS for some restricted cases. We conclude in Section 6.

2. PRELIMINARIES

For a maximization problem, an algorithm provides an α-***approximation*** if it outputs a solution of value at least opt/α, where opt is the value of the optimal solution. An (α, β)-***approximation*** algorithm for HGRAPHPACKING is an algorithm that returns a (potentially infeasible) solution of value at least opt/α such that each hyperedge f contains at most $\max(\#(f), \beta)$ vertices of the solution.

2.1 LP Relaxation and the Rounding Scheme

We consider the following natural LP relaxation for the HGRAPHPACKING problem. For each vertex v, we have a variable x_v with the interpretation that x_v is 1 if v is selected, and 0 otherwise. For each hyperedge f, we have a constraint that enforces that the number of vertices of f that are selected is at most the capacity of f.

HYPERGRAPH-LP :

$$\max \sum_{v \in \mathsf{V}} w_v x_v$$
$$\sum_{v \in f} x_v \le \#(f) \qquad \forall f \in \mathsf{E}$$
$$0 \le x_v \le 1 \qquad \forall v \in \mathsf{V}.$$

The ***energy*** of a subset $X \subseteq \mathsf{V}$ is $\mathcal{E}(X) = \sum_{v \in X} x_v$. In the following, $\mathcal{E}$ denotes the ***energy*** of the LP solution; that is $\mathcal{E} = \mathcal{E}(\mathsf{V}) = \sum_{v \in \mathsf{V}} x_v$. Note that the energy is at most the number of vertices of the hypergraph. Also, we assume that $\mathcal{E} \ge 1$ (which is always true since all the capacities are at least one).

The minimum capacity of a packing instance is a useful measure of how hard the instance is; formally, the ***minimum capacity*** of a given instance G is

$$\nu = \nu(\mathsf{G}) = \min_{f \in \mathsf{E}} \#(f). \qquad (1)$$

Let $\mathsf{G} = (\mathsf{V}, \mathsf{E})$ be a hypergraph, and let $X \subseteq \mathsf{V}$ be a subset of its vertices. The sub-hypergraph of G ***induced*** by X is $\mathsf{G}_X = \left(X, \left\{ f \cap X \,\middle|\, f \in \mathsf{E} \right\}\right)$.

Definition 2.1 *Let $\mathsf{G} = (\mathsf{V}, \mathsf{E})$ be a hypergraph. For any integer k, let $F_k(\cdot)$ denote the function*

$$F_k(t) = \max_{X \subseteq \mathsf{V}, |X| \le t} \left| \left\{ f \,\middle|\, f \in \mathsf{E} \text{ and } |X \cap f| = k+1 \right\} \right|;$$

that is, $F_k(t)$ is the maximum number of hyperedges of size $k+1$ of a sub-hypergraph of G that is induced by a subset of at most t vertices. We say that G has the **bounded growth** *property if the following conditions are satisfied:*

(A) *There exists a non-decreasing function* $\gamma(\cdot)$ *such that* $F_k(t) \leq 2^{O(k)} t\gamma(t)$ *for any* k *and* t.

(B) *There exists a constant* c *such that* $F_k(xt) \leq cF_k(t)$ *for any* t, k *and* x *such that* $1 \leq x \leq 2$.

This notion of bounded growth is a hereditary property of the hypergraph, and it is somewhat similar to the bounds on the size of set systems with bounded VC dimension. Hypergraphs with bounded growth arise naturally in geometric settings.

Running Example 2.2 *To keep the presentation accessible, we interpret this algorithm for instances of* PACKREGIONS *in which the regions are disks. Specifically, we are given a weighted set of disks* $\mathcal{D}$ *and set of points* P *with capacities. The hypergraph has a vertex for each disk in* $\mathcal{D}$ *and a hyperedge for each point* $\mathsf{p} \in \mathsf{P}$*; the hyperedge* f_{p} *consists of the vertices corresponding to all disks of* $\mathcal{D}$ *that contain* p.

In this case, the mysterious quantity $F_k(t)$ *(see Definition 2.1) is bounded by the number of faces in an arrangement of* t *disks that have depth exactly* $k+1$*. Since the union complexity of* t *disks is linear, a standard application of the Clarkson technique implies that* $F_k(t) = O(kt)$*. Thus in this case we have* $\gamma(t) = O(1)$.

3. APPROXIMATE PACKING FOR HYPERGRAPHS

In this section, we present the algorithm for computing a packing for a given hypergraph $\mathsf{G} = (\mathsf{V}, \mathsf{E})$. We assume that $|\mathsf{E}|$ is polynomial in $|\mathsf{V}|$ and that G has the bounded growth property introduced in Definition 2.1 (properties which both hold for hypergraphs arising out of natural geometric settings). Let x be an optimal solution to the HYPERGRAPH-LP relaxation described in Section 2.1.

3.1 The algorithm

We round the fractional solution to an integral solution using a standard randomized rounding with alteration approach. The first step is to choose an appropriate ordering of the vertices. We will see later how to choose a good ordering; for now, we assume that we are given the ordering. The rounding then proceeds in two phases, the ***selection phase*** and the ***alteration phase***. In the selection phase, we pick a random sample $\mathcal{C}$ of the vertices by selecting each vertex v independently at random with probability x_v/Δ, where Δ is a parameter that we will determine later. In the alteration phase, we pick a subset of $\mathcal{C}$ as follows: We consider the sampled vertices in the order chosen and we add the current vertex to our solution if the resulting solution remains feasible. We say that a vertex is ***selected*** if it is present in the sample, and we say that it is ***accepted*** if it is present in the solution. The main insight is that we can take advantage of the bounded growth property of the hypergraph to show that there is an ordering such that each vertex is accepted with constant probability, provided that it is selected. This will immediately imply that the algorithm achieves an $O(\Delta)$-approximation.

The main challenge is to prove that a good ordering for the alteration phase exists, that is an ordering such that we accept each selected vertex with constant probability. We now proceed to give such a proof. This proof will suggest a natural $O(n^{C+O(1)})$ time brute force algorithm to actually compute this good ordering, where C is the maximum capacity of an edge in the given instance. In Section 3.3 we show how one can improve the dependence on C and make the running time polynomial in the input size.

3.1.1 *Constructing a good ordering*

Before we describe how to construct a good ordering of the vertices, it is useful to understand what will force a vertex to be rejected in the alteration phase. With this goal in mind, consider an ordering of the vertices. Let $\mathcal{C}$ be a sample of the vertices in V such that each vertex v is in $\mathcal{C}$ independently at random with probability x_v/Δ. Let v be a vertex in $\mathcal{C}$. When we consider v in the alteration phase, we will reject v if and only if there exists a hyperedge f of capacity $\#(f)$ such that f contains v and we have already accepted $\#(f)$ vertices of f. The event that we already accepted $\#(f)$ vertices of f is difficult to analyze. However, as we will see, we can settle for a more conservative analysis that upper bounds the probability that v is rejected, given that *all* of the vertices in $\mathcal{C}$ that appear before v in the ordering are accepted. (In the alteration phase, it is possible that not all vertices in $\mathcal{C}$ that appear before v will be accepted, but this can only help us.) Since we are only interested in the event that $\mathcal{C}$ contains $k+1$ vertices — the vertex v and k other vertices that appear before v in the ordering — that are contained in a hyperedge of capacity k, only the set of vertices that appear before v in the sample matter, and not the actual ordering of the vertices. With this observation in mind, we define a k***-conflict*** to be a set of $k+1$ vertices that are contained in a hyperedge of capacity k. In the following, $\mathcal{H}_k$ denotes the set of all k-conflicts, and $\mathcal{H} = \cup_k \mathcal{H}_k$ denotes the set of all conflicts. We are interested in the probability of the event that all of the vertices of a k-conflict, h, are present in the sample, and we refer to this probability as the Δ***-potential*** of the conflict, $\rho_\Delta(h)$. For the analysis it will also be useful to define the unscaled version of this quantity that is the probability that all the vertices of a conflict are present given that we sampled each vertex with probability x_v instead of x_v/Δ. We refer to this quantity as simply the ***potential*** of the conflict, $\rho(h)$. Formally, we have

$$\rho_\Delta(h) = \prod_{v \in h} \frac{x_v}{\Delta} \qquad \text{and} \qquad \rho(h) = \prod_{v \in h} x_v.$$

Another quantity of interest is the expected number of conflicts in which a vertex v participates, given that v is in the sample. We refer to this quantity as the Δ***-resistance*** of a vertex v in a set of vertices $X \subseteq \mathsf{V}$, and we use $\eta_\Delta(v, X)$ to denote it:

$$\eta_\Delta(v, X) = \frac{\Delta}{x_v} \sum_{h \in \mathcal{H}, h \subseteq X, v \in h} \rho_\Delta(h).$$

The ordering.

Note that, if the Δ-resistance of v with respect to the set X of vertices that come before it in the ordering is small, the probability of rejecting v is also small. This suggests that the vertex with least resistance (with respect to V) should be the last vertex in the ordering. This gives us the following algorithm for constructing an ordering: We compute the vertex of least resistance and put it last in our ordering (i.e., it is v_n). We then recursively consider the remaining vertices and we compute an ordering for them.

In the following, we assume for simplicity that the resulting ordering is $v_1, \ldots, v_n$.

Note that computing the resistance of a vertex by brute force takes $O\left(n^{C+O(1)}\right)$ time, where C is the maximum capacity of a hyperedge, and therefore this algorithm is not efficient. We give a polynomial time algorithm for constructing the ordering in Section 3.3.

3.2 Analysis

Our main insight is that, if the hypergraph satisfies the bounded growth property defined in Definition 2.1, then for any set $X \subseteq \mathsf{V}$ there exists a vertex $v \in X$ such that $\eta_\Delta(v, X) \leq 1/4$. We prove this below in Section 3.2.2 (see Lemma 3.8). This proof requires that we set $\Delta = \alpha\gamma(\mathcal{E})^{1/\nu}$, where α is some sufficiently large constant. As such, in the remainder of this section we assume $\Delta = \alpha\gamma(\mathcal{E})^{1/\nu}$.

We now show that given $\eta_\Delta(v, X) \leq 1/4$, proving the quality of approximation of the algorithm is straightforward.

Lemma 3.1 *Let $\mathcal{C}$ and $\mathcal{O}$ be the set of vertices that were selected and accepted by the algorithm, respectively. For each i, we have* $\mathbf{Pr}\left[v_i \in \mathcal{O} \,\middle|\, v_i \in \mathcal{C}\right] \geq 3/4$.

Proof: Let $X_i = \langle v_1, \ldots, v_i \rangle$. Note that, if we selected v_i, we rejected v_i in the alteration phase only if v_i participates in a conflict with some of the vertices in $\{v_1, \ldots, v_{i-1}\} \cap \mathcal{C}$. Let Z_i be the number of conflicts of X_i that contain v_i and are realized in $\mathcal{C}$, i.e., $h \subseteq \mathcal{C}$. In the following, we show that the probability that Z_i is non-zero is at most $1/4$, which implies the lemma.

Consider a k-conflict $h = \{v_{j_1}, \ldots, v_{j_k}, v_i\}$, where each vertex of h is in X_i and h contains v_i. The probability that all of the vertices of h are selected, given that v_i is selected, is equal to $\frac{x_{j_1}}{\Delta} \cdot \frac{x_{j_2}}{\Delta} \cdots \frac{x_{j_k}}{\Delta} = \frac{\Delta}{x_i}\rho_\Delta(h)$. Therefore we have

$$\mathbf{E}\left[Z_i \,\middle|\, v_i \in \mathcal{C}\right] = \sum_{h \in \mathcal{H}, h \subseteq X_i, v_i \in h} \frac{\Delta}{x_i}\rho_\Delta(h) = \eta_\Delta(v_i, X_i) \leq \frac{1}{4},$$

where the last inequality follows from Lemma 3.8 and the fact that v_i is the vertex of minimum resistance in X_i. Thus

$$\mathbf{Pr}\left[v_i \notin \mathcal{O} \,\middle|\, v_i \in \mathcal{C}\right] \leq \mathbf{Pr}\left[Z_i > 0 \,\middle|\, v_i \in \mathcal{C}\right] \leq \mathbf{E}\left[Z_i \,\middle|\, v_i \in \mathcal{C}\right] \leq \frac{1}{4}.$$

Therefore, if v_i is selected, it is accepted with probability at least $3/4$. ■

Corollary 3.2 *The total expected weight of the set of vertices output by the algorithm is $\Omega\left(\mathrm{opt}/\gamma(\mathcal{E})^{1/\nu}\right)$, where* opt *is the weight of the optimal solution, and ν is the minimum capacity of the given instance.*

Proof: By Lemma 3.1, for each vertex $v \in \mathsf{V}$, we have

$$\begin{aligned}\mathbf{Pr}\left[v \in \mathcal{O}\right] &= \mathbf{Pr}\left[(v \in \mathcal{O}) \cap (v \in \mathcal{C})\right] \\ &= \mathbf{Pr}\left[v \in \mathcal{O} \,\middle|\, v \in \mathcal{C}\right] \mathbf{Pr}\left[v \in \mathcal{C}\right] \\ &\geq \frac{3}{4}\mathbf{Pr}\left[v \in \mathcal{C}\right] \geq \frac{3x_v}{4\Delta},\end{aligned}$$

where $\Delta = O\left(\gamma(\mathcal{E})^{1/\nu}\right)$. By linearity of expectation, we have that the expected weight of the generated solution is at least

$$\sum_{v \in \mathsf{V}} \frac{3x_v}{4\Delta} w_v = \Omega\left(\frac{\sum_v x_v w_v}{\gamma(\mathcal{E})^{1/\nu}}\right) = \Omega\left(\frac{\mathrm{opt}}{\gamma(\mathcal{E})^{1/\nu}}\right),$$

as $\sum_v x_v w_v$ is the value of the fractional LP solution, which is bigger than (or equal to) the weight of the optimal solution. ■

3.2.1 *On the expected number of realized conflicts*

To analyze the algorithm we need to understand how conflicts might form during its execution, and show that the damage of such conflicts to the generated solution is limited. To this end, for a subset $X \subseteq \mathsf{V}$, consider the quantity

$$F_k(t) = \max_{A \subseteq X, |A| \leq t} \left|\left\{ f \,\middle|\, f \in \mathsf{E} \text{ and } |A \cap f| = k+1 \right\}\right|.$$

(The reader may recall that this is the function from Definition 2.1 of bounded growth, which is a hereditary property.) This is the maximum number of k-conflicts that can be realized by a subset of t vertices from X. The quantity of interest in the following is $\sum_{h \in \mathcal{H}_k} \rho(h)$, as it is the expected number of conflicts that would be realized if we sampled according to the LP solution. Our goal is to prove that $\sum_{h \in \mathcal{H}_k} \rho(h)$ is bounded by a function of $F_k(\mathcal{E}(X))$ (i.e. a function of the maximum number of k-conflicts that can be realized by a subset of X whose size is the energy of the LP).

With this goal in mind, we let R be a random sample of X such that each vertex $v \in X$ is in R independently at random with probability $x_v/2$. We stress that R is a random sample that we use for the purposes of defining a quantity $\mathcal{M}$ (i.e., the expected number of conflicts realized in R), and it should not be confused with the random sample $\mathcal{C}$ that is used by the algorithm. In the following, we bound $\mathcal{M}$ from above in Lemma 3.3 and from below in Lemma 3.4. Putting these two bounds together imply the desired bound on $\sum_{h \in \mathcal{H}_k} \rho(h)$.

A conflict $h \in \mathcal{H}$ is ***realized*** in R if there is a hyperedge $f \in \mathsf{E}$ such that $h = f \cap \mathsf{R}$ and $|h| = \#(f) + 1$.

The following is similar in spirit to the Clarkson technique (a similar but simpler argument was used by Chan and Har-Peled [8]).

Lemma 3.3 *The expected number of k-conflicts realized in R is $\mathcal{M} = O(F_k(\mathcal{E}(X)))$, where R is a random sample of X such that each vertex $v \in X$ is in R independently at random with probability $x_v/2$.*

Proof: Each k-conflict h that is realized corresponds to a hyperedge f with capacity k such that $h = f \cap \mathsf{R}$. Additionally, two realized conflicts that are distinct correspond to different hyperedges. Therefore the number of k-conflicts that are realized in R is at most the number of hyperedges f such that the capacity of f is k and $|f \cap \mathsf{R}| = k+1$. It follows from the definition of $F_k(\cdot)$ that the number of k-conflicts is at most $F_k(|\mathsf{R}|)$. Therefore it suffices to upper bound the expected value of $F_k(|\mathsf{R}|)$.

Note that $\mathbf{E}[|\mathsf{R}|] = \mathcal{E}(X)/2$. We have

$$\begin{aligned}\mathbf{E}\Big[F_k(|\mathsf{R}|)\Big] &\leq \sum_{t=0}^{\infty} \mathbf{Pr}\Big[|\mathsf{R}| \geq t\frac{\mathcal{E}(X)}{2}\Big] F_k\Big((t+1)\frac{\mathcal{E}(X)}{2}\Big)\\ &\leq \sum_{t=0}^{\infty} 2^{-(t+1)/2} F_k\Big((t+1)\frac{\mathcal{E}(X)}{2}\Big)\\ &\leq \sum_{t=0}^{\infty} 2^{-(t+1)/2} c^{O(\log t)} F_k\Big(\mathcal{E}(X)\Big)\\ &= O\Big(F_k(\mathcal{E}(X))\Big),\end{aligned}$$

since G has the bounded growth property (see Definition 2.1), and by the Chernoff inequality (we use here implicitly that $\mathcal{E}(X) \geq 1$). ■

Lemma 3.4 *For each k-conflict h, the probability that h is realized in R is at least $\rho(h)/2(2e)^k$. Therefore the expected number of k-conflicts realized in R is $\mathcal{M} = \Omega\left(\left(\sum_{h\in\mathcal{H}_k, h\subseteq X}\rho(h)\right)/(2e)^k\right)$.*

Proof: Let $f \in \mathsf{E}$ be a hyperedge with capacity k that generated the conflict h. Since x is a feasible solution for the LP, we have that $\sum_{v\in f-h} x_v \leq \sum_{v\in f} x_v \leq \#(f) = k$. Clearly, the conflict h is realized if we pick all the vertices of h, and none of the vertices of $f-h$, and the probability of that event is

$$\begin{aligned}\prod_{v\in h}\frac{x_v}{2}\prod_{v\in f-h}\Big(1-\frac{x_v}{2}\Big) &\geq \frac{1}{2^{k+1}}\prod_{v\in h}x_v\prod_{v\in f-h}\exp(-x_v)\\ &= \frac{\rho(h)}{2^{k+1}}\cdot\exp\left(-\sum_{v\in f-h}x_v\right) \geq \frac{\rho(h)}{2(2e)^k},\end{aligned}$$

In the first line we used the inequality $1 - x_v/2 \geq \exp(-x_v)$, which holds since $x_v \leq 1$. ■

Putting the above two lemmas together, we get the following.

Lemma 3.5 *For any non-negative integer k we have* $\sum_{h\in\mathcal{H}_k, h\subseteq X}\rho(h) = O\Big((2e)^k F_k(\mathcal{E}(X))\Big)$.

Running Example 3.6 *In our running example, we have that the expected number of k-conflicts that are being realized by a random sample (sampling more or less according to the LP values) is $\sum_{h\in\mathcal{H}_k}\rho(h) = O\Big((2e)^k k\mathcal{E}\Big)$. This is a hefty quantity, but the key observation is that if we sample according to the LP values scaled down by a large enough constant, then the probability of such a conflict to be realized drops exponentially with k. In particular, for a sufficiently large constant, the expected number of realized k-conflicts in such a sample is going to be $\leq \mathcal{E}/\left(10\cdot 2^k\right)$. Intuitively, this implies that such conflicts can only cause the algorithm to drop very few vertices during the rounding stage, thus guaranteeing a good solution.*

3.2.2 Resistance is futile, if you pick the right vertex

In the following, we consider a subset X of the vertices and we show that there exists a vertex $v \in X$ whose Δ-resistance $\eta_\Delta(v, X)$ is at most $1/4$. Recall that $\mathcal{H}_k$ is the set of all k-conflicts involving vertices in V. We can rewrite the Δ-resistance of v in X as

$$\begin{aligned}\eta_\Delta(v,X) &= \frac{\Delta}{x_v}\sum_{h\in\mathcal{H}, h\subseteq X, v\in h}\rho_\Delta(h)\\ &= \frac{1}{x_v}\sum_{k=\nu}^{\infty}\frac{1}{\Delta^k}\sum_{h\in\mathcal{H}_k, h\subseteq X, v\in h}\rho(h).\end{aligned}$$

As shown in Lemma 3.5, we can relate the total potential of the conflicts of $\mathcal{H}_k$ that are contained in X to the maximum number of k-conflicts contained in a set of at most $\mathcal{E}(X)$ vertices, where $\mathcal{E}(X) = \sum_{v\in X} x_v$.

Recall that the hypergraph has the bounded growth property (see Definition 2.1) and this property is hereditary. Therefore the function $F_k(\cdot)$ in the lemma above has the two properties described in Definition 2.1 and we get the following corollary.

Corollary 3.7 *We have $\sum_{h\in\mathcal{H}_k, h\subseteq X}\rho(h) = O(2^{ck}\mathcal{E}(X)\,\gamma(\mathcal{E}(X))$, where c is a constant.*

We can now use Corollary 3.7 to complete the proof of Lemma 3.8 as follows.

Lemma 3.8 *Suppose that the hypergraph G satisfies the bounded growth property (see Definition 2.1). Let $\Delta = \alpha\,\gamma(\mathcal{E})^{1/\nu}$, where $\alpha > 0$ is a sufficiently large constant and ν is the minimum capacity of the given instance (see Eq. (1)). Then, for any set $X \subseteq \mathsf{V}$, there exists a vertex $v \in X$ such that $\eta_\Delta(v, X) \leq 1/4$.*

Proof: Let $T = \sum_{v\in X} x_v\eta_\Delta(v, X)$. The quantity $T/\mathcal{E}(X)$ is the weighted average of the resistances of the vertices in X, where the weight of a vertex v is $x_v/\mathcal{E}(X)$. Therefore it suffices to show that $T \leq \mathcal{E}(X)/4$, since the minimum resistance is at most the weighted average. We have

$$\begin{aligned}T &= \sum_{k=\nu}^{\infty}\frac{1}{\Delta^k}\sum_{v\in X}\sum_{\substack{h\in\mathcal{H}_k\\ h\subseteq X\\ v\in h}}\rho(h) = \sum_{k=\nu}^{\infty}\frac{k+1}{\Delta^k}\sum_{\substack{h\in\mathcal{H}_k\\ h\subseteq X}}\rho(h)\\ &= \sum_{k=\nu}^{\infty}\frac{k+1}{\Delta^k}O\Big(2^{ck}\mathcal{E}(X)\gamma(\mathcal{E}(X))\Big)\\ &\leq \mathcal{E}(X)\cdot\underbrace{\beta\sum_{k=\nu}^{\infty}\left(\frac{2^c}{\Delta}\right)^k(k+1)\gamma(\mathcal{E}(X))}_{=S},\end{aligned}$$

by Corollary 3.7, where β is some constant. Since $\Delta = \alpha\gamma(\mathcal{E})^{1/\nu}$, we have

$$\begin{aligned}S &= \sum_{k=\nu}^{\infty}\beta\left(\frac{2^c}{\alpha}\right)^k(k+1)\left(\frac{1}{\gamma(\mathcal{E})}\right)^{k/\nu}\gamma(\mathcal{E}(X))\\ &\leq \sum_{k=\nu}^{\infty}\beta\left(\frac{2^c}{\alpha}\right)^k(k+1)\left(\frac{1}{\gamma(\mathcal{E})}\right)^{k/\nu}\gamma(\mathcal{E}) \leq \frac{1}{4},\end{aligned}$$

In the second to last inequality, we have used the fact that $\gamma(\cdot)$ is non-decreasing. The last inequality follows if we pick α to be a sufficiently large constant. Therefore $T \leq \mathcal{E}(X)/4$, and the lemma follows. ■

3.3 Improving the running time

In Section 3, we described an algorithm that constructs an ordering of the vertices by repeatedly finding the vertex of least resistance with respect to the set of remaining vertices. Computing the resistance of a vertex by brute force takes $O(n^{C+O(1)})$ time, where C is the maximum capacity of an edge in E. However, for our analysis to go through, we only need to find a vertex that is safe with respect to the set of remaining vertices; informally, a vertex v is safe if the probability that it participates in a conflict with a random sample of the remaining vertices is smaller than some constant (that is strictly smaller than one), where each remaining vertex u is included in the sample with probability x_u/Δ. In this section we show that there is a sampling algorithm that finds a safe vertex with high probability and its running time is polynomial in the maximum capacity C.

Lemma 3.9 *Computing a good ordering of the vertices can be done in polynomial time. Namely, the algorithm of Section 3 can be implemented in polynomial time.*

Proof: To get the same quality of approximation we do not need to take the vertex of least resistance in each round (of computing the ordering), but merely a vertex that is "safe." More precisely, let X be the current set of vertices, let v be a vertex of this set, and let R be a random sample of X in which each vertex u is included with probability x_u/Δ (also we force v to be in R). We say that v is ***violated*** in R if v is contained in a hyperedge f such that the number of vertices of f that are in R is larger than its capacity $\#(f)$. Let $\mu(v, X)$ denote the probability that v is violated in R. Note that $\mu(v, X)$ is a (conservative) upper bound on the probability that v is rejected by our rounding algorithm if we started with an ordering in which $X \setminus \{v\}$ is the set of all vertices that come before v. Therefore, in order for our rounding to succeed, in each round we only need to find a vertex v for which the probability $\mu(v, X)$ is low, where X is the set of all vertices that still need to be ordered at the beginning of the round. (We remark that it follows from the argument of Lemma 3.8 that, for any set X, there is a vertex v for which $\mu(v, X) \leq 1/4$.)

Now we are ready to describe how to construct an ordering for our algorithm. Let X be the set of vertices that still need to be ordered. As we will see shortly, for each vertex $v \in X$, we can compute an estimate $\overline{\mu}(v, X)$ of the probability $\mu(v, X)$. We pick the vertex v with minimum estimated probability $\overline{\mu}(v, X)$, we make v the last vertex (in the ordering of X) and we recursively order $X \setminus \{v\}$.

We can compute the estimates $\overline{\mu}(v, X)$ in polynomial time as follows. Fix a vertex v. Let ψ be a sufficiently large polynomial in n. We pick ψ independent random samples of X (again, forcing v to be in each of these samples); in each random sample, each vertex u is included with probability x_u/Δ. We set $\overline{\mu}(v, X)$ to be the fraction of the samples in which the vertex v is violated. Using a standard argument based on the Chernoff inequality, we can show that our estimates are very close with high probability, and therefore our rounding algorithm achieves the required approximation with high probability as well; we omit the easy but tedious details. ■

3.4 The result

Theorem 3.10 *Let $\mathsf{G} = (\mathsf{V}, \mathsf{E})$ be a hypergraph with a weight function $w(\cdot)$ on the vertices and a capacity function $\#(\cdot)$ on the edges, such that $|\mathsf{E}|$ is polynomial in $|\mathsf{V}|$ and G satisfies the bounded growth property (see Definition 2.1). Then we can compute in polynomial time a subset $X \subseteq \mathsf{V}$ of vertices such that no hyperedge f contains more than its capacity $\#(f)$ vertices of X, and such that in expectation, the total weight of the output set is $\Omega\left(\mathrm{opt}/\gamma(\mathcal{E})^{1/\nu}\right)$, where opt is the weight of the optimal solution, $\mathcal{E}$ is the energy of the LP, ν is the minimum capacity of the given instance, and $\gamma(\cdot)$ is the growth function of Definition 2.1.*

Consider an integer constant $\phi > 0$, and observe that one can always relax the capacity constraints of a given instance of HGraphPacking by replacing all capacities smaller than ϕ by ϕ. Theorem 3.10 thus implies the following.

Corollary 3.11 *Given an instance of HGraphPacking, with the bounded growth property, one can compute in polynomial time a $\left(O\left(\gamma(\mathcal{E})^{1/\phi}\right), \phi\right)$-approximation to the optimal solution.*

4. APPLICATIONS

Using our main result (Theorem 3.10), we get several approximation algorithms for the packing problems mentioned in the introduction. We present some of these results here.

4.1 Packing regions with low union complexity

Let $\mathcal{D}$ be a set of n weighted regions in the plane, and let the maximum union complexity of $m \leq n$ objects of $\mathcal{D}$ be $\mathsf{U}(m) = m\mathsf{u}(m)$. We assume that (i) $\mathsf{U}(n)/n = \mathsf{u}(n)$ is a non-decreasing function, and (ii) there exists a constant c, such that $\mathsf{U}(xr) \leq c\,\mathsf{U}(r)$, for any r and $1 \leq x \leq 2$. We are also given a set of points P, where each point $\mathsf{p} \in \mathsf{P}$ is assigned a positive integer $\#(\mathsf{p})$ which is the capacity of p.

We are interested in solving PackRegions (Problem 1.1) for $\mathcal{D}$ and P. Consider the hypergraph G obtained by creating a vertex for each region and a hyperedge for each subset of regions containing a given point of P. Here, $F_k(t)$ is bounded by the number of faces in the arrangement of t regions of depth exactly $k+1$. The number of such faces can be bounded by the union complexity by a standard application of the Clarkson technique [11, 12].

Lemma 4.1 *Consider a set of regions in the plane such that the boundary of every pair intersects a constant number of times. The number of faces of depth at most $k+1$ in the arrangement of any subset of these regions of size t is $O\left(k^2\mathsf{U}(t/k)\right)$.*

Plugging this bound into Theorem 3.10 yields the following result.

Theorem 4.2 *Let $\mathcal{D}$ be a set of m weighted regions in the plane such that the union complexity of any t of them is $\mathsf{U}(t) = t\mathsf{u}(t)$. Let P be a set of n points in the plane, where there is a capacity $\#(\mathsf{p})$ associated with each point $\mathsf{p} \in \mathsf{P}$. There is a polynomial time algorithm that computes a subset $\mathcal{O} \subseteq \mathcal{D}$ of regions such that no point $\mathsf{p} \in \mathsf{P}$ is contained in more than $\#(\mathsf{p})$ regions of $\mathcal{O}$. Furthermore, in expectation,* *the total weight of the output set is* $\Omega\Big(\mathrm{opt}/\mathsf{u}(\mathcal{E})^{1/\nu}\Big)$, *where* opt *is the weight of the optimal solution,* $\mathcal{E}$ *is the energy of the LP solution and* ν *is the minimum capacity of the given instance.*

Alternatively, for any integer constant ϕ*, one can get a* $\Big(O\left(\mathsf{u}(\mathcal{E})^{1/\phi}\right), \phi\Big)$*-approximation to the optimal solution for the given instance.*

The following results follow from the theorem above.

Corollary 4.3 *(A) The union complexity of pseudo-disks and fat triangles of similar size is linear; that is,* $\mathsf{U}(t) = O(t)$*. Therefore we get an* $O(1)$*-approximation for* PACKREGIONS *if the regions are fat triangles of similar size, disks, or pseudo-disks.*

(B) The union complexity of fat triangles is $O(n \log^* n)$ *[1]. This results in an* $O(\log^* n)$*-approximation for* PACKREGIONS *if the regions are (arbitrary) fat triangles.*

(C) Consider a set of regions in the plane such that any pair of them intersects a constant number of times (e.g., a set of arbitrary triangles). In this case, $\mathsf{U}(t) = O(t^2)$ *and* $\mathsf{u}(t) = O(t)$*. Therefore, for any integer constant* $\phi > 0$*, we get an* $\Big(O\Big(\mathcal{E}^{1/\phi}\Big), \phi\Big)$*-approximation for instances of* PACKREGIONS *on such regions.*

4.2 Packing halfspaces, rays and disks

Problem 4.4 *(A)* PACKHALFSPACES*: Given a weighted set of halfspaces* $\mathcal{S}$ *and a set of points* P *with capacities in* $\mathbb{R}^3$*, find a maximum weight subset* $\mathcal{O}$ *of* $\mathcal{S}$ *so that, for each point* p*, the number of halfspaces of* $\mathcal{O}$ *that contains* p *is at most* $\#(\mathsf{p})$*.*

(B) PACKRAYSINPLANES*: Given a weighted set of vertical rays* $\mathcal{R}$ *and a set of planes* $\mathcal{H}$ *with capacities in* $\mathbb{R}^3$*, find a maximum weight subset* $\mathcal{O}$ *of* $\mathcal{R}$ *so that, for each plane* h*, the number of rays of* $\mathcal{O}$ *that intersect* h *is at most* $\#(\mathsf{h})$*.*

(C) PACKPOINTSINDISKS*: Given a set* $\mathcal{D}$ *of disks with capacities and a weighted set* P *of points, find a maximum weight subset* $\mathcal{O}$ *of the points so that each disk* $r \in \mathcal{D}$ *contains at most* $\#(r)$ *points of* $\mathcal{O}$*.*

Since the union complexity of halfspaces in three dimensions is linear, Theorem 3.10 (along with the 3d analogue of Lemma 4.1) implies the following.

Corollary 4.5 *One can compute, in polynomial time, a constant factor approximation to the optimal solution of the* PACKHALFSPACES *problem.*

Standard point/plane duality implies that the same result holds for the dual problem. Namely, a point (a, b, c) gets mapped to the plane $z = ax + by - c$ and a plane $z = ax + by + c$ gets mapped to the point $(a, b, -c)$. Also, a point lies below a given plane if and only if the dual point of the plane lies below the dual plane of the point. As such, the dual of an instance of PACKHALFSPACES is an instance of PACKRAYSINPLANES (and vice versa).

Thus, Corollary 4.5 implies the following.

Corollary 4.6 *One can compute, in polynomial time, a constant factor approximation to the optimal solution of the* PACKRAYSINPLANES *problem.*

Observe that an instance of PACKPOINTSINDISKS can be lifted into an instance of PACKRAYSINPLANES, by the standard lifting $f(x, y) = (x, y, x^2 + y^2)$, which maps points and disks in the plane to halfspaces and points in three dimensions [13].

Corollary 4.7 *One can compute, in polynomial time, a constant factor approximation to the optimal solution to the* PACKPOINTSINDISKS *problem.*

5. HARDNESS OF APPROXIMATION

5.1 Packing same size fat triangles into points

Here we show that PACKREGIONS (Problem 1.2) does not have a PTAS, even if the regions have unit weight and their union complexity is linear. We show that the problem is APX-hard using a reduction from the maximum bounded 3-dimensional matching problem. Since maximum bounded 3-dimensional matching is APX-complete [21], this will imply the claim (unless P = NP).

Theorem 5.1 *Unless P = NP there is no PTAS for* PACKREGIONS *(Problem 1.2) even if the regions are unweighted, in the plane, and have linear union complexity. In particular, this holds if the regions are fat triangles of similar size. (See Corollary 4.3 (A) for the matching approximation algorithm.)*

See the full version of the paper [14] for the proof.

5.2 Packing points into fat triangles

Lemma 5.2 *There is an approximation-preserving reduction from the* INDEPENDENT SET *problem in general graphs to the* PACKPOINTS *problem. In particular, for instances of the problem* PACKPOINTS *in which the regions are fat triangles with unit capacities and the points are unweighted, no approximation better than* $\Omega(n^{1-\varepsilon})$ *is possible in polynomial time, for any constant* $\varepsilon > 0$*, unless P = NP.*

Proof: Consider an instance of the INDEPENDENT SET problem, namely a graph $G = (V, E)$. Let $n = |V|$. Place n distinct points on the unit circle (arbitrarily) and map every vertex of V to a unique point of the resulting set of points P. For every edge $uv \in E$, consider the segment $\mathsf{p}_u\mathsf{p}_v$, where p_u and p_v are the points corresponding to u and v in P. We construct a fat triangle containing $\mathsf{p}_u\mathsf{p}_v$ by connecting p_u, p_v, and a third vertex in the interior of the unit disk; this can always be done so as to achieve roughly 2-fatness. We add this triangle to our set of regions $\mathcal{D}$, and assign it capacity one.

Clearly, solving the resulting instance $(\mathsf{P}, \mathcal{D})$ of PACKPOINTS is equivalent to solving the INDEPENDENT SET problem for G. The claim now follows from the hardness results known for the INDEPENDENT SET problem [20]. ■

6. CONCLUSIONS

In this paper, we presented a general framework for approximating geometric packing problems with non-uniform constraints. We then applied this framework in a systematic fashion to get improved algorithms for specific instances of this problem, many of which required additional non-trivial ideas. There are several special cases of this problem for which we currently do not know any useful approximation; for example, the special case of packing axis-parallel boxes into points, in which the boxes are in four dimensions is still wide open. Making some progress on these special cases is an interesting direction for future work.

Acknowledgments

The authors thank Timothy Chan, Chandra Chekuri, and Esther Ezra for several useful discussions.

7. REFERENCES

[1] B. Aronov, M. de Berg, E. Ezra, and M. Sharir. Improved bound for the union complexity of locally fat objects in the plane. manuscript, 2011.

[2] B. Aronov, M. de Berg, E. Ezra, and M. Sharir. Improved bound for the union of fat objects in the plane. manuscript, 2011.

[3] B. Aronov, E. Ezra, and M. Sharir. Small-size ε-nets for axis-parallel rectangles and boxes. *SIAM J. Comput.*, 39(7):3248–3282, 2010.

[4] P. Cappanera. A survey on obnoxious facility location problems. Technical report, University of Pisa, 1999.

[5] P. Chalermsook and J. Chuzhoy. Maximum independent set of rectangles. In *Proc. 20th ACM-SIAM Sympos. Discrete Algorithms*, pages 892–901, 2009.

[6] T. M. Chan. Polynomial-time approximation schemes for packing and piercing fat objects. *J. Algorithms*, 46(2):178–189, 2003.

[7] T. M. Chan and S. Har-Peled. Approximation algorithms for maximum independent set of pseudo-disks. In *Proc. 25th Annu. ACM Sympos. Comput. Geom.*, pages 333–340, 2009. cs.uiuc.edu/~sariel/papers/08/w_indep.

[8] T. M. Chan and S. Har-Peled. Approximation algorithms for maximum independent set of pseudo-disks. *CoRR*, abs/1103.1431, 2011.

[9] C. Chekuri, K. L. Clarkson., and S. Har-Peled. On the set multi-cover problem in geometric settings. In *Proc. 25th Annu. ACM Sympos. Comput. Geom.*, pages 341–350, 2009.

[10] M. Chlebík and J. Chlebíková. The complexity of combinatorial optimization problems on d-dimensional boxes. *SIAM J. Discrete Math.*, 21(1):158–169, 2007.

[11] K. L. Clarkson. Applications of random sampling in computational geometry, II. In *Proc. 4th Annu. ACM Sympos. Comput. Geom.*, pages 1–11, 1988.

[12] K. L. Clarkson and P. W. Shor. Applications of random sampling in computational geometry, II. *Discrete Comput. Geom.*, 4:387–421, 1989.

[13] M. de Berg, O. Cheong, M. van Kreveld, and M. H. Overmars. *Computational Geometry: Algorithms and Applications*. Springer-Verlag, 3rd edition, 2008.

[14] A. Ene, S. Har-Peled, and B. Raichel. Geometric packing under non-uniform constraints. *CoRR*, abs/1107.2949, 2011.

[15] T. Erlebach, K. Jansen, and E. Seidel. Polynomial-time approximation schemes for geometric intersection graphs. *SIAM J. Comput.*, 34(6):1302–1323, 2005.

[16] E. Ezra, B. Aronov, and M. Sharir. Improved bound for the union of fat triangles. In *Proc. 16th ACM-SIAM Sympos. Discrete Algorithms*, pages 1778–1785, 2011.

[17] J. Fox and J. Pach. Computing the independence number of intersection graphs. In *Proc. 22nd ACM-SIAM Sympos. Discrete Algorithms*, pages 1161–1165, 2011.

[18] E. Grant and T. M. Chan. Exact algorithms and apx-hardness results for geometric set cover. In *Proc. 23rd Canad. Conf. Comput. Geom.*, 2011.

[19] S. Har-Peled. Being fat and friendly is not enough. *CoRR*, abs/0908.2369, 2009.

[20] J. Hastad. Clique is hard to approximate within $n^{1-\varepsilon}$. *Acta Mathematica*, pages 105–142, 1999.

[21] V. Kann. Maximum bounded 3-dimensional matching is max snp-complete. *Inform. Process. Lett.*, 37:27–35, January 1991.

[22] N. H. Mustafa and S. Ray. Improved results on geometric hitting set problems. *Discrete Comput. Geom.*, 44(4):883–895, 2010.

[23] A. Srinivasan. New approaches to covering and packing problems. In *Proc. 12th ACM-SIAM Sympos. Discrete Algorithms*, pages 567–576, 2001.

[24] P. G. Szabó, M. Cs. Markót, T. Csendes, E. Specht, L. G. Casado, and I. Garcäa. *New Approaches to Circle Packing in a Square: With Program Codes (Springer Optimization and Its Applications)*. Springer-Verlag New York, Inc., 2007.

[25] A. Tamir. Obnoxious facility location on graphs. *SIAM J. Discrete Math.*, 4(4):550–567, 1991.

[26] K. Varadarajan. Weighted geometric set cover via quasi-uniform sampling. In *Proc. 42nd Annu. ACM Sympos. Theory Comput.*, pages 641–648, 2010.

[27] D. P. Williamson and D. B. Shmoys. *The Design of Approximation Algorithms*. Cambridge University Press, 2011.

Computing Cartograms with Optimal Complexity *

Md. Jawaherul Alam †
University of Arizona
Tucson, AZ, USA
mjalam@cs.arizona.edu

Therese Biedl ‡
University of Waterloo
Waterloo, ON, Canada
biedl@uwaterloo.ca

Stefan Felsner §
Technische Universität Berlin
Berlin, Germany
felsner@math.tu-berlin.de

Michael Kaufmann ¶
Universität Tübingen
Tübingen, Germany
mk@informatik.uni-tuebingen.de

Stephen G. Kobourov †
University of Arizona
Tucson, AZ, USA
kobourov@cs.arizona.edu

Torsten Ueckerdt §
Technische Universität Berlin
Berlin, Germany
ueckerdt@math.tu-berlin.de

ABSTRACT

In a rectilinear dual of a planar graph vertices are represented by simple rectilinear polygons, while edges are represented by side-contact between the corresponding polygons. A rectilinear dual is called a cartogram if the area of each region is equal to a pre-specified weight. The complexity of a cartogram is determined by the maximum number of corners (or sides) required for any polygon. In a series of papers the polygonal complexity of such representations for maximal planar graphs has been reduced from the initial 40 to 34, then to 12 and very recently to the currently best known 10. Here we describe a construction with 8-sided polygons, which is optimal in terms of polygonal complexity as 8-sided polygons are sometimes necessary. Specifically, we show how to compute the combinatorial structure and how to refine it into an area-universal rectangular layout in linear time. The exact cartogram can be computed from the area-universal layout with numerical iteration, or can be approximated with a hill-climbing heuristic.

We also describe an alternative construction of cartograms for Hamiltonian maximal planar graphs, which allows us to directly compute the cartograms in linear time. Moreover, we prove that even for Hamiltonian graphs 8-sided rectilinear polygons are necessary, by constructing a non-trivial lower bound example. The complexity of the cartograms can be reduced to 6 if the Hamiltonian path has the extra property that it is one-legged, as in outer-planar graphs. Thus, we have optimal representations (in terms of both polygonal complexity and running time) for Hamiltonian maximal planar and maximal outer-planar graphs.

*Research initiated at Dagstuhl 10461 on Schematization

†Funded in part by NSF grants CCF-0545743 and CCF-1115971

‡Research supported by NSERC.

§Funded in part: EUROGIGA project GraDR & DFG Fe 340/7-2

¶Partially supported by EUROGIGA project GraDR

SCG'12, June 17–20, 2012, Chapel Hill, North Carolina, USA.

Categories and Subject Descriptors

F.2.2 [**Analysis of Algorithms and Problem Complexity**]: Nonnumerical Algorithms and Problems—*computations on discrete structures, geometrical problems and computations*; G.2.2 [**Discrete Mathematics**]: Graph Theory—*graph algorithms*

General Terms

Algorithms, Theory

Keywords

planar graphs, geometric representations, contact graphs, cartograms

1. INTRODUCTION

There is a large body of work about representing planar graphs as contact graphs, where vertices are represented by geometrical objects with edges corresponding to two objects touching in some specified fashion. Typical classes of objects might be curves, line segments, or polygons. An early result is Koebe's 1936 theorem [20] that all planar graphs can be represented by touching disks.

In this paper, we consider contact representations of planar graphs, with vertices represented by simple interior-disjoint polygons and adjacencies represented by non-trivial contacts (shared boundary) between the corresponding polygons. We are specifically interested in rectilinear weighted version where the vertices are represented by simple (axis-aligned) rectilinear polygons. This type of a representation is known as a *rectilinear dual* of the input graph.

In the weighted version, the input is a planar graph $G = (V, E)$ along with a weight function $w : V(G) \to \mathbb{R}^+$ that assigns a weight to each vertex of G. A rectilinear dual is called a *cartogram* if the area of each region is equal to the pre-specified weight of the corresponding vertex. Such representations have practical applications in cartography [26], geography [30], sociology [17], VLSI layout, floor-planning [24] etc. Other applications are in visualization of relational data, where using the adjacency of regions to represent edges in a graph can lead to a more compelling visualization than just drawing a line segment between two points [5].

For both rectilinear duals (unweighted) and cartograms (weighted) it is often desirable, for aesthetic, practical and cognitive reasons, to limit the *polygonal complexity* of the representation, measured by the number of sides (or by the number of corners). Similarly, it is also desirable to minimize the unused area in the representation, also known as *holes* in floor-planning and VLSI layouts. A given rectilinear dual is *area-universal* if it can realize a cartogram with any pre-specified set of weights for the vertices of the graph without disturbing the underlying adjacencies or increasing the polygonal complexity. With these considerations, we study the problem of constructing area-universal rectilinear duals and show how to compute cartograms with worst-case optimal polygonal complexity and without any holes.

1.1 Related Work

In our paper and in most of the papers cited here, "planar graph" refers to an inner-triangulated planar graph with a simple outer-face; the former restriction is required if at most three rectilinear polygons are allowed to meet in a point and the latter restriction is customary to achieve that the union of all the polygons in the representation is a rectangle.

Rectilinear duals (unweighted) were first studied in graph theoretic context, and then with renewed interest in the context of VLSI layouts and floor planning. It is known that 8 sides are sometimes necessary and always sufficient [15, 23, 36]. The case when the rectilinear polygons are restricted to rectangles has been of particular interest and there are several (independent) characterizations of the class of planar graphs that allows such *rectangular duals* [21, 22, 32]. A summary of the history and the state of the art in the rectangle contact graphs literature can be found in [5].

In the above results on rectilinear and rectangular duals, the areas of the polygons are not considered; i.e., these results deal with the unweighted version of the problem. The weighted version dates back to 1934 when Raisz described rectangular cartograms [26]. Algorithms by van Kreveld and Speckmann [33] and Heilmann *et al.* [16] yield representations with touching rectangles but the adjacencies may be disturbed and there may also be a small distortion of the weights. Recently, Eppstein *et al.* [14] characterized the class of planar graphs with area-universal rectangular duals. Given an area-universal rectangular dual and a weight function the construction of the actual cartogram can be accomplished using a result by Wimer *et al.* [35], which in turn requires numerical iteration.

The result of Eppstein *et al.* above is restricted to planar graphs that have rectangular duals. Going back to the more general rectilinear duals, leads to a series of papers where the main goal has been to reduce the polygonal complexity while respecting all areas and adjacencies. De Berg *et al.* initially showed that 40 sides suffice [9]. This was later improved to 34 sides [19]. In a recent paper [4] the polygonal complexity was reduced to 12 sides and even more recently to 10 sides [1]. But in practice for many maps the average region complexity is closer to 5 [10], so the natural question is to improve this further.

Side contact representations of planar graphs have also been studied without the restriction to rectilinear polygons. In the unweighted case 6-sided polygons are sometimes necessary and always sufficient [13]. The constructive upper bound relies on convex 6-sided polygons. In the weighted version, where the area of each polygon is prescribed, examples are known for which polygons with 7 sides are necessary [31]. This lower bound is matched by constructive upper bound of 7 sides if holes are allowed [2]. In the same paper it is shown that even allowing arbitrarily high polygonal complexity and holes of arbitrary size, there exist examples with prescribed areas which cannot be represented with convex polygons. If holes are not allowed then the best previously known polygonal complexity is 10, and it is achieved with rectilinear polygons [1].

1.2 Our Results

Recall that the known lower bound on the polygonal complexity even for unweighted rectilinear duals is 8 while the best known upper bound is 10. Here we present the first construction that matches the lower bound. The main difference to all previous papers is that we do not create the cartogram directly from the graph and the area function. Instead, we first create an 8-sided rectilinear dual layout of the graph with a very simple construction. Then we argue that this layout is area-universal, i.e., any are assignment can be realized by deforming the rectilinear dual layout. With this approach, both the construction of the rectilinear dual layout and the argument that a cartogram exists are much simpler than in previous paper that achieved a complexity of 10 or more. The exact cartogram can be computed from the area-universal rectangular layout with numerical iteration, or can be approximated with a hill-climbing heuristic.

For Hamiltonian maximal planar graphs we have an alternative construction which allows us to directly compute cartograms with 8-sided rectilinear polygons in linear time. Moreover, we prove that 8-sided rectilinear polygons are necessary by constructing a non-trivial lower bound example. If the Hamiltonian path has the extra property that it is one-legged, then we can reduce the polygonal complexity and realize cartograms with 6-sided polygons. This can be used to obtain 6-sided cartograms of maximal outer-planar graphs. Thus we have optimal (in terms of both polygonal complexity and running time) representations for Hamiltonian maximal planar and maximal outer-planar graphs.

2. PRELIMINARIES

A *planar graph* $G = (V, E)$ is one that has a drawing without crossings in the plane. A *plane graph* is a planar graph with a fixed planar embedding, defined via a cyclic ordering of the edges around each vertex. It splits the plane into connected regions called *faces*; the unbounded region is the *outer-face* and all other faces are *interior faces*. A planar (plane) graph is *maximal* if no edge can be added to it without violating planarity. Thus each face of a maximal plane graph is a triangle. A *Hamiltonian cycle* in a graph G is a simple cycle containing all the vertices of G. A graph is *Hamiltonian* if it contains a Hamiltonian cycle.

A set P of closed simple interior-disjoint polygons with an isomorphism $\mathcal{P} : V \to P$ is a *polygonal contact representation* of a graph $G = (V, E)$ if for any two vertices $u, v \in V$ the boundaries of $\mathcal{P}(u)$ and $\mathcal{P}(v)$ share a non-empty line-segment if and only if $(u, v) \in E$. ($\mathcal{P}(u)$ and $\mathcal{P}(v)$ may share

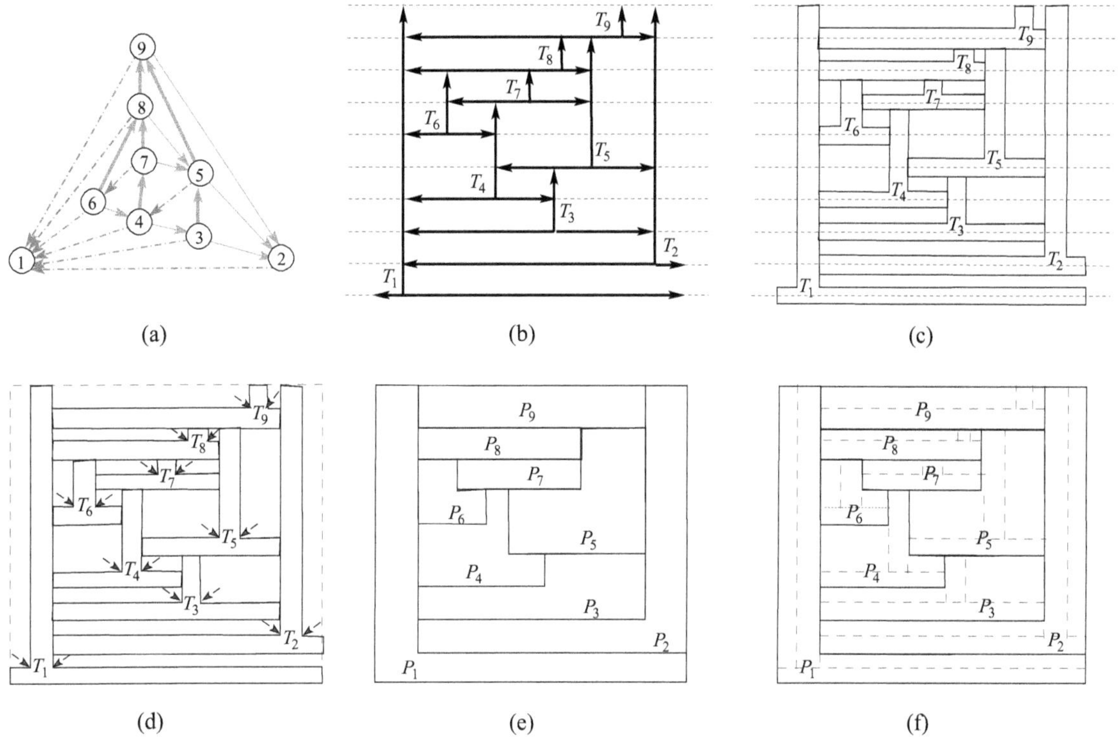

Figure 1: Construction of a rectilinear dual of a maximal planar graph with 8-sided polygons.

multiple line-segments). Such a representation is a *rectilinear dual* of the input graph if the polygons are rectilinear. In the weighted version the input is the graph G, along with a weight function $w : V(G) \to \mathbb{R}^+$ assigning a weight to each vertex. A rectilinear dual is a *cartogram* if the area of each polygon is equal to the assigned weight of the corresponding vertex. The *complexity of a polygon* is the number of sides in it. A common objective is to realize a given graph and a set of weights, using polygons with minimal complexity.

2.1 Canonical Orders & Schnyder Realizers

Next we briefly summarize the concepts of a "canonical order" of a planar graph [12] and that of a "Schnyder realizer" [29]. Let $G = (V, E)$ be a maximal plane graph with outer vertices u, v, w in clockwise order. Then we can compute in linear time [8] a *canonical order* or *shelling order* of the vertices $v_1 = u$, $v_2 = v$, v_3, ..., $v_n = w$, which is defined as one that meets the following criteria for every $4 \leq i \leq n$.

- The subgraph $G_{i-1} \subseteq G$ induced by v_1, v_2, ..., v_{i-1} is biconnected, and the boundary of its outer-face is a cycle C_{i-1} containing the edge (u, v).
- The vertex v_i is in the exterior face of G_{i-1}, and its neighbors in G_{i-1} form an (at least 2-element) subinterval of the path $C_{i-1} - (u, v)$.

A *Schnyder realizer* of G is a partition of the interior edges of G into the sets $\mathcal{S}_1$, $\mathcal{S}_2$ and $\mathcal{S}_3$ of directed edges so that for each interior vertex v, the following conditions hold:

- v has out-degree exactly one in each of $\mathcal{S}_1$, $\mathcal{S}_2$ and $\mathcal{S}_3$,
- the counterclockwise order of edges incident to v is: entering $\mathcal{S}_1$, leaving $\mathcal{S}_2$, entering $\mathcal{S}_3$, leaving $\mathcal{S}_1$, entering $\mathcal{S}_2$, leaving $\mathcal{S}_3$.

Schnyder proved that any maximal plane graph has a Schnyder realizer and it can be computed in $O(n)$ time [29]. The first condition implies that $\mathcal{S}_i$, for $i = 1, 2, 3$ defines a tree rooted at exactly one exterior vertex and containing all the interior vertices where the edges are directed towards the root. Denote by $\Phi_k(v)$ the parent of vertex v in tree T_k. The following well-known lemma shows a profound connection between canonical orders and Schnyder realizers.

LEMMA 2.1. *Let G be a maximal plane graph. Then the following conditions hold.*

(a) A canonical order of the vertices of G defines a Schnyder realizer of G, where the outgoing edges of a vertex v are to its first and last predecessor (where "first" is with respect to the clockwise order around v), and to its highest-numbered successor.

(b) A Schnyder realizer with trees S_1, S_2, S_3 defines a canonical order, which is a topological order of the acyclic graph $\mathcal{S}_1^{-1} \cup \mathcal{S}_2^{-1} \cup \mathcal{S}_3$, where $\mathcal{S}_k^{-1}$ is the tree $\mathcal{S}_k$ with the direction of all its edges reversed.

3. CARTOGRAMS WITH 8-GONS

In this section we show that 8-sided polygons are always sufficient and sometimes necessary for a cartogram of a maximal planar graph. Our algorithm for constructing 8-sided area-universal rectilinear duals has three main phases. In the first phase we create a contact representation of the graph G, where each vertex of G is represented by an upside-down **T**, i.e., a horizontal segment and a vertical segment. Fig. 1(a)-(b) show a maximal planar graph and its contact representation using **T**'s, where the three ends of each **T** are marked with arrows. In the second phase we make both the horizontal and vertical segments of each **T** into thin polygons with λ thickness for some $\lambda > 0$. We then have a contact representation of G with T-shaped polygons; see Fig. 1(c). In the third phase we remove all the unused area in the representation by assigning each (rectangular) hole to one of the polygons adjacent to it; see Fig. 1(d). We show that the resulting representation is an area-universal rectilinear dual of G with polygonal complexity 8, as illustrated in Fig. 1(e).

3.1 Contact Representation with T's

Our contact representation with **T**'s is similar to the approach described by de Fraysseix *et al.* [11].

Let G be a planar graph. As mentioned earlier, we assume that G is inner-triangulated with a simple outer-face. If need be, we can add two vertices (which we later choose as v_1 and v_2) and connect them to the outer-face to ensure that the graph is maximal. Now let $v_1, v_2, v_3, \ldots, v_n$ be a canonical order of the vertices in G with corresponding Schnyder trees $\mathcal{S}_1$, $\mathcal{S}_2$ and $\mathcal{S}_3$ rooted at v_1, v_2 and v_n. Add to $\mathcal{S}_1$ the edges (v_2, v_1), (v_n, v_1) oriented towards v_1 and add to $\mathcal{S}_2$ the edge (v_n, v_2) oriented towards v_2. In what follows, we sometimes identify vertex v_i with its canonical label i.

We assign to each vertex v_i the **T**-shape T_i consisting horizontal and vertical segments h_i and b_i, respectively. Begin by placing T_1 and T_2 so that h_1 is placed at $y = 1$, h_2 is placed at $y = 2$, the topmost points of both b_1 and b_2 have y-coordinate $n+1$ and the leftmost point of the h_2 touches b_1. Next the algorithm iteratively constructs the contact representation by defining T_k so that h_k is placed at $y = k$ and the topmost point of b_k has y-coordinate $\Phi_3(k)$ for $3 \leq k < n$. After the k-th step of the algorithm we have a contact representation of G_k, and we maintain the invariant that the order of the vertical segments with non-empty parts in the half-plane $y > k$ corresponds to the same circular order of the vertices along $C_k - (v_1, v_2)$.

Consider inserting T_k for v_k. The neighbors $v_{k_1}, v_{k_2}, \ldots, v_{k_d}$ of v_k in G_{k-1} form a subinterval of $C_{k-1} - (v_1, v_2)$ and hence the corresponding vertical segments are also in the same order in the half-plane $y > k-1$ of the representation of G_{k-1}. Since $v_k = \Phi_3(v_{k_i})$ for $1 < i < d$ (Lemma 2.1), the topmost points of the corresponding vertical segments have y-coordinate k. As v_{k_1} and v_{k_d} are the parents of v_k in $\mathcal{S}_1$ and $\mathcal{S}_2$, the x-coordinates of b_{k_1} and b_{k_d} define the x-coordinates of the two endpoints of h_k. Let these coordinates be x_l and x_r; then h_k is placed between the two points (x_l, k), (x_r, k) and b_k is placed between the two points (x_m, k), $(x_m, \Phi_3(k))$ with $x_l + 1 < x_m < x_r - 1$. Finally for $k = n$, we place T_n so that h_n touches b_1 to the left, b_2 to the right and the topmost point of b_n has y-coordinate $n+1$.

We note that this representation can be computed in linear time in a way such that all coordinates are integers. We do this by pre-computing a topological order π of $\mathcal{S}_1^{-1} \cup \mathcal{S}_2$; then h_k is the segment $[\pi(\Phi_1(k)), \pi(\Phi_2(k))] \times k$ and b_k is the segment $\pi(k) \times [k, \Phi_3(k)]$.

3.2 λ-Fattening of T_i's

Let Γ' be the contact representation of G using **T**'s obtained above. In this phase of the algorithm, we "fatten" **T**'s so that each vertex is represented by a T-shaped polygon. We replace each horizontal segment h_i by an axis-aligned rectangle H_i which has the same width as h_i, and whose top (bottom) side is $\lambda/2$ above (below) h_i, for some $0 < \lambda$, as illustrated in Fig. 2(a). Similarly, we replace each vertical segment b_i by an axis-aligned rectangle B_i which has the same height as b_i and whose left (right) side is $\lambda/2$ to the left (right) of b_i. We call this process *λ-fattening* of T_i. Note that this process creates intersections of H_i with B_i, $B_{\Phi_1(i)}$ and $B_{\Phi_2(i)}$ and intersection of B_i with $H_{\Phi_3(i)}$. We remove these intersections by replacing H_i by $H_i - B_{\Phi_1(i)} - B_{\Phi_2(i)}$ and replacing B_i by $B_i - H_i - H_{\Phi_3(i)}$. The resulting layout is a contact representation Γ'' of G where each vertex v_i of G is represented by the T-shaped polygon $H_i \cup B_i$.

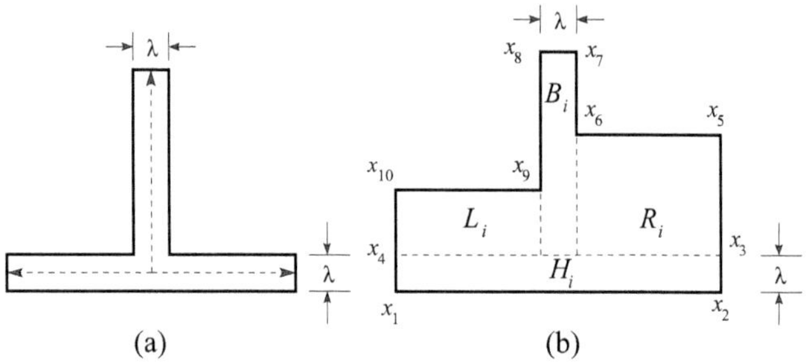

Figure 2: (a) λ-fattening of T, and (b) subdividing a T-shaped polygon into four rectangles.

3.3 Removing unused area

In this step, we begin with the λ-fat T-shaped polygonal layout, Γ'', from above and assign each (rectangular) hole to a polygon adjacent to it. We start by placing an axis-aligned rectangle of minimum size that encloses Γ''. This creates five new bounded holes. Note that all these holes are rectangles, and each of them is bounded at the bottom by H_i for some vertex v_i. We assign each hole to this vertex. This assigns at most two holes to each vertex v_i: one hole L_i to the left of B_i, and one hole R_i to the right of B_i. Now for each vertex v_i, define $P_i = T_i \cup L_i \cup R_i$. It is easy to see that P_i is an 8-sided rectilinear polygon since the left side of L_i has the same x-coordinate as the left side of H_i and the right side of R_i has the same x-coordinate as the right side of H_i. Thus we have a rectilinear dual, Γ, of G where each vertex v_i is represented by P_i.

Note that the coordinates of P_i could be computed directly, not going through T-shapes and λ-fattening, using the values $\Phi_k(v_i)$ for $k = 1, 2, 3$ and a topological order π of $\mathcal{S}_1^{-1} \cup \mathcal{S}_2$. In particular, we can compute the representation in linear time, and all coordinates are integers of size $O(n)$.

3.4 Area-Universality

A rectilinear dual Γ is *area-universal* if any assignment of areas to its polygons can be realized by a combinatorially equivalent layout. Eppstein *et al.* [14] studied this concept for the case when all the polygons are rectangles and the outer-face boundary is also a rectangle (which they call a *rectangular layout*). They gave a characterization of area-universal rectangular layouts using the concept of "maximal line-segment". A *line-segment* in a layout is the union of inner edges forming a consecutive part of a straight-line. A line-segment not contained in any other line-segment is *maximal*. A maximal line-segment s is called *one-sided* if it forms a full side of at least one rectangular face, or in other words, if the perpendicular line segments that attach to its interior are all on one side of s.

LEMMA 3.1. *[14] A rectangular layout is area-universal if and only if each maximal segment is one-sided.*

No such characterization is known when some faces are not rectangles. Still we can use the characterization from Lemma 3.1 to show that the rectilinear dual obtained by the algorithm from the previous section is area-universal, with the following Lemma.

LEMMA 3.2. *Let Γ be the rectilinear dual obtained by the above algorithm. Then Γ is area-universal.*

PROOF. To show the area-universality of Γ, we divide all the polygons in Γ into a set of rectangles such that the resulting rectangular layout is area-universal. Specifically, we divide each polygon P_i into four rectangles H_i B_i, L_i and R_i (as defined in the previous subsection) by adding three auxiliary segments: one horizontal and two vertical, as illustrated in Fig. 2(b). Any horizontal segment s not on the bounding box belongs to some H_i (either top or bottom), and by expanding it to both sides, it ends at $B_{\Phi_1(v_i)}$ on the left and $B_{\Phi_2(i)}$ on the right. So s is one-sided since it is a side of H_i. Any vertical segment s not on the bounding box belongs to some B_i (either left or right), and by expanding it to both sides it ends at H_i on the bottom and $H_{\Phi_3(i)}$ on the top. So s is one-sided since it is a side of B_i.

Now given an assignment of areas $w : V \to \mathbb{R}^+$ to the vertices of G, we split $w(v_i)$ arbitrarily into four parts and assign the four values to its four associated rectangles. Since Γ^* is area-universal, there exists a rectilinear dual of G that is combinatorially equivalent to Γ where these areas are realized. Fig. 1(f) illustrates the rectangular layout obtained from the rectilinear dual in Fig. 1(e). □

So for any area-assignment, the rectilinear dual that we found can be turned into a combinatorially equivalent one that respects the area requirements. This proves our main result for maximal planar graphs. Omitting v_1 and v_2 from the drawing still results in a cartogram where the union of all polygons is a rectangle, so the result also holds for all inner-triangulated planar graphs with a simple outer-face.

Recall that the lower bound on the complexity of polygons in any rectilinear dual (and hence in any cartogram) is 8, as proven by Yeap and Sarrafzadeh [36]. The algorithm described in this section, thus leads to our main theorem.

THEOREM 3.1. *Eight-sided polygons are always sufficient and sometimes necessary for a cartogram of an inner triangulated planar graph with a simple outer-face.*

3.5 Feature Size and Supporting Line Set

In addition to optimal polygonal complexity, we point out here a practical feature of the 8-sided area-universal rectilinear layout constructed with our algorithm. Earlier constructions, e.g., [4, 9], often rely on "thin connectors" to maintain adjacencies, whereas our construction does not. Moreover, we have the freedom to choose how to divide the area assigned to any vertex v_i among the four rectangles associated with it. This flexibility makes it possible to achieve other desired properties. In particular, we show in the followings that the *minimum feature size* (i.e., the smallest distance between non-adjacent edges) can be made as large as $1/2\sqrt{A}\cdot\min_{v\in V(G)} w(v)$ (where A is the sum of the weights), and that this is worst-case optimal.

Choose W and H such that $W \times H = A = \sum_{v\in V(G)} w(v)$. We are interested in cartograms within a rectangle of width W and height H. Define $w_{min} = min_{v\in V(G)} w(v)$. Recall that each vertex v_i is represented by the union of at most four rectangles $H_i \cup B_i \cup R_i \cup L_i$, with H_i and B_i nonempty. We can distribute the weight assigned to v_i arbitrarily among them. In particular, we can assign zero areas to the rectangles L_i and R_i and split the weight of v_i into two equal parts, assigning them to H_i and B_i, respectively. In this layout each original vertex is represented by rectangles H_i and B_i whose union is some fattened T or L, and all the necessary contacts remain. Hence we can use this simplified layout to produce the cartogram.

The distribution of the weight of v_i in equal parts to H_i and B_i allows to bound the feature size. The height and width of each rectangle are bounded by H and W, respectively. Its weight is at least $w_{min}/2$. Therefore, the height and width of each rectangle is at least $\frac{w_{min}}{2\max\{W,H\}}$. Thus the minimum feature size is at least $\frac{w_{min}}{2\max\{W,H\}}$. This is worst-case optimal, as the polygon with the smallest weight might need to reach from left to right and from top to bottom in the representation. We may choose $W = H = \sqrt{A}$ so that the minimum feature size is $\frac{w_{min}}{2\sqrt{A}}$. Furthermore the rectangular layout based on only H_i and B_i yields a cartogram with at most $2n$ supporting lines, instead of the $3n$ supporting lines in the cartogram based on four rectangles per vertex.

3.6 Computing the Cartogram

The proof of Lemma 3.1 implies an algorithm for computing the final cartogram. Splitting the T-shaped polygons into four rectangles and distributing the weights yields an area-universal rectangular dual. This combinatorial structure has to be turned into an actual cartogram, i.e., into a layout respecting the given weights. Wimer *et al.* [35] gave a formulation of the problem combining flows and quadratic equations. Eppstein *et al.* [14] indicated that a solution can be found with a numerical iteration. Alternate methods also exist, based on non-linear programming [28], geometric programming [25], and convex programming [7]. Heuristic hill-climbing schemes converge much quicker and can be used in practice, at the expense of small errors [6, 18, 34].

We implemented the entire algorithm, along with a force-directed heuristic to compute the final cartogram. We treat each region as a rectilinear "room" containing an amount of "air" equal to the weight assigned to the corresponding vertex. We then simulate the natural phenomenon of air pressure applied to the "walls", which correspond to the line segment borders in our layout. At each iteration, we consider the segment that feels the maximum pressure and let it move in the appropriate direction.

For each vertex v_i of G, the polygon P_i contains air with volume $w(v_i)$. If the area of P_i is A_i, then the pressure applied to each of the walls surrounding P_i is given by $\mathcal{P}(v_i) = \frac{w(v_i)}{A_i}$. In Section 3.4, we saw that the maximal segments of the layout are the two horizontal and the two vertical segment associated with each polygon. For each polygon, the horizontal segment other than the base is entirely inside the polygon, hence it feels no "pressure" on it. For each of the other three segments s for the polygon P_i, the "inward force" it feels is given by $\mathcal{F}(s) = \sum_{v_j\in V(s)}[\mathcal{P}(v_j)l_j] - \mathcal{P}(v_i)l_i$. Here $V(s)$ is the set of vertices other that v_i whose corresponding polygon touches the segment s and l_i (resp. l_j) denotes

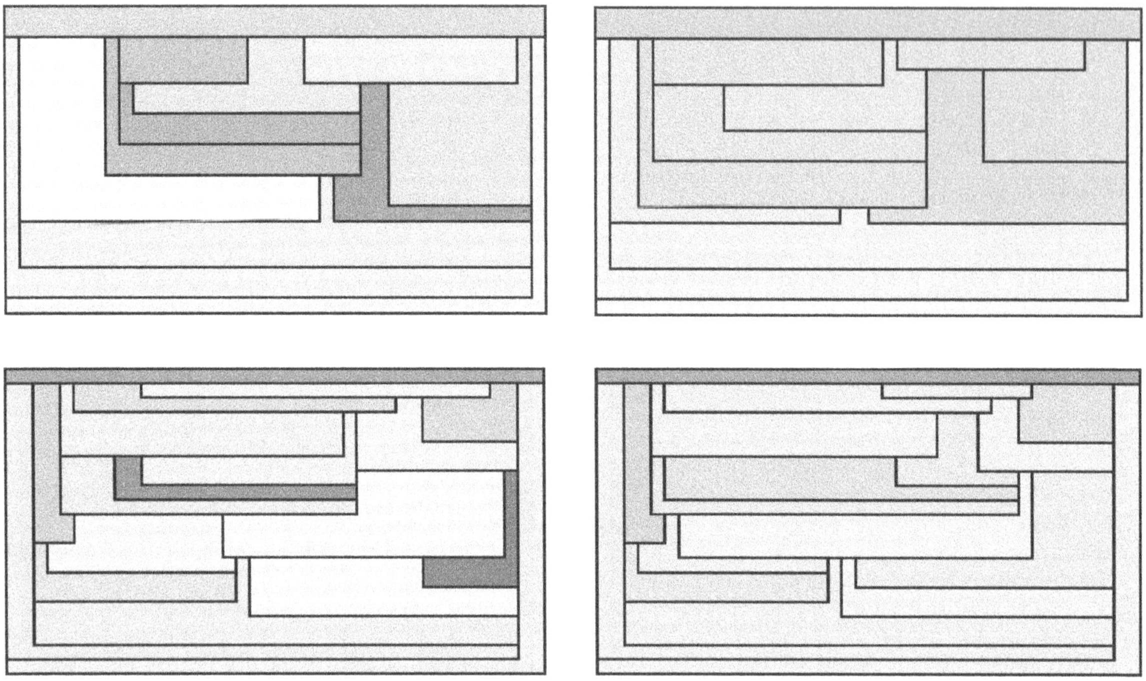

Figure 3: Input-output pairs: weights are assigned at random in the range $[10, 100]$, and the cartographic error in the output is less than 1%. The colors indicate air-pressure: a gray region has almost the desired area; the lighter a region is, the more it needs to shrink; the darker a regions is, the more it needs to grow.

the length of s that is shared with P_i (resp. P_j). At each iteration, we consider the segment that feels the maximum pressure and let it move in the appropriate direction. This process is guaranteed to converge with the desired cartogram according to [18], but in practice we stop when the (cartographic) errors fall below a certain threshold. Some sample input-output pairs are shown in Fig. 3; more examples and movies showing the gradual transformation can be found at `www.cs.arizona.edu/~mjalam/optocart`.

We ran a few simple experiments to test the heuristic for time and accuracy. In the first experiment we generated 5 graphs on n vertices with each integer n in the range $[10, 50]$ and assigned 5 random weight distributions with weights in the range $[10, 100]$. Next we ran the heuristic until the "cartographic error" dropped below 1% and recorded the average time. The *cartographic error* for a cartogram of a planar graph $G = (V, E)$ is defined as in [33]: $max_{v \in V}(|A(v) - w(v)|/w(v))$, where $w(v)$ denotes the weight assigned to a vertex v and $A(v)$ denotes the area of the polygon representing v. All the averages were below 50 milliseconds, which confirms that good solutions can be found very quickly in practice; see Fig 4(a). In the second experiment we fixed the time allowed and tested the quality of the cartograms obtained within the time limit. Specifically, we generated 5 graphs of n vertices with each integer n in the range $[10, 50]$ and assigned 5 random weight distributions with weights in the range $[10, 100]$. We allowed the program to run for 1 millisecond and recorded the average cartographic error. Even with this small time limit, the average cartographic error was under 2.5%; see Fig. 4(b). All of the experiments were run on an Intel Core i3 machine with a 2.2GHz processor and 4GB RAM.

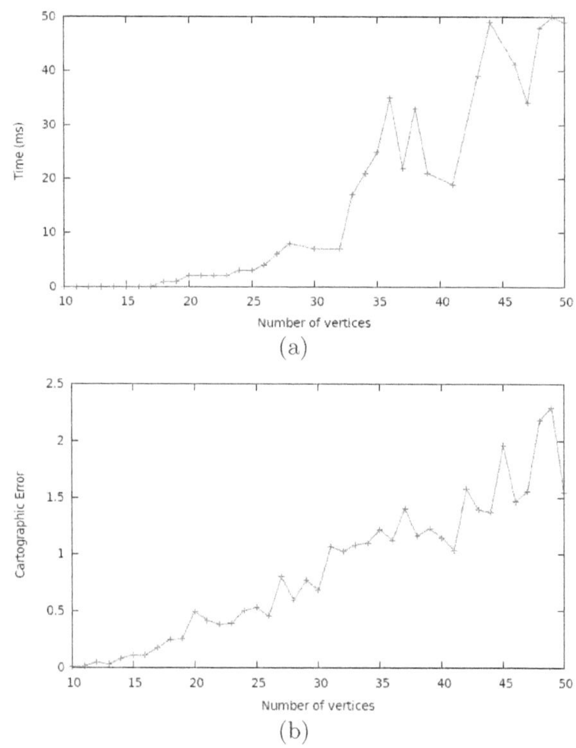

Figure 4: Experimental results for graphs with 10-50 vertices. Each sample point corresponds to 25 graphs. (a) Plotting the average time it takes to reach cartographic error of 1%, (b) Plotting the average cartographic error achieved in 1 millisecond.

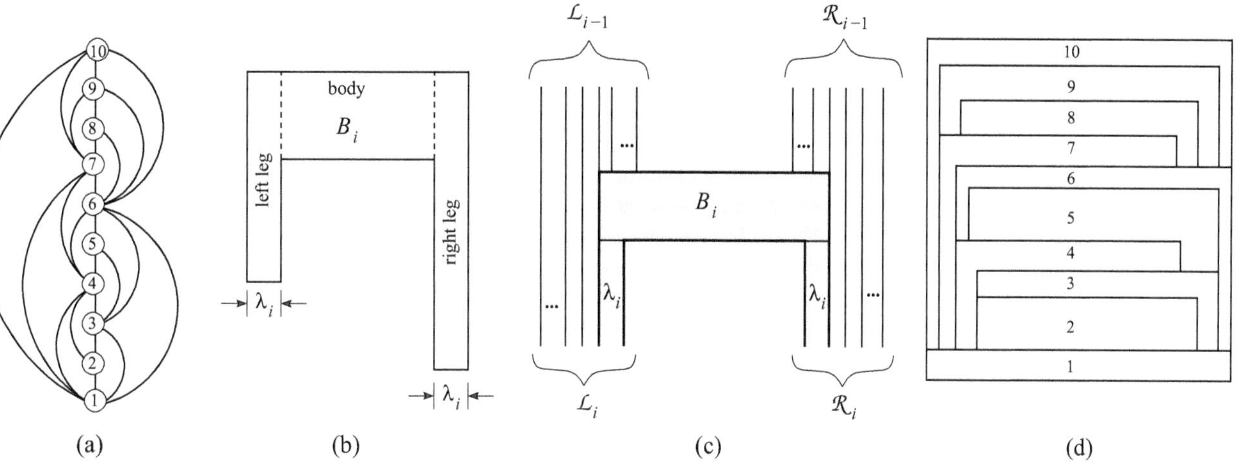

Figure 5: (a) A Hamiltonian maximal planar graph G, (b) an 8-sided polygon for vertex i, (c) illustration for the algorithm to construct a cartogram of G, (d) a cartogram of G with 8-sided polygons.

4. HAMILTONIAN GRAPHS

In this section we show that 8-sided polygons are always sufficient and sometimes necessary for a cartogram of a *Hamiltonian* maximal planar graph. We first give a direct linear-time construction with 8-sided regions without relying on numerical iteration or heuristics, as in the previous section. We then prove that this is optimal by showing that 8 sides are necessary, with a non-trivial lower bound example.

4.1 Sufficiency of 8-sided Polygons

Let $v_1, \ldots, v_n$ be a Hamiltonian cycle of a maximal planar graph G. Consider a plane embedding of G with the edge (v_1, v_n) on the triangular outer-face. The Hamiltonian cycle splits the plane graph G into two outer-planar graphs which we call the *left graph* G_l and *right graph* G_r. Edges on the Hamiltonian cycle belong to both graphs. The naming is with respect to a planar drawing Γ of G in which the vertices $v_1, \ldots, v_n$ are placed in increasing order along a vertical line, and the edges are drawn as y-monotone circular arcs with leftmost edge (v_1, v_n); see Fig. 5(a).

LEMMA 4.1. *Let $G = (V, E)$ be a Hamiltonian maximal planar graph with a weight function $w : V \to \mathbb{R}^+$. A cartogram with 8-sided polygons can be computed in linear time.*

PROOF. Let $v_1, \ldots, v_n$ be a Hamiltonian cycle and Γ be the drawing defined above with (v_1, v_n) on the outer-face. Suppose R is a rectangle of width W and height H where $W \times H = A = \sum_{v \in V} w(v)$. Each vertex v_i will be represented as the union of three rectangles, the *left leg*, the *body* B_i, and *right leg* of v_i. We set the width of the legs to $\lambda_i = w(v_i)/(2H + W)$; see Fig. 5(b).

Our algorithm places vertices $v_1, \ldots, v_n$ in this order, and also reserves vertical strips for legs of all vertices that have earlier neighbors. More precisely, let $\mathcal{L}_j$ be all vertices v_k with an edge (v_i, v_k) in G_l for which $i \le j < k$. Similarly define $\mathcal{R}_j$ with respect to edges in G_r. In the drawing Γ, $\mathcal{L}_j$ are those vertices above v_j for which the horizontal ray left from v_j crosses an incident edge.

We place vertices $v_1, \ldots, v_j$ with the following invariant: The horizontal line through the top of B_j intersects, from left to right: (a) a vertical strip of width λ_k for each $v_k \in \mathcal{L}_j$, in descending order, (b) a non-empty part of the top of B_j, and (c) a vertical strip of width λ_k for each $v_k \in \mathcal{R}_j$, in ascending order. We start by placing B_1 as a rectangle that spans the bottom of R. At the left and right end of the top of B_1, we reserve vertical strips of width λ_k for each vertex in $\mathcal{L}_1$ and $\mathcal{R}_1$, respectively.

To place B_i, $i > 1$, first locate the vertical strips reserved for v_i in previous steps (since $v_i \in \mathcal{L}_{i-1}$ and $i \in \mathcal{R}_{i-1}$, there always are such strips, though they may have started only at the top of B_{i-1}). Since vertical strips are in descending/ascending order, the strips for v_i are the innermost ones. Let B_i be a rectangle just above B_{i-1} connecting these strips. Choose the height of B_i so large that it, together with the left and right leg inside the strips, has area $w(v_i)$; we discuss the appropriate height below.

Finally, at the top left of the polygon of v_j we reserve a new vertical strip of width λ_k for each vertex k in $\mathcal{L}_i - \mathcal{L}_{i-1}$. Similarly reserve strips for vertices in $\mathcal{R}_i - \mathcal{R}_{i-1}$. By planarity, the vertices in $\mathcal{L}_i - \mathcal{L}_{i-1}$ must have smaller indices than vertices in $\mathcal{L}_{i-1}$, and so this can be done such that the order required for the invariant is respected, see Fig. 5(c).

Clearly this algorithm takes linear time and constructs 8-gons of the correct area. To see that it creates contacts for all edges, consider an edge (v_i, v_k) with $i < k$ in G_l (edges in G_r are similar.) By definition $k \in \mathcal{L}_i$. If $v_k \in \mathcal{L}_i - \mathcal{L}_{i-1}$, then we reserved a vertical strip for v_k when placing v_i. This vertical strip is used for the left leg of v_k, which touches v_i. Otherwise ($v_k \notin \mathcal{L}_i - \mathcal{L}_{i-1}$) we have $v_k \in \mathcal{L}_{i-1}$. When v_{i-1} was placed, there was a vertical strip for v_k. There was also a vertical strip for $v_i \in \mathcal{L}_{i-1}$. These two strips must be adjacent, because by planarity (and edge (v_i, v_k)) there can be no vertex v_j with $i < j < k$ in $\mathcal{L}_{i-1}$. So these strips create a contact between the two left legs of v_i and v_k.

We now discuss the choice of $\lambda_i = w(v_i)/(2H + W)$. Each leg of v_i has height $\le H$ and width λ_i, hence area $\le H\lambda_i$. Then B_i has area $\ge w(v_i) - 2H\lambda_i$ and width $\le W$, hence height $\ge \frac{w(v_i) - 2H\lambda_i}{W} = \lambda_i$. It follows that B_i has positive height. Also all vertical strips fit: after placing B_i, we have a strip of width λ_k for each vertex $v_k \in (\mathcal{L}_i \cup \mathcal{R}_i)$, and these strips use width $\sum_{v_k \in (\mathcal{L}_i \cup \mathcal{R}_i)} \frac{w(v_k)}{2H+W} \le \frac{2\sum_{v_k \in V - \{v_i\}} w(v_k)}{2H+W}$ $< \frac{2A}{2H} - \frac{2w(v_i)}{2H+W} = W - 2\lambda_i$. Hence B_i has width $> 2\lambda_i$ and the polygon of v_i has minimum feature size λ_i. $\square$

Fig. 5(d) illustrates a cartogram for graph in Fig. 5(a), obtained by the above algorithm. This algorithm also gives a minimum feature size for the cartogram: $\min_{v_i \in V}\{\lambda_i\} = \frac{w_{min}}{2H+W}$, where $w_{min} = \min_{v \in V}\{w(v_i)\}$. Choosing $W = \sqrt{2A}$, $H = \sqrt{A/2}$, yields the minimum feature size $\frac{w_{min}}{2\sqrt{2}\sqrt{A}}$.

4.2 Necessity of 8-sided Polygons

While it was known that 8-sided polygons are necessary for general planar graphs [27], here we show that this necessity holds even for Hamiltonian maximal planar graphs.

LEMMA 4.2. *Consider the Hamiltonian maximal planar graph $G = (V, E)$ in Fig. 6(a). Define $w(j) = w(l) = D$ and $w(v) = \delta$ for $v \in V \setminus \{j, l\}$, where $D \gg \delta$. Then any cartogram of G with respect to w contains an 8-sided polygon.*

PROOF. Assume for a contradiction that G admits a cartogram Γ with respect to w such that each polygon $\{P(v)\}$ in Γ representing the vertex v has complexity at most 6. Observe that if $\{u, v, x\}$ is a *separating triangle* in G, i.e., three mutually adjacent vertices whose removal disconnect the graph, then the region R_{uvw} used for the inside of the separating triangle contains at least one reflex corner of the polygon $P(u)$, $P(v)$, or $P(x)$. The 5-vertex set $\{a, c, e, g, i\}$ in G is the union of the five separating triangles $\{a, c, g\}$, $\{a, c, e\}$, $\{c, e, g\}$, $\{a, e, i\}$, and $\{e, g, i\}$ with disjoint interiors. Since all the polygons in Γ are either 4-sided or 6-sided, the union of the polygons for these five vertices has at most five reflex corners and hence each of the five separating triangles above contains the only reflex corner of the polygon for a, c, e, g, or i. In particular, the outer boundary of R_{aei} contains exactly one reflex corner from one of $P(a)$, $P(e)$ and $P(i)$, hence it is a rectangle, say 1234. By symmetry, assume that the reflex corner of $P(i)$ is *not* used for R_{aei}.

The 4-vertex set $\{a, i, k, m\}$ is the disjoint union of three separating triangles $\{a, k, m\}$, $\{k, i, m\}$, $\{i, a, m\}$, containing l, j and n, respectively, in the interior. The reflex corner of $P(i)$ is not used for R_{aei}, so, it cannot be used for any of these triangles. Thus each of $P(j)$, $P(l)$, $P(n)$ contains exactly one reflex corner from $P(a)$, $P(k)$ and $P(m)$. In particular, $R_{aei} = 1234$ must contain the reflex corner of $P(a)$. We also conclude that $P(j)$, $P(l)$, $P(n)$ are rectangles, since there are no additional reflex corners to accommodate additional convex corners from $P(j)$, $P(l)$ and $P(n)$.

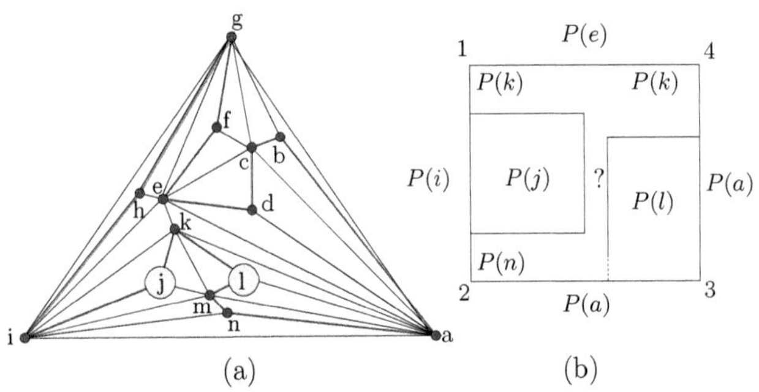

Figure 6: (a) A maximal planar Hamiltonian graph with a weight function that requires at least one 8-sided polygon in any cartogram. (b) Illustration for the proof of Lemma 4.2.

Assume the naming in Fig. 6(b) is such that edge 12 belongs to $P(i)$, edges 23 and 34 belong to $P(a)$ and edge 41 belongs to $P(e)$. By the adjacencies, $P(k)$ must occupy corners 1 and 4 and $P(n)$ must occupy corner 2, while corner 3 (which is the reflex corner of $P(a)$) could belong to n or l.

Now consider the rectangles $P(j)$ and $P(l)$. If D is sufficiently big, then these two rectangles each occupy almost half of rectangle 1234. Therefore, either their x-range or their y-range must overlap. Assume their y-range overlaps, the other case is similar. Which polygon should occupy the area horizontally between $P(j)$ and $P(l)$? It cannot be k, because $P(k)$ contains corners 1 and 2 and hence would obtain 2 reflex angles from $P(j)$ and $P(l)$. So it must be $P(m)$, since n is not adjacent to j and l. But $P(m)$ must also separate $P(n)$ from both $P(j)$ and $P(l)$. Regardless of whether n or l occupies corner 3, this is not possible without two reflex vertices for m. Thus some polygon must have 8 sides. □

Lemma 4.1 and Lemma 4.2 yield the following theorem.

THEOREM 4.1. *Eight-sided polygons are always sufficient and sometimes necessary for a cartogram of a Hamiltonian maximal planar graph.*

5. CARTOGRAMS WITH 6-GONS

Here we study cartograms with rectilinear 6-gons. We first note that these are easily constructed for outer-planar graphs. Then we generalize this technique to a class of maximal planar Hamiltonian graphs.

5.1 Maximal Outer-planar Graphs

Our algorithm from Lemma 4.1 naturally gives drawings of maximal outer-planar graphs that use 6-sided polygons. Another linear-time algorithm for constructing a cartogram of a maximal outer-planar graph with 6-sided rectilinear polygons is also described in [1], however, our construction based on Lemma 4.1 is much simpler. Any maximal outer-planar graph G can be made into a maximal Hamiltonian graph by duplicating G and gluing the copies together at the outer-face such that $G_l = G = G_r$. (This graph has double edges, but the algorithm in Lemma 4.1 can handle double edges as long as one copy is in the left and one in the right graph.) Create the drawing based on Lemma 4.1 with all vertices having double the weight, and cut it in half with a vertical line. This gives a drawing of G with 6-sided rectilinear polygons as desired.

5.2 One-Legged Hamiltonian Cycles

We now aim to find more maximal Hamiltonian graphs which have cartograms with 6-sided polygons. In a Hamiltonian cycle $v_1, \ldots, v_n$, call vertex v_j *two-legged* if it has a neighbor v_i^l in G_l with $i^l < j - 1$ and also a neighbor v_i^r in G_r with $i_r < j - 1$. Call a Hamiltonian cycle *one-legged* if none of its vertices is two-legged. In the construction from Lemma 4.1, the polygon of v_j obtains a reflex vertex on both sides only if it has a neighbor below v_{j-1} on both sides, or in other words, if it is two-legged. Hence we have:

LEMMA 5.1. *Let $G = (V, E)$ be a maximal planar graph with a one-legged Hamiltonian cycle and let $w : V \to R^+$ be a weight function. Then a cartogram with 6-sided polygons can be computed in linear time.*

It is a natural question to characterize graphs with such Hamiltonian cycles. The following lemma gives a nice relation between a one-legged Hamiltonian cycle, a canonical ordering and a Schnyder realizer of a graph.

LEMMA 5.2. *Let $v_1, \ldots, v_n$ be a Hamiltonian cycle in a maximal plane graph G with (v_1, v_n) on the outer triangle. Define $w_i := v_{n-i+1}$. Then these are equivalent:*

(a) The Hamiltonian cycle is one-legged.

(b) For $i = 2, \ldots, n$, edge (v_{i-1}, v_i) is an outer edge of the embedded graph G_i induced by $v_1, v_2, \ldots, v_i$.

(c) v_{n-1} is an outer vertex and vertex v_i has at least two neighbors with a larger index for $i = 1, \ldots, n-2$.

(d) $w_1, \ldots, w_n$ is a canonical ordering for G.

(e) G admits a Schnyder realizer $(\mathcal{S}_1, \mathcal{S}_2, \mathcal{S}_3)$ in which w_1, w_2 and w_n are the roots of $\mathcal{S}_1$, $\mathcal{S}_2$ and $\mathcal{S}_3$, respectively and every inner vertex is a leaf in $\mathcal{S}_1$ or $\mathcal{S}_2$.

PROOF. (a) $\iff$ (b): For $i = 2, \ldots, n$ we argue that a vertex v_i is one-legged if and only if (b) holds for i. Indeed, (v_{i-1}, v_i) is an inner edge in G_i if and only if there are outer edges (v_i, v_j) and (v_i, v_k) with $j, k < i-1$ in G_l and G_r, respectively. But then v_i is two-legged by definition.

(b) $\iff$ (c): Since v_n is an outer vertex and $G_n = G$, (b) holds for $i = n$ if and only if v_{n-1} is an outer vertex. For $i = 2, \ldots, n-1$ we argue that (b) holds for i if and only if (c) holds for $i-1$. Let v_i^l, respectively v_i^r, denote the third vertex in the inner facial triangle containing the edge (v_{i-1}, v_i) in G_l, respectively G_r. (Both triangles exist, since (v_{i-1}, v_i) is an inner edge in G.) Now (v_{i-1}, v_i) is an inner edge in G_i if and only if both, v_i^l and v_i^r, have a smaller index than v_{i-1}, which in turn holds if and only if the index of *every* neighbor of v_{i-1}, other than v_i, is less than $i-1$.

(c) $\implies$ (d): By (c) $\{w_1, w_2, w_n\} = \{v_n, v_{n-1}, v_1\}$ is the outer triangle of G. Moreover, $\tilde{G}_3$, induced by v_n, v_{n-1}, v_{n-2}, is a triangle, hence its outer boundary is a simple cycle C_3 with the edge (w_1, w_2). Thus the first condition of canonical ordering is met for $i = 4$. Assuming (c) and the first condition for $i = 4, \ldots, n-1$, we show that the second and first condition hold for i and $i+1$, respectively. Finally, the second condition holds for $i = n$ since w_n is an outer vertex.

Note that w_i is in the exterior face of $\tilde{G}_{i-1}$ since w_n lies in the exterior face and the path $w_i, \ldots, w_n$ is disjoint from vertices in $\tilde{G}_{i-1}$ and the embedding is planar. By (c) w_i has at least two neighbors in $\tilde{G}_{i-1}$. If the neighbors do not form a subinterval of the path $C_{i-1} \setminus (w_1, w_2)$, there is a non-triangular inner face in $\tilde{G}_i$, containing a vertex w_j with $j > i$ in its interior. Then the path $w_j, \ldots, w_n$, which is disjoint from $\tilde{G}_i$, starts and ends in an interior and the exterior face of $\tilde{G}_i$, respectively, contradicting planarity. Thus the second condition of canonical ordering is satisfied for i. Moreover $\tilde{G}_i$ is inner-triangulated, has a simple outer cycle C_i containing the edge (w_1, w_2). Thus, the first condition holds for $i+1$.

(d) $\implies$ (c): Since $w_1, \ldots, w_n$ is a canonical ordering, (w_1, w_2) is an outer edge. In particular, $w_2 = v_{n-1}$ is an outer vertex. Clearly v_1 has at least two neighbors and every neighbor has a larger index, i.e., (c) holds for $i = 1$. Moreover, by the second condition of a canonical ordering every vertex $v_i = w_{n-i+1}$, for $i = 2, \ldots, n-2$, has at least two neighbors in $\tilde{G}_{n-i} = G \setminus G_i$, which is the subgraph induced by $v_n, \ldots, v_{i+1}$.

(d) $\implies$ (e): Consider the Schnyder realizer $(\mathcal{S}_1, \mathcal{S}_2, \mathcal{S}_3)$ of G defined by the canonical order $w_1, \ldots, w_n$ according to Lemma 2.1. For $i = 3, \ldots, n-1$ the outer cycle C_i of $\tilde{G}_i$ consists of the edge (w_1, w_2), the $w_i w_1$-path P_1 in $\mathcal{S}_1$, and the $w_i w_2$-path P_2 in $\mathcal{S}_2$. Due to the counterclockwise order of edges in a Schnyder realizer, no vertex on P_1, respectively P_2, has an incoming inner edge in $\tilde{G}_i$ in $\mathcal{S}_2$, respectively $\mathcal{S}_1$. Thus considering only edges in $\tilde{G}_i$ every outer vertex in $\tilde{G}_i$, different from w_1, w_2, is a leaf in $\mathcal{S}_1$ or $\mathcal{S}_2$. When in the canonical ordering vertex w_{i+1} is attached to $\tilde{G}_i$, some vertices on C_i become inner vertices of $\tilde{G}_{i+1}$. Every inner edge in $\tilde{G}_{i+1}$, which was not an edge in $\tilde{G}_i$ is in $\mathcal{S}_3$. Thus every inner vertex in $\tilde{G}_i$ is a leaf in either $\mathcal{S}_1$ or $\mathcal{S}_2$.

(e) $\implies$ (d): Consider a canonical ordering $w_1, w_2, \ldots, w_n$ of G defined by the Schnyder realizer $(\mathcal{S}_1, \mathcal{S}_2, \mathcal{S}_3)$ according to Lemma 2.1. Then $\{w_1, w_2, w_3\}$ is a triangle, hence C_3 consists of the edge (w_1, w_2), the $w_3 w_1$-path P_1 in $\mathcal{S}_1$, and the $w_3 w_2$-path P_2 in $\mathcal{S}_2$. For $i = 4, \ldots, n$, w_i is attached to $\tilde{G}_{i-1}$. If w_{i+1} has no edge to w_i then the outgoing edge of w_i in $\mathcal{S}_1$ or $\mathcal{S}_2$ is connected an inner vertex in the $w_i w_2$-path or $w_i w_1$-path, respectively. But this vertex would then have an incoming edge in both $\mathcal{S}_1$ and $\mathcal{S}_2$ – a contradiction. □

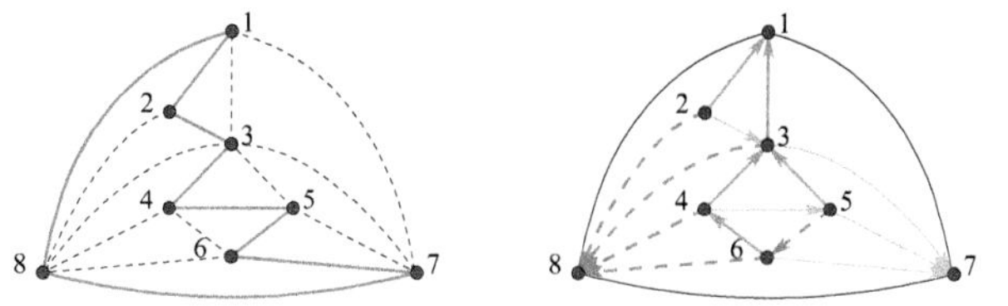

Figure 7: A graph with a one-legged Hamiltonian cycle and the corresponding Schnyder realizer.

Fig. 7 shows a one-legged Hamiltonian cycle, corresponding canonical ordering, and Schnyder realizer in a graph.

Once we have a one-legged Hamiltonian cycle, we can build a 6-sided cartogram via Lemma 5.1 in linear time. Alternately we could obtain a Schnyder wood, where every vertex is a leaf in $\mathcal{S}_1$ or $\mathcal{S}_2$, and hence obtain a 6-sided cartogram via the algorithm in Section 3. However, we prefer the construction of Lemma 5.1 due to its linear runtime.

Not every Hamiltonian maximal planar graph admits a one-legged Hamiltonian cycle; for example, the graph in Fig. 6 does not even admit a cartogram with 6-gons. However, we believe that some non-trivial subclasses of Hamiltonian maximal planar graphs are also one-legged Hamiltonian. In particular, we have the following conjecture:

CONJECTURE 5.1. *Every 4-connected maximal planar graph has a one-legged Hamiltonian cycle.*

Note that by Lemma 5.2, the conjecture is equivalent to asking whether every 4-connected maximal planar graph has a Hamiltonian cycle that induces a canonical ordering.

6. CONCLUSION AND OPEN PROBLEMS

We presented a cartogram construction for maximal planar graphs with optimal polygonal complexity. For the precise realization of the actual cartogram this approach requires numerical iteration. Even though the simple heuristic works well in practice, a natural open problem is whether everything can be computed with an entirely combinatorial linear-time approach. We also presented an entirely different linear-time construction for Hamiltonian maximal planar graphs and showed that the resulting 8-sided cartograms are optimal. Finally, we showed that if the graph admits a one-legged Hamiltonian cycle (for example outer-planar graphs), only 6 sides are needed. It remains to identify larger classes of planar graphs which are one-legged Hamiltonian and thus have 6-sided cartograms. We conjecture that 4-connected maximal planar graphs have this property.

All of the constructions in this paper yield area-universal rectilinear duals with optimal polygonal complexity. While Eppstein *et al.* [14] characterized area-universal *rectangular* layouts, a similar characterization remains an open problem for general area-universal *rectilinear* layouts.

Is it NP-hard to test whether a graph with a given area-assignment has a k-sided cartogram? Note that if a rectilinear layout of the graph is given, then it is NP-hard to test whether the layout can be deformed into a cartogram, even if all faces have at most 8 sides [3]. Does the freedom to choose the rectilinear layout help?

7. REFERENCES

[1] M. J. Alam, T. C. Biedl, S. Felsner, A. Gerasch, M. Kaufmann, and S. G. Kobourov. Linear-time algorithms for hole-free rectilinear proportional contact graph representations. In *ISAAC*, pp. 281–291, 2011.

[2] M. J. Alam, T. C. Biedl, S. Felsner, M. Kaufmann, and S. G. Kobourov. Proportional contact representations of planar graphs. In *Graph Drawing*, pp. 26–38, 2012.

[3] T. C. Biedl and B. Genc. Complexity of octagonal and rectangular cartograms. In *CCCG*, pp. 117–120, 2005.

[4] T. C. Biedl and L. E. R. Velázquez. Orthogonal cartograms with few corners per face. In *WADS*, pp. 98–109, 2011.

[5] A. L. Buchsbaum, E. R. Gansner, C. M. Procopiuc, and S. Venkatasubramanian. Rectangular layouts and contact graphs. *ACM Trans. on Algorithms*, 4, 2008.

[6] I. Cederbaum. Analogy between VLSI floorplanning problems and realisation of a resistive network. *IEE Circuits, Devices and Systems*, 139(1):99–103, 1992.

[7] T. Chen and M. K. H. Fan. On convex formulation of the floorplan area minimization problem. In *Symposium on Physical Design*, pp. 124–128, 1998.

[8] M. Chrobak and T. H. Payne. A linear-time algorithm for drawing a planar graph on a grid. *Information Processing Letters*, 54(4):241–246, 1995.

[9] M. de Berg, E. Mumford, and B. Speckmann. On rectilinear duals for vertex-weighted plane graphs. *Discrete Mathematics*, 309(7):1794–1812, 2009.

[10] M. de Berg, E. Mumford, and B. Speckmann. Optimal bsps and rectilinear cartograms. *Int. J. Computational Geometry and Applications*, 20(2):203–222, 2010.

[11] H. de Fraysseix, P. O. de Mendez, and P. Rosenstiehl. On triangle contact graphs. *Combinatorics, Probability and Computing*, 3:233–246, 1994.

[12] H. de Fraysseix, J. Pach, and R. Pollack. How to draw a planar graph on a grid. *Combinatorica*, 10(1):41–51, 1990.

[13] C. A. Duncan, E. R. Gansner, Y. F. Hu, M. Kaufmann, and S. G. Kobourov. Optimal polygonal representation of planar graphs. *Algorithmica*, 63(3):672–691, 2012.

[14] D. Eppstein, E. Mumford, B. Speckmann, and K. Verbeek. Area-universal rectangular layouts. In *SoCG*, pp. 267–276, 2009.

[15] X. He. On floor-plan of plane graphs. *SIAM J. Computing*, 28(6):2150–2167, 1999.

[16] R. Heilmann, D. A. Keim, C. Panse, and M. Sips. Recmap: Rectangular map approximations. In *INFOVIS*, pp. 33–40, 2004.

[17] D. House and C. Kocmoud. Continuous cartogram construction. In *Visualization*, pp. 197–204, 1998.

[18] T. Izumi, A. Takahashi, and Y. Kajitani. Air-pressure model and fast algorithms for zero-wasted-area layout of general floorplan. *IEICE Trans. Fundam. Electron., Commun. and Comp. Sci.*, E81–A:857–865, 1998.

[19] A. Kawaguchi and H. Nagamochi. Orthogonal drawings for plane graphs with specified face areas. In *TAMC 2007*, pp. 584–594, 2007.

[20] P. Koebe. Kontaktprobleme der konformen Abbildung. *Berichte über die Verhandlungen der Sächsischen Akademie der Wissenschaften zu Leipzig. Math.-Phys. Klasse*, 88:141–164, 1936.

[21] K. Koźmiński and E. Kinnen. Rectangular duals of planar graphs. *Networks*, 15:145–157, 1985.

[22] S. M. Leinwand and Y.-T. Lai. An algorithm for building rectangular floor-plans. In *Design Automation Conference*, pp. 663–664, 1984.

[23] C.-C. Liao, H.-I. Lu, and H.-C. Yen. Compact floor-planning via orderly spanning trees. *J. Algorithms*, 48:441–451, 2003.

[24] J. Michalek, R. Choudhary, and P. Papalambros. Architectural layout design optimization. *Engineering Optimization*, 34(5):461–484, 2002.

[25] T.-S. Moh, T.-S. Chang, and S. L. Hakimi. Globally optimal floorplanning for a layout problem. *IEEE Trans. on Circuits and Systems*, 43(9):713–720, 1996.

[26] E. Raisz. The rectangular statistical cartogram. *Geographical Review*, 24(3):292–296, 1934.

[27] I. Rinsma. Nonexistence of a certain rectangular floorplan with specified area and adjacency. *Environment and Planning*, 14:163–166, 1987.

[28] E. Rosenberg. Optimal module sizing in VLSI floorplanning by nonlinear programming. *Methods and Models of Operations Research*, 33:131–143, 1989.

[29] W. Schnyder. Embedding planar graphs on the grid. In *SODA*, pp. 138–148, 1990.

[30] W. Tobler. Thirty five years of computer cartograms. *Annal. Assoc. Amer. Geographers*, 94:58–73, 2004.

[31] T. Ueckerdt. *Geometric Representations of Graphs with low Polygonal Complexity*. PhD thesis, Technische Universität Berlin, 2011.

[32] P. Ungar. On diagrams representing graphs. *J. London Mathematical Society*, 28:336–342, 1953.

[33] M. J. van Kreveld and B. Speckmann. On rectangular cartograms. *Comput. Geom.*, 37(3):175–187, 2007.

[34] K. Wang and W.-K. Chen. Floorplan area optimization using network analogous approach. In *IEEE International Symposium on Circuits and Systems*, pp. 167–170, 1995.

[35] S. Wimer, I. Koren, and I. Cederbaum. Floorplans, planar graphs, and layouts. *IEEE Trans. on Circuits and Systems*, 35(3):267–278, 1988.

[36] K.-H. Yeap and M. Sarrafzadeh. Floor-planning by graph dualization: 2-concave rectilinear modules. *SIAM J. Computing*, 22:500–526, 1993.

Approximate Bregman Near Neighbors in Sublinear time: Beyond the triangle inequality *

[Extended Abstract]†

Amirali Abdullah
University of Utah
amirali@cs.utah.edu

John Moeller
University of Utah
moeller@cs.utah.edu

Suresh Venkatasubramanian
University of Utah
suresh@cs.utah.edu

ABSTRACT

In this paper, we give the first provably *approximate nearest neighbor* (ANN) algorithms for Bregman divergences over bounded domain. These process queries in $O(\log n)$ time for fixed dimensions. We also obtain poly-log$n$ bounds for a more abstract class of distance measures (containing Bregman divergences) which satisfy certain structural properties. Both of these bounds apply to the regular asymmetric Bregman divergences as well as their symmetrized versions. Our first algorithm resolves a query for a d-dimensional $(1+\varepsilon)$ ANN in $O\left(\left(\frac{\log n}{\varepsilon}\right)^{O(d)}\right)$ time and $O\left(n\log^{d-1} n\right)$ space and holds for generic μ-defective distance measures satisfying a *reverse triangle inequality*. Our second algorithm is more specific in analysis to the Bregman divergences and uses a further structural constant, the maximum ratio of second derivatives over each dimension of our domain (c_0). This allows us to locate a $(1+\varepsilon)$-ANN in $O(\log n)$ time and $O(n)$ space, where there is a further $(c_0)^d$ factor in the big-Oh for the query time.

Categories and Subject Descriptors

F.2.2 [**Nonnumerical Algorithms and Problems**]: [Geometrical problems and computations, Sorting and searching]

Keywords

ACM proceedings, Approximate Nearest Neighbor, Bregman divergences

1. INTRODUCTION

The nearest neighbor problem is one of the most extensively studied problems in data analysis. The past 20 years has seen tremendous research into the problem of computing near neighbors efficiently as well as approximately in different kinds of metric spaces.

*This research was supported in part by the NSF under grant CCF-0953066

†A full version of this paper is available at `http://arxiv.org/abs/1108.0835`

An important application of the nearest-neighbor problem is in querying content databases (images, text, and audio databases, for example). In these applications, the notion of similarity is based on a distance metric that arises from information-theoretic or other considerations. Popular examples include the Kullback-Leibler divergence [16], the Itakura-Saito distance [20] and the Mahalanobis distance [27]. These distance measures are examples of a general class of divergences called the *Bregman divergences* [9], and this class has received much attention in the realm of machine learning, computer vision and other application domains.

Bregman divergences possess a rich geometric structure but are not metrics in general, and are not even symmetric in most cases! While the geometry of Bregman divergences has been studied from a combinatorial perspective and for clustering, there have been no algorithms with provable guarantees for the fundamental problem of nearest-neighbor search. This is in contrast with extensive *empirical* study of the problem [10, 33, 34, 36, 37].

In this paper we present the first provably approximate nearest-neighbor (ANN) algorithms for Bregman divergences. Our first algorithm processes queries in $O(\log^d n)$ time using $O(n\log^d n)$ space and only uses general properties of the underlying distance function (which includes Bregman divergences as a special case). The second algorithm processes queries in $O(\log n)$ time using $O(n)$ space and exploits structural constants associated specifically with Bregman divergences. An interesting feature of our algorithms is that they extend the "ring-tree + quad-tree" paradigm for ANN searching beyond Euclidean distances and metrics of bounded doubling dimension to distances that might not even be symmetric or satisfy a triangle inequality.

1.1 Overview of Techniques

One approach [35] for near-neighbor search in low dimensions is to build a quad-tree-like data structure to process queries. Since the quad tree cells reduce in size by a constant factor at each stage, the triangle inequality can be used to infer that we never need to expand cells that are smaller than a fraction of the true nearest neighbor distance in order to get a good approximation to the nearest neighbor, i.e. we only need to explore our tree upto a certain depth. Packing arguments then limit the number of cells we need explore at each level of our tree, in order to obtain a query running time that is a function only of the desired error and the spread. The terms involving the spread are eliminated by finding an initial crude approximation to the nearest neighbor.

Since the resulting depth to explore is bounded by the logarithm of the ratio of the cell sizes, any c-approximation of the nearest neighbor results in a depth of $O(\log(c/\varepsilon))$. A standard data structure that yields such a crude bound is the *ring tree* [26].

Unfortunately, these methods (which work also for doubling metrics [14, 26, 7]) require two key properties: the existence of the triangle inequality, as well as packing bounds for fitting small-radius balls into large-radius balls. Bregman divergences in general are not symmetric and do not even satisfy a directed triangle inequality! We note in passing that such problems do not occur for the *exact* nearest neighbor problem in constant dimension: this problem reduces to point location in a Voronoi diagram, and Bregman Voronoi diagrams possess the same combinatorial structure as Euclidean Voronoi diagrams [8].

Reverse Triangle Inequality. The first observation we make is that while Bregman divergences do not satisfy a triangle inequality, they satisfy a weak *reverse triangle inequality*: along a line, the sum of lengths of two contiguous intervals is always *less* than the length of the union. This immediately yields a packing bound: intuitively, we cannot pack too many disjoint intervals in a larger interval because their sum would then be too large, violating the reverse triangle inequality.

μ-defectiveness. The second idea is to allow for a *relaxed* triangle inequality. We do so by defining a distance measure to be *μ-defective* w.r.t. a given domain if there exists a fixed $\mu \geq 1$ such that for all triples of points (x,y,z), we have $|D(x,y) - D(x,z)| \leq \mu D(y,z)$. This notion was first employed by Farago *et al* [19] for an algorithm based on optimizing average case complexity of ANN search.

A different natural way to relax the triangle inequality would be to show there exists a fixed $\mu < 1$ such that for all triples (x,y,z), the inequality $D(x,y) + D(y,z) \geq \mu D(x,z)$. In fact, this is the notion of *μ-similarity* used by Ackermann *et al* [3] to *cluster* data under a Bregman divergence. However, this version of a relaxed triangle inequality is too weak for the nearest-neighbor problem, as we see in Figure 1.

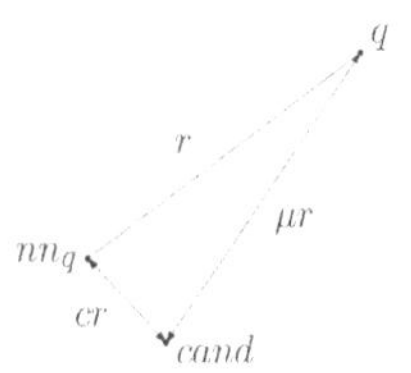

Figure 1: The ratio $\frac{D(q,\mathbf{cand})}{D(q,\mathbf{nn}_q)} = \mu$, regardless of c

Let q be a query point, cand a point in P such that $D(q,\text{cand})$ is known and nn_q is the nearest neighbor to q. The principle of grid related machinery is that for $D(q,\text{nn}_q)$ and $D(q,\text{cand})$ sufficiently large, and $D(\text{cand},\text{nn}_q)$ suffici the same lines asently small, we can verify that $D(q,\text{cand})$ is a $(1+\varepsilon)$ nearest neighbor, i.e. we can short-circuit our grid. Figure 1 illustrates a case where this short-circuit may not be valid for μ-similarity. Note that μ-similarity is satisfied here for any $c < 1$. Yet the ANN quality of cand, i.e., $\frac{D(q,\text{cand})}{D(q,\text{nn}_q)}$, need not be better than μ even for arbitrarily close nn_q and cand! This demonstrates the difficulty of naturally adapting the Ackermann notion of μ-similarity to finding a $1+\varepsilon$ nearest neighbor.

Unfortunately, Bregman divergences in general do not satisfy μ-defectiveness for any size domain or value of μ! One of our technical contributions is demonstrating in Section 4 that the square root of Bregman divergences does satisfy this property with μ depending on the boundedness of the domain and choice of divergence.

A Generic Approximate Near-Neighbor Algorithm.
After establishing that Bregman divergences satisfy the reverse triangle inequality and μ-defectiveness (Section 4), we first show (Section 6) that *any* distance measure satisfying the reverse triangle inequality, μ-defectiveness, and some mild technical conditions admits a ring-tree-based construction to obtain a weak near neighbor. However, applying it to a quad-tree construction creates a problem. The μ-defectiveness of a distance measure means that if we take a unit length interval and divide it into two parts, all we can expect is that each part has length between $1/2$ and $1/(\mu+1)$. This implies that while we may have to go down to level $\lceil \log_2 \ell \rceil$ to guarantee that all cells have side length $O(\ell)$, some cells might have side length as little as $\ell^{\log_2(\mu+1)}$, weakening packing bounds considerably.

We deal with this problem in two ways. For Bregman divergences, we can exploit geometric properties of the associated convex function ϕ (see Section 3) to ensure that cells at a fixed level have bounded size (Section 8); this is achieved by reexamining the second derivative ϕ''. For more general abstract distances that satisfy the reverse triangle inequality and μ-defectiveness, we instead construct a portion of the quad tree "on the fly" for each query (Section 7). While this is expensive, it still yields polylog(n) bounds for the overall query time in fixed dimensions. Both of these algorithms rely on packing/covering bounds that we prove in Section 5. An important technical point is that for exposition and simplicity, we initially work with the *symmetrized* Bregman divergences (of the form $D_{s\phi}(x,y) = D_\phi(x \mid y) + D_\phi(y \mid x)$), and then extend these results to general Bregman divergences (Section 9). We note that the results for symmetrized Bregman divergences might be interesting in their own right, as they have also been used in applications [33, 34, 32, 30].

2. RELATED WORK

Approximate nearest-neighbor algorithms come in two flavors: the high dimensional variety, where all bounds must be polynomial in the dimension d, and the constant-dimensional variety, where terms exponential in the dimension are permitted, but query times must be sublinear in n. In this paper, we focus on the constant-dimensional setting. The idea of using ring-trees appears in many works [25, 26, 23], and a good exposition of the general method can be found in Har-Peled's textbook [35, Chapter 11].

The Bregman distances were first introduced by Bregman [9]. They are the unique divergences that satisfy certain axiom systems for distance measures [17], and are key players in the theory of information geometry [5]. Bregman distances are used extensively in machine learning, where they have been used to unify boosting with different loss functions [15] and unify different mixture-model density estimation problems [6]. A first study of the algorithmic geometry of Bregman divergences was performed by Nielsen, Nock and Boissonnat [8]. This was followed by a series of papers analyzing the behavior of clustering algorithms under Bregman divergences [3, 2, 1, 28, 13].

Many heuristics have also been proposed for spaces endowed with Bregman divergences. Nielsen and Nock [31] developed a Frank-Wolfe like iterative scheme for finding minimum enclosing balls under Bregman divergences. Cayton [10] proposed the first nearest-neighbor search strategy for Bregman divergences, based on a clever primal-dual branch and bound strategy. Zhang *et al* [37] developed another prune-and-search strategy they argue is more scalable and uses operations better suited to use within a standard database system. For good broad reviews of near neighbor search in theory and practice, the reader is referred to the books by Har-Peled [35], Samet [24] and Shakhnarovich *et al* [29].

3. DEFINITIONS

In this paper we study the approximate nearest neighbor problem for distance functions D: Given a point set P, a query point q, and an error parameter ε, find a point $\mathrm{nn}_q \in P$ such that $D(\mathrm{nn}_q, q) \leq (1+\varepsilon)\min_{p\in P} D(p,q)$. We start by defining general properties that we will require of our distance measures. In what follows, we will assume that the distance measure D is *reflexive*: $D(x,y) = 0$ iff $x = y$.

Definition 3.1 (Monotonicity). *Let $M \subset \mathbb{R}$, $D : M \times M \to \mathbb{R}$ be a distance function, and let $a,b,c \in M$ where $a < b < c$. If the following are true for any such choice of a, b, and c: $0 \leq D(a,b) < D(a,c)$, $0 \leq D(b,c) < D(a,c)$, and $D(x,y) = 0$ iff $x = y$, then we say that D is* monotonic.

For a general distance function $D : M \times M \to \mathbb{R}$, where $M \subset \mathbb{R}^d$, we say that D is monotonic if it is monotonic when restricted to any subset of M parallel to a coordinate axis.

Definition 3.2 (Reverse Triangle Inequality). *Let M be a subset of $\mathbb{R}$. We say that a monotone distance measure $D : M \times M \to \mathbb{R}$ satisfies a* reverse triangle inequality *or RTI if for any three elements $a \leq b \leq c \in M$, $D(a,b) + D(b,c) \leq D(a,c)$*

Definition 3.3 (μ-defectiveness). *Let D be a symmetric monotone distance measure satisfying the reverse triangle inequality. We say that D is* μ-defective *with respect to domain M if for all $a,b,q \in M$,*

$$|D(a,q) - D(b,q)| < \mu D(a,b) \tag{3.1}$$

For an asymmetric distance measure D, we define left and right sided μ-defectiveness respectively as $|D(q,a) - D(q,b)| < \mu D(a,b)$ and $|D(a,q) - D(b,q)| < \mu D(b,a)$.

Two technical notes.

The distance functions under consideration are typically defined over $\mathbb{R}^d$. We will assume in this paper that the distance D is *decomposable*: roughly, that $D((x_1,\ldots,x_d),(y_1,\ldots,y_d))$ can be written as $g(\sum_i f(x_i,y_i))$, where g and f are monotone. This captures all the Bregman divergences that are typically used (with the exception of the Mahalanobis distance). We will also need to compute the diameter of an axis parallel box of side length ℓ. Our results hold as long as the diameter of such a box is $O(\ell d^{O(1)})$: note that this captures standard distances like those induced by norms, as well as decomposable Bregman divergences. In what follows, we will mostly make use of the *square root* of a Bregman divergence, for which the diameter of a box is $\ell(\mu+1)d^{\frac{1}{2}}$ or $\ell d^{\frac{1}{2}}$, and so without loss of generality we will use this in our bounds.

Bregman Divergences.

Let $\phi : M \subset \mathbb{R}^d \to \mathbb{R}$ be a *strictly convex* function that is differentiable in the relative interior of M. The *Bregman divergence* D_ϕ is defined as $D_\phi(x,y) = \phi(x) - \phi(y) - \langle \nabla\phi(y), x-y\rangle$

In general, D_ϕ is asymmetric. A *symmetrized* Bregman divergence by averaging is $D_{s\phi}(x,y) = \frac{1}{2}(D_\phi(x,y) + D_\phi(y,x))$

An important subclass of Bregman divergences are the *decomposable* Bregman divergences. Suppose ϕ has domain $M = \prod_{i=1}^d M_i$ and can be written as $\phi(x) = \sum_{i=1}^d \phi_i(x_i)$, where $\phi_i : M_i \subset \mathbb{R} \to \mathbb{R}$ is also strictly convex and differentiable in relint(S_i). Then $D_\phi(x,y) = \sum_{i=1}^d D_{\phi_i}(x_i,y_i)$ is a *decomposable* Bregman divergence.

Most commonly used Bregman divergences are decomposable: [11, Chapter 3] illustrates some of the commonly used ones, including the Euclidean distance, the KL-divergence, and the Itakura-Saito distance. In this paper we will hence limit ourselves to considering decomposable distance measures. We note that due to the primal-dual-based equality of $D_\phi(a,b)$ and $D_{\phi^*}(b^*,a^*)$, for our results on the asymmetric Bregman divergence we need only consider right-sided μ-defective distance measures.

4. STRUCTURAL PROPERTIES OF BREGMAN DIVERGENCES

The previous section defined key properties that we desire of a distance function D. The Bregman divergences (or modifications thereof) satisfy the following properties, as can be shown by direct computation.

Lemma 4.1. *Any one-dimensional Bregman divergence is monotonic.*

Lemma 4.2. *Any one-dimensional Bregman divergence satisfies the reverse triangle inequality. Let $a \leq b \leq c$ be three points in the domain of D_ϕ. Then it holds that $D_\phi(a,b) + D_\phi(b,c) \leq D_\phi(a,c)$ and that $D_\phi(c,b) + D_\phi(b,a) \leq D_\phi(c,a)$. The analaogous claim also holds for $D_{s\phi}$ and $\sqrt{D_{s\phi}}$.*

Note that this lemma can be extended similarly by induction to any series of n points between a and c. Further, using the relationship between $D_\phi(a,b)$ and the "dual" distance $D_{\phi^*}(b^*,a^*)$, we can show that the reverse triangle inequality holds going "left" as well: $D_\phi(c,b) + D_\phi(b,a) \leq D_\phi(c,a)$. These two separate reverse triangle inequalities together yield the result for $D_{s\phi}$. The corresponding proof for $\sqrt{D_{s\phi}}$ merely requires some algebraic manipulation.

For the remainder of the claims in this section, the proofs are straightforward but tedious and hence we consign them to the extended version. Now while the Bregman divergences satisfy both monotonicity and the reverse triangle inequality, they are not μ-defective with respect to *any* domain! An easy example of this is ℓ_2^2, which is also a Bregman divergence. A surprising fact however is that $\sqrt{D_{s\phi}}$ and $\sqrt{D_\phi}$ do satisfy μ-defectiveness (with μ depending on the bounded size of our domain). While we were unable to show precise bounds for μ in terms of the domain, the values are small. For example, for the symmetrized KL-divergence on the simplex where each coordinate is bounded between 0.1 and 0.9, μ is 1.22. If each coordinate is between 0.01 and 0.99,then μ is 2.42.

Lemma 4.3. *Given any interval $I = [x_1 x_2]$ on the real line, there exists a finite μ such that $\sqrt{D_{s\phi}}$ is μ-defective with respect to I, and $\sqrt{D_\phi}$ is both left and right-sided μ-defective with respect to I.*

We note that the result for $\sqrt{D_\phi}$ uses the following weak symmetry between $D_\phi(a,b)$ and $D_\phi(b,a)$ over a bounded interval $I \subset \mathbb{R}$.

Lemma 4.4. *Given a Bregman divergence D_ϕ and a bounded interval $I \subset \mathbb{R}$, $\sqrt{D_\phi(a,b)}/\sqrt{D_\phi(b,a)}$ is bounded by a constant c_0 $\forall a,b \in I$ where c_0 depends on the choice of divergence and interval.*

Lemma 4.5. *Given any interval $I = [x_1 x_2]$ on the real line, there exists finite μ such that $\sqrt{D_\phi}$ is left-sided and right-sided μ-defective with respect to I.*

Our results extend to d dimensions naturally, in that if M is a domain such that $\sqrt{D_{s\phi}}$ and $\sqrt{D_\phi}$ are μ-defective with respect to the projection of M onto each coordinate axis, then $\sqrt{D_{s\phi}}$ and $\sqrt{D_\phi}$ are μ-defective with respect to all of M.

Lemma 4.6. *Consider three points, $a = (a_1 \ldots a_d)$, $b = (b_1 \ldots b_d)$, $q = (q_1 \ldots, q_d)$ such that:*
$|\sqrt{D_{s\phi}(a_i,q_i)} - \sqrt{D_{s\phi}(b_i,q_i)}| < \mu\sqrt{D_{s\phi}(a_i,b_i)}$, $\forall i \in [1..d]$, *then it follows that* $\left|\sqrt{D_{s\phi}(a,q)} - \sqrt{D_{s\phi}(b,q)}\right| < \mu\sqrt{D_{s\phi}(a,b)}$.

Similarly, if $|\sqrt{D_\phi(a_i,q_i)} - \sqrt{D_\phi(b_i,q_i)}| < \mu\sqrt{D_\phi(a_i,b_i)}, \forall i \in [1..d]$, *then* $\left|\sqrt{D_\phi(a,q)} - \sqrt{D_\phi(b,q)}\right| < \mu\sqrt{D_\phi(b,a)}$.

5. PACKING AND COVERING BOUNDS

The aforementioned key properties (monotonicity, the reverse triangle inequality, decomposability, and μ-defectiveness) can be used to prove packing and covering bounds for a distance measure D. We now present some of these bounds.

Lemma 5.1 (Interval packing). *Given a monotone distance measure D satisfying the reverse triangle inequality, an interval $[ab]$ s.t. $D(a,b)=s$ and a collection of disjoint intervals I intersecting $[ab]$, where $\forall [xx'] \in I, D(x,x') \geq \ell$. Then $|I| \leq \frac{s}{\ell}+2$.*

Proof. Let I' be the intervals of I that are totally contained in $[ab]$. The combined length of all intervals in I' is at most $|I'|\ell$, but by the reverse triangle inequality, their total length cannot exceed s, so $|I'| \leq \frac{s}{\ell}$. There can be only two members of I not in I', so $|I| \leq \frac{s}{\ell}+2$. □

Corollary 5.1. *Given two points $a \leq b$ on the line s.t. $D(a,b)=s$, we can construct a packing of $[ab]$ by $r \leq \frac{1}{\varepsilon}$ intervals $[x_i x_{i+1}]$, $1 \leq i \leq r$ such that $D(a,x_0)=D(x_i,x_{i+1})=\varepsilon s$, $\forall i$ and $D(x_r,b) \leq \varepsilon s$. Here D is a monotone distance measure satisfying the reverse triangle inequality.*

Recall here that D_ϕ, $D_{s\phi}$ and $\sqrt{D_{s\phi}}$ satisfy the conditions of Lemma 5.1 and Corollary 5.1 as they satisfy an RTI and are decomposable. However, since $\sqrt{D_\phi}$ may not satisfy the reverse triangle inequality, we can instead prove a weaker packing bound on $\sqrt{D_\phi}$ by using D_ϕ.

Lemma 5.2 (Weak interval packing). *Given a distance measure $\sqrt{D_\phi}$,an interval $[ab]$ s.t. $\sqrt{D_\phi}(a,b)=s$ and a set of disjoint intervals I intersecting $[ab]$ where $I=\{[xx'] \mid [xx'], \sqrt{D_\phi}(x,x') \geq \ell\}$. Then $|I| \leq \frac{s^2}{\ell^2}+2$. Such a set of intervals can be explicitly constructed.*

The above bounds can be generalized to higher dimensions to provide packing bounds for balls and cubes (which we define below) with respect to a monotone, decomposable distance measure.

Definition 5.1. *Given a collection of d intervals $[a_i,b_i]$, such that $D(a_i,b_i)=s$ where $1 \leq i \leq d$, the* cube *in d dimensions is defined as $\prod_{i=i}^{d}[a_i b_i]$ and is said to have side length s.*

Lemma 5.3. *Given a d dimensional cube B_1 of side length s under distance measures D_ϕ, $D_{s\phi}$ and $\sqrt{D_{s\phi}}$, we can cover it with at most ε^d cubes of side length* exactly *εs. In the case of $\sqrt{D_\phi}$, we can cover it with at most ε^{2d} cubes of side length εs.*

Proof. Note that $D_{s\phi}$, D_ϕ, and $\sqrt{D_{s\phi}}$ satisfy conditions of Corollary 5.1. Hence we can construct a gridding of at most $\frac{1}{\varepsilon}$ points in each dimension spaced εs apart. We then take a product over all d dimensions, and the lemma follows trivially. For $\sqrt{D_\phi}$, we refer to Lemma 5.2 and follow the same procedure, gridding by at most $\frac{1}{\varepsilon^2}$ points in each dimension, spaced εs apart. □

Lemma 5.4. *Consider a ball B of radius s and center C with respect to a distance measure D. Then in the case of $D_{s\phi}$ and D_ϕ it can be covered with $\frac{2^d}{\varepsilon^d}$ balls of radius $d\varepsilon s$. In the case of $\sqrt{D_{s\phi}}$, B can be covered with $\frac{2^d}{\varepsilon^d}$ balls of radius $\sqrt{d}\varepsilon s$. And for $\sqrt{D_\phi}$, B can be covered by $\frac{2^d}{\varepsilon^{2d}}$ balls of radius $\sqrt{d}\varepsilon s$.*

Proof. Deferred to the expanded version. □

6. COMPUTING A ROUGH APPROXIMATION

Armed with our packing and covering bounds, we now describe how to compute a $O(\log n)$ rough approximate nearest-neighbor on our point set P under $\sqrt{D_\phi}$, which we will use in the next section to find the $(1+\varepsilon)$-approximate nearest neighbor. The technique we use is based on ring separators. Ring separators are a fairly old concept in geometry, notable appearances of which include the landmark paper by Indyk and Motwani [25]. Our approach here is heavily influenced by Har-Peled and Mendel [23], and by Krauthgamer and Lee [26], and our presentation is along the template of the textbook by Har-Peled [35, Chapter 11].

We note here that the constant of $d^{d/2}$ which appears in our final bounds for storage and query time is specific to $\sqrt{D_{s\phi}}$. However, an argument on the same lines will yield a constant of $d^{O(d)}$ for any generic μ-defective, symmetric RTI-satisfying decomposable distance measure such that the diameter of a cube of side length 1 is bounded by $d^{O(1)}$.

Let $B(m,r)$ denote the ball of radius r centered at m, and let $B'(m,r)$ denote the complement (or exterior) of $B(m,r)$. A *ring* R is the difference of two concentric balls: $R=B(m,r_2)\setminus B(m,r_1), r_2 \geq r_1$. We will often refer to the larger ball $B(m,r_2)$ as B_{out} and the smaller ball as B_{in}. We use $P_{\text{out}}(R)$ to denote the set $P \cap B_{\text{out}}$, and similarly use $P_{\text{in}}(R)$ as $P \cap B_{\text{in}}$, where we may drop the reference to R when the context is obvious. A *t-ring separator* $R_{P,c}$ on a point set P is a ring such that $\frac{n}{c} < |P_{\text{in}}| < (1-\frac{1}{c})n$, $\frac{n}{c} < |P_{\text{out}}| < (1-\frac{1}{c})n$, $r_2 \geq (1+t)r_1$ and $B_{\text{out}} \setminus B_{\text{in}}$ is empty. A t-ring tree is a binary tree obtained by repeated dispartition of our point set P using a t-ring separator.

Note that later on in this section, we will abuse this notation slightly by using ring-separators where the annulus is not actually empty, but we will bound the added space complexity and tree depth that we introduce. Finally, denote the minimum sized ball containing at least $\frac{n}{c}$ points of P by $B_{\text{opt},c}$; its radius is denoted by $r_{\text{opt},c}$.

We demonstrate that for any point set P, a ring separator exists and can always be computed efficiently. Applying this "separator" recursively on our point structure yields a ring-tree structure for searching our point set. Before we proceed further, we need to establish some properties of disks under a μ-defective distance. Lemma 6.1 is immediate from the definition of μ-defectiveness, Lemma 6.2 is similar to one obtained by Har-Peled and Mazumdar [22] and the idea of repeating points in both children of a ring-separator derives from a result by Har-Peled and Mendel [23].

Lemma 6.1. *Let D be a μ-defective distance, and let $B(m,r)$ be a ball with respect to D. Then for any two points $x,y \in B(m,r)$, $D(x,y) < (\mu+1)r$.*

Lemma 6.2. *Given a constant $1 \leq c \leq n$, we can compute in $O(nc)$ randomized time a $\mu+1$ approximation to the smallest radius ball containing $\frac{n}{c}$ points.*

Proof. We let S be a random sample from P, generated by choosing every point of P with probability $\frac{c}{n}$. Next, compute for every $p \in S$, the smallest disk centered at p containing c points. By median selection, this can be done in $O(n)$ time and since $E(|S|)=c$, this gives us the expected running time of $O(nc)$. Now, let r' be the minimum radius computed. Note that by lemma 6.1, if $|S \cap B_{\text{opt},c}| > 0$ then we have that $r' \leq (\mu+1)r_{opt}$. But since $B_{\text{opt},c}$ contains $\frac{n}{c}$ points, we can upper bound the probability of failure as the probability that we do not select any of the $\frac{n}{c}$ points in B_{opt} in our sample. Hence $Pr(|S \cap B_{\text{opt},c}| > 0) = 1-(1-\frac{c}{n})^{\frac{n}{c}} \geq 1-\frac{1}{e}$. □

We can now use Lemma 6.2 to construct our ring-separator.

Lemma 6.3. *For arbitrary t s.t. $1 < t < n$, we can construct a $\frac{1}{t}$-ring separator $R_{P,c}$ in $O(n)$ expected time on a point set P by repeating points.*

Proof. Using Lemma 6.2, we compute a ball $S = B(m, r_1)$ (where $m \in P$) containing $\frac{n}{c}$ points such that $r_1 \leq (\mu+1) r_{\text{opt},c}$ where c is a constant to be set. Consider the ball $\bar{S} = B(m, 2r_1)$. We shall argue that there must be $\frac{n}{c}$ points of P in $\bar{S}'$, for careful choices of c. As described in Lemma 5.4, $\bar{S}$ can be covered by 2^d hypercubes of side length $2r_1$, the union of which we shall refer to as H. Set $L = (\mu+1)\sqrt{d}$. Imagine a partition of H into a grid, where each cell is of side-length $\frac{r_1}{L}$ and hence of diameter at most $\Delta(\frac{r_1}{L}, d) = \frac{r_1}{\mu+1} \leq r_{\text{opt},c}$. A ball of radius $r_{\text{opt},c}$ on any corner of a cell will contain the entire cell, and so it will contain at most $\frac{n}{c}$ points, by the definition of $r_{\text{opt},c}$.

By Lemma 5.3 the grid on H has at most $2^d(2r_1/\frac{r_1}{L})^d = (4(\mu+1)\sqrt{d})^d$ cells. Set $c = 2(4(\mu+1)\sqrt{d})^d$. Then we have that $\bar{S} \subset H$ contains at most $\frac{n}{c}(4(\mu+1)\sqrt{d})^d = \frac{n}{2}$ points. Since the inner ball S contains at least $\frac{n}{c}$ points, and the outer ball $\bar{S}$ contains at most $\frac{n}{2}$ points, hence the annulus $\bar{S} \setminus S$ contains at most $\frac{n}{2} - \frac{n}{c}$ points. Now, divide $\bar{S} \setminus S$ into t rings of equal width, and by the pigeonhole principle at least one of these rings must contain at most $O(\frac{n}{t})$ points of P. Now let the inner ball corresponding to this ring be B_{in} and the outer ball be B_{out} and add these points to *both* children. Even for $t = 1$, each child contains at most $\frac{n}{2} + (\frac{n}{2} - \frac{n}{c}) = (1 - \frac{1}{c})n$ points. Also, the thickness of the ring is bounded by $\frac{2r_1 - r_1}{t}/2r_1 = \frac{1}{2t}$, i.e. it is a $O(\frac{1}{t})$ ring separator. Finally, we can check in $O(n)$ time if the randomized process of Lemma 6.2 succeeded simply by verifying the number of points in the inner and outer ring. □

Lemma 6.4. *Given any point set P, we can construct a $O(\frac{1}{\log n})$ ring-separator tree T of depth $O(d^{\frac{d}{2}}(\mu+1)^d \log n)$.*

Proof. Repeatedly partition P by Lemma 6.2 into P_{in}^{v} and P_{out}^{v} s.t. $\mathbf{v}$ is the parent node. Store only the point $\text{rep}_v = m \in P$ explicitly in node $\mathbf{v}$, the center of the ball $B(m, r_1)$. We continue this partitioning until we have nodes with only a single point contained in them. Since each child contains at least $\frac{n}{c}$ points (by proof of Lemma 6.3), each subset reduces by a factor of at least $1 - \frac{1}{c}$ at each step, and hence the depth of the tree is logarithmic. We recall that in Lemma 6.3, $c = O(d^{\frac{d}{2}}(\mu+1)^d)$. Hence the depth x can be obtained by solving for $n(1-\frac{1}{c})^x = 1$ to get that $x = O\left(d^{\frac{d}{2}}(\mu+1)^d \log n\right)$. □

Finally, we verify that the storage space required is not excessive.

Lemma 6.5. *To construct a $O(\frac{1}{\log n})$ ring-separator tree requires $O(n)$ storage and $O(d^{\frac{d}{2}}(\mu+1)^d n \log n)$ time.*

Proof. By Lemma 6.4 the depth bounds on our tree still hold. For storage, we have to bound the total number of points in our data structure after repetition, let us say P_R. Since each node corresponds to a splitting of P_R,there may be only $O(P_R)$ nodes and total storage. Note from the proof of Lemma 6.3, for a node containing x points, at most an additional $\frac{x}{\log n}$ may be duplicated in the two children. To bound this over each level of our tree, we sum across each node to obtain that the number of points T_i at the i-th level, as $T_i = T_{i-1}\left(1 + \frac{1}{\log T_{i-1}}\right)$ Note also by Lemma 6.4, the tree depth is $O(\log n)$ or bounded by $k \log n$ where k is a constant. Hence we only need to bound the storage at the level $i = O(\log n)$. We solve the recurrence, noting that $T_0 = n$ and $T_i > n$ for all i and hence $T_i < T_{i-1}(1 + \frac{1}{\log n})$. Thus the recurrence works out to:

$$T_i < n\left(1 + \frac{1}{\log n}\right)^{O(\log n)} < n\left(\left(1 + \frac{1}{\log n}\right)^{\log n}\right)^k < n(e^k), \tag{6.1}$$

where k and hence e^k is a constant and the main algebraic step is that $(1+\frac{1}{x})^x < e$. This proves that the number of points, and hence our storage complexity is $O(n)$. Multiplying the depth by $O(n)$ for computing the smallest ball across nodes on each level, gives us the time complexity of $O(n \log n)$. We note that other tradeoffs are available for different values of approximation quality (t) and construction time / query time. □

Algorithm and Quality Analysis.

Let best_q be the best candidate for nearest neighbor to q found so far and $D_{\text{near}} = D(\text{best}_q, q)$. Let nn_q be the exact nearest neighbor to q from point set P and $D_{\text{exact}} = D(\text{nn}_q, q)$ be the exact nearest neighbor distance. Finally, let **curr** be the tree node currently being examined by our algorithm, and rep_{curr} be a representative point $p \in P$ of **curr**. By convention r_v represents the radius of the *inner* ball associated with a node $\mathbf{v}$, and within each node $\mathbf{v}$ we store $\text{rep}_v = m_v$, which is the center of $B_{\text{in}}(m_v, r_v)$. The node associated with the inner ball B_{in} is denoted by $\mathbf{v}_{\text{in}}$ and the node associated with B_{out} is denoted by $\mathbf{v}_{\text{out}}$.

Lemma 6.6. *Given a t-ring tree T for a point set with respect to a μ-defective distance D, we can find a $O(\mu + \frac{2\mu^2}{t})$ nearest neighbor in $O((\mu+1)^d d^{\frac{d}{2}} \log n)$ time.*

Proof. Our search algorithm is a binary tree search. Whenever we reach node $\mathbf{v}$, if $D(\text{rep}_v, q) < D_{\text{near}}$ set $\text{best}_q = \text{rep}_v$ and $D_{\text{near}} = D(\text{rep}_v, q)$ as our current nearest neighbor and nearest neighbor distance respectively. Our branching criterion is that if $D(\text{rep}_v, q) < (1+\frac{t}{2}) r_v$, we continue search in $\mathbf{v}_{\text{in}}$, else we continue the search in $\mathbf{v}_{\text{out}}$. Since the depth of the tree is $O(\log n)$ by Lemma 6.4, this process will take $O(\log n)$ time.

Now, let $\mathbf{w}$ be the first node such that $\text{nn}_q \in \mathbf{w}_{\text{in}}$ but we searched in $\mathbf{w}_{\text{out}}$, or vice-versa. After examining rep_w, $D_{\text{near}} \leq D(\text{rep}_w, q)$ and D_{near} can only decrease at each step. An upper bound on $D(q, \text{rep}_w)/D(q, \text{nn}_q)$ yields a bound on the quality of the approximate nearest neighbor produced. In the first case, suppose $\text{nn}_q \in \mathbf{w}_{\text{in}}$, but we searched in $\mathbf{w}_{\text{out}}$. Then $D(\text{rep}_w, q) > \left(1+\frac{t}{2}\right) r_w$ and $D(\text{rep}_w, \text{nn}_q) < r_w$. Now μ-defectiveness implies that $\mu D(q, \text{nn}_q) > D(\text{rep}_w, q) - D(\text{rep}_w, \text{nn}_q)$, so we have $D(q, \text{nn}_q) > \frac{t}{2\mu} r_w$. And for the upper bound on $D(\text{rep}_w, q)/D(q, \text{nn}_q)$, apply μ-defectiveness to conclude $D(\text{rep}_w, q) - D(q, \text{nn}_q) < \mu D(\text{nn}_q, \text{rep}_w)$, which now yields

$$\frac{D(\text{rep}_w, q)}{D(q, \text{nn}_q)} < 1 + \mu \frac{r_w}{D(q, \text{nn}_q)} < 1 + \mu \frac{r_w}{\frac{t}{2\mu} r_w} = 1 + 2\frac{\mu^2}{t} \tag{6.2}$$

We now consider the other case. Suppose $\text{nn}_q \in \mathbf{w}_{\text{out}}$ and we search in $\mathbf{w}_{\text{in}}$ instead. By construction we must have $D(\text{rep}_w, q) < \left(1+\frac{t}{2}\right) r_w$ and $D(\text{rep}_w, \text{nn}_q) > (1+t) r_w$. Again, μ-defectiveness yields $D(q, \text{nn}_q) > \frac{t}{2\mu} r_w$. Now we can simply take the ratios of the two: $\frac{D(\text{rep}_w, q)}{D(q, \text{nn}_q)} < \frac{(1+\frac{t}{2}) r_w}{\frac{t}{2\mu} r_w} = \mu + \frac{2\mu}{t}$. Taking an upper bound of the approximation provided by each case, the ring tree provides us a $\mu + 2\frac{\mu^2}{t}$ approximation. □

Corollary 6.1. *Setting $t = \frac{1}{\log n}$, we can find a $O(\mu + 2\mu^2 \log n)$ approximate nearest neighbor to a query point q in $O(d^{\frac{d}{2}}(\mu + 1)^d \log(n))$ time, using a $O(\frac{1}{\log n})$ ring separator tree.*

Proof. By Lemma 6.4, Lemma 6.3 and Lemma 6.6. Note that we are slightly abusing notation in Lemma 6.3, in that the separating ring obtained there is not empty of points of P as originally stipulated. However remember that if nn_q is in the ring, then nn_q repeats in *both* children and cannot fall off the search path. Hence we can "pretend" the ring is empty as in our analysis in Lemma 6.6. □

7. OVERALL ALGORITHM

We give now our overall algorithm for obtaining a $1+\varepsilon$ nearest neighbor in $O\left(\frac{1}{\varepsilon^d}\log^{2d} n\right)$ query time. This section is tailored towards $\sqrt{D_{s\varphi}}$, but the general bounds will hold for any decomposable, symmetric, monotone distance measure that satisfies an RTI.

7.1 Preprocessing

We first construct an improved ring-tree R on our point set P in $O(n\log n)$ time as described in Lemma 6.5, with ring thickness $O(\frac{1}{\log n})$. We then compute an efficient orthogonal range reporting data structure on P in $O(n\log^{d-1} n)$ time, such as that described in [4] by Afshani *et al.* We note the main result we need:

Lemma 7.1. *We can compute a data structure on P that requires $O(n\log^{d-1} n)$ storage (and same construction time), such that given an arbitrary axis parallel box B we can determine in $O(\log^d n)$ query time a point $p \in P \cap B$ if $|P \cap B| > 0$*

7.2 Query handling

Given a query point q, we use R to find $q_{\text{rough}} \in P$ in $O(\log n)$ time such that $D_{\text{rough}} = D(q, q_{\text{rough}}) \leq (1+\mu^2\log n)D(q, nn_q)$. We can now use Lemma 5.4 to find a family F of 2^d cubes of side length exactly D_{rough} such that they cover $B(q, D_{\text{rough}})$. We use our range reporting structure to find a point $p \in P$ for all non-empty cubes in F in a total of $O(2^d \log^d n)$ time. These points act as representatives of the cubes for what follows. Note that nn_q must necessarily be in one of these cubes, and hence there must be a $(1+\varepsilon)$-nearest neighbor $q_{\text{approx}} \in P$ in some $G \in F$. To locate this q_{approx}, we construct a quadtree [35, Chapter 11] [18] for repeated bisection and search on each $G \in F$.

Algorithm 1 describes the overall procedure. We call the collection of all cells produced during the procedure a *quadtree*. We borrow the presentation in Har-Peled's book [35] with the important qualifier that we construct our quadtree at runtime. The terminology here is as introduced earlier in section 6.

Lemma 7.2. *Algorithm 1 will always return a $(1+\varepsilon)$-approximate nearest neighbor.*

Proof. Let best_q be the point returned by the algorithm at the end of execution. By the method of the algorithm, for all points p for which the distance is directly evaluated, we have that $D(\text{best}_q, q) < D(p,q)$. The terminology here is as in section 6. We look at points p which are *not* evaluated during the running of the algorithm, i.e. we did not expand their containing cells. But by the criterion of the algorithm for not expanding a cell, it must be that $D(\text{best}_q, q)(1-\frac{\varepsilon}{2}) < D(p,q)$. For $\varepsilon < 1$, this means that $(1+\varepsilon)D(p,q) > D(\text{best}_q, q)$ for any $p \in P$, including nn_q. So best_q is indeed a $1+\varepsilon$ approximate nearest neighbor. □

Algorithm 1 QueryApproxNN(P, **root**, q)

```
Instantiate a queue Q containing all cells from F along with their
representatives and enqueue root
Let D_near = D(rep_root, q), best_q = rep_root
repeat
  Pull off the head of the queue and place it in curr.
  if D(rep_curr, q) < D(best_q, q) then
    Let best_q = rep_curr, D_near = D(best_q, q)
    Bisect curr according to procedure of Lemma 7.3; place
    the result in {G_i}.
    for all G_i do
      As described in 7.3, check if G_i is non-empty by passing
      it to our range reporting structure, which will also return
      us some p ∈ P if G_i is not empty.
      Also check if G_i may contain a point closer than (1 −
      ε/2)D_near to q. (This may be done in O(d) time for each
      cell, given the coordinates of the corners.)
      if G_i is non-empty AND has a close enough point to q
      then
        Let rep_{G_i} = p
        Enqueue G_i
      end if
    end for
  end if
until Q is empty
Return best_q
```

We must analyze the time complexity of a single iteration of our algorithm, namely the complexity of a subdivision of a cube G and determining which of the 2^d subcells of G are non-empty.

Lemma 7.3. *Let G be a cube with maximum side length s and G_i its subcells produced by bisecting along each side of G. For all non-empty subcubes G_i of G, we can find $p_i \in P \cap G_i$ in $O(2^d \log^d n)$ total time complexity, and the maximum side length of any G_i is at most $\frac{s}{2}$.*

Proof. Note that G is defined as a product of d intervals. For each interval, we can find an approximate bisecting point in $O(1)$ time(see extended version) and by the RTI each subinterval is of length at most $\frac{s}{2}$. This leads to an $O(d)$ cost to find a bisection point for all intervals, which define $O(2^d)$ subcubes or children.

We pass each subcube of G to our range reporting structure. By lemma 7.1, this takes $O(\log^d n)$ time to check emptiness or return a point $p_i \in P$ contained in the child, if non-empty. Since there are $O(2^d)$ non-empty children of G, this implies a cost of $2^d(\log^d n)$ time incurred.

Checking each of the non-empty subcubes G_i to see if it may contain a point closer than $(1-\frac{\varepsilon}{2})D_{\text{near}}$ to q takes a further $O(d)$ time per cell or $O(d2^d)$ time. □

We now bound the number of cells that will be added to our search queue. We do so indirectly, by placing a lower bound on the maximum side length of all such cells.

Lemma 7.4. *Algorithm 1 will not add the children of node **C** to our search queue if the maximum side length of **C** is less than $\frac{\varepsilon D(q, nn_q)}{2\mu\sqrt{d}}$.*

Proof. Let $\Delta(\mathbf{C})$ represent the diameter of cell $\mathbf{C}$. By construction, we can expand $\mathbf{C}$ only if some subcell of $\mathbf{C}$ contains a point p such that $D(p,q) \leq (1-\frac{\varepsilon}{2})D_{\text{near}}$. Note that since $\mathbf{C}$ is examined, we have $D_{\text{near}} \leq D(\text{rep}_C, q)$. Now assuming we expand $\mathbf{C}$, then we must have:

$$\mu\Delta(\mathbf{C}) > D(\text{rep}_C, q) - D(p,q) \geq D_{\text{near}} - (1-\frac{\varepsilon}{2})D_{\text{near}} = \frac{\varepsilon}{2}D_{\text{near}}, \quad (7.1)$$

so $\varepsilon D_{\text{near}}/(2\mu) < \Delta(\mathbf{C})$. Note that $D(\text{rep}_C, q) \leq D_{\text{near}}$, that by definition $D(q, \text{nn}_q) \leq D_{\text{near}}$, and that $\Delta(\mathbf{C}) \leq \sqrt{d}s$, where s is the maximum side length of $\mathbf{C}$. Making the appropriate substitutions yields our required bound. □

Given the bound on quadtree depth (Lemma 7.4), and using the fact that at most 2^{xd} nodes are expanded at level x, we have:

Lemma 7.5. *Given a cube G of side length D_{rough}, we can compute a $(1+\varepsilon)$-nearest neighbor in $O\left(\frac{1}{\varepsilon^d}2^d\mu^d d^{\frac{d}{2}}\left(\frac{D_{rough}}{D(q,nn_q)}\right)^d \log^d n\right)$ time.*

Proof. Consider a quadtree search from q on a cube G of side length D_{rough}. By lemma 7.4, our algorithm will not expand cells with all side lengths smaller than $\frac{\varepsilon D(q,\text{nn}_q)}{2\mu\sqrt{d}}$. But since the side length reduces by at least half in each dimension upon each split, all side lengths are less than this value after $x = \log\left(D_{\text{rough}}/\frac{\varepsilon D(q,\text{nn}_q)}{2\mu\sqrt{d}}\right)$ repeated bisections of our root cube.

Noting that $O(\log^d n)$ time is spent at each node by lemma 7.3, and that at the x-th level the number of nodes expanded is 2^{xd}, we get a final bound of $O\left(\frac{1}{\varepsilon^d}2^d\mu^d d^{\frac{d}{2}}\left(\frac{D_{\text{rough}}}{D(q,\text{nn}_q)}\right)^d \log^d n\right)$ time. □

Substituting $D_{\text{rough}} = \mu^2 D(q, \text{nn}_q)\log n$ in Lemma 7.5 gives us a bound of $O\left(2^d \frac{1}{\varepsilon^d}\mu^{3d}d^{\frac{d}{2}}\log^{2d} n\right)$. This time is per cube of F that covers $B(q, D_{\text{rough}})$. Noting that there are 2^d such cubes gives us a final time complexity of $O\left(2^{2d}\frac{1}{\varepsilon^d}\mu^{3d}d^{\frac{d}{2}}\log^{2d} n\right)$. For the space complexity of our run-time queue, observe that the number of nodes in our queue increases only if a node has more than one non-empty child, i.e., there is a split of our n points. Since our point set mdiameteray only split n times, this gives us a bound of $O(n)$ on the space complexity of our queue.

8. LOGARITHMIC BOUNDS, WITH FURTHER ASSUMPTIONS.

For a given $D_{s\phi}$, let $c_0 = \max_{i\in[1..d]}\sqrt{\frac{\max_x \phi_i''(x)}{\min_y \phi_i''(y)}}$. We conjecture that $c_0 = \Theta(\mu)$ although we cannot prove it. In particular, we show that if we assume a bounded c_0 (in addition to μ), we can obtain a $1+\varepsilon$ nearest neighbor in time $O(\log n + (\frac{1}{\varepsilon})^d)$ time for $\sqrt{D_{s\phi}}$. We do so by constructing a *Euclidean* quadtree T on our set in preproccessing and using c_0 and μ to express the bounds obtained in terms of $\sqrt{D_{s\phi}}$.

We will refer to the Euclidean distance l_2 as D_e and note first the following key relation between $\sqrt{D_{s\phi}}$ and D_e.

Lemma 8.1. *Suppose we are given a interval $I = [x_1 x_2] \subset \mathbb{R}$ s.t. $x_1 < x_2$, $D_e(x_1, x_2) = r_e$, and $\sqrt{D_{s\phi}(x_1,x_2)} = r_\phi$. Suppose we divide I into m subintervals of equal length with endpoints $x_1 = a_0, a_1, \ldots a_{m-1}, a_m = x_2$, where $a_i < a_{i+1}$ and $D_e(a_i, a_{i+1}) = r_e/m$, $\forall i \in [0..m-1]$. Then $\frac{r_\phi}{c_0 m} \leq \sqrt{D_{s\phi}(a_i,a_{i+1})} \leq \frac{c_0 r_\phi}{m}$.*

Proof Sketch: We can relate $\sqrt{D_{s\phi}}$ to D_e via the Taylor expansion of $\sqrt{D_{s\phi}}$: $\sqrt{D_{s\phi}(a,b)} = \sqrt{\phi''(\bar{x})}D_e(a,b)$ for some $\bar{x} \in [ab]$. Combining this with c_0 yields $\frac{\min_i \sqrt{D_{s\phi}(a_i,a_{i+1})}}{\sqrt{D_{s\phi}(x_1,x_2)}} \geq \frac{D_e(a_i,a_{i+1})}{c_0 D_e(x_1,x_2)} = \frac{1}{c_0 m}$ and $\frac{\max_i \sqrt{D_{s\phi}(a_i,a_{i+1})}}{\sqrt{D_{s\phi}(x_1,x_2)}} \leq c_0\frac{D_e(a_i,a_{i+1})}{D_e(x_1,x_2)} = \frac{c_0}{m}$. □

Corollary 8.1. *If we recursively bisect an interval $I = [x_1 x_2] \subset \mathbb{R}$ s.t. $D_e(x_1,x_2) = r_e$ and $\sqrt{D_{s\phi}(x_1,x_2)} = r_\phi$ into 2^i equal subintervals (under D_e), then $\frac{r_\phi}{c_0 2^i} \leq \sqrt{D_{s\phi}(a_k,a_{k+1})} \leq \frac{c_0 r_\phi}{2^i}$ for any of the subintervals $[a_k a_{k+1}]$ so obtained. Hence after $\log\frac{c_0 r_\phi}{x}$ subdivisions, all intervals will be of length at most x under $\sqrt{D_{s\phi}}$. Also, given a cube of initial side length r_ϕ, after $\log\frac{c_0 r_\phi}{x}$ repeated bisections (under D_e) the diameter will be at most $\sqrt{d}x$ under $\sqrt{D_{s\phi}}$.*

We find the smallest enclosing Bregman cube of side length s that bounds our point set, and then construct our compressed Euclidean quadtree in preprocessing. Corollary 8.1 gives us that for cells formed at the i-th level of decomposition, the side length under $\sqrt{D_{s\phi}}$ is between $\frac{s}{c_0 2^i}$ and $\frac{c_0 s}{2^i}$. Refer to these cells formed at the i-th level as L_i.

Lemma 8.2. *Given a ball B of radius r under $\sqrt{D_{s\phi}}$, let $i = \log\frac{s}{c_0 r}$. Then $|L_i \cap B| \leq O(2^d)$ and the side length of each cell in L_i is between r and ${c_0}^2 r$ under $\sqrt{D_{s\phi}}$. We can also explicitly retrieve the quadtree cells corresponding to $|L_i \cap B|$ in $O(2^d \log n)$ time.*

Proof. Note that for cells in L_i, we have side lengths under $\sqrt{D_{s\phi}}$ between $\frac{s}{c_0 2^i}$ and $\frac{c_0 s}{2^i}$ by Corollary 8.1. Substituting $i = \log\frac{s}{c_0 r}$, these cells have side length between r and ${c_0}^2 r$ under $\sqrt{D_{s\phi}}$. By the reverse triangle inequality and a similar argument to Lemma 5.4, we get our required bound for $|L_i \cap B|$. In preconstruction of our quadtree T we maintain for each dimension the corresponding interval quadtree T_k, $\forall k \in [1..d]$. Observe this incurs at most $O(n)$ storage, with d in the big-Oh. For retrieving the actual cells $|L_i \cap B|$, we first find the $O(1)$ intervals from level i in each T_k that may intersect B. Taking a product of these, we get $O(2^d)$ cells which are a superset of the canonical cells $L_i \subset T$. Each cell may be looked up in $O(\log n)$ time from the compressed quadtree [35] so our overall retrieval time is $O(2^d \log n)$. □

Given query point q, we first obtain in $O(\log n)$ time with our ring-tree a rough $O(n)$ ANN q_{rough} s.t. $D_{\text{rough}} = \sqrt{D_{s\phi}(q, q_{\text{rough}})} = \mu^2 n\sqrt{D_{s\phi}(q,\text{nn}_q)}$. By Lemma 8.2, we have $O(2^d)$ quadtree cells intersecting $B(q, \sqrt{D_{s\phi}(q, q_{\text{rough}})})$.

Let us call this collection of cells Q. We then carry out a quadtree search on each element of Q. Note that we expand only cells which may contain a point nearer to query point q than the current best candidate. We bound the depth of our search using μ-defectiveness similar to Lemma 7.4:

Lemma 8.3. *We will not expand cells of diameter less than $\frac{\varepsilon\sqrt{D_{s\phi}(q,nn_q)}}{2\mu}$ or cells whose side-lengths w.r.t. $\sqrt{D_{s\phi}}$ are less than $\frac{\varepsilon\sqrt{D_{s\phi}(q,nn_q)}}{2\mu\sqrt{d}}$.*

For what follows, refer to our *spread* as $\beta = \frac{D_{\text{rough}}}{\sqrt{D_{s\phi}(q,\text{nn}_q)}}$.

Lemma 8.4. *We will only expand our tree to a depth of $k = \log(2{c_0}^3\mu\beta\sqrt{d}/\varepsilon)$.*

Proof. Using Lemma 8.3 and Corollary 8.1, each cell of Q will be expanded only to a depth of $k = \log\left(c_0{c_0}^2 D_{\text{rough}}/\frac{\varepsilon\sqrt{D_{s\phi}(q,\text{nn}_q)}}{2\mu\sqrt{d}}\right)$. This gives us a depth of $\log(2{c_0}^3\mu\beta\sqrt{d}/\varepsilon)$. □

Lemma 8.5. *The number of cells examined at the i-th level is* $n_i = O\left(2^d\left(\mu^d d^{\frac{d}{2}} c_0^{4d} + (\frac{2^i c_0}{\beta})^d\right)\right)$.

Proof. Recalling that the cells of Q start with side length at most $c_0^2 D_{\text{rough}}$, at the i-th level the diameter of cells is at most $\frac{c_0^3\sqrt{d}D_{\text{rough}}}{2^i}$ under $\sqrt{D_{s\phi}}$, by Corollary 8.1. Hence by μ-defectiveness, there must be some point examined by our algorithm at distance at most $D_{\text{best}} = \sqrt{D_{s\phi}(q,\text{nn}_q)} + \frac{\mu c_0^3\sqrt{d}D_{\text{rough}}}{2^i}$. Note that our algorithm will only expand cells within this distance of q.

The side-length of a cell $\mathbf{C}$ at this level is at least $\Delta(\mathbf{C}) = \frac{D_{\text{rough}}}{c_0 2^i}$. Applying the packing bounds from Lemma 5.3, and the fact that $(a+b)^d < 2^d(a^d+b^d)$, the number of cells expanded is at most

$$n_i = \left(\frac{D_{\text{best}}}{\Delta(\mathbf{C})}\right)^d = O\left(2^d\left(\mu^d d^{\frac{d}{2}} c_0^{4d} + \left(\frac{c_0 2^i}{\beta}\right)^d\right)\right). \quad \square$$

Finally we add the n_i to get the total number of nodes explored:

$$\sum_i n_i = O\left(2^d \mu^d d^{\frac{d}{2}} c_0^{4d} \log(2c_0{}^3\mu\beta\sqrt{d}/\varepsilon) + 2^{2d} c_0^{4d}\mu^d d^{\frac{d}{2}}/\varepsilon^d\right).$$

Recalling that $\beta = \frac{D_{\text{rough}}}{\sqrt{D_{s\phi}(q,\text{nn}_q)}} = \mu^2 n$, substituting back and ignoring lower order terms, the time complexity is

$$O\left(2^d\mu^d d^{\frac{d}{2}} c_0^{4d}\log n + 2^{2d}c_0^{4d}\mu^d d^{\frac{d}{2}}/\varepsilon^d\right).$$

Accounting for the 2^d cells in Q that we need to search, this adds a further 2^d multiplicative factor. We note that compressed Euclidean quadtrees can be built in $O(n\log n)$ time and require $O(n)$ space [35], which matches our bound for the ring-tree search phase of our algorithm requiring $O(n\log n)$ time and $O(n)$ space.

9. THE GENERAL CASE: ASYMMETRIC DIVERGENCES

Without loss of generality we will focus on the *right-sided* nearest neighbor: given a point set P, query point q and $\varepsilon > 0$, find $x \in P$ that approximates $\min_{p\in P} D(p,q)$ to within a factor of $(1+\varepsilon)$. Since a Bregman divergence is not in general μ-defective, we will consider instead $\sqrt{D_\phi}$: by monotonicity and with an appropriate choice of ε, the result will carry over to D_ϕ.

We list three issues that have to be resolved to complete the algorithm. Firstly, because of asymmetry, we cannot bound the diameter of a quadtree cell $\mathbf{C}$ of side length s by $s\sqrt{d}$. However as the proof of Lemma 5.4 shows, we can choose a *canonical corner* of a cell such that a (directed) ball of radius $s\sqrt{d}$ centered at that corner covers the cell. By μ-defectiveness, we can now conclude that the diameter of $\mathbf{C}$ is at most $(\mu+1)s\sqrt{d}$ (note that this incurs an extra factor of $\mu+1$ in all expressions). Secondly, while $\sqrt{D_\phi}$ satisfies μ-defectiveness (unlike D_ϕ) the opposite is true for the reverse triangle inequality, which is satisfied by D_ϕ but not $\sqrt{D_\phi}$. This requires the use of a weaker packing bound based on Lemma 5.2, introducing dependence in $1/\varepsilon^2$ instead of $1/\varepsilon$. And last, asymmetry implies the need for care in the use of directionality when proving our bounds.

With some small adjustments, similar bounds can be obtained for more generic asymmetric, monotone, decomposable and μ-defective distance measures satisfying packing bounds. The left-sided asymmetric nearest neighbor can be determined analogously.

Finally, given a bounded domain D, we have that $\sqrt{D_\phi}$ is left-sided μ-defective for some μ_L and right sided μ-defective for some μ_R (see extended version for detailed proof). For what follows, let $\mu = \max(\mu_L,\mu_R)$ and describe D as simply μ-defective.

Most of the proofs here mirror their counterparts in Sections 6 and 7.

9.1 Asymmetric ring-trees

Since we focus on *right*-near-neighbors, all balls and ring separators referred to will use *left-balls*, i.e. balls $B(m,r) = \{x \mid D(m,x) < r\}$. As in Section 6, we will design a ring-separator algorithm and use that to build a ring-separator tree.

Lemma 9.1. *Let D be a μ-defective distance, and let $B(m,r)$ be a left-ball with respect to D. Then for any two points $x,y \in B(m,r)$, $D(x,y) < (\mu+1)r$.*

As in Lemma 6.2 we can construct (in $O(nc)$ expected time) a $(\mu+1)$-approximate left-ball enclosing $\frac{n}{c}$ points. This in turn yields a ring-separator construction, and from it a ring tree with an extra $(\mu+1)^d d^{\frac{d}{2}}$ factor in depth as compared to symmetric ring-trees, due to the weaker packing bounds for $\sqrt{D_\phi}$.

We note that the asymptotic bounds for ring-tree storage and construction time follow from purely combinatorial arguments and hence are unchanged for $\sqrt{D_\phi}$. Once we have the ring-tree, we can use it as before to identify a rough near-neighbor for a query q; once again, exploiting μ-defectiveness gives us the desired approximation guarantee for the result. As before, all proofs are deferred to the extended version.

Lemma 9.2. *Given any constant $1 \le c \le n$, we can compute in $O(nc)$ randomized time a left-ball $B(m,r')$ such that $r' \le (\mu+1)r_{opt,c}$ and $B(m,r') \cap P \ge \frac{n}{c}$.*

Lemma 9.3. *There exists a constant c (which depends only on d and μ), such that for any d-dimensional point set P and any μ-defective distance D, we can find a $O(\frac{1}{\log n})$ left-ring separator $R_{P,c}$.*

We proceed now to the construction of our ring-tree using the basic ring-separator structure of Lemma 9.3.

Lemma 9.4. *Given any point set P, we can construct a $O(\frac{1}{\log n})$ left ring-separator tree T of depth $O(d^d(\mu+1)^{2d}\log n)$.*

Note that Lemma 9.4 also serves to bound the query time of our data structure. We need only bound the approximation quality. The derivation is similar to Lemma 6.6, but with some care about directionality.

Lemma 9.5. *Given a t-ring tree T for a point set with respect to a right-sided μ-defective distance D, we can find a $O(\mu+\frac{2\mu^2}{t})$ nearest neighbor $O((\mu+1)^{2d}d^d\log n)$ time.*

Lemma 9.5, Lemma 9.3 and Lemma 9.4 together give us our required result.

9.2 Asymmetric quadtree decomposition

As in Section 7, we use the approximate near-neighbor returned by the ring-separator-tree query to progressively expand cells, using a subdivide-and-search procedure similar to Algorithm 1. A key difference is the weaker packing bounds on cell bisection.

Lemma 9.6. *Let G be a cube with maximum side length s and G_i its subcells produced by partitioning each side of G into two equal intervals. For all non-empty subcubes G_i of G, we can find $p_i \in P \cap G_i$ in $O(2^d \log^d n)$ total time complexity, and the maximum side length of any G_i is at most $\frac{s}{\sqrt{2}}$.*

Proof. Note that G is defined as a product of d intervals. For each interval, we can find an approximate bisecting point in $O(1)$ time.

Here the bisection point x of interval $[ab]$ is such that $\sqrt{D_\phi(a,x)} = \sqrt{D_\phi(x,b)}$. By resorting to the RTI for D_ϕ, we get that $D_\phi(a,x) + D_\phi(x,b) < s^2$ and hence $D_\phi(a,x) = D_\phi(x,b) < \frac{s^2}{2}$ which implies $\sqrt{D_\phi(a,x)} = \sqrt{D_\phi(x,b)} < \frac{s}{\sqrt{2}}$. The rest of our proof follows as in Lemma 7.3. □

We now bound the number of cells that will be added to our search queue. We do so indirectly, by placing a lower bound on the maximum side length of all such cells, and note that for the asymmetric case we get an additional factor of $\frac{1}{\mu+1}$.

Lemma 9.7. *Algorithm 1 will not add the children of node **C** to our search queue if the maximum side length of **C** is less than* $\frac{\varepsilon D(nn_q,q)}{2\mu(\mu+1)\sqrt{d}}$.

The remaining analysis is similar to Section7, the full proof of which we defer to the expanded version.

Theorem 9.1. *We can compute a* $(1+\varepsilon)$ *right-sided nearest neighbor to q in* $O\left(2^{3d}\frac{1}{\varepsilon^{2d}}\mu^{6d}(\mu+1)^{2d}d^d\log^{3d}n\right)$ *time.*

Logarithmic bounds for Asymmetric Bregman divergences.

We now extend our logarithmic bounds from Section 8 to asymmetric Bregman divergence $\sqrt{D_\phi}$. First note that the following Lemma goes through by identical argument to Lemma 8.1.

Lemma 9.8. *Suppose we are given an interval* $I = [x_1x_2] \subset \mathbb{R}$ *s.t.* $x_1 < x_2$, $D_e(x_1,x_2) = r_e$ *and* $\sqrt{D_\phi(x_1,x_2)} = r_\phi$. *Suppose we divide I into m subintervals of equal length with endpoints* $x_1 = a_0 < a_1 < \ldots < a_{m-1} < a_m = x_2$ *where* $D_e(a_i,a_{i+1}) = r_e/m$, *for all* $i \in [0..m-1]$. *Then* $\frac{r_\phi}{c_0 m} \leq \sqrt{D_\phi(a_i,a_{i+1})} \leq \frac{c_0 r_\phi}{m}$.

Corollary 9.1. *If we recursively bisect an interval* $I = [x_1x_2] \subset \mathbb{R}$ *s.t.* $D_e(x_1,x_2) = r_e$ *and* $\sqrt{D_\phi(x_1,x_2)} = r_\phi$ *into* 2^i *equal subintervals (under* D_e*), then* $\frac{r_\phi}{c_0 2^i} \leq \sqrt{D_\phi(a_k,a_{k+1})} \leq \frac{c_0 r_\phi}{2^i}$ *for any of the subintervals* $[a_k a_{k+1}]$ *so obtained. Hence after* $i = \lceil \log \frac{c_0 r_\phi}{x} \rceil$ *subdivisions, all intervals will be of length at most x.*

We now construct a compressed Euclidean quad tree as before, modifying the Section 8 analysis slightly to account for the weaker packing bounds for $\sqrt{D_\phi}$ and the extra $\mu+1$ factor on the diameter of a cell.

Theorem 9.2. *Given an asymmetric decomposable Bregman divergence* D_ϕ *that is* μ*-defective over a domain with associated* c_0 *as in Section 8, we can compute a* $(1+\varepsilon)$*-approximate right-near-neighbor in time* $O\left((\mu+1)^d d^{\frac{d}{2}}\log n + \left(\frac{2c_0{}^4(\mu+1)\mu^3\sqrt{d}}{\varepsilon}\right)^d\right)$.

We note now a slightly modified packing bound due to $\sqrt{D_\phi}$ not having a direct RTI. It follows from the Lagrange form and full proof is deferred to the expanded version.

Lemma 9.9. *Given an interval* $[x_1x_2] \subset \mathbb{R}$ *s.t.* $\sqrt{D_\phi(x_1,x_2)} = r > 0$, *and intervals with endpoints* $a_0 < a_1 < \ldots < a_{m-1} < a_m$, *s.t. for all* $i \in [0..m-1]$, $\sqrt{D_\phi(a_i,a_{i+1})} \geq l$, *at most* $O(\frac{c_0 r}{l})$ *such intervals intersect* $[x_1x_2]$.

Corollary 9.2. *Given a ball B of radius r under* $\sqrt{D_\phi}$, *there can be at most* $c_0^d(\frac{r}{l})^d$ *disjoint cubes that can intersect B where each cube has side length at least l under* $\sqrt{D_\phi}$.

As before, we find the smallest enclosing Bregman cube of side length s that encloses our point set, and then construct a compressed Euclidean quad-tree in pre-processing. Let L_i denote the cells at the i-th level.

Lemma 9.10. *Given a ball B of radius r under* $\sqrt{D_\phi}$, *let* $i = \log\frac{s}{c_0 r}$. *Then* $|L_i \cap B| \leq O(c_0{}^d)$ *and the side length of each cell in* L_i *is between r and* $c_0{}^2 r$ *under* $\sqrt{D_{s\phi}}$. *We can also explicitly retrieve the quadtree cells corresponding to* $|L_i \cap B|$ *in* $O(c_0{}^d \log n)$ *time.*

We first obtain in $O(\log n)$ time with our asymmetric ring-tree an $O(n)$ ANN q_{rough} to query point q, such that $\sqrt{D_\phi(q_{\text{rough}},q)} = O\left(\mu^2 n\sqrt{D_\phi(\text{nn}_q,q)}\right)$. We then use Lemma 9.10 to get $O(c_0{}^d)$ cells of our quadtree that intersect right ball $B\left(q,\sqrt{D_\phi(q_{\text{rough}},q)}\right)$.

Let us call this collection of cells as Q. We then carry out a quadtree search on each element of Q. Note that we expand only cells which may contain a point nearer to query point q than the current best candidate. We bound the depth of our search using μ-defectiveness similar to Lemma 8.4.

Lemma 9.11. *We need only expand cells of diameter greater than* $\frac{\varepsilon\sqrt{D_\phi(nn_q,q)}}{2\mu}$

Proof. By μ-defectiveness, similar to Lemma 7.4. □

Corollary 9.3. *We will not expand cells where the length of each side is less than* $x = \frac{\varepsilon\sqrt{D_\phi(nn_q,q)}}{2\mu(\mu+1)\sqrt{d}}$

Let the spread be $\beta = \frac{D_{\text{rough}}}{\sqrt{D_{s\phi}(\text{nn}_q,q)}} = O(\mu^2 n)$. The remaining analysis is similar to Section 8, so we simply state the Lemmas with the proofs deferred to the expanded version.

Lemma 9.12. *We will only expand our tree to a depth of* $k = \log(2c_0^3\mu(\mu+1)\beta\sqrt{d}/\varepsilon)$.

Lemma 9.13. *The number of cells expanded at the i-th level is* $n_i = O\left(2^d(\mu^d d^{\frac{d}{2}} c_0^{5d} + (\frac{c_0^2 2^i}{\beta})^d)\right)$.

The total number of nodes explored is

$$O(2^d\mu^d(\mu+1)^d c_0^{5d}\log(2c_0^3\mu\beta\sqrt{d}/\varepsilon) + 2^{2d}c_0^{5d}\mu^d(\mu+1)^d d^{\frac{d}{2}}/\varepsilon^d)$$

or $O\left(2^d\mu^d(\mu+1)^d c_0^{5d}\log n + 2^{2d}c_0^{5d}\mu^d(\mu+1)^d d^{\frac{d}{2}}/\varepsilon^d\right)$, substituting back for β and ignoring smaller terms. Recalling that there are c_0^d cells in Q adds a further c_0^d multiplicative factor.

10. FURTHER WORK

A major open question is whether bounds independent of μ-defectiveness can be obtained for the complexity of ANN-search under Bregman divergences. As we have seen, traditional grid based methods rely heavily on the triangle inequality, and there are technical difficulties in adapting other methods such as cone decompositions [12] or approximate Voronoi diagrams [21]. We expect that we will need to exploit geometry of Bregman divergences more substantially.

11. ACKNOWLEDGEMENTS

We thank Sariel Har-Peled and anonymous reviewers for helpful comments.

12. REFERENCES

[1] Ackermann, M., and Blömer, J. Bregman clustering for separable instances. In *SWAT* (2010), pp. 212–223.

[2] Ackermann, M., Blömer, J., and Scholz, C. Hardness and non-approximability of bregman clustering problems. Tech. rep., ECCC, 2011. http://eccc.hpi-web.de/report/2011/015/.

[3] Ackermann, M. R., and Blömer, J. Coresets and approximate clustering for Bregman divergences. In *Proc. 20th SODA* (2009), pp. 1088–1097.

[4] Afshani, P., Arge, L., and Larsen, K. D. Orthogonal range reporting in three and higher dimensions. In *Proc. IEEE FOCS* (2009), pp. 149–158.

[5] Amari, S., and Nagaoka, H. *Methods of Information Geometry*, vol. 191 of *Translations of Mathematical monographs*. Oxford University Press, 2000.

[6] Banerjee, A., Merugu, S., Dhillon, I. S., and Ghosh, J. Clustering with bregman divergences. *J. Mach. Learn. Res. 6* (December 2005), 1705–1749.

[7] Beygelzimer, A., Kakade, S., and Langford, J. Cover trees for nearest neighbor. In *Proc. 23rd ICML* (New York, NY, USA, 2006), ACM, pp. 97–104.

[8] Boissonnat, J.-D., Nielsen, F., and Nock, R. Bregman voronoi diagrams. *Discrete and Computational Geometry 44* (2010), 281–307. 10.1007/s00454-010-9256-1.

[9] Bregman, L. M. The Relaxation Method of Finding the Common Point of Convex Sets and Its Application to the Solution of Problems in Convex Programming. *USSR Computational Mathematics and Mathematical Physics 7* (1967), 200–217.

[10] Cayton, L. Fast nearest neighbor retrieval for bregman divergences. In *Proc. 25th ICML* (New York, NY, USA, 2008), ACM, pp. 112–119.

[11] Cayton, L. *Bregman proximity search*. PhD thesis, University of California, San Diego, 2009.

[12] Chan, T. M. Approximate nearest neighbor queries revisited. *Discrete and Computational Geometry 20* (1998), 359–373. 10.1007/PL00009390.

[13] Chaudhuri, K., and McGregor, A. Finding metric structure in information theoretic clustering. In *COLT '08* (2008).

[14] Cole, R., and Gottlieb, L.-A. Searching dynamic point sets in spaces with bounded doubling dimension. In *Proc. 38th STOC* (2006), pp. 574–583.

[15] Collins, M., Schapire, R., and Singer, Y. Logistic regression, adaboost and bregman distances. *Machine Learning 48*, 1 (2002), 253–285.

[16] Cover, T. M., and Thomas, J. A. *Elements of information theory*. Wiley-Interscience, New York, NY, USA, 1991.

[17] Csiszár, I. I-Divergence Geometry of Probability Distributions and Minimization Problems. *The Annals of Probability 3* (1975), 146–158.

[18] Eppstein, D., Goodrich, M. T., and Sun, J. Z. The skip quadtree: a simple dynamic data structure for multidimensional data. In *Proc 21st. SoCG* (2005), pp. 296–305.

[19] Farago, A., Linder, T., and Lugosi, G. Fast nearest-neighbor search in dissimilarity spaces. *Pattern Analysis and Machine Intelligence, IEEE Transactions on 15*, 9 (sep 1993), 957 –962.

[20] Gray, R., Buzo, A., Gray Jr, A., and Matsuyama, Y. Distortion measures for speech processing. *Acoustics, Speech and Signal Processing, IEEE Transactions on 28*, 4 (Aug 1980), 367–376.

[21] Har-Peled, S. A replacement for Voronoi diagrams of near linear size. In *Proc. 42nd FOCS* (2001), pp. 94–105.

[22] Har-Peled, S., and Mazumdar, S. Fast algorithms for computing the smallest k-enclosing circle. *Algorithmica 41* (2005), 147–157. 10.1007/s00453-004-1123-0.

[23] Har-Peled, S., and Mendel, M. Fast construction of nets in low dimensional metrics, and their applications. In *Proc. 21st SoCG* (2005), pp. 150–158.

[24] Hjaltason, G. R., and Samet, H. Index-driven similarity search in metric spaces (survey article). *ACM Trans. Database Syst. 28* (December 2003), 517–580.

[25] Indyk, P., and Motwani, R. Approximate nearest neighbors: towards removing the curse of dimensionality. In *Proc. 30th STOC* (1998), pp. 604–613.

[26] Krauthgamer, R., and Lee, J. R. The black-box complexity of nearest-neighbor search. *Theor. Comput. Sci. 348*, 2 (Dec. 2005), 262–276.

[27] Mahalanobis, P. C. On the generalised distance in statistics. *Proc. National Institute of Sciences in India 2*, 1 (1936), 49–55.

[28] Manthey, B., and Röglin, H. Worst-case and smoothed analysis of k-means clustering with bregman divergences. In *Algorithms and Computation*, Y. Dong, D.-Z. Du, and O. Ibarra, Eds., vol. 5878 of *Lecture Notes in Computer Science*. Springer Berlin / Heidelberg, 2009, pp. 1024–1033.

[29] Moraleda, J., Shakhnarovich, Darrell, T., and Indyk, P. Nearest-neighbors methods in learning and vision. theory and practice. *Pattern Analysis and Applications 11* (2008), 221–222. 10.1007/s10044-007-0076-8.

[30] Nielsen, F., and Boltz, S. The burbea-rao and bhattacharyya centroids. *CoRR abs/1004.5049* (2010).

[31] Nielsen, F., and Nock, R. On the smallest enclosing information disk. *Information Processing Letters 105*, 3 (2008), 93 - 97.

[32] Nielsen, F., and Nock, R. Sided and symmetrized bregman centroids. *IEEE Trans. Inf. Theor. 55* (June 2009), 2882–2904.

[33] Nielsen, F., Piro, P., and Barlaud, M. Bregman vantage point trees for efficient nearest neighbor queries. In *Proceedings of the 2009 IEEE international conference on Multimedia and Expo* (Piscataway, NJ, USA, 2009), ICME'09, IEEE Press, pp. 878–881.

[34] Nielsen, F., Piro, P., and Barlaud, M. Tailored bregman ball trees for effective nearest neighbors. In *In European Workshop on Computational Geometry* (2009).

[35] Sariel-Har-Peled. *Geometric Approximation Algorithms*. AMS, 2011. `http://goo.gl/pLiEO`.

[36] Spellman, E., and Vemuri, B. Efficient shape indexing using an information theoretic representation. In *Image and Video Retrieval*, W.-K. Leow, M. Lew, T.-S. Chua, W.-Y. Ma, L. Chaisorn, and E. Bakker, Eds., vol. 3568 of *Lecture Notes in Computer Science*. Springer Berlin / Heidelberg, 2005, pp. 590–590.

[37] Zhang, Z., Ooi, B. C., Parthasarathy, S., and Tung, A. K. H. Similarity search on bregman divergence: towards non-metric indexing. *Proc. VLDB Endow. 2* (August 2009), 13–24.

The Point-Set Embeddability Problem for Plane Graphs*

Therese Biedl†
David R. Cheriton School of Computer Science
University of Waterloo
Waterloo, ON N2L 1A2, Canada
biedl@uwaterloo.ca

Martin Vatshelle‡
Department of Informatics
University of Bergen
Bergen, Norway
vatshelle@ii.uib.no

ABSTRACT

In this paper, we study the Point-set embeddability-problem, i.e., given a planar graph and a set of points, is there a mapping of the vertices to the points such that the resulting straight-line drawing is planar? It was known that this problem is NP-hard if the embedding can be chosen, but becomes polynomial for triangulated graphs of treewidth 3. We show here that in fact it can be answered for all planar graphs with a fixed combinatorial embedding that have constant treewidth and constant face-degree.

We also prove that as soon as one of the conditions is dropped (i.e., either the treewidth is unbounded or some faces have large degrees), Point-set embeddability with a fixed embedding becomes NP-hard. The NP-hardness holds even for a 3-connected planar graph with constant treewidth, triangulated planar graphs, or 2-connected outer-planar graphs.

Categories and Subject Descriptors

F.2.2 [**Theory of Computation**]: Analysis of Algorithms and Problems Complexity; G.2 [**Mathematics of Computing**]: Discrete Mathematics—*Graph Theory*; I.3.5 [**Computing Methodologies**]: Computer Graphics—*Computational Geometry and Object Modeling*

General Terms

Algorithms, Theory

Keywords

Graph drawing, point-set embedding, carving width

*The authors would like Dimitrios Thilikos for helpful input.

†Research supported by NSERC and by the Ross and Muriel Cheriton fellowship.

‡Research done while visiting the University of Waterloo.

1. INTRODUCTION

Planar graph drawing is the art of drawing a graph in the plane such that no two edges cross. It has been known for many decades that any planar graph can be drawn with straight lines without crossing [27, 10, 24]. In fact, one of the first problems studied in the graph drawing community was how to achieve a straight-line drawing such that coordinates of points are small integers. Multiple constructions were given to achieve an $O(n) \times O(n)$-grid for a planar graph with n vertices [21, 11].

One line of research in this field has worked on restricting the placement of the vertices further. In particular, is it possible to restrict the placement of vertices to a given set of points? This is the so-called *Point-set embeddability* problem: Given a planar graph G and a set S of points in the plane, test whether there exists a mapping of the vertices of G onto the point-set S that results in a planar straight-line drawing of G. This problem is the topic of the present paper.

Existing results It has been discovered multiple times [15, 8, 5] that Point-set embeddability is easy for outer-planar graphs, presuming the points are in general position and the drawing that we want is the "standard" one (with all points on the outer-face). Namely, *any* outer-planar graph can be embedded in *any* set of n points in general position so that the drawing is planar.

On the other hand, for arbitrary planar graphs a point-set embedding is not possible for all point sets. Cabello [7] proved that it is NP-hard to decide whether a planar graph can be embedded in a given set of points, even if the graph is 2-outer-planar. However, his proof crucially uses that the embedding of the graph is not fixed; NP-hardness remained open for 3-connected graphs until very recently, when it was shown (in work done independently from the present paper) by Durocher and Mondal [9].

Since then, many variations of Point-set embeddability have been investigated. For example, Badent et al. studied version with further restrictions as to which vertices may be mapped to which points [2], and Di Giacomo et al. studied the situation where only a subgraph is restricted to points [14]. Directed versions have also been considered, see e.g. [13]. Other papers considered Point-set embeddability for special graph classes. Nishat et al. [18] and Moosa and Rahman [16] gave algorithms to test Point-set embeddability in triangulated planar graphs of treewidth 3.

Our results Our investigation was motivated by the last two papers. If Point-set embeddability is polynomial for triangulated graphs of treewidth 3, then what can be said about triangulated graphs of treewidth k, for some constant k? We show that Point-set embeddability is polynomial for all these, and we can even relax the triangulation restriction. Specifically we show:

> Point-set embeddability is polynomial for any graph of constant treewidth and constant face-degree if the embedding of the graph must be respected.

On the other hand, if the embedding can be chosen freely, then Cabello's result [7] shows NP-hardness even for graphs of constant treewidth (his constructed graphs have treewidth 5) and for constant face-degree (only one of his faces has degree greater than 4, and this can be overcome by modifying his construction a bit.)

This leaves the case of graphs that have faces with arbitrarily large degree, or that have large treewidth, but the embedding is fixed. (The NP-hardness construction of Durocher and Mondal [9] has unbounded treewidth and unbounded face-degrees.) We prove that Point-set embeddability is NP-hard for

- 2-connected outer-planar graphs (which have constant treewidth) if the embedding is fixed,
- 3-connected planar graphs (which have a unique embedding) of constant treewidth,
- triangulated planar graphs (which have a unique embedding).

So the first two items show that even with constant treewidth Point-set embeddability is NP-hard if the face-degrees are unbounded, while the third item shows that even with constant face-degrees Point-set embeddability is NP-hard if the treewidth is unbounded. So for our polynomial result, we truly need both bounded treewidth and bounded face-degrees.

Our paper is organized as follows. After giving definitions, we present the algorithm for graphs of constant treewidth and face-degrees in Section 3. We prove the NP-hardness of Point-set embeddability in various scenarios in Section 4 and conclude with open problems.

2. BACKGROUND

We assume familiarity with graph-theoretic concepts such as graphs, degrees, and connectivity. Throughout this paper, let $G = (V, E)$ denote a graph with n vertices and m edges. G is always assumed to be *planar*, i.e., it can be drawn in the plane without crossing. In such a planar drawing, the connected pieces that remain when removing the drawing are called *faces*, with the unbounded piece called *outer-face*. All our input graphs are assumed to be simple, i.e., to have neither loops nor multiple edges. Any simple planar graph (even if disconnected) has $m \leq 3n - 3$ edges.

Combinatorial embeddings: To describe drawings and faces without using geometry, we need two ingredients. A *rotational system* assigns to each vertex of G a circularly ordered list of the edges of G incident to v. A planar drawing *respects the rotational system* if for every vertex the order in the rotational system is the counter-clockwise (ccw) order of incident edges in the drawing. We assume throughout this paper that a rotational system of G has been fixed in such a way that it can be respected by some planar drawing.

Such a rotational system determines all planar drawings of G (up to choice of the outer-face and topological deformations) if G is connected. If G is not connected, then we need more information to determine its drawing uniquely. We define two kinds of *angle* in a graph. One kind of angle is an ordered triple $\{e_1, v, e_2\}$ such that e_1 and e_2 both end at v, and the ccw order at v contains e_1 followed by e_2. If v has degree 1 then e_1 and e_2 are the same edge. In a planar drawing, the area ccw between e_1 and e_2 and v belongs to some face f; we say that f is *incident to this angle.* The other kind of angle occurs at a vertex that has degree 0, and is described simply by the vertex itself. In any drawing, the face that is adjacent to this vertex is called the face *incident to this angle.*

A *combinatorial embedding* of a graph now consists of a planar graph with a rotational system, a finite set F, one special element o of F, and a mapping of angles to elements of F. We say that a planar drawing *respects the combinatorial embedding* if it respects the rotational system, its faces are in 1-1 correspondence with F, with the outer-face mapped to o, and the face incident to each angle is as prescribed by the mapping. In this paper we always assume that a combinatorial embedding is fixed and only consider drawings that respect it. Therefore, we will often use geometric terms (such as faces and outer-face) interchangeably with combinatorial terms (such as F and o.)

The *degree* $\deg(f)$ of a face f is the number of angles to which f is incident. If the edges incident to f form a simple cycle (as they do for 2-connected graphs), then the degree of a face equals the number of edges (or number of vertices) that are incident to f. But the degree may be higher than that if there are cut-vertices or vertices of degree 0 incident to f.

If G is 2-connected, then the boundary of each face f is a simple cycle and hence there is a natural order of vertices around each face f of G. If G has cut-vertices or is disconnected, then for some faces the boundary is not simple. To be able to handle this, we assume that every face f stores a fixed order $\rho(f)$ of vertices that are adjacent to f. The precise order of $\rho(f)$ is not important as long as it has been fixed.

A *triangulated graph* is a planar graph for which all faces have degree 3. Such a graph is 3-connected and has a combinatorial embedding (up to choice of the outer-face and reversing all vertex orders.)

Drawings and Point-set embeddability: In this paper, we only consider straight-line drawings, i.e., drawings where edges are straight-line segments between the points assigned to their endpoints. Therefore, we can describe a *proper drawing* of a planar graph G by giving only the mapping from vertices to points, and demand that this mapping satisfies: (a) No two vertices are mapped to the same point. (b) The interior of the line-segment of each edge contains no points of other vertices or edges. (c) The rotational system and outer-face are respected. (d) The face-angle incidences are respected. We will also consider drawings of a subgraph G' of G, and consider them to be proper if the conditions are satisfied for all elements of G' that also exist in G. So in particular a proper drawing of G' means a drawing that satisfies (a) and (b) above, respects the induced rotational system, and respects the face-angle incidences for every face f and angle α of G' that also existed in G.

Given a mapping of the n vertices to points, we can check

in $O(n \log n)$ time whether this mapping corresponds to a proper drawing; the bottleneck being the time for testing intersections of line segments and finding the face-incidences for disconnected components.

The *Point-set embeddability* problem is defined formally as follows. Given a planar graph $G = (V, E)$ and a set S of points in the plane, where we may assume $|S| \geq n = |V|$. Is there a mapping of V to S such that the resulting drawing of G is proper? Note that we allow S to have more points than necessary to draw G. We also do not require S to be in general position. In our paper, we will only consider the Point-set embeddability problem for graphs for which the combinatorial embedding is given, and we are only interested in proper drawings that respect this combinatorial embedding.

Crucial for solving Point-set embeddability on some graphs will be to consider the dual graph. For a graph G with a combinatorial embedding, let the *dual graph* G^* consist of assigning a vertex v_f to each face $f \in F$. For each edge e adjacent to two faces f_1, f_2, add an edge from v_{f_1} to v_{f_2} in the dual graph. The dual graph (at least in our definition) ignores all vertices of degree 0; our algorithm will take care of them separately since they are included in the list $\rho(f)$ for the face f that they are incident to. The dual graph is always connected even if G was not. It may have loops or multiple edges even if G was simple.

Treewidth: The treewidth of a graph is a graph parameter that has proved very useful both in theoretical studies (for the Graph Minor Theorem [19]) and for algorithmic development and fixed-parameter tractable algorithms (see e.g. [3].) We will not give the (somewhat complicated) definition of treewidth as we do not use it directly.

Computing the treewidth of a graph is NP-hard in general [1], but can be done in polynomial time if the treewidth is a constant [4]. Graphs of treewidth t are also called *partial t-trees*. Any graph of treewidth 2 is planar, and is also called a *series-parallel graph*. A special case of series-parallel graphs are *outer-planar graphs*, which can be drawn such that all vertices are on the outer-face.

3. ALGORITHMS

In this section, we give an algorithm for Point-set embeddability if some graph parameters are small. Recall that the O^*-notation is used for asymptotic bounds that hide polynomial terms.

THEOREM 3.1. *Let G be a planar graph with a fixed combinatorial embedding. If the dual graph G^* of G has treewidth at most t and maximum degree at most Δ, then Point-set embeddability can be solved for G on any point set S in $O^*(|S|^{1.5(t+1)\Delta})$ time and space.*

We first give an outline of the proof. Rather than using the treewidth directly, we use the treewidth and maximum degree to bound another graph parameter of the dual graph G^* known as the *carving-width* and defined precisely below. Translating back to G, the carving-width roughly corresponds to the number of edges needed on a cycle to separate the faces of G into parts (and recursively in the subgraphs). Since the carving-width is bounded, we can brute-force enumerate all possible ways in which such a cycle could be mapped to the given point set S. This therefore gives rise to a dynamic programming approach. In the base case, for each face f of G find all possible sets of points in S that could possibly hold the boundary of f. Then in the recursive steps consider all possible ways of gluing faces together to obtained a bigger subgraph; this is polynomial if the carving-width is a constant.

3.1 Carving-width

The carving-width of a graph (not necessarily planar) was first introduced by [22] and can be defined as follows:

DEFINITION 3.2 (CARVING-WIDTH). *Let $G = (V, E)$ be a graph. A* carving-decomposition *of width k is a rooted binary tree T and a bijection δ between V and the leaves of T such that for any arc a of T, there are at most k edges e in G that* cross *arc a, i.e., the endpoints of e are mapped to different components of $T - a$. The carving-width of G is the minimum width over all carving-decompositions.*

The carving-width is a generalization of the *cut-width*, where we require the tree T to be a caterpillar, see [25, 26] for an overview. Computing the carving-width of a graph is NP-hard in general, but becomes polynomial if the graph is planar [22]. (It is also fixed-parameter tractable in the carving-width [25].)

Any graph of treewidth t has carving-width at least $\frac{1}{3}t$ [25]. Also, for graphs without loops the carving-width is at least Δ, since the arc from the leaf that stores vertex v is crossed by $\deg(v)$ edges. Vice versa, it is known that the carving-width of G is at most $b \cdot \Delta$, where b is the so-called *branch-width* of G [17]. The branch-width in turn is bounded by $t + 1$ [20], so we get:

LEMMA 3.3. *A graph of treewidth t and maximum degree Δ has carving-width at most $(t + 1) \cdot \Delta$.*

3.2 Embedding one face

We now solve Point-set embeddability for any graph G for which the dual graph G^* has a carving decomposition T of width at most k; this proves Theorem 3.1 by the previous lemma. We solve Point-set embeddability using a bottom-up dynamic programming algorithm in T. The leaves of T are in bijection with the vertices of G^*, hence the faces of G. Let f be one such face of G. Recall that $\rho(f)$ is some enumeration of the vertices of f, and $|\rho(f)| \leq \deg(f) \leq \Delta$.

Now consider any ordered set π of $|\rho(f)|$ points in S. If we map vertices in $\rho(f)$ to points of π in this order, then this gives one possible way of drawing the graph G_f that is formed by face f. Of course, for most sets π, this will not be a proper drawing.[1] By testing this for each π, we hence compute and store,

$$M_b(f, \pi) = \begin{cases} 1 & \text{if mapping } \rho(f) \text{ to } \pi \text{ in this order} \\ & \text{gives a proper drawing of } G_f \\ 0 & \text{otherwise} \end{cases}$$

The computation time for all these values $M_b(.,.)$ is $O(n \cdot \Delta \log \Delta \cdot |S|^{\Delta})$, since there are $O(n)$ faces f, at most $\Delta! \binom{|S|}{\Delta} \leq |S|^{\Delta}$ ordered sets of length $|\rho(f)| \leq \deg(f) \leq \Delta$, and we spend $O(\Delta \log \Delta)$ time to test whether the drawing of f is proper.

[1] We could improve the run-time of our algorithm by only considering point sets for which the simple cycles that form the boundary of f are crossing-free. A set of s points has at most $O(86.89^s)$ crossing-free spanning cycles [23], so this would eliminate many candidates (but still be exponential.)

3.3 Subgraphs and boundaries

Now we consider some larger subgraphs of G. To explain how to combine such subgraphs, we need a definition. Let F' be a subset of faces of G. The *boundary* B of F' is the set of all edges e for which the face on one side belongs to F' and the face on the other side does not belong to F'. We say that a vertex *belongs to the boundary* of F' if one of its incident edges is on the boundary of F'.

We specifically need boundaries for sets of faces defined by the carving decomposition T of G^*. For every node $v \in T$, let $F(v)$ be all vertices of G^* (hence faces of G) mapped to a leaf below v in T. Let $B(v)$ be the boundary of $F(v)$. By definition of boundary and $F(v)$, any edge in $B(v)$ has one side on a face below v in T and the other side on a face not below v in T. Therefore, the dual of any edge in $B(v)$ crosses the arc from v to its parent in T. By definition of carving-width, therefore $|B(v)| \leq k$.

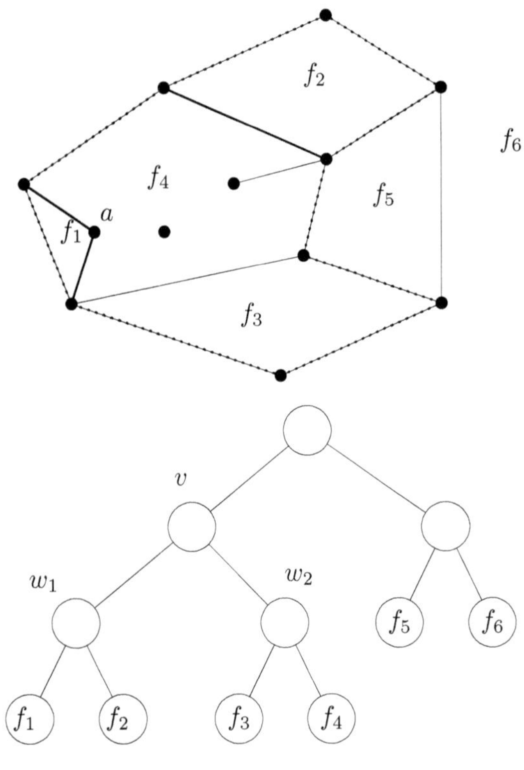

Figure 1: An example of a planar graph. Face f_4 has degree 9 (this includes the isolated vertex.) For the given carving decomposition of the dual graph. the boundary $B(v)$ is shown dotted; the duals of these edges cross the arc from v to its parent. Set $E_\sigma(v)$ is shown bold, and $V_\sigma(v) = \{a\}$.

Similarly as for faces, we assume that for every node v of T we have fixed an order $\rho(v)$ of the vertices that are on the boundary $B(v)$. We also need the notation $G(v)$ for the graph formed by faces $F(v)$, i.e., the graph consisting of all vertices and edges that are on at least one face in $F(v)$.

We now aim to compute the following function for all nodes v of T and all ordered sets π of $|\rho(v)|$ points of S:

$$M(v, \pi) = \begin{cases} 1 & \text{if mapping } \rho(v) \text{ to } \pi \text{ in this order gives a proper drawing of } B(v), \\ & \text{and it can be expanded to a proper drawing of } G(v) \text{ on } S. \\ 0 & \text{otherwise} \end{cases}$$

Note that in the previous subsection we computed a very similar function for any node ℓ that is a leaf of T (hence corresponds to a face f.) To compute $M(\ell, \pi)$ for a leaf ℓ, we only need to find all entries $M_b(f, \pi')$ that are 1 and set $M(\ell, \pi) = 1$ for the induced ordered list π (observe that $\rho(\ell)$ is a subset of $\rho(f)$.) This takes $O(\Delta \cdot s^\Delta)$ time per leaf, since $\deg(f) \leq \Delta$.

3.4 Combining subgraphs

Now consider a non-leaf v in the carving decomposition T and let w_1 and w_2 be its children. We can use the entries in $M(w_1, \pi_1)$ and $M(w_2, \pi_2)$ to compute the entries for $M(v, \pi)$ as follows. Let $E_\sigma(v)$ be all edges in $B(w_1) \cap B(w_2)$, therefore $B(v) = B(w_1) \cup B(w_2) - E_\sigma(v)$. Let $V_\sigma(v)$ be all those vertices in $\rho(w_1) \cup \rho(w_2)$ that are not in $\rho(v)$. A vertex in $V_\sigma(v)$ hence "disappears" from the boundary, which is possible only if it has at least two incident edges in $E_\sigma(v)$; therefore $|V_\sigma(v)| \leq |E_\sigma(v)|$.

Assume we have an ordered set π of $|\rho(v)|$ points and want to compute $M(v, \pi)$. First test whether assigning $\rho(v)$ to π draws $B(v)$ properly. If so, then try all possible ways to expand π by adding an ordered list π_σ of points for the vertices in $V_\sigma(v)$. There are $|S|^{|V_\sigma(v)|}$ many ways of doing this, and we test all of them. For each one of them, assigning $\rho(v) \cup V_\sigma(v)$ to $\pi \cup \pi_\sigma$ defines the induced assignments of $\rho(w_1)$ and $\rho(w_2)$ to ordered lists π_1 and π_2. Test whether $M(w_1, \pi_1) = 1 = M(w_2, \pi_2)$. It is not hard to see that if this holds, then combining the drawings of $G(w_1)$ and $G(w_2)$ gives a proper drawing of $G(v)$, so we set $M(v, \pi) = 1$. If we do not succeed for any π_σ, then set $M(v, \pi) = 0$.

3.5 Putting it all together

To finish the proof of the theorem, we now do the obvious dynamic programming in the tree T: Compute $M(\ell, \pi)$ for all leaves ℓ and all ordered lists π of $\rho(\ell)$. Then go bottom-up in the tree and compute $M(v, \pi)$ for every node v using the values of its two children. Note that for the root r the boundary $B(r)$ is empty. Hence $B(r, \emptyset)$ will tell us whether the whole graph G can be embedded in the given point set.

It remains to analyze the running time. Constructing the dual graph and computing its carving-decomposition is polynomial (and for all but the smallest values of k it is dominated by the other operations.) For every leaf ℓ with corresponding face f we spend $O(\Delta \log \Delta \cdot s^\Delta)$ time to compute $M_b(f, .)$, and then again as much to compute $M(\ell, .)$.

Now consider a node v higher up in a tree. Given one fixed ordered set of points π, we can test whether this gives a proper drawing of the boundary in $O(k \log k)$ time since $|\rho(v)| \leq |B(v)| \leq k$. Computing π_1 and π_2 takes $O(k)$ time. There are $O(|S|^{|V_\sigma(v)|})$ ways to choose π_σ, and for each one of them we look up two values.[2] So we spend $O(k \log k + |S|^{|V_\sigma(v)|})$ time for one ordered set π of length

[2] We presume $M(w, \pi)$ is stored so that these values can be looked up in constant time. More space-saving data structures are possible, but they have larger lookup time.

$\rho(v)$. Summing over all such sets gives run-time

$$O\left(|S|^{|\rho(v)|}\cdot\left(k\log k+|S|^{|V_\sigma(v)|}\right)\right).$$

To bound this, we need to bound $|\rho(v)|+|V_\sigma(v)|\leq|B(v)|+|E_\sigma(v)|$. Notice that $|B(v)|,|B(w_1)|$ and $|B(w_2)|$ all are bounded by the carving width k of the dual graph. Also, $|B(w_1)|+|B(w_2)|=|B(v)|+2|E_\sigma(v)|$, therefore $2|B(v)|+2|E_\sigma(v)|\leq k+|B(v)|+2|E_\sigma(v)|=k+|B(w_1)|+|B(w_2)|\leq 3k$. So we spent

$$O\left(k\log k\cdot|S|^{|B(v)|+|E_\sigma(v)|}\right)=O\left(k\log k\cdot|S|^{1.5k}\right)$$

time per node v. Since T has $O(n)$ nodes and $k\leq\Delta(t+1)$, therefore the total run-time is

$$O\left(n\Delta(t+1)\log(\Delta(t+1))\cdot|S|^{1.5\Delta(t+1)}\right),$$

which finishes the proof of Theorem 3.1.

COROLLARY 3.4. *Point-set embeddability is polynomial for any triangulated graph G of bounded treewidth.*

PROOF. Since G is triangulated, the dual graph has maximum degree 3. Since G has bounded treewidth, so does the dual [6]. The result then holds by the previous theorem. □

4. NP-HARDNESS

In this section, we show that Point-set embeddability is NP-hard. This was already shown by Cabello [7] and Durocher and Mondal [9], but our proof holds even for 3-connected graphs with constant treewidth, triangulated planar graphs, or 2-connected outer-planar graphs with a fixed combinatorial embedding. We first give a construction for disconnected graphs, and then make it increasingly more connected.

Our reduction is from the *3-partition* problem defined as follows: Given $3n$ positive integers $a_1,\ldots,a_{3n}$ with $\sum_{i=1}^{3n}a_i=n\cdot B$, partition the numbers into n groups such that the numbers within each group sum to B. It is well-known that 3-partition is strongly NP-hard [12].

4.1 Disconnected graphs

So assume we are given an instance $\{a_1,\ldots,a_{3n}\}$ of 3-partition, with $\sum_{i=1}^{3n}a_i=n\cdot B$. We explain the construction of S and G simultaneously. The convex hull of S consists of $4n+8$ points placed on an oval (see also Figure 2), with the leftmost point at $(0,0)$ and the rightmost point at $(2n+4,0)$. Let $\varepsilon>0$ be a small constant. On the lower hull, for $j=1,\ldots,n+1$ we have points at x-coordinates $2j-\varepsilon,2j+\varepsilon$; we call these b_j and d_j. On the upper hull, we have points at x-coordinates $2j+1-\varepsilon,2j+1+\varepsilon$, for $j=1,\ldots,n$; we call these b'_j and d'_j. We also have points at x-coordinates ε and $2n+4-\varepsilon$ on both upper and lower hull.

The outer-face of graph G has $4n+8$ vertices in a cycle. Any mapping of G on the point set S that respects the outer-face must place the outer-face of G on the convex hull of S.

Next, add a point p_a near $(\varepsilon/2,0)$. In the graph, pick three arbitrary consecutive vertices c_1,c_2,c_3 of the outer-face and add edge (c_1,c_3). Add a new vertex v_a inside the resulting triangle $\{c_1,c_2,c_3\}$ and connect it to c_1 and c_2. Thus in any realization of G on S, vertices c_1,c_2,c_3 must be mapped to consecutive points on the convex hull such that the triangle $\{c_1,c_2,c_3\}$ contains a point. We will not add any such points *except* p_a. Therefore, in any realization c_2 must be mapped to $(0,0)$, which fixes the mapping of the entire outer-face of G to the points of the convex hull of S.

We next add $n+1$ *lower teeth*. For $j=1,\ldots,n+1$, add $M+1$ points on the vertical line with x-coordinate $2j$, for M sufficiently large (we will discuss this choice below.) The topmost point (called p_j) is above the y-axis. All other points are below the y-axis, and "sufficiently far away from the convex hull", i.e., not inside any triangle defined by three consecutive points on the convex hull. [3]

For $j=1,\ldots,n+1$ let vertices b_j,d_j in G be those that are mapped to points b_j,d_j in S (by the above the mapping of the outer-face is fixed.) We introduce a new vertex t_j that is adjacent to b_j and d_j. Inside the triangle $\{t_j,b_j,d_j\}$ place an M-path (a path with M vertices) and make one end adjacent to t_j and the other to b_j. We can realize this by placing t_j and the M-path vertices at p_j and the M points below it. Vice versa, assume we have an mapping of the graph on the point set. Then t_j must be placed so that the triangle $\{b_j,t_j,d_j\}$ contains M points. But the base of this triangle has width $\approx 2\varepsilon$, so these M points would have to be within distance $\approx\varepsilon$ of a line that passes through the line segment (b_j,d_j). We will not add such points except near the vertical line that defined the jth lower tooth. So t_j must be placed on the jth lower tooth (and it must be at p_j so that $\{b_j,t_j,d_j\}$ has M points inside.)

We complete the point set by placing, for $j=1,\ldots,n$, the *jth blob* consisting of B points on (or near) the vertical line with x-coordinate $2j+1$. All these points have negative y-coordinate and are sufficiently far away from the convex hull. We complete the graph by adding $3n$ paths inside the face adjacent to $t_1,\ldots,t_{n+1}$. The ith path has a_i vertices (we call it the *a_i-path.*)

If the instance of 3-partition has a solution, then we can realize the graph on the point set by drawing these a_i-paths on the blobs according to the partition into groups. Vice versa, assume G can be realized on the point sets. Then all points not in the blobs are occupied by the outer-face and the teeth as discussed above. Any two blobs are separated by the line segment (b_j,t_j). So by planarity no two points on two different blobs can have an edge between them. Therefore, each of the $3n$ paths of G must be entirely realized within one blob. Since there are B points in a blob, this gives a partition of $a_1,\ldots,a_n$ into groups summing to B, as desired.

Note that our construction used exactly $|V|$ points for S. But NP-hardness holds even for a point set with $s\geq|V|$ points, for any s, as follows: Replace point p_a (the one near $(\varepsilon/2,0)$) by $s-|V|+1$ points that all are within the triangle $\{c_1,c_2,c_3\}$. Then v_a must be mapped to one of these points, and the triangle $\{c_1,c_2,c_3\}$ enforces that the other $s-|V|$ points cannot be used anywhere in the construction.

4.2 Connected outer-planar graphs

To make the graph connected, we first introduce lots of cycles (with the a_i-paths held between them), and more teeth to force those cycles into specific locations. Figure 3 illustrates the following construction.

First we add n *upper teeth*, one at each edge (b'_j,d'_j). This is done exactly as for lower teeth, except that the points are placed on the vertical line with x-coordinate $2j+1$ ($j=$

[3]There is no need for these points to be exactly on the line, as long as the triangle $\{p_j,b_j,d_j\}$ contains them all. In particular, points could be put in general position.

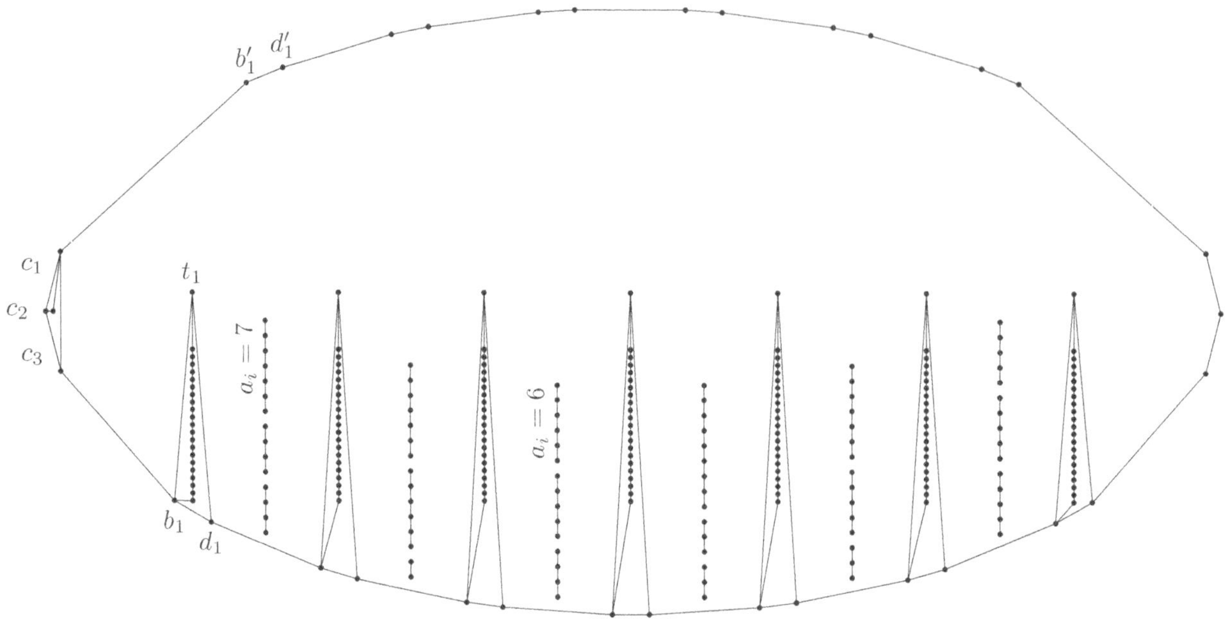

Figure 2: An example of NP-completeness construction for disconnected graphs with $n = 6$ and $B = 15$.

$1, \dots, n$), with the bottommost point below the tips of the lower teeth.

Next, we add more points near what we call the *vertical anchor lines*, which have x-coordinates 1 and $2n + 3$. We place M points on each of these vertical lines. We also place (above, below, and outside) a triplet of points that encloses the M points. M is larger than the total number of points that are not on teeth or anchor lines (including points on the layers that will be defined below); $M \in \Theta(n^3 + nB)$ suffices.

Likewise, in the graph we define two triangles, each of which has an M-path inside that connects to two vertices of the triangle. In any realization, each of these two *anchor triangles* hence must contain points from exactly one of the two anchor lines, since points from the teeth are already occupied by other elements of the graph, M is large enough, and we cannot use points from both anchor lines without the triangle intersecting the teeth.

Finally, for some $\ell \geq (3n + 1)(4n + 9)$, we add ℓ *layers*, which are cycles of length $4n + 8$. These are all inside each other; the outermost cycle is in the face adjacent to all teeth, and the innermost cycle is around the two anchor triangles. After every $4n + 9$ such layers, we place one of the paths for a_i, $i = 1, \dots, 3n$. To make everything connected, we connect each a_i-path to some vertex on each of the adjacent layers, and connect two consecutive layers without an a_i-path between them by adding one arbitrary edge. Finally we connect the outermost layer by adding an edge to t_1, and connect the innermost layer to the anchor triangles at diametrically opposed vertices of the layer. This finishes the construction of the connected graph.

To finish the construction of the point set, we must add more points for the layers as follows: The vertical line at a lower tooth obtains 2ℓ additional points, all above the tooth. The vertical line at an upper tooth obtains 2ℓ additional points, all below the tooth. The vertical line at an anchor triangle obtains 2ℓ additional points, of which ℓ are above and ℓ are below. Finally we also add ℓ points on (or near) the x-axis to the left of the left anchor-triangle, and ℓ points on the x-axis to the right of the right anchor-triangle. This finishes the construction of the point set.

Assume G is embedded in this point set. This fixates the place of the upper and lower teeth (the points for the layers cannot help, since the outer-face vertices are fixed and the triangle $\{b_j, d_j, t_j\}$ must have exactly M points in it.) Also, each anchor triangle must enclose points on one anchor line, so one must be left of all teeth and the other right of all teeth.

In the fixed combinatorial embedding, any layer has both anchor triangles *inside*. But in the drawing any layer must avoid all teeth by planarity. So each layer must zig-zag from left to right between the teeth, go around the anchor triangle, then zig-zag from right to left between the teeth, then go around the other anchor triangle. So each of the ℓ layers uses up two points below each upper tooth, two points above each lower tooth, and one point each below, above and outside of each anchor triangle. This uses up all the points, except for B points below each of the upper teeth. Hence each a_i-path (for $i = 1, \dots, 3n$) must be placed entirely below one of the upper teeth, thereby giving a solution to 3-partition.

Vice versa, if 3-partition has a solution, then the outer-face, teeth and layers can be placed as explained above. Each layer should use the outermost point not used by an earlier layer or an a_i-path. Each a_i-path is placed on the next a_i points of the blob that corresponds to its assignment in the solution to 3-partition. It remains to argue that the edges connecting layers can be drawn. Observe that such an edge could be drawn between corresponding points of the routes (i.e., horizontally or vertically), or one can draw it slanted to the next point of the next layer's route. There are $4n+9$ layers between a_i-paths and each layer has $4n+8$ vertices, so by shifting one or zero units forward with each

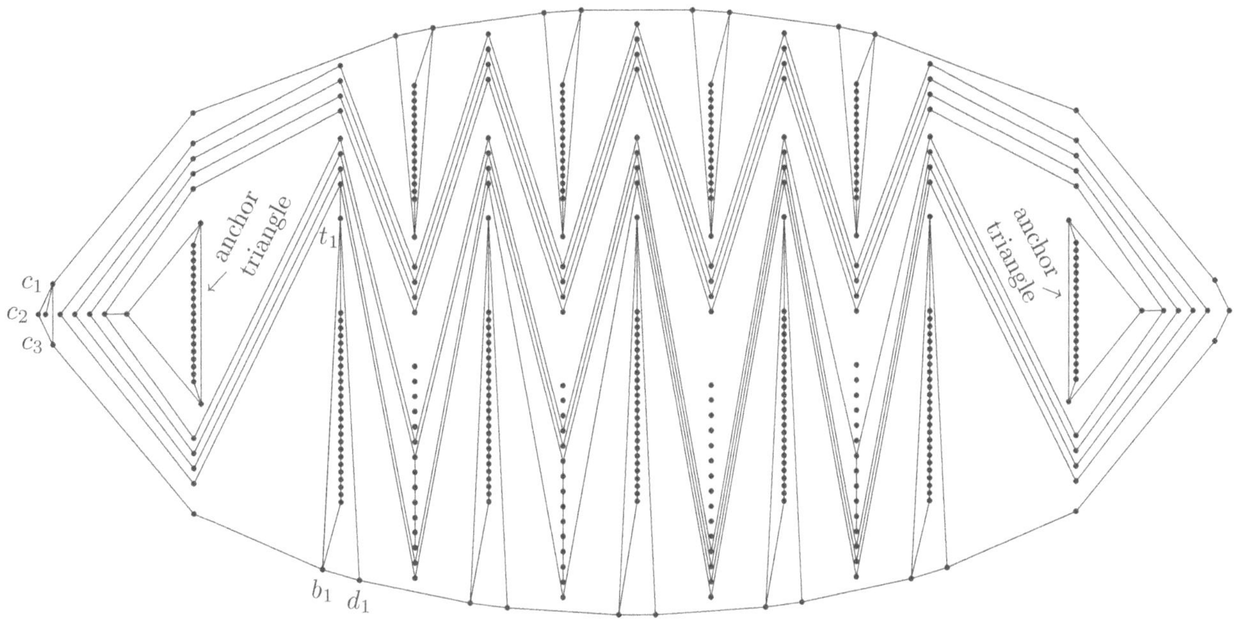

Figure 3: The connected graph construction for $n = 4$. Many of the 325 layers are missing. We added a few extra points on the outer-face for legibility.

layer, we can move to any other position for the next a_i-path. The first and last $4n + 9$ layers can be used to shift routes so that the connections to the outer-face (resp. anchor triangles) can be obtained regardless of the blob that contains a_1 and a_{3n}.

So if 3-partition has a solution, then we can realize G in S, which finishes the NP-hardness proof for connected graphs with a fixed planar embedding. One can easily see that our construction is outer-planar in a different embedding; see also Figure 4.

4.3 Higher connectivity

We now give three different ways of how to connect layers further, depending on which properties we want to achieve.

4.3.1 Outer-planar 2-connected graphs

The graph as constructed is already outer-planar, and the only cut-vertices occur at the connections between two consecutive layers. To make the graph 2-connected, we add a second edge between two consecutive layers (and also from the outermost layer to the outer-face and from the innermost layer to the anchor triangles.) This edge is added so that the two connections between layers end at two adjacent vertices of the layers. The resulting graph then is still outer-planar as illustrated in Figure 4. Also, we still have the choice of drawing edges between corresponding vertices of the layers (i.e., horizontally or vertically), or "shifting over" by one unit, so the argument why a solution to 3-partition can be used to create a proper drawing remains valid.

4.3.2 Inner triangulated graphs

If we add more edges, then we can make all inner faces triangles as follows:

- Fix one arbitrary vertex of the outermost layer to be placed at the leftmost point of the outermost route.

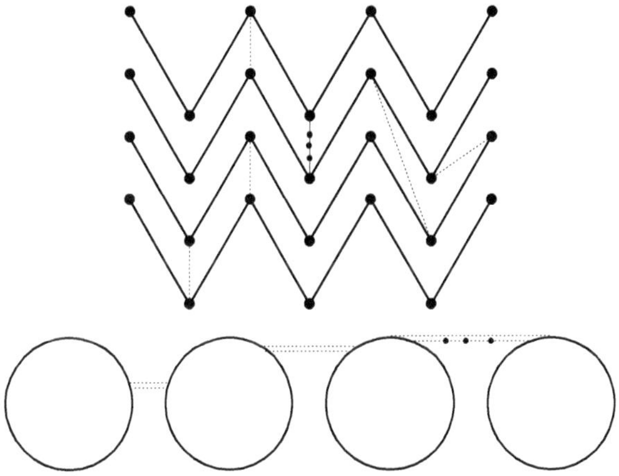

Figure 4: A close-up (not to scale) of making the graph 2-connected, and a sketch of the combinatorial embedding that achieves outer-planarity.

This does not affect embeddability of G, since there are enough layers left to shift the a_1-path to any of the blobs.

Then triangulate any face outside the outermost layer in such a way that the edges can be drawn without creating crossings. In other words, in the fixed drawing of the outer-face, teeth and outermost layer, triangulate the polygons that represent faces and add the corresponding edges to G.

- Similarly fix the innermost layer and triangulate the faces inside it.

- Let the vertices of two consecutive layers be $v_1, \ldots, v_N$ and $w_1, \ldots, w_N$, respectively. If these layers have no

a_i-path between them, then triangulate the faces between the layers by adding the zig-zag path $v_1, w_1, v_2, w_2, \ldots, v_N, w_N, v_1$. This still allows to shift the route one over in one direction: Either we draw all (v_i, w_i) vertically or horizontally and all (w_i, v_{i+1}) slanted, or vice versa. See also the bottom two layers in Figure 5.

- If the two layers have an a_i-path between them, then don't add edge (v_2, w_2); instead connect the a_i-path to v_2 and w_2, and connect all vertices on it to v_1 and w_3. The a_i-path has to be drawn vertically for this to be realizable, but we can add this restriction since there are enough layers between a_i-paths to reach the next blob already earlier.

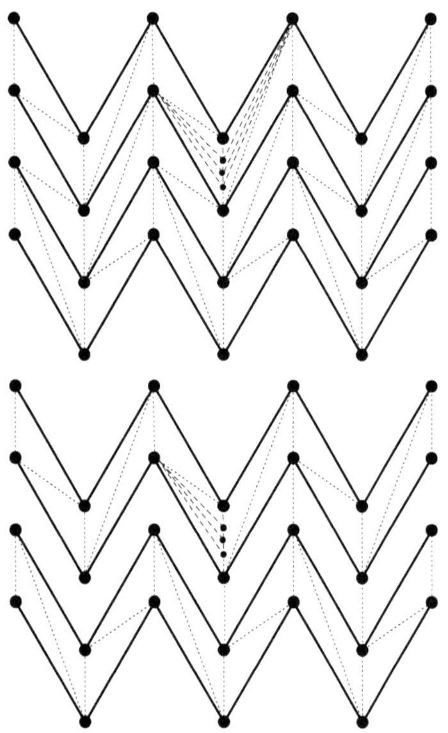

Figure 5: Making the graph (left) inner triangulated and (right) 3-connected with constant pathwidth.

Observe that most vertices of the graph have constant maximum degrees. Exceptions occur only at vertices that are adjacent to all the vertices of a K-path (where $K = M$ or $K = a_i$). We can avoid these high degrees by choosing points within each blob more carefully: Rather than placing points on a line, we place them in two not-quite parallel lines that bend away from each other, forming a)(shape. We can then triangulate the interior of the K-path leading to degree at most 7 at all vertices. We sketch this placement in Figure 6 (which is not to scale; the points must be much closer to a line to ensure that the layers can leave exactly a_i points free between them.)

Triangulated graphs: The previous construction gave an inner triangulated graph. We can triangulate the outer-face as well by adding more points and expanding the graph. Let D be the number of vertices on the outer-face. While $D > 40$, add into the outer-face a cycle of length $\lceil D/2 \rceil$ and

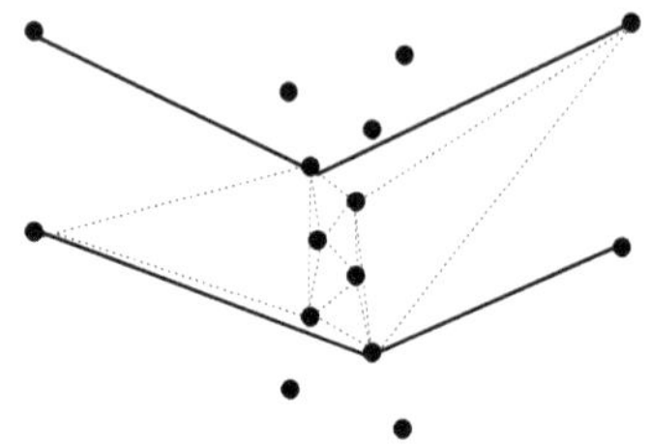

Figure 6: With carefully chosen points near a line, we can add edges within K-paths to triangulate while keeping the maximum degree small.

connect each vertex on it to 2 or 3 vertices of the previous outer-face. Repeat (with the new value of D) until for the first time we have $D \leq 40$ and therefore $D > 20$. Let D_L be this last value of D. Observe that all added vertices have degree at most 7; see also Figure 7. Now to close off, add a triangle T and connect each vertex on it to $\approx D_L/3 + 1$ vertices of the previous outer-face. The vertices of the triangle have degree at most $\lceil D_L/3 \rceil + 1 \leq 15$.

With each cycle added to the graph, we add a corresponding set of points so that they form the convex hull and the cycles can be drawn on them. So if 3-partition has a solution, then there exists a proper drawing of the graph on the point set.

Vice versa, assume that the graph can be drawn on the point set. Since the graph is triangulated, any face could theoretically be drawn on the three points C that form the convex hull of the point set. However, removing C leaves a point set with D_L points on the convex hull, so we can only draw a triangle on C such that removing the triangle from the graph leaves a face with D_L vertices. A degree argument shows that this means that triangle T must be on the convex hull: All vertices not on T have degree at most 7, and hence for any triangle $T' \neq T$ the degrees of the three vertices sum to at most $2(D_L + 1)/3 + 9$ (as can be seen with a case analysis on D_L mod 3.) Therefore removing T' leaves a face of degree at most $2(D_L + 1)/3 + 6 < D_L$ since $D_L > 20$.

So triangle T is drawn on the convex hull. Removing the drawing of T leaves a point set whose convex hull is exactly the size of the face created by removing T. So this fixes the next cycle to be drawn on the convex hull, and so on. So the added cycles exactly use up the added points, and the remaining drawing gives a solution to 3-partition as in the previous part. So the problem remains hard for triangulated graphs with maximum degree 15.

3-connected graphs with small treewidth: The construction of the inner triangulated graph has unbounded treewidth since the layers and connections between them form a $\Omega(n) \times \Omega(n^2)$-grid, which has treewidth $\Omega(n)$. This is to be expected (recall that Point-set embeddability is polynomial if the treewidth is constant and the graph is triangulated).

We show now that the treewidth can be constant if we replace "triangulated" by "3-connected." To give an upper bound on the treewidth, it suffices to analyze the *pathwidth* defined as follows: A graph has pathwidth k if its vertices can be enumerated as $v_1, \ldots, v_n$ such that for any $1 \leq i \leq n$, at most k vertices in $\{v_1, \ldots, v_i\}$ have a neighbour in

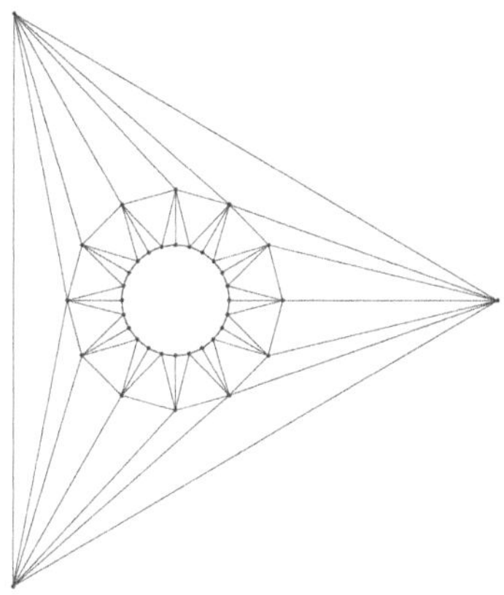

Figure 7: Triangulation of the outer face; the length of the cycles is not to scale.

$\{v_{i+1}, \dots, v_n\}$. Any graph of pathwidth k also has treewidth at most k, but not vice versa.

If two layers $v_1, \dots, v_N$ and $w_1, \dots, w_N$ are connected by an alternating path, then the subgraph induced by them is 3-connected and has pathwidth 4 (use $v_1, w_1, \dots, v_N, w_N$ as enumeration; then only v_1, w_1, v_i, w_i or only v_1, w_1, w_i, v_{i+1} have later neighbours for any i). If two layers have an a_i-path between them, then adding the edges between them as described above also gives a subgraph of pathwidth 4 (use w_1, v_2, v_3, the vertices of the a_i-path, $w_2, w_3, v_4, w_4, \dots, v_N, w_N, v_1$ as enumeration). Similarly the graphs outside the outer-most layer and inside the anchor triangles can be made to have constant pathwidth by triangulating suitably.

The main idea is now to *not* triangulate the faces between every other pair of consecutive layers. Instead, we use only three edges between three consecutive vertices on each of the layers; see also Figure 5. This means that the pathwidth of the resulting graph is at most 3 more than the pathwidth of the components, which was constant. So the resulting graph has constant pathwidth. By choosing points for K-paths as before, we can even achieve that it has constant maximum degree. We leave exact determination of pathwidth and maximum degree to the reader; we estimate both to at most 7.

All our reductions are polynomial in n and B, and we have hence proved:

THEOREM 4.1. *Point-set embeddability is NP-hard, even in the following situations:*

- *For a 2-connected outer-planar graph if the combinatorial embedding must be respected.*
- *For a triangulated planar graph with maximum degree 15.*
- *For a 3-connected planar graph with constant pathwidth and maximum degree.*

5. CONCLUSION

In this paper, we studied the Point-set embeddability problem, especially for the case where the combinatorial embedding of the graph is fixed. We showed that the problem is polynomial if the treewidth and the face-degrees are bounded, but becomes NP-hard as soon as one of these conditions is dropped. NP-hardness holds even for the special case of a 2-connected outer-planar graph (with a fixed combinatorial embedding, which doesn't have all vertices on the outer-face), or a triangulated planar graph (which hence has only one combinatorial embedding), or a 3-connected graph with constant treewidth (which also has only one combinatorial embedding.)

A few open problems remain:

- Our algorithm is polynomial if the treewidth and the face-degrees are constant, but it is not fixed-parameter tractable in these parameters. Does a fixed-parameter tractable algorithm exist, or is the problem $W[1]$-hard (say, with respect to the carving width of the dual graph)?
- What is the right kind of graph parameter to characterize when Point-set embeddability is polynomial if the combinatorial embedding is not fixed? It is not hard to see that the graph used in the reduction of Cabello [7] has treewidth 3, and one can modify it so that it has constant face-degrees. So these parameters alone do not suffice. Under what conditions (other than outer-planarity) is Point-set embeddability polynomial when the combinatorial embedding can be chosen?

6. REFERENCES

[1] S. Arnborg, D.G. Corneil, and A. Proskurowski. Complexity of finding embeddings in a k-tree. *SIAM J. Alg. Disc. Meth.*, 8:277–284, 1987.

[2] M. Badent, E. Di Giacomo, and G. Liotta. Drawing colored graphs on colored points. *Theor. Comput. Sci.*, 408(2-3):129–142, 2008.

[3] H. Bodlaender. Treewidth: algorithmic techniques and results. In *Mathematical Foundations of Computer Science (MFCS 1997)*, volume 1295 of *Lecture Notes in Computer Science*, pages 19–36. Springer-Verlag, 1997.

[4] H. L. Bodlaender and Ton Kloks. Efficient and constructive algorithms for the pathwidth and treewidth of graphs. *J. Algorithms*, 21(2):358–402, 1996.

[5] P. Bose. On embedding an outer-planar graph in a point set. *Computational Geometry: Theory and Applications*, 23(3):303–312, 2002.

[6] V. Bouchitté, F. Mazoit, and I. Todinca. Chordal embeddings of planar graphs. *Discrete Mathematics*, 273(1-3):85–102, 2003.

[7] S. Cabello. Planar embeddability of the vertices of a graph using a fixed point set is NP-hard. *Journal of Graph Algorithms and Applications*, 10(2):353–363, 2006.

[8] N. Castañeda and J. Urrutia. Straight line embeddings of planar graphs on point sets. In *Canadian Conference on Computational Geometry (CCCG '96)*, pages 312–318, 1996.

[9] S. Durocher and D. Mondal. On the hardness of point-set embeddability. In *6th Workshop on Algorithms and Computation (WALCOM 2012)*, volume 7157 of *Lecture Notes in Computer Science*, pages 148–159. Springer, 2012.

[10] I. Fáry. On straight line representation of planar graphs. *Acta Scientiarum Mathematicarum (Szeged)*, 11:229–233, 1948.

[11] H. de Fraysseix, J. Pach, and R. Pollack. How to draw a planar graph on a grid. *Combinatorica*, 10:41–51, 1990.

[12] M.R. Gary and D.S. Johnson. *Computers and Intracatablity: A Guide to the Theory of NP-completeness*. Freeman, 1979.

[13] M. Geyer, M. Kaufmann, T. Mchedlidze, and A. Symvonis. Upward point-set embeddability. In *Theory and Practice of Computer Science (SOFSEM 2011)*, volume 6543 of *Lecture Notes in Computer Science*, pages 272–283. Springer, 2011.

[14] E. Di Giacomo, W. Didimo, G. Liotta, H. Meijer, and S. Wismath. Constrained point-set embeddability of planar graphs. *Intl. Journal of Computational Geometry and Applications*, 20(5):577–600, 2010.

[15] P. Gritzmann, B. Mohar, J. Pach, and R. Pollack. Embedding a planar triangulation with vertices at specified positions. *American Mathematic Monthly*, 98:165–166, 1991.

[16] T.M. Moosa and M. Sohel Rahman. Improved algorithms for the point-set embeddability problem for plane 3-trees. In *COCOON*, volume 6842 of *Lecture Notes in Computer Science*, pages 204–212. Springer, 2011.

[17] N. Nestoridis and D. Thilikos. Square roots of minor closed graph classes, 2011. Presented at EuroComb'11.

[18] R.I. Nishat, D. Mondal, and Md. S. Rahman. Point-set embeddings of plane 3-trees. *Comput. Geom.*, 45(3):88–98, 2012.

[19] N. Robertson and P. Seymour. Graph minors. XVIII. Tree-decompositions and well-quasi-ordering. *J. Combin. Theory Ser. B*, 89(1):77–108, 2003.

[20] N. Robertson and P.D. Seymour. Graph minors. X. Obstructions to tree-decomposition. *J. Comb. Theory, Ser. B*, 52(2):153–190, 1991.

[21] W. Schnyder. Embedding planar graphs on the grid. In *ACM-SIAM Symposium on Discrete Algorithms (SODA '90)*, pages 138–148, 1990.

[22] P. D. Seymour and R. Thomas. Call routing and the ratcatcher. *Combinatorica*, 14(2):217–241, 1994.

[23] M. Sharir and E. Welzl. On the number of crossing-free matchings, (cycles, and partitions). In *Symposium on Discrete Algorithms (SODA 2006)*, pages 860–869. ACM Press, 2006.

[24] S. Stein. Convex maps. In *American Mathematical Society*, volume 2, pages 464–466, 1951.

[25] D. M. Thilikos, M. Serna, and H. L. Bodlaender. Cutwidth I: A linear time fixed parameter algorithm. *Journal of Algorithms*, 56:1–24, 2005.

[26] D. M. Thilikos, M. Serna, and H. L. Bodlaender. Cutwidth II: Algorithms for partial ω-trees of bounded degree. *Journal of Algorithms*, 56:25–49, 2005.

[27] K. Wagner. Bemerkungen zum Vierfarbenproblem. *Jahresbericht der Deutschen Mathematiker-Vereinigung*, 46:26–32, 1936.

Algorithmic Complexity of Finding Cross-Cycles in Flag Complexes

Michał Adamaszek*
Mathematics Institute and DIMAP
University of Warwick
Coventry, CV4 7AL, UK
aszek@mimuw.edu.pl

Juraj Stacho†
Mathematics Institute and DIMAP
University of Warwick
Coventry, CV4 7AL, UK
stacho@cs.toronto.edu

ABSTRACT

A cross-cycle in a flag simplicial complex K is an induced subcomplex that is isomorphic to the boundary of a cross-polytope and that contains a maximal face of K. A cross-cycle is an efficient way to define a non-zero class in the homology of K. For an independence complex of a graph G, a cross-cycle is equivalent to a combinatorial object: *induced matching containing a maximal independent set.*

We study the complexity of finding cross-cycles in independence complexes. We show that in general this problem is NP-complete when input is a graph whose independence complex we consider. We then focus on the class of chordal graphs, where, as we show, cross-cycles detect all of homology of the independence complex. As our main result, we present a polynomial time algorithm for detecting a cross-cycle in the independence complex of a chordal graph. Our algorithm is based on the geometric intersection representation of chordal graphs and has an efficient implementation.

As a corollary, we obtain polynomial time algorithms for such topological properties as contractibility or simple-connectedness of independence complexes of chordal graphs. These problems are undecidable for general independence complexes.

We further prove that even for chordal graphs, it is NP-complete to decide if there is a cross-cycle of a given cardinality, and hence, if a particular homology group of the independence complex is nontrivial. As a corollary we obtain that computing homology groups of arbitrary simplicial complexes given as a list of facets is NP-hard.

Keywords

computational topology, independence complex, homology groups, chordal graphs

*Supported by the Centre for Discrete Mathematics and its Applications (DIMAP), EPSRC award EP/D063191/1.

†Supported by EPSRC grant EP/I01795X/1.

Categories and Subject Descriptors

F.2.2 [**Analysis of algorithms and problem complexity**]: Nonnumerical algorithms and problems—*Computations on discrete structures, Geometrical problems and computations*

1. INTRODUCTION

The purpose of this paper is to investigate the algorithmic complexity of detecting certain non-trivial homology classes in flag simplicial complexes and to explore the consequences for deciding their topological properties. There is a certain efficient construction that defines a class in the homology of a flag complex and, at the same time, delivers a proof that this class is non-trivial. We describe it now. Let S^0 be the zero-sphere, that is, the simplicial complex consisting of two disjoint vertices. The k-fold join

$$O_k = \underbrace{S^0 * \cdots * S^0}_{k}$$

is a complex with $2k$ vertices, combinatorially equivalent to the boundary of the cross-polytope and homeomorphic to the $(k-1)$-sphere.

Definition. A cross-cycle (of size k) in a flag simplicial complex K is an induced subcomplex of K isomorphic to O_k and such that it contains a maximal face of K.

A cross-cycle is an embedded sphere $S^{k-1} \subseteq K$; hence, it defines a homology class in $\widetilde{H}_{k-1}(K)$. This class must be non-zero as its representing cycle contains a maximal face, and so it cannot be hit by a differential (see Section 2.2 for details). The name cross-cycle refers to the fact that the sphere is isomorphic to the boundary of a cross-polytope.

Cross-cycles have been used to construct homology classes in a number of contexts [1, 4, 15, 25]. They appear as the main contribution to the homology of the clique complexes of random geometric graphs [17]. Moreover, they are the minimal models, in the sense that for a flag complex K every non-zero homology class in $\widetilde{H}_{k-1}(K)$ must be supported on at least $2k$ vertices and, if the support size is exactly $2k$, then the class must be given by an embedded O_k, see [16]. This leads to our interest in their algorithmic properties.

There are two equivalent view-points of flag complexes. Geometers prefer to think of them as clique complexes of graphs that encode some incidence relation, e.g. proximity in a metric space (the Vietoris-Rips complex). We will follow the combinatorial topologists' approach via indepen-

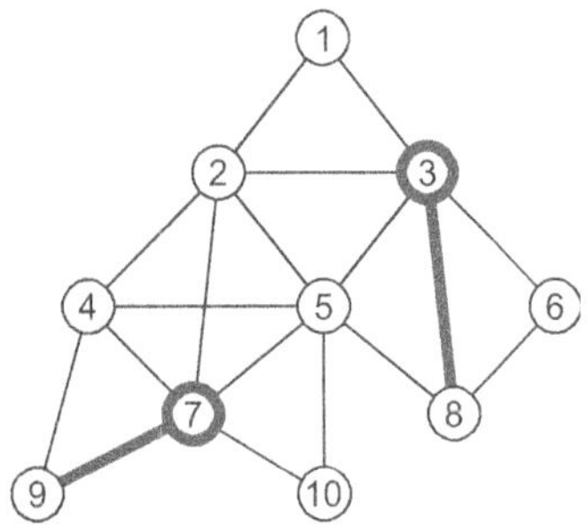

Figure 1: Example graph with an induced matching containing a maximal independent set.

dence complexes of graphs. For a graph G, the *independence complex* $I(G)$ is the simplicial complex whose vertices are the vertices of G and whose faces are the independent sets of G. Speaking about independence, rather than clique complexes, is a matter of choice, because the independence complex of a graph is precisely the clique complex of its complement.

Note that O_k is the independence complex of the disjoint union of k edges. This immediately leads to a characterization of cross-cycles in $I(G)$ in terms familiar to graph theorists.

Observation. If G is a graph then a cross-cycle of size k in $I(G)$ determines, and is determined, by an induced matching M of size k in G such that the vertex set of M contains a subset D that is a maximal independent set of G. For simplicity we shall say *an induced matching containing a maximal independent set*.

Note that D must contain exactly one vertex from each edge of M, and D is a dominating set of G (see Section 2.1). For example, in the graph G in Fig. 1, $M = \{(7,9),(3,8)\}$ is an induced matching containing an independent set $D = \{7,3\}$, and D is maximal in G. This defines a cross-cycle and therefore a non-trivial homology class in $\widetilde{H}_1(I(G))$.

Note that as a clique complex, O_k is the clique complex of rather more involved object, namely the complete k-partite graph $K_{2,\ldots,2}$ with all parts of size 2. We hope that this convinces the reader of the simplicity and usefulness of the representation via independence complexes.

For our algorithmic analysis we always assume that a flag complex is presented as $I(G)$ and the graph G is the input.

Our first observation is that finding a cross-cycle is hard.

THEOREM 1. *Given a graph G, it is NP-complete to decide if G has an induced matching containing a maximal independent set. Consequently, it is NP-complete to decide if $I(G)$ has a cross-cycle.*

We then concentrate on chordal graphs. A graph is *chordal* if every induced cycle has length three (cf. Section 2.1 for other characterizations). Chordal graphs are a very classical class of graphs [3, 11], and quite a lot is known about their independence complexes as well. In particular, they are vertex-decomposable [29] and therefore homotopy equivalent to wedges of spheres (also proved in [27, 19]). Moreover, every wedge of spheres arises, up to homotopy, as an independence complex of a chordal graph [19]. Induced matchings in chordal graphs are directly related to algebraic invariants of edge ideals [14]. Another reason to study chordal graphs in this context is that for this family of graphs, cross-cycles detect all of the homology of the independence complex.

THEOREM 2. *For a chordal graph G and any $k \geq 0$, the homology group $\widetilde{H}_{k-1}(I(G))$ is non-trivial if and only if G has an induced matching of size k containing a maximal independent set.*

Theorem 2 establishes a strong connection between the homotopy type of $I(G)$ and the combinatorics of induced matchings of a chordal graph G. The next two results describe quite precisely how much of the homotopical information can be recovered from G in polynomial time.

THEOREM 3. *There is an $O(|E(G)|^2)$ time algorithm that decides, for a chordal graph G, if G has an induced matching containing a maximal independent set.*

THEOREM 4. *Given a chordal graph G and an integer k, it is NP-complete to decide if G has an induced matching of size k containing a maximal independent set.*

By Theorem 2 these results have immediate topological corollaries.

THEOREM 5. *For a chordal graph G one can decide in polynomial time*

- *if $I(G)$ is contractible,*
- *if $I(G)$ is simply-connected.*

THEOREM 6. *Given a chordal graph G and integer k, it is NP-complete to decide if the group $\widetilde{H}_{k-1}(I(G))$ is non-trivial.*

This has a more general consequence for the hardness of calculating homology groups of arbitrary flag complexes.

THEOREM 7. *The following problems are NP-hard.*

- *Given a flag complex K, represented by its 1-skeleton, and an integer k, decide if $\widetilde{H}_k(K) = 0$.*
- *Given any simplicial complex K, represented by the list of maximal faces, and an integer k, decide if $\widetilde{H}_k(K) = 0$.*

The second part answers [18, Prob.33]. Theorem 7 is not surprising, as there are no known methods of calculating $\widetilde{H}_k(K)$ without enumerating in some way or other the (exponentially many) k-faces. There is much research on practical algorithms and their performance (see [30] and the references therein). However, we are not aware of any previous proof that the problems are in fact hard. See [24] for a recent result about the hardness of computing the Euler characteristic.

Also the positive results break some barriers. First, note that for an arbitrary graph G the problems of deciding if $I(G)$ is simply-connected or contractible are both undecidable [13]. Moreover, all previous work on topological features of chordal graphs [29, 7, 19, 27] made use exclusively of the existence of simplicial vertices (see Section 2.1). Our algorithm of Theorem 5 makes essential use of the geometric intersection representation of chordal graphs and their tree models. Finally, we remark that a straightforward application of Lemma 11 yields an algorithm that solves the problem of Theorem 3 in linear time for forests. This improves the result of [20] and answers a question of [6]. If G is a forest we obtain a complete information about the homotopy type of $I(G)$, since it is known to be homotopy equivalent to a single sphere or contractible [6].

Finally, one can ask for which natural subclasses of chordal graphs also the dimensions of non-trivial homology groups can be recovered in polynomial time. We prove such results for interval graphs and (more generally) for chordal graphs of bounded leafage (see Section 5 for definitions).

THEOREM 8. *There is an $O(k \cdot |V(G)|^3)$ time algorithm that decides, for an interval graph G and integer k, if G has an induced matching of size k containing a maximal independent set.*

THEOREM 9. *There is a $O(k^{\ell-1}|V(G)|^{\ell+1})$ time algorithm that decides, for an chordal graph G of leafage ℓ and integer k, if G has an induced matching of size k containing a maximal independent set.*

The rest of the paper is set out as follows. In Sections 2.1 and 2.2, we provide the necessary prerequisites from graph theory and combinatorial topology. In Section 3, we prove Theorem 2, thus establishing the bridge between topology and combinatorics, and use that bridge to deduce the topological statements (Theorems 5, 6 and 7) from the corresponding combinatorial results. After that, we work only with graphs. The main algorithm from Theorem 3 is presented in Section 4, its extensions (Theorems 8, 9) are described in Section 5 and finally, in Section 6, we prove the hardness results (Theorems 1 and 4).

2. PRELIMINARIES

2.1 Graph theory

A graph $G = (V, E)$ with vertex set $V(G)$ and edge set $E(G) \subseteq V(G) \times V(G)$ is always finite, undirected, with no loops or parallel edges. We write uv for the edge $(u, v) \in E(G)$. The *neighbourhood* of a vertex $v \in V(G)$ is defined as $N(v) = \{u \mid uv \in E(G)\}$. We write $N[v]$ for $N(v) \cup \{v\}$.

For a set $X \subseteq V(G)$, we define $N[X] = \bigcup_{v \in X} N[v]$ and write $N(X)$ for $N[X] \setminus X$. We denote by $G[X]$ the subgraph of G *induced* on X, i.e. the graph $G[X] = (X, E \cap X \times X)$. Further, we write $G - X$ for the graph $G[V(G) \setminus X]$.

A *clique* of G is a set of pairwise adjacent vertices of G. An *independent set* of G is a set of pairwise non-adjacent vertices of G. A set $X \subseteq V(G)$ is *dominating* if $N[X] = V(G)$.

A vertex $v \in V(G)$ is *simplicial* if $N(v)$ is a clique of G. A set of edges of G is a *matching* if no two edges in the set share a common endpoint.

A matching M of G is an *induced matching* if the subgraph of G induced by the endpoints of the edges in M contains only the edges in M. In other words, a matching of G is induced if it occurs as an induced subgraph of G.

A graph is *chordal* if it does not contains an induced cycle of length four or longer. By [5], every chordal graph has a simplicial vertex. Removing it yields a smaller chordal graph, which gives rise to the so-called *perfect elimination ordering*. Such orderings only exist for chordal graphs [11].

A *tree model* of a graph $G = (V, E)$ consists of a tree T, called *a host tree*, and a collection of subtrees of T, one for each vertex u of G, denoted by $\{T_u\}_{u \in V}$, with the property that $uv \in E$ if and only if $V(T_u) \cap V(T_v) \neq \emptyset$. For clarity, we shall use capital letters $X, Y, \ldots$ for the vertices of T and call them *nodes*. The following is a well-known fact [10].

THEOREM 10. *[10] A graph is chordal if and only if it has a tree model.*

We remark that there is a linear time algorithm [11, 23] to determine whether an input graph G is chordal, and if so, to construct a tree model of G.

For further notation and terminology we invite the reader to consult [11, 28] and particularly [3, Chap.9] or [11, Chap.4] for the properties of chordal graphs.

2.2 Combinatorial algebraic topology

We now recall some basic notions of algebraic topology and homology theory. An excellent concise introduction is [26, Chap.0.3]. More details can be found in [21].

A *simplicial complex* K with vertex set V is a collection of subsets of V such that if $B \in K$ and $A \subseteq B$ then $A \in K$. The elements of K are called *faces*. For a face $\sigma \in K$ we denote by $|\sigma|$ its cardinality (number of vertices) and by $\dim \sigma = |\sigma| - 1$ its dimension. We think of K as of the triangulation of a topological space (namely the geometric realization of K). The *join* of two simplicial complexes K and L with disjoint vertex sets is the complex $K * L = \{\sigma \cup \tau : \sigma \in K, \tau \in L\}$. It is also convenient to treat the empty simplicial complex (with no vertices and the unique face $\emptyset$) as the (-1)-dimensional sphere S^{-1}.

A simplicial complex K is called *flag* if every minimal non-face of K has dimension 1. This is equivalent to saying that K is the clique complex of its 1-dimensional skeleton or the independence complex of the complement of that skeleton.

Two topological spaces X, Y are *homotopy equivalent* or *have the same homotopy type*, which we denote $X \simeq Y$ if there are continuous maps $f : X \to Y$ and $g : Y \to X$ such that $fg \sim \mathrm{id}_Y$ and $gf \sim \mathrm{id}_X$ where $\sim$ denotes the relation of *homotopy* between maps [21]. Loosely speaking it means that one can continuously deform one map to the other. This is a relaxation of the notion of homeomorphism. A space is *contractible* if it is homotopy equivalent to a point.

We now proceed to define simplicial homology. An *oriented face* of K is a face σ together with an equivalence class of orderings of the vertices of σ, where two orderings are equivalent if they differ by an even permutation. We will denote by $[v_0, \ldots, v_p]$ the oriented face corresponding to the face $\sigma = \{v_0, \ldots, v_p\}$ together with the ordering $v_0 < \cdots < v_p$. Let $C_p(K)$ be the free abelian group with basis consisting of the oriented faces of dimension p modulo the relations $\sigma_1 + \sigma_2 = 0$ whenever σ_1 and σ_2 are oriented faces corresponding to the same face σ but with opposite orientations. The elements of $C_p(K)$ are called *p-chains* and $C_p(K) = 0$ when $p \leq -2$. There are maps $\partial_p : C_p(K) \to C_{p-1}(K)$ defined on the basis elements via

$$\partial_p[v_0, \ldots, v_p] = \sum_{i=0}^{p} (-1)^i [v_0, \ldots, v_{i-1}, v_{i+1}, \ldots, v_p].$$

Then ∂_p extend to group homomorphisms and $\partial_p \partial_{p+1} = 0$. The elements of $\ker \partial_p \subseteq C_p(K)$ are called *cycles* and the elements of $\mathrm{im}\, \partial_{p+1} \subseteq C_p(K)$ are *boundaries*. The *p-th reduced homology group of K* is the quotient $\widetilde{H}_p(K) = \ker \partial_p / \mathrm{im}\, \partial_{p+1}$. If K is finite, it is a finitely generated abelian group whose rank is called the *p-th reduced Betti number*

$\widetilde{b}_p(K)$. The homology groups are an invariant of the homotopy type of the space.

We can now describe the homology classes defined by cross-cycles in flag complexes. First of all, suppose

$$M = \{(v_{1,1}, v_{1,2}), (v_{2,1}, v_{2,2}), \ldots, (v_{k,1}, v_{k,2})\}$$

is an induced matching of size k in a graph G. Then we define a $(k-1)$-chain $\alpha_M \in C_{k-1}(I(G))$ by

$$\alpha_M = \sum_{\epsilon_1,\ldots,\epsilon_k \in \{1,2\}} (-1)^{\epsilon_1+\cdots+\epsilon_k}[v_{1,\epsilon_1}, v_{2,\epsilon_2}, \ldots, v_{k,\epsilon_k}].$$

Since M is an induced matching, each $\{v_{1,\epsilon_1}, \ldots, v_{k,\epsilon_k}\}$ is an independent set in G, and the definition makes sense. We leave it to the reader to verify $\partial_{k-1}\alpha_M = 0$. So α_M is a cycle and it determines a homology class $[\alpha_M] \in \widetilde{H}_{k-1}(I(G))$.

Note that if $i_M : O_k \hookrightarrow I(G)$ is the inclusion of the cross-cycle defined by M, then $[\alpha_M]$ is just the image of the fundamental class in $\widetilde{H}_{k-1}(O_k) = \widetilde{H}_{k-1}(S^{k-1})$ under the induced map in homology.

Now assume that M contains a maximal independent set D of G, which we can without loss of generality assume to be $D = \{v_{1,1}, v_{2,1}, \ldots, v_{k,1}\}$. Then the face $[v_{1,1}, v_{2,1}, \ldots, v_{k,1}]$ is maximal in $I(G)$, and so it does not appear in the image of $\partial_k : C_k(I(G)) \to C_{k-1}(I(G))$. It follows that the homology class $[\alpha_M] \in \widetilde{H}_{k-1}(I(G))$ represented by α_M is non-zero.

Unfortunately it is not always the case that $\widetilde{H}_{k-1}(I(G))$ is generated by the classes $[\alpha_M]$ over all induced matchings M of size k. For instance, for the cycle C_5 on 5 vertices, it is easy to see that $I(C_5) = S^1$, but C_5 does not even have induced matchings of size two.

In the next section, we show that the situation is much better for chordal graphs.

3. HOMOLOGY FOR CHORDAL GRAPHS

In this section, we prove that if G is chordal then all of the homology of $I(G)$ can be detected using cross-cycles.

Recall that for a sequence of topological spaces $X_1, \ldots, X_n$ the *wedge sum* (or one-point union) $\bigvee_{i=1}^n X_i$ is the space obtained from the disjoint union of the X_i's by choosing a sequence of distinguished points $x_i \in X_i$ and identifying them to a single point. If $n = 0$, we declare the wedge sum of an empty family to be the space consisting of one point. For a simplicial complex K, the *suspension* is $\Sigma K = S^0 * K$. These constructions are well-behaved with respect to reduced homology: we have $\widetilde{H}_k(\bigvee_{i=1}^n X_i) = \bigoplus_{i=1}^n \widetilde{H}_k(X_i)$ and $\widetilde{H}_k(\Sigma K) = \widetilde{H}_{k-1}(K)$. In fact, these properties are usually taken as the axioms of a reduced homology theory.

We are going to use the following result of Engström [7].

LEMMA 11. *[7] If G is any graph and v is a simplicial vertex, then there is a homotopy equivalence*

$$I(G) \simeq \bigvee_{u \in N(v)} \Sigma I(G \setminus N[u]).$$

This gives a method of recursively computing the homotopy type of $I(G)$ when G is chordal. For example, for the graph in Fig. 1, choosing $v = 1$ yields

$$I(G) \simeq \Sigma I\big(G[6,8,9,10]\big) \vee \Sigma I\big(G[4,7,9,10]\big).$$

The first graph has isolated vertices, so its independence complex is a cone, hence contractible. In the other graph, $v = 10$ is a simplicial vertex of degree 1. We conclude that $I(G[4,7,9,10]) \simeq \Sigma I(\emptyset) = \Sigma S^{-1} = S^0$, and so $I(G) \simeq S^1$.

This naive method leads only to an exponential time algorithm. However, it provides a connection to graph theory. For chordal graphs, this connection is summarized by Theorem 2 whose proof we now present.

PROOF OF THEOREM 2. The 'if' part follows from the discussion of cross-cycles in the previous section. To prove the 'only if' part we use induction on the size of G.

If $V(G) = \emptyset$, then $I(G) = S^{-1}$ and $\widetilde{H}_{-1}(S^{-1}) = \mathbb{Z}$. In this case the empty matching of size 0 satisfies the requirements.

Thus suppose that G is a chordal graph with at least one vertex and $\widetilde{H}_{k-1}(I(G)) \neq 0$. Let v be any simplicial vertex of G (which exists by [5]). By Lemma 11 there is a splitting

$$\widetilde{H}_{k-1}(I(G)) = \bigoplus_{u \in N(v)} \widetilde{H}_{k-2}(I(G \setminus N[u])).$$

It follows that there exists a vertex $u \in N(v)$ such that $\widetilde{H}_{k-2}(I(G \setminus N[u])) \neq 0$. The graph $G \setminus N[u]$ is chordal, so by induction it has an induced matching M' of size $k-1$ containing a maximal independent set D'.

Now define a new pair (M, D) of size k in G by setting

$$M = M' \cup \{vu\}, \qquad D = D' \cup \{u\}.$$

We easily see that M is an induced matching in G and D is an independent set which is maximal in G.

This completes the proof. □

Remark. A more careful analysis of this argument shows that, in fact, slightly more is true. We leave the proof to the reader, as we do not need the full strength of the next result for our algorithmic applications.

PROPOSITION 12. *If G is chordal and $k \geq 0$ then the homology group $\widetilde{H}_{k-1}(I(G))$ is generated by the classes $[\alpha_M]$ as M runs through all induced matchings of size k containing maximal independent sets.*

We close this section by deducing Theorems 5, 6 and 7 from their combinatorial counterparts.

PROOF OF THEOREMS 5 AND 6. Since for a chordal graph G the space $I(G)$ has the homotopy type of a wedge sum of spheres [29], it is contractible if and only if all its homology groups vanish. Therefore, the first part of Theorem 5 follows directly from Theorems 3 and 2. The second part is a consequence of the fact that simply-connectedness of $I(G)$ is equivalent to the vanishing of $\widetilde{H}_1(I(G))$, which by Theorem 2 is equivalent to G not having an induced matching of size two containing a maximal independent set. Clearly this condition can be checked in polynomial time. Finally, Theorem 6 follows by combining Theorems 4 and 2. □

PROOF OF THEOREM 7. The first statement follows directly from Theorem 6. To prove the second statement we describe a reduction from the NP-complete problem of Theorem 6. Suppose we have a chordal graph G with n vertices. Let K_G be the simplicial complex with vertices $V(G)$ whose maximal faces are the "complements" of edges in G, that is

$$\begin{aligned} \sigma \in K_G &\iff G[\overline{\sigma}] \text{ contains an edge} \\ &\iff \overline{\sigma} \text{ is not independent in } G \end{aligned}$$

where $\overline{\sigma}$ denotes $V(G) \setminus \sigma$.

Note that K_G is the Alexander dual [2] of $I(G)$ and that the list of its maximal faces of K_G has size polynomial in the size of G. Next, we have

$$\widetilde{H}_k(I(G)) = \widetilde{H}^k(I(G)) = \widetilde{H}_{n-k-3}(K_G)$$

where the second equality is Alexander duality [2, Thm. 1.1] and the first holds because $I(G)$ has the homotopy type of a wedge of spheres. This reduces the problem of Theorem 6 to the problem of computing the homology of K_G. □

4. ALGORITHM FOR CHORDAL GRAPHS

4.1 Introduction and outline

We now embark on proving Theorem 3. We present a polynomial time algorithm checking if a chordal graph G has an induced matching M containing a maximal independent set D. For simplicity, we call the pair (M, D) a *solution* to G.

Below, we first describe necessary definitions and preliminary steps before we present the proof as Section 4.2; the proofs of technical lemmas are deferred to Section 4.3.

Let G be a chordal graph. Recall that one can construct a tree model of G in linear time [23]. Thus, for the rest of this section, we shall assume that we have a fixed tree model of G and all subsequent considerations are always with respect to this model.

Namely, let a tree T with a collection $\{T_u\}_{u \in V(G)}$ of subtrees of T be a tree model of G. We consider T rooted at some node, and we direct all edges of T away from the root (Fig. 2). For $X, Y \in V(T)$, we write $X \preceq Y$ if there is in T a directed path from Y to X. Observe that $\preceq$ is a partial order. We write $X \prec Y$ if $X \preceq Y$ and $X \neq Y$. If XY is an edge of T oriented from Y to X, we say that Y is the *parent* of X, and X is a *child* of Y. If $X \prec Y$, we say that Y is an *ancestor* of X, and X is a *descendant* of Y.

Notation. For $u \in V(G)$, $\mathsf{top}(u)$ denotes the maximum element of $V(T_u)$ with respect to $\preceq$.

The following is a simple consequence of the definition of a tree model.

FACT 13. *If* $uv \in E(G)$, *then* $\mathsf{top}(u) \succeq \mathsf{top}(v) \in V(T_u)$ *or* $\mathsf{top}(v) \succeq \mathsf{top}(u) \in V(T_v)$.

To decide whether a solution to G exists, it suffices to consider particular type of solutions.

Definition. A solution (M, D) to G is *canonical* if every edge uv in M is such that $u \in D$ if $\mathsf{top}(u) \succ \mathsf{top}(v)$ and $v \in D$ if $\mathsf{top}(u) \prec \mathsf{top}(v)$.

For example, the solution $(\{(7,9),(3,8)\},\{7,3\})$ to the graph presented (with its tree model) in Fig.2 is canonical, because $\mathsf{top}(7) = C \succ D = \mathsf{top}(9)$ and $\mathsf{top}(3) = A \succ F = \mathsf{top}(8)$.

The following is an important result which is ultimately a consequence of Lemma 11. (For proof, see Section 4.3.)

LEMMA 14. *If there exists a solution to* G, *then there exist a canonical solution to* G *of the same size.*

Our algorithm is based on dynamic programming on T that tries to find canonical solutions for subgraphs of G and then combines these solutions to obtain a solution to G if one exists. In particular, we focus on subgraphs induced by vertices whose subtrees lie completely below some node of T.

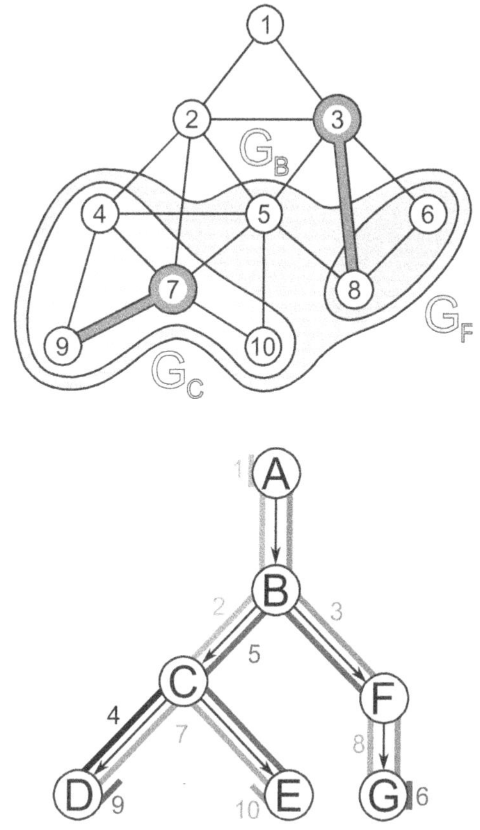

Figure 2: Example chordal graph G and its rooted tree model.

Notation. For $X \in V(T)$, G_X denotes the subgraph of G induced on all vertices v with $\mathsf{top}(v) \preceq X$.

If X is the root of T, then $G_X = G$. We distinguish the following special type of solutions to G_X.

Definition. A solution (M, D) to G_X is *rooted* if there exists $u \in D$ such that $\mathsf{top}(u) = X$.

Notation. $\mathcal{S}$ denotes the set of all nodes $X \in V(T)$ such that there exists a solution to G_X.

Notation. $\mathcal{R}$ denotes the set of all nodes $X \in V(T)$ such that there exists a rooted canonical solution to G_X.

Let us explain the notation and our strategy using the example in Fig. 2. We want to know whether the root of T is in $\mathcal{S}$. To find out, we recursively find the nodes that admit a rooted solution (the set $\mathcal{R}$). For example, $C \in \mathcal{R}$ since the subgraph G_C has a solution $(\{(7,9)\},\{7\})$. On the other hand, $B \notin \mathcal{R}$ since for any choice of an edge $(5, v)$ in G_B, the graph $G_B \setminus (N[5] \cup N[v])$ has an isolated vertex, and thus no solution. Still, G_B has a non-rooted solution combined from the rooted solutions of the disjoint subgraphs G_C and G_F.

Finally for $G = G_A$, we try the edge $(3,8)$ and see that $G_A \setminus (N[3] \cup N[8]) = G_C$. We then check that the rooted solution to G_C together with the edge $(3,8)$ indeed form a solution to G_A. This implies that $A \in \mathcal{S}$ as we wanted.

In the rest of this section, we state three technical lemmas that explain how we can compute the sets $\mathcal{R}$ and $\mathcal{S}$. Namely, Lemma 16 explains how to obtain a solution to G_X (if one exists) by combining rooted canonical solutions for descendants of X, and Lemma 17 provides a way to determine if a rooted solution exists to G_X. (For proofs see Section 4.3.)

LEMMA 15. *Let $X \in V(T)$ and suppose that there exists a canonical solution (M, D) to G_X. Then*

(i) if $Y = \mathsf{top}(u)$ for some $u \in D$, then $Y \in \mathcal{R}$, and

(ii) if $Y \preceq X$ and $Y \notin \bigcup_{u \in D} V(T_u)$, then $Y \in \mathcal{S}$.

LEMMA 16. *Let $X \in V(T)$. Define $\mathcal{R}_X = \{Y \mid Y \preceq X,\ Y \in \mathcal{R}\}$ and let $\mathcal{R}_X^*$ denote the set of maximal elements of $\mathcal{R}_X$ with respect to $\preceq$. For each $Y \in \mathcal{R}_X^*$, let (M_Y, D_Y) be a rooted canonical solution to G_Y, and define*

$$M^* = \bigcup_{Y \in \mathcal{R}_X^*} M_Y \qquad D^* = \bigcup_{Y \in \mathcal{R}_X^*} D_Y$$

Then, if there exists a solution to G_X, then (M^*, D^*) is a canonical solution to G_X.

LEMMA 17. *Let $Z \in V(T)$, and let $uv \in E(G)$ be an edge such that $\mathsf{top}(u) = Z$ and $\mathsf{top}(v) \preceq Z$.*

Define $\mathcal{L}_{uv} = \{Y \mid Y \preceq Z,\ Y \notin V(T_u) \cup V(T_v)\}$ and let $\mathcal{L}_{uv}^*$ be the set of maximal elements of $\mathcal{L}_{uv}$ with respect to $\preceq$.

For each $X \in \mathcal{L}_{uv}^*$, define $\mathcal{R}_X^*$ just like in Lemma 16, and for each $Y \in \mathcal{R}_X^*$, let (M_Y, D_Y) be a rooted canonical solution to G_Y. Finally, define

$$M_{uv}^* = \{uv\} \cup \bigcup_{X \in \mathcal{L}_{uv}^*} \bigcup_{Y \in \mathcal{R}_X^*} M_Y$$

$$D_{uv}^* = \{u\} \cup \bigcup_{X \in \mathcal{L}_{uv}^*} \bigcup_{Y \in \mathcal{R}_X^*} D_Y$$

Then, if there exists a canonical solution (M, D) to G_Z such that $uv \in M$ and $u \in D$, then (M_{uv}^*, D_{uv}^*) is a rooted canonical solution to G_Z.

With these observations we can now present our algorithm.

4.2 Proof of Theorem 3

We proceed in two steps. In the first phase, we process all nodes X of T from leaves to the root and decide if $X \in \mathcal{R}$. For every node X for which the answer is 'yes', we also store (for later use) a rooted canonical solution to G_X.

In the second phase, we check if the root of T is in $\mathcal{S}$. This determines whether or not there is a solution to G.

The recursive step of the first phase works as follows. Suppose we are at a node $Z \in V(T)$ and that all descendants of Z have already been processed. We try every edge $uv \in E(G)$ such that $\mathsf{top}(u) = Z$ and $\mathsf{top}(v) \preceq Z$. We construct the sets M_{uv}^* and D_{uv}^* as described in Lemma 17. Note that we are able to construct these sets because all descendants of Z have already been processed. Then we test if D_{uv}^* is a dominating set of G_Z. If so, we declare that there exists a rooted canonical solution (M_{uv}^*, D_{uv}^*) to G_Z. If we fail for every possible choice of uv, then we declare that there is no rooted canonical solution to G_Z. The correctness of this procedure is guaranteed by Lemma 17.

We then proceed with the second phase and test if there exists a canonical solution to G. To do so we construct the sets M^* and D^* as described in Lemma 16 for $X =$ root of T and test if D^* is a dominating set of G. If so, we declare that G has a solution given by (M^*, D^*). If not, we declare that no solution for G exists. The correctness of this step is guaranteed by Lemma 16.

To analyze the running time, denote $n = |V(G)|$ and $m = |E(G)|$. Each time we process a node Z, we test a subset of edges of G_Z and each edge uv of G is tested this way exactly once during the whole run of the algorithm. To construct M_{uv}^* and D_{uv}^* it suffices to search through the descendants of Z in T. This clearly takes at most $O(n)$ time, since both the sets M_{uv}^* and D_{uv}^* have no more than n elements and also T contains at most n nodes. Afterwards, we test if D_{uv}^* is a dominating set of G_Z. This can be done directly in time $O(m)$ by exploring the neighbourhood of every vertex in $V(G_Z) \setminus D_{uv}^*$. The same applies to the construction and testing of M^* and D^* in the second phase. Thus, altogether, the total complexity is $O(m^2)$ which concludes the proof. □

4.3 Proofs of technical lemmas

Before the proofs, we need to state an easy observation by reinterpreting Lemma 11 in the language of graphs. Namely, we observe that one can always tweak a solution (M, D) so that M contains an edge where one endpoint is a prescribed simplicial vertex while the other endpoint is in D.

We summarize it below, followed by a simple consequence.

FACT 18. *Let G be a graph and v be a simplicial vertex of G. Suppose that G has an induced matching M containing a maximal independent set D.*

(i) If $v \in D$, then $uv \in M$ for some $u \in N(v)$ and we define $M' = M$ and $D' = D \setminus \{v\} \cup \{u\}$.

(ii) If $v \notin D$, then $u \in D$ and $uw \in M$ for some $u \in N(v)$ and $w \in N(u)$ (possibly $w = v$), and we define $M' = M \setminus \{uw\} \cup \{uv\}$ and $D' = D$.

Then M' is an induced matching of G, and D' is a maximal independent set of G where M' contains D'.

FACT 19. *Let G be a graph with a simplicial vertex v.*

(i) If (M, D) is a solution to G and there exists $u \in N(v)$ with $uv \in M$ and $u \in D$, then $(M \setminus \{uv\}, D \setminus \{u\})$ is a solution to $G - N[u]$.

(ii) If $u \in N(v)$ and (M', D') is a solution to $G - N[u]$ then $(M' \cup \{uv\}, D' \cup \{u\})$ is a solution to G.

Using these observations, we can now prove Lemma 14.

PROOF OF LEMMA 14. The proof is by induction on G. Suppose that there exists a solution (M, D) to G. Let v be a vertex of G such that $\mathsf{top}(v)$ is a minimal element of the set $\{\mathsf{top}(u) \mid u \in V(G)\}$ with respect to $\preceq$. We claim that v is a simplicial vertex of G. Indeed, otherwise there would be non-adjacent vertices $u, w \in N(v)$. By Lemma 13 and the minimality of v, we conclude that $\mathsf{top}(v) \in T_u$ and $\mathsf{top}(v) \in T_w$. But then $\mathsf{top}(v) \in T_u \cap T_w$ implying $uw \in E(G)$, a contradiction. Thus v is indeed a simplicial vertex of G.

So by Lemma 18, we may assume that $uv \in M$ and $u \in D$ for some vertex $u \in N(v)$. By Lemma 19(i) the pair $M' = M \setminus \{uv\}$ and $D' = D \setminus \{u\}$ is a solution to $G' = G - N[u]$.

Consider a tree model of G' by taking T as the host tree and the collection $\{T_x\}_{x \in V(G) \setminus N[u]}$ as subtrees corresponding to the vertices of G'. By induction, there exists a canonical solution (M'', D'') to G' with respect to this tree model with $|M''| = |M'|$. In particular, each $xy \in M''$ satisfies $x \in D''$ if $\mathsf{top}(x) \succ \mathsf{top}(y)$, and $y \in D''$ if $\mathsf{top}(x) \prec \mathsf{top}(y)$.

Define $M^* = M'' \cup \{uv\}$ and $D^* = D'' \cup \{u\}$. It follows that (M^*, D^*) is a solution to G. By Lemma 13 and the minimality of v, we also conclude $\mathsf{top}(v) \preceq \mathsf{top}(u)$. Thus (M^*, D^*) is actually a canonical solution to G, and $|M^*| = |M''| + 1 = |M'| + 1 = |M|$ as required. □

In the rest of this section, we prove the three technical lemmas needed for the recursive step of our algorithm.

Proof of Lemma 15. Consider a node $Y \preceq X$ satisfying either the assumption of (i) or the assumption of (ii). Define $M' = M \cap E(G_Y)$ and $D' = D \cap V(G_Y)$. We show that (M', D') is a canonical solution to G_Y which will imply that $Y \in \mathcal{S}$ and the claim will follow.

Since (M, D) is a solution to G_X, it follows that M' is an induced matching of G_Y, and D' is an independent set of G_Y. We show that M' contains D'. Suppose otherwise, namely that there exists $u \in D'$ that is not an endpoint of any edge in M'. Since $u \in D$, there exists v such that $uv \in M$. Thus $v \notin V(G_Y)$ by our assumption about u. Hence, from the definition of G_Y, we obtain $\mathsf{top}(u) \preceq Y$ and $\mathsf{top}(v) \not\preceq Y$. This, with Lemma 13, implies that $\mathsf{top}(v) \succ \mathsf{top}(u)$. But then (M, D) is not a canonical solution to G_X.

It remains to show that D' is a maximal independent set of G_Y. Suppose otherwise, namely that there exists a vertex $v \in V(G_Y) \setminus N[D']$. Observe that $v \notin D$ by the definition of D'. Recall that (M, D) is a solution to G_X, and so there exists $w \in D$ such that $vw \in E(G)$. This implies $w \notin D'$, and so $w \notin V(G_Y)$. Thus, the definition of G_Y yields $\mathsf{top}(v) \preceq Y$ and $\mathsf{top}(w) \not\preceq Y$. Therefore, by Lemma 13, we conclude that $\mathsf{top}(w) \succ \mathsf{top}(v) \in V(T_w)$. This shows that $\mathsf{top}(w) \succeq Y \succeq \mathsf{top}(v)$ which implies $Y \in V(T_w)$, since T_w is a connected subgraph of T and $\mathsf{top}(v) \in V(T_w)$.

We now conclude that Y does not satisfy the assumption of (ii), since $Y \in V(T_w)$ and $w \in D$. So, Y must satisfy the assumption of (i), namely that there exist $u \in D$ with $\mathsf{top}(u) = Y$ and $u \neq w$, since $\mathsf{top}(w) \succ Y$. Therefore, $uw \in E(G)$, since $\mathsf{top}(u) = Y \in V(T_w)$. But this is impossible, since D is an independent set of G_X and $u, w \in D$.

This proves that $Y \in \mathcal{S}$. Finally, we observe that if Y satisfies the assumption of (i), then the vertex $u \in D$ such that $Y = \mathsf{top}(u)$ also belongs to D' which shows that $Y \in \mathcal{R}$.

That concludes the proof. □

Proof of Lemma 16. By Lemma 14, we may assume that there is a canonical solution (M, D) to G_X. We prove that (M^*, D^*) is also a canonical solution to G_X.

First, observe that for all distinct $Y, Y' \in \mathcal{R}^*_X$, there is no edge in G with one endpoint in G_Y and another endpoint in $G_{Y'}$. Indeed, any such edge uv where $u \in V(G_Y)$ and $v \in V(G_{Y'})$ must satisfy $\mathsf{top}(u) \preceq Y$ and $\mathsf{top}(v) \preceq Y'$. That contradicts Lemma 13 as Y and Y' are incomparable by $\preceq$.

This shows that M^* is indeed an induced matching, and D^* is an independent set of G_X. Also, if (M^*, D^*) is a solution to G_X, then it is a canonical solution, since (M_Y, D_Y) is a canonical solution to G_Y for all $Y \in \mathcal{R}^*_X$.

Thus, it remains to show that D^* is a maximal independent set of G_X. Suppose otherwise, namely that there exists a vertex $v \in V(G_X) \setminus N[D^*]$.

Consider a node $Y \in \mathcal{R}^*_X$. Observe that $v \notin V(G_Y)$ because (M_Y, D_Y) is a solution to G_Y. Next, recall that (M_Y, D_Y) is a rooted canonical solution of G_Y. Thus, there exists $w \in D_Y$ with $\mathsf{top}(w) = Y$. Suppose that v is adjacent to some $u \in V(G_Y)$. We conclude $\mathsf{top}(u) \preceq Y$ and $\mathsf{top}(v) \not\preceq Y$ by the definition of G_Y. Hence, by Lemma 13, $\mathsf{top}(v) \succ \mathsf{top}(u) \in V(T_v)$ which implies $\mathsf{top}(v) \succeq \mathsf{top}(w) \succeq \mathsf{top}(u)$. So $\mathsf{top}(w) \in V(T_v)$, since T_v is connected. But now $vw \in E(G)$ and $w \in D^*$ contradicting our assumption that v is not adjacent to any vertex of D^*.

This proves that for every $Y \in \mathcal{R}^*_X$, the vertex v is neither in G_Y nor is adjacent to some vertex in G_Y. Suppose that $v \in D$ and recall that (M, D) is a canonical solution to G_X. This implies by Lemma 15 that $\mathsf{top}(v) \in \mathcal{R}_X$. Thus, since $\mathcal{R}^*_X$ is the set of maximal elements of $\mathcal{R}_X$ with respect to $\preceq$, there must exist $Y \in \mathcal{R}^*_X$ such that $Y \succeq \mathsf{top}(v)$. Thus $v \in V(G_Y)$ by the definition of G_Y, which is a contradiction to the above. So we conclude that $v \notin D$ and we recall that D is a maximal independent set of G_X. Thus, there exists $u \in D$ such that $uv \in E(G)$. Similarly as before, we conclude that $\mathsf{top}(u) \in \mathcal{R}_X$ by Lemma 15. Thus, there exists $Y \in \mathcal{R}^*_X$ such that $Y \succeq \mathsf{top}(u)$. Consequently, $u \in V(G_Y)$ by the definition of G_Y. But this is impossible, since $uv \in E(G)$ and v is not adjacent to any vertex in G_Y as shown above.

This shows that D^* is indeed a maximal independent set of G_X and completes the proof. □

Proof of Lemma 17. First, note similarly as in the proof of Lemma 16 that M^*_{uv} is an induced matching and D^*_{uv} is an independent set of G. This follows from the fact that the elements of $\mathcal{L}^*_{uv}$ are pairwise incomparable by $\preceq$, and so the elements in $\bigcup_{X \in \mathcal{L}^*_{uv}} \mathcal{R}^*_X$ are pairwise incomparable by $\preceq$.

Now, let (M, D) be a canonical solution to G_Z with $uv \in M$ and $u \in D$. For contradiction, suppose that (M^*_{uv}, D^*_{uv}) is not a solution to G_Z. It means that there exists a vertex $w \in V(G_Z)$ that is not adjacent to any vertex in D^*_{uv}.

Consider $X \in \mathcal{L}^*_{uv}$, and define $M^*_X = \bigcup_{Y \in \mathcal{R}^*_X} M_Y$ and $D^*_X = \bigcup_{Y \in \mathcal{R}^*_X} D_Y$. Note, by the definition of $\mathcal{L}^*_{uv}$, that the parent of X belongs to $V(T_u) \cup V(T_v)$. Since (M, D) is a canonical solution to G_Z with $uv \in M$, X necessarily satisfies one of the assumptions in Lemma 15 which allows us to conclude that $X \in \mathcal{S}$. In other words, there exists a solution to G_X. Thus, by Lemma 16, (M^*_X, D^*_X) is also a solution to G_X. In particular, D^*_X is a maximal independent set of G_X. This shows that w is not a vertex of G_X, since $D^*_{uv} \supseteq D^*_X$ and we assumed that $D^*_{uv} \cup \{w\}$ is an independent set of G.

We conclude that $\mathsf{top}(w) \not\preceq X$ for all $X \in \mathcal{L}^*_{uv}$. This implies that $\mathsf{top}(w) \notin \mathcal{L}_{uv}$. Recall that $w \in V(G_Z)$ which means $\mathsf{top}(w) \preceq Z$. So $\mathsf{top}(w) \in V(T_u) \cup V(T_v)$ by the definition of $\mathcal{L}_{uv}$.

Now, recall that (M, D) is a solution to G_Z and so D is a maximal independent set of G_Z. Thus there exist $x \in D$ such that $xw \in E(G)$. Since $u \in D$ but also $u \in D^*_{uv}$, we conclude that $x \neq u$. This implies that $\mathsf{top}(x) \in \mathcal{L}_{uv}$. Hence, there exists $X \in \mathcal{L}^*_{uv}$ such that $\mathsf{top}(x) \preceq X$, and so, by Lemma 15, we conclude that $\mathsf{top}(x) \in \mathcal{R}_X$. Thus there exists $Y \in \mathcal{R}^*_X$ such that $\mathsf{top}(x) \preceq Y$. Recall that (M_Y, D_Y) is a rooted canonical solution to G_Y. Thus, there exists $y \in D_Y$ such that $\mathsf{top}(y) = Y$. Recall that $\mathsf{top}(w) \not\preceq X$ by the above, and so $\mathsf{top}(w) \not\preceq Y$. This implies, by Lemma 13, that $\mathsf{top}(x) \preceq \mathsf{top}(y) \preceq \mathsf{top}(w)$ and $\mathsf{top}(x) \in V(T_w)$. Thus $\mathsf{top}(y) \in V(T_w)$ since T_w is connected, which implies that y is adjacent to w. But then $y \in D_Y \subseteq D^*_{uv}$ which contradicts our assumption that $D^*_{uv} \cup \{w\}$ is an independent set of G_Z.

So no such w exists which proves that D^*_{uv} is indeed a maximal independent set of G_Z. That concludes the proof. □

5. EXTENSIONS AND SMALL CASES

In this section, we briefly examine a modification of our algorithm from Section 4 allowing us to find exact-size solutions in special cases. This will prove Theorems 8 and 9.

We shall use the notation from Section 4. Again, we have a chordal graph G and a tree model for G consisting of a tree T and subtrees $\{T_u\}_{u\in V(G)}$. Similarly, we define the partial order $\preceq$ on $V(T)$, and the subgraphs G_X.

The following two technical lemmas are analogues of Lemmas 16 and 17 for deciding if there exists a solution of size k. Their proofs are straightforward, and similar to the proofs of Lemmas 16 and 17. (We omit further details.) After the lemmas, we discuss how they imply Theorems 8 and 9.

Notation. $\mathcal{R}^{(k)}$ denotes the set of nodes $X \in V(T)$ such that there is a rooted canonical solution to G_X of size k, and $\mathcal{R}^{(k)}_X$ for $X \in V(T)$ denotes the set $\{Y \mid Y \preceq X,\ Y \in \mathcal{R}^{(k)}\}$.

LEMMA 20. *Let $X \in V(T)$. Suppose that there is a solution to G_X of size k. Then there exist positive integers $k_1, k_2, \ldots, k_t$ and nodes $Y_1, \ldots, Y_t$ where $Y_i \in \mathcal{R}^{(k_i)}_X$ for all $i \in \{1, \ldots, t\}$ such that*

(i) $k_1 + k_2 + \cdots + k_t = k$,

(ii) $Y_1, Y_2, \ldots, Y_t$ *are pairwise incomparable by* $\preceq$, *and*

(iii) if (M_{Y_i}, D_{Y_i}) is any rooted canonical solution to G_{Y_i} of size k_i for each $i \in \{1, \ldots, t\}$, and

$$M^* = \bigcup_{i=1}^{t} M_{Y_i} \qquad D^* = \bigcup_{i=1}^{t} D_{Y_i}$$

then (M^*, D^*) is a canonical solution to G_X of size k.

Notation. For $uv \in E(G)$ with $\mathsf{top}(u) \succeq \mathsf{top}(v)$, we denote by $\mathcal{R}^{(k)}_{uv}$ the set $\mathcal{R}^{(k)}_{\mathsf{top}(u)} \setminus \big(V(T_u) \cup V(T_v)\big)$.

LEMMA 21. *Let $X \in V(T)$, and let $uv \in E(G)$ be such that $\mathsf{top}(u) = X$ and $\mathsf{top}(v) \preceq X$. Suppose that there is a canonical solution (M, D) to G_X with $uv \in M$ and $u \in D$. Then there exist positive integers $k_1, k_2, \ldots, k_t$ and nodes $Y_1, \ldots, Y_t$ where $Y_i \in \mathcal{R}^{(k_i)}_{uv}$ for all $i \in \{1, \ldots, t\}$ such that*

(i) $k_1 + k_2 + \cdots + k_t = k - 1$,

(ii) $Y_1, Y_2, \ldots, Y_t$ *are pairwise incomparable by* $\preceq$, *and*

(iii) if (M_{Y_i}, D_{Y_i}) is any rooted canonical solution to G_{Y_i} of size k_i for each $i \in \{1, \ldots, t\}$, and

$$M^* = \{uv\} \cup \bigcup_{i=1}^{t} M_{Y_i} \qquad D^* = \{u\} \cup \bigcup_{i=1}^{t} D_{Y_i}$$

then (M^*, D^*) is a rooted canonical solution to G_X of size k.

We now look at how these lemmas can be applied in special subclasses of chordal graphs.

First, we illustrate this on *interval* graphs. These are defined as the intersection graphs of intervals of the real line. Every interval graph G is chordal and, in particular, the intersection model of intervals allows us to choose a tree model of G such that $\preceq$ is a total order. In particular, this means that the largest set of pairwise incomparable nodes of T with respect to $\preceq$ is of size one. It follows that we can compute the sets $\mathcal{R}^{(k)}$ in polynomial time as it suffices to apply Lemmas 20 and 21 with $t \in \{0, 1\}$. In particular, this proves Theorem 8. (For complexity analysis see below.)

More generally, we consider chordal graphs of bounded *leafage*. The leafage $\ell(G)$ of a chordal graph G is defined as the smallest integer ℓ such that G has a tree model whose host tree has exactly ℓ nodes of degree one. It was first defined in [22] and it can be computed in polynomial time by the algorithm of [12]. The same algorithm also constructs a tree model of G with said property.

Thus we may assume that T has exactly $\ell = \ell(G)$ nodes of degree one. We shall consider T rooted at one of these nodes (of degree one) and define $\preceq$ accordingly. Similarly to the case of interval graphs, this implies that the largest set of pairwise incomparable nodes of T with respect to $\preceq$ has size $\ell - 1$. Thus to compute the sets $\mathcal{R}^{(k)}$ it suffices to apply Lemmas 20 and 21 for all $t \in \{0, 1, \ldots, \ell - 1\}$ which again yields a polynomial time algorithm, as discussed below.

For the complexity analysis, let $n = |V(G)|$. To find a solution of size k, we compute k sets $\mathcal{R}^{(1)}, \ldots, \mathcal{R}^{(k)}$ each of which has at most n elements. For $j \in \{1 \ldots k\}$, to find out if $Z \in \mathcal{R}^{(j)}$, we apply Lemma 21. This involves choosing integers $k_1, \ldots, k_t$ and a set of t nodes, constructing the sets M^* and D^*, and testing if this constitutes a solution to G_Z. There are $\binom{j-t}{t-1} = O(j^t)$ choices for the integers and $O(n^t)$ choices for the nodes. The construction and testing of (M^*, D^*) takes $O(n^2)$ time, assuming we are storing partial solutions just like in Section 4. In total, for all j and all t, it takes $\sum_{j=0}^{k}\sum_{t=0}^{\ell-1} O(j^t n^{t+2}) = O(k^{\ell-1} n^{\ell+1})$ time to compute the sets $\mathcal{R}^{(1)}, \ldots, \mathcal{R}^{(k)}$. By the same token, the complexity of Lemma 20 applied to the root is also $O(k^{\ell-1} n^{\ell+1})$ once we have the sets $\mathcal{R}^{(1)}, \ldots, \mathcal{R}^{(k)}$. This proves Theorem 9.

Finally, for Theorem 8, we note that interval graphs coincide precisely with the case $\ell = 2$. That ends the proof.

6. HARDNESS

Here, we prove the hardness results of Theorems 1 and 4.

6.1 Proof of Theorem 1

The problem is clearly in NP. To prove that it is also NP-hard we construct a reduction from 3SAT which is NP-hard [9]. Consider an instance $\mathcal{I}$ to 3SAT, namely clauses $C_1, \ldots, C_m$ using variables $v_1, \ldots, v_n$. By standard arguments, we may assume that no variable appears twice in the same clause (positively or negatively). We write $\overline{v_i}$ for the negation of v_i. From $\mathcal{I}$ we construct a graph $G_\mathcal{I}$ as follows. The vertex set of $G_\mathcal{I}$ is $\{c_1, \ldots, c_m,\ v_1, \ldots, v_n,\ \overline{v_1}, \ldots, \overline{v_n}\}$. It contains edges $v_1\overline{v_1}, \ldots, v_n\overline{v_n}$ and there is an edge between v_i (resp. $\overline{v_i}$) and c_j if the literal v_i (resp. $\overline{v_i}$) appears in the clause C_j. Finally, the set $\{c_1, \ldots, c_m\}$ forms a clique.

We prove that $\mathcal{I}$ has a satisfying truth assignment if and only if $G_\mathcal{I}$ has an induced matching containing a maximal independent set. If $\varphi : \{v_1, \ldots, v_n\} \to \{\mathsf{true}, \mathsf{false}\}$ is a satisfying truth assignment, we define $M = \{v_1\overline{v_1}, \ldots, v_n\overline{v_n}\}$ and $D = \{v_i \mid \varphi(v_i) = \mathsf{true}\} \cup \{\overline{v_i} \mid \varphi(v_i) = \mathsf{false}\}$. By construction, M is an induced matching containing D, and D is an independent set and is maximal because every clause vertex c_j is adjacent to a literal that evaluates to true. Conversely, if $G_\mathcal{I}$ has an induced matching M containing a maximal independent set D, we first observe that $D \cap \{c_1, \ldots, c_m\} = \emptyset$ and exactly one of v_i, $\overline{v_i}$ belongs to D for each i. We assign true to all literals that are in D, and then observe that this constitutes a satisfying truth assignment.

6.2 Proof of Theorem 4

Since the problem of Theorem 4 (hardness of exact-size induced matching containing a maximal independent set in a chordal graph) is used to prove Theorem 7 (hardness of computing homology for arbitrary simplicial complexes), we present its proof in detail.

To prove Theorem 4, we first prove a somewhat stronger result about maximal independent sets in chordal graphs. Finding a smallest or largest such set is possible in polynomial time in chordal graphs [8, 11]. We prove that finding such a set of exact cardinality is NP-hard.

LEMMA 22. *It is NP-complete to decide, for a given chordal graph G and integer k, whether or not G has a maximal independent set of size exactly k.*

PROOF. We construct a reduction from the following problem: given a graph H and an integer ℓ, decide if H has a dominating set of size ℓ. This problem is NP-complete since the corresponding minimization problem (minimum size dominating set) is NP-hard (cf. [9, Prob.GT2]).

Consider an instance to this problem, a graph $H = (V, E)$ and an integer ℓ. Construct the following graph G:

$$\begin{aligned} V(G) =& \{v_0, v_1, \ldots, v_6 \mid v \in V\} \\ E(G) =& \{u_0v_0 \mid u, v \in V,\ u \neq v\} \cup \{u_0v_1 \mid uv \in E\} \cup \\ & \{v_0v_1, v_1v_2, v_1v_3, v_1v_4, v_1v_5, v_2v_6 \mid v \in V\} \end{aligned}$$

Let $V_0 = \{v_0 \mid v \in V\}$. Note that V_0 is a clique of G. From this, it is not difficult to see that G is a chordal graph.

We show that the following two conditions are equivalent:

- H has a dominating set of cardinality ℓ,
- G has a maximal independent set of cardinality $4|V| - 2\ell$.

This will yield the polynomial-time reduction and the proof.

First, if $D \subseteq V$ is a dominating set in H with $|D| = \ell$, then

$$\{v_1, v_6 \mid v \in D\} \cup \{v_2, v_3, v_4, v_5 \mid v \notin D\}$$

is a maximal independent set in G of size

$$2\ell + 4(|V| - \ell) = 4|V| - 2\ell.$$

For the converse, let $D \subseteq V(G)$ be a maximal independent set in G of size $|D| = 4|V| - 2\ell$. For any vertex v, we have

- if $v_1 \in D$, then $v_6 \in D$ and $v_2, v_3, v_4, v_5 \notin D$,
- if $v_1 \notin D$, then $v_3, v_4, v_5 \in D$ and exactly one of v_2, v_6 is in D.

This yields: $$|D \cap \{v_1, \ldots, v_6\}| = \begin{cases} 2 & \text{if } v_1 \in D, \\ 4 & \text{if } v_1 \notin D. \end{cases} \qquad (\star)$$

Now, let $D' = \{v \in V \mid v_1 \in D\}$. By $(\star)$ we can conclude

$$|D| = |D \setminus V_0| + |D \cap V_0| = 2|D'| + 4(|V| - |D'|) + |D \cap V_0|$$

Since V_0 is a clique of G, and D is an independent set of G, we conclude $|D \cap V_0| \leq 1$. However, $|D \cap V_0| = 1$ implies that the cardinality of D is odd, which is not the case.

We must conclude $D \cap V_0 = \emptyset$, and thus, $4|V| - 2\ell = |D| = 2|D'| + 4(|V| - |D'|)$. From this, we obtain $|D'| = \ell$. Moreover, D' is a dominating set of H, as otherwise some vertex $v_0 \in V_0$ could be used to enlarge D to a bigger independent set in G. That completes the proof. □

We are now ready to prove Theorem 4 by reduction from the exact cardinality independent dominating set problem in chordal graphs (Lemma 22). Consider a chordal graph $G = (V, E)$ and an integer k. We construct a graph G' from G by substituting an edge for every vertex of G. Namely,

$$\begin{aligned} V(G') =& \{v_1, v_2 \mid v \in V\} \\ E(G') =& \{v_1v_2 \mid v \in V\} \cup \\ & \{u_1v_1, u_1v_2, u_2v_1, u_2v_2 \mid uv \in E\} \end{aligned}$$

Note that G' is a chordal graph, since chordal graphs are closed under the operation of replacing a vertex by a clique.

If $D \subseteq V$ is a maximal independent set in G then

$$M' = \{v_1v_2 \mid v \in D\}, \quad D' = \{v_1 \mid v \in D\}$$

is an induced matching containing a maximal independent set in G' of the same cardinality as D.

Conversely, if (M', D') is a solution to G', then by setting

$$D = \{v \in V \mid v_1 \in D' \text{ or } v_2 \in D'\}$$

we obtain a maximal independent set of G of the same size.

This concludes the proof of Theorem 4. □

Open Problem. Note that our hardness results do not address the problem of finding *minimum* or *maximum* size cross-cycles in the case of chordal graphs. By Theorem 2 these are equivalent to calculating the connectivity and homological dimension of $I(G)$ for a chordal graph G.

We suspect that these two problems are also hard, but we were not able to prove this.

7. REFERENCES

[1] M. Adamaszek, *Special cycles in independence complexes and superfrustration in some lattices*, arxiv/1109.4914

[2] A. Björner, M. Tancer, *Note: Combinatorial Alexander Duality – A Short and Elementary Proof*, Discrete & Computational Geometry 42(4): 586-593 (2009)

[3] J. A. Bondy, U. S. R. Murty, *Graph theory*, Graduate Texts in Mathematics, 244, Springer, New York, 2008

[4] P. Csorba, *Subdivision yields Alexander duality on independence complexes*, Electr. J. Comb. 16(2) (2009), #R11

[5] G. A. Dirac, *On rigid circuit graphs*, Abhandlungen aus dem Mathematischen Seminar der Universität Hamburg, 25 (1961) 71-76

[6] R. Ehrenborg, G. Hetyei. *The topology of the independence complex*, Eur. J. Comb. 27(6): 906-923 (2006)

[7] A. Engström, *Complexes of directed trees and independence complexes*, Discrete Math. 309 (2009), 3299-3309

[8] M. Farber, *Independent domination in chordal graphs*, Operations Research Letters 1 (1982) 134-138

[9] M. R. Garey, D. S. Johnson, Computers and Intractability: A Guide to the Theory of NP-Completeness, W. H. Freeman, 1979

[10] F. Gavril, *The intersection graphs of subtrees in trees are exactly the chordal graphs*, Journal of Combinatorial Theory B 16 (1974), 47-56

[11] M. C. Golumbic, Algorithmic Graph Theory and Perfect Graphs (2nd edition), North Holland, 2nd ed., 2004

[12] M. Habib, J. Stacho, *Polynomial-time algorithm for the leafage of chordal graphs*, Algorithms - ESA 2009, Lecture Notes In Computer Science 5757 (2009) 290-300

[13] W. Haken, *Connections between topological and group theoretical decision problems*, In: Boone, Cannonito and Lyndon (1973), 427-441.

[14] H.T. Hà, A. Van Tuyl, *Monomial ideals, edge idcals of hypergraphs, and their graded Betti numbers*, J. Algebraic Combin. 27 (2008), no. 2, 215-245,

[15] J. Jonsson, *Certain Homology Cycles of the Independence Complex of Grids*, Discrete & Computational Geometry 43(4): 927-950 (2010)

[16] M. Kahle, *Topology of random clique complexes*, Discrete Math. 309(6): 1658-1671 (2009)

[17] M. Kahle, E. Meckes, *Limit theorems for Betti numbers of random simplicial complexes*, `arxiv/1009.4130`

[18] V. Kaibel, M. E. Pfetsch, *Some Algorithmic Problems in Polytope Theory* in Algebra, Geometry, and Software Systems 2003: 23-47

[19] K. Kawamura, *Independence complexes of chordal graphs* Discrete Math. 310 (2010), 2204-2211

[20] K. Kawamura, *Homotopy types of independence complexes of forests*, Contr. to Disc. Math. 5 (2010), 67-75

[21] D. Kozlov, *Combinatorial Algebraic Topology*, Algorithms and Computation in Mathematics, Vol. 21, Springer-Verlag Berlin Heidelberg 2008

[22] I.-J. Lin, T. A. McKee, D. B. West, *The leafage of a chordal graph*, Discussiones Mathematicae Graph Theory 18 (1998) 23-48

[23] D. J. Rose, R. E. Tarjan, G. S. Leuker, *Algorithmic aspects of vertex elimination on graphs*, SIAM Journal on Computing 5 (1976) 266-283

[24] B. H. Roune, E. Sáenz de Cabezón, *Complexity and Algorithms for Euler Characteristic of Simplicial Complexes*, `arxiv/1112.4523`

[25] J. Shareshian, M. L. Wachs, *Torsion in the matching complex and chessboard complex*, Adv. Math. 212 (2007), no. 2, 525-570.

[26] R. P. Stanley, *Combinatorics and commutative algebra*, Progress in Mathematics 41, Birkhauser, 1983

[27] A. Van Tuyl, R. H. Villarreal, *Shellable graphs and sequentially Cohen-Macaulay bipartite graphs*, J. Combin. Theory Ser. A 115 (2008), no. 5, 799-814

[28] D. B. West, Introduction to Graph Theory, Prentice Hall, 2nd ed., 2000.

[29] R. Woodroofe, *Vertex decomposable graphs and obstructions to shellability*, Proc. Amer. Math. Soc. 137 (2009), No. 10, 3235-3246

[30] A. Zomorodian, *The tidy set: a minimal simplicial set for computing homology of clique complexes*, Symposium on Computational Geometry 2010, 257-266

Counting Crossing-Free Structures

Victor Alvarez*
Fachrichtung Informatik
Universität des Saarlandes
66041 Saarbrücken, Germany
alvarez@cs.uni-saarland.de

Karl Bringmann
Max-Planck-Institut für Informatik
66123 Saarbrücken, Germany
kbringma@mpi-inf.mpg.de

Radu Curticapean
Fachrichtung Informatik
Universität des Saarlandes
66041 Saarbrücken, Germany
curticapean@cs.uni-saarland.de

Saurabh Ray
Max-Planck-Institut für Informatik
66123 Saarbrücken, Germany
saurabh@mpi-inf.mpg.de

ABSTRACT

Let P be a set of n points in the plane. A crossing-free structure on P is a straight-edge planar graph with vertex set in P. Examples of crossing-free structures include triangulations of P, and spanning cycles of P, also known as polygonalizations of P, among others. There has been a large amount of research trying to bound the number of such structures. In particular, bounding the number of triangulations spanned by P has received considerable attention. It is currently known that *every* set of n points has at most $O(30^n)$ and at least $\Omega(2.43^n)$ triangulations. However, much less is known about the algorithmic problem of counting crossing-free structures of a given set P. For example, no algorithm for counting triangulations is known that, on all instances, performs faster than enumerating all triangulations. In this paper we develop a general technique for computing the number of crossing-free structures of an input set P. We apply the technique to obtain algorithms for computing the number of triangulations and spanning cycles of P. The running time of our algorithms is upper bounded by $n^{O(k)}$, where k is the number of *onion layers* of P. In particular, we show that our algorithm for counting triangulations is not slower than $O(3.1414^n)$. Given that there are several well-studied configurations of points with at least $\Omega(3.464^n)$ triangulations, and some even with $\Omega(8^n)$ triangulations, our algorithm is the first to asymptotically outperform any enumeration algorithm for such instances. In fact, it is widely believed that any set of n points must have at least $\Omega(3.464^n)$ triangulations. If this is true, then our algorithm is strictly sub-linear in the number of triangulations counted. We also show that our techniques are general enough to solve the *restricted triangulation counting problem*, which we prove to be $W[2]$-hard in the parameter k. This implies a "no free lunch" result: In order to be fixed-parameter tractable, our general algorithm must rely on additional properties that are specific to the considered class of structures.

*Partially Supported by CONACYT-DAAD of México.

SCG'12, June 17–20, 2012, Chapel Hill, North Carolina, USA.

Categories and Subject Descriptors

F.2.2 [**Theory of Computation**]: Analysis of Algorithms and Problem Complexity—*Nonnumerical Algorithms and Problems*

General Terms

Algorithms, Theory

Keywords

Counting, Triangulations, Onion Layers, Parameterized Complexity

1. INTRODUCTION

Let P be a set of n points on the plane. A crossing-free structure defined by P, or on P, is a straight-edge planar graph whose vertex set is P. Examples of such crossing-free structures are triangulations, spanning cycles, matchings, spanning trees, etc. One can naturally ask: what are upper and lower bounds on the number of such structures over all possible sets of n points on the plane. Or given P, how can the number of such geometric objects be computed. The search for bounds has spawned a large amount of research over almost 30 years, starting with an upper bound of 10^{13n} on the number of crossing-free graphs on every set of n points, see [1]. This bound implies that the size of *each* class of crossing-free structures can be upper-bounded by c^n, with $c \in \mathbb{R}$ depending on the particular class. Since then, research has focused on tightening the upper and lower bounds on c. For example, in the case of spanning cycles, it is currently known that $4.6 \leq c \leq 54.55$, see [2] for the upper bound and [3] for the lower bound. Thus, every set of n points has at least $\Omega(4.6^n)$ and at most $O(54.55^n)$ spanning cycles. For triangulations, [4] provides the bound $c \leq 30$, and [5] provides $c \geq 2.4$. The interested reader can visit [6, 7] for an up-to-date list of bounds on other classes of crossing-free structures. The references therein gather the modern history of all listed bounds.

The second question on crossing-free structures, which was mentioned above, is of algorithmic flavor since we consider the problem of *computing* the number of crossing-free structures of a particular class for a *given* input set P. This problem is closely related to that of sampling crossing-free structures of the class uniformly at random, that is, if P spans, say t spanning cycles, we want to sample every spanning cycle with probability $1/t$. A first approach to the counting problem would be to produce *all* elements of the class, using methods for enumeration (see e.g. [14]), and then simply count the number of elements. This has the obvious disadvantage that the total time spent will be, at best, linear in the number of elements counted. By the first part, this number is always exponential in the input size. Thus an important question is whether we can count crossing-free structures of a given class in time sub-linear in the number of elements counted. Only for the class of plane graphs this is known to be possible [15], while for the class of all triangulations there are algorithms that are reported to be faster than enumeration [16, 13], but that have no theoretical runtime guarantees.

In this paper we focus on counting the elements of two particular classes of crossing-free structures defined over a given set P. The first class, $\mathcal{F}_T$, is the class of triangulations of P. The second class, $\mathcal{F}_C$, is the class of spanning cycles of P, also known as polygonalizations of P.

To state the main results of our paper, we need to define the notion of an *onion layer* of P. The formal definition will be given in the next section, but intuitively, k onion layers mean k nested convex sets. Thus every point set has at least one onion layer, so $k \geq 1$.

THEOREM 1. *Let P be a set of n points with k onion layers. The number of triangulations of P can be computed in time $O(k^2 g(n/k)^n)$, where $g(x) = ((x^3 + 3x^2 + 2x + 2)/2)^{\frac{1}{x}}$. Since $k \leq \lceil \frac{n}{3} \rceil$, this bound never exceeds $O(3.1414^n)$. Note that it can also be bounded by $n^{O(k)}$, which is polynomial for constant k.*

THEOREM 2. *Let P be as before. The number of spanning cycles of P can be computed in $n^{O(k)}$ time.*

For $k = O(1)$, our results yield algorithms that have polynomial running time in n, regardless of the cardinalities $|\mathcal{F}_C|$ and $|\mathcal{F}_T|$. Thm. 2 gives in particular a partial answer to Problem 16 of [8].

As stated before, for every set of n points, $|\mathcal{F}_T|$ can be lower-bounded by $\Omega(2.4^n)$, but it is widely believed that this bound can be improved to $\Omega(\sqrt{12}^n) \approx \Omega(3.464^n)$. If this stronger bound is true, the algorithm of Thm. 1 would count triangulations in time $O(3.1414^n) = o(|\mathcal{F}_T|)$, thus answering one of the aforementioned open questions in the positive. The general layout of our algorithms is similar to the one found in [17] where these ideas have been used for optimization problems.

The running times of both algorithms can be stated as $n^{f(k)}$, for some function f that does not depend on n. With regard to parameterized complexity, it is natural to ask if these runtimes can be improved to $f(k) \cdot n^{O(1)}$, thus proving that our problems belong to FPT, the class of fixed-parameter tractable problems. However, our techniques are general enough to solve harder problems, such as the *restricted triangulation counting problem*: Given P and a set of admissible edges E as input, count the triangulations of P that use only edges from E. We prove the following hardness result:

THEOREM 3. *The restricted triangulation counting problem is $W[2]$-hard if the parameter is considered to be k, the number of onion layers of P. This result even holds for the problem of deciding the* existence *of a restricted triangulation.*

The book [9] is a standard reference for parameterized complexity theory and defines the classes FPT and $W[2]$. For now, it suffices to say that the separation $FPT \neq W[2]$ is widely believed and indicates that we may have to exploit the particular structure of the problems in order to obtain fixed-parameter tractable algorithms for counting crossing-free structures.

The rest of the paper is organized as follows: In Section 2, we introduce our general framework. In Section 3 we will formally introduce the terminology and definitions needed to prove Thm. 1. In Sections 4 and 5, we prove Thm. 2 and 3. We finish in Section 6 with some conclusions.

2. GENERAL FRAMEWORK

In this section, we describe the central ideas of our counting algorithms. This section aims at an intuitive description and will thus abstain from technical details, which will be given in the next two sections, where the ideas are demonstrated with two concrete applications.

Let P be a set of n points and suppose that we want to count triangulations of P. A set S of non-crossing edges is called a *separator* if the union of edges in S splits conv(P) into at least two regions. Suppose that there exists a set of separators $\mathcal{S}$ with the following properties: (1) every triangulation T of P contains a unique separator $S \in \mathcal{S}$, and (2) we can enumerate the members of $\mathcal{S}$. With a set of separators $\mathcal{S}$, the triangulations of P can be counted as follows: For each $S \in \mathcal{S}$, let R_1^S, R_2^S,..., R_t^S be the regions of conv$(P) \setminus S$. Recursively compute the number of triangulations N_i^S of each R_i^S. The number of triangulations containing S is then $N^S = \prod_{i=1}^t N_i^S$, and the total number of triangulations of P is simply $\sum_{S \in \mathcal{S}} N^S$. Of course, in the recursion, a set of separators is required in each R_i^S, and the efficiency of the algorithm depends heavily on the choice of $\mathcal{S}$. One well-known family of separators $\mathcal{S}$ for triangulations is the set of T-paths of a set of points, see [13]. In Section 3, we introduce another set of separators for counting triangulations.

For now, let us move on to a slightly more complicated counting problem. Suppose we want to count crossing-free matchings spanned by the point set. Since any matching can be completed to a triangulation by adding edges, we could try the technique used for counting triangulations. Take any set $\mathcal{S}$ of separators. For each $S \in \mathcal{S}$, count the matchings in triangulations containing S, and finally add this up over all $S \in \mathcal{S}$. In any matching M that can be completed to a triangulation containing S, each vertex in S is either unmatched, or it is matched to a vertex within some R_i^S, or it is matched to another vertex in S. We can *annotate* each separator S with this information. When counting, for each $S \in \mathcal{S}$, we iterate over all annotations of S, and take care to be consistent with the current annotation when recursing into the subproblems.

This simple algorithm fails because some matchings M could contain several, say $s_M > 1$, separators and would thus fool our algorithm to count M exactly s_M times. This is no problem if $s_M = s$ were constant over all matchings, but we are not aware of any set of separators $\mathcal{S}$ with this property.

However, there is a different way to ensure that each matching is counted exactly once: we embed each matching M into a unique triangulation $T \supseteq M$. Given a family $\mathcal{S}$ of separators for the triangulations of P, we associate a unique $S \in \mathcal{S}$ to each matching. For concreteness, let us associate to each M the constrained delaunay triangulation (CDT) Δ^M *constrained* to contain M. Under standard non-degeneracy assumptions, there is a unique CDT for any given set of mandatory edges. We revise our algorithm as follows: Whenever we recurse, in each subproblem we only count matchings M with $S \subseteq \Delta^M$. If this last condition can be satisfied locally in each subproblem, i.e., choices in one subproblem do not depend on choices in others, we are done. While not every $\mathcal{S}$ admits such a locality condition, some do as we will see later.

The annotations required for counting *matchings* are not very complicated, but for many other counting problems, they are. An example of more involved annotations is given in Section 4, where we consider the problem of counting spanning cycles.

The techniques we described are fairly general and can be applied to several counting problems. The choice of separator structures and annotations depends on the specific problem and affects the efficiency of the algorithm. We start now by demonstrating the technique for counting triangulations.

3. COUNTING TRIANGULATIONS

Let P be a set of n points on the plane. Let conv(P) denote the convex hull of P and let ∂conv(P) denote its boundary. We define the onion layers of P as follows: the first onion layer $P^{(1)}$ of P is $P \cap \partial\text{conv}(P)$. For $i > 1$, the ith onion layer $P^{(i)}$ of P is defined inductively as $P \cap \partial\text{conv}(P \setminus \bigcup_{j=1}^{i-1} P^{(j)})$. By "number of onion layers of P", we mean the number of non-empty onion layers of P. For any $p \in P$, let $\ell(p)$ denote the index of the onion layer to which p belongs. Let us label the points $p \in P$ with distinct labels in $\{1, \ldots, n\}$ such that if $\ell(p) < \ell(q)$ then p also receives a label smaller than q. This is clearly possible. Figure 1 shows the onion layers of a set of 17 points and the labels assigned to them. From now on we will refer to the points of P by their labels *i.e.*, we will think of P as the set $\{1, \ldots, n\}$ and when we say "$p \in P$", we will mean the point with label p.

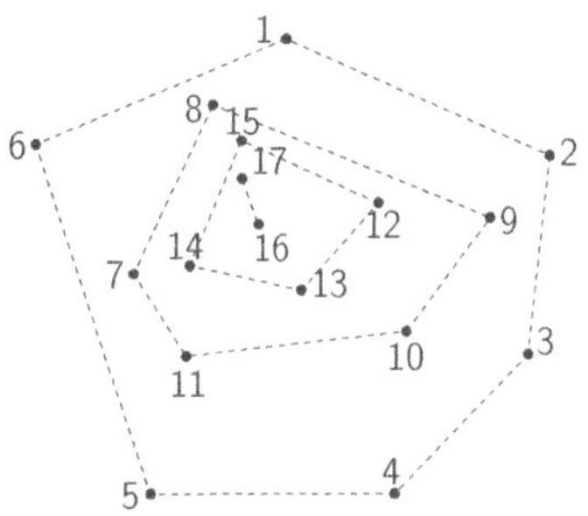

Figure 1: Four onion layers.

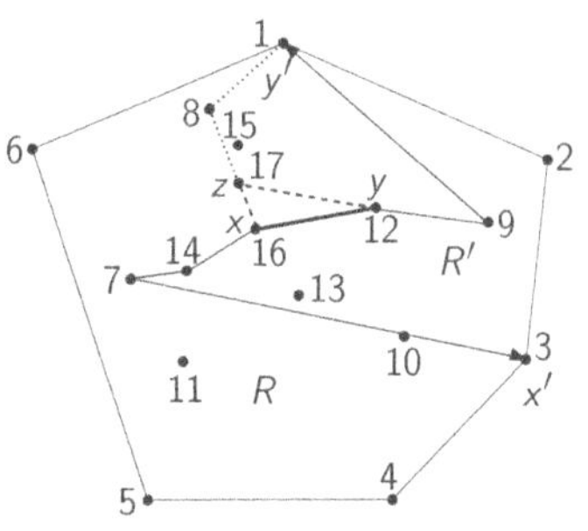

Figure 2: R and R' are the *sn*-regions (x, y).

Let T be any triangulation of P. For $p \in P \setminus P^{(1)}$, let $sn_T(p)$ be the smallest neighbor of p in T. Observe that any such point p has at least one neighbor q such that $\ell(q) < \ell(p)$ and therefore $sn_T(p) < p$. If $p \in P^{(1)}$, we set $sn_T(p) = p$. When T is clear from context, we will just write $sn(p)$ instead of $sn_T(p)$.

We denote by sn-path$_T(p)$ the unique path $p = a_0, a_1, \ldots, a_m$ in T such that for each $0 \leq i < m$, we have that $a_{i+1} = sn(a_i)$ and $sn(a_m) = a_m$. We will also direct this path from a_0 towards a_m and call this the direction of *descent* since $\ell(\cdot)$ decreases along the path. Note that any sn-path consists of at most one point from each onion layer and precisely one point from the first onion layer.

Let (p, q) be some edge in T and suppose that sn-path(p) ends at $p' \in P^{(1)}$ and sn-path(q) ends in $q' \in P^{(1)}$. There are two paths in T from p' to q' along ∂conv(P), one in the clockwise direction and the other in the counter-clockwise direction. Each of these paths along with the edge (p, q) and the two sn-paths starting at p and q respectively, defines a region within conv(P). We call these two regions the sn-regions of (p, q). See Figure 2. Given any sn-region R, we refer to the number of triangles in any triangulation of R as the *size* of R. This is well defined since the number of triangles is the same regardless of the triangulation chosen.

Let ab be an edge on ∂conv(P). Observe that in any triangulation, conv(P) is one of the sn-regions of (a, b), the other region being empty. In any triangulation T of P, there is precisely one triangle apb that the edge ab belongs to. Let R_{ap} be the sn-region of (a, p) that does not contain apb and similarly let R_{pb} be the sn-region of (p, b) that does not contain apb.

3.1 The algorithm

The main idea of our algorithm is as follows. We can easily enumerate all the points p such that the triangle apb appears in some triangulation. This is just the set Q of points p such that the triangle apb is free of other points of P. For every element p of Q, suppose that we can enumerate the sn-paths ρ of p over all triangulations of P. For ever pair (p, ρ), let $\mathcal{T}_{p,\rho}$ be the set of triangulations that contain the triangle apb and in which ρ is the sn-path of p. If, for each such pair that we can obtain, we can compute $|\mathcal{T}_{p,\rho}|$, then we are done, since each triangulation of P must contain precisely one pair (p, ρ), adding the numbers over all pairs gives us the total number of triangulations.

Let us fix a pair (p, ρ) for which we would like to compute $|\mathcal{T}_{p,\rho}|$. The pair already defines the regions R_{ap} and R_{pb} for all triangulations in $\mathcal{T}_{p,\rho}$. Observe that any triangulation in $\mathcal{T}_{p,\rho}$ contains a triangulation T_{ap} of R_{ap} and a triangulation T_{pb} of R_{pb}, each of which satisfy the following ***sn*-constraint**: for each edge (q, r) in ρ there is no edge (q, s) in the triangulation (either T_{ap} or T_{pb}) such that $s < r$. Furthermore, putting together any pair of triangulations T_{ap} and T_{pb}, each satisfying the constraint, and the triangle apb gives a triangulation in $\mathcal{T}_{p,\rho}$. This observation follows from the fact that ρ is an sn-path of p in any triangulation of $\mathcal{T}_{p,\rho}$ and allows us to separately compute the number of (sn-constraint satisfying) triangulations N_{ap} of R_{ap} and N_{pb} of R_{pb} whose product gives $|\mathcal{T}_{p,\rho}|$.

The numbers N_{ap} and N_{pb} are computed recursively. We will maintain that at any point in the recursion we are dealing with an sn-region of some edge. This is certainly true in the beginning since we start with an sn-region of the edge ab and also in the next step since we recurse on sn-regions defined by the edges (a, p) and (p, b) respectively. At any point, let us say that we are dealing with an sn-region R defined by an edge (x, y) and let ρ_x and ρ_y be the sn-paths starting at x and y respectively.

We recurse almost exactly as we did before: we enumerate the set of points z such that the triangle xzy lies within R and is free of other points of P contained in R, see Figure 2. Furthermore, we ensure that if z happens to be a point in either ρ_x or ρ_y, and (z, t) is an edge in that sn-path, then both x and y are bigger than t. This way, we do not violate the sn-constraint. For each such z, we enumerate the portions of sn-paths starting at z that lie within R. See Figure 2. Each such path splits the region R into regions R_{xz} and R_{zy} which are sn-regions defined by (x, z) and (z, y) respectively. Each of the regions R_{xz} and R_{zy} have sizes smaller than R. Recall that the size of a region is the number of triangles required to triangulate it. The recursion bottoms out when the size is ≤ 1 - in which case we know that there is exactly one triangulation. Note that even though we enumerate only the portions of the sn-paths of z that lie within R, these portions implicitly define the entire sn-path of z. This is because such a portion either ends at a point on the first onion layer in which case it is the entire sn-path, or at a point w on either ρ_x or ρ_y. The direction of descent along that sn-path, starting at w, is then the remaining portion of the sn-path of z.

One detail is still missing. How do we enumerate the portions of the sn-paths of z that belong to at least one triangulation of R? Well, we will not do it, instead, we enumerate a superset of paths which are *descending* in the sense that they start at z and each successive point is in a strictly smaller layer (layer with a smaller index). Again, we only enumerate the portion of such paths that lie inside R since the rest is implicitly defined. For any descending path that does not really belong to any triangulation of R, at least one of the regions R_{xz} or R_{zy} has no triangulations satisfying the sn-constraint. This will be detected somewhere down the recursion where we will not be able find any z satisfying the sn-constraint. At that point, we return 0 as the number of triangulations. Thus the algorithm still works. There is one other ingredient that we add for efficiency: memoization. Whenever we compute the number of triangulations of a certain sn-region that satisfy the sn-constraint dictated by the sn-paths defining the region, we store it in a hash table (or any other data structure). Consider a *call graph* in which each node represents an sn-region and there is a directed edge from a region R to a region R' if from R we make a recursive call to R'. The number of egdes in this graph is an upper bound on the running time of the algorithm since, because of memoization, no edge is *traversed* more than once.

We will now prove an upper bound on the number of edges in the call graph. Each call from a region R to a region R' can be charged to a triple of descending paths - two defining R and a third that, along with a triangle, splits R into two regions, one of which is R'. The triples (ρ_1, ρ_2, ρ_3) that are produced in the algorithm have the property that once two paths merge in the direction of descent, they never split again. This is ensured by the fact that we only enumerate the portions of the third descending path within the region R and the rest is implicitly defined, as noted before. Let ρ_2' be the portion of ρ_2 that does not have any point in common with ρ_1, and let ρ_3' be the portion of ρ_3 that does not have any point in common with either ρ_1 or ρ_2. The descending paths ρ_1, ρ_2' and ρ_3' are vertex disjoint, and along with some additional information they completely describe ρ_1, ρ_2 and ρ_3. The additional information that is required is whether, and where, ρ_2 merges with ρ_1, and whether, and where, ρ_3 merges with one of the other paths. If P has k onion layers, then each descending path has length at most k and therefore there are at most k ways that ρ_2' may merge with ρ_1, and at most $2k$ ways ρ_3' may merge with one of ρ_1 or ρ_2. Therefore, if U is an upper bound on the number of triples of vertex disjoint descending paths, then $2k^2U$ is an upper bound on the number of triples (ρ_1, ρ_2, ρ_3) as described above, and hence also an upper bound on the running time of the algorithm.

3.2 Number of vertex-disjoint triples of descending paths

Each descending path uses at most one vertex from every onion layer. Let $n_i = |P^{(i)}|$ be the size of the ith onion layer. Let us count how many ways there are for any triple of paths to use at most one vertex each from this layer. There is one way for each of the paths to skip this onion layer. There are n_i ways of choosing one point among the n_i which may then be used by any of the paths. This gives $3n_i$ ways for the three paths. There are $\binom{n_i}{2}$ ways to choose two points, and any two of the paths may use them. This gives $6\binom{n_i}{2}$ ways for the three paths. Finally there are $\binom{n_i}{3}$ ways of choosing three points, and there are three (not six) ways for the three paths to use one of these vertices. This is because these paths are non-crossing planar curves, and therefore the clockwise order of these paths along any $\partial \text{conv} P^{(i)}$ that intersects all three of them is the same for each i. The overall number of ways in which at most three points can be used from the ith layer is therefore $f(n_i)$, where $f(x) = 1 + 3x + 6\frac{x(x-1)}{2} + 3\frac{x(x-1)(x-2)}{6}$.

The number of triples of vertex disjoint descending paths is therefore at most $U = \prod_{i=1}^{k} f(n_i)$. Since each n_i is a positive integer, and the function $f(\cdot)$ is log-concave, as can be checked, for $x \geq 1$, the above product is maximized when each n_i is equal to n/k. This give an upper bound of $f(n/k)^k = g(n/k)^n$, where $g(x) = f(x)^{1/x}$. Now, $g(x)$ is maximized for some value of x between 0 and 1 and is a decreasing function for $x \geq 1$. Since each onion layer except the kth one must have at least three points, we have $U = O(g(3)^n)$. The fact that the kth onion layer may have less than three points makes only a difference of a constant factor. Therefore the running time of our algorithm is $O(k^2 g(3)^n) = O(3.1414^n)$. This concludes the proof of Thm. 1.

Often the number of onion layers can be much smaller than the maximum possible $\lceil n/3 \rceil$. For example, Dalal [10] has shown that if n points are chosen uniformly at random from a disk, then the expected number of onion layers of the resulting point set is $\Theta(n^{2/3})$.

4. COUNTING SPANNING CYCLES

In this section we give an algorithm for counting crossing-free spanning cycles of a point set P. We start by defining a suitable family of separator.

4.1 Triangular Path

We assume again that P has k onion layers. For every point $p \in P$ (on layer $P^{(i)}$ which is not the first layer) we fix in advance a ray ρ_p which emanates from p and does not intersect the interior of $\text{conv}(P^{(i)})$.

For any triangulation T of P there is a unique triangle $\Delta_p = p, q_1, q_2$ adjacent to p and intersecting ρ_p. Let q_p be the smaller of q_1 and q_2. Clearly q_p lies in a layer lower than the one containing p. Let $p_0, p_1, \ldots, p_r$ be the sequence so that $p_0 = p$, $p_{i+1} = q_{p_i}$, $\forall\, 0 \leq i < k$, and p_r lies on the first layer. We call $P_p(T) := \bigcup_i \Delta_{p_i}$ the *triangular path* of p w.r.t. T and we call p_r the *last point* of $P_p(T)$. See Figure 3.

$P_p(T)$ is uniquely defined for any triangulation. Moreover, for distinct triangulations T_1 and T_2, $P_p(T_1)$, $P_p(T_2)$ are either identical or they intersect properly: let i be the first position where $\Delta_{p_i}(T_1) \neq \Delta_{p_i}(T_2)$, then those two triangles intersect, as they both are adjacent to p, intersect ρ_p and have interiors free of points in P.

Before we continue, we describe constrained Delaunay triangulations, which will be needed in our algorithm.

4.2 Constrained Delaunay Triangulation

A constrained Delaunay triangulation (CDT) Δ^S of S is a Delaunay triangulation constrained to include a given set of *mandatory edges* S. More formally, it is the triangulation T containing S such that no edge e in $T \setminus S$ is flippable in the following sense: let Δ_1, Δ_2 be triangles sharing e. The edge e is flippable if and only if $\Delta = \Delta_1 \cup \Delta_2$ is convex, and replacing e with the other diagonal of Δ increases the smallest angle of the triangulation of Δ.

The CDT is unique if the points are non-cocircular, which is a reasonable assumption for counting crossing-free structures, as they do not change when perturbing the point set slightly.

We are now ready to describe our algorithm.

4.3 The Algorithm

Instead of counting spanning cycles, we count *rooted and oriented* spanning cycles. Given any cycle, we make it rooted by designating a *starting vertex*, and we make it oriented by assigning an *orientation*- clockwise or counter-clockwise. We then number the vertices in the cycle from 1 to n, beginning at the starting vertex, and continuing along the assigned direction. We also direct the edges along this direction. This way, each spanning cycle is counted exactly $2n$ times. At the end we divide by $2n$ to get the required number. In the remainder, we use the term HamCycle for rooted an oriented spanning cycles.

Given a HamCycle H let Δ^H be the CDT of H. We annotate Δ^H as follows:

- each vertex v of Δ^H is annotated with $(pos_v, prev_v, next_v)$, where pos_v is the number assigned to v in H, $prev_v$ is the vertex lying immediately before v in H, and $next_v$ is the vertex lying immediately after v in H.
- each edge e in Δ^H is annotated with a bit b_e that indicates whether e belongs to H or not.

The annotated Δ^H will be denoted by $\bar{\Delta}^H$. Let S be a separator contained in Δ^H that splits conv(P) into regions $R_1, \ldots R_t$. Separator S inherits the following information from $\bar{\Delta}^H$: each vertex $v \in S$ inherits pos_v from $\bar{\Delta}^H$. If $prev_v$ and $next_v$ are present in S, then this information is also inherited. If $prev_v$ is absent in S then v is annotated with the index i, $1 \leq i \leq t$, of the region R_i that $prev_v$ falls in. The same holds for $next_v$. Each edge e of S carries the annotation it has in $\bar{\Delta}^H$. The separator S of Δ^H thus annotated will be denoted by $\bar{\Delta}^H_S$.

We say that an annotated constrained Delaunay triangulation is *legal* if and only if it is identical to $\bar{\Delta}^H$, for some HamCycle H. Since there is a one-to-one correspondence between HamCycles and legal constrained Delaunay triangulations, our goal is to count the latter.

Our algorithm is essentially the same as for counting triangulations: instead of sn-paths we use annotated triangular paths. We start with an edge ab on ∂conv(P), and enumerate the set of points p such that the triangle apb is free of other points of P. For each such p, the triangle apb along with the triangular path starting at p forms a separator, see Figure 4. We enumerate such separators and all possible annotations for each one of them. Each such annotated separator splits conv(P) into smaller regions in which we recurse. In each recursive sub-problem we count (legal) annotated constrained Delaunay triangulations consistent with the annotated separator.

The reason for which we use triangular paths instead of simple sn-paths is the following: no edge in a separator, formed by a triangular path, lies on the boundary of more than one sub-problem. This allows us to verify flippability of edges separately in each sub-problem. If an edge belonged to more than one sub-problem, then the flippability of this edge would depend on the choices made in each sub-problem, thus introducing dependency between these sub-problems.

As in the case for counting triangulations, we use *memoization*. The running time as before is dominated by the number of triples of

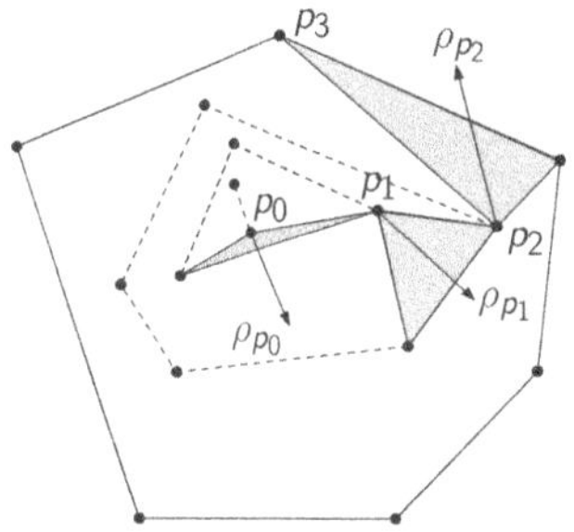

Figure 3: Triangular path P_p starting in onion layer $P^{(4)}$. Onion layers are shown in dashed. P_p can be extended to a triangulation T, in such a case P_p will be unique for T.

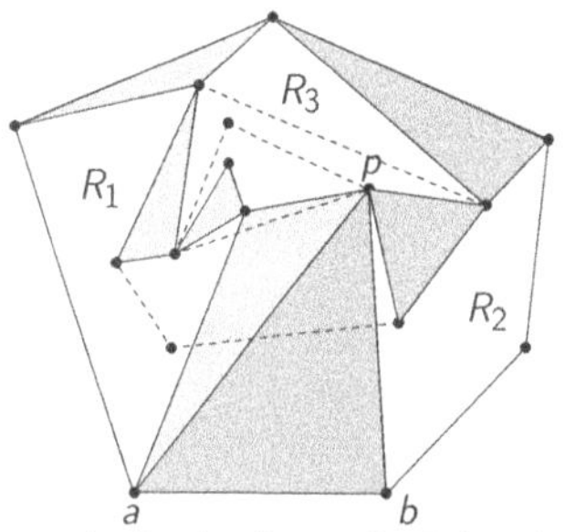

Figure 4: In the first call of the algorithm, the triangular path shown in dark gray is created. It divides the problem into regions $R_1 \cup R_3$ and R_2. A call for the latter creates the triangular path shown in light gray. Annotations are not shown for simplicity.

annotated triangular paths. The size of each triangular path is $O(k)$, thus there are at most $n^{O(k)}$ triangular paths. There are at most $n^{O(k)}$ annotations per triangular path, as can be easily checked. Hence there are $n^{O(k)}$ annotated triangular paths, and $n^{O(k)}$ triples of annotated triangular paths. The overall running time is thus $n^{O(k)}$, which concludes the proof of Thm. 2.

5. EXISTENCE OF RESTRICTED TRIANGULATIONS IS W[2]-HARD

Let P be a set of n points with k onion layers and let E be some set of "admissible" edges spanned by P. We say that a triangulation T of P is *restricted* w.r.t E if $T \subseteq E$. In this section, we consider the following *restricted triangulation existence problem*: On input (P, E), decide whether there exists a triangulation of P that is restricted w.r.t. E. This also defines a counting problem in the natural way, and the existence problem is trivially reducible to the counting problem.

The restricted triangulation existence problem was proven to be NP-complete in [12, 11]. We observed that both reductions are actually parsimonious, implying $\#P$-completeness of the restricted triangulation *counting* problem. So far all reductions involving restricted triangulations rely heavily on the ability to specify a particular set E as part of the input. If E is instead fixed to the set of *all* edges spanned by P, we obtain the problem of counting *all* tri-

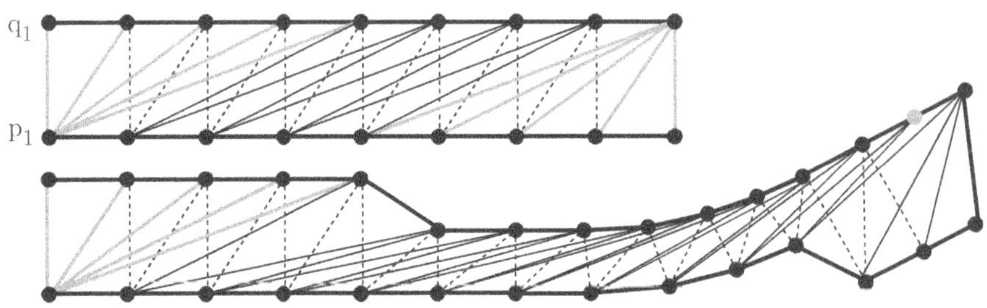

Figure 5: **(Top)** A pipe with 2 states and $l = 9$. Thick black edges constitute F, thick grey edges constitute S, thin edges are zig-zag L_1, and dashed edges are zig-zag L_0. This pipe is also a wire. **(Bottom)** A strechted and bent wire with a terminal.

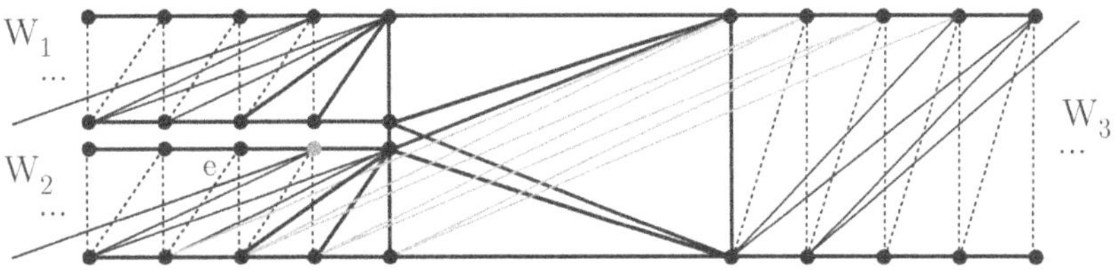

Figure 6: The or-gadget. The grey edges from W_2 to W_3 are "transfer edges". An analogous set of edges is also present from W_1 to W_3, but suppressed in this figure to improve legibility.

angulations of P, which we believe to be $\#P$-complete. Note that the corresponding existence problem is trivial in this case.

In this section, we parameterize the restricted triangulation counting/existence problems by k, the number of onion layers of P. Observe that the counting algorithm presented in Section 3 can be easily adapted to solve the counting problem, and thus the existence problem, in time $n^{O(k)}$. This, along with Thm. 3, means that our algorithmic framework is general enough to solve a $W[2]$-hard problem and therefore does not admit a general improvement to runtimes of the form $f(k) \cdot n^{O(1)}$ under the widely-believed assumption $FPT \neq W[2]$. Still, it might be possible to count crossing-free structures of particular classes (other than restricted triangulations) in time $f(k) \cdot n^{O(1)}$ by exploiting the structure of the class.

In the remainder of this section, we turn our attention to the parameterized existence problem to prove Thm. 3. Our proof is by reduction from the *parameterized hitting set problem*, which is proven to be $W[2]$-hard in [9]. An instance A of this problem is formed by numbers $n, m, k \in \mathbb{N}$, along with sets $S_1, \dots, S_m \subset [n]$, where k is considered as a parameter, and $[n] := \{0, \dots, n-1\}$. The output to A is "yes" iff there is a set $H \subseteq [n]$ of size at most k, such that $H \cap S_i \neq \emptyset$ for every $1 \leq i \leq m$.

In our reduction, several gadgets are used to transform an instance A of the hitting set problem to an instance $G_A = (P, E)$ of the restricted triangulation existence problem. The reduction is an *fpt-reduction* in the sense of [9], that is, it maps every instance A with parameter k to an instance G_A with $O(k)$ onion layers. Each gadget is given by a set of points with $O(1)$ onion layers, along with a set of admissible edges. The basic gadget is the *pipe*, shown in Fig. 5:

DEFINITION 1. *A **pipe** Q with n states and length $l > 8(n-1)$ consists of points $p_1 \dots p_l$, $q_1 \dots q_l$ with $p_t = (t, 0)$, $q_t = (t, 1)$ and a set $S \cup F \cup L_0 \cup \dots \cup L_{n-1}$ of admissible edges. The individual sets are defined as follows:*

*For $i \geq 1$, the **zig-edges** $a_{i,t}$ and the **zag-edges** $b_{i,t}$ are defined by $a_{i,t} = \{p_t, q_{t+4i}\}$ and $b_{i,t} = \{q_{t+4i}, p_{t+1}\}$. For $i = 0$, we define other zig- and zag-edges by $a_{0,t} = \{p_t, q_{t+1}\}$ and $b_{0,t} = \{q_{t+1}, p_{t+1}\}$. For $i \in [n]$, we define the **zig-zag line** $L_i = \{a_{i,1}, \dots, a_{i,l-w}, b_{i,1}, \dots, b_{i,l-w-1}\}$ with $w = 1$ for $i = 0$, and $w = 4i$ else.*

*Next, we add **completion edges** $S = \{\{p_1, q_{1+t}\} \mid 0 \leq t \leq 4(n-1)\} \cup \{\{p_{l-t}, q_l\} \mid 0 \leq t \leq 4(n-1)\}$. Finally, we add **frame edges** $F = \{\{p_i, p_{i+1}\} \mid i < l\} \cup \{\{q_i, q_{i+1}\} \mid i < l\}$.*

It is clear that in any triangulation T of a pipe Q, exactly one zig-zag L_i, for $i \in [n]$, is contained, since different zig-zags lines cross. If $L_i \subseteq T$, we say that Q "carries" the value i in T. The edges of the sets S and F allow to complete a triangulation of a pipe once a zig-zag is chosen. Note that $F \subseteq T$ holds for every valid triangulation T. The same is not true for S.

A pipe with n states will always be "horizontal", i.e., it will not turn in any other direction. This is required for the final point set to feature a bounded number of onion layers.

In our construction, we will also require vertical connections between pipes. These are obtained by **wires**, which are pipes with two states. Since they feature only two states, wires can be stretched by arbitrary factors, and bent by arbitrary angles, while increasing their length only by a constant additive term. This is shown in Fig. 5. For wires, we relabel the values 0 and 1 by *false* and *true* respectively.

The remaining gadgets are specified and defined as follows:

An or-gadget is connected to input wires W_1, W_2, and an output wire W_3, as shown in Fig. 6. (1) If one of W_1 or W_2 carries *true*, then W_3 may carry *true*. (2) If W_3 carries *true*, then at least one of W_1 or W_2 must carry *true*.

A terminal-gadget can be attachted to a wire W, replacing its "end part" as exemplified in the bottom part of Fig. 5. It admits a valid triangulation iff W carries *true*

A tester-gadget for i at t is connected to a pipe Q, between $a_{i,t}$ and $b_{i,t}$, and has an output wire W, see Fig. 7a. (1) If Q carries i, W may carry *true*. (2) If W carries *true*, Q must carry i.

A crossing-gadget allows an input wire V to intersect a pipe Q, leaving it as an output wire W. The value carried by Q is not

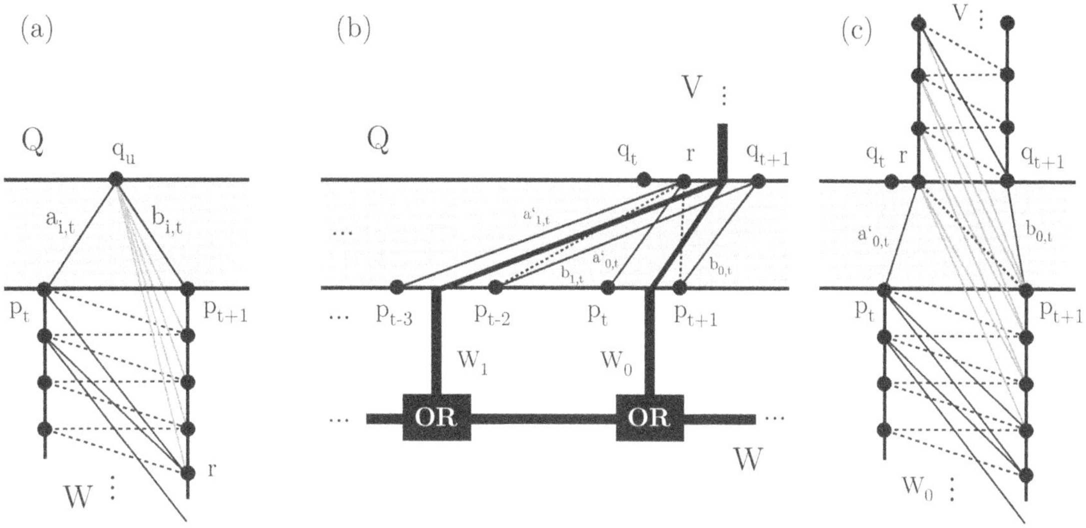

Figure 7: **(a)** The tester-gadget for i at t. Q is modified by shifting, for $k > 0$, all p_{t+k} and q_{u+k} to the right until the triangle (r, p_{t+1}, q_u) is oriented counterclockwise. **(b)** A crossing between pipe Q and wire W. **(c)** Details at W_0.

influenced by V. (1) If V carries *true*, then W may carry *true*. (2) If W carries *true*, then V must also carry *true*.

As shown in Fig. 7b, V enters the crossing-gadget from the top. Between points q_t and q_{t+1}, a new point r is added. For all p_u adjacent to q_{t+1} via some a-edge $a_{i,u}$, the edge $a_{i,u}$ is replaced by $a'_{i,u} = \{p_u, r\}$ and $\{p_{u+1}, r\}$. Latter edge is shown dashed in Fig. 7b.

In the area between $a'_{i,u}$ and $b_{i,u}$, the wire V is prolonged to W_i, as shown in Fig. 7c for $i = 0$. All wires W_i are stretched by a factor α. This α is chosen small enough to ensure that, for all i, the grey edges shown in Fig. 7c do not intersect either $a'_{i,u}$ or $b_{i,u}$.

Finally, the wires $W_0, \dots, W_{n-1}$ are connected to a chain of or-gadgets, as shown in Fig. 7b.

We now prove that each of the gadgets fulfills its specifications.

Or-gadget: (1) Assume wlog that W_2 carries *true* in some triangulation T. We construct a triangulation as follows: Add the transfer edges at W_2 to T, as shown in Fig. 6. Then triangulate W_3 by L_{true}. Triangulate the middle part of the or-gadget by thick edges. If W_1 carries *false*, this already suffices, else additionally include the thick edges in W_1.

(2) If W_3 carries *true*, the transfer edges from either W_1 or W_2, say W_2, must be present in the triangulation. If W_2 carried *false*, it can include L_{false} only up to edge e, since all following edges intersect transfer edges. But then, the grey point fails to be part of a triangle.

Terminal: If W carries *true*, the terminal is triangulated. If W carries *false*, the grey point in Fig. 5 fails to be in a triangle.

Tester: (1) Assume Q carries i in some triangulation T: Since no grey edges in the tester intersect L_i, they can be added to T, and W can thus be triangulated by L_{true}.

(2) Given some triangulation T, in which W carries *true*, all grey edges must be present in T. But for all $j \neq i$, there is an edge $c \in L_j$ in Q that intersects both $a_{i,t}$ and $b_{i,t}$ and all grey edges, and therefore $c \notin T$. Then $L_j \nsubseteq T$ for all $j \neq i$, forcing $L_i \subseteq T$.

Crossing: (1) Let T be a triangulation in which Q carries i and V carries *true*. Include the grey edges at W_i, triangulate W_i to carry *true* and leave the or-gadget chain with *true* at W.

(2) Assume W carries *true* in a triangulation T. Then, by correctness of the or-gadget, some wire W_i must carry true. But then, the grey edges shown in Fig. 7c must be present at W_i. Thus, V must carry true.

A triangulation is also possible if V and W carry *false*: Assume Q carries i in T and would thus contain the edge $a_{i,u} = \{p_u, q_{t+1}\}$, for some u, if the crossing were not present. Instead of $a_{i,u}$, we include $a'_{i,u} = \{p_u, r\}$ and $\{p_{u+1}, r\}$ into T. This ensures that the parallelogram $(p_u, p_{u+1}, q_{t+1}, r)$ has the diagonal $\{p_{u+1}, r\}$, shown dashed in Fig. 7b and Fig. 7c.

To describe how the gadgets fit together, recall that an instance A of the parameterized hitting set problem is formed by numbers $n, m, k \in \mathbb{N}$, along with sets $S_1, \dots, S_m \subseteq [n]$, where k is the parameter. We create, in polynomial time, an instance $G_A = (P, E)$ of size $poly(n, m)$ that has $O(k)$ onion layers and admits a triangulation w.r.t. E iff A admits a hitting set of size $\leq k$. The mapping $A \mapsto G_A$ will clearly be polynomial-time computable, and thus an fpt-reduction.

In the construction, we start with parallel pipes $Q_1, \dots, Q_k$ of n states each, and of length polynomial in m and n, as shown in Fig. 8. Pipe Q_i lies above pipe Q_{i+1}. Let Q_i be a pipe, $1 \leq i \leq k$, and let $S_j = \{s_{j,1}, \dots s_{j,t}\} \subseteq [n]$ be a set of instance A. We define the *stripe* $B_{i,j}$ as a set of t testers attached to Q_i that check if Q_i carries any of the values of set S_j. The stripe $B_{i+1,j}$ will lie in the same vertical slab as $B_{i,j}$. The testers of $B_{i,j}$ are connected to a chain of or-gadgets that lies between pipes Q_i, Q_{i+1}. For $i < k$, the output of the last or-gadget in $B_{i,j}$ is carried to $B_{i+1,j}$ by a crossing-gadget, see Fig. 8. For $i = k$, the last or-gadget in $B_{i,j}$ is connected to a terminal-gadget instead.

The *block* $\mathcal{B}_j$ is the union of the stripes $B_{1,j}, \dots, B_{k,j}$. The blocks $\mathcal{B}_1, \dots, \mathcal{B}_m$ are arranged horizontally in such a way that the points in stripes $B_{i,1}, \dots, B_{i,m}$, with $1 \leq i \leq k$, are aligned by their y-coordinates.

Finally, P is defined to be the set of points of all the gadgets involved. To define the set E of edges, we first include the admissible

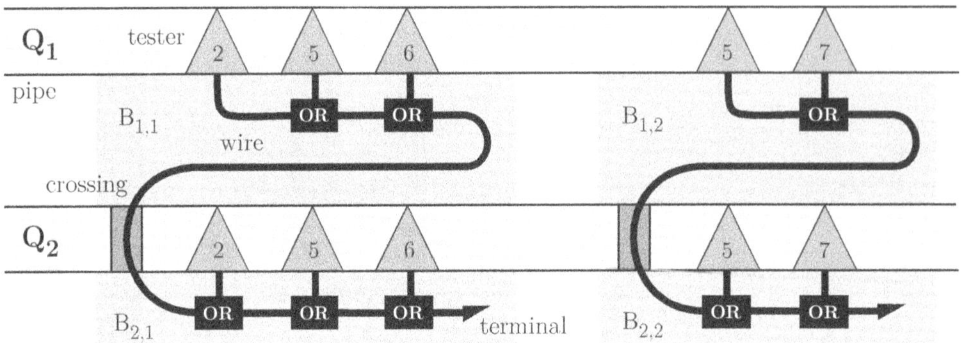

Figure 8: Instance G_A produced from instance A of the parameterized hitting set problem with $n = 8, k = 2$ and $S_1 = \{2, 5, 6\}, S_2 = \{5, 7\}$.

edges from all gadgets involved. Then, the empty spaces between gadgets are triangulated arbitrarily, and these edges are added to E. Setting $G_A = (P, E)$, we can now complete the proof of Thm. 3 by proving the following lemma:

LEMMA 1. *G_A has $O(k)$ onion layers and admits a triangulation iff A admits a hitting set of size $\leq k$.*

Proof. Consider the number of different y-coordinates of P. This is an upper bound for the number of onion layers of P. The pipes contribute $2k$ different y-coordinates. Every other gadget features $O(1)$ different y-coordinates. Each wire can be stretched and bent with $O(1)$ extra length, and gives $O(1)$ different y-coordinates. Since the points in stripes $B_{i,1}, \ldots, B_{i,m}$ are aligned by their y-coordinates, each set $C_i := B_{i,1} \cup \ldots \cup B_{i,m}$ has $O(1)$ different y-coordinates. This totals to $2k + O(k) = O(k)$ different y-coordinates for all points in P.

Given a hitting set $H = \{x_1, \ldots, x_k\}$, we construct a triangulation that uses only edges from E: For every $i \leq k$, make Q_i carry x_i. For every $j \leq m$, pick some $h(j) \in \{1, \ldots, k\}$ with $x_{h(j)} \in S_j$. In stripe $B_{h(j),j}$, triangulate the output wire of the tester for $x_{h(j)}$ to carry *true*. Triangulate all following or-gadgets to output *true*, possibly passing crossing gadgets, until the terminal of $\mathcal{B}_t$ is reached, which can then be triangulated.

On the other hand, the values $H = \{x_1, \ldots, x_k\}$ carried by the pipes $Q_1, \ldots, Q_k$ in any valid triangulation of G_A form a hitting set: Since every terminal is triangulated, the wire of every block $\mathcal{B}_j$ must carry *true* at some place. Thus, the output of some or-gadget in $\mathcal{B}_j$ must carry *true*. Consider the first or-gadget that fulfills this, and say it lies in stripe $B_{h(j),j}$. It must be connected to a tester that outputs *true*. This implies $x_{h(j)} \in S_j$ and $H \cap S_j \neq \emptyset$. ■

6. CONCLUSIONS

Some of the important open questions at this moment are: (1) is it true that every set of n points contains at least $\Omega(3.464^n)$ triangulations? (2) Is it possible to compute the number of triangulations, and of spanning cycles, of a given point set P in polynomial time? These questions look at the moment very challenging.

It would be very interesting to practically compare our algorithm for counting triangulations with the ones presented in [13, 16]. While we suspect that our algorithm will be fast in practice, it remains to be verified experimentally.

7. ACKNOWLEDGMENT

We thank Raimund Seidel for valuable feedback and interesting discussions.

8. REFERENCES

[1] M. Ajtai, V. Chvátal, M.M. Newborn, E. Szemerédi, *Crossing-free subgraphs*, Annals Discrete Math. **12**:9–12, 1982.

[2] M. Sharir, A. Sheffer, E. Welzl, *Counting Plane Graphs: Perfect Matchings, Spanning Cycles, and Kasteleyn's Technique*, Manuscript, 2011.

[3] A. García, M. Noy, J. Tejel, *Lower bounds on the number of crossing-free subgraphs of K_n*, Computational Geometry: Theory and Applications **16**:211–221, 2000.

[4] M. Sharir, A. Sheffer, *Counting Triangulations of Planar Point Sets*, Electr. J. Comb. **18:1**:P70, 2011.

[5] M. Sharir, A. Sheffer, E. Welzl, *On Degrees in Random Triangulations*, Proc. 26th ACM Symp. on Computational Geometry, 297–306, 2010.

[6] A. Sheffer, *Number of plane graphs: http://www.cs.tau.ac.il/~sheffera/counting/PlaneGraphs.html*

[7] E. Demaine, *Simple Polygonizations: http://erikdemaine.org/polygonization/*

[8] *http://maven.smith.edu/ orourke/TOPP/P16.html#Problem.16*

[9] J. Flum, M. Grohe, *Parameterized Complexity Theory*, Springer, ISBN 978-3-540-29952-3, 2006.

[10] K. Dalal, *Counting the onion*, Random Struct. Algorithms, **24**:2:155–165, 2004.

[11] A. Schulz, *The Existence of a Pseudo-triangulation in a given Geometric Graph*, Proc. 22nd European Workshop on Computational Geometry, 17–20, 2006.

[12] E. L. Lloyd, *On triangulations of a set of points in the plane*, Proc. 18th Annual Symp. on Foundations of Computer Science, 228–240, 1977.

[13] Oswin Aichholzer, *The path of a triangulation*, Proceedings of the fifteenth Annual Symposium on Computational Geometry - SoCG '99. 14–23, 1999.

[14] N. Katoh, S. Tanigawa, *Fast enumeration algorithms for non-crossing geometric graphs*, Proceedings of the twenty-fourth Annual Symposium on Computational Geometry - SoCG '08. 328–337, 2008.

[15] A. Razen, E. Welzl, *Counting Plane Graphs with Exponential Speed-Up*, Lecture Notes in Computer Science **6570**:36–46, 2011.

[16] S. Ray, R. Seidel, *A simple and less slow method for counting triangulations and for related problems*, Proceedings of the 20th European Workshop on Computational Geometry - EuroCG '04. 2004.

[17] E. Anagnostou, D. Corneil, *Polynomial-time instances of the minimum weight triangulation problem*, Computational Geometry: Theory and Applications, **3:5**:247–259, 1993.

A Fast and Simple Surface Reconstruction Algorithm*

Siu-Wing Cheng
Dept. Comp. Sci. Eng.
HKUST, Hong Kong
scheng@cse.ust.hk

Jiongxin Jin
Dept. Comp. Sci. Eng.
HKUST, Hong Kong
jamesjjx@cse.ust.hk

Man-Kit Lau
Dept. Comp. Sci. Eng.
HKUST, Hong Kong
lmkaa@cse.ust.hk

ABSTRACT

We present an algorithm to reconstruct a surface from a dense sample. Given n sample points, it runs in $O(n \log n)$ time, which is optimal in the pointer machine model. The only existing $O(n \log n)$-time algorithm due to Funke and Ramos uses some sophisticated data structures for the key task of extracting a locally uniform subsample. Our algorithm is based on a variant of the standard octree, and it is much simpler. We built a prototype, which runs an implementation of our algorithm to extract a locally uniform subsample, invokes Cocone to reconstruct a surface from the subsample, and adds back the samples points absent from the subsample via edge flips. The subsample extraction step is very fast and effective. In our experiments with some non-uniform samples, our prototype gives a 51% to 68% speedup from using Cocone alone. Even for locally uniform samples, our prototype is usually much faster.

Categories and Subject Descriptors

F.2.2 [**Analysis of Algorithms and Problem Complexity**]: Nonnumerical Algorithms and ProblemsGeometrical problems and computations; G.2.1 [**Discrete Mathematics**]: Combinatorics

General Terms

Algorithms, Theory, Experimentation

Keywords

Surface reconstruction, Delaunay triangulation, octree

1. INTRODUCTION

The problem of reconstructing surfaces in $\mathbb{R}^3$ from a set of sample points has been extensively studied in computer graphics and visions (e.g. [1, 16, 24]). Amenta and Bern [2] and Amenta, Choi, Dey, and Leekha [3] developed the first provably correct surface reconstruction algorithms. When the sample is dense enough, the output approximates the unknown surfaces well in the sense that their normal deviation and Hausdorff distance are small, and they are homeomorphic. More results on surface reconstruction can be found in [4, 7, 17]. Let n denote the number of input sample points. All algorithms with theoretical guarantees mentioned above compute the 3D Delaunay triangulation or Voronoi diagram. The 3D Voronoi diagram can be constructed in $O(n^2)$ time [12], and the fastest output-sensitive algorithm runs in $O((n+f)\log^2 n)$ time, where f is the output size [11]. If the smooth surface is generic and the sample is uniform, Attali, Boissonnat and Lieutier [6] showed that the 3D Voronoi diagram has $O(n \log n)$ complexity; however, in the case of non-generic smooth surfaces, Erickson [21] proved that the 3D Voronoi diagram can have $\Omega(n\sqrt{n})$ complexity even if the sampling is uniform.

Given a locally uniform sample, Dey, Funke and Ramos [18] showed that the Cocone algorithm [3] can be made to run in $O(n \log n)$ time. This modified Cocone algorithm employs an approximate nearest neighbor searching data structure. For locally uniform samples, Dumitriu, Funke, Kutz, and Milosavljević [19, 20] developed a reconstruction algorithm that requires very little geometry (only the distances among the sample points). The robustness is thus greatly enhanced.

For a general dense sample, Funke and Ramos [22] extended the Cocone algorithm and improved the running time to $O(n \log n)$, which is optimal in the pointer machine model. The key task is to extract a locally uniform subsample from the input points. Unfortunately, the algorithm employs several sophisticated data structures such as the well-separated pair decomposition, approximate directional nearest neighbor searching, and approximate range searching.

Our main result is an algorithm to extract a locally uniform subsample from a dense input sample without using any sophisticated data structure. After building a variant of the standard octree in $O(n \log n)$ time, the extraction works in $O(n)$ time by traversing the octree. Then, reconstructing from the subsample takes $O(n \log n)$ time using the modified Cocone algorithm [18], and adding back the sample points not in the subsample can be done in $O(n \log n)$ time as shown by Funke and Ramos [22]. The underlying surfaces are assumed to be smooth, but they do not have to be generic. In reconstructing from the subsample, the octree also allows us to avoid the approximate nearest neighbor searching data structure employed in the modified Cocone algorithm. We built a prototype, which runs an implementa-

*Research supported by the Research Grant Council, Hong Kong, China (project no. 612109).

SCG'12, June 17–20, 2012, Chapel Hill, North Carolina, USA.

tion of our subsample extraction algorithm, invokes Cocone to reconstruct a surface from the subsample, and adds back the remaining samples points via edge flips. Our experiments show that the subsample extraction step is very fast and effective. For non-uniform samples, our prototype is 51% to 68% faster than running Cocone alone. For locally uniform samples, our prototype is 8% to 41% faster, and most of the time is spent on running Cocone to reconstruct the surfaces from the extracted subsample.

We use $d(\cdot, \cdot)$ to denote the distance between two geometric objects. Let $B(c, r)$ denote the ball centered at a point c with radius r. Let $\Sigma \subset \mathbb{R}^3$ be a smooth compact two-dimensional manifold without boundary. The manifold Σ may have multiple connected components, and we call each component a *surface*. The *medial axis* is the closure of the set of points in $\mathbb{R}^3$ that have two or more closest points in Σ. The *local feature size* of a point $x \in \Sigma$, denoted $\text{lfs}(x)$, is the distance between x and the medial axis. The local feature size is 1-Lipschitz, i.e., for every pair of points $x, y \in \Sigma$, $\text{lfs}(x) \le \text{lfs}(y) + d(x, y)$. We use $\mathbf{n}_p$ to denote the outward unit normal at a point $p \in \Sigma$. We assume that the input sample P is an *ε-sample* of Σ, i.e., $d(x, P) \le \varepsilon\text{lfs}(x)$ for every point $x \in \Sigma$. For any subset $U \subset \Sigma$ of discrete points, we call U *locally uniform*[1] if $|B(p, \beta r_p) \cap U| = O(1)$ for every point $p \in U$, where $\beta > 1$ is a constant fixed a priori, and r_p is the radius of the largest ball that is centered at some point in Σ, contains p in its boundary, and does not contain any point in U in its interior.

The following results from the literature will be useful.

LEMMA 1.1 ([3, 5, 17, 23]). *Let p, q and r be any three points in Σ. The following hold for sufficiently small ε.*

(i) *If $d(p, q) \le \varepsilon\text{lfs}(p)$, the acute angle between the support lines of $\mathbf{n}_p$ and pq is at least $\pi/2 - O(\varepsilon)$, and $\angle(\mathbf{n}_p, \mathbf{n}_q) = O(\varepsilon)$.*

(ii) *If the circumradius of pqr is at most $\varepsilon\text{lfs}(p)$, the acute angle between the support lines of $\mathbf{n}_p$ and the normal of pqr is $O(\varepsilon)$.*

(iii) *For every point x on the tangent plane at p, if $d(p, x) \le \varepsilon\text{lfs}(p)$, then $d(x, \Sigma) = O(\varepsilon d(p, x))$.*

2. OVERVIEW

We follow the same three high-level steps in the algorithm of Funke and Ramos [22].

1. Select a subset $U \subseteq P$ such that U is a locally uniform $O(\varepsilon)$-sample of Σ.

2. For each sample point $p \in U$, estimate the surface normal at p and construct the incident Delaunay triangles (with respect to U) whose dual Voronoi edges intersect the cocone at p.[2] Then, traverse all such triangles to extract a surface triangulation M' as described in [3]. This traversal is a simple $O(n)$-time search.

3. Construct the final surface triangulation M by computing the Delaunay triangulation of P restricted to M'.

Step 1 is the hardest. All aforementioned sophisticated algorithmic tools are employed for this purpose. Our first contribution is a variant of the standard octree that allows us to perform step 1 in $O(n \log n)$ time.

In step 2, we cannot compute the 3D Delaunay triangulation of U because this would take more than $O(n \log n)$ time. Since U is locally uniform, the candidate triangles can only connect p to other sample points within a distance βr_p from p for some appropriate constant $\beta > 1$ and there are $O(1)$ such points. Therefore, by computing the 3D Delaunay triangulation of p and these $O(1)$ points, step 2 takes only $O(1)$ time for each $p \in U$, provided that we can estimate the support line of $\mathbf{n}_p$ and identify the sample points within a distance βr_p from p.

Consider step 3. The triangulation M' is a provably good reconstruction. The only downside is that it does not include all points in P. As shown in [22], the remaining points in P can be added by computing the Delaunay triangulation of P restricted to M', which is the subset of Delaunay triangles (with respect to P) whose dual Voronoi edges intersect M'. For every triangle τ in M', the vertices of Delaunay triangles restricted to τ must be at distance $\rho\gamma_\tau$ or less from every vertex p of τ, where γ_τ is the circumradius of τ, and ρ is a known constant. Since τ is a Delaunay triangle with respect to U with the circumcenter very close to the surface, the circumradius γ_τ is at most slightly more than r_p for any vertex p of τ. By setting $\beta = 2\rho$, we can use the points in $P \cap B(p, \beta r_p)$ to compute the Delaunay triangulation restricted to τ, which reduces to computing the 3D convex hull of $|P \cap B(p, \beta r_p)|$ points by standard geometric transformations. Each point in P participates in the restricted Delaunay computation for only $O(1)$ triangles in M' by the local uniformity of U. Thus, the overall running time is $O(n \log n)$, provided that the sample points of P at distance βr_p or less from each point $p \in U$ are available.

Our octree allows us to estimate the surface normal at each sample point in P, and to collect the point sets $U \cap B(p, \beta r_p)$ and $P \cap B(p, \beta r_p)$ for each $p \in U$ as needed by steps 2 and 3 above.

The following observation motivates our approach to extract a locally uniform sample.

LEMMA 2.1. *Let R be a compact subset of $\mathbb{R}^3$. Suppose that R contains at least one sample point in P. Then one of the following is true for a small enough ε:*

(i) *$d(R, P \setminus R) \le 4\varepsilon\text{lfs}(p)$ for some sample point $p \in P \cap R$.*

(ii) *For each surface $\mathcal{S}$ in Σ, either R contains $P \cap \mathcal{S}$, or R is disjoint from $P \cap \mathcal{S}$.*

Proof. Suppose that (ii) does not hold for some surface $\mathcal{S}$ in Σ, i.e., $P \cap \mathcal{S}$ has some points in R and some points not in R. Let $\mathcal{U}_\mathcal{S} = \bigcup_{p \in P \cap \mathcal{S}} B(p, 1.5\varepsilon\text{lfs}(p))$. For a small enough ε, $\mathcal{S} \subseteq \mathcal{U}_\mathcal{S}$, so $\mathcal{U}_\mathcal{S}$ is connected. The connectedness of $\mathcal{U}_\mathcal{S}$ implies the existence of $p, q \in P \cap \mathcal{S}$ such that $p \in R$, $q \notin R$, and $d(p, q) \le 1.5\varepsilon\text{lfs}(p) + 1.5\varepsilon\text{lfs}(q)$. As $\text{lfs}(q) \le \text{lfs}(p) + d(p, q)$, we have $d(R, P \setminus R) \le d(p, q) \le \frac{3\varepsilon}{1-1.5\varepsilon}\text{lfs}(p) \le 4\varepsilon\text{lfs}(p)$. ◻

In the sequel, R will be some octree cells with a high concentration of sample points, and Lemma 2.1 helps us to decide whether R contains a surface or the sample points in R should be decimated to achieve local uniformity.

[1] Our definition is weaker than those in [18, 22].

[2] The cocone at p is the set of points $x \in \mathbb{R}^3$ such that the acute angle between px and the support line of the estimated normal at p is at least $\frac{\pi}{2} - \theta$ for some constant $\theta \in (0, \pi/8)$. The choice of θ depends on the target ε for which the surface reconstruction algorithm works.

Consider the simple case that there is only one surface in Σ. Our idea is to start with a smallest axes-aligned bounding cube of P and build an octree by repeated splitting. The splitting of octree cells continue until for every leaf cell C, the sample points in $P \cap C$ concentrate in an octree cube C' of one-eighth the side length of C, called the core of C. Since there is only one surface in Σ, Lemma 2.1(ii) cannot hold with $R = C'$. Then, Lemma 2.1(i) applies with $R = C'$. In the simplest situation that C' does not touch the boundary of C, by the octree alignment, the distance between the boundaries of C' and C is at least the side length of C', and this implies that the side lengths of C' and C are $O(\varepsilon)$ times lfs. Thus, the side length of C is a local length scale that reflects the local sampling density. We smooth the different local length scales induced by different octree leaf cells, and use these scales to prune P to a locally uniform subsample U. In general, the core C' of a leaf cell C can touch the boundary of C, and it may be in contact with the cores of some other leaf cells. We need to group the cores into clusters, and apply Lemma 2.1 with R equal to a cluster.

When there are several surfaces in Σ, some sample points may concentrate in a cluster X because they lie on some surfaces in Σ that lie (almost) entirely inside X. It is wrong to decimate the sample points in X. Fortunately, Lemma 2.1(i) does not hold with $R = X$. Then Lemma 2.1(ii) implies that the sample points in X can be treated as the input for an independent instance of the surface reconstruction problem, which can be solved recursively. Of course, nobody can foretell whether Lemma 2.1(i) holds or not. We first recursively reconstruct from the sample points in X, if the reconstruction for X is successful, we ignore the sample points in X; otherwise, it is safe to decimate the sample points in X.

We describe the octree construction in Section 3 and the full algorithm in Section 4.

3. OCTREE DECOMPOSITION

3.1 The first octree

We describe our octree decomposition in any fixed dimension d. Examples in 2D are used to illustrate the concepts.

The side length of a box C in $\mathbb{R}^d$ is denoted by ℓ_C. Take a smallest axis-aligned bounding box of the input point set P. It is the root and the only leaf cell of the initial octree. Without loss of generality, assume that this root cell is $[0,1]^d$. A box $C \subseteq [0,1]^d$ is *canonical* if $\ell_C = 2^{-i}$ for some integer $i \geq 0$, and the coordinates of C's corners are integral multiples of 2^{-i}. A canonical box C is *splittable* if $P \cap C$ is not contained in a single canonical box with side length $2^{-d}\ell_C$. Two canonical boxes are *neighbors* if their interiors are disjoint and their boundaries are in contact. The octree is grown by invoking the following two rules alternately.

- *Splitting rule:* As long as there is a splittable leaf cell C, split C into 2^d octree cells of side length $\frac{1}{2}\ell_C$ and make them the children of C in the octree.
- *Balancing rule:* As long as there are two neighboring leaf cells C and C' such that $\ell_C > 2\ell_{C'}$, split C as described in the splitting rule.

The splitting rule is applied first until no leaf cell is splittable. The balancing rule is then applied until the tree is *balanced*, i.e. every two neighboring leaf cells differ in side lengths by at most a factor two. We apply the two rules alternately until the tree stops growing, which must happen because, in the worst case, every leaf cell contains a single point and is non-splittable, and then the tree will stop growing after another round of balancing.

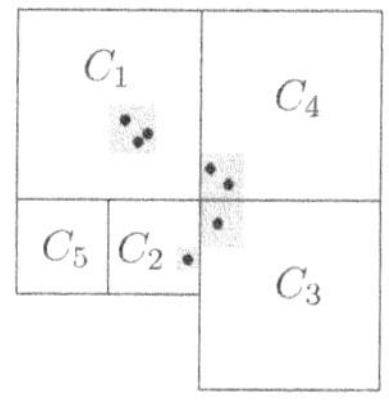

Figure 1: Five leaf cells are shown. C_5 is empty. C_2 is a neighbor of C_1, C_3, C_4, and C_5. Their cores are shown in gray: $\text{core}(C_1)$ alone is a cluster, and $\text{core}(C_2) \cup \text{core}(C_3) \cup \text{core}(C_4)$ is another cluster.

Let T_P be the resulting octree. We add pointers between neighboring cells in T_P that have the same side length. These pointers enable us to access the cells of similar sizes near a given cell. A cell is *empty* if it contains no point in P. For every leaf cell C, since C is not splittable, $P \cap C$ is contained in a canonical box with side length $2^{-d}\ell_C$ inside C. We call it the *core* of C, denoted $\text{core}(C)$. Two cores are connected if their boundaries are in contact. Each connected component of cores is called a *cluster*. An isolated core is a cluster by itself. See Figure 1.

Lemma 3.1 below proves some properties of the octrees. It is stated in a more general setting as we need it to describe the recursive octree construction on the clusters in T_P.

Consider an octree T_Q built for a subset $Q \subseteq P$ by the procedure above. (A canonical box C is splittable if $Q \cap C$ is not contained in a single canonical box of side length $2^{-d}\ell_C$, and a cell in T_Q is empty if it contains no point in Q.) We say that P *conforms* to T_Q if:

- for every cell C in T_Q, either $Q \cap C = \emptyset$ or $Q \cap C = P \cap C$, and
- if $Q \neq P$, then $d(Q, P \setminus Q)$ is at least 2^{-d-2} times the side length of the root cell of T_Q.

LEMMA 3.1. *Let T_Q be the octree built for some $Q \subseteq P$ by applying the splitting and balancing rules alternately. If P conforms to T_Q, the properties below are satisfied by T_Q.*

(i) *The side lengths of two neighboring leaf cells differ by at most a factor two.*

(ii) *Every pair of leaf cells whose cores are in the same cluster are neighbors. Hence, each cluster consists of at most 2^d cores, and the side lengths of two cores in the same cluster differ by at most a factor two.*

(iii) *Let C be a non-empty leaf cell. Let X denote the cluster that contains $\text{core}(C)$. If $Q \cap X$ and $Q \setminus X$ contain sample points from the same surface, then, for a small enough ε, $\ell_C \leq 2^d 20\varepsilon \text{lfs}(p)$ for any $p \in Q \cap C$.*

Proof. Property (i) follows from the octree construction. Let X be a cluster and let $\widehat{X}$ be the set of leaf cells whose cores are in X. We show that every two cells in $\widehat{X}$ are neighbors and then (ii) follows.

Let $\widehat{K}$ be a maximal subset of $\widehat{X}$ such that the cores of cells in $\widehat{K}$ are connected and every two cells in $\widehat{K}$ are neighbors.

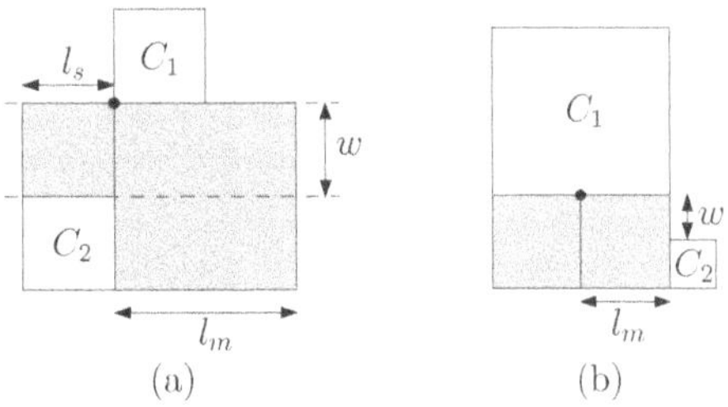

Figure 2: The shaded boxes are cells in $\bar{K}$. In (a), these cells share the lower left vertex of C_1. In (b), C_2 is at L_∞-distance l_m from the left shaded box.

Let K be set of cores of cells in $\widehat{K}$. If $\widehat{K} = \widehat{X}$, we are done. Pick $C_2 \in \widehat{X} \setminus \widehat{K}$ such that core(C_2) is connected to a core in K. Pick $C_1 \in \widehat{K}$ that maximizes $w = d_\infty(C_1, C_2)$. We have $w > 0$ by the maximality of $\widehat{K}$. Take an axes-aligned slab that has width w and separates C_1 and C_2. Let $\bar{K} \subseteq \widehat{K}$ be the subset of cells that overlap with this slab. Let l_s and l_m be the side lengths of the smallest and largest cells in $\bar{K}$, respectively. Observe that $|\bar{K}| \leq 2^{d-1}$; Figure 2(a) shows an example that 2^{d-1} cells in $\bar{K}$ share a common vertex of C_1.

We claim that $w \geq l_m/2$. Let C be the largest cell in $\bar{K}$. If C_2 is not a neighbor of C, then $w \geq d_\infty(C, C_2) \geq \ell_C/2 = l_m/2$. Suppose that C_2 is a neighbor of C. Then $\ell_{C_2} \geq \ell_C/2$. Also, C_1 and C are neighbors as they belong to $\widehat{K}$, so $\ell_{C_1} \geq \ell_C/2$. The octree alignment implies that $d_\infty(C_1, C_2) \geq \min\{\ell_{C_1}, \ell_{C_2}\} \geq \ell_C/2 = l_m/2$.

If $|\bar{K}| < 2^{d-1}$ or $w \geq l_m$, then since K is connected, the gap between C_1 and C_2 must be bridged by the cores of cells in $\bar{K}$; however, $d_\infty(C_1, C_2) \leq |\bar{K}| 2^{-d} l_m < w$, a contradiction. The remaining case is that $|\bar{K}| = 2^{d-1}$ and $w = l_m/2$. If $l_s < l_m$, then $l_s = l_m/2$, but then $d_\infty(C_1, C_2) \leq (|\bar{K}| - 1)2^{-d} l_m + 2^{-d} l_s < l_m/2 \leq w$, a contradiction again. If $l_s = l_m$, then C_2 is at L_∞-distance $l_m > w$ from some cell in $\bar{K}$ as illustrated in Figure 2(b), contradicting the choice of C_1 to maximize $d_\infty(C_1, C_2)$.

Consider (iii). If $Q \cap X$ and $Q \setminus X$ contain sample points from the same surface $\mathcal{S}$, then $P \cap X$ and $P \setminus X$ contain sample points from $\mathcal{S}$ too as P conforms to T_Q. Lemma 2.1(i) implies that $d(X, P \setminus X) \leq 4\varepsilon\text{lfs}(q)$ for some $q \in P \cap X = Q \cap X$. Then, using the fact that the distance between X and any core not in X is at least $\frac{1}{4}\ell_{\text{core}(C)}$, we can derive that $\ell_C \leq 2^d 20\varepsilon\text{lfs}(p)$ for every $p \in Q \cap C$. ◻

3.2 The other octrees by recursion

If $\Sigma \subset \mathbb{R}^3$ is known to be connected, then T_P suffices. Otherwise, we build an octree in $\mathbb{R}^3$ for each cluster of T_P in a recursive manner. In general, after building an octree T_Q for a subset $Q \subseteq P$, for each cluster X in T_Q that contains more than one sample point, we recursively build $T_{Q\cap X}$ by initializing an octree of $O(1)$ levels, followed by alternate splitting and balancing. The initialization ensures that no cell of $T_{Q\cap X}$ contains sample points from different cores in X, which helps to achieve the $O(n \log n)$ running time as explained in the footnote referenced near the end of the proof of Lemma 3.4. The initialization works as follows.

By Lemma 3.1(ii), there are at most eight cores in X. Label them C_i for $i = 1, 2, \cdots$. Let l be the side length of the largest core in X. Lemma 3.1(ii) also implies that every core C_i has side length l or $l/2$. For each C_i, let $\widehat{C}_i$ be the canonical cube (with respect to T_Q) such that $\ell_{\widehat{C}_i} = l$. $\widehat{C}_i$ is contained in the leaf cell of T_Q whose core is C_i, and hence $\widehat{C}_i$ and $\widehat{C}_j$ have disjoint interiors for $i \neq j$. We construct the root cell C_X of $T_{Q\cap X}$ and the initial octree as follows.

If X contains only one core, set C_X to be a smallest axis-aligned bounding cube of $Q \cap X$, and C_X alone is the initial octree. If X contains more than one core, set C_X to be the smallest axis-aligned cube such that $\ell_{C_X} = 2^k l$ for some integer k, and each $\widehat{C}_i$ is a canonical cube with respect to C_X. Since X has at most eight cores, $\ell_{C_X} \leq 8l$. Split C_X repeatedly until each leaf cell has size l. This gives the initial octree, which has $O(1)$ levels. Each $\widehat{C}_i$ is a leaf cell of the initial tree, so a leaf cell contains either no point in in $Q \cap X$ or points from exactly one C_i. Notice that $d(Q \cap X, Q \setminus X)$ is at least half of the side length of the smallest core in X, which is at least $\frac{1}{32}\ell_{C_X}$.

The initialization takes $O(1)$ time. Afterwards, we split and balance alternately to obtain the final octree $T_{Q\cap X}$. Pointers are added between neighboring octree cells in $T_{Q\cap X}$ with the same side length. We then recurse on the clusters in $T_{Q\cap X}$ that contain more than one sample point. If P conforms to T_Q, then P conforms to $T_{Q\cap X}$ for every cluster X in T_Q. Inductively, P conforms to all octrees constructed in the recursion, and all octrees satisfy Lemma 3.1 for $d = 3$.

We omit the proofs of the next two technical lemmas.

LEMMA 3.2. *There are two or more clusters in $T_{Q\cap X}$, or the sample points in some core in X are divided into two or more leaf cells of $T_{Q\cap X}$.*

LEMMA 3.3. *Let $m = 3^3$. Let C be a cell in T_Q that is created after the initialization. Let D be the cube with the same center as C and side length $3\ell_C$. If no splittable cell in T_Q that intersects D has side length in the range $[2^{-m}\ell_C, 2\ell_C]$, then the subtree rooted at C has at most $m - 1$ levels.*

LEMMA 3.4. *The total size of the octrees is $O(n)$. The total construction time is $O(n \log n)$.*

Proof. Consider T_Q for some $Q \subseteq P$. Let C be a cell in T_Q that is not in the initial octree. If the subtree rooted at C has more than 3^3 levels, by Lemma 3.3, there is a splittable cell with side length $\Theta(\ell_C)$ at distance $\Theta(\ell_C)$, so we can charge C to it. Each splittable cell is charged $O(1)$ times, which implies a total charge of $O(n_Q + 1)$, where n_Q is the number of the splittable cells in T_Q. The cells of T_Q not covered by the $O(n_Q + 1)$ charge are at $O(1)$ levels below the root. The total size of all octrees is thus $O(n)$ by Lemma 3.2.

We use a trick in [10] to get the $O(n \log n)$ running time. First, sort P three times by the x-, y-, and z-coordinates. Store the three sorted lists at the root of T_P. Testing whether a cell C in T_P is splittable can be done in $O(1)$ time by maintaining the minimum and maximum x-, y-, and z-coordinates of the points in C. To split C, we divide C into two equal halves A_1 and A_2 by a plane perpendicular to the x-axis. We split the list sorted by x-coordinates in $\min\{|P \cap A_1|, |P \cap A_2|\}$ time by moving from both ends of the list towards the middle. Suppose that $|P \cap A_1| \leq |P \cap A_2|$. Remove the points of $P \cap A_1$ from the three sorted lists, and create three sorted lists for A_1. This takes $O(|P \cap A_1|)$ time. After the removal of $P \cap A_1$, the modified sorted lists for C are the sorted lists for A_2, and we only store reference pointers to them at A_2. Divide A_1 into four cells directly in $O(|P \cap A_1|)$

time. For A_2, we repeat the above to split it by planes in the other two directions. In the end, there is exactly one child C' of C with reference pointers to the final sorted lists left at C. The total time spent is $O(|P \cap (C \setminus C')|)$. We charge each point in $C \setminus C'$ $O(1)$ amount of work. Since every child of C other than C' has at most $|P \cap C|/2$ points, a point can be charged only $O(\log n)$ times. Similar analysis applies to the recursive calls.[3] The total time is thus $O(n \log n)$. ◻

4. SURFACE RECONSTRUCTION

We call RECONSTRUCT(T_P) to perform the surface reconstruction. Given T_Q for some $Q \subseteq P$, RECONSTRUCT(T_Q) performs the following steps.

First, for every cluster X in T_Q, call RECONSTRUCT($T_{Q \cap X}$) recursively, which reconstructs the surfaces whose sample points are completely contained in $T_{Q \cap X}$, and then the sample points on these surfaces are removed from T_Q.

Second, call TRIM(T_Q), which returns yes iff the remaining points in T_Q are complete samples of some surfaces. If not, since the root cell of T_Q is induced by a cluster in the "parent octree", there is a large gap between points in this cluster and other points from the same surfaces. This gap detection is done in a bottom-up manner in T_Q so that if a non-empty leaf cell C in T_Q is near a gap on the estimated tangent plane for points in C that is large compared with ℓ_C, then parent(C) is made a leaf cell. In the end, TRIM(T_Q) returns yes iff the final pruned tree $\overline{T}_Q$ has more than one node.

Third, if TRIM(T_Q) returns no, RECONSTRUCT(T_Q) just returns; otherwise, we proceed to reconstruct the surfaces with sample points in $\overline{T}_Q$. The pruning by TRIM ensures that all non-empty cells in $\overline{T}_Q$ have the right sizes with respect to the local neighborhood. Thus, we can traverse $\overline{T}_Q$ to smooth the side lengths of non-empty leaf cells by pruning the tree further. Finally, one point is picked from each non-empty leaf cell to form a locally uniform sample. We then perform steps 2 and 3 in Section 2 to reconstruct the surfaces.

We give the details of the second and third steps in the following subsections.

4.1 Tree trimming

We have already recursively reconstructed surfaces whose sample points are contained in the clusters of T_Q, and these sample points have been removed. Thus, if C is a non-empty leaf cell of T_Q, then C does not contain a complete sample, meaning that $\ell_C = O(\varepsilon \mathrm{lfs}(x))$ for every $x \in Q \cap C$ by Lemma 3.1(iii). We first estimate the tangent plane for a point $p \in Q \cap C$: find two sample points q and r near p such that $\angle qpr$ is neither too large nor too small, and then the support plane of pqr is a good estimate by Lemma 1.1(ii). If the tangent plane estimation fails, we make parent(C) a leaf cell. Otherwise, we proceed to detect if there is a big gap on the estimated tangent plane. If there is a big gap, we also make parent(C) a leaf cell. The pseudocode of TRIM is given below. Let $R(x, l)$ be the axis-aligned cube with center x and side length l.

[3]Since every leaf of the initial octree for a cluster X in T_Q contains sample points from one core in X, we can keep reference pointers at such a leaf to the sorted lists by x-, y-, and z-coordinates stored at an appropriate cell in T_Q.

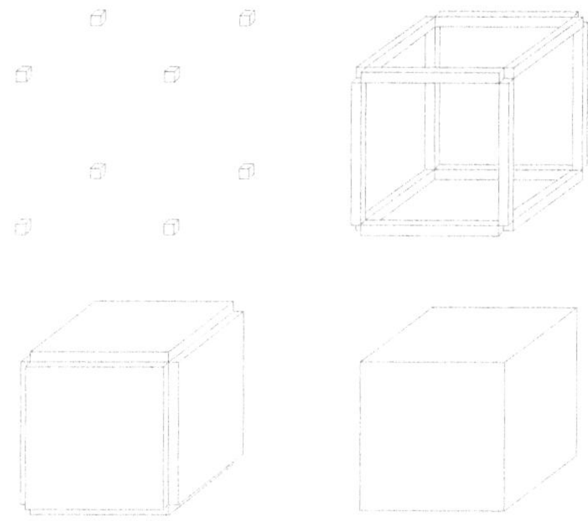

Figure 3: The upper left figure is $G_{i,0}$. The upper right figure is $G_{i,1}$. The lower left figure is $G_{i,2}$, which consists of six slabs and has a rectangular hole in the middle. The lower right figure is $G_{i,3}$, which fits into the hole in $G_{i,2}$.

TRIM(T_Q)

1. $\overline{T}_Q := T_Q$. Initialize L to be the list containing all non-empty leaf cells in $\overline{T}_Q$.
2. If L is empty, return yes; otherwise, extract a cell C from L (in any order).
3. If C is the root of $\overline{T}_Q$, return no.
4. Let c denote the center of C. Partition $R(c, 5\ell_C)$ into a set $\{R_i : 1 \le i \le 5^3\}$ of canonical cubes of side length ℓ_C. For each non-empty R_i, pick an arbitrary sample point $p_i \in Q \cap R_i$. Let p denote the pick of C.
5. Let $\theta = \arccos(0.97)$. If there exist two points p_i and p_j such that $\angle p_i p p_j \in [\theta, \pi - \theta]$, the normal of $p_i p p_j$ serves as the estimated surface normal at every point in $Q \cap C$. Otherwise, make parent(C) a leaf of $\overline{T}_Q$, add parent(C) to L, remove the descendants of parent(C) from L, and go to step 2.
6. Let H be the plane through p and orthogonal to the estimated normal. Let $\mathcal{U}$ be the union of all empty R_i. If there exists $x \in H$ such $R(x, \frac{1}{8}\ell_C) \subseteq \mathcal{U}$, then make parent($C$) a leaf cell of $\overline{T}_Q$, add it to L, remove its descendants from L, and go to step 2.

In step 6 of TRIM, the empty cube $R(x, \frac{1}{8}\ell_C)$ is the "big gap" indicating that C is too small. The checking of whether $\mathcal{U}$ contains an empty $R(x, \frac{1}{8}\ell_C)$ for some $x \in H$ can be performed in $O(1)$ time as follows. Divide each empty R_i into canonical cubes of side length $\frac{1}{16}\ell_C$ and put them into four groups $G_{i,m}$ for $m \in [0, 3]$ such that $G_{i,m}$ is the union of the cubes that intersect an m-dimensional face of R_i but not any face of R_i with dimension less than m. Figure 3 shows an example. $G_{i,3}$ is one cube in the interior of R_i with side length $\frac{14}{16}\ell_C$, $G_{i,2}$ consists of six rectangular slabs in contact with the facets of R_i, $G_{i,1}$ consists of twelve rectangular rods in contact with the edges of R_i, and $G_{i,0}$ consists of eight cubes in contact with the vertices of R_i. Assume that the plane H intersects an empty R_i. Then, H intersects a connected component K of $G_{i,m}$ for some $m \in [0, 3]$. There exists $x \in H \cap K$ such that $R(x, \frac{1}{8}\ell_C) \subseteq \mathcal{U}$ iff the

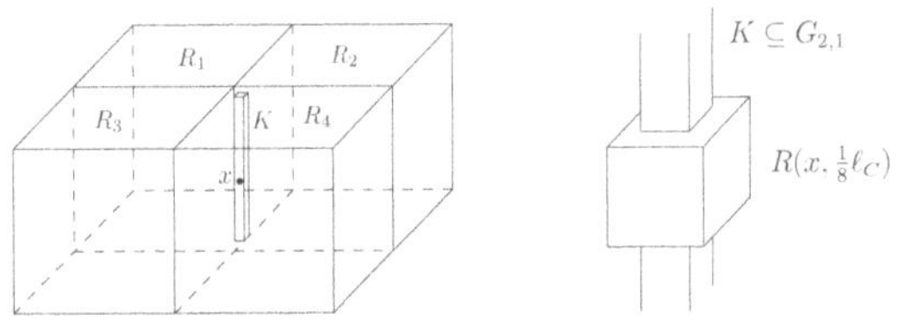

Figure 4: The point x lies in a rod K in $G_{2,1}$. R_1, R_2, R_3 and R_4 are the four cubes in contact of K. Since the base side length of K is $\ell_C/16$, the box $R(x, \frac{1}{8}\ell_C)$ must intersect the all these four cubes. Moreover, it is contained in the union of these four cubes. Thus, $R(x, \frac{1}{8}\ell_C) \subseteq \mathcal{U}$ iff these four cubes belong to $\mathcal{U}$.

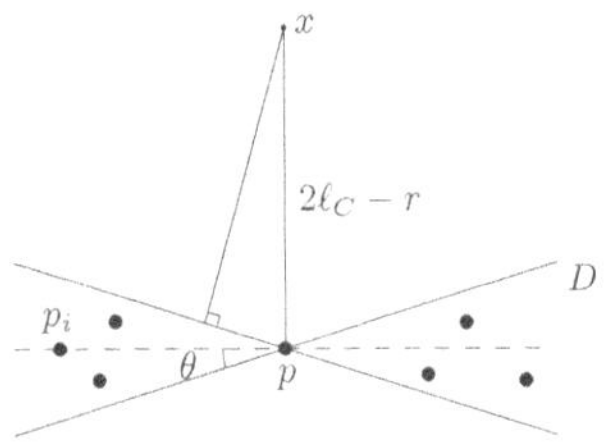

Figure 5: The distance between x and any pick is at least $d(x,p)\cos\theta$.

set $\{R_j \subseteq \mathcal{U} : R_j$ touches or intersects $K.\}$ contains 2^{3-m} cubes. Figure 4 shows an example. So step 6 of TRIM takes $O(1)$ time for each C. We can show that that C is indeed small if an empty cube $R(x, \frac{1}{8}\ell_C)$ is found. (Lemma 4.2.)

LEMMA 4.1. *Suppose that $\ell_C = O(\varepsilon \text{lfs}(q))$ for any point $q \in Q \cap C$. If the normal estimation in step 5 of* TRIM *fails for C, then $\ell_C \le 80\varepsilon\text{lfs}(q)$ for any point $q \in Q \cap C$.*

Proof. Let p be the sample point in C that is picked for the normal estimation. For $1 \le i \le 5^3$, let R_i be the canonical cubes used in estimating the surface normals at the samples points in C. We have picked at most one sample point p_i from each R_i. There are two cases.

Consider a cube $R_i \ne C$ with pick p_i. Take a point $x \in \Sigma$ such that px is perpendicular to pp_i and $d(p,x) = 2\ell_C - r$, where $r = 0.1\ell_C$. Such x must exist as $d(p,x) = O(\varepsilon\text{lfs}(p))$. Let D be the double cone with p as the apex, the line through p and p_i as the axis, and aperture 2θ, where $\theta = \arccos(0.97)$. See Figure 5. Since we cannot find a pick p_j such that $\angle p_i p p_j \in [\theta, \pi - \theta]$, the double cone D must contain the sample point p_j picked from each non-empty R_j. Thus, $d(x,p_j) \ge d(p,x)\cos\theta = (2\ell_C - r)\cos\theta$ for all R_j. Any sample point in $Q \cap R_j$ is at distance $\sqrt{3}\ell_C$ or less from the pick p_j. Therefore, the distance from x to any sample point of Q in any R_j (including C) is at least $d(x,p_j) - \sqrt{3}\ell_C \ge (2\ell_C - r)\cos\theta - \sqrt{3}\ell_C > r$.

The distance between any point in $B(x,r)$ and the cell C is at most $r + d(p,x) = 2\ell_C$, implying that $B(x,r)$ lies inside $R(c, 5\ell_C)$, where c is the center of C. Suppose that the ball $B(x,r)$ does not contain any sample point in $P \cap R(c, 5\ell_C)$. Then we have $\varepsilon\text{lfs}(x) > r = 0.1\ell_C$. For any sample point q in C, $d(x,q) \le d(x,p) + d(p,q) = O(\ell_C) = O(\varepsilon\text{lfs}(x))$. By the Lipschitz condition, $\text{lfs}(x) \le \text{lfs}(q) + d(x,q) \le (1 + O(\varepsilon))\text{lfs}(q)$. Therefore, for a small enough ε, $\ell_C \le 10\varepsilon\text{lfs}(x) < 15\varepsilon\text{lfs}(q)$.

Suppose that $B(x,r) \cap P \ne \emptyset$. Since P conforms to T_Q, $B(x,r) \cap Q = \emptyset$ and $B(x,r) \subset R(c, 5\ell_C)$, we have $\ell_C = \Omega(d(Q, P \setminus Q)) = \Omega(\ell_{C_Q})$, where C_Q is the side length of root cell of T_Q. Then $\ell_{C_Q} = O(\ell_C) = O(\varepsilon\text{lfs}(q))$ for any $q \in Q \cap C$, and, hence, Q cannot contain a complete sample from some surface. For a non-root cell C, $\ell_C \le \frac{1}{2}\ell_{C_Q}$, which is at most $80\varepsilon\text{lfs}(q)$ for any point $q \in Q \cap C$ by Lemma 3.1(iii).

The other possibility is that we cannot pick any sample point p_i in any $R_i \ne C$. In this case, $R(c, 5\ell_C) \setminus C$ contains no sample point in P, where c is the center of C. We can pick any point $x \in \Sigma$ such that $d(p,x) = 2\ell_C - r$ and repeat the analysis in the previous paragraph. ◻

LEMMA 4.2. *Suppose that $\ell_C \le 180\varepsilon\text{lfs}(q)$ for every point $q \in Q \cap C$. If $\mathcal{U}$ contains an empty $R(x, \frac{1}{8}\ell_C)$ for some $x \in H$, then $\ell_C \le 80\varepsilon\text{lfs}(q)$ for every sample point q in C.*

Proof. Let p be the pick for C. We have $d(p,x) = O(\ell_C) = O(\varepsilon\text{lfs}(p))$. Let x' be the projection of x in the tangent plane at p. Since the plane H makes an $O(\varepsilon)$ angle with the tangent plane at p, we have $d(x,x') = O(\varepsilon d(p,x))$, and $d(p,x') \le d(p,x) + d(x,x') = (1 + O(\varepsilon))d(p,x)$.

Let $\tilde{x}$ be the nearest point in Σ to x'. By Lemma 1.1(iii), $d(x',\tilde{x}) = O(\varepsilon d(p,x')) = O(\varepsilon d(p,x))$. It follows that

$$d(x,\tilde{x}) \le d(x,x') + d(x',\tilde{x}) = O(\varepsilon d(p,x)), \text{ and}$$
$$d(p,\tilde{x}) \le d(p,x) + d(x,\tilde{x}) = (1 + O(\varepsilon))d(p,x).$$

Since x is in $R(c, 5\ell_C)$, we have $d(p,x) = O(\ell_C)$, which is $O(\varepsilon\text{lfs}(p))$ by the assumption of the lemma. Thus, $\text{lfs}(\tilde{x}) \le \text{lfs}(p) + d(p,x) \le (1 + O(\varepsilon))\text{lfs}(p)$. Notice that

$$B(\tilde{x}, \tfrac{1}{16}\ell_C - d(x,\tilde{x})) \subset B(x, \tfrac{1}{16}\ell_C) \subset R(x, \tfrac{1}{8}\ell_C).$$

So $B(\tilde{x}, \frac{1}{16}\ell_C - d(x,\tilde{x}))$ does not contain any sample point in Q. Suppose that it does not contain any sample point in P either. Then $\varepsilon\text{lfs}(\tilde{x}) \ge \frac{1}{16}\ell_C - d(x,\tilde{x})$, implying that $\ell_C \le 16(\varepsilon\text{lfs}(\tilde{x}) + d(x,\tilde{x})) \le (16 + O(\varepsilon))\varepsilon\text{lfs}(p)$. For any sample point q in C, since $d(p,q) \le \sqrt{3}\ell_C = O(\varepsilon\text{lfs}(q))$, we have $\text{lfs}(p) = (1 + O(\varepsilon))\text{lfs}(q)$. Hence, $\ell_C \le 16(1 + O(\varepsilon))^2\varepsilon\text{lfs}(q) \le 20\varepsilon\text{lfs}(q)$ for a small enough ε.

Suppose that $B(\tilde{x}, \frac{1}{16}\ell_C - d(x,\tilde{x}))$ contains some point in P. Since P conforms to T_Q and this ball lies in $R(c, 5\ell_C)$, ℓ_C is at least some constant fraction of the side length of the root cell of T_P. By the same argument in the proof of Lemma 4.1, $\ell_C \le 80\varepsilon\text{lfs}(q)$ for any point $q \in Q \cap C$. ◻

Lemmas 4.1 and 4.2 imply that if a leaf cell C is trimmed, then $\ell_C \le 80\varepsilon\text{lfs}(q)$ for every sample point q in C. So $\ell_{\text{parent}(C)} = 2\ell_C \le 160\varepsilon\text{lfs}(q)$ for every sample point q in C. For each sample point q' in parent(C) but not in C, we have $d(q,q') \le 2\sqrt{3}\ell_C = O(\varepsilon\text{lfs}(q))$. Thus, $\text{lfs}(q) = (1 + O(\varepsilon))\text{lfs}(q')$ and $\ell_{\text{parent}(C)} \le 160\varepsilon\text{lfs}(q) \le 160(1+O(\varepsilon))\varepsilon\text{lfs}(q')$, which is at most $180\varepsilon\text{lfs}(q')$ for a small ε. Therefore, Lemmas 4.1 and 4.2 are applicable to parent(C) when we repeat the normal estimation and trimming at parent(C).

Lemma 4.3 below shows several properties of the final tree $\overline{T}_Q$. First, each non-empty leaf cell is small relative to the local feature sizes of the sample points in it. Second, they are not too small relative to the local sampling. The most important property is that $\overline{T}_Q$ includes a partial sample on some surface in Σ if and only if TRIM(T_Q) returns no.

LEMMA 4.3. *Let $\overline{T}_Q$ be the final trimmed T_Q.*

(i) *Trimming T_Q to $\overline{T}_Q$ takes $O(|T_Q|)$ time. For every non-empty leaf cell C in $\overline{T}_Q$ and every sample point q in C, $\ell_C \le 180\varepsilon\text{lfs}(q)$.*

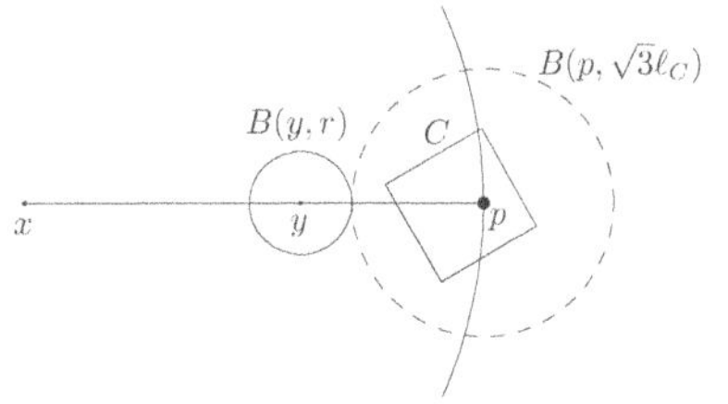

Figure 6: Proof of Lemma 4.3(ii).

(ii) *Let x be a point in Σ. Let p be the nearest point in $\overline{T}_Q$ to x. If $d(p,x) = O(\varepsilon \mathrm{lfs}(x))$, then the non-empty leaf cell C in $\overline{T}_Q$ that contains p satisfies $d(p,x) < 2\ell_C$, unless $\overline{T}_Q$ has only one node.*

(iii) *The sample points in $\overline{T}_Q$ include a partial sample of a surface iff* TRIM(T_Q) *returns no.*

Proof. The correctness of (i) follows from the description of the algorithm and the previous discussion on the side length of parent(C) after trimming a leaf cell C.

Consider (ii). Suppose that $d(p,x) \geq 2\ell_C$. We will show that C is the only node in $\overline{T}_Q$. Refer to Figure 6. Let y be the point on the segment px such that $d(p,y) = \sqrt{3}\ell_C + r$, where $r = (1-\sqrt{3}/2)\ell_C$. The distance between any point in $B(y,r)$ and C is at most $r + d(p,y) = 2r + \sqrt{3}\ell_C = 2\ell_C$. So $B(y,r) \subset R(c, 5\ell_C)$, where c is the center of C.

Let y' be the nearest point of y in the plane H through p perpendicular to the estimated surface normal at p. Since H makes an $O(\varepsilon)$ angle with the tangent plane at p, we have $d(y,y') = O(\varepsilon d(p,y)) = O(\varepsilon \ell_C)$, which is less than $r - (\sqrt{3}/16)\ell_C$ for a small enough ε. So

$$R\Big(y', \frac{1}{8}\ell_C\Big) \subset B\Big(y', \frac{\sqrt{3}}{16}\ell_C\Big) \subset B\Big(y, \frac{\sqrt{3}}{16}\ell_C + d(y,y')\Big),$$

which is inside $B(y,r) \subset R(c, 5\ell_C)$.

Recall the partition of $R(c, 5\ell_C)$ into the set $\{R_i : 1 \leq i \leq 5^3\}$ of canonical cubes of side length ℓ_C. If R_i intersects $R(y', \frac{1}{8}\ell_C)$, it also intersects $B(y,r)$, which implies that

$$R_i \subset B(y, \sqrt{3}\ell_C + r) = B(y, d(p,y)) \subset B(x, d(p,x)).$$

Since p is the nearest point of x in Q, R_i does not contain any point in Q. So $R(y', \frac{1}{8}\ell_C)$ is contained in the union of the empty R_i. But then this should have caused the leaf cell C to be trimmed unless C is the only node in $\overline{T}_Q$. This completes the proof of (ii).

Consider (iii). Suppose that TRIM(T_Q) returns no. In this case, $\overline{T}_Q$ contains the root node only. Then by (i), the root has side length at most $180\varepsilon \mathrm{lfs}(q)$ for any sample point q in $\overline{T}_Q$. It is too small for the sample points in $\overline{T}_Q$ to contain a complete sample of any surface in Σ.

As for the other direction of the claim, suppose that the sample points in $\overline{T}_Q$ include a partial sample of some surface in Σ. Then, Q is a proper subset of P, and $Q = Q_0 \cap X$ for some cluster X of some octree T_{Q_0}. By construction, the root cell C_X of $\overline{T}_Q$ has side length proportional to the side lengths of the cores in X. Since the sample points in $\overline{T}_Q$ include a partial sample, a surface in Σ contains a sample point in X and a sample point outside X. Moreover, the distance between X and any sample point not in X is at least half the side length of the smallest core in X. Then, Lemma 3.1(iii) implies that $\ell_{C_X} = O(\varepsilon \mathrm{lfs}(q))$ for every sample point $q \in Q$.

Since ℓ_{C_X} is so small and there are points in Σ at distance $\mathrm{lfs}(q)$ from q, we can find a point $x \in \Sigma$ such that $d(x, C_X) = 2\ell_{C_X}$. Then, $d(q,x) \leq d(x, C_X) + \sqrt{3}\ell_{C_X} = (2+\sqrt{3})\ell_{C_X} = O(\varepsilon \mathrm{lfs}(q))$. Combining this with $\mathrm{lfs}(q) \leq \mathrm{lfs}(x) + d(q,x)$, we get $d(q,x) = O(\varepsilon \mathrm{lfs}(x))$. Take an arbitrary leaf cell C of $\overline{T}_Q$. It is contained in C_X as C_X is the root cell. The choice of x implies that $d(x,C) \geq d(x, C_X) = 2\ell_{C_X} \geq 2\ell_C$. So by (ii), $\overline{T}_Q$ has only one node, i.e., TRIM(T_Q) returns no. ◻

4.2 Subsample and reconstruction

Suppose that TRIM(T_Q) returns yes. We call the following procedure EXTRACT to traverse $\overline{T}_Q$ and obtain a locally uniform subsample U_Q. Let $\alpha = 3\beta$, where β is the constant used in the definition of a locally uniform sample. (Refer to Section 1 for the definition.) Recall that $R(x,l)$ denotes the axis-aligned cube with center x and side length l.

EXTRACT($\overline{T}_Q$)

1. Initialize $\widehat{T}_Q = \overline{T}_Q$. Start a breadth-first search of $\widehat{T}_Q$ at its root to run the following steps on the non-empty leaf cells in non-increasing order of their sizes.
2. Let C be the next non-empty leaf cell of $\widehat{T}_Q$ encountered in the breadth-first search. (Non-leaf cells and empty cells are skipped.) Let c be the center of C.
3. For each octree cell C' in $\widehat{T}_Q$ (not necessarily leaf cells) such that $\ell_{C'} = \frac{1}{2}\ell_C$ and C' intersects $R(c, 2\alpha\ell_C)$, we discard the subtree rooted at C' to make C' a leaf cell of $\widehat{T}_Q$ and regard the sample points in the descendants of C' as the sample points in C'.
4. Go to step 2 if the breadth-first search of $\widehat{T}_Q$ has not finished.
5. Return the sample U_Q that consists of an arbitrary sample point from each non-empty leaf cell of $\widehat{T}_Q$.

EXTRACT($\overline{T}_Q$) runs in $O(|\overline{T}_Q|)$ time if step 3 takes $O(1)$ time. In step 3, there are only $O(\alpha^3) = O(1)$ such cells C', and they can be found in $O(1)$ time using the pointers between neighboring octree cells of the same side length and the parent-child pointers.

LEMMA 4.4. EXTRACT($\overline{T}_Q$) *computes in $O(|\overline{T}_Q|)$ time a locally uniform $O(\varepsilon)$-sample $U_Q \subseteq Q$ of the surfaces in Σ that contain the sample points in $\widehat{T}_Q$.*

Proof. The running time follows from the above discussion. It remains to show that U_Q is a locally uniform $O(\varepsilon)$-sample.

We first prove that U_Q is an $O(\varepsilon)$-sample. Take any point x on a surface that contains a sample point in $\widehat{T}_Q$. We need to identify a point in U_Q at distance $O(\varepsilon \mathrm{lfs}(x))$ from x.

Since TRIM(T_Q) succeeds, by Lemma 4.3(iii), Q is a complete sample for some surfaces in Σ. So Q is an ε-sample for these surfaces. Let p be a sample point in Q such that $d(p,x) \leq \varepsilon \mathrm{lfs}(x)$. If p is included in U_Q, we are done. Assume not. Let C_1 be the non-empty leaf cell in $\widehat{T}_Q$ that contains p. Let $(C_k, C_{k-1}, \ldots, C_1)$ be the maximal sequence of octree cells of T_Q such that for $i \in [2,k]$, the processing of C_i made C_{i-1} a non-empty leaf cell of $\widehat{T}_Q$ during the execution of EXTRACT. Notice that C_k is a non-empty leaf cell of the input octree $\overline{T}_Q$. By step 3 of EXTRACT, for $i \in [2,k]$, $\ell_{C_{i-1}} = \frac{1}{2}\ell_{C_i}$, and the distance between any two points in C_{i-1} and C_i is at most $\sqrt{3}\alpha\ell_{C_i} + \frac{\sqrt{3}}{2}\ell_{C_i} + \sqrt{3}\ell_{C_{i-1}} = \sqrt{3}(\alpha+1)\ell_{C_i}$. This implies that the distance between any two points in C_1 and C_k is at most $\sum_{i=2}^{k}\sqrt{3}(\alpha+1)\ell_{C_i} < 2\sqrt{3}(\alpha+1)\ell_{C_k}$.

Let q be the sample point in C_1 that is included in U_Q. Then, $d(p,q) < 2\sqrt{3}(\alpha+1)\ell_{C_k} = O(\varepsilon\text{lfs}(q))$ by Lemma 4.3(i). Thus, $d(q,x) \leq d(p,x) + d(p,q) \leq \varepsilon\text{lfs}(x) + O(\varepsilon\text{lfs}(q))$. Combining this with $\text{lfs}(q) \leq \text{lfs}(x) + d(q,x)$ gives $\text{lfs}(q) \leq (1+O(\varepsilon))\text{lfs}(x)$, and therefore $d(q,x) = O(\varepsilon\text{lfs}(x))$, i.e., U_Q is an $O(\varepsilon)$-sample.

To prove the local uniformity of U_Q, for every sample point $p \in U_Q$, we need to show that $|B(p,\beta d(p,x)) \cap U_Q| = O(1)$ for all point $x \in \Sigma$ such that $B(x,d(p,x))$ is empty. Since $B(x,d(p,x))$ is empty, p is the nearest sample point in U_Q to x. We have $d(p,x) = O(\varepsilon\text{lfs}(x))$ as U_Q is an $O(\varepsilon)$-sample. Let C and C' be the non-empty leaf cells in $\widehat{T}_Q$ and $\overline{T}_Q$, respectively, that contain p. Notice that $C' \subseteq C$ as $\overline{T}_Q$ is a refinement of $\widehat{T}_Q$. By Lemma 4.3(ii), $d(p,x) \leq 2\ell_{C'} \leq 2\ell_C$. Consider an arbitrary non-empty leaf cell C_1 of $\widehat{T}_Q$ that intersects $B(p,\beta d(p,x))$. Let c be the center of C. We have $d(c,C_1) < \beta d(p,x) + d(p,c) \leq \beta d(p,x) + \frac{\sqrt{3}}{2}\ell_C < (2\beta+1)\ell_C$, which is less than $\alpha\ell_C$ because $\beta > 1$ and $\alpha = 3\beta$. This implies that C_1 intersects $R(c,2\alpha\ell_C)$. We conclude that $\ell_{C_1} \geq \frac{1}{2}\ell_C$, because otherwise step 3 of EXTRACT would have removed C_1. Then, a packing argument shows that only $O(1)$ non-empty leaf cells of $\widehat{T}_Q$ can intersect $B(p,\beta d(p,x))$, meaning that $|B(p,\beta d(p,x)) \cap U_Q| = O(1)$ as U_Q contains one sample point from each non-empty leaf cell of $\widehat{T}_Q$. ◻

Steps 2 and 3 in Section 2 describe how to reconstruct surface(s) from a locally uniform $O(\varepsilon)$-sample and how to add the sample points not in the subsample to the reconstruction.

To reconstruct from U_Q in $O(|U_Q|)$ time in step 2, for each sample point $p \in U_Q$, we need the cocone at p and the sample points of U_Q at distance at most βr_p from p. The cocone at p can be obtained from the estimated normal at p. Let C be the leaf cell in $\widehat{T}_Q$ containing p. The proof of Lemma 4.4 reveals that $B(p,\beta r_p) \subset R(c,\alpha\ell_C)$, where c is the center of C. Every cell that intersects $R(c,\alpha\ell_C)$ has side length at least $\frac{1}{2}\ell_C$ by the working of EXTRACT. Therefore, we can find the leaf cells in $\widehat{T}_Q$ that intersect $R(c,\alpha\ell_C)$ using the pointers between neighboring equal-sized cells and the parent-child pointers, and then report the sample points of U_Q in them in $O(1)$ time. The set of reported sample points contains all points in $B(p,\beta r_p) \cap U_Q$.

In step 3, we need to collect relevant sample points of P to compute the Delaunay triangulation restricted to each triangle τ in the initial mesh M'. Let p be a vertex of τ. Suppose that the surface containing p is initially reconstructed by the call RECONSTRUCT(T_Q) and that C is the leaf cell in the octree $\widehat{T}_Q$ that contains p. Let c be the center of C. As shown above, $B(p,\beta r_p) \subset R(c,\alpha\ell_C)$, and we can retrieve the leaf cells of $\widehat{T}_Q$ intersecting $R(c,\alpha\ell_C)$ in $O(1)$ time. In Section 2, we have shown, that by setting β an appropriate constant, all the relevant sample points needed for the local restricted Delaunay triangulation computation are in $B(p,\beta r_p)$. We use more points by including all the points of P in the leaf cells of $\widehat{T}_Q$ that intersect $R(c,\alpha\ell_C)$. The Delaunay triangulation restricted to τ can be computed in $O(n_\tau \log n_\tau)$ time, where $n_\tau = |P \cap R(c,\alpha\ell_C)|$. We claim that each sample point can only participate in a constant number of computations, so the total running time is $O(n \log n)$. The sample points in a leaf cell C_1 will be collected only when a triangle in M' has a vertex in some leaf cell C_2 such that C_1 intersects $R(c_2,\alpha\ell_{C_2})$, where c_2 is the center of C_2. EXTRACT ensures that $\ell_{C_1} \geq \frac{1}{2}\ell_{C_2}$, so C_2 intersects $R(c_1,2\alpha\ell_{C_1})$. Then, EXTRACT ensures that $\ell_{C_2} \geq \frac{1}{2}\ell_{C_1}$, so there are $O(1)$ such C_2. Thus, the sample points in C_1 only participate in $O(1)$ computations.

THEOREM 4.1. *Let P be an ε-sample of a two-dimensional smooth compact manifold in $\mathbb{R}^3$ for a sufficiently small ε. A surface triangulation homeomorphic to Σ and approximating the geometry of Σ can be computed in $O(|P|\log|P|)$ time.*

5. EXPERIMENTAL RESULTS

Our implementation differs slightly from the preceding description. In steps 4 and 5 of TRIM, we use $R(c,3\ell_C)$ instead of $R(c,5\ell_C)$, pick two points from each of the canonical cubes of side length ℓ_C in $R(c,3\ell_C)$, and then use PCA to estimate the surface normal. The PCA-based approach improves the robustness of the surface normal estimation. In step 6 of TRIM, we perform an extra test before trimming a cell C. We use the surface normal estimated in step 5 of TRIM to form a cocone at $p \in C$ of angular radius $\pi/12$. We trim C only if the original step 5 tells us to and this cocone at p contains all sample points in $R(c,3\ell_C)$. We need this cocone check because some test cases do not meet the sampling requirement of the algorithm. If the cocone at p excludes some sample point in $R(c,3\ell_C)$, the surface around p is curvy with respect to the local sampling density, so we should not trim C. In step 3 of EXTRACT, we set $\alpha = 2$. In step 5 of EXTRACT, for each non-empty leaf cell C in $\widehat{T}_P$, if C is also a leaf cell in $\overline{T}_P$, then sample one point in C as described in EXTRACT; otherwise, we sample one point from each child of C in $\overline{T}_P$ instead of C alone. This gives us a denser subsample to facilitate the subsequent surface reconstruction. This implementation still runs in $O(n \log n)$ time.

We run Cocone to reconstruct the first surface from the subsample. Afterwards, the sample points absent from the subsample are added back as follows. For each remaining sample point p, we find the nearest triangle t in the current reconstruction, connect p to the three sides of t to split t into three triangles, and then apply edge flips until no new edge produced is flippable [14]. We compile our code using `g++-4.1.2` with `O2` flag, and obtain the binary of Cocone from its author's webpage. The experiments were run on a Dell Optiplex 745 with a 2.4GHz CPU and 4GB RAM.

Table 1 shows the results of our experiments on locally non-uniform samples. Figure 7 shows the samples and models used. For each model, we randomly pick a subset of triangles, sample extra points in those triangles and their neighboring triangles, and then include the vertices of the model to form a locally non-uniform input sample. The white patches in Figures 1(a) and 1(d) are the resulting clusters of sample points. Refer to the top table in Table 1. Our prototype is 51% to 68% faster than running Cocone alone. The columns `In` and `Sn` show that the subsamples are significantly more locally uniform than the input samples. The columns `Crad` and `Srad` show that the average circumradii in Cocone's output and our prototype's output are similar. We also experimented with some locally uniform samples.

The results are shown in the bottom table in Table 1. Our prototype is 8% faster than running Cocone alone for Buddha, and 22% to 41% faster for other models. Note that most of the time is spent on running Cocone to reconstruct the surfaces from the extracted subsample.

6. DISCUSSION

Suppose that the sample points are perturbed with a noise magnitude $O(\varepsilon^2)$ times the local feature sizes. We may estimate the surface normals using PCA as in our prototype. The three eigenvalues should tell us whether the estimated surface normal is good. Similar analysis has been conducted before for detecting manifold dimension from noisy samples [15]. So if the manifold is known to be connected (meaning no recursive calls), our approach should be able to cope with noise. However, if the manifold contains multiple surfaces, our approach may fail, as the recursive calls try to reconstruct the smallest surface first, and the noise may collaborate to form a tiny spurious surface.

Our approach should also work for a smooth compact k-manifold in $\mathbb{R}^d$, assuming that k is known. It seems likely that our results can be extended to $\mathbb{R}^d$, in which case one can produce the octree in $O(n \log n)$ time, where the hidden constant is exponential in $O(d)$. One should be able to use PCA to estimate the normal space. After obtaining a locally uniform subsample, several algorithms can be applied to reconstruct the k-manifold [13, 9, 8].

7. REFERENCES

[1] M. Alexa, J. Behr, D. Cohen-Or, S. Fleishman, D. Levin, and C. T. Silva. Point set surfaces. *Proceedings of the IEEE Conference on Visualization*, 2001, 21–28.

[2] N. Amenta, M. Bern. Surface reconstruction by Voronoi filtering. *Discrete and Computational Geometry*, 22 (1999), 481–504.

[3] N. Amenta, S. Choi, T. K. Dey and N. Leekha. A simple algorithm for homeomorphic surface reconstruction. *International Journal on Computational Geometry and Applications*, 12 (2002), 125–141.

[4] N. Amenta, S. Choi and N. Kolluri. The power crust, unions of balls, and the medial axis transform. *Computational Geometry: Theory and Applications*, 19 (2001), 127–153.

[5] N. Amenta and T. K. Dey. Normal variation with adaptive feature size. A note as an erratum to Lemma 2 in [17].

[6] D. Attali, J.-D. Boissonnat, and A. Lieutier. Complexity of the Delaunay triangulation of points on surfaces: the smooth case. *Proceedings of the 19th Annual Symposium on Computational Geometry, 2003, 201–210.*

[7] J.-D. Boissonnat and F. Cazals. Smooth surface reconstruction via natural neighbor interpolation of distance functions. *Proceedings of the 16th Annual Symposium on Computational Geometry*, 2000, 223–232.

[8] J.D. Boissonnat and A. Ghosh. Manifold reconstruction using tangential Delaunay complexes. *Proceedings of the 26th Annual Symposium on Computational Geometry*, 2010, 324–333.

[9] J.D. Boissonnat, L. Guibas, and S.Y. Oudot. Manifold reconstruction in arbitrary dimensions using witness complexes. *Proceedings of the 23rd Annual Symposium on Computational Geometry*, 2007, 194–203.

[10] P. B. Callahan and S. R. Kosaraju. A decomposition of multidimensional point sets with applications to k-nearest-neighbors and n-body potential Fields. *Journal of ACM*, 42 (1995), 67–90.

[11] T.M. Chan, J. Snoeyink, and C.-K. Yap. Primal dividing and dual pruning: output-sensitive construction of four-dimensional polytopes and three-dimensional Voronoi diagrams. *Discrete and Computational Geometry*, 1997, 433–454.

[12] B. Chazelle. An optimal convex hull algorithm in any fixed dimension. *Discrete and Computational Geometry*, 10 (1993), 377–409.

[13] S.-W. Cheng, T.K. Dey, and E.A. Ramos. Manifold reconstruction from point samples. *Proceedings of the 16th Annual ACM-SIAM Symposium on Discrete Algorithms*, 2005, 1018–1027.

[14] S.-W. Cheng and J. Jin. Edge flips and deforming surface meshes. *Proceedings of the 27th Annual Symposium on Computational Geometry*, 2011, 331–340.

[15] S.-W. Cheng, Y. Wang, and Z. Wu. Provable dimension detection using principal component analysis. *International Journal of Computational Geometry and Applications*, 18 (2008), 414–440.

[16] B. Curless and M. Levoy. A volumetric method for building complex models from range images. *SIGGRAPH*, 1996, 306–312.

[17] T. K. Dey. *Curve and surface reconstruction: Algorithms with mathematical analysis.* Cambridge University Press, New York, 2006.

[18] T. K. Dey, S. Funke and E. A. Ramos. Surface Reconstruction in Almost Linear Time under Locally Uniform Sampling. *European Workshop on Computational Geometry*, Berlin, March 2001.

[19] D. Dumitriu, S. Funke, M. Kutz, and N. MilosavljevIć. On the locality of extracting a 2-manifold in R^3. *Proceedings of the 11th Scandinavian Workshop on Algorithm Theory*, 2008, 270–281.

[20] D. Dumitriu, S. Funke, M. Kutz, and N. MilosavljevIć. How much geometry it takes to reconstruct a 2-manifold in $\mathbb{R}^3$. *ACM Journal of Experimental Algorithmics*, 14 (2009), 2.2:1–2.2:17.

[21] J. Erickson. Nice point sets can have nasty Delaunay triangulations. *Proceedings of the 17th Annual Symposium on Computational Geometry*, 2001, 96–105.

[22] S. Funke and E.A. Ramos. Smooth-surface reconstruction in near-linear time. *Proceedings of the 13th Annual ACM-SIAM Symposium on Discrete Algorithms*, 2002, 781–790.

[23] J. Giesen and U. Wagner. Shape dimension and intrinsic metric from samples of manifolds. *Discrete and Computational Geometry*, 32 (2004), 245–267.

[24] H. Hoppe, T. DeRose, T. Duchamp, J. McDonald, and W. Stuetzle. Surface reconstruction from unorganized points. *SIGGRAPH*, 1992, 71–78.

Non-uniform	I	Cocone	In	Crad	S	Prune+Build	Sn	Total	Srad
Torus	348820	224s	101	23	70038	7s+44s	36	74s	23
Bunny	210646	100s	589	36	29242	4s+14s	28	33s	36
Armadillo	411777	216s	171	23	112593	9s+59s	31	93s	23
Buddha	512524	299s	261	19	143021	11s+79s	72	121s	19
Dragon	452679	256s	1177	22	80989	9s+41s	69	81s	23
Uniform	**I**	**Cocone**	**In**	**Crad**	**S**	**Prune+Build**	**Sn**	**Total**	**Srad**
Torus	57600	36s	5	74	53956	2s+25s	13	28s	74
Bunny	35286	20s	21	116	23503	1s+11s	23	13s	117
Armadillo	115897	60s	24	52	73706	3s+37s	41	45s	54
Buddha	142534	77s	62	42	112076	4s+62s	60	71s	43
Dragon	126359	66s	308	50	58439	3s+30s	84	39s	52

Table 1: `I`: input size; `Cocone`: Cocone's running time on the input; `In`: maximum no. of sample points within $1.5R_v$ from a vertex v, where R_v is the maximum circumradius among its incident triangles in the Cocone output; `Crad`: average circumradius in the Cocone output; `S`: subsample size; `Prune+Build`: time to extract the subsample + time to reconstruct from it using Cocone; `Sn`: maximum no. of subsample points within $1.5S_v$ from a vertex v, where S_v is the maximum circumradius among its incident triangles after running Cocone on the subsample; `Total`: prototype's running time; `Srad`: average circumradius in the prototype's output.

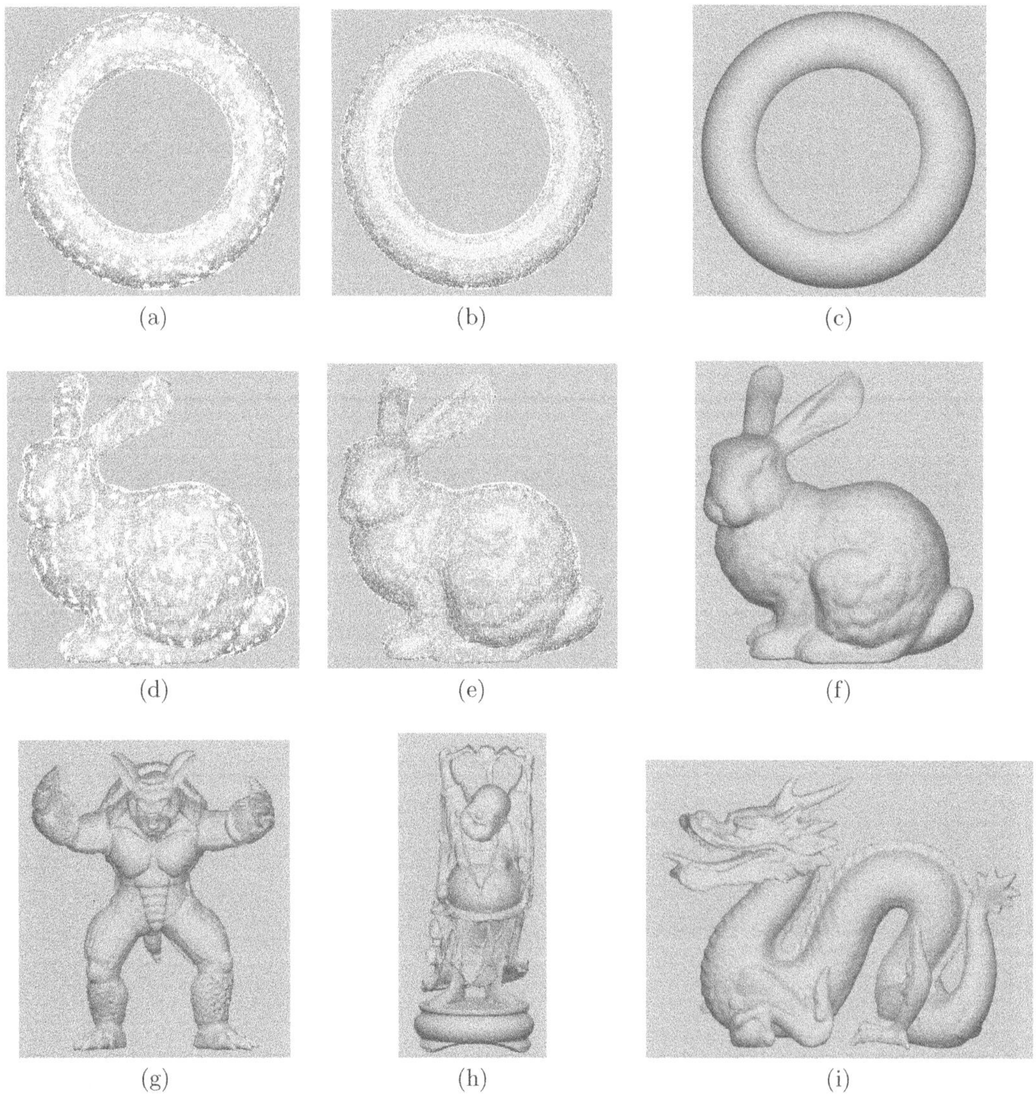

Figure 7: (a) and (d) show the input non-uniform samples. (b) and (e) show the subsamples extracted. (c) and (f)–(i) show the surfaces reconstructed by our prototype.

Protein Molecules in 3D: Complex Behavior from Simple Local Geometry plus Weak Interactions

Jane S. Richardson
Duke University Medical Center
Department of Biochemistry
Durham, North Carolina USA
jsr@kinemage.biochem.duke.edu

David C. Richardson
Duke University Medical Center
Department of Biochemistry
Durham, North Carolina USA
dcr@kinemage.biochem.duke.edu

ABSTRACT

Protein structures in 3D are large. elegant, complex, and subtle enough that their prediction and understanding still largely eludes us. In contrast, their local building blocks are quite simple: to a good approximation their connectivity is a tree graph of backbone and sidechains, and the atoms are connected with specific, near-invariant covalent bond-length and bond-angle geometry. However, the connections are mostly tetrahedral, with essentially no right angles and thus no independent, orthogonal degrees of freedom. Also, the non-local interactions such as van der Waals, hydrophobicity, and hydrogen bonds are very much weaker than the covalent bonds and produce the compact, semi-rigid "native" structure only collectively by their great numbers. These two factors are major causes of the cooperativity of protein folding, allostery, and function, and help guide the evolution of a protein sequence to achieve just one or a few equilibrium 3D structures.

We will explore several ways in which geometrical analysis – or sometimes even a geometrical metaphor – has illuminated some facet of protein folding, conformation, or mobility. A final message is that the currently most successful algorithms in structure prediction and design attack the complexity by global energetics evaluated using unabashed brute-force computation. There may yet be opportunity for developing more efficient solutions by hierarchical buildup of cleaner and more intuitive geometrical understanding.

Categories and Subject Descriptors

J.3 [**Life and Medical Sciences**]: Biology and genetics – *structural biology*.

General Terms

Experimentation, Measurement, Design.

Keywords

Biological macromolecules, molecular conformation, molecular motions, non-orthogonal parameters.

SCG'12, June 17–20, 2012, Chapel Hill, North Carolina, USA.
ACM 978-1-4503- 1299-8/12/06.

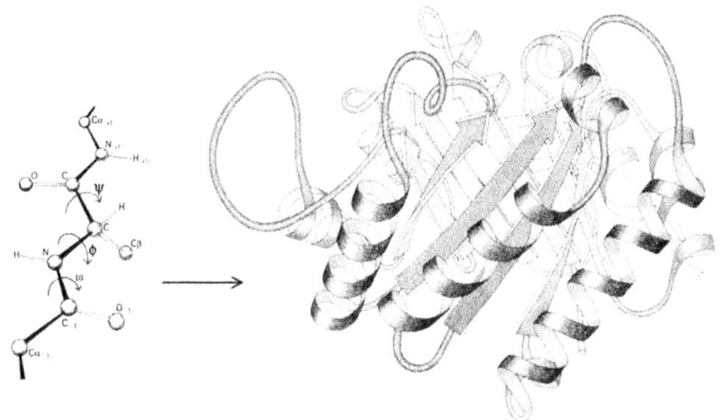

Figure 1. Simple backbone geometry (left) leads to complex protein folds such as triose phosphate isomerase (ribbon).

1. BIOGRAPHY

Jane and Dave Richardson are Professors of Biochemistry at Duke University whose research on protein and RNA structure has covered x-ray crystallography, structural bioinformatics, protein design, graphical representations (both hand-drawn ribbons [1] and computer graphics [2]), and the use of all-atom contacts to validate and improve experimental structures [3]. More details at http://en.wikipedia.org/wiki/Jane_Richardson. The validation site is http://molprobity.kinemage.biochem.duke.edu. The above and other open license figures are available on Wikimedia Commons at http://commons.wikimedia.org/wiki/User:Dcrjsr.

2. REFERENCES

[1] Richardson, J.S. 1981. The Anatomy and Taxonomy of Protein Structure. *Advances in Protein Chem.* 34, 167-339. ISBN 0-12-034234-0. Available on-line (annotated) at http://kinemage.biochem.duke.edu/teaching/anatax.

[2] Richardson, D.C., and Richardson, J.S. 1992. The Kinemage: A Tool for Scientific Illustration. *Protein Sci.* 1, 3-9.

[3] Chen, V.B., Arendall, W.B. III, Headd, J.J., Keedy, D.A., Immormino, R.M., Kapral, G.J., Murray, L.W., Richardson, J.S., and Richardson, D.C. 2010. MolProbity: All-atom structure validation for macromolecular crystallography. *Acta Crystallogr.* D66,12-21.

Higher-Order Erdős–Szekeres Theorems*

Marek Eliáš
Department of Applied Mathematics
Charles University, Malostranské nám. 25
118 00 Praha 1, Czech Republic
eliam6am@ss1000.ms.mff.cuni.cz

Jiří Matoušek
Department of Applied Mathematics and
Institute of Theoretical Computer Science (ITI)
Charles University, Malostranské nám. 25
118 00 Praha 1, Czech Republic, and
Institute of Theoretical Computer Science
ETH Zurich, 8092 Zurich, Switzerland
matousek@kam.mff.cuni.cz

ABSTRACT

Let $P = (p_1, p_2, \ldots, p_N)$ be a sequence of points in the plane, where $p_i = (x_i, y_i)$ and $x_1 < x_2 < \cdots < x_N$. A famous 1935 Erdős–Szekeres theorem asserts that every such P contains a monotone subsequence S of $\lceil\sqrt{N}\,\rceil$ points. Another, equally famous theorem from the same paper implies that every such P contains a convex or concave subsequence of $\Omega(\log N)$ points.

Monotonicity is a property determined by pairs of points, and convexity concerns triples of points. We propose a generalization making both of these theorems members of an infinite family of Ramsey-type results. First we define a $(k+1)$-tuple $K \subseteq P$ to be *positive* if it lies on the graph of a function whose kth derivative is everywhere nonnegative, and similarly for a *negative* $(k+1)$-tuple. Then we say that $S \subseteq P$ is *kth-order monotone* if its $(k+1)$-tuples are all positive or all negative.

We investigate quantitative bound for the corresponding Ramsey-type result (i.e., how large kth-order monotone subsequence can be guaranteed in every N-point P). We obtain an $\Omega(\log^{(k-1)} N)$ lower bound ($(k-1)$-times iterated logarithm). This is based on a quantitative Ramsey-type theorem for *transitive colorings* of the complete $(k+1)$-uniform hypergraph (these were recently considered by Pach, Fox, Sudakov, and Suk).

For $k = 3$, we construct a geometric example providing an $O(\log\log N)$ upper bound, tight up to a multiplicative constant. As a consequence, we obtain similar upper bounds for a Ramsey-type theorem for *order-type homogeneous* subsets in $\mathbb{R}^3$, as well as for a Ramsey-type theorem for hyperplanes in $\mathbb{R}^4$ recently used by Dujmović and Langerman.

Categories and Subject Descriptors

G.2 [**Discrete Mathematics**]: Combinatorics

*Both authors supported by the ERC Advanced Grant No. 267165.

Keywords

Combinatorial geometry, Erdős–Szekeres theorem, Ramsey theory, order type

1. INTRODUCTION

In this paper we mainly consider sets $P = \{p_1, p_2, \ldots, p_N\}$ of points in the plane, where $p_i = (x_i, y_i)$. We always assume that no two of the x-coordinates coincide, and unless stated otherwise, we also assume that the p_i are numbered so that $x_1 < x_2 < \cdots < x_N$ (the same also applies to subsets of P, which we will enumerate in the order of increasing x-coordinates).

Two theorems of Erdős and Szekeres. Among simple results in combinatorics, only few can compete with the following one in beauty and usefulness:

THEOREM 1.1. (ERDŐS AND SZEKERES ON MONOTONE SUBSEQUENCES [ES35]) *For every positive integer n, among every $N = (n-1)^2 + 1$ points $p_1, \ldots, p_N \in \mathbb{R}^2$ as above, one can always choose a* monotone subset *of at least n points, i.e., indices $i_1 < i_2 < \cdots < i_n$ such that either $y_{i_1} \le y_{i_2} \le \cdots \le y_{i_n}$ or $y_{i_1} \ge y_{i_2} \ge \cdots \ge y_{i_n}$.*

See, for example, Steele [Ste95] for a collection of six nice proofs and some applications. For many purposes, it is more natural to view the above theorem as a purely combinatorial result about permutations, but here we prefer the geometric formulation (which is also similar to the one in the original Erdős–Szekeres paper).

Another result of the same paper of Erdős and Szekeres is the following well-known gem in discrete geometry:[1]

THEOREM 1.2. (ERDŐS–SZEKERES ON CONVEX/CONCAVE CONFIGURATIONS [ES35]) *For every positive integer n, among every $N = \binom{2n-4}{n-2} + 1 \approx 4^n/\sqrt{n}$ points $p_1, \ldots, p_N \in \mathbb{R}^2$ as above, one can always choose a* convex configuration *or a* concave configuration *of n points, i.e., indices $i_1 < i_2 < \cdots < i_n$ such that the slopes of the segments $p_{i_j}p_{i_{j+1}}$, $j = 1, 2, \ldots, n-1$, are either monotone nondecreasing or monotone nonincreasing.*

See, e.g., [MS00, Mat02] for proofs and surveys of developments around this result.

[1]Somewhat unfortunately, the name Erdős–Szekeres theorem refers to Theorem 1.1 in some sources and to Theorem 1.2 or similar statements in other sources.

k-general position. To simplify our forthcoming discussion, at some places it will be convenient to assume that the considered point sets are in a "sufficiently general" position. Namely, we define a set P to be in *k-general position* if no $k+1$ points of P lie on the graph of a polynomial of degree at most $k-1$. In particular, 1-general position requires that no two y-coordinates coincide, and 2-general position means the usual general position, i.e., no three points collinear.

kth-order monotone subsets. Here we propose a view of Theorems 1.1 and 1.2 as the first two members in an infinite sequence of Ramsey-type results about planar point sets.[2]

In Theorem 1.1, monotonicity of a subset is a property of *pairs* of points of the subset, and actually, it suffices to look at pairs of consecutive points. Similarly, convexity or concavity of a configuration in Theorem 1.2 is a property of triples, and again it is enough to look at consecutive triples.

In the former case, we are considering the slope of the segment determined by a pair of points, which can be thought of as the first derivative. In the latter case, a triple is convex iff its points lie on the graph of a smooth convex function, i.e., one with nonnegative second derivative everywhere.

With this point of view, it is natural to define a $(k+1)$-tuple $K \subseteq P$ to be *positive* if it lies on the graph of a function whose k-th derivative (exists and) is everywhere nonnegative, and similarly for a *negative* $(k+1)$-tuple (in Section 2, we will provide several other, equivalent characterizations of these properties). Then we say that an arbitrary subset $S \subseteq P$ is *kth-order monotone* if its $(k+1)$-tuples are all positive or all negative.

First-order monotonicity is obviously equivalent to monotonicity as in Theorem 1.1, and second-order monotonicity is equivalent to convexity/concavity as in Theorem 1.2. We will also see (Lemma 2.5) that, to certify kth-order monotonicity, it is enough to consider all $(k+1)$-tuples of *consecutive* points.

Let us remark that every $(k+1)$-tuple K is positive or negative, and moreover, if K is in k-general position, it cannot be both positive and negative (Corollary 2.3). We will write $\mathrm{sgn}(K) = +1$ if K is positive and $\mathrm{sgn}(K) = -1$ if K is negative.

Ramsey's theorem, quantitative bounds, and transitive colorings. Using the just mentioned facts, one can immediately derive a Ramsey-type theorem for kth-order monotone subsets from Ramsey's theorem.

PROPOSITION 1.3. *For every k and n there exists N such that every N-point planar set in k-general position contains an n-point kth-order monotone subset.*

Proof. We recall Ramsey's theorem (for two colors; see, e.g., Graham, Rothschild, and Spencer [GRS90]): for every ℓ and n there exists N such that for every coloring of the set $\binom{X}{\ell}$ of all ℓ-element subsets of an N-element set X there exists an n-element *homogeneous* set $Y \subseteq X$, i.e., a subset in which all ℓ-tuples have the same color. The smallest N for which the claim holds is usually denoted by $R_\ell(n)$.

In our case, we set $X = P$ and color each $(k+1)$-tuple $K \subseteq P$ with the color $\mathrm{sgn}(K) \in \{\pm 1\}$. Then homogeneous subsets are exactly kth-order monotone subsets. □

Let us denote by $\mathrm{ES}_k(n)$ the smallest value of N for which the claim in this proposition holds. We have $\mathrm{ES}_1(n) \le (n-1)^2+1$ and $\mathrm{ES}_2(n) \le \binom{2n-4}{n-2}+1$ according to Theorems 1.1 and 1.2, respectively; moreover, these inequalities actually hold with equality [ES35]. Our main goal is to estimate the order of magnitude of $\mathrm{ES}_k(n)$ for $k \ge 3$.

The above proof gives $\mathrm{ES}_k(n) \le R_{k+1}(n)$. However, for $k=1$, and most likely for all k, the order of magnitude of $R_{k+1}(n)$ is much larger than that of $\mathrm{ES}_k(n)$. Indeed, considering k fixed and n large, the best known lower and upper bounds of $R_{k+1}(n)$ are of the form[3] $R_2(n) = 2^{\Theta(n)}$ and, for $k \ge 2$,

$$\mathrm{twr}_k(\Omega(n^2)) \le R_{k+1}(n) \le \mathrm{twr}_{k+1}(O(n)),$$

where the tower function $\mathrm{twr}_k(x)$ is defined by $\mathrm{twr}_1(x) = x$ and $\mathrm{twr}_{i+1}(x) = 2^{\mathrm{twr}_i(x)}$. It is widely believed that the upper bound is essentially the truth. This belief is supported by known bounds for more than two colors, where the lower bound for $(k+1)$-tuples is also a tower of height $k+1$; see Conlon, Fox, and Sudakov [CFS11] for a recent improvement and more detailed overview of the known bounds.

The coloring of the $(k+1)$-tuples in the above proof of Proposition 1.3 is not arbitrary. In particular, it has a property we call *transitivity* (see Lemma 2.5). Transitive colorings were introduced earlier in the recent preprint Fox et al. [FPSS11, Section 6], under the same name.

To define a transitive coloring in general, we need to consider a hypergraph whose vertex set is linearly ordered; w.l.o.g. we can identify it with the set $[N] := \{1, 2, \ldots, N\}$. A coloring $c\colon \binom{[N]}{\ell} \to [m]$ is *transitive* if, for every $i_1, \ldots, i_{\ell+1} \in [N]$, $i_1 < \cdots < i_{\ell+1}$, whenever the ℓ-tuples $\{i_1, \ldots, i_\ell\}$ and $\{i_2, \ldots, i_{\ell+1}\}$ have the same color, then *all* ℓ-element subsets of $\{i_1, \ldots, i_{\ell+1}\}$ have the same color. Let $R_\ell^{\mathrm{trans}}(n)$ denote the Ramsey number for transitive colorings, i.e., the smallest N such that any transitive coloring of the complete ℓ-uniform hypergraph on $[N]$ contains an n-element homogeneous subset. We have the following bound.[4]

THEOREM 1.4. *For $k = 1, 2$, we have $R_{k+1}^{\mathrm{trans}}(n) = \mathrm{ES}_k(n)$, and for every fixed $k \ge 3$,*

$$\mathrm{ES}_k(n) \le R_{k+1}^{\mathrm{trans}}(n) \le \mathrm{twr}_k(O(n)).$$

We note that Fox et al. [FPSS11] proved the slightly weaker upper bound $R_{k+1}^{\mathrm{trans}}(n) \le \mathrm{twr}_k(O(n \log n))$.

The proof of Theorem 1.4 is given in Section 3. The inequality $\mathrm{ES}_k(n) \le R_{k+1}^{\mathrm{trans}}(n)$ is clear (since every N-point set in k-general position provides a transitive coloring of $\binom{[N]}{k+1}$). The upper bounds for $R_2^{\mathrm{trans}}(n)$ and $R_3^{\mathrm{trans}}(n)$ follow by translating the proofs of Theorem 1.1 and 1.2 to the

[2]There is also a (trivial) 0th member, namely, the statement that in every P, at least half of the points either have all y-coordinates nonnegative or have or all y-coordinates nonpositive.

[3]We employ the usual asymptotic notation for comparing functions: $f(n) = O(g(n))$ means that $|f(n)| \le C|g(n)|$ for some C and all n, where C may depend on parameters declared as constants (in our case on k); $f(n) = \Omega(g(n))$ is equivalent to $g(n) = O(f(n))$; and $f(n) = \Theta(g(n))$ means that both $f(n) = O(g(n))$ and $f(n) = \Omega(g(n))$.

[4]By inspecting the proof of the next theorem, it is easy to verify that the transitivity condition is not used in full strength—it suffices to assume only that the subsets obtained by omitting one of i_2, i_3 have the same color.

setting of transitive colorings almost word by word, and they are contained in [FPSS11]. The upper bound on $R_{k+1}^{\mathrm{trans}}(n)$ is then obtained by induction on k, with $k = 3$ as the base case, following one of the usual proofs of Ramsey's theorem.

A set with no large third-order monotone subsets. For $k \leq 2$, the numbers $\mathrm{ES}_k(n)$ (and thus $R_{k+1}^{\mathrm{trans}}(n)$) are known exactly. Our perhaps most interesting result is an asymptotically matching lower bound for $\mathrm{ES}_3(n)$.

THEOREM 1.5. *For all $n \geq 2$ we have $R_4^{\mathrm{trans}}(2n+1) \geq \mathrm{ES}_3(2n+1) \geq 2^{2^{n-1}} + 1$. Consequently, $\mathrm{ES}_3(n) = 2^{2^{\Theta(n)}}$.*

The proof is given in Section 4. A Ramsey function with known doubly exponential growth seems to be rare in geometric Ramsey-type problems (a notable example is a result of Valtr [Val04]).

Order types. Here we change the setting from the plane to $\mathbb{R}^d$ and we consider an ordered sequence $P = (p_1, p_2, \ldots, p_N)$ in $\mathbb{R}^d$. This time we do *not* assume the first coordinates to be increasing. For simplicity, we assume P to be in general position, which now means that no $d+1$ points of P lie on a common hyperplane.

We recall that *order type* of P specifies the orientation of every $(d+1)$-tuple of points of P, and it this way, it describes purely combinatorially many of the geometric properties of P. More formally, the order type of P is the mapping $\chi\colon \binom{[N]}{d+1} \to \{-1, +1\}$, where for a $(d+1)$-tuple $I = \{i_1, \ldots, i_{d+1}\}$, $i_1 < i_2 < \cdots < i_{d+1}$,

$$\chi(I) := \operatorname{sgn}\det M(p_{i_1}, p_{i_2}, \ldots, p_{i_{d+1}}),$$

where $M(q_1, \ldots, q_{d+1})$ is the $(d+1) \times (d+1)$ matrix whose jth column is $(1, q_j)$, i.e., 1 followed by the vector of the d coordinates of q_j. See, e.g., Goodman and Pollack [GP93] or [Mat02] for more background about order types.

From Ramsey's theorem for $(d+1)$-tuples, we can immediately derive a Ramsey-type result for order types: for every d and n there exists N such that every N-point sequence contains an n-point subsequence in which all the $(d+1)$-tuples have the same orientation (we call such a subsequence *order-type homogeneous*). Let us write $\mathrm{OT}_d(n)$ for the smallest such N.

In Section 5 we first observe that, by simple and probably well known considerations, $\mathrm{OT}_1(n) = (n-1)^2 + 1$ and $\mathrm{OT}_2(n) = 2^{\Theta(n)}$. For $d \geq 3$, the best upper bound for $\mathrm{OT}_d(n)$ we are aware of is the one from the Ramsey argument above, i.e., $\mathrm{OT}_d(n) \leq R_{d+1}(n) \leq \mathrm{twr}_{d+1}(O(n))$. In particular, for $\mathrm{OT}_3(n)$ this upper bound is triply exponential; in Section 5 we prove a doubly exponential lower bound.

PROPOSITION 1.6. *For all d and n, $\mathrm{OT}_d(n) \geq \mathrm{ES}_d(n)$. In particular, $\mathrm{OT}_3(n) = 2^{2^{\Omega(n)}}$.*

A Ramsey-type result for hyperplanes. Let us consider a finite set H of hyperplanes in $\mathbb{R}^d$ in general position (every d intersecting at a single point). Let us say that H is *one-sided* if $V(H)$, the vertex set of the arrangement of H, lies completely on one side of the coordinate hyperplane $x_d = 0$.

Let $\mathrm{OSH}_d(n)$ be the smallest N such that every set H of N hyperplanes in $\mathbb{R}^d$ in general position contains a one-sided subset of n hyperplanes. Ramsey's theorem for d-tuples immediately gives $\mathrm{OSH}_d(n) \leq R_d(n)$ (a d-tuple gets color $+1$ if its intersection has a positive last coordinate, and color -1 otherwise).

Matoušek and Welzl [MW92] observed that, actually, $\mathrm{OSH}_2(n) = \mathrm{ES}_1(n) = (n-1)^2 + 1$, and applied this in a range-searching algorithm. Recently Dujmović and Langerman [DL11] used the existence of $\mathrm{OSH}_d(n)$ (essentially Lemma 9 in the arXiv version of their paper) to prove several interesting results, such as a ham-sandwich and centerpoint theorems for hyperplanes.

In Section 5 we show that lower bounds for kth-order monotone subsets in the plane can be translated into lower bounds for OSH_d.

PROPOSITION 1.7. *We have $\mathrm{OSH}_d(n) \geq \mathrm{ES}_{d-1}(n)$, and in particular, $\mathrm{OSH}_3(n) = 2^{\Omega(n)}$ and*[5] *$\mathrm{OSH}_4(n) = 2^{2^{\Omega(n)}}$.*

The lower bounds for $\mathrm{OSH}_d(n)$ can also be translated into lower bounds in the theorems of Dujmović and Langerman. For example, in their ham-sandwich theorem, we have d collections $H_1, \ldots, H_d$ of hyperplanes in $\mathbb{R}^d$, each of size N, and we want a hyperplane g such that in each H_i, we can find disjoint subsets A_i, B_i of n hyperplanes each such $V(A_i)$ lies on one side of g and $V(B_i)$ on the other side.

To derive a lower bound for the smallest necessary N, we fix d affinely independent points $p_1, \ldots, p_d$ in the $x_d = 0$ hyperplane, and a set H of N hyperplanes in general position with no one-sided subset of size n. We let H_i be an affinely transformed copy of H such that all of $V(H_i)$ lies very close to p_i. Then every potential ham-sandwich hyperplane g for these H_i has to be almost parallel to the $x_d = 0$ hyperplane, and thus there cannot be A_i, B_i of size n for all i.

The work of Fox et al. While preparing a draft of the present paper, we learned about a recent preprint of Fox, Pach, Sudakov, and Suk [FPSS11]. They investigated various combinatorial and geometric problems inspired by Theorems 1.1 and 1.2, and as was mentioned above, among others, they introduced transitive colorings,[6] but mainly they studied a related but different Ramsey-type quantity: let $N_\ell(q, n)$ be the smallest integer N such that, for every coloring of $\binom{[N]}{\ell}$ with q colors, there exists an n-element $I = \{i_1, \ldots, i_n\} \subseteq [N]$, $i_1 < \cdots < i_n$, inducing a *monochromatic monotone path*, i.e., such that all the ℓ-tuples of the form $\{i_j, i_{j+1}, \ldots, i_{j+\ell-1}\}$, $j = 1, 2, \ldots, n-\ell+1$, have the same color.

They note that $R_\ell^{\mathrm{trans}}(n) \leq N_\ell(2, n)$, and they obtained the following bounds for $N_\ell(2, n)$: $N_2(2, n) = \mathrm{ES}_1(n)$, $N_3(2, n) = \mathrm{ES}_2(n)$, and for every fixed $k \geq 3$,

$$\mathrm{twr}_k(\Omega(n)) \leq N_{k+1}(2, n) \leq \mathrm{twr}_k(O(n \log n)).$$

As we mentioned after Theorem 1.4, this also yields an upper bound for $R_{k+1}^{\mathrm{trans}}(n)$ only slightly weaker than the one in that theorem.

Open problems.

1. We have obtained reasonably tight bounds for $\mathrm{ES}_3(n)$, but the gaps are much more significant for $\mathrm{ES}_k(n)$ with $k \geq 4$. According to the cases $k = 1, 2, 3$, one may

[5] An exponential lower bound for OSH_3 was known to the authors of [MW92], and perhaps to others as well, but as far as we know, it hasn't appeared in print.

[6] With still another geometric source of such colorings besides the Erdős–Szekeres theorems, namely, noncrossing convex bodies in the plane

guess that $\mathrm{ES}_k(n)$ is of order $\mathrm{twr}_k(\Theta(n))$, and thus that stronger lower bounds are needed, but a possibility of a better upper bound shouldn't also be overlooked. This question looks both interesting and challenging.

2. A perhaps more manageable task might be a better lower bound for $R_k^{\mathrm{trans}}(n)$, $k \geq 4$. A natural approach would be to imitate the Stepping-Up Lemma used for lower bounds for the Ramsey numbers $R_k(n)$ (see, e.g., [CFS11]). But so far we have not succeeded in this, since even if we start with a transitive coloring of k-tuples, we could not guarantee transitivity for the coloring of $(k+1)$-tuples.

3. As for order-type homogeneous sequences, for $\mathrm{OT}_3(n)$ we have the lower bound of $2^{2^{\Omega(n)}}$, but upper bound only $\mathrm{twr}_4(O(n))$ directly from Ramsey's theorem. It seems that the colorings given by the order type are not transitive in any reasonable sense, and we have no good guess of which of the upper and lower bounds should be closer to the truth. Similar comments apply to the problem with one-sided subsets of planes in $\mathbb{R}^3$ (concerning $\mathrm{OSH}_3(n)$), and the higher-dimensional cases are even more widely open.

4. Another interesting question is whether $n \log n$ can be replaced by n in the upper bound for the quantity $N_\ell(2,n)$ considered by Fox et al. [FPSS11].

5. In our definition of kth-order positivity, every $(k+1)$-tuple of points should lie on the graph of a function with a nonnegative kth derivative, and different functions can be used for different $(k+1)$-tuples. In an earlier version of this paper, we conjectured that, assuming k-general position, a single function should suffice for all $(k+1)$-tuples; in other words, that every kth-order monotone finite set finite set in k-general position lies on a graph of a k-times differentiable function $f\colon \mathbb{R} \to \mathbb{R}$ whose kth derivative is everywhere nonnegative or everywhere nonpositive.

   However, Rote [Rot12] disproved this for $k = 3$ (while the cases $k = 1, 2$ do hold, as is not hard to check). With his kind permission, we reproduce his example at the end of Section 2.

   Naturally, this opens up interesting new questions: How can one characterize point sets lying on the graph of a function whose kth derivative is positive everywhere? Is there a Ramsey-type theorem for such sets, and if yes, how large is the corresponding Ramsey function?

2. ON THE DEFINITION OF kTH-ORDER MONOTONICITY

Here we provide several equivalent characterizations of kth-order monotonicity of planar point sets and some of their properties. First we recall several known results.

Divided differences and Newton's interpolation. Let $p_1, p_2, \ldots, p_{k+1}$ be points in the plane, $p_i = (x_i, y_i)$, where the x_i are all distinct (but not necessarily increasing). We recall that the *kth divided difference* $\Delta_k(p_1, p_2, \ldots, p_{k+1})$ is defined recursively as follows:

$$\Delta_0(p_1) := y_1$$

$$\Delta_k(p_1, p_2, \ldots, p_{k+1}) := \frac{\Delta_{k-1}(p_2, p_3, \ldots, p_{k+1}) - \Delta_{k-1}(p_1, p_2, \ldots, p_k)}{x_{k+1} - x_1}.$$

For example, $\Delta_1(p_1, p_2)$ equals the slope of the line p_1p_2. In general, the kth divided difference is related to the kth derivative as follows (see, e.g., [Phi03, Eq. 1.33]; note that the case $k = 1$ is the Mean Value Theorem):

LEMMA 2.1 (CAUCHY). *Let the points $p_1, \ldots, p_{k+1}$, $a := x_1 < x_2 < \cdots < b := x_{k+1}$, lie on the graph of a function f such that the kth derivative $f^{(k)}$ exists everywhere on the interval (a, b). Then there exists $\xi \in (a, b)$ such that*

$$\Delta_k(p_1, \ldots, p_{k+1}) = \frac{f^{(k)}(\xi)}{k!}.$$

We will also need the following result (see, e.g., [Phi03, Eq. 1.11–1.19]).

LEMMA 2.2 (NEWTON'S INTERPOLATION). *Let $p_1, \ldots, p_{k+1} \in \mathbb{R}^2$ be points with distinct x-coordinates (here we need not assume that the x-coordinates are increasing). Then the unique polynomial f of degree at most k whose graph contains $p_1, \ldots, p_{k+1}$ is given by*

$$f(x) = \sum_{i=1}^{k+1} \left(\Delta(p_1, \ldots, p_i) \prod_{j=1}^{i-1} (x - x_j) \right)$$

In particular, the coefficient of x^k is $\Delta(p_1, \ldots, p_{k+1})$, and it equals $f^{(k)}(x)/k!$ (which is a constant function).

We recall that a $(k+1)$-tuple $K = \{p_1, \ldots, p_{k+1}\}$ was defined to be positive if it is contained in the graph of a function having a nonnegative kth derivative everywhere. We obtain the following equivalent characterization:

COROLLARY 2.3. *A $(k+1)$-tuple $K = \{p_1, \ldots, p_{k+1}\}$ is positive iff $\Delta_k(p_1, \ldots, p_{k+1}) \geq 0$ (and similarly for a negative $(k+1)$-tuple). If K is in k-general position, we have $\operatorname{sgn} K = \operatorname{sgn} \Delta_k(p_1, \ldots, p_{k+1})$.*

Proof. If K is contained in the graph of f with $f^{(k)} \geq 0$ everywhere, then $\Delta_k(p_1, \ldots, p_{k+1}) \geq 0$ by Lemma 2.1.

Conversely, if $\Delta_k(p_1, \ldots, p_{k+1}) \geq 0$, then by Lemma 2.2, the unique polynomial of degree at most k whose graph contains K is the required function with nonnegative kth derivative.

If, moreover, K is in k-general position, then

$$\Delta_k(p_1, \ldots, p_{k+1}) \neq 0,$$

and so K cannot be both kth-order positive and kth-order negative by Lemma 2.1. □

We will also need the following criterion for the sign of a $(k+1)$-tuple.

LEMMA 2.4. *Let $K = \{p_1, p_2, \ldots, p_{k+1}\}$ be a $(k+1)$-tuple of points in k-general position, $x_1 < \cdots < x_{k+1}$, let $i \in [k+1]$, and let f_i be the (unique) polynomial of degree at most $k-1$ whose graph passes through the points of $K \setminus \{p_i\}$. Then $\operatorname{sgn} K = (-1)^{k-i}$ if p_i lies below the graph of f_i, and $\operatorname{sgn} K = (-1)^{k+1-i}$ if p_i lies above the graph.*

Let f be the polynomial of degree at most k passing through all of K. We use Newton's interpolation (Lemma 2.2), but with the points reordered so that p_i comes last, and we get that

$$f(x) = f_i(x) + \Delta_k(p_1, \ldots, p_{i-1}, p_{i+1}, \ldots, p_{k+1}, p_i) \cdot \prod_{j \in [k+1] \setminus \{i\}} (x - x_j).$$

Using this with $x = x_i$, we get

$$\begin{aligned} \operatorname{sgn}(y_i - f_i(x_i)) &= \operatorname{sgn}(f(x_i) - f_i(x_i)) \\ &= \operatorname{sgn} \Delta_k(p_1, \ldots, p_{i-1}, p_{i+1}, \ldots, p_{k+1}, p_i) \\ &\quad \cdot \operatorname{sgn} \prod_{j \in [k+1] \setminus \{i\}} (x_i - x_j). \end{aligned}$$

Divided differences are invariant under permutations of the points (as can be seen, e.g., from Lemma 2.2, since the interpolating polynomial does not depend on the order of the points), and so $\operatorname{sgn} \Delta_k(p_1, \ldots, p_{i-1}, p_{i+1}, \ldots, p_{k+1}, p_i) = \operatorname{sgn} K$. Finally, the product $\prod_{j \in [k+1] \setminus \{i\}} (x_i - x_j)$ has $k+1-i$ negative factors, thus its sign is $(-1)^{k+1-i}$, and the lemma follows. □

It remains to prove transitivity.

LEMMA 2.5. *Let $P = \{p_1, \ldots, p_N\}$ be a point set in k-general position. Then the 2-coloring of $(k+1)$-tuples $K \in \binom{P}{k+1}$ by their sign is transitive.*

Proof. We consider a $(k+2)$-tuple $L = \{p_1, \ldots, p_{k+2}\}$ with $\operatorname{sgn}\{p_1, \ldots, p_{k+1}\} = \operatorname{sgn}\{p_2, \ldots, p_{k+2}\} = +1$, and we fix $i \in \{2, \ldots, k+1\}$. Let $f_{i,k+2}$ be the polynomial of degree at most $k-1$ passing through $L \setminus \{p_i, p_{k+2}\}$, and similarly for $f_{1,k+2}$. Our goal is to show that $f_{i,k+2}(x_{k+2}) < y_{k+2}$, since this gives $\operatorname{sgn}(L \setminus \{p_i\}) = +1$ by Lemma 2.4.

Since $\operatorname{sgn}(L \setminus \{p_1\}) = +1$, we have $f_{1,k+2}(x_{k+2}) < y_{k+2}$ (Lemma 2.4 again), and so it suffices to prove $f_{i,k+2}(x_{k+2}) < f_{1,k+2}(x_{k+2})$.

Let us consider the polynomial $g := f_{1,k+2} - f_{i,k+2}$; as explained above, our goal is proving $\operatorname{sgn} g(x_{k+2}) = +1$. To this end, we first determine $\operatorname{sgn} g(x_1)$: We have $f_{i,k+2}(x_1) = y_1$ and $\operatorname{sgn}(y_1 - f_{1,k+2}(x_1)) = (-1)^k$ (using $\operatorname{sgn}(L \setminus \{p_1\}) = +1$ and Lemma 2.4). Hence $\operatorname{sgn} g(x_1) = (-1)^{k-1}$.

Next, we observe that g is a polynomial of degree at most $k-1$, and it vanishes at $x_2, \ldots, x_{i-1}, x_{i+1}, \ldots, x_{k+1}$. These are $k-1$ distinct values; thus, they include all roots of g, and each of them is a simple root. Consequently, g changes sign $(k-1)$-times between x_1 and x_{k+2}. Hence, finally, $\operatorname{sgn} g(x_{k+2}) = (-1)^{k-1} \operatorname{sgn} g(x_1) = +1$ as claimed. □

Rote's example. Fig. 1 shows a 6-point set $P = \{p_1, \ldots, p_6\}$ in 3-general position (no four points on a parabola). It is easy to check 3rd-order positivity using Lemma 2.4: By transitivity, it suffices to look at 4-tuples of consecutive points. For $p_1, \ldots, p_4$ we use the parabola through p_1, p_2, p_3 (which actually degenerates to the x-axis); for $p_2, \ldots, p_5$ we use the dashed parabola through p_2, p_3, p_4 (which is very close to the x-axis in the relevant region); and for $p_3, \ldots, p_6$, the parabola through p_4, p_5, p_6 (drawn full).

It remains to check that P does not lie on the graph of a function f with $f^{(3)} \geq 0$ everywhere. Assuming for contradiction that there is such an f, we consider the point $q := (x_0, f(x_0))$, where x_0 is such that the full parabola is below the x-axis at x_0. For the 4-tuple $\{p_1, p_2, p_3, q\}$ to be positive, q has to lie above the x-axis, but the 4-tuple $\{q, p_4, p_5, p_6\}$ is positive only if q lies below the parabola through p_4, p_5, p_6—a contradiction.

3. UPPER BOUNDS ON THE RAMSEY NUMBERS FOR TRANSITIVE COLORINGS

In this section we prove Theorem 1.4. As we mentioned in the remark following that theorem, it suffices to establish the case $k \geq 3$.

Thus, we want to prove that $R_{k+1}^{\text{trans}}(n) \leq \operatorname{twr}_k(C_k n)$ for all n and for every $k \geq 3$, with suitable constants C_k depending on k. As the base of the induction we use $R_3^{\text{trans}}(n) \leq 4^n$, which, as was remarked earlier, follows by imitating the proof of Theorem 1.2.

Thus, let $k \geq 3$ be fixed, let n be given, and let us set $M := R_k^{\text{trans}}(n)$. We will prove that

$$R_{k+1}^{\text{trans}}(n) \leq N := 2^{M^k}. \tag{1}$$

Theorem 1.4 then follows from this recurrence and from the fact that $2^{\operatorname{twr}_{k-1}(n)^k} \leq \operatorname{twr}_k(kn)$ for $k \geq 3$, which is easy to check.

To prove (1), we follow an inductive proofs of Ramsey's theorem going back to Erdős and Rado [ER52]. Let $\chi\colon \binom{[N]}{k+1} \to \{1, 2\}$ be an arbitrary transitive 2-coloring. We set $A_{k-1} := \{1, 2, \ldots, k-1\}$ and $X_{k-1} := [N] \setminus A_{k-1}$. For $i = k, k+1, \ldots, M$ we will inductively construct sets $A_i, X_i \subseteq [N]$ such that

(i) $A_i < X_i$ (i.e., all elements of A_i precede all elements of X_i);

(ii) $|A_i| = i$ and $|X_i| \geq |X_{i-1}|/2^{M^{k-1}}$; and

(iii) the color of a $(k+1)$-tuple whose first k elements all belong to A_i does not depend on its last element; in other words, for $K \in \binom{A_i}{k}$ and $x, y \in A_i \cup X_i$ with $K < \{x, y\}$, we have $\chi(K \cup \{x\}) = \chi(K \cup \{y\})$.

For the inductive step, suppose that A_i and X_i have already been constructed. We let x_i be the smallest element of X_i, we set $A_{i+1} := A_i \cup \{x_i\}$, and we write $X_i' := X_i \setminus \{x_i\}$.

Let us call two elements $x, y \in X_i'$ *equivalent* if we have, for every $K \in \binom{A_{i-1}}{k-1}$, $\chi(K \cup \{x_i, x\}) = \chi(K \cup \{x_i, y\})$. There are $\binom{i}{k-1}$ possible choices of K, and hence there are at most $2^{\binom{i}{k-1}} < 2^{M^{k-1}}$ equivalence classes. We choose $X_{i+1} \subseteq X_i'$ as the largest equivalence class. Then (i), (iii) obviously hold for A_{i+1} and X_{i+1}, and we have $|X_{i+1}| \geq (|X_i| - 1)/(2^{M^{k-1}} - 1) \geq |X_i|/2^{M^{k-1}}$ (since $i \leq M$ and thus we have $|X_i| \geq N/(2^{M^{k-1}})^{i-1} = 2^{M^k - (i-1)M^{k-1}} \geq 2^{M^{k-1}}$). This finishes the inductive construction of A_i and X_i.

In this way, we construct the sets $A := A_M$ and X_M (note that $|X_M| \geq 1$ by (ii)). Let x be the first element of X_M, and let us define a 2-coloring $\chi^*\colon \binom{A}{k} \to \{1, 2\}$ of the k-tuples of A by $\chi^*(K) := \chi(K \cup \{x\})$.

We claim that, crucially, χ^* is transitive (which is not entirely obvious). So we consider elements $a_1 < a_2 < \cdots < a_{k+1}$ of A, and we suppose that

$$\chi^*(\{a_1, \ldots, a_k\}) = \chi^*(\{a_2, \ldots, a_{k+1}\}) =: c.$$

We want to show that $\chi^*(\{a_1, \ldots, a_{k+1}\} \setminus \{a_i\}) = c$ for every $i = 2, 3, \ldots, k$. We have $c = \chi^*(\{a_1, \ldots, a_k\}) =$

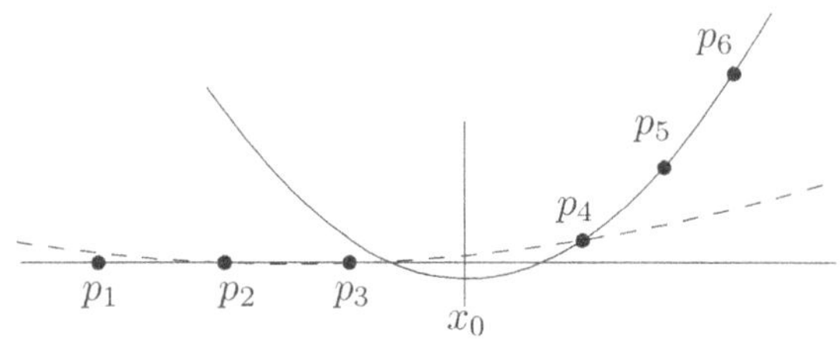

Figure 1: Rote's example: a 6-point 3rd-order positive set in 3-general position that does not lie on the graph of any function with nonnegative 3rd derivative.

$\chi(\{a_1,\dots,a_k,x\}) = \chi(\{a_1,\dots,a_{k+1}\})$ (by definition and by the independence of χ of the last element), and $c = \chi^*(\{a_2,\dots,a_{k+1}\}) = \chi(\{a_2,\dots,a_{k+1},x\})$. Next we use the transitivity of χ on the $(k+2)$-tuple $(a_1,\dots,a_{k+1},x)$, obtaining

$$\chi(\{a_1,\dots,a_{k+1},x\}\setminus\{a_i\}) = c = \chi^*(\{a_1,\dots,a_{k+1}\}\setminus\{a_i\})$$

as needed.

Now we can apply the inductive hypothesis to A, which yields an n-element subset of A homogeneous w.r.t. χ^*, and this subset is homogeneous w.r.t. χ as well, finishing the proof of Theorem 1.4. □

4. A LOWER BOUND FOR ES_3

Here we prove Theorem 1.5, a lower bound for $ES_3(2n+1)$. We proceed by induction on n; the goal is to construct a set P_n of $N := 2^{2^{n-1}}$ points with no $(2n+1)$-point third-order monotone subset. The induction starts for $n = 2$ with an arbitrary P_2 of size $2^{2^1} = 4$.

In the inductive step, given P_n, we will construct P_{n+1} so that $|P_{n+1}| = |P_n|^2$; then the bound on the size of P_n clearly holds.

We may assume that $P = P_n$ is in 3-general position (this can always be achieved by a small perturbation). By an affine transformation we also make sure that $P \subset [1,2]\times[0,1]$; or actually, $P \subset [1,1.9]\times[0,1]$ so that there is some room for perturbation. Moreover, there is a small $\delta > 0$ such that if P' is obtained from P by moving each point arbitrarily by at most δ, then P' is still in 3-general position, the order of the points of P' along the x-axis is the same as that for P, and the sign of every 4-tuple in P' is the same as the sign of the corresponding 4-tuple in P.

The construction. The construction of P_{n+1} from $P = P_n$ as above proceeds in the following steps.

1. We choose a sufficiently large number $A = A(P)$ (the requirements on it will be specified later), and we set $\varepsilon := 1/A^2$.

2. For every point $p \in P$, let Q_p be the image of P under the affine map that sends the square $[1,2]\times[0,1]$ to the axis-parallel rectangle of width ε, height ε^2, and with the lower left corner at p; see Fig. 2.

3. Let $\psi_p(x) = Ax^2 + C_p$ be a quadratic function, where A is as above and C_p is chosen so that $\psi_p(x(p)) = 0$ (where $x(p)$ is the x-coordinate of p). Let $\check{Q}_p$ be the set obtained by "adding ψ_p to Q_p", i.e., by shifting each point $(x,y) \in Q_p$ vertically upwards by $\psi_p(x)$.

We set $P_{n+1} := \bigcup_{p\in P}\check{Q}_p$. We call the $\check{Q}_p$ the *clusters* of P_{n+1}.

First we check that each cluster $\check{Q}_p$ lies close to p.

LEMMA 4.1. *Each $\check{Q}_p$ is contained in an $O(\sqrt{\varepsilon})$-neighborhood of p.*

Proof. Writing $p = (x_0, y_0)$, the set Q_p obviously lies in the 2ε-neighborhood of p, and the maximum amount by which a point of Q_p was translated upwards is at most

$$\psi_p(x_0+\varepsilon) = A\left((x_0+\varepsilon)^2 - x_0^2\right) = A(2x_0\varepsilon + \varepsilon^2) = O(\sqrt{\varepsilon}).$$

□

Here is a key property of the construction.

LEMMA 4.2 (SLOPE LEMMA). *Let λ be a parabola passing through three points of P_{n+1} that belong to three different clusters, or a line passing through two points of different clusters. Let μ be a parabola passing through three points of a single cluster $\check{Q}_p$, or a line passing through two such points. Then the maximum slope (first derivative) of λ on the interval $[1,2]$ is smaller than the minimum slope of μ on $[1,2]$, provided that A was chosen sufficiently large.*

Proof. Clearly, the maximum slope of any such λ can be bounded from above by some finite number depending only on P but not on A. Thus, it suffices to show that, with A large, for every μ as in the lemma, the minimum slope is bounded from below by A.

First let us assume that μ is a parabola passing through three points of $\check{Q}_p$, where $p = (x_0, y_0)$, let $\tilde{\mu}$ be the parabola passing through the corresponding three points of P, and let the equation of $\tilde{\mu}$ be $y = ax^2 + bx + c$.

By the construction of $\check{Q}_p$, the affine map transforming P to Q_p sends a point with coordinates (x,y) to the point $(\varepsilon(x-1)+x_0, \varepsilon^2 y + y_0)$. Calculation shows that the image of $\tilde{\mu}$ under this affine map has the equation $y = ax^2 + (2a\varepsilon + b\varepsilon - 2ax_0)x + c'$, where the value of the absolute term c' need not be calculated since it doesn't matter. Hence the minimum slope of this curve on $[1,2]$ is bounded from below by $-(8|a| + 4|a|\varepsilon + 2|b|\varepsilon + 8|a|)$. Finally, μ is obtained by adding $\psi_p(x) = Ax^2 + C_p$ to this curve, and the minimum slope of ψ_p on $[1,2]$ is at least $2A$.

Next, let μ be a line passing through two points $q, r \in \check{Q}_p$. Let us choose another point $s \in \check{Q}_p$ and consider the parabola μ' through q, r, s. By the Mean Value Theorem, the slope of μ equals the slope of μ' at some point between q and r, and the latter is at least A by the above. The lemma is proved. □

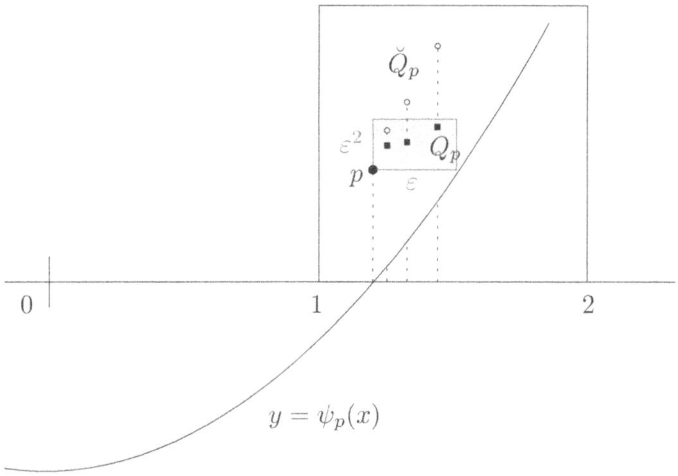

Figure 2: A schematic illustration of the construction of P_{n+1}.

Let $K = \{p_1, p_2, p_3, p_4\} \subseteq P_{n+1}$ be a 4-tuple, $p_i = (x_i, y_i)$, $x_1 < \cdots < x_4$. We assign a *type* to K, which is an ordered partition of 4 given by the distribution of K among the clusters; for example, K has type $1+1+2$ if the first point p_1 lies in some $\check{Q}_p$, p_2 lies in $\check{Q}_{p'}$ for $p' \neq p$, and $p_3, p_4 \in \check{Q}_{p''}$, $p'' \neq p, p'$.

The next lemma shows that the sign K is determined by its type. We provide a complete classification, although we will not use all of the types in the subsequent proof.

LEMMA 4.3. *Let $K = \{p_1, p_2, p_3, p_4\} \subseteq P_{n+1}$ be a 4-tuple. If K is of type $1+1+1+1$ or 4, then the sign of K is the same as that of the corresponding 4-tuple in P. Otherwise, the sign of K is determined by its type as follows:*

- *for types $3+1$ and $1+3$ it is -1;*
- *for types $1+1+2$ and $2+1+1$ it is $+1$;*
- *for type $1+2+1$ it is -1; and*
- *for type $2+2$ it is $+1$.*

Proof. Since the transformation that converts P into $\check{Q}_p$ preserves the types of 4-tuples, the statement for type 4 is clear. The statement for type $1+1+1+1$ follows since, by Lemma 4.1, K is obtained by a sufficiently small perturbation of the corresponding 4-tuple in P (this gives one of the lower bounds on A, since we need the bound in Lemma 4.1 to be smaller than the δ considered at the beginning of our description of the construction).

The statements for the remaining types are obtained by simple application of the slope lemma (Lemma 4.2) together with Lemma 2.4. Namely, for type $3+1$, we get that the parabola through p_1, p_2, p_3 lies above p_4 (by comparing its slope to the slope of the line p_3p_4); see Fig. 3. For type $1+3$ we similarly get that p_1 lies above the parabola through p_2, p_3, p_4, and so the sign is -1 in both of these cases.

For type $1+1+2$, the segment p_3p_4 is steeper than the parabola through $p_1p_2p_3$, and so the sign is $+1$. Similarly for type $2+1+1$ we get that p_1 lies below the parabola through p_2, p_3, p_4, which again gives sign $+1$. For type $1+2+1$, p_3 lies above the parabola through p_1, p_2, p_4, giving sign -1. Finally, for type $2+2$, the segment p_1p_2 is steeper than p_2p_3, thus the parabola through p_1, p_2, p_3 is concave, and hence its slope at p_3 and after it is no larger than the slope of the segment p_2p_3. Thus, p_4 lies above this parabola and the sign is $+1$ as claimed. □

Finishing the proof of Theorem 1.5. It remains to show that P_{n+1} contains no $(2n+3)$-point third-order monotone subset.

For contradiction, suppose that $M \subseteq P_{n+1}$ is such a $(2n+3)$-point subset. Let $2n+3 = n_1 + n_2 + \cdots + n_s$ be the type of M (i.e., M has $n_i \geq 1$ points in the ith leftmost cluster it intersects). By the inductive assumption we have $s \leq 2n$ and $n_i \leq 2n$ for all i.

Let $n_a = \max_i n_i$ and $n_b = \max_{i \neq a} n_i$ be the two largest among the n_i. For convenience, let us assume $a < b$; the case $a > b$ is handled symmetrically. We distinguish three cases.

First, if $n_a \geq 3$ and $n_b \geq 2$, then we can select 4-tuples of types $3+1$ and $2+2$ from the corresponding two clusters, which have different signs, and so M is not homogeneous.

Second, if $n_a \geq 3$ and $n_b = 1$, then we have at least three n_i equal to 1 (since $n_a \leq 2n$), and at least two of them lie on the same side of the cluster corresponding to n_a, say to the right of it. Then we can select 4-tuples of types $3+1$ and $2+1+1$, again of opposite signs.

Third, if $n_a = 2$, then there are at least two other clusters of size 2. From these three 2-element clusters, we can select 4-tuples of types 2+2 and 1+2+1, again of opposite signs.

This exhausts all possibilities ($n_a = 1$ cannot happen, because $s \leq 2n$), and Theorem 1.5 is proved. □

5. ORDER TYPES AND ONE-SIDED SETS OF HYPERPLANES

First we substantiate the two claims made above Proposition 1.6, concerning OT_1 and OT_2. For $d = 1$, an order-type homogeneous sequence in $\mathbb{R}^1$ is just a monotone sequence of real numbers, so $\mathrm{OT}_1(n) = (n-1)^2 + 1$ by Theorem 1.1.

In a similar spirit, it is easy to check that a planar order-type homogeneous sequence corresponds to the vertices of a convex n-gon, enumerated in a clockwise or counterclock-

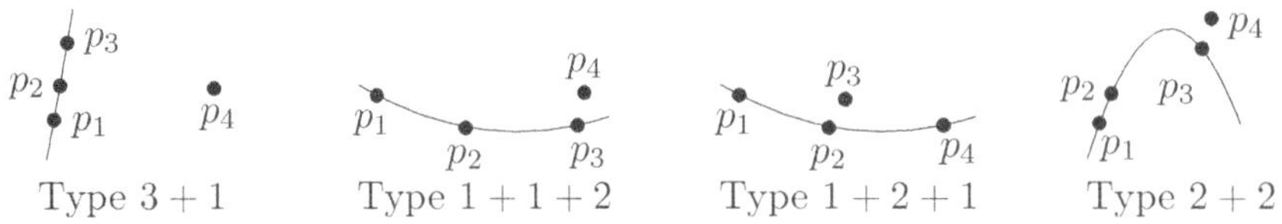

Figure 3: Determining the signs of 4-tuples by type.

wise order. Thus, $\mathrm{OT}_2(n) \geq \mathrm{ES}_2(\lceil n/2 \rceil) = 2^{\Omega(n)}$. On the other hand, given any N-point sequence, we can first select a subsequence of $\lceil \sqrt{N} \rceil$ points with increasing or decreasing x-coordinates, and then we select a convex or concave configuration from it. Thus, by Theorem 1.2, we have $\mathrm{OT}_2(n) = 2^{O(n)}$.

Proof of Proposition 1.6. For a point $p = (x, y) \in \mathbb{R}^2$, we define the point $\tilde{p} := (x, x^2, \ldots, x^{d-1}, y) \in \mathbb{R}^d$.

To prove that $\mathrm{ES}_d(n) \leq \mathrm{OT}_d(n)$, we consider a set $P = \{p_1, \ldots, p_N\} \subset \mathbb{R}^2$ in d-general position, $p_i = (x_i, y_i)$, where $N = \mathrm{ES}_d(n) - 1$ and $x_1 < \cdots < x_N$, with no dth-order monotone subset of n points. It suffices to prove that the sequence $\tilde{P} := (\tilde{p}_1, \tilde{p}_2, \ldots, \tilde{p}_N)$ has no n-point order-type homogeneous subsequence. This follows from the next lemma.

LEMMA 5.1. *For every $(d+1)$-tuple $(p_1, \ldots, p_{d+1})$ of points in $\mathbb{R}^2$, $x_1 < \cdots < x_{d+1}$, we have $\mathrm{sgn}(\{p_1, \ldots, p_{d+1}\}) = \mathrm{sgn}\det M(\tilde{p}_1, \tilde{p}_2, \ldots, \tilde{p}_{d+1})$, where $M(q_1, \ldots, q_{d+1})$ is the matrix from the definition of order type above Proposition 1.6.*

Proof. By Lemma 2.2 and Corollary 2.3, the sign of $\{p_1, \ldots, p_{d+1}\}$ equals the sign of the coefficient a_d of the unique polynomial $f(x) = \sum_{j=0}^{d} a_j x^j$ of degree at most d whose graph passes through the points $p_1, \ldots, p_{d+1}$.

The vector $a = (a_0, \ldots, a_d)$ can be expressed as the solution of the linear system $Va = y$, where $y = (y_1, \ldots, y_{d+1})$ and V is the *Vandermonde matrix* with $v_{ij} = x_i^{j-1}$, $i, j = 1, 2, \ldots, d+1$. By Cramer's rule, we obtain

$$a_d = \frac{\det W}{\det V},$$

where W stands for the matrix V with the last column replaced with the vector y. As is well known, $\det V = \prod_{1 \leq i < j \leq d+1} (x_j - x_i)$, and since $x_1 < \cdots < x_{d+1}$, we have $\det V > 0$. Thus, $\mathrm{sgn}\, a_d = \mathrm{sgn} \det W$. Finally, we have

$$W = \begin{pmatrix} 1 & x_1 & x_1^2 & \ldots & x_1^{d-1} & y_1 \\ \vdots & \vdots & \vdots & \vdots & \vdots & \vdots \\ 1 & x_{d+1} & x_{d+1}^2 & \ldots & x_{d+1}^{d-1} & y_{d+1} \end{pmatrix}$$
$$= M(\tilde{p}_1, \tilde{p}_2, \ldots, \tilde{p}_{d+1})^T.$$

The lemma follows, and Proposition 1.6 is proved. □

Proof of Proposition 1.7. The proof is very similar to the previous one. This time we start with a set $P = \{p_1, \ldots, p_N\} \subset \mathbb{R}^2$ in $(d-1)$-general position, $p_i = (x_i, y_i)$, where $N = \mathrm{ES}_{d-1}(n) - 1$ and $x_1 < \cdots < x_N$, with no $(d-1)$th-order monotone subset of n points. We define a collection $H = \{h_1, \ldots, h_N\}$ of N hyperplanes in $\mathbb{R}^d$, where h_i is given by

$$h_i = \left\{ (\xi_1, \ldots, \xi_d) \in \mathbb{R}^d : \sum_{j=1}^{d} x_i^{j-1} \xi_j = y_i \right\}.$$

The intersection point $\xi = (\xi_1, \ldots, \xi_d)$ of, say, $h_1, \ldots, h_d$ is the solution of the linear system $V\xi = y$, where V is the $d \times d$ Vandermonde matrix this time, $v_{ij} = x_i^{j-1}$. Cramer's rule then gives that the dth coordinate ξ_d, whose sign we are interested in, equals $(\det W)/(\det V)$, where W is obtained from V by replacing the last column with y.

As we saw in the proof of Proposition 1.6, $(\det W)/(\det V)$ also expresses the leading coefficient in the polynomial of degree $d-1$ passing through $p_1, \ldots, p_d$, and thus its sign equals $\mathrm{sgn}\, \Delta_{d-1}(p_1, \ldots, p_d)$. It follows that one-sided subsets of H precisely correspond to $(d-1)$st-order monotone subsets in P, and the proposition is proved. □

Acknowledgment

We would like to thank János Pach for kindly discussing some of the results of Fox et al. [FPSS11] with us. We also thank Günter Rote for informing us about about his refutation of our conjecture and for permission to present it in this paper.

6. REFERENCES

[CFS11] D. Conlon, J. Fox, and B. Sudakov. An improved bound for the stepping-up lemma. *Discrete Applied Mathematics*, 2011. In press.

[DL11] V. Dujmović and S. Langerman. A center transversal theorem for hyperplanes and applications to graph drawing. In *Proc. 27th ACM Symposium on Computational Geometry*, pages 117–124, 2011. Full version arXiv:1012.0548.

[ER52] P. Erdős and R. Rado. Combinatorial theorems on classifications of subsets of a given set. *Proc. London Math. Soc.*, 3:417–439, 1952.

[ES35] P. Erdős and G. Szekeres. A combinatorial problem in geometry. *Compositio Math.*, 2:463–470, 1935.

[FPSS11] J. Fox, J. Pach, B. Sudakov, and A. Suk. Erdős–Szekeres-type theorem for monotone paths and convex bodies. Arxiv preprint 1105.2097v1, 2011. *Proc. London Math. Soc.*, in press.

[GP93] J.E. Goodman and R. Pollack. Allowable sequences and order types in discrete and computational geometry. In J. Pach, editor, *New Trends in Discrete and Computational Geometry*, volume 10 of *Algorithms and Combinatorics*, pages 103–134. Springer, Berlin etc., 1993.

[GRS90] R.L. Graham, B.L. Rothschild, and J. Spencer. *Ramsey Theory*. J. Wiley & Sons, New York, 1990.

[Mat02] J. Matoušek. *Lectures on Discrete Geometry*. Springer, New York, 2002.

[MS00] W. Morris and V. Soltan. The Erdős–Szekeres problem on points in convex position—a survey. *Bull. Amer. Math. Soc., New Ser.*, 37(4):437–458, 2000.

[MW92] J. Matoušek and Emo Welzl. Good splitters for counting points in triangles. *J. Algorithms*, 13:307–319, 1992.

[Phi03] George M. Phillips. *Interpolation and approximation by polynomials.* Springer, Berlin etc., 2003.

[Rot12] G. Rote. Private communication, February 2012.

[Ste95] M. J. Steele. Variations on the monotone subsequence theme of Erdős and Szekeres. In *D. Aldous et al., editors, Discrete Probability and Algorithms, IMA Volumes in Mathematics and its Applications 72*, pages 111–131. Springer, Berlin etc., 1995.

[Val04] P. Valtr. Open caps and cups in planar point sets. *Discr. Comput. Geom.*, 37:365–567, 2004.

Add Isotropic Gaussian Kernels at Own Risk: More and More Resilient Modes in Higher Dimensions

Herbert Edelsbrunner*
IST Austria
Klosterneuburg, Austria
edels@ist.ac.at

Brittany Terese Fasy†
Dept of Computer Science
Duke University
Durham, North Carolina
brittany@cs.duke.edu

Günter Rote
Institut für Informatik
Freie Universität Berlin
Berlin, Germany
rote@inf.fu-berlin.de

ABSTRACT

It has been an open question whether the sum of finitely many isotropic Gaussian kernels in $n \geq 2$ dimensions can have more modes than kernels, until in 2003 Carreira-Perpiñán and Williams exhibited $n+1$ isotropic Gaussian kernels in $\mathbb{R}^n$ with $n+2$ modes. We give a detailed analysis of this example, showing that it has exponentially many critical points and that the resilience of the extra mode grows like $\sqrt{n}$. In addition, we exhibit finite configurations of isotropic Gaussian kernels with superlinearly many modes.

Categories and Subject Descriptors

I.4.7 [**Image Processing and Computer Vision**]: Feature measurement—*size and shape*; G.3 [**Probability and Statistics**]: Distribution functions; I.6 [**Simulation and Modeling**]: General

Keywords

Unit Gaussian kernels, diffusion, convolution, standard simplex, critical points, modes, resilience.

1. INTRODUCTION

The diffusion of chemical substances, such as hormones, and of physical quantities, such as temperature, is a general phenomenon. Assuming a uniform medium, the process is described by the solution to the heat equation. In Euclidean space, solving this equation is synonymous to convolving with a Gaussian kernel. This is also a popular computational method, in particular in computer vision, where the 1-parameter family of convolutions of a given image is known as its *scale space*; see [13, 18]. A one-dimensional Gaussian kernel is known as a *normal density function* in probability [8].

We are interested in the quantitative analysis of diffusion and Gaussian convolution. In particular, we study the evolution of the critical points of a function that is convolved with a progressively wider Gaussian kernel. If the function is 1-dimensional, from $\mathbb{R}$ to $\mathbb{R}$, then Gaussian convolution does not create new critical points [1, 9, 16, 19]. As a consequence, the diffusion of m point masses (a sum of m Dirac delta functions) cannot have more than m modes (local maxima); see [2, 4, 17]. For two- or higher-dimensional functions, this is no longer true; see [12] for a two-dimensional function for which diffusion temporarily increase the number of modes and [7, 15] for a mathematical analysis of the unfolding events that cause this effect. It has been observed that these events are rare in practice [10, 11] and it has been confirmed that the ability to create critical points with non-negligible persistence deteriorates rapidly [6]. It is also known that $n+1$ point masses can be arranged in $\mathbb{R}^n$ so that diffusion creates $n+2$ modes during a non-empty time interval; see [5].

The contribution of this paper is a strengthening of the cautionary voice on using Gaussian convolution in dimensions beyond 1. In particular, we give a detailed analysis of the sum of $n+1$ identical isotropic Gaussian kernels placed at the vertices of a regular n-simplex in $\mathbb{R}^n$. We prove that all critical points lie on the symmetry axes of the n-simplex, and we characterize their indices, confirming the $(n+2)$-nd mode at the barycenter as the sole extra mode. While the extra mode seems fragile, we show that the interval of widths during which it exists grows like the square root of the dimension. It thus seems likely that the phenomenon of extra modes is more prevalent in higher dimensions. Providing additional evidence, we construct finite configurations of isotropic Gaussian kernels with superlinearly many modes.

Outline. Section 2 provides background on Gaussian kernels and the geometry of regular simplices. Section 3 analyzes the sum of kernels placed at the vertices of a regular simplex, characterizes its critical points, estimates the resilience of the extra mode, and exhibits configurations with superlinearly many modes. Section 4 concludes this paper.

2. BACKGROUND

Our results depend on 1-dimensional Gaussian kernels and n-dimensional regular simplices. We study these two topics in two subsections.

2.1 Curve Analysis

In this subsection, we introduce Gaussian kernels and discuss some of their fundamental properties.

Gaussian kernels and derivatives. We call a real-valued function of the form $g(x) = We^{-C\|x-z\|^2}$ an $(n+1)$*-dimensional Gaussian kernel*, where W and C are real constants, $z \in \mathbb{R}^{n+1}$ is a point and $\|x - z\|$ denotes the Euclidean distance between the two points. We call the kernel *normalized* if it integrates to 1, in which

*... and Departments of Computer Science and of Mathematics, Duke University, Durham, North Carolina; Delaunay Lab of Discrete and Computational Geometry, Yaroslavl' State University, Russia; Geomagic, Research Triangle Park, North Carolina.

†... and IST Austria (Institute of Science and Technology Austria), Klosterneuburg, Austria.

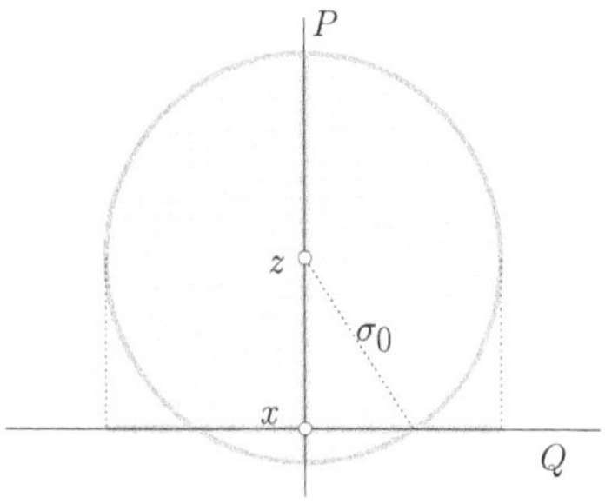

Figure 1: A unit Gaussian kernel with center z in $\mathbb{R}^2$ represented by the disk with radius $\sigma_0 = 1/\sqrt{2\pi}$. The lines P and Q define 1-dimensional sections.

case it can be written as:

$$g(x) \quad = \quad \frac{1}{(2\pi\sigma^2)^{\frac{n+1}{2}}} \cdot e^{-\frac{\|x-z\|^2}{2\sigma^2}}.$$

We call $z \in \mathbb{R}^{n+1}$ the *center* or *mean* and $\sigma > 0$ the *width* (or *standard deviation* if $n = 0$). Finally, g is *unit* if it is normalized with height 1. Independent of the dimension, the width and the formula of the unit Gaussian kernel are:

$$\begin{aligned} \sigma_0 &= 1/\sqrt{2\pi}, \\ g(x) &= e^{-\pi\|x-z\|^2}, \end{aligned}$$

which will simplify our computations. For the case $n = 0$, we can write formulas of its first two derivatives:

$$g'(x) \quad = \quad [-2\pi x] \cdot g(x), \tag{1}$$

$$g''(x) \quad = \quad [4\pi^2x^2 - 2\pi] \cdot g(x). \tag{2}$$

Note that g'' is negative in the interior of $[-\sigma_0, \sigma_0]$ and positive outside this interval.

We write g_z for the unit Gaussian kernel with center z. It can be transformed into every other Gaussian kernel by a translation (to change the center), a scaling of the domain (to change the width), and a scaling of the range (to change the height). A Gaussian kernel is concave at every point in the interior of the closed ball with center z and radius σ, and it fails to be concave at every point outside this ball. This ball is a convenient illustration of the kernel; see Figure 1.

Balanced sums. Consider the sum of two unit Gaussian kernels. For symmetry, we choose their centers at distance $z \geq 0$ to the left and right of the origin. As proven in [3], $G = g_{-z} + g_z$ has either 1 or 3 critical points and no other number is possible. More specifically, G has 1 maximum iff $z \leq \sigma_0$ and G has 2 maxima and 1 minimum iff $z > \sigma_0$. We present our own proof of this result, as we need the concepts it uses.

Since $G' = g'_{-z} + g'_z$, a point $x \in \mathbb{R}$ is a critical point of G iff the graphs of $p = -g'_{-z}$ and $q = g'_z$ intersect above x. Since g' is an odd function, the number of intersections (counted with multiplicities) must be odd, and it is visually plausible that the number can only be 1 or 3. To be sure, we introduce the *ratio function*, $r = \frac{p}{q} - 1$, with formula

$$r(x) \quad = \quad \frac{z+x}{z-x} \cdot e^{-4\pi z \cdot x} - 1; \tag{3}$$

see Figure 2. Setting $r(x) = 0$ gives us the intersections of p and q and thus the critical points of G. For simple geometric reasons, the roots of r are necessarily in $[-z, z]$. Independent of z, we have $r(0) = 0$. To see whether there are additional roots, we take

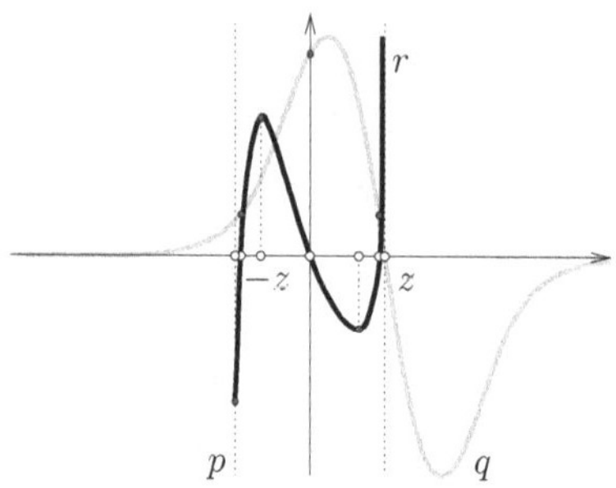

Figure 2: The graphs of the derivatives of g_{-z} and of g_z intersect above the roots of the (black) ratio function.

the derivative:

$$r'(x) \quad = \quad \frac{2z - 4\pi z(z^2 - x^2)}{(z-x)^2} \cdot e^{-4\pi z \cdot x}.$$

Setting $r'(x) = 0$, we have $x^2 = z^2 - \frac{1}{2\pi}$, which has two real solutions if $z > \sigma_0$ and no solution if $z < \sigma_0$. Observing that $r(-z) = -1$ and $r(z)$ goes to infinity as x approaches z, we conclude that r has 1 root if $z \leq \sigma_0$ and 3 roots if $z > \sigma_0$. As anticipated, G has 1 critical point if $z \leq \sigma_0$ and it has 3 non-degenerate critical points if $z > \sigma_0$.

Unbalanced sums. Next, we study sums $G_w = g_{-z} + wg_z$, where $w \geq 0$ is the *weight* of the second term. The number of critical points of G_w is at most 3, but in contrast to the balanced case, it can also be 2, as we will see. More specifically, [2] gives necessary and sufficient conditions for all three cases (1, 2, or 3 critical points), but they are not as easy to state as in the balanced case. As before, we present our own proof since we need the concepts it uses.

Recalling that the critical points of G_w correspond to the intersections between the graphs of $p = -g'_{-z}$ and $wq = wg'_z$, we

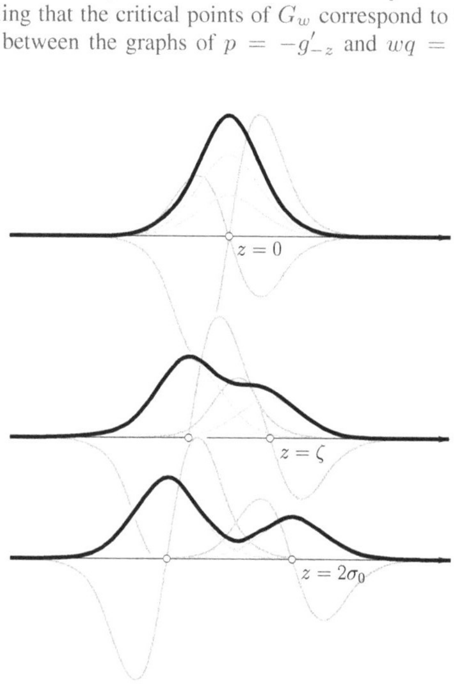

Figure 3: From top to bottom: the sum of a unit kernel and half a unit kernel. The kernels are blue, their sum is black, their derivatives (plus-minus) are pink, and the trajectories of the critical points, drawn over the interval from 0 to $2\sigma_0$, are yellow.

consider their ratio function, which is

$$r_w = \frac{p}{wq} - 1 = \frac{r}{w} - \left(1 - \frac{1}{w}\right).$$

Its derivative is $r_w' = \frac{r'}{w}$, which has at most 2 roots. Similarly, $r_w(-z) = -1$ and $r_w(z)$ goes to infinity as x approaches z, so we see that r_w has at most 3 roots. It follows that G_w has at most 3 critical points, just like G.

A new phenomenon is the possibility of 2 critical points. To see when this case arises, we set $w = r(x) + 1$ and note that for this choice of weight $r_w(x) = 0$. In words, p and wq intersect above x and, equivalently, x is a critical point of G_w. If x has the additional property of being critical for r, the intersection between p and wq is degenerate. As computed above, the critical points of r_w are given by $x^2 = z^2 - \sigma_0^2$. Let $x_1 = -\sqrt{z^2 - \sigma_0^2}$ and $x_2 = \sqrt{z^2 - \sigma_0^2}$ be the two solutions, and note that x_1 gives a weight $w_1 = r(x_1) + 1$ that is larger than 1, while x_2 gives a weight $w_2 = r(x_2) + 1$ between 0 and 1. We call w_1 and w_2 the *transition weights* for z, remembering that they exist iff $z \geq \sigma_0$.

We illustrate the case analysis by moving the centers and observing how the shape of G_w changes. We get qualitatively the same behavior for every positive weight. Fixing the weight to $w = \frac{1}{2}$, let 2ζ be the distance between the two centers for which w is a transition weight. Starting with $-z = z = 0$, we see the evolution sketched in Figure 3. For $z < \zeta$, the function G_w has only one maximum. At $z = \zeta$, we have a degenerate critical point forming a shoulder on the right. For $z > \zeta$, this shoulder turns into a min-max pair. This is a generic event in the 1-parameter evolution of a Morse function, known as an anti-cancellation.

2.2 Simplex Design

In this subsection, we design a sum of Gaussian kernels in $\mathbb{R}^{n+1}$ that has the symmetry group of the regular n-simplex. We begin with a geometric study of the simplex, whose shape properties will play a central role in our design.

Standard simplex. A convenient model is the *standard n-simplex*, defined as the convex hull of the $n+1$ unit coordinate vectors in $\mathbb{R}^{n+1}$: $\Delta^n = \text{conv}\{e_0, e_1, \ldots, e_n\}$. Each subset of $k+1$ vectors defines a *k-face* of Δ^n, which is itself a standard k-simplex. The *barycenter* of Δ^n is the point whose $n+1$ coordinates are all equal to $\frac{1}{n+1}$. Let $0 \leq k \leq \ell$ with $k + \ell = n - 1$ and consider a k-face and the unique disjoint ℓ-face. Their barycenters use complementary subsets of the $n+1$ coordinates, which makes it easy to compute the distance between them as

$$D_{k,\ell} = \sqrt{\frac{1}{k+1} + \frac{1}{\ell+1}}. \qquad (4)$$

For example, the height of the n-simplex, which is defined as the distance between a vertex to the opposite $(n-1)$-face, is $D_{0,n-1} = \sqrt{(n+1)/n}$. Similarly, we can compute the radius of the circumsphere:

$$R_n = \sqrt{\frac{n}{n+1}}. \qquad (5)$$

Sections. A *k-section* of a Gaussian kernel is the restriction to a k-dimensional plane P. Assuming a unit Gaussian kernel, we can write this as $g_z|_P(x) = g_z(y) \cdot g_y(x)$, where y is the orthogonal projection of z onto P and g_y is the k-dimensional unit Gaussian kernel with center $y \in P$. We call $g_z(y)$ the *weight*, noting that it is equal to the integral of g_z over P and we call $g_z|_P$ a *weighted* unit Gaussian kernel. Importantly, we see that the section has the same width as the original kernel. It is a unit kernel itself iff P passes through z. In this case, $g_z|_P(x)$ is the weight of the $(n-k)$-section defined by the plane Q that intersects P orthogonally at x; see Figure 1. Iterating this construction, we can write the $(n+1)$-dimensional unit Gaussian kernel as a product of 1-dimensional kernels:

$$g_z(x) = \prod_{i=0}^{n} g(x_i - z_i), \qquad (6)$$

where the x_i and z_i are the Cartesian coordinates of x and z. In words, the high-dimensional unit kernel can be *separated* into mutually orthogonal one-dimensional unit kernels.

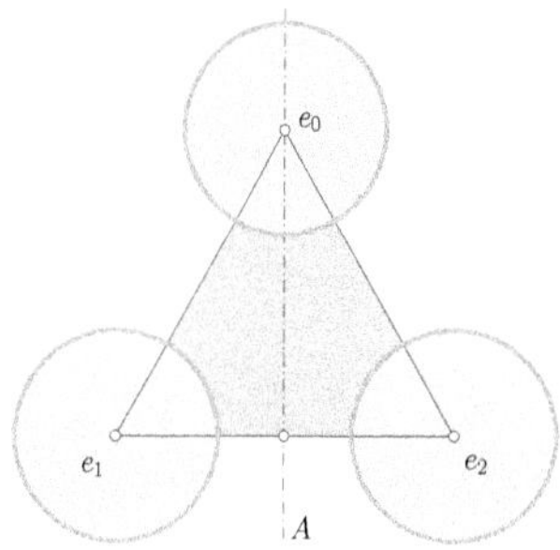

Figure 4: A 2-design, which is the sum of 3 unit Gaussian kernels placed at the vertices of an equilateral triangle.

Standard design. We turn Δ^n into a function by placing a unit Gaussian kernel at every vertex. Writing g_i for the kernel with center e_i, we get $G = g_0 + g_1 + \ldots + g_n$, which we call an *$n$-design*; see Figure 4. Its symmetry group is that of the n-simplex with the additional reflection across the n-dimensional plane spanned by the n-simplex. It is therefore isomorphic to $\Sigma_{n+1} \oplus \mathbb{S}^0$, where Σ_{n+1} is the symmetric group on $n+1$ elements. To argue about the symmetries, we use lines that connect barycenters of complementary faces of the n-simplex. We call these lines *axes*. Suppose for example that $0 \leq k \leq \ell$ are integers with $k + \ell = n - 1$, and that A is the axis that connects the barycenter of the k-face spanned by e_0 to e_k with the barycenter of the ℓ-face spanned by e_{k+1} to e_n. Except for the barycenter of Δ^n, every point x of A has two distinct barycentric coordinates, one with multiplicity $k+1$ and the other with multiplicity $\ell + 1$. The orbit of x has therefore size $\binom{n+1}{k+1}$. Recall that the section defined by the axis is the restriction of the function to the line: $G|_A$. The restrictions of $g_0, g_1, \ldots, g_k$ to A are all identical, namely a weighted 1-dimensional unit Gaussian kernel whose center is the barycenter of the k-face. Similarly, the restrictions of $g_{k+1}, \ldots, g_{n-1}, g_n$ are all identical, and we can write the 1-section as the sum of two kernels:

$$G|_A = (k+1) \cdot g_0|_A + (\ell + 1) \cdot g_n|_A, \qquad (7)$$

which are one-dimensional weighted unit Gaussian kernels, with weights $(k+1)g(R_k)$ and $(\ell+1)g(R_\ell)$, and distance $D_{k,\ell}$ between their centers.

We are interested in changing the widths of the $(n+1)$-dimensional kernels uniformly. Equivalently, we scale the n-simplex by moving the centers of the unit Gaussian kernels closer to or further from each other, without changing their widths and heights. To do this, we introduce the *scaled n-design*, $G_s = g_{se_0} + g_{se_1} + \ldots + g_{se_n}$. Here, we call s the *scale factor*, and we write $s\Delta^n$ for the *scaled n-simplex* whose vertices are the se_i. We are interested in the evolution of the critical points in the 1-parameter family of scaled n-design $G_s : \mathbb{R}^{n+1} \to \mathbb{R}$, as s goes from zero to infinity.

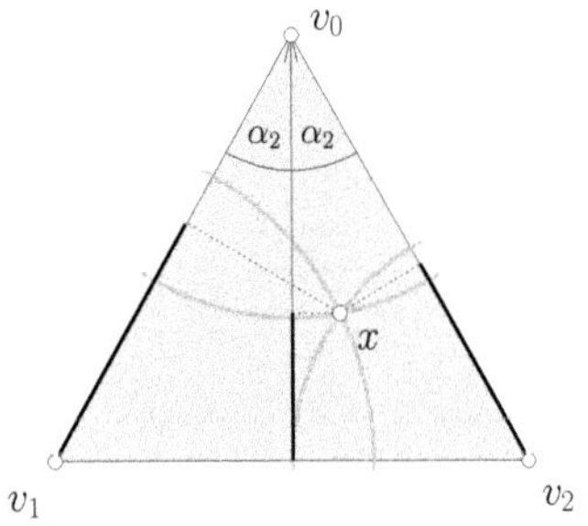

Figure 5: The radii of the three circles are the distance coordinates of the point x. The orthogonal projections onto the vertical axis and the left and right edges of the triangle give x_0, x_{01}, and x_{02}.

3. ANALYSIS

We begin this section by proving that all critical points lie on the axes of the n-simplex. Thereafter, we analyze each axis, characterizing for which scales we see 1, 2, or 3 critical points. To decide which of the maxima are modes, we analyze the n-sections orthogonal to the axes. As it turns out, all modes lie on axes that pass through vertices of the n-simplex. Most interesting is the critical point at the barycenter, which changes from unique mode during an initial interval of scales, to $(n+2)$-nd mode during a non-empty intermediate interval, to a saddle of index one during a final interval. We call the length of the intermediate interval the resilience of the extra mode and show that it grows like the square root of the dimension. Finally, we construct sums of isotropic Gaussian kernels with a superlinear number of modes.

3.1 Lines of Critical Points

In this subsection, we note that all critical points of a scaled n-design lie on the axes of the scaled n-simplex. We begin by introducing coordinates that are more natural for the n-design, and we show how they relate to the barycentric coordinates.

Distance coordinates. Write $v_i = se_i$, for $0 \leq i \leq n$, and let x be a point of the corresponding scaled n-simplex $s\Delta^n$. Setting $r_i = \|x - v_i\|$, we note that x is uniquely defined by the vector of $n+1$ distances since x lies on the hyperplane spanned by $\{v_i\}$. We express this by writing $x = (r_0, r_1, \ldots, r_n)_D$, and by calling the r_i the *distance coordinates* of x. Recall that $(x_0, x_1, \ldots, x_n)_B$ is the representation of the same point in barycentric coordinates. We are interested in computing the barycentric from the distance coordinates via the Coordinate Transformation below.

COORDINATE TRANSFORMATION. *For $0 \leq i \leq n$, the ith barycentric coordinate is given by:*

$$x_i = \frac{1}{n+1} + \frac{1}{2(n+1)s^2}\left(\sum_{j=0}^{n} r_j^2 - (n+1)r_i^2\right). \quad (8)$$

PROOF. Let $(r_0, r_1, \ldots, r_n)_D$ be the distance coordinates of a point x in the scaled n-simplex $s\Delta^n$. Let $i \neq j$ and consider the edge connecting v_i with v_j, recalling that v_i and v_j are two vertices of $s\Delta^n$. The length of the edge is $s\sqrt{2}$. Let x be a point of $s\Delta^n$ and set x_{ij} to the distance between v_j and the orthogonal projection of x onto the edge, normalized by dividing with $s\sqrt{2}$; see Figure 5. We first show that

$$x_{ij} = \frac{1}{2} + \frac{1}{4s^2}(r_j^2 - r_i^2). \quad (9)$$

Indeed, if $x = (1-t)v_j + tv_i$, then we have $r_j = s\sqrt{2}t$ and $r_i = s\sqrt{2}(1-t)$. Furthermore, $x_{ij} = t$, which agrees with the equation we get by plugging the values of r_i and r_j into (9). Realizing that $r_j^2 - r_i^2$ is constant along hyperplanes orthogonal to the edge, we get (9) for all points of the n-simplex.

For the next step, let b_i be the barycenter of the $(n-1)$-face complementary to v_i and y be the orthogonal projection of x onto the edge between v_i and v_j. Set α_n to the angle between the edges that connect v_i to v_j and to b_i. Because $s\Delta^n$ is regular, this angle does not depend on the choice of i and j. Suppose x lies on the latter edge, which connects v_i and b_i. Then $x = (1-x_i)b_i + x_i v_i$ and we have two expressions for $\cos\alpha_n$. Setting these two expression equal we arrive at

$$\frac{D_{0,n-1}}{\sqrt{2}} = \frac{\sqrt{2}(1-x_{ij})}{D_{0,n-1}(1-x_i)},$$

for every $0 \leq j \leq n$ and $j \neq i$. Adding the n equations gives

$$nD_{0,n-1}^2(1-x_i) = 2n - \sum_{i \neq j} x_{ij}. \quad (10)$$

Similar to before, we notice that the two sides of the equation are constant along hyperplanes orthogonal to the axis defined by v_i and b_i. Hence, (10) holds for all points x of the scaled n-simplex. It remains to plug (4) and (9) into (10), which gives

$$(n+1)(1-x_i) = n + \frac{1}{2s^2}\left(nr_i^2 - \sum_{i \neq j} r_j^2\right).$$

The equation simplifies to the claimed equation. □

Non-zero gradients. Recall that $G_s : \mathbb{R}^{n+1} \to \mathbb{R}$ is the scaled n-design formed by taking the sum of the $n+1$ unit Gaussian kernels whose centers are the vertices of $s\Delta^n$. We use the Coordinate Transformation to show that G_s has no critical points away from the axes of the scaled n-simplex:

AXES LEMMA. *Every critical point of G_s lies on an axis of the scaled n-simplex $s\Delta^n$.*

PROOF. Recall that a point x belongs to an axis of $s\Delta^n$ iff it has at most two distinct barycentric coordinates. We will show that if x has three distinct barycentric coordinates then the gradient of G_s at x is non-zero. Writing $f_i = g_{se_i}$, we obtain

$$\nabla G_s(x) = -2\pi G_s(x) \cdot x + 2\pi \sum_{i=0}^{n} f_i(x) \cdot v_i.$$

Setting the gradient to zero, we solve for x

$$x = \sum_{i=0}^{n} \frac{f_i(x)}{G_s(x)} \cdot v_i. \quad (11)$$

We will show that Equation (11) can hold only if x has at most two distinct barycentric coordinates. To this end, we write x in distance coordinates: $x = (r_0, r_1, \ldots, r_n)_D$. Similar to the barycentric coordinates, x lies on an axis of $s\Delta^n$ iff there are at most two distinct distance coordinates. Transforming x into barycentric coordinates, we have $x = (x_0, x_1, \ldots, x_n)_B$, in which

$$x_i = \frac{1}{n+1} + \frac{1}{2(n+1)s^2}\left(\sum_{j=0}^{n} r_j^2 - (n+1)r_i^2\right),$$

for $0 \leq i \leq n$. Assume now that x does not lie on any of the axes. If follows there are three distinct distance coordinates: $r_k < r_\ell < r_m$. Subtracting the m-th barycentric coordinate from the other two , we obtain

$$x_k - x_m = \frac{1}{2s^2}(r_m^2 - r_k^2), \quad (12)$$

$$x_\ell - x_m = \frac{1}{2s^2}(r_m^2 - r_\ell^2). \quad (13)$$

Assuming a zero gradient, the barycentric coordinates of x have the form given in (11). Hence, $x_k - x_m$ and $x_\ell - x_m$ are equal to

$$\frac{f_k(x) - f_m(x)}{G_s(x)} = \frac{e^{-\pi r_k^2} - e^{-\pi r_m^2}}{G_s(x)}, \quad (14)$$

$$\frac{f_\ell(x) - f_m(x)}{G_s(x)} = \frac{e^{-\pi r_\ell^2} - e^{-\pi r_m^2}}{G_s(x)}, \quad (15)$$

respectively. Since the right hand sides of (12) and (14) are equal, as well as the right hand sides of (13) and (15), we have

$$\frac{r_m^2 - r_k^2}{e^{-r_m^2} - e^{-r_k^2}} = \frac{r_m^2 - r_\ell^2}{e^{-r_m^2} - e^{-r_\ell^2}}.$$

But this is impossible because $r_m^2 - r_k^2 > r_m^2 - r_\ell^2$, by assumption, and because the function $f(t) = e^{-t}$ is strictly convex. □

3.2 One-Dimensional Sections

The restriction of G_s to an axis of $s\Delta^n$ is a sum of two weighted Gaussian kernels. This sum has two maxima for a range of scale factors, which we now analyze.

Transitions. Recall that the n-design consists of $n+1$ unit Gaussian kernels placed at the vertices of the standard n-simplex. Consider the 1-section defined by the line that connects the barycenter of a k-face with the barycenter of the complementary ℓ-face, with $k+\ell = n-1$, and vary the construction by scaling the design with $s \geq 0$. We call a value a *transition* if the number of critical points of the 1-section changes as s passes the value. It is easy to compute the transition for $k = \ell = \frac{n-1}{2}$ because the corresponding 1-section is balanced for all scale factors s. The distance between the two centers is $sD_{\ell,\ell} = 2s/\sqrt{n+1}$ and we find the transition by setting the distance equal to $2\sigma_0$, which gives s equal to

$$U_n = \sqrt{\frac{n+1}{2\pi}}. \quad (16)$$

Consider next the case $k < \ell$. Equation (7) gives the weights of the two kernels in the decomposition of the 1-section as $(k+1)g(sR_k)$ and $(\ell+1)g(sR_\ell)$. Using (5) and taking the ratio, the *weight function* is computed by the following function:

$$\omega_{k,\ell}(s) = \frac{\ell+1}{k+1} \cdot e^{-\pi s^2(\frac{1}{k+1} - \frac{1}{\ell+1})}. \quad (17)$$

We compare this with the two *transition functions*, which we get by setting $z = \frac{s}{2}D_{k,\ell}$ and plugging the two solutions of $x^2 = z^2 - \sigma_0^2$ into the formula for $r(x)+1$, which we get from (3). This gives

$$\tau_{k,\ell}(s) = \frac{z - \sqrt{z^2 - \sigma_0^2}}{z + \sqrt{z^2 - \sigma_0^2}} \cdot e^{4\pi z\sqrt{z^2 - \sigma_0^2}}, \quad (18)$$

$$v_{k,\ell}(s) = \frac{z + \sqrt{z^2 - \sigma_0^2}}{z - \sqrt{z^2 - \sigma_0^2}} \cdot e^{-4\pi z\sqrt{z^2 - \sigma_0^2}}. \quad (19)$$

Note that $v_{k,\ell}(s) = 1/\tau_{k,\ell}(s)$. We find the first transition, $T_{k,\ell}$, by solving $\omega_{k,\ell}(s) = \tau_{k,\ell}(s)$, and the second transition, U_n, by solving $\omega_{k,\ell}(s) = v_{k,\ell}(s)$. Appendix A will prove that both transitions are well defined, also showing that the second transition depends on n but not on k and ℓ and is given by (16) in all cases. While we have no analytic expression for $T_{k,\ell}$, we will derive one for an upper bound in Section 3.4.

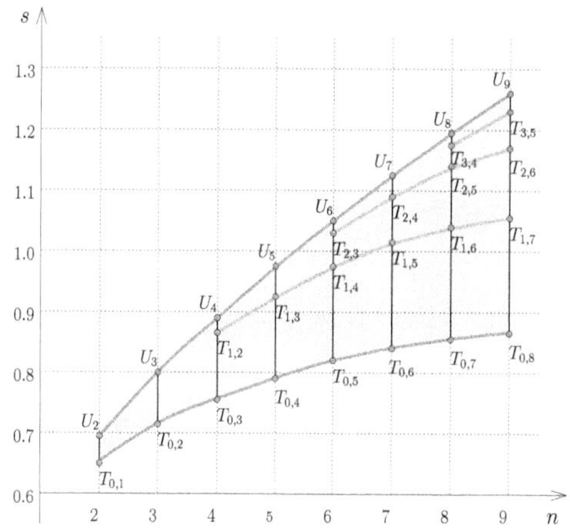

Figure 6: The vertical interval bounded from below by $T_{0,\ell}$ and above by U_n contains the scale factors for which an extra mode appears.

Section evolution. We follow the 1-section defined by an axis of the n-simplex as the scale factor, s, goes from 0 to infinity. By construction, we have qualitative changes at the transitions, which we now summarize.

1-SECTION LEMMA. *Let $0 \leq k \leq \ell$ with $k+\ell = n-1$, and let A be the axis passing through the barycenters of a k-face and its complementary ℓ-face of $s\Delta^n$. Then $G_s|_A$ has one maximum whenever $s < T_{k,\ell}$, and two maxima whenever $T_{k,\ell} < s$ and $s \neq U_n$.*

Indeed, the double intersection is responsible for the special evolution of the 1-section. In particular, we go from one maximum for $s < T_{k,\ell}$ to two maxima for $T_{k,\ell} < s < U_n$, of which one is the barycenter of the n-simplex. After the second transition at the double intersection, we still have two maxima, but now the separating minimum is the barycenter of the n-simplex. Figure 6 shows all transition scale factors in a single picture for small values of k and ℓ. First, we look at $T_{k,\ell}$ for a fixed value of n. We observe that $T_{k,\ell}$ increases with growing k. This implies that for constant n, the axes defined for small values of k spawn a second maximum earlier than do axes defined by large values of k. For $k = \ell$, the two transitions coincide and the corresponding 1-section does not witness the extra maximum at all. Second, we fix k and observe that $T_{k,\ell}$ increases with growing ℓ. This implies that the k-faces of a low-dimensional simplex spawn second maxima earlier than do the k-faces in high-dimensional simplices.

Next, we look at the second transition, U_n. As we have observed, it depends only on n. This implies that all 1-sections lose the maximum at the barycenter at the same scale factor. For $k = \ell$, the two transitions coincide, so the interval collapses.

3.3 n-Dimensional Sections

In this subsection, we show that most maxima of the 1-sections are not modes. We begin with the analysis of the barycenter of $s\Delta^n$, which belongs to every axis of the scaled n-simplex.

Barycenter of n-simplex. The n-design has the symmetry group of the n-simplex, which implies that the barycenter, $b_G \in s\Delta^n$, is a critical point of G_s. Indeed, if b_G is not a critical point then it has a non-zero gradient, which contradicts the symmetry.

BARYCENTER LEMMA. *Let $n \geq 1$. Then the barycenter of $s\Delta^n$ is a mode of G_s for $s < U_n$ and it is a saddle of index 1 for $s > U_n$.*

PROOF. We compute the Hessian of G_s at b_G by taking partial derivatives with respect to the Cartesian basis of $\mathbb{R}^{n+1}$. Because of the symmetry, we have

$$d = \frac{\partial^2 G_s}{\partial x_0^2}(b_G) = \frac{\partial^2 G_s}{\partial x_i^2}(b_G),$$
$$c = \frac{\partial^2 G_s}{\partial x_0 \partial x_1}(b_G) = \frac{\partial^2 G_s}{\partial x_i \partial x_j}(b_G),$$

for all $0 \leq i \leq n$ and all $j \neq i$. The characteristic polynomial of the Hessian is therefore

$$\det \begin{bmatrix} d-\xi & c & \dots & c \\ c & d-\xi & \dots & c \\ \vdots & \vdots & \ddots & \vdots \\ c & c & \dots & d-\xi \end{bmatrix}. \qquad (20)$$

Its roots are the eigenvalues of the Hessian, which are $\xi = d + nc$, with multiplicity one, and $\xi = d - c$, with multiplicity n. For simple geometric reasons, $d+nc$ is negative and corresponds to the eigenvector in the diagonal direction of $\mathbb{R}^{n+1}$. To compute $d-c$, it suffices to consider just one 1-section through the barycenter, and we choose the line, B, that passes through the vertex $v_0 = se_0$ and the barycenter of the complementary $(n-1)$-face, which we denote as b_0. The distances from the barycenter of the n-simplex are $\|b_G - v_0\| = sR_n$ and $\|b_G - b_0\| = \frac{s}{n}R_n$. Furthermore, the common distance of the vertices $v_i = se^i$ from B is $\|v_i - b_0\| = sR_{n-1}$, for $1 \leq i \leq n$. Plugging these distances into (2), we compute the second derivatives at b_G of the 1-sections defined by the $n+1$ unit Gaussian kernels as

$$[4\pi^2 s^2 R_n^2 - 2\pi] \cdot e^{-\pi s^2 R_n^2}, \qquad (21)$$
$$[4\pi^2 \frac{s^2}{n^2} R_n^2 - 2\pi] \cdot e^{-\pi s^2 (R_{n-1}^2 + R_n^2/n^2)}, \qquad (22)$$

where the first line applies for $i = 0$ and the second line for $1 \leq i \leq n$. Note that $R_{n-1}^2 + R_n^2/n^2 = R_n^2$ and $R_n^2(1+\frac{1}{n}) = 1$. Adding (21) and n times (22), we obtain the second derivative of the sum of Gaussian kernels as

$$d - c = [4\pi^2 s^2 - 2\pi(n+1)] \cdot e^{-\pi s^2 R_n^2}.$$

It has the same sign as $s^2 - \frac{n+1}{2\pi}$. This implies that b_G is a maximum of G_s for $s < U_n$ and a saddle of index one for $s > U_n$ as claimed. ☐

We note here that the barycenter is an index-1 saddle for $s > U_n$, as opposed to a minimum, because we place the n-simplex in $\mathbb{R}^{n+1}$. At the transition, when $s = U_n$, the barycenter of the n-simplex is a degenerate critical point.

Orthogonal sections. We generalize the analysis of the barycenter. Let $1 \leq k \leq \ell$ with $k + \ell = n - 1$, and consider a k-face of the n-simplex as well as the complementary ℓ-face. Writing G_s as the sum of the $f_i = g_{se_i}$, for $0 \leq i \leq n$, we assume that the centers of f_0 to f_k span the k-face, and that the centers of f_{k+1} to f_n span the ℓ-face. Hence, $G_s = K_s + L_s$, where $K_s = \sum_{i=0}^{k} f_i$ and $L_s = \sum_{i=k+1}^{n} f_i$. Writing b_K and b_L for the barycenters of the two faces, we let A be the axis defined by $A(t) = (1-t)b_K + tb_L$. We are interested in the Hessian of G_s at $x = A(t)$. For symmetry reasons, it has at most four distinct eigenvalues, each a second derivative along pairwise orthogonal lines. One line is the axis, another is orthogonal to the n-simplex, a third line is parallel to the k-face, and a fourth line is parallel to the ℓ-face. The latter two eigenvalues have multiplicity k and ℓ. We write κ for the length parameter along the third line and λ for the length parameter along the fourth line.

n-SECTION LEMMA. *Let $1 \leq k \leq \ell$ with $k + \ell = n - 1$. The second derivatives of G_s at $x = A(t)$ along lines parallel to the complementary k- and ℓ-faces of $s\Delta^n$ are*

$$\frac{\partial^2 G_s}{\partial \kappa^2}(x) = -2\pi G_s(x) + 4\pi^2 s^2 f_0(x), \qquad (23)$$
$$\frac{\partial^2 G_s}{\partial \lambda^2}(x) = -2\pi G_s(x) + 4\pi^2 s^2 f_n(x). \qquad (24)$$

PROOF. Recall $G_s = \sum_{i=0}^{n} f_i$ and $f_i(x) = e^{-\pi\|x - se_i\|^2}$. The derivative with respect to the i-th coordinate direction is

$$\frac{\partial G_s}{\partial x_i}(x) = -2\pi x_i G_s(x) + 2\pi s f_i(x).$$

Deriving again, with respect to the same and a different coordinate direction, we have

$$\frac{\partial^2 G_s}{\partial x_i^2}(x) = [-2\pi + 4\pi^2 x_i^2] G_s(x) - 4\pi^2 (2sx_i - s^2) f_i(x), \qquad (25)$$
$$\frac{\partial^2 G_s}{\partial x_i \partial x_j}(x) = 4\pi^2 [x_i x_j G_s(x) - s x_i f_j(x) - s x_j f_i(x)]. \qquad (26)$$

The point at which we take the second derivative has only two distinct coordinates, $\frac{(1-t)s}{k+1}$, repeated $k+1$ times, and $\frac{ts}{\ell+1}$, repeated $\ell + 1$ times. We can therefore substitute x_0 and x_1 for any two among the first $k + 1 \geq 2$ coordinate directions, and we can substitute x_n and x_{n-1} for any two among the last $\ell + 1 \geq 2$ coordinate directions. The Hessian at the point x is

$$H(x) = \begin{bmatrix} d & \dots & c & \gamma & \dots & \gamma \\ \vdots & \ddots & \vdots & \vdots & \ddots & \vdots \\ c & \dots & d & \gamma & \dots & \gamma \\ \gamma & \dots & \gamma & D & \dots & C \\ \vdots & \ddots & \vdots & \vdots & \ddots & \vdots \\ \gamma & \dots & \gamma & C & \dots & D \end{bmatrix},$$

where

$$d = \frac{\partial^2 G_s}{\partial x_0^2}(x), \quad c = \frac{\partial^2 G_s}{\partial x_0 \partial x_1}(x), \qquad (27)$$
$$D = \frac{\partial^2 G_s}{\partial x_n^2}(x), \quad C = \frac{\partial^2 G_s}{\partial x_n \partial x_{n-1}}(x), \qquad (28)$$
$$\gamma = \frac{\partial^2 G_s}{\partial x_0 \partial x_{n-1}}(x). \qquad (29)$$

We get the eigenvalues as the roots of the characteristic polynomial, which we find by subtracting the variable ξ from each diagonal element and taking the determinant, as in (20). In particular, $d - c$ is the k-fold eigenvalue that corresponds to the k-face, and $D - C$ is the ℓ-fold eigenvalue that corresponds to the ℓ-face. Plugging (25) and (26) into (27) and (28), we arrive at

$$d - c = -2\pi G_s(x) + 4\pi^2 s^2 f_0(x),$$
$$D - C = -2\pi G_s(x) + 4\pi^2 s^2 f_n(x).$$

These are the two claimed second derivatives of (23) and (24). ☐

Sign change. A point $x = A(t)$ is a mode of $G_s : \mathbb{R}^{n+1} \to \mathbb{R}$ iff it is a maximum of the 1-section defined by A as well as of the n-section defined by H_t. Focusing on the latter, we compute the values of the parameter t at which the second derivatives with respect to κ and with respect to λ vanish. Beginning with κ, we set (23) to zero and find

$$[4\pi^2 s^2 - 2\pi(k+1)] f_0(x) \quad = \quad 2\pi(\ell+1) f_n(x). \quad (30)$$

We note that the natural logarithm of $f_n(x)/f_0(x)$ is $-\pi$ times the following difference of squared distances:

$$\|x - se_n\|^2 - \|x - se_0\|^2 = 2s^2 \cdot \frac{(\ell+1) - t(n+1)}{(k+1)(\ell+1)}. \quad (31)$$

Plugging (31) into (30) gives us

$$\frac{2\pi s^2 - (k+1)}{\ell+1} \quad = \quad e^{-2\pi s^2 \frac{(\ell+1)-t(n+1)}{(k+1)(\ell+1)}}.$$

Solving this equation, we get t as a function of the scale parameter. We call this function t_K. Doing the symmetric computations for λ, we find a second function $t_L : \mathbb{R} \to \mathbb{R}$, both defined by

$$t_K(s) \;=\; \frac{k\ell+n}{2\pi s^2(n+1)} \cdot \ln \frac{2\pi s^2 - k - 1}{\ell+1} + \frac{\ell+1}{n+1}, \quad (32)$$

$$t_L(s) \;=\; \frac{k\ell+n}{2\pi s^2(n+1)} \cdot \ln \frac{k+1}{2\pi s^2 - \ell - 1} + \frac{\ell+1}{n+1}. \quad (33)$$

For example, for $s = U_n$, we get $t_K = t_L = \frac{\ell+1}{n+1}$, which is consistent with the Barycenter Lemma, where $s = \frac{\ell+1}{n+1}$ is identified as the scale factor at which the barycenter of the n-simplex changes from a maximum to a minimum. Note also that t_K is undefined for $s = U_k$, and t_L is undefined for $s = U_\ell$.

Chandelier. To get a feeling for the situation, we draw the trajectories of the critical points of G_s, and in particular those of the modes. We call this set in $\mathbb{R}^{n+1} \times \mathbb{R}$ the *chandelier* of the 1-parameter family of functions. Letting s increase from bottom to top, Figure 7 sketches the chandelier for $n = 1, 2$. The most prominent fea-

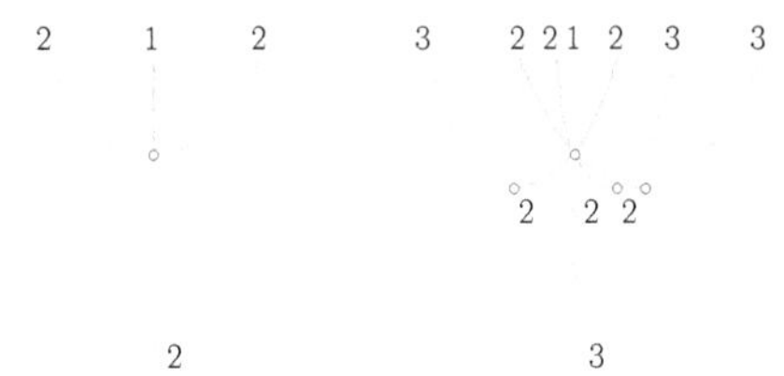

Figure 7: The chandelier for $n = 1$ on the the left and for $n = 2$ on the right. Each curve is labeled by the index of its critical points.

ture is the *base point*, which we use to decompose the chandelier into curves. Two of these curves are vertical, both swept out by the barycenter of the n-simplex, which changes from index $n+1$ to index 1 when it passes through the base point. For each curve, we consider the *height function* defined by mapping $(x, s) \in \mathbb{R}^{n+1} \times \mathbb{R}$ to s, and we further subdivide so that the height function is injective. In other words, we cut each curve at the local minima and maxima of the height function. The benefit of this subdivision is that now each curve is swept out by a critical point of G_s with constant index. While the total number of curves in the chandelier grows exponentially with the dimension, the number of curves that correspond to modes grows only by one for each dimension. To count the curves, we compute the number of complementary face pairs of the n-simplex:

$$p_n \quad = \quad \frac{1}{2} \sum_{k=0}^{n-1} \binom{n+1}{k+1} \quad = \quad 2^n - 1.$$

For each pair, two branches emanate from the base point. Adding the vertical line, we count $2p_n + 1 = 2^{n+1}$ branches. For each complementary face pair with $0 \le k < \ell$, the height function of one of the two corresponding branches has a local minimum and is therefore subdivided into two curves. The number of local minima is

$$l_n \quad = \quad \begin{cases} p_n & \text{if } n \text{ is even,} \\ p_n - \frac{1}{2}\binom{n+1}{(n+1)/2} & \text{if } n \text{ is odd.} \end{cases}$$

The total number of curves is therefore $2p_n + 2 + l_n$. Of these, only $n + 2$ correspond to modes.

Indices. The index swept out by a curve in the chandelier is easy to determine numerically, but at this time, we lack analytic proofs. We fist state the result and second explain the numerical evidence that supports it.

$0 \le k < \ell$**:** There are $\binom{n+1}{k+1}$ complementary face pairs of k- and ℓ-faces. Besides the barycenter, the corresponding axes witness critical points of index $\ell+2$ and $\ell+1$ for $s \in (T_{k,\ell}, U_n)$ and two critical points of index $k+2$ for $s > U_n$.

$k = \ell = \frac{n-1}{2}$**:** There are $\frac{1}{2}\binom{n+1}{k+1}$ complementary pairs of k-faces. Besides the barycenter, the corresponding axes witness two critical points of index $k+2$ for $s > T_{k,k} = U_n$.

To explain the numerical evidence, we consider $t_K(s)$ and $t_L(s)$, which are given by (32) and (33). We make t_K injective by restricting it to the range $[0, \frac{\ell+1}{n+1}]$ and we make t_L injective by restricting it to the range $[\frac{\ell+1}{n+1}, 1]$; see Figure 8, which plots the inverses of the restricted functions. Drawing the horizontal line for a value of s, we note that the portion below the graphs of t_K and t_L consists of the points x at which the n-section orthogonal to the axis has a maximum at x We see these graphs for even and odd values of n in Figure 8. For each scale factor s, there is either one or two modes witnessed by $A_{k,\ell}$, drawn in cyan in Figure 8. We notice empirically that the mode at the barycenter, given by $t = \frac{\ell+1}{n+1}$, is the only mode under the piecewise defined curve for $0 < k < \ell$. This means that the only mode is at the barycenter and the other critical points are saddles of index $\ell+1$ and $\ell+2$.

3.4 Resilient Modes

We have seen that the sum of Gaussian kernels can have extra modes. In this subsection, we study their significance, showing that they last for an interval of scale factors whose length increases with the dimension.

Balancing scales. To get started, we need more information on the transition at which the extra minima appear. We get an upper bound on $T_{k,\ell}$ by studying the scale factor at which the weights of the two 1-dimensional kernels in the decomposition of G_s restricted to a relevant axis are balanced. For $k = \ell$, the two 1-dimensional kernels in the decomposition are always balanced. For $k < \ell$, the *balancing scale factor* is

$$B_{k,\ell} \quad = \quad \sqrt{\frac{\ln(\ell+1) - \ln(k+1)}{\pi\left(\frac{1}{k+1} - \frac{1}{\ell+1}\right)}}. \quad (34)$$

Indeed, recomputing the weights gives $(k+1)g(B_{k,\ell}R_k) = (\ell+1)g(B_{k,\ell}R_\ell)$. Similar to $T_{k,\ell}$ and U_n, the balancing scale factor increases with respect to k and ℓ. Numerically, we observed that $B_{k,\ell}$ is not very different, but consistently larger than $T_{k,\ell}$. We prove that this relationship is not accidental.

TRANSITION LEMMA. *We have $T_{k,\ell} < B_{k,\ell} < U_n$ for all integers $0 \le k < \ell$ with $k + \ell = n - 1$.*

PROOF. We prove the claim indirectly, by showing that $s = B_{k,\ell}$ gives two maxima in the 1-section along any axis connecting

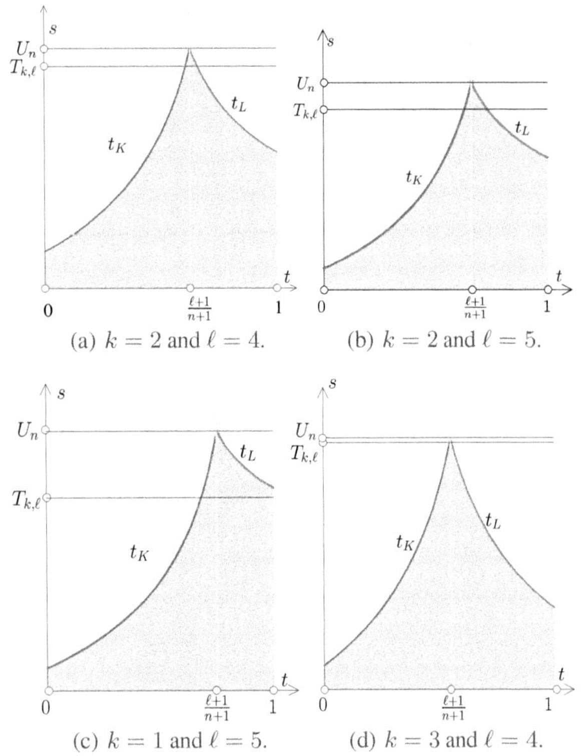

(a) $k = 2$ and $\ell = 4$. (b) $k = 2$ and $\ell = 5$.

(c) $k = 1$ and $\ell = 5$. (d) $k = 3$ and $\ell = 4$.

Figure 8: Graphs of t_K and t_L for select values of k and ℓ.

the barycenter of a k-face with the barycenter of the complementary ℓ-face. For balanced weights, we have two maxima iff the centers of the two 1-dimensional kernels are further apart than twice the width; see Section 2. To prove the latter property, we compute

$$\frac{B_{k,\ell} D_{k,\ell}}{2\sigma_0} = \sqrt{\frac{n+1}{2(\ell-k)} \cdot \ln \frac{\ell+1}{k+1}}, \quad (35)$$

using (4), (1), and (34). Recall the logarithmic inequality:

$$\frac{x}{1+\frac{x}{2}} < \ln(1+x),$$

for $x > 0$. Setting $x = \frac{\ell-k}{k+1}$, we see that the right hand side of (35) exceeds 1 for all choices of $0 \leq k < \ell$. This implies that we have two maxima along the axis, which implies that the balancing scale factor lies between the first and second transitions, as claimed. ☐

Resilience. We define the *resilience* of a mode as the length of the interval of scale values at which it exists. This definition is not satisfactory for a general 1-parameter family of smooth functions; however, it will suffice in our context, in which we know enough about the modes to follow them through the family parametrized by the scale s. Specifically, we have a single mode for $0 \leq s \leq T_{0,n-1}$, and we have $n+1$ modes for $U_n \leq s$. The picture is more interesting in the interval $T_{0,n-1} < s < U_n$, in which we have $n+2$ modes. One of these modes is the barycenter of the n-simplex, and we study the resilience of this *extra* mode. The upper endpoint of the interval is defined in (16), and an upper bound for the lower endpoint is given in the Transition Lemma, with the definition of the bound in (34):

$$T_{0,n-1} < \sqrt{\frac{\ln n}{\pi(1-1/n)}}.$$

As n goes to infinity, U_n grows roughly like the square root of n, and $T_{0,n-1}$ grows roughly like the square root of the logarithm of n. The gap between the two widens, so that the resilience of the mode at the barycenter of the n-simplex grows roughly like $\sqrt{n}$; see Figure 6.

Summary. We are now ready to summarize the findings in regard to the critical points and the modes of the 1-parameter family of functions $G_s : \mathbb{R}^{n+1} \to \mathbb{R}$. For values $s < T_{0,n-1}$, we have a single critical point with index $n+1$. Thereafter, we pick up $2\binom{n+1}{k+1}$ critical points at every $T_{k,\ell}$, for $0 \leq k < \ell$, until we accumulated $2l_n + 1$ critical points right before reaching U_n. The barycenter has index $n+1$, and the other critical points come in pairs, with indices $\ell + 2$ and $\ell + 1$, for $\frac{n-1}{2} < \ell \leq n - 1$. For $U_n < s$, we have $2p_n + 1$ critical points. The barycenter has index 1, and the other critical points come in pairs with indices $\ell + 2$ and $k + 2$, for $\frac{n-1}{2} \leq \ell \leq n - 1$. As a sanity check, we consider the Euler-Poincaré formula, which states that the alternating sum of critical points is equal to the Euler characteristic of $\mathbb{R}^{n+1}$:

$$\sum_{i=0}^{n+1} (-1)^i c_i = (-1)^{n+1}, \quad (36)$$

where c_i counts the critical points with index i. We also write $c = \sum_{i=0}^{n+1} c_i$. Trivially, (36) holds in the first case. Thereafter, we pick up the critical points in pairs whose contribution to the alternating sum cancel, so (36) is maintained. Finally, for $U_n < s$, we have a bijection between the critical points and the faces of the n-simplex such that the index is $n+1$ minus the dimension of the face. Since the n-simplex is a closed ball, its Euler characteristic is 1, which again implies (36). We thus have a complete description of the critical points of the n-design as the scale factor increases from zero to infinity.

MAIN THEOREM. *Let $n \geq 1$ and consider the sum of $n+1$ unit Gaussian kernels placed at the vertices of the scaled standard n-simplex, $s\Delta^n$.*

(1) For $s < T_{0,n-1}$, we have 1 critical point which is also a mode.

(2) For $T_{0,n-1} < s < U_n$, we have gradually more critical points after passing each $T_{k,\ell}$, until we accumulate $2l_n + 1$ critical points right before U_n. Of these critical points, $n+2$ are modes, and they exist during the entire interval.

(3) For $U_n < s$, we have $2p_n + 1$ critical points, of which $n+1$ are modes.

The resilience of the extra mode in Case (2) is $U_n - T_{0,n-1}$, which grows like $\sqrt{n}$.

3.5 Many Modes

In this subsection, we construct a finite configuration of isotropic Gaussian kernels with a superlinear number of modes. While there is a family of such constructions, it will suffice to explain one.

Products of simplices. The basic building block of our construction is the standard 2-simplex. Let the dimension be $3n$ and write the $3n$-dimensional Euclidean space as the Cartesian product of n 3-dimensional planes: $\mathbb{R}^{3n} = H_1 \times H_2 \times \ldots \times H_n$, in which H_i is spanned by the three coordinate vectors $e_{3i-2}, e_{3i-1}, e_{3i}$, for $1 \leq i \leq n$. Let Δ_i^2 be the standard 2-simplex in H_i, with vertices $v_{i0} = e_{3i-2}$, $v_{i1} = e_{3i-1}$, $v_{i2} = e_{3i}$. Correspondingly, we write $g_{ij} : H_i \to \mathbb{R}$ for the 3-dimensional unit Gaussian kernel with center v_{ij}, for $0 \leq j \leq 2$, and $G_i : H_i \to \mathbb{R}$ defined by

$$G_i(x) = g_{i0}(x) + g_{i1}(x) + g_{i2}(x)$$

for their sum. Next, we construct a $3n$-dimensional sum of Gaussian kernels by taking products. To begin, we let $P \subseteq \mathbb{R}^{3n}$ be the largest subset of points whose orthogonal projection to H_i is $\{v_{i0}, v_{i1}, v_{i2}\}$, for $1 \leq i \leq n$. This is the set of 3^n points formed by taking the Cartesian product of the n triplets of points. For each point $p \in P$, let $f_p : \mathbb{R}^{3n} \to \mathbb{R}$ be the unit Gaussian kernel with center p. Adding these kernels, we get $F : \mathbb{R}^{3n} \to \mathbb{R}$, defined by

$$F(x) = \sum_{p \in P} f_p(x).$$

To understand F, we recall that f_p can be written as the product of $3n$ 1-dimensional unit Gaussian kernels; see (6). Collecting the terms in sets of three, we can write

$$f_p(x) = \prod_{i=1}^{n} g_{ij}(x),$$

where j is chosen such that v_{ij} is the orthogonal projection of p onto H_i. Substituting the sum of the three kernels for the singletons, we obtain

$$F(x) = \prod_{i=1}^{n} G_i(x).$$

In words, the sum of the 3^n $3n$-dimensional unit Gaussian kernels is the product of n sums of three 3-dimensional unit Gaussian kernels.

Counting modes. We arrive at the final construction by reintroducing the scale factor, writing $F_s : \mathbb{R}^{3n} \to \mathbb{R}$ for the product of the $G_{is} : H_i \to \mathbb{R}$, where G_{is} is of course the sum of the three unit Gaussian kernels with centers se_{3i-2}, se_{3i-1}, se_{3i}. We have seen in Section 3 that s can be chosen such that G_{is} has 4 modes. Since F_s is the product of the G_{is}, its sets of modes is the largest subset of $\mathbb{R}^{3n}$ whose orthogonal projection to H_i is the set of four modes of G_{is}, for $1 \leq i \leq n$. Its size is $4^n = 3^{(1+\log_3 \frac{4}{3})n} > 3^{1.261n}$. This shows that the number of modes is roughly the number of kernels to the power 1.261.

There is an entire family of similar constructions. The one presented here neither maximizes the number nor the resilience of the extra modes. Indeed, we can increase the exponent by improving the ratio of modes over kernels in each H_i (see [5]), and we can improve the resilience by using higher-dimensional simplices.

4. DISCUSSION

The main contribution of this paper is a cautionary message about the sum of Gaussian kernels. Giving a detailed analysis of the construction studied in [5], we show that there is indeed only one extra mode, but that its resilience increases like the square root of the dimension. We also exhibit configurations of finitely many identical isotropic Gaussian kernels whose sums have superlinearly many modes. We thus give precisely quantified contradictions to our intuition that diffusion erodes and eliminates local density maxima.

The results in this paper raise a number of questions. How stable are the extra maxima? Our analysis in Section 3.4 answers the question when the perturbation is the diffusion of density. How robust are they under moving individual kernels or changing their weights? Related to this question, we ask about the probability of extra modes for randomly placed Gaussian kernels in Euclidean or other spaces. Carreira-Perpiñán and Williams report that their computerized searches in $\mathbb{R}^2$ did not turn up any extra modes [5], but what if we did similar experiments in three and higher dimensions? Finally, it would be interesting to determine the persistence of the extra modes; see [6] for a recent related study. In other words, how large is the difference in function value between an extra mode and the highest saddle?

5. ACKNOWLEDGMENTS

This research is partially supported by the National Science Foundation (NSF) under grant DBI-0820624, by the European Science Foundation under the Research Networking Programme, and the Russian Government Project 11.G34.31.0053.

6. REFERENCES

[1] J. BABAUD, A. P. WITKIN, W. BAUDIN AND R. O. DUDA. Uniqueness of the Gaussian kernel for scale-space filtering. *IEEE Trans. Pattern Anal. Mach. Intel.* **8** (1986), 26–33.

[2] J. BEHBOODIAN. On the modes of a mixture of two normal distributions. *Technometrics* **12** (1970), 131–139.

[3] P. J. BURKE. Solution of Problem 4616 [1954, 718], proposed by A. C. Cohen, Jr. *Amer. Math. Monthly* **63** (1956), 129.

[4] M. CARREIRA-PERPIÑÁN AND C. WILLIAMS. On the number of modes of a Gaussian mixture. In *Scale-Space Methods in Computer Vision*, Lecture Notes in Computer Science **2695**, 2003, 625–640.

[5] M. CARREIRA-PERPIÑÁN AND C. WILLIAMS. An isotropic Gaussian mixture can have more modes than components. Rept. EDI-INF-RR-0185, School of Informatics, Univ. Edinburgh, Scotland, 2003.

[6] C. CHEN AND H. EDELSBRUNNER. Diffusion runs low on persistence fast. In "Proc. 13th Internat. Conf. Comput. Vision, 2011", 423–430.

[7] J. DAMON. Local Morse theory for solutions to the heat equation and Gaussian blurring. *J. Diff. Equations* **115** (1995), 368–401.

[8] W. FELLER. *An Introduction to Probability Theory and Its Applications, Vol. I.* Wiley, New York, 1950.

[9] J. J. KOENDERINK. The structure of images. *Biol. Cybern.* **50** (1984), 363–370.

[10] A. KUIJPER AND L. M. J. FLORACK. The application of catastrophe theory to image analysis. Rept. UU-CS-2001-23, Dept. Comput. Sci., Utrecht Univ., Utrecht, the Netherlands, 2001.

[11] A. KUIJPER AND L. M. J. FLORACK. The relevance of non-generic events in scale space models. *Internat. J. Comput. Vision* **57** (2004), 67–84.

[12] L. M. LIFSHITZ AND S. M. PIZER. A multiresolution hierarchical approach to image segmentation based on intensity extrema. *IEEE Trans. Pattern Anal. Mach. Intell.* **12** (1990), 529–540.

[13] T. LINDEBERG. *Scale-Space Theory in Computer Vision.* Kluwer, Dortrecht, the Netherlands, 1994.

[14] J. MILNOR. *Morse Theory.* Princeton Univ. Press, New Jersey, 1963.

[15] J. RIEGER. Generic evolutions of edges on families of diffused greyvalue surfaces. *J. Math. Imaging Vision* **5** (1995), 207–217.

[16] S. J. ROBERTS. Parametric and non-parametric unsupervised cluster analysis. *Pattern Recognition* **30** (1997), 261–272.

[17] B. W. SILVERMAN. Using kernel density estimates to investigate multimodality. *J. Royal Stat. Soc., B* **43** (1981), 97–99.

[18] A. P. WITKIN. Scale-space filtering. *In* "Proc. 8th Internat. Joint Conf. Artif. Intell., 1983", 1019-1022.

[19] A. L. YUILLE AND T. A. POGGIO. Scaling theorems for zero crossings. *IEEE Trans. Pattern Anal. Mach. Intel.* **8** (1986), 15–25.

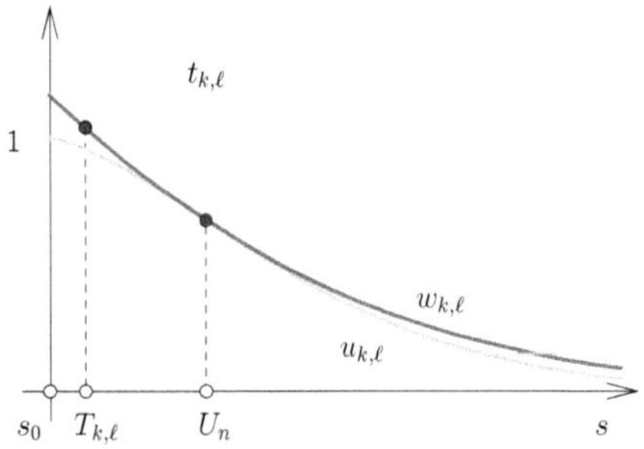

Figure 9: The graphs of the weight function and the two transition functions. The intersections define the transitions.

Appendix A

In this appendix, we give a detailed analysis of the intersections between the weight function, $\omega_{k,\ell}$, and the two transition functions, $\tau_{k,\ell}$ and $\upsilon_{k,\ell}$, all introduced in Section 3.2. We recall that the two transitions, $T_{k,\ell}$ and U_n, are the solutions to $\omega_{k,\ell}(s) = \tau_{k,\ell}(s)$ and to $\omega_{k,\ell}(s) = \upsilon_{k,\ell}(s)$ respectively. We will find that both transitions are well defined, and the second transition depends on n but not on the choice of k and ℓ, as claimed in Section 3.2.

Curve analysis. We discuss the graphs of the three functions to convey a feeling for how they intersect.To begin, we note that $s_0 = 2\sigma_0/D_{k,\ell}$ is the smallest scale factor for which the transition functions are defined. Writing $z = \frac{s}{2}D_{k,\ell}$, we have $z^2 - \sigma_0^2 = 0$ for $s = s_0$. It follows that $\tau_{k,\ell}(s_0) = \upsilon_{k,\ell}(s_0) = 1$. To compare this with the weight function at the same value, we compute

$$\omega_{k,\ell}(s_0) = \frac{\ell+1}{k+1} \cdot e^{-\frac{2(\ell-k)}{n+1}},$$

$$\ln \omega_{k,\ell}(s_0) = \ln \frac{\ell+1}{k+1} - \frac{2(\ell-k)}{n+1}. \quad (37)$$

We interpret the right hand side of (37) as the difference between the area below the graph of $\frac{1}{x}$, for $k+1 \leq x \leq \ell+1$, and the area below the line that touches the graph at the midpoint of the interval, again for $k+1 \leq x \leq \ell+1$. Since $\frac{1}{x}$ is a convex function, the second area is smaller, which implies $\ln \omega_{k,\ell}(s_0) > 0$ and therefore $\omega_{k,\ell}(s_0) > 1$. This implies (38), which is the first of the two pairs of inequalities that describe the relation between the three functions on the left and the right:

$$\tau_{k,\ell}(s_0) = \upsilon_{k,\ell}(s_0) < \omega_{k,\ell}(s_0), \quad (38)$$

$$\upsilon_{k,\ell}(s) < \omega_{k,\ell}(s) < \tau_{k,\ell}(s), \quad (39)$$

as in Figure 9. To compare the functions on the right, for sufficiently large s, we look at the exponents. Writing $z = \frac{s}{2}D_{k,\ell}$, as before, the exponent of $\omega_{k,\ell}(s)$ is $-4\pi z^2 \frac{\ell-k}{n+1}$, which is clearly smaller than the exponent of $\tau_{k,\ell}(s)$, which is $4\pi z\sqrt{z^2-\sigma_0^2}$. Less obvious is the comparison with the exponent of $\upsilon_{k,\ell}(s)$, which is $-4\pi z\sqrt{z^2-\sigma_0^2}$. After dividing by $-4\pi z$ and squaring, we obtain $z^2(\ell-k)^2/(n+1)^2 < z^2 - \sigma_0^2$. It follows that the exponent of $\omega_{k,\ell}(s)$ is larger than that of $\upsilon_{k,\ell}(s)$, for s large which implies (39).

From (38) and (39), we conclude that the number of intersections between the graphs of $\omega_{k,\ell}$ and $\tau_{k,\ell}$ (with multiplicity)is odd. Since both functions are monotonic, with slopes of opposite signs, we have exactly one intersection. It follows that $T_{k,\ell}$ is well defined.

Double intersection. Similarly, we use (38) and (39) to conclude that the number of intersections between the graphs of $\omega_{k,\ell}$ and $\upsilon_{k,\ell}$ (again counting with multiplicity) is even. We will establish that there is only one double intersection, namely at $s = U_n$. To prove that U_n is a solution to the equation $\omega_{k,\ell}(s) = \upsilon_{k,\ell}(s)$, we write $A = (\ell+1)+(k+1)$ and $B = (\ell+1)-(k+1)$, noting that $A^2 - B^2 = 4(k+1)(\ell+1)$. Setting

$$s = \sqrt{\frac{n+1}{2\pi} + \frac{y}{2\pi(n+1)}},$$

and recalling that $A = n+1$, we have

$$z = \sqrt{\frac{A^2+y}{2\pi(A^2-B^2)}},$$

$$\sqrt{z^2-\sigma_0^2} = \sqrt{\frac{B^2+y}{2\pi(A^2-B^2)}},$$

$$\frac{z+\sqrt{z^2-\sigma_0^2}}{z-\sqrt{z^2-\sigma_0^2}} = \frac{\sqrt{A^2+y}+\sqrt{B^2+y}}{\sqrt{A^2+y}-\sqrt{B^2+y}},$$

$$-4\pi z\sqrt{z^2-\sigma_0^2} = -\frac{2\sqrt{A^2+y}\sqrt{B^2+y}}{A^2-B^2}.$$

Using the definitions of the weight function in (17) and the second transition function in (19), we arrive at

$$\omega_{k,\ell}(s) = \frac{(A+B)^2}{A^2-B^2} \cdot e^{-\frac{2AB+2\frac{B}{A}y}{A^2-B^2}},$$

$$\upsilon_{k,\ell}(s) = \frac{[\sqrt{A^2+y}+\sqrt{B^2+y}]^2}{A^2-B^2} \cdot e^{-\frac{2\sqrt{A^2+y}\sqrt{B^2+y}}{A^2-B^2}}.$$

Clearly, $\omega_{k,\ell}(s) = \upsilon_{k,\ell}(s)$ if $y = 0$, which shows that $s = U_n$ is indeed a solution to the equation. We continue by showing that for small but non-zero y, we have $\omega_{k,\ell}(s) > \upsilon_{k,\ell}(s)$. This is equivalent to showing that the natural logarithm of $\upsilon_{k,\ell}(s)$ over $\omega_{k,\ell}(s)$ is smaller than 0. Equivalently, LHS < RHS, where

$$\text{LHS} = \ln \frac{\sqrt{A^2+y}+\sqrt{B^2+y}}{A+B},$$

$$\text{RHS} = \frac{\sqrt{A^2+y}\sqrt{B^2+y} - AB - \frac{B}{A}y}{A^2-B^2}.$$

To prove this inequality for small values of y, we use the Taylor expansions of the square root and the natural logarithm functions:

$$\sqrt{A^2+y} = A + \frac{y}{2A} - \frac{y^2}{8A^3} + \frac{y^3}{16A^5} - \ldots,$$

$$\ln(1+x) = x - \frac{x^2}{2} + \frac{x^3}{3} - \frac{x^4}{4} + \ldots.$$

With this, we can re-write the two sides of the inequality: LHS $= l_1 y + l_2 y^2 + \ldots$, and RHS $= r_1 y + r_2 y^2 + \ldots$. Computing the coefficients, in turn, we find

$$l_1 = \frac{1}{2AB} = r_1,$$

$$l_2 = -\frac{A^2+B^2}{8A^3B^3} < -\frac{A^2-B^2}{8A^3B^3} = r_2.$$

In words, $s = U_n$ is a double solution,and $\omega_{k,\ell}(s) > \upsilon_{k,\ell}(s)$ for values of s chosen in a small neighborhood but different from U_n.

On the Expected Complexity of Voronoi Diagrams on Terrains*

Anne Driemel[†]
Dept. Inform. Comput. Sci.
Utrecht University
The Netherlands
anne@cs.uu.nl

Sariel Har-Peled[‡]
Dept. Computer Science
University of Illinois
201 N. Goodwin Avenue
Urbana, IL, 61801, USA
sariel@uiuc.edu

Benjamin Raichel[§]
Dept. Computer Science
University of Illinois,
Urbana-Champaign
201 N. Goodwin Avenue
Urbana, IL, 61801, USA
raichel2@uiuc.edu

ABSTRACT

We investigate the combinatorial complexity of geodesic Voronoi diagrams on polyhedral terrains using a probabilistic analysis. Aronov *et al.* [AdBT08] prove that, if one makes certain realistic input assumptions on the terrain, this complexity is $\Theta(n+m\sqrt{n})$ in the worst case, where n denotes the number of triangles that define the terrain and m denotes the number of Voronoi sites. We prove that under a relaxed set of assumptions the Voronoi diagram has expected complexity $O(n+m)$, given that the sites have a uniform distribution on the domain of the terrain (or the surface of the terrain). Furthermore, we present a worst-case construction of a terrain which implies a lower bound of $\Omega(nm^{2/3})$ on the expected worst-case complexity if these assumptions on the terrain are dropped.

Categories and Subject Descriptors

F.2.2 [**ANALYSIS OF ALGORITHMS AND PROBLEM COMPLEXITY**]: Nonnumerical Algorithms and Problems—*Geometrical problems and computations*; E.2 [**DATA STORAGE REPRESENTATIONS**]: Object representation

General Terms

Algorithms, Theory

* The current full version of this paper can be found online [DHPR11]

[†]Work by AD has been supported by the Netherlands Organisation for Scientific Research (NWO) under RIMGA (Realistic Input Models for Geographic Applications).

[‡]Work on this paper by SH has been partially supported by a NSF AF award CCF-0915984.

[§]Work on this paper was partially supported by a NSF AF award CCF-0915984.

SCG'12, June 17–20, 2012, Chapel Hill, North Carolina, USA.

1. INTRODUCTION

Voronoi diagrams on terrains are a basic geometric data-structure that have a variety of applications in many areas: geographic information science (GIS) [AJT01, DG08, PA06, CG10], robot motion planning [TS89], mesh generation [KWR97] and image analysis [SF04, WDB+08] to name a few. The geodesic Voronoi diagram of point sites on a polyhedral terrain is a subdivision of the surface into cells, according to the set of sites, such that every cell contains exactly the surface points which are closest to the site that is associated with the cell. Here, the distance is measured by the length of the shortest path on the terrain. It is tempting to believe that in practice – that is, given that the terrain is well-behaved – the complexity of such a geodesic Voronoi diagram should be linear, because of its similarity to the Euclidean Voronoi diagram of point sites in the plane.

However, in the worst case, this complexity can be much higher, even if one makes certain realistic assumptions on the shape of the terrain. Indeed, Aronov *et al.* [AdBT08] show that the worst-case complexity is $\Theta(n + m\sqrt{n})$ for a certain class of well-behaved terrains, where n is the number triangles that define the terrain and m is the number of Voronoi sites. This shows that assuming realistic input indeed brings the complexity down, but it is still far from being linear. They conjecture that, in order to prove a linear bound, one needs to make further assumptions on how the sites are distributed. Our purpose in this paper is to study the complexity of a geodesic Voronoi diagram if we assume that the sites are being chosen randomly from the terrain.

Previous work.

Analyzing the expected complexity of geometric structures for random inputs has a long history in computational geometry. See for instance the work of Rényi and Sulanke [RS63] and Raynaud [Ray70] on the complexity of convex hulls of random points. Weil and Wieacker give an overview of related results in [WW93]. Naturally, Voronoi diagrams (and their counterpart, Delaunay triangulations) have been analyzed in this way as they are a fundamental data-structure, used in fields such as mesh generation [Rup95], surface reconstruction [Dey11], molecular biology, and many others. It is well-known that the size of the Voronoi diagram of point sites in $\mathbb{R}^3$ is quadratic in the worst-case, however it is near-linear in most practical situations. To address this dichotomy, people have investigated the complexity when the point sites are: (i) generated by a random processes, (ii) well spaced, (iii) have bounded spread, or (iv) were sampled from surfaces according to curvature. See [DEG08] and references therein for more information on such work. In particular, there is a vast amount of work on *Poisson Voronoi Diagrams* (*PVD*). Here, the domain has a density associated with it (say, the area). The probability of n points to appear in an area of measure μ has, as the name suggests, a Poisson distribution parameterized by the area. Similarly, the distribution of points selected into disjoint areas is independent. Poisson Voronoi diagrams are used in many areas, such as physics, biology, animal ecology, and others. See [OBSC00, JN04] and references therein. However, this work does not seem to have considered geodesics at all.

In this paper, our main interest is in the complexity of geodesic Voronoi diagrams on polyhedral terrains.

Moet *et al.* [MvKvdS08] were the first to study this complexity using a set of parameterized assumptions that describe realistic terrains. In this approach, one assumes that a certain property, for example, the maximum slope of the terrain, can be bounded by a constant independent of the input size. This allows one to avoid certain worst-case configurations which are highly unlikely to occur in practice. Instead, the analysis is confined to classes of well-behaved inputs and consequently this method is described as using *realistic input models*. Moet also did an experimental validation of the used parameters [Moe08] and confirmed that the parameters indeed behave like constants on realistic terrains. The realistic input models introduced in this work have also been adopted by subsequent papers. As such, Aronov *et al.* [AdBT08] improved the bounds given by Moet *et al.* and showed that (i) the bisector between two sites has worst-case complexity $\Theta(n)$, (where n denotes the number of triangles of the terrain) if the triangulation is low density and the lifted triangles have bounded slope; and (ii) that the worst-case combinatorial complexity of the Voronoi diagram is $\Theta(n + m\sqrt{n})$, (where m denotes the number of sites) if in addition the triangles are of similar size and the aspect ratio of the domain is bounded. The realistic assumptions made in these papers are described in more detail in the next section.

Finally, note that Schreiber and Sharir [SS08] showed how to compute an implicit representation of the geodesic Voronoi diagram on the surface of a convex polyhedron, in time $O((n+m)\log(n+m))$, so that the site closest to a query point can be reported in time $O(\log(n+m))$. Schreiber [Sch07] also extended their method for single-source shortest paths to the case of non-convex polyhedra using several realistic input models. Naturally, these analyses do not inform about the complexity of the explicit Voronoi diagram.

Our results.

We study the expected complexity of geodesic Voronoi diagrams on terrains. To this end, we use realistic input assumptions on the terrain, and sample the sites uniformly at random from the domain of the terrain. See Section 2 for the exact definitions. In Section 3 we show that under these assumptions the complexity of the geodesic Voronoi diagram is indeed linear. That is, we show that the complexity is bounded by $O(n+m)$, where n is the complexity of the terrain, and m is the number of sites being randomly picked. The constants in the asymptotic analysis depend on how well-behaved the terrain is, which is formalized using the input models described in the next section. See Theorem 3.3 for the exact result. In Section 4 we analyze the expected complexity if these assumptions on the shape of the terrain are dropped. In particular, in Theorem 4.12 we show a lower bound of $\Omega\left(nm^{2/3}\right)$. This lower bound, in a sense, justifies the input assumptions made previously, since it implies that the randomness assumption by itself is not sufficient if we want the geodesic Voronoi diagram to have a low complexity. The construction that leads to this lower bound is intricate and requires a careful balancing of the variance of the distances of the sampled sites, and how closely they can be packed together.

Organization.

In Section 2 we introduce the concepts used. Section 3 contains the poofs of the upper bound on the complexity of the Voronoi diagram. The lower bound is proved in Section 4. Finally, we conclude in Section 5.

2. PRELIMINARIES

2.1 Voronoi Diagrams on Terrains

A polyhedral terrain $\mathcal{T}$ is defined by a triangulation Δ of n vertices V in $\mathbb{R}^2$, a convex domain $D \subseteq \mathbb{R}^2$ which contains V, and a height function on these vertices. The ***surface*** of the terrain is defined by the triangles of Δ lifted according to this height function. We refer to $\mathcal{T}$ simply as a ***terrain*** and we denote the set of edges of the triangulation with E. For simplicity of exposition we restrict our discussion to the case where D is the unit square, however, our results can be easily extended to the more general case of convex regions with bounded aspect ratio. For two points $\mathsf{q}, \mathsf{s} \in D$, we denote their euclidean distance in the (x, y)-plane with $\|\mathsf{q}-\mathsf{s}\|$. When q and s are lifted to the surface of $\mathcal{T}$, we define their geodesic distance to be the length of the shortest path connecting them that is constrained to lie in the surface of $\mathcal{T}$, and we denote this value by $\mathsf{d}_{\mathcal{T}}(\mathsf{q}, \mathsf{s})$.

The (geodesic) ***Voronoi diagram*** of a set of m points on $\mathcal{T}$ (which are called ***sites***) is a planar subdivision of the surface of $\mathcal{T}$, where every cell of the subdivision is associated with exactly one site, and such that for any point in the cell the associated site is the closest site, where the distances are measured using the geodesic distance. We denote the Voronoi diagram with $\mathcal{V}or(\mathsf{P})$, where P denotes the set of sites, and we call a cell of the subdivision a ***Voronoi cell***. The ***bisector*** between two sites q and s on the surface of $\mathcal{T}$ is defined as the set of points p, such that p has the same distance to q and s. The Voronoi diagram can be represented as the structured set of curves and straight lines which delineate the Voronoi cells and which are subsets of the bisectors between these points. We call a point which is incident to at least three cells a ***Voronoi vertex*** and we call each maximally connected subset of the bisector incident to two Voronoi cells a ***Voronoi edge*** (note that two cells can have multiple edges between them). Usually one assumes general position of the sites so that no two sites are equidistant from a terrain vertex, which ensures that bisectors are one-dimensional and that the Voronoi cells subdivide the terrain surface without overlap, see also [AdBT08]. In our case the sites are randomly sampled, and so general position is implied.

Since a terrain $\mathcal{T}$ is defined by a height function over a domain D, there is a natural bijection between points of $\mathcal{T}$ and points of D. Hence, the various objects defined in the previous paragraph can be viewed either in $\mathcal{T}$ or in D. Generally in the paper we shall refer to these objects by their projection in D, unless otherwise stated.

2.2 Input Model

The main idea of ***realistic input models*** is to parametrize certain properties of the input, which are suspected to capture contrived configurations leading to high complexities or running times. In cases where there exists a high discrepancy between the theoretical bounds and the complexities observed in practice, it is often useful to analyze the complexities not only as a function of the input size, but also with respect to these parameters. This sometimes leads to more informative asymptotic bounds. As such, the realistic input assumptions do not only distinguish between "good" and "bad" input, instead they enable a more differentiated view on which inputs are 'better' or 'worse'.

In this paper, we use the following realistic input model. A set of line segments is ***λ-low density*** if and only if the number of edges that intersect an arbitrary ball, which are longer than the radius of the ball, is smaller than λ. Low density has been used in the analysis of many different geometric problems, see [dBKSV02] for an overview.

To model a ***realistic terrain*** we adopt the realistic assumptions made in [AdBT08]. According to these assumptions, there exist constants λ and ξ independent of n, such that

(i) the set of edges of the triangulation, is a λ-*low density* set, and

(ii) any line segment embedded in the lifted triangulation has slope at most ξ.

We now state some useful facts that follow from these assumptions. For notational ease in the rest of the paper we define the constant $\beta = \sqrt{1+\xi^2}$.

First, the number of pairs of objects from two low density sets, which intersect each other is linear in the total number of objects in these sets. This can be easily verified, the idea is to charge each intersecting pair to the smaller of the two objects.

Fact 2.1 ([AdBT08]). *Let A be a set of n objects with λ-low density and let B be a set of m objects with ϕ-low density, then the number of pairs of objects $(u, v) \in A \times B$, such that u intersects v is $O(\lambda m + \phi n)$.*

Second, since the slope is bounded by a constant, the geodesic distance is the same as the euclidean distance up to a constant factor, and similarly, geodesic disks have an area that is approximately the same as that of planar disks in this case.

Fact 2.2 ([AdBT08]). *For any two points $\mathsf{q}, \mathsf{s} \in D$, we have that $\|\mathsf{q} - \mathsf{s}\| \le d_{\mathcal{T}}(\mathsf{q}, \mathsf{s}) \le \beta \|\mathsf{q} - \mathsf{s}\|$.*

Lemma 2.3. *Let D be a geodesic disk of radius r on the surface of a terrain with bounded slope ξ and let A denote its area. We have that $\pi(r/\beta)^2 \le A \le \pi\beta r^2$.*

Proof: Let c be the center of D and let c_p be its projection. Let $D_{r/\beta}$ and D_r be the planar disks with center c_p and radius r/β and r, respectively. Let $T_{r/\beta}$ and T_r denote the portion of the terrain that lies directly above $D_{r/\beta}$ and D_r, respectively.

Clearly the projection of D is contained in D_r and so if we can bound the area of T_r then this will bound the area of D. Observe that D_r consists of a set of triangles from Δ, which have been clipped at the boundary of D_r. It is a well-known fact that if we lift any such (clipped) triangle up to the terrain then its area can increase by at most a factor of β. Therefore the total area of T_r is at most $\pi\beta r^2$.

Now consider the disk $D_{r/\beta}$. First observe that $T_{r/\beta}$ must have area at least as large as that of $D_{r/\beta}$. Second, note that D must contain $T_{r/\beta}$ (since by Fact 2.2 the distance between any point $T_{r/\beta}$ and c is at most $\beta(r/\beta) = r$). Therefore the area of D is at least $\pi(r/\beta)^2$. ∎

2.3 Complexity of the Voronoi Diagram

The complexity of the Voronoi diagram is measured by the complexity of the structured set of curves and line segments that delineate the Voronoi cells. This set consists of pieces of bisectors and it can be characterized as follows. Again, we adopt the definitions used in [AdBT08].

For most of the points on a bisector, the shortest path to either site will be unique. If the shortest path is not unique, we call p a ***breakpoint***. The breakpoints partition the bisector into a set of curved pieces which we call ***chords***. The ***combinatorial complexity*** of the Voronoi diagram is now defined as the sum of (i) the number of Voronoi vertices, (ii) the number of breakpoints of Voronoi edges, and (iii) the number of intersections of the chords of Voronoi edges with the triangulation of the terrain.

We continue with some useful facts and lemmas used in the analysis of the complexity. First, it was observed by Moet *et al.* that the number of breakpoints of the Voronoi diagram is bounded by n, since each of them can be attributed to a terrain vertex. To see why this is true, imagine walking along the bisector while sweeping the shortest paths to either site from the current position. The points on the bisector, where the edge sequence of the shortest path changes, are the breakpoints.

Fact 2.4 ([MvKvdS08]). *Given a terrain $\mathcal{T}$ which is defined by a triangulation with n vertices, the number of breakpoints of any Voronoi diagram on $\mathcal{T}$ is smaller than or equal to n.*

Furthermore, we will use the following result by Aronov *et al.* [AdBT08].

Lemma 2.5 ([AdBT08]). *Given two points q and s, the set of chords that form the bisector of q and s on $\mathcal{T}$ (projected to the (x, y)-plane) is $O(\xi)$-low density.*

We remark that Aronov *et al.* use this result to show that the bisector has linear complexity. However, note that this result does not imply that the overall set of chords of the Voronoi diagram is low density, which would imply a linear complexity for the whole diagram. Consider for example the situation, where all the sites lie close to each other on a straight line, and all the triangles of the terrain surface are coplanar. In this example, the bisectors are pairwise parallel lines, which extend from one side of the domain to the other and could therefore lead to a quadratic complexity Voronoi diagram by intersecting many triangles of the terrain.

Finally, we observe that the number of Voronoi vertices and edges is linear in the number of sites, as the following lemma and corollary testify. This fact was also observed by Aronov *et al.*, we include an independent proof for the sake of completeness.

Lemma 2.6. *Let $\mathcal{T}$ be a terrain and let* P *be a set of m points. Then the number of Voronoi edges and Voronoi vertices of $\mathcal{V}or(\mathsf{P})$ is $O(m)$.*

Proof: For $m \leq 2$ the claim is clearly true, since we have at most one bisector, which contributes exactly one Voronoi edge. For $m > 2$, we argue as follows.

First, observe that the cells in this Voronoi diagram are connected. Indeed, consider a point p that belongs to the interior of the cell of $\mathsf{s} \in \mathsf{P}$. Consider the shortest path π from p to s, and consider any point $\mathsf{q} \in \pi$. If q is closer to some other site t than to s, then we have that

$$\begin{aligned} \mathsf{d}_{\mathcal{T}}(\mathsf{p},\mathsf{s}) &= \mathsf{d}_{\mathcal{T}}(\mathsf{p},\mathsf{q}) + \mathsf{d}_{\mathcal{T}}(\mathsf{q},\mathsf{s}) \\ &\geq \mathsf{d}_{\mathcal{T}}(\mathsf{p},\mathsf{q}) + \mathsf{d}_{\mathcal{T}}(\mathsf{q},\mathsf{t}) \geq \mathsf{d}_{\mathcal{T}}(\mathsf{p},\mathsf{t}), \end{aligned}$$

but this is a contradiction to p being in the interior of the cell of s.

Now, consider the dual graph $\mathcal{G}$ of the graph formed by the Voronoi vertices and Voronoi edges. In this graph, every vertex corresponds to a Voronoi cell and every face corresponds to a Voronoi vertex. Note that we can derive a geometric embedding of this graph by using the sites as vertices and picking an arbitrary point on each Voronoi edge and connecting it by its shortest path to either site to form a (possibly curved) edge between two vertices.

It is well-known that a cell in the Voronoi diagram might not be simply connected. Indeed, consider a mountain surrounded by a plane. If we place a site s on the top of the mountain, and a site t at the bottom of the mountain (and the mountain slope is large enough) then the Voronoi cell of s would be completely surrounded by the cell of t and thus would create a 'hole' in this cell. In $\mathcal{G}$, the two vertices that correspond to the cells of s and t would be connected by an edge, which is incident to only one face in $\mathcal{G}$.

Furthermore, it is known that the dual graph can have multiple edges between two sites. To see this, again, place s on the top of the mountain and place two sites t and r at the bottom, such that the bisector between t and r intersects the mountain. The boundary between the cells of t and r would contain two Voronoi edges from the same bisector.

However, the dual graph $\mathcal{G}$ is planar and connected and as such its Euler characteristic is 2. Hence $v - e + f = 2$, where e denotes the number of edges, f the number of faces and v the number of vertices of $\mathcal{G}$.

Now, by definition, every Voronoi vertex is incident to at least three Voronoi cells. Therefore, every face of $\mathcal{G}$ is incident to at least three edges of $\mathcal{G}$. Since $\mathcal{G}$ is planar, every edge of $\mathcal{G}$ is incident to at most two faces of $\mathcal{G}$. Hence, we have that $3f \leq 2e$, which implies that $3f \leq 2(v+f-2)$, and therefore it holds that $f \leq 2v - 2 = 2m - 2$. It follows that the number of Voronoi vertices is in $O(m)$. Applying Euler's formula again, we obtain that also the number of Voronoi edges is in $O(m)$. ■

Corollary 2.7. *Let $\mathcal{T}$ be a terrain and let* P *be a set of m points. Let $\square$ be a sub-square which intersects k Voronoi cells in their projection. The number of Voronoi edges which intersect $\mathcal{V}or(\mathsf{P}) \cap \square$ in their projection is in $O(k)$.*

3. UPPER BOUND

We prove the following lemma first in the planar case and then extend it to terrains with bounded slope. The bounded expected complexity then follows by examining the number of intersections of the chords with the terrain triangulation in an $\sqrt{m} \times \sqrt{m}$ grid.

Lemma 3.1. *Let* P *be a set of m points, sampled uniformly at random from a unit square, and let $\square$ be a sub-square contained in the unit square of side length $1/\sqrt{m}$. Then the expected number of points in* P *that contribute to $\mathcal{V}or(\mathsf{P}) \cap \square$ is $O(1)$.*

Proof: We place a sequence of exponentially growing disks centered at the center point of $\square$. Let $r_i = \frac{1}{\sqrt{2m}} 2^i$, for $i = 0, \ldots, k = \lceil \lg \sqrt{2m} \rceil$ (i.e. r_0 is the radius of the circumscribed circle of $\square$). Let d_i be the disk of radius r_i, which is clipped to the unit square and let $\mathcal{R}_i = \mathsf{d}_i \setminus \mathsf{d}_{i-1}$, for $i = 1, \ldots, k$.

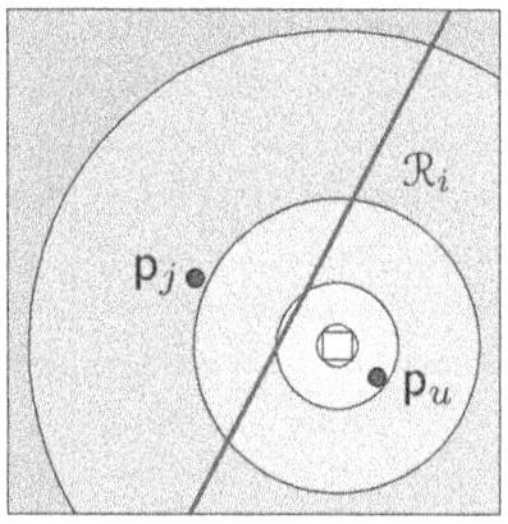

Let the points in P be labeled $\mathsf{p}_1, \ldots, \mathsf{p}_m$. Observe that the expected number of points from P that fall into d_2 is $(\pi r_2^2)m = \frac{16\pi m}{2m} = 8\pi = O(1)$. Hence we do not need to worry about their contribution to $\mathcal{V}or(\mathsf{P}) \cap \square$. Otherwise, we claim that a point p_j which falls into $\mathcal{R}_i$ for $i > 2$, can only contribute to $\mathcal{V}or(\mathsf{P}) \cap \square$ if d_{i-2} contains no points of P. Assume for the sake of contradiction that the Voronoi cell of p_j intersects $\square$, and there exists some point p_u in P that lies in d_{i-2}. By construction, we know p_j has distance greater than $(r_{i-1} - r_0)$ to any point in d_0 and p_u has distance at most $(r_{i-2} + r_0)$ to any point in d_0. Hence we have that $\mathsf{d}(\mathsf{p}_j, \mathsf{d}_0) > r_{i-1} - r_0 = 2r_{i-2} - r_0 \geq r_{i-2} + r_0 \geq \mathsf{d}(\mathsf{p}_u, \mathsf{d}_1)$ for $i > 2$ and as such every point in d_0 is strictly closer to p_u than p_j, where $\mathsf{d}(\mathsf{p}, X) = \min_{\mathsf{q} \in X} \|\mathsf{p} - \mathsf{q}\|$.

Hence it is sufficient to bound the expected number of points p_j which fall into an annulus $\mathcal{R}_i$ such that d_{i-2} is empty. For p_j we define the indicator variable X_i^j which is equal to 1 if and only if $\mathsf{p}_j \in \mathcal{R}_i$, and the indicator variable Y_i^j which is equal to 1 if and only if no other point falls into d_{i-2}. Hence a given point p_j can contribute to $\mathcal{V}or(\mathsf{P}) \cap \square$ if and only if $X_i^j Y_i^j = 1$ for some value of i. Now, we know that

$$\begin{aligned} \mathbf{Pr}\Big[X_i^j = 1\Big] &\leq \pi r_i^2 - \pi r_{i-1}^2 \\ &= \frac{\pi}{2m}(2^{2i} - 2^{2(i-1)}) \\ &= \frac{3\pi}{2m} 4^{i-1}, \end{aligned}$$

and

$$\begin{aligned} \mathbf{Pr}\Big[Y_i^j = 1\Big] &\leq \left(1 - \frac{1}{4}\pi r_{i-2}^2\right)^{m-1} \\ &\leq \exp\Big(-\frac{\pi}{4} r_{i-2}^2 (m-1)\Big) \\ &= \exp\left(-\frac{\pi 4^{i-3}(m-1)}{2m}\right) \\ &\leq \exp(-4^{i-3}). \end{aligned}$$

where a factor of 1/4 was added in the bound for Y_i^j due to boundary effects that might arise from the position of $\square$

in the unit square. Hence the number of points that can affect $\mathcal{V}or(\mathsf{P}) \cap \square$ is bounded by $\sum_j \sum_{i>2} X_i^j Y_i^j$, for which in expectation we have,

$$\begin{aligned}\mathbf{E}\left[\sum_j \sum_{i>2} X_i^j Y_i^j\right] &= \sum_j \sum_{i>2} \mathbf{E}\left[X_i^j\right] \mathbf{E}\left[Y_i^j\right] \\ &\le \sum_j \sum_{i>2} \frac{3\pi}{2m} 4^{i-1} e^{-4^{i-3}} \\ &= \frac{3\pi}{2} \sum_{i>2} 4^{i-1} e^{-4^{i-3}} = O(1),\end{aligned}$$

by linearity of expectation and the independence of X_i^j and Y_i^j for all i and j. ■

Lemma 3.2. *Let $\mathcal{T}$ be a terrain with bounded slope ξ. Let P be a random sample of m points, either sampled uniformly from $\mathcal{T}$, or uniformly from the unit square and then lifted vertically up to $\mathcal{T}$. Let $\square$ be a sub-square contained in the unit square of side length $1/\sqrt{m}$. Then the expected number of points in P that contribute to the portion of the Voronoi diagram of P on $\mathcal{T}$ that lies above $\square$ is $O(\beta^5)$.*

Proof: We adapt the proof of Lemma 3.1. To accommodate for the larger distances on the surface, we increase the radii of the disks slightly.

So, let $r_i = \frac{1}{\sqrt{2m}}(2\beta+1)^i$, for $i = 1, \ldots, k$. Let d_i be the disk (in the plane) of radius r_i centered at the center of $\square$. And let $\mathcal{R}_i = \mathsf{d}_i \setminus \mathsf{d}_{i-1}$, for $i = 1, \ldots, k$, as defined above. For points $\mathsf{p}, \mathsf{q} \in \mathsf{d}_{i-2}$ and $\mathsf{s} \in \mathcal{R}_i = \mathsf{d}_i \setminus \mathsf{d}_{i-1}$, we have that

$$\begin{aligned}\mathsf{d}_{\mathcal{T}}(\mathsf{p},\mathsf{q}) &\le \beta\|\mathsf{p}-\mathsf{q}\| \le \beta 2 r_{i-2} \\ &\le r_{i-1} - r_{i-2} < \|\mathsf{q}-\mathsf{s}\| \le \mathsf{d}_{\mathcal{T}}(\mathsf{q},\mathsf{s}).\end{aligned}$$

As such, as before, for a point in $\mathsf{P} \cap \mathcal{R}_i$ to affect the Voronoi diagram on $\square$ requires that d_{i-2} is empty of any points of P.

Now, consider the case that the points are sampled from the terrain. Define X_i^j and Y_i^j as in Lemma 3.1. Note that the area of the terrain can only increase from one by lifting the individual triangles. We have that

$$\begin{aligned}\mathbf{Pr}\left[X_i^j = 1\right] &\le \frac{\text{area on terrain of } \mathcal{R}_i}{\text{area of terrain}} \\ &\le \beta\left(\pi r_i^2 - \pi r_{i-1}^2\right) \\ &\le \frac{\pi\beta}{2m}\left((2\beta+1)^{2i} - (2\beta+1)^{2i-2}\right) \\ &\le \frac{\pi\beta}{2m}(2\beta+1)^{2i}.\end{aligned}$$

Similarly, since only a quarter of d_{i-2} might be in the terrain, the probability of this area is bounded from below by $\pi r_{i-2}^2/(4\beta)$. Plugging this into the analysis of Lemma 3.1, we have

$$\begin{aligned}\mathbf{Pr}\left[Y_i^j = 1\right] &\le \left(1 - \frac{\pi r_{i-2}^2}{4\beta^2}\right)^{m-1} \\ &\le \exp\left(-\frac{\pi r_{i-2}^2}{4\beta^2}(m-1)\right) \\ &\le \exp\left(-(2\beta+1)^{2i-5}\right).\end{aligned}$$

As before, we thus have

$$\begin{aligned}&\mathbf{E}\left[\sum_j \sum_{i>2} X_i^j Y_i^j\right] \\ &\le \sum_{j=1}^{m} \sum_{i>2} \frac{\pi\beta}{2m}(2\beta+1)^{2i} \cdot \exp\left(-(2\beta+1)^{2i-5}\right) \\ &= O(1).\end{aligned}$$

The only missing component is bounding the expected number of points of $\mathsf{P} \cap \mathsf{d}_2$, as they can affect the Voronoi diagram in $\square$. Arguing as above, this quantity is bounded by $m\beta \cdot (\text{area of } \mathsf{d}_2) \le m\beta\pi r_2^2 \le \frac{\beta}{2}(2\beta+1)^4 = O(\beta^5)$.

Note that if we drop the factors of β in the bounds for X_i^j, Y_i^j, and the area of d_2, then the above becomes a proof for the case when we sample from the unit square ■

Theorem 3.3. *Let $\mathcal{T}$ be a terrain. Let P be a random sample of m points, either sampled uniformly from the surface of $\mathcal{T}$, or uniformly from the domain and then lifted vertically up to the surface. The expected combinatorial complexity of $\mathcal{V}or(\mathsf{P})$ is in $O(\xi\beta^5\lambda(n+m))$.*

Proof: As described in Section 2.3, the combinatorial complexity of the Voronoi diagram is the sum of the number of breakpoints of Voronoi edges, the number of Voronoi vertices and the number of intersections of triangulation edges with chords of Voronoi edges. By Fact 2.4 the number of breakpoints is bounded by $O(n)$, and by Lemma 2.6 the number of Voronoi vertices is bounded by $O(m)$.

It remains to bound the number of intersections of the set of chords with the triangulation. To this end, we place a grid on the domain of the terrain, such that the side length of each grid cell is $l = 1/\sqrt{m}$ and we obtain $M = O(m)$ grid cells which together cover the domain of the terrain. Now, let $\mathsf{C}_1, \ldots, \mathsf{C}_M$ denote these grid cells. Consider the grid cell C_i and the set of chords of the Voronoi diagram which intersect this grid cell, let this set be B_i. Similarly, let E_i denote the subset of edges of the triangulation, which intersect C_i. Since we assumed that the triangulation is λ-low density, also E_i is a λ-low density set. By Fact 2.5 we have that the set of chords, which originate from the same Voronoi edge (and therefore from the same bisector) form an $O(\xi)$-low density set. Let k_i denote the number of Voronoi edges that contribute chords to B_i, we have that B_i is a $O(\xi k_i)$-low density set. By Fact 2.1, the number of intersections between objects of E_i and objects of B_i is in $O(\xi k_i |\mathsf{E}_i| + \lambda|\mathsf{B}_i|)$.

Now, in order to bound the overall number of intersections, let $\mathsf{B}_i^{>l}$ denote the subset of chords which are longer than l, similarly, let $\mathsf{B}_i^{\le l}$ denote the chords in B_i which have length smaller or equal to l and let $\mathsf{E}_i^{\le l}$ and $\mathsf{E}_i^{>l}$ be defined analogously. By the above analysis, we have that there exists some constant c_1, such that it holds for the overall number of intersections χ,

$$\begin{aligned}\chi &\le c_1 \sum_{i\ge 1}^{M} (\xi k_i |\mathsf{E}_i| + \lambda|\mathsf{B}_i|) \\ &= c_1 \sum_{i\ge 1}^{M} \left(\xi k_i\left(\left|\mathsf{E}_i^{>l}\right| + \left|\mathsf{E}_i^{\le l}\right|\right) + \lambda\left(\left|\mathsf{B}_i^{>l}\right| + \left|\mathsf{B}_i^{\le l}\right|\right)\right).\end{aligned}$$

By the definition of low density sets, we have that $\left|\mathsf{E}_i^{>l}\right| = O(\lambda)$ and $\left|\mathsf{B}_i^{>l}\right| = O(\xi k_i)$, since they intersect the bounding ball of the grid cell C_i, which has radius $O(l)$. Therefore, it must be that there exists a constant c_2 such that,

$$\chi \leq c_1 \left(\sum_{i \geq 1}^{M} \xi k_i \left| \mathsf{E}_i^{\leq l} \right| + \lambda \left| \mathsf{B}_i^{\leq l} \right| + c_2 \lambda \xi k_i \right)$$
$$\leq c_1 \left(\sum_{i \geq 1}^{M} \xi k_i \left| \mathsf{E}_i^{\leq l} \right| + c_2 \lambda \xi k_i \right) + c_1 \lambda 4 \left| \mathsf{B} \right| ,$$

where the last inequality follows from the fact that any chord in $\mathsf{B}_i^{\leq l}$ can intersect at most four grid cells, since the grid cells have side length equal to l (similarly any edge in $\mathsf{E}_i^{\leq l}$ can intersect at most three grid cells).

Finally, note that the number of Voronoi cells that are expected to intersect a grid cell in their projection is bounded by $O(\beta^5)$ by Lemma 3.2. Thus by Corollary 2.7 we have that $\mathbf{E}[k_i] = O(\beta^5)$. Therefore in expectation,

$$\mathbf{E}[\chi] \leq c_1 \, \mathbf{E} \left[\sum_{i \geq 1}^{M} \left(\xi k_i \left| \mathsf{E}_i^{\leq l} \right| + c_2 \lambda \xi k_i \right) \right] + c_1 \lambda 4 \left| \mathsf{B} \right|$$
$$= c_1 \sum_{i \geq 1}^{M} \left(\xi \, \mathbf{E}[k_i] \left| \mathsf{E}_i^{\leq l} \right| + c_2 \lambda \xi \, \mathbf{E}[k_i] \right) + c_1 \lambda 4 \left| \mathsf{B} \right|$$
$$\leq c_3 \xi \beta^5 \lambda \left(\sum_{i \geq 1}^{M} \left(\left| \mathsf{E}_i^{\leq l} \right| + 1 \right) + 4 \left| \mathsf{B} \right| \right)$$
$$\leq c_3 \xi \beta^5 \lambda (3 \left| \mathsf{E} \right| + M + 4 \left| \mathsf{B} \right|)$$

for some constant c_3 (note that we used the fact that $\left| \mathsf{E}_i^{\leq l} \right|$ is independent of the random sampling). Furthermore, observe that by Lemma 2.6 the overall number of Voronoi edges is $O(m)$. Recall that every Voronoi edge is broken up by breakpoints into chords. Every breakpoint increases the number of chords by one. Using Fact 2.4, it follows that the overall number of chords $|\mathsf{B}|$ is in $O(n + m)$. Since the number of edges of the triangulation $|\mathsf{E}|$ is $O(n)$, we conclude that $\mathbf{E}[\chi] = O(\xi \beta^5 \lambda (n + m))$. ■

4. LOWER BOUND

In this section we show that if we drop the assumptions on the terrain, then the expected worst-case complexity of the resulting geodesic Voronoi diagram can be $\Omega\left(nm^{2/3}\right)$ if the sites are sampled uniformly at random from the unit square.

In the following we will refer to the walls of the unit square defined by $x = 0$, $x = 1$, $y = 0$, and $y = 1$ as the ***west, east, south, and north walls***, respectively.

Example 4.1. We start with the easy case of planar map and points in the unit square. Refer to Figure 1(i).

First place m points in a column near and parallel to the west wall of the unit square such that the spacing between each adjacent pair of points is $\Theta(1/m)$ (we assume $m = O(n)$). The planar map consists of n vertical lines near the east wall of the unit square that extend from the north wall to the south wall. Now, the boundaries of the Voronoi cells of these points extend from the west to the east wall and are parallel to the north and south walls, and hence the complexity of the overlay of the Voronoi diagram with the planar map is $\Theta(nm)$, since it is an $n \times m$ grid.

The later constructions use this as their starting point.

4.1 Farming – an $\Omega(n\sqrt{m})$ example

4.1.1 Construction

The height function used in the following construction of a terrain has (essentially) only two values, zero and h. The areas between a part of the terrain that is of height zero and of height h consist of very narrow and steep boundary regions. This intermediate boundary would have a very small measure in the projection, and the reader can think of it as having measure zero. Moreover, h is chosen to be sufficiently large so that no point at height zero can affect the Voronoi diagram at height h. One can therefore view the following terrain construction as a flat unit square, where we have cut out or "forbidden" areas (that have height 0). Therefore, for the sake of simplicity of exposition, an area being constructed is flat, at height h, and the adjacent forbidden area is at height zero. Our main building blocks will be farms. We define a ***farm*** to be a square of side length $1/(c\sqrt{m})$. Intuitively, farms are part of the terrain which with constant probability (the constant will depend on c) will receive at least one point from the random sample (i.e. a farm takes the place of a point from the example in Exercise 4.1). We define the ***diameter*** of a farm to be the quantity $\delta = \sqrt{2}/(c\sqrt{m})$ (that is, the distance of the furthest two points in a farm).

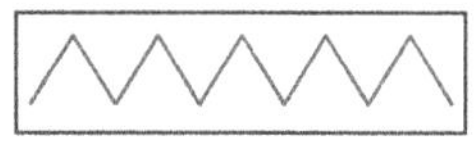

We now define a sequence of ridges to take the place of the planar map from Exercise 4.1 (i.e. in expectation we would like the Voronoi diagram to look like a grid over the ridges). Formally, let a ***sequence of ridges*** of length n be a sequence of $2n$ rectangles, $r_1, \ldots, r_{2m}$ such that the east edge of r_i is the same as the west edge of r_{i+1} for $i = 1, \ldots, 2n - 1$, r_i has a slope of 45° for odd i and $-45°$ for even i, all rectangles extend from the north to south walls of the unit square, and the geodesic distance from the west edge of r_1 to the east edge of r_{2n} is $1/(c2^n)$ (which is $O(1/2^m)$ since we assumed $m = O(n)$). Refer to the figure above to the right for a side view of the ridges.

The construction of the $\Omega(n\sqrt{m})$ example is as follows. Place $\Theta(\sqrt{m})$ farms from north to south along the west wall of the unit square, with $2/(c\sqrt{m})$ spacing in between each adjacent pair. Next build a sequence of $\Theta(n)$ ridges near (and parallel to) the east wall of the unit square. Then connect each farm directly to the westmost ridge by creating a line parallel to the north and south walls connecting the south-east corner of the farm to the first ridge. See Figure 1(ii). We refer to such a line as ***road***. The roads stay at height h and to the left and right of a road, the height drops to zero as described earlier.

4.1.2 Analysis

Definition 4.2. *The point at which a farm connects to its road is its* **entrance** *(i.e. the south-east corner of the farm), and the point at which the road connects to the westmost ridge is its* **exit**. *We say that the point (from the random sample of m points) that is closest to the entrance for some farm, is that farm's* **dominating point**. *Let p and e be the dominating point and exit, respectively, of some farm. We say that another point q from the random sample that is contained in another farm and such that $\mathsf{d}_{\mathcal{T}}(q, e) < \mathsf{d}_{\mathcal{T}}(p, e)$,*

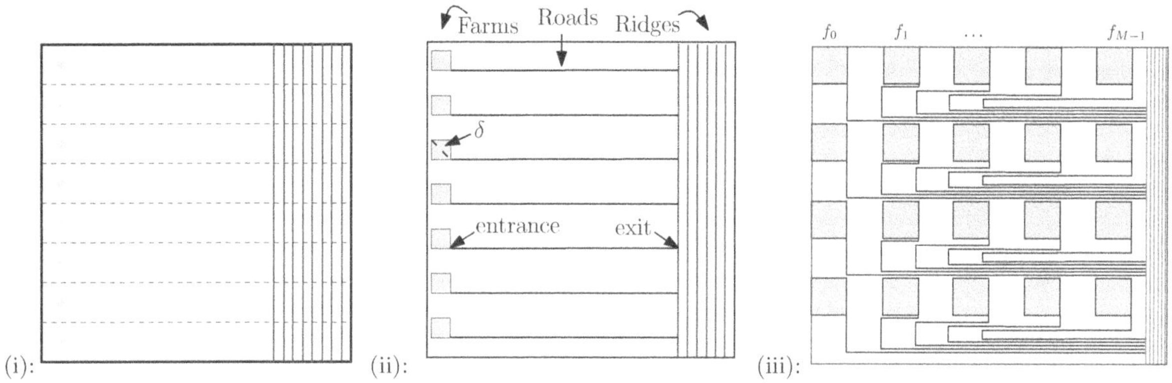

Figure 1: (i) Starting point of the construction (ii) $\Theta(\sqrt{m})$ farms, (iii) $\Theta(m^{2/3})$ farms.

eliminates *(the Voronoi cell of) p, where $d_{\mathcal{T}}(p,e)$ denotes the shortest path on the terrain from p to e. If there are no points which eliminate a given dominating point, then the dominating point is* **alive**.

Lemma 4.3. *For the construction of the terrain described above, if one picks uniform at random m points in the unit square, their induced geodesic Voronoi diagram on this terrain has complexity $\Omega(n\sqrt{m})$.*

Proof: The area of each farm is $\Theta(1/m)$ and hence a sample of m points picked uniformly at random from the unit square, will have at least one point with constant probability in each farm. Moreover, since we constructed $\Theta(\sqrt{m})$ farms, this implies that in expectation $\Theta(\sqrt{m})$ farms will receive at least one point. Furthermore, the width of the sequence of ridges and roads was chosen such that the probability that either receives a point is exponentially small (and hence in the following we assume they do not receive any point).

Now consider a farm which received at least one point, and let p be its dominating point. Observe that the Voronoi cell of p contains the entire road connecting this farm to the ridges, and its Voronoi cell extends all the way to the eastmost edge of the sequence of ridges, and hence will be of complexity $\Omega(n)$. Indeed, by our construction, only a point from another farm can prevent the Voronoi cell of p from reaching the eastmost ridge. However, the spacing of the farms was chosen to prevent this. In the worst case p is in the north-west corner of its farm, and an adjacent farm has a point q at the south-east corner. Let l be the length of a road. Let e_p (resp. e_q) be the exit of the farm containing p (resp q). Now consider the geodesic shortest path connecting e_p to the eastmost ridge. Every point on this segment is in distance at most $\delta + l + 1/(c2^n)$ from p. However, the closest point on this segment from q is at a distance of at least $l + 2/(c\sqrt{m}) \geq \delta + l + 1/(c2^n)$, and so q cannot prevent the Voronoi cell of p from reaching all the way to the eastmost ridge.

Therefore, in expectation, we have that $\Theta(\sqrt{m})$ farms have a point whose Voronoi cell extends all the way across the sequence of ridge, which gives a Voronoi diagram that in expectation has complexity $\Omega(n\sqrt{m})$. ∎

4.2 Industrial farming – an $\Omega\left(nm^{2/3}\right)$ example

The challenge in improving the example above is that the distance of a dominating point to the exit of a farm has too much variance (i.e., $\sqrt{1/m}$). Since there does not seem to be a way to decrease the variance directly, we still connect all the farms to the ridges, and carefully argue about the expected complexity of the generated Voronoi diagram.

4.2.1 Construction

In the following we assume that $m = O(n)$.

We set the side length of each square farm to be $1/\sqrt{m}$ and construct a sequence of $\Theta(n)$ ridges near the east wall of the unit square. We will place an $M \times M$ grid of farms inside the unit square, where $M = \lfloor\sqrt{m}/4\rfloor$. Specifically, the spacing between columns (which extend from north to south) will be $1/\sqrt{m}$ and the spacing between rows (which extend from west to east) will be $2/\sqrt{m}$. The grid starts in the north-west corner of the unit square.

We now describe the connecting roads from the farms to the ridges. The following construction of the roads will ensure that the length of each road is the same and that the distance between adjacent exits on the ridges is at least $1/m$. These two properties will be sufficient for the analysis in the next section to go through.

Consider a given row of farms. Number the farms in this row $f_0, \ldots, f_{M-1}$ in increasing order of their distance to the west wall. Every farm has dimensions $1/\sqrt{m} \times 1/\sqrt{m}$, and the spacing between two consecutive farms in a row is $1/\sqrt{m}$. As such, the x coordinate of the entrance of the ith farm is $x_i = (2i+1)/\sqrt{m}$ (as before, the entrance to each road will be at the south-east corner of the farm). The directions the ith farm's road goes from entrance to exit is described as follows:

(a) south for a distance of $\alpha_{i,a} = i/m$,
(b) west for a distance of $\alpha_{i,b} = (x_i + \alpha_{i,a})/2$,
(c) south for a distance of $\alpha_{i,c} = 1/\sqrt{m} - 2\alpha_{i,a}$, and
(d) east for a distance of $\alpha_{i,d} = w - (x_i - \alpha_{i,b})$ all the way to the first ridge, where w is the distance from the west wall to the first ridge.

This layout is sketched in Figure 1(iii). Note that the spacing in this figure only approximately matches the description.

Sanity checks.

The road of the ith farm starts at x coordinate x_i, goes west for a distance of $\alpha_{i,b}$ and east for a distance of $\alpha_{i,d}$. Observe that the x coordinate of the exit of this road is $x_i - \alpha_{i,b} + \alpha_{i,d} = x_i - \alpha_{i,b} + w - (x_i - \alpha_{i,b}) = w$.

Observe that the ith farm will connect to the ridges in north to south distance $\alpha_{i,a} + \alpha_{i,c} = 1/\sqrt{m} - i/m$ from the southern boundary of the row of farms. That is, adjacent farms in the row have exits in distance i/m apart along the first ridge (exits are $\Theta(1/\sqrt{m})$ apart between rows). Furthermore, each road is of the same length. Indeed, let r_i be the length of the road for the ith farm to its exit. We have that

$$r_i = \alpha_{i,a} + \underbrace{\frac{x_i + \alpha_{i,a}}{2}}_{\alpha_{i,b}} + \underbrace{\frac{1}{\sqrt{m}} - 2\alpha_{i,a}}_{\alpha_{i,c}} + \underbrace{w - \left(x_i - \frac{x_i + \alpha_{i,a}}{2}\right)}_{\alpha_{i,d}}$$
$$= \frac{1}{\sqrt{m}} + w.$$

That is, all the roads have exactly the same length.

4.2.2 Competing farms

We now prove that in expectation $\Theta\left(m^{2/3}\right)$ dominating points will have their Voronoi cells reach all the way to the east wall across the sequence of ridges.

Observation 4.4. Let p and e_p be the dominating point and exit, respectively, of some farm f. Let q be a point which is in some other farm f' with exit e_q. If f' is i farms away (in the north to south order of the exits of the farms along the first ridge) then $\|e_p - e_q\| = i/m$ and hence q eliminates p if and only if $\mathsf{d}_{\mathcal{T}}(q, e_p) = \mathsf{d}_{\mathcal{T}}(q, e_q) + i/m < \mathsf{d}_{\mathcal{T}}(p, e_p)$.

Next, we prove that each dominating point is alive with probability $\Omega\left(1/m^{1/3}\right)$, using the following helper lemmas.

Lemma 4.5. *Let X be a positive random variable with expected value μ. We then have that $\mathbf{E}\left[e^{-X}\right] \geq e^{-2\mu}/2$*

Proof: Markov's inequality implies that $\mathbf{Pr}[X < 2\mu] = 1 - \mathbf{Pr}[X \geq 2\mu] \geq 1 - \mu/2\mu = 1/2$. Therefore, by the definition of expectation, we get $\mathbf{E}\left[e^{-X}\right] \geq \mathbf{Pr}[X \geq 2\mu] * 0 + \mathbf{Pr}[X < 2\mu] * e^{-2\mu} \geq e^{-2\mu}/2$. ■

Lemma 4.6. *Let f be a farm, and let $r(f)$ be the random variable that is the distance of the closest site (that falls into this farm) from the farm entrance (if there is no site in this farm, we set $r(f) = \sqrt{2/m}$). Then, for any distance s, we have $\mathbf{Pr}\left[r(f) \leq s\right] \leq ms^2\pi/4$, where equality holds for $0 \leq s \leq 1/\sqrt{m}$.*

Proof: Recall that the entrance of a farm is at the south-east corner. Hence a point in the farm which is in distance at most s from the entrance must fall into the intersection of the farm with a circle of radius s whose center is at the entrance of the farm, see figure.

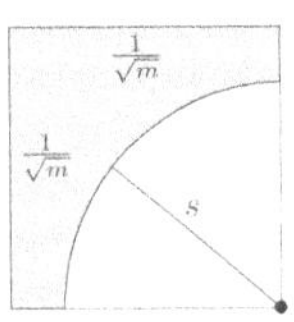

Therefore, if the radius of the circle is less than the side length of the square, i.e. $s \leq 1/\sqrt{m}$, then the intersection is a quarter disk and so the area is exactly $\pi s^2/4$. Otherwise, the top and left portions of the quarter disk will be needed to be clipped to the farm and so the area is $\leq \pi s^2/4$. Now, the probability of the ith site to fall into this disk is $\leq \pi s^2/4$, and since we sample m sites (independently), by the union bound the claim follows. ■

Lemma 4.7. *Let p and e be the dominating point and exit, respectively, of a farm f. Let $r = \mathsf{d}_{\mathcal{T}}(p, e)$. Let f_i be a farm which is i farms away from f (either in north or in south direction), and let X_i be the number of points which fell into f_i. Let α_{X_i} denote the probability that no point from f_i eliminates p (see Definition 4.2). Then $\alpha_{X_i} \geq \exp(-m(r - i/m)^2 X_i/2)$.*

Proof: Let $q_1, \ldots, q_{X_i}$ be the X_i points that fall into farm f_i. Let e_i be the exit of f_i, and let $d_j = \mathsf{d}_{\mathcal{T}}(q_j, e_i)$, for $j = 1, \ldots, X_i$. Arguing as in Lemma 4.6, we have that $\mathbf{Pr}[d_j \leq s] \leq s^2\pi/4$ for all j. By Observation 4.4, a point q_j eliminates p if and only if $d_j < r - i/m$. For $j \neq l$, whether or not q_j or q_l kill p are independent events and hence

$$\alpha_{X_i} = \prod_{j=1}^{X_i} \mathbf{Pr}\left[d_j \geq r - i/m\right]$$
$$\geq \left(1 - \frac{(r - i/m)^2 \frac{\pi}{4}}{\text{area}(f)}\right)^{X_i}$$
$$\geq \exp\left(-m(r - i/m)^2 \frac{\pi X_i}{2}\right).$$
■

Lemma 4.8. *Let p and e be the dominating point and exit, respectively, of some farm f. Let $r = \mathsf{d}_{\mathcal{T}}(p, e)$. Let X_i (resp. Y_i), for $i = 1, \ldots, \lfloor rm \rfloor$, denote the number of points which fall into the farm which is i farms to the north (resp. south), from f in the order of the exits along the first ridge. Then the probability that p is not eliminated given the values of X_i and Y_i for all i, is at least*

$$\exp\left(-m \sum_{i=1}^{\lfloor rm \rfloor} (r - i/m)^2 \pi (X_i + Y_i)/2\right)$$

Proof: First note that for $i' > \lfloor rm \rfloor$, we have that $i'/m \geq (\lfloor rm \rfloor + 1)/m > r$. Namely no point from a farm i' farms away can kill p, and hence we can ignore such farms.

Given the value X_i and Y_i for all i, whether a farm contains a point which eliminates p is independent from whether any other farm contains a point which eliminates p. Therefore, by Lemma 4.7,

$$\mathbf{Pr}\left[p \text{ is alive} \,\middle|\, X_1, \ldots, X_{\lfloor rm \rfloor}, Y_1, \ldots, Y_{\lfloor rm \rfloor}\right]$$
$$= \left(\prod_{i=1}^{\lfloor rm \rfloor} \alpha_{X_i}\right)\left(\prod_{i=1}^{\lfloor rm \rfloor} \alpha_{Y_i}\right)$$
$$\geq \left(\prod_{i=1}^{\lfloor rm \rfloor} \exp(-m(r - i/m)^2 \pi X_i/2)\right)$$
$$\cdot \left(\prod_{i=1}^{\lfloor rm \rfloor} \exp(-m(r - i/m)^2 \pi Y_i/2)\right)$$
$$= \exp\left(-m \sum_{i=1}^{\lfloor rm \rfloor} (r - i/m)^2 \pi (X_i + Y_i)/2\right)$$
■

Lemma 4.9. *Let p and e be the dominating point and exit, respectively, of some farm f, and let $r = \mathsf{d}_{\mathcal{T}}(p, e)$. Then the probability that p is alive is $\geq \frac{1}{2}\exp\left(-2r^3m^2\right)$.*

Proof: Let X_i and Y_i, for $i = 1, \ldots, \lfloor rm \rfloor$, be random variables equal to the number of points which fall into the farm which is i farms to the north or south, respectively, from f in the order of the exits along the first ridge (note that if there is no farm i farms to the north or south, then $X_i = 0$ or $Y_i = 0$, respectively).

Lemma 4.8 tells us that

$$\mathbf{Pr}\left[p \text{ is alive} \,\middle|\, X_1, \ldots, X_{\lfloor rm \rfloor}, Y_1, \ldots, Y_{\lfloor rm \rfloor}\right] \geq e^{-T},$$

where $T = \sum_{i=1}^{\lfloor rm \rfloor}(r - i/m)^2\pi(X_i + Y_i)/2$. Since the area of each farm is $1/m$, We know that $\mathbf{E}[X_i] \leq 1$ and $\mathbf{E}[Y_i] \leq 1$ for $i = 1, \ldots, \lfloor rm \rfloor$ ("≤ 1" is used instead of "$= 1$" since there might not be a farm at that distance). Therefore,

$$\begin{aligned}\mathbf{E}[T] &= \mathbf{E}\left[\sum_{i=1}^{\lfloor rm \rfloor} m(r - i/m)^2\pi(X_i + Y_i)/2\right] \\ &= \sum_{i=1}^{\lfloor rm \rfloor} m(r - i/m)^2\pi(\mathbf{E}[X_i] + \mathbf{E}[Y_i])/2 \\ &\leq \sum_{i=1}^{\lfloor rm \rfloor} m(r - i/m)^2 \leq m(r^2m) = r^3m^2\end{aligned}$$

By Lemma 4.5,

$$\begin{aligned}\mathbf{Pr}\left[p \text{ is alive}\right] &\geq \mathbf{E}\left[e^{-T}\right] \\ &\geq \frac{1}{2}\exp(-2\,\mathbf{E}[T]) \\ &\geq \frac{1}{2}\exp\left(-2r^3m^2\right).\end{aligned}$$

■

Observation 4.10. Lemma 4.9 implies that if $r \leq 1/m^{2/3}$ then

$$\mathbf{Pr}[p \text{ is alive}] \geq \frac{1}{2}\exp\left(-2r^3m^2\right) \geq \frac{1}{2e^2}.$$

Lemma 4.11. *The probability that a farm f gives rise to a Voronoi cell that is not eliminated is $\geq \pi/(8e^2m^{1/3})$.*

Proof: By the equality in Lemma 4.6 for short distances, we know that $\mathbf{Pr}[r(f) \leq s] = ms^2\pi/4$, for $0 \leq s \leq 1/\sqrt{m}$. Let p be the dominating point of f. By Observation 4.10 and Lemma 4.6, we have

$$\begin{aligned}\mathbf{Pr}[p \text{ is alive}] &\geq \mathbf{Pr}\left[\left(r(f) \leq m^{-2/3}\right) \cap (p \text{ is alive})\right] \\ &= \mathbf{Pr}\left[\,p \text{ is alive} \,\middle|\, r(f) \leq m^{-2/3}\right] \\ &\quad\cdot \mathbf{Pr}\left[r(f) \leq m^{-2/3}\right] \\ &\geq \frac{1}{2e^2}\mathbf{Pr}\left[r(f) \leq m^{-2/3}\right] \\ &\leq \frac{1}{2e^2}\cdot\frac{\pi m}{4m^{4/3}} \\ &= \frac{\pi}{8e^2m^{1/3}}\end{aligned}$$

■

Theorem 4.12. *In expectation, the Voronoi diagram will be of complexity $\Omega\left(nm^{2/3}\right)$.*

Proof: Every farm which receives a point from the random sample has a dominating point. Since $\Theta(m)$ farms were built, and each farm receives one point in expectation, the expected number of dominating points is $\Theta(m)$. Therefore, by Lemma 4.11, the expected number of alive dominating points is $\Omega\left(m * (\pi/(8e^2m^{1/3}))\right) = \Omega\left(m^{2/3}\right)$.

Given that a dominating point is alive, the probability that its Voronoi cell does not reach the eastmost ridge is negligible. Therefore, in expectation, if there are $\Omega\left(m^{2/3}\right)$ alive dominating points and $\Theta(n)$ ridges then the complexity of the Voronoi diagram will be $\Omega\left(nm^{2/3}\right)$. ■

5. CONCLUSIONS

We investigated the expected combinatorial complexity of geodesic Voronoi diagrams on polyhedral terrains in two settings where the sites are being picked randomly. Usually, such random settings are the great simplifier – for example, the expected complexity of the convex hull of n points picked uniformly in the unit square is $O(\log n)$ – but in our case the situation is considerably more subtle.

We proved that the expected complexity is linear if one assumes low density and bounded slope and the domain of the terrain is a unit square. On the other hand, we described a worst-case construction of a terrain which implies a super-linear lower bound on the expected complexity if these assumptions are dropped. This implies that the probabilistic analysis alone does not yield a linear complexity.

There are still many interesting open question for further research. In particular, if we relax the realistic input assumptions, is the expected complexity still linear or can one show other lower bounds? One could, for instance allow the terrain to have a constant number of triangles, where the slope is unbounded, or drop the steepness assumption completely. Consider the farming layout to the right, where the sequence of ridges have been replaced by a recursive low density construction. One problem with this construction is that the low density assumption requires the ridges to have a non-negligible area, which could catch points from the random sample.

6. REFERENCES

[AdBT08] B. Aronov, M. de Berg, and S. Thite. The complexity of bisectors and Voronoi diagrams on realistic terrains. In *Proc. 16th Annu. European Sympos. Algorithms*, pages 100–111, 2008.

[AJT01] H. Alani, C. B. Jones, and D. Tudhope. Voronoi-based region approximation for geographical information retrieval with gazetteers. *Int. J. of Geo. Inf. Sci.*, 15(4):287–306, 2001.

[CG10] R. Cavalli and S. Grigolato. Influence of characteristics and extension of a forest road network on the supply cost of forest woodchips. *J. Forest Res.*, 15:202–209, 2010. 10.1007/s10310-009-0170-4.

[dBKSV02] M. de Berg, M. J. Katz, A. F. van der Stappen, and J. Vleugels. Realistic input models for geometric algorithms. *Algorithmica*, 34:81–97, 2002.

[DEG08] O. Devillers, J. Erickson, and X. Goaoc. Empty-ellipse graphs. In *Proc. 19th ACM-SIAM Sympos. Discrete Algorithms*, SODA '08, pages 1249–1257, 2008.

[Dey11] T.K. Dey. *Curve and Surface Reconstruction: Algorithms with Mathematical Analysis.* Cambridge Monographs on Applied and Computational Mathematics. Cambridge University Press, 2011.

[DG08] M. T. Dickerson and M. T. Goodrich. Two-site Voronoi diagrams in geographic networks. In *Proc. 16th ACM SIGSPATIAL int. conf. on Adv. geo. inf. sys.*, GIS '08, pages 59:1–59:4. ACM, 2008.

[DHPR11] Anne Driemel, Sariel Har-Peled, and Benjamin Raichel. On the expected complexity of voronoi diagrams on terrains. *CoRR*, abs/1112.0695, 2011.

[JN04] F. Jarai-Szabo and Z. Neda. On the size-distribution of poisson voronoi cells. *eprint arXiv:cond-mat/0406116*, June 2004.

[KWR97] R. Kunze, F.-E. Wolter, and T. Rausch. Geodesic Voronoi diagrams on parametric surfaces. In *Proc. of Comp. Graph. Int.*, pages 230–237, 1997.

[Moe08] E. Moet. *Computation and Complexity of Visibility in Geometric Environments.* PhD thesis, Utrecht University, 2008.

[MvKvdS08] Esther Moet, Marc J. van Kreveld, and A. Frank van der Stappen. On realistic terrains. *Comput. Geom. Theory Appl.*, 41(1-2):48–67, 2008.

[OBSC00] A. Okabe, B. Boots, K. Sugihara, and S. N. Chiu. *Spatial tessellations: Concepts and applications of Voronoi diagrams.* Probability and Statistics. Wiley, 2nd edition edition, 2000.

[PA06] J.N. Portela and M.S. Alencar. Cellular network as a multiplicatively weighted voronoi diagram. In *Consum. Comm. Net. Conf.*, volume 2, pages 913—917, 2006.

[Ray70] H. Raynaud. Sur l'enveloppe convex des nuages de points aleatoires dans R^n. *J. Appl. Probab.*, 7:35–48, 1970.

[RS63] A. Rényi and R. Sulanke. Über die konvexe Hülle von n zufällig gewählten Punkten I. *Z. Wahrsch. Verw. Gebiete*, 2:75–84, 1963.

[Rup95] J. Ruppert. A Delaunay refinement algorithm for quality 2-dimensional mesh generation. *J. Algorithms*, 18(3):548–585, 1995.

[Sch07] Y. Schreiber. Shortest paths on realistic polyhedra. In *Proc. 23rd Annu. ACM Sympos. Comput. Geom.*, pages 74–83. ACM, 2007.

[SF04] G. Stylianou and G. Farin. Crest lines for surface segmentation and flattening. *IEEE Trans. Vis. Comp. Graphics*, 10(5):536–544, 2004.

[SS08] Y. Schreiber and M. Sharir. An optimal-time algorithm for shortest paths on a convex polytope in three dimensions. *Discrete Comput. Geom.*, 39:500–579, 2008.

[TS89] O. Takahashi and R.J. Schilling. Motion planning in a plane using generalized Voronoi diagrams. *IEEE Trans. Robot. Auto.*, 5(2):143–150, apr 1989.

[WDB+08] O. Weber, Y. S. Devir, A. M. Bronstein, M. M. Bronstein, and R. Kimmel. Parallel algorithms for approximation of distance maps on parametric surfaces. *ACM Trans. Graph.*, 27:104:1–104:16, November 2008.

[WW93] W. Weil and J. A. Wieacker. Stochastic geometry. In P. M. Gruber and J. M. Wills, editors, *Handbook of Convex Geometry*, volume B, chapter 5.2, pages 1393–1438. North-Holland, 1993.

On Klee's Measure Problem for Grounded Boxes

Hakan Yıldız
Department of Computer Science
University of California, Santa Barbara
Santa Barbara, CA 93106, USA
hakan@cs.ucsb.edu

Subhash Suri
Department of Computer Science
University of California, Santa Barbara
Santa Barbara, CA 93106, USA
suri@cs.ucsb.edu

ABSTRACT

A well-known problem in computational geometry is Klee's measure problem, which asks for the volume of a union of axis-aligned boxes in d-space. In this paper, we consider Klee's measure problem for the special case where a 2-dimensional orthogonal projection of all the boxes has a common corner. We call such a set of boxes *2-grounded* and, more generally, a set of boxes is *k-grounded* if in a k-dimensional orthogonal projection they share a common corner.

Our main result is an $O(n^{(d-1)/2} \log^2 n)$ time algorithm for computing Klee's measure for a set of n 2-grounded boxes. This is an improvement of roughly $O(\sqrt{n})$ compared to the fastest solution of the general problem. The algorithm works for k-grounded boxes, for any $k \geq 2$, and in the special case of $k = d$, also called the *hypervolume indicator problem*, the time bound can be improved further by a $\log n$ factor. The key idea of our technique is to reduce the d-dimensional problem to a semi-dynamic *weighted volume* problem in dimension $d-2$. The weighted volume problem requires solving a combinatorial problem of maintaining the *sum of ordered products*, which may be of independent interest.

Categories and Subject Descriptors

F.2.2 [**Analysis of Algorithms and Problem Complexity**]: Nonnumerical Algorithms and Problems—*Geometrical problems and computations*

General Terms

Algorithms, Theory

Keywords

Grounded boxes, hypervolume indicator, Klee's measure, sum of ordered products, weighted volume

SCG'12, June 17–20, 2012, Chapel Hill, North Carolina, USA.

1. INTRODUCTION

Klee's Measure Problem [14] is a classical problem in computational geometry, dating back to 1970s: Given a set of n axis-aligned boxes in d-space, compute the volume of their union. The first non-trivial solution, with running time $O(n^{d-1} \log n)$, was given by Bentley using his space-sweep approach [3], which was quickly improved to $O(n^{d-1})$ time by van Leeuwen and Wood [16]. Several years later, Overmars and Yap [15] made a breakthrough and achieved the bound of $O(n^{d/2} \log n)$, which essentially remains unbeaten more than 20 years later [8]. All the improvements during these intervening years have come in the form of reducing the log factor (the best bound achieved by Chan [8]) or for cubes, unit hypercubes and "fat" boxes [1, 2, 5, 7, 9]. The problem is known to be $\#P$-hard when the dimension is part of the input [6], so the exponential dependence on the dimension seems unavoidable.

In this paper, we present a new result for Klee's measure under the assumption that a 2-dimensional orthogonal projection of all the boxes has a common corner. We call such a collection of boxes *2-grounded.* (When the boxes share a common corner in a k-dimensional projection, for $1 \leq k \leq d$, we call the set *k-grounded.* Naturally, the algorithm for 2-grounded boxes work trivially for k-grounded, where $k \geq 2$.) Our approach has the following general form. We transform the d-dimensional problem into one of maintaining its $(d-1)$-dimensional cross-section with a sweeping plane. The sweeping is a standard step used in algorithms for the Klee's measure, but instead of solving the cross-section problem directly in $(d-1)$-space, we exploit the grounding property and transform the problem into a $(d-2)$-dimensional *weighted volume* problem. In this problem, each box has a non-negative weight and, in computing the volume, the contribution of each point in space equals the weight of the heaviest box containing it. Solving the weighted volume problem efficiently is the main result of our paper. In particular, we show that the d-dimensional weighted volume can be maintained under insertion at the amortized cost of $O(n^{(d-1)/2} \log^2 n)$, which leads to an $O(n^{(d-1)/2} \log^2 n)$ algorithm for the d-dimensional Klee's Measure Problem on 2-grounded boxes. In solving the weighted volume problem, we also introduce and solve a combinatorial problem of maintaining the *sum of ordered products*, which may be of independent interest.

Besides their intrinsic theoretical interest as a special case of the general Klee's problem, the k-grounded boxes also arise frequently in applications when some of the coordinate axes have a natural "start" or "end" position for ranges. For instance, if one of the axes represents time recording the duration of some event, then the origin marks the natural start point, and therefore a source of grounding on that axis. Similarly, in sensor databases, physical attributes such as temperature, humidity etc. naturally have a common "lower bound," and the measured quantity is really the deviation from this "grounded" value. In these cases, as long as two or more dimensions are grounded, our algorithm leads to an improved bound for computing the volume of the union.

Finally, if the boxes are d-grounded, namely they all share a common corner, our bound improves by a $\log n$ factor. This particular case of Klee's measure is also known as the *hypervolume indicator*, which is a metric frequently used in multi-objective optimization and evolutionary computing [10,12,17]. This particular special case has been studied extensively on its own, as well as a special case of unit cubes. Nevertheless, our new algorithm achieves a new improved bound for the following dimensions: $d = 4, 5, 6$.

The paper is organized in seven sections. In Section 2, we explain how to reduce the d-dimensional problem to a $(d-2)$-dimensional weighted volume problem. In Sections 3 and 4, we solve a special case of the weighted volume involving *halfspaces*, which turns out to be the key problem. In Section 5, we show how the general weighted problem can be solved based on the solution of the halfspace problem. In Section 6, we briefly mention our results for the hypervolume indicator. In Section 7, we conclude with a summary and discussion.

2. DIMENSION REDUCTION: SWEEPING AND WEIGHTING

The classical approach of Overmars and Yap [15] solves the d-dimensional Klee's Measure Problem by reducing it to a dynamic $(d-1)$-dimensional problem. Our key idea is to reduce the d-dimensional problem to a $(d-2)$-dimensional *weighted* volume problem with *insert-only* updates. We reduce one dimension by plane sweep, and another by converting the unweighted problem to a weighted problem. We begin with the high level ideas behind the plane sweep, and the weighting.

Reducing a Dimension by Plane Sweep. Consider a set of axis-aligned boxes $\mathcal{B} = \{B_1, \ldots, B_n\}$ in d-space, for which we want to compute the volume of their union. Without loss of generality, we assume that all the boxes are contained in the positive orthant $\mathbb{R}^d_+$—otherwise, we can divide the problem into 2^d groups, one for each quadrant. We say that a box B is *grounded* with respect to dimension k if B's extent along the kth dimension has the form $(0, B^k)$, where $B^k > 0$ is the length of B along dimension k. Supposing that all boxes in $\mathcal{B}$ are grounded with respect to dimension d, we can compute the volume of their union as follows. We sort the boxes in $\mathcal{B}$ in the descending order of their dth coordinate. Without loss of generality, let the resulting order be $B_1, B_2, \ldots, B_n$, meaning that $B_1^d > B_2^d > \cdots > B_n^d > 0$. We further assume that $B_{n+1}^d = 0$. Then, it is easy to see that the total volume covered by $\mathcal{B}$ is given by the formula

$$\sum_{i=1}^{n} V_{i,d-1} \times (B_i^d - B_{i+1}^d),$$

where $V_{i,d-1}$ is the volume of the set $\{B_1, \ldots, B_i\}$ *projected*

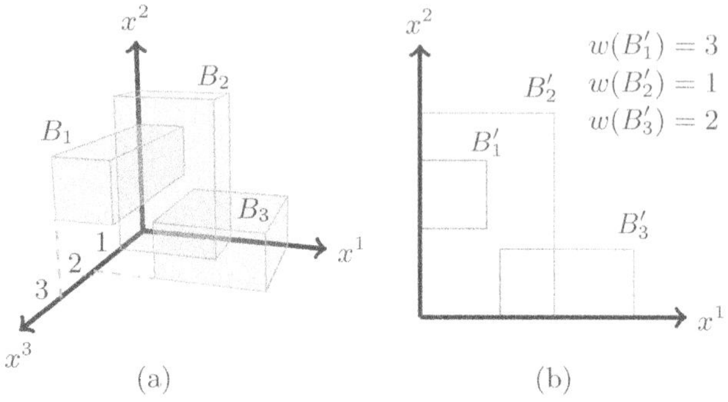

Figure 1: (a) A set of 3 boxes B_1, B_2, B_3 in 3-space, grounded with respect to the x^3 axis, and (b) their 2-dimensional weighted projections.

onto the plane $\mathbf{x}^d = 0$.[1] This is easily seen by visualizing a plane parallel to $\mathbf{x}^d = 0$, sweeping the space from B_1^d to $B_{n+1}^d = 0$, and by observing that the intersection of this plane with the boxes of $\mathcal{B}$ is constant between two consecutive coordinates in the sorted list, determined entirely by the boxes that precede B_i. If we write $V_{i,d-1}$ for the $(d-1)$-dimensional volume of this intersection, then the d-dimensional volume that lies between B_i^d and B_{i+1}^d is precisely $V_{i,d-1} \times (B_i^d - B_{i+1}^d)$.

The $(d-1)$-dimensional slice changes only when the sweeping plane encounters a new box B_i, and thus we can compute the d-dimensional volume of n boxes by maintaining the $(d-1)$-dimensional volume of their projections subject to n insert-only updates. If the *amortized* update and query cost is $O(F(n))$, then the d-dimensional problem can be solved in time $O(nF(n))$ plus the preprocessing cost including the initial sorting.

Reducing a Dimension by Weighting. Another conceptual way to reduce the dimension of the volume problem is the following. Assuming again that the boxes of $\mathcal{B}$ are grounded with respect to the dth dimension, we orthogonally project each box B_i in $\mathcal{B}$ onto the plane $\mathbf{x}^d = 0$, obtaining a $(d-1)$-dimensional box, denoted B'_i, and assign a *weight* $w(B'_i) = B_i^d$ to this projected box. The *weighted volume* of B'_i is defined as $w(B'_i)$ times the (ordinary, $(d-1)$-dimensional) volume of B'_i. See Figure 1 for illustration.

This transformation converts the set $\mathcal{B}$ into a set $\mathcal{B}' = \{B'_1, B'_2, \ldots, B'_n\}$ of $(d-1)$-dimensional weighted boxes. For each point $\mathbf{x} \in \mathbb{R}^{d-1}_+$ on the plane $\mathbf{x}^d = 0$, assign to it the weight of the heaviest box containing $\mathbf{x}$. That is,

$$w(\mathbf{x}) = \max\{w(B'_i) \mid \mathbf{x} \in B'_i\}$$

Then the volume covered by the boxes in $\mathcal{B}$ is given by the integral

$$\int_{\mathbf{x} \in \mathbb{R}^{(d-1)}} w(\mathbf{x})\, d\mathbf{x}$$

We call this expression the *weighted volume* of the union of the set $\mathcal{B}'$. Intuitively, this expression weights each point $\mathbf{x}$ on the projection plane by the maximum *height* of a box in $\mathcal{B}$ that contains it, thus correctly computing the volume contribution of each box in $\mathcal{B}$.

[1] $\mathbf{x}^d$ denotes the dth coordinate of point $\mathbf{x}$.

Joint Sweep and Weighting for 2-grounded Boxes. Suppose $\mathcal{B}$ is a set of boxes, grounded along dimensions d and $d-1$, and that their orthogonal projection onto the dimensions $(d-1)$ and d has origin as the common corner. We can now apply both of the above dimension-reduction techniques simultaneously, as follows. We sweep the space along the dth dimension, which requires us to dynamically maintain the $(d-1)$-dimensional volume of the set of boxes intersecting the sweep plane. The updates are *insert only* because boxes are only inserted, and never deleted. Depending on that the boxes are grounded with respect to $(d-1)$th dimension, we maintain the dynamic $(d-1)$-dimensional volume by converting it to a *weighted* $(d-2)$-dimensional volume.

Thus, the d-dimensional Klee's Measure Problem is transformed into a $(d-2)$ dimensional problem of maintaining the weighted volume under insertion of boxes. Solving this problem efficiently is the main contribution of this paper, and the focus of the next three sections. In fact, the crux of the problem proves to be the following special case:

Under insert-only updates, maintain the weighted volume of a set of axis-parallel halfspaces.

In reality, we require the weighted volume of axis-parallel *strips*, but generalizing the halfspace solution to strips is relatively straightforward, incurring an extra $\log n$ factor in complexity. In the interest of simplicity, therefore, we focus on halfspace in the next two sections, and return to the strips only in Section 4.4. For ease of presentation, we first describe the halfspace algorithm in two dimensions (Section 3), and then its generalization to higher dimensions. (Section 4). Section 5 describes how to solve the general weighted volume by combining the strips problem with a space partition technique.

3. WEIGHTED VOLUME OF HALFSPACES: THE PLANAR CASE

We have a set $\mathcal{H}$ of n axis-parallel weighted halfplanes in 2-space, each containing the origin, and a positive orthant axis-aligned rectangle R anchored at the origin. For a point $\mathbf{p}$ in R, define the weight of $\mathbf{p}$ with respect to a subset $\mathcal{H}' \subseteq \mathcal{H}$, denoted $w(\mathbf{p}, \mathcal{H}')$, as the weight of the heaviest halfplane in $\mathcal{H}'$ that contains $\mathbf{p}$; if no such halfplane exists, then the weight is zero. The weighted volume (area) of $\mathcal{H}'$ over R is $\int_{\mathbf{p}\in R} w(\mathbf{p}, \mathcal{H}')\, d\mathbf{p}$. Our goal is to maintain the weighted volume as $\mathcal{H}'$ undergoes insert-only updates. We will show a data structure with amortized cost of $O(\log n)$ per insertion. The set of axis-aligned halfplanes in $\mathcal{H}'$ is naturally divided into two groups: vertical and horizontal. The vertical halfplanes have the form $0 \le \mathbf{x}^1 \le a$, and the horizontal ones have the form $0 \le \mathbf{x}^2 \le b$. We maintain the weight distribution imposed by these two classes separately, and then show how to compute the joint weight implicitly and efficiently.

3.1 Vertical and Horizontal Gradients

The intersection of R with a halfplane H is a "strip," either vertical or horizontal, containing the origin. Let us focus on the vertical halfplanes of $\mathcal{H}'$, and consider the *partition* they induce on R where each point of R is "claimed" by the maximum weight halfplane containing it. This partition is a sequence of vertical strips in which each strip belongs entirely to one halfplane, each halfplane contributes at most one strip, and the strips are ordered in *descending* weight order from left to right. This follows because all halfplanes contain the origin, and a larger weight halfplane completely overrides the smaller weight halfplane to its left. Visually, the resulting structure looks like a "waterfall", and for ease of reference, we call it the *vertical gradient.*[2] Similarly, the horizontal halfplanes, considered in isolation, induce a *horizontal gradient* of strips ordered in descending order of weights from bottom to top. (Figure 2(a) shows an example, where the vertical gradient consists of three strips of respective weights $w_7 > w_4 > w_1$, and respective widths 4.5, 3.5, and 5.)

The gradients give a nice and compact representation of the weight structure imposed by the vertical and the horizontal halfplanes *separately*. In order to compute the weighted volume of $\mathcal{H}'$ over R, however, we need to *intersect* the two gradients. An explicit intersection entails $\Omega(n^d)$ complexity in d-dimensions, and so the key is to perform this intersection *implicitly*.

Let w_i be the weight of the ith halfplane $H_i \in \mathcal{H}$, and assume that the halfplanes are ordered in the increasing order of their weights. Without loss of generality, we assume that the weights w_i's are distinct. We maintain two arrays, A_1 and A_2, storing the *widths* of the strips contributed by vertical and horizontal halfplanes, respectively. Both the arrays have size n, whose ith entries correspond to the halfplane H_i. The entry $A_1[i]$ or $A_2[i]$ stores the width of the strip contributed by H_i: if H_i is vertical, it contributes to $A_1[i]$, otherwise to $A_2[i]$. (A halfplane contributes to neither gradient if it is dominated by heavier halfplanes, in which case both the entries are zero.) (See Figure 2(b) for an illustration.)

Let L_1, L_2 be the dimensions of the rectangular region R. Then it is easy to see that $\sum_i A_1[i] \le L_1$ and $\sum_i A_2[i] \le L_2$. (The vertical strips whose widths populate A_1 are disjoint, and their sum cannot exceed the width of R.) Furthermore, *considering each gradient in isolation*, the weighted volume contribution of H_i is precisely $w_i \times (A_1[i] \cdot L_2 + A_2[i] \cdot L_1)$: this follows because only one of $A_1[i]$ or $A_2[i]$ can be nonzero, and so this term is precisely the weighted area of the rectangular strip of H_i.

The next problem is to determine how much of each gradient is claimed by the other. We do that in the next subsection, but first let us consider how to maintain the arrays A_1 and A_2 under *insertion of new halfplanes*. Consider an update to A_1; the horizontal case A_2 is entirely symmetric. When a vertical halfplane H, with weight w, is to be inserted, we first determine its *rank*, namely, the *index* i for the weight w in the ordered sequence $w_1, \ldots, w_n$, which can be done in $O(\log n)$ time. Suppose the rank of H is i, namely, $H \equiv H_i$. Two changes occur in the gradient structure: (1) some of the vertical strips that overlap with H_i are deleted—in particular, those with weights less than w_i, and (2) the strip containing the vertical line defining H_i "shrinks" in width. We can locate the strip to be shrunk in logarithmic time by maintaining a binary search tree on the strips, keyed by their position, and then appropriately update its array value. Starting with that strip, we then traverse the list of vertical strips to the left, deleting each as long as their weights are less than w_i. These strips are deleted also from the search tree, in logarithmic time per

[2]This name is inspired by color gradients, which are images with decreasing color intensity in one direction.

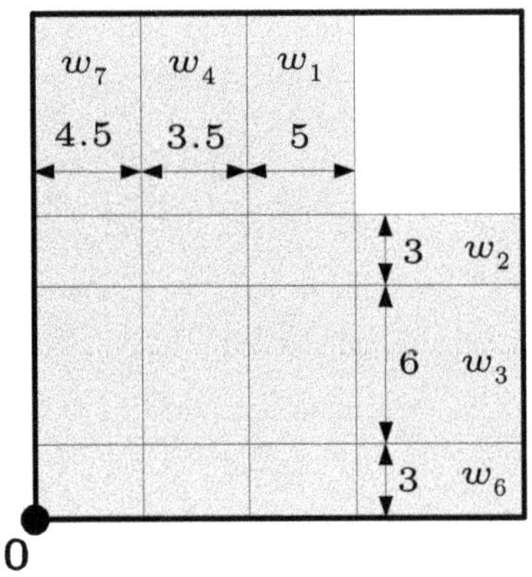

(a)

i	$A_1[i]$	$A_2[i]$
1	5	0
2	0	3
3	0	6
4	3.5	0
5	0	0
6	0	3
7	4.5	0

(b)

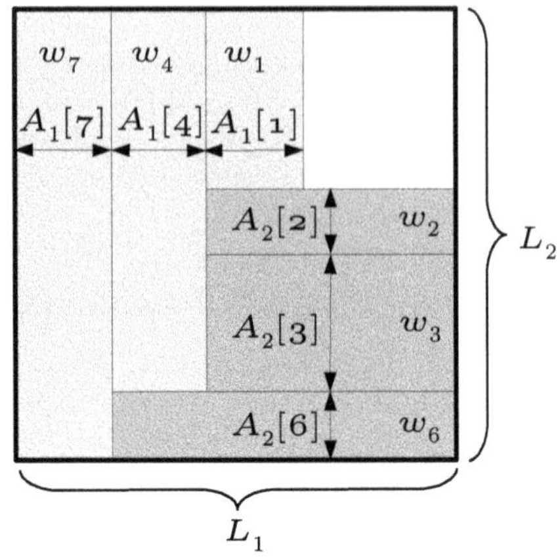

(c)

Figure 2: (a) Illustrating the gradient structures. The vertical gradient has three strips, with weights in descending order $w_7 > w_4 > w_1$, and widths $4.5, 3.5$ and 5, respectively. (b) The array representation of the two gradients shown in (a). The halfplane H_5 is either not inserted yet or covered by halfplanes with higher weights, and so its array entries are zero. (c) Illustrating the "intersection" of the two gradients. The "light gray" (resp., "dark gray") section shows the portion where the weight of the vertical (resp., horizontal) gradient dominates. Recall that the halfplane with larger index has higher weight than the one with smaller index.

strip. (We note that once a strip is deleted from the gradient, it is never reinserted.) We can easily compute the width of the strip produced by H_i and write this value to $A_1[i]$, leading to the following lemma.

LEMMA 1. *The arrays A_1 and A_2 representing the vertical and horizontal gradients can be updated in $O(\log n)$ amortized time after the insertion of a halfplane. The number of array entries whose values change is $O(1)$ amortized.*

3.2 Intersecting Gradient Volumes via Partial Sums

The key problem now is to deduce the correct weighted volume by overlapping the two gradients. Any point that is covered by both the vertical and the horizontal gradients should only be counted once, and receive the weight of the heavier halfplane containing it. In particular, let V_1 denote the weighted volume *contributed* by the vertical gradient: this is the weighted sum over the points where the vertical halfplanes prevail. Similarly, V_2 is the contribution of the horizontal gradient. We claim that V_1 has the following form:

$$V_1 = \sum_{i=1}^{n}\left(w_i \times A_1[i] \times \left(L_2 - \sum_{j=i+1}^{n} A_2[j]\right)\right)$$

This follows because $\sum_{i=1}^{n} w_i \cdot A_1[i] \cdot L_2$ is the weighted volume *without* considering the horizontal gradient, and the subtracted term is precisely what the horizontal gradient claims away from vertical strips. More precisely, for the halfplane H_i, the portion claimed by the horizontal gradient involves only those strips whose weight is larger than w_i, and if such a strip has "thickness" $A_2[j]$, then the area claimed by the vertical strip away from H_i is $A_1[i] \cdot A_2[j]$. (See Figure 2(c).) By subtracting the total over all such horizontal strips leaves the portion that H_i contributes to the final weighted volume. The complementary term V_2 has the similar form:

$$V_2 = \sum_{i=1}^{n}(w_i \times A_2[i] \times (L_1 - \sum_{j=i+1}^{n} A_1[j]))$$

We need to maintain these weighted volumes under insert-only updates, which we do by using a dynamic partial sums structure. Recall that the *partial sums* problem is the following:

> Maintain an array $\mathcal{A}$ of size n under an intermixed sequence of *update* and *query* operations, where $update(i, \Delta)$ changes $A[i]$ to $A[i] + \Delta$, and $query(k)$ reports the partial sum $\sum_{i=1}^{k} A[i]$.

Using a balanced binary tree, one can easily support these operations in $O(\log n)$ time each using linear space. The partial sums structure allows us to maintain our weighted volumes V_1 and V_2 as the gradient structures change due to insertion of new halfplanes. In particular, when a halfplane insertion changes $A_1[i]$ by Δ, the value V_1 changes by

$$w_i \times \Delta \times (L_2 - \sum_{j=i+1}^{n} A_1[j])$$

This requires computing the partial sum $\sum_{i=j+1}^{n} A_1[j]$, which is easily done by computing the partial sum $query(j)$, and subtracting it from $query(n)$, in $O(\log n)$ time. On the other hand, if the halfplane H changes the entry $A_2[j]$, then the change in V_1 is $-(\Delta \times \sum_{i=1}^{j-1} w_i \times A_1[i])$. By maintaining a partial sums structure for the array $A[i] = w_i \times A_1[i]$, this update can also be implemented in logarithmic time. In summary, the disjoint weighted volume contributions V_1 and V_2 of the two gradients can be maintained at the cost of $O(\log n)$ time per update, and we have the following key result.

THEOREM 1. *The weighted volume of a set of axis-aligned halfplanes in a rectangle R can be maintained in $O(\log n)$*

amortized time per insertion, with a linear-space data structure.

4. MULTIDIMENSIONAL WEIGHTED VOLUME

In dimension $d > 2$, the basic idea is the same: we organize the halfspaces in d independent gradient structures, which can be updated efficiently when a new halfspace is added. The key difference from the two-dimensional problem arises in how we compute the final weighted volume by intersecting the d gradients structures. Unlike the planar case, where we only needed to compute partial sums, the higher dimensional problems involves a more complex *sum of ordered products*. We discuss the details of these sums in the next two subsections, but first let us express the general form of the d-dimensional gradients.

We have a set $\mathcal{H}$ of n axis-parallel weighted halfspaces, each containing the origin, in d-space. We also have an orthogonal region R in the positive quadrant, anchored at the origin. We wish to maintain the weighted volume of a subset $\mathcal{H}' \subseteq \mathcal{H}$ over R, under insertions of halfspaces from $\mathcal{H}$. We maintain d gradient structures, where the jth structure, denoted G_j, is formed by the halfspaces normal to the jth axis, for $j = 1, 2, \ldots, d$. The gradient G_j consists of strips, in descending weight order, along the positive $\mathbf{x}^j$ direction. We represent G_j as an array A_j of size n, where the entry $A_j[i]$ contains the width of the strip contributed by the halfspace H_i to the gradient G_j, with weight w_i. The array indices are arranged in the increasing weight order of the halfspaces. Each of these d arrays can be maintained at the amortized cost of $O(\log n)$, per insertion of a halfspace, requiring modifications to a constant (amortized) number of array entries, using the technique of the previous section.

4.1 Intersecting the Gradients

Let us now consider how to maintain the weighted volume of the current set of halfspaces $\mathcal{H}'$, given the d gradient arrays. Write the weighted volume as the sum $V_1 + \cdots + V_d$, where V_j is the contribution by G_j to the total volume. We show how to maintain V_1; the others are analogous. In the absence of the other gradients, the weighted volume induced by a strip in G_1 is simply the product of its weight, its width and the $(d-1)$-dimensional volume of its projection on the plane $\mathbf{x}^1 = 0$. Formally, the weighted volume of ith strip is $w_i \times A_1[i] \times \prod_{\alpha=2}^{d} L_\alpha$, where L_α is the length of R along the αth coordinate axis. But some of this volume is *lost* to other gradients because of their higher weight. In particular, for a nonempty set $S = \{\alpha_1, \ldots, \alpha_m\} \subseteq \{2, \ldots, d\}$, let $Loss_i(S)$ be the weighted volume of the ith strip in G_1 that is claimed by all of the gradients in the set $\{G_{\alpha_1}, \ldots, G_{\alpha_m}\}$. In other words, $Loss_i(S)$ is the intersection of the weighted volumes claimed by the gradients $G_{\alpha_1}, \ldots, G_{\alpha_m}$. We can write $Loss_i(S)$ as $w_i \times A_1[i]$ times the $(d-1)$-dimensional volume on the plane $\mathbf{x}^1 = 0$ that is covered by the intersection of heavier subsections of the gradients $G_{\alpha_1}, \ldots, G_{\alpha_m}$. Formally, we write

$$Loss_i(S) = w_i \times A_1[i] \times \prod_{\alpha \in \{2,\ldots,d\} \setminus S} L_\alpha \times \sum_{i < j_{\alpha_1} \le n} A_{\alpha_1}[j_{\alpha_1}] \times \cdots \times \sum_{i < j_{\alpha_m} \le n} A_{\alpha_m}[j_{\alpha_m}]$$

Let $Loss(S)$ be the portion of G_1's weighted volume (over all of its strips) that is claimed by the intersection of the set of gradients $\{G_{\alpha_1}, \ldots, G_{\alpha_m}\}$. Then, we clearly have

$$\begin{aligned} &Loss(S) \\ &= \sum_{1 \le i \le n} Loss_i(S) \\ &= \sum_{1 \le i \le n} \Bigg(w_i \times A_1[i] \times \prod_{\alpha \in \{2,\ldots,d\} \setminus S} L_\alpha \times \\ &\quad \sum_{i < j_{\alpha_1} \le n} A_{\alpha_1}[j_{\alpha_1}] \times \cdots \times \sum_{i < j_{\alpha_m} \le n} A_{\alpha_m}[j_{\alpha_m}] \Bigg) \\ &= \prod_{\alpha \in \{2,\ldots,d\} \setminus S} L_\alpha \times \sum_{1 \le i < j_{\alpha_1}, j_{\alpha_2}, \ldots, j_{\alpha_m} \le n} \Bigg(w_i \times \\ &\quad A_1[i] \times A_{\alpha_1}[j_{\alpha_1}] \times \cdots \times A_{\alpha_m}[j_{\alpha_m}] \Bigg) \end{aligned}$$

The volume contributed by G_1, namely, V_1 can be written as the quantity *not* claimed by any of the other gradients. Using the inclusion-exclusion principle, we get

$$V_1 = \prod_{2 \le \alpha \le d} L_\alpha \times \sum_{1 \le i \le n} w_i \times A_1[i] \; - \sum_{S \subseteq \{2,\ldots,d\} \wedge S \neq \emptyset} (-1)^{|S|+1} Loss(S)$$

In this expression, the product involving L_α's is a constant, while the sum $\sum_{1 \le i \le n} w_i \times A_1[i]$ is easily maintained. The non-trivial term is the summation involving $Loss$, and we observe that it contains $(2^{d-1} - 1)$ inner terms of the following general form:

$$C \times \sum_{1 \le i_1 < i_2, i_3, \ldots, i_k \le n} \mathcal{A}_1[i_1] \times \cdots \times \mathcal{A}_k[i_k],$$

where C is constant, the number of indices k is at most d, and each array $\mathcal{A}_j[\,]$ corresponds to either a gradient array or the array $\mathcal{A}[i] = w_i \times A_1[i]$. We further decompose each term of the above form into $(k-1)!$ terms by grouping the inner terms of the summation by the ordering of their indices $i_2, \ldots, i_k$. This yields at most $(d-1) \times (d-1)!$ terms of the following form:[3]

$$C \times \sum_{1 \le i_1 < \cdots < i_k \le n} \mathcal{A}_1[i_1] \times \cdots \times \mathcal{A}_k[i_k].$$

Let us call this expression a *sum of ordered products*. All we need now is a data structure that can maintain the sum of ordered products efficiently as array entries are modified. We show in the following subsection (Section 4.2) how to maintain the sum of ordered products in $O(\log n)$ time per update and $O(1)$ time per query, which is sufficient for the following key theorem for the weighted volume of halfspaces in any fixed dimension d.

THEOREM 2. *The weighted volume of a set of axis-aligned halfspaces in a rectangular box R in d dimensions can be*

[3]By distinct weight assumption, all inner terms that contain equal indices are 0, and so we can safely ignore them.

maintained in $O(\log n)$ amortized time per insertion, with a linear-space data structure.

4.2 Maintaining the Sum of Ordered Products

Given d arrays $\mathcal{A}_i$, $i = 1, 2, \ldots, d$, each of size n, we want to efficiently maintain the following sum of ordered products, under updates to individual array entries:

$$\sum_{1 \le i_1 < \ldots < i_d \le n} \mathcal{A}_1[i_1] \times \ldots \times \mathcal{A}_d[i_d]$$

For simplicity, let us assume that n is a power of two. For $0 \le j \le \log n$ and $1 \le k \le n/2^j$, let $S(j,k)$ denote the set of consecutive integers in the range $[(k-1)2^j + 1, k2^j]$. Observe that $S(0,k) = \{k\}$ and $S(\log n, 1) = \{1, \ldots, n\}$. Moreover, $S(j,k)$ is the concatenation of $S(j-1, 2k-1)$ and $S(j-1, 2k)$, for $j \ge 1$. Conceptually, S represents a hierarchically ordered binary partition of the set $\{1, \ldots, n\}$ into singleton integers. If the partition is viewed as a tree, then $S(j,k)$ refers to the kth node from the left at the jth level and it is the parent of $S(j-1, 2k-1)$ and $S(j-1, 2k)$.

Let $T(j,k,l,r)$ denote the following sum of ordered products

$$\sum_{i_l < \ldots < i_r \ \wedge \ i_l, \ldots, i_r \in S(j,k)} \mathcal{A}_l[i_l] \times \ldots \times \mathcal{A}_r[i_r],$$

where $0 \le j \le \log n$, $1 \le k \le n/2^j$ and $1 \le l \le r \le d$. We observe that $T(\log n, 1, 1, d)$ is the sum of ordered products that we want to maintain. Additionally, $T(0,k,l,l) = \mathcal{A}_l[k]$ and $T(0,k,l,r) = 0$ for $l \ne r$. For $j \ge 1$, we can write $T(j,k,l,r)$ in terms of $T()$ values whose first parameter is $(j-1)$, as shown in the following lemma.

LEMMA 2. *For $j \ge 1$, we have the following recurrence:*

$$T(j,k,l,r) = T(j-1, 2k-1, l, r) + T(j-1, 2k, l, r) + \sum_{l \le c < r} \Big(T(j-1, 2k-1, l, c) \times T(j-1, 2k, c+1, r) \Big)$$

PROOF. See Appendix A. □

We maintain $T()$ values in a table: for each valid selection of (j,k,l,r), the table has an entry storing $T(j,k,l,r)$. Then, a query can be answered in $O(1)$ time by reporting $T(\log n, 1, 1, d)$. Moreover, it is easy to show that the table has $O(n)$ entries. When an array entry $A_l[k]$ is updated, we update the table entry for $T(0,k,l,l)$ and the table entries for all $T()$ values that are dependent on $T(0,k,l,l)$ through the recurrence relation given in Lemma 2. Notice that updating an entry takes constant time, assuming d is a constant. It can be easily seen that for a constant value of j, there are at most $d + \binom{d}{2}$ table entries dependent on $T(0,k,l,l)$. Thus, assuming d is a constant, $O(\log n)$ entries are updated in total. This leads to the following theorem.

THEOREM 3. *The sum of ordered products of d arrays of size n can be maintained with an update time of $O(\log n)$, query time of $O(1)$, using a linear-space data structure.*

4.3 Higher Order Partial Sums and Sum of Ordered Products

The reader will notice that the partial sum problem utilized in the halfplane solution is replaced by sum of ordered products in higher dimensions, suggesting a link between the two problems. In fact, the sum of ordered products can be viewed as an *iterated* or *higher-order* generalization of the classical partial sum problem.

Given an array $\mathcal{A}$ of n numbers, the basic *partial sum* problem, addresses the following query operation: report $query(k) = \sum_{i=1}^{k} \mathcal{A}[i]$. There are n different partial sum queries, for $1 \le k \le n$, and one can view them as forming another array $\mathcal{A}'[k] = query(k)$. We can then ask the partial sum problem on $\mathcal{A}'$, which could be considered the *second order* partial sum of $\mathcal{A}$. An iterated application of this process leads to higher order partial sums, with the following general form.

> Given an array $\mathcal{A}$ of size n, its *kth partial sum of order d*, denoted $P_d(k)$, is defined recursively as follows:
>
> $$P_d(k) = \begin{cases} \sum_{i=1}^{k} \mathcal{A}[i] & \text{if } d = 1 \\ \sum_{i=1}^{k} P_{d-1}(i) & \text{if } d > 1 \end{cases}$$

Considering the d arrays $\mathcal{A}_1, \ldots, \mathcal{A}_d$ of size n, the following definition is a further generalization of the iterated partial sums problem for $\mathcal{A}_1$:

> Given d arrays $\mathcal{A}_1, \ldots, \mathcal{A}_d$ of size n, *their kth weighted partial sum*, denoted $W_d(k)$, is defined as follows:
>
> $$W_d(k) = \begin{cases} \sum_{i=1}^{k} \mathcal{A}_1[i] & \text{if } d = 1 \\ \sum_{i=1}^{k} \mathcal{A}_d[i] \times W_{d-1}(i) & \text{if } d > 1 \end{cases}$$

The following lemma shows that the sum of ordered products is actually a weighted partial sum.[4]

LEMMA 3. *For d arrays $\mathcal{A}_1, \ldots, \mathcal{A}_d$,*

$$W_d(n) = \sum_{1 \le i_1 \le \cdots \le i_d \le n} \mathcal{A}_1[i_1] \times \cdots \times \mathcal{A}_d[i_d]$$

PROOF. By induction. The lemma clearly holds for $d = 1$. For $d > 1$,

$$\begin{aligned} W_d(n) &= \sum_{i=1}^{n} \mathcal{A}_d[i] \times W_{d-1}(i) \\ &= \sum_{i=1}^{n} \Big(\mathcal{A}_d[i] \times \sum_{1 \le i_1 \le \cdots \le i_{d-1} \le i} \mathcal{A}_1[i_1] \times \cdots \times \mathcal{A}_{d-1}[i_{d-1}] \Big) \\ &= \sum_{1 \le i_1 \le \cdots \le i_d \le n} \mathcal{A}_1[i_1] \times \cdots \times \mathcal{A}_d[i_d] \end{aligned}$$

□

Our sum of ordered products structure can be easily used to maintain weighted (or iterated) partial sums of d arrays $\mathcal{A}_1, \ldots, \mathcal{A}_d$. Let $\mathcal{A}_{d+1}$ be an additional array such that $\mathcal{A}_{d+1}[i] = 0$ for all i. Then, the weighted partial sum $W_d(i)$ can be obtained by simply incrementing $\mathcal{A}_{d+1}[i]$ by 1 and then querying for the sum of ordered products of the set $\{\mathcal{A}_1, \ldots, \mathcal{A}_{d+1}\}$.

[4]For simplicity of presentation only, we allow the indices in the sum of products to be equal. The original sum of products with strictly ordered indices can be easily reduced to this form by shifting arrays.

4.4 Weighted Volume of Axis-parallel Strips

The weighted volume algorithm given in Section 5 requires us to maintain the weighted volume of a structure known as a *trellis*, which consists of axis-parallel strips in the form $a \leq \mathbf{x}^k \leq b$, rather than halfspaces of the form $\mathbf{x}^k \leq b$. Our technique for maintaining the volume of the halfspaces also works for trellises, by maintaining separate arrays for strips orthogonal to each axis. We utilize a standard data structure known as a *segment tree* to maintain each of these arrays. In particular, the segment tree used to maintain the array for the strips orthogonal to the ith axis stores the projections of these strips on the ith coordinate axis as one-dimensional *weighted* intervals. Note that the segment tree stores each interval in $O(\log n)$ of its nodes, which conceptually partitions the interval into $O(\log n)$ subintervals. (See [4] for details on segment trees.)

The length dominated by each interval (which corresponds to the array entry of its strip) can then be maintained in $O(\log n)$ amortized time per strip insertion. Briefly, this is achieved by deleting the subintervals that are subsumed (both in extent and in weight) by the inserted interval during insertions. The implication of this policy is an invariant that along a root-to-leaf path in the tree, the interval entries are stored in ascending weight order. It then becomes trivial to compute the contribution of each interval to the array via a recursive relation on the nodes. We defer the details of this algorithm to Section 5, where we apply it to a "binary space partition tree" rather than a segment tree.

Each strip insertion affects $O(\log n)$ array entries amortized. This follows from the fact that each inserted interval is stored on $O(\log n)$ nodes in the tree, with $O(\log n)$ total ancestors. The array contribution of the lower weight intervals stored in these $O(\log n)$ ancestors are possibly modified, whereas the contributions of the lower weight intervals in all remaining nodes are *deleted*, for which we charge to their corresponding insertions.

Each array entry change is coupled with an $O(\log n)$ time update in our sums of ordered products structure. It follows that we can maintain the weighted volume of strips in $O(\log^2 n)$ amortized time per insertion. We can now state the following theorem.

THEOREM 4. *The weighted volume of a set of axis-aligned strips in a rectangular box R in d dimensions can be maintained in $O(\log^2 n)$ amortized time per insertion, with a linear-space data structure.*

We finally note that we also need an elimination operation as part of the main algorithm described in Section 5. In this elimination operation, we delete all strips with weights less than a given weight w. This can be easily achieved by doing a binary search on the weights to identify the strips to delete and then deleting the corresponding intervals from the segment trees. As usual, we can charge the deletions to the corresponding insertions, thus this operation can be achieved in $O(\log^2 n)$ amortized time as well.

5. DYNAMIC WEIGHTED VOLUME FOR ARBITRARY BOXES

We now argue that the weighted volume of n arbitrary boxes in d-space, can be maintained under insert-only updates in $O(n^{(d-1)/2} \log^2 n)$ amortized time. For simplicity, we describe a detailed solution only for the two-dimensional case. The extension of this solution to higher dimensions is straightforward as we mention later.

Let $\mathcal{B}$ be a set of n weighted boxes in 2-space. Our objective is to maintain the weighted volume of a dynamic set of boxes that undergoes insertions from $\mathcal{B}$. We propose a structure that achieves this in $O(\sqrt{n} \log^2 n)$ amortized time per update.

The main idea of our solution is to partition the space into rectangular regions such that the set of boxes, $\mathcal{B}$, forms a set of axis-parallel strips inside each region. This enables us to use the specialized structure defined in Section 4.4 to efficiently maintain the overall weighted volume. To form this partition, we follow a classical technique by Overmars and Yap [15].

The partition is formed as follows. We draw a vertical line through each $\sqrt{n}$-th vertical boundary of the boxes along the first coordinate (horizontal) axis. As a result, the space is divided into $\Theta(\sqrt{n})$ slabs, each of which contain $\Theta(\sqrt{n})$ boundaries. We then individually divide each slab further by drawing a horizontal line through: (1) each box corner in the slab and, (2) each $\sqrt{n}$-th horizontal boundary of boxes passing through the slab. Consequently, each slab is partitioned into $\Theta(\sqrt{n})$ final regions. We call these final regions *cells*. The following lemma summarizes the properties of the partition and its proof can be found in [15].

LEMMA 4. *The partition has the following properties.*

- *There are $O(n)$ cells.*
- *Any box intersects $O(\sqrt{n})$ cells on its boundary.*
- *Each cell intersects the boundary of $O(\sqrt{n})$ boxes in $\mathcal{B}$.*
- *The boxes of $\mathcal{B}$ contain no corner in the interior of any cell.*

The last property implies that $\mathcal{B}$ forms a set of strips inside each cell. As the first key part of our algorithm, we maintain, for each cell C, the weighted volume on C contributed *only* by the boxes that intersect C on their boundary. We do this by utilizing an instance of the strips structure defined in Section 4.4 for each cell. By Lemma 4, there are $O(\sqrt{n})$ boxes that intersects any cell on its boundary, thus, the size of each strip structure is $O(\sqrt{n})$. Also, during the insertion of a box B, we can update all strip structures in $O(\sqrt{n} \log^2 n)$ amortized time, because B intersects $O(\sqrt{n})$ cells on its boundary and updating the strip structure of each of these cells takes $O(\log^2 n)$ time. We note that the cells lying on the boundary of B can be efficiently obtained by using the partition tree that we describe below.

Clearly, the weighted volumes maintained in the cells exclude the contributions of the boxes that *entirely* contain the cells. The second key part of our algorithm is to include these contributions so that the overall weighted volume is correctly maintained. To do this efficiently, we utilize a *binary space partition tree*. In particular, we maintain a balanced binary tree in which every node v is associated with a rectangular region of the space R_v, such that for all internal nodes v with children v_l and v_r, R_v is divided into R_{v_l} and R_{v_r} by an axis-aligned hyperplane. Moreover, the root is associated with the whole space and each leaf is associated with one cell of our partition bijectively. It is trivial to show that such a balanced space partition tree can be constructed in $O(n \log n)$ time.

When a box B is inserted to the structure, we store an entry for B at every node v such that B subsumes R_v but not $R_{parent(v)}$. This storage scheme conceptually partitions B into a set of maximal fragments, where each fragment is the intersection $B \cap R_v$ for a node v that B is stored. Note that this fragmentation excludes the sections of B that partially overlap the cells, which are already stored in the corresponding strip structures. It can easily be shown that B is stored in $O(\sqrt{n}\log n)$ nodes in the tree and these nodes can be traversed in $O(\sqrt{n}\log n)$ steps. [15]

For efficient weighted volume maintenance, we remove any box fragments that are entirely contained by boxes of higher weight. In particular, when we insert a box entry B to a node v, we delete all box entries B' stored below v such that $w(B') < w(B)$. Note that the deleted fragments have no contribution to the weighted volume because they are contained by a heavier box, and thus they are safe to be deleted. Similarly, for each cell that a newly inserted box contains, we delete all lower weight boxes in the corresponding strip structure. Finally, we keep only the highest weight box at each node. The result of this policy is an invariant that for any particular box B at a node v, all boxes stored above v have lower weights while all boxes stored below (including the strip structures) have higher weights. Consequently, the weighted volumes of the strip structures directly contribute to the overall. Moreover, we can write the weighted volume contribution of a box B stored at node v as

$$w(B) \times (|R_v| - A(v))$$

where $|R_v|$ is the ordinary volume (area) of region R_v and $A(v)$ is the ordinary volume of the union of the boxes stored below v (including the strip structures). Note that $A(v)$ equals the volume that is claimed from B by higher weight fragments, which are all stored below v. The maintenance of $A(v)$ for leaf nodes can be done trivially due to the simplicity of the strips arrangement [15]. For each internal node v with children v_l and v_r, $A(v)$ can be written recursively in terms of $A(v_l)$ and $A(v_r)$, and thus can easily be updated during the tree traversals. (See [15] for details.)

The overall weighted volume is easily maintained as the sum of weighted volume contributions of each strip structure and each box entry in the tree. It remains to show that removing lower weight box entries during insertions do not increase the overall amortized cost of $O(\sqrt{n}\log^2 n)$ per insertion. Recall that the tree traversal performed to insert a box takes $O(\sqrt{n}\log n)$ steps. To efficiently delete the lower weight boxes after an insertion, we maintain at each node v, the weight of the lowest weight box stored below v. This makes it possible to directly traverse to the box entries to be deleted. The additional traversals performed to delete the entries can be charged to the traversals which inserted these entries. Specifically, the traversal of within a subtree that contains entries of a lower weight box B is charged to the traversal of the same subtree during the insertion of B. Thus, the amortized cost of a box insertion remains $O(\sqrt{n}\log^2 n)$.

One can easily prove that the partition tree and the strip structures can be constructed in $O(n\sqrt{n}\log n)$ time and consume $O(n\sqrt{n})$ space. This yields to the following theorem.

THEOREM 5. *There exists a data structure that maintains the weighted volume of n 2-dimensional boxes in $O(\sqrt{n}\log^2 n)$ amortized time per insertion. This structure can be constructed in $O(n\sqrt{n}\log n)$ time and consumes $O(n\sqrt{n})$ space.*

The partition technique of [15] is generalized to higher dimensions as given in the following lemma.

LEMMA 5. *Given a set $\mathcal{B}$ of n axis-aligned boxes in d-space, one can partition the space into cells such that*

- *There are $O(n^{d/2})$ cells.*
- *Any box intersects $O(n^{(d-1)/2})$ cells on its boundary.*
- *Each cell intersects the boundary of $O(n^{(d-1)/2})$ boxes in $\mathcal{B}$.*
- *No cell contains a $(d-2)$-dimensional facet of any box in its interior.*

By applying the partition tree algorithm together with d-dimensional strip structures, we deduce the following result.

THEOREM 6. *We can maintain the weighted volume of a set of n d-dimensional boxes in $O(n^{(d-1)/2}\log^2 n)$ amortized time per box insertion. This data structure can be constructed in $O(n^{(d+1)/2}\log n)$ time using $O(n^{(d+1)/2})$ space.*

Computing the Klee's measure for 2-grounded boxes in d-space requires n box insertions into a $(d-2)$-dimensional structure, for the total complexity of $O(n \cdot n^{(d-3)/2}\log^2 n) = O(n^{(d-1)/2}\log^2 n)$, giving us the main result of our paper.

THEOREM 7. *Klee's measure for n 2-grounded boxes in d-space can be computed in worst-case time $O(n^{(d-1)/2}\log^2 n)$ and space $O(n^{(d-1)/2})$.*

6. KLEE'S MEASURE FOR d-GROUNDED BOXES

When the boxes are d-grounded, namely, they are all anchored at a common corner, the time complexity of our algorithm improves by a factor of $\log n$. This follows from the fact that the Overmars-Yap partition, when applied on d-grounded boxes, yields to regions that contain half-spaces rather than strips. By Theorem 2, we can maintain weighted volume of each region in $O(\log n)$ time per update (improving on $O(\log^2 n)$ for strips), reducing the running time of our algorithm to $O(n^{(d-1)/2}\log n)$. The d-grounded case of the problem is also known as the hypervolume indicator problem, which is utilized in evolutionary computing often to assess the quality of multi-objective optimization algorithms. Several techniques in the computational geometry literature can be used solve the problem efficiently. The current best bounds with respect to the number of dimensions are as follows:

- For $d \leq 3$, $O(n\log n)$, based on space-sweep [11].
- For $d = 4$, $O(n^{3/2}\text{polylog } n)$ by Chan [7], using reduction to unit-cubes.
- For $d = 5$, $O(n^2\text{polylog } n)$ by Kaplan et al. [13].
- For $d \geq 6$, $O(n^{(d+2)/3})$ by Bringmann [5], using reduction to fat-boxes.

Compared to the above results, our algorithm is faster in dimensions 4, 5 and 6, improving the bound for computing the hypervolume indicator in these dimensions, and using a simpler approach.

7. CONCLUSION

We have proposed a new method for computing Klee's measure on grounded boxes, which includes the hypervolume indicator as a special case. In particular, we obtained a bound of $O(n^{(d-1)/2} \log^2 n)$ for the k-grounded problem for $2 \leq k < d$, which is an improvement of roughly $\sqrt{n}$ over the general Klee's bound. Our technique also leads to a faster algorithm for the hypervolume indicator, which is a special case of the grounded boxes, in dimensions 4, 5 and 6. Given the long and distinguished history of Klee's measure problem, where all the previous improvements have been limited to cube-like boxes, the grounded boxes offer an interesting new direction to pursue.

8. REFERENCES

[1] P. Agarwal. An Improved Algorithm for Computing the Volume of the Union of Cubes. In *Proceedings of the 26th Annual Symposium on Computational Geometry*, pages 230–239, 2010.

[2] P. Agarwal, H. Kaplan, and M. Sharir. Computing the Volume of the Union of Cubes. In *Proceedings of the 23rd Annual Symposium on Computational Geometry*, pages 294–301, 2007.

[3] J. L. Bentley. Solutions to Klee's rectangle problems. *Unpublished manuscript*, 1977.

[4] M. de Berg, O. Cheong, M. van Kreveld, and M. Overmars. *Computational Geometry: Algorithms and Applications*. Springer-Verlag, 2008.

[5] K. Bringmann. Klee's Measure Problem on Fat Boxes in Time $O(n^{(d+2)/3})$. In *Proceedings of the 26th Annual Symposium on Computational Geometry*, pages 222–229, 2010.

[6] K. Bringmann and T. Friedrich. Approximating the volume of unions and intersections of high-dimensional geometric objects. In *Proceedings of 19th International Symposium on Algorithms and Computation*, pages 436–447, 2008.

[7] T. Chan. Semi-online Maintenance of Geometric Optima and Measures. In *Proceedings of the 13th Annual ACM-SIAM Symposium on Discrete Algorithms*, pages 474–483, 2002.

[8] T. M. Chan. A (slightly) faster algorithm for Klee's measure problem. *Computational Geometry*, 43(3):243–250, 2010.

[9] L. Chew, D. Dor, A. Efrat, and K. Kedem. Geometric Pattern Matching in d-Dimensional Space. *Discrete and Computational Geometry*, 21(2):257–274, 1999.

[10] M. Fleischer. The measure of pareto optima. applications to multi-objective metaheuristics. In *Evolutionary Multi-Criterion Optimization, Second International Conference*, pages 519–533, 2003.

[11] C. Fonseca, L. Paquete, and M. López-Ibáñez. An Improved Dimension-Sweep Algorithm for the Hypervolume Indicator. In *IEEE Congress on Evolutionary Computation*, pages 1157–1163, 2006.

[12] S. Huband, P. Hingston, L. While, and L. Barone. An evolution strategy with probabilistic mutation for multi-objective optimisation. In *The 2003 Congress on Evolutionary Computation*, volume 4, pages 2284–2291, 2003.

[13] H. Kaplan, N. Rubin, M. Sharir, and E. Verbin. Counting colors in boxes. In *Proceedings of the 18th Annual ACM-SIAM Symposium on Discrete Algorithms*, pages 785–794, 2007.

[14] V. Klee. Can the measure of $\cup_1^n [a_i, b_i]$ be computed in less than $O(n \log n)$ steps? *American Mathematical Monthly*, 84(4):284–285, 1977.

[15] M. H. Overmars and C.-K. Yap. New upper bounds in Klee's measure problem. *SIAM Journal on Computing*, 20(6):1034–1045, 1991.

[16] J. van Leeuwen and D. Wood. The measure problem for rectangular ranges in d-space. *Journal of Algorithms*, 2(3):282–300, 1981.

[17] E. Zitzler and S. Künzli. Indicator-based selection in multiobjective search. In *Proceedings of 8th International Conference on Parallel Problem Solving from Nature*, pages 832–842, 2004.

APPENDIX

A. PROOF OF LEMMA 2

A member of $S(j,k)$ is either a member of $S(j-1,2k-1)$ or $S(j-1,2k)$. We can, therefore, decompose $T(j,k,l,r)$ into sums of products based on the sets to which the indices belong, as follows:

$$T(j,k,l,r) = \sum_{i_l<\ldots<i_r \ \wedge \ i_l,\ldots,i_r \in S(j-1,2k-1)} A_l[i_l] \times \ldots \times A_r[i_r]$$

$$+ \sum_{i_l<\ldots<i_r \ \wedge \ i_l,\ldots,i_r \in S(j-1,2k)} A_l[i_l] \times \ldots \times A_r[i_r]$$

$$+ \sum_{l\leq c<r} \left(\sum_{i_l<\ldots<i_r \ \wedge \ i_l,\ldots,i_c \in S(j-1,2k-1) \ \wedge \ i_{c+1},\ldots,i_r \in S(j-1,2k)} A_l[i_l] \times \ldots \times A_r[i_r] \right)$$

By the distributive property of multiplication over addition, we get

$$T(j,k,l,r) = \sum_{i_l<\ldots<i_r \ \wedge \ i_l,\ldots,i_r \in S(j-1,2k-1)} A_l[i_l] \times \ldots \times A_r[i_r]$$

$$+ \sum_{i_l<\ldots<i_r \ \wedge \ i_l,\ldots,i_r \in S(j-1,2k)} A_l[i_l] \times \ldots \times A_r[i_r]$$

$$+ \sum_{l\leq c<r} \left(\begin{array}{c} \sum_{i_l<\ldots<i_c \ \wedge \ i_l,\ldots,i_c \in S(j-1,2k-1)} A_l[i_l] \times \ldots \times A_c[i_c] \\ \times \\ \sum_{i_{c+1}<\ldots<i_r \ \wedge \ i_{c+1},\ldots,i_r \in S(j-1,2k)} A_{c+1}[i_{c+1}] \times \ldots \times A_r[i_r] \end{array} \right)$$

This expression is equivalent to

$$\begin{aligned} T(j,k,l,r) &= T(j-1,2k-1,l,r) \ + \ T(j-1,2k,l,r) \\ &+ \sum_{l\leq c<r} (T(j-1,2k-1,l,c) \times T(j-1,2k,c+1,r)) \end{aligned}$$

The lemma follows.

How to Walk Your Dog in the Mountains with No Magic Leash*

Sariel Har-Peled†
Dept. Computer Science
University of Illinois,
Urbana-Champaign
201 N. Goodwin Avenue
Urbana, IL , 61801, USA
sariel@uiuc.edu

Amir Nayyeri
Dept. Computer Science
University of Illinois,
Urbana-Champaign
201 N. Goodwin Avenue
Urbana, IL, 61801, USA
nayyeri2@uiuc.edu

Mohammad Salavatipour‡
Dept. of Computing Science
University of Alberta
mreza@cs.ualberta.ca

Anastasios Sidiropoulos
Toyota Technological Institute
at Chicago
tasos@ttic.edu

ABSTRACT

We describe a $O(\log n)$-approximation algorithm for computing the homotopic Frechét distance between two polygonal curves that lie on the boundary of a triangulated topological disk. Prior to this work, algorithms where known only for curves on the Euclidean plane with polygonal obstacles.

A key technical ingredient in our analysis is a $O(\log n)$-approximation algorithm for computing the minimum height of a homotopy between two curves. No algorithms were previously known for approximating this parameter. Surprisingly, it is not even known if computing either the homotopic Frechét distance, or the minimum height of a homotopy, is in NP.

Categories and Subject Descriptors

F.2.2 [**ANALYSIS OF ALGORITHMS AND PROBLEM COMPLEXITY**]: Nonnumerical Algorithms and Problems – *Geometrical problems and computations*

General Terms

Algorithms, Theory

Keywords

Frechét distance, Approximation algorithms

*The current full version of this paper can be found online [19].

†Work on this paper was partially supported by a NSF AF award CCF-0915984.

‡Part of this work was done while visiting Toyota Technological Institute at Chicago.

SCG12, June 17–20, 2012, Chapel Hill, North Carolina, USA.

1. INTRODUCTION

Comparing the shapes of curves – or sequenced data in general – is a challenging task that arises in many different contexts. The ***Frechét distance*** and its variants (e.g. dynamic time-warping [22]) have been used as a similarity measure in various applications such as matching of time series in databases [23], comparing melodies in music information retrieval [26], matching coastlines over time [24], as well as in map-matching of vehicle tracking data [4, 28], and moving objects analysis [5, 6]. See [16, 1, 2] for algorithms for computing the Frechét distance.

Informally, for a pair of such curves $f, g : [0,1] \to \mathcal{D}$, for some ambient space $(\mathcal{D}, \mathsf{d})$, their Frechét distance is the minimum length leash needed to traverse both curves in sync. To this end, imagine a person traversing f starting from $f(0)$, and a dog traversing g starting from $g(0)$, both traveling along these curves without ever moving backwards. Then, the Frechét distance is the infimum over all possible traversals, of the maximum distance between the person and the dog. Specifically, given a bijective continuous parameterization $\phi : [0,1] \to [0,1]$, the ***width*** of this reparameterization, i.e., the longest leash needed by this reparameterization, is $\mathrm{width}(\phi) = \sup_{x \in [0,1]} \mathsf{d}\Big(f(x), g(\phi(x))\Big)$. As such, the Frechét distance between f and g is defined to be

$$\mathsf{d}_{\mathcal{F}}(f,g) = \inf_{\phi:[0,1]\to[0,1]} \mathrm{width}(\phi),$$

where ϕ ranges over all orientation-preserving homeomorphisms.

While this distance makes complete sense when the underlying distance is the Euclidean metric, it becomes less useful if the distance function is more interesting. For example, imagine walking a dog in the woods. The leash might get tangled as the dog and the person walk on two different sides of a tree. Since the Frechét distance cares only about the distance between the two moving points, the leash would "magically" jump over the tree.

Homotopic Frechét distance.

To address this shortcoming, a natural extension of the above notion called ***homotopic Frechét distance*** was in-

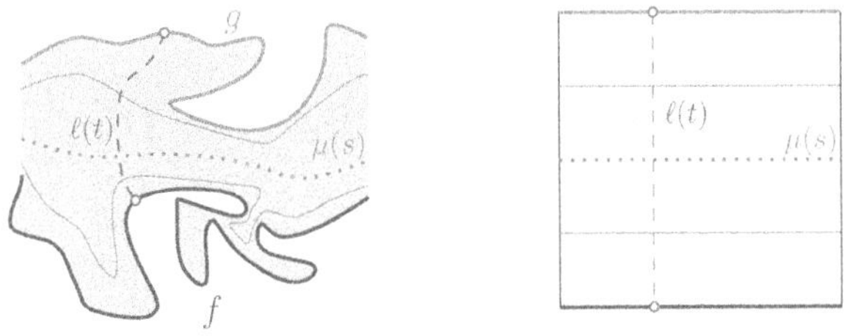

(i) (ii)

Figure 1: (i) Two curves f and g, and (ii) the parametrization of their homotopic Frechét distance.

troduced by Chambers *et al.* [7]. Informally, revisiting the above person-dog analogy, we consider the infimum over all possible traversals of the curves, but this time, we require that the person is connected to the dog via a leash, i.e. a curve that moves continuously over time. Furthermore, one keeps track of the leash during the motion, where the purpose is to minimize the maximum leash length needed.

To this end, consider a continuous mapping $\psi : [0,1]^2 \to \mathcal{D}$. For parameters $s, t \in [0,1]$ consider the one dimensional functions $\ell_t(y) = \psi(t,y)$ and $\mu_s(x) = \psi(x,s)$. The functions $\ell(y) \equiv \ell_t(y)$ and $\mu(x) \equiv \mu_s(x)$ are parametrized curves that are the natural restrictions of ψ to one dimension, by the x and y coordinates, respectively. We require that $\mu(0) = f$ and $\mu(1) = g$. The ***homotopic width*** of ψ is $\mathrm{width}(\psi) = \max_{t\in[0,1]} \|\ell(t)\|$, and the ***homotopic Frechét distance*** between f and g is

$$\mathsf{d}_{\mathcal{H}}(f,g) = \inf_{\psi:[0,1]^2\to\mathcal{D}} \mathrm{width}(\psi),$$

where the infimum is over all such mappings, and $\|\cdot\|$ denotes the length of a curve.

Clearly, $\mathsf{d}_{\mathcal{H}}(f,g) \geq \mathsf{d}_{\mathcal{F}}(f,g)$; in fact, $\mathsf{d}_{\mathcal{H}}(f,g)$ can be arbitrary larger than $\mathsf{d}_{\mathcal{F}}(f,g)$. We remark that $\mathsf{d}_{\mathcal{H}}(f,g) = \mathsf{d}_{\mathcal{F}}(f,g)$ for any pair of curves in the Euclidean plane, as we can always pick the leash to be a straight line segment between the person and the dog. In other words, the map ψ in the definition of $\mathsf{d}_{\mathcal{H}}$ can be obtained from the map ψ in the definition of $\mathsf{d}_{\mathcal{F}}$ via an appropriate affine extension. However, this is not true for general ambient spaces, where the leash might have to pass over obstacles, hills, etc.

The homotopic Frechét distance is refereed to as the ***morphing width*** of f and g, and it bounds how far a point on f has to travel to its corresponding point in g under the morphing of ψ [13]. The length of $\mu(s)$ is the ***height of the morph at time*** s, and the ***height*** of such a morphing is $\mathsf{height}(\mu) = \max_{s\in[0,1]} |\mu(s)|$. The ***homotopy height*** between f and g bounded by $\ell(0)$ and $\ell(1)$ is

$$\mathsf{h}(f, g, \ell(0), \ell(1)) = \inf_{\mu} \mathsf{height}(\mu),$$

where μ varies over all possible morphings between f and g, such that each curve in μ has one end on $\ell(0)$ and one end in $\ell(1)$. See Figure 1 for an example. Note that if we do not constraint the endpoints of the curves during the morphing to stay on $\ell(0)$ and $\ell(1)$, the problem of computing the minimum height homotopy is trivial. One can contract f to a point, send it to g from the shortest (f,g)-path, and then expand it to g. To keep the notation simple, we use $\mathsf{h}(f,g)$ when f and g have common endpoints.

Intuitively, the homotopy height measures how long the curve has to become as it deforms from f to g, and it was introduced by Chambers and Letscher [8, 9]. Observe that if we are given the starting and ending leashes $\ell(0)$ and $\ell(1)$ then the homotopy height of f and g, is the homotopic Frechét distance between $\ell(0)$ and $\ell(1)$.

Here, we are interested in the problems of computing the homotopic Frechét distance and the homotopic height between two simple polygonal curves that lie on the boundary of an arbitrary triangulated topological disk.

Why are these measures interesting?.

For the sake of the discussion here, assume that we know the starting and ending leash of the homotopy between f and g. The region bounded by the two curves and these leashes, form a topological disk, and the mapping realizing the homotopic Frechét distance is a mapping of the unit square to this disk $\mathcal{D}$. This mapping specifies how to sweep over $\mathcal{D}$ in a geometrically "efficient" way (especially if the leash does not sweep over the same point more than once), so that the leash (i.e., the sweeping curve) is never too long [13]. As a concrete example, consider the two curves as enclosing several mountains between them on the surface – computing the homotopic Frechét distance corresponds to deciding which mountains to sweep first and in which order.

Furthermore, this mapping can be interpreted as surface parameterization [15, 27] and can thus be used in applications such as texture mapping [3, 25]. In the texture mapping problem, we wish to find a continuous and invertible mapping from the texture, usually a two-dimensional rectangular image, to the surface.

Another interesting interpretation is when f is a closed curve, and g is a point. Interpreting f as a rubber band in a 3d model, the homotopy height between f and g here is the minimum length the rubber band has to be so that it can be collapsed to a point (here, the rubber band stays on the surface as this is happening). In particular, a short closed curve with large homotopic height to any point in the surface is a "neck" in the 3d model.

To summarize, these measures seem to provide us with a fundamental understanding of the structure of the given surface/model.

Previous work.

Chambers *et al.* [7] gave a polynomial time algorithm to compute the homotopic Frechét distance between two polygonal curves on the Euclidean plane with polygonal obstacles. Chambers and Letscher [8, 9] introduced the notion of minimum homotopy height, and proved structural properties for the case of a pair of paths on the boundary of a topological disk. We remark that in general, it is not known whether the optimum homotopy has polynomially long description. In particular, it is not known whether the problem is in NP.

The problem of computing the (standard) Frechét distance between curves has been considered by Alt and Godau [2], who gave a polynomial time algorithm. Eiter and Mannila [14] studied the easier discrete version of this problem. Computing the Frechét distance between surfaces [17], appears to be a much more difficult task, and its complexity is poorly understood. The problem has been shown to be NP-hard by Godau [18], while the best algorithmic result is due to Alt and Buchin [1], who showed that it is upper semi-computable.

Efrat *et al.* [13] considered the Frechét distance inside a simple polygon as a way to facilitate sweeping it efficiently. They also used the Frechét distance with the underlining geodesic metric as a way to get a morphing between two curves. For recent work on the Frechét distance, see [11, 12, 20, 10] and references therein.

Our results.

In this paper, we consider the problems of computing the homotopic Frechét distance and the homotopy height between two simple polygonal curves that lie on the boundary of a triangulated topological disk $\mathcal{D}$ that is composed of n triangles.

We give a polynomial time $O(\log n)$-approximation algorithm for computing the homotopy height between f and g. Our algorithm to compute an approximate homotopy between f and g is via a simple, yet delicate divide and conquer algorithm.

We use the homotopy height algorithm as an ingredient for a $O(\log n)$-approximation algorithm for the homotopic Frechét distance problem. Intuitively, our algorithm for homotopic Frechét distance works as follows. We first approximately guess the optimum (i.e. $\mathsf{d}_{\mathcal{H}}(f,g)$). Using this guess, we classify parts of $\mathcal{D}$ as "obstacles", i.e. regions over which a short leash cannot pass. Let D' be the punctured disk obtained from $\mathcal{D}$ after removing these obstacles. The isotopy class of any leash is determined by the set of punctures that are on its left side. Observe that the leashes of the optimum solution belong to the same isotopy class. We describe a greedy algorithm to pick an isotopy class out of exponential number of choices, s.t. the homotopic Frechét distance constrained inside it is a constant factor of the homotopic Frechét distance in $\mathcal{D}$. Then, we use an extended version of our homotopy height algorithm to compute the homotopic Frechét distance.

The $O(\log n)$ factor shows up in the homotopic Frechét distance algorithm only because it uses the homotopy height as a subroutine. Thus, any constant factor approximation algorithm for the homotopy height problem implies a constant factor approximation algorithm for the homotopic Frechét distance.

As a warm-up exercise and in order to simplify the presentation we first consider the discrete version of the homotopy height problem in Section 2.1. This is how Chambers and Letscher formulated the problem. Later, in Section 2.2, we describe an algorithm to approximately find the shortest homotopy in continuous settings. In Section 3, we address the homotopic Frechét distance, both discrete and continuous. Further basic definitions are provided in the full version [19].

2. APPROXIMATING THE HEIGHT OF A HOMOTOPY

In this section we give an approximation algorithm for finding a homotopy of minimum height in a topological disc $\mathcal{D}$, whose boundary is defined by two walks L and R that share their end-points s and t. We start with the discrete case, i.e. when the disk is a triangulated edge-weighted planar graph and then generalize it to the continuous case. We then use this algorithm as a subroutine (in the next section) in our algorithm for the minimum homotopic Frechét distance problem.

2.1 The discrete case

To start, let us assume we are given an embedded planar graph G all of whose faces (except possibly the outer face) are triangles. Let $s,t \in \partial\mathsf{G}$ and L and R be the two non-crossing (s,t)-walks on $\partial\mathsf{G}$ in counter-clockwise and clockwise order, respectively. We use $\mathcal{D}$ to denote the topological disk enclosed by $\mathsf{L} \cup \mathsf{R}$. We refer to vertices of G (inside or on the boundary of $\mathcal{D}$) as vertices of $\mathcal{D}$. Our goal is to find a minimum height homotopy from L to R of non-crossing walks. Informally, the homotopy is defined by a sequence of walks, where every two consecutive walks differ by either a triangle, or an edge (being traversed twice). For a formal definition, see [19].

Lemma 2.1. *Let x and y be vertices of G that are at distance ρ. Then any homotopy between L and R has height at least ρ.*

Proof: Fix a homotopy of height δ. This homotopy contains an s-t walk ω that passes through x, and an s-t walk χ that passes through y. We have, by the triangle inequality, that $\rho \leq \|\omega[s,x]\| + \|\chi[s,y]\|$, and $\rho \leq \|\omega[x,t]\| + \|\chi[y,t]\|$. Therefore, $\rho \leq (\|\omega\| + \|\chi\|)/2 \leq \delta$, as required. ■

Lemma 2.2. *Suppose d_1 is the maximum distance of a vertex of G from either of L or R, d_2 is the largest edge weight, and let $d = \max\{d_1, d_2\}$. Furthermore, let $\mathcal{D}$, L, R, and d be defined as above. Then any homotopy between L and R has height at least d.*

Proof: For every homotopy between L and R, and for every edge e, there exists a walk in the homotopy that passes through e. Therefore, the height of the homotopy is at least d_2. Moreover, the height is at least d_1 by Lemma 2.1. ■

Here we present an algorithm which finds an (L,R)-homotopy of height at most $\|\mathsf{L}\| + \|\mathsf{R}\| + O(d \log n)$.

Lemma 2.3. *Let $\mathcal{D}$ be an edge-weighted triangulated topological disk with n faces such that its boundary is formed by two walks L and R that share endpoints s and t. Then, one can compute, in $O(n \log n)$ time, a homotopy from L to R of height at most $\|\mathsf{L}\| + \|\mathsf{R}\| + O(d \log n)$, where d is the largest among (a) the maximum distance of a vertex of $\mathcal{D}$ from either of L and R, and (b) the maximum edge weight.*

Proof: Let $f(\|\mathsf{L}\| + \|\mathsf{R}\|, d, n)$ denote the maximum height of such a homotopy. We will show that $f(u,d,n) = u + O(d \log n)$.

The base case $n = 0$ is easy. Indeed, if we have two edges (u,v) and (v,u) consecutive in R (or in L) we can retract these two edges. By repeating this we arrive at both L and R being identical, and we are done. The case $n = 1$ is handled in a similar fashion. After one face flip, the problem reduces to the case $n = 0$. As such, $f(\|\mathsf{L}\|+\|\mathsf{R}\|, d, 1) \leq \|\mathsf{L}\|+\|\mathsf{R}\|+d$.

For $n > 1$, compute for each vertex of G its shortest path to L, and consider the set of edges $\mathcal{E}$ used by all these shortest paths. Clearly, these shortest paths can be chosen so that $\mathsf{L} \cup \mathcal{E}$ form a tree. We consider each edge of R to be "thick" and have two sides (i.e., we think about these edges as being corridors with thickness). If $\mathcal{E}$ uses an edge of R then it uses the inner copy of this edge, while R uses the outer side. Similarly, we will consider each original vertex of R to be two vertices (one inside and other one on the boundary R). The set $\mathcal{E}$ would use only the inner vertices of R, while R would use only the outer vertices. To keep the graph triangulated we also arbitrarily triangulate inside each thick edge of R by adding corridor edges. Observe that, each corridor edge either connects two copies of a single vertex (thus has weight zero) or copies of two neighbors on R (and so has the same weight as the original edge).

Clearly, if we cut $\mathcal{D}$ along the edges of $\mathcal{E}$, what remains is a simple triangulated polygon (it might have "thin" corridors along the edges of R). One can find a diagonal uv such that each side of the diagonal contains at least $\lceil n/3 \rceil$ triangles of G (and at most $(2/3)n$). (Here, we count only the "real" triangles of G – we consider the faces of the thin corridors of the edges of R to have weight 0.) Observe that, because the faces inside corridors have weight zero, we can ensure that if uv is a corridor edge then u and v are copies of the same vertex. We use this property in the following case analysis.

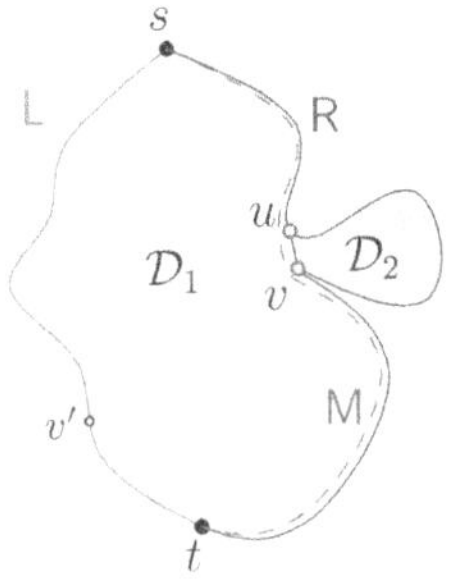

(A) Consider the case that u and v are both vertices of R. In this case, let $\mathsf{R}[u,v]$ be the portion of R in between u and v, and let $\mathcal{D}_2$ be the disk having $\mathsf{R}[u,v] \cup uv$ as its outer boundary. Let $\mathcal{D}_1$ be the disk $\mathcal{D} \backslash \mathcal{D}_2$. Let $\mathsf{M} = \mathsf{R}[s,u] \cup uv \cup \mathsf{R}[v,t]$.

Clearly, the distance of any vertex of $\mathcal{D}_1$ from L is at most d. By induction, there is a homotopy of height $f(\|\mathsf{L}\|+\|\mathsf{M}\|, d, (2/3)n)$ from L to M. Similarly, the distance of any vertex of $\mathcal{D}_2$ from uv is smaller than its distance to L. As such, by induction, there is a homotopy between uv and $\mathsf{R}[u,v]$ of height at most $f(\|\mathsf{R}[u,v]\| + d, d, (2/3)n)$. Clearly, we can extend this to a homotopy of M to R of height

$$\|\mathsf{R}[s,u]\| + f(\|\mathsf{R}[u,v]\| + d, d, (2/3)n) + \|\mathsf{R}[v,t]\|.$$

Putting these two homotopies together results in the desired homotopy from L to R.

(B) If v is a vertex of $\mathcal{E}$ and u is a vertex of R. As such, v is an inner vertex of R (that belongs to $\mathcal{E}$) and u is an outer vertex of R. Recall that we can assume that v and u are inner and outer copies of a same vertex of R. Let π_v be the shortest path in $\mathcal{D}$ from v to L, and let v' be its endpoint on L.

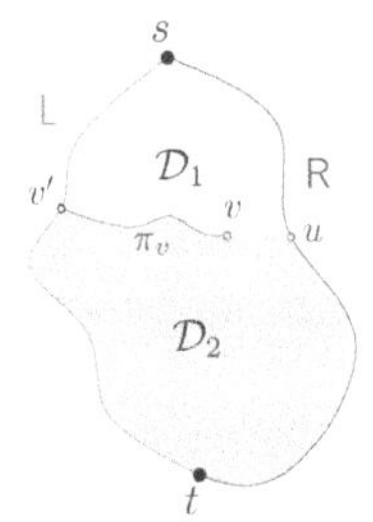

Consider the disk $\mathcal{D}_1$ having the "left" boundary $\mathsf{L}_1 = \mathsf{L}[s,v'] \cup \pi_v \cup vu$ and $\mathsf{R}_1 = \mathsf{R}[s,u]$ as its "right" boundary. This disk contains at most $(2/3)n$ triangles, and by induction, it has a homotopy of height $f(\|\mathsf{L}_1\| + \|\mathsf{R}_1\|, d, (2/3)n)$. To see why we can apply the recursion, observe that u and v are copies of the same vertex of R. That is, all shortest paths of vertices inside $\mathcal{D}_1$ to L are completely inside $\mathcal{D}_1$. Particularly, the distance of all vertices in $\mathcal{D}_1$ to L_1 are at most d.

Similarly, the topological disk $\mathcal{D}_2$ with the left boundary $\mathsf{L}_2 = uv \cup \pi_v \cup \mathsf{L}[v',t]$ and the right boundary $\mathsf{R}_2 = \mathsf{R}[u,t]$ has a homotopy of height $f(\|\mathsf{L}_2\| + \|\mathsf{R}_2\|, d, (2/3)n)$.

Starting with L, extending a tendril from v' to v, from v to u, and then applying the homotopy to first half of this walk (i.e., L_1) to move to R_1, and then the homotopy of $\mathcal{D}_2$ to the second part, results in a homotopy of L to R of height

$$\max\begin{pmatrix} \|\mathsf{L}\| + 2d, \\ f\Big(\|\mathsf{L}_1\| + \|\mathsf{R}_1\|, d, (2/3)n\Big) + \|\mathsf{L}_2\|, \\ \|\mathsf{R}_1\| + f\Big(\|\mathsf{L}_2\| + \|\mathsf{R}_2\|, d, (2/3)n\Big) \end{pmatrix}.$$

If the first number is the maximum, we are done. Otherwise, the above value is at most $f(\|\mathsf{L}\| + \|\mathsf{R}\| + 2d, d, 2/3n)$.

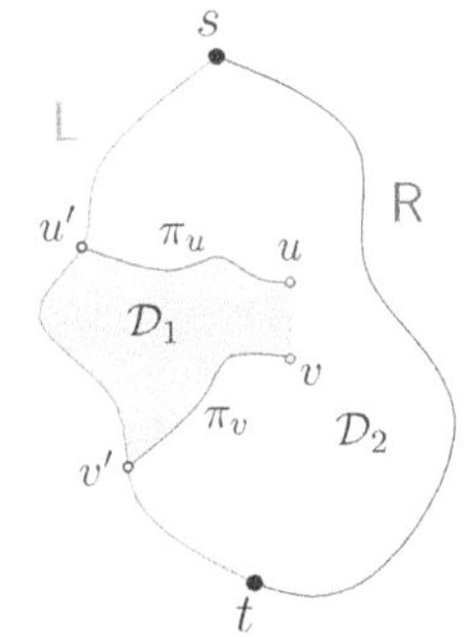

(C) Here we handle the case that u and v are both vertices of $\mathsf{L} \cup \mathcal{E}$. Then, as before, let u' and v' be the closest points on L to u and v, respectively. Now, let π_u (resp. π_v) be the shortest path from u (resp. v) to u' (resp. v').

Consider the disk $\mathcal{D}_1$ having $\mathsf{L}_1 = \mathsf{L}[u',v']$ as left boundary, and $\mathsf{R}_1 = \pi_u \cup uv \cup \pi_v$ as right boundary. This disk contains between $n/3$ and $2n/3$ triangles of the original surface. The distance of any vertex of $\mathcal{D}_1$ to L_1 (when restricted to $\mathcal{D}_1$) is at most d, and as such by induction, there is a homotopy from L_1 to R_1 of height $\alpha = f(\|\mathsf{L}_1\| + \|\mathsf{R}_1\|, d, (2/3)n) \leq f(\|\mathsf{L}[u',v']\| + 3d, d, (2/3)n)$. This yields a homotopy of height $\alpha_1 = \|\mathsf{L}[s,u']\| + \alpha + \|\mathsf{L}[v',t]\|$, from L to $\mathsf{L}_2 = \mathsf{L}[s,u'] \cup \pi_u \cup uv \cup \pi_v \cup \mathsf{L}[v,t]$. It is straight forward to check that $\alpha_1 \leq f(\|L\| + 3d, d, (2/3)n)$.

Next, let $\mathcal{D}_2$ be the disk with its left boundary being L_2 and its right boundary being $\mathsf{R}_2 = R$. Observe, that as before, the maximum distance of any vertex of $\mathcal{D}_2$ to L_2 is at most d. As before, by induction, there is a homotopy form L_2 to R_2 of height $\alpha_2 = f(\|\mathsf{L}_2\| + \|\mathsf{R}_2\|, d, (2/3)n)$. Since $\|\mathsf{L}_2\| \leq \|\mathsf{L}\| + 3d$, we have $\alpha_2 \leq f(\|\mathsf{L}\| + \|\mathsf{R}\| + 3d, d, (2/3)n)$.

In all cases the length of the homotopy is at most

$$f(\|\mathsf{L}\| + \|\mathsf{R}\| + 3d, d, (2/3)n)$$

Now, it is easy to verify that the solution to the recursion $f(u,d,n)$ that complies with all the above inequalities is $f(u,d,n) = u + O(d \log n)$, as desired.

We can compute the shortest path tree in linear time using the algorithm of Henzinger *et al.* [21]. The separating edge can also be found in linear time using DFS. So, the running time for a graph with n faces is $T(n) = T(n_1) + T(n_2) + O(n)$, where $n_1 + n_2 = n$ and $n_1, n_2 \leq 2/3n$. It follows that $T(n) = O(n \log n)$. ■

Remark 2.4. In the algorithm of Lemma 2.3, it is not necessary that we have the shortest paths from L to all the vertices of $\mathcal{D}$. Instead, it is sufficient if we have a tree structure that provides paths from any vertex of $\mathcal{D}$ to L of distance at most d in this tree. We will use this property in the continuous case, where recomputing the shortest path tree is relatively expensive.

2.2 The continuous case

In this section we extend the arguments to the continuous case. Here we are given a piecewise linear triangulated topological disk, $\mathcal{D}$, with n triangles. The boundary of $\mathcal{D}$ is composed of two paths L and R with shared endpoints s and t. Observe that the distance of any point x in $\mathcal{D}$ from L and R is not longer than the homotopy height as there is a (s,t)-path that contains x. Here, we build a homotopy of height $\|\mathsf{L}\| + \|\mathsf{R}\| + O(d \log n)$, where d is the maximum distance of any point in $\mathcal{D}$ from either L or R.

We use the following observations (see [19] for details):

(A) The shortest path from a set of $O(n)$ edges to the whole surface can be computed in $O(n^3 \log n)$ time.

(B) A shortest path (i.e., a geodesic) intersects a face along a segment and it locally looks like a segment if the adjacent faces are rotated to be coplanar.

2.2.1 *Homotopy height if edges are short*

Here, we assume that the longest edge in $\mathcal{D}$ has length at most $2d$, where d is the maximum distance for any point of $\mathcal{D}$ from either L or R.

As in the discrete case, let $\mathcal{E}$ be the union of all the shortest paths from the vertices of $\mathcal{D}$ to L (as before, we treat the edges and vertices of R as having infinitesimal thickness). For a vertex v of $\mathcal{D}$, its shortest path π_v is a polygonal path that crosses between faces (usually) in the middle of edges (it might also go to a vertex, merge with some other shortest paths and then follow a common shortest path back to L). In particular, each such shortest path might intersect a face of $\mathcal{D}$ along a single segment. As such, the polygon resulting from cutting $\mathcal{D}$ along $\mathcal{E}$, call it P, is a polygon that has complexity $O(n^2)$. A face of P is a hexagon, a pentagon, a quadrilateral, or a triangle. However, it has at most 3 edges that are portions of the edges of $\mathcal{D}$. We say the degree of a face is i if it has i edges that are portions of the edges of $\mathcal{D}$. Observe that, each triangle of $\mathcal{D}$ is now decomposed into a set of faces. Obviously, each triangle of $\mathcal{D}$ contains at most one face of degree 3. Overall, there are $O(n)$ faces of degree 3 in P.

Now consider C^*, the dual of the graph that is inside the polygon (ignore the edges on the boundary). More precisely, C^* has a vertex corresponding to each face inside the polygon P. Two vertices of C^* are adjacent if and only if their corresponding faces share a portion of an edge of $\mathcal{D}$ (this shared edge is a diagonal of the polygon resulting from the cutting). Since the maximum degree of C^* is 3, there is an edge that is a good separator. We use this edge in a similar fashion to the proof of Lemma 2.3, except that in the recursion we avoid recomputing the shortest paths (i.e., we use the old shortest paths and distances computed in the original disk), see Remark 2.4. So, we compute the shortest paths once in the beginning in $O(n^3 \log n)$ time. Then, in each step we can find the separator in $O(n^2)$ time. Namely, the total time spent on computing the separators is $T(n) = T(n_1) + T(n_2) + O(n^2)$, where $n_1 + n_2 = O(n^2)$ and $n_1, n_2 \leq (2/3)(n_1 + n_2)$; that is, $T(n) = O(n^2 \log n)$. As such, the total running time is dominated by the computation of the shortest paths.

The proof of Lemma 2.3 then goes through literally in this case. Since all the edges have length at most $2d$, by assumption, we get the following.

Lemma 2.5. *Let $\mathcal{D}$ be a topological disk with n faces where every face is a triangle (here, the distance between any two points on the triangle is their Euclidean distance). Furthermore, the boundary of $\mathcal{D}$ is formed by two walks* L *and* R *(that share two endpoints). Let d be the maximum distance of any point of $\mathcal{D}$ from either* L *or* R*. Furthermore, assume that all edges of $\mathcal{D}$ have length at most $2d$. Then, one can compute a continuous homotopy from* L *to* R *of height $\leq \|\mathsf{L}\| + \|\mathsf{R}\| + O(d \log n)$ in $O(n^3 \log n)$ time.*

2.2.2 *Homotopy height if there are long edges*

Algorithm.

For any two points in $\mathcal{D}$ consider a shortest path π connecting them. The signature of π is the ordered sequence of edges (crossed or used) and vertices used by π, see [19]. For a point $\mathsf{p} \in \mathsf{R}$, let $\mathsf{s_L}(\mathsf{p})$ denote the signature of the shortest path from p to L. The signature $\mathsf{s_L}(\mathsf{p})$ is well defined in R except for a finite set of *medial* points, where there are two (or more) distinct shortest paths from L to p. In particular, let Π_R be the set of all shortest paths from any medial point on R to L. Observe that, the medial points are the only points that the signature of the shortest path from R to L changes in any non-degenerate triangulation.

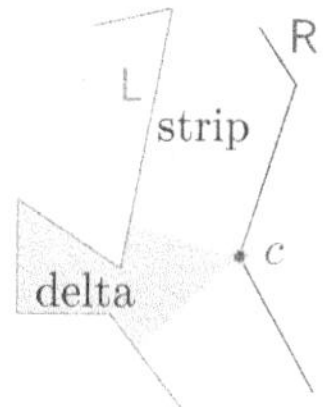

Cutting $\mathcal{D}$ along the paths of Π_R breaks $\mathcal{D}$ into ***corridors***. If the intersection of a corridor with R is a point (resp. segment) then it is a ***delta*** (resp. ***strip***). In a strip C, all the shortest paths to L from the points in the interior of the segment $C \cap \mathsf{R}$ have the same signature. Intuitively, strips have a natural way to morph from one side to the other. We further break each delta into chunks and pockets.

So, consider a delta C with an apex c (i.e., the point of R on the boundary of C). For a point $x \in \mathsf{L} \cap C$, we define its signature (in relation to C), to be the signature of the shortest path from x to c (restricted to lie inside C). Again, we partition $\mathsf{L} \cap C$ into maximum intervals that have the same signature. If a newly created region has a single intersection point with both L and R, then it is a ***pocket***, otherwise, it is a ***chunk***.

Consider such a chunk C. Its intersection with L is a segment, and its intersection with R is a point (i.e., the apex c of the delta). Observe that the distance of any point of $\mathsf{L} \cap C$ to c is at most $2d$, and this provides a natural way to morph $\mathsf{L} \cap C$ to c. A pocket, on the other hand, is a topological disk that its intersections with L and R are both single points, and the two boundary paths between these intersections are of length $\leq 2d$. Pockets are handled by using the recursive scheme developed for the discrete case.

Applying the above partition scheme to all the deltas results in breaking $\mathcal{D}$ into strips, chunks and pockets. Next, order the resulting regions according to their order along L, and transform each one of them at time, such that starting with L we end up with R.

(A) **Morphing a chunk/strip** S: Let $\sigma_\mathsf{L} = \mathsf{L} \cap S$ and $\sigma_\mathsf{R} = \mathsf{R} \cap S$. There is a natural homotopy from $\pi_t \cup \sigma_\mathsf{L}$ to $\sigma_\mathsf{R} \cup \pi_b$.

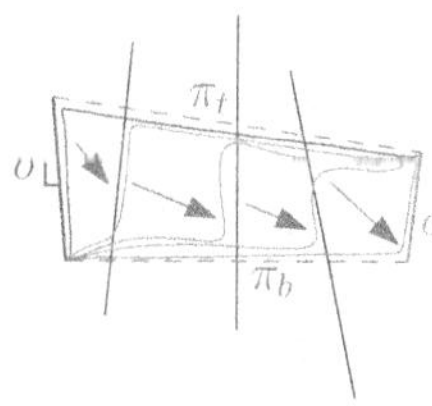

The strip/chunk S has no vertex of $\mathcal{D}$ in its interior, and as such it is formed by taking planar quadrilaterals and gluing them together along common edges. Observe that by the triangle inequality, all such edges of any of these quadrilaterals are of length $\leq \max(\|\sigma_\mathsf{L}\|, \|\sigma_\mathsf{R}\|) + 4d$. It is now easy to check that we can collapse each such quadrilateral in turn to get the required homotopy. Since each of π_t and π_b is composed of two shortest paths, there is a linear number of such quadrilaterals, and each collapse can be done in constant time. See figure for an example.

(B) **Morphing a pocket**: A pocket has perimeter $4d$, and there is a point on its boundary, such that the distance of any point in it to this base point is at most $2d$. By the the triangle inequality, we have that for a topological disk $\mathcal{D}$, such that all the points of $\mathcal{D}$ are in distance at most $2d$ from some point c. Then the longest edge in $\mathcal{D}$ has length at most $4d$. As such, all the edges inside a pocket can not be longer than $4d$. We can now apply Lemma 2.5 to such a pocket. This results in the desired homotopy.

Analysis. The shortest paths from R to L can be computed in $O(n^3 \log n)$ time. The shortest paths inside a delta to its apex can be computed in $O(n^2 \log n)$. Since there is a linear number of deltas, the total running time for building the strips is $O(n^3 \log n)$.

Lemma 2.6. *The number of paths in* Π_R *is* $O\Big(|\mathsf{V}(\mathcal{D})|\Big)$, *where* $\mathsf{V}(\mathcal{D})$ *is the set of vertices of* $\mathcal{D}$.

Proof: Let $\{\zeta_1, \zeta_2, \ldots, \zeta_k\}$ be the paths in Π_R sorted by the order of their endpoints along R. Observe that these paths are geodesics and such one can assume that they are interior disjoint. Now, if $l_i \in L$ and $r_i \in R$ are the endpoints of ζ_i, for $i = 1, \ldots, k$, then these endpoints are sorted along their respective curves. In particular, let $\mathcal{D}_i$ be the disk having $\mathsf{L}[s, l_i] \cup \zeta_{i+1} \cup \mathsf{R}[s, r_i]$ for boundary. We have that $\mathcal{D}_1 \subseteq \mathcal{D}_2 \subseteq \cdots \subseteq \mathcal{D}_k$. The signatures of ζ_i and ζ_{i+2} must be different as otherwise they would be consecutive. Furthermore, because of the inclusion property, if an edge or a vertex of $\mathcal{D}$ intersects ζ_i but does not intersect ζ_{i+1} then, it can not intersect any later path. As such, every other path in Π_R can be charged to vertices or edges that are added or removed from the signature of the respective path. Since an edge or a vertex can be added at most once, and deleted at most once, this implies the desired bound on the number of paths. ■

Arguing as in Lemma 2.6, we have that the total number of parts (i.e., chunks, corridors and pockets) generated by the above decomposition is $O(|\mathsf{V}(\mathcal{D})|)$.

Lemma 2.7. *Consider a strip or a chunk S generated by the above partition of $\mathcal{D}$. Let $\sigma_\mathsf{L} = \mathsf{L} \cap S$ and $\sigma_\mathsf{R} = \mathsf{R} \cap S$. Let π_t and π_b be the top and bottom paths forming the two sides of S that do not lie on R or L.*

(A) We have $\|\pi_b\| \leq 2d$ and $\|\pi_t\| \leq 2d$.

(B) If $\|\sigma_\mathsf{L}\| > 0$ or $\|\sigma_\mathsf{R}\| > 0$ then there is no vertex of $\mathcal{D}$ in the interior of S.

(C) If $\|\sigma_\mathsf{L}\| > 0$ or $\|\sigma_\mathsf{R}\| > 0$ then there is a homotopy from $\pi_t \cup \sigma_\mathsf{L}$ to $\sigma_\mathsf{R} \cup \pi_b$ of height $\max(\|\sigma_\mathsf{L}\|, \|\sigma_\mathsf{R}\|) + 4d$. This homotopy can be computed in linear time.

Proof: (A) If the strip was generated by the first stage of partitioning then the claim is immediate.

Otherwise, consider a delta C with an apex c. For any point $x \in \mathsf{L} \cap C$ we claim that there is a path of length $\leq 2d$ to c. Indeed, consider the shortest path π_x from x to R in $\mathcal{D}$. If this path goes to c the claim holds immediately. Otherwise, the shortest path (that has length at most d) must cross either the top or bottom shortest path forming the boundary of C that are emanating from c. We can now modify π_x, so that after its intersection point with this shortest path, it follows it back to c. Clearly, the resulting shortest path has length at most $2d$ and lies inside the resulting chunk.

(B) Indeed, the boundary paths π_t and π_b have the same signature (formally, they are the limit paths of same signature). Since $\mathcal{D}$ is non-degenerate, if there was any vertex in the middle, then the path on one side of the vertex, and the path on the other side of the vertex can not possibly have the same signature.

(C) Immediate from the algorithm description. ■

We thus get the following result.

Theorem 2.8. *Suppose that we are given a piecewise linear triangulated surface with the topology of a disk, such that its boundary is formed by two walks L and R. Then, there is a continuous homotopy from L to R of height $\leq \|\mathsf{L}\| + \|\mathsf{R}\| + O(d \log n)$. This homotopy can be computed in $O\big(n^3 \log n\big)$ time.*

3. COMPUTING THE HOMOTOPIC FRECHÉT DISTANCE

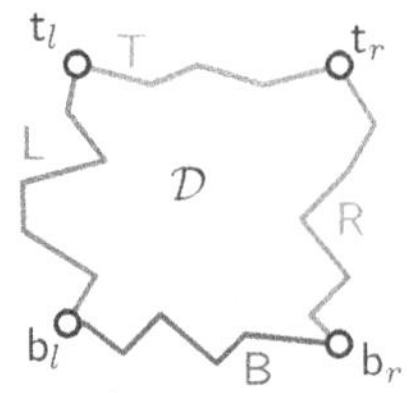

In this section, fix $\mathcal{D}$ to be a triangulated topological disk with n faces. Let the boundary of $\mathcal{D}$ be composed of T, R, B and L, four internally disjoint walks appearing in clockwise order along the boundary. Also, let $\mathsf{t}_l = \mathsf{L} \cap \mathsf{T}$, $\mathsf{b}_l = \mathsf{L} \cap \mathsf{B}$, $\mathsf{t}_r = \mathsf{R} \cap \mathsf{T}$ and $\mathsf{b}_r = \mathsf{R} \cap \mathsf{B}$.[1] See figure on the right.

3.1 Approximating the Regular Frechét Distance

Let $\mathsf{d}_\mathcal{F}(\mathsf{T}, \mathsf{B})$ (resp. $\mathsf{d}_\mathcal{H}(\mathsf{T}, \mathsf{B})$) be the regular (resp. homotopic) Frechét distance between T and B (when restricted to $\mathcal{D}$). Clearly, $\mathsf{d}_\mathcal{F}(\mathsf{T}, \mathsf{B}) \leq \mathsf{d}_\mathcal{H}(\mathsf{T}, \mathsf{B})$. The following lemma implies that the Frechét distance can be approximated within a constant factor.

Lemma 3.1. *Let $\mathcal{D}$, n, T and B be as above. Then, for the continuous case, one can compute, in $O\big(n^3 \log n\big)$ time, reparametrizations of T and B of width $\leq 2\delta$, where $\delta = \mathsf{d}_\mathcal{F}(\mathsf{T}, \mathsf{B})$.*

Proof: Let Π be the set of shortest paths from all points of T to the curve B. As in the proof of Lemma 2.6, let Π_T be the set of all shortest paths from medial points on T to B. Arguing as in Lemma 2.6, we have that the set Π_T is composed of a linear number of paths. The paths in Π_T do not cross and so partition $\mathcal{D}$ into a set of regions. Each region is bounded by a portion of T, a portion of B and two paths in Π_T. A region is a *delta* if the two paths of Π_T in its boundary share a single endpoint (on T), it is a ***pocket*** if they share two endpoints (one on T and one on B), and it is *strip* if they share no endpoints.

Obviously, the (endpoints of the) paths in Π covers all of T. The paths in Π also cover all of B except for the bases of deltas. Now, for each delta we compute the set of all shortest paths from the vertices of its base to its apex inside the delta. Let Π_B be the set of all such paths in all deltas.

[1]We use the same notation to argue about both the discrete problem and the continuous problem.

Clearly, the union of Π_B and Π_T is a set of non-crossing paths whose endpoints cover all the vertices of T and B.

The shortest path from any point of T to B is at most δ. So, all paths in Π have length at most δ. Similarly, the shortest path from a point of B to T is at most δ. Now, consider a delta C with apex c. Let b be a point on the base of C (and so on B). The shortest path π_b from b to T has length at most δ. Let x be the first point that π_b intersects a boundary path of C, π_C. Now, $\pi_b[b, x] \cdot \pi_C[x, c]$ has length at most 2δ and it is inside C. We conclude that all paths in Π_B have length at most 2δ.

The paths in $\Pi_B \cup \Pi_T$ decompose $\mathcal{D}$ into strips and corridors. The left and right portions of a strip is of length at most 2δ, and its top and bottom sides have as such Frechét distance at most 2δ from each other. Similarly, the leash can jump over a pocket from the left leash to the right leash. Doing this to all corridors and pockets, results in reparametrizations of L and R such that their maximum length of a leash for these reparametrizations are at most 2δ. This implies that the Frechét distance is at most 2δ, and we have an explicit reparametrization that realizes this distance.

As for the running time, in $O(n^3 \log n)$ time, one can compute all shortest paths from T to the whole surface. Then, one can, in $O(n^2 \log n)$ time, compute the shortest paths inside each of the linear number of deltas. It follows that the total running time is $O(n^3 \log n)$. ■

Lemma 3.2. *Let $\mathcal{D}$ be a triangulated topological disk with n faces, and* T *and* B *be two internally disjoint walks on the boundary of $\mathcal{D}$. Then, one can compute, in $O(n)$ time, reparametrizations of* T *and* B *that approximate the discrete Frechét distance between* T *and* B. *The computed reparametrizations have width $\leq 3\delta$, where δ is the Frechét distance between* T *and* B.

Proof: First, compute the set of shortest paths, $\Pi_T = \{\pi_1, \pi_2, \cdots, \pi_k\}$, from vertices of T to the path B. Now, let π_i and π_{i+1} be two consecutive paths, that is the endpoints of π_i and π_{i+1}, a_i and a_{i+1}, are adjacent vertices on T. For all $1 \leq i < k$, we add the paths $\pi_i^+ = (a_i, a_{i+1}) \cdot \pi_{i+1}$ to the set Π_T to obtain Π_T^+. Observe that each path in Π_T^+ has length at most 2δ; it is composed of zero or one edge of T and a shortest path from a vertex of T to B. Further, Π_T^+ partitions the graph into regions, similar to the continuous case. Now for each vertex of B that is not an endpoint of a path in Π_T^+, we compute the shortest path inside its region to T. Because the region is bounded by paths of length at most 2δ, the length of such a shortest path is at most 3δ. If Π_B is the set of all such shortest paths, then $\Pi_T^+ \cup \Pi_B$ is a leash sequence of height at most 3δ.

We use the algorithm of Henzinger *et al.* [21] to compute the shortest paths from T in linear time. Since all regions are disjoint, we can compute all the shortest paths inside different regions in $O(n)$ time, as well. ■

3.2 Homotopic Frechét distance if there are no mountains

The following lemma implies that if all the vertices in $\mathcal{D}$ are not too far from the two curves, then one can transform the Frechét distance into the continuous variant (i.e., without jumps in the leash).

Lemma 3.3. *Let $\mathcal{D}$ be a triangulated topological disk with n faces, and* T *and* B *be two internally disjoint walks on the boundary of $\mathcal{D}$. Further, assume for all* $\mathsf{p} \in \mathcal{D}$, p*'s distance to both* T *and* B *is $\leq x$. Then, one can compute reparametrizations of* T *and* B *of width $O(x \log n)$. The running time is $O(n^4 \log n)$ for the continuous case. For the discrete case the running time is $O(n \log n)$.*

In particular, if $x = O(\mathsf{d}_{\mathcal{H}}(\mathsf{T}, \mathsf{B}))$ then this is an $O(\log n)$-approximation to the optimal homotopic Frechét distance.

Proof: Using the algorithm of Lemma 3.1 (or Lemma 3.2 for the discrete case) we compute (not necessarily continuous) reparametrization of T and B of (regular) Frechét distance $\leq \delta$, where $\delta = O(x)$.

Note that the leash movement $s(\cdot)$ associated with this reparametrizations, is not required to deform continuously in s. For a given time $t \in [0, 1]$, let $s^-(t) = \lim_{t' \to t^-} s(t')$ and $s^+(t) = \lim_{t' \to t^+} s(t')$. By definition, s is discontinuous at t if and only if $s^-(t) \neq s^+(t)$. In the discrete case, let $s^-(t) = s(t)$ and $s^+(t) = s(t+1)$. In this case, we say s is discontinuous at t if $s^-(t)$ and $s^+(t)$ are more than one flip operation apart. In both cases, we also say that s has a *gap* at t.

Observe that a gap at time t can be filled by attaching an $(s^-(t), s^+(t))$-homotopy to s at time t. Observe that all the vertices inside the disc with boundary $s^-(t) \cup s^+(t)$ have distance $O(x)$ to T and B and so to $s^-(t)$ and $s^+(t)$, so Lemma 2.8 and Lemma 2.3 imply that an $(s^-(t), s^+(t))$-homotopy with height $O(x \log n)$ can be computed.

Suppose that s has a gap at t. In the discrete case, the disc with boundary $s^-(t) \cup s^+(t)$ contains at least one face. In the continuous case, that disc contains at least one vertex. Thus, in both cases, there are $O(n)$ gaps.

We start with s and fill in all linear number of gaps to obtain a continuous leash sequence of height $O(x \log n)$.

In the discrete case, the initial leash sequence can be computed in $O(n)$ time. Since the gaps are disjoint, they can be filled in overall $O(n \log n)$.

In the continuous case, the initial leash sequence can be computed in $O(n^3 \log n)$ time. Each of the $O(n)$ number of gaps can be filled in $O(n^3 \log n)$ time. ■

3.3 A Decision Procedure for the Homotopic Frechét distance in the presence of mountains

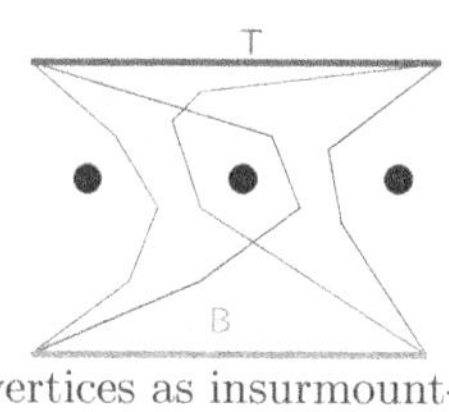

For a parameter $\tau \geq 0$, a vertex $v \in \mathsf{V}(\mathcal{D})$ is *τ-tall* if and only if its distance to T or B is larger than τ (intuitively τ is a guess for the value of $\mathsf{d}_{\mathcal{H}}(\mathsf{T}, \mathsf{B})$). Here, we consider the case where there are τ-tall vertices. Intuitively, one can think about tall vertices as insurmountable mountains. Thus, to find a good homotopy between T and B, we have to choose which "valleys" to use (i.e., what homotopy class the solution we compute belongs to if we think about tall vertices as punctures in the disk). As a concrete example, consider the figure on the right, where there are three tall vertices, and two possible solutions are being shown. As suggested by the figure, we have to make a combinatorial decision: Which tall vertices are going to be to the "left" of computed homotopy, and which tall vertices are going to be on the other side.

In the discrete case, we subdivide each edge in the beginning so that if an edge has length $> 2\tau$, then the vertex inserted in the middle of it is τ-tall. Observe that, if $\tau \geq \mathsf{d}_{\mathcal{H}}(\mathsf{T}, \mathsf{B})$ then no leash of the optimum homotopic motion can afford to contain a τ-tall vertex. We use M^{τ} to denote the set of all τ-tall vertices in $\mathsf{V}(\mathcal{D})$.

Now, let ω and ω' be two walks connecting points on T and B. We say that ω and ω' are ***isotopic*** in $\mathcal{D} \setminus \mathsf{M}^{\tau}$ if and only if they are homotopic in $\mathcal{D} \setminus \mathsf{M}^{\tau}$ after contracting T and B. Consequently, the disc with boundary $\mathsf{T} \cup \mathsf{B} \cup \omega \cup \omega'$ contains no tall vertices if ω and ω' are isotopic.

Definition 3.4. *Given a subset $X \subseteq \mathsf{M}^{\tau}$, consider a path ζ from T to B, such that X is contained in one side of $\mathcal{D} \setminus \zeta$ (i.e., cutting $\mathcal{D}$ along ζ breaks it into two connected components), and $\mathsf{M}^{\tau} \setminus X$ is contained in the other side. The set of all such paths is the* **isotopy class** *of X. For each such X, the isotopy class of X is a τ-isotopy class.*

Let $\pi_{\mathrm{L},\mathsf{h}}$ (resp. $\pi_{\mathrm{R},\mathsf{h}}$) be the **left geodesic** *(resp.* **right geodesic***) of an isotopy class h; that is, $\pi_{\mathrm{L},\mathsf{h}}$ denotes the shortest path in h from t_l to b_l (resp. from t_r to b_r).*

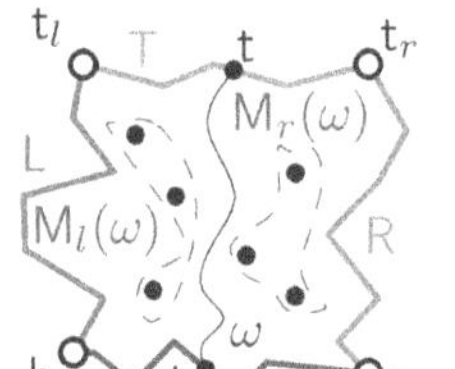

Let ω be any walk in h from $\mathsf{b} \in \mathsf{B}$ to $\mathsf{t} \in \mathsf{T}$. We define the ***left tall set*** of h, denote $\mathsf{M}_l(\mathsf{h}) = \mathsf{M}_l(\omega)$ to be the set of all τ-tall vertices to the left of ω; inside the disc with boundary $\mathsf{L} \cup \mathsf{T}[\mathsf{t}_l, \mathsf{t}] \cup \omega \cup \mathsf{B}[\mathsf{b}_l, \mathsf{b}]$, where L is the "left" portion of the boundary of $\mathcal{D}$, having endpoints t_l and b_l. We similarly define the ***right tall set*** of h, $\mathsf{M}_r(\mathsf{h}) = \mathsf{M}_r(\omega)$, to be the set of all τ-tall vertices to the right of ω. See figure on the right. Note that the sets $\mathsf{M}_l(\mathsf{h})$ and $\mathsf{M}_r(\mathsf{h})$ do not depend on the particular choice of ω, since all paths in h are in the same isotopy class.

We say that h is ***τ-extendable*** from the left if and only if $\|\pi_{\mathrm{L},\mathsf{h}}\| \leq \tau$ and there is an isotopy class h', such that $\|\pi_{\mathrm{L},\mathsf{h}'}\| \leq \tau$ and $\mathsf{M}_l(\mathsf{h}) \subset \mathsf{M}_l(\mathsf{h}')$. In particular, h is ***τ-saturated*** if it is not τ-extendable and $\|\pi_{\mathrm{L},\mathsf{h}}\| \leq \tau$.

3.3.1 On the left and right geodesics

Lemma 3.5. *Let h be a τ-saturated isotopy class, where $\tau \geq \mathsf{d}_{\mathcal{H}}(\mathsf{T}, \mathsf{B})$. Then, $\|\pi_{\mathrm{R},\mathsf{h}}\| \leq 4\tau$.*

Proof: Let $\mathsf{h}_{\mathrm{opt}}$ be the isotopy class of the leashes in the optimum solution. Of course, no leash in the optimum solution contains a τ-tall vertex, and therefore, all leashes in the optimal solution are isotopic.

Since h is saturated, the set $\mathsf{M}_l(\mathsf{h})$ is not a proper subset of $\mathsf{M}_l(\mathsf{h}_{\mathrm{opt}})$. If $\mathsf{M}_l(\mathsf{h}) = \mathsf{M}_l(\mathsf{h}_{\mathrm{opt}})$ then $\mathsf{h} = \mathsf{h}_{\mathrm{opt}}$, and in particular $\|\pi_{\mathrm{R},\mathsf{h}}\| = \|\pi_{\mathrm{R},\mathsf{h}_{\mathrm{opt}}}\| \leq \tau$.

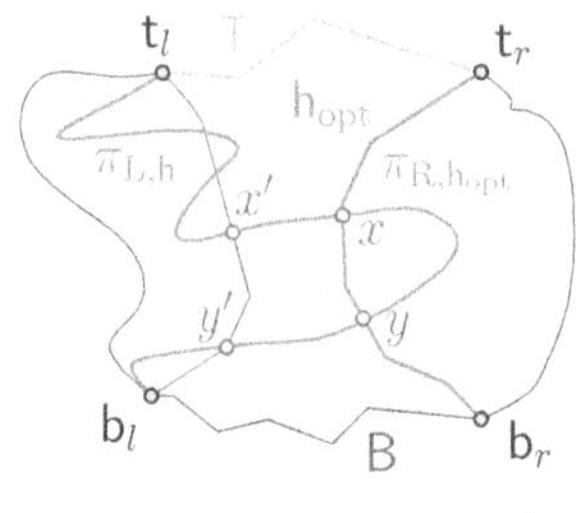

Otherwise, the set $\mathsf{M}_l(\mathsf{h}) \cap \mathsf{M}_r(\mathsf{h}_{\mathrm{opt}})$ is not empty. It follows that $\pi_{\mathrm{L},\mathsf{h}}$ crosses $\pi_{\mathrm{R},\mathsf{h}_{\mathrm{opt}}}$.

Let x be the first intersection point between $\pi_{\mathrm{L},\mathsf{h}}$ and $\pi_{\mathrm{R},\mathsf{h}_{\mathrm{opt}}}$, as one traverses $\pi_{\mathrm{L},\mathsf{h}}$ from t_l to b_l. Let x' be the last intersection point of $\pi_{\mathrm{L},\mathsf{h}}[\mathsf{t}_l, x]$ with $\pi_{\mathrm{L},\mathsf{h}_{\mathrm{opt}}}$. Similarly, y is the last intersection point between $\pi_{\mathrm{L},\mathsf{h}}$ and $\pi_{\mathrm{R},\mathsf{h}_{\mathrm{opt}}}$, and y' is the first intersection of $\pi_{\mathrm{L},\mathsf{h}}[y, \mathsf{b}_l]$ and $\pi_{\mathrm{L},\mathsf{h}_{\mathrm{opt}}}$. Observe that the interiors of $\pi_{\mathrm{L},\mathsf{h}}[x', x]$ and $\pi_{\mathrm{L},\mathsf{h}}[y, y']$ does not intersect the curves $\pi_{\mathrm{L},\mathsf{h}_{\mathrm{opt}}}$ and $\pi_{\mathrm{R},\mathsf{h}_{\mathrm{opt}}}$.

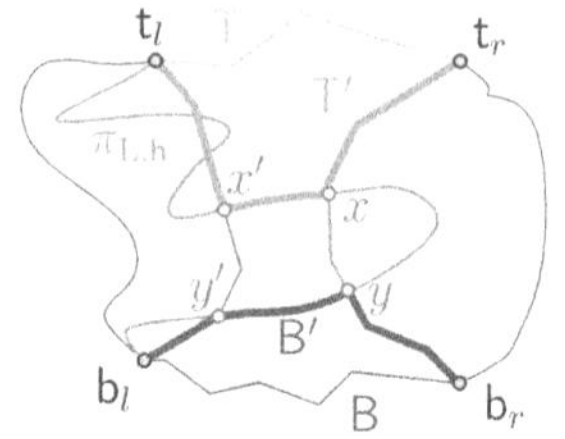

As the curves $\pi_{\mathrm{L},\mathsf{h}}$ and $\pi_{\mathrm{R},\mathsf{h}}$ are isotopic (by definition), the disk with the boundary $\mathsf{T} \cdot \pi_{\mathrm{L},\mathsf{h}} \cdot \mathsf{B} \cdot \pi_{\mathrm{R},\mathsf{h}}$ does not contain any tall vertex, and $\mathsf{T} \cdot \pi_{\mathrm{L},\mathsf{h}} \cdot \mathsf{B}$ is homotopic to $\pi_{\mathrm{R},\mathsf{h}}$.

Consider the following walk $\mathsf{T}' = \pi_{\mathrm{R},\mathsf{h}_{\mathrm{opt}}}[\mathsf{t}_r, x] \cdot \pi_{\mathrm{L},\mathsf{h}}[x, x'] \cdot \pi_{\mathrm{L},\mathsf{h}_{\mathrm{opt}}}[x', \mathsf{t}_l]$. The walk T' is homotopic to T. Similarly, $\mathsf{B}' = \pi_{\mathrm{L},\mathsf{h}_{\mathrm{opt}}}[\mathsf{b}_l, y'] \cdot \pi_{\mathrm{L},\mathsf{h}}[y', y] \cdot \pi_{\mathrm{R},\mathsf{h}_{\mathrm{opt}}}[y, \mathsf{b}_r]$ is homotopic to B. It follows that $\pi_{\mathrm{R},\mathsf{h}}$ is homotopic to $\mathsf{T}' \cdot \pi_{\mathrm{L},\mathsf{h}} \cdot \mathsf{B}'$. As $\pi_{\mathrm{R},\mathsf{h}}$ is the shortest path in its homotopy class with these endpoints, it follows that

$$
\begin{aligned}
\|\pi_{\mathrm{R},\mathsf{h}}\| &\leq \|\mathsf{T}' \cdot \pi_{\mathrm{L},\mathsf{h}} \cdot \mathsf{B}'\| \\
&\leq \|\pi_{\mathrm{L},\mathsf{h}}\| + \left(\|\pi_{\mathrm{L},\mathsf{h}_{\mathrm{opt}}}\| + \|\pi_{\mathrm{L},\mathsf{h}}\| + \|\pi_{\mathrm{R},\mathsf{h}_{\mathrm{opt}}}\|\right) \leq 4\tau,
\end{aligned}
$$

as T' and B' are disjoint, and $\mathsf{T}' \cup \mathsf{B}' \subseteq \pi_{\mathrm{R},\mathsf{h}_{\mathrm{opt}}} \cup \pi_{\mathrm{L},\mathsf{h}_{\mathrm{opt}}} \cup \pi_{\mathrm{L},\mathsf{h}}$. ■

Lemma 3.6. *Let h be a τ-isotopy class, such that we have $\max(\|\pi_{\mathrm{L},\mathsf{h}}\|, \|\pi_{\mathrm{R},\mathsf{h}}\|) \leq x$, where $x \geq \tau \geq \mathsf{d}_{\mathcal{H}}(\mathsf{T}, \mathsf{B})$. Let $\mathcal{D}'$ be the disk with boundary $\mathsf{T} \cdot \pi_{\mathrm{R},\mathsf{h}} \cdot \mathsf{B} \cdot \pi_{\mathrm{L},\mathsf{h}}$. Then, all the points inside $\mathcal{D}'$ are closer than $O(x)$ to both T and B in $\mathcal{D}'$.*

Proof: We first consider the continuous case.

By the definition of τ-isotopy, the disk $\mathcal{D}'$ has no τ-tall vertices. Furthermore, by the definition of x, we have that the distance of any point on T to B, restricted to paths in $\mathcal{D}'$ is at most δ_1, where $\delta_1 = x + \mathsf{d}_{\mathcal{F}}(\mathsf{T}, \mathsf{B}) \leq 2x$. Indeed, the shortest path from any point on T to B in $\mathcal{D}$, either stays inside $\mathcal{D}'$, or alternatively intersects either $\pi_{\mathrm{L},\mathsf{h}}$ or $\pi_{\mathrm{R},\mathsf{h}}$.

We can now deploy the decomposition of $\mathcal{D}'$ into strips and pockets, as done in Section 2.2.2. Every strip is being swept by a leash of length $\leq \delta_2 = 2\delta_1 \leq 4x$ (the factor two is because a strip might rise out of a delta), and as such the claim trivially holds for points inside a strip.

Every pocket P has perimeter of length at most $\|\partial \mathsf{P}\| \leq \delta_3 = 2\delta_2 = 8x$ (the perimeter also contains two points of T and B and they are in distance at most δ_2 from each other in either direction along the perimeter).

So consider such a pocket P. Since $\mathcal{D}'$ contains no τ-tall vertices, P does not contain any tall vertex. Let e be an edge in P (or a subedge if it intersects the boundary of P). The two endpoints of e are in P, and such an endpoint is either a (not tall) vertex or it is contained in $\partial \mathsf{P}$. In either case, these endpoints are in distance at most x from $\partial \mathsf{P}$, and as such they are in distance at most $\delta_4 = 2x + \|\partial \mathsf{P}\|/2 = 2x + \delta_2 \leq 6x$ from each other (inside P). We conclude that $\|\mathsf{e}\| \leq \delta_4$, and as such, any point in e is in distance at most $\delta_6 = \|\mathsf{e}\|/2 + x + \delta_2 \leq 3x + x + 8x \leq 12x$ from T and B.

Now, consider any point p in P, and consider the face F that contains it. Since the surface is triangulated, F is a triangle. Clipping $\triangle$ to P results in a planar region F' that has perimeter $\leq \delta_5 = 3\delta_4 + \|\partial \mathsf{P}\| \leq 3 \cdot 6x + \delta_3 \leq (18+8)x \leq 26x$ (note, that an edge might be fragmented into several subedges, but the furthest two points along a single edge is at most δ_4 using the same argumentation as above). As such, the furthest a point of P can be from an edge of P is at most $\delta_7 = \delta_5/2\pi \leq 5x$. As such, the maximum distance of a point of P from either T or B (inside $\mathcal{D}'$) is at most $\delta_6 + \delta_7 \leq 12x + 5x = 17x$.

The discrete case is easier. Any edge of length $\geq 2\tau$ was split, by introducing a middle vertex, which must be τ-tall.

As such, there are no long edges using by any curves in the same isotopic class. A simplified version of the above proof now implies the claim. ■

3.3.2 *The decision algorithm*

The proof of the following three lemmas, can be found in the full version of the paper [19].

Lemma 3.7. *Let $\mathcal{D}, n, \mathsf{T}, \mathsf{L}, \mathsf{B}, \mathsf{R}, \mathsf{t}_l, \mathsf{b}_l, \tau$ be as above, and let $X \subseteq \mathsf{V}(\mathcal{D})$ be the set of τ-tall vertices. Consider the shortest path ζ (between t_l and b_l) that belongs to any homotopy class h such that $X \subseteq \mathsf{M}_l(\mathsf{h})$. Then, path ζ can be computed in $O(n^4 \log n)$ (resp. $O(n^2 \log n)$) time in the continuous (resp. discrete) case.*

Lemma 3.8. *Let $\mathcal{D}$ be a triangulated topological disk with n faces, and T and B be two internally disjoint walks on $\mathcal{D}$'s boundary. Given $\tau > 0$, one can compute a τ-saturated isotopy class in $O(n^5 \log n)$ (resp. $O(n^3 \log n)$) time the continuous (resp. discrete) case.*

Lemma 3.9. *Let $\mathcal{D}$ be a triangulated topological disk with n faces, and T and B be two internally disjoint walks on the boundary of $\mathcal{D}$. Given a real number $x > 0$, one can either:*

(A) Compute a homotopy from T to B of width $O(x \log n)$.

(B) Return that $x < \mathsf{d}_{\mathcal{H}}(\mathsf{T}, \mathsf{B})$.

The running time of this procedure is $O(n^5 \log n)$ (resp. $O(n^3 \log n)$) in the continuous (resp. discrete) case.

Proof: Assume $x \geq \delta_H = \mathsf{d}_{\mathcal{H}}(\mathsf{T}, \mathsf{B})$, and we use x as a guess for this value δ_H. Using Lemma 3.8, one can compute a x-saturated isotopy class, h. Lemma 3.5 implies that both $\pi_{\mathrm{L},\mathsf{h}}$ and $\pi_{\mathrm{R},\mathsf{h}}$ are at most $4x$. Let $\mathcal{D}' \subseteq \mathcal{D}$ be the disc with boundary $\mathsf{T} \cup \cup\pi_{\mathrm{L},\mathsf{h}} \cup \mathsf{B} \cup \pi_{\mathrm{R},\mathsf{h}}$. By Lemma 3.6, all vertices in $\mathcal{D}'$ are in distance $O(x)$ from T and B (i.e., there are no $O(x)$-tall vertices in $\mathcal{D}'$). Finally, Lemma 3.3 implies that a continuous leash sequence of height $O(x \log n)$ between T and B, inside $\mathcal{D}'$, can be computed.

Thus, if x is larger than $\mathsf{d}_{\mathcal{H}}(\mathsf{T}, \mathsf{B})$ then this algorithm returns the desired approximation. Otherwise, it fails only if $x < \mathsf{d}_{\mathcal{H}}(\mathsf{T}, \mathsf{B})$. In the case of such failure, return that x is too small. ■

3.4 A strongly polynomial algorithm for the Frechét distance

Identifying the tall vertices.

Observe that using the algorithm of Lemma 3.9, we can decide given a candidate value δ_H for $\mathsf{d}_{\mathcal{H}}(\mathsf{T}, \mathsf{B})$ if it is too large, too small, or leads to the desired approximation. Indeed, if the algorithm returns an approximation of values $O(\delta_H \log n)$ but fails for $\delta_H/2$, we know it is the desired approximation.

So, compute for each vertex $v \in \mathsf{V}(\mathcal{D})$ its tallness; that is α_v would be the maximum distance of v to either T or B. Sort these values, and using binary search, compute the vertex w, with the minimum value α_w, such that Lemma 3.9 returns a parametrization with homotopic Frechét distance $O(\alpha_w \log n)$. If the algorithm of Lemma 3.9 returns that α_w/n is too small of a guess, then $[\alpha_w/n, \alpha_w \log n]$ contains δ_H. In this case, we can use binary search to find an interval $[\gamma/2, \gamma]$ that contains δ_H and use Lemma 3.9 to obtain the desired approximation. Similarly, if v is the tallest vertex shorter than w, then we can assume that $\alpha_v n$ is too small of a guess, otherwise we are again done as $[\alpha_v, \alpha_v n]$ contains δ_H.

As such, in the following, we know that the desired distance δ_H lies in interval $[x, y]$ where $x = \alpha_v n$ and $y = \alpha_w/n$, and for every vertex u of $\mathcal{D}$ it holds that (i) $\alpha_u < x/n$, or (ii) $\alpha_v > yn$. Naturally, we consider all the vertices that satisfy (ii) as tall vertices, by setting $\tau = 2x/n$. In the following, let M denote the set of these τ-tall vertices.

Rough approximation via greedy addition of vertices.

For a vertex $v \in \mathsf{V}(\mathcal{D})$, define $\mathrm{cost}(v)$ to be the length of the shortest path between t_l and b_l that has v on its left side. Similarly, for a set of vertices $X \subseteq \mathsf{V}(\mathcal{D})$, let $\mathrm{Cost}(X)$ be the length of the shortest path between t_l and b_l that has X on its left side. For a specific v or X, one can compute $\mathrm{cost}(v)$ and $\mathrm{cost}(X)$ by invoking the algorithm of Lemma 3.7 once.

The following is easy to verify.

Lemma 3.10. *For $X, Y \subseteq \mathsf{V}(\mathcal{D})$, we have $\mathrm{Cost}(X \cup Y) \leq \mathrm{Cost}(X) + \mathrm{Cost}(Y)$.*

Consider the algorithm that starts with $X_0 = \emptyset$. In the ith iteration, the algorithm computes the vertex $v_i \in \mathsf{M} \setminus X_{i-1}$, such that $\mathrm{Cost}(X_{i-1} \cup \{v_i\})$ is minimized, and set $X_i = X_{i-1} \cup \{v_i\}$. Let h_i be the isotopy class having X_i on its left side, and $\mathsf{M} \setminus X_i$ on its right side.

Lemma 3.11. *The cheapest homotopic Frechét parametrization among $\mathsf{h}_1, \ldots, \mathsf{h}_n$ has width $O(\mathsf{d}_{\mathcal{H}}(\mathsf{T}, \mathsf{B})\, n \log n)$.*

Proof: Consider the set Y that is the subset of tall vertices on the left side of the optimal solution. Let i be the first index such that $Y \subseteq X_i$ and $Y \not\subseteq X_{i-1}$. Let v be any vertex in $Y \setminus X_{i-1}$. We have that

$$\begin{aligned}\mathrm{Cost}(X_i) &\leq \mathrm{Cost}(X_{i-1} \cup \{v\}) \leq \mathrm{Cost}(X_{i-1}) + \mathrm{cost}(v)\\ &\leq \mathrm{Cost}(X_{i-1}) + \mathrm{Cost}(Y)\\ &\leq \mathrm{Cost}(X_{i-2}) + 2\mathrm{Cost}(Y) \leq \cdots \leq i\mathrm{Cost}(Y)\\ &\leq n\mathrm{Cost}(Y).\end{aligned}$$

Now, setting $\tau = \mathrm{Cost}(X_i)$, it follows that the set X_i is τ-saturated. Applying Lemma 3.5, implies that $\|\pi_{\mathrm{R},\mathsf{h}_i}\| \leq 4\tau$. Observe, that the disk defined by T, $\pi_{\mathrm{L},\mathsf{h}_i}$, B, $\pi_{\mathrm{R},\mathsf{h}_i}$ can not contain any tall vertex (by construction).

Now, plugging this into Lemma 3.3 implies the homotopic Frechét width of h_i (starting with $\pi_{\mathrm{L},\mathsf{h}_i}$ and ending up with $\pi_{\mathrm{R},\mathsf{h}_i}$) is $O(\tau \log n)$, which implies the claim since $\mathrm{Cost}(X_i) \leq n\mathrm{Cost}(Y) \leq n\mathsf{d}_{\mathcal{H}}(\mathsf{T}, \mathsf{B})$. ■

The algorithm.

We approximate the homotopic Frechét width of each one of the classes $\mathsf{h}_1, \ldots, \mathsf{h}_n$. We identify the value x with the smallest width, and we do a binary search in the interval $[x/n^2, x]$ for the homotopic Frechét distance.

Theorem 3.12. *Let $\mathcal{D}$ be a topological disk that is triangulated and has n faces, and T and B be two internally disjoint walks on the boundary of $\mathcal{D}$. One can compute a homotopic Frechét parametrization of width $O(\mathsf{d}_{\mathcal{H}}(\mathsf{T}, \mathsf{B}) \log n)$, where $\mathsf{d}_{\mathcal{H}}(\mathsf{T}, \mathsf{B})$ is the homotopic Frechét distance between T and B in $\mathcal{D}$.*

The running time of this procedure is $O(n^6 \log n)$ (resp. $O(n^4 \log n)$) in the continuous (resp. discrete) case.

Proof: The algorithm requires $O(n^2)$ calls to Lemma 3.7, which takes $O(n^6 \log n)$ (resp. $O(n^4 \log n)$) time in the continuous (resp. discrete) case. Then, the algorithm requires Lemma 3.3 to compute the homotopic Frechét distance of the classes $\mathsf{h}_1, \ldots, \mathsf{h}_n$. The algorithm also performs $O(\log n)$ calls to the algorithm of Lemma 3.9. ■

4. CONCLUSIONS

We presented a $O(\log n)$ approximation algorithm for approximating the homotopy height and the homotopic Frechét distance between curves on piecewise linear surfaces. It seems quite believable that the approximation quality can be further improved, and we leave this as the main open problem of our work. Since our algorithm works both for the continuous and discrete cases, it seems natural to conjecture that this algorithm should work more general surfaces and metrics.

5. REFERENCES

[1] H. Alt and M. Buchin. Semi-computability of the Frechét distance between surfaces. In *Proc. 21st Euro. Workshop on Comput. Geom.*, 2005.

[2] H. Alt and M. Godau. Computing the Fréchet distance between two polygonal curves. *Internat. J. Comput. Geom. Appl.*, 5:75–91, 1995.

[3] Chakib Bennis, Jean-Marc Vézien, Gérard Iglésias, and André Gagalowicz. Piecewise surface flattening for non-distorted texture mapping. In Thomas W. Sederberg, editor, *Proc. SIGGRAPH '91*, volume 25, pages 237–246, 1991.

[4] S. Brakatsoulas, D. Pfoser, R. Salas, and C. Wenk. On map-matching vehicle tracking data. In *Proc. 31st VLDB Conference*, pages 853–864, 2005.

[5] K. Buchin, M. Buchin, and J. Gudmundsson. Detecting single file movement. In *Proc. 16th ACM SIGSPATIAL Int. Conf. Adv. GIS*, pages 288–297, 2008.

[6] K. Buchin, M. Buchin, J. Gudmundsson, Maarten L., and J. Luo. Detecting commuting patterns by clustering subtrajectories. In *Proc. 19th Annu. Internat. Sympos. Algorithms Comput.*, pages 644–655, 2008.

[7] E. W. Chambers, E. C. de Verdière, J. Erickson, S. Lazard, F. Lazarus, and S. Thite. Walking your dog in the woods in polynomial time. In *Proc. 24th Annu. ACM Sympos. Comput. Geom.*, pages 101–109. ACM, 2008.

[8] E. W. Chambers and D. Letscher. On the height of a homotopy. In *Proc. 21st Canad. Conf. Comput. Geom.*, 2009.

[9] E. W. Chambers and D. Letscher. Erratum for on the height of a homotopy. `http://mathcs.slu.edu/~chambers/papers/hherratum.pdf`, 2010.

[10] A. F. Cook, A. Driemel, S. Har-Peled, J. Sherette, and C. Wenk. Computing the Fréchet distance between folded polygons. In *Proc. 12th Workshop Algorithms Data Struct.*, pages 267–278, 2011.

[11] A. F. Cook and C. Wenk. Geodesic Fréchet distance inside a simple polygon. *ACM Trans. Algo.*, 7:9:1–9:19, 2010.

[12] A. Driemel, S. Har-Peled, and C. Wenk. Approximating the Fréchet distance for realistic curves in near linear time. In *Proc. 26th Annu. ACM Sympos. Comput. Geom.*, pages 365–374, 2010.

[13] A. Efrat, L. J. Guibas, S. Har-Peled, J. S.B. Mitchell, and T.M. Murali. New similarity measures between polylines with applications to morphing and polygon sweeping. *Discrete Comput. Geom.*, 28:535–569, 2002.

[14] T. Eiter and H. Mannila. Computing discrete Frechét distance. Tech. Report CD-TR 94/64, Christian Doppler Lab. Expert Sys., TU Vienna, Austria, 1994.

[15] M. S. Floater. Parameterization and smooth approximation of surface triangulations. *Comput. Aided Geom. Design*, 14(4):231–250, 1997.

[16] M. Frechét. Sur quelques points du calcul fonctionnel. *Rendic. Circ. Mat. Palermo*, 22:1–74, 1906.

[17] M. Frechét. Sur la distance de deux surfaces. *Ann. Soc. Polonaise Math.*, 3:4–19, 1924.

[18] M. Godau. *On the complexity of measuring the similarity between geometric objects in higher dimensions*. PhD thesis, Free University of Berlin, 1999.

[19] S. Har-Peled, A. Nayyeri, M. Salavatipour, and A. Sidiropoulos. How to walk your dog in the mountains with no magic leash. `http://valis.cs.uiuc.edu/~sariel/papers/11/homotopy/`, 2012.

[20] S. Har-Peled and B. Raichel. The Fréchet distance revisited and extended. In *Proc. 27th Annu. ACM Sympos. Comput. Geom.*, pages 448–457, 2011. `http://www.cs.uiuc.edu/~sariel/papers/10/frechet3d/`.

[21] M. R. Henzinger, P. Klein, S. Rao, and S. Subramanian. Faster shortest-path algorithms for planar graphs. *J. Comput. Sys. Sci.*, 55:3–23, August 1997.

[22] E. J. Keogh and M. J. Pazzani. Scaling up dynamic time warping to massive dataset. In *Proc. of the Third Euro. Conf. Princip. Data Mining and Know. Disc.*, pages 1–11, 1999.

[23] M.S. Kim, S.W. Kim, and M. Shin. Optimization of subsequence matching under time warping in time-series databases. In *Proc. ACM symp. Applied comput.*, pages 581–586, 2005.

[24] A. Mascret, T. Devogele, I. Le Berre, and A. Hénaff. Coastline matching process based on the discrete Fréchet distance. In *Proc. 12th Int. Sym. Spatial Data Handling*, pages 383–400, 2006.

[25] D. Piponi and G. Borshukov. Seamless texture mapping of subdivision surfaces by model pelting and texture blending. In *Proc. SIGGRAPH 2000*, pages 471–478, August 2000.

[26] J. Serrà, E. Gómez, P. Herrera, and X. Serra. Chroma binary similarity and local alignment applied to cover song identification. *IEEE Transactions on Audio, Speech & Language Processing*, 16(6):1138–1151, 2008.

[27] Alla Sheffer and Eric de Sturler. Surface parameterization for meshing by triangulation flattening. In *Proc. 9th International Meshing Roundtable*, pages 161–172, 2000.

[28] C. Wenk, R. Salas, and D. Pfoser. Addressing the need for map-matching speed: Localizing global curve-matching algorithms. In *Proc. 18th Int. Conf. Sci. Statis. Database Manag.*, pages 879–888, 2006.

Tracing Compressed Curves in Triangulated Surfaces*

Jeff Erickson†
Department of Computer Science
University of Illinois, Urbana-Champaign

Amir Nayyeri
Department of Computer Science
University of Illinois, Urbana-Champaign

ABSTRACT

A simple path or cycle in a triangulated surface is *normal* if it intersects any triangle in a finite set of arcs, each crossing from one edge of the triangle to another. We describe an algorithm to "trace" a normal curve in $O(\min\{X, n^2 \log X\})$ time, where n is the complexity of the surface triangulation and X is the number of times the curve crosses edges of the triangulation. In particular, our algorithm runs in polynomial time even when the number of crossings is exponential in n. Our tracing algorithm computes a new cellular decomposition of the surface with complexity $O(n)$; the traced curve appears as a simple path or cycle in the 1-skeleton of the new decomposition.

We apply our abstract tracing strategy to two different classes of normal curves: abstract curves represented by *normal coordinates*, which record the number of intersections with each edge of the surface triangulation, and simple *geodesics*, represented by a starting point and direction in the local coordinate system of some triangle. Our normal-coordinate algorithms are competitive with and conceptually simpler than earlier algorithms by Schaefer, Sedgwick, and Štefankovic [COCOON 2002, CCCG 2008] and by Agol, Hass, and Thurston [*Trans. AMS* 2005].

Categories and Subject Descriptors: F.2.2 [Analysis of Algorithms and Problem Complexity]: Nonnumerical Algorithms and Problems—*Computations on discrete structures, Geometrical problems and computations*

Keywords: computational topology, normal coordinates, geodesics

1. INTRODUCTION

Curves on abstract surfaces are usually represented by describing the interaction between the curve and a decomposition of the surface into elementary pieces. For example, given a triangulation of the surface, any sufficiently well-behaved curve can be described by listing the sequence of edges of the triangulation that the curve crosses, in order along the curve. (See Section 2 for more formal definitions.) This *intersection sequence* identifies the curve up to a continuous deformation that avoids the vertices. We call a subpath of a curve between two consecutive edge crossings an *elementary segment*.

For *simple* curves, however, there are several more compact representations. For example, given a triangulation of the surface, any sufficiently well-behaved simple curve can be described by listing the number of elementary segments connecting each pair of edges in each triangle. These numbers are called the *normal coordinates* of the curve [15, 19]. Any vector of normal coordinates identifies a unique simple curve (again up to continuous deformation), because there is only one way to fill each triangle with the correct number of elementary segments without intersection. The normal coordinate representation is remarkably compact; only $O(n \log(X/n))$ bits are needed to list the normal coordinates of a curve with X crossings on a triangulated surface with complexity n. Several algorithms in two- and three-dimensional topology owe their efficiency to the compactness of the normal-coordinate representation [1, 6, 7, 16, 35, 36, 38, 40, 44].

Schaefer *et al.* [35, 38, 44] consider several algorithmic questions about normal curves, such as computing the number of components of a curve, deciding whether two given curves are isotopic, and computing algebraic and geometric intersection numbers of pairs of curves. Classical algorithms for these problems require explicit intersection sequences as input. By connecting normal coordinates with grammar-based text compression [22, 24, 34] and word equations [31, 32, 33], Schaefer *et al.* developed algorithms whose running times are polynomial in the bit complexity of the normal coordinates. These algorithms rely on a complex algorithm of Plandowski and Rytter [31] to compute compressed solutions of word equations. We are unaware of any precise time analysis, but as Plandowski and Rytter's algorithm uses a nested sequence of quadratic- and cubic-time reductions, its running time is quite high. Štefankovic [44] described simpler linear-time algorithms for some of these problems, by reducing them to an algorithm of Robson and Deikert [32, 33] to solve word equations with a certain special structure. Some of the problems considered by Schaefer *et al.* can also be solved in polynomial time using the orbit-counting algorithm of Agol *et al.* [1], which was designed for normal *surfaces* in triangulated 3-manifolds.

Other compact representations of curves on surfaces include weighted train tracks [4, 5, 13, 14, 30], Dehn-Thurston coordinates (with respect to a fixed pants decomposition of the surface) [9, 13, 14, 29, 46], and compressed intersection sequences [38, 44].

*This work was partially supported by NSF grant CCF 09-15519. See http://www.cs.uiuc.edu/~jeffe/pubs/tracing.html for the most recent version of this paper.
†Portions of this work were done while this author was visiting IST Austria.

SCG'12, June 17–20, 2012, Chapel Hill, North Carolina, USA.

Our Results

We propose an alternate strategy to efficiently compute with curves on surfaces. Instead of using complex compression techniques to avoid unpacking the intersection sequence of the input curve, our algorithms *modify the underlying cellular decomposition* of the surface so that the curve has a small *explicit* description with respect to the new decomposition. Specifically, given the normal coordinates of a curve γ on a triangulated surface with n edges, we compute a new cellular decomposition of the surface with complexity $O(n)$, called a *street complex*, such that γ is a simple path or cycle in the 1-skeleton. After reviewing some background terminology, we formally define the street complex in Section 2; see Figure 2 for an example.

At a high level, our algorithm simply *traces* the curve, continuously updating the street complex to reflect the portion of the curve traced so far. A naïve implementation of our tracing strategy runs in $O(X)$ time, where X is the total number of edge crossings; each time the curve enters a triangle by crossing an edge, we can easily determine in $O(1)$ time which of the other two edges of the triangle to cross next. The main result of this paper is a tracing algorithm that runs in $O(n^2 \log X)$ time, an exponential improvement over the naïve algorithm for any fixed surface triangulation.

Our new algorithm relies on two simple ideas. First, we observe that for typical curves, most of the decisions made by the brute-force tracing algorithm are redundant. If a curve enters a triangle Δ between two older elementary segments that leave Δ through the same edge, the new elementary segment must also leave Δ through that edge. The street complex allows us to skip these redundant decisions.

Second, even with redundant decisions filtered out, the naïve algorithm may repeat the same series of crossings many times in a row when the input curve contains a *spiral* [28, 37, 39]. Our algorithm detects spirals as they occur, quickly determines the depth of the spiral (the number of repetitions), and then skips ahead to the first crossing after the spiral. We describe our generic tracing algorithm in Section 3 and analyze its running time in Section 4.

The street complex allows us to answer several fundamental topological questions about simple curves using elementary algorithms. For example, to determine whether a curve represented by normal coordinates is connected, we can trace one component of the curve, and then check whether the number of edge crossings we encountered is equal to the sum of the normal coordinates. To determine whether a connected normal curve is contractible, we can trace the curve and then apply a $O(n)$-time depth-first search in the dual of the resulting street complex [12].

In particular, our tracing algorithm can be adapted to solve many of the problems considered by Schaefer *et al.* [35] in $O(n^2 \log X)$ time. The resulting algorithms are significantly faster and simpler than both the corresponding word-equation algorithms of Schaefer *et al.* and the orbit-counting algorithm of Agol *et al.* [1]. In Section 5, as an illustrative example, we describe an algorithm to classify the components of a normal curve into isotopy classes in $O(n^2 \log X)$ time; our algorithm returns the number of isotopy classes and the number of components in each class. More examples will appear in the full paper.

Our technique has two disadvantages compared to earlier results. First, for a few of the problems we consider, such as computing the number of components of a normal curve, our algorithms are slower by a factor of n than those described by Štefankovic in his PhD thesis [44]. Second, our algorithm is limited to curves on surfaces; whereas, the orbit-counting and word-equation algorithms are dimension-agnostic.

In Section 6, we sketch and extension of our tracing algorithm to simple *geodesic* paths on piecewise-linear surfaces. Here, the input surface is presented as a set of n triangles, each with its own local Euclidean coordinate system, with some pairs of equal-length edges identified; the geodesic path is specified by a starting point and a direction, in the local coordinate system of one of the triangles. In particular, we do *not* assume that the input surface is embedded (or embeddable) in any Euclidean space. As an example application, we sketch an algorithm to find the first point of self-intersection on a geodesic path in $O(n^2 \log X)$ time.

Due to space limitations, many proofs and low-level details are omitted from this version of the paper, especially in Section 6.

2. BACKGROUND

We begin by recalling several standard definitions from combinatorial topology; for further background, see Edelsbrunner and Harer [11] or Stillwell [45].

2.1 Surfaces, Curves, and Isotopy

A ***surface*** (more formally, a *2-manifold with boundary*) is a Hausdorff space in which every point has an open neighborhood homeomorphic to either the plane $\mathbb{R}^2$ or the closed halfplane $\{(x, y) \mid x \geq 0\}$. We consider only compact, connected, *orientable* surfaces in this paper. The set of points in a surface Σ with halfplane neighborhoods is the ***boundary*** of the surface, denoted $\partial\Sigma$; the boundary is homeomorphic to a finite set of disjoint circles.

A ***simple cycle*** in a surface Σ is (the image of) a continuous injective map $\gamma\colon S^1 \to \Sigma$. A ***simple path*** is (the image of) a continuous injective map $\pi\colon [0, 1] \to \Sigma$; a ***simple arc*** α is a simple path whose endpoints $\alpha(0)$ and $\alpha(1)$ lie on the boundary $\partial\Sigma$. An arc is ***properly embedded*** if it intersects $\partial\Sigma$ only at its endpoints; similarly, a cycle is properly embedded if it avoids $\partial\Sigma$ entirely. A ***properly embedded curve*** is a finite collection of disjoint, properly embedded arcs and cycles. We emphasize that properly embedded curves are not necessarily connected.

An ***isotopy*** between two cycles γ and γ' is a continuous map $h\colon [0, 1] \times S^1 \to \Sigma$ such that $h(0, \cdot) = \gamma$ and $h(1, \cdot) = \gamma'$, and $h(t, \cdot)$ is a properly embedded cycle for all $t \in [0, 1]$. Similarly, an isotopy between two arcs α and α' is a continuous map $h\colon [0, 1] \times [0, 1] \to \Sigma$ such that $h(0, \cdot) = \alpha$ and $h(1, \cdot) = \alpha'$, and $h(t, \cdot)$ is a properly embedded arc for all $t \in [0, 1]$. The definition of isotopy easily extends to properly embedded curves with multiple components. Two curves are ***isotopic***, or in the same ***isotopy class***, if there is an isotopy between them. A simple cycle or arc is ***contractible*** if it is isotopic to a point. The ***genus*** of a surface is the maximum number of disjoint simple cycles that can be removed without disconnecting the surface.

2.2 Triangulations and Normal Curves

An ***embedding*** of a graph G on a surface Σ is a function mapping the vertices of G to distinct points in Σ and the edges of G to paths in Σ that are simple and disjoint except at common endpoints. The ***faces*** of the embedding are maximal subsets of Σ that are disjoint from the image of the graph. An embedding is ***cellular*** if every face is homeomorphic to an open disk; in particular, $\partial\Sigma$ must be the image of a set of disjoint cycles in G. A ***triangulation*** of Σ is a cellularly embedded graph in which a walk around the boundary of any face has length three. Equivalently, a triangulation expresses Σ as a set of disjoint triangles with certain pairs of edges identified; the ***1-skeleton*** of the resulting cell complex is the induced graph of vertices and edges.

We assume that our input surfaces are presented as triangulations, either as a set of triangles and gluing rules, or as an abstract graph with a rotation system [26]. We do *not* assume that triangulations are simplicial complexes. That is, triangulations may contain parallel edges and loops; two triangles may share more than a single vertex or a single edge; and the same triangle may be incident to a vertex or edge more than once.

Let T be a triangulation of a surface Σ with n triangular faces. A properly embedded curve γ in Σ is ***normal*** with respect to T if (1) γ avoids the vertices of T; (2) every intersection between γ and an edge of T is transverse; and (3) the intersection of γ with any face of T is a finite set of disjoint ***elementary segments***: simple paths whose endpoints lie on distinct sides of the triangle.

A ***normal isotopy*** between two normal curves is an isotopy h such that $h(t, \cdot)$ is a normal curve for all t. Two curves are ***normal isotopic***, or in the same ***normal isotopy class***, if there is a normal isotopy between them. The following lemma, essentially due to Kneser [19], follows easily from Euler's formula.

Lemma 2.1. *Let γ be a normal curve on a triangulated surface with n triangles, genus g, and b boundary cycles. The components of γ fall into at most $O(g + b)$ isotopy classes and at most $O(n)$ normal isotopy classes.*

A normal cycle is ***trivial*** if it bounds a disk in Σ containing a single vertex of T. We call a normal curve ***reduced*** if no component of γ is trivial and no two components of γ are normal isotopic. Lemma 2.1 immediately implies that any reduced normal curve has at most $O(n)$ components.

Any normal curve can be identified, up to normal isotopy, by two different vectors of $O(n)$ non-negative integers. There are three types of elementary segments within any face Δ, each separating one corner of Δ from the other two; the ***corner coordinates*** of γ list the number of elementary segments of each type in each face of T. The ***edge coordinates*** of γ list the number of times γ intersects each edge of T. We collectively refer to the corner and edge coordinates of a curve as its ***normal coordinates***.[1] Given either coordinate representation, it is easy to compute the other in $O(n)$ time. The ***total crossing number*** of a normal curve is the sum of its edge coordinates.

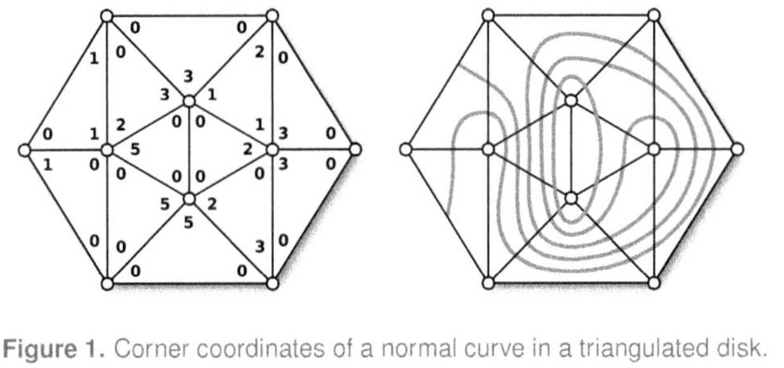

Figure 1. Corner coordinates of a normal curve in a triangulated disk.

2.3 Piecewise-Linear Surfaces

A ***piecewise-linear*** surface is a 2-manifold, possibly with boundary, constructed from a finite number of closed *Euclidean* polygons by identifying pairs of equal-length edges. The interiors of the constituent polygons are called *faces* of the surface; without loss of generality, we assume that all faces are triangles. The *vertices* and *edges* of the surface are the equivalence classes of vertices and edges of the polygons, respectively. The simplest example of a piecewise-linear surface is the boundary of a convex polyhedron in $\mathbb{R}^3$. However, most piecewise-linear surface cannot be embedded in *any* Euclidean space so that every face is flat; consider, for example, the *flat torus* obtained by identifying opposite sides of the unit square. Our algorithms assume only that the surface is orientable and that each face has its own local coordinate system; affine transformations between the local coordinate systems of neighboring faces can be derived from the gluing rules.

A ***geodesic*** is a path that is *locally* as short as possible; for any point x in a geodesic γ, a sufficiently small neighborhood of γ around x is a shortest path. If γ is a geodesic in a piecewise-linear surface Σ, any subpath of γ that lies entirely within a face of Σ is a straight line segment. Similarly, a subpath of γ that crosses an edge of Σ from one face A to another face B is a line segment in the polygon obtained by *unfolding* A and B into a common planar coordinate system [10, 23]. A geodesic is ***simple*** if it does not self-intersect.

2.4 Ports, Blocks, Junctions, and Streets

The intersections between any normal curve γ and the edges of any triangulation T partition γ into ***elementary segments*** and partition the edges of T into segments called ***ports***. The ***overlay graph*** $T \| \gamma$ is the cellularly embedded graph whose edges are these elementary segments and ports. Every vertex of $T \| \gamma$ is either a vertex of T or an intersection point of γ and some edge of T. Every face of $T \| \gamma$ is a subset of some face Δ of T. We call each face a ***junction*** if it is incident to all three sides of its containing face Δ, and a ***block*** if it is incident to only two sides of Δ; these are the only two possibilities. Each face of T contains exactly one junction. Each block is bounded either by two elementary segments and two ports, or by one elementary segment and two ports that share a vertex of T.

We call a port ***redundant*** if it separates two blocks; because each face of T contains exactly one junction, each edge of T contains at most two non-redundant ports. Removing all the redundant ports from the overlay graph $T \| \gamma$ merges contiguous sets of blocks into ***streets***. Each street is either a single disk with exactly two non-redundant ports on its boundary (called the ***ends*** of the street), a disk bounded by a trivial component of γ, or an annulus bounded by two normal isotopic components of γ. In particular, if γ is reduced, all streets are of the first type. For any reduced normal curve γ, The ***street complex*** $S(T, \gamma)$ is the complex of streets and junctions in the overlay $T \| \gamma$. See Figure 2 for an example.

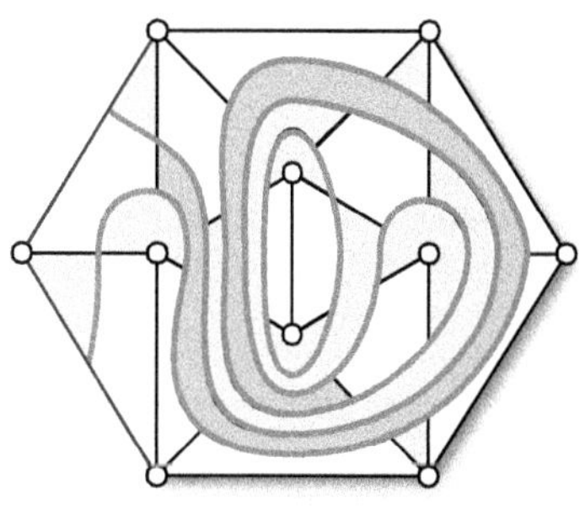

Figure 2. A street complex. Unshaded faces are junctions; shaded faces are streets; one street is shaded darker (green) for emphasis.

By construction, the components of any reduced normal curve γ appear as disjoint paths and cycles in the 1-skeleton of the street complex. Although the complexity of the overlay graph $T \| \gamma$ can be arbitrarily large, the street complex $S(T, \gamma)$ is at most a constant factor more complex than the original triangulation T.

[1] Schaefer *et al.* [35, 38, 44] refer to the edge coordinates as "normal coordinates", but the standard coordinate system for normal surfaces [15] is a generalization of corner coordinates.

Lemma 2.2. *Let T be a surface triangulation with n triangles. For any* ***reduced*** *normal curve γ in Σ, the street complex $S(T,\gamma)$ has complexity $O(n)$.*

Proof: The triangulation T trivially has at most $3n$ vertices and at most $3n$ edges. Each interior edge of T contains at most two non-redundant ports, so $S(T,\gamma)$ has $O(n)$ interior vertices. Each boundary vertex of $S(T,\gamma)$ is either a boundary vertex of T or an endpoint of one of the $O(n)$ components of γ, so $S(T,\gamma)$ has $O(n)$ boundary vertices. Each vertex of $S(T,\gamma)$ is either a vertex of T or has degree at most 4, so $S(T,\gamma)$ has $O(n)$ edges. Each non-redundant port is an end of at most one street, so $S(T,\gamma)$ has $O(n)$ streets. Finally, $S(T,\gamma)$ has exactly n junctions, one in each triangle of T. □

Our restriction to reduced curves is convenient for two reasons. First, the number of components in a non-reduced curve can be arbitrarily larger than n. Second, the street complex of any non-reduced curve contains faces that are not open disks but open annuli, bounded either by two components of γ or by one component of γ and a vertex of T. Fortunately, as we argue in Section 5, it is easy to avoid tracing trivial components or more than one component in the same normal isotopy class.

The ***crossing length*** of a street is the number of constituent blocks plus one, or equivalently, the minimum number of edge intersections of a curve traversing the street from end to end. To simplify our analysis, we regard any port between two junctions, as well as any boundary port incident to a junction, as a street with crossing length 1. The sum of the crossing lengths of the streets in any street complex $S(T,\gamma)$ is the total crossing number of γ plus the number of edges in T.

Any normal curve γ' that is disjoint from γ subdivides each port in $S(T,\gamma)$ into smaller ports, each street in $S(T,\gamma)$ into narrower "blocks", and each junction in $S(T,\gamma)$ into blocks and exactly one smaller junction. Removing all redundant ports from the overlay $S(T,\gamma)\|\gamma'$ gives us the street complex $S(T,\gamma\cup\gamma')$. The intersection of γ' with any street or junction in $S(T,\gamma)$ is a set of elementary arcs. There are three types of elementary arcs within any junction, each connecting two of the junction's three ports. The ***junction coordinates*** of γ' list the number of elementary arcs of each type in each junction of $S(T,\gamma)$. Similarly, the ***street coordinates*** of γ' list the number of such arcs within each street of $S(T,\gamma)$. The normal coordinates of a curve γ are just the junction and street coordinates of γ in the trivial street complex $S(T,\varnothing)$.

Our tracing strategy must handle normal curves that are partially drawn on the surface; we slightly extend our definitions to include such curves. A ***normal path*** in Σ is a simple path whose endpoints lie in the interior of edges of T and that can be extended to a normal curve on Σ. Each endpoint of a normal path is the endpoint of exactly two ports; we call the union of these two ports a ***fork***; for most purposes, we can think of a fork as a degenerate junction.

3. TRACING CONNECTED CURVES

In this section, we describe our algorithm to trace *connected* normal curves. Given a triangulation T of an orientable surface Σ and the corner and edge coordinates of a connected normal curve γ, our tracing algorithm computes the street complex $S(T,\gamma)$. We extend our algorithm to arbitrary *reduced* curves in Section 4, and we consider arbitrary normal curves in Section 5.

Our algorithm maintains a normal subpath π of γ that is growing at one end, along with the street complex $S(T,\pi)$ and the junction and street coordinates of the complementary subpath $\gamma\setminus\pi$. If γ is an arc, we simply trace it from one endpoint to the other. If γ is a cycle, we start the trace at an arbitrary intersection point with an edge of T; this intersection point splits the edge into a fork.

3.1 Steps

In each ***step*** of our algorithm, we extend the path π through one junction or fork, and then through one street, updating both the street complex and the junction and street coordinates. We call the streets that contain the moving endpoint of π the left and right ***active*** streets.

Suppose π is about to enter a junction. We call the streets adjacent to the junction but not the endpoint of π the *left exit* and the *right exit*. Suppose the local junction coordinates are a, b, and c, and the active street coordinates are l and r, as shown in Figure 3. These coordinates satisfy the equation $l + r + 1 = a + c$, so either $l < a$ or $r < c$. If $l < a$, we extend π through the junction and through its left exit into the next junction; the left active street grows to the end of the left exit, and the left exit becomes the new right active street. We call this case a ***left turn***; the symmetric case $r < c$ is called a ***right turn***. In either case, we update the street and junction coordinates as shown in Figure 3. A similar case analysis applies when π crosses a fork, as shown in Figure 4.

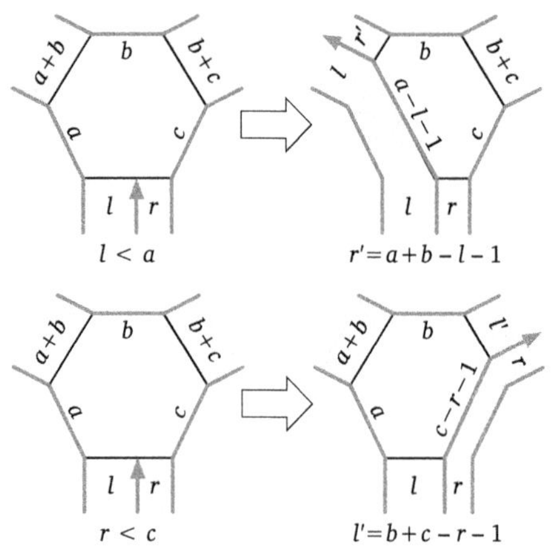

Figure 3. Tracing a curve through a junction.

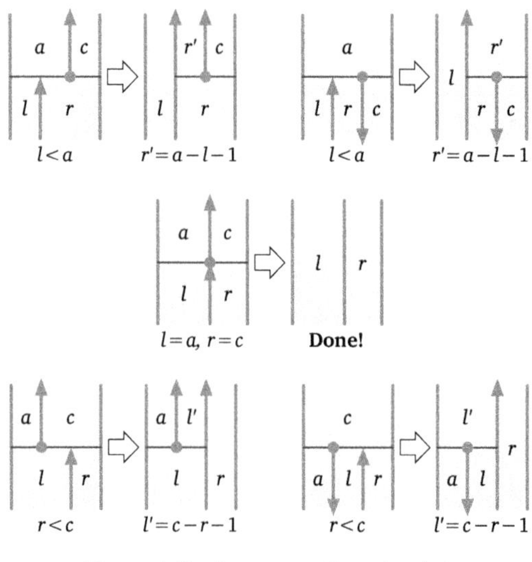

Figure 4. Tracing a curve through a fork.

In all cases, each step makes one active street longer, replaces the other active street, and makes the new active street narrower. All necessary operations for a single step—comparing and updating the junction and street coordinates and updating the street complex—can be performed in $O(1)$ time. The tracing algorithm ends when π hits either the boundary of Σ or the starting point of the trace.

3.2 Phases and Spirals

Unfortunately, executing each step by brute force is not necessarily efficient. To improve the brute-force algorithm, we more coarsely partition the tracing process into ***phases***. Each phase is a maximal sequence of either left turns or right turns. Every step in a phase consisting of left turns extends the same left active street; similarly, every step in a right phase extends the same right active street. In either case, each phase extends a single ***active street***.

During each phase, we maintain a sequence of *directed* streets and junctions traversed in that phase. If the growing path π ever enters a street for the second time, in the same direction, during the same phase, then π has entered a ***spiral***. The reentered street must be the first street traversed during the current phase; for the remainder of the phase, π repeatedly traverses the same sequence of directed streets and junctions. The ***length*** of a spiral is the total number of streets it traverses, counted with multiplicity, and the ***depth*** of the spiral is the number of times we repeat the entire sequence of directed streets and junctions. If the spiral has length ℓ and traverses m distinct directed streets, then the depth of the spiral is $\lceil \ell/m \rceil - 1$.

Instead of tracing the spiral step by step, we compute the depth of the spiral directly in $O(m)$ time as follows. Let $J_0, J_1, \ldots, J_{m-1}$ be the junction coordinates modified during the first iteration of the spiral. Let w denote the ***width*** of the active street, defined as the corresponding street coordinate plus 1. The depth of the spiral is $d = \min_i \lfloor J_i/w \rfloor$, and the spiral ends at the first junction whose coordinate J_i is smaller than dw. Once we compute d, we can update the street complex $S(T, \pi)$ and the appropriate street and junction coordinates in $O(m)$ time.

The crude upper bound $m = O(n)$ immediately implies that each phase of our tracing algorithm can be executed in $O(n)$ time. We bound the number of phases, as a function of the total crossing number of the traced curve, in Section 4.

3.3 History

For some applications of our tracing algorithm, it is useful to maintain the ***history*** of the street complex, which records the evolution of each street during the algorithm's execution. For each phase, the history records which street is active for the entire phase, the sequence of m distinct directed streets traversed during the phase, and the length ℓ of the phase. The resulting history is equivalent to a context-free grammar whose terminals are the edges of T, whose variables are the directed streets extended in each phase of the tracing algorithm and whose productions have the following forms:

$$S_a \to S_b (S_{i_0} S_{i_1} \cdots S_{i_{m-1}})^d S_{i_0} S_{i_1} \cdots S_{i_{(\ell-1) \bmod m}}$$
$$\overline{S}_a \to \overline{S}_{i_{(\ell-1) \bmod m}} \cdots \overline{S}_{i_1} \overline{S}_{i_0} (\overline{S}_{i_{m-1}} \cdots \overline{S}_{i_1} \overline{S}_{i_0})^d \overline{S}_b.$$

Each non-terminal S_i represents a single string, recording the intersection sequence of the street at the end of some phase, and $\overline{S}_i$ is the reversal of S_i. In the example above, S_a is the intersection sequence of the active street just *after* the phase ends; S_b is the intersection sequence of the active street just *before* the phase begins; $i_0, i_1, \ldots, i_{m-1}$ are the indices of the directed streets traversed during the first iteration of the spiral. We may have $S_{i_j} = \overline{S}_{i_k}$ for some indices j and k, but otherwise the indices i_j are distinct.

This context-free grammar can be transformed into Chomsky normal form by replacing each production in the form above with $O(m + \log d)$ productions of the form $A \to BC$. (Chomsky normal form grammars whose non-terminals generate exactly one string are often called *straight-line programs*.) Thus, our history data structure is essentially equivalent to the *compressed intersection sequence* constructed by Schaefer *et al.* [38, 44]. We analyze the complexity of our history data structure and the resulting compressed intersection sequence in the next section.

4. ANALYSIS

We now bound the running time of our tracing algorithm as functions of n, the number of triangles in the surface, and X, the total crossing number of the traced curve. We assume that $X = \Omega(n^2)$, since otherwise our analysis yields a time bound worse than the trivial bound $O(n + X)$. Throughout our analysis, we let N denote the number of streets in the evolving street complex. Because we only trace *reduced* normal curves, Lemma 2.2 implies that $N = \Theta(n)$.

In Section 4.1, we bound the time required to trace a *connected* normal curve; our argument can be viewed as a generalization of Lamé's classical analysis of Euclid's GCD algorithm in terms of Fibonacci numbers [20, 43].[2] We extend our analysis to curves with multiple components in Section 4.2 and to the complexity of compressed intersection sequences in Section 4.3.

4.1 Abstract Tracing

In each phase, the crossing length of the active street increases by the sum of the crossing lengths of the other traversed streets, counted with appropriate multiplicity. The algorithm AbstractTrace, shown in Figure 5, abstractly models this growth.

```
AbstractTrace(N):
  for j ← 1 to N
      x[j] ← 1
  a ← 1
  while not done
      choose an integer m ∈ [N]
      choose an integer ℓ ≥ m
      choose a vector (i_0, i_1, ..., i_{m-1}) ∈ [n]^m
      d ← ⌈ℓ/m⌉ − 1
      for j ← 0 to m − 1
          x[a] ← x[a] + d · x[i_j]
      for j ← 0 to (ℓ − 1) mod m
          x[a] ← x[a] + x[i_j]
      a ← i_{(ℓ−1) mod m}
```

Figure 5. Our abstract tracing algorithm.

AbstractTrace maintains an array $x[1 .. N]$ of positive integers, corresponding to the crossing lengths of the streets maintained in our tracing algorithm, along with the index a of the current active street. Each iteration of the outer loop of AbstractTrace models a phase of our tracing algorithm. The inner loops update the crossing length $x[a]$ of the active street as the curve traverses a spiral of length ℓ and depth d, containing m distinct streets whose indices are in the vector $(i_0, i_1, \ldots, i_{m-1})$. The last street traversed in the current phase becomes the active street for the next phase. For purposes of analysis, we assume that the parameters ℓ, m, and $(i_0, i_1, \ldots, i_{m-1})$, as

[2]This connection is not a coincidence; for tracing geodesics on the flat torus, our algorithm actually reduces to Euclid's algorithm. Euclid's algorithm is explicitly used in the orbit-counting algorithm of Agol *et al.* [1]. See also related results by Moeckel [25] and Series [41, 42] on encoding (infinite) geodesics in surfaces of constant curvature by continued fractions.

well as the termination condition for the outer loop, are determined *adversarially* rather than being determined by the topology of the curve.

To prove an upper bound on the number of phases of AbstractTrace, we can assume conservatively that $m = 1$ in every phase; equivalently, we can ignore the contribution to the active street's crossing length from all but the last street in every spiral. Thus, we consider the simpler algorithm SimpleTrace shown in Figure 6. The new variable δ is the number of times the last street in the spiral is traversed; specifically, $\delta = d$ if ℓ/m is an integer and $\delta = d + 1$ otherwise. The other new variable Δ is used only in the analysis.

```
SimpleTrace(N):
  for j ← 1 to N
      x[j] ← 1
  Δ ← 0
  a ← 1
  while not done
      choose an index i ∈ [N]
      choose an integer δ ≥ 1
      x[a] ← x[a] + δ · x[i]
      Δ ← Δ + lg(δ + 1)
      a ← i
```

Figure 6. A simplified tracing algorithm for analysis.

Lemma 4.1. *At the end of each iteration of* SimpleTrace, *we have* $\Delta \le 2\sum_{i=1}^{N} \lg x[i]$.

Proof: Consider the potential function $\Phi := 2\sum_{i=1}^{N} \lg x[i] - \lg x[a]$. Initially we have $\Phi = 0$. There are two cases to consider, depending on whether $x[a]$ is smaller or larger than $x[i]$ at the start of each iteration of the loop.

- If $x[a] \le x[i]$, then the assignment $x[a] \leftarrow x[a] + \delta \cdot x[i]$ increases Φ by at least $\lg(\delta + 1)$, and the assignment $a \leftarrow i$ does not decrease Φ.
- If $x[a] \ge x[i]$, then the assignment $x[a] \leftarrow x[a] + \delta \cdot x[i]$ does not decrease Φ, and the assignment $a \leftarrow i$ increases Φ by at least $\lg(\delta + 1)$.

In both cases, Φ increases by at least $\lg(\delta + 1)$ in each iteration. It immediately follows by induction that $\Delta \le \Phi \le 2\sum_{i=1}^{N} \lg x[i]$ at the end of every iteration. □

Lemma 4.2. AbstractTrace(N) *runs for at most* $2L = O(N \log X)$ *phases, where* L *is the final value of* $\sum_{i=1}^{N} \lg x[i]$ *and* X *is the final value of* $\sum_{i=1}^{N} x[i]$.

Proof: To maximize the number of phases, we assume $\ell = 1$ in every phase. This assumption allows us to simplify the execution to an instance of SimpleTrace where $\delta = 1$ in every phase, and therefore Δ is simply the number of phases. Lemma 4.1 implies that the algorithm terminates after at most $2L$ phases. The parameter L is maximized as a function of n and X when $x[i] = X/N$ for all i. (Our assumption that $X = \Omega(n^2)$ implies that $\log(X/N) = \Theta(\log X)$.) □

Corollary 4.3. AbstractTrace(N) *runs in* $O(NL) = O(N^2 \log X)$ *time, where* L *is the final value of* $\sum_{i=1}^{N} \lg x[i]$ *and* X *is the final value of* $\sum_{i=1}^{N} x[i]$.

Theorem 4.4. *Let* Σ *be a surface composed of* n *triangles, and let* γ *be a* ***connected*** *normal curve in* Σ *with total crossing number* X. *Given the normal coordinates of* γ, *we can trace* γ *in* $O(n^2 \log X)$ *time.*

In the full paper, we prove that the upper bounds reported in Lemma 4.2 and Corollary 4.3 are tight in the worst case, at least when X and n are sufficiently large. However, we optimistically conjecture that the upper bound in Theorem 4.4 is *not* tight.

4.2 Tracing Reduced Curves

Now we consider the more general case where γ is a *reduced* curve, possibly with more than one component. (For the application we describe in Section 5, this is the most general case we need to consider.) Our tracing algorithm requires little modification to handle these curves; we simply trace the components one at a time, in arbitrary order. Each component refines the street complex defined by the previous components. Lemma 2.2 immediately implies that the resulting algorithm runs in $O(n^3 \log X)$ time, but this time bound can be improved with more careful analysis, as follows.

Theorem 4.5. *Let* Σ *be a surface composed of* n *triangles, and let* γ *be a* ***reduced*** *normal curve in* Σ *with total crossing number* X. *Given the normal coordinates of* γ, *we can trace every component of* γ *in* $O(n^2 \log X)$ *time.*

Proof: Consider the effect of ending one component and starting another on the vector of crossing lengths modeled by the array $x[1..N]$ in SimpleTrace. When we begin tracing a new cycle component, we split some street into three smaller streets by introducing a fork; one of the three new streets becomes the active street for the first phase of the new component. This update can be modeled in SimpleTrace by adding the following lines to the beginning of the main loop:

```
if starting a cycle:
    choose an index i ∈ [N]
    choose an integer y ∈ [x[i]]
    x[i] ← x[i] − y + 1
    x[N + 1] ← y
    x[N + 2] ← y
    N ← N + 2
    a ← N + 1
```

When we finish tracing a cycle component, we merge the four streets adjacent to the initial fork into two longer streets; see the bottom center of Figure 3. This update can be modeled in SimpleTrace by adding the following lines:

```
if ending a cycle:
    choose an index j ∈ [N]
    choose an index k ∈ [N]
    x[j] ← x[j] + x[N − 1]
    x[k] ← x[k] + x[N]
    N ← N − 2
```

Similarly, when we begin tracing a new arc component, we split some street into two narrower streets; this update can be modeled in SimpleTrace by adding the following lines:

```
if starting an arc:
    choose an index i ∈ [N]
    x[N + 1] ← x[i]
    N ← N + 1
    a ← N + 1
```

No additional changes are necessary when we end an arc component.

Altogether, ending one component and starting a new one decreases the potential function Φ by at most $O(\log X)$. An easy modification of the proof of Lemma 4.1 now implies that after each iteration of SimpleTrace, we have $\Delta \le 2\sum_{i=1}^{N} \lg x[i] + O(t \log X)$, where t is the number of components we have completely traced so far. Lemma 2.1 implies that any reduced normal curve has $O(n)$ components. We conclude that SimpleTrace executes at most $O(n \log X)$ phases; each phase trivially requires $O(n)$ time. □

4.3 Compression: Logarithmic Spiral Cost

After tracing a disconnected normal curve γ, our history data structure can be transformed into a straight-line program such that the intersection sequence of any component of γ is the language of some non-terminal. For each phase of the tracing algorithm, the straight-line program contains $O(m + \log d)$ productions, where m is the number of distinct streets traversed in that phase and d is the depth of that phase's spiral. We have already proved that the sum of the $O(m)$ terms, over all phases, is at most $O(n^2 \log X)$. The sum of the $O(\log d)$ terms is within a constant factor of the variable Δ computed by SIMPLETRACE (now *without* the simplifying assumption that $\delta = 1$) and thus is at most $O(n \log X)$ by Lemma 4.1.

Theorem 4.6. *Let Σ be a triangulated surface with n triangles, and let γ be a* ***reduced*** *normal curve in Σ with total crossing number X. Given the normal coordinates of γ, we can compute a straight-line program of length $O(n^2 \log X)$ that encodes the intersection sequence of every component of γ in $O(n^2 \log X)$ time.*

5. COUNTING ISOTOPY CLASSES

Our tracing algorithm implies efficient solutions for several problems involving normal curves represented by their normal coordinates. As an illustrative example, we describe an algorithm to compute the number of distinct isotopy classes of components of a given normal curve, as well as the number of components in each isotopy class. The input to our algorithm is a surface triangulation T with n triangles, and the edge and corner coordinates a normal curve γ with total crossing length X. Our algorithm runs in $O(n^2 \log X)$ time, significantly improving the large-polynomial-time solution of Schaefer *et al.* [35].

We begin by counting and deleting the trivial components of γ. The number of trivial components that separate any vertex v from the other vertices is just the minimum of the corner coordinates incident to v. Thus, we can easily count all trivial components and delete them from γ, by reducing the appropriate normal coordinates, in $O(n)$ time. We separately record the number of trivial cycles and the number of trivial arcs with endpoints on each boundary cycle.

Next, for each non-trivial *normal* isotopy class, we trace one component of γ in that class, and then count and remove all other components in that class, as follows. Suppose we have already traced components $\gamma_1, \ldots, \gamma_{i-1}$. Let $\hat{\gamma}_{<i}$ denote the reduced normal curve $\gamma_1 \cup \cdots \cup \gamma_{i-1}$, and let $\gamma_{\geq i}$ denote the union of all components of γ *not* normal-isotopic to any component of $\hat{\gamma}_{<i}$. In particular, we have $\hat{\gamma}_{<1} = \varnothing$ and $\gamma_{\geq 1} = \gamma$. By assumption, we have computed the street complex $S(T, \hat{\gamma}_{<i})$ as well as the street and junction coordinates of $\gamma_{\geq i}$. Let x be the leftmost intersection point between $\gamma_{\geq i}$ and some non-redundant port p in $S(T, \hat{\gamma}_{<i})$, and let γ_i denote the component of $\gamma_{\geq i}$ that contains x. We trace γ_i through $S(T, \hat{\gamma}_{<i})$ to produce the street complex $S(T, \hat{\gamma}_{<(i+1)})$, along with the street and junction coordinates of $\gamma_{\geq i} \setminus \gamma_i$. The number of other components of γ that are normal isotopic to γ_i is the minimum of the junction coordinates just to the right of γ_i in the new street complex $S(T, \hat{\gamma}_{<(i+1)})$. Thus, we can easily count these components and reduce the appropriate street and junction coordinates in $O(n)$ time, thereby computing the street and junction coordinates of $\gamma_{\geq(i+1)}$.

Theorem 4.5 implies that the total time spent tracing all components γ_i is $O(n^2 \log X)$. Lemma 2.1 implies that there are at most $O(n)$ normal-isotopy classes of components in γ, so the total time spent counting and removing parallel components of γ is only $O(n^2)$.

Finally, let $\hat{\gamma}$ be the union of the traced components γ_i. Each traced component γ_i is a simple arc or cycle in the 1-skeleton of the final street complex $S(T, \hat{\gamma})$. Thus, we can compute the Euler characteristic of every component of $\Sigma \setminus \hat{\gamma}$ in $O(n)$ time using a depth-first search in the dual graph of $S(T, \hat{\gamma})$ [12]; in particular, we can identify which components of $\Sigma \setminus \hat{\gamma}$ are disks ($\chi = 1$) and annuli ($\chi = 0$). Then it is straightforward to cluster the components of $\hat{\gamma}$ into isotopy classes in $O(n)$ time. We can also compute the number of components of γ in each isotopy class in $O(n)$ time by adding the sizes of the appropriate normal-isotopy classes.

Theorem 5.1. *Let Σ be a triangulated surface with n triangles, and let γ be a normal curve in Σ with total crossing length X, represented by its normal coordinates. We can compute the number of isotopy classes of components of γ and the number of components in each isotopy class in $O(n^2 \log X)$ time.*

The algorithm of Schaefer *et al.* [35] to identify isotopy classes actually computes the normal coordinates of one component in each isotopy class. If we require the same output representation, we can recover the normal coordinates of any component of $\hat{\gamma}$ (or the union of any subset of components) in $O(n^2 \log X)$ time, essentially by running the tracing algorithm backward; details will appear in the full paper. Lemma 2.1 implies that the overall time to compute the normal coordinates of every component of $\hat{\gamma}$ is $O((g + b)n^2 \log X)$, which is still significantly faster than Schaefer *et al.*

6. TRACING GEODESICS

Finally, we extend our tracing algorithm to simple geodesic paths on piecewise-linear triangulated surfaces. The input to our algorithm is a piecewise-linear surface Σ (specified by triangles and gluing rules), along with a starting point p and a direction vector v in the local coordinate system of some face of Σ that contains p. As an example application, we sketch an algorithm to trace the geodesic path γ that starts at p in direction v up to its first point of self-intersection.

To simplify our exposition, we assume that both the surface Σ and the direction vector v are *generic*; thus, the geodesic γ does not intersect any vertex of Σ but does eventually intersect itself.[3] Our algorithm requires only minor changes to handle degenerate inputs. We emphasize that even for generic inputs, the total crossing number of γ is not bounded *a priori* by *any* function of n.

Any simple geodesic path γ that starts and ends on edges of the triangulation is a *normal* path. Thus, the street complex of γ is well-defined and has complexity $O(n)$ by Lemma 2.2. Moreover, each of the $O(n)$ faces of the street complex is isometric to a convex polygon with at most six sides; in particular, every street is either a triangle or a convex quadrilateral. (The street complex may have $\Omega(n)$ vertices on the boundary of a single street, but at most four are true vertices of that street.)

Although we do not know the normal coordinates of γ in advance, we can still easily decide in constant time at each step of our tracing algorithm whether a geodesic γ entering a junction leaves through its left exit, leaves through its right exit, or hits an earlier segment of γ. Geometrically, this decision is equivalent to a ray-shooting query in a convex polygon with at most six sides. Thus, we can easily adapt the brute force tracing algorithm to the geodesic setting. However, to achieve a running time of $O(n^2 \log X)$, we require a new algorithm to efficiently compute the depth of a geodesic spiral.

[3] If the total angle around every vertex of Σ is an integer multiple of π, almost every geodesic path can be extended infinitely without self-intersection. Examples of such surfaces include the boundary of a regular tetrahedron, the flat torus, and the *eierlegende Wollmilchsau* [17].

6.1 Annular Ray Shooting

We reduce computing spiral depth to the following ***annular ray shooting*** problem, which may be of independent interest: Given a ray ρ on a piecewise-linear annulus A, how many times does ρ wrap around A before hitting the boundary? Specifically, we are given a triangulated simple polygon P in the plane, with two edges e_0 and e_1 of equal length, and a ray ρ that starts on e_0 and points into P. The annulus A is obtained from P by identifying e_0 and e_1. Equivalently, e_0 and e_1 are *portals*; when the ray exits P at any point on e_1, it immediately reenters P through the corresponding point on e_0 at the same incidence angle [27, 48, 49]. Our goal is to determine the number of times ρ crosses the portal(s) before hitting a non-portal edge of P.

The naïve solution to this requires $O(n + t^* \log n)$ time, where t^* is the output of the query: Preprocess P in $O(n)$ time, and then perform $t^* + 1$ ray-shooting queries, each in $O(\log n)$ time [18]. Our algorithm improves this naïve bound exponentially.

Lemma 6.1. *The annular ray-shooting problem in a piecewise-linear annulus with n vertices can be solved in $O(n + \log t^*)$ time and space, where t^* is the output value.*

Let $\tau: \mathbb{R}^2 \to \mathbb{R}^2$ denote the rigid motion (either translation or rotation) that maps e_0 onto e_1. For each integer $i > 0$, let $\tau^i = \tau \circ \tau^{i-1}$, and let $e_i = \tau(e_{i-1}) = \tau^i(e_0)$. Similarly, let $P_0 = P$, and for any integer $i > 0$, let $P_i = \tau(P_{i-1}) = \tau^i(P_0)$; by construction, the segment e_i is an edge of both P_{i-1} and P_i. Finally, for any non-negative integer t, let $P_{<t}$ denote the region obtained by gluing the polygons $P_0, P_1, \ldots, P_{t-1}$ along corresponding edges e_j. Although $P_{<t}$ is a topological disk, it is not in general a simple polygon, as different copies of P may overlap.

The output t^* of the annular ray-shooting problem is the maximum of all integers t such that ρ intersects edges e_0 and e_t but no other edge of $P_{<t}$. Our algorithm finds t^* using a standard unbounded search strategy due to Bentley and Yao [3]. We emphasize that our algorithm does not actually construct $P_{<t^*}$ or any significant portion thereof, but instead computes on the fly only the vertices and edges required for the search.

To simplify our presentation, we assume that the edge e_0 is vertical, the polygon P lies locally to the right of e_0, and that the transformation τ is either a translation or a counterclockwise rotation by angle $\theta > 0$. Thus, $P_{<k}$ grows to the right and possibly curves upward as the parameter k increases. We also assume that $t^* \geq 1$, since it is easy to solve the ray shooting problem in $O(n)$ time otherwise.

The *funnel* of P is the union of all shortest paths from points in e_0 to points in e_1. The funnel is bounded by the shortest paths in P between corresponding endpoints of the two portals; we call these shortest paths A ("above") and B ("below"); see Figure 7. Our assumption that $t^* \geq 1$ implies that ρ reaches e_1 without intersecting either A or B; thus, A and B are disjoint convex chains [2, 47]. Because P is already triangulated, it is straightforward to construct its funnel in $O(n)$ time [8, 21, 47]. From now on, we assume implicitly that P is equal to its funnel.

Let $a_0, a_1, \ldots, a_\alpha$ denote the vertices of A in order from left to right, and let $b_0, b_1, \ldots, b_\beta$ denote the vertices of B in order from left to right. Thus $e_0 = a_0 b_0$ and $e_1 = a_\alpha b_\beta$; we also have $\alpha + \beta \leq n + 2$. For any indices i and j, let $a_{i,j} = \tau^j(a_i)$ denote the vertex of P_j corresponding to a_i, and let $b_{i,j} = \tau^j(b_i)$ denote the vertex of P_j corresponding to b_i. In particular, we have $a_{0,j} = a_{\alpha,j-1}$ and $b_{0,j} = b_{\beta,j-1}$ for every positive integer j.

We define two families of segments that connect adjacent copies of A and B. For any index j, let c_j ("the jth ceiling") denote the lower

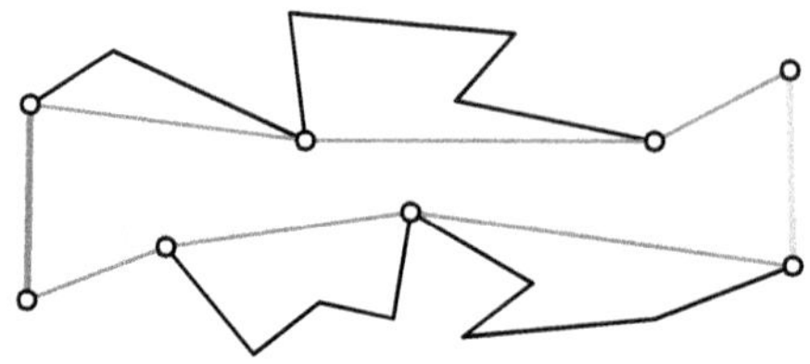

Figure 7. The funnel between the portals of P.

common tangent of A_{j-1} and A_j. By symmetry, there are indices l and r such that $c_j = a_{l,j-1} a_{r,j}$ for every index j. Similarly, let f_j ("the jth floor") denote the upper common tangent of B_{j-1} and B_j. By symmetry, there are indices p and q such that $f_j = b_{p,j-1} b_{q,j}$ for every index j. Because τ is either a translation or a counterclockwise rotation, we have $r \leq l$ and $p \leq q$; thus, segments c_j and c_{j+1} are interior-disjoint, but segments f_j and f_{j+1} intersect inside P_j. See Figure 8. In the degenerate case where τ is a translation, we have $l = r$ and $p = q$.

Our algorithm applies a standard unbounded search strategy of Bentley and Yao [3] to find the smallest value of t that satisfies at least one of the following conditions:

(A) Segment c_t contains a point below ρ.

(A') The slope of segment c_t is larger than the slope of ρ.

(B) Segment f_t contains a point above ρ.

If $t > t^* + 1$, then at least one of these three conditions must be true. On the other hand, if $t < t^*$, then conditions (A) and (B) do not hold, and condition (A') holds only if ρ does not intersect *any* upper chain A_j. In particular, condition (A') is necessary to detect the situation where ρ crosses some upper chain A_t twice between the ceiling endpoints $a_{r,t}$ and $a_{l,t}$.

Starting with the estimate $t = 1$, our algorithm repeatedly doubles t until at least one of these three conditions is satisfied, and then performs a binary search for the critical value of t. When the unbounded search ends, there are three cases to consider, depending on which conditions are satisfied.

(A) Suppose c_t contains a point below ρ but c_{t-1} does not. Then ρ must hit either A_{t-1} or A_t; that is, either $t^* = t - 1$ or $t^* = t$. We can distinguish between these two cases in $O(n)$ time by brute force.

(A') Suppose c_{t-1} and c_t both lie above ρ, and the slope of ρ lies between the slopes of c_{t-1} and c_t. Then either ρ hits A_{t-1}, or ρ does not intersect any upper chain A_j. Again, we can distinguish between these two cases in $O(n)$ time by brute force. If ρ hits A_{t-1}, we return $t^* = t - 1$. Otherwise, we perform a second unbounded search without termination conditions (A) and (A').

(B) Finally, if f_t contains a point above ρ but f_{t-1} does not, then ρ must hit either B_{t-1} or B_t. Again, we can distinguish between these two cases in $O(n)$ time by brute force.

If the critical value of t satisfies more than one termination condition, we perform the relevant computation for all satisfied conditions and return the smallest value found.

An important subtlety in the algorithm is that computing the coordinates of c_t or f_t from scratch requires $\Theta(\log t)$ time. However, by computing an array of $O(\log t^*)$ transformations of the form τ^{2^i} during the doubling phase of the unbounded search, we can compute the coordinates of the appropriate segments c_t or f_t in $O(1)$ time in each iteration of the search.

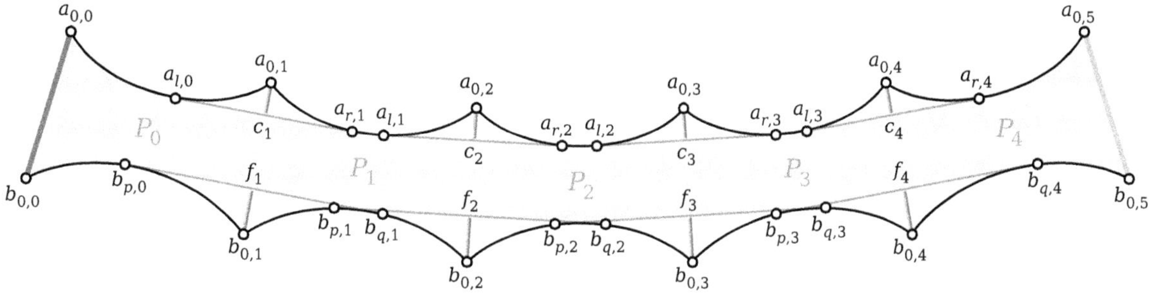

Figure 8. Floors and ceilings in the universal cover of the annulus.

6.2 Self-Collisions and Directed Streets

To compute the depth of a geodesic spiral, we actually require an algorithm for the following minor modification of the annular ray-shooting problem: Given a polygon P with equal-length portal edges and a ray ρ, we need to determine how many times ρ crosses the portal before hitting either a non-portal edge of P *or some earlier point in* ρ. Equivalently, we need to compute the largest integer t such that ρ does not intersect $A_{<t}$ or $B_{<t}$ *and* does not cross the ray $\tau^{-1}(\rho)$ inside $P_{<t}$. To solve this modified problem, we add a fourth termination condition to our unbounded search algorithm:

(*C*) The point $\rho \cap \tau^{-1}(\rho)$ lies inside the convex quadrilateral $\text{conv}\{e_{t-1}, e_t\}$.

Adding this condition increase the running time of the unbounded search algorithm by only a small constant factor.

Another important subtlety is that a geodesic may traverse a street in both directions during a single phase of the tracing algorithm; in this case, the simple self-intersection test described in the previous paragraph may fail. To avoid this possibility, we partition each street into two ***lanes*** by drawing a ***median*** geodesic segment between vertices of the triangular faces at each end of the street, as shown in Figure 9. The median segments also partition the junctions into smaller convex polygons of constant complexity. Straightforward case analysis implies that the geodesic crosses the median of a street only in the last step of a phase, and thus traverses each lane in only one direction during each phase. Informally, the geodesic "drives on the right" during right-turning phases and "drives on the left" during left-turning phases.

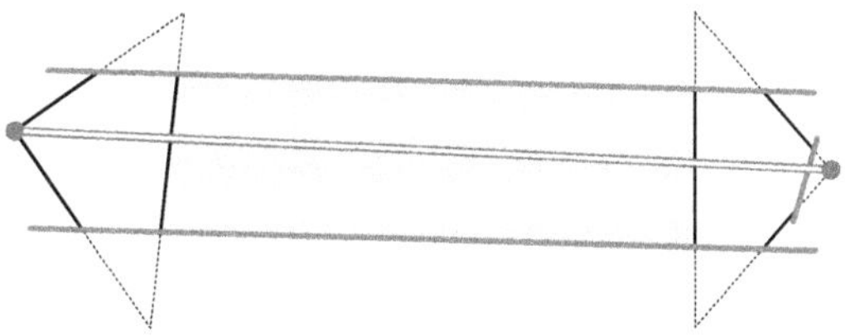

Figure 9. Splitting a geodesic street into two one-way lanes

Finally, to compute the depth of a spiral, we can unfold the m relevant *lanes* and *junction fragments* into a simple polygon P with complexity $O(m)$, and then invoke our modified annular ray-shooting algorithm.

Lemma 6.2. *The depth d of a geodesic spiral through m distinct directed streets can be computed in $O(m + \log d)$ time.*

Corollary 4.3 now immediately implies that our geodesic tracing algorithm runs in $O(n^2 \log X)$ time, where X is the number of times the geodesic crosses an edge of the original triangulation. The output of the tracing algorithm is the final street complex $S(T, \gamma)$, along with the location of the first self-intersection point x, in the local coordinates of the face of $S(T, \gamma)$ that contains x. If necessary, we can locate x in the original triangulation T in $O(n^2 \log X)$ additional time by running the tracing algorithm backward and maintaining the location of x is the devolving street complex. We defer the remaining details to the full paper.

Theorem 6.3. *Let Σ be a triangulated piecewise-linear surface with n triangles. Given the starting point and direction of a geodesic γ in Σ, we can compute the first self-intersection point in γ in $O(n^2 \log X)$ time, where X is the number of edges γ crosses before it self-intersects.*

References

[1] Ian Agol, Joel Hass, and William P. Thurston. The computational complexity of knot genus and spanning area. *Trans. Amer. Math. Soc.* 358(9):3821–3850, 2006. ArXiv:math/0205057.

[2] David Avis, Teren Gum, and Godfried T. Toussaint. Visiblity between two edges of a simple polygon. *Vis. Comput.* 2:342–357, 1986.

[3] Jon Louis Bentley and Andrew Chi-Chih Yao. An almost optimal algorithm for unbounded searching. *Inform. Proc. Lett.* 5:82–87, 1976.

[4] Joan S. Birman and Caroline Series. Geodesics with bounded intersection number on surfaces are sparsely distributed. *Topology* 24(2):217–225, 1985.

[5] Joan S. Birman and Caroline Series. Algebraic linearity for an automorphism of a surface group. *J. Pure Appl. Algebra* 52:227–275, 1988.

[6] Benjamin A. Burton. The complexity of the normal surface solution space. *Proc. 26th Ann. Symp. Comput. Geom.*, 201–209, 2010.

[7] Benjamin A. Burton and Melih Ozlen. Computing the crosscap number of a knot using integer programming and normal surfaces. *ACM Trans. Math. Software* (to appear), 2012. ArXiv:1107.2382.

[8] Bernard Chazelle. A theorem on polygon cutting with applications. *Proc. 23rd Ann. IEEE Symp. Found. Comput. Sci.*, 339–349, 1982.

[9] Max Dehn. Die Gruppe der Abbildungsklassen (Das arithmetische Feld auf Flächen). *Acta Mathematica* 69:135–206, 1938.

[10] Erik D. Demaine and Joseph O'Rourke. *Geometric Folding Algorithms: Linkages, Origami, Polyhedra.* Cambridge Univ. Press, 2007.

[11] Herbert Edelsbrunner and John Harer. *Computational Topology: An Introduction*. American Mathematical Society, 2010.

[12] Jeff Erickson and Sariel Har-Peled. Optimally cutting a surface into a disk. *Discrete Comput. Geom.* 31(1):37–59, 2004.

[13] Albert Fathi, François Laudenbach, and Valentin Poénaru. *Travaux de Thurston sur les surfaces*. Astérisque 66-67. Soc. Math. de France, 1979. Séminaire Orsay. English translation in [14].

[14] Albert Fathi, François Laudenbach, and Valentin Poénaru. *Thurston's Work on Surfaces*. Mathematical Notes. Princeton Univ. Press, 2011. Translated by Djun Kim and Dan Margalit. English translation of [13].

[15] Wolfgang Haken. Theorie der Normalflächen: Ein Isotopiekriterium für den Kreisknoten. *Acta Mathematica* 105:245–375, 1961.

[16] Joel Hass, Jeffrey C. Lagarias, and Nicholas Pippenger. The computational complexity of knot and link problems. *J. ACM* 46(2):185–211, 1999.

[17] Frank Herrlich and Gabriela Schmithüsen. An extraordinary origami curve. *Math. Nachr.* 281(2):219–237, 2008.

[18] John Hershberger and Subhash Suri. A pedestrian approach to ray shooting: Shoot a ray, take a walk. *J. Algorithms* 18(3):403–431, 1995.

[19] Helmuth Kneser. Geschlossene Flächen in dreidimensionalen Mannigfaltigkeiten. *Jahresbericht Deutschen Math.-Verein.* 38:248–260, 1930.

[20] Gabriel Lamé. Note sur la limite du monbre des divisions dand la recherche du plus grand commun divisur entre dex nombres entiers. *Compt. Rend. Acad. Sci., Paris* 19:857–870, 1844.

[21] Der-Tsai Lee and Franco P. Preparata. Euclidean shortest paths in the presence of rectilinear barriers. *Networks* 14:393–410, 1984.

[22] Yury Lifshits. Processing compressed texts: A tractabiliity border. *Proc. 18th Ann. Symp. Combin. Pattern Matching*, 228–240, 2007. Lecture Notes Comput. Sci. 4850, Springer-Verlag.

[23] Lazar Aronovich Lyusternik. *Shortest Paths: Variational Problems*. Popular Lectures Math. 13. Pergamon Press, 1964. Translated and adapted from the Russian by P. Collins and Robert B. Brown.

[24] Masamichi Miyazaki, Ayumi Shinohara, and Masayuki Takeda. An improved pattern matching algorithm for strings in terms of straight-line programs. *J. Discrete Algorithms [Hermes]* 1(1):187–204, 2000.

[25] Richard Moeckel. Geodesics on modular surfaces and continued fractions. *Ergodic Theory Dynam. Sys.* 2:69–83, 1982.

[26] Bojan Mohar and Carsten Thomassen. *Graphs on Surfaces*. Johns Hopkins Univ. Press, 2001.

[27] Nuclear Monkey Software. Narbacular Drop. Video game, 2005.

[28] János Pach and Géza Tóth. Recognizing string graphs is decidable. *Discrete Comput. Geom.* 28(4):593–606, 2001.

[29] Robert C. Penner. The action of the mapping class group on curves in surfaces. *L'Enseignment Mathématique* 30:39–55, 1984.

[30] Robert C. Penner and John L. Harer. *Combinatorics of Train Tracks*. Annals of Math. Studies 125. Princeton Univ. Press, 1992.

[31] Wojciech Plandowski and Wojciech Rytter. Application of Lempel-Ziv encodings to the solution of word equations. *Proc. 25th Int. Conf. Automata Lang. Prog.*, 731–742, 1998. Lecture Notes Comput. Sci. 1443, Springer-Verlag.

[32] John M. Robson and Volker Diekert. On quadratic word equations. *Proc. 16th Ann. Conf. Theoretical Aspects Comput. Sci.*, 217–226, 1999. Lecture Notes Comput. Sci. 1563, Springer-Verlag.

[33] John M. Robson and Volker Diekert. Quadratic word equations. *Jewels are Forever, Contributions on Theoretical Computer Science in Honor of Arto Salomaa*, 314–326, 1999. Springer-Verlag.

[34] Wojciech Rytter. Application of Lempel-Ziv factorization to the approximation of grammar-based compression. *Theoret. Comput. Sci.* 302:211–222, 2003.

[35] Marcus Schaefer, Eric Sedgwick, and Daniel Štefankovič. Algorithms for normal curves and surfaces. *Proc. 8th Int. Conf. Comput. Combin.*, 370–380, 2002. Lecture Notes Comput. Sci. 2387, Springer-Verlag.

[36] Marcus Schaefer, Eric Sedgwick, and Daniel Štefankovič. Recognizing string graphs in NP. *J. Comput. System Sci.* 67(2):365–380, 2003.

[37] Marcus Schaefer, Eric Sedgwick, and Daniel Štefankovič. Spiraling and folding: The topological view. *Proc. 19th Ann. Canadian Conf. Comput. Geom.*, 73–76, 2007.

[38] Marcus Schaefer, Eric Sedgwick, and Daniel Štefankovič. Computing Dehn twists and geometric intersection numbers in polynomial time. *Proc. 20th Canadian Conf. Comput. Geom.*, 111–114, 2008. Full version: Tech. Rep. 05–009, Comput. Sci. Dept., DePaul Univ., April 2005, ⟨http://facweb.cs.depaul.edu/research/techreports/abstract05009.htm⟩.

[39] Marcus Schaefer, Eric Sedgwick, and Daniel Štefankovič. Spiraling and folding: The word view. *Algorithmica* 60(3):609–626, 2011.

[40] Saul Schleimer. Sphere recognition lies in NP. Preprint, July 2004. ArXiv:math/0407047.

[41] Caroline Series. The modular surface and continued fractions. *J. London Math. Soc.* 31:69–80, 1985.

[42] Caroline Series. Geometrical Markov coding of geodesics on surfaces of constant negative curvature. *Ergodic Theory Dynam. Sys.* 6(4):601–625, 1986.

[43] Jeffrey Shallit. Origins of the analysis of the Euclidean algorithm. *Hist. Math.* 21:401–419, 1994.

[44] Daniel Štefankovič. *Algorithms for simple curves on surfaces, string graphs, and crossing numbers*. Ph.D. thesis, Dept. Comput. Sci., Univ. Chicago, June 2005.

[45] John Stillwell. *Classical Topology and Combinatorial Group Theory*, 2nd edition. Graduate Texts in Mathematics 72. Springer-Verlag, 1993.

[46] William P. Thurston. On the geometry and dynamics of diffeomorphisms of surfaces. *Bull. Amer. Math. Soc.* 19(2):417–431, 1988. Circulated as a preprint in 1976.

[47] Godfried T. Toussaint. Shortest path solves edge-to-edge visibility in a polygon. *Patt. Recog. Letters* 4(3):165–170, 1986.

[48] Valve Corporation. Portal. Video game, 2007.

[49] Andy Wachowski and Larry Wachowski. *Matrix Revolutions*. Warner Bros., 2003. Motion picture.

Testing Graph Isotopy on Surfaces

Éric Colin de Verdière
Département d'informatique
École normale supérieure, CNRS
Paris, France
Eric.Colin.de.Verdiere@ens.fr

Arnaud de Mesmay
Département d'informatique
École normale supérieure
Paris, France
Arnaud.de.Mesmay@ens.fr

ABSTRACT

We investigate the following problem: Given two embeddings G_1 and G_2 of the same abstract graph G on an orientable surface S, decide whether G_1 and G_2 are isotopic; in other words, whether there exists a continuous family of embeddings between G_1 and G_2.

We provide efficient algorithms to solve this problem in two models. In the first model, the input consists of the arrangement of G_1 (resp., G_2) with a fixed graph cellularly embedded on S; our algorithm is linear in the input complexity, and thus, optimal. In the second model, G_1 and G_2 are piecewise-linear embeddings in the plane minus a finite set of points; our algorithm runs in $O(n^{3/2} \log n)$ time, where n is the complexity of the input.

The graph isotopy problem is a natural variation of the homotopy problem for closed curves on surfaces and on the punctured plane, for which algorithms have been given by various authors; we use some of these algorithms as a subroutine.

As a by-product, we reprove the following mathematical characterization, first observed by Ladegaillerie (1984): Two graph embeddings are isotopic if and only if they are homotopic and congruent by an oriented homeomorphism.

Categories and Subject Descriptors

F.2.2 [**Analysis of Algorithms and Problem Complexity**]: Nonnumerical algorithms and problems—*Computations on discrete structures; Geometric problems and computations*; G.2.2 [**Discrete Mathematics**]: Graph Theory—*Graph algorithms*; I.3.5 [**Computer Graphics**]: Computational geometry and object modeling—*Geometric algorithms, languages and systems*

General Terms

Algorithms, Performance, Theory

Keywords

Computational topology, topological graph theory, surface, embedded graph, isotopy, mapping class group

1. INTRODUCTION

Deciding whether a curve in a space can be deformed to another one is a fundamental topological task. Since the problem is undecidable in general, even in two-dimensional simplicial complexes and four-manifolds [30, p. 242–247], one has to put restrictions on the ambient space. The case of surfaces has been much investigated recently (see references below), since it is at the same time non-trivial, interesting, and tractable, and since the underlying mathematics, notably in the field of combinatorial group theory, are well-understood.

This paper also studies deformations on surfaces, but instead of curves, we consider graphs drawn on surfaces: Given two embeddings (crossing-free drawings) of a graph G on a surface, can we deform one continuously to the other without introducing intersections between edges during the process?

As a motivating special case, consider a finite set of obstacle points P in the plane and a graph G embedded in $\mathbb{R}^2 \setminus P$ in two different ways, G_1 and G_2. Does there exist a "morph" between G_1 and G_2 (possibly moving the vertices and bending the edges) that avoids passing over any obstacle? To the best of our knowledge, no algorithm is known for this purpose. This is relevant for morphing applications: To compute a morphing between two images, it is helpful to first build a deformation between compatible graphs representing the most salient features of the images. In such applications, it is sometimes desirable to add some topological requirements on the morphing, e.g., to force some area of the deforming image to always cover a fixed point of the plane during the deformation. Such requirements can be encoded using obstacle points, since a face of the graph containing an obstacle point has to contain it during the whole deformation.

Another motivation for this problem comes from geographic information systems and map simplification. When simplifying a road network, it is crucial that the features of the map (cities, mountains) stay on the same side of the roads. This can be tested by inputting these features as obstacle points and testing whether the road networks obtained before and after straightening are isotopic; see, e.g., Cabello et al. [7].

More generally, assume that we have a triangulated surface in $\mathbb{R}^3$, and two embeddings G_1 and G_2 of the same graph G on that surface: Each graph G_i is encoded by its combinatorial arrangement with the triangulation. Can we continuously move G_1 to G_2? In this setting, the graphs G_1 and G_2 might represent textures on the surface, and the question is whether one can continuously move one texture so that it coincides with the other.

1.1 Related Work

A *homotopy* between two paths is a continuous family of paths with the same endpoints between them. A *homotopy* between two cycles (closed curves) is a continuous family of cycles between them[1]. Dey and Guha [13], and later Lazarus and Rivaud [26], study the problem of deciding whether two paths or cycles on a surface S are homotopic. Both problems can be solved in optimal linear time if the input curves are represented as walks in a graph embedded on S. Cabello et al. [7] give efficient algorithms for testing homotopy of paths in the special case where the surface S is a punctured plane (a plane minus a finite set of obstacle points) and the input paths are represented by polygonal paths in the plane.

A related problem, to which a large body of research is devoted, is that of computing shortest homotopic paths or cycles. It is first studied by Hershberger and Snoeyink [22] for a triangulated surface where the vertices lie on the boundary, and revisited by Efrat et al. [14] and Bespamyatnikh [2] for paths in a punctured plane. These algorithms, in particular, allow to decide whether two paths are homotopic, and have a better complexity than the algorithm by Cabello et al. [7] in some cases. Results on the shortest homotopic path/cycle problem are also known for combinatorial surfaces [10, 11]: it is solvable in polynomial time [9].

An *isotopy* between two simple curves (paths or cycles without self-intersections) is a homotopy that keeps the deforming curve simple at each step. If two simple curves are isotopic, they are homotopic. The converse does not hold in general [19], though it holds for simple paths with endpoints on the boundary and for simple cycles not bounding a disk [16]; in these cases, it is equivalent to test homotopy or isotopy. More generally, an *isotopy with fixed vertices* between two graph embeddings is a continuous family of graph embeddings between them, where the vertices are not allowed to move. It is known that one can compute shortest graph embeddings with fixed vertices in polynomial time [8, 9], and similar techniques allow to test whether two graph embeddings are isotopic with fixed vertices in polynomial time.

Two homeomorphisms from S to S are *isotopic* if there is a continuous family of homeomorphisms connecting them. The *mapping class group* of a surface (without boundary) S is, roughly, the set of isotopy classes of all orientation-preserving homeomorphisms from S to S. See, e.g., Farb and Margalit [18] for a recent and exhaustive survey on this topic. Although we use little of this vast theory, it is quite connected to our problem: If G_1 and G_2 are cellularly embedded on S, a homeomorphism of S that maps G_1 to G_2 represents a unique element of the mapping class group, and our problem amounts to testing whether this element is the identity. This problem has already been tackled in a more general setting [4, 31], but these algorithms take a different input from ours and their complexities are exponential.

[1]This is sometimes referred to as *free* homotopy in textbooks, in opposition to *fixed point homotopy*, which we will not use.

1.2 Our Results

In this paper, we study the problem of deciding whether two embeddings G_1 and G_2 of the same abstract graph G on a surface S are isotopic; in other words, whether there exists a continuous family of embeddings of G (possibly moving the vertices of G) connecting G_1 with G_2.

In more detail, the input of the algorithm is a description of the surface S, the graph G, and the graph embeddings G_1 and G_2. All surfaces are assumed to be compact, connected, and orientable; they may have boundary. An embedding maps each vertex (or edge, or halfedge) of G to the corresponding feature on the surface. In particular, the correspondence between the vertices (or edges, or halfedges) of G_1 and G_2 is given.

Our algorithmic results come in two flavors, depending on the model used. In the general model, G_1 and G_2 are graph embeddings of G on an arbitrary surface S. To represent them, we use a model similar to a *cross-metric surface* [9]: a graph embedding is pictured by the intersections it forms with a given cellularly embedded graph. More specifically, we assume that S has a fixed graph H cellularly embedded on it (i.e., each face of H is homeomorphic to a disk; for example, H may be a triangulation of S); the input of the algorithm is the combinatorial map of the arrangement of G_1 with H (resp., G_2 with H)[2]. We emphasize that the input does not consider the crossings betweeen G_1 and G_2. We give a linear-time (and thus optimal) algorithm to decide whether G_1 and G_2 are isotopic:

Theorem 1. *Let S be an orientable surface, possibly with boundary. Let H be a fixed graph cellularly embedded on S. Let G_1 and G_2 be two graph embeddings of the same graph G on S, each in general position with respect to H. Given the combinatorial map of the arrangement of G_1 with H (resp., G_2 with H), of complexity k_1 (resp., k_2), we can determine whether G_1 and G_2 are isotopic in $O(k_1+k_2)$ time.*

Let us emphasize that the surface S is not fixed in this result; the constant in the $O(\cdot)$ notation does not depend on S.

We also study the complexity of the problem in the case where S is the plane minus a finite set P of obstacle points, and G_1 and G_2 are piecewise-linear graph embeddings of G in $\mathbb{R}^2 \setminus P$. In this case, the input is the point set P together with the embeddings G_1 and G_2, where each edge of each embedding is represented as a polygonal path. (Again, the embeddings G_1 and G_2 may intersect arbitrarily.) In this setting, we prove the following result.

Theorem 2. *Let P be a set of p points in the plane, and let G_1 and G_2 be two piecewise-linear graph embeddings of the same graph G in $\mathbb{R}^2 \setminus P$, of complexities (number of segments) k_1 and k_2 respectively. We can determine whether G_1 and G_2 are isotopic in $\mathbb{R}^2 \setminus P$ in time $O(n^{3/2}\log n)$ time, where n is the total size of the input. In more detail, the running time is, for any $\varepsilon > 0$,*

$$O\Big((k_1+p)\log(k_1+p) + (k_2+p)\log(k_2+p) + \min\Big\{(k_1+k_2)p \ ,\ p^{1+\varepsilon} + (k_1+k_2)\sqrt{p}\log p\Big\}\Big).$$

Let us emphasize that the isotopy is a continuous family of topological embeddings; one may assume that all these embeddings are piecewise-linear, but we claim no upper bound on their complexities.

[2]i.e., the graph obtained by overlaying G_i and H.

1.3 Overview of the Techniques

If two graph embeddings G_1 and G_2 of the same graph G are isotopic, then clearly (1) there is an oriented homeomorphism of the surface that maps G_1 to G_2; and (2) if γ is a cycle in G (possibly with repeated vertices), then its images in G_1 and G_2 are homotopic.

It was shown by Ladegaillerie [25] that such necessary conditions are, in fact, sufficient. However, the second condition is not algorithmic, since there are infinitely many cycles in G. A close inspection of Ladegaillerie's proof reveals that $O(g+b)$ pairs of cycles need to be tested for homotopy, where g and b denote the genus and number of boundary components of the surface. However, in Ladegaillerie's construction, the complexity of the family of cycles is not explicitly given. With some work, and using some of our techniques, one might be able to obtain an explicit algorithm using his construction, but the running time would certainly be larger, by at least an additional $O(g)$ factor (since Ladegaillerie's decomposition contains a pants decomposition), if not more.

Using a very different method, we reprove Ladegaillerie's characterization in a strengthened form (Theorem 10 below): we provide an explicit set of cycles Γ in G of linear overall complexity that have to be tested for homotopy. Our algorithmic results follow, since one can perform efficiently the homeomorphism test (this essentially amounts to checking equality of two combinatorial maps), as well as the test of homotopy between two given cycles (using known algorithms or variants of them [13, 26, 7]).

We note that it is not straightforward to find a suitable set Γ in G of overall linear complexity satisfying the above condition. In particular, a natural candidate for Γ would be the set of all facial cycles in G_1 (and thus in G_2). However, Figure 1 shows that the condition is not fulfilled, even in the case where the surface is the sphere with four punctures.

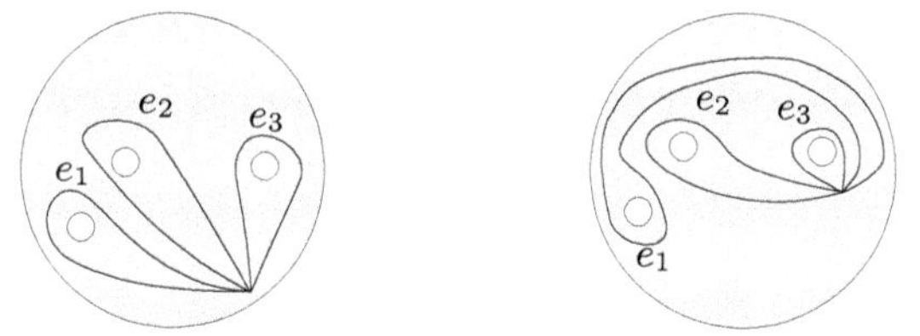

Figure 1: Two embeddings G_1 and G_2 of a one-vertex graph with three loop edges on the sphere with four punctures. These two embeddings are not isotopic, although there exists an oriented homeomorphism sending one to the other, and the four cycles following the boundaries of the faces are homotopic in G_1 and G_2.

The paper is organized as follows. After some preliminaries (Section 2), we characterize the existence of an isotopy between two arrangements of *stable* cycles, containing no disk with less than four vertices on its boundary (Section 3). We then prove the aforementioned strengthened form of Ladegaillerie's result (Section 4); the idea for the proof is to compute from G_1 and G_2 two arrangements of stable curves in their tubular neighborhoods, which are isotopic if and only if G_1 and G_2 are isotopic, and to apply the characterization for stable curves. Finally, we deduce our main theorems in Section 5. Due to space restrictions, some of the proofs were omitted, we refer the reader to the full version available on the webpages of the authors.

2. PRELIMINARIES

Here, we review some topological notions. We refer the reader to Hatcher [21], Stillwell [30], or any previous paper on the same topic [9] for standard background in algebraic topology used in the present paper, and Farb and Margalit [18] for the connection between covering spaces, hyperbolic geometry, and surface homeomorphisms.

Curves and graphs on surfaces. Henceforth, S is a compact, connected, orientable surface with genus g and b boundary components; its ***Euler characteristic*** is $\chi(S) = 2 - 2g - b$. A ***path***, respectively a ***cycle***, on a surface S is a continuous map $p : [0,1] \to S$, respectively a continuous map $\gamma : S^1 \to S$, where S^1 is the unit circle. An ***arc*** is a path intersecting the boundary of S exactly at its endpoints. Paths and cycles are ***simple*** if they are one-to-one. We emphasize that, according to our definition, a cycle in a graph is a closed walk *possibly with repeated vertices* (in contrast to the graph-theoretic notion).

An ***embedding*** of a graph G on a surface S is, informally, a crossing-free drawing of G on S. A graph embedding is ***cellular*** if its ***faces***, namely, the connected components of the complement of the image of the graph, are homeomorphic to open disks. If G is embedded cellularly on a surface S and if we denote by v, e, and f the respective numbers of its vertices, edges and faces, we have $v - e + f = \chi(S) = 2 - 2g - b$.

A (finite) family of cycles on S is ***in general position*** if the cycles are in the interior of S, there are finitely many (self-)intersection points, and each intersection is a crossing between exactly two pieces of cycles. Similarly, a (finite) family of graph embeddings on S is ***in general position*** if all embeddings are in the interior of S, there are finitely many intersection points between two different embeddings, and each intersection is a crossing of the interiors of exactly two edges.

The simultaneous drawing of a family of cycles or of graph embeddings in general position on S gives a set of vertices (original vertices or intersection points) and edges (connecting two such vertices); this graph is called the ***arrangement*** of the family. (Here we also allow edges that are simple cycles without any vertex on them.)

Isotopy. Two embeddings G_0 and G_1 of the same abstract graph G on S are ***isotopic*** if there is a continuous family of embeddings $(G_t)_{t\in[0,1]}$ between G_0 and G_1. In more detail, the data of an embedding is given by the choice of a point for each vertex and a path connecting the appropriate vertices for each edge (with some conditions asserting that no crossing occurs); a family (G_t) of embeddings is continuous if all these maps vary continuously over t. (The vertices may, in particular, move.)

Two homeomorphisms h_0 and h_1 from S to S are ***isotopic*** if there is a continuous family of homeomorphisms $(h_t)_{t\in[0,1]}$ connecting them. A homeomorphism isotopic to the identity is called an ***(ambient) isotopy***[3] of the surface.

It actually holds that two embeddings of a graph on S are isotopic if and only if there exists an ambient isotopy

[3]This is not exactly the standard definition, but a slight adaptation more suitable for our paper.

of S mapping one to the other (provided the two graph embeddings are in the interior of S); this follows from tools in differential topology (e.g., vector fields); see for instance Farb and Margalit [18, Proposition 1.11] or Hirsch [23, Theorem 1.3]. However, our proof also implies the statements of Theorems 1 and 2 in which graph isotopy is replaced with ambient isotopy, and actually *reproves* the aforementioned fact.

We will need the following basic lemma, whose (simple) proof can be found, e.g., in Farb and Margalit [18, Lemma 2.1].

LEMMA 3 (ALEXANDER'S LEMMA). *Let D be a disk and $h : D \to D$ be a homeomorphism fixed on the boundary of D. Then h is isotopic to the identity.*

Combinatorial maps for non-cellular embeddings. To each embedding G_1 of an abstract graph G on S corresponds a ***combinatorial map***, storing the combinatorial information of that embedding. For cellularly embedded graphs, this notion is well-known, and there are several essentially equivalent data structures available to store such combinatorial maps on surfaces [24], like the gem representation [15, 27] and the halfedge data structure. All these data structures have linear complexity in the number of edges of the graph. Given a graph G and two combinatorial maps of cellular embeddings G_1 and G_2 of G on S, one can easily decide whether there exists a homeomorphism of S mapping G_1 to G_2: this amounts to checking that the two combinatorial maps are isomorphic, and this takes linear time in the complexities of the combinatorial maps.

We can easily extend the notion of combinatorial map for possibly non-cellular graph embeddings. Essentially, this boils down to keeping track of the genus of each face and of the number of boundary components of the surface within each face. Furthermore, one can orient a combinatorial map by giving an orientation to an arbitrary face. Given two such oriented combinatorial maps of (possibly non-cellularly) embedded graphs G_1 and G_2, one can, in linear time, decide whether there exists an oriented homeomorphism of S taking G_1 to G_2. We refer to Appendix A for details.

Covering spaces and hyperbolic geometry. If S has a negative Euler characteristic and has no boundary, it can be provided with a hyperbolic metric; this naturally induces a hyperbolic metric on its universal cover $\widetilde{S}$, which is then isometric to the open hyperbolic disk $\boldsymbol{D_2}$. This open disk is classically compactified with a boundary ∂D_2 [18, Chapter 1], and a cycle on S lifts into an arc in $D_2 \cup \partial D_2$. We call its intersections with the boundary the ***endpoints*** of the lift or, according to the orientation of the lift, its ***source*** and its ***target***. Two lifts in D_2 have common endpoints if and only if they stay at a bounded distance from each other. If S has boundary, $\widetilde{S}$ is isometric to a totally geodesic subspace of D_2.

It is folklore [6, Lemma 1.6.5] that if two cycles on a surface S are homotopic, they have lifts in D_2 with the same endpoints. We also recall that on a hyperbolic surface, every cycle is homotopic to a unique geodesic [18, Proposition 1.3].

3. ISOTOPIES OF STABLE FAMILIES OF CURVES

Let Γ be a family of cycles in general position on S. A ***k-gon*** in Γ, for $k \geq 1$, is an open disk on S whose boundary is formed by exactly k subpaths of Γ. A ***0-gon*** is a disk whose boundary is a single simple cycle in Γ. In general, k-gons may contain and may be crossed by other pieces of the arrangement of Γ; if this is not the case, we say that the k-gon is ***empty***.

We say that Γ is ***stable*** if its arrangement contains no *empty* k-gon for $k \leq 3$. In this section, we prove the following result.

THEOREM 4. *Let S be an orientable surface and let $\Gamma_1 = (\gamma_{1,1}, \ldots, \gamma_{1,n})$ and $\Gamma_2 = (\gamma_{2,1}, \ldots \gamma_{2,n})$ be two* stable *families of cycles on S in general position such that:*

1. *there exists an oriented homeomorphism h of S mapping each curve $\gamma_{1,j}$ of Γ_1 to the corresponding curve $\gamma_{2,j}$ of Γ_2, and*
2. *each curve of Γ_1 is homotopic to the corresponding curve of Γ_2.*

Then there is an ambient isotopy of S mapping each curve of Γ_1 to the corresponding curve of Γ_2.

We note that conversely, if there exists an ambient isotopy of S mapping Γ_1 to Γ_2, then conditions (1) and (2) are satisfied. Let us also emphasize that, when we write that a homeomorphism h *maps* a curve $\gamma_{1,j}$ to another curve $\gamma_{2,j}$, we really mean that it is the case pointwise, namely $h \circ \gamma_{1,j}(t) = \gamma_{2,j}(t)$ for each $t \in S^1$; similarly for an ambient isotopy.

The remaining part of this section is devoted to the proof of this theorem if the surface S has negative Euler characteristic. The proof for the remaining surfaces uses slightly different tools, and we refer the reader to the full version.

3.1 Basic Consequences of Euler's Formula

The following lemmas are simple consequences of Euler's formula.

LEMMA 5. *Assume just for this lemma that S is a sphere. Let G be a connected (hence cellularly embedded) graph on S, such that every vertex has degree four. Then there are at least three faces in G with degree smaller than four.*

LEMMA 6. *Let Γ be a stable family of cycles in a surface S. Then no k-gon can exist in Γ for $k \leq 3$, and no cycle in Γ is null-homotopic.*

3.2 Following the Geodesics

Since we assume that the Euler characteristic of S is negative, we can endow S with a hyperbolic metric and identify $\widetilde{S}$ with a subset of the open hyperbolic disk. Recall that no cycle in Γ_1 (or Γ_2) is null-homotopic, by Lemma 6. Thus, for each j, let g_j be the unique geodesic homotopic to $\gamma_{1,j}$ (and $\gamma_{2,j}$). (Some g_j's may be identical.)

The following result uses a proposition by de Graaf and Schrijver [12, Proposition 13], the proof of which is based on a slight refinement of Ringel's theorem [29] for arrangements of curves in the disk.

PROPOSITION 7. *For $i = 1, 2$, for each $\varepsilon > 0$, up to replacing Γ_i with its image by an ambient isotopy of S, we*

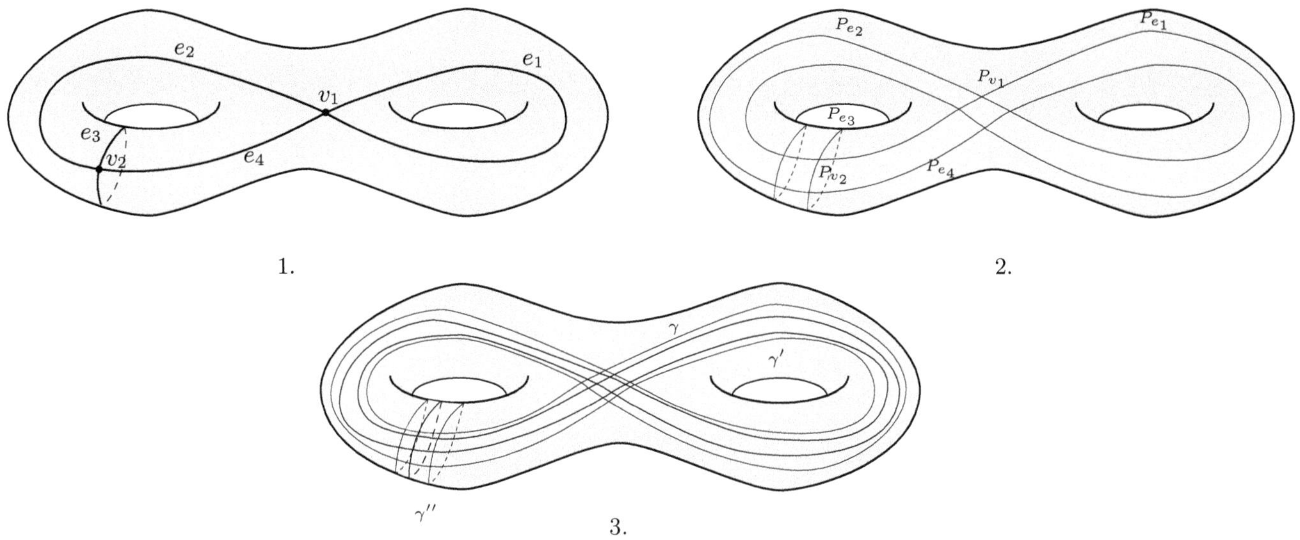

Figure 2: 1. Two geodesics on a double torus, and the associated graph. 2. The corridors associated to them. 3. Cycles in the corridors. In vertex polygons, they intersect in a grid-like pattern.

may assume that each cycle $\gamma_{i,j}$ has a lift that belongs to an ε-neighborhood of a lift of g_j.

PROOF. It follows from the aforementioned result by de Graaf and Schrijver [12, Proposition 13] that the cycles $\gamma_{i,j}$ in Γ_i can be moved through Reidemeister moves and ambient isotopies into cycles of which lifts are ε-close to some lift of a geodesic α_j. Since lifts of $\gamma_{i,j}$ and α_j are ε-close, they share the same endpoints and thus α_j is the unique geodesic homotopic to $\gamma_{i,j}$, which shows that $\alpha_j = g_j$. As Γ_i is stable, no Reidemeister move *at all* is possible, so the only possible moves are actually isotopies of the surface. □

So henceforth we assume that Γ_1 and Γ_2 satisfy the conclusion of the above proposition.

The union of the geodesics g_j forms a graph, possibly with simple cycles without vertex, on the surface S; we denote by E and V its edges and vertices. (Simple cycles without vertex are considered as closed edges.) Each vertex has even degree, and each geodesic arriving at a vertex from an edge leaves it via the opposite edge.

Now, following ideas by de Graaf and Schrijver [12], we introduce a polygonal decomposition of an ε-neighborhood of the graph (V, E). To each edge $e \in E$, we associate an ***edge polygon*** P_e (actually, a quadrilateral), and to each vertex $v \in V$ of degree $2d$, we associate a ***vertex polygon*** P_v with $2d$ sides, such that each edge $e = uv$ lies in the interior of $P_e \cup P_u \cup P_v$ and v lies in the interior of P_v, and such that the union of all the polygons form a tubular neighborhood of the cycles g_i containing their ε-neighborhoods.[4] We can assume that all the polygons are mutually disjoint, except for P_e and P_u for u a vertex incident to e, which share an edge.

Let us call a *corridor* the polygonal neighborhood of a single geodesic g_i, that is $\bigcup P_e \cup \bigcup P_u$ for all e and u in g_i. Every cycle in Γ_i belongs to a single corridor. Examples of corridors are pictured in Figure 2.

[4]In the case where a geodesic coincides with a boundary of the surface, one of the edges of each polygon of that geodesic actually lies on this boundary. Also, for simple cycles without vertex, we introduce an edge polygon with two opposite sides glued together.

The following proposition is proved using Lemmas 5 and 6 and by standard bigon flipping arguments.

PROPOSITION 8. *For $i = 1, 2$, up to replacing Γ_i with its image by an ambient isotopy of S, we may assume that:*

- *each maximal piece of a cycle in Γ_i within a polygon is simple, and has its endpoints on opposite sides of the polygon;*
- *two such pieces cross at most once; moreover, if they cross, then the four endpoints of these two pieces are all in different sides of the polygon (in particular, the polygon is a vertex polygon).*

PROOF. If a maximal piece of Γ_i within a polygon is non-simple, it forms a 1-gon, which is impossible (Lemma 6). If two pieces cross twice, they form a 2-gon, which is impossible for the same reason.

If such a piece has its endpoints on two different sides of a polygon that are not opposite in that polygon, that polygon is a vertex polygon, and the corresponding cycle does not belong to a single corridor, which is impossible by construction.

If such a piece p has its endpoints on the same side s of a polygon, without loss of generality assume that the disk D bounded by s and p contains no other piece with both endpoints on s. Thus the pieces inside D are simple, pairwise disjoint (otherwise these two pieces would form a 3-gon with p), and connect s to p; so with an isotopy of the surface, we can push p across s, decreasing the total number of intersections between the cycles and the sides of the polygon. After finitely many such operations, no such piece p exists.

There only remains to prove that there cannot be any (self-)intersection among cycles in the same corridor C. We distinguish two cases:

- If C contains at least one vertex polygon, a cycle from another corridor C' crosses all cycles in C; if there were a (self-)intersection in C, there would be a 3-gon.

- On the other hand, if the corridor C contains no vertex polygon, then C is an annulus. Consider the arrangement of the cycles in Γ, and assume there is a crossing. One connected component of this arrangement is a graph where all vertices have degree four. It follows that, on the sphere obtained by attaching two disks on the boundaries of C, that graph has three faces of degree at most three (Lemma 5). At least one of these faces does not contain one of these two disks; that face is a k-gon for $k \leq 3$, contradicting Lemma 6. □

We can now assume that Γ_1 and Γ_2 satisfy the conclusion of Proposition 8. It follows that all the arcs of Γ_i in a given edge polygon P_e are simple, disjoint, and belong to different cycles. Moreover, each polygon P_v is actually a quadrilateral, because otherwise three arcs in P_v coming from six different sides would cross inside P_v, yielding a 3-gon, which is impossible (Lemma 6). Finally, within each polygon P_v, the arcs intersect in a grid-like fashion, as in Figure 2.

3.3 A Technical Result on Corridors

Henceforth, let us choose an arbitrary orientation on S. Let C be a corridor, oriented in the direction of its geodesic g; let C_l and C_r be the left and right boundaries of C, respectively.

Let Γ_1^C be the subfamily of cycles in Γ_1 that belong to C, and let Γ_2^C be its image by h. Recall that Γ_i^C has no crossing in an edge polygon P_e. For $i = 1, 2$, the *ordering* of Γ_i^C along C is defined as follows: Consider an arc in an edge polygon P_e, going from C_l to C_r, crossing each cycle in Γ_i^C exactly once, and record the index of the cycles in Γ_i encountered, in this order along the arc. By construction, this ordering does not depend on the choice of the polygon and of the arc.

LEMMA 9. *The orderings of Γ_1^C and Γ_2^C are the same.*

PROOF. We first claim that the oriented homeomorphism $h : S \to S$ lifts to an oriented homeomorphism $\widetilde{h} : \widetilde{S} \to \widetilde{S}$. Indeed, if we denote by π the projection $\pi : \widetilde{S} \to S$ and apply the lifting theorem [21, Proposition 1.33] to $\pi \circ h$, we get[5] a continuous map $\widetilde{h} : \widetilde{S} \to \widetilde{S}$ satisfying $\pi \circ h = \widetilde{h} \circ \pi$. Let $x \in \widetilde{S}$ and $y = \widetilde{h}(x)$. Similarly, we lift h^{-1} to a continuous map $\widetilde{h^{-1}}$ such that $\widetilde{h^{-1}}(y) = x$. Then $\widetilde{h^{-1}}$ and $\widetilde{h}$ are inverse continuous maps on $\widetilde{S}$, so $\widetilde{h}$ is a homeomorphism. Furthermore, $\widetilde{h}$ is oriented because h is oriented.

Let $\widetilde{g}$ be the lift of a geodesic inside a lift $\widetilde{C}$ of C. The homeomorphism $\widetilde{h}$ maps $\widetilde{g}$ into a possibly different lift of C. However, up to composing $\widetilde{h}$ with a deck transformation, we may assume that $\widetilde{h}$ maps $\widetilde{g}$ into $\widetilde{C}$. Furthermore, $\widetilde{g}$ and its image by $\widetilde{h}$ have the same orientation in $\widetilde{C}$, because otherwise the endpoints of each lift of Γ_1^C in $\widetilde{C}$ would be exchanged under $\widetilde{h}$, which is not the case since h preserves the homotopy classes of the cycles in Γ_1^C. Therefore, $\widetilde{h}$ maps $\widetilde{g}$ to a path in $\widetilde{C}$ with the same source and target. Since $\widetilde{h}$ is oriented, the orderings of the cycles in $\widetilde{\Gamma}_1^C$ and $\widetilde{\Gamma}_2^C$, from left to right in $\widetilde{C}$, are the same. It follows that they are also the same in C. □

[5]The hypotheses are trivially fulfilled since the universal cover has trivial fundamental group.

3.4 End of Proof

PROOF OF THEOREM 4. We can build an isotopy of S that moves the endpoints of the arcs of Γ_1 in an edge polygon P_e to the corresponding endpoints of Γ_2: This follows from the fact that within P_e, the orderings of the arcs are the same in Γ_1 and Γ_2, which is proved in Lemma 9. Now, within P_e, the arcs of Γ_1 are simple and pairwise disjoint; thus, there exists a homeomorphism from P_e to P_e that is the identity on its boundary and sends the arcs of Γ_1 inside P_e to the corresponding arcs of Γ_2. Therefore, by Alexander's lemma, the arcs of Γ_1 can be mapped to the corresponding arcs of Γ_2 by an isotopy of P_e.

Now, within each vertex polygon P_v, the endpoints of the arcs of Γ_1 and Γ_2 coincide; moreover, they form combinatorially isomorphic arrangements of arcs (namely, grids) inside P_v. The same argument as above shows that an isotopy of P_v sends the arcs of Γ_1 to the corresponding arcs of Γ_2. □

4. ISOTOPIES OF GRAPH EMBEDDINGS

In this section, we prove the following result.

THEOREM 10. *Let G_1 and G_2 be two graph embeddings of a graph G on an orientable surface S. Assume that there is an oriented homeomorphism h of S mapping G_1 to G_2. There exists a family Γ of cycles in G such that the following holds: If, for each cycle γ in Γ, the images of γ in G_1 and G_2 are homotopic, then there exists an ambient isotopy of S taking G_1 to G_2.*

Furthermore, the cycles in Γ use each edge of G at most twice in total and, given only the combinatorial map of G_1 on S, one can compute the cycles of Γ in linear time in the complexity of that combinatorial map.

Note that, by the homeomorphism condition, the combinatorial maps of G_1 and G_2 on S have to be the same.

Conversely, if G_1 and G_2 are isotopic (in particular, if there is an ambient isotopy taking G_1 to G_2), there must exist an oriented homeomorphism mapping one to the other, and the images of any cycle of G in G_1 and G_2 are homotopic. Therefore, we reprove Ladegaillerie's result [25] stated in the introduction.

The difficulty of the construction resides in the fact that the families Γ_1 and Γ_2 must have small complexity. In a sense, Γ_1 forms a topological decomposition of the tubular neighborhood of G_1 using cycles. However, all known topological decompositions of surfaces made of cycles (like pants decompositions [10], octagonal decompositions [9], or systems of loops [17]) have worst-case complexity $\Omega(gn)$, where n is the complexity of the surface; our construction has linear size in the complexity of the object studied. We suspect that our construction can be useful for other purposes as well.

4.1 Preprocessing Step

For the proof of Theorem 10, we assume for simplicity of exposition that G_1 and G_2 are known. It is immediate to check that, actually, only the combinatorial map of G_1 on S is needed.

PROPOSITION 11. *Without loss of generality, we may assume that (1) G_1 has a single face, or none of its faces is a disk, and (2) G_1 has no vertex of degree zero or one.*

Intuitively, the proof is simple: Whenever e is an edge of G bounding two different faces, at least one of which is a disk, removing e in G_1 and G_2 does not change whether G_1 and G_2 are isotopic. Similarly, removing vertices of degree zero or one does not harm.

4.2 Construction of the Stable Family

Henceforth, we assume that G_1 satisfies the properties of Proposition 11.

PROPOSITION 12. *In linear time, we can construct a family of cycles Γ in G such that:*

- *each edge of G is used at most twice by all the cycles in Γ;*
- *there exists a* stable *family Γ_1 on S whose cycles are homotopic (by an arbitrarily small perturbation) to the cycles in the images of Γ in G_1;*
- *G_1 does not meet the interior of the faces of the arrangement of Γ_1 that are not disks.*

PROOF. It is actually simpler to explain the construction of Γ_1 first; see Figure 3 for an example. For simplicity of notation, we let $G_1 := (V, E)$. Recall that the *cyclomatic number* of a connected graph is the minimum number of edges one needs to delete to obtain a tree. Equivalently, it equals its number of edges minus its number of vertices plus one.

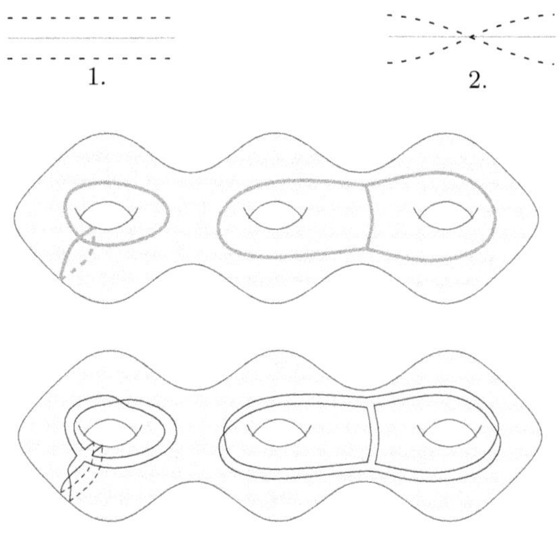

Figure 3: Top: A crossover along an edge of $E' \setminus E''$. Middle: An embedded graph G_1 on a genus 3 surface. Bottom: The corresponding family Γ_1.

Let (V', E') be a connected component of (V, E). By Proposition 11, we can assume that each vertex has degree at least two.

- If (V', E') has cyclomatic number one, then it must be a single cycle. In this case, we add that cycle to Γ_1.
- Otherwise, (V', E') has cyclomatic number at least two. We add to Γ_1 the cycles that are the boundaries of a tubular neighborhood of (V', E'). Furthermore, let E'' be the edge set of a spanning tree of (V', E'). For each edge $e \in E' \setminus E''$, we introduce a "crossover" as in Figure 3 (top) on the two pieces of Γ_1 that run along edge e: instead of locally having two pieces of cycles that run along edge e without touching it, we now have two pieces of cycles in the neighborhood of edge e that cross at a single interior point of e. Of course, this operation may change the number of cycles of Γ_1 and create self-intersections.[6] See Figure 3.

We now prove that Γ_1 is a stable family. If S is a sphere, then, by the conclusion of Proposition 11, (V, E), and therefore also Γ_1, are empty, in which case the conclusion is obviously true; so we can assume that S is not a sphere. Let f be a face of the arrangement of the curves in Γ_1. By construction, there are two cases:

- The face f belongs to a single face f' of (V, E): In this case, f and f' have the same topology. If these are disks, then (V, E) is a cut graph (by virtue of Proposition 11), and f' is its unique face. In particular, S has empty boundary, and (V, E) is connected and has cyclomatic number $4g$. Therefore, the number of crossings of Γ_1 along the boundary of f is exactly $4g \geq 4$. To conclude, f is not a disk with at most three crossings of Γ_1 along its boundary.
- The face f encloses a spanning tree (V', E'') of a connected component (V', E'). By construction, f is a disk, and $|E' \setminus E''| \geq 2$. Each edge $e \in E' \setminus E''$ corresponds to a single crossing between cycles of Γ_1, and this crossing appears twice along the boundary of f; so f has $2|E' \setminus E''| \geq 4$ crossings of Γ along its boundary.

Hence Γ_1 is a stable family. It follows from the construction that the computation of Γ_1 takes linear time in the complexity of the combinatorial map of G_1. The cycles in Γ_1 have been constructed in a tubular neighborhood of the graph G_1, so we can deduce, again in linear time, a family of cycles Γ in G whose images under G_1 are slight perturbations of Γ_1. All these cycles use each edge of G at most twice. Furthermore, also by construction, G_1 is included in Γ_1 and in the faces of Γ_1 that are disks. □

4.3 End of Proof of Theorem 10

We now conclude the proof of Theorem 10.

We first apply Proposition 12 to G_1, obtaining a family of cycles Γ on G; then we test whether the images of these cycles in G_1 and G_2 are pairwise homotopic. Assume it is the case. After a homotopy, these cycles correspond, respectively, to stable families Γ_1 and $\Gamma_2 := h(\Gamma_1)$; therefore, each cycle in Γ_1 is homotopic to the corresponding cycle in Γ_2. It follows from Theorem 4 that there is an isotopy of S taking Γ_1 to Γ_2. So (up to composing with this isotopy) we can assume that $\Gamma_1 = \Gamma_2$. Moreover, each face of Γ_1 is mapped to itself, because h is an oriented homeomorphism.

In particular, $G_1 \cap \Gamma_1 = G_2 \cap \Gamma_2$. In every disk of $S \setminus \Gamma_1 = S \setminus \Gamma_2$, the oriented homeomorphism h is the identity on the boundary; therefore, by Alexander's lemma, it is isotopic to the identity (the isotopy being fixed on the boundary). This gives us an isotopy between G_1 and G_2 on every such disk. By gluing these isotopies together along their boundaries, we get an isotopy of S mapping G_1 to G_2, because G_1 and G_2 are included in the closures of the faces of $\Gamma_1 = \Gamma_2$ that are disks (Proposition 12). This concludes the proof of Theorem 10.

[6] If G_1 is a system of loops, the arrangement of Γ_1 coincides with the dual of the so-called "radial graph" by Lazarus and Rivaud [26]. Our construction is, however, more general.

5. ALGORITHMS

For Theorems 1 and 2, it suffices to be able to test the existence of an oriented homeomorphism, and of homotopies between the cycles in Γ, as computed by Theorem 10, in the indicated amount of time. We prove Theorems 1 and 2 in Sections 5.1 and 5.2, respectively.

5.1 Surfaces: Proof of Theorem 1

Homeomorphism test. For the case of graphs on surfaces, the existence of an oriented homeomorphism that maps G_1 to G_2 can be checked in $O(k_1 + k_2)$ time. Indeed, let us choose an arbitrary orientation on S; this induces an orientation of the combinatorial map of H, hence an orientation of the combinatorial map of the arrangement of G_i and H, for $i = 1, 2$. Computing the number of boundary components of S in each face, as well as the genus of each face (using the Euler characteristic), and "erasing" the graph H in both arrangements gives us oriented combinatorial maps for each G_i. They are isomorphic if and only if there exists an oriented isomorphism of S between G_1 and G_2, and this can be checked in linear time (refer to Appendix A).

Homotopy tests. The homotopy tests can also be performed in $O(k_1 + k_2)$ time. Indeed, recall that the input of the algorithm consists of the combinatorial maps of the arrangement of G_1 and H on one hand, and of G_2 and H on the other hand, where H is a fixed cellular graph embedding; k_1 and k_2 denote the complexities of these two maps. In $O(k_1 + k_2)$ total time, we can compute the cyclically ordered list of edges of H crossed by each cycle of Γ in the embeddings G_1 and G_2. This gives us a set of pairs of cycles in the dual graph H^* of H that have to be tested for homotopy. The total complexity of these cycles is $O(k_1 + k_2)$, and H^* has complexity $O(k_1 + k_2)$ as well. Lazarus and Rivaud [26] prove that, after a preprocessing linear in the complexity of the cellular graph H^*, one can test homotopy of cycles in H^* in time linear in the complexities of these cycles. An earlier paper by Dey and Guha [13] claims a similar result, except for some low-genus surfaces, but Lazarus and Rivaud point out some problems in their proof. Also, both papers address only the case of surfaces without boundary, but the case of surfaces with boundary is easier, as the fundamental group is free; alternatively, homotopy tests for cycles on surfaces with boundary can be performed using an algorithm for surfaces without boundary by first attaching a handle to each boundary component, which does not change the outcomes of the homotopy tests.

This concludes the proof of Theorem 1.

5.2 Punctured Plane: Proof of Theorem 2

For the algorithm in the punctured plane model, let B be the union of the set P of obstacles and of the four corners of a bounding box containing all input points in its interior. We give an algorithm to test the existence of an oriented homeomorphism, and two algorithms to test homotopy between cycles; depending on the ratio between $k_1 + k_2$ and p, one is faster than the other.

Homeomorphism test. To test whether there exists an oriented homeomorphism of the plane that maps G_1 to G_2 in $\mathbb{R}^2 \setminus P$, we compute the oriented combinatorial map of G_1 in the punctured plane using a sweep-line algorithm for $G_1 \cup P$ in $O((k_1 + p)\log(k_1 + p))$ time [1]. Then we apply the same procedure with G_2 instead of G_1, and check that the two resulting oriented combinatorial maps are isomorphic.

First algorithm for homotopy tests. We transform the input into the surface model. For this purpose, we compute a triangulation T of B in $O(p \log p)$ time (for example, a Delaunay triangulation). We can then easily determine the arrangement of G_1 with T in $O(k_1 p)$ time, because each segment in G_1 has $O(p)$ crossings with T; that arrangement has complexity $O(k_1 p)$. We can apply the same procedure to G_2. After a slight modification of T that does not affect its complexity, we may assume that T is a triangulation of the bounding box minus a set of small triangular obstacles. We can then test homotopy of cycles in time linear in the number of their crossings with T, either by computing and comparing their cyclically reduced crossing words with T (since the fundamental group is a free group) or by applying the algorithm by Lazarus and Rivaud. This takes $O((k_1 + k_2)p)$ time.

Second algorithm for homotopy tests. To get a subquadratic running time in the input size, we improve the homotopy test by adapting an algorithm by Cabello et al. [7, Section 4]. Their algorithm tests homotopy for paths in the punctured plane, not homotopy of cycles; however, it can be modified to handle this case also. More precisely, we show below that, after $O(p^{1+\varepsilon})$ preprocessing time (for any $\varepsilon > 0$), one can test homotopy of two (possibly non-simple) cycles γ_1 and γ_2 of complexities m_1 and m_2, respectively, in $O((m_1 + m_2)\sqrt{p}\log p)$ time.

The main idea is to replace the triangulation T in the first algorithm above with a cellular decomposition of B that has nicer properties: each line in the plane crosses at most $O(\sqrt{p})$ segments of that decomposition. Cabello et al. [7] show how to compute such a cellular decomposition in $O(p^{1+\varepsilon})$ time. The two input cycles γ_1 and γ_2 cross this decomposition $O((m_1+m_2)\sqrt{p})$ times. Then, computing the cyclically ordered lists of edges of the decomposition crossed by these two cycles takes $O((m_1 + m_2)\sqrt{p}\log p)$ time using ray shooting, as done also in the paper by Cabello et al. We conclude using the same method as in the first algorithm, with the cellular decomposition in place of the triangulation. This proves Theorem 2.

6. CONCLUSION

We have given an optimal algorithm to test whether two graph embeddings of the same graph on an orientable surface are isotopic, where the vertices are allowed to move. On the other hand, the case of non-orientable surfaces remains open. Our characterization heavily relies on testing the existence of an oriented homeomorphism; it is not clear how to adapt this test in the non-orientable case.

It is also interesting to test isotopy of graphs, where some vertices of the graph are required to be fixed. Also, if we allow the input graph embeddings to meet the boundary, there is a difference between ambient isotopy fixing the boundary, ambient isotopy that allows to move the boundary points, and graph isotopy. It appears that all these cases are tractable using simple variations of our algorithm, and we will develop this in the full version of this paper.

Regarding the case of the punctured plane, the most obvious open question is to improve the running time of the algorithm. There is hope that Bespamyatnikh's $O(n^{4/3} \times \text{polylog}(n))$-time algorithm for testing path homotopy [3,

Theorem 7] extends to the homotopy test of cycles, which would immediately improve the complexity of our algorithm accordingly.

Also, we only test the existence of a topological isotopy: the edges are allowed to bend during the deformation. It is easy to see that, in the presence of obstacles, the existence of a topological isotopy between two straight-line embeddings does not imply the existence of a straight-line isotopy, in contrast to the case without obstacle [20, 5]. Could it be that, in such a situation, there exists a straight-line isotopy after splitting each edge in two (or a constant number of) segments? Computing such an isotopy efficiently may be a not easy task, but related techniques [28] might apply.

Finally, in both the surface model and the punctured plane model, computing shortest graph embeddings within a given isotopy class would be very interesting, and would generalize known results for computing shortest paths within a given homotopy class [22, 14, 2, 10, 11, 9], even though we suspect the problem to be much harder.

Acknowledgment

We thank Francis Lazarus and Julien Rivaud for sending us an advance copy of their paper [26], which inspired us for the construction of the stable family in Section 4.

7. REFERENCES

[1] J. L. Bentley and Th. Ottmann. Algorithms for reporting and counting geometric intersections. *IEEE Transactions on Computing*, 28(9):643–647, Sept. 1979.

[2] S. Bespamyatnikh. Computing homotopic shortest paths in the plane. *Journal of Algorithms*, 49(2):284–303, 2003.

[3] S. Bespamyatnikh. Encoding homotopy of paths in the plane. In *LATIN 2004: Theoretical Informatics, 6th Latin American Symposium*, volume 2976 of *Lecture Notes in Computer Science*, pages 329–338. Springer-Verlag, 2004.

[4] M. Bestvina and M. Handel. Train-tracks for surface homeomorphisms. *Topology*, 34(1):109–140, 1995.

[5] R. H. Bing and M. Starbird. Linear isotopies in E^2. *Transactions of the American Mathematical Society*, 237:205–222, 1978.

[6] P. Buser. *Geometry and spectra of compact Riemann surfaces*, volume 106 of *Progress in Mathematics*. Birkhäuser, 1992.

[7] S. Cabello, Y. Liu, A. Mantler, and J. Snoeyink. Testing homotopy for paths in the plane. *Discrete & Computational Geometry*, 31:61–81, 2004.

[8] É. Colin de Verdière. *Raccourcissement de courbes et décomposition de surfaces*. PhD thesis, Université Paris 7, Dec. 2003. English translation available at `http://www.di.ens.fr/~colin/03these.html`.

[9] É. Colin de Verdière and J. Erickson. Tightening nonsimple paths and cycles on surfaces. *SIAM Journal on Computing*, 39(8):3784–3813, 2010.

[10] É. Colin de Verdière and F. Lazarus. Optimal system of loops on an orientable surface. *Discrete & Computational Geometry*, 33(3):507–534, 2005.

[11] É. Colin de Verdière and F. Lazarus. Optimal pants decompositions and shortest homotopic cycles on an orientable surface. *Journal of the ACM*, 54(4):Article 18, 2007.

[12] M. de Graaf and A. Schrijver. Making curves minimally crossing by Reidemeister moves. *Journal of Combinatorial Theory, Series B*, 70(1):134–156, 1997.

[13] T. K. Dey and S. Guha. Transforming curves on surfaces. *Journal of Computer and System Sciences*, 58:297–325, 1999.

[14] A. Efrat, S. G. Kobourov, and A. Lubiw. Computing homotopic shortest paths efficiently. *Computational Geometry: Theory and Applications*, 35:162–172, 2006.

[15] D. Eppstein. Dynamic generators of topologically embedded graphs. In *Proceedings of the 14th Annual ACM-SIAM Symposium on Discrete Algorithms (SODA)*, pages 599–608, 2003.

[16] D. B. A. Epstein. Curves on 2-manifolds and isotopies. *Acta Mathematica*, 115:83–107, 1966.

[17] J. Erickson and K. Whittlesey. Greedy optimal homotopy and homology generators. In *Proceedings of the 16th Annual ACM-SIAM Symposium on Discrete Algorithms (SODA)*, pages 1038–1046, 2005.

[18] B. Farb and D. Margalit. *A primer on mapping class groups*. Princeton University Press, 2011.

[19] C. D. Feustel. Homotopic arcs are isotopic. *Proceedings of the AMS*, 17:891–896, 1966.

[20] C. Gotsman and V. Surazhsky. Guaranteed intersection-free polygon morphing. *Computers and Graphics*, 25(1):67–75, 2001.

[21] A. Hatcher. *Algebraic topology*. Cambridge University Press, 2002. Available at `http://www.math.cornell.edu/~hatcher/`.

[22] J. Hershberger and J. Snoeyink. Computing minimum length paths of a given homotopy class. *Computational Geometry: Theory and Applications*, 4:63–98, 1994.

[23] M. W. Hirsch. *Differential topology*, volume 33 of *Graduate Texts in Mathematics*. Springer-Verlag, 1994. Corrected reprint of the 1976 original.

[24] L. Kettner. Using generic programming for designing a data structure for polyhedral surfaces. *Computational Geometry: Theory and Applications*, 13:65–90, 1999.

[25] Y. Ladegaillerie. Classes d'isotopie de plongements de 1-complexes dans les surfaces. *Topology*, 23(3):303–311, 1984.

[26] F. Lazarus and J. Rivaud. On the homotopy test on surfaces. arXiv:1110.4573, 2011.

[27] S. Lins. Graph-encoded maps. *Journal of Combinatorial Theory, Series B*, 32:171–181, 1982.

[28] A. Lubiw and M. Petrick. Morphing planar graph drawings with bent edges. *Electronic Notes in Discrete Mathematics*, 31:45–48, 2008.

[29] G. Ringel. Teilungen der Ebene durch Geraden oder topologische Geraden. *Mathematische Zeitschrift*, 64:79–102 (1956), 1955.

[30] J. Stillwell. *Classical topology and combinatorial group theory*. Springer-Verlag, New York, second edition, 1993.

[31] I. Takarajima. A combinatorial representation of curves using train tracks. *Topology and its Applications*, 106:169–198, 2000.

APPENDIX

A. COMBINATORIAL MAPS FOR NON-CELLULAR EMBEDDINGS

To each embedding G_1 of an abstract graph G on S corresponds a ***combinatorial map***, storing the combinatorial information of that embedding. For graphs cellularly embedded, this notion is well-known, and there are several essentially equivalent data structures available to store such combinatorial maps on surfaces [24], like the gem representation [15, 27] and the halfedge data structure. All these data structures have linear complexity in the number of edges of the graph. The gem representation stores four *flags* per edge; intuitively, if the edge is oriented, two flags close to its head, one to its left and one to its right, and similarly for the tail. More formally, a flag represents an incidence between a vertex, an edge, and a vertex of the embedding. Three involutions allow to move from flag to an "incident" flag in the graph: The first one i_f keeps the same vertex-edge incidence and moves to the opposite face, i_v keeps the same edge-face incidence and moves to the opposite vertex, and the last one i_e keeps the same vertex-face incidence and moves to the opposite edge. Also, each flag has a pointer to the underlying vertex and edge of G.

Two combinatorial maps of the same graph G are ***isomorphic*** if there is a bijection φ between their sets of flags F_1 and F_2, commuting with the three involutions, and such that the underlying vertex (resp., edge) of a flag f in F_1 is the same as that of $\varphi(f)$. It is not hard to test whether two combinatorial maps representing two embeddings G_1 and G_2 of the same abstract graph G on S are isomorphic in linear time. Furthermore, we have the following lemma.

LEMMA 13. *The combinatorial maps representing two embeddings G_1 and G_2 of the same abstract graph G on S are isomorphic if and only if there exists a homeomorphism of S mapping G_1 to G_2.*

PROOF. The isomorphism of combinatorial maps extends naturally to a homeomorphism between the tubular neighborhoods of the graphs, and since all the faces are disks, this homeomorphism extends once again to the whole surface S. □

We adapt the gem representation (Figure 4) to handle graphs that are not cellularly embedded, by adding the following information. (For simplicity of exposition, we only consider graphs without isolated vertices; it is not hard to extend the data structure to handle this case.) First, each face may be incident to several cycles of G, and stores a list containing one flag of each such cycle. Conversely, each flag has a pointer to the face it belongs to. Also, a face may have non-zero genus and contain some boundary components of S, so we store this genus and number of boundary components of S.

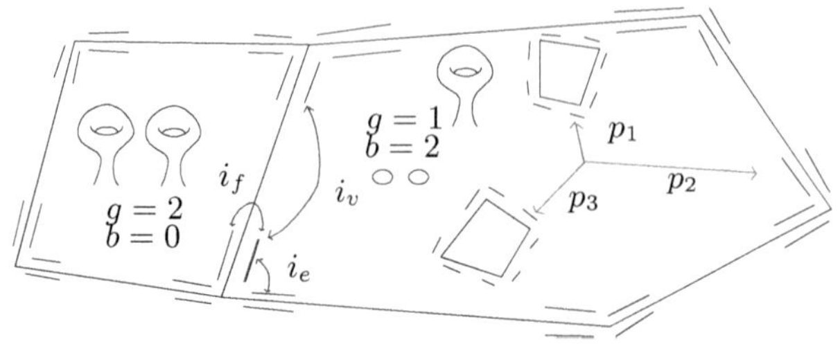

Figure 4: The extended combinatorial map for a face of the graph, with the three involutions i_1, i_2 and i_3 for a given flag (in bold), and the pointers p_1, p_2 and p_3 between the face and one flag for each incident cycle.

Two such ***extended combinatorial maps*** are ***isomorphic*** if, in addition to the conditions above for standard combinatorial maps, the corresponding faces have the same genus and the same number of boundary components of the surface. As in the cellular case, one can check whether two extended combinatorial maps of two embeddings G_1 and G_2 of G are isomorphic in linear time in their complexity. And as in the standard case, we have the following lemma.

LEMMA 14. *Two extended combinatorial maps of G_1 and G_2 are isomorphic if and only if there exists a homeomorphism of the surface mapping G_1 to G_2.*

PROOF. The proof is as straightforward as the one of Lemma 13. The isomorphism extends to a homoeomorphism between the neighborhoods of the graphs G_1 and G_2, and since each face of G_1 has the same topology as the corresponding one in G_2, this homeomorphism extends naturally to the whole surface. □

Our construction does not depend on whether S is orientable. If S is oriented, a flag has a natural ***orientation***, depending on whether we turn clockwise or counterclockwise around the vertex of the flag when starting on the edge of the flag, close to the vertex, and moving towards the face of the flag. Furthermore, each of the three involutions reverses the orientation. An *orientation* of an extended combinatorial map assigns an orientation "clockwise" or "counterclockwise" to each flag such that each involution reverses the orientation of the flag. There exists an ***orientation-preserving isomorphism*** between two oriented extended combinatorial maps if they are isomorphic as extended combinatorial maps and the bijection between the flags preserves the orientation. This can also be done in linear time. Equivalently, there exists an oriented homeomorphism of the surface mapping one graph embedding to the other.

On Laplacians of Random Complexes

Anna Gundert*   Uli Wagner*†
Institut für Theoretische Informatik
ETH Zürich
CH-8092 Zürich, Switzerland
{anna.gundert,uli}@inf.ethz.ch

ABSTRACT

Eigenvalues associated to graphs are a well-studied subject. In particular the spectra of the adjacency matrix and of the Laplacian of random graphs $G(n,p)$ are known quite precisely. We consider generalizations of these matrices to simplicial complexes of higher dimensions and study their eigenvalues for the Linial–Meshulam model $X^k(n,p)$ of random k-dimensional simplicial complexes on n vertices. We show that for $p = \Omega(\log n/n)$, the eigenvalues of both, the higher-dimensional adjacency matrix and the Laplacian, are a.a.s. sharply concentrated around two values.

In a second part of the paper, we discuss a possible higher-dimensional analogue of the Discrete Cheeger Inequality. This fundamental inequality expresses a close relationship between the eigenvalues of a graph and its combinatorial expansion properties; in particular, spectral expansion (a large eigenvalue gap) implies edge expansion.

Recently, a higher-dimensional analogue of edge expansion for simplicial complexes was introduced by Gromov, and independently by Linial, Meshulam and Wallach and by Newman and Rabinovich. It is natural to ask whether there is a higher-dimensional version of Cheeger's inequality. We show that the most straightforward version of a higher-dimensional Cheeger inequality fails: for every $k > 1$, there is an infinite family of k-dimensional complexes that are spectrally expanding (there is a large eigenvalue gap for the Laplacian) but not combinatorially expanding.

Categories and Subject Descriptors: G.2.2 [Discrete Mathematics]: Graph Theory—*Hypergraphs*

Keywords: Eigenvalues, Laplacians, Hypergraphs, Simplicial Complexes, Random Complexes, Measure Concentration.

*Research supported by the Swiss National Science Foundation (SNF Projects 200021-125309 and 200020-138230).

†Partially supported by a Research Membership (within the program *Quantitative Geometry*) at MSRI, Berkeley, CA, August–October, 2012.

1. INTRODUCTION

Eigenvalues of graphs are a classical and well-studied subject, which goes back to a fundamental paper of Kirchhoff [46], in which he used the combinatorial graph Laplacian to analyze electrical networks and formulated his celebrated *Matrix-Tree Theorem* for the number of spanning trees of a graph (which includes, as the special case of the complete graph, Cayley's [9] famous formula n^{n-2} for the number of labeled trees on n vertices).

The eigenvalues of a graph G encode many important properties of G, in particular regarding connectivity and *expansion* properties of G (the mixing rate of a random walk on G) as well as other *quasirandomness* properties of G. Because of this, eigenvalues of graphs also play a major role in the design and analysis of algorithms, including heuristic and approximation algorithms for hard graph partitioning problems (*spectral partitioning*) and Markov Chain Monte Carlo approximation algorithms for hard counting problems. We cannot hope to survey the relevant literature here and refer the reader to the survey articles and monographs [12, 43, 49, 39, 50, 18, 63] for background and further references.

Adjacency Matrix and Laplacians of Graphs

We recall the three $(n \times n)$-matrices commonly associated with a graph[1] $G = (V, E)$ on n vertices. The *adjacency matrix* $A = A(G) \in \{0,1\}^{V \times V}$ has entries defined by $A_{u,v} = 1$ iff $\{u,v\} \in E$. The *combinatorial Laplacian* is defined as $L = L(G) := D - A$, where $D = D(G) \in \mathbb{R}^{V \times V}$ is the diagonal matrix with entries $D_{v,v} = \deg_G(v)$, the *degrees* of the vertices. Both of these are *symmetric* matrices.

A symmetric real $(n \times n)$-matrix has a multiset of n real eigenvalues, called its *spectrum*, and $\mathbb{R}^n$ has an orthonormal basis of corresponding eigenvectors.

The eigenvalues of A and of L turn out to be quite sensitive to the maximum and minimum degree of G. For graphs with very nonuniform degree distributions, it is often more convenient to consider the *normalized Laplacian*, which is defined as $\Delta = \Delta(G) := D^{-1}L = I - D^{-1}A$, where $I \in \mathbb{R}^{V \times V}$ is the identity matrix.[2]

[1]Throughout this paper, we will assume that G is simple, i.e., we do not consider loops or multiple edges.

[2]Strictly speaking, D^{-1} is defined only if there are no isolated vertices, i.e., if $\deg_G(v) > 0$ for all $v \in V$, which will be the case of primary interest to us. If there are isolated vertices, we adopt the convention that $D^{-1}_{v,v} = 0$ whenever $\deg_G(v) = 0$ and retain the definition $\Delta = D^{-1}L$. (The second equation $\Delta = I - D^{-1}A$ no longer holds in this case, since Δ has zero diagonal entries at isolated vertices.)

The normalized Laplacian is not symmetric but corresponds to a self-adjoint operator on $\mathbb{R}^n$ with respect to a weighted inner product (see Section 2) and so also has n real eigenvalues. Both versions of the Laplacian are *positive semidefinite* relative to their respective inner products and so have nonnegative eigenvalues, typically listed in increasing order $\lambda_1(L) \leq \ldots \leq \lambda_n(L)$ and $\lambda_1(\Delta) \leq \ldots \leq \lambda_n(\Delta)$. The "all-1" vector $\mathbf{1} = (1, \ldots, 1)^T$ satisfies $L\mathbf{1} = \Delta\mathbf{1} = 0$, hence $\lambda_1(L) = \lambda_1(\Delta) = 0$, which is called the *trivial eigenvalue.*

The graph G is connected iff $\lambda_2(L) > 0$ iff $\lambda_2(\Delta) > 0$. More generally, the multiplicity of 0 as an eigenvector of either Laplacian equals the number of connected components of G, and if G is connected, then the second eigenvalue λ_2 of either Laplacian controls the *edge expansion* of the graph (see the discussion below). It is known that the spectrum of the normalized Laplacian Δ is contained in the interval $[0, 2]$, and that $\lambda_n(\Delta) = 2$ iff G has a nontrivial *bipartite* connected component [12, Lemma 1.7]. Moreover, if G has no isolated vertices then $\lambda_{n-1}(\Delta) \geq \frac{n}{n-1}$. The spectrum of the normalized Laplacian also controls the *mixing rate* of the random walk on the graph, see, e.g., [12, Section 1.5].

For the adjacency matrix, the eigenvalues are typically listed in *decreasing* order as $\mu_1(A) \geq \ldots \geq \mu_n(A)$.

Regular Graphs and Ramanujan Graphs

If G is *d-regular*, i.e., $\deg_G(v) = d$ for all $v \in V$ (where d may depend on n), then $L = d \cdot I - A = d \cdot \Delta$, and so the spectra of A, L, and Δ are equivalent (up to scaling and linear shifts): $\lambda_i(L) = d \cdot \lambda_i(\Delta)$ and $\mu_i(A) = d - \lambda_i(L)$, $1 \leq i \leq n$. In particular, $\mu_1(A) = d$, $\mu_2(A) < d$ iff G is connected, and $\mu_n(A) = -d$ iff G has a nontrivial bipartite connected component.

Define $\mu(G) := \max\{\mu_2(A), |\mu_n(A)|\}$. For the normalized Laplacian, this means that all nontrivial eigenvalues $\lambda_i(\Delta)$, $i > 1$, lie in the interval $[1 - \mu(G)/d, 1 + \mu(G)/d]$. It is not hard to show that $\mu(G) \geq \sqrt{d \cdot (n-d)/(n-1)}$ for every d-regular graph (see, e.g., [39, Claim 2.8]), hence $\mu(G) \geq \Omega(\sqrt{d})$ for $d \leq 0.99n$, say. For constant d, one has the sharper *Alon-Boppana bound* $\mu(G) \geq 2\sqrt{d-1} \cdot (1 - O(1/\log^2 n))$, see [59, 28].

A d-regular graph G is called a *Ramanujan graph* if it meets this bound for the spectral gap, i.e., if $\mu(G) \leq 2\sqrt{d-1}$. It is a deep result due to Lubotzky, Phillips and Sarnak [53] and independently to Margulis [55] that for every fixed number d with $d-1$ prime, there exist Ramanujan graphs on n vertices for infinitely many n (and moreover, these graphs can be explicitly constructed).

Eigenvalues of Random Graphs

Let $G(n, p)$ be the binomial random graph on n vertices, for which every edge is included independently with probability $p = p(n)$, and let $d = p(n-1) \sim np$ be the expected average degree.

For the adjacency matrix $A = A(G(n,p))$, it is known that $\mu_1(A(G(n,p))) \geq d$ *asymptotically almost surely* (a.a.s.), i.e., with probability tending to 1 as $n \to \infty$. Moreover, for $p \geq C \cdot \log n/n$, it is known that a.a.s. $\mu(G(n,p)) = O(\sqrt{d})$.

Sometimes, (e.g., in [13, 12, 15]) a slightly different matrix is referred to as the normalized Laplacian, namely $\mathscr{L} := I - D^{-1/2}AD^{-1/2}$. Assuming that there are no isolated vertices, Δ and $\mathscr{L}$ have the same spectra, since $\Delta x = \lambda x$ for some $\lambda \in \mathbb{R}$ and $x \in \mathbb{R}^V$ iff $\mathscr{L}y = \lambda y$, where $y = D^{1/2}x$.

Füredi and Komlós [32] showed this for constant p, and their method of proof can be extended fairly straightforwardly to p as small as roughly $(\log n)^6/n$ (cf. the discussion in [26]). Feige and Ofek [26] extended this to values of p as small as $C \cdot \log n/n$. They used methods of Friedman, Kahn, and Szemerédi [31], who proved that $\mu(G) = O(\sqrt{d})$ holds a.a.s. for *random d-regular graphs* (with constant d). Relating the spectrum of $\Delta(G(n,p))$ to that of $A(G(n,p))$, Coja-Oghlan [15] proved the corresponding result for the normalized Laplacian.

We summarize these results as follows (it will be important later that the error probability is not just $o(1)$ but can be made as small as n^{-c} for any $c > 0$):

THEOREM 1 ([26, 15]). *Suppose* $0.99 \geq p \geq C \cdot \log n/n$ *for a suitable constant* C *and let* $d = p(n-1)$. *Then, for every* $c > 0$ *there exists* $c' > 0$ *such that the following statements hold with probability at least* $1 - n^{-c}$*:*

(i) $\mu(G(n,p)) \leq c' \cdot \sqrt{d}$;

(ii) $1 - \frac{c'}{\sqrt{d}} \leq \lambda_2(\Delta(G(n,p)) \leq \ldots \leq \lambda_n(\Delta(G(n,p)) \leq 1 + \frac{c'}{\sqrt{d}}$.

One type of application of such results is the analysis of spectral heuristics for algorithms that deal with random instances of NP-hard graph partitioning and related problems, see the discussions in [26, 15].

We remark that both parts of the theorem can be extended to very sparse random graphs $G(n,p)$ with $p = \Theta(1/n)$ (for which they fail to hold as stated) by passing to a suitable large *core subgraph*, see [26, 15]. Moreover, analogous results are also known for other random graph models, including random d-regular graphs (see above) and random graphs with prescribed expected degree sequences [13, 16].

Higher-dimensional Laplacians

Eckmann [25] introduced a generalization of the graph Laplacian L to higher-dimensional simplicial complexes X to study discrete boundary value problems on such complexes.

More precisely, let X be a finite simplicial complex and let $C^i(X; \mathbb{R})$, $i \in \mathbb{Z}$, be the vector space of i-dimensional simplicial cochains with real coefficients (we refer to Section 2 for the necessary definitions). Eckmann defines three linear operators $L_i^{\text{down}}(X)$, $L_i^{\text{up}}(X)$ and $L_i(X) = L_i^{\text{down}}(X) + L_i^{\text{up}}(X)$ on the space $C^i(X; \mathbb{R})$ and proves a discrete analogue of *Hodge theory* [37], which implies, in particular, that the subspace $\mathcal{H}_i(X) := \ker L_i(X)$ of so-called *harmonic cochains* on X is isomorphic to $\widetilde{H}^i(X; \mathbb{R})$, the i-th reduced cohomology.

In the case of a 1-dimensional simplicial complex (graph) G, $L_0^{\text{up}}(G)$ coincides with the usual graph Laplacian $L(G)$ discussed previously.

Subsequently, combinatorial Laplacians were applied in a variety of contexts. Dodziuk [19] and Dodziuk and Patodi [21] showed how the continuous Laplacian of a Riemannian manifold can be approximated by the combinatorial Laplacians of a suitable sequence of successively finer triangulations of the manifold.

Kalai [44] used combinatorial Laplacians to prove a higher-dimensional generalization of Cayley's formula for the number of labeled trees, and further results in this direction, including a generalization of the Matrix-Tree Theorem, were obtained in [1, 24]. For further combinatorial applications, see, e.g., [30, 29, 47, 22]. For further background and references regarding combinatorial Laplacians, see also [40].

We will mostly work with a normalized version of the Laplacian, $\Delta_i(X) = \Delta_i^{\text{down}}(X) + \Delta_i^{\text{up}}(X)$ (see Section 2 for the definition). Again, for graphs, $\Delta_0^{\text{up}}(G)$ agrees with the normalized graph Laplacian $\Delta(G)$ discussed above.

Random Complexes

Linial and Meshulam [51] introduced a higher-dimensional analogue of the binomial random graph model $G(n,p)$. By definition, the random k-dimensional complex $X^k(n,p)$ has n vertices, a *full* $(k-1)$*-skeleton* (i.e., every subset of k of fewer vertices form a face of the complex), and every $(k+1)$-element set of vertices is taken as a k-face independently with probability p, which may be constant or, more generally, a function $p(n)$ depending on n.

This model has been studied quite extensively, and the *threshold probabilities* for a number of basic topological properties of $X^k(n,p)$ have been determined quite precisely:

(i) the sharp threshold for vanishing of the $(k-1)$-st cohomology $H^{k-1}(X^k(n,p);\mathbb{G})$ with coefficients in any fixed finite group $\mathbb{G}$ is at $p = \frac{k \log n}{n}$, see [51, 57];

(ii) for $k = 2$, the threshold for vanishing of the fundamental group $\pi_1(X^2(n,p))$ is roughly at $p = 1/\sqrt{n}$, see [7];

(iii) the thresholds for vanishing of the top-dimensional homology $H_k(X^k(n,p))$ (any group of coefficients) and for collapsibility onto the $(k-1)$-skeleton are at $p = \Theta(1/n)$, see [6, 14, 17, 48];

(iv) the threshold for embeddability of $X^k(n,p)$ into $\mathbb{R}^{2k}$ is at $p = \Theta(1/n)$, see [62].

Our first result is a higher-dimensional analogue of Theorem 1. Let Δ_{k-1}^{up} and A_{k-1} denote the normalized up-Laplacian and the adjacency matrix of a k-dimensional complex X, respectively (see Section 2.3 for the precise definitions). Both matrices have rows and columns indexed by the $(k-1)$-faces of X; we assume that X has n vertices and a full $(k-1)$-skeleton, so the matrices have dimension $\binom{n}{k} \times \binom{n}{k}$. A_{k-1} has entries in $\{0, \pm 1\}$, and $(A_{k-1})_{F,G} = \pm 1$ (with appropriate signs) iff $F \cup G$ is a k-face of X. For random complexes $X^k(n,p)$ the spectra of $\Delta_{k-1}^{\text{down}}$ and Δ_{k-1} are essentially determined by that of Δ_{k-1}^{up} (see Section 2.3). Hence, we focus on the latter.

THEOREM 2. *For all $c > 0$ and $k \geq 1$ there exists a constant $C = C(c,k) > 0$ with the following property: Assume $p \geq C\frac{\log(n)}{n}$ and let*[3] *$d := p(n-k)$. Then there exist $\gamma_A = O(\sqrt{d \log n})$ and $\gamma_\Delta = O(1/\sqrt{d})$ such that the following statements hold with probability at least $1 - n^{-c}$:*

(i) *The largest $\binom{n-1}{k-1}$ eigenvalues of $A_{k-1}(X^k(n,p))$ lie in the interval $[d - \gamma_A, d + \gamma_A]$, and the remaining $\binom{n-1}{k}$ eigenvalues lie in the interval $[-\gamma_A, +\gamma_A]$.*

(ii) *The smallest $\binom{n-1}{k-1}$ eigenvalues of $\Delta_{k-1}^{\text{up}}(X^k(n,p))$ are (trivially) zero, and the remaining $\binom{n-1}{k}$ eigenvalues lie in the interval $[1 - \gamma_\Delta, 1 + \gamma_\Delta]$. In particular, $\tilde{H}^{k-1}(X^k(n,p);\mathbb{R}) = 0$.*

[3]Thus, d is the expected *degree* of any $(k-1)$-face F in $X^k(n,p)$, i.e., the expected number of k-faces incident to F.

For the eigenvalues of A_{k-1}, the theorem only yields concentration in intervals of width $O(\sqrt{d \log n})$ around the typical eigenvalues, as opposed to the bound $O(\sqrt{d})$ in Theorem 1. This seems to be an artifact of the proof presented here, which uses techniques due to Oliveira [60], who obtained analogous, slightly suboptimal bounds of $\sqrt{d \log n}$ for the case of $G(n,p)$. Work in progress indicates that the methods in [26] can be adapted to higher dimensions and that the factor of $\sqrt{\log n}$ is, in fact, superfluous. This will be discussed in detail in the full version.

For the normalized Laplacian, we achieve the sharper concentration in an interval of width $O(1/\sqrt{d})$ by a different method, namely by applying a fundamental estimate due to Garland [34] (see the discussion below), which reduces the higher-dimensional problem to estimates for the eigenvalues of random graphs, i.e., to Theorem 1.

Garland's Estimate

In his seminal paper [34], Garland proves a conjecture of Serre's regarding the cohomology of certain groups. As a technical lemma of independent interest, Garland proves a lower bound for the nontrivial eigenvalues of the normalized Laplacian Δ_i^{up} of a finite simplicial complex. Qualitatively speaking, Garland's estimate says the following (see Section 4 for the precise statement): If X is k-dimensional and if the *links* of all $(k-2)$-dimensional faces of X are sufficiently spectrally expanding (more precisely, these links are graphs, and we require $\lambda_2(\Delta(\text{lk}(F,X))) > \frac{k-1}{k}$ for all $(k-2)$-faces F) then also X is *spectrally expanding* in the sense that all the nontrivial eigenvalues of $\Delta_{k-1}^{\text{up}}(X)$ are bounded away from zero, and in particular, $\tilde{H}^1(X;\mathbb{R}) = 0$.

For a combinatorial application of Garland's ideas (to clique complexes of graphs) see [2]. Garland's estimate was subsequently further strengthened and extended. In particular, Żuk [64] proved that if a 2-dimensional complex X satisfies $\lambda_2(\Delta(\text{lk}(v,X))) > 1/2$ for all vertex links, then the fundamental group of X has *Kazhdan's Property (T)*.

Edge Expansion and Cheeger's Inequality

For a graph of arbitrary density, its *edge expansion* can be defined as follows. Let $\varepsilon > 0$ be a parameter. We say that $G = (V,E)$ is ε-edge expanding if for every $S \subseteq V$,

$$\frac{|E(S, V \setminus S)|}{|E|} \geq \varepsilon \cdot \frac{\min\{|S|, |V \setminus S|\}}{|V|}, \tag{1}$$

where $E(S, V \setminus S) = \{\{u,v\} \in E : u \in S, v \in V \setminus S\}$ is the set of edges across the cut $(S, V \setminus S)$. Moreover, we call the best possible constant ε the *edge expansion* of G and denote it by $\varepsilon(G)$.[4] For a survey of the numerous applications of graph expansion in theoretical computer science and connections to other branches of mathematics, we refer to [39].

As mentioned above, the edge expansion of a graph is controlled by the second-smallest eigenvalue of its Laplacian. Here, we state this fact in its simplest form, for d-regular graphs (due to Dodziuk [20], Alon and Milman [4, 3]; Cheeger [10] proved an analogous result for Laplacians on Riemannian manifolds.):

[4]Note that (1) is equivalent, to the more common condition that $|E(S, V \setminus S)| \geq \frac{\varepsilon}{2} \cdot d \cdot |S|$ for all $S \subseteq V$ with $|S| \leq |V|/2$, where $d = 2|E|/|V|$ is the average degree. Thus, $\varepsilon(G) = 2h(G)$, where $h(G) := \min\{\frac{|E(S,V \setminus S)|}{d|S|} : S \subseteq V, |S| \leq |V|/2\}$ is the (normalized) *Cheeger constant* of G.

THEOREM 3 (DISCRETE CHEEGER INEQUALITY). *Let $G = (V, E)$ be a d-regular graph, and let $\lambda_2 = \lambda_2(\Delta(G))$ be the second-smallest eigenvalue of its normalized Laplacian. Then the edge expansion $\varepsilon(G)$ satisfies*

$$\lambda_2 \leq \varepsilon(G) \leq \sqrt{8\lambda_2}.$$

The inequality on the left-hand side is proved fairly easily by expressing the characteristic function $\mathbf{1}_S \in \mathbb{R}^V$ of a subset $S \subseteq V$ as a linear combination of eigenvectors of the Laplacian Δ. We will refer to this as "*the easy part of Cheeger's inequality.*" The harder part is the inequality on the right-hand side. For a short proof see, e.g., [5].

We remark that even the easy part of Cheeger's inequality is very useful. For instance, essentially all explicit constructions of constant-degree expanders [54, 33, 53, 55, 61] prove a lower bound on the edge expansion of the constructed graphs by analyzing their eigenvalues.

Higher-Dimensional Expansion

Recently, a higher-dimensional analogue of edge-expansion of graphs, which we refer to as *combinatorial expansion*, was introduced by Gromov [36], and independently by Linial, Meshulam and Wallach [51, 57] and by Newman and Rabinovich [58]. The precise definition will be given in Section 2. (For further related results, see, also [27, 45, 56, 23].)

It is natural to ask whether there is a higher-dimensional analogue of Cheeger's inequality; this question was raised explicitly, e.g., by Dotterrer and Kahle [23]. As our second result we show, by a simple probabilistic construction, that the most straightforward attempt at a higher-dimensional Cheeger Inequality fails, even for the "easy part". In higher dimensions, *spectral expansion* (an eigenvalue gap for the Laplacian) does not imply combinatorial expansion:

THEOREM 4. *For every $k > 1$ and $\varepsilon > 0$ there is an infinite family of k-dimensional complexes Y such that all nontrivial eigenvalues of $\Delta^{\text{up}}_{k-1}(Y)$ are $1 \pm o(1)$ (hence, Y is spectrally expanding) but every Y contains a* large *nontrivial cohomology class $[a] \in \tilde{H}^{k-1}(Y; \mathbb{Z}_2)$ of normalized Hamming weight $\|[a]\| \geq \frac{1}{2} - o(1)$ (so, Y is not combinatorially expanding; in particular, $\tilde{H}^{k-1}(Y; \mathbb{Z}_2) \neq 0$).*

For a graph G and any abelian group $\mathbb{G}$, $\tilde{H}^0(G; \mathbb{G}) = 0$ iff G is connected. In higher dimensions, however, it is well-known that the vanishing of a cohomology group may depend on the choice of coefficients. A basic example for this is the real projective plane $\mathbb{R}P^2$ for which $\tilde{H}^1(\mathbb{R}P^2; \mathbb{R}) = 0$ but $\tilde{H}^1(\mathbb{R}P^2; \mathbb{Z}_2) = \mathbb{Z}_2$. In general, $\tilde{H}^1(Y; \mathbb{G}) = 0$ iff Y is ε-expanding, with respect to a given norm on $\mathbb{G}$-cochains, for some small $\varepsilon > 0$ that may depend on Y. Thus, the point of Theorem 4 is that there is an infinite family of examples that are not combinatorially expanding and whose spectral expansion is bounded away from zero.

Related Work

Chung [11] studies a higher Laplacian for hypergraphs that is closely related[5] to the combinatorial Laplacian $L_{k-1} = L^{\text{up}}_{k-1} + L^{\text{down}}_{k-1}$. In [11, Section 7], she proves a somewhat weaker concentration result for eigenvalues of random hypergraphs, namely, essentially, that for *constant* p and any $\varepsilon > 0$, the eigenvalues of $L_{k-1}(X^k(n, p))$ are concentrated in an interval of width $O(n^{1/2+\varepsilon})$. She also states, without proof, that the proof methods for random graphs can be extended to yield the sharp bound of $O(\sqrt{pn})$.

The probabilistic construction of the examples in Theorem 4 is well-known in the study of quasirandomness for hypergraphs, see, e.g., the discussion in [35, Section 5]. In [11, Section 8], it is asserted, again without proof, that the eigenvalues of the combinatorial Laplacian of these examples are concentrated in an interval of width $O(\sqrt{n})$, but we are not aware of a proof appearing in the literature.

After submitting the original manuscript of the present paper, we became aware of a preprint by Hoffman, Kahle and Paquette [38], who prove closely related results. Specifically, following the basic approach of [31], they show that for any $\varepsilon > 0$ and $p \geq (k + \varepsilon)\frac{\ln n}{n}$, $\lambda_2(\Delta(G(n, p))) > 1/2$ with probability $1 - o(n^{1-k})$ (thus, compared to the known results, they trade precise information about the constant factor in front of $\ln n/n$ for weaker concentration). Using Żuk's result mentioned above, they obtain as an immediate corollary that for $p \geq (2 + \varepsilon)\frac{\ln n}{n}$, the random 2-complex $X^2(n, p)$ a.a.s. has Property (T).

In another recent preprint, Lu and Peng [52] study a rather different kind of Laplacian for random complexes. Specifically, given a k-dimensional complex X on a vertex set V and a parameter $s \leq \frac{k+1}{2}$, they consider an auxiliary weighted graph on the vertex set $\binom{V}{s}$ in which $I, J \in \binom{V}{s}$ are connected by an edge of weight w if $I \cap J = \emptyset$ and I and J are contained in precisely w common k-faces of X. Lu and Peng study the normalized Laplacian of this auxiliary weighted graph. However, this Laplacian seems to capture the topology of X only in a limited way. For instance, in the case $k = 2$ and $s = 1$, any two 2-dimensional complexes on n vertices that have a complete 1-skeleton and are d-regular (every edge is contained in d triangles) yield the same auxiliary graph, even though the topologies of these complexes (as measured by real cohomology groups and the usual Laplacian, say) may be very different.

2. PRELIMINARIES

2.1 Simplicial Complexes and Cohomology

A (finite, abstract) *simplicial complex* X is a finite set system that is closed under taking subsets, i.e. $F \subseteq G \in X$ implies $F \in X$. The sets in X are called *simplices* or *faces* of X. The *dimension* of a face F is $\dim(F) := |F| - 1$. We denote the set of i-dimensional faces of X by X_i. The dimension of X is the maximum dimension of any of its faces. The 0-dimensional faces are called *vertices*. Formally, these are singletons (one-element sets) but in this context we will usually identify the singleton $\{v\}$ with its unique element v.

A k-dimensional simplicial complex is *pure* if all maximal simplices in X have dimension k. We define the *degree* of a face F as $\deg(F) = |\{G \in X_k : F \subseteq G\}|$. The *link* of F in X is $\text{lk}(F, X) := \{G \in X : F \cup G \in X, F \cap G = \emptyset\}$. We denote by K^k_n the *complete k-dimensional complex* on n vertices, i.e. $K^k_n = \{F \subseteq [n] : |F| \leq k + 1\}$.

Orientations and Incidence Numbers

Throughout we assume that we have fixed a linear ordering on the vertex set $V := X_0$ of X, and we consider the faces of X with the orientations given by the order of their vertices.

[5] One difference is that Chung's Laplacian operates not just on cochains, i.e., skew-symmetric functions on oriented simplices, but on arbitrary real-valued functions.

Formally, consider an i-simplex $F = \{v_0, v_1, \ldots, v_i\} \in X_i$, where $v_0 < v_1 < \ldots < v_i$. For an $(i-1)$-simplex $G \in X_{i-1}$, we define the *oriented incidence number* $[F:G]$ by setting $[F:G] := (-1)^j$ if $G \subseteq F$ and $F \setminus G = \{v_j\}$, $0 \le j \le i$, and $[F:G] := 0$ if $G \not\subseteq F$. In particular, for every vertex $v \in X_0$ and the unique empty face $\emptyset \in X_{-1}$, we have $[v:\emptyset] = 1$.

Cohomology

Let X be a finite simplicial complex and let $\mathbb{G}$ be an Abelian group (we will mostly be concerned with the cases $\mathbb{G} = \mathbb{Z}_2$ and $\mathbb{G} = \mathbb{R}$, respectively). We denote by $C^i(X;\mathbb{G})$ the group $\mathbb{G}^{X_i}$ of functions from X_i to $\mathbb{G}$, which are called *i-dimensional cochains of X with coefficients in* $\mathbb{G}$. In particular, since $\emptyset$ is the unique empty face of X, we have $C^{-1}(X;\mathbb{G}) \cong \mathbb{G}$. It is convenient to define $C^i(X;\mathbb{G}) := 0$ for $i < -1$ or $i > \dim X$. If $\mathbb{G}$ is a field, the characteristic functions e_F of faces $F \in X_i$ form a basis of the vector space $C^i(X;\mathbb{G})$. They are called *elementary cochains*.

The *coboundary map* $\delta_i \colon C^i(X;\mathbb{G}) \to C^{i+1}(X,\mathbb{G})$ is the linear map given by

$$(\delta_i f)(F) := \sum_{G \in X_i} [F:G] \cdot f(G)$$

for $f \in C^i(X;\mathbb{G})$, $-1 \le i < \dim X$, and $\delta_i = 0$ otherwise.

It is an easy but central observation that the composition $\delta_i \circ \delta_{i-1} = 0$, which means that $B^i(X;\mathbb{G}) := \operatorname{im} \delta_{i-1} \subseteq Z^i(X;\mathbb{G}) := \ker \delta_i$. The elements of $B^i(X;\mathbb{G})$ and $Z^i(X;\mathbb{G})$ are called i-dimensional *coboundaries* and *cocycles*, respectively. Since $B^i(X;\mathbb{G}) \subseteq Z^i(X;\mathbb{G})$, we can form the quotient group $\tilde{H}^i(X;\mathbb{G}) := Z^i(X;\mathbb{G})/B^i(X;\mathbb{G})$, the i-th (reduced) *cohomology group* of X with coefficients in $\mathbb{G}$.

2.2 Norms on Cochains and Expansion

We now describe a very general definition of *expansion* for simplicial complexes, which was introduced in [36] (with a slightly different normalization and under the name *inverse (co)filling norm*).

Let X be a finite simplicial complex. Assume that every cochain group $C^i(X;\mathbb{G})$ is equipped with a *pseudonorm* $\|\cdot\|$, taking real values and satisfying $\|f\| = \|-f\|$ and $\|f+g\| \le \|f\| + \|g\|$ for all $f, g \in C^i(X;\mathbb{G})$. We will focus on the following two cases.

1. **$\mathbb{R}$-cochains with weighted ℓ_2-norm:** Assume that we are given a *weight function* w with nonnegative real values on the simplices of X. Define by $\langle f, g\rangle := \sum_{F \in X_i} w(F) f(F) g(F)$ a weighted inner product on $C^i(X;\mathbb{R})$. Observe that the inner products obtained in this way are characterized by the condition that the elementary cochains be pairwise orthogonal. We then consider the corresponding weighted ℓ_2-norm $\|f\| = \|f\|_2 := \sqrt{\langle f, f\rangle}$.

2. **$\mathbb{Z}_2$-cochains with weighted Hamming norm:** Let w be as before and define the *weighted Hamming norm* on $C^i(X;\mathbb{Z}_2)$ by $\|f\| := \sum_{F \in X_i : f(F)=1} w(F)$.

The idea is to define a notion of i-dimensional expansion that provides lower bounds for the norm of the coboundary $\delta_{i-1}(f) \in C^i(X;\mathbb{G})$ of $(i-1)$-dimensional cochains $f \in C^{i-1}(X;\mathbb{G})$. However, we cannot define such a lower bound in terms of the norm $\|f\|$ of f, since the set $B^{i-1}(X;\mathbb{G})$ is always contained in the kernel of the coboundary operator $\delta = \delta_{i-1}$. Thus, the right comparison measure is the *distance* of a cochain f from this *trivial part of the kernel*. That is, we define, for $f \in C^{i-1}(X;\mathbb{G})$,

$$\|[f]\| := \min\{\|f + \delta_{i-2} g\| : g \in C^{i-2}(X;\mathbb{G})\}.$$

Coboundary Expansion for Arbitrary Coefficients

Suppose every cochain group $C^i(X;\mathbb{G})$ is equipped with a pseudonorm $\|\cdot\|$ as above. We say that X is *ε-expanding in dimension i* (with respect to $\mathbb{G}$ and the given norm) if

$$\|\delta f\| \ge \varepsilon \cdot \|[f]\|. \tag{2}$$

for all $f \in C^{i-1}(X;\mathbb{G})$. The best possible ε is called the i-dimensional expansion of X. Note that, in particular, $\tilde{H}^{i-1}(X;\mathbb{G}) = 0$ if X has i-dimensional expansion $\varepsilon > 0$.

Combinatorial Expansion

Now we focus on the case of $\mathbb{Z}_2$-coefficients. Define a weight function by $w(F) := 1/|X_i|$ for $F \in X_i$ (whenever $|X_i| > 0$). In this setting, the normalized Hamming weight of a $\mathbb{Z}_2$-cochain $f \in C^{i-1}(X;\mathbb{Z}_2)$ is just the number of faces in the support of f divided by the number of all $(i-1)$-faces of X.

If X is is ε-expanding in dimension i with respect to this norm, we also say that X is *combinatorially ε-expanding in dimension i*.

Note that in the case $i = 1$ of graphs, there are just two 0-dimensional cochains, namely the constant functions $\mathbf{0}$ and $\mathbf{1}$ on the set $V = X_0$ of vertices. Moreover, a 0-dimensional cochain $f \in C^0(X;\mathbb{Z}_2)$ is in bijective correspondence with its support $S = \{v \in V : f(v) = 1\} \subseteq V$, and $\|[f]\| = \frac{\min\{|S|, |V\setminus S|\}}{|S|}$. Thus, 1-dimensional combinatorial expansion corresponds precisely to the definition (1) of edge expansion discussed in the introduction.

A basic observation in this context is that complete complexes are combinatorially expanding in all dimensions. This was observed independently by Gromov [36], Linial, Meshulam and Wallach [51, 57] and Newman and Rabinovich [58]:

PROPOSITION 5. *The complete complex K_n^k has i-dimensional combinatorial expansion 1 for all $i \in \{0, 1, \ldots, k\}$.*

From this, standard Chernoff bounds immediately imply that a.a.s., $X^k(n,p)$ is combinatorially expanding in dimension k and $H^{k-1}(X^k(n,p);\mathbb{Z}_2) = 0$ if $p > C \log n / n$ for a suitable constant C. Much of the work in [51, 57] is devoted to refining this argument to obtain the optimal constant $C = k$ for the threshold.

Dotterrer and Kahle [23] prove results analogous to Proposition 5 for some other complexes, specifically for skeleta of crosspolytopes and for complete multipartite complexes. They also explicitly raise the question whether there is some higher-dimensional analogue of Cheeger's inequality. The most straightforward attempt at such an inequality would be to relate combinatorial expansion and eigenvalue gaps of higher-dimensional Laplacians, which we discuss next.

2.3 Laplacians and Adjacency Matrixes

Adjacency matrices

For a finite k-dimensional simplicial complex X and $0 \le i \le k-1$ we define the *adjacency matrix* $A_i = A_i(X)$ by

$$(A_i(X))_{F,G} = \begin{cases} -[F \cup G : F][F \cup G : G] & \text{if } F \sim G, \\ 0 & \text{otherwise,} \end{cases}$$

where $F, G \in X_i$ and we write $F \sim G$ if F and G share a common $(i-1)$-face and $F \cup G \in X_{i+1}$. Note that $A_0(X)$ agrees with the adjacency matrix of the graph (X_0, X_1) because $[\{u,v\}:u][\{u,v\}:v] = -1$ for all vertices $u, v \in X_0$. The motivation for the signs in higher dimensions will hopefully become clear later on.

Weighted Laplacians

Following the exposition in [40], we begin by defining a general weighted Laplacian. Suppose we are given a nonnegative weight function w on the faces of a finite simplicial complex X and that the spaces $C^i(X;\mathbb{R})$ are equipped with the weighted inner product and the corresponding weighted ℓ_2-norm as described above.

The elementary cochains e_F, $F \in X_i$, form an orthogonal basis of $C^i(X;\mathbb{R})$. With respect to these bases, the coboundary map $\delta_i \colon C^i(X;\mathbb{R}) \to C^{i+1}(X;\mathbb{R})$ is given by the following $|X_{i+1}| \times |X_i|$-matrix (for which we abuse notation and again use the symbol δ):

$$(\delta_i(X))_{F,G} = \begin{cases} [F:G] & \text{if } G \subsetneq F, \\ 0 & \text{otherwise.} \end{cases}$$

Consider the *transpose map* $\delta_i^* \colon C^{i+1}(X;\mathbb{R}) \to C^i(X;\mathbb{R})$ of $\delta_i(X)$ with respect to the given inner product. This transpose is determined by the condition that $\langle \delta_i^* f, g\rangle = \langle f, \delta_i g\rangle$ for all $f \in C^{i+1}(X;\mathbb{R})$ and $g \in C^i(X;\mathbb{R})$. More explicitly,

$$(\delta_i^* f)(G) = \sum_{F \in X_{i+1}} \frac{w(F)}{w(G)} [F:G] f(F)$$

for $f \in C^{i+1}(X;\mathbb{R})$ and $G \in X_i$.

For example, in the case of unit weights $w(F) = 1$ for all $F \in X$, we get the standard inner product on $C^i(X;\mathbb{R})$, and $\delta_i^* = \partial_{i+1}$ coincides with the usual *boundary map* given on elementary cochains by $\partial_{i+1}(e_F) = \sum_{G \in X_i} [F:G] e_G$, $F \in X_{i+1}$.

In general, for arbitrary weights w on X, we define the *weighted Laplacian* by

$$\mathcal{L}_i^{\text{down}} := \delta_{i-1}\delta_{i-1}^*, \qquad \mathcal{L}_i^{\text{up}} := \delta_i^*\delta_i, \qquad \mathcal{L}_i := \mathcal{L}_i^{\text{down}} + \mathcal{L}_i^{\text{up}}.$$

Note that all three maps $\mathcal{L}_i^{\text{down}}, \mathcal{L}_i^{\text{up}}, \mathcal{L}_i$ are *self-adjoint* and *positive semidefinite* (with respect to the given weighted inner product) linear operators on $C^i(X;\mathbb{R})$.

In general, setting $\mathcal{H}_i = \mathcal{H}_i(X;\mathbb{R}) := \ker \mathcal{L}_i = \ker \mathcal{L}_i^{\text{down}} \cap \ker \mathcal{L}_i^{\text{up}} = \ker \delta_{i-1}^* \cap Z^i(X;\mathbb{R})$, one gets a *Hodge decomposition* of $C^i(X;\mathbb{R})$ into pairwise orthogonal subspaces

$$C^i(X;\mathbb{R}) = \mathcal{H}_i \oplus B^i(X;\mathbb{R}) \oplus \operatorname{im}(\delta_i^*), \tag{3}$$

(see [25, 40]); in particular, $\mathcal{H}_i \cong H^i(X;\mathbb{R})$.

Spectra of $\mathcal{L}_i^{\text{up}}$ and Spectral Expansion

By the Hodge decomposition (3) the spectrum of $\mathcal{L}_i$ is determined by the spectra of $\mathcal{L}_i^{\text{down}}$ and $\mathcal{L}_i^{\text{up}}$. Furthermore, for any linear map A, the spectra of AA^* and A^*A differ only in the multiplicity of 0; in particular, this holds for the spectra of $\mathcal{L}_i^{\text{up}}$ and $\mathcal{L}_{i+1}^{\text{down}}$. Thus, it is enough to consider the eigenvalues of $\mathcal{L}_i^{\text{up}}$.

Observe that, trivially, $B^i(X;\mathbb{R}) \subseteq \ker \mathcal{L}_i^{\text{up}}$. Thus, every $f \in B^i(X;\mathbb{R})$ is an eigenvector of $\mathcal{L}_i^{\text{up}}$ with eigenvalue zero. We call these the trivial eigenvectors of $\mathcal{L}_i^{\text{up}}$ and the trivial part of its spectrum. Thus, the nontrivial eigenvalues of $\mathcal{L}_i^{\text{up}}$ are, by definition, the eigenvalues of the restriction of $\mathcal{L}_i^{\text{up}}$ to the orthogonal complement (with respect to the given weighted inner product) $(B^i(X;\mathbb{R}))^{\perp}$.

By the variational definition of eigenvalues, the minimal nontrivial eigenvalue of $\mathcal{L}_i^{\text{up}}$ is given by

$$\min_{f \perp B^i(X;\mathbb{R})} \frac{\langle \mathcal{L}_i^{\text{up}} f, f\rangle}{\langle f, f\rangle} = \min_{f \perp B^i(X;\mathbb{R})} \frac{\|\delta_i f\|^2}{\|f\|^2}.$$

Thus, we see that the minimal nontrivial eigenvalue of $\mathcal{L}_i^{\text{up}}$ is at least ε^2 iff X has $(i+1)$-dimensional expansion at least ε with respect to the given weighted ℓ_2-norms on real cochains. In this case, we will also say that X is *spectrally expanding* in dimension i.

Combinatorial Laplacians

The combinatorial Laplacian $L_i = L_i^{\text{down}} + L_i^{\text{up}}$ corresponds to the special case of the standard inner product $\langle f, g\rangle = \sum_{f \in X_i} f(F)g(F)$, that is, the case of *unit weights* $w(F) = 1$ for all $F \in X$. Thus, $L_i^{\text{up}} = L_i^{\text{up}}(X) = \partial_{i+1}\delta_i$.

Recall that the matrix corresponding to the coboundary map δ_i with respect to the orthogonal basis of elementary cochains is, by abuse of notation, also denoted by $\delta_i = \delta_i(X)$, and its transpose δ_i^T corresponds to the boundary map ∂_{i+1}. The combinatorial Laplacian L_i^{up} can be expressed as the matrix $\delta_i^T \delta_i$.

We can now motivate the signs in the definition of the adjacency matrix $A_i(X)$: Recall that for a graph G the combinatorial Laplacian satisfies $L(G) = D(G) - A(G)$. If we let $D_i(X)$ denote the diagonal matrix with entry $D_{i\,F,F} = \deg(F)$ for $F \in X_i$, we also have $L_i^{\text{up}}(X) = D_i(X) - A_i(X)$.

Normalized Laplacians

Suppose that X is a pure k-dimensional simplicial complex. The normalized Laplacian $\Delta_i = \Delta_i^{\text{down}} + \Delta_i^{\text{up}}$ is the special case of the weighted Laplacian obtained by taking the weight function $w(F) := \deg(F)$. That is, the corresponding weighted inner product is $\langle f, g\rangle = \sum_{F \in X_i} \deg(F) f(F) g(F)$.

Let δ_i^* be the adjoint of δ_i with respect to this weighted inner product. Thus,

$$(\delta_i^* f)(G) = \sum_{F \in X_{i+1}} \frac{\deg(F)}{\deg(G)} [F:G] f(F).$$

Note that we have $\deg(F) > 0$ for every $F \in X$, since we assume that X is pure. The normalized Laplacian is then $\Delta_i^{\text{up}} = \Delta_i^{\text{up}}(X) = \delta_i^* \delta_i$.

With respect to the basis of elementary cochains, Δ_i^{up} corresponds to the matrix $D_i^{-1}\delta_i^T D_{i+1}\delta_i$. As $\deg(F) = 1$ for any k-face $F \in X_k$, we can write Δ_{k-1}^{up} as the matrix $D_{k-1}^{-1} L_{k-1}^{\text{up}} = I - D_{k-1}^{-1} A_{k-1}$.

Eigenvalues of the Complete Complex

As an example we consider the spectra of the three matrices for the complete complex K_n^k.

LEMMA 6. *The eigenvalues of the combinatorial Laplacian $L_{k-1}^{\text{up}}(K_n^k)$ are 0 with multiplicity $\binom{n-1}{k-1}$ and n with multiplicity $\binom{n-1}{k}$. The weighted Laplacian $\Delta_{k-1}^{\text{up}}(K_n^k)$ has eigenvalues 0 with multiplicity $\binom{n-1}{k-1}$ and $\frac{n}{n-k}$ with multiplicity $\binom{n-1}{k}$. The eigenvalues of $A_{k-1}(K_n^k)$ are $n-k$ with multiplicity $\binom{n-1}{k-1}$ and $-k$ with multiplicity $\binom{n-1}{k}$.*

PROOF. Because K_n^k is $(n-k)$-regular, it suffices to show the statement for the spectrum of $L_{k-1}^{\text{up}}(K_n^k)$.

The following equality is contained implicitly in [44] and follows from a straightforward calculation using the matrix representations of the Laplacians:

$$L^{\mathrm{up}}_{k-1}(K^k_n) + L^{\mathrm{down}}_{k-1}(K^k_n) = nI.$$

Any non-zero element of $\ker L^{\mathrm{down}}_{k-1}(K^k_n) = \ker \delta^*_{k-2}(K^k_n)$ is hence an eigenvector of L^{up}_{k-1} with eigenvalue n. Naturally, any non-zero element of $\ker L^{\mathrm{up}}_{k-1}(K^k_n) = B^{k-1}(K^k_n)$ is an eigenvector of L^{up}_{k-1} with eigenvalue 0.

To see that these two spaces have dimensions $\binom{n-1}{k}$ and $\binom{n-1}{k-1}$, respectively, one can check that bases are given by $\{\delta^*_{k-1}e_G : 1 \in G \in \binom{[n]}{k+1}\}$ and $\{\delta_{k-2}e_F : 1 \notin F \in \binom{[n]}{k-1}\}$. $\square$

3. THE ADJACENCY MATRIX

In this section we prove the first part of Theorem 2:

THEOREM 7. *For all $c > 0$ and $k \geq 1$ there exist constants $C = C(c,k) > 0$ and $c' > 0$ with the following property: Assume $p \geq C\frac{\log n}{n}$ and let $d := p(n-k)$ and $\gamma_A := c'\sqrt{d \log n}$. Then, with probability at least $1-n^{-c}$, the largest $\binom{n-1}{k-1}$ eigenvalues of $A_{k-1}(X^k(n,p))$ lie in $[d-\gamma_A, d+\gamma_A]$, and the remaining $\binom{n-1}{k}$ eigenvalues are contained in the interval $[-pk-\gamma_A, -pk+\gamma_A]$.*

The bound for the nontrivial eigenvalues of $\Delta^{\mathrm{up}}_{k-1}(X^k(n,p))$ are considered in Section 5 (Theorem 13).

Chernoff-type Bounds

A basic result on the concentration of binomially distributed random variables is the well-known Chernoff bound (see, e.g., [41, Theorem 1]).

THEOREM 8 (CHERNOFF BOUND). *Let X be a binomially distributed random variable with parameters n and p. Then for any $t \geq 0$*

$$\Pr[X \geq \mathbb{E}X+t] \leq e^{-\frac{t^2}{2(np+t/3)}} \text{ and } \Pr[X \leq \mathbb{E}X-t] \leq e^{-\frac{t^2}{2np}}.$$

In [60] Oliveira proves a concentration result for matrix martingales which has the following Chernoff-type inequality for matrices as a corollary. He uses this to obtain a concentration result for the eigenvalues of the adjacency matrix and the normalized Laplacian for any random graph model where potential edges appear independently, with possibly different probabilities. We will apply his techniques to the adjacency matrix of random complexes.

THEOREM 9 ([60, COROLLARY 7.1]). *Let independent, mean-zero random real symmetric matrices $X_1, X_2, \ldots, X_N$ be given that are bounded, i.e., a.a.s. $\|X_i\| \leq M$ almost surely for some $M > 0$ and every $1 \leq i \leq N$, then for all $t \geq 0$,*

$$\Pr\left[\left\|\sum_{i=1}^{n} X_i\right\| \geq t\right] \leq 2me^{-\frac{t^2}{8\sigma^2+4Mt}},$$

where $\sigma^2 = \lambda_{\max}\left(\sum_{i=1}^N \mathbb{E}[X_i^2]\right)$.

Here, for a symmetric matrix M with eigenvalues λ_i, $1 \leq i \leq m$, $\|M\| := \max_i |\lambda_i|$ denotes the *spectral radius*, and $\lambda_{\max}(M) := \max_i \lambda_i$ denotes the largest eigenvalue of M.

Following the strategy in [60], we show that the eigenvalues of the adjacency matrix of the random k-dimensional complex $X^k(n,p)$ are concentrated around those of a certain "typical" matrix: Let $A^{\mathrm{typ}}_{k-1,n,p} = pA_{k-1}(K^k_n)$ (we mostly suppress k and n from the notation and write A^{typ}_p), which can be thought of as the adjacency matrix of the weighted complete k-dimensional complex on n vertices where each k-dimensional face has weight p. Lemma 6 directly implies:

LEMMA 10. *The eigenvalues of $A^{\mathrm{typ}}_{k-1,n,p}$ are $p(n-k)$ with multiplicity $\binom{n-1}{k-1}$ and $-pk$ with multiplicity $\binom{n-1}{k}$.*

PROOF OF THEOREM 7. Let $c > 0$ and $k \geq 1$. To improve readability we write A_p for $A_{k-1}(X^k(n,p))$.

The key step in the proof is a decomposition of the difference $A_p - A^{\mathrm{typ}}_p$ as a sum of independent matrices X_H, each of which has only a constant number of non-zero entries and such that we can estimate the "variance" $\lambda_{\max}(\sum_H \mathbb{E}[X_H^2])$.

First observe that for any complex X the adjacency matrix $A_{k-1}(X)$ can be written as a sum of matrices corresponding to the k-faces of X: $A_{k-1}(X) = \sum_{H \in X_k} A_H$ where

$$(A_H)_{F,G} = \begin{cases} -[H:F][H:G] & \text{if } F,G \subsetneq H \text{ and } F \neq G, \\ 0 & \text{otherwise.} \end{cases}$$

For $H \in X_k$, the matrix A_H can be obtained from the adjacency matrix $A_{k-1}(K^k_{k+1})$ of a k-simplex by adding zero rows and columns corresponding to $(k-1)$-faces that are not contained in H. Furthermore, define independent indicator random variables I_H, for each $H \in \binom{[n]}{k+1}$, where $I_H = 1$ with probability p and $I_H = 0$ with probability $1-p$. Then $A_p = \sum_{H \in \binom{[n]}{k+1}} I_H A_H$ and $A^{\mathrm{typ}}_p = \sum_{H \in \binom{[n]}{k+1}} pA_H$. Hence,

$$A_p - A^{\mathrm{typ}}_p = \sum_{H \in \binom{[n]}{k+1}} X_H \text{ where } X_H = (I_H - p)A_H$$

In order to apply Theorem 9 to the sum of the X_H we check that the assumptions are fulfilled:

- Since the I_H are independent, the X_H are as well. Because $\mathbb{E}[I_H] = p$, we also have mean zero.
- To check the second condition, first consider $\|A_H\|$. The spectrum of A_H is that of $A_{k-1}(K^k_{k+1})$ plus additional zeros. By Lemma 6 $A_{k-1}(K^k_{k+1})$ has eigenvalues ± 1, hence $\|X_H\| \leq \|A_H\| \leq 1$.

Thus, we can apply Theorem 9. To interpret the result we compute $\sigma^2 = \lambda_{\max}(\sum_H \mathbb{E}[X_H^2])$. A simple computation shows that $A_H^2 = kD_H - (k-1)A_H$ where

$$(D_H)_{F,G} = \begin{cases} 1 & \text{if } F,G \subsetneq H \text{ and } F = G, \\ 0 & \text{otherwise.} \end{cases}$$

Thus, $\mathbb{E}[X_H^2] = p(1-p)(kD_H - (k-1)A_H)$ and hence

$$\sum_{H \in \binom{[n]}{k+1}} \mathbb{E}[X_H^2] = p(1-p)(k(n-k)I - (k-1)A_{k-1}(K^k_n)).$$

By Lemma 6, $A_{k-1}(K^k_n)$ has eigenvalues $n-k$ and $-k$. As any eigenvector of $A_{k-1}(K^k_n)$ is also an eigenvector of I we get that $\sigma^2 = p(1-p)k(n-1) \leq 2kd$.

If we now choose $C > 0$ with $d \geq 2(d\log(2\binom{n}{k}/n^{-c}))^{1/2}$, then for $p \geq C\log(n)/n$, we can apply Theorem 9 with $t = 8k(d\log(2\binom{n}{k}/n^{-c}))^{1/2} \leq 4kd$ to get

$$\Pr[\|A_p - A^{\mathrm{typ}}_p\| \geq t] \leq 2\tbinom{n}{k}e^{-\frac{64k^2 d\log(2\binom{n}{k}/n^{-c})}{64k^2 d}} = n^{-c}.$$

A straightforward calculation using the variational definition of eigenvalues shows $\|A-B\| \geq \max_i |\lambda_i(A)-\lambda_i(B)|$ for symmetric matrices A and B with spectra $\lambda_1(A), \ldots, \lambda_n(A)$ and $\lambda_1(B), \ldots, \lambda_n(B)$. Hence, this shows that with probability $1-n^{-c}$ the eigenvalues of $A_{k-1}(X^k(n,p))$ are concentrated around those of $A_{k-1}(X_{k,p}^{\mathrm{typ}})$ with a radius of at most $8k(d\log(2\binom{n}{k}/n^{-c}))^{1/2} = O(\sqrt{d\log(n)})$. □

4. GARLAND'S ESTIMATE REVISITED

In [34] Garland studies the normalized Laplacian $\Delta_i^{\mathrm{up}}(X)$. His main result regards a conjecture of Serre's on the cohomology of certain groups. As a technical lemma, he proves a bound for the nontrivial eigenvalues of $\Delta_i^{\mathrm{up}}(X)$ in terms of the eigenvalues of the Laplacian on links of lower-dimensional faces (see also [8] for a very clear exposition).

We state the result for the case of $\Delta_{k-1}^{\mathrm{up}}(X)$ and the links of $(k-2)$-dimensional faces $F \in X_{k-2}$. In this case, $\mathrm{lk}\, F = \mathrm{lk}(F, X)$ is a graph on $n-k+1$ vertices and the normalized Laplacian $\Delta_0^{\mathrm{up}}(\mathrm{lk}\, F)$ agrees with the usual normalized graph Laplacian $\Delta(\mathrm{lk}\, F)$.

Theorem 11 ([34], see also [8, Theorem 1.5,1.6]). *Let X be a pure k-dimensional complex. Denote by $\langle,\rangle$ the weighted inner product on $C^{k-1}(X;\mathbb{R})$ that is defined by $\langle f,g\rangle = \sum_{F\in X_i} \deg(F) f(F) g(F)$. Assume for all $F \in X_{k-2}$*

$$\lambda_{\min} \leq \lambda_2(\Delta(\mathrm{lk}\, F)) \leq \lambda_{n-k+1}(\Delta(\mathrm{lk}\, F)) \leq \lambda_{\max}.$$

Then for all $f \in B^{k-1}(X)^{\perp}$ (where the orthogonal complement is taken with respect to the weighted inner product)

$$(1+k\lambda_{\min}-k)\langle f,f\rangle \leq \langle \Delta_{k-1}^{\mathrm{up}} f, f\rangle \leq (1+k\lambda_{\max}-k)\langle f,f\rangle.$$

Hence, all nontrivial eigenvalues of $\Delta_{k-1}^{\mathrm{up}}(X)$ on $B^{k-1}(X)^{\perp}$ lie in $[1+k\lambda_{\min}-k, 1+k\lambda_{\max}-k]$.

We remark that Garland only states the lower bound. The upper bound follows directly from the proof, which we reproduce here in our notation. The main idea of the proof is to present the normalized Laplacian as a sum of matrices each of which has non-zero entries only on the link of some $(k-2)$-face. These matrices then correspond to the Laplacians of the links.

For a pure k-dimensional simplicial complex X, fix a face $F \in X_{k-2}$ of dimension $k-2$. Let ρ_F be the diagonal $|X_{k-1}| \times |X_{k-1}|$-matrix defined by

$$(\rho_F)_{G,H} = \begin{cases} 1 & \text{if } G=H \text{ and } F \subset G, \\ 0 & \text{otherwise.} \end{cases}$$

We set $\Delta_{k-1}^{\mathrm{up},F} := \rho_F \Delta_{k-1}^{\mathrm{up}} \rho_F$. For $f \in C^{k-1}(X)$ we define $f_F \in C^0(\mathrm{lk}\, F)$ by $f_F(\{u\}) = [F\cup\{u\} : F] f(F\cup\{u\})$. Straightforward calculations yield the following statements:

Lemma 12. *a) $\sum_{F\in X_{k-2}} \Delta_{k-1}^{\mathrm{up},F} = \Delta_{k-1}^{\mathrm{up}} + (k-1)I$.*

b) For $u,v \in V(\mathrm{lk}\, F)$ let $F_u = F\cup\{u\}$ and $F_v = F\cup\{v\}$. Then $(\Delta_{k-1}^{\mathrm{up},F})_{F_u,F_v} = [F_u:F][F_v:F](\Delta(\mathrm{lk}\, F))_{u,v}$. So, for $f \in C^{k-1}(X)$, $\langle \Delta_{k-1}^{\mathrm{up},F} f, f\rangle = \langle \Delta(\mathrm{lk}\, F) f_F, f_F\rangle$.

c) If $f \in B^{k-1}(X)^{\perp}$ then $f_F \in \mathbf{1}^{\perp}$.

These can easily be combined to prove Garland's estimate:

Proof of Theorem 11. Let $f \in B^{k-1}(X)^{\perp}$. Then

$$\langle \sum_{F\in X_{k-2}} \Delta_{k-1}^{\mathrm{up},F} f, f\rangle = \sum_{F\in \mathcal{F}_f} \langle \Delta(\mathrm{lk}\, F) f_F, f_F\rangle,$$

where $\mathcal{F}_f = \{F \in X_{k-2} | F \subset G \text{ for some } G \text{ with } f(G) \neq 0\}$. Now, since $f \in B^{k-1}(X)^{\perp}$, we have $f_F \in \mathbf{1}^{\perp}$ and $f_F \neq 0$ for $F \in \mathcal{F}_f$. As furthermore $\sum_{F\in\mathcal{F}_f} \langle f_F, f_F\rangle = k\langle f,f\rangle$,

$$k\lambda_{\min}\langle f,f\rangle \leq \langle \sum_{F\in X_{k-2}} \Delta_{k-1}^{\mathrm{up},F} f, f\rangle \leq k\lambda_{\max}\langle f,f\rangle.$$

By Lemma 12 we have furthermore

$$\langle \Delta_{k-1}^{\mathrm{up}} f, f\rangle = \langle \sum_{F\in X_{k-2}} \Delta_{k-1}^{\mathrm{up},F} f, f\rangle - (k-1)\langle f,f\rangle,$$

which finishes the proof. □

5. NORMALIZED LAPLACIAN

In this section, we prove the second part of Theorem 2, the bounds on the eigenvalues of the normalized Laplacian. The idea of the proof is to use Theorem 11 to reduce the statement on eigenvalues of $X^k(n,p)$ to a statement on the links of $(k-2)$-faces. Since for every $(k-2)$-face F, the link $\mathrm{lk}(F, X^k(n,p))$ is a random graph with the same distribution as $G(n-k+1,p)$, we can then apply results on the eigenvalues of Laplacians of random graphs.

Theorem 13. *For all $c>0$ and $k\geq 1$ there exist constants $C = C(k,c) > 0$ and $c' > 0$ with the following property: Assume $0.99 \geq p \geq C\log(n)/n$ and let $d := p(n-k)$. Then with probability $1-n^{-c}$ all nontrivial eigenvalues of $\Delta_{k-1}^{\mathrm{up}}(X^k(n,p))$ lie in the interval $[1-\frac{c'}{\sqrt{d}}, 1+\frac{c'}{\sqrt{d}}]$.*

Observe that $B^{k-1}(K_n^k) \subseteq \ker \Delta_{k-1}^{\mathrm{up}}(X^k(n,p))$ because $X^k(n,p)$ has a complete $(k-1)$-skeleton, so the multiplicity of 0 as an eigenvalue of $\Delta_{k-1}^{\mathrm{up}}(X^k(n,p))$ is at least $\binom{n-1}{k-1}$.

Proof of Theorem 13. Let $C>0$. For simplicity, we write $\mathrm{lk}\, F$ instead of $\mathrm{lk}(F, X^k(n,p))$, where $F \in \binom{[n]}{k-1}$. For every such F, the $\mathrm{lk}(F)$ is a random graph with distribution $G(\nu,p)$ with $\nu = n-k+1$. By Theorem 1 there exist constants $C>0$ and $c'>0$ such that with probability at least n^{-c-k+1} the nontrivial eigenvalues of $\Delta(G(\nu,p))$ are contained in the interval $[1-c'/(k\sqrt{d}), 1+c'/(k\sqrt{d})]$ for $0.99 \geq p \geq C\cdot \log n/n$.

For every $(k-1)$-face $G \in \binom{[n]}{k}$ of $X^k(n,p)$, the random variable $\deg(G)$ is binomially distributed with parameters $(n-k)$ and p. By making C slightly larger, if necessary, we can ensure that for $p \geq C\cdot\log n/n$, the complex $X^k(n,p)$ is pure with probability at least n^{-c-k+1}.

Now, assume that $0.99 \geq p \geq C\cdot \log n/n$. For any pure complex, we can apply Theorem 11. Hence, the probability that the nontrivial part of the spectrum of $\Delta_{k-1}^{\mathrm{up}}(X^k(n,p))$ lies in the interval $[1-c'/\sqrt{d}, 1+c'/\sqrt{d}]$ is bounded below by the probability that $X^k(n,p)$ is pure and that the nontrivial eigenvalues of $\Delta(\mathrm{lk}\, F)$ are contained in $[1-c'/(k\sqrt{d}), 1+c'/(k\sqrt{d})]$ for all $F \in \binom{[n]}{k-1}$. Applying a union bound and the considerations above we see that this probability is at least $1-n^{-c}$. □

Remark 1. Note that that the preceding proof works for any random distribution $\mathcal{X}_k(n,p)$ on k-dimensional simplicial complexes with n vertices and complete $(k-1)$-skeleton with the property that the link $\mathrm{lk}(F, \mathcal{X}_k(n,p))$ of every $F \in \binom{[n]}{k-1}$ is a random graph with distribution $G(n-k+1,p)$.

6. SPECTRAL VS. COMBINATORIAL

In this section, we prove Theorem 4. As mentioned in the introduction, the examples are obtained by a simple probabilistic construction: Denote by $Y^k(n,p)$ the random k-dimensional simplicial complex with vertex set $V = [n]$ and complete $(k-1)$-skeleton obtained as follows: Randomly choose $a\colon \binom{V}{k} \to \mathbb{Z}_2$ by setting $a(F) = 1$ with probability p and $a(F) = 0$ otherwise, independently for each $F \in \binom{V}{k}$. (Thus, the support of a has the same distribution as the $(k-1)$-faces of the Linial-Meshulam random complex $X^{k-1}(n,p)$.) For every $G \in \binom{V}{k+1}$, we add G as a k-face to $Y^k(n,p)$ iff G contains an *even* number of $(k-1)$-faces F with $a(F) = 0$. Note that, by construction, a is a $\mathbb{Z}_2$*-cocycle* in the complex $Y^k(n,p)$, i.e., $a \in Z^{k-1}(Y^k(n,p); \mathbb{Z}_2)$. For simplicity, we concentrate on the case $p = 1/2$ from now on and write Y instead of $Y^k(n, 1/2)$.

For any fixed $b \in C^{k-1}(Y; \mathbb{Z}^2) = \mathbb{Z}_2^{\binom{V}{k}}$, the expected normalized Hamming distance between b and the randomly chosen a equals $1/2$. Since there are fewer than $2^{\binom{n}{k-1}}$ *coboundaries* $b \in B^{k-1}(Y; \mathbb{Z}^2)$ and $\binom{n}{k}$ independent random choices for the entries of $a \in \mathbb{Z}^2$, a straightforward application of Chernoff plus a union bound implies that, a.a.s., a has normalized Hamming distance $1/2 - o(1)$ from any coboundary, i.e., $\|[a]\| \geq 1/2 - o(1)$. In particular, a.a.s. $\tilde{H}^{k-1}(Y, \mathbb{Z}_2) \neq 0$.

Note that for $G \in \binom{V}{k+1}$, the probability that G is a k-face of Y equals $1/2$. However, in contrast to the model $X^k(n, 1/2)$, the decisions for different k-faces that share some $(k-1)$-face are not independent. Nevertheless, we can still easily analyze the links of $(k-2)$-faces in Y:

Lemma 14. *For every $(k-2)$-face $H \in Y_{k-2} = \binom{V}{k-1}$, the random graph $\mathrm{lk}(H,Y)$ has the distribution $G(n-k+1, 1/2)$.*

Proof. Let $U := V \setminus H$. For $e \in \binom{U}{2}$, consider the event that $e \in \mathrm{lk}(H,Y)$, i.e., that $H \cup e \in Y$. We need to show that these events are mutually independent. To see this, choose and fix, for each $e \in \binom{U}{2}$, an arbitrary $(k-1)$-simplex F_e with $e \subseteq F_e \subseteq H \cup e$; we call these the "*undecided*" $(k-1)$-simplices, and let $\mathcal{D} := \binom{V}{k} \setminus \{F_e \colon e \in \binom{U}{2}\}$ be the set of remaining, "*decided*" $(k-1)$-simplices. Note that, by construction, each k-simplex of the form $H \cup e$, $e \in \binom{U}{2}$, contains exactly one undecided $(k-1)$-simplex F_e and that these are pairwise distinct. Fix a map $r\colon \mathcal{D} \to \mathbb{Z}_2$ and condition upon the event that r is the restriction of a to $\mathcal{D}$. For each $e \in \binom{U}{2}$, we have $e \in \mathrm{lk}(H,Y)$ iff $a(F_e) = \sum_{F \in \mathcal{D}, F \subset H \cup e} r(F)$. For a fixed r, the (conditional) probability of this happening is $1/2$, and the values $a(F_e)$ are mutually independent since the F_e are pairwise distinct. Thus, for any set of edges $e_1, \ldots, e_\ell \in \binom{U}{2}$ and for any fixed r, we get the conditional probability $\Pr[\forall i : e_i \in \mathrm{lk}(H,Y) \mid a|_{\mathcal{D}} = r] = (1/2)^\ell$. Since this holds for all choices of r, it also holds unconditionally, which proves the lemma. □

By this lemma and Remark 1, we can proceed as in the proof of Theorem 13 to show that there exists a constant $c > 0$ such that a.a.s. the nontrivial part of the spectrum of $\Delta^{\mathrm{up}}_{k-1}(Y)$ lies in the interval $[1 - \frac{c}{\sqrt{n-k}}, 1 + \frac{c}{\sqrt{n-k}}]$. This completes the proof of Theorem 4.

7. CONCLUDING REMARKS

We have given a brief survey of higher-dimensional expansion properties of simplicial complexes and shown concentration results for the eigenvalues of the Laplacian and of the adjacency matrix of random complexes in the Linial–Meshulam model. The bounds for the Laplacian are essentially tight, while the bounds for the adjacency matrix still contain an extra $\sqrt{\log n}$ factor, which we are confident can be removed. Details will be presented in the full version.

We have also shown that the most straightforward attempt to formulate a higher-dimensional analogue of Cheeger's inequality fails: In higher dimensions, a large eigenvalue gap for the Laplacian does not imply combinatorial expansion. The difficulties as well as the counterexamples are closely related to the ones that arise in the study of regularity and quasirandomness for (*dense*) hypergraphs. Currently we are investigating two basic approaches to a higher-dimensional Cheeger inequality. The first is to relate combinatorial expansion not to an eigenvalue gap for the Laplacian, but to other quasirandomness properties, specifically to *octahedral quasirandomness* and small *higher uniformity norms* (see, e.g., [35]). The second is to formulate combinatorial expansion in higher dimensions as a combinatorial optimization problem and to study semidefinite relaxations thereof. We plan to discuss this in more detail in the full version.

Acknowledgements

The second author is grateful to Roy Meshulam for helpful discussions during which, in particular, he learned about Garland's results. We would also like to thank the referees of this paper for their helpful comments.

8. REFERENCES

[1] R. M. Adin. Counting colorful multi-dimensional trees. *Combinatorica*, 12(3):247–260, 1992.

[2] R. Aharoni, E. Berger, and R. Meshulam. Eigenvalues and homology of flag complexes and vector representations of graphs. *Geom. Funct. Anal.*, 15(3):555–566, 2005.

[3] N. Alon. Eigenvalues and expanders. *Combinatorica*, 6(2):83–96, 1986.

[4] N. Alon and V. D. Milman. λ_1, Isoperimetric inequalities for graphs, and superconcentrators. *J. Combin. Theory Ser. B*, 38(1):73–88, 1985.

[5] N. Alon, O. Schwartz, and A. Shapira. An elementary construction of constant-degree expanders. *Combin. Probab. Comput.*, 17(3):319–327, 2008.

[6] L. Aronshtam, N. Linial, T. Luczak, and R. Meshulam. Collapsibility and vanishing of top homology in random simplicial complexes. Preprint, `arXiv:1010.1400v2`, 2010.

[7] E. Babson, C. Hoffman, and M. Kahle. The fundamental group of random 2-complexes. *J. Amer. Math. Soc.*, 24(1):1–28, 2011.

[8] A. Borel. Cohomologie des certains groupes discrets et Laplacien p-adique. In *Séminaire Bourbaki, Exposé 437*, volume 431 of *Lecture Notes in Math.*, pages 12–35. Springer, 1973.

[9] A. Cayley. A theorem on trees. *Quart. J. Pure Appl. Math.*, 23:376–378, 1889. Collected Mathematical Papers Vol. 13, Cambridge Univ. Press, 1897, 26–28.

[10] J. Cheeger. A lower bound for the smallest eigenvalue of the Laplacian. In *Problems in analysis (Papers dedicated to Salomon Bochner, 1969)*, pages 195–199. Princeton Univ. Press, Princeton, N. J., 1970.

[11] F. Chung. *The Laplacian of a Hypergraph*, volume 10 of *DIMACS Ser. Discrete Math. Theoret. Comput. Sci.*, pages 21–36. Amer. Math. Soc., 1993.

[12] F. Chung. *Spectral graph theory*, volume 92 of *CBMS Reg. Conf. Ser. Math.* Amer. Math. Soc., 1997.

[13] F. Chung, L. Lu, and V. Vu. The spectra of random graphs with given expected degrees. *Internet Math.*, 1(3):257–275, 2004.

[14] D. C. Cohen, M. Farber, and T. Kappeler. The homotopical dimension of random 2-complexes. Preprint, arXiv:1005.3383v1, 2010.

[15] A. Coja-Oghlan. On the Laplacian eigenvalues of $G_{n,p}$. *Combin. Probab. Comput.*, 16(6):923–946, 2007.

[16] A. Coja-Oghlan and A. Lanka. The spectral gap of random graphs with given expected degrees. *Electron. J. Combin.*, 16(1):R138, 2009.

[17] A. Costa, M. Farber, and T. Kappeler. Topology of random 2-complexes. Preprint, arXiv:1006.4229v2, 2010.

[18] D. Cvetković, P. Rowlinson, and S. Simić. *An introduction to the theory of graph spectra*, volume 75 of *London Math. Soc. Stud. Texts*. Cambridge Univ. Press, Cambridge, 2010.

[19] J. Dodziuk. Finite-difference approach to the Hodge theory of harmonic forms. *Amer. J. Math.*, 98(1):79–104, 1976.

[20] J. Dodziuk. Difference equations, isoperimetric inequality and transience of certain random walks. *Trans. Amer. Math. Soc.*, 284(2):787–794, 1984.

[21] J. Dodziuk and V. K. Patodi. Riemannian structures and triangulations of manifolds. *J. Indian Math. Soc. (N.S.)*, 40(1-4):1–52 (1977), 1976.

[22] X. Dong and M. L. Wachs. Combinatorial Laplacian of the matching complex. *Electron. J. Combin.*, 9(1):Research Paper 17, 11, 2002.

[23] D. Dotterrer and M. Kahle. Coboundary expanders. Preprint, arXiv:1012.5316, 2010.

[24] A. M. Duval, C. J. Klivans, and J. L. Martin. Simplicial matrix-tree theorems. *Trans. Amer. Math. Soc.*, 361(11):6073–6114, 2009.

[25] B. Eckmann. Harmonische Funktionen und Randwertaufgaben in einem Komplex. *Comment. Math. Helv.*, 17:240–255, 1945.

[26] U. Feige and E. Ofek. Spectral techniques applied to sparse random graphs. *Random Structures Algorithms*, 27(2):251–275, 2005.

[27] J. Fox, M. Gromov, V. Lafforgue, A. Naor, and J. Pach. Overlap properties of geometric expanders. Preprint, arXiv:1005.1392, 2010.

[28] J. Friedman. Some geometric aspects of graphs and their eigenfunctions. *Duke Math. J.*, 69(3):487–525, 1993.

[29] J. Friedman. Computing Betti numbers via combinatorial Laplacians. *Algorithmica*, 21(4):331–346, 1998.

[30] J. Friedman and P. Hanlon. On the Betti numbers of chessboard complexes. *J. Algebraic Combin.*, 8(2):193–203, 1998.

[31] J. Friedman, J. Kahn, and E. Szemerédi. On the second eigenvalue of random regular graphs. In *Proc. 21st Ann. ACM Symp. Theor. Comput.*, STOC '89, pages 587–598, New York, 1989. ACM.

[32] Z. Füredi and J. Komlós. The eigenvalues of random symmetric matrices. *Combinatorica*, 1(3):233–241, 1981.

[33] O. Gabber and Z. Galil. Explicit constructions of linear-sized superconcentrators. *J. Comput. System Sci.*, 22(3):407–420, 1981.

[34] H. Garland. p-adic curvature and the cohomology of discrete subgroups of p-adic groups. *Ann. of Math. (2)*, 97(3):375–423, 1973.

[35] W. T. Gowers. Quasirandomness, counting and regularity for 3-uniform hypergraphs. *Combin. Probab. Comput.*, 15(1-2):143–184, 2006.

[36] M. Gromov. Singularities, expanders and topology of maps. Part 2: From combinatorics to topology via algebraic isoperimetry. *Geom. Funct. Anal.*, 20(2):416–526, 2010.

[37] W. V. D. Hodge. *The Theory and Applications of Harmonic Integrals*. Cambridge Mathematical Library. Cambridge Univ. Press, Cambridge, 1989.

[38] C. Hoffman, M. Kahle, and E. Paquette. A sharp threshold for Kazhdan's property (T). Preprint, arXiv:1201.0425, 2012.

[39] S. Hoory, N. Linial, and A. Wigderson. Expander graphs and their applications. *Bull. Amer. Math. Soc. (N.S.)*, 43(4):439–561 (electronic), 2006.

[40] D. Horak and J. Jost. Spectra of combinatorial Laplace operators on simplicial complexes. Preprint, arXiv:1105.2712v3, 2011.

[41] S. Janson. On concentration of probability. In B. Bollobás, editor, *Contemporary Combinatorics*, volume 10 of *Boly. Soc. Math. Stud.*, pages 289–301. Springer, 2002.

[42] S. Janson, T. Łuczak, and A. Ruciński. *Random Graphs*. Wiley-Intersci. Ser. in Discrete Math. Optim. Wiley, 2000.

[43] M. Jerrum. *Counting, sampling and integrating: algorithms and complexity*. Lectures Math. ETH Zürich. Birkhäuser, Basel, 2003.

[44] G. Kalai. Enumeration of Q-acyclic simplicial complexes. *Israel J. Math.*, 45:337–351, 1983.

[45] R. N. Karasev. A Simpler Proof of the Boros-Füredi-Bárány-Pach-Gromov Theorem. *Discrete Comput. Geom.*, 47(3):492–495, 2012.

[46] G. Kirchhoff. Ueber die Auflösung der Gleichungen, auf welche man bei der Untersuchung der linearen Vertheilung galvanischer Ströme geführt wird. *Ann. Phys.*, 148(12):497–508, 1847.

[47] W. Kook, V. Reiner, and D. Stanton. Combinatorial Laplacians of matroid complexes. *J. Amer. Math. Soc.*, 13(1):129–148, 2000.

[48] D. Kozlov. The threshold function for vanishing of the top homology group of random d-complexes. *Proc. Amer. Math. Soc.*, 138(12):4517–4527, 2010.

[49] M. Krivelevich and B. Sudakov. Pseudo-random graphs. In *More sets, graphs and numbers*, volume 15 of *Bolyai Soc. Math. Stud.*, pages 199–262. Springer, 2006.

[50] D. A. Levin, Y. Peres, and E. L. Wilmer. *Markov chains and mixing times*. Amer. Math. Soc., 2009.

[51] N. Linial and R. Meshulam. Homological connectivity of random 2-complexes. *Combinatorica*, 26(4):475–487, 2006.

[52] L. Lu and X. Peng. Loose Laplacian spectra of random hypergraphs. Preprint, arXiv:1109.3433, 2011.

[53] A. Lubotzky, R. S. Phillips, and P. Sarnak. Ramanujan graphs. *Combinatorica*, 8(3):261–277, 1988.

[54] G. A. Margulis. Explicit constructions of expanders. *Problemy Peredači Informacii*, 9(4):71–80, 1973.

[55] G. A. Margulis. Explicit group-theoretic constructions of combinatorial schemes and their applications in the construction of expanders and concentrators. *Problemy Peredachi Informatsii*, 24(1):51–60, 1988.

[56] J. Matoušek and U. Wagner. On Gromov's method of selecting heavily covered points. Preprint, arXiv:1102.3515v1, 2011.

[57] R. Meshulam and N. Wallach. Homological connectivity of random k-dimensional complexes. *Random Structures Algorithms*, 34(3):408–417, 2009.

[58] I. Newman and Y. Rabinovich. On multiplicative λ-approximations and some geometric applications. In *Proc. 23rd Ann. ACM-SIAM Symp. Discrete Algorithms*. SIAM, 2011.

[59] A. Nilli. On the second eigenvalue of a graph. *Discrete Math.*, 91(2):207–210, 1991.

[60] R. I. Oliveira. Concentration of the adjacency matrix and of the Laplacian in random graphs with independent edges. Preprint, arXiv:0911.0600v2, 2010.

[61] O. Reingold, S. Vadhan, and A. Wigderson. Entropy waves, the zig-zag graph product, and new constant-degree expanders. *Ann. of Math. (2)*, 155(1):157–187, 2002.

[62] U. Wagner. Minors in random and expanding hypergraphs. In *Proc. 27th Ann. Symp. Comput. Geom. (SoCG)*, pages 351–360, 2011.

[63] X.-D. Zhang. The Laplacian eigenvalues of graphs: a survey. Preprint, arXiv:1111.2897v1, 2011.

[64] A. Żuk. La propriété (T) de Kazhdan pour les groupes agissant sur les polyèdres. *C. R. Acad. Sci. Paris*, 323:453–458, 1996.

Approximating Low-Dimensional Coverage Problems

Ashwinkumar Badanidiyuru*
Dept. Computer Science
Cornell University
Ithaca, NY
ashwinkumarbv@gmail.com

Robert Kleinberg†
Dept. Computer Science
Cornell University
Ithaca, NY
rdk@cs.cornell.edu

Hooyeon Lee
Dept. Computer Science
Stanford University
Stanford, CA
haden.lee@stanford.edu

ABSTRACT

We study the complexity of the maximum coverage problem, restricted to set systems of bounded VC-dimension. Our main result is a fixed-parameter tractable approximation scheme: an algorithm that outputs a $(1-\epsilon)$-approximation to the maximum-cardinality union of k sets, in running time $O(f(\epsilon, k, d) \cdot poly(n))$ where n is the problem size, d is the VC-dimension of the set system, and $f(\epsilon, k, d)$ is exponential in $(kd/\epsilon)^c$ for some constant c. We complement this positive result by showing that the function $f(\epsilon, k, d)$ in the running-time bound cannot be replaced by a function depending only on (ϵ, d) or on (k, d), under standard complexity assumptions.

We also present an improved upper bound on the approximation ratio of the greedy algorithm in special cases of the problem, including when the sets have bounded cardinality and when they are two-dimensional halfspaces. Complementing these positive results, we show that when the sets are four-dimensional halfspaces neither the greedy algorithm nor local search is capable of improving the worst-case approximation ratio of $1-1/e$ that the greedy algorithm achieves on arbitrary instances of maximum coverage.

Categories and Subject Descriptors

F.2.2 [**Analysis of Algorithms and problem complexity**]: Nonnumerical Algorithms and Problems—*Geometrical problems and computations*

General Terms

Algorithms, Theory

*Supported by NSF grants AF-0910940 and IIS-0905467.

†Supported by NSF grants AF-0910940, CCF-0643934, and IIS-0905467, AFOSR grant FA9550-09-1-0100, a Microsoft Research New Faculty Fellowship, a Google Research Grant, and an Alfred P. Sloan Foundation Fellowship.

SCG'12, June 17–20, 2012, Chapel Hill, North Carolina, USA.

Keywords

Fixed-parameter tractable, Low VC-dimension

1. INTRODUCTION

The MAXIMUM COVERAGE problem is one of the classical NP-hard combinatorial optimization problems. An instance of MAXIMUM COVERAGE is specified by a triple $(U, \mathcal{R}, k)$ where U is a finite set, $\mathcal{R}$ is a collection of subsets of U, and k is a positive integer. The objective is to output a k-tuple of elements of $\mathcal{R}$ such that their union contains as many elements as possible. (In the *weighted* version of the problem, elements $x \in U$ have non-negative weights $w(x)$ and the objective is to maximize the combined weight of elements in the union.) A very natural greedy algorithm for MAXIMUM COVERAGE chooses k sets sequentially, where each new set is chosen to maximize the number (or combined weight) of elements covered by the new set but not by any of the preceding ones. It has been known for decades that this algorithm has an approximation factor of $(1-\frac{1}{e})$ [6]; in fact, it is a special case of the greedy algorithm for maximizing a monotone submodular function subject to a cardinality constraint [15], and the algorithm's approximation factor remains $(1-\frac{1}{e})$ even in this more general case. It was shown by Feige [8] that this approximation factor is the best possible, even for the unweighted MAXIMUM COVERAGE problem, unless P=NP.

One of the reasons the greedy algorithm for MAXIMUM COVERAGE is so widely studied is that it has innumerable applications: originally introduced by Cornuejols, Fisher, and Nemhauser [5] to model the problem of locating bank accounts in multiple cities to maximize "float," it was subsequently applied in databases [10], social networks [12], sensor placement [13], information retrieval [18, 20], and numerous other areas. A prototypical application arises in the information retrieval setting, when considering the problem of assembling a list of k documents to satisfy the information needs of as many users as possible. Equating every document with the set of users whom it satisfies, we see that this information retrieval problem is modeled by the MAXIMUM COVERAGE problem.

Given the extremely broad applicability of MAXIMUM COVERAGE problems, it is natural to wonder whether the approximation ratio of $1-\frac{1}{e}$ is the strongest theoretical guarantee one can hope for. Feige's hardness result eliminates the possibility of obtaining a better worst-case approximation ratio in polynomial-time, but the problem instances arising in applications are unlikely to resemble worst-case instances of MAXIMUM COVERAGE. Is it possible to identify broad classes of MAXIMUM COVERAGE instances (hopefully resembling those that arise in practice) such that the greedy algorithm provably achieves an approximation factor better than $1 - \frac{1}{e}$ on these instances? If not, can one design a different polynomial-time algorithm with an improved approximation factor? These are the questions that inspired our paper.

Let us reconsider the problem of assembling a top-k list of documents, mentioned above, in light of these questions. At least two aspects of this application distinguish it from an arbitrary instance of MAXIMUM COVERAGE.

(1) The value k is quite small compared to n, the input size. A typical instance might involve processing a list of thousands or millions of documents to extract a list of $k = 10$ top choices.

(2) The set system $(U, \mathcal{R})$ is likely to have a "low-dimensional" structure. For example, a natural model of users' preferences might assume that there are $d \ll n$ topics, a document contains a mix of topics described by a vector in $\mathbb{R}^d_+$, and the user's information need is satisfied if the dot product of this vector with another vector in $\mathbb{R}^d_+$ (describing the mix of topics the user seeks to read about) exceeds some threshold.

Is the approximation ratio of the greedy algorithm better than $1 - \frac{1}{e}$ under these circumstances? If not, is there some other algorithm that is significantly better?

We answer the first question negatively and the second one affirmatively. More precisely, for $d \geq 4$ we show that the greedy algorithm's approximation ratio in this special case is no better than its worst-case approximation ratio of $1 - \left(1 - \frac{1}{k}\right)^k \approx 1 - \frac{1}{e}$, but that there is an algorithm with running time $O(f(\epsilon, k, d) \cdot poly(n))$ whose approximation factor is $(1 - \epsilon)$, for some function $f(\epsilon, k, d)$. (Of course, for *very* small values of k a trivial brute-force search over all collections of k sets in $\mathcal{R}$ finds an exactly optimal solution in $O(n^{k+1})$ time, but a fixed-parameter algorithm whose running time is exponential in k but quadratic in n is vastly faster when $k = 10$ and $n = 10^6$, for instance.)

The following subsection describes our contributions in more detail.

1.1 Our contributions

Our main contribution is a fixed-parameter approximation scheme (fpt-AS) for the MAXIMUM COVERAGE problem, parameterized by the number of sets k, the approximation parameter ϵ, and the VC-dimension of the set system, d. Letting n denote the problem size — i.e. the sum of cardinalities of all the sets in $\mathcal{R}$ — the approximation scheme has running time $O(f(\epsilon, k, d) \cdot poly(n))$, where $f(\epsilon, k, d) = \exp\left(\tilde{O}(k^2 d\epsilon^{-5})\right)$[1]. The algorithm, which is presented in Section 3, is based on three ingredients. First, set systems of bounded VC-dimension have bounded-size ϵ-approximations (see Section 3 for definitions) and there is even a deterministic algorithm to find them in linear time [4]. Second, this means it is easy to design a fpt-AS for the special case of MAXIMUM COVERAGE in which the set system has bounded VC-dimension *and* the optimum solution covers a constant fraction of the elements. Third, the general case can be reduced to this special case by an intricate non-deterministic algorithm, which can then be made deterministic at the cost of blowing up the running time by a factor that is exponential in $k^2 d\epsilon^{-5}$, but independent of n.

In Section 5 we show that various aspects of this result cannot be improved, under standard complexity assumptions. First, the function $f(\epsilon, k, d)$ cannot be replaced by a function depending polynomially on k unless $P = NP$. Second, it cannot be replaced by a function depending polynomially on $1/\epsilon$ unless $P = W[1]$. (The question of whether the exponential dependence on d can be eliminated is intriguing, but it is unlikely to be easily resolvable since fixed-parameter complexity theory lacks machinery analogous to the PCP Theorem for proving $W[1]$-hardness of approximation.) Furthermore, these hardness results apply even in some very simple cases: MAXIMUM COVERAGE with set systems of VC-dimension 2, or with halfspaces in dimension 4, or with rectangles in dimension 2. Moreover, in all three of these special cases, the greedy algorithm fails to achieve an approximation factor better than $1 - \frac{1}{e}$.

These negative results about the greedy algorithm are counterbalanced by some positive results that we present in Section 4. We identify a parameter of the problem instance — the *covering multiplicity*, denoted by r — such that the greedy algorithm's approximation factor is never worse than $1 - (1 - \frac{1}{r})^r$. The covering multiplicity satisfies $r \leq k$, and when the inequality is strict this improves upon the worst-case approximation bound for the greedy algorithm. For problem instances whose sets have cardinality at most r, the covering multiplicity is bounded by r, and for instances in which the sets are two-dimensional halfspaces the covering multiplicity is bounded by 2, implying that the greedy algorithm is a $\frac{3}{4}$-approximation in the latter case.

1.2 Related work

As mentioned above, the MAXIMUM COVERAGE problem was introduced, and the greedy algorithm first analyzed, by Cornuejols et al. in [5]. This work was subsequently generalized to the context of submodular functions by Nemhauser et al. [15]. A matching hardness of approximation for MAXIMUM COVERAGE was obtained by Feige [8], in a paper that also settled the approximation hardness of the closely related SET COVER problem, establishing that the greedy algorithm achieves the optimal approximation ratio (up to lower order terms) for both problems.

The approximability of special cases of SET COVER and MAXIMUM COVERAGE was subsequently investigated in numerous papers. For example, the MAXIMUM VERTEX COVERAGE problem is the special case of MAXIMUM COVERAGE in which U is the edge set of a graph and every set in $\mathcal{R}$ is the set of edges incident to one vertex of that graph. This special case of the problem was shown to be APX-hard by Petrank [17]. A landmark paper by Ageev and Sviridenko [1] introduced the technique of *pipage rounding* and used it to give a (non-greedy) polynomial-time $\frac{3}{4}$-approximation algorithm for MAXIMUM VERTEX COVERAGE; more generally, they gave a polynomial-time algorithm with approximation factor $\left(1 - (1 - \frac{1}{k})^k\right)$ for the special case in which every element of U belongs to at most k sets in $\mathcal{R}$.

Computational geometers have intensively studied special cases of SET COVER, or the dual problem of HITTING SET, when the set system is defined geometrically, e.g. by rectangles, disks, or halfspaces. A seminal paper by Bronniman and Goodrich [3] introduced a multiplicative-weights

[1] $\tilde{O}$ hiding log factors

method for approximating Hitting Set, and applied this method to design constant-factor approximation algorithms for various classes of bounded VC-dimensional set systems, e.g. disks in the plane. The weighted case of these problems turns out to be much more challenging; see [9, 19]. A breakthrough paper by Mustafa and Ray [11] presented a new method to analyze local search algorithms for geometric hitting set problems, thereby proving that local search yields a PTAS for many interesting special cases such as three-dimensional halfspaces.

The study of fixed-parameter approximation schemes — and fixed-parameter approximation algorithms more generally — is still in its youth. An excellent survey by Marx [14] includes an fpt-AS for Maximum Vertex Coverage (also known as Partial Vertex Cover), a problem which is a special case of bounded VC-dimensional Maximum Coverage. Thus, one consequence of our algorithm in Section 3 is an alternative fpt-AS for Partial Vertex Cover, although the techniques underlying our algorithm are very different from those in Marx's algorithm.

2. PRELIMINARIES

An instance of the Maximum Coverage problem is specified by a finite set U, a collection of subsets $\mathcal{R}$, and a positive integer k. We will assume that the input is specified by simply listing the elements of U and those of each set in $\mathcal{R}$; thus, the problem size is $n = \sum_{R \in \mathcal{R}} |R|$. In the Weighted Maximum Coverage problem, we are also given a function $w : U \to \mathbb{R}_+$; the weight of a set $S \subseteq U$ is defined to be $w(S) = \sum_{x \in S} w(x)$ and the goal is to output a k-tuple of elements of $\mathcal{R}$ whose union has maximum weight. We will denote this maximum by $OPT(U)$.

For $A \subseteq U$, we will use the notation $\mathcal{R}|_A$ to denote the collection of all subsets $B \subseteq A$ of the form $B = A \cap R$, where $R \in \mathcal{R}$. The set A is *shattered* by $\mathcal{R}$ if $\mathcal{R}|_A$ is equal to 2^A, the collection of all subsets of A. The VC-dimension of $(U, \mathcal{R})$ is the cardinality of the largest set that is shattered by $\mathcal{R}$. If $(U, \mathcal{R})$ has VC-dimension d and $A \subseteq U$, it is known that $|\mathcal{R}|_A|$ is bounded above by $O\left(|A|^d\right)$.

Our focus will be on Maximum Coverage problems such that $(U, \mathcal{R})$ has bounded VC-dimension. Among these, two special cases of particular interest are Maximum Halfspace Coverage— in which U is a subset of $\mathbb{R}^d$ and each of the sets in $\mathcal{R}$ is the intersection of a halfspace with U — and Maximum Rectangle Coverage, in which U is again a subset of $\mathbb{R}^d$ and each of the sets in $\mathcal{R}$ is obtained by intersecting an axis-parallel rectangle with U.

3. FIXED-PARAMETER APPROXIMATION SCHEME

In this section, we work with the *unweighted* Maximum Coverage problem. Following Chazelle and Matoušek, we assume that $\mathcal{R}$ is represented by a *subsystem oracle of dimension d*, defined as follows.

Definition 3.1 ([4]). *A subsystem oracle of dimension d for a set system $(U, \mathcal{R})$ is an algorithm which, given a subset $A \subseteq U$, returns a list of all sets in $\mathcal{R}|_A$ in time $O\left(|A|^{d+1}\right)$; the number of sets belonging to this list must also be bounded above by $O\left(|A|^d\right)$.*

The following fact is obvious but useful: a subsystem oracle of dimension d for $(U, \mathcal{R})$ also constitutes a subsystem oracle of dimension d for $(V, \mathcal{R}|_V)$, for every subset $V \subseteq U$.

We define a set $A \subseteq U$ to be an ϵ-approximation of $(U, \mathcal{R})$ if the inequality

$$\left| \frac{|A \cap R|}{|A|} - \frac{|R|}{|U|} \right| \leq \epsilon$$

holds for all $R \in \mathcal{R}$. A crucial ingredient of our approximation scheme is an algorithm, due to Chazelle and Matoušek [4], that computes an ϵ-approximation of cardinality $O\left(d\epsilon^{-2}\log(d/\epsilon)\right)$ for $(U, \mathcal{R})$ in time $O\left(d^{3d}\epsilon^{-2d}\log^d(d/\epsilon)n\right)$, given a subsystem oracle of dimension d for $(U, \mathcal{R})$.

Let $\mathcal{R}^{\cup k}$ denote the collection of all sets $R_1 \cup \cdots \cup R_k$ such that $R_1, \ldots, R_k \in \mathcal{R}$. To apply Chazelle and Matoušek's algorithm, we will need a subsystem oracle for $\mathcal{R}^{\cup k}$. The existence of such an oracle is ensured by the following lemma.

Lemma 3.2. *If $(U, \mathcal{R})$ has a subsystem oracle of dimension d, then for all $k > 0$, $(U, \mathcal{R}^{\cup k})$ has a subsystem oracle of dimension kd.*

Proof. The proof is by induction on k, the base case $k = 1$ being trivial. Given subsystem oracles for $(U, \mathcal{R})$ and $(U, \mathcal{R}^{\cup k-1})$ of dimensions d and $(k-1)d$, respectively, the following simple algorithm constitutes a subsystem oracle for $(U, \mathcal{R}^{\cup k})$. First, we use the given two subsystem oracles to generate lists $\mathcal{Q}_1$ and $\mathcal{Q}_{k-1}$, consisting of all sets in $\mathcal{R}|_A$ and $\mathcal{R}^{\cup k-1}|_A$, respectively. Letting $a = |A|$, the induction hypothesis implies that $|\mathcal{Q}_1| = O(a^d)$ and $|\mathcal{Q}_{k-1}| = O(a^{(k-1)d})$, and that the two lists are generated in time $O(a^{d+1})$ and $O(a^{(k-1)d+1})$, respectively. Now, for every pair $B_1 \in \mathcal{Q}_1$ and $B_{k-1} \in \mathcal{Q}_{k-1}$, we form the set $B = B_1 \cup B_{k-1}$ and add it to $\mathcal{Q}_k$. There are $O(a^{kd})$ such pairs, and for each pair the union can be computed in $O(a)$ time, so the algorithm runs in time $O(a^{kd+1})$, as desired. □

As an easy consequence, we derive that the Maximum Coverage problem has a fpt-AS when $(U, \mathcal{R})$ has a bounded-dimensional subsystem oracle *and* the optimum is a constant fraction of $|U|$.

Lemma 3.3. *For any constants $c, \delta > 0$, consider the* Maximum Coverage *problem, restricted to set systems $(U, \mathcal{R})$ having a subsystem oracle of dimension d and satisfying $OPT(U) \geq c|U|$. This special case of the* Maximum Coverage *problem has a $\left(\frac{c-2\delta}{c}\right)$-approximation algorithm with running time bounded by $O(d^{3kd}k^{3kd}\delta^{-2kd-2}\log^{kd+1}(kd/\delta)n)$.*

Proof. The set system $(U, \mathcal{R}^{\cup k})$ has a subsystem oracle of dimension kd, so it is possible to compute a set $A \subseteq U$ which is a δ-approximation to $(U, \mathcal{R}^{\cup k})$, in time $O(d^{3kd}k^{3kd}\delta^{-2kd}\log^{kd}(kd/\delta)n)$. Furthermore, the cardinality of A is $O(kd\delta^{-2}\log(kd/\delta))$. We can solve the Maximum Coverage problem for the set system $(A, \mathcal{R}|_A)$ by brute force. First we call the subsystem oracle to obtain a list of all the sets in $\mathcal{R}|_A$; there are at most $O(k^d d^d \delta^{-2d}\log^d(kd/\delta))$ such sets. Then we enumerate all k-tuples of sets in this list, compute their union, and output the k-tuple whose union has the largest cardinality. Computing the union of k sets requires $O(k|A|) = O(k^2 d\delta^{-2}\log(kd/\delta))$ time, and multiplying this by the number of k-tuples we obtain an overall running time of $O(k^{kd+2}d^{kd+1}\delta^{-2kd-2}\log^{kd+1}(kd/\delta))$.

Let $R_1, \ldots, R_k$ be sets in $\mathcal{R}$ whose restrictions to A constitute an optimal solution of the Maximum Coverage problem for $(A, \mathcal{R}|_A)$. Let $S_1, \ldots, S_k$ be an optimal solution of the MAXIMUM COVERAGE problem for $(U, \mathcal{R})$. We have

$$\frac{|(R_1 \cup \cdots \cup R_k) \cap A|}{|A|} \geq \frac{|(S_1 \cup \cdots \cup S_k) \cap A|}{|A|}$$
$$\frac{|R_1 \cup \cdots \cup R_k|}{|U|} \geq \frac{|S_1 \cup \cdots \cup S_k|}{|U|} - 2\delta$$
$$\frac{|R_1 \cup \cdots \cup R_k|}{|S_1 \cup \cdots \cup S_k|} \geq 1 - 2\delta\left(\frac{|U|}{|S_1 \cup \cdots \cup S_k|}\right)$$
$$\geq 1 - \frac{2\delta}{c} = \frac{c-2\delta}{c}$$

where the first line follows from the construction of $R_1, \ldots, R_k$, the second line follows because A is an δ-approximation for $(U, \mathcal{R}^{\cup k})$, and the third line follows from our assumption that $|S_1 \cup \cdots \cup S_k| = OPT(U) \geq c|U|$. $\square$

For the remainder of this section, we work on eliminating the assumption that $OPT(U) \geq c|U|$. Our plan of attack is to perform a preprocessing step that extracts a subset $V \subseteq U$ such that $OPT(V) \geq (1-\epsilon/3)OPT(U)$ and $OPT(V) \geq c|V|$, for a constant $c = c(\epsilon, k)$ depending only on ϵ and k. Then we will run the algorithm from Lemma 3.3 on $(V, \mathcal{R}|_V)$, using an appropriate choice of $\delta = \delta(\epsilon, k)$, to obtain a $(1-\epsilon)$-approximation to $OPT(U)$.

To design and analyze the preprocessing algorithm that constructs V, we must first define a new problem that we call CONSTRAINED MAXIMUM COVERAGE and analyze a simple greedy algorithm for the problem.

DEFINITION 3.4. *An instance of the* CONSTRAINED MAXIMUM COVERAGE *problem is specified by a universe U and k collections of sets $\mathcal{R}_1, \ldots, \mathcal{R}_k \subseteq 2^U$. A solution of the problem is specified by designating a k-tuple of sets $R_1, \ldots, R_k$ such that $R_i \in \mathcal{R}_i$ for $i = 1, \ldots, k$. The objective is to maximize $|R_1 \cup \cdots \cup R_k|$.*

The greedy algorithm for CONSTRAINED MAXIMUM COVERAGE *selects $R_1, R_2, \ldots, R_k$, in that order, by choosing R_1 to be the maximum-cardinality set in $\mathcal{R}_1$ and, for $i > 2$, choosing R_i to be the set in $\mathcal{R}_i$ that maximizes $|R_i \setminus (R_1 \cup \cdots \cup R_{i-1})|$.*

Note that MAXIMUM COVERAGE is the special case of CONSTRAINED MAXIMUM k-COVERAGE in which $\mathcal{R}_1 = \cdots = \mathcal{R}_k$, and that the greedy algorithm specializes, in that case, to the familiar greedy algorithm for maximum k-coverage. The approximation ratio of the greedy algorithm for CONSTRAINED MAXIMUM k-COVERAGE is not equal to $1 - \frac{1}{e}$ in general; in fact it is equal to $\frac{1}{2}$. However, for our purposes the following property of the greedy algorithm will be more crucial to the analysis.

LEMMA 3.5. *Given an instance of the* CONSTRAINED MAXIMUM COVERAGE *problem, let $R_1, \ldots, R_k$ be the sets selected by the greedy algorithm and let $S_1, \ldots, S_k$ be any other solution. Let $\mathbf{R} = R_1 \cup \cdots \cup R_k$ and $\mathbf{S} = S_1 \cup \cdots \cup S_k$. For every $\delta > 0$, at least one of the following two alternatives holds.*

1. $|\mathbf{R}| \geq (1-\delta)|\mathbf{S}|$.
2. $|\mathbf{S} \setminus \mathbf{R}| < (1-\delta)|\mathbf{S}|$.

PROOF. We will construct a one-to-one mapping from $\mathbf{S} \setminus \mathbf{R}$ into $\mathbf{R}$. This suffices to prove the lemma, since either $|\mathbf{S} \setminus \mathbf{R}| < (1-\delta)|\mathbf{S}|$ or $|\mathbf{S} \setminus \mathbf{R}| \geq (1-\delta)|\mathbf{S}|$, and in the latter case our one-to-one mapping will certify that

$$|\mathbf{R}| \geq |\mathbf{S} \setminus \mathbf{R}| \geq (1-\delta)|\mathbf{S}|.$$

To construct the one-to-one mapping, partition $\mathbf{S} \setminus \mathbf{R}$ into k sets $T_1, T_2, \ldots, T_k$, where $T_i = S_i \setminus (\mathbf{R} \cup S_1 \cup S_2 \cup \cdots \cup S_{i-1})$. Note that T_i is a subset of $S_i \setminus (R_1 \cup \cdots \cup R_{i-1})$, hence

$$|T_i| \leq |S_i \setminus (R_1 \cup \cdots \cup R_{i-1})| \leq |R_i \setminus (R_1 \cup \cdots \cup R_{i-1})|,$$

where the second inequality follows from the definition of the greedy algorithm. This means that there is a one-to-one mapping from T_i to $R_i \setminus (R_1 \cup \cdots \cup R_{i-1})$. Combining these one-to-one mappings gives us the desired one-to-one mapping from $\mathbf{S} \setminus \mathbf{R} = \coprod_{i=1}^k T_i$ into $\mathbf{R} = \coprod_{i=1}^k R_i \setminus (R_1 \cup \cdots \cup R_{i-1})$. $\square$

We now describe and analyze a non-deterministic algorithm to solve MAXIMUM COVERAGE on a set system $(U, \mathcal{R})$, given a subsystem oracle of dimension d; later we will make the algorithm deterministic. The algorithm proceeds in a sequence of phases numbered $1, \ldots, p = \lceil \frac{6}{\epsilon} \ln\left(\frac{3}{\epsilon}\right) \rceil$. In each phase q, it chooses a k-tuple of sets $R_1^q, \ldots, R_k^q$. Let

$$\mathbf{R}^q = \bigcup_{i=1}^{q} \bigcup_{j=1}^{k} R_j^i.$$

In phase q, the algorithm computes a set A^q, of cardinality $O(kd\epsilon^{-2}p^2 \log(kdp/\epsilon))$, which is an $(\epsilon/6p)$-approximation to $(\mathbf{R}^{q-1}, \mathcal{R}^{\cup k})$. It non-deterministically guesses a sequence of k subsets $B_1^q, \ldots, B_k^q \subseteq A^q$ and defines set systems $\mathcal{R}_1^q, \ldots, \mathcal{R}_k^q$ as

$$\mathcal{R}_i^q = \{R \in \mathcal{R} \mid R \cap A^q = B_i^q\}, \quad i = 1, \ldots, k.$$

It then selects the sets $R_1^q, \ldots, R_k^q$ using the greedy algorithm for CONSTRAINED MAXIMUM COVERAGE, applied to the universe $U \setminus \mathbf{R}^{q-1}$ with set systems $\mathcal{R}_1^q, \ldots, \mathcal{R}_k^q$. After repeating this process for $p = \lceil \frac{6}{\epsilon} \ln\left(\frac{3}{\epsilon}\right) \rceil$ phases, it defines $V = \mathbf{R}^p = \bigcup_{i=1}^p \bigcup_{j=1}^k R_j^i$. Setting $c = 1/p$ and

$$\delta = \frac{\epsilon}{6p}\left(1 + \frac{\epsilon}{3}\right)^{-1},$$

so that $(c-2\delta)/c \geq 1-\epsilon/3$, it runs the algorithm of Lemma 3.3 on the set system $(V, \mathcal{R})$ to find a $(1-\epsilon/3)$-approximation to the optimum of the MAXIMUM COVERAGE problem for $(V, \mathcal{R})$.

We aim to prove that there exists an execution of this non-deterministic algorithm that yields a $(1-\epsilon)$-approximation to the optimum of the MAXIMUM COVERAGE problem for $(U, \mathcal{R})$. If our algorithm produces a set V satisfying $OPT(V) \geq c|V| = |V|/p$ and $OPT(V) \geq (1-\epsilon/3)OPT(U)$, then Lemma 3.3 ensures that we finish up by producing a $(1-\epsilon/3)$-approximation to $OPT(V)$, which will also be a $(1-\epsilon/3)^2 \geq (1-\epsilon)$-approximation to $OPT(U)$. Proving that $OPT(V) \geq |V|/p$ is easy: $V = \mathbf{R}^p$ is the union of p sets $\mathbf{R}^q \setminus \mathbf{R}^{q-1}$, each of which has cardinality at most $OPT(V)$ since it can be covered by the k sets $R_1^q, \ldots, R_k^q$.

To prove that there exists an execution yielding a set V such that $OPT(V) \geq (1-\epsilon/3)OPT(U)$, we use Lemma 3.5. Let $S_1, \ldots, S_k$ denote an optimal solution of the MAXIMUM COVERAGE problem for $(U, \mathcal{R})$. Consider the execution in which the algorithm's choice of B_i^q is equal to $S_i \cap A^q$ for every q, i. There are two cases to consider. First, suppose that exists a phase q such that

$$|(R_1^q \cup \cdots \cup R_k^q) \setminus \mathbf{R}^{q-1}| \geq \left(1 - \frac{\epsilon}{6}\right)|(S_1 \cup \cdots \cup S_k) \setminus \mathbf{R}^{q-1}|. \quad (1)$$

Recall that $\mathcal{R}_i^q = \{R \in \mathcal{R} \mid R \cap A^q = B_i^q\}$, and that we are assuming $B_i^q = A^q \cap S_i$. Hence, we have $R_i^q \cap A^q = S_i \cap A^q$

for all i and, consequently, $(R_1^q \cup \cdots \cup R_k^q) \cap A^q = (S_1 \cup \cdots \cup S_k) \cap A^q$. Using the fact that A^q is an $(\epsilon/6p)$-approximation for $(\mathbf{R}^{q-1}, \mathcal{R}^{\cup k})$, we now obtain

$$|(R_1^q \cup \cdots \cup R_k^q) \cap \mathbf{R}^{q-1}| \geq |(S_1 \cup \cdots \cup S_k) \cap \mathbf{R}^{q-1}| - \frac{\epsilon}{6p}|\mathbf{R}^{q-1}|. \quad (2)$$

Letting $\mathbf{S}$ denote $S_1 \cup \cdots \cup S_k$, we sum (1) and (2) to obtain

$$\begin{aligned} |R_1^q \cup \cdots \cup R_k^q| &\geq (1 - \tfrac{\epsilon}{6})\,|\mathbf{S} \setminus \mathbf{R}^{q-1}| + |\mathbf{S} \cap \mathbf{R}^{q-1}| - \tfrac{\epsilon}{6p}|\mathbf{R}^{q-1}| \\ &= |\mathbf{S}| - \frac{\epsilon}{6}|\mathbf{S} \setminus \mathbf{R}^{q-1}| - \frac{\epsilon}{6p}|\mathbf{R}^{q-1}|. \quad (3) \end{aligned}$$

Now, as above, $\mathbf{R}^{q-1}$ can be partitioned into sets $\mathbf{R}^i \setminus \mathbf{R}^{i-1}$, $(i = 1, \ldots, q-1)$, each having cardinality at most $OPT(U) = |\mathbf{S}|$. The number of pieces of the partition is $q - 1 < p$, so $\frac{1}{p}|\mathbf{R}^{q-1}| \leq |S|$. Substituting this back into (3), we obtain (4) as desired.

$$\begin{aligned} OPT(V) \geq & \quad |R_1^q \cup \cdots \cup R_k^q| \\ \geq & \quad (1 - \tfrac{\epsilon}{6} - \tfrac{\epsilon}{6})\,|S| = (1 - \tfrac{\epsilon}{3})\,OPT(U), \quad (4) \end{aligned}$$

Finally, there remains the case that (1) is not satisfied by any q. Then Lemma 3.5 implies that

$$|(S_1 \cup \cdots \cup S_k) \setminus \mathbf{R}^q| < \left(1 - \frac{\epsilon}{6}\right) |(S_1 \cup \cdots \cup S_k) \setminus \mathbf{R}^{q-1}| \quad (5)$$

for all q. Combining (5) for $q = 1, \ldots, p$, we get that

$$|(S_1 \cup \cdots \cup S_k) \setminus \mathbf{R}^p| < \left(1 - \frac{\epsilon}{6}\right)^p |S_1 \cup \cdots \cup S_k| \leq \frac{\epsilon}{3}|S_1 \cup \cdots \cup S_k|,$$

which implies that

$$|(S_1 \cup \cdots \cup S_k) \cap \mathbf{R}^p| > \left(1 - \frac{\epsilon}{3}\right) |S_1 \cup \cdots \cup S_k|,$$

and hence $OPT(V) \geq (1 - \epsilon/3)OPT(U)$ since $V = \mathbf{R}^p$.

To turn the non-deterministic algorithm into a deterministic one, we simply run every possible execution of the non-deterministic algorithm and output the best answer. An execution of the non-deterministic algorithm is determined by the choice of sets B_i^q, $(1 \leq q \leq p, 1 \leq i \leq k)$. Recall that B_i^q must be a subset of A^q and that $|A^q| = O(kd\epsilon^{-2}p^2 \log(kdp/\epsilon))$. Hence if $N(k, d, \epsilon)$ denotes the number of executions of the non-deterministic algorithm, it satisfies

$$\begin{aligned} N(k, d, \epsilon) &= \prod_{q=1}^{p} \prod_{i=1}^{k} 2^{|A_q|} \\ \log N(k, d, \epsilon) &\leq pk \cdot O(kd\epsilon^{-2}p^2 \log(kdp/\epsilon)) \\ &= O(k^2 d\epsilon^{-2}p^3 \log(kdp/\epsilon)) \\ &= \tilde{O}(k^2 d\epsilon^{-5}) \end{aligned}$$

Each iteration runs in time $O(g(k, d, \epsilon) \cdot n)$ where $\log g(k, d, \epsilon) = O(kd \log(kd/\epsilon))$. Hence, the algorithm's overall running time is $O(f(k, d, \epsilon) \cdot n)$ where

$$\log f(k, d, \epsilon) = \log N(k, d, \epsilon) + \log g(k, d, \epsilon) = \tilde{O}(k^2 d\epsilon^{-5}).$$

In deriving this bound on the algorithm's running time, we have assumed that $(U, \mathcal{R})$ has a subsystem oracle of dimension d. If we instead assume that $(U, \mathcal{R})$ has VC dimension d and is represented in the input by simply listing all the elements of $\mathcal{R}$, the running time increases by a factor of n. This is because the trivial implementation of a subsystem oracle — computing $\mathcal{R}_A$ by enumerating each set of $\mathcal{R}$ and intersecting it with A — has running time $O(|A|^{d+1}n)$, n times slower than the bound required by the definition of a subsystem oracle.

4. BOUNDED COVERING MULTIPLICITY

In this section we show that the greedy algorithm gives a $1 - (1 - 1/r)^r$-approximate solution when the *covering multiplicity* of the set system is at most r.

DEFINITION 4.1. *An instance of the maximum coverage problem $(U, \mathcal{R}, k)$ has* covering multiplicity r *if for every k-tuple of sets $a_1, \ldots, a_k \in \mathcal{R}$ there exists an optimal solution $(o_1, \ldots, o_k)$ of the maximum coverage problem, with union denoted by OPT, such that each of the sets $a_i \cap OPT$ for $1 \leq i \leq k$ is contained in the union of r elements of $\{o_1, \ldots, o_k\}$.*

One of the interesting special cases which satisfies this property is when the cardinality of every set in $\mathcal{R}$ is bounded by r. In Appendix A we prove that it is also satisfied (with $r = 2$) when $U \subset \mathbb{R}^2$ and $\mathcal{R}$ consists of halfspaces in $\mathbb{R}^2$.

Let $g_1, g_2, \ldots, g_k$ be the k sets choosen by the greedy algorithm in the order that they are choosen. Let w be the coverage function and $o_1, o_2, \ldots, o_k$ be the k sets choosen by OPT.

THEOREM 4.2. *Greedy algorithm is a $1 - (1 - 1/r)^r$ approximation algorithm for* MAXIMUM COVERAGE *with* covering multiplicity r.

COROLLARY 4.3. *Greedy algorithm is a $1 - (1 - 1/r)^r$ approximation algorithm for* MAXIMUM COVERAGE *with each set having cardinality at most r.*

4.1 Reduction to a special case

For simplifying the analysis we first argue that we can consider the following special case without loss of generality. We take the problem instance on which the greedy algorithm (which we henceforth abbreviate as "greedy") has a given approximation ratio and convert it into a special instance with no better approximation ratio. Then it is enough to analyze the special case.

- The sets chosen by greedy are different from the optimal sets. This assumption can be made as we can just duplicate the sets. Note that this does not change the covering multiplicity.
- The sets chosen by greedy are disjoint. This is because if one defines a new problem instance with $\tilde{g}_i = g_i \setminus (\cup_{j=1}^{i-1} g_j)$ then the values of the optimal solution and the greedy solution are unchanged. Note that this step uses the fact that the sets chosen by greedy do not belong to the optimal solution. Also note that this does not change the covering multiplicity since we are not modifying any sets in the optimal solution.
- Let $o_1, o_2, \ldots, o_k$ be any optimal solution such that each set $g_i \cap OPT$ $(1 \leq i \leq k)$ is contained in the union of r elements of $\{o_1, \ldots, o_k\}$. We can assume that these sets o_i are pairwise disjoint. This is because we can define a new problem instance in which every point belonging to two or more of the sets in $\{o_1, \ldots, o_k\}$ is assigned to one of those sets and deleted from the others. The values of the greedy and optimal solutions are unchanged. To preserve the property that each set $g_i \cap OPT$ $(1 \leq i \leq k)$ is contained in the union of r

elements of $\{o_1, \ldots, o_k\}$, we simply ensure that every element of g_i is assigned to one of those r sets, for all i. This is possible due to our previous assumption that the sets $g_1, \ldots, g_k$ are disjoint.

4.2 Simple case

Consider the simple case $k = t \cdot r$ for some integer t. We will prove the approximation for this special case to get some intuition. We will do it in steps.

- Let $x_i = \sum_{j=(i-1)\cdot t+1}^{i \cdot t} w(g_i)$.
- Let $o_1, o_2, \ldots, o_k$ be the optimal sets in decreasing order of $w(o_i)$.
- Note that $w(g_1) \geq w(o_1)$, $w(g_2) \geq w(o_{r+1})$, $w(g_3) \geq w(o_{2r+1}), \ldots, w(g_t) \geq w(o_{(t-1)\cdot r+1})$. These inequalities use the fact that the covering multiplicity is r and the sets $o_1, \ldots, o_k$ are disjoint. Now summing the t terms we get $\sum_{i=1}^{t} w(g_i) \geq \sum_{i=1}^{t} w(o_{(i-1)r+1}) \geq \frac{1}{r} w(OPT)$.
- Repeating the above step on the residual problem we get $\sum_{i=t+1}^{2t} w(g_i) \geq \frac{OPT - x_1}{r}$. Similarly we get the following series of equations.

$$\forall 1 \leq l \leq r-1, \sum_{i=l\cdot t+1}^{(l+1)t} w(g_i) \geq \frac{OPT - \sum_{i=1}^{l} x_i}{r} \qquad (6)$$

- Multiplying (6) by $(1-1/r)^{r-l-1}$ and summing we get $\sum_{i=1}^{i=k} w(g_i) \geq \left(1 - \left(1 - \frac{1}{r}\right)^r\right) w(OPT)$.

4.3 General case

Let $k = t \cdot r + q$ for some $0 \leq q \leq r-1$. We will use the following lemma in the proof.

LEMMA 4.4. *$\forall 1 \leq l \leq r, 0 \leq z \leq q \leq r$ we have $r \cdot \sum_{m=0}^{z} \binom{l-1}{m}\binom{r-l}{q-m-1} \geq q \cdot \sum_{m=0}^{z} \binom{l-1}{m}\binom{r-l+1}{q-m}$*

PROOF. Consider $p(z) = \frac{\sum_{m=0}^{z} \binom{l-1}{m}\binom{r-l}{q-m-1}}{\sum_{m=0}^{z} \binom{l-1}{m}\binom{r-l+1}{q-m}}$. Consider a random process in which q out of r bins are chosen uniformly at random (without replacement) and a ball is added to each one of the q bins. Now $p(z)$ represents the conditional probability that a ball is in the lth bin, given that at most z bins from the first $l-1$ are chosen. One can easily see that this function should be a decreasing function of z and hence $p(z) \geq p(q) = \frac{q}{r}$. □

Consider r bins and arrange the k greedy sets in the r bins with each bin having either t or $t+1$ greedy sets. Let bin 1 have the first t or $t+1$ sets, bin 2 have the second t or $t+1$ sets, and so on. Let σ be one such arrangement. We will apply inequalities similar to the simpler case. Let $x_i^\sigma = \sum_{g_j \in bin_i} w(g_j)$. Let $\sigma(l)$ denote the number of sets in the first l bins. Let x_{qmin}^t be the residual value of the qth minimum set among the optimal sets after the first t greedy sets are choosen. Note that x_{qmin}^t is a decreasing function of t. Let $B(t)$ denote the set of bins with t sets and $B(t+1)$ denote the set of bins with $t+1$ sets.

- Consider bin l with $t+1$ items. Then $\sum_{g_i \in bin_l} w(g_i) \geq \frac{OPT - \sum_{i=1}^{l-1} x_i^\sigma + (r-q) x_{qmin}^{\sigma(l-1)}}{r}$. This inequality is proved similar to inequality 6.
- Consider bin l with t items. Then $\sum_{g_i \in bin_l} w(g_i) \geq \frac{OPT - \sum_{i=1}^{l-1} x_i^\sigma - q \cdot x_{qmin}^{\sigma(l-1)}}{r}$. This inequality is proved similar to inequality 6.

Multiplying the equation corresponding to bin l with $(1 - 1/r)^{r-l}$ and summing we get equation 7 in Figure 1.[2]

Now taking the average over all arrangements σ we get equation 8 in Figure 1. Now using the fact that x_{qmin}^t is a decreasing function of t and Lemma 4.4 we get $w(greedy) \geq (1 - (1 - 1/r)^r) w(OPT)$.

5. LOWER BOUNDS

This section considers three different low-dimensional restrictions of the MAXIMUM COVERAGE problem: set systems of VC-dimension 2, halfspaces in $\mathbb{R}^4$, and axis-parallel rectangles in $\mathbb{R}^2$. In each case, we show that the problem is APX-hard and that the greedy algorithm's approximation ratio, restricted to that special case, is no better than its worst-case approximation ratio, $1 - \frac{1}{e}$.

All of these lower bounds are based on the MAXIMUM VERTEX COVERAGE problem, the special case of MAXIMUM COVERAGE in which each element of U belongs to exactly two sets in $\mathcal{R}$. In this special case, we can identify $\mathcal{R}$ with the vertex set of a graph G, and U with its edge set, such that the endpoints of the edge corresponding to $x \in U$ are the vertices that correspond to the two sets containing x. Thus, MAXIMUM VERTEX COVERAGE can be defined as the problem of choosing k vertices of a graph to maximize the number of edges they cover. The problem is known to be APX-hard [17] and it is known that the approximation ratio of the greedy algorithm, specialized to MAXIMUM VERTEX COVERAGE, is no better than in the general case [6]. In fact, the following lemma shows that the performance of the greedy algorithm does not improve when we further specialize to bipartite instances of MAXIMUM VERTEX COVERAGE.

LEMMA 5.1. *For any $\epsilon > 0$, there exist instances of* MAXIMUM VERTEX COVERAGE *in which the graph is bipartite, the instance has a vertex cover of size k, but the output of the greedy algorithm covers only $1 - \frac{1}{e} + \epsilon$ fraction of the edges.*

PROOF. We construct a bipartite graph with edge set $E = \{1, \ldots, kN\}$ (for some sufficiently large N) and vertex set $U \cup W$, where $k = |U| \leq |W|$. We refer to U and W as the *left* and *right* vertex sets, respectively.

Define a sequence of positive integers $n_0, n_1, n_2, \ldots$ by the formula $n_i = \left\lceil N \cdot \left(1 - \frac{1}{k}\right)^i \right\rceil + 1$ and let $s_i = \sum_{j=0}^{i-1} n_j$ denote the sequence of partial sums, interpreting s_0 to be 0. If $r = \min\{i \mid s_i \geq kN\}$ then $W = \{w_1, \ldots, w_r\}$, while $U = \{u_1, \ldots, u_k\}$. The right endpoint of edge j is the unique w_i such that $s_{i-1} < j \leq s_i$, while the left of endpoint of j is the unique u_i such that $i \equiv j \pmod{k}$.

By construction, $|U| = k$ and U is a vertex cover. Each element of U has exactly N elements. However, the greedy algorithm instead chooses vertices $w_1, \ldots, w_k$. To prove this by induction, observe that after choosing $w_1, \ldots, w_i$, the number of remaining uncovered edges is less than $kN\left(1 - \frac{1}{k}\right)^i$, and these edges are consecutively numbered. Each element of U covers a congruence class of edges, and therefore it covers fewer than $N\left(1 - \frac{1}{k}\right)^i + 1$ of the remaining edges, whereas

[2]The two-column format of this paper necessitated using a separate figure to typeset these equations.

$$w(greedy) \geq (1-(1-1/r)^r w(OPT) + \sum_{bin_l \in B(t+1)} \frac{r-q}{r} \cdot (1-1/r)^{r-l} x_{qmin}^{\sigma(l-1)} - \sum_{bin_l \in B(t)} \frac{q}{r} \cdot (1-1/r)^{r-l} x_{qmin}^{\sigma(l-1)} \quad (7)$$

$$\begin{aligned} w(greedy) &\geq (1-(1-1/r)^r w(OPT) + \textstyle\sum_l \frac{(1-1/r)^{r-l}}{\binom{r}{q}} \left(\frac{r-q}{r} \sum_{w=0}^{l-1} \binom{l-1}{w}\binom{r-l}{q-w-1} x_{qmin}^{(l-1)t+w} - \frac{q}{r} \sum_{w=0}^{l-1} \binom{l-1}{w}\binom{r-l}{q-w} x_{qmin}^{(l-1)t+w} \right) \\ &\geq (1-(1-1/r)^r w(OPT) + \sum_l \frac{(1-1/r)^{r-l}}{r \cdot \binom{r}{q}} \left(\sum_{w=0}^{l-1} (((r-q)\binom{l-1}{w}\binom{r-l}{q-w-1} - q\binom{l-1}{w}\binom{r-l}{q-w})x_{qmin}^{(l-1)t+w} \right) \\ &\geq (1-(1-1/r)^r w(OPT) + \sum_l \frac{(1-1/r)^{r-l}}{r \cdot \binom{r}{q}} \left(\sum_{w=0}^{l-1} ((r\binom{l-1}{w}\binom{r-l}{q-w-1} - q\binom{l-1}{w}\binom{r-l+1}{q-w})x_{qmin}^{(l-1)t+w} \right) \quad (8) \end{aligned}$$

Figure 1: Performance of greedy

w_{i+1} covers n_i edges and $n_i \geq N\left(1-\frac{1}{k}\right)^i + 1$. It follows that the greedy algorithm chooses w_{i+1} and this completes the induction step.

The number of edges covered by $w_1, \ldots, w_k$ is bounded above by $2k + N\sum_{i=0}^{k-1}\left(1-\frac{1}{k}\right)^i = 2k + kN\left[1-\left(1-\frac{1}{k}\right)^k\right]$. For k, N sufficiently large, this is less than $\left(1-\frac{1}{e}+\epsilon\right)kN$. □

THEOREM 5.2. *Each of the following special cases of* MAXIMUM COVERAGE *is APX-hard:*
(a) Set systems of VC-dimension $d \geq 2$.
(b) Halfspaces in $\mathbb{R}^d$, $d \geq 4$.
(c) Rectangular ranges in $\mathbb{R}^d$, $d \geq 2$.
Furthermore, the worst-case approximation ratio of the greedy algorithm, when restricted to any of these special cases, is $1-\frac{1}{e}$.

PROOF. Recall that MAXIMUM VERTEX COVERAGE can be defined as the instance of MAXIMUM COVERAGE in which every $x \in U$ belongs to exactly two sets in $\mathcal{R}$. Any such set system $(U, \mathcal{R})$ has VC-dimension at most 2: indeed, if $\mathcal{R}$ shatters a three-element set $\{x, y, z\}$ then there exist sets $R_1, \ldots, R_4$ in $\mathcal{R}$ whose intersections with $\{x, y, z\}$ are the sets $\{x\}$, $\{x, y\}$, $\{x, z\}$, $\{x, y, z\}$, respectively, and consequently x belongs to at least four distinct sets in $\mathcal{R}$. Thus, we see that MAXIMUM COVERAGE restricted to set systems of VC-dimension d includes MAXIMUM VERTEX COVERAGE as a special case, as long as $d \geq 2$. Part (a) of the theorem now follows from the fact that MAXIMUM VERTEX COVERAGE is APX-hard [17] and from Lemma 5.1.

To prove Part (b) we again show that MAXIMUM VERTEX COVERAGE is a special case. To do so, consider any graph with vertex set $V = \{v_1, \ldots, v_n\}$ and associate to each vertex $v_t \in V$ the vector $b_t = (t, t^2, t^3, t^4, 0, \ldots, 0) \in \mathbb{R}^d$. Define a halfspace $\mathbf{h}_t \subset \mathbb{R}^d$ by the inequality $v_t \cdot x \geq 1$. For every edge (v_r, v_s) we construct a vector $y_{rs} \in \mathbb{R}^d$ that belongs to $\mathbf{h}_r \cap \mathbf{h}_s$ but not to $\mathbf{h}_t$ for any $t \neq i, j$. The construction is as follows. First, write the polynomial $(z-r)^2(z-s)^2$ in the form $\sum_{i=0}^4 a_i z^i$, and then put

$$y_{rs} = -\frac{1}{a_0}(a_1, a_2, a_3, a_4, 0, \ldots, 0) \in R^d.$$

The inequality $y_{rs} \cdot v_t \geq 1$ can be rewritten as $-\sum_{i=1}^4 a_i t^i \geq a_0$ (using the fact that $a_0 = r^2 s^2 > 0$) and it follows that the inequality is satisfied only when $(t-r)^2(t-s)^2 \leq 0$, i.e. only when $t \in \{r, s\}$. Thus, the set system defined by the vectors $\{y_{rs}\}$ and the halfspaces $\{\mathbf{h}_t\}$ is identical to the MAXIMUM VERTEX COVERAGE instance defined by G.

To prove Part (c), we specialize to rectangular ranges in $\mathbb{R}^2$. (The case of rectangular ranges in $\mathbb{R}^d$, $d > 2$ follows *a fortiori.*) To begin with, we observe that every *bipartite* instance of MAXIMUM VERTEX COVERAGE can be represented using axis-parallel rectangles in $\mathbb{R}^2$. The construction is as follows. If we label the vertices of the bipartite graph as $\{u_1, \ldots, u_p, w_1, \ldots, w_q\}$ such that every edge has one endpoint in $\{u_1, \ldots, u_p\}$ and the other endpoint in $\{w_1, \ldots, w_q\}$, then we can represent edge (u_i, w_j) using the point $(2i, 2j) \in \mathbb{R}^2$. Vertex u_i is represented by the rectangle $[2i-1, 2i+1] \times [1, 2q+1]$ and vertex w_j by the rectangle $[1, 2p+1] \times [2j-1, 2j+1]$. This construction, combined with Lemma 5.1, suffice to show that the greedy algorithm has worst-case approximation ratio $1-\frac{1}{e}$ when specialized to rectangular ranges in $\mathbb{R}^2$. To prove APX-hardness, we need to use a different reduction that is based on bounded-degree graphs rather than bipartite graphs. We use the following theorem from [16]: there exists a constant Δ such that VERTEX COVER, restricted to graphs of maximum degree Δ, is APX-hard.

For any graph G, create an instance of MAXIMUM COVERAGE as follows. Assuming that G has vertex set $\{v_1, \ldots, v_n\}$ and edge set $\{e_1, \ldots, e_m\}$. For each edge e_k with endpoints v_i, v_j, the set $U \subset \mathbb{R}^2$ contains the three points $(6k-2, 2i)$, $(6k, 0)$, $(6k+2, 2j)$. These $3m$ points constitute the entire set U. The rectangles in $\mathcal{R}$ are as follows. For each edge e_k there are two rectangles $\mathbf{r}_1(e_k) = [6k-3, 6k+1] \times [-1, n+1]$ and $\mathbf{r}_2(e_k) = [6k-1, 6k+3] \times [-1, n+1]$. For each vertex v_i there is one rectangle $\mathbf{r}(v_i) = [0, 6m+3] \times [2i-1, 2i+1]$. If G has a vertex cover C of size s, then there is a set of $m+s$ rectangles in $\mathcal{R}$ that cover all the points in U: we take rectangle $\mathbf{r}(v_i)$ for each $v_i \in C$, this covers at least one of the points $(6k-2, 2i)$, $(6k+2, 2j)$ for each edge e_k and the remaining two points corresponding to that edge can be covered using either $\mathbf{r}_1(e_k)$ or $\mathbf{r}_2(e_k)$. Conversely, if U can be covered by $m+s$ elements of $\mathcal{R}$, then the covering must have a subcollection of m rectangles that contains one of the rectangles $\mathbf{r}_1(e_k), \mathbf{r}_2(e_k)$ for each k. Let T be the subset of U that is not covered by this subcollection, and let C be the set of all vertices v_i such that T contains a point whose y-coordinate is $2i$. It is easy to see that C is a vertex cover of G, and $|C| \leq s$.

Now let $\epsilon > 0, \Delta < \infty$ be chosen such that it is NP-hard to distinguish between graphs of maximum degree Δ having a vertex cover of size s (henceforth, *yes instances*) and those having no vertex cover of size less than $(1+\epsilon)s$ (*no instances*). If G is a yes instance, then the corresponding MAXIMUM COVERAGE instance with parameter $k = m + s$ has optimum value $3m$. If G is a no instance, then the corresponding *MaximumCoverage* instance with parameter $k = m + s$ has optimum value at most $3m - \epsilon s$. Indeed, if there exist $m + s$ rectangles that cover more than $3m - \epsilon s$ points, then it is trivial to find fewer than $m + (1 + \epsilon)s$ rectangles that cover all $3m$ points, which is impossible if G is a no instance. If it possible for a graph with m edges and maximum degree Δ to have a vertex cover of size s then $s \geq m/\Delta$. Thus, we have shown that it a yes instance of VERTEX COVER maps to a MAXIMUM COVERAGE instance whose optimum value is $3m$ while a no instance maps to one whose optimum value is at most $(3 - \epsilon/\Delta)m$, implying the claimed APX-hardness. □

An immediate corollary of Theorem 5.2 is the following statement, which justifies that in our fixed-parameter algorithm, the super-polynomial dependence of the running time on k and $1/\epsilon$ is unavoidable.

COROLLARY 5.3. *Suppose that the* MAXIMUM COVERAGE *problem, specialized to instances with VC-dimension d, has a $(1-\epsilon)$-approximation algorithm running in time $O(f(\epsilon, k, d) \cdot \text{poly}(n))$, for every ϵ, k. If $P \neq NP$, then $f(\epsilon, k, d)$ must be super-polynomial in k. If $P \neq W[1]$, then $f(\epsilon, k, d)$ must be super-polynomial in ϵ^{-1}. In fact, both of these statements hold even if we restrict to $d = 2$.*

PROOF. Stating that $f(\epsilon, k, 2)$ must be super-polynomial in k is a restatement of the APX-hardness of MAXIMUM COVERAGE in VC-dimension 2, which is Part (a) of Theorem 5.2. To prove that $f(\epsilon, k, 2)$ must be super-polynomial in ϵ^{-1}, we observe that MAXIMUM COVERAGE, specialized to instances with VC-dimension 2, is a generalization of the $W[1]$-hard PARTIAL VERTEX COVER problem, and that approximating the optimum of PARTIAL VERTEX COVER within a factor of $(1 - \epsilon)$, for $\epsilon < 1/|E|$, is equivalent to solving it exactly. □

6. OPEN QUESTIONS

We leave several interesting open questions.

- Give an algorithm to maximize a submodular function subject to cardinality constraint with approximation ratio $1 - \epsilon$ and running time $exp(k, \epsilon) \cdot poly(n)$ or prove its impossibility.
- Improve the running time of our algorithm for sets with bounded VC-dimension.
- Give an algorithm better than $1 - (1 - 1/r)^r$ approximation when the cardinality of each set is bounded by r. Such an algorithm could have a running time exponential in r.
- Resolve the approximability of max-coverage on three-dimensional halfspaces. We conjecture that local search is a PTAS for the problem. Appendix A.2 presents a proof of the two-dimensional version of this conjecture.

7. REFERENCES

[1] Alexander A. Ageev and Maxim I. Sviridenko. Pipage rounding: A new method of constructing algorithms with proven performance guarantee. *J. Combinatorial Optimization*, 8:307–328, 2004.

[2] Ashwinkumar Badanidiyuru, Robert Kleinberg, and Hooyeon Lee. Approximating low-dimensional coverage problems. *CoRR*, abs/1112.0689, 2011.

[3] Hervé Bronniman and Michael T. Goodrich. Almost optimal set covers in finite vc-dimension. *Discrete Comp. Geom.*, 14(4):463–479, 1995.

[4] Bernard Chazelle and Jiří Matoušek. On linear-time deterministic algorithms for optimization problems in fixed dimension. *J. Algorithms*, 21:579–597, 1996.

[5] Gerard Cornuejols, Marshall L. Fisher, and George L. Nemhauser. Location of bank accounts to optimize float: An analytic study of exact and approximate algorithms. *Management Science*, 23(8):789–810, 1977.

[6] Gerard Cornuejols, George L. Nemhauser, and Laurence A. Wolsey. Worst-case and probabilistic analysis of algorithms for a location problems. *Operations Research*, 28:847–858, 1980.

[7] Tomás Feder and Daniel Greene. Optimal algorithms for approximate clustering. In *Proc. 20th Annual ACM Symposium on Theory of Computing (STOC)*, pages 434–444, 1988.

[8] Uriel Feige. A threshold of $\ln n$ for approximating set cover. *J. ACM*, 45:634–652, 1998.

[9] Sariel Har-Peled and Mira Lee. Weighted geometric set cover problems revisited, 2008. manuscript.

[10] Venky Harinarayan, Anand Rajaraman, and Jeffrey D. Ullman. Implementing data cubes efficiently. In H. V. Jagadish and Inderpal Singh Mumick, editors, *SIGMOD Conference*, pages 205–216. ACM Press, 1996.

[11] Nabil H. Hustafa and Saurabh Ray. Improved results on geometric hitting set problems. *Discrete Comp. Geom.*, 44(4):883–895, 2010.

[12] David Kempe, Jon M. Kleinberg, and Éva Tardos. Maximizing the spread of influence through a social network. In Lise Getoor, Ted E. Senator, Pedro Domingos, and Christos Faloutsos, editors, *KDD*, pages 137–146. ACM, 2003.

[13] Andreas Krause. *Optimizing Sensing: Theory and Applications*. PhD thesis, Carnegie Mellon University, December 2008.

[14] Dániel Marx. Parameterized complexity and approximation algorithms. *The Computer Journal*, 51(1):60–78, 2008.

[15] George L. Nemhauser, Laurence A. Wolsey, and Marshall L. Fisher. An analysis of approximations for maximizing submodular set functio ns i. *Mathematical Programming*, 14, 1978.

[16] Christos H. Papadimitriou and Mihalis Yannakakis. Optimization, approximation, and complexity classes. *J. Computer and System Sciences*, 43:425–440, 1991.

[17] Erez Petrank. The hardness of approximations: Gap location. *Computational Complexity*, 4:133–157, 1994.

[18] F. Radlinski, R. Kleinberg, and T. Joachims. Learning diverse rankings with multi-armed bandits. In *International Conference on Machine Learning (ICML)*, pages 784–791, 2008. First presented at NIPS07 Workshop on Machine Learning for Web Search.

[19] Kasturi R. Varadarajan. Weighted geometric set cover

via quasi-uniform sampling. In Leonard J. Schulman, editor, *STOC*, pages 641–648. ACM, 2010.

[20] Yisong Yue and T. Joachims. Predicting diverse subsets using structural SVMs. In *International Conference on Machine Learning (ICML)*, pages 271–278, 2008.

APPENDIX

A. TWO-DIMENSIONAL HALFSPACES

Theorem 5.2 rules out the possibility of designing a PTAS for MAXIMUM COVERAGE specialized to halfspaces in $\mathbb{R}^d$ for $d \geq 4$ (unless P=NP) and it likewise rules out the possibility of proving an approximation ratio better than $1 - \frac{1}{e}$ for the greedy algorithm. But in very low dimensions, the situation is different. When $d = 1$, it is easy to see that the greedy algorithm itself always computes an optimal solution. When $d = 2$, a dynamic programming algorithm due to Har-Peled and Lee [9] computes an optimal solution in polynomial time. (The algorithm given in that paper is for SET COVER rather than MAXIMUM COVERAGE, but a trivial modification of their algorithm solves MAXIMUM COVERAGE.) Despite the existence of a polynomial-time algorithm for two-dimensional MAXIMUM HALFSPACE COVERAGE, it is interesting to investigate the approximation ratio of some other archetypical algorithms for this problem, especially since this investigation may shed light on the approximability of three-dimensional MAXIMUM HALFSPACE COVERAGE, which is NP-hard [7] and hence the two-dimensional dynamic programming algorithm is unlikely to generalize. In this section, we show that when $d = 2$ the greedy algorithm has approximation ratio 3/4, and there is a natural local search algorithm yielding a PTAS.

A.1 Analysis of the greedy algorithm

To analyze the greedy algorithm for MAXIMUM HALFSPACE COVERAGE in two dimensions we prove that the *covering multiplicity* r of the problem instance is 2. Then by Theorem 4.2 we have that greedy algorithm is a factor $\frac{3}{4}$ approximation algorithm for MAXIMUM HALFSPACE COVERAGE in two dimensions.

LEMMA A.1. *In two dimensions, the* covering multiplicity *for* MAXIMUM HALFSPACE COVERAGE *is* 2.

PROOF. The proof is a series of simple observations and can be found in the full version at [2]. □

THEOREM A.2. *The greedy algorithm for two-dimensional* MAXIMUM HALFSPACE COVERAGE *has approximation ratio 3/4.*

PROOF. The proof follows easily from Lemma A.1 and Theorem 4.2. An example showing that the analysis is tight can be found in the full version at [2]. □

A.2 A PTAS via local search

If $(U, \mathcal{R}, k)$ is an instance of MAXIMUM COVERAGE and $S = \{R_1, \ldots, R_k\}$ is a solution, define a t-swap to be the operation of transforming this solution into another solution $S' = \{R'_1, \ldots, R'_k\}$ such that there are at most t sets belonging to L but not L', and vice-versa. If $(U, \mathcal{R}, k)$ is a two-dimensional instance of MAXIMUM HALFSPACE COVERAGE and $S = \{R_1, \ldots, R_k\}$ is a solution, define $area(S) \subseteq \mathbb{R}^2$ to be the set $\bigcup_{i=1}^{k} \mathbf{h}_i$, where $\mathbf{h}_i$ is the halfspace corresponding to R_i.

In this section we analyze the following local search algorithm. We assume an *unweighted* instance of two-dimensional MAXIMUM HALFSPACE COVERAGE, i.e. an instance in which each element has weight 1.

1. Start with a arbitrary k-tuple of sets S.
2. While possible do a t-swap to improve the number of elements covered.
3. If there exists a 1-swap to obtain a solution S' such that $area(S) \subseteq area(S')$ and $area(S) \neq area(S')$ then perform this 1-swap and go to step 2. Otherwise terminate the algorithm.

It is simple to see that step 3 does not run for more than n times without the solution improving because once a set is deleted from S in step 3, the only event that can re-insert it is a t-swap in step 2.

We will prove that this local search algorithm achieves an approximation ratio of $2t/(2t+1)$. (This implies that we can obtain a PTAS with running time $n^{O(1/\epsilon)}$ by setting $t = 1/\epsilon$.) The proof of the approximation ratio is in two steps. We first assume the existence of a certain chain decomposition and prove that this implies a $2t/(2t+1)$ approximation ratio. Then we construct such a chain decomposition.

A.2.1 *Approximation ratio assuming chain decomposition*

For succinctness, we will refer to the sets in the optimal solution and in the output of the local search algorithm as *opt sets* and *local sets*, respectively. Opt sets will be denoted by o_i and local sets by ℓ_i. If L_i is a subcollection of the local sets, we will frequently use the notation $v(L_i)$ to denote the set of elements covered by L_i but not by any of the other local sets, i.e.

$$v(L_i) = \left(\bigcup_{\ell_j \in L_i} \ell_j \right) - \left(\bigcup_{\ell_j \notin L_i} \ell_j \right).$$

Let the opt sets and local sets be grouped into chains $C_1, \ldots, C_l$ such that the following properties are satisfied.

- In a chain C_i, the local and opt sets alternate (cyclically).
- Let $\ldots o_1 \ell_1 o_2 \ldots \ell_t o_{t+1} \ldots$ be a portion of any chain (cyclically). Define $L = \{\ell_1, \ell_2, \ldots, \ell_t\}$, $O_1 = \{o_1, o_2, \ldots, o_t\}$ and $O_2 = \{o_2, o_3, \ldots o_{t+1}\}$. Then $v(L) \cap OPT \subseteq v(L) \cap (O_1 \cup O_2)$.

Consider L_i and O_i each having same number of sets and at most t sets. We derive some equations based on local optimality in equation 9 in Figure 2. Then, we find sets used in equation (9) and then add these equations to get the desired result.

- Consider chain C_i. If number of local sets in C_i is $\leq t$, then let L_i be the collection of all local sets in C_i and let O_i be the collection of all opt sets in C_i. Make $2t$ such copies, i.e. the same equation will be used $2t$ times in the proof.
- If chain C_i chain has more than t local sets, then let L_i be any t consecutive local sets, and let O_i be the opt sets in the chain which are shifted from L_i by 1 either

$$\begin{aligned} w(v(L_i)-OPT)+w(v(L_i)\cap OPT) &\geq w(O_i-local)+w(O_i\cap v(L_i)) \\ \Rightarrow w(v(L_i)-OPT)+w(v(L_i)\cap(OPT-O_i)) &\geq w(O_i-local) \end{aligned} \quad (9)$$

$$\begin{aligned} 2t\cdot w(LOC-OPT)+w(LOC\cap OPT) &\geq 2t\cdot w(OPT-LOC) \\ \Rightarrow 2t\cdot w(LOC-OPT)+(2t+1)\cdot w(LOC\cap OPT) &\geq 2t\cdot w(OPT) \\ \Rightarrow (2t+1)\cdot w(LOC-OPT)+(2t+1)\cdot w(LOC\cap OPT) &\geq 2t\cdot w(OPT) \\ \Rightarrow (2t+1)w(LOC) &\geq 2t\cdot w(OPT) \\ \Rightarrow w(LOC) &\geq \tfrac{2t}{2t+1}\cdot w(OPT) \end{aligned} \quad (10)$$

Figure 2: Conditions for local optimum

clockwise or counterclockwise. Note that a particular choice for L_i appears twice since there are two options for O_i.

Here are some properties of the above decomposition.

1. Each ℓ_i belongs to $2t$ of the sets L_j
2. Each o_i belongs to $2t$ of the sets O_j
3. Let L_i be associated with O_j and O_k. Then $v(L_i)\cap OPT \subseteq v(L_i)\cap(O_j\cup O_k)$. This is just a restatement of the assumed property of the chain decomposition.

Based on the above properties we derive the the final inequality. Sum the equation (9) over all L_i,O_i. Then we bound each term in the sum. Let OPT denote the set of elements covered by the opt sets, and let LOC denote the set of elements covered by the local sets.

- $\sum w(v(L_i)-OPT)\leq 2t\cdot w(LOC-OPT)$. This is due to property 1.
- $\sum w(O_i-LOC)\geq 2t\cdot w(OPT-LOC)$. This is due to property 2. Note the difference in the direction of inequalities.
- $\sum w(v(L_i)\cap(OPT-O_i))\leq w(LOC\cap OPT)$. This is due to property 3.

From the above three equations and equation (9) we get the final necessary equation 10 in Figure 2.

A.2.2 *Obtaining a chain decomposition*

We argue about some properties of two-dimensional MAXIMUM HALFSPACE COVERAGE instances, based on which we get some associations. Consider the optimal solution which has the maximum number of sets in common with the output of the local search algorithm.

1. For each ℓ_i we have that $\exists o_j, o_l$ such that $\ell_i\cap OPT\subseteq o_j\cup o_l$. It is simple to see that if this is not true then we can change the optimal solution so that the number of sets in common with the local optimum increases. Now associate ℓ_i to the corresponding o_j and o_l. Let $A^{\text{init}}(o_i)$ be the local sets associated with o_i and $A^{\text{init}}(\ell_i)$ be the opt sets associated with ℓ_j.
2. For each o_i we have that $\exists \ell_j,\ell_t\in A^{\text{init}}(o_i)$ such that $o_i\cap A^{\text{init}}(o_i)\subseteq \ell_j\cup\ell_t$. This is true due to the different form of local search used. If no such ℓ_j,ℓ_t existed, we could change the local optimum to increase the area. Now if o_i has more than local two sets associated with it, then among them choose two ℓ_j,ℓ_t such that $o_i\cap A^{\text{init}}(o_i)\subseteq\ell_j\cup\ell_t$ and keep the association and remove the rest of the associations for o_i. Let the new associations be called $A^{\text{new}}(\ell_i)$ and $A^{\text{new}}(o_j)$.
3. Note that in step 2 we remove some associations. Hence it might no longer be true that $\ell_i\cap OPT\subseteq A^{\text{new}}(\ell_i)$. But also note that it is still true that $o_i\cap A^{\text{init}}(o_i)\subseteq A^{\text{new}}(o_i)$.
4. Note that due to step 1,2 we have that each local set is associated to at most two opt sets and each opt set is associated to at most two local sets.
5. Now form maximal alternating pseudo chains such that a pseudo chain is a list of alternating local and optimal sets. Additionally each local set has the association to its adjacent sets and each opt set has association to its adjacent sets.
6. Now there are three kinds of pseudo chains depending on their endpoints. They are either $l-l$ or $o-o$ or $o-l$ (here $l-l$ means a chain starting with a local set and ending with a local set).
7. merge $l-l$ pseudo chains arbitrarily with $o-o$ to get only $o-l$ chains.

The $o-l$ chains thus obtained are the chains we desired in the analysis. It is left to be proven that this decomposition satisfies the properties needed.

- By construction the local and opt sets alternate (cyclically) in a chain C_i.
- Let $...o_1\ell_1 o_2...\ell_t o_{t+1}...$ be a portion of any chain (cyclically). Define $L=\{\ell_1,\ell_2,...,\ell_t\}$, $O_1=\{o_1,o_2,...,o_t\}$ and $O_2=\{o_2,o_3,...o_{t+1}\}$. Then we need to prove that $v(L)\cap OPT\subseteq v(L)\cap(O_1\cup O_2)$. The proof is by contradiction. i.e. let $x\in v(L)\cap OPT$ but $x\notin O_1\cup O_2$. Then $x\in\ell_c\in L$. We follow through the associations.
  - Consider the initial association. Then by its property $o_j\in A^{\text{init}}(\ell_c)$ such that $x\in o_j$.
  - If $o_j\in A^{\text{new}}(\ell_c)$ then o_j is adjacent to ℓ_c in the chain and hence $o_j\in\{o_1,o_2,...,o_{t+1}\}$ which is a contradiction to the fact that $x\notin O_1\cup O_2$.
  - If $o_j\notin A^{\text{new}}(\ell_c)$ then in step 2 of associations $o_j\cap A^{\text{init}}(o_j)\subseteq A^{\text{new}}(o_j)$. Hence $\ell_d\in A^{\text{new}}(o_j)$ such that $x\in\ell_d$ and o_j and ℓ_d are adjacent in some chain. If $\ell_d\in\{\ell_1,\ell_2,...,\ell_t\}$ then $o_j\in\{o_1,o_2,...,o_{t+1}\}$ and we arrive at a contradiction that $x\notin O_1\cup O_2$. Otherwise $x\notin v(L)$ and we still arrive at a contradiction.

Improved Range Searching Lower Bounds

Kasper Green Larsen*
MADALGO†
Department of Computer Science
Aarhus University, Aarhus, Denmark
larsen@cs.au.dk

Huy L. Nguyễn‡
Center for Computational Intractability
Department of Computer Science
Princeton University, Princeton, NJ, USA
hlnguyen@cs.princeton.edu

ABSTRACT

In this paper we present a number of improved lower bounds for range searching in the pointer machine and the group model. In the pointer machine, we prove lower bounds for the approximate simplex range reporting problem. In approximate simplex range reporting, points that lie within a distance of $\varepsilon \cdot \text{diam}(s)$ from the border of a query simplex s, are free to be included or excluded from the output, where $\varepsilon \geq 0$ is an input parameter to the range searching problem. We prove our lower bounds by constructing a hard input set and query set, and then invoking Chazelle and Rosenberg's [CGTA'96] general theorem on the complexity of navigation in the pointer machine.

For the group model, we show that input sets and query sets that are hard for range reporting in the pointer machine (i.e. by Chazelle and Rosenberg's theorem), are also hard for dynamic range searching in the group model. This theorem allows us to reuse decades of research on range reporting lower bounds to immediately obtain a range of new group model lower bounds. Amongst others, this includes an improved lower bound for the fundamental problem of dynamic d-dimensional orthogonal range searching, stating that $t_q t_u = \Omega((\lg n / \lg \lg n)^{d-1})$. Here t_q denotes the query time and t_u the update time of the data structure. This is an improvement of a $\lg^{1-\delta} n$ factor over the recent lower bound of Larsen [FOCS'11], where $\delta > 0$ is a small constant depending on the dimension.

*Supported in part by MADALGO and in part by a Google Fellowship in Search and Information Retrieval. In addition, this research was done while the author was visiting Princeton University. This visit was supported by an EliteForsk travel scholarship from the Danish Ministry of Science and by a grant from Aarhus University's Research Foundation.

†Center for Massive Data Algorithmics, a Center of the Danish National Research Foundation.

‡Supported in part by CCF 0832797 and a Gordon Wu fellowship.

SCG'12, June 17–20, 2012, Chapel Hill, North Carolina, USA.

Categories and Subject Descriptors

F.2.2 [**Nonnumerical Algorithms and Problems**]: Geometrical problems and computations

General Terms

Theory

Keywords

Range searching, lower bounds, pointer machine, group model

1. INTRODUCTION

Range searching is one of the most fundamental topics in computational geometry. Here the goal is to represent an input set of geometric objects, i.e. maintain a data structure, such that given a query range, one can efficiently compute various statistics over the input objects intersecting the query. The input objects are typically points in d-dimensional space and the most common query ranges are axis-aligned rectangles, simplices, halfspaces and balls. The type of information computed when answering a query includes counting the number of input objects in the intersection (range counting), reporting the k objects in the intersection (range reporting), determining whether the intersection is empty (range emptiness) and computing the sum of a set of weights assigned the input objects in the intersection (weighted range counting and range searching in the semi-group and group model).

In this paper we present improved lower bounds for a number of range searching problems. In the pointer machine model, we present the first lower bounds for approximate simplex range reporting. Secondly, we present a new powerful theorem for proving lower bounds for dynamic range searching in the group model. This theorem allows us to reuse decades of research in pointer machine range reporting lower bounds to immediately obtain a range of improved group model lower bounds.

1.1 Range Reporting in the Pointer Machine

The pointer machine model was introduced in 1979 by Tarjan [18]. In this model, a range reporting data structure is represented by a directed graph. Each node of the graph may store either an input object or some auxiliary data. The nodes have constant out-degrees and one node is designated as the root. When answering a query, the data structure starts by reading the root node. The data structure then examines the contents of that node and if the node stores an input object, the data structure may choose to report that object if it intersects the query range. Following that, it either terminates or selects an edge leaving the root and retrieves the node pointed to by that edge. This process continues, where at each step, the data structure selects an edge leaving one of the previously seen nodes and retrieves the node pointed to by that edge. When the process terminates, we require that all input objects that intersects the query range have been reported, i.e. each reported input object must be stored in at least one of the explored nodes. Thus a data structure in the pointer machine model, is a data structure where all memory accesses are through pointers and random accesses are disallowed.

The space of a pointer machine data structure is defined as the number of nodes in the corresponding graph, and the query time is the number of nodes explored when answering a query.

While the pointer machine model is somewhat constrained compared to the popular word-RAM model, there are several motivations for studying the complexity of range searching in this model. First and foremost, we can prove polynomially high and often very tight lower bounds in this model. This stands in sharp contrast to the highest query time lower bound for any static data structure problem in the word-RAM (or cell probe model), which is a mere $\Omega(\lg n/\lg\lg n)$, see e.g. [17]. Additionally, most word-RAM range reporting upper bounds are really pointer-based, or can easily be implemented without random accesses with a small overhead, typically at most an $O(\lg n)$ multiplicative cost in the query time and/or space. Thus pointer machine lower bounds indeed shed much light on the complexity of range reporting problems.

Previous Results.

With only a few exceptions (see [1, 3]), lower bounds for range reporting in the pointer machine model have all been proved by appealing to a theorem initially due to Chazelle [5] and later refined by Chazelle and Rosenberg [10]. Since our lower bounds also rely on this theorem, we introduce it in the following: First, let P be a set of input objects to a range searching problem and let $\mathcal{R}$ be a set of query ranges. Then we say that $\mathcal{R}$ is *(t, h)-favorable* if

1. $|R \cap P| \geq t$ for all $R \in \mathcal{R}$.
2. $|R_1 \cap R_2 \cap \cdots \cap R_h \cap P| = O(1)$ for all sets of h different queries $R_1, \ldots, R_h \in \mathcal{R}$.

Letting k denote the output size of a query, Chazelle and Rosenberg proved that

THEOREM 1 (CHAZELLE AND ROSENBERG [10]). *Let P be a set of n input objects to a range searching problem and $\mathcal{R}$ a set of m query ranges. If $\mathcal{R}$ is $(\Omega(t_q), h)$-favorable, then any pointer machine data structure for P with query time $t_q + O(k)$ must use space $\Omega(mt_q/h)$.*

This theorem completely reduces the task of proving pointer machine range reporting lower bounds to a geometric problem of constructing a hard input and query set. In the case of simplex range reporting, Chazelle and Rosenberg [10] used this theorem to prove that any pointer machine data structure with query time $t_q + O(k)$ for d-dimensional simplex range reporting must use space $\Omega((n/t_q)^{d-\varepsilon})$, where $\varepsilon > 0$ is an arbitrarily small constant. Thus to obtain a query time as inefficient as $O(\sqrt{n} + k)$ for simplex range reporting in $\mathbb{R}^{10}$, one needs space almost $\Omega(n^5)$. This is extremely prohibitive for most natural applications.

Because of the high space/time requirement for most exact range reporting problems, researchers also look at the approximate version of the question: for a query range s, if an input object is within distance $\varepsilon \cdot \mathrm{diam}(s)$ from the boundary of s, the data structure has the freedom to arbitrarily include or exclude it from the output. There is an extensive literature in this area but the result most relevant to us, is the data structure by Arya, da Fonseca, and Mount [4] for approximate simplex range searching in the semi-group model. Their data structure answers queries in $O(\lg n + \lg(1/\varepsilon) + 1/\varepsilon^{d-1-u})$ time and uses $O(n/\varepsilon^{\frac{u^2 d}{d-1}})$ space for any integer $u \geq 0$. While their data structure is for semi-group range searching, they briefly describe a data structure for reporting with the same space and query time, except for an additional linear term in the number of reported points. We are not aware of any previous lower bounds for approximate range reporting for any kind of ranges despite the long history of the problem.

Since favorable query sets play an important role in our group model results, we also mention a number of previous favorable query set constructions. All these results were used to prove pointer machine range reporting lower bounds by invoking Theorem 1. Chazelle [5] (or alternatively [6]), showed that one can construct a $(t, 2)$-favorable set of $\Theta(n/t \cdot (\lg n/\lg t)^{d-1})$ queries for *orthogonal range reporting* in d-dimensional space. Henceforth, $t = \Omega(1)$ is an adjustable parameter. We note that the input to orthogonal range reporting is a set of n points and the queries are axis-aligned rectangles. For the "dual" problem of d-dimensional *rectangle stabbing* (the input is n axis-aligned rectangles and a query asks to report all rectangles containing a query point), Afshani, Arge and Larsen [2] showed that one can construct a $(t, 2)$-favorable set of $n/t \cdot 2^{\Theta\left(\frac{\lg n}{t^{1/(d-1)}}\right)}$ queries. For *line range reporting* (the input consists of n two-dimensional points and a query asks to report all points on a query line), a classical construction of Erdös (see e.g. [14]) shows that one can construct a $(\Theta(n^{1/3}), 2)$-favorable set of n query lines. Finally, for *convex polytope intersection reporting* in $\mathbb{R}^3$ (the input is an n-vertex convex polytope and a query asks to report all edges of the polytope intersecting a query plane), Chazelle and Liu [8] showed that one can construct a $(t, 2)$-favorable set of $\Theta(n^2/t^3)$ query planes.

1.2 Range Searching in the Group Model

The group model was introduced in 1982 by Fredman [11]. In this model, each input object to a range searching problem is assigned a weight from a commutative group and the goal is to preprocess the input into a collection of precomputed group elements, such that one can efficiently compute the group sum of the weights assigned to all input objects intersecting a query range. In this paper, we focus on dynamic range searching in the group model. Here a data structure must also support updating the weights assigned to the input objects (while the set of input objects is fixed).

Since the group model was first introduced, there has been two slightly different definitions of dynamic data structures in the group model. These types of data structures have been named *weakly oblivious* and *oblivious* data structures, respectively. The lower bounds we prove apply to *oblivious* data structures:

Oblivious Data Structures.

An *oblivious* data structure in the group model (see [11] or [13]), is a dynamic data structure with no understanding of the particular group in question, i.e. it can only access and manipulate weights through black-box addition and subtraction. Thus from the data structure's point of view, each precomputed group element is just a linear combination over the weights assigned to the input objects. When answering a query, such a data structure adds and subtracts a subset of these linear combinations (another linear combination) to finally yield the linear combination summing exactly the weights assigned to the input objects intersecting the query range. When given an update request, the data structure simply re-evaluates every precomputed group element for which the weight of the updated input object occurs with non-zero coefficient in the corresponding linear combination.

We define the query time of an oblivious data structure as the number of precomputed group elements used when answering a query, and the update time is defined as the number of linear combinations that need to be re-evaluated when updating the weight of an input object. For a more formal definition, see Section 2.

We note that weakly oblivious data structures differ from oblivious data structures, in that they are allowed slightly more elaborate update procedures and therefore lower bounds proved for weakly oblivious data structures also apply to oblivious data structures. For a definition of weakly oblivious data structures, we refer the reader to [13].

Previous Results.

Proving high lower bounds in the group model has until very recently remained a great barrier. When Fredman [11] defined the model, he proved a lower bound of $\Omega(n \lg n)$ over a sequence of n updates and queries to the *partial sums* problem (which is the special case of 1-d orthogonal range searching where the points have coordinates $0, \dots, n-1$). This bound holds for oblivious data structures. Following that, Fredman and Saks [12] proved an $\Omega(n \lg n / \lg \lg n)$ lower bound for the same problem, which does however hold for weakly oblivious data structures.

Chazelle [6, 7] later proved lower bounds for *offline* range searching in the group model. He also proved a lower bound of $\Omega(n \lg n)$ for offline two-dimensional *halfspace range searching* [7]. The input to the offline halfspace range searching problem is a set of n query halfspaces and n input points, each assigned a weight from a commutative group, and the goal is to compute for every query halfspace, the group sum of the weights assigned to the points contained therein. In [6] he considered offline two-dimensional orthogonal range searching and proved a lower bound of $\Omega(n \lg \lg n)$. Both these lower bounds were established using a general theorem for proving lower bounds for offline range searching in the group model. Letting A denote the *incidence matrix* corresponding to the input set of points and queries (i.e. the matrix with one row for each query R_i and one column for each point p_j, such that entry $a_{i,j}$ is 1 if $p_j \in R_i$ and it is 0 otherwise), Chazelle showed that for any $1 \leq k \leq n$, if λ_k denotes the k'th largest eigenvalue of A, then the offline problem requires $\Omega(k \cdot \lg(\lambda_k))$ time [7]. Thus proving offline group model lower bounds was reduced to constructing input and query sets where the incidence matrix has large eigenvalues.

The next step forward was by Pătraşcu and Demaine [16], who improved the lower bound of Fredman and Saks to $\Omega(n \lg n)$, i.e. they matched the initial bound of Fredman, but also for weakly oblivious data structures. Following that, Pătraşcu [15] considered two-dimensional orthogonal range searching and proved an $\Omega(\lg n / \lg(s/n + \lg n))$ query time lower bound for the static case. Here s denotes the space usage of the data structure. He also showed how to extend this static lower bound to give a query time lower bound of $\Omega((\lg n / \lg(\lg n + t_u))^2)$ for dynamic data structures with update time t_u. This bound holds for weakly oblivious data structures.

None of the above bounds exceed $\Omega((\lg n / \lg \lg n)^2)$ per query, and this barrier was not overcome until Larsen's [13] recent results. Larsen moved past this barrier by introducing one additional restriction on oblivious data structures: Recall that an oblivious data structure stores group elements corresponding to linear combinations over the weights assigned to the input objects, and it answers queries by again computing linear combinations over the precomputed group elements. Larsen defined the *multiplicity* of an oblivious data structure as the largest absolute value of any coefficient in these linear combinations, and noted that all known upper bounds use only coefficients amongst $\{-1, 0, 1\}$, i.e. they have multiplicity 1. With this definition, he demonstrated a connection between oblivious data structures and the *discrepancy* of the corresponding range searching problem. The discrepancy, disc, of a range searching problem with query ranges $\mathcal{R}$, is the maximum over all sets P of n input objects, of the deviation of the best 2-coloring of the objects in P from an even coloring, i.e.

$$\text{disc} = \max_{P:|P|=n} \min_{\chi:P \to \{-1,+1\}} \max_{R \in \mathcal{R}} |\sum_{p \in R \cap P} \chi(p)|.$$

More precisely, Larsen's main theorem states that $t_q t_u = \Omega(\text{disc}^2 / \Delta^4 \lg n)$, where t_q is the worst case query time, t_u the worst case update time and Δ is the multiplicity of the data structure. Plugging in the vast amount of discrepancy lower bounds, this relation immediately yielded a whole range of polynomially high group model lower bounds for data structures with constant multiplicity. The two bounds most important to our work (stated here for constant multiplicity), is a lower bound of $t_q t_u = \Omega(n^{1/3} / \lg n)$ for line range searching and a bound of $t_q t_u = \Omega(\lg^{d-2+\mu(d)} n)$ for d-dimensional orthogonal range searching, where $\mu(d) > 0$ is a small but strictly positive function of d.

1.3 Our Results

In the following two paragraphs, we present our new results for range searching in the group model and approximate range reporting in the pointer machine.

Our Group Model Results.

In Section 2, we present an alternative to Larsen's connection to discrepancy theory. Our relation allows us to almost immediately translate range reporting lower bounds in the pointer machine to dynamic group model lower bounds for data structures with bounded multiplicity. More specifically, let P be a set of n input objects to a range searching problem and $\mathcal{R}$ a set of m query ranges over P. We say that $\mathcal{R}$ is *strongly* (t, h)*-favorable* if

1. $|R \cap P| = \Theta(t)$ for all $R \in \mathcal{R}$.

2. $|R_1 \cap R_2 \cap \cdots \cap R_h \cap P| = O(1)$ for all sets of h different queries $R_1, \ldots, R_h \in \mathcal{R}$.

3. $|\{R \in \mathcal{R} \mid p \in R\}| = O(mt/n)$ for all $p \in P$.

Thus a favorable query set is strongly favorable, if in addition, all query ranges contain roughly equally many input objects and all input objects are contained in roughly equally many query ranges. With this definition, we prove the following result

THEOREM 2. *Let P be a set of n input objects to a range searching problem and $\mathcal{R}$ a set of $m \leq n$ query ranges over P. If $\mathcal{R}$ is strongly $(t, 2)$-favorable, then any oblivious data structure for the range searching problem must have $t_q t_u = \Omega(m^2 t/n^2 \Delta^4)$ on the input set P. Here t_q denotes the worst case query time, t_u the worst case update time and Δ the multiplicity of the data structure. For $m = \Theta(n)$ and $\Delta = O(1)$, this bound simplifies to $t_q t_u = \Omega(t)$.*

Fortunately, all the favorable query sets described in Section 1.1 are also strongly favorable. By adjusting the parameter t (in Section 1.1) such that the number of queries in the favorable query sets is $m = \Theta(n)$ and $m \leq n$, we immediately obtain the following lower bounds (listed here for constant multiplicity):

For d-dimensional orthogonal range searching, we get a lower bound of

$$t_q t_u = \Omega((\lg n / \lg \lg n)^{d-1}).$$

Compared with the result of Larsen [13], this is an improvement of a $\lg^{1-\mu(d)} n/(\lg \lg n)^{d-1} = \Omega(\lg^{1-\delta} n)$ factor, where $\delta > 0$ is some small constant. For rectangle stabbing, we get a matching bound either by a reduction or from the construction of Afshani, Arge and Larsen [2]. This again improves over Larsen's result [13] by a $\lg^{1-\delta} n$ factor. For line range searching, we get a lower bound of $t_q t_u = \Omega(n^{1/3})$, which is an improvement of a $\lg n$ factor over [13]. Additionally, we get a lower bound for convex polytope intersection searching in $\mathbb{R}^3$ of $t_q t_u = \Omega(n^{1/3})$ using the result of Chazelle and Liu [8].

Our proof of Theorem 2 is based on carefully bounding the eigenvalues of the incidence matrix corresponding to P and $\mathcal{R}$. In fact, we prove the following stronger theorem during our establishment of Theorem 2:

THEOREM 3. *Let P be a set of n input objects to a range searching problem, $\mathcal{R}$ a set of m query ranges over P and A the corresponding incidence matrix. Then for every $3 \leq k \leq n$, any oblivious data structure for the range searching problem must have $t_q t_u = \Omega(\lambda_k k^2/mn\Delta^4)$ on the input set P. Here λ_k denotes the k'th largest eigenvalue of $A^T A$, t_q the worst case query time of the data structure, t_u the worst case update time and Δ the multiplicity of the data structure.*

This theorem can be considered a complement to Chazelle's theorem [7] for establishing lower bounds on offline range searching in the group model, however our dependence on λ_k is exponentially better than the one of Chazelle. Using this theorem, we also obtain a lower bound of $t_q t_u = n^{\Omega(1/\lg \lg n)}$ for orthogonal range searching in non-constant dimension $d = \Omega(\lg n / \lg \lg n)$ (see Section 2 for a proof).

Our Pointer Machine Results.

In Section 3, we give the first lower bound for approximate range reporting in the pointer machine model. Specifically, we show that for any fixed constant $\delta > 0$, any data structure for approximate simplex range reporting that answers a simplex query s in $t_q + O(k)$ time, where k is the number of points either inside s or within a distance $\varepsilon \cdot \mathrm{diam}(s)$ from the boundary of s, needs space at least $n\varepsilon^{1+\delta-d}/t_q^{1+\delta}$. This bound matches the upper bound from [4] up to a factor $\varepsilon^{\delta d/(1+\delta)}$ when the space is $\Theta(n)$, but unfortunately there is still a rather large gap between the two bounds in the rest of the tradeoff.

2. THE GROUP MODEL

In this section, we present the proofs of Theorem 2 and Theorem 3, relating range reporting in the pointer machine and dynamic range searching in the group model. Before starting our proof, we present an equivalent but more formal definition of oblivious data structures (see [13] or [11]):

Oblivious Data Structures.

An oblivious data structure for a range searching problem, is a factorization of each incidence matrix A, corresponding to a set n input objects P and a set of m query ranges $\mathcal{R}$, into two matrices Q and D such that $Q \cdot D = A$.

The *data matrix* $D \in \mathbb{Z}^{s \times n}$ represents the precomputed group sums stored by the data structure on input P. Each of the s rows is interpreted as a linear combination over the weights assigned to the n input objects, and we think of the data structure as maintaining the corresponding group sums when given an assignment of weights to the input objects.

The *query matrix* $Q \in \mathbb{Z}^{m \times s}$ specifies the query algorithm. It has one row for each query R in $\mathcal{R}$, and we interpret this row as a linear combination over the precomputed group elements, denoting which precomputed elements to add and subtract when answering the query R on input set P.

The *worst case query time* of an oblivious data structure is defined as the maximum number of non-zero entries in a row of Q over all input sets P and query sets $\mathcal{R}$. The *worst case update time* is similarly defined as the maximum number of non-zero entries in a column of D over all P. The *space* of the data structure is the maximum number of columns in Q (equivalently number of rows in D) over all P and $\mathcal{R}$. Finally, we define the *multiplicity* as the largest absolute value of an entry in D and Q over all P and $\mathcal{R}$.

We refer the reader to [13] or [11] for more intuition on this definition.

2.1 Proofs of Theorem 2 and Theorem 3

In light of the above definition, proving lower bounds for oblivious data structures boils down to arguing when an incidence matrix cannot be factored into two sparse matrices Q and D. The key insight here is, that for sparse matrices Q and D with bounded coefficients, the product QD must have small singular values (i.e. $(QD)^T QD$ has small eigenvalues). Thus if A has large singular values, Q and D cannot be sparse if $QD = A$. This is precisely the intuition behind our proof of Theorem 3:

Proof of Theorem 3.

Let P, $\mathcal{R}$ and A be as in Theorem 3. Furthermore, let $QD = A$ be the factorization of A provided by an oblivious data structure, where Q is an $m \times s$ matrix such that each row has at most t_q non-zero entries and where D is an $s \times n$ matrix where each column has at most t_u non-zero entries. Finally, let Δ be the multiplicity of the oblivious data structure, i.e. any coefficient in Q and D is bounded in absolute value by Δ. Now let $U(D)\Sigma(D)V(D)^T$ be the singular value decomposition of D. Here $U(D)$ and $V(D)$ are unitary matrices and $\Sigma(D)$ is a diagonal matrix where the diagonal entries equals the singular values of D, i.e. if we let $\gamma_i(D^TD) \geq 0$ denote the i'th largest eigenvalue of the $n \times n$ positive semi-definite matrix D^TD, then the i'th diagonal entry of $\Sigma(D)$ is $\sigma_{i,i}(D) = \sqrt{\gamma_i(D^TD)}$. Similarly, let $U(Q)\Sigma(Q)V(Q)^T$ be the singular value decomposition of Q. Letting $\gamma_i(Q^TQ) \geq 0$ denote the i'th largest eigenvalue of Q^TQ, we have that the i'th diagonal entry of $\Sigma(Q)$ is $\sigma_{i,i}(Q) = \sqrt{\gamma_i(Q^TQ)}$. Letting $d_{i,j}$ denote entry (i,j) in D, it now follows from D^TD being square and real that

$$\sum_i \gamma_i(D^TD) = \mathrm{tr}(D^TD) = \sum_{i,j} d_{i,j}^2 \leq t_u\Delta^2 n,$$

where we used that the coefficients of D are bounded in absolute value by Δ. Similarly, we have $\sum_i \gamma_i(Q^TQ) \leq t_q\Delta^2 m$. Finally since $\gamma_i(D^TD)$ and $\gamma_i(Q^TQ)$ are non-negative for all i, we conclude that $\gamma_{\lfloor k/2 \rfloor}(D^TD) = O(t_u\Delta^2 n/k)$ and $\gamma_{\lceil k/2 \rceil - 1}(Q^TQ) = O(t_q\Delta^2 m/k)$.

Our last step is to bound from above the eigenvalues of $(QD)^TQD$. Letting $\gamma_k((QD)^TQD)$ denote the k'th largest eigenvalue of $(QD)^TQD$, we get from the Courant-Fischer characterization of eigenvalues that

$$\gamma_k((QD)^TQD) = \min_{S:\dim(S)\geq n-k+1} \max_{x\in S:\|x\|_2=1} \|QDx\|_2^2,$$

i.e. $\gamma_k((QD)^TQD)$ equals the minimum over all subspaces S of $\mathbb{R}^n$ of dimension at least $n-k+1$, of the maximum square of the stretch of a unit length vector x when multiplying with QD. We thus aim to find a subspace S of dimension at least $n-k+1$, such that every unit vector in S is scaled as little as possible when multiplied with QD. We choose the subspace S consisting of all vectors x, for which $\langle v_i(D), x\rangle = 0$ for $i = 1, \dots, \lfloor k/2 \rfloor$ and $\langle v_i(Q), Dx\rangle = 0$ for $i = 1, \dots, \lceil k/2 \rceil - 1$. Here $\langle \cdot, \cdot \rangle$ denotes the standard inner product, $v_i(D)$ denotes the i'th column vector of $V(D)$ and $v_i(Q)$ denotes the i'th column vector of $V(Q)$. Clearly $\dim(S) \geq n-k+1$. Now let $x \in S$ be a unit length vector and consider first the product $Dx = U(D)\Sigma(D)V(D)^Tx$. Since x is orthogonal to the first $\lfloor k/2 \rfloor$ row vectors in $V(D)^T$ and since $\sigma_{i,i}(D) = O(\sqrt{t_u n/k}\Delta)$ for $i \geq \lfloor k/2 \rfloor$, we get that $\|\Sigma(D)V(D)^Tx\|_2 = O(\sqrt{t_u n/k}\Delta)$. Since $U(D)$ is unitary, this implies $\|Dx\|_2 = O(\sqrt{t_u n/k}\Delta)$. Finally, since Dx is orthogonal to the first $\lceil k/2 \rceil - 1$ row vectors of $V(Q)^T$, we conclude $\|QDx\|_2^2 = O(t_q t_u \Delta^4 mn/k^2)$. But $A^TA = (QD)^TQD$ and thus it must hold that $t_q t_u = \Omega(\lambda_k k^2/mn\Delta^4)$. This completes the proof of Theorem 3.

Proof of Theorem 2.

Let P and $\mathcal{R}$ be as in Theorem 2, i.e. $\mathcal{R}$ is a strongly $(t,2)$-favorable set of queries. Furthermore, let A be the $m \times n$ incidence matrix corresponding to P and $\mathcal{R}$, where $m \leq n$. Our proof is based on lower bounding the eigenvalues of $M = A^TA$, and then applying Theorem 3. We lower bound these eigenvalues using the following theorem of Chazelle and Lvov:

THEOREM 4 (CHAZELLE AND LVOV [9]). *Let A be an $m\times n$ real matrix where $m \leq n$ and let $M = A^TA$. Then M has at least*

$$\frac{n}{16\,\mathrm{tr}(M^2)n/9\,\mathrm{tr}(M)^2 - 7/9}$$

eigenvalues that are greater than or equal to $\mathrm{tr}(M)/4n$.

To use Theorem 4, we bound $\mathrm{tr}(M)$ and $\mathrm{tr}(M^2)$. The first is easily seen to be $\mathrm{tr}(M) = \sum_{R\in\mathcal{R}}|R| = \Omega(mt)$ and the latter is bounded by

$$\begin{aligned}
\mathrm{tr}(M^2) &= \sum_{R_1\in\mathcal{R}}\sum_{R_2\in\mathcal{R}} |R_1 \cap R_2|^2 \\
&= \sum_{R\in\mathcal{R}} |R|^2 + \sum_{R_1\in\mathcal{R}}\ \sum_{R_2\in\mathcal{R}|R_1\neq R_2} |R_1 \cap R_2|^2 \\
&= O(mt^2) + \\
&\quad \sum_{p\in P}\ \sum_{R_1\in\mathcal{R}|p\in R_1}\ \sum_{R_2\in\mathcal{R}|p\in R_2\wedge R_1\neq R_2} |R_1\cap R_2| \\
&= O(mt^2) + \sum_{p\in P}\ \sum_{R_1\in\mathcal{R}|p\in R_1}\ \sum_{R_2\in\mathcal{R}|p\in R_2\wedge R_1\neq R_2} O(1) \\
&= O(mt^2) + \sum_{p\in P} O((mt/n)^2) \\
&= O(mt^2).
\end{aligned}$$

Plugging these values into Theorem 4, we conclude that $M = A^TA$ has $\Omega(nm^2t^2/mt^2n) = \Omega(m)$ eigenvalues greater than $\Omega(mt/n)$. Finally invoking Theorem 3, we get that $t_q t_u = \Omega(mt/n \cdot m^2/mn\Delta^4) = \Omega(m^2t/n^2\Delta^4)$, which completes the proof of Theorem 2.

Implications.

As already mentioned in Section 1.2, we obtain a number of lower bounds from Theorem 2 by reusing favorable query sets constructed for proving pointer machine range reporting lower bounds. Thus in the following, we only mention our proof of the lower bound for orthogonal range reporting in non-constant dimension $d = \Omega(\lg n/\lg\lg n)$.

For non-constant dimensions $d = \Omega(\lg n/\lg\lg n)$, Chazelle and Lvov [9] showed one can construct a set of n points and n query rectangles such that the corresponding incidence matrix A satisfies $\mathrm{tr}(A^TA) = n^{1+\Omega(1/\lg\lg n)}$ and $\mathrm{tr}((A^TA)^2) = O(\mathrm{tr}(A^TA)^2/n)$. From Theorem 4, this means that A^TA has $\Omega(n)$ eigenvalues that are greater than $n^{\Omega(1/\lg\lg n)}$. The result follows immediately from Theorem 3.

3. APPROXIMATE SIMPLEX RANGE REPORTING

In this section we show our lower bounds for approximate simplex range reporting in constant dimensions. The idea of the proof is as follows. From the result by Chazelle and Rosenberg [10], we know that thin *slab* (everything contained between two parallel hyperplanes) queries are hard for range reporting. Therefore, we consider a set of simplex queries that approximate thin slabs and then appeal to Theorem 1 and the properties of thin slabs shown in [10]. First, before proving the lower bound, we formally define the notion of approximations for range reporting.

DEFINITION 1. *Given a query range R, define*

$$R^+ = \{x \mid \mathrm{dist}(x,R) \leq \varepsilon\,\mathrm{diam}(R)\}$$

and $R^- = \{x \in R | \operatorname{dist}(x, \partial R) \geq \varepsilon \operatorname{diam}(R)\}$ *where* ∂R *is the boundary of* R.

For any given query range R*, a data structure for approximate range reporting should report all points inside* R^- *and none of the point outside* R^+.

We now define the approximate analog of *favorable* query sets.

Definition 2. *Let* P *be a set of points and* $\mathcal{R}$ *a set of queries. Then* $\mathcal{R}$ *is* (t,h)*-approximate favorable for* P *if*

1. *The set* $\{R^- | R \in \mathcal{R}\}$ *is* (t,h)*-favorable for* P.
2. *For any* $R \in \mathcal{R}$, $|P \cap R^+| = O(|P \cap R^-|)$.

We get the following corollary of Theorem 1.

Corollary 1. *Let* P *be a set of points and* $\mathcal{R}$ *a* (t_q, h)*-approximate favorable set of queries for* P*. If for any query* $R \in \mathcal{R}$*, a data structure can report all points in* $P \cap R^-$ *and none outside* $P \cap R^+$ *in time* $O(t_q + |R^+ \cap P|)$*, then the space used by the data structure is* $\Omega(|\mathcal{R}| t_q / h)$.

Proof. Since $|R^+ \cap P| = O(R^- \cap P)$, the data structure satisfies the condition of Theorem 1 with respect to the query set $\{R^- | R \in \mathcal{R}\}$ and the corollary follows immediately. □

Now we construct our set $\mathcal{R}$ of (t_q, h)-approximate favorable queries for d-dimensional approximate simplex range reporting. First select a set of $m = \varepsilon^{1+\delta-d}/t_q^{1+\delta}$ uniform random points in $[0,1]^{d-1}$, for an arbitrary fixed constant $\delta > 0$. Then map each point $(x_1, \ldots, x_{d-1})$ to $\Theta(1/\varepsilon)$ points using the map

$$(x_1, \ldots, x_{d-1}) \to 10\varepsilon d^3 i(x_1 + 1, \ldots, x_{d-1} + 1, 2)$$

where i ranges over integers such that $1/2 \leq 20\varepsilon d^3 i \leq 3/4$. For each new point p, consider the slab $H_{p,2\varepsilon d^2} = \{x \in [0,1]^d | \, |\langle x, p\rangle - |p|^2| \leq 2\varepsilon d^2 |p|\}$. The intersection of each such slab and $[0,1]^{d-1} \times \mathbb{R}$ forms a parallelotope I_p. Let $I'_p = H_{p,4\varepsilon d^2} \cap [-2\varepsilon d^2, 1 + 2\varepsilon d^2]^{d-1} \times \mathbb{R}$. Let $u_{p,0}$ be a vertex of I'_p and $u_{p,1}, \ldots, u_{p,d}$ be its neighboring vertices of I'_p. Let R_p be the simplex whose vertices are $u_{p,0}$ and $u_{p0} + d(u_{p,i} - u_{p,0})$ for $i = 1, \ldots, d$. The query set $\mathcal{R}$ is the set of all R_p's constructed above. Also define the parallelotope $O_p = H_{p,10\varepsilon d^3} \cap [-2d, 2d]^{d-1} \times \mathbb{R}$. First, we show some properties of the constructed queries.

Lemma 1. *For any query* R_p *in the query set* $\mathcal{R}$, $I_p \subset R_p^- \subset R_p \subset R_p^+ \subset O_p$.

Proof. First notice that the diameter of I'_p is bounded by $2d$ so the diameter of R_p is bounded by $2d^2$.

Now we prove the first containment. Notice that every vertex of I'_p can be written as a convex combination of vertices of R_p so $I'_p \subset R_p$. The distance from any point in I_p to $\partial I'_p$ is at least $2\varepsilon d^2 \geq \varepsilon \operatorname{diam}(R_p)$ so $I_p \subset R_p^-$.

Finally we prove the last containment. Since $I'_p \subset H_{p,4\varepsilon d^2} \cap [-2\varepsilon d^2, 1 + 2\varepsilon d^2]^{d-1} \times \mathbb{R}$, by construction, $R_p \subset H_{p,8\varepsilon d^3} \cap [-3d/2, 3d/2]^{d-1} \times \mathbb{R}$. Thus, $R_p^+ \subset H_{p,10\varepsilon d^3} \cap [-2d, 2d]^{d-1} \times \mathbb{R} \subset O_p$. □

We have the following lemma from [10], adapted to our parameters.

Lemma 2. (Chazelle and Rosenberg [10] (Lemma 3.1)) *The constructed query set* $\mathcal{R}$ *satisfies*

1. $\mathcal{R}$ *is a set of size* $\Theta(m/\varepsilon) = \Theta(\varepsilon^{\delta-d}/t_q^{1+\delta})$.
2. I_p *and* O_p *have intersection with* $[0,1]^d$ *of volume* $\Theta(\varepsilon)$ *for any* $R_p \in \mathcal{R}$.
3. *For any* $k = \lg m$ *different* O_p*'s, there exists a subset* $\{O_{p_1}, O_{p_2}, \ldots, O_{p_d}\}$ *of size* d *such that* $O_{p_1} \cap \cdots \cap O_{p_d} \cap [0,1]^d$ *has volume* $O(\varepsilon^d m (\lg m)^{d-2}) = O(\varepsilon^{1+\delta}(\lg(1/\varepsilon))^{d-2}/t_q^{1+\delta})$.

Applying this lemma, we show the following.

Lemma 3. *Consider* $t_q = \omega(\lg(1/\varepsilon))$*. Choose* t_q/ε *points* $P = \{p_1, \ldots, p_{t_q/\varepsilon}\}$ *independently and uniformly in* $[0,1]^d$*. With probability* $1 - o(1)$*, the query set* $\mathcal{R}$ *is* $(t_q, O(\lg(1/\varepsilon)))$*-approximate favorable for the point set* P.

Proof. Since the volume of the intersection between I_p and $[0,1]^d$ is $\Theta(\varepsilon)$, each point p_i is included in I_p with probability $\Theta(\varepsilon)$. By the Chernoff bound, with probability $1 - \exp(\Omega(t_q))$, the number of points in I_p is within a factor 2 of its expectation of $\Theta(t_q)$. Similarly with probability $1 - \exp(\Omega(t_q))$, the number of points in O_p is within a factor 2 of its expectation of $\Theta(t_q)$. Therefore, if $t_q = \omega(\lg(1/\varepsilon))$, then by the union bound, with probability $1 - o(1)$, $|I_p \cap P| = \Theta(t_q)$ and $|O_p \cap P| = \Theta(t_q)$ for all queries Q_p.

Next we apply the following lemma from [10].

Lemma 4. (Chazelle and Rosenberg [10] (Lemma 3.4)) *With probability* $1 - o(1)$*, for all distinct* $p_1, \ldots, p_k$ *where* $k = \lceil \lg m \rceil = O(\lg 1/\varepsilon)$*, we have* $|P \cap O_{p_1} \cap \ldots O_{p_k} \cap [0,1]^d| = O(1)$.

By the above lemma and the fact that $R_{p_i} \subset O_{p_i}$, we have $|P \cap R_{p_1}^- \cap \ldots \cap R_{p_k}^- \cap [0,1]^d| = O(1)$. Thus, with probability $1 - o(1)$, $\mathcal{R}$ is $(t_q, O(\lg(1/\varepsilon)))$-approximate favorable. □

Theorem 5. *Consider a function* $t_q = \omega(\lg(1/\varepsilon))$*. For any data structure that can answer any query* $R \in \mathcal{R}$ *on the point set* P *in time* $O(t_q + |R^+ \cap P|)$*, the space it uses is at least* $n\varepsilon^{1+\delta-d}/t_q^{1+\delta}$ *for any arbitrary constant* $\delta > 0$.

Proof. Consider the disjoint union of $n\varepsilon/t_q$ hypercubes and the corresponding queries constructed above. The union of the query sets is still $(t_q, \lg(1/\varepsilon))$-approximate favorable. Thus, by corollary 1, the space needed is at least $n\varepsilon/t_q \cdot (\varepsilon^{\delta-d}/t_q^{1+\delta} \cdot t_q/\lg(1/\varepsilon)) = n\varepsilon^{1+\delta-d}/(t_q^{1+\delta}\lg(1/\varepsilon))$ for any $\delta > 0$. □

4. CONCLUSION

In this paper, we presented a new theorem for proving lower bounds on dynamic range searching in the group model. This theorem allows us to almost immediately translate lower bounds for range reporting in the pointer machine to lower bounds in the group model. However, as in the recent result of Larsen [13], the lower bounds obtained are conditioned on data structures having low multiplicity. Echoing Larsen, we believe that proving polynomially high lower bounds, that are independent of the multiplicity, remains one of the most important open problems in the group model. This seems to require a radically different approach. A more modest goal would be to develop a technique that allows for different types of query time and update time tradeoffs than the linear tradeoffs presented here and in [13], even under the assumption of constant multiplicity.

We also presented the first lower bounds for approximate range reporting in the pointer machine. Tightening the gap between the upper and lower bound for approximate simplex range reporting remains an interesting open problem. We believe both are subject to improvement.

5. REFERENCES

[1] P. Afshani. Improved pointer machine and I/O lower bounds for simplex range reporting and related problems. In *Proc. 28th ACM Symposium on Computational Geometry*, 2012.

[2] P. Afshani, L. Arge, and K. D. Larsen. Orthogonal range reporting: Query lower bounds, optimal structures in 3d, and higher dimensional improvements. In *Proc. 26th ACM Symposium on Computational Geometry*, pages 240–246, 2010.

[3] P. Afshani, L. Arge, and K. G. Larsen. Higher-dimensional orthogonal range reporting and rectangle stabbing in the pointer machine model. In *Proc. 28th ACM Symposium on Computational Geometry*, 2012.

[4] S. Arya, G. D. da Fonseca, and D. M. Mount. Tradeoffs in approximate range searching made simpler. In *Proceedings of the XXI Brazilian Symposium on Computer Graphics and Image Processing (SIBGRAPI)*, pages 237–244, 2008.

[5] B. Chazelle. Lower bounds for orthogonal range searching: I. the reporting case. *Journal of the ACM*, 37(2):200–212, 1990.

[6] B. Chazelle. Lower bounds for off-line range searching. In *Proc. 27th ACM Symposium on Theory of Computation*, pages 733–740, 1995.

[7] B. Chazelle. A spectral approach to lower bounds with applications to geometric searching. *SIAM Journal on Computing*, 27:545–556, 1998.

[8] B. Chazelle and D. Liu. Lower bounds for intersection searching and fractional cascading in higher dimension. In *Proc. 33rd ACM Symposium on Theory of Computation*, pages 322–329, 2001.

[9] B. Chazelle and A. Lvov. A trace bound for the hereditary discrepancy. In *Proc. 16th ACM Symposium on Computational Geometry*, pages 64–69, 2000.

[10] B. Chazelle and B. Rosenberg. Simplex range reporting on a pointer machine. *Computational Geometry: Theory and Applications*, 5:237–247, January 1996.

[11] M. L. Fredman. The complexity of maintaining an array and computing its partial sums. *Journal of the ACM*, 29:250–260, January 1982.

[12] M. L. Fredman and M. Saks. The cell probe complexity of dynamic data structures. In *Proc 21st ACM Symposium on Theory of Computation*, pages 345–354, 1989.

[13] K. G. Larsen. On range searching in the group model and combinatorial discrepancy. In *Proc. 52nd IEEE Symposium on Foundations of Computer Science*, 2011.

[14] J. Pach and P. K. Agarwal. *Combinatorial geometry.* Wiley-Interscience series in discrete mathematics and optimization. Wiley, 1995.

[15] M. Pătraşcu. Lower bounds for 2-dimensional range counting. In *Proc. 39th ACM Symposium on Theory of Computation*, pages 40–46, 2007.

[16] M. Pătraşcu and E. D. Demaine. Logarithmic lower bounds in the cell-probe model. *SIAM Journal on Computing*, 35:932–963, April 2006.

[17] M. Pătraşcu and M. Thorup. Higher lower bounds for near-neighbor and further rich problems. *SIAM Journal on Computing*, 39(2):730–741, 2010.

[18] R. E. Tarjan. A class of algorithms that require nonlinear time to maintain disjoint sets. *Journal of Computer and System Sciences*, 18:110–127, 1979.

An Output-sensitive Algorithm for Computing Projections of Resultant Polytopes

Ioannis Z. Emiris*
Department of Informatics & Telecommunications
University of Athens, Athens, Greece
emiris@di.uoa.gr

Vissarion Fisikopoulos*
Department of Informatics & Telecommunications
University of Athens, Athens, Greece
vissarion@di.uoa.gr

Christos Konaxis**
Archimedes Center for Modeling,
Analysis & Computation (ACMAC)
University of Crete, Heraklio, Greece
ckonaxis@acmac.uoc.gr

Luis Peñaranda*
Department of Informatics & Telecommunications
University of Athens, Athens, Greece
lpenaranda@di.uoa.gr

ABSTRACT

We develop an incremental algorithm to compute the Newton polytope of the resultant, aka resultant polytope, or its projection along a given direction. The resultant is fundamental in algebraic elimination and in implicitization of parametric hypersurfaces. Our algorithm exactly computes vertex- and halfspace-representations of the desired polytope using an oracle producing resultant vertices in a given direction. It is output-sensitive as it uses one oracle call per vertex. We overcome the bottleneck of determinantal predicates by hashing, thus accelerating execution from 18 to 100 times. We implement our algorithm using the experimental CGAL package `triangulation`. A variant of the algorithm computes successively tighter inner and outer approximations: when these polytopes have, respectively, 90% and 105% of the true volume, runtime is reduced up to 25 times. Our method computes instances of 5-, 6- or 7-dimensional polytopes with 35K, 23K or 500 vertices, resp., within 2hr. Compared to tropical geometry software, ours is faster up to dimension 5 or 6, and competitive in higher dimensions.

*This research was partially funded by the University of Athens' Special Account of Research Grants n° 10812. Partial support from project "Computational Geometric Learning", which acknowledges the financial support of the Future and Emerging Technologies (FET) programme within the 7th Framework Programme for research of the European Commission, under FET-Open grant number: 255827.

**The research leading to these results has received funding from the European Union's 7th Framework Programme (FP7-REGPOT-2009-1) under grant agreement n° 245749; most of the work in this paper was done at University of Athens, supported by "Computational Geometric Learning".

Categories and Subject Descriptors

F.2.2 [**Analysis of Algorithms and Problem Complexity**]: Nonnumerical Algorithms and Problems—*Geometrical problems and computations*; G.4 [**Mathematical Software**]: Algorithm design and analysis.

General Terms

Algorithms, Design, Experimentation, Theory.

Keywords

General Dimension, Convex Hull, Regular Triangulation, Secondary Polytope, Resultant, CGAL Implementation, Experimental Complexity.

1. INTRODUCTION

Given pointsets $A_0, \dots, A_n \subset \mathbb{Z}^n$, we define the pointset

$$\mathcal{A} := \bigcup_{i=0}^{n} (A_i \times \{e_i\}) \subset \mathbb{Z}^{2n}, \qquad (1)$$

where $e_0, \dots, e_n$ form an affine basis of $\mathbb{R}^n$: e_0 is the zero vector, $e_i = (0, \dots, 0, 1, 0, \dots, 0), i = 1, \dots, n$. Clearly, $|\mathcal{A}| = |A_0| + \cdots + |A_n|$, where $|\cdot|$ denotes cardinality. By Cayley's trick (Prop. 2) the regular tight mixed subdivisions of the Minkowski sum $A_0 + \cdots + A_n$ are in bijection with the regular triangulations of $\mathcal{A}$, which are the vertices of the *secondary polytope* $\Sigma(\mathcal{A})$.

The *Newton polytope* of a polynomial is the convex hull of its *support*, i.e. the exponent vectors of monomials with nonzero coefficient. It subsumes the notion of degree for sparse multivariate polynomials by providing more information, unless the polynomial is completely dense. Given $n + 1$ polynomials in n variables, with fixed supports A_i and symbolic coefficients, their *sparse (or toric) resultant* $\mathcal{R}$ is a polynomial in these coefficients which vanishes exactly when the polynomials have a common root (Def. 1). The resultant is the most fundamental tool in elimination theory, and is instrumental in system solving; it is also important in changing representation of parametric hypersurfaces.

The Newton polytope of the resultant $N(\mathcal{R})$, or *resultant polytope*, is the object of our study, especially when some of the input coefficients are not symbolic, in which case we seek a projection of the resultant polytope. The lattice points in $N(\mathcal{R})$ yield a superset of the support of $\mathcal{R}$; this reduces implicitization [14, 30] and computation of $\mathcal{R}$ to sparse interpolation (Sect. 2). The number of coefficients of the $n+1$ polynomials ranges from $O(n)$ to $O(n^d d^n)$, where d bounds their total degree. In system solving and implicitization, one computes $\mathcal{R}$ when all but $O(n)$ of the coefficients are specialized to constants, hence the need for resultant polytope projections.

The resultant polytope is a Minkowski summand of $\Sigma(\mathcal{A})$. For its construction, we exploit an equivalence relation defined on the secondary vertices, where the classes are in bijection with the vertices of the resultant polytope. This yields an oracle producing a resultant vertex in a given direction, thus avoiding to compute $\Sigma(\mathcal{A})$, which has much more vertices than $N(\mathcal{R})$. Although there exist efficient software for $\Sigma(\mathcal{A})$ [28], it is useless in computing resultant polytopes. For instance, in implicitizing parametric surfaces with < 100 input terms, we compute the Newton polytope of the equations in $<$ 1sec, which includes all common instances in geometric modeling, whereas $\Sigma(\mathcal{A})$ is intractable.

Our main contribution is twofold. First, we design an output-sensitive algorithm for computing the Newton polytope of $\mathcal{R}$, or of specializations of $\mathcal{R}$. The algorithm computes both vertex (V) and halfspace (H) representations, which is important for the targeted applications. Its incremental nature implies that we also obtain a triangulation of the polytope, which may be useful for enumerating its lattice points in subsequent applications. The complexity is proportional to the number of output vertices and facets; the overall cost is dominated by computing as many regular triangulations of $\mathcal{A}$ (Thm. 10). We work in the space of the projected $N(\mathcal{R})$ and revert to the high-dimensional space of $\Sigma(\mathcal{A})$ only if needed. Our algorithm readily extends to computing $\Sigma(\mathcal{A})$, projections of $\Sigma(\mathcal{A})$ and, more generally, any polytope that can be efficiently described by a vertex oracle. A variant of our algorithm computes successively tighter inner and outer approximations: typically, these polytopes have, respectively, 90% and 105% of the true volume, while runtime is reduced up to 25 times. This may lead to an approximation algorithm.

Second, we describe an efficient implementation based on CGAL [7] and the experimental package `triangulation`. Our method computes instances of 5-, 6- or 7-dimensional polytopes with 35K, 23K or 500 vertices, respectively, in $<$ 2hr. Our code is faster up to dimensions 5 or 6, and competitive in higher dimensions, to a method computing $N(\mathcal{R})$ via tropical geometry and based on the `Gfan` library [21]. Moreover, our code in the critical step of computing convex hulls, uses `triangulation` which compared to state-of-the-art software `lrs`, `cdd`, and `polymake`, is the fastest together with `polymake`. We factor out repeated computation by reducing the bulk of our work to a sequence of determinants: this is often the case in high-dimensional geometric computing. Here, we exploit the nature of our problem to capture the similarities of the predicates, and hash the computed minors which are needed later, to speedup subsequent determinants. This is of independent interest and can be used to improve other related problems such as convex hull and triangulation computations.

Let us review previous work. Sparse (or toric) elimination theory was introduced in [18]. They show that $N(\mathcal{R})$, for two univariate polynomials with k_0+1, k_1+1 monomials, has $\binom{k_0+k_1}{k_0}$ vertices and, when both $k_i \geq 2$, it has k_0k_1+3 facets. In [29, Sec.6] is proven that $N(\mathcal{R})$ is 1-dimensional iff $|A_i| = 2$, for all i, the only planar $N(\mathcal{R})$ is the triangle, whereas the only 3-dimensional ones are the tetrahedron, the square-based pyramid, and the polytope of two univariate trinomials; we compute an instance of the latter (Fig. 2(b)). Following [29, Thm.6.2], the 4-dimensional polytopes include the 4-simplex, some $N(\mathcal{R})$ obtained by pairs of univariate polynomials, and those of 3 trinomials, which we can investigate with our code: the maximal such polytope we have computed has f-vector $(22, 66, 66, 22)$ (Fig. 2(c)).

In [26] they describe all Minkowski summands of $\Sigma(A)$. In [27] is defined an equivalence class over $\Sigma(A)$ vertices having the same mixed cells. The classes map in a many-to-1 fashion to resultant vertices; our algorithm exploits a stronger equivalence relationship. Tropical geometry is a polyhedral analogue of algebraic geometry which gives alternative ways of recovering resultant polytopes [21] and Newton polytopes of implicit equations [30]. Sect. 5 discusses it and shows our approach is faster up to dimensions 5, 6 and competitive in higher dimensions.

We focus on sequences of determinantal predicates. For determinants, the record bit complexity is $O(n^{2.697})$ [23]. Methods exist for the sign of general determinants, e.g. [6]. We compared linear algebra libraries LinBox [11] and Eigen [19], which seem most suitable in dimension > 100 and medium-high dimensions, respectively, whereas CGAL provides the most efficient determinant computation for the dimensions to which we focus.

The roadmap of the paper follows: Sect. 2 describes the combinatorics of resultants, and the following section presents our algorithm. Sect. 4 overcomes the bottleneck of Orientation predicates. Sect. 5 discusses the implementation, experiments, and comparison with other software. We conclude with future work.

2. RESULTANT POLYTOPES AND THEIR PROJECTIONS

We introduce tools from combinatorial geometry [25, 31] to describe resultants [9, 18]. Let $\mathrm{vol}(\cdot) \in \mathbb{N}$ denote normalized Euclidean volume, and $(\mathbb{R}^m)^\times$ the linear m-dimensional functionals.

Let $\mathcal{A} \subset \mathbb{R}^d$ be a pointset whose convex hull is of dimension d. For any triangulation T of $\mathcal{A}$, define vector $\phi_T \in \mathbb{R}^{|\mathcal{A}|}$ with coordinate

$$\phi_T(a) = \sum_{\sigma \in T: a \in \sigma} \mathrm{vol}(\sigma), \qquad a \in \mathcal{A}, \tag{2}$$

summing over all simplices σ of T having a as a vertex; $\Sigma(\mathcal{A})$ is the convex hull of ϕ_T for all triangulations T. Let $\mathcal{A}^w$ denote pointset $\mathcal{A}$ lifted to $\mathbb{R}^{d+1}$ via a generic lifting function w in $(\mathbb{R}^{|\mathcal{A}|})^\times$. *Regular triangulations* of $\mathcal{A}$ are obtained by projecting the upper (or lower) hull of $\mathcal{A}^w$ back to $\mathbb{R}^d$.

PROPOSITION 1. *[18] The vertices of $\Sigma(\mathcal{A})$ correspond to the regular triangulations of $\mathcal{A}$, while its face lattice corresponds to the poset of regular polyhedral subdivisions of $\mathcal{A}$, ordered by refinement. A lifting vector produces regular triangulation T (resp. a regular polyhedral subdivision of $\mathcal{A}$) iff it lies in the normal cone of vertex ϕ_T (resp. of the corresponding face) of $\Sigma(\mathcal{A})$. The dimension of $\Sigma(\mathcal{A})$ is $|\mathcal{A}|-d-1$.*

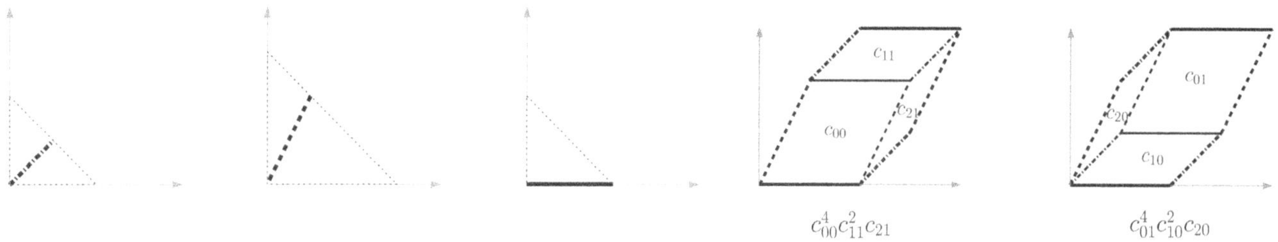

Figure 1: Let $f_0 := c_{00} - c_{01}x_1x_2$, $f_1 := c_{10} - c_{11}x_1x_2^2$, $f_2 := c_{20} - c_{21}x_1^2$. Left: Newton polygons and those of fully dense polynomials of same total degree (dashed). Right: There are 2 regular tight mixed subdivisions of the Minkowski sum giving the extreme monomials of $\mathcal{R} = -c_{00}^4c_{11}^2c_{21} + c_{01}^4c_{10}^2c_{20}$.

Let $A_0, \ldots, A_n$ be subsets of $\mathbb{Z}^n$, $P_0, \ldots, P_n \subset \mathbb{R}^n$ their convex hulls, and $P = P_0 + \cdots + P_n$ their Minkowski sum. A *Minkowski (maximal) cell* of P is any full-dimensional convex polytope $B = \sum_{i=0}^n B_i$, where each B_i is a convex polytope with vertices in A_i. Minkowski cells $B, B' = \sum_{i=0}^n B_i'$ intersect properly when $B_i \cap B_i'$ is a face of both and their Minkowski sum descriptions are compatible, i.e. coincide on the common face. A *mixed subdivision* of P is any family of Minkowski cells which partition P and intersect properly. A cell is *i-mixed* or *v_i-mixed*, if it is the Minkowski sum of n one-dimensional segments from P_j, $j \neq i$, and some vertex $v_i \in P_i$.

Mixed subdivisions contain *faces* of all dimensions between 0 and n, the maximum dimension corresponding to cells. Every face of a mixed subdivision of P has a unique description as Minkowski sum of $B_i \subset P_i$. A mixed subdivision is *regular* if it is obtained as the projection of the upper (or lower) hull of the Minkowski sum of lifted polytopes $P_i^{w_i} := \{(p_i, w_i(p_i)) \mid p_i \in P_i\}$, for lifting $w_i : P_i \to \mathbb{R}$. If the lifting function $w := (w_0 \ldots, w_n)$ is sufficiently generic, then the mixed subdivision is *tight*, and $\sum_{i=0}^n \dim B_i = \dim \sum_{i=0}^n B_i$, for every cell. Given $A_0, \ldots, A_n$ and the affine basis $\{e_0, \ldots, e_n\}$ of $\mathbb{R}^n$, we define the Cayley pointset $\mathcal{A} \subset \mathbb{Z}^{2n}$ as in equation (1).

Proposition 2. *[Cayley trick] [18] There exist bijections between: the regular tight mixed subdivisions of P and the regular triangulations of $\mathcal{A}$; the tight mixed subdivisions of P and the triangulations of $\mathcal{A}$; the mixed subdivisions of P and the polyhedral subdivisions of $\mathcal{A}$.*

The family $A_0, \ldots, A_n \subset \mathbb{Z}^n$ is *essential* if they jointly affinely span $\mathbb{Z}^n$ and every subset of cardinality $j, 1 \leq j < n$, spans a space of dimension $\geq j$. It is straightforward to check this property algorithmically and, if it does not hold, to find an essential subset. In the sequel, the input $A_0, \ldots, A_n \subset \mathbb{Z}^n$ is supposed to be essential.

Definition 1. *Let $A_0, \ldots, A_n \subset \mathbb{Z}^n$ be an essential family, and $f_0, \ldots, f_n \in \mathbb{C}[x_1, \ldots, x_n]$ polynomials with these supports and symbolic coefficients $c_{ij}, i = 0, \ldots, n, j = 1, \ldots, |A_i|$, i.e. $f_i = \sum_{a \in A_i} c_{ij}x^a$. Their sparse (or toric) resultant is a polynomial in $\mathbb{Z}[c_{ij} : i = 0, \ldots, n, j = 1, \ldots, |A_i|]$, defined up to sign, which vanishes iff $f_0 = f_1 = \cdots = f_n = 0$ has a common root in $(\mathbb{C}^*)^n$, $\mathbb{C}^* = \mathbb{C} \setminus \{0\}$.*

For $n = 1$, the resultant is named after Sylvester. For linear systems, it equals the determinant of the $(n+1) \times (n+1)$ coefficient matrix. The discriminant of a polynomial $F(x_1, \ldots, x_n)$ is given by the resultant of $F, \partial F/\partial x_1, \ldots, \partial F/\partial x_n$.

The Newton polytope $N(\mathcal{R})$ of the resultant is a lattice polytope called *resultant polytope*: the resultant has $|\mathcal{A}| = \sum_{i=0}^n |A_i|$ variables, hence $N(\mathcal{R})$ lies in $\mathbb{R}^{|\mathcal{A}|}$, though of smaller dimension (Prop. 4). The monomials corresponding to vertices of $N(\mathcal{R})$ are the extreme resultant monomials. For a sufficiently generic lifting $w \in (\mathbb{R}^{|\mathcal{A}|})^\times$, the w-extreme resultant monomial, whose exponent vector maximizes inner product with w, is recovered from the regular tight mixed subdivision S of P induced by w. Let T be the regular triangulation corresponding, via the Cayley trick, to S, and $\rho_T \in \mathbb{N}^{|\mathcal{A}|}$ the exponent of the w-extreme monomial. Then,

$$\rho_T(a) = \sum_{\substack{a-\text{mixed} \\ \sigma \in T : a \in \sigma}} \operatorname{vol}(\sigma) \in \mathbb{N}, \qquad a \in \mathcal{A}, \tag{3}$$

where simplex σ is a-mixed iff the corresponding cell is a-mixed in S. Note that, $\rho_T(a) \in \mathbb{N}$, since it is a sum of volumes of mixed cells σ, each of them equal to the *mixed volume* of a set of *lattice* polytopes, the Minkowksi summands of σ:

$$\operatorname{vol}(\sigma) = MV(\sigma_0, \ldots, \sigma_n), \quad \sigma = \sigma_0 + \cdots + \sigma_n, \ \sigma_i \subset \mathcal{A}_i,$$

where MV denotes the mixed volume function. Now, $N(\mathcal{R})$ is the convex hull of all ρ_T vectors [18, 29].

There exists a many-to-1 surjection from vertices of $\Sigma(\mathcal{A})$ to those of $N(\mathcal{R})$. One defines an *equivalence relationship* on all regular tight mixed subdivisions, where equivalent subdivisions yield the same vertex in $N(\mathcal{R})$. Thus, equivalent vertices of $\Sigma(\mathcal{A})$ correspond to the same resultant vertex. Consider $w \in (\mathbb{R}^{|\mathcal{A}|})^\times$ lying in the union of outer-normal cones of equivalent vertices of $\Sigma(\mathcal{A})$. They correspond to a resultant vertex whose outer-normal cone contains w; this defines a w-extremal resultant monomial. If w is non-generic, it specifies a sum of extremal monomials in $\mathcal{R}$, i.e. a face of $N(\mathcal{R})$ (Fig. 2(a),(b)).

We compute some orthogonal projection of $N(\mathcal{R})$, denoted Π, in $\mathbb{R}^m$:

$$\pi : \mathbb{R}^{|\mathcal{A}|} \to \mathbb{R}^m : N(\mathcal{R}) \to \Pi, \ m \leq |\mathcal{A}|.$$

By reindexing, this is the subspace of the first m coordinates, so $\pi(\rho) = (\rho_1, \ldots, \rho_m)$. It is possible that none of the coefficients c_{ij} is specialized, hence $m = |\mathcal{A}|$, π is trivial, and $\Pi = N(\mathcal{R})$. Assuming the specialized coefficients take sufficiently generic values, Π is the Newton polytope of the corresponding specialization of $\mathcal{R}$. The following is used for preprocessing.

Lemma 3. *[21, Lem.3.20] If $a_{ij} \in A_i$ corresponds to a specialized coefficient of f_i, and lies in the convex hull of the*

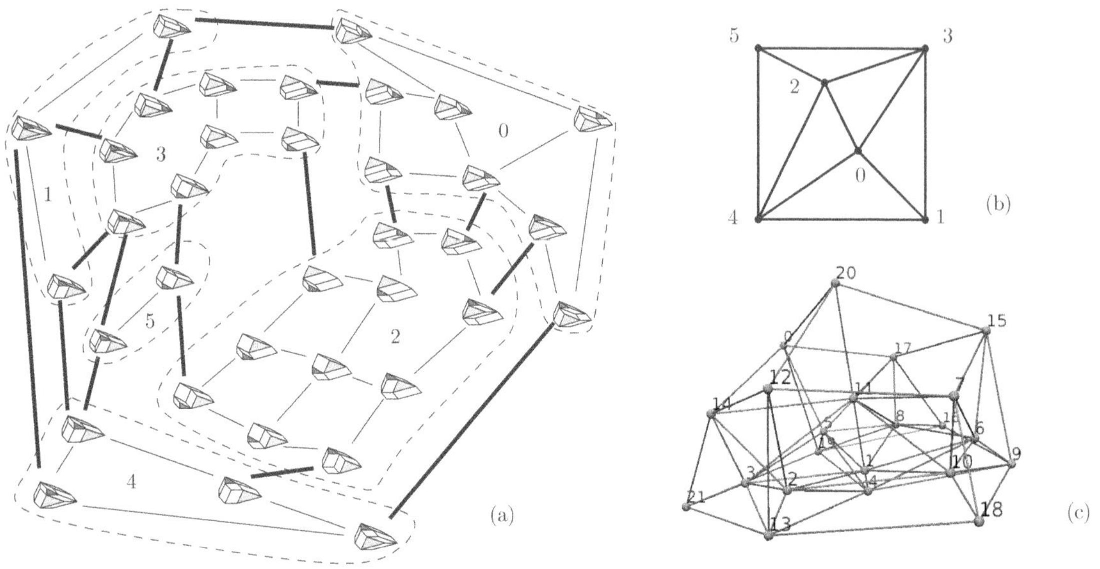

Figure 2: (a) Σ polytope of two triangles (dark, light grey) and one segment $A_0 = \{(0,0),(1,2),(4,1)\}$, $A_1 = \{(0,1),(1,0)\}$, $A_2 = \{(0,0),(0,1),(2,0)\}$; vertices correspond to mixed subdivisions of the Minkowski sum $A_0 + A_1 + A_2$ and edges to flips between them (b) $N(\mathcal{R})$, whose vertices correspond to the dashed classes of Σ. Bold edges of Σ, called cubical flips, map to edges of $N(\mathcal{R})$ (c) 4-dimensional $N(\mathcal{R})$ of 3 generic trinomials with f-vector $(22,66,66,22)$; figure made with `polymake`**.**

other points in A_i corresponding to specialized coefficients, then removing a_{ij} from A_i does not change the Newton polytope of the specialized resultant.

PROPOSITION 4. *[18] $N(\mathcal{R})$ is a Minkowski summand of $\Sigma(\mathcal{A})$, of dimension $|\mathcal{A}| - 2n - 1$.*

Let us describe the $2n+1$ hyperplanes in whose intersection lies $N(\mathcal{R})$. For this, let M be the $(2n+1) \times |\mathcal{A}|$ matrix whose columns are the points in the A_i, where each $a \in A_i$ is followed by the i-th unit vector in $\mathbb{N}^{n+1}$. Then, the inner product of any coordinate vector of $N(\mathcal{R})$ with row i of M is: constant, for $i = 1,\ldots,n$, and known, and depends on i, for $i = n+1,\ldots,2n+1$ [18, Prop.7.1.11]. This implies that one obtains an isomorphic polytope when projecting $N(\mathcal{R})$ along $2n+1$ points in $\mathcal{A}$ which affinely span $\mathbb{R}^{2n}$; this is possible because of the assumption of essential family. Having computed the projection, we obtain $N(\mathcal{R})$ by computing the missing coordinates as the solution of a linear system: we write the aforementioned inner products as $M[X\,V]^T = C$, where C is a known matrix and $[X\,V]^T$ is a transposed $(2n+1) \times u$ matrix, expressing the partition of the coordinates to unknown and known values, where u is the number of $N(\mathcal{R})$ vertices. If the first $2n+1$ columns of M correspond to specialized coefficients, $M = [M_1\, M_2]$, where submatrix M_1 is of dimension $2n+1$ and invertible, hence $X = M_1^{-1}(C - M_2 B)$.

We focus on 3 applications. First, we interpolate the resultant in all coefficients, thus illustrating an alternative method for computing resultants.

EXAMPLE 1. *Let $f_0 = a_2x^2 + a_1x + a_0$, $f_1 = b_1x^2 + b_0$, with supports $A_0 = 2,1,0$, $A_1 = 1,0$. Their (Sylvester) resultant is a polynomial in a_2, a_1, a_0, b_1, b_0. Our algorithm computes its Newton polytope with vertices $(0,2,0,1,1)$, $(0,0,2,2,0)$, $(2,0,0,0,2)$; it contains 4 points, corresponding to 4 potential monomials $a_1^2b_1b_0$, $a_0^2b_1^2$, $a_2a_0b_1b_0$, $a_2^2b_0^2$. There is a parameterization of resultants [24], which yields: $a_2 = (2t_1+t_2)t_3^2t_4$, $a_1 = (-2t_1 - 2t_2)t_3t_4$, $a_0 = t_2t_4$, $b_1 = -t_1t_3^2t_5$, $b_0 = t_1t_5$, where the t_i's are parameters. We substitute these expressions to the monomials, evaluate at 4 sufficiently random t_i's, and obtain a matrix whose kernel vector $(1,1,-2,1)$ yields $\mathcal{R} = a_1^2b_1b_0 + a_0^2b_1^2 - 2a_2a_0b_1b_0 + a_2^2b_0^2$.* □

Second, consider system solving by the rational univariate representation of roots [3]. Given $f_1,\ldots,f_n \in \mathbb{C}[x_1,\ldots,x_n]$, define an overconstrained system by adding $f_0 = u_0 + u_1x_1 + \cdots + u_nx_n$ with symbolic u_i's. Let coefficients $c_{ij}, i \geq 1$, take specific values, and suppose the roots of $f_1 = \cdots = f_n = 0$ are isolated, denoted $r_i = (r_{i1},\ldots,r_{in})$. Then the u-resultant is $\mathcal{R}_u = a\prod_{r_i}(u_0 + u_1r_{i1} + \cdots + u_nr_{in})^{m_i}$, $a \in \mathbb{C}^*$, where m_i is the multiplicity of r_i. Computing $\mathcal{R}_u$ is the bottleneck; our method computes (a superset of) $N(\mathcal{R}_u)$.

The last application comes from geometric modeling, where $y_i = f_i(x)$, $i = 0,\ldots,n$, $x = (x_1,\ldots,x_n) \in \Omega \subset \mathbb{R}^n$, defines a parametric hypersurface. Many applications require the equivalent implicit representation $F(y_1,\ldots,y_n) = 0$. This amounts to eliminating x, so it is crucial to compute the resultant when coefficients are specialized except the y_i's. Our approach computes a polytope that contains the Newton polytope of F, thus reducing implicitization to interpolation [14]. In particular, we compute the polytope of surface equations within 1sec, assuming < 100 terms in parametric polynomials, which includes all common instances in geometric modeling, whereas the corresponding $\Sigma(\mathcal{A})$ are intractable.

3. ALGORITHMS AND COMPLEXITY

This section analyzes our exact and approximate algorithms for computing orthogonal projections of polytopes whose vertices are defined by an *oracle*. This oracle computes the vertex of the polytope which is extremal in a given direction w. If there are more than one vertices extremal in w it computes exactly one of these. Moreover, we define such an oracle for the vertices of orthogonal projections Π of $N(\mathcal{R})$ which results to algorithms for computing Π without computing $N(\mathcal{R})$. Finally, we analyze the asymptotic complexity of these algorithms.

Given pointset V, reg_subdivision(V, w) computes the regular subdivision of its convex hull by projecting the upper hull of V lifted by w, and conv(V) computes the H-representation of its convex hull. The oracle $\text{Vtx}(\mathcal{A}, w, \pi)$ computes a point in Π, extremal in the direction w, by refining the output of reg_subdivision($\mathcal{A}, \widehat{w}$), $\widehat{w} = (w, \vec{0}) \in (\mathbb{R}^{|\mathcal{A}|})^{\times}$, into a regular triangulation T of $\mathcal{A}$, then returning $\pi(\rho_T)$. It is clear that, triangulation T constructed by Vtx, is regular and corresponds to some secondary vertex ϕ_T which maximizes the inner product with $\widehat{w}$.

LEMMA 5. *All points computed by Vtx are vertices of Π.*

PROOF. Let $v = \pi(\rho_T) = \text{Vtx}(\mathcal{A}, w, \pi)$. We first prove that v lies on $\partial\Pi$. Point ρ_T of $N(\mathcal{R})$ is a Minkowski summand of vertex ϕ_T of $\Sigma(\mathcal{A})$ extremal wrt $\widehat{w}$, hence ρ_T is extremal wrt $\widehat{w}$. Since $\widehat{w}$ is perpendicular to projection π, ρ_T projects to a point in $\partial\Pi$. The same argument implies that every vertex ϕ'_T, where T' is a triangulation refining the subdivision produced by $\widehat{w}$, corresponds to a resultant vertex $\rho_{T'}$ s.t. $\pi(\rho_{T'})$ lies on the same face of Π with ρ_T, hence also lies on $\partial\Pi$.

Now we prove that v is a vertex of Π. Let w be such that the face f of $N(\mathcal{R})$ extremal wrt $\widehat{w}$ contains a vertex ρ_T which projects to relint($\pi(f)$), where relint($\cdot$) denotes relative interior. Then, if $\text{Vtx}(\mathcal{A}, w, \pi)$ returns $\pi(\rho_T)$, this is a point on $\partial\Pi$ but not a vertex of Π. We resolve such degeneracies by adding an infinitesimal generic perturbation vector to w, thus obtaining w_p. Since the perturbation is arbitrarily small, $\widehat{w_p}$ shall be normal to a vertex of f, extremal wrt w, but projecting to a vertex of $\pi(f)$. This equivalent to computing a triangulation refining the regular subdivision of $\mathcal{A}$ induced by $\widehat{w}$, whose normal cone's projection is full (m) dimensional. The perturbation can be implemented in Vtx, without affecting any other parts of the algorithm. In practice, our implementation does avoid degenerate cases. □

The *initialization algorithm* computes an inner approximation of Π in both V- and H-representations (denoted Q, Q^H, resp.), and triangulated. First, it calls $\text{Vtx}(\mathcal{A}, w, \pi)$ for $w \in W \subset (\mathbb{R}^m)^{\times}$; set W is either random or contains, say, vectors in the $2m$ coordinate directions. Then, it updates Q by adding $\text{Vtx}(\mathcal{A}, w, \pi)$ and $\text{Vtx}(\mathcal{A}, -w, \pi)$, where w is normal to hyperplane $H \subset \mathbb{R}^m$ containing Q, as long as either of these points lies outside H. We stop when these points do no longer increase $\dim(Q)$.

LEMMA 6. *The initialization algorithm computes $Q \subseteq \Pi$ s.t. $\dim(Q) = \dim(\Pi)$.*

Incremental Alg. 1 computes both V- and H-representations of Π and a triangulation of Π, given an inner approximation Q, Q^H of Π computed at initialization. A hyperplane H is *legal* if it is a supporting hyperplane to a facet of Π, otherwise it is *illegal*. At every step of Alg. 1, we compute $v = \text{Vtx}(\mathcal{A}, w, \pi)$ for a supporting hyperplane H of a facet of Q with normal w. If $v \notin H$, it is a new vertex thus yielding a tighter *inner approximation* of Π by inserting it to Q, i.e. $Q \subset \text{CH}(Q \cup v) \subseteq \Pi$, where CH($\cdot$) denotes convex hull. This happens when the preimage $\pi^{-1}(f) \subset N(\mathcal{R})$ of the facet f of Q defined by H, is not a Minkowski summand of a face of $\Sigma(\mathcal{A})$ having normal $\widehat{w}$. Otherwise, there are two cases: either $v \in H$ and $v \in Q$, thus the algorithm simply decides hyperplane H is legal, or $v \in H$ and $v \notin Q$, in which case the algorithm again decides H is legal but also inserts v to Q.

The algorithm computes Q^H from Q, then iterates over the new hyperplanes to either compute new vertices or decide they are legal, until no increment is possible, which happens when all hyperplanes are legal. Alg. 1 ensures that each normal w to a hyperplane supporting a facet of Q is used only *once*, by storing all used w's in a set W. When a new normal w is created, the algorithm checks if $w \notin W$, then calls $\text{Vtx}(\mathcal{A}, w, \pi)$ and updates $W \leftarrow W \cup w$. If $w \in W$ then the same or a parallel hyperplane has been checked in a previous step of the algorithm. It is straightforward that w can be safely ignored; Lem. 7 formalizes the latter case.

LEMMA 7. *Let H' be a hyperplane supporting a facet constructed by Alg. 1, and $H \neq H'$ an illegal hyperplane at a previous step. If H', H are parallel then H' is legal.*

Algorithm 1: ComputeΠ $(A_0, \ldots, A_n, \pi)$

Input : essential $A_0, \ldots, A_n \subset \mathbb{Z}^n$ processed by lem. 3,
projection $\pi : \mathbb{R}^{|\mathcal{A}|} \to \mathbb{R}^m$,
H-, V-repres. Q^H, Q; triang. T_Q of $Q \subseteq \Pi$.
Output: H-, V-repres. Q^H, Q; triang. T_Q of $Q = \Pi$.

```
A ← ⋃_0^n (A_i × e_i)    // Cayley trick
H_illegal ← ∅
foreach H ∈ Q^H do H_illegal ← H_illegal ∪ {H}
while H_illegal ≠ ∅ do
    select H ∈ H_illegal and H_illegal ← H_illegal \ {H}
    w is the outer normal vector of H
    v ← Vtx(A, w, π)
    if v ∉ H ∩ Q then
        Q^H_temp ← conv(Q ∪ {v})
        foreach (d − 1)-face f ∈ T_Q with f ⊂ ∂H do
            T_Q ← T_Q ∪ {faces of conv(f, v)}
        foreach H' ∈ {Q^H \ Q^H_temp} do
            H_illegal ← H_illegal \ {H'}//H'separates Q, v
        foreach H' ∈ {Q^H_temp \ Q^H} do
            H_illegal ← H_illegal ∪ {H'} // new h/plane
        Q ← Q ∪ {v}
        Q^H ← Q^H_temp
return Q, Q^H, T_Q
```

The next lemma formulates the termination criterion of our algorithm.

LEMMA 8. *Let vertex v be computed by $Vtx(\mathcal{A}, w, \pi)$, where w is normal to a supporting hyperplane H of Q, then $v \notin H \Leftrightarrow H$ is not a supporting hyperplane of Π.*

PROOF. Let $v = \pi(\rho_T)$, where T is a triangulation refining subdivision S in Vtx. It is clear that, since $v \in \partial\Pi$ extremal wrt w, if $v \notin H$ then H cannot be a supporting hyperplane of Π. Conversely, let $v \in H$. By the proof of Lem. 5, every other vertex $\pi(\rho'_T)$ on the face of $N(\mathcal{R})$ is extremal wrt w, hence lies on H, thus H is a supporting hyperplane of Π. □

We now bound the *complexity* of our algorithm. Beneath-and-beyond, given a k-dimensional polytope with l vertices, computes its H-representation and a triangulation in $O(k^5 l t^2)$, where t is the number of full-dimensional faces (cells) [22]. Let $|\Pi|, |\Pi^H|$ be the number of vertices and facets of Π.

LEMMA 9. *Alg. 1 computes at most* $|\Pi| + |\Pi^H|$ *regular triangulations of* $\mathcal{A}$.

PROOF. The steps of Alg. 1 increment Q. At every such step, and for each supporting hyperplane H of Q with normal w, we compute one vertex of Π, by Lem. 5. If H is illegal, this vertex is unique because H separates the set of (already computed) vertices of Q from the set of vertices of $\Pi \backslash Q$ which are extremal wrt w, hence, an appropriate translate of H also separates the corresponding sets of vertices of $\Sigma(\mathcal{A})$ (Fig. 3). This vertex is never computed again because it now belongs to Q. The number of regular triangulations of $\mathcal{A}$ yielding vertices is thus bounded by $|\Pi|$.

For a legal hyperplane of Q, we compute one vertex of Π that confirms its legality; the regular triangulation of $\mathcal{A}$ yielding this vertex is accounted for by the legal hyperplane. Since every normal to a hyperplane of Q is used only once in Alg. 1 (by the earlier discussion concerning the set W of all used normals), the statement follows. □

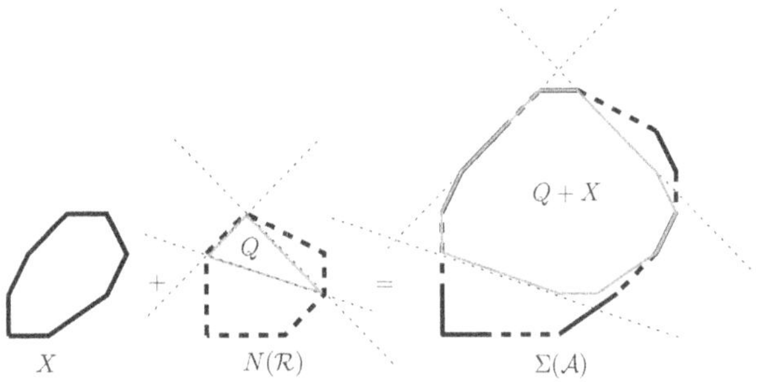

Figure 3: Lem. 9: each illegal hyperplane of Q with normal w, separates the already computed vertices of Π (here equal to $N(\mathcal{R})$) from new ones, extremal wrt w. X is a polytope s.t. $X + N(\mathcal{R}) = \Sigma(\mathcal{A})$.

Let the size of a triangulation be the number of its cells. Let $s_{\mathcal{A}}$ denote the size of the largest triangulation of $\mathcal{A}$ computed by Vtx, and s_Π that of Π computed by Alg. 1. In Vtx, the computation of a regular triangulation reduces to a convex hull, computed in $O(n^5|\mathcal{A}|s_{\mathcal{A}}^2)$; for ρ_T we compute Volume for all cells of T in $O(s_{\mathcal{A}} n^3)$. The overall complexity of Vtx becomes $O(n^5|\mathcal{A}|s_{\mathcal{A}}^2)$. Alg. 1 calls, in every step, Vtx to find a point on $\partial\Pi$ and insert it to Q, or conclude that a hyperplane is legal. By Lem. 9 it executes Vtx as many as $|\Pi| + |\Pi^H|$ times, in $O((|\Pi| + |\Pi^H|)n^5|\mathcal{A}|s_{\mathcal{A}}^2)$, and computes the H-representation of Π in $O(m^5|\Pi|s_\Pi^2)$. Now we have, $|\mathcal{A}| \leq (2n+1)s_{\mathcal{A}}$ and as the input $|\mathcal{A}|, m, n$ grows large we can assume that $|\Pi| \gg |\mathcal{A}|$ and thus s_Π dominates $s_{\mathcal{A}}$. Moreover, $s_\Pi(m+1) \geq |\Pi^H|$. Now, let $\widetilde{O}(\cdot)$ imply that polylogarithmic factors are ignored.

THEOREM 10. *The time complexity of Alg. 1 to compute* $\Pi \subset \mathbb{R}^m$ *is* $O(m^5|\Pi|s_\Pi^2 + (|\Pi| + |\Pi^H|)n^5|\mathcal{A}|s_{\mathcal{A}}^2)$, *which becomes* $\widetilde{O}(|\Pi|s_\Pi^2)$ *when* $|\Pi| \gg |\mathcal{A}|$.

This implies our algorithm is output sensitive. Its experimental performance confirms this property, see Sect. 5.

We have proven that oracle Vtx (within our algorithm) has two important properties:

1. Its output is a vertex of the target polytope (Lem. 5)
2. When the direction w is normal to an illegal facet, then the vertex computed by the oracle is computed once (Lem. 9)

The algorithm can easily be generalized to incrementally compute any polytope P if the oracle associated with the problem satisfies property (1); if it satisfies also property (2), then the computation can be done in $O(|P| + |P^H|)$ oracle calls, where $|P|$, $|P^H|$ denotes the number of vertices and number of facets of P, respectively. For example, if the described oracle returns $\pi(\phi_T)$ instead of $\pi(\rho_T)$, it can be used to compute orthogonal projections of secondary polytopes.

The algorithm readily yields an approximate variant: for each supporting hyperplane H, we use its normal w to compute v =Vtx$(\mathcal{A}, w, \pi)$. Instead of computing a convex hull, now simply take the hyperplane parallel to H through v. The set of these hyperplanes defines a polytope $Q_o \supseteq \Pi$, i.e. an *outer approximation* of Π. Thus, we have an approximation algorithm by stopping Alg. 1 when vol(Q)/vol(Q_o) achieves a user-defined threshold. Then, vol(Q)/vol(Π) is bounded by the same threshold. This algorithm is up to 25 times faster than the deterministic version. Of course, vol(Q) is available by our incremental convex hull algorithm. However, vol(Q_o) is the critical step; we plan to examine algorithms that update (exactly or approximately) this volume.

When all hyperplanes of Q are checked, knowledge of legal hyperplanes accelerates subsequent computations of Q^H, although it does not affect its worst-case complexity. Specifically, it allows us to avoid checking legal facets against new vertices.

4. HASHING OF DETERMINANTS

This section discusses methods to avoid duplication of computations by exploiting the nature of the determinants appearing in the inner loop of our algorithm. Our algorithm computes many regular triangulations, which are typically dominated by the computation of determinants. Similar techniques can be used to improve determinantal predicates in incremental convex hull computations [13].

Consider the $2n \times |\mathcal{A}|$ matrix with the points of $\mathcal{A}$ as columns. Define P as the extension of this matrix by adding lifting values $\widehat{w}$ as the last row. We use the Laplace (or cofactor) expansion along the last row for computing the determinant of the square submatrix formed by any $2n+1$ columns of P; wlog these are the first $2n+1$ columns $a_1, \ldots, a_{2n+1}$. Let $\langle a_1, \ldots, a_{2n+1}\rangle \setminus i$ be the vector $\langle a_1, \ldots, a_{2n+1}\rangle$ without its i-th element; $P_{\langle a_1,\ldots,a_{2n+1}\rangle \setminus i}$ is the $(2n) \times (2n)$ matrix obtained from the $2n$ first elements of the columns whose indices are in $\langle a_1, \ldots, a_{2n+1}\rangle \setminus i$.

Orientation is the sign of the determinant of $P^{hom}_{\langle a_1,\ldots,a_{2n+2}\rangle}$, constructed by columns $a_1, \ldots, a_{2n+2}$ when we add $\vec{1} \in \mathbb{R}^{2n+2}$ as last row. Computing a regular subdivision is a long sequence of such predicates with different a_i's. We expand along the next-to-last row which contains the lifting values and compute the determinants $|P_{\langle a_1,\ldots,a_{2n+2}\rangle \setminus \{i\}}|$ for $i \in \{1, \ldots, 2n+2\}$. Another predicate is Volume, used by Vtx. It equals the determinant of $P^{hom}_{\langle a_1,\ldots,a_{2n+1}\rangle}$, constructed by columns $a_1, \ldots, a_{2n+1}$ when we replace its last row by $\vec{1} \in \mathbb{R}^{2n+1}$.

Our contribution consists in maintaining a hash table with the computed minors, which will be reused at subsequent steps of the algorithm. We store all minors of sizes between 2 and $2n$. For Orientation, they are independent of w and once computed they are stored in the hash table. The main advantage of our scheme is that, for a new w, the only change in P are m (nonzero) coordinates in the last row, hence computing the new determinants can be done by reusing hashed minors. This also saves time from matrix constructions.

Laplace expansion has complexity $O(n!)$, which decreases as long as minors do not need to be recomputed.

LEMMA 11. *Using hashing of determinants, the complexity of the Orientation and Volume predicates is $O(n)$ and $O(1)$, respectively, if all minors have already been computed.*

Many determinant algorithms modify the input matrix; this makes necessary to create a new matrix and introduces a constant overhead on each minor computation. Computing with Laplace expansion, while hashing the minors of smaller size, performs better than state-of-the-art algorithms, in practice. Experiments in Sect. 5 show that our algorithm with hashed determinants outperforms the version without hash. For $m = 3$ and $m = 4$, we experimentally observed that the speedup factor is between 18 and 100; Fig. 4(b) illustrates the second case.

We looked for a hashing function that takes as input a vector of integers and returns an integer that minimizes collisions. We considered many different hash functions, including some variations of the well-known FNV hash [15]. We obtained the best results with the implementation of Boost Hash [20], which shows fewer collisions than the other tested functions.

The drawback is the amount of storage, which is in $O(n!)$. The hash table can be cleared at any moment to limit memory consumption, at the cost of dropping all previously computed minors. Finding a heuristics to clear the hash table according to the number of times each minor was used would decrease the memory consumption, while keeping running times low. In our experiments, we obtained a good trade-off between efficiency and memory consumption by clearing the entire hash table when it contains 10^6 minors. Last column of Table 1 shows that the memory consumption of our algorithm is related to $|A|$ and $dim(\Pi)$.

It is possible to exploit the structure of the above $(2n) \times (2n)$ minor matrices. Let M be such a matrix, with columns corresponding to points of $A_0, \ldots, A_n$. Let $f : \{1, \ldots, 2n\} \to \{0, \ldots n\}$ return i, given the column index. After column permutations, we split M into 4 $n \times n$ submatrices A, B, D, I, where I is the identity and D is a permutation matrix. This follows from the fact that the bottom half of every column in M has at most one 1, and the last n rows of M contain at least one 1 each, unless $\det M = 0$, which is easily checked. Now, $\det M = \pm \det(B - AD)$, with AD constructed in $O(n)$. Hence, the computation of $(2n) \times (2n)$ minors is asymptotically equal to computing an $n \times n$ determinant. This only decreases the constant within the asymptotic bound. A simple implementation of this idea is not faster than Laplace expansion in the dimensions that we currently focus. However, this idea should be valuable in higher dimensions.

5. IMPLEMENTATION AND EXPERIMENTS

We implemented Alg. 1 in C++ to compute Π; our code can be obtained from `http://respol.sourceforge.net`. All timings are on an Intel Core i5-2400 3.1GHz, with 6MB L2 cache and 8GB RAM, running 64-bit Debian GNU/Linux. We rely on CGAL, using mainly a preliminary version of package `triangulation` under development, for both regular triangulations, as well as for the V- and H-representation of Π.

We first compare 4 state-of-the-art exact convex hull packages, `triangulation` [4] implementing [8], and `beneath-and-beyond (bb)` in `polymake` [17]; double description implemented in `cdd` [16], and `lrs` implementing reverse search [1], to compute Π actually extending the work in [2] for the new class of polytopes Π. The first was shown to be faster in computing Delaunay triangulations in ≤ 6 dimensions [4]. The last 3 are run through `polymake`, where we have ignored the time to load the data. We test all packages in an offline version. We first compute the V-representation of Π using our implementation and then we give this as an input to the convex hull packages that compute the H-representation of Π. Moreover, we test `triangulation` by inserting points in the order that Alg. 1 computes them, while improving the point location of these points since we know by the execution of Alg. 1 one facet to be removed (online version). The experiments show that `triangulation` and `bb` are faster than `lrs`, which outperforms `cdd`. Furthermore, the online version of `triangulation` is 2.5 times faster than its offline counterpart due to faster point location (Table 1).

An advantage of `triangulation` is that it maintains a polytope whose boundary and interior are triangulated. This is useful when the oracle $\mathrm{Vtx}(\mathcal{A}, w, \pi)$ needs to refine the regular subdivision of $\mathcal{A}$ which is obtained by projecting the upper hull of the lifted pointset $\mathcal{A}^{\widehat{w}}$ (Sect. 3). In fact this refinement is attained by a placing triangulation, i.e., by computing the projection of the upper hull of the placing triangulation of $\mathcal{A}^{\widehat{w}}$. This is implemented in two steps:

Step 1. compute the placing triangulation of the last $|\mathcal{A}| - m$ points in the order they appear in $\mathcal{A}^{\widehat{w}}$ (they all have height zero),

Step 2. place the first m points of $\mathcal{A}^{\widehat{w}}$ in T_0 in the order they appear in $\mathcal{A}$.

Step 1 is performed only once at the beginning of the algorithm, whereas Step 2 is performed every time we check a new w. The order of placing the points in Step 2 only matters if w is not generic; otherwise, w already produces a triangulation of the m points, so any placing order results in this triangulation.

We can formulate this 2-step construction using a single lifting. Let $c > 0$ be a sufficiently large constant, $a_i \in \mathcal{A}$, $q_i \in \mathbb{R}$, $q_i > c\,q_{i+1}$, for $i = 1, \ldots, |\mathcal{A}|$. Define lifting $h : \mathcal{A} \to \mathbb{R}^2$, where $h(a_i) = (w_i, q_i)$, for $i = 1, \ldots, m$, and $h(a_i) = (0, q_i)$, for $i = m + 1, \ldots, |\mathcal{A}|$. Then, projecting the upper hull of $\mathcal{A}^h$ to $\mathbb{R}^{2n}$ yields the triangulation of $\mathcal{A}$ obtained by the 2-step construction.

This is the implemented method; although different from the perturbation in the proof of Lem. 5, it is more efficient because of the reuse of triangulation T_0 in Step 1 above. Moreover, our experiments show that it always validates the two conditions in Sect. 3.

Fixing the dimension of the triangulation at compile time results in $< 1\%$ speedup. We also tested a filtered kernel

$dim(\Pi)$	$\|\mathcal{A}\|$	# of Π vertices	time (seconds) respol	triang/on	triang/off	bb	cdd	lrs	respol mem (Mb)
3	2490	318	85.03	0.07	0.10	0.07	1.20	0.10	37
4	27	830	15.92	0.71	1.08	0.50	26.85	3.12	46
4	37	2852	97.82	2.85	3.91	2.29	335.23	39.41	64
5	15	510	11.25	2.31	5.57	1.22	47.87	6.65	44
5	18	2584	102.46	13.31	34.25	9.58	2332.63	215.22	88
5	24	35768	4610.31	238.76	577.47	339.05	> 1hr	> 1hr	360
6	15	985	102.62	20.51	61.56	28.22	610.39	146.83	2868
6	19	23066	6556.42	1191.80	2754.30	> 1hr	> 1hr	> 1hr	6693
7	12	249	18.12	7.55	23.95	4.99	6.09	11.95	114
7	17	500	302.61	267.01	614.34	603.12	10495.14	358.79	5258

Table 1: Total time and memory consumption of our code (`respol`) and time comparison of online version of `triangulation` (`triang/on`) and offline versions of all convex hull packages for computing the H-representation of Π.

input	m	3	3	4	4	5	5
	$\|\mathcal{A}\|$	200	490	20	30	17	20
approxim. algorithm	# of Q vertices	15	11	63	121	> 10hr	> 10hr
	$\text{vol}(Q)/\text{vol}(\Pi)$	0.96	0.95	0.93	0.94	> 10hr	> 10hr
	$\text{vol}(Q_o)/\text{vol}(\Pi)$	1.02	1.03	1.04	1.03	> 10hr	> 10hr
	time	0.15	0.22	0.37	1.42	> 10hr	> 10hr
uniformly random	$\|Q\|$	34	45	123	207	228	257
	random vectors	606	576	613	646	977	924
	$\text{vol}(Q)/\text{vol}(\Pi)$	0.93	0.99	0.94	0.90	0.90	0.90
	$\text{vol}(Q_o)/\text{vol}(\Pi)$	1.05	1.01	1.02	1.02	1.03	> 10hr
	time	5.61	12.78	1.1	4.73	8.41	16.9
random variant	# of Q vertices	26	21	102	380	341	544
	$\text{vol}(Q)/\text{vol}(\Pi)$	0.92	0.90	0.80	0.92	0.80	0.81
	$\text{vol}(Q_o)/\text{vol}(\Pi)$	1.02	1.04	1.14	1.03	1.08	1.10
	time	0.16	0.31	1.54	23.66	59.87	211.50
exact alg.	# of Π vertices	98	133	416	1296	1674	5093
	time	2.03	5.87	3.72	25.97	51.54	239.96

Table 2: Typical results on experiments computing Q, Q_o^H using the approximation algorithm and the random vectors procedures (the uniform and its variant); we stop the approximation algorithm when $\frac{\text{vol}(Q)}{\text{vol}(Q_o)} > 0.9$; results on random procedures is the average values over 10 independent experiments; "> 10hr" means computation of vol(Q_o) takes > 10hr on `polymake`.

with a similar time speedup. On the other hand, **`triangulation`** is expected to implement incremental high-dimensional regular triangulations wrt to a lifting, faster than the above method [10]. Moreover, we use a modified version of **`triangulation`** in order to benefit from our hashing scheme. Therefore, all cells of the triangulated facets of Π have the same normal vector and we use a structure (**`STL set`**) to maintain the set of unique normal vectors, thus computing only one regular triangulation per triangulated facet of Π.

We perform an *experimental analysis* of our algorithm. We design experiments parameterized on: the total number of input points $|\mathcal{A}|$, the dimension n of pointsets A_i, and the dimension of projection m. First, we examine our algorithm on random inputs for implicitization and u-resultants, where $m = n + 1$, while varying $|\mathcal{A}|, n$. We fix $\delta \in \mathbb{N}$ and select random points on the δ-simplex to generate dense inputs, and points on the $(\delta/2)$-cube to generate sparse inputs. For *implicitization* the projection coordinates correspond to point $a_{i1} = (0, \ldots, 0) \in A_i$. For $n = 2$ the problem corresponds to implicitizing surfaces: when $|\mathcal{A}| < 60$, we compute the polytopes in < 1sec (Fig. 4(a)). When computing the *u-resultant* polytope, the projection coordinates correspond to $A_0 = \{(1, \ldots, 0), \ldots, (0, \ldots, 1)\}$. For $n = 2$, when $|\mathcal{A}| < 500$, we compute the polytopes in < 1sec (Fig. 4(a)).

By using the *hashing determinants* scheme we gain a 18× speedup when $n = 2, m = 3$. For $m = 4$ we gain a larger speedup; we computed in < 2min an instance where $|\mathcal{A}| = 37$ and would take > 1hr to compute otherwise. Thus, when the dimension and $|\mathcal{A}|$ becomes larger, this method allows our algorithm to compute instances of the problem that would be intractable otherwise, as shown for $n = 3, m = 4$ (Fig. 4(b)).

We confirm experimentally the *output-sensitivity* of our algorithm. First, our algorithm always computes vertices of Π either to extend Π or to legalize a facet. We experimentally show that our algorithm has a subexponential behaviour wrt to both input and output (Fig. 4(c), 4(d)) and its output is subexponential wrt the input.

We explore the *limits* of our implementation. By bounding runtime to < 2hr, we compute instances of 5-, 6-, 7-dimensional Π with 35K, 23K, 500 vertices, resp. (Table 1).

We also compare with the implementation of [21], which is based on **`Gfan`** library. They develop two algorithms to compute projections of $N(\mathcal{R})$. Assuming $\mathcal{R}$ defines a hypersurface, their methods computes a union of (possibly over-

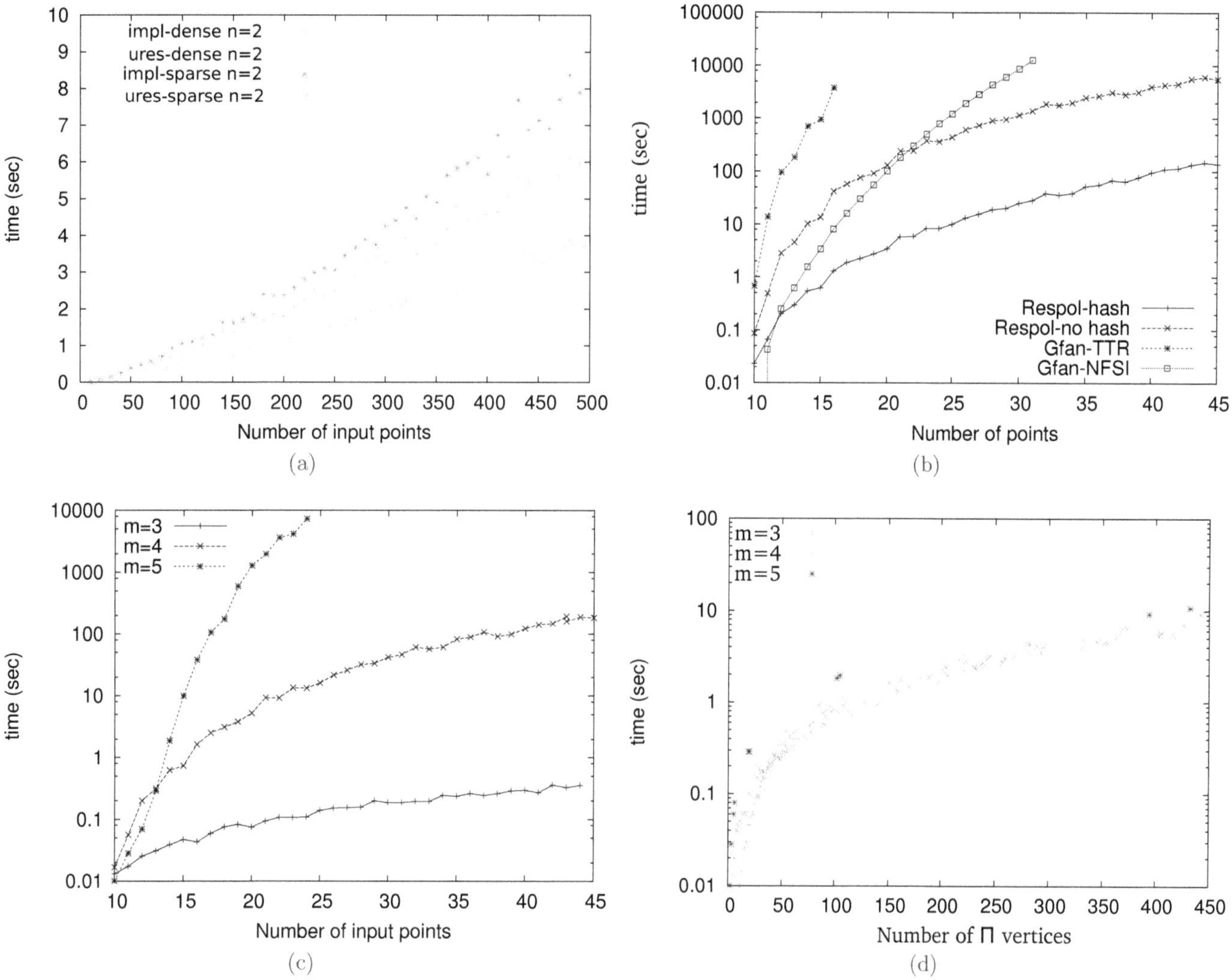

Figure 4: (a) Implicitization and u-resultants for $n = 2, m = 3$ (b) Comparison of `respol` (hashing and not hashing determinants) and `Gfan` (traversing tropical resultants and computing normal fan from stable intersection) for $m = 4$ (c) Performance of Alg 1 for $m = 3, 4, 5$ as a function of input (d) Performance of Alg 1 as a function of its output; y-axis in (b),(c),(d) are in logarithmic scale.

lapping) cones, along with their multiplicities [21, Thm.2.9]. From this intermediate result they construct the normal cones to the resultant vertices. We compare with the best timings of `Gfan` methods using the examples and timings of [21]: our method is faster for $m < 7, |\mathcal{A}| < 16$ and competitive (up to 2 times slower) when $m = 5, |\mathcal{A}| = 20$ or $m = |\mathcal{A}| = 15$. We run extensive experiments for $n = 3$, considering implicitization, where $m = 4$ and our method, with and without using hashing, is much faster than any of the two algorithms based on `Gfan` (Fig. 4(b)). However, for $n = 4, m = 5$ the beta version of `Gfan` we run was not stable and always crashes when $|\mathcal{A}| > 13$.

We analyze the computation of inner and outer *approximations* Q and Q_o^H. We test the variant of Sect. 3 by stopping it when $\frac{\text{vol}(Q)}{\text{vol}(\Pi)} > 0.9$. In the experiments, the number of Q vertices is $< 15\%$ of the Π vertices, thus the speedup is up to 25 times faster than the exact algorithm and $\frac{\text{vol}(Q_o^H)}{\text{vol}(\Pi)} < 1.04$ (Table 2). The bottleneck of this computation is the computation of $\text{vol}(Q_o^H)$, which is in H-representation. We use `lrs` through `polymake` in every step to compute $\text{vol}(Q_o^H)$ because we are lacking of an implementation that given a polytope P in H-representation, its volume and a halfspace H, computes the volume of the intersection of P and H. Note that we don't count this computation time in total time.

Next, we study procedures that compute only the V-representation of Q. For this, we count how many *random vectors* uniformly distributed on the m-sphere are needed to obtain $\frac{\text{vol}(Q)}{vol(\Pi)} > 0.9$. This procedure runs up to 10 times faster than the exact algorithm. Its drawback is that it does not provide guarantees for $\frac{\text{vol}(Q)}{\text{vol}(\Pi)}$ because we do not know Π in advance. To overcome this, we test a variant procedure which starts with a sequence of random vectors and produces vectors that are affine combinations of the existing ones. It stops when the produced vectors do not give a new vertex for Q. Even if it computes Q s.t. $\frac{\text{vol}(Q)}{\text{vol}(\Pi)} > 0.9$ up to 10 times faster than the exact algorithm, on average it computes Q s.t. $\frac{\text{vol}(Q)}{\text{vol}(\Pi)} > 0.8$ in similar runtimes (Table 2).

6. FUTURE WORK

As shown, `polymake`'s convex hull algorithm is competitive, thus one may use it for implementing our algorithm. On the other hand, `triangulation` is expected to include [10] fast enumeration of all regular triangulations for a given (non generic) lifting, in which case Π may be extended by more than one (coplanar) vertices.

The resultant edges are associated to flips on mixed cells [29]. We studied them algorithmically [12] and may revisit them to generate all neighbors of a vertex of $N(\mathcal{R})$.

Any incremental convex hull algorithm has a worst-case super-polynomial *total time complexity* [5] in the number of input points and output facets. The question is whether there is a polynomial total time incremental algorithm for Π. For this we study the structure of $N(\mathcal{R})$ through experiments; it appears to be an exciting family of polytopes.

7. ACKNOWLEDGMENTS

We thank O. Devillers and S. Hornus for discussions on `triangulation`, and A. Jensen and J. Yu for discussions and for sending us a beta version of their code.

8. REFERENCES

[1] D. Avis. lrs: A revised implementation of the reverse search vertex enumeration algorithm. In *Polytopes: Combinatorics & Computation*, volume 29 of *Oberwolfach Seminars*, pages 177–198. 2000.

[2] D. Avis, D. Bremner, and R. Seidel. How good are convex hull algorithms? *Comput. Geom.: Theory & Appl.*, 7:265–301, 1997.

[3] S. Basu, R. Pollack, and M.-F. Roy. *Algorithms in real algebraic geometry.* Springer-Verlag, Berlin, 2003.

[4] J.-D. Boissonnat, O. Devillers, and S. Hornus. Incremental construction of the Delaunay triangulation and the Delaunay graph in medium dimension. In *Proc. SoCG*, pages 208–216, 2009.

[5] D. Bremner. Incremental convex hull algorithms are not output sensitive. In *Proc. 7th Intern. Symp. Algorithms and Comput.*, pages 26–35, London, UK, 1996. Springer.

[6] H. Brönnimann, I.Z. Emiris, V. Pan, and S. Pion. Sign determination in Residue Number Systems. *Theor. Comp. Science, Spec. Issue on Real Numbers & Computers*, 210(1):173–197, 1999.

[7] CGAL: Computational geometry algorithms library. http://www.cgal.org.

[8] K.L. Clarkson, K. Mehlhorn, and R. Seidel. Four results on randomized incremental constructions. *Comput. Geom.: Theory & Appl.*, 3:185–121, 1993.

[9] D. Cox, J. Little, and D. O'Shea. *Using Algebraic Geometry.* Number 185 in GTM. Springer, New York, 2nd edition, 2005.

[10] O. Devillers, 2011. Personal communication.

[11] J.-G. Dumas, T. Gautier, M. Giesbrecht, P. Giorgi, B. Hovinen, E. Kaltofen, B. D. Saunders, W. J. Turner, and G. Villard. Linbox: A generic library for exact linear algebra. In *Proc. Intern. Congress Math. Software*, pages 40–50, Beijing, 2002.

[12] I.Z. Emiris, V. Fisikopoulos, and C. Konaxis. Regular triangulations and resultant polytopes. In *Proc. Europ. Works. Comput. Geometry*, Dortmund, 2010.

[13] I.Z. Emiris, V. Fisikopoulos, and L. Peñaranda. Optimizing the computation of sequences of determinantal predicates. In *Proc. Europ. Workshop Computat. Geometry*, Assisi, Italy, 2012.

[14] I.Z. Emiris, T. Kalinka, and C. Konaxis. Implicitization using predicted support. In *Proc. Intern. Works. Symbolic-Numeric Computation*, 2011.

[15] G. Fowler, L.C. Noll, and P. Vo. Fowler-Noll-Vo hash. www.isthe.com/chongo/tech/comp/fnv, 1991.

[16] K. Fukuda. cdd and cdd+ Home Page. ETH Zürich. www.ifor.math.ethz.ch/~fukuda/cdd_home/, 2008.

[17] E. Gawrilow and M. Joswig. Polymake: an approach to modular software design in computational geometry. In *Proc. Annual ACM Symp. Computational Geometry*, pages 222–231. ACM Press, 2001.

[18] I.M. Gelfand, M.M. Kapranov, and A.V. Zelevinsky. *Discriminants, Resultants and Multidimensional Determinants.* Birkhäuser, Boston, 1994.

[19] G. Guennebaud, B. Jacob, et al. Eigen v3. http://eigen.tuxfamily.org, 2010.

[20] D. James. Boost functional library. www.boost.org/libs/functional/hash, 2008.

[21] A. Jensen and J. Yu. Computing tropical resultants. *arXiv:math.AG/1109.2368v1*, 2011.

[22] M. Joswig. Beneath-and-beyond revisited. In M. Joswig and N. Takayama, editors, *Algebra, Geometry, and Software Systems*, Mathematics and Visualization. Springer, Berlin, 2003.

[23] E. Kaltofen and G. Villard. On the complexity of computing determinants. *Computational Complexity*, 13:91–130, 2005.

[24] M.M. Kapranov. Characterization of A-discriminantal hypersurfaces in terms of logarithmic Gauss map. *Math. Annalen*, 290:277–285, 1991.

[25] J.A. De Loera, J. Rambau, and F. Santos. *Triangulations: Structures for Algorithms and Applications*, volume 25 of *Algorithms and Computation in Mathematics.* Springer, 2010.

[26] T. Michiels and R. Cools. Decomposing the secondary Cayley polytope. *Discr. Comput. Geometry*, 23:367–380, 2000.

[27] T. Michiels and J. Verschelde. Enumerating regular mixed-cell configurations. *Discr. Comput. Geometry*, 21(4):569–579, 1999.

[28] J. Rambau. TOPCOM: Triangulations of point configurations and oriented matroids. In *Proc. Intern. Congress Math. Software*, pages 330–340, 2002.

[29] B. Sturmfels. On the Newton polytope of the resultant. *J. Algebraic Combin.*, 3:207–236, 1994.

[30] B. Sturmfels and J. Yu. Tropical implicitization and mixed fiber polytopes. In *Software for Algebraic Geometry*, volume 148 of *IMA Volumes in Math. & its Applic.*, pages 111–131. Springer, New York, 2008.

[31] G.M. Ziegler. *Lectures on Polytopes.* Springer, 1995.

Counting Plane Graphs: Perfect Matchings, Spanning Cycles, and Kasteleyn's Technique*

Micha Sharir
School of Computer Science
Tel Aviv University
Tel Aviv 69978, Israel
Courant Inst. of Math. Sci.
New York University
New York, NY 10012, USA
michas@tau.ac.il

Adam Sheffer
School of Computer Science
Tel Aviv University
Tel Aviv 69978, Israel
sheffera@tau.ac.il

Emo Welzl
Institute for Theoretical
Computer Sci., ETH Zürich
CH-8092 Zürich, Switzerland
welzl@inf.ethz.ch

ABSTRACT

We derive improved upper bounds on the number of crossing-free straight-edge spanning cycles (also known as Hamiltonian tours and simple polygonizations) that can be embedded over any specific set of N points in the plane. More specifically, we bound the ratio between the number of spanning cycles (or perfect matchings) that can be embedded over a point set and the number of triangulations that can be embedded over it. The respective bounds are $O(1.8181^N)$ for cycles and $O(1.1067^N)$ for matchings. These imply a new upper bound of $O(54.543^N)$ on the number of crossing-free straight-edge spanning cycles that can be embedded over any specific set of N points in the plane (improving upon the previous best upper bound $O(68.664^N)$). Our analysis is based on a weighted variant of Kasteleyn's linear algebra technique.

Categories and Subject Descriptors

F.2.2 [**Nonnumerical Algorithms and Problems**]: Geometrical problems and computations; G.2.2 [**DISCRETE MATHEMATICS**]: Graph Theory

*Work on this paper by the first two authors was partially supported by Grant 338/09 from the Israel Science Fund and by the Israeli Centers of Research Excellence (I-CORE) program, (Center No. 4/11). Work by Micha Sharir was also supported by NSF Grant CCF-08-30272, by Grant 2006/194 from the U.S.-Israel Binational Science Foundation, and by the Hermann Minkowski–MINERVA Center for Geometry at Tel Aviv University. Emo Welzl acknowledges support from the EuroCores/EuroGiga/ComPoSe SNF grant 20GG21_134318/1. Part of the work on this paper was done at the Centre Interfacultaire Bernoulli (CIB), EPFL, Lausanne, during the Special Semester on Discrete and Computational Geometry, Fall 2010, and supported by the Swiss National Science Foundation.

SCG'12, June 17–20, 2012, Chapel Hill, North Carolina, USA.

Keywords

Spanning cycles, perfect matchings, Kasteleyn's Technique, Triangulations.

1. INTRODUCTION

In this paper we consider the problem of bounding the number of all crossing-free straight-edge spanning cycles that can be embedded over a specific set of points in the plane. That is, given a set S of N labeled points in the plane, we consider the number of spanning cycles that have a straight-edge planar embedding over S. We rely on Kasteleyn's linear-algebra technique [13], and on edge-flipping techniques that were developed in a previous paper by the authors [8]. No familiarity with [8] is necessary, since we re-introduce all the notions that we require from it. We now give a detailed and more formal definition of the problem.

A *planar graph* is a graph that can be embedded in the plane in such a way that its vertices are embedded as points and its edges are embedded as Jordan arcs that connect the respective pairs of points and can meet only at a common endpoint. A *crossing-free straight-edge graph* is a plane embedding of a planar graph such that its edges are embedded as non-crossing straight line segments. In this paper, we only consider crossing-free straight-edge graphs, and we also assume that the points of the vertex set S are in general position, that is, no three points are collinear. (For upper bounds on the number of graphs, this involves no loss of generality, because the number of graphs can only grow when a degenerate point set is slightly perturbed into general position.) For simplicity, we sometimes refer to such graphs as *plane graphs*.

We focus on upper bounding the maximal number of plane *spanning cycles* (also known as *Hamiltonian cycles*, *Hamiltonian tours*, and *simple polygonizations*) that can be embedded over a fixed set of points in the plane. For a set S of points in the plane, we denote by $\mathcal{C}(S)$ the set of all crossing-free straight-edge spanning cycles of S, and put $\mathsf{sc}(S) := |\mathcal{C}(S)|$. Moreover, we let $\mathsf{sc}(N) = \max_{|S|=N} \mathsf{sc}(S)$. The main goal of this paper is thus to obtain improved upper bounds on $\mathsf{sc}(N)$.

There are many similar variants of this problem, such as bounding the number of plane forests, spanning trees, triangulations, and general plane graphs. Recent work on some of these variants can be found in [1, 8, 19], and we try to keep a comprehensive list of the up-to-date upper and lower bounds in a dedicated webpage[1]. It seems that the case of spanning cycles is the most popular one, already considered in [2, 3, 4, 5, 6, 15, 21] and many other works. Moreover, spanning cycles were the first case for which bounds were published, namely the bounds $3/20 \cdot 10^{N/3} \leq \mathsf{sc}(N) \leq 2 \cdot 6^{N-2} \cdot (\lfloor N/2 \rfloor)!$ in [15]. A brief history of the steady progress on bounding the number of spanning cycles can be found in a dedicated webpage by Erik Demaine[2]. Currently, the best known lower bound is $\mathsf{sc}(N) = \Omega(4.642^N)$, due to García, Noy, and Tejel [6], and the previous upper bound is $\mathsf{sc}(N) = O(68.664^N)$ by Dumitrescu *et al.* [5]. We derive the improved bound $\mathsf{sc}(N) = O(54.543^N)$.

These problems have also been studied from an algorithmic point of view, deriving algorithms for enumeration or counting of the plane graphs (or other graph types) that can be embedded over a given point set (such as in [12, 17]). The combinatorial upper bounds are useful for analyzing the running times of such algorithms, and also to answer questions such as "how many bits are required to represent a triangulation (or any other kind of plane graphs)?".

Our bound (as do some of the previous bounds) relies on *triangulations*. A triangulation of a set S of N points in the plane is a maximal plane graph on S (that is, no additional straight edges can be inserted without crossing some of the existing edges). For a set S of points in the plane, we denote by $\mathcal{T}(S)$ the set of all triangulations of S, and put $\mathsf{tr}(S) := |\mathcal{T}(S)|$. Moreover, we let $\mathsf{tr}(N) = \max_{|S|=N} \mathsf{tr}(S)$. Currently, the best known bounds for $\mathsf{tr}(N)$ are $\mathsf{tr}(N) < 30^N$ [19], and $\mathsf{tr}(N) = \Omega(8.65^N)$ [5].

The upper bound by Dumitrescu *et al.* [5] is obtained by proving that for every set S of N points in the plane $\mathsf{sc}(S) = O\left(2.2888^N\right) \cdot \mathsf{tr}(S)$. This has sharpened an earlier bound of Buchin *et al.* [4], who showed that every triangulation T of S contains at most $30^{N/4} \approx 2.3404^N$ spanning cycles (i.e., cycles whose edges belong to T), implying[3] that $\mathsf{sc}(S) < 2.3404^N \cdot \mathsf{tr}(S)$. Combining the above ratio with the bound $\mathsf{tr}(N) < 30^N$ directly implies the asserted bound. We derive our bound in a similar manner, showing that $\mathsf{sc}(S) = O\left(1.8181^N\right) \cdot \mathsf{tr}(S) = O(54.5430^N)$.

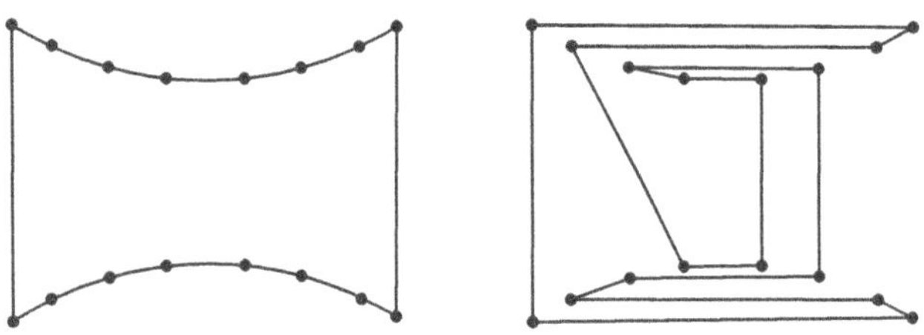

Figure 1: Two spanning cycles embedded over a double chain point configuration.

In spite of our improved bound, we strongly believe, and conjecture, that for every point set S (of size at least N_0, for some sufficiently large constant N_0) one has $\mathsf{sc}(S) < \mathsf{tr}(S)$, and perhaps even a much sharper ratio holds. The best lower bound for this ratio that we know of is obtained from the *double chain configuration*, presented in [6] (and depicted in Figure 1). It is shown in [6] that when S is a double chain configuration with N vertices, $\mathsf{tr}(S) = \Theta^*\left(8^N\right)$ and $\mathsf{sc}(S) = \Omega^*\left(4.64^N\right)$.[4] Thus, in this case, $\mathsf{sc}(S)/\mathsf{tr}(S) = \Omega^*(0.58^N)$. (It is stated in [1], albeit without proof, that $\mathsf{sc}(S) = O(5.61^N)$, so this example supports our conjecture.)

In Section 2 we go over the preliminaries required for our analysis. These include, among others, the edge-flip techniques used in [8]. Section 3 derives the bound $\mathsf{sc}(S) = O\left(12^{N/4}\right) \cdot \mathsf{tr}(S) = O\left(1.8613^N\right) \cdot \mathsf{tr}(S)$ for any set S of N points in the plane. As part of this derivation, we describe Kasteleyn's technique for counting perfect matchings and present a new way of applying it. The more advanced analysis, deriving the improved bound $\mathsf{sc}(S) = O\left(10.9247^{N/4}\right) \cdot \mathsf{tr}(S) = O\left(1.8181^N\right) \cdot \mathsf{tr}(S)$, is presented in Section 5. In Section 4 we use the same methods to prove an upper bound on the ratio between the number of plane perfect matchings and the number of triangulations, showing that $\mathsf{pm}(S) = O(1.1067^N) \cdot \mathsf{tr}(S)$ (where $\mathsf{pm}(S)$ is the number of crossing-free straight-edge matchings that can be embedded over the point set S).

While this paper constitutes a significant improvement over previous bounds, it is only one stepping stone towards the goal of establishing a sharp bound on $\mathsf{sc}(N)$, or of at least showing that $\mathsf{sc}(N) < \mathsf{tr}(N)$, as conjectured above. The interest in this paper, in our opinion, is in the technique that it employs, where it combines recent results on edge flippability in triangulations [8] with the beautiful (and fairly old) technique of Kasteleyn [13, 14] that applies tools from linear algebra to derive upper bounds on the number of perfect matchings in planar graphs. Kasteleyn's technique has already been used recently in [4] for deriving bounds on $\mathsf{sc}(N)$, but the application in this paper is different, as it handles edge-weighted planar graphs. Instead of bounding the number of perfect matchings, it bounds the sum of their weights, where the weight of a matching is the product of the weights of its edges. This enhanced version allows us to "push" the technique much further and obtain our improved bounds. We hope that this enhanced tool will lead to further results in this area.

2. PRELIMINARIES

In this section we establish some notations and lemmas that are required for the following sections.

Given two plane graphs G and H over the same point set S, if every edge of G is also an edge of H, we write $G \subseteq H$.

Hull edges and vertices (resp., interior edges and vertices) of a graph embedded on a point set S are those that are part of the boundary of the convex hull of S (resp., not part of the convex hull boundary).

Given a set S of N points in the plane, we denote by h the number of hull vertices of S, and put $n = N - h$, which is the number of interior vertices of S.

2.1 The support of a graph

Let us denote by $\mathsf{sc}_\Delta(N)$ the maximal number of plane spanning cycles that can be contained in any fixed triangulation of a set of N points in the plane. Moreover, denote

[1] `http://www.cs.tau.ac.il/~sheffera/counting/PlaneGraphs.html` (version of March 2012).

[2] `http://erikdemaine.org/polygonization/` (version of March 2012).

[3] The implication comes from the fact that every spanning cycle, and in fact every plane graph, is contained in at least one triangulation; see Section 2.

[4] In the notations $O^*()$, $\Theta^*()$, and $\Omega^*()$, we neglect polynomial factors.

the set of spanning cycles contained in a triangulation T by $\mathcal{C}(T)$, so $\mathsf{sc}_\Delta(N) = \max_{|S|=N,\ T\in\mathcal{T}(S)} |\mathcal{C}(T)|$.

Any spanning cycle (or, for that matter, any plane graph) is contained in at least one triangulation. Therefore, we can upper bound the number of spanning cycles of a set S of N points in the plane by going over every triangulation $T \in \mathcal{T}(S)$ and counting the number of spanning cycles contained in T. This implies the bound $\mathsf{sc}(N) \leq \mathsf{tr}(N)\cdot\mathsf{sc}_\Delta(N)$. Applying the bounds $\mathsf{tr}(N) < 30^N$ from [19] and $\mathsf{sc}_\Delta(N) \leq 30^{N/4}$ from [4], we obtain $\mathsf{sc}(N) < 30^{5N/4} \approx 70.21^N$.

This bounding method seems rather weak since it potentially counts some spanning cycles many times. For example, consider a spanning cycle of the double-chain configuration consisting of two convex chains facing each other, as depicted in the left-hand side of Figure 1. García, Noy, and Tejel [6] show that such a spanning cycle is contained in $\Theta^*(8^N)$ triangulations of its point set. Therefore, the above method will count this spanning cycle $\Theta^*(8^N)$ times. In this case the above analysis method will be grossly over-counting because, as stated in [1], this point set has only $O(5.61^N)$ spanning cycles.

In order to deal with this inefficiency, we define the notion of *support* (the same notion was also used in [5, 8, 19, 20, 22]). Given a plane edge graph G embedded over a set S of points in the plane, we say that G has a support of x if G is contained in (exactly) x triangulations of S; we write $\mathsf{supp}(G) = x$. Notice that

$$\mathsf{sc}(S) = \sum_{T\in\mathcal{T}(S)} \sum_{C\in\mathcal{C}(T)} \frac{1}{\mathsf{supp}(C)}, \qquad (1)$$

because every spanning cycle C contributes exactly one to the right-hand side of the equation (it appears in $\mathsf{supp}(C)$ terms of the first sum, and contributes $1/\mathsf{supp}(C)$ in every appearance). We will use (1) to obtain better upper bounds for $\mathsf{sc}(N)$, by showing that, on average, $\mathsf{supp}(C)$ is large.

2.2 Ps-flippable edges

An edge in a triangulation is said to be *flippable*, if its two incident triangles form a convex quadrilateral. A flippable edge can be *flipped*, that is, removed from the graph of the triangulation and replaced by the other diagonal of the resulting quadrilateral. Such an operation is depicted in Figure 2(a), where the edge ce can be flipped to the edge ad.

In [8], we present the concept of *pseudo simultaneously flippable edges* (or *ps-flippable edges*, for short). Given a triangulation T, we say that a subset F of its edges is a set of ps-flippable edges if the edges of F are diagonals of interior-disjoint convex polygons (whose boundaries are also parts of T). For example, in Figure 2(b), the three dashed edges form a set of ps-flippable edges, since they are diagonals of interior-disjoint convex quadrilateral and convex pentagon (another set of ps-flippable edges, in a different triangulation, is depicted in Figure 2(c)).

Ps-flippable edges are related to *convex decompositions*. A convex decomposition of a point set S is a crossing-free straight-edge graph D on S such that (i) D includes all the hull edges, (ii) each bounded face of D is a convex polygon, and (iii) no point of S is isolated in D. See Figure 2(c) for an illustration. For additional information about convex decompositions, see, for example, [9]. Notice that if T is a triangulation that contains D, the edges of $T \setminus D$ form a set of ps-flippable edges, since they are the diagonals of

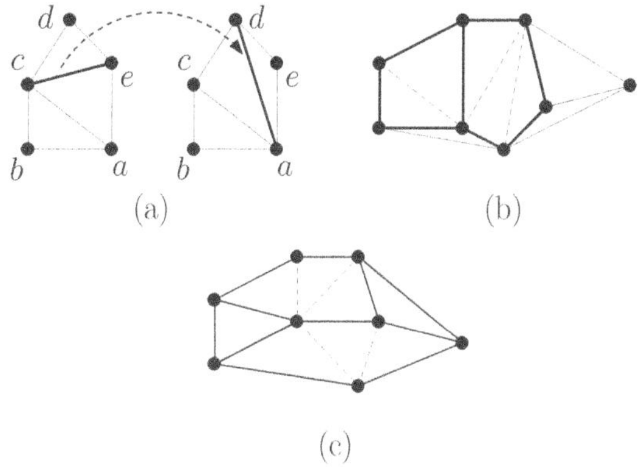

Figure 2: (a) The edge ce can be flipped to the edge ad. (b) A set of three (dashed) ps-flippable edges which are diagonals of interior-disjoint convex quadrilateral and convex pentagon. (c) A convex decomposition with four bounded cells obtained by removing the dashed edges from the triangulation; here too the dashed edges are diagonals of pairwise disjoint convex faces, and form a set of ps-flippable edges.

the interior-disjoint convex polygons of D (again, consider the dashed edges in Figure 2(c) for an illustration). Thus, finding a large set of ps-flippable edges in a triangulation T is equivalent to finding a convex decomposition with a small number of faces (or edges) in T.

In [8], we prove the two following lemmas.

LEMMA 2.1. *Every triangulation T over a set of N points in the plane contains a set F of $N/2 - 2$ ps-flippbale edges. Also, there are triangulations with no larger sets of ps-flippable edges.*

LEMMA 2.2. *Consider a triangulation T, a set F of $N/2-2$ ps-flippable edges in T, and a graph $G \subseteq T$. If G does not contain j edges from F then $\mathsf{supp}(G) \geq 2^j$.*

Proof sketch. Consider the set $F' = F \setminus G$ of j ps-flippable edges. The convex faces of $T \setminus F'$ can be triangulated in at least 2^j ways (the actual number, which is a product of Catalan numbers [23, Section 5.3], attains this minimum when every edge of F' is a diagonal of a distinct quadrangular face of $T \setminus F'$), and each of the resulting triangulations contains G. See [8] for more details. ∎

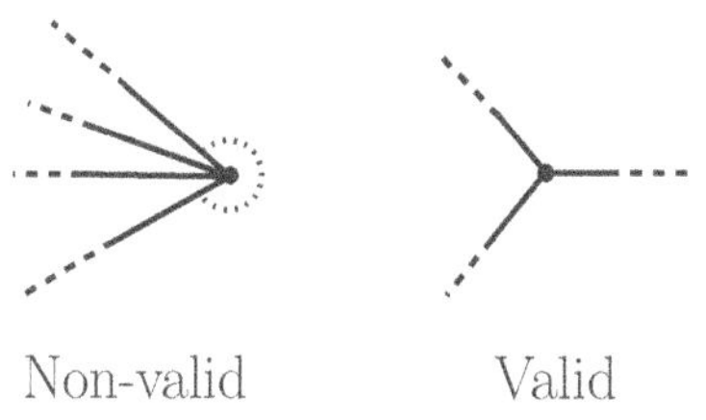

Figure 3: A vertex is valid if and only if it is not a reflex vertex of any face.

We now describe another property of convex decompositions (not discussed in [8]). Consider a set S of points in the plane and a crossing-free straight-edge graph G embedded on S. We say that an interior point $p \in S$ has a *valid triple of edges in G* if there exist three points $a, b, c \in S$ such that p is contained in the convex hull of $\{a, b, c\}$ and the edges ap,

bp, and cp belong to G. To simplify the notation, we refer to vertices with valid triples as *valid* (with respect to G), and to the other interior vertices as *non-valid*. See Figure 3 for an illustration.

LEMMA 2.3. *Let S be a set of points in the plane and let G be a crossing-free straight-edge graph over S that contains all the edges of the convex hull of S. Then G is a convex decomposition of S if and only if every interior vertex of S is valid with respect to G.*

Proof. An interior vertex v is a reflex vertex of some face of G if and only if v is non-valid (see Figure 3). The lemma follows by observing that G is a convex decomposition if and only if no bounded face of G has a reflex vertex. ◻

Figure 4: In a point set of an even size, every spanning cycle is the union of two edge-disjoint perfect matchings.

2.3 Spanning cycles and perfect matchings

Our analysis, as most of the previous works dealing with the number spanning cycles, heavily relies on the number of plane perfect matchings on S (for example, see [4, 5, 21]). To see the connection between the two problems, notice that if $|S|$ is even, every spanning cycle C is the union of two edge-disjoint perfect matchings on S; namely, the matching consisting of the even-indexed edges of C, and the matching consisting of the odd-indexed edges. An illustration of this property is depicted in Figure 4. Denote by $\mathcal{M}(S)$ the set of all plane perfect matchings on S, and put $\mathsf{pm}(S) = |\mathcal{M}(S)|$. We also set $\mathsf{pm}(N) = \max_{|S|=N} \mathsf{pm}(S)$. Hence, a simple upper bound on $\mathsf{sc}(S)$ is $\mathsf{pm}(S)^2$. In general, the union of two edge-disjoint perfect matchings is not always a spanning cycle, but it is a cover of S by vertex-disjoint even-sized cycles.

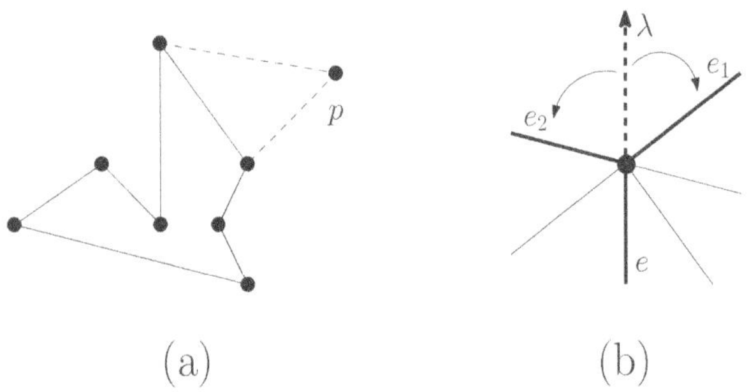

Figure 5: (a) We can always connect a new vertex p outside the convex hull of S to two endpoints of some edge of the spanning cycle. (b) The set $\{e, e_1, e_2\}$ is a valid triple of edges.

To deal with point sets of odd size, we use the following lemma:

LEMMA 2.4. *Let $c > 1$ be a constant such that every set S of an even number of points in the plane satisfies $\mathsf{sc}(S) = O(c^{|S|})$. Then $\mathsf{sc}(S) = O(c^{|S|})$ also holds for sets S of an odd number of points.*

Proof. Consider a set S of N points in the plane, where N is odd. Pick a new point p outside the convex hull of S, and put $S' = S \cup \{p\}$. Let C be a plane spanning cycle of S. Then there exists an edge $e = vu$ of C such that p can be connected to the two endpoints u, v of e without crossing C (e.g., see Figure 5(a)). Indeed, this is a projective variant of the property, noted in [7], that every finite collection of non-crossing straight segments in the plane contains a segment e such that no other segment lies vertically above any point of e (see also [16, Section 8.7]). By replacing e with the edges vp and pu, we obtain a crossing-free spanning cycle of S'. This implies that we can map every spanning cycle of S to a distinct spanning cycle of S', and thus, $\mathsf{sc}(S) \le \mathsf{sc}(S')$. The lemma then follows since $\mathsf{sc}(S') = O(c^{N+1}) = O(c^N)$. ◻

Bounding the number of perfect matchings on S within a fixed triangulation T can be done by the beautiful linear-algebra technique of Kasteleyn [13], described in detail in [14, Section 8.3]; see Section 3 for more details. Buchin *et al.* [4] have used this technique to show that any triangulation T of S contains at most $6^{N/4}$ perfect matchings, and at most $30^{N/4} \approx 2.3403^N$ spanning cycles. We also note that Sharir and Welzl [21] showed that $\mathsf{pm}(S) = O\left(10.05^N\right)$, completely bypassing the approach of counting matchings (or other graphs) *within* a triangulation. While this bound is fairly small, it does not seem to be useful for obtaining a good bound on $\mathsf{sc}(N)$. For example, using the inequality $\mathsf{sc}(N) \le \mathsf{pm}(N)^2$, noted above, only gives the rather weak bound $\mathsf{sc}(N) = O(101.01^N)$.

3. AN UPPER BOUND ON THE NUMBER OF SPANNING CYCLES

In this section we first review an enhanced variant of Kasteleyn's technique and then use it to derive an upper bound on $\mathsf{sc}(S)$.

THEOREM 3.1. *For any set S of N points in the plane,*

$$\mathsf{sc}(S) = O\left(12^{N/4}\right) \cdot \mathsf{tr}(S) = O\left(1.8613^N\right) \cdot \mathsf{tr}(S).$$

This bound is slightly weaker than the one stated in the introduction, but its proof is considerably simpler; the improved bound is derived in Section 5.

Proof. First, by Lemma 2.4, we may assume that N is even. Consider a triangulation T of S. As already observed, every spanning cycle contained in T is the union of two edge-disjoint perfect matchings contained in T. Given a plane graph G, we denote by $\mathcal{M}(G)$ the set of all perfect matchings that are contained in G. Recalling (1), we have

$$\mathsf{sc}(S) \le \sum_{T \in \mathcal{T}(S)} \sum_{\substack{M_1, M_2 \in \mathcal{M}(T) \\ M_1, M_2 \text{ edge-disjoint}}} \frac{1}{\mathsf{supp}(M_1 \cup M_2)}.$$

(The inequality comes from the fact that not every pair M_1, M_2 of matchings, as in the sum, necessarily yields a spanning cycle.) Let us fix the "first" perfect matching $M_1 \subset T$; as mentioned above, Buchin *et al.* [4] prove that $|\mathcal{M}(T)| \le 6^{N/4}$, so there are at most $6^{N/4}$ choices of M_1. Next, we construct a convex decomposition D such that $M_1 \subset D \subset T$, as follows. We start with M_1 and add all the missing hull edges; let us denote the resulting graph as D'. By Lemma 2.3, it suffices to add edges to D' so as to ensure that every interior point $p \in S$ is connected in D to (at least) three points $a, b, c \in S$, such that p is inside the convex hull of $\{a, b, c\}$. Every interior vertex p of S has degree 1 in D', so we start by setting $D := D'$, and then, for each interior point $p \in S$, we add to D two additional edges of T adjacent to p, so as to create a valid triple. To do so, let e be the edge of D' (that is, of M_1) incident to p, and let λ be the ray emanating from p in the opposite direction. Let e_1 (resp., e_2) be the first edge of T incident to p encountered in clockwise (resp., counterclockwise) direction from λ; see Figure 5(b). Then, as is easily checked, $\{e, e_1, e_2\}$ is a valid triple of edges, and we add e_1, e_2 to D. After applying this step to each interior point p, the resulting graph D is indeed a convex decomposition of S.

We denote by F the set of edges that are in T but not in D. The edges of F are diagonals of interior-disjoint convex polygons, and thus F is a set of ps-flippable edges. By Euler's formula, the triangulation T contains $3N-2h-3$ interior edges, and D contains at most $2n+N/2$ interior edges (at most $N/2$ edges of M_1 and at most $2n$ added edges to form n valid triples). Therefore,

$$|F| \geq 3N - 2h - 3 - (2n + N/2) = N/2 - 3.$$

Remark. Note the strength of this bound: Lemma 2.1 has a rather involved proof, given in [8], and it yields a set of $N/2-2$ ps-flippable edges in the entire triangulation. In contrast, here we get the same number (minus 1) after we remove from T an arbitrary perfect matching, with a considerably simpler analysis. Thus the significance of the analysis in [8] (giving the proof of Lemma 2.1) is only for triangulations which contain no perfect matching on S. For example, any triangulation with more than $N/2$ interior vertices of degree 3 cannot contain a perfect matching, since, as is easily checked, two interior vertices of degree 3 cannot share an edge.

Without loss of generality, we assume that F consists of *exactly* $N/2-3$ edges. We now proceed to bound the number of ways to choose the second matching M_2 while taking the supports of the resulting graphs $M_1 \cup M_2$ into account. Since M_1 and M_2 have to be edge-disjoint, we can remove the $N/2$ edges of M_1 from T, and remain with a subgraph T' that has fewer than $5N/2$ edges. Next, we define a weight function μ over the edges of T', such that every edge in F has a weight of 1 and every other edge has a weight of 1/2. We define the weight $\mu(M_2)$ of a perfect matching $M_2 \subset T'$ as the product of the weights of its edges. Therefore, if M_2 contains exactly j edges of F, then $\mu(M_2) = (1/2)^{N/2-j}$. Moreover, for such a matching M_2, we have $|F \setminus M_2| = N/2 - 3 - j$. Clearly, $F \setminus M_2$ is also a set of ps-flippable edges, none of which belongs to $M_1 \cup M_2$. We thus have

$$\frac{1}{\mathsf{supp}(M_1 \cup M_2)} \leq \frac{1}{2^{N/2-3-j}} = 8\mu(M_2),$$

which implies that, given a specific triangulation T and a specific perfect matching $M_1 \subset T$,

$$\sum_{M_2 \in \mathcal{M}(T')} \frac{1}{\mathsf{supp}(M_1 \cup M_2)} \leq 8 \sum_{M_2 \in \mathcal{M}(T')} \mu(M_2), \qquad (2)$$

with $T' = T \setminus M_1$, as above.

Kasteleyn's technique: An enhanced version. We now apply an extension of Kasteleyn's technique to estimate the sum in the right-hand side of (2). Here is a brief overview of the technique being used (where instead of the original technique, we apply a weighted extension of it). Given an oriented graph[5] $\vec{G} = (V, E)$ with no anti-parallel edges and a weight function μ over the edges, we define the following weighted adjacency matrix $B_{\vec{G},\mu} = (b_{ij})_{N \times N}$ of $(\vec{G}, \mu)$,

$$b_{ij} = \begin{cases} \mu(e), & \text{if } e = (i,j) \in E, \\ -\mu(e), & \text{if } e = (j,i) \in E, \\ 0, & \text{otherwise} \end{cases}$$

(where $N = |V|$, and the rows and columns of $B_{\vec{G},\mu}$ correspond to an arbitrary fixed enumeration of the vertices).

An easy extension of Kasteleyn's theorem states that every planar graph G can be oriented into some digraph $\vec{G}$ such that, for any real-valued weight function μ on its edges, we have

$$\left(\sum_{M \in \mathcal{M}(G)} \mu(M) \right)^2 = \left| \mathsf{det}\left(B_{\vec{G},\mu} \right) \right| \qquad (3)$$

(recall that $\mu(M) = \prod_{e \in M} \mu(e)$). In the "pure" form of Kasteleyn's theorem $\mu \equiv 1$ (i.e., G is unweighted) and the left-hand side is just the squared number of perfect matchings in G. A detailed presentation of Kasteleyn's theorem can be found in [14, Section 8.3]. The extension (3) to weighted graphs is given in Exercise 8.3.9 therein.

We denote by b_i the column vectors of B, for $1 \leq i \leq N$, and estimate the above determinant using Hadamard's inequality

$$\left| \mathsf{det}\left(B_{\vec{G},\mu} \right) \right| \leq \prod_{i=1}^{N} \|b_i\|_2. \qquad (4)$$

Applying the above machinery to our plane graph T' (i.e., using (3) and (4)), with the edge weights μ as defined above, we have

$$\begin{aligned} \sum_{M_2 \in \mathcal{M}(T')} \mu(M_2) &= \sqrt{\left|\mathsf{det}(B_{\vec{T'},\mu})\right|} \leq \sqrt{\prod_{i=1}^{N} \|b_i\|_2} \\ &= \left(\prod_{i=1}^{N} \|b_i\|_2^2 \right)^{1/4} \leq \left(\frac{1}{N} \sum_{i=1}^{N} \|b_i\|_2^2 \right)^{N/4} \\ &= \left(\frac{2}{N} \sum_{e \in T'} \mu(e)^2 \right)^{N/4} \end{aligned} \qquad (5)$$

(where we have used the arithmetic-geometric mean inequality and the fact that every edge of $\vec{T'}$ has two corresponding matrix entries). We note that the bound $6^{N/4}$ on the number of perfect matchings in a triangulation T is obtained in [4] by applying the unweighted version of Kasteleyn's theorem to the entire T. In this case $\sum_{e \in T} \mu(e)^2$ is the number of edges of T, which is at most $3N$, and the bound follows.

By noting that

$$|T' \setminus F| \leq 5N/2 - (N/2 - 3) = 2N + 3,$$

[5] We follow here the notation used in [14] to denote a digraph obtained from an underlying undirected graph by giving each of its edges an orientation.

and combining this with (2) and (5), we obtain

$$\sum_{M_2 \in \mathcal{M}(T')} \frac{1}{\mathsf{supp}(M_1 \cup M_2)} \leq 8 \cdot \left(\frac{2}{N} \sum_{e \in T'} \mu(e)^2 \right)^{N/4}$$

$$\leq 8 \cdot \left(\frac{2}{N} \cdot \left(1^2 \cdot (N/2 - 3) + \left(\frac{1}{2}\right)^2 \cdot (2N+3) \right) \right)^{N/4}$$

$$= O\left(2^{N/4}\right). \qquad (6)$$

Recalling once again that a triangulation contains at most $6^{N/4}$ perfect matchings [4] (that is, there are $6^{N/4}$ ways of choosing M_1), and combining this with (6), we obtain

$$\mathsf{sc}(S) \leq \sum_{T \in \mathcal{T}(S)} \sum_{\substack{M_1, M_2 \in \mathcal{M}(T) \\ M_1, M_2 \text{ edge-disjoint}}} \frac{1}{\mathsf{supp}(M_1 \cup M_2)}$$

$$\leq \sum_{T \in \mathcal{T}(S)} 6^{N/4} \cdot O\left(2^{N/4}\right) = O\left(12^{N/4}\right) \cdot \mathsf{tr}(S),$$

as asserted. □

As already noted, by applying a more complex analysis, we will obtain in Section 5 a slightly better bound.

4. PERFECT MATCHINGS AND TRIANGULATIONS

In this section we apply the machinery of the previous section to derive an upper bound on the ratio between the number of plane perfect matchings and the number of triangulations. As already mentioned in Section 2, Kasteleyn's technique implies that a triangulation of a set of N points can contain at most $6^{N/4}$ perfect matchings (see [4]). This implies that for every set S of N points in the plane, $\mathsf{pm}(S) \leq 6^{N/4} \cdot \mathsf{tr}(S) \approx 1.5651^N \cdot \mathsf{tr}(S)$. We will improve this bound, using lower bounds on the supports of perfect matchings, in a manner similar to that in Section 3.

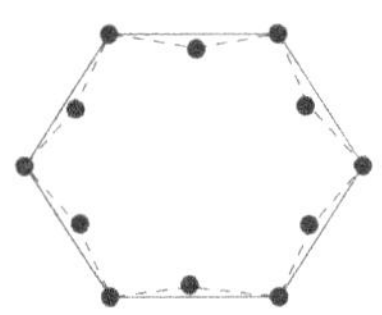

Figure 6: A set of 12 points in a double circle configuration.

Before proceeding, we note the following lower bound on the ratio $\mathsf{pm}(S)/\mathsf{tr}(S)$. Let S be a *double circle configuration*, depicted in Figure 6, consisting of N points (see [10] for a precise definition). An inclusion-exclusion argument implies that $\mathsf{tr}(S) = 12^{N/2} \approx 3.464^N$ (see [10, 18]). Moreover, Aichholzer *et al.* [1] proved that $\mathsf{pm}(S) = \Theta^*(2.2^N)$. Therefore, in this case, $\mathsf{pm}(S)/\mathsf{tr}(S) \approx \Theta^*(0.635^N)$.

We now present an improved upper bound for this ratio.

THEOREM 4.1. *For any set S of N points in the plane,*

$$\mathsf{pm}(S) \leq 8 \cdot (3/2)^{N/4} \cdot \mathsf{tr}(S) = O(1.1067^N) \cdot \mathsf{tr}(S).$$

Proof. The exact value of $\mathsf{pm}(S)$ is

$$\mathsf{pm}(S) = \sum_{T \in \mathcal{T}(S)} \sum_{M \in \mathcal{M}(T)} \frac{1}{\mathsf{supp}(M)}. \qquad (7)$$

Consider a triangulation $T \in \mathcal{T}(S)$ and a perfect matching $M \subseteq T$. As shown in the proof of Theorem 3.1, there exists a set of $N/2 - 3$ ps-flippable edges in $T \setminus M$. Therefore, the support of M is at least $2^{N/2-3}$. Combining this with (7) implies

$$\mathsf{pm}(S) \leq \sum_{T \in \mathcal{T}(S)} \sum_{M \in \mathcal{M}(T)} \frac{1}{2^{N/2-3}} \leq \sum_{T \in \mathcal{T}(S)} \frac{6^{N/4}}{2^{N/2-3}}$$

$$= 8 \cdot (3/2)^{N/4} \cdot \mathsf{tr}(S).$$

□

As already mentioned above, this does not imply a new bound on $\mathsf{pm}(N)$, since Sharir and Welzl [21] showed that $\mathsf{pm}(S) = O\left(10.05^N\right)$, bypassing the approach of counting matchings *within* a triangulation. We are not aware of any construction for which $\mathsf{pm}(S) \geq \mathsf{tr}(S)$, and offer the conjecture that there exists a constant $c < 1$ such that $\mathsf{pm}(S) = O(c^{|S|} \cdot \mathsf{tr}(S))$ for every finite set S of points in the plane. (See also the stronger conjecture concerning spanning cycles, made in the introduction.)

5. AN IMPROVED BOUND

In this section we present a more complex analysis for the number of spanning cycles, obtaining a slightly better bound than the one presented in Section 3. The analysis has three parts, each presented in a separate subsection.

Let us denote the number of interior vertices of degree 3 in the triangulation T as $v_3(T)$, and the number of flippable edges in T as $\mathsf{flip}(T)$. In Subsection 5.1 we give an upper bound for $\sum_{C \in \mathcal{C}(T)} \frac{1}{\mathsf{supp}(C)}$ that depends on $v_3(T)$. In Subsection 5.2 we give an alternative upper bound that depends on $\mathsf{flip}(T)$. Finally, in Subsection 5.3 we combine these two bounds to obtain

$$\mathsf{sc}(N) = O(1.8181^N) \cdot \mathsf{tr}(N) = O(54.543^N).$$

5.1 A $v_3(T)$-sensitive bound

In this subsection we derive the following bound, which is a function of N and $v_3(T)$.

LEMMA 5.1. *Let T be a triangulation over a set S of $N \geq 6$ points in the plane, such that N is even and S has a triangular convex hull; also, let $v_3(T) = tN$. Then*

$$\sum_{C \in \mathcal{C}(T)} \frac{1}{\mathsf{supp}(C)} < 8 \left(\frac{3}{2^t} \left(\frac{(2-t)(2-t/2)}{(1-t)^2} \right)^{1-t} \right)^{N/4}.$$

Proof. As before, we treat every spanning cycle as the union of two edge-disjoint perfect matchings $M_1, M_2 \in \mathcal{M}(T)$. We start by bounding the number of ways to choose the first perfect matching M_1. For this, we use the standard variant of Kasteleyn's technique, with the weight function $\mu \equiv 1$ (i.e., the underlying graph G is unweighted).

Recall the inequality $\sum_{M \in \mathcal{M}(T)} \mu(M) \leq \left(\prod_{i=1}^N \|b_i\|_2^2\right)^{1/4}$ obtained in Equation (5), where the b_i's (for $1 \leq i \leq N$) are the column vectors of the (signed) adjacency matrix of the oriented graph $\vec{T}$. Substituting $\mu \equiv 1$, the left hand side becomes the number of perfect matchings in T, and the squared l_2-norm of each column vector is the degree of the vertex corresponding to that column. Since every column that corresponds to a vertex of degree 3 has a squared norm of 3, the product of the squared norms of these columns is $3^{v_3(T)} = 3^{tN}$.

For the remaining $N - v_3(T)$ columns, we use, as in Section 3, the arithmetic-geometric mean inequality to bound the product of their squared norms (as in Equation (5)). This yields the bound

$$\left(\frac{X}{N - v_3(T)}\right)^{(N-v_3(T))/4} = \left(\frac{X}{N(1-t)}\right)^{(N(1-t))/4}, \quad (8)$$

where X is the sum of the degrees of all vertices other than those counted in $v_3(T)$. The sum of the degrees over the vertices of any specific triangulation is smaller than $6N$, and the sum of the degrees of the interior degree-3 vertices in T is $3v_3(T)$. Therefore, we have

$$X < 6N - 3v_3(T) = 3N(2-t). \quad (9)$$

Combining (8), (9), and the product of the squared norms that correspond to interior vertices of degree 3, implies that the number of ways to choose M_1 is less than

$$\left(3^t \cdot \left(\frac{3N(2-t)}{N(1-t)}\right)^{1-t}\right)^{N/4} = \left(3 \cdot \left(\frac{2-t}{1-t}\right)^{1-t}\right)^{N/4}. \quad (10)$$

Next, let us fix a specific perfect matching $M_1 \in \mathcal{M}(T)$. As shown at the beginning of the proof of Theorem 3.1, there exists a set F of $N/2 - 3$ ps-flippable edges in T, none of which belongs to M_1.

We continue as in the proof of Theorem 3.1, by assigning a weight of 1 to the edges of F and a weight of 1/2 to the rest of the edges of $T \setminus M_1$, recalling (2), and then applying Kaseteleyn's technique to bound the sum

$$\sum_{M_2 \in \mathcal{M}(T')} \frac{1}{\mathsf{supp}(M_1 \cup M_2)} \le 8 \sum_{M_2 \in \mathcal{M}(T \setminus M_1)} \mu(M_2) \le 8 \left(\prod_{i=1}^{N} \|b_i'\|_2^2\right)^{1/4},$$

where b_i' are the column vectors of the oriented weighted adjacency matrix of $T \setminus M_1$.

An interior vertex v of degree 3 in T has only two edges adjacent to it in $T \setminus M_1$, both not in F (since an edge adjacent to an interior vertex of degree 3 cannot be flippable). Therefore, the squared norm of a matrix column that corresponds to such a vertex is $(1/2)^2 + (1/2)^2 = 1/2$, and the product of the squared norms of all such columns is $1/2^{v_3(T)} = 1/2^{tN}$.

For the remaining $N - v_3(T)$ columns, we may once again use the arithmetic-geometric mean inequality to obtain a bound similar to the one in (8). Namely

$$\left(\frac{Y}{N(1-t)}\right)^{(N(1-t))/4}, \text{ where } Y = \sum \|b_i'\|_2^2,$$

and the sum is over the $N - v_3(T)$ vertices of $T \setminus M_1$ which are not of degree 3 in T. Each such vertex contributes to Y the sum of the squared weights of its incident edges. The estimate for Y will therefore be different, since (i) some of the edges of T were removed, and (ii) the weight function μ is not identically 1. The edges of F have remained and still have a weight of 1 each, so they contribute at most $2 \cdot (N/2 - 3) \cdot 1 < N$ to Y. Every other edge contributes $2 \cdot 1/4 = 1/2$ if it is not incident to an interior vertex of degree 3 in T, and 1/4 otherwise. Since a triangulation has fewer than $3N - 3$ edges, there are fewer than $2N$ edges in $T \setminus \{F \cup M_1\}$, and we get

$$\begin{aligned} X < N + (2N - 2v_3(T)) \cdot \frac{1}{2} + 2v_3(T) \cdot \frac{1}{4} &= 2N - \frac{v_3(T)}{2} \\ &= N(2 - t/2). \end{aligned}$$

By combining this with the rest of the squared norms and with the present version of (8), we have

$$\begin{aligned} \sum_{M_2 \in \mathcal{M}(T \setminus M_1)} \frac{1}{\mathsf{supp}(M_1 \cup M_2)} &< 8 \left(\frac{1}{2^t}\left(\frac{N(2-t/2)}{N(1-t)}\right)^{1-t}\right)^{\frac{N}{4}} \\ &= 8\left(\frac{1}{2^t} \cdot \left(\frac{2-t/2}{1-t}\right)^{1-t}\right)^{N/4}. \quad (11) \end{aligned}$$

Finally, to complete the proof, we combine (10) and (11), and obtain

$$\begin{aligned} \sum_{C \in \mathcal{C}(T)} \frac{1}{\mathsf{supp}(C)} &\le \sum_{\substack{M_1, M_2 \in \mathcal{M}(T) \\ M_1, M_2 \text{ edge-disjoint}}} \frac{1}{\mathsf{supp}(M_1 \cup M_2)} \\ &< \left(3 \cdot \left(\frac{2-t}{1-t}\right)^{1-t}\right)^{N/4} \cdot 8 \left(\frac{1}{2^t} \cdot \left(\frac{2-t/2}{1-t}\right)^{1-t}\right)^{N/4} \\ &= 8\left(\frac{3}{2^t}\left(\frac{(2-t)(2-t/2)}{(1-t)^2}\right)^{1-t}\right)^{N/4}. \end{aligned}$$

□

Remark. Notice that in the worst case (i.e., when $t = 0$) we obtain the same asymptotic value as in our initial bound of $12^{N/4}$. Similarly, the bound in (10) becomes $6^{N/4}$ when $t = 0$, as in Buchin *et al.* [4].

5.2 A $\mathsf{flip}(T)$-sensitive bound

Hurtado, Noy, and Urrutia [11] proved that $\mathsf{flip}(T) \ge N/2 - 2$, and that this bound is tight in the worst case (the upper bound is also implied by Lemma 2.1; see also [8]). In this subsection we obtain a bound on $\sum_{C \in \mathcal{C}(T)} \frac{1}{\mathsf{supp}(C)}$ as a function of $\mathsf{flip}(T)$, which improves our initial bound of $12^{N/4}$ when $\mathsf{flip}(T)$ is larger than $N/2$ by some positive fraction of N.

We define c_{gon} as the maximum real number satisfying the following property. Every simple polygon P that has a triangulation T_P with k of its diagonals flippable and with $l \le k$ of these diagonals forming a ps-flippable set, has at least $2^l c_{\text{gon}}^{k-l}$ triangulations. Notice that the triangulations under consideration, including T_P, are triangulations of the polygon P, and not of its vertex set. Note also that we can have the equality $l = k$ only when P is convex.

LEMMA 5.2. *$x \le c_{gon} \le 5/4$ where $x \approx 1.17965$ is the unique real root of the polynomial $1 + 4x^2 - 4x^3$. That is, the left inequality means that every simple polygon P that has a triangulation T_P with k of its diagonals flippable and with $l \le k$ of these diagonals forming a ps-flippable set, has at least $2^l x^{k-l}$ triangulations.*

For a proof of this lemma, see the full version of the paper.[6] Next, we show how to use c_{gon} (or rather, its lower bound x) and $\mathsf{flip}(T)$ to upper bound $\sum_{C \in \mathcal{C}(T)} \frac{1}{\mathsf{supp}(C)}$.

[6] `http://arxiv.org/abs/1109.5596`

LEMMA 5.3. *Consider a triangulation T with $\mathsf{flip}(T) = N/2-3+\kappa N$, for some $\kappa \geq 0$, and let x be the constant in Lemma 5.2. Then*

$$\sum_{C\in\mathcal{C}(T)} \frac{1}{\mathsf{supp}(C)} < 8\left(\frac{(3+(\gamma^2-1)(\kappa+1/2))(4+(x^2-1)\kappa)}{x^{4\kappa}}\right)^{N/4},$$

where

$$\gamma = x\cdot e^{-\frac{x^2-1}{4(4+(x^2-1)\kappa)}}.$$

Proof. Once again, we treat every spanning cycle as the union of a pair of edge-disjoint perfect matchings $M_1, M_2 \in \mathcal{M}(T)$, and use Kasteleyn's technique (as presented in Section 3) to bound the number of such pairs. We start by fixing some perfect matching $M_1 \in \mathcal{M}(T)$ and denote the number of flippable edges of T that are in M_1 as $\mathsf{flip}_T(M_1)$. As shown in the proof of Theorem 3.1, there is a set of at least $N/2-3$ ps-flippable edges in $T\setminus M_1$. We restrict our attention to a set F of *exactly* $N/2-3$ ps-flippable edges in $T\setminus M_1$.

For analyzing the complementary matchings M_2, we define a weight function $\mu(\cdot)$ on the edges of $T\setminus M_1$, such that

$$\mu(e) = \begin{cases} 2, & \text{if } e\in F,\\ x, & \text{if } e\notin F \text{ is flippable},\\ 1, & \text{if } e \text{ is not flippable.}\end{cases}$$

Notice that any spanning cycle partitions the convex hull of its point set into interior-disjoint simple polygons. The support of the spanning cycle is the product of the number of triangulations of each of these polygons. For a fixed choice of M_2 (and of M_1), denote by $P_1,\ldots,P_m$ the polygons in the partition produced by $M_1\cup M_2$ (assuming that $M_1\cup M_2$ is indeed a spanning cycle). For each i, let k_i be the number of flippable diagonals of P_i, and let l_i be the number of those diagonals (among the k_i flippable ones) that belong to F. If M_2 uses $\mathsf{flip}_T(M_2)$ flippable edges of $T\setminus M_1$, l of which are in F, then $\sum_{i=1}^m k_i = \mathsf{flip}(T)-\mathsf{flip}_T(M_1)-\mathsf{flip}_T(M_2)$ and $\sum_{i=1}^m l_i = |F|-l = N/2-3-l$. Applying Lemma 5.2 to each P_i and multiplying the resulting bounds, we obtain a total of at least $2^{\sum l_i}x^{\sum k_i - \sum l_i}$ triangulations. Hence,

$$\begin{aligned}\mathsf{supp}(M_1\cup M_2) &\geq (2/x)^{\sum l_i}x^{\sum k_i}\\ &= (2/x)^{N/2-3-l}x^{\mathsf{flip}(T)-\mathsf{flip}_T(M_1)-\mathsf{flip}_T(M_2)}.\end{aligned}$$

Next, notice that $\mu(M_2) = 2^l x^{\mathsf{flip}_T(M_2)-l}$, so we have

$$\begin{aligned}\mathsf{supp}(M_1\cup M_2) &\geq \frac{2^{N/2-3}x^{\mathsf{flip}(T)-\mathsf{flip}_T(M_1)-(N/2-3)}}{\mu(M_2)}\\ &= \frac{2^{N/2-3}x^{\kappa N-\mathsf{flip}_T(M_1)}}{\mu(M_2)}. \qquad (12)\end{aligned}$$

By combining (12) with Kasteleyn's method, we obtain

$$\begin{aligned}\sum_{\substack{M_2\in\mathcal{M}(T\setminus M_1)\\ M_1\cup M_2\in\mathcal{C}(T)}} \frac{1}{\mathsf{supp}(M_1\cup M_2)} &\leq \sum_{M_2\in\mathcal{M}(T\setminus M_1)} \frac{\mu(M_2)}{2^{N/2-3}x^{\kappa N-\mathsf{flip}_T(M_1)}}\\ &\leq \frac{1}{2^{N/2-3}x^{\kappa N-\mathsf{flip}_T(M_1)}}\cdot\left(\frac{2}{N}\sum_{e\in T\setminus M_1}\mu(e)^2\right)^{N/4}. \qquad (13)\end{aligned}$$

To bound the sum in the parentheses, we notice that $T\setminus M_1$ contains exactly $N/2-3$ edges of F, exactly $\kappa N - \mathsf{flip}_T(M_1)$ flippable edges not in F, and fewer than $2N-(\kappa N-\mathsf{flip}_T(M_1))$ non-flippable edges. Therefore,

$$\begin{aligned}\frac{2}{N}\sum_{e\in T\setminus M_1}\mu(e)^2 &< \frac{2}{N}\Big(1^2\cdot(2N-(\kappa N-\mathsf{flip}_T(M_1))) \qquad (14)\\ &\quad + x^2\cdot(\kappa N-\mathsf{flip}_T(M_1)) + 2^2\cdot N/2\Big)\\ &= 8+2(x^2-1)\kappa - 2(x^2-1)\cdot\mathsf{flip}_T(M_1)/N\\ &= (8+2(x^2-1)\kappa)\left(1-\frac{2(x^2-1)}{8+2(x^2-1)\kappa}\cdot\frac{\mathsf{flip}_T(M_1)}{N}\right)\\ &\leq (8+2(x^2-1)\kappa)\cdot e^{-\frac{x^2-1}{4+(x^2-1)\kappa}\cdot\frac{\mathsf{flip}_T(M_1)}{N}}\\ &= (8+2(x^2-1)\kappa)\cdot(\gamma/x)^{4\cdot\mathsf{flip}_T(M_1)/N}, \qquad (15)\end{aligned}$$

where in the penultimate inequality we applied $1-u\leq e^{-u}$ (which holds for $u\geq 0$). Combining (13) and (15), we get

$$\begin{aligned}\sum_{C\in\mathcal{C}(T)}\frac{1}{\mathsf{supp}(C)} &\leq \sum_{\substack{M_1,M_2\in\mathcal{M}(T)\\ M_1,M_2\text{ edge-disjoint}\\ M_1\cup M_2\in\mathcal{C}(T)}} \frac{1}{\mathsf{supp}(M_1\cup M_2)}\\ &\leq \sum_{M_1\in\mathcal{M}(T)} \frac{\left((8+2(x^2-1)\kappa)\cdot(\gamma/x)^{4\cdot\mathsf{flip}_T(M_1)/N}\right)^{N/4}}{2^{N/2-3}x^{\kappa N-\mathsf{flip}_T(M_1)}}\\ &= 8\left(\frac{8+2(x^2-1)\kappa}{4x^{4\kappa}}\right)^{N/4}\sum_{M_1\in\mathcal{M}(T)}\gamma^{\mathsf{flip}_T(M_1)}. \qquad (16)\end{aligned}$$

To bound the sum in (16), we once again use Kasteleyn's technique. This time, we define a weight function $\nu(\cdot)$ over the edges of T, such that every flippable edge gets a weight of γ, and every other edge a weight of 1. Notice that, in this manner, $\nu(M_1) = \gamma^{\mathsf{flip}_T(M_1)}$ for every $M_1\in\mathcal{M}(T)$. We thus have

$$\begin{aligned}\sum_{M_1\in\mathcal{M}(T)}\gamma^{\mathsf{flip}_T(M_1)} &\leq \left(\frac{2}{N}\sum_{e\in T}\nu(e)^2\right)^{N/4}\\ &< \left(\frac{2}{N}\left(\gamma^2\cdot\mathsf{flip}(T)+1\cdot(3N-\mathsf{flip}(T))\right)\right)^{N/4}\\ &< \left(6+\frac{2}{N}(\gamma^2-1)(N/2+\kappa N)\right)^{N/4}\\ &< \left(6+2(\gamma^2-1)(\kappa+1/2)\right)^{N/4}. \qquad (17)\end{aligned}$$

Finally, combining (16) and (17) implies the assertion of the lemma. $\square$

Note that in the worst case, when $\kappa = 0$, the bound becomes $O\left((10+2\gamma^2)^{N/4}\right)$. For $k=0$, we have $\gamma = x\cdot e^{-(x^2-1)/16}$, and it is easy to verify that $\gamma > 1$ for $1 < x \leq$

5/4. So the bound is actually asymptotically *worse* than our initial bound of $12^{N/4}$, and it continues to be worse when κ is sufficiently small. As the next subsection shows, in this case the v_3-dependent bound from Subsection 5.1 becomes small and can be used instead.

5.3 Integration

In this subsection we combine the results from the two previous subsections to obtain an improved bound for $\mathsf{sc}(N)$. This is done by deriving a connection between $v_3(T)$ and $\mathsf{flip}(T)$. We start by presenting a generalization of Lemma 2.4.

LEMMA 5.4. *Let $c > 1$ be a constant such that every set S of an even number of points in the plane and a triangular convex hull satisfies $\mathsf{sc}(S) = O(c^{|S|})$. Then $\mathsf{sc}(S) = O(c^{|S|})$ also holds for every other finite point set S in the plane.*

Proof. Consider a point set S. If S has an even number of points, we pick a new point p outside the convex hull of S, and put $S' = S \cup \{p\}$. As mentioned in the proof of Lemma 2.4, inserting an additional vertex outside the convex hull of the point set can only increase the number of spanning cycles. If S has an odd number of points, we put $S' = S$. Notice that, either way, S' has an odd number of points. Let Δabc be a large triangle containing S' in its interior, and let $S'' = S' \cup \{a, b, c\}$. Again, since inserting an additional vertex outside the convex hull of the point set can only increase the number of spanning cycles, we have $\mathsf{sc}(S') \le \mathsf{sc}(S'')$. Since S'' has an even number of points and a triangular convex hull, $\mathsf{sc}(S) \le \mathsf{sc}(S') \le \mathsf{sc}(S'') = O(c^{N+4}) = O(c^{|S|})$. □

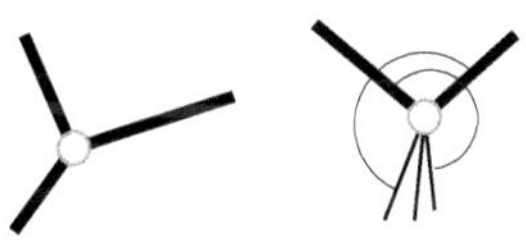

Figure 7: Separable edges.

We also require the notion of *separable edges*, as presented in [20]. Consider a point set S, a triangulation $T \in \mathcal{T}(S)$, and an interior point $p \in S$. We call an edge e incident to p in T a *separable edge at* p if it can be separated from the other edges incident to p by a line through p. An equivalent condition is that the two angles between e and its clockwise and counterclockwise neighboring edges (around p) sum up to more than π. We observe the easy following properties (see Figure 7 for an illustration).

(S0) No edge is separable at both vertices induced by its endpoints.

(S1) If p has degree 3 in T, every edge incident to it is separable (recall that p is an interior point).

(S2) If p has degree at least 4 in T, at most two incident edges can be separable at w.

(S3) If p has degree at least 4 in T and there are two edges separable at p, then they must be consecutive in the order around it.

We are now ready for the main theorem of the section.

THEOREM 5.5. *For any set S of N points in the plane,*

$$\mathsf{sc}(S) = O\left(10.9247^{N/4}\right) \cdot \mathsf{tr}(S) = O\left(1.8181^{N}\right) \cdot \mathsf{tr}(S).$$

Proof. By Lemma 5.4, we may assume that N is even and that S has a triangular convex hull. Recall that

$$\mathsf{sc}(S) = \sum_{T \in \mathcal{T}(S)} \sum_{C \in \mathcal{C}(T)} \frac{1}{\mathsf{supp}(C)}.$$

We sort the triangulations in the first sum according to the number of interior vertices of degree 3 that they contain, and get

$$\mathsf{sc}(S) = \sum_{i=0}^{(2N+1)/3} \sum_{\substack{T \in \mathcal{T}(S) \\ v_3(T)=i}} \sum_{C \in \mathcal{C}(T)} \frac{1}{\mathsf{supp}(C)}. \tag{18}$$

(The fact that $v_3(T) \le (2N+1)/3$ for every triangulation T is established, e.g., in [22].) Given a triangulation T with $v_3(T) = i$, we can use Lemma 5.1 to bound $\sum_{C \in \mathcal{C}(T)} \frac{1}{\mathsf{supp}(C)}$. However, when $v_3(T)$ is small, the improvement in Lemma 5.1 is not significant. In this case we will use instead the bound in Lemma 5.3 which, as we now proceed to show, becomes significant when $v_3(T)$ is small.

Consider a triangulation $T \in \mathcal{T}(S)$. Since S has a triangular convex hull, T contains $3N - 9$ interior edges. Notice that an interior edge e is flippable if and only if e is not separable at either of its endpoints (this property is equivalent to e being a diagonal of a convex quadrilateral). From the above properties of separable edges, we have

$$\mathsf{flip}(T) \ge \overbrace{3N-9}^{\text{Interior edges}} - 3 \cdot \overbrace{v_3(T)}^{\text{Interior vertices of degree 3}} - 2 \cdot \overbrace{(N - v_3(T) - 3)}^{\text{Other interior vertices}} = N - 3 - v_3(T).$$

To find for which values of i it is better to use Lemma 5.1, and for which values it is better to use Lemma 5.3, we define $t = v_3(T)/N$ and

$$\kappa = \frac{\mathsf{flip}(T) - (N/2 - 3)}{N} \ge \frac{(N - tN - 3) - (N/2 - 3)}{N} = 1/2 - t,$$

and solve the equation

$$8\left(\frac{3}{2^t}\left(\frac{(2-t)(2-t/2)}{(1-t)^2}\right)^{1-t}\right)^{N/4} = 8\left(\frac{(3 + (\gamma^2 - 1)(\kappa + 1/2))(4 + (x^2 - 1)\kappa)}{x^{4\kappa}}\right)^{N/4}$$

with $\kappa = 1/2 - t$, where $x \approx 1.17965$ and $\gamma = x \cdot e^{-\frac{x^2-1}{4(4+(x^2-1)\kappa)}}$; this will determine the threshold where the two bounds coincide. That is, we need to solve the equation (again, with $\kappa = 1/2 - t$)

$$\frac{3}{2^t}\left(\frac{(2-t)(2-t/2)}{(1-t)^2}\right)^{1-t} = \frac{(3 + (\gamma^2 - 1)(\kappa + 1/2))(4 + (x^2 - 1)\kappa)}{x^{4\kappa}}.$$

For this, we use the Wolfram Mathematica software [24], and obtain the solution $t \approx 0.1072$. Moreover, it is easily shown that for $i \geq 0.1072N$ the bound from Lemma 5.1 is smaller, and for $i \leq 0.1072N$ the bound from Lemma 5.3 is smaller. In fact, these bounds, in their appropriate usage, are all dominated by the common bound for $t \approx 0.1072$. This, together with (18), Lemma 5.1, and Lemma 5.3 imply the asserted bound. □

By combining Theorem 5.5 with the bound $\mathsf{tr}(N) < 30^N$ [19], we obtain:

COROLLARY 5.6. $\mathsf{sc}(N) = O\left(54.543^N\right)$.

6. REFERENCES

[1] O. Aichholzer, T. Hackl, C. Huemer, F. Hurtado, H. Krasser, and B. Vogtenhuber, On the number of plane geometric graphs, *Graphs and Combinatorics* **23**(1) (2007), 67–84.

[2] M. Ajtai, V. Chvátal, M.M. Newborn, and E. Szemerédi, Crossing-free subgraphs, *Annals Discrete Math.* **12** (1982), 9–12.

[3] S. Akl, A lower bound on the maximum number of crossing-free Hamilton cycles in a rectilinear drawing of K_n, *Ars Combinat.* **7** (1979), 7–18.

[4] K. Buchin, C. Knauer, K. Kriegel, A. Schulz, and R. Seidel, On the number of cycles in planar graphs, *Proc. 17th Computing and Combinatorics Conf.*, Lecture Notes Comput. Sci., Vol 4598, Springer-Verlag, Berlin, 2007, 97–107.

[5] A. Dumitrescu, A. Schulz, A. Sheffer, and Cs.D. Tóth, Bounds on the maximum multiplicity of some common geometric graphs, *Proc. 28th Symp. Theoret. Aspects Comput. Sci.* (2011), 637–648.

[6] A. García, M. Noy, and J. Tejel, Lower bounds on the number of crossing-free subgraphs of K_N, *Comput. Geom. Theory Appl.* **16**(4) (2000), 211–221.

[7] L.J. Guibas and F.F. Yao, On translating a set of rectangles, *Proc. 12th ACM Symp. on Theory of Computing* (1980), 154–160.

[8] M. Hoffmann, M. Sharir, A. Sheffer, Cs.D. Tóth, and E. Welzl, Counting plane graphs: Flippability and its applications, *Proc. 12th Symp. on Algs. and Data structs.* (2011), 524–535. An extended version that also includes A. Schulz will appear in the book "Thirty Esssays in Geometric Graph Theory" (ed. J. Pach).

[9] K. Hosono, On convex decompositions of a planar point set, *Discrete Math.* **309** (2009), 1714–1717.

[10] F. Hurtado and M. Noy, Counting triangulations of almost-convex polygons, *Ars Combinat.* **45** (1997), 169–179.

[11] F. Hurtado, M. Noy, and J. Urrutia, Flipping edges in triangulations, *Discrete Comput. Geom.* **22** (1999), 333–346.

[12] N. Katoh and S. Tanigawa, Fast enumeration algorithms for non-crossing geometric graphs, *Proc. 24th ACM Symp. on Comput. Geom.* (2008), 328–337.

[13] P. Kasteleyn, Graph theory and crystal physics, in *Graph Theory and Theoretical Physics*, Academic Press, London, 1967, 43–110.

[14] L. Lovász and M. Plummer, *Matching Theory*, North Holland, Budapest-Amsterdam, 1986.

[15] M. Newborn and W.O.J. Moser, Optimal crossing-free Hamiltonian circuit drawings of K_n, *J. Combinat. Theory Ser. B* **29** (1980), 13–26.

[16] J. O'Rourke, *Computational Geometry in C*, 2nd edition, Cambridge University Press, New York, 1998.

[17] A. Razen and E. Welzl, Counting crossing-free geometric graphs with exponential speed-up, *Rainbow of Computer Science — Dedicated to Hermann Maurer on the Occasion of His 70th Birthday*, Lecture Notes Comput. Sci., Vol 6570, Springer-Verlag, Berlin, 2011, 36–46.

[18] F. Santos and R. Seidel, A better upper bound on the number of triangulations of a planar point set, *J. Combinat. Theory* A **102**(1) (2003), 186–193.

[19] M. Sharir and A. Sheffer, Counting triangulations of planar point sets, *Electr. J. Comb.* **18**(1) (2011), P. 70. Also in arXiv:0911.3352v2.

[20] M. Sharir, A. Sheffer, and E. Welzl, On degrees in random triangulations of point sets, *J. Combinat. Theory* A **118** (2011), 1979–1999.

[21] M. Sharir and E. Welzl, On the number of crossing-free matchings (cycles, and partitions), *SIAM J. Comput.* **36**(3) (2006), 695–720.

[22] M. Sharir and E. Welzl, Random triangulations of planar point sets, *Proc. 17th Ann. ACM-SIAM Symp. on Discrete Algorithms* (2006), 860–869.

[23] R.P. Stanley, *Enumerative Combinatorics*, vol. **2**, Cambridge University Press, Cambridge, 1999.

[24] Wolfram Research, Inc., Mathematica, Version 7.0.1, Champaign, IL, 2009.

Mappings and Meshes

Christopher Bishop
Stony Brook University
Stony Brook, NY, USA
bishop@math.sunysb.edu

ABSTRACT

In my talk I will attempt to draw some connections between complex analysis and computational geometry, particularly between conformal mappings, hyperbolic geometry, the medial axis and optimal meshing. The Riemann mapping theorem says that there is a conformal (angle preserving) map of the unit disk, $\mathbb{D}$, to the interior Ω of any simple n-gon. How much work is needed to compute this map?

Theorem 1: *[1] We can compute the conformal map* $f : \mathbb{D} \to \Omega$ *to within error* ϵ *in time* $O(n \cdot \log \frac{1}{\epsilon} \log\log \frac{1}{\epsilon})$.

The proof uses an iteration that converges quadratically to f, but it needs a good initial guess that is close to the correct answer. The medial axis allows us to quickly construct a map $\Omega \to \mathbb{D}$ that is close to Riemann's map with precise estimates. This gives our starting point, and the time estimates in the theorem depend on time needed to compute the medial axis (linear by work of Chin, Snoeyink and Wang).

Conversely, the proof of Theorem 1 uses ideas from analysis that may be of interest in CG. For example, the dome of a planar domain Ω is the surface in $\mathbb{R}^3_+$ given by the upper envelope of all hemispheres whose base on $\mathbb{R}^2$ is a medial axis disk of Ω. Domes arise in the study of 3-manifolds, and a fundamental result of Dennis Sullivan about convex sets in hyperbolic 3-space lies at the heart of Theorem 1.

The proof of Theorem 1 also introduces the thick-thin decomposition of a polygon, inspired by the thick-thin decomposition of a hyperbolic manifold. The thin parts correspond to "long, narrow" pieces of the polygon (but not in the obvious way; the precise definition uses a conformal invariant called extremal length). This decomposition plays in important role in proving:

Theorem 2: *[2] Any simple* n*-gon has a* $O(n)$ *quadrilateral mesh with all angles* $\leq 120°$ *and all new angles* $\geq 60°$.

The sharp upper bound is due to Bern and Eppstein; the lower bound is the novel part here. The result is also true for PSLGs with at most n edges and vertices, although simple examples show the $O(n)$ must be replaced by $O(n^2)$:

Theorem 3: *[4] Every PSLG has an* $O(n^2)$ *quadrilateral mesh with all angles* $\leq 120°$ *and all new angles* $\geq 60°$. *Only* $O(n/\epsilon)$ *angles satisfy* $|\theta - 90°| > \epsilon$, *for any* $\epsilon > 0$.

Adding diagonals to the quadrilaterals gives a triangulation with all angles $\leq 120°$, improving results of S. Mitchell (157.5°) and Tan (132°). Applying thick-thin decompositions directly to triangulations gives an even better result: for any $\epsilon > 0$, a PSLG has a $O(n^2)$ triangulation with angles $\leq 90° + \epsilon$ (but the constant blows up as $\epsilon \searrow 0$). For $\epsilon = 0$, a more intricate construction shows:

Theorem 4: *[3] Every PSLG has a* $O(n^{2.5})$ *nonobtuse triangulation.*

No polynomial bound was previously known, and I strongly suspect $n^{2.5}$ can be improved to n^2 (known sharp). The triangles may be taken to be either all acute or all right. Some consequences of Theorem 4 and its proof include

- Every PSLG has a $O(n^{2.5})$ conforming Delaunay triangulation (improves $O(n^3)$ by Edelsbrunner and Tan).
- Any triangulation of an n-gon has a $O(n^2)$ nonobtuse refinement (improves $O(n^4)$ by Bern and Eppstein).

Most of the algorithms I will describe seem too complex to implement (at least for me), and are strongly 2-dimensional, but I hope the audience will have some suggestions for getting around these problems.

BIO

I graduated from Michigan State, did Part III at Cambridge and was a Ph.D. student of Peter W. Jones at the University of Chicago. I am currently a professor of mathematics at Stony Brook University working in geometric function theory and its connections to potential theory, hyperbolic geometry, quasiconformal maps, probability, dynamics and (most recently) computational geometry. My path from classical analysis to meshing is described in the essay [5]. See the 'papers' link at `http://www.math.sunysb.edu/~bishop`.

1. REFERENCES

[1] C. J. Bishop. Conformal mapping in linear time. *Discrete Comput. Geom.*, 44(2):330–428, 2010.

[2] C. J. Bishop. Optimal angle bounds for quadrilateral meshes. *Discrete Comput. Geom.*, 44(2):308–329, 2010.

[3] C. J. Bishop. Nonobtuse triangulation of PSLGs. 2011.

[4] C. J. Bishop. Quadrilateral meshes for PSLGs. 2011.

[5] C. J. Bishop. A random walk in analysis. In A. Cordoba, J. Fernandez, and P. Fernandez, editors, *All That Math: portraits of mathematicians as young readers.* Revisita Matematica Iberoamericana, 2011.

Categories & Subject Descriptors: G.0. Mathematics of Computing: General

Keywords: conformal maps, Schwarz-Christoffel formula, hyperbolic geometry, medial axis, quadrilateral meshes, nonobtuse triangulation, acute triangulation, Delaunay triangulation

SCG12, June 17–20, 2012, Chapel Hill, North Carolina, USA.
ACM 978-1-4503-1299-8/12/06.

Cubic Plane Graphs on a Given Point Set

Jens M. Schmidt*
MPI für Informatik
Saarbrücken, Germany
jens.schmidt@mpi-inf.mpg.de

Pavel Valtr†
Dept. of Applied Math. and Institute for
Theoretical Comp. Sci. (ITI)
Charles University
Prague, Czech Republic

ABSTRACT

Let P be a set of $n \geq 4$ points in the plane that is in general position and such that n is even. We investigate the problem whether there is a cubic plane straight-line graph on P. No polynomial-time algorithm is known for this problem. Based on a reduction to the existence of certain diagonals of the boundary cycle of the convex hull of P, we give the first polynomial-time algorithm; the algorithm is constructive and runs in time $O(n^3)$. We also show which graph structure can be expected when there is a cubic plane graph on P; e. g., if P admits a 2-connected cubic plane graph, we show that P admits also a 2-connected cubic plane graph that contains the boundary cycle of P. The algorithm extends to checking P on admitting a 2-connected cubic plane graph.

Categories and Subject Descriptors

F.2.2 [**Nonnumerical Algorithms and Problems**]: Computations on discrete structures, Geometrical problems and computations; G.2 [**Graph Theory**]: Graph algorithms

Keywords

planar embedding, point set, cubic graphs

*Part of the research by J. M. Schmidt was done at FU Berlin.

†Research by P. Valtr was partially supported by project CE-ITI (GACR P2020/12/G061) of the Czech Science Foundation

1. INTRODUCTION

Let P be a point set in the plane that is in *general position*, i. e., that does not contain three points on a line. A *straight-line embedding* of a graph $G = (V, E)$ is an injective function π: $V \to \mathbb{R}^2$ such that for any two distinct edges ab and cd the straight line segments $\pi(a)\pi(b)$ and $\pi(c)\pi(d)$ are internally disjoint (i. e., they may only intersect at their endpoints). Let P *admit* a graph $G = (V, E)$ if $|P| = |V|$ and there is a straight-line embedding that maps V to P; we also say that G *is on* P. Thus, P can only admit plane graphs.

We are interested in classifying the point sets P that admit at least one simple plane graph G with a given additional property, e. g., being k-connected, k-edge-connected or k-regular. The graph G is not part of the input: it suffices to find any graph G on P with the desired properties. Using Euler's formula, none of these properties can exist for $k \geq 6$, so we focus on $k \in \{0, \ldots, 5\}$. In addition, there are no k-regular graphs for $k = 1, 3, 5$ when n is odd, since every graph must have an even number of odd vertices. We will assume in these cases that n is even. Since we are dealing only with simple graphs, we can further assume $n > k$ throughout the paper. If not stated otherwise, all graphs are assumed to be simple and plane, but not necessarily connected. All point sets are assumed to be in general position.

For $k \in \{0, 1\}$, it is easy to see that every point set admits a 0-connected 0-regular as well as a connected graph, but a 1-regular graph can only be found when n is even (see Table 1). For $k = 2$, every point set P admits a 2-regular 2-connected (and thus also 2-edge-connected) graph, as there is always a plane cycle on P [4].

For $k = 3$, Dey et al. [2] give a construction proving that there is a 3-connected graph on P if and only if $n > 3$ and P is not in convex position (the same characterization holds for 3-edge-connected graphs). García et al. [3] investigate how many edges are sufficient to allow a 3-connected graph on P. Let h be the number of points on the convex hull boundary of P. If P is not in convex position, they give a construction of a 3-connected graph on P that has $max(\frac{3}{2}n, n+h-1)$ edges; the same construction on minimality constraints holds for 3-edge-connected graphs. They also prove that this number is minimal for any 3-connected and for any 3-edge-connected graph on P.

As a corollary, $h \leq \frac{n}{2} + 1$ implies the existence of a 3-regular graph on P. García et al. show in addition that there is a 3-regular graph on P (not necessarily 3-connected, but still 2-connected) when $\frac{n}{2} + 1 \leq h \leq \frac{3}{4}n$ [3, Theorem 4]. While this gives a characterization of the point sets admitting 3-regular graphs for $h \leq \frac{3}{4}n$, the problem remained

k	Necessary and sufficient conditions for a k-connected plane graph	k-edge-connected plane graph	k-regular plane graph	Minimum number of edges in a k-(edge-)connected plane graph on P
0	none	none	none	0
1	none	none	n even	$n-1$
2	none	none	none	n
3	P not in convex position	P not in convex position	? (known for $h \leq \frac{3}{4}n$)	$max(\frac{3}{2}n, n+h-1)$
4	? (known for $h=3$)	?	?	?
5	?	?	?	?

Table 1: Conditions on P that are both necessary and sufficient for the existence of a k-connected, k-edge-connected and k-regular plane graph, respectively, on a point set P in general position, where $|P| = n > k$.

open for higher values of h. Examples show that the existence of a 3-regular graph is then not any more dependent on only h and n [3]. We give a characterization for all values of h, which leads to the first polynomial time algorithm that computes a 3-regular graph on P if it exists; the running time is $O(n^3)$. We also show that the existence of a 2-connected cubic graph on P implies that there is also a 2-connected cubic graph on P that contains the boundary cycle of the convex hull of P.

In contrast to $k \leq 3$, very little is known about the case $k \in \{4, 5\}$. There exist point sets that are neither in convex position nor admit 4-connected graphs. For the special case of $h = 3$, Dey et al. [2] could characterize the point sets that admit 4-connected graphs.

Preliminaries.

A graph is *cubic* if it is 3-regular. Let P be a set of $n \geq 4$ points in the plane in general position with n even. Let $ch(P)$ denote the boundary cycle of the convex hull of P and let a (combinatorial) edge in $\binom{P}{2}$ be a *diagonal* of P if it joins two non-consecutive points in $ch(P)$. Let H be the set of points in $ch(P)$, $h := |H|$, and let $I = P \setminus H$ be the set of inner points in P, $i := |I|$.

Let D be a set of non-crossing diagonals of P. We call the bounded regions in which $ch(P)$ and D subdivides the plane *faces induced by* D; let $F(D)$ be the set of faces induced by D. For every induced face $f \in F(D)$, let V_f be the set of endpoints of diagonals on the boundary of f, let H_f be the set of points on the boundary of f that are not in V_f and let I_f be the set of points strictly inside f. Moreover, let $i_f := |I_f|$, $h_f := |H_f|$, $v_f := |V_f|$, $N_f := H_f \dot{\cup} I_f$ and $n_f := |N_f|$.

For a vertex $v \in V(G)$ and $X \subseteq V(G)$ of a graph G, let $G[X]$ be the subgraph of G that is *vertex-induced* by X. For a simple cubic plane graph G on P and an induced face f of $D(C_G)$, let $G[f] = G[N_f \cup V_f]$.

2. NECESSARY CONDITIONS FOR CUBIC GRAPHS

Every simple cubic plane graph G on P contains a (possibly empty) set D of non-crossing diagonals on P. This way, G naturally defines a set of induced faces. Instead of working with concrete cubic graphs, we will consider only their set of diagonals and, for any non-diagonal edge e, the induced face that contains e. It will turn out that it is only important how many non-diagonal edges in an induced face f are incident to a vertex $p \in P$; the precise neighbors of p in f are not crucial. We model this by representing non-diagonal edges with *half-edges*, i. e., with the two parts generated by cutting an edge.

Definition 1. A *diagonal configuration* $\mathcal{C}$ on P consists of a set $D(C)$ of pairwise non-crossing diagonals of P and a multiset of half-edges on the points of P such that, for every vertex $p \in P$,

- p has degree 3 in C (counting diagonals and half-edges) and
- each half-edge on p is assigned to a face that is induced by $D(C)$ and contains p.

Any simple cubic plane graph G on P determines a unique diagonal configuration C_G by cutting every non-diagonal edge e into two half-edges such that both half-edges are assigned to the induced face that contained e. We list necessary conditions on graphs and diagonal configurations to allow cubic graphs on P. García et al. proved the following theorem.

Theorem 2 ([3], implicitly)**.** *Let P be a set of $n \geq 4$ points in general position such that n is even. If $h \leq \frac{3}{4}n$, there is an algorithm with running time $O(n^3)$ that constructs a simple cubic 2-connected plane graph on P that contains $ch(P)$. If $h = n$, P does not admit a simple cubic plane graph.*

We can therefore focus on the case $h > \frac{3}{4}n$; then, every cubic plane graph on P must contain at least one diagonal.

Lemma 3. *Let $h > \frac{3}{4}n$. For every simple cubic (not necessarily connected) plane graph G on P, $D(C_G) \neq \emptyset$. Moreover, $|D(C_G)| \geq 2(h - \frac{3}{4}n) = \frac{h-3i}{2}$.*

Proof. Every vertex v in $H \setminus V(D(C_G))$ has at least one neighbor in I, as v can have at most two neighbors in H and has degree 3. Let s be the number of edges in G that join vertices of H with vertices of I. Then $h - 2|D(C_G)| \leq s$. However, $s \leq 3i$, because every inner vertex can be adjacent to at most 3 vertices in H. It follows that $|D(C_G)| \geq \frac{h-3i}{2} = 2(h - \frac{3}{4}n)$. If $h > \frac{3}{4}n$, then the right-hand side is positive, which concludes the proof.

To establish further conditions on diagonal configurations for cubic graphs, we classify the vertices in each induced face f. A vertex in V_f can be one of the following types with respect to f (see Figure 1).

Definition 4. Let C be a diagonal configuration on P and let $f \in F(D(C))$. We call a vertex $v \in V_f$ *sated*, *balanced* or *hungry* (*in* f) if its number of incident diagonals and half-edges in f is 1, 2 or 3, respectively.

A vertex in V_f that is not balanced is also called *unbalanced*. Let V_f^0, V_f^+ and V_f^- be the set of *balanced*, *hungry* and *sated* vertices in V_f, respectively. For a sated vertex $v \in V_f^-$, let w be the unique neighbor of v in $ch(P)$ that is contained in f. If w is also sated in f, v is called *matched* (with w); otherwise, v is *unmatched*. Let V_f^{-m} and V_f^{-u} be the set of *matched* and *unmatched* vertices in V_f, respectively. Let $v_f^0 := |V_f^0|$, $v_f^+ := |V_f^+|$, $v_f^- := |V_f^-|$, $v_f^{-m} := |V_f^{-m}|$ and $v_f^{-u} := |V_f^{-u}|$.

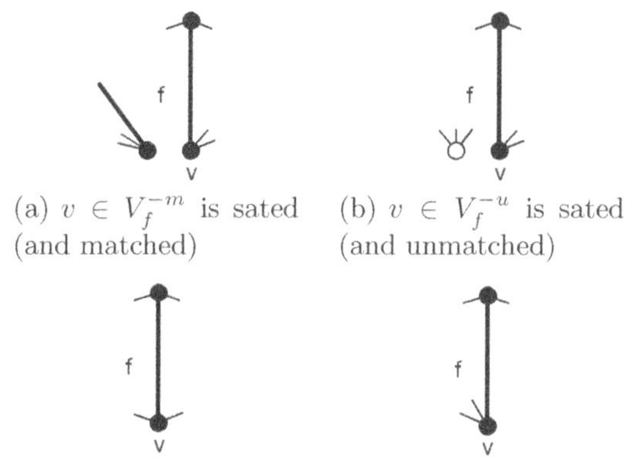

(a) $v \in V_f^{-m}$ is sated (and matched)

(b) $v \in V_f^{-u}$ is sated (and unmatched)

(c) $v \in V_f^0$ is balanced

(d) $v \in V_f^+$ is hungry

Figure 1: Diagonal configurations that contain balanced and unbalanced vertices v (with respect to the induced face f).

We now assign each induced face f the following integer value $\Delta(f)$ in order to prove necessary conditions on these values for cubic graphs.

Definition 5. For a diagonal configuration C on P and each induced face $f \in F(D(C))$, let $\Delta(f) = 3i_f - h_f - v_f^+ - v_f^{-u}$.

It is useful to imagine $h_f + v_f^+ + v_f^{-u}$ as the number of edges between inner and boundary vertices of f that are necessary for any cubic graph on P. Note that an unmatched vertex requires such an edge indirectly, as it forces its non-sated neighbor to require one additional such edge.

Lemma 6. *Let G be a simple cubic plane graph on P. For every face $f \in F(D(C_G))$, $\Delta(f) \geq 0$.*

PROOF. Let f be a face in $F(D(C_G))$. We count the number s_f of edges in G that are incident with exactly one vertex in I_f. Each of the v_f^+ hungry vertices in V_f is incident to at least one such edge. There is no diagonal of $H_f \cup V_f$ in G, as otherwise this diagonal would be a diagonal of P due to $H_f \cup V_f \subseteq H$ and contradict f to be an induced face. Therefore, every vertex v in H_f has at most two neighbors in $H_f \cup V_f$ and, hence, at least one neighbor in I_f. Each unmatched vertex $v \in V_f$ has a unique neighbor w in $ch(P) \cap H_f$ such that $vw \notin E(G)$. This forces w to be incident with an additional edge to a vertex in I_f for each such vertex v. It follows that $h_f + v_f^+ + v_f^{-u} \leq s_f$. Since $s_f \leq 3i_f$, $\Delta(f) \geq 0$.

Let a set of diagonals be *disjoint* if no two of them contain a common end vertex. Note that $D(C_G)$ in Lemma 6 does not have to be disjoint. However, if G contains $ch(P)$, all diagonals are disjoint and consist of balanced vertices only. This gives the following corollary from Lemma 6.

Corollary 7. Let G be a simple cubic plane graph on P that contains $ch(P)$. Then $h_f \leq \frac{3}{4} n_f$ for every induced face $f \in F(D(C_G))$.

We show a necessary parity condition for $\Delta(f)$.

Lemma 8. *Let G be a simple cubic plane graph on P. Then $\Delta(f)$ is even for every induced face $f \in F(D(C_G))$.*

PROOF. Consider the graph $G[f]$ and its vertex set $I_f \cup H_f \cup V_f^+ \cup V_f^{-u} \cup V_f^{-m} \cup V_f^0$. By definition, the degree in $G[f]$ of all vertices in V_f^0 is even while the degree of all other vertices in $G[f]$ is odd. As every graph has an even number of odd-degree vertices, $i_f + h_f + v_f^+ + v_f^{-u} + v_f^{-m}$ must be even. However, v_f^{-m} is even, as matched vertices come in pairs. Thus, $i_f + h_f + v_f^+ + v_f^{-u}$ is even and it follows that $3i_f - h_f - v_f^+ - v_f^{-u} = \Delta(f)$ is even.

3. CONSTRUCTIONS

We give sufficient conditions for diagonal configurations to admit cubic graphs. The following result of Tamura and Tamura will be used.

Lemma 9 (Tamura and Tamura [6]). *For points $p_1, p_2, \ldots, p_n$ in general position in the plane and any assignment of degrees from $\{1, 2, \ldots, n-1\}$ to the points such that the sum of degrees is $2n-2$, there is a plane tree with these prescribed degrees on $p_1, p_2, \ldots, p_n$. Moreover, the plane tree can be constructed in time $O(n \log n)$ [1, 3].*

We call an induced face f *empty* if $i_f = h_f = 0$.

Lemma 10. *Let C be a diagonal configuration without unmatched vertices such that for every non-empty induced face $f \in F(D(C))$, $\Delta(f)$ is even and $0 \leq \Delta(f) < 2i_f$. Then there is a simple cubic plane graph G on P with $D(C_G) = D(C)$. If $D(C)$ is additionally disjoint, there is a simple cubic 2-connected plane graph G on P with $D(C_G) = D(C)$ that contains the boundary cycle of P.*

PROOF. The proof for the first claim builds on a construction given in [3], but avoids the creation of additional diagonals. We omit the proof in this extended abstract, but state an important structure that is used. Let G' be the graph that consists of $D(C)$ and of all edges in $ch(P)$ that do not join a matched vertex pair. In each induced face $f \in F(D(C))$, we augment G' by a collection L of plane trees, each on at least three vertices, such that (1) the union of all trees is plane, (2) every vertex v in $I_f \cup H_f \cup V_f^+$ is contained in exactly one tree $T \in L$ and (3) v has degree 3 in T if $v \in I_f$ and degree 1 in T if $v \in H_f \cup V_f^+$. This constructs a cubic graph. Note that the precondition of having no unmatched vertex is crucial, as otherwise boundary vertices of f may need more than one incident edge to the interior of f.

Remark. Intuitively, the precondition $\Delta(f) < 2i_f$ in Lemma 10 avoids that we have to build cycles in the induced face f. If $\Delta(f) \geq 2i_f$, the desired sum of degrees for L would exceed the sum of degrees of every forest in f. The precondition is tight for the statement of Lemma 10, as there are counterexamples even for $\Delta(f) = 2i_f$ ($\Leftrightarrow i_f = h_f + v_f^+$): E. g., in Figure 2, each of the vertices a, b and c has to be adjacent with exactly one of the 3 black vertices for a cubic graph in face f. Thus, a, b and c must form a triangle, which forces every possible edge from b to the boundary to induce a crossing.

For the special case that $\Delta(f) = 0$ for every face f in Lemma 10, each tree will be a $K_{1,3}$. We give an efficient algorithm to construct the trees for this case.

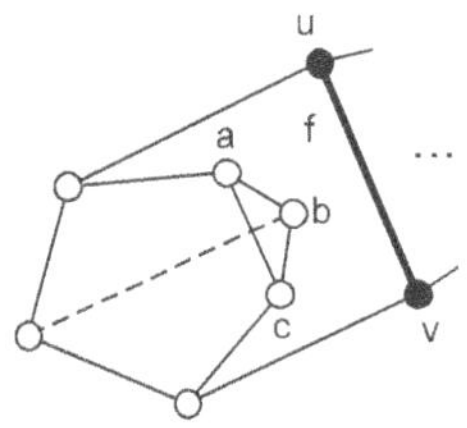

Figure 2: An induced face f with $\Delta(f) = 2i_f = 6$ that does not admit a simple cubic plane graph.

Lemma 11. *Let C be a diagonal configuration without unmatched vertices and $\Delta(f) = 0$ for every induced face $f \in F(D(C))$. There is a $O(n \log n)$ algorithm that constructs a simple cubic plane graph G on P with $D(C_G) = D(C)$ and no edge that joins two points of I.*

Proof. We create a graph G' and, for every induced face $f \in F(D(C))$, a collection L of trees that satisfy the properties (1)–(3), as described in the proof of Lemma 10. As $\Delta(f) = 0$, the number of vertices in $H_f \cup V_f^+$ for every face f is exactly $3i_f$ and every such vertex needs exactly one additional incident edge to the interior of f. Thus, L must consist of i_f trees, each of which is a $K_{1,3}$. We take two point sets of equal size: one is $Z := H_f \cup V_f^+$ and the other set Y is generated from I_f by replacing each point $p \in I_f$ with three points that are in a sufficiently small ϵ-neighborhood of p.

We need to compute a plane matching between Y and Z. Clearly, it can be assumed that $Y \cup Z$ is in general position. We compute a ham-sandwich cut for Y and Z in time $O(n)$ [5]. Iterating the computation for the subsets of $Y \cup Z$ that are contained in each side of the cut, respectively, will terminate with cells containing exactly one point of Y and one point of Z. Joining the two points for every cell by an edge constructs L, giving a total running time of $O(n \log n)$.

4. REDUCTION TO DIAGONAL CONFIGURATIONS

We have shown that a suitable diagonal configuration allows to construct a cubic graph on P. If we can show that every cubic graph on P implies the existence of such a suitable diagonal configuration, we have reduced the problem of computing cubic graphs to diagonal configurations. We will give an efficient algorithm for finding such a diagonal configuration in the next section. Let $h > \frac{3}{4}n$ due to Theorem 2.

Let C_G be the diagonal configuration of a simple cubic plane graph G on P. We will transform C_G to the desired diagonal configuration by iteratively applying two operations. In each step, we maintain a diagonal configuration C, which is initialized with C_G. Note that this does not transform the graph G itself.

Operation 1. Let $\Delta(f) > 0$ for an induced face f of $D(C)$ (see, e. g., any graph in Figure 1). Cut a diagonal vw on the boundary of f into two (non-diagonal) half-edges. Both half-edges are assigned to the newly generated induced face z.

Operation 2. Let $v \in V_f^{-u}$ for an induced face f of $D(C)$ (see, e. g., Figures 3(b) and 3(c)). Cut the unique diagonal vw on the boundary of f into two (non-diagonal) half-edges. Both half-edges are assigned to the newly generated induced face z.

For being able to construct a cubic graph from a diagonal configuration, we need that $\Delta(f) \geq 0$ and $\Delta(f)$ is even for every induced face f. We prove that the above operations preserve these properties of C in every step.

Lemma 12. *Operations 1 and 2 preserve for every induced face f' that $\Delta(f') \geq 0$ and that $\Delta(f')$ is even.*

Proof. Consider the diagonal configuration C before an Operation 1. As $D(C) \neq \emptyset$ due to $h > \frac{3}{4}n$, at least one diagonal vw exists. Let g be the induced face of $D(C)$ that is separated from f by vw. According to Lemmas 6 and 8, $\Delta(f) \geq 2$ and $\Delta(g) \geq 0$. We check the effects of Operation 1 on $\Delta(z)$ in dependence of the type of v. By symmetry, the same effects hold for w. Figure 3 lists the five possible configurations for vertex types of v in f and g; note that this covers all cases, as v cannot be hungry or sated in both faces, respectively. We illustrate the effect on $\Delta(z)$ for the case $v \in V_f^{-m}$ and $v \in V_g^+$ (see Figure 3(e)).

Since $v \in V_f^{-m} \cap V_g^+$, the contribution of v to $\Delta(g)$ in C is -1, while the contribution of v to $\Delta(f)$ is 0, as matched vertices do not influence Δ. Applying Operation 1 replaces these two contributions by one value dependent on z. If v has been incident to only one diagonal of D, v will be a new vertex in H_z (and, thus, in N_z) after performing Operation 1, causing $\Delta(z)$ to decrease by 1. Otherwise, v will be in V_z^+, which also decreases $\Delta(z)$ by 1. Additionally, $\Delta(z)$ decreases by another 1 in both cases, since the vertex in f that was formerly matched to v is now unmatched. In total, the effect of v on $\Delta(z)$ is $1 - 2 = -1$. Table 2 lists the effects for the other four cases.

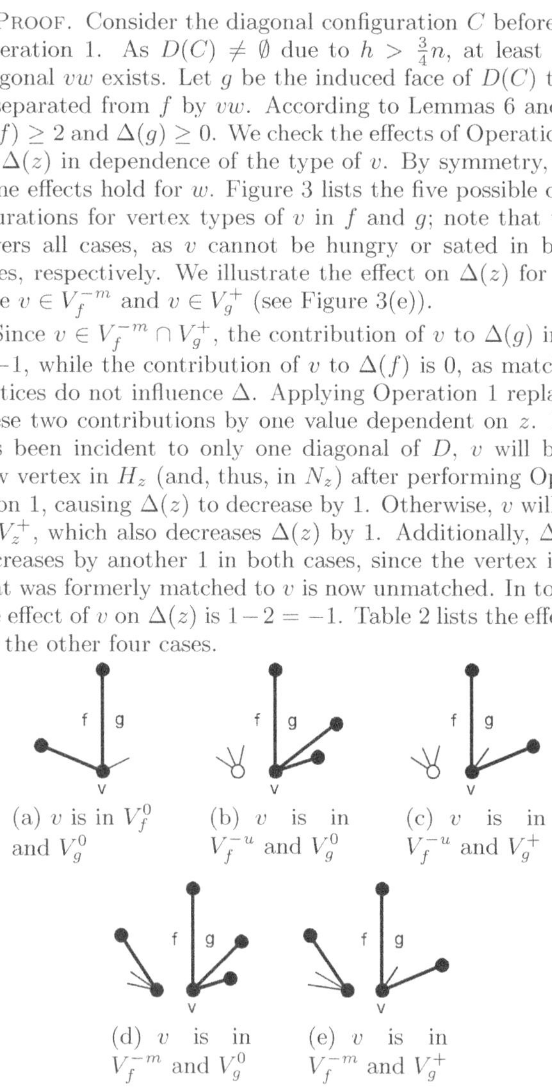

(a) v is in V_f^0 and V_g^0

(b) v is in V_f^{-u} and V_g^0

(c) v is in V_f^{-u} and V_g^+

(d) v is in V_f^{-m} and V_g^0

(e) v is in V_f^{-m} and V_g^+

Figure 3: Possible initial configurations for v.

According to Table 2, $\Delta(z) = \Delta(f) + \Delta(g) + x$ with $x \in \{-2, 0, 2\}$ after applying Operation 1, since exactly two

$v \in$	$\Delta(f)$	$v \in$	$\Delta(g)$	$v \in$	$\Delta(z)$	effect on $\Delta(z)$	Fig.
V_f^0	$+0$	V_g^0	$+0$	$H_z \cup V_z^+$	-1	-1	3(a)
V_f^{-u}	-1	V_g^0	$+0$	V_z^0	$+0$	$+1$	3(b)
V_f^{-u}	-1	V_g^+	-1	$H_z \cup V_z^+$	-1	$+1$	3(c)
V_f^{-m}	$+0$	V_g^0	$+0$	V_z^0	$-1^{(\star)}$	-1	3(d)
V_f^{-m}	$+0$	V_g^+	-1	$H_z \cup V_z^+$	$-2^{(\star)}$	-1	3(e)

Table 2: The five possible configurations of a diagonal end vertex v before Operation 1. Columns 2 and 4 depict the contribution of v to the given value. Column 6 depicts the contribution of v to $\Delta(z)$ after Operation 1. For entries marked with $^{(\star)}$, the sated vertex in f that was matched to v is not matched anymore in z.

vertices change. This implies $\Delta(z) \geq 0$ and that $\Delta(z)$ is still even.

Consider the diagonal configuration C before an Operation 2. As $v \in V_f^{-u}$, v must be either contained in V_g^0 or in V_g^+ (see Figures 3(b) and 3(c)). In both cases, the effect of v on $\Delta(z)$ is $+1$, according to Table 2. Thus, after applying Operation 2, $\Delta(z) = \Delta(f) + \Delta(g) + x$ with $x \in \{0, 2\}$ and the claim follows.

We are now able to state our main structural results.

Theorem 13. *The following statements are equivalent:*

(1) *P admits a simple cubic plane graph G.*

(2) *$h \leq \frac{3}{4}n$ or there is a diagonal configuration C on P such that $\Delta(f) = 0$ for every induced face $f \in F(D(C))$ and no vertex is unmatched.*

PROOF. The proof for *(2) ⇒ (1)* follows directly from Theorem 2 for $h \leq \frac{3}{4}n$ and from Lemma 11 for $h > \frac{3}{4}n$. We prove *(1) ⇒ (2)*. Assume that $h > \frac{3}{4}n$. Then $D(C_G) \neq \emptyset$ with Lemma 3 and we can iteratively apply (any of the) Operations 1 and 2 on C_G as long as possible until the process terminates with a diagonal configuration C'. Note that the application of each operation decreases the number of diagonals by one; thus, $D(C') \subseteq D(C_G)$. Due to Lemma 12, every induced face $f \in F(D(C'))$ satisfies $\Delta(f) = 0$ and no vertex can be unmatched.

Corollary 14. If $h > \frac{3}{4}n$, every simple cubic plane graph on P implies the existence of a simple cubic plane graph on P that does neither contain an unmatched vertex nor an edge joining two vertices in I.

Theorem 15. *The following statements are equivalent:*

(1) *P admits a simple cubic 2-connected plane graph G.*

(2) *P admits a simple cubic 2-connected plane graph G' such that G' contains $ch(P)$ and the diagonals $D(C_{G'})$ are disjoint.*

(3) *$h \leq \frac{3}{4}n$ or there is a diagonal configuration C on P such that there are no unbalanced vertices and $h_f = \frac{3}{4}n_f$ for each induced face $f \in F(D(C))$.*

PROOF. The proof for *(2) ⇒ (1)* is immediate. We prove *(3) ⇒ (2)*. In case that $h \leq \frac{3}{4}n$, Theorem 2 settles the claim. Let $h > \frac{3}{4}n$. Then $D(C)$ must be disjoint, as every vertex v that is end point of two diagonals would contradict that v is balanced in every induced face. With $h_f = \frac{3}{4}n_f$ and $d_f^+ = d_f^- = 0$ for every induced face f, $\Delta(f) = 0$ follows. Applying the construction of Lemma 11 yields the desired graph G' and ensures $ch(P) \subseteq G'$, as $D(C)$ is disjoint.

It remains to prove *(1) ⇒ (3)*. We can assume $h > \frac{3}{4}n$. Then G contains at least one diagonal with Lemma 3. As any unbalanced in C_G would imply G to contain a cut vertex, there are no unbalanced vertices in C_G. Iteratively applying Operation 1 on C_G results in a diagonal configuration that satisfies $\Delta(f) = 0$ for every induced face f. As Operation 1 does not introduce new unbalanced vertices, $h_f = \frac{3}{4}n_f$ for every f, which gives the claim.

The properties we get from not being able to apply Operation 1 are valid also for cubic plane graphs that contain the least possible number of diagonals for P. This implies for every face f that every vertex in I_f has to be joined to exactly three vertices in H_f. We get the following corollary.

Corollary 16. If $h > \frac{3}{4}n$, every simple 2-connected cubic plane graph on P that contains the least possible number of diagonals contains $ch(P)$.

If we insist on 2-connected cubic graphs that contain $ch(P)$, the number of diagonals in these graphs is completely determined by n and h. Given n and h, the number of diagonals is $2h - \frac{3}{2}n$.

Corollary 17. If $h > \frac{3}{4}n$, every simple cubic 2-connected plane graph on P implies the existence of a simple cubic 2-connected plane graph on P such that $ch(P) \subseteq G$, $D(C_G)$ is disjoint, $|D(C_G)| = 2h - \frac{3}{2}n$ and there is no edge in G that joins two vertices in I.

One could be tempted to prove that every point set P admitting a cubic plane graph also admits a cubic 2-connected plane graph. However, this is not true because of the following counterexample.

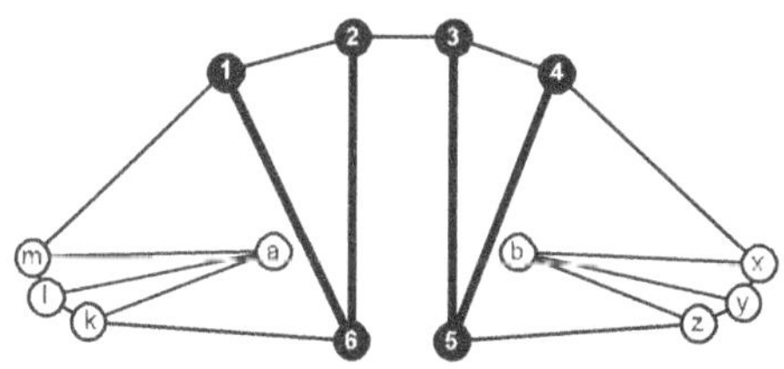

Figure 4: A point set that admits no simple 2-connected cubic plane graph.

Lemma 18. *There is no simple 2-connected cubic plane graph on the point set P in Figure 4.*

PROOF. Assume to the contrary there is such a graph G. Since $h > \frac{3}{4}n$, we may assume with Corollary 17 that G contains $ch(P)$ and exactly 3 disjoint diagonals. For every diagonal Z with an end vertex in $\{x, y, z, k, l, m\}$, let $hp(Z)$ be the open halfplane defined by Z that contains not a. Note that $hp(Z)$ does not contain b either. Assuming there is a diagonal ending at a vertex in $\{x, y, z, k, l, m\}$, we choose one such diagonal Z with a minimal number of points in $hp(Z)$. As $hp(Z)$ is non-empty, but contains no inner vertex, every vertex in $hp(Z)$ has degree 2, contradicting the cubicness of G. This leaves the six remaining candidates $\{1, \ldots, 6\}$ for an end vertex of a diagonal. No diagonal can join two vertices of $\{1, \ldots, 4\}$, as these vertices form a path in $ch(P)$. Thus, there can be only two disjoint diagonals that have end vertices 5 and 6, respectively, contradicting that there have to be 3 disjoint diagonals.

With a bit more effort, one can show that the simple cubic plane graph shown in Figure 4 is the only one possible. Lemma 18 gives rise to the question whether we can expect connectivity on a point set P admitting a simple cubic plane graph. As we could not find an analogue counterexample that distinguishes 0- and 1-connectivity, this is still an open problem.

5. THE ALGORITHMS

In this section we describe two algorithms. We first describe an algorithm for finding a 2-connected cubic plane graph on a given point set P (if it exists). By Theorem 15, it suffices to look for a plane graph G such that it contains all edges of $ch(P)$ and all faces induced by the set of the diagonals of G satisfy $\Delta(f) = 0$. Then we outline a similar, very technical algorithm for finding a cubic plane graph (not necessarily connected) on a given point set P (if it exists). In both cases we use a dynamic-programming approach. For a pair of points a, b, let $R(a, b)$ be the closed halfplane to the right of the line ab (oriented from a to b). For a point $x \in H$, let x^+ and x^- be the points of H counterclockwise and clockwise of x, respectively.

Let T be the set of ordered pairs of distinct non-neighboring vertices of $ch(P)$. During the first algorithm, for each pair $(a, b) \in T$, we compute $d(a, b)$, which is the maximum number of pairwise disjoint diagonals of P such that all these diagonals lie in $R(a, b)$ and all faces induced by them satisfy $\Delta(f) = 0$, with the possible exception for the unique face intersecting the complement of $R(a, b)$.

For $(a, b) \in T$, let $H(a, b)$ be the set of points of H lying in the interior of $R(a, b)$, let $I(a, b)$ be the set of points of I lying in $R(a, b)$, and let $\Delta(a, b) := 3|I(a, b)| - |H(a, b)|$. When computing the numbers $d(a, b)$, we make use of the following observation:

Observation 19. *(i) Let $(a, b) \in T$. Let C be a set of disjoint diagonals in $R(a, b)$, which does not include the diagonal ab. Let f be the unique face induced by C, which contains the segment ab and some points to the right of the segment ab. Then there is a $c \in H(a, b) \cup \{a\}$ such that the face f also contains the pair c, c^+ of adjacent vertices of $ch(P)$. Consequently, every diagonal of C lies entirely in one of the closed halfspaces $R(a, c)$ and $R(c^+, b)$.*

(ii) Let $(a, b) \in T$. Let C be a set of disjoint diagonals in $R(a, b)$, containing the diagonal ab. Then all the other diagonals of C lie in $R(a^+, b^-)$.

Algorithm for finding a 2-connected cubic graph

If n is odd or $n \leq 3$, there is no cubic graph on P; we therefore assume that n is even and $n \geq 4$. If $h \leq \frac{3}{4}n$, we can find a 2-connected cubic plane graph on P in time $O(n^3)$ due to Theorem 2. Let $h > \frac{3}{4}n$. In particular, $h > 3$.

For all pairs $(a, b) \in T$ with $|H(a, b)| \leq 2$, no diagonal in $H(a, b)$ can exist and we set $d(a, b) := 0$. For technical reasons, we also set $d(a, a^+) := d(a, a) := 0$ for each $a \in H$. The values $d(a, b)$ for pairs $(a, b) \in T$ with $|H(a, b)| \geq 3$ are computed in the order of increasing value of $|H(a, b)|$. Thus, we first compute $d(a, b)$ for all pairs $(a, b) \in T$ with $|H(a, b)| = 3$, then for all pairs $(a, b) \in T$ with $|H(a, b)| = 4$, etc.. For each pair $d(a, b)$, we proceed using the following recursion rule.

For a given $(a, b) \in T$ with $|H(a, b)| \geq 3$, let

$$d := \max\{d(a, c) + d(c^+, b) : c \in H(a, b) \cup \{a\}\}.$$

The number d equals $d(a, b)$, unless $d(a, b)$ is witnessed only by diagonal sets containing the diagonal ab. If $d = d(a^+, b^-)$ and $\Delta(a, b) + 2d = 0$, we can add the diagonal ab to the $d(a^+, b^-)$ diagonals, as $\Delta(f) = 0$ for the induced face f in $R(a, b)$ that contains ab. In this case, we set $d(a, b) := d + 1$. Otherwise, we set $d(a, b) := d$. Note that no diagonal in $R(a, b)$ is counted twice by taking $d(a, c) + d(c^+, b)$ in the recursion, as $R(a, c)$ and $R(c^+, b)$ are disjoint. Due to Observation 19, $d(a, b)$ is indeed the maximum number of pairwise disjoint diagonals of P such that all these diagonals lie in $R(a, b)$ and all faces induced by them satisfy $\Delta(f) = 0$, with the possible exception for the unique face intersecting the complement of $R(a, b)$.

After computing $d(a, b)$ for all pairs $(a, b) \in T$ in this way, we check if there is a pair $(a, b) \in T$ satisfying

$$2(d(a, b) + d(b, a)) \geq h - 3i.$$

The computation of $d(a, b)$ for each pair $(a, b) \in T$ takes time $O(n)$. Clearly, all the computation so far can be done in time $O(n^3)$.

Lemma 20. *If $h > 3$ then P admits a 2-connected cubic plane graph if and only if $2(d(a, b) + d(b, a)) \geq h - 3i$ for some pair $(a, b) \in T$.*

PROOF. If P admits a 2-connected cubic plane graph, then by Theorem 13, there is a diagonal configuration C on P such that $\Delta(f) = 0$ for every induced face $f \in F(D(C))$ and no vertex is unmatched. Choose any diagonal ab lying in C. By the definition of $d(a, b)$, the number of diagonals of C contained in $R(a, b)$ is at most $d(a, b)$. Similarly, the number of diagonals of C contained in $R(b, a)$ is at most $d(b, a)$. Since each diagonal of C lies in $R(a, b)$ or in $R(b, a)$, we obtain that $d(a, b) + d(b, a) \geq |C|$. Consequently, by Lemma 3, $2(d(a, b) + d(b, a)) \geq |2|C| \geq h - 3i$.

On the other hand, suppose now that $2(d(a, b) + d(b, a)) \geq h - 3i$ for some pair $(a, b) \in T$. Consider the two sets, D_1 and D_2, of diagonals witnessing the values of $d(a, b)$ and $d(b, a)$, respectively. If the diagonal ab lies in both of them, then each face induced by the set $D_1 \cup D_2$ satisfies $\Delta(f) = 0$ and we can find a 2-connected cubic plane graph on P due to Lemma 11. Suppose now that the diagonal ab does not lie in $D_1 \cap D_2$. If it lies neither in D_1 nor in D_2 then $\Delta(f) = 0$ holds for each face induced by the set $D_1 \cup D_2$, with the possible exception for the face containing the segment ab. If the diagonal ab lies exactly in one of the sets D_1 and D_2, then

$\Delta(f) = 0$ holds for each face induced by the set $D_1 \cup D_2$, with the possible exception for one face adjacent to the diagonal ab. Due to Lemma 3, removing consequently $\frac{h-3i}{2} - (d(a,b) + d(b,a))$ diagonals at the boundary of the unique face not satisfying $\Delta(f) = 0$ results in a set of diagonals inducing only faces satisfying $\Delta(f) = 0$. Then, due to Lemma 11, there is a 2-connected cubic plane graph on P.

Thus, if there is no pair $(a,b) \in T$ satisfying $2(d(a,b) + d(b,a)) \geq h - 3i$ then there is no 2-connected cubic plane graph on P. Otherwise, if we find a pair $(a,b) \in T$ satisfying $2(d(a,b) + d(b,a)) \geq h - 3i$, then we use the following recursive procedure which finds diagonals of two sets witnessing the values of $d(a,b)$ and $d(b,a)$, respectively.

Procedure Split(x,y);
If $d(x,y) = d(x^+, y^-) + 1$ and $\Delta(x,y) + 2d(x^+, y^-) = 0$, we add the diagonal xy to D and run Split(x^+, y^-). Otherwise, if $d(x,y) > 0$, we find a $c \in H(x,y) \setminus \{y^-\}$ such that

$$d(x,y) = d(x,c) + d(c^+, y)$$

and then run (for one such c) procedures Split(x,c) and Split(c^+, y).
End of Procedure Split.

The procedure Split follows the way in which the numbers $d(a,b)$ were computed. Therefore, if we first set $D := \emptyset$ and then run procedures Split(a,b) and Split(b,a), we obtain a set D of diagonals such that each face induced by D satisfies $\Delta(f) \geq 0$. This can be done in time $O(n^2)$. We then apply Operation 1 as long as $\Delta(f) > 0$ for some of the induced faces. We obtain a set D' of diagonals such that $\Delta(f) = 0$ holds for each induced face. Then we find pairwise non-intersecting 3-stars in each face covering all inner points in the face according to Lemma 11. These stars in all faces together with the diagonals of D' and with the edges of $ch(P)$ form a 2-connected cubic plane graph on P. The whole algorithm can be done in time $O(n^3)$.

Algorithm for finding any cubic graph

If 2-connectivity is not required then the algorithm is similar but much more involved than the above algorithm for finding 2-connected cubic plane graph. No substantially new ideas are needed, therefore we only give a sketch of the algorithm.

An *H-edge* is an edge connecting a pair of vertices of H.

In its main part, the algorithm computes special quantities $z(a,b,s,t,j)$, where $a, b \in H$ with $a \neq b$, s and t are contained in $\{0,1,2,3\}$ and $j \in \{0,1\}$. In fact, not all the quantities $z(a,b,s,t,j)$ are computed. Some of them do not make sense, e. g., for parity reasons. The quantity $z(a,b,s,t,j)$, whenever it exists, is equal to the maximum size of a set S of H-edges lying in $R(a,b)$ such that the following three conditions hold:

(i) No four edges of S have a common vertex,

(ii) S contains the diagonal ab if and only if $j = 1$, and

(iii) the diagonals lying in S are exactly the diagonals of some "nearly" diagonal configuration on the point set $P \cap R(a,b)$, meaning that the configuration satisfies all the properties of a diagonal configuration, except that the degrees of a and b are $3 - s$ and $3 - t$, respectively. Moreover, the degrees in each face allow to connect the half-edges into edges such that 3-stars are formed and each 3-star connects at least one vertex of H with a vertex of I.

Note that j is an indicator of the diagonal ab (Condition (ii)) and s and t are "free" degrees (half-edges) to be used for edges going outside of $R(a,b)$.

When computing the numbers $z(a,b,s,t,j)$, we make use of the following analogue of Observation 19:

Observation 21. *Let $(a,b) \in T$. Let C be a set of non-crossing diagonals in $R(a,b)$, which may include the diagonal ab. Let f be the unique face induced by C, which contains the segment ab and some points to the right of the segment ab. Then there is a $c \in H(a,b)$ contained in the face f. Consequently, every diagonal of C other than ab lies entirely in one of the closed halfspaces $R(a,c)$ and $R(c,b)$.*

At the beginning of the algorithm, for each $a \in H$, we set $z(a, a^+, 2, 2, 1) := 1$ and $z(a, a^+, 3, 3, 0) := 0$. All the other values $z(a, a^+, s, t, j)$ are not defined. Then, in the main part of the algorithm, if $z(a,b,s,t,j)$ with $|H(a,b)| \geq 1$ exists then, due to Observation 21, it is determined as

$$\max(j + z(a,c,s',t',j') + z(c,b,s'',t'',j'')),$$

where the maximum is taken over all $c \in H(a,b), s', s'', t', t'', j', j''$, for which the following five conditions hold:

1. $z(a,c,s',t',j')$ and $z(c,b,s'',t'',j'')$ exist,
2. $s + j \leq s'$ (degree condition at a),
3. $t + j \leq t''$ (degree condition at b),
4. $s'' + t' \geq 3$ (degree condition at c),
5. if $j = 1$, $\Delta'(a,b) + 2z(a,c,s',t',j') + 2z(c,b,s'',t'',j'') = 0$, where $\Delta'(a,b) = 3i - 3h - (3 - s - j) - (3 - t - j)$.

The last condition ensures that we find a diagonal configuration inducing faces with $\Delta(f) \geq 0$. After applying Operations 1 and 2 as long as it is needed, we obtain a diagonal configuration with no unmatched vertex and with $\Delta(f) = 0$ for each induced face. Then we can complete the construction of a plane cubic graph similarly as in the previous algorithm. Again, the whole algorithm runs in time $O(n^3)$. This can be checked in a similar way as for the first algorithm.

6. FINAL REMARKS

The natural open problems left are to extend the structural results of this paper to 4-regular and 5-regular plane graphs and to find polynomial-time algorithms for recognizing the point sets that admit such graphs. As discussed before, it is also open whether there are point sets that admit a cubic plane graph but no connected cubic plane graph. Shortly before submitting the final version of this paper the authors have found a proof that no such point sets exist, i. e., if P admits a cubic plane graph it admits also a connected one.

7. REFERENCES

[1] P. Bose, M. McAllister, and J. Snoeyink. Optimal algorithms to embed trees in a point set. *J. Graph Algorithms Appl.*, 1, 1997.

[2] T. K. Dey, M. B. Dillencourt, S. K. Ghosh, and J. M. Cahill. Triangulating with high connectivity. *Comput. Geom. Theory Appl.*, 8(1):39–56, 1997.

[3] A. García, F. Hurtado, C. Huemer, J. Tejel, and P. Valtr. On triconnected and cubic plane graphs on given point sets. *Comput. Geom. Theory Appl.*, 42:913–922, 2009.

[4] M. Gemignani. On finite subsets of the plane and simple closed polygonal paths. *Mathematics Magazine*, 39(1):38–41, 1966.

[5] C.-Y. Lo, J. Matoušek, and W. Steiger. Algorithms for ham-sandwich cuts. *Discrete & Computational Geometry*, 11:433–452, 1994.

[6] A. Tamura and Y. Tamura. Degree constrained embedding into points in the plane. *Information Processing Letters*, 44:211–214, 1992.

Multinerves and Helly Numbers of Acyclic Families

Éric Colin de Verdière
Département d'informatique
École normale supérieure
CNRS
Paris, France
eric.colin.de.verdiere@ens.fr

Grégory Ginot
UPMC
Institut de Mathématiques
de Jussieu
Paris, France
ginot@math.jussieu.fr

Xavier Goaoc
Project-team VEGAS
INRIA Nancy Grand-Est
LORIA
Nancy, France
goaoc@loria.fr

ABSTRACT

The nerve of a family of sets is a simplicial complex that records the intersection pattern of its subfamilies. Nerves are widely used in computational geometry and topology, because the nerve theorem guarantees that the nerve of a family of geometric objects has the same topology as the union of the objects, if they form a good cover.

In this paper, we relax the good cover assumption to the case where each subfamily intersects in a disjoint union of possibly several homology cells, and we prove a generalization of the nerve theorem in this framework, using spectral sequences from algebraic topology. We then deduce a new topological Helly-type theorem that unifies previous results of Amenta, Kalai and Meshulam, and Matoušek. This Helly-type theorem is used to (re)prove, in a unified way, bounds on transversal Helly numbers in geometric transversal theory.

Categories and Subject Descriptors: G.2.1 [Mathematics of Computing]: Discrete Mathematics—*Combinatorics*

Keywords: acyclic family, combinatorics, geometric transversal theory, good cover, Helly's theorem, multinerve, nerve, spectral sequence, topology

1. INTRODUCTION

The *nerve* of a family F of sets is a combinatorial structure that encodes the intersection patterns of subfamilies of F. More precisely, it is an abstract simplicial complex with vertex set F where $G \subseteq F$ forms a simplex if and only if the intersection of the subfamily G is non-empty.

The notion of nerve is widely used in computational geometry (see discussion below), primarily because, under adequate assumptions, it provides a faithful combinatorial description of the topology of the union of the sets. Indeed, if the family is geometric, the *nerve theorem*, due to Borsuk [8], states that the nerve is homotopy equivalent to the union of the sets in the family, provided that the family is a *good cover*: the intersection of any subfamily is either empty or contractible.

In the nerve theorem, the requirement that the family be a good cover is rather strong and not always adequate. Here, we consider *acyclic* families of sets, where every subfamily intersects in finitely many connected components, each of which is a homology cell (in particular, a set that is contractible is a homology cell). The contributions of our paper are threefold:

1. We define an analogue of the nerve, called the *multinerve*, that is suitable for general acyclic families, and we prove that this combinatorial structure enjoys an analogue of the nerve theorem.

2. Furthermore, we demonstrate the usefulness of this construction by proving a new *topological Helly-type theorem* for acyclic families. Helly-type theorems are results of the following form: Given a family F of sets, if every subfamily of cardinality at most h has non-empty intersection, then the whole family has non-empty intersection. In such a case, F is said to have *Helly number* at most h. Our theorem generalizes previous results of Amenta [3], Kalai and Meshulam [33], and Matoušek [36].

3. Finally, this result is used to (re)prove, in a unified way, bounds on transversal Helly numbers in *geometric transversal theory*.

Before describing our results in more detail, we review various incarnations of nerves in discrete and computational geometry.

Nerves in discrete and computational geometry.

While the study of unions and intersections of basic shapes is a classical topic in discrete and computational geometry, the methods used to study these structures have recently evolved toward a greater integration of geometric, combinatorial, and topological arguments. Several of them can be traced back to the classical notion of nerve:

- Standard objects relate to or are particular cases of nerve complexes: (a) the *intersection graph* of a family of sets is the 1-dimensional skeleton of the nerve of that family, (b) the nerve of a family of geometric objects is the smallest abstract simplicial complex containing their *dual range space* (seen as a hypergraph), and (c) given a point set P and a parameter $t > 0$, the *Čech complex* $C(P, t)$ used in computational topology [11, 17, 6] is no other than the nerve of the family of balls of radius t centered in the points of P. In the latter case, the nerve theorem justifies that the Čech complex is a topologically faithful construction. Also the nerve theorem ensures that the restricted Delaunay triangulation of an object has the same topology as that object, under mild assumptions [21].

- Helly's theorem [29] and its topological generalization [30] have implications in combinatorial optimization, e.g., they allow to bound the combinatorial dimension of some LP-type problems [2], or in reconstruction, e.g., to define the maximum cardinality of a *blocker* [5]. Helly's topological theorem states that the Helly number of a good cover in $\mathbb{R}^d$ is at most $d+1$ (see also Debrunner [18]). For open sets, this follows easily from the nerve theorem. Indeed, if F is an open good cover in $\mathbb{R}^d$ such that[1] $\bigcap_F = \emptyset$ and $\bigcap_G \neq \emptyset$ for all proper subfamilies $G \subsetneq F$, then $\mathcal{N}(F)$ is the boundary of a simplex with $|F|$ vertices; hence, $\mathcal{N}(F)$ has non-zero homology in dimension $|F|-2$. On the other hand, $\mathcal{N}(F)$ has the same homology as $\bigcup_F$, which, as an open set in $\mathbb{R}^d$, has zero homology in dimension d and larger. So $|F|-2$ is at most $d-1$. It follows that if all subsets of F of size $d+1$ have non-empty intersection then $\bigcap_F$ is non-empty.

- The inclusion-exclusion formula asserts that for any family F of subsets of $\mathbb{R}^d$ we have
$$\mathbf{1}_{\bigcup_F} = \sum_{\emptyset \subsetneq G \subseteq F} (-1)^{|G|+1} \mathbf{1}_{\bigcap_G},$$
where $\mathbf{1}_X$ denotes the indicator function of set X. This allows, for instance, to evaluate integral quantities such as volumes for $\bigcup_F$ (a fundamental question in bioinformatics) from the values of that quantity for $\bigcap_G$, where $G \subseteq F$. This identity has a topological flavor to it: for any point $p \in \mathbb{R}^d$, if F_p denotes the subset of F of sets that contain p, then
$$\sum_{\emptyset \subsetneq G \subseteq F} (-1)^{|G|+1} \mathbf{1}_{\bigcap_G}(p) = \sum_{\emptyset \subsetneq G \subseteq F_p} (-1)^{|G|+1}$$
which equals zero if $p \notin \bigcup_F$ and the Euler characteristic of the simplex with $|F_p|$ vertices, namely one, if $p \in \bigcup_F$. This point of view allows to characterize and compute, for certain families of objects, more compact inclusion-exclusion formulas using tools such as the nerve theorem, as shown by Attali and Edelsbrunner [4].

Previous work on Helly numbers.

Previous bounds on Helly numbers of families of non-connected sets come in two different flavors. On the one hand, one can start with a "ground" family $\mathcal{H}$ whose Helly number is bounded and consider families F such that the intersection of any subfamily $G \subseteq F$ is a disjoint union of at most r elements of $\mathcal{H}$. When $\mathcal{H}$ is closed under intersection and *non-additive* (that is, the union of two disjoint elements of $\mathcal{H}$ is never an element of $\mathcal{H}$) the Helly number of

[1]Here and in the rest of the paper we use $\bigcap_A$ and $\bigcup_A$ to denote, respectively, $\bigcap_{x \in A} x$ and $\bigcup_{x \in A} x$.

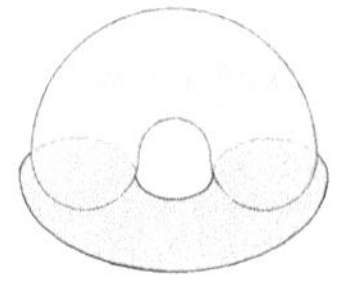

Figure 1. A family of two sets that is not a good cover but satisfies the hypotheses of our theorem.

F can be bounded by r times the Helly number of $\mathcal{H}$. This was conjectured (and proven for $r=2$) by Grünbaum and Motzkin [27] and a proof of the general case was recently published by Eckhoff and Nischke [20], building on ideas of Morris [38]. Direct proofs were also given by Amenta [3] in the case where $\mathcal{H}$ is a finite family of compact convex sets in $\mathbb{R}^d$ and by Kalai and Meshulam [33] in the case where $\mathcal{H}$ is a good cover in $\mathbb{R}^d$. On the other hand, Matoušek [36] and Alon and Kalai [1] showed, independently, that if F is a family of sets in $\mathbb{R}^d$ such that the intersection of any subfamily is the union of at most r (possibly intersecting) convex sets, then the Helly number of F can be bounded from above by some function of r and d. Matoušek also gave a topological analogue [36, Theorem 2]: he bounds from above (again, by a function of r and d) the Helly number of families of sets in $\mathbb{R}^d$ assuming that the intersection of any subfamily has at most r connected components, each of which is $(\lceil d/2 \rceil - 1)$-connected, that is, has its ith homotopy group vanishing for $i \leq \lceil d/2 \rceil - 1$.

Our results in more detail.

While powerful generalizations of nerves were developped in algebraic topology, their statements require heavy formalism with elaborated algebraic structures (spectral sequences, in particular), which prove difficult to manipulate for the non-expert. The first contribution of our paper is the definition of an analogue of the nerve, called the *multinerve*, and a generalization of the nerve theorem to the multinerve, that is relatively easy to state and can be used as a black box in applications. The multinerve is not a simplicial complex, but a more general *simplicial poset*. The proof of our "multinerve theorem" is based on Leray's acyclic cover theorem (see [9, 34, 10, 23, 22] for instance) and Čech complexes from algebraic topology.[2]

Our second contribution is a topological Helly-type theorem. We prove that if F is a family of open sets in $\mathbb{R}^d$ such that every subfamily intersects in at most r connected components, each of which is a homology cell, then the Helly number of F is at most $r(d+1)$. We obtain this Helly-type theorem by revisiting the proof of Kalai and Meshulam and replacing a certain construction by the multinerve, demonstrating the effectiveness of this new tool. This generalizes previous results of Amenta [3] and Kalai and Meshulam [33]: Figure 1 shows a family for which our result applies with $r=2$, but where the results of Amenta and of Kalai and Meshulam do not (as the family of connected components is not a good cover). Matoušek proved [36] that the Helly numbers of the families we consider were bounded; we provide a tight, explicit bound. Finally, our result and the Eckhoff-Morris-Nischke theorem [20] do not seem to imply

[2]Let us emphasize that here we mean a *Čech complex* in the sense of algebraic toplogy, which is a more general object than what the eponymous structure generally used in computational topology.

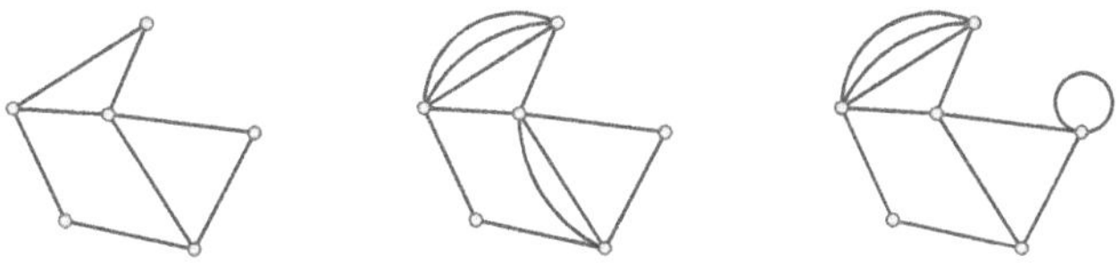

Figure 2. One-dimensional simplicial complex (left) vs simplicial poset (middle). Simplicial sets (right) are even more general structures.

one another, but to be distinct generalizations of the Kalai-Meshulam theorem.

Also, an interesting feature of our result is that it generalizes to families living in rather general topological spaces, not only $\mathbb{R}^d$. Our bound on Helly numbers, when applied in a k-dimensional submanifold embedded in $\mathbb{R}^d$, depends only on k and is oblivious to d. Since Helly numbers arise in the context of surface reconstruction as the maximum size of blockers [5], this could be a useful tool to overcome the "curse of dimensionality", a major issue in high-dimensional manifold reconstruction and related areas.

As a last contribution, we use our new Helly-type theorem to reprove or sharpen, in a unified way, bounds on Helly numbers in *geometric transversal theory*. While most bounds were previously known, they were obtained through ad hoc, and sometimes tedious, methods. In essence, it turns out that the set of lines piercing a given family of objects in $\mathbb{R}^d$ is, in many cases, a union of disjoint contractible sets in line space; our Helly-type result is thus suitable for bounding the Helly numbers of such sets of transversals.

Due to space limitations, we restrict the presentation in this conference version to the main ideas and proof outlines, and refer to the complete version [15] for full details.

2. MULTINERVES

In this section we recall the definitions of simplicial complexes and simplicial posets and introduce the multinerve, a simplicial poset that refines the notion of nerve.

2.1 Simplicial Complexes and Simplicial Posets

At the combinatorial level, an (abstract) *simplicial complex* X over a set of *vertices* V is a non-empty family of subsets of V closed under taking subsets; in particular, $\emptyset$ belongs to every simplicial complex. An element σ of X is a *simplex*; its *dimension* is the cardinality of σ minus one; a d-simplex is a simplex of dimension d.

Intuitively, a simplicial partially ordered set (simplicial poset for short) is a set of simplices with an incidence relation; a d-simplex still has $d+1$ distinct vertices; however, in contrast to simplicial complexes, there may be several simplices with the same vertex set, but no two can be incident to the same higher-dimensional simplex (see Figure 2).

Formally, let X be a finite set and $\preceq$ a partial order on X; we also say that $(X, \preceq)$ is a partially ordered set, or that X is a *poset* to save breath. Let $[\alpha, \beta] = \{\tau \in X \mid \alpha \preceq \tau \preceq \beta\}$ denote the *segment* defined by α and β. (Similarly, we will denote by $(\alpha, \cdot]$ the set $\{\tau \in X \mid \alpha \prec \tau\}$, and so on.) A map $\varphi : X \to Y$ between two posets $(X, \preceq^X)$ and $(Y, \preceq^Y)$ is *monotone* if it preserves the order: for any $\sigma, \tau \in X$ $\sigma \preceq^X \tau \Rightarrow \varphi(\sigma) \preceq^Y \varphi(\tau)$. An *isomorphism* of posets is a monotone bijection between them. For an overview of basic properties of simplicial posets see, e.g., Stanley [42].

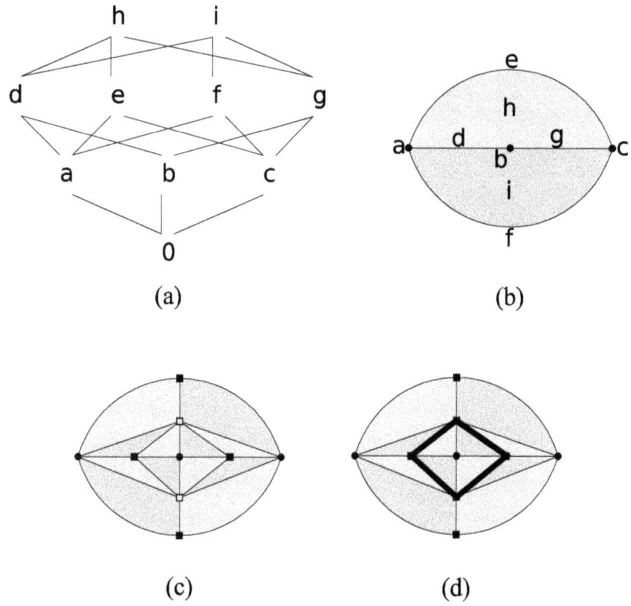

Figure 3. (a) A simplicial poset X. (b) The geometric realization of X. (c) The geometric realization of $\mathrm{sd}(X)$, which also equals $\dot{D}_X(0)$. (d) $\dot{D}_X(b)$ is a 1-dimensional simplicial complex that is a cycle of length four (in black bold lines).

A poset X is a *simplicial poset* if it satisfies two conditions (see Figure 3(a)). First, X must have a *least element* 0, that is $0 \preceq \sigma$ for any $\sigma \in X$. Second, for any $\sigma \in X$, there must exist some integer d such that the lower segment $[0, \sigma]$ is isomorphic to $2^{\{0,\ldots,d\}}$, the poset of faces of a d-simplex partially ordered by the inclusion; d is then called the *dimension* of σ. The elements of X are called its *simplices* and the simplices of dimension 0 (i.e. that only dominate 0) are its *vertices*. If $\tau \preceq \sigma$ we also say that τ is *contained in* (or *a face of*) σ. Simplicial posets lie in-between simplicial complexes and the more general notion of *simplicial sets* used in algebraic topology [37, 24]. The simplices of a simplicial complex, ordered by inclusion, form a simplicial poset (with $\emptyset$ as least element). The converse is not always true: the one-dimensional simplicial complexes are precisely the graphs without loops or multiple edges, while the one-dimensional simplicial posets correspond to the graphs without loops but possibly with multiple edges (refer back to Figure 2). If τ is a simplex of a simplicial poset with set of vertices V, the map that associates to any face of τ the set of vertices of that face is a bijection between $[0, \tau]$ and 2^V. There may, however, exist several simplices with the same set of vertices, but no two of them can be faces of one and the same simplex. Just as for simplicial complexes, every simplicial poset has an associated topological space, its *geometric realization*; see Figure 3(b).

2.2 Multinerves: Definition and Properties

Let F be a finite family of subsets of a topological space. We define the *multinerve* $\mathcal{M}(F)$ of F as:

$$\mathcal{M}(F) = \left\{ (C, A) \;\middle|\; \begin{array}{l} A \subseteq F, \\ C \text{ is a connected component of } \bigcap_A \end{array} \right\}.$$

By convention, we put $\bigcap_\emptyset = \bigcup_F$, and in particular, $(\bigcup_F, \emptyset)$ belongs to $\mathcal{M}(F)$. We turn $\mathcal{M}(F)$ into a poset by equipping

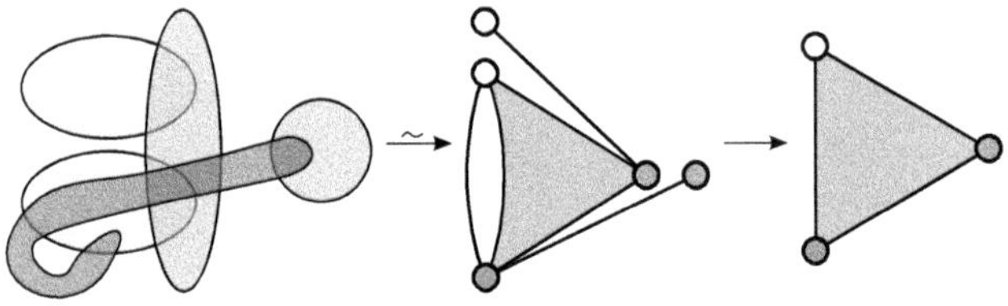

Figure 4. Left: A family F of subsets of $\mathbb{R}^2$. Middle: Its multinerve $\mathcal{M}(F)$. Right: Its nerve $\mathcal{N}(F)$.

it with the partial order:

$$(C', A') \preceq (C, A) \Leftrightarrow C \supseteq C' \text{ and } A \subseteq A'.$$

To get an intuition, it does not harm to assume that, whenever A and A' are different subsets of F, the connected components of $\bigcap_A$ and of $\bigcap_{A'}$ are different. Under this assumption, $\mathcal{M}(F)$ can be identified with the set of all connected components of the intersection of any subfamily of F, equipped with the opposite of the inclusion order. See Figure 4 for an example.

LEMMA 1. *The poset $\mathcal{M}(F)$ is simplicial.*

PROOF. The projection on the second coordinate identifies any lower segment $[(\bigcup_F, \emptyset), (C, A)]$ with the simplex 2^A. Indeed, let $A' \subseteq A$ and let $C' \subseteq \bigcup_F$. The lower segment $[(\bigcup_F, \emptyset), (C, A)]$ contains (C', A') if and only if C' is the connected component of $\bigcap_{A'}$ containing C. Moreover, by definition, $\mathcal{M}(F)$ contains a least element, namely $(\bigcup_F, \emptyset)$. The statement follows. □

Intuitively, $\mathcal{M}(F)$ is an "expanded" version of $\mathcal{N}(F)$: while $\mathcal{N}(F)$ has one simplex for each non-empty intersecting subfamily, $\mathcal{M}(F)$ has one simplex for each *connected component* of an intersecting subfamily.

3. MULTINERVE THEOREM

In this section, we state a "homological multinerve theorem" (Theorem 3): under certain conditions on a family F, the multinerve $\mathcal{M}(F)$ captures the homology of $\bigcup_F$. Before, we explain how the notion of homology for simplicial complexes extends, in the natural way, to simplicial posets.

3.1 Homology of Simplicial Posets

The *homology* of a simplicial poset can be defined as a direct extension of simplicial homology for simplicial complexes (or, alternatively, as the singular homology of the geometric realization of the simplicial poset, but we deliberately avoid this point of view here). Let X be a simplicial poset and assume chosen an ordering on the set of vertices of X. For $n \geq 0$, let $C_n(X)$ be the $\mathbb{Q}$-vector space with basis the set of simplices of X of dimension exactly n. If σ is an n-dimensional simplex, the lower segment $[0, \sigma]$ is isomorphic to the poset of faces of a standard n-simplex $2^{\{0,\dots,n\}}$; here we choose the isomorphism so that it preserves the ordering on the vertices. Thus, we get $n+1$ faces $d_i(\sigma) \in X$ (for $i = 0, \dots, n$), each of dimension $n-1$: namely, $d_i(\sigma)$ is the (unique) face of σ whose vertex set is mapped to $\{0, \dots, n\} \setminus \{i\}$ by the above isomorphism. Extending the maps d_i by linearity, we get the *face operators* $d_i : C_n(X) \to C_{n-1}(X)$. Let $d = \sum_{i=0}^{n} (-1)^i d_i$ be the linear map $C_n(X) \to C_{n-1}(X)$ (which is defined for any $n \geq 0$). The fact that $d \circ d = 0$ is easy and follows from the same argument as for simplicial complexes since it is computed inside the vector space generated by $[0, \sigma]$ which is isomorphic to a standard simplex. The (simplicial) nth homology group $H_n(C_\bullet(X), d)$ is defined as the quotient vector space of the kernel of $d : C_n(X) \to C_{n-1}(X)$ by the image of $d : C_{n+1}(X) \to C_n(X)$. The simplicial homology of a simplicial poset coincides with the homology of its geometric realization.

We emphasize that, in this paper, we only consider homology over $\mathbb{Q}$. In the sequel, we denote by $H_i(O)$ the ith $\mathbb{Q}$-homology group of O (whether O is a simplicial poset or a topological space), and by $\tilde{H}_i(O)$ the corresponding *reduced* homology group (see, e.g., Hatcher [28]). A homology cell is a set O such that $\tilde{H}_i(O) = 0$ for each i; in particular, every contractible space is a homology cell.

3.2 Multinerve Theorem

Let Γ be a locally arc-wise connected topological space. If F is a good cover of Γ then the multinerve of F is simply its nerve, and the nerve theorem asserts that $\bigcup_F$ and $\mathcal{M}(F)$ have the same homotopy type. We prove a homological version of the nerve theorem that is valid for multinerves of *acyclic families*, where a family is *acyclic* if for any non-empty subfamily $G \subseteq F$, each connected component of the intersection of the elements of G is a $\mathbb{Q}$-homology cell.

THEOREM 2. *Let F be a family of open sets in a locally arc-wise connected topological space Γ. If F is* acyclic, *then $\tilde{H}_\ell(\mathcal{M}(F)) \cong \tilde{H}_\ell(\bigcup_F)$ for any non-negative integer ℓ.*

Any good cover is acyclic, as a contractible space is a homology cell, but the reverse is not true as (a) acyclicity allows a subfamily to intersect in more than one connected component, and (b) there exist homology cells that are not contractible. See Figures 1 and 4 for examples of acyclic families that are not good covers. In fact, we slightly strengthen Theorem 2 by allowing intersections of few elements of the family to have non-zero homology in low dimension; this variant will be useful for our applications to geometric transversal theory. Specifically, we call a family F of subsets of Γ *acyclic with slack s* if for every non-empty subfamily $G \subseteq F$ and every $i \geq \max(1, s - |G|)$ we have $\tilde{H}_i(\bigcap_G) = 0$. Here is the strengthened version of Theorem 2.

THEOREM 3 (HOMOLOGICAL MULTINERVE THEOREM). *Let F be a family of open sets in a locally arc-wise connected topological space Γ. If F is* acyclic with slack s *then $\tilde{H}_\ell(\mathcal{M}(F)) \cong \tilde{H}_\ell(\bigcup_F)$ for any non-negative integer $\ell \geq s$.*

We only provide a sketch of proof, which is independent from the rest of the paper and can safely be skipped.

SKETCH OF PROOF OF THEOREMS 2 AND 3. The homology of topological spaces can be computed as the homology of standard chain complexes [28]. (Recall that a chain complex is a graded vector space $E_\bullet = \oplus_n E_n$ and a differential, that is, a linear map $d : E_n \to E_{n-1}$ such that $d \circ d = 0$, and that its homology groups $H_n(C_\bullet(X), d)$ are the quotients of the kernels of $d : E_n \to E_{n-1}$ by the images of $d : E_{n+1} \to E_n$.)

The key idea of the proof is to compute the homology of the union $\bigcup_F$ using a peculiar chain complex, the *Čech complex* $\check{C}_\bullet(F)$, that depends not only on $\bigcup_F$ but on the way the family F covers that union. Intuitively, $\check{C}_\bullet(F)$ takes into account the nerve of F as well as the topology of the intersections of subfamilies. Roughly, the vector space of $\check{C}_\bullet(F)$ is

$$\check{C}_\bullet(F) = \bigoplus_n \Big(\bigoplus_{p+|G|=n+1} C_p(\textstyle\bigcap_G) \Big)$$

where $C_p(X))$ is the degree-p subspace of a standard[3] chain complex computing the homology of the topological space X. The differential $d : \check{C}_\bullet(F) \to \check{C}_{\bullet-1}(F)$ is obtained from the differentials of the chain complexes $C_\bullet(\bigcap_G)$ using an analogue of the inclusion-exclusion principle. We refer to the full version for a formal definition, or to standard textbooks [10, 23] (or [9] for an analogue in de Rham cohomology).

A key feature of the Čech complex is that its homology (and thus the one of the topological space $\bigcup_F$) can be computed by *successive approximations*, starting by the direct sum of the homology of the topological spaces $\bigcap_G$ for all subfamilies; then the differential on the Čech complex will induce another differential (of combinatorial nature) on this direct sum whose homology groups will be a new approximation of the homology of the Čech complex (and thus of the space $\bigcup_F$). This new approximation also inherits a differential so that we can compute a third approximation, and so on. These approximation eventually yields the homology of the Čech complex.

This idea is made mathematically rigorous by *spectral sequences*. A spectral sequence is a sequence $(E^r_{\bullet,\bullet})_{r\geq 0}$ of bigraded vector spaces $E^r_{p,q}$ $(p, q \geq 0)$ equipped with a differential $d^r : E^r_{p,q} \to E^r_{p-r,q+r-1}$ such that $E^{r+1}_{\bullet,\bullet} \cong H_\bullet(E^r_{\bullet,\bullet}, d^r)$ is the homology of $E^r_{\bullet,\bullet}$ with respect to the differential d^r. The term $E^r_{\bullet,\bullet}$ is called the r-page of the spectral sequence. See Figure 5 for an example. Note that, for degree reasons, $d^r : E^r_{p,q} \to E^r_{p-r,q+r-1} = 0$ for $r > p$. It follows that the terms $E^r_{p,q}$ *stabilize*, that is, $E^r_{p,q} \cong E^{r+1}_{p,q} \cong \cdots$ for r large enough. We write $E^\infty_{p,q}$ for the stabilized groups $E^\infty_{p,q} \cong \cdots \cong E^{p+2}_{p,q} \cong E^{p+1}_{p,q}$. The homology of the Čech complex can be read off from the stabilized groups; precisely there is a linear isomorphism of vector spaces $H_n(\check{C}(F)) \cong \bigoplus_{p+q=n} E^\infty_{p,q}$ for every $n \geq 0$. Let us note that the fact that the Čech complex can be computed by such a spectral sequence follows from the observation that it is endowed with a natural *filtration* given by the cardinality of the subfamilies $G \subseteq F$; namely, for $q \geq 0$, we have vector spaces $\big(\check{C}_\bullet(F)\big)^p$ obtained from $\check{C}_\bullet(F)$ by considering only families G such that $|G| \leq p + 1$. Then the page E^0 of the spectral sequence is just $E^0_{p,q} := \big(\check{C}_{p+q}(F)\big)^p / \big(\check{C}_{p+q}(F)\big)^{p-1}$ equipped with the differential induced by d on the quotient spaces. When the family consists of only two open sets, our computation of the homology of a space using the above spectral sequence boils down to the standard Mayer-Vietoris exact sequence [28, 41].

A *key property* of spectral sequences is the following result: assume that there exists integers N and $r \geq 0$ such that

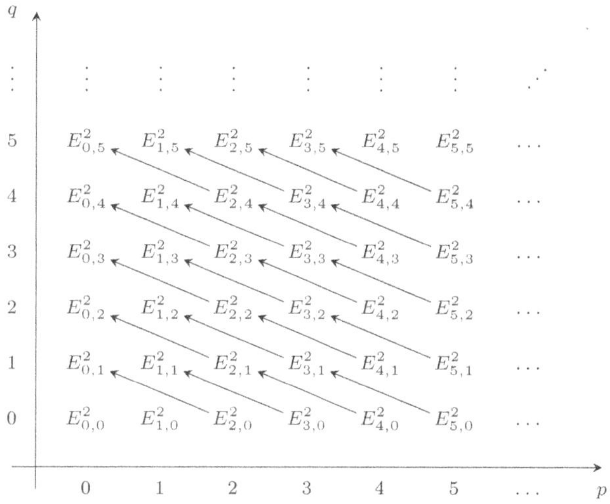

Figure 5. General form of the E^2-page of a spectral sequence. The arrows show the differential d^2.

$E^r_{p,q} = 0$ whenever $p + q \geq N$. Then $H_k(\check{C}_\bullet(F), d) = 0$ for $k \geq N$ as well.

We now explain what the spectral sequence for the Čech complex looks like with our assumptions on the family. First, the E^1-page is given by $E^1_{p,q} = \bigoplus_{|G|=p+1} H_q(\bigcap_G)$. The differential $d^1 : E^1_{p,q} \to E^1_{p-1,q}$ can be computed by formulae analogous to the one giving the differential of a simplicial poset. In fact, restricting to degree $q = 0$, $(E^1_{p,0}, d^1)$ is precisely $\big(C_p(\mathcal{M}(F)), d\big)$ and thus the component of the E^2-page in degree zero, $E^2_{p,0}$, is $H_p\big(\mathcal{M}(F)\big)$, the homology of the multinerve (viewed as a simplicial poset).

Thus for general covers, the E^2-page consists of bigraded vector spaces $E^2_{p,q}$ where $E^2_{p,0}$ is the homology of the multinerve $\mathcal{M}(F)$ associated to the cover.

In the case of an acyclic cover, all the homology groups $H_q(\bigcap_G)$ vanish for $q > 0$ and thus all the differentials d^r $(r \geq 2)$ as well. Hence $E^1_{p,q>0} = E^2_{p,q>0} = 0$ and the spectral sequence stabilizes immediately at page E^2. It follows that the homology of the Čech complex is the homology of the multinerve, which proves Theorem 2.

In the case of a cover that is acyclic with slack s, by assumption, for any $q \geq \max(1, s - p - 1)$ and any subfamily $G \subseteq F$ with $|G| = p + 1$, we have $H_q(\bigcap_G) = 0$ and thus $E^1_{p,q} = E^2_{p,q} = 0$ for $q \geq \max(1, s - p - 1)$ (see Figure 6). It follows that, for $p + q \geq s$, the spectral sequence stabilizes[4] at the page E^2, *i.e.*, $H_n(\check{C}(F)) \cong \bigoplus_{p+q=n} E^\infty_{p,q} = \bigoplus_{p+q=n} E^2_{p,q} = E^2_{n,0} = H_n(\mathcal{M}(F))$. Hence the homology of the Čech complex in degree greater or equal to s is the homology of the multinerve, which proves Theorem 3. □

4. HELLY NUMBERS OF ACYCLIC FAMILIES

The main result of this section is a Helly-type theorem for acyclic families.

[3] To prove that the Čech complex has the same homology as the topological space $\bigcup_F$ given by the union, a technical point here is that one needs to use "nice enough" chain complexes computing homology, for instance as given by cosheaf theory. However this technical point can be harmlessly disregarded since it has no influence when computing the E^1 and onward pages of the spectral sequence below.

[4] However, for $p + q < s$, it is in general necessary to compute higher order differential d^r to get some of the lower dimensional homology groups.

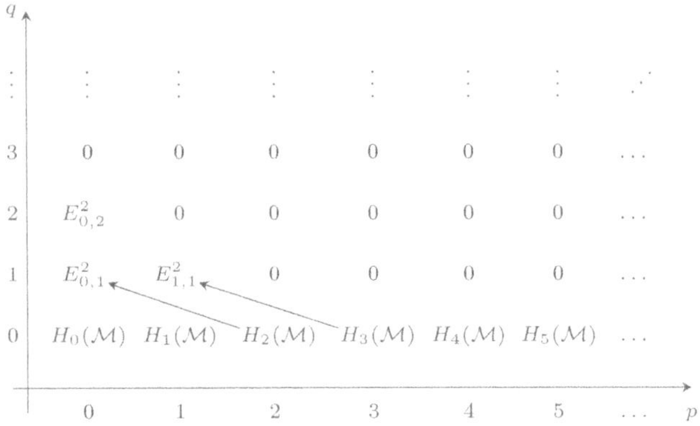

Figure 6. E^2-page of the Čech complex spectral sequence when F is acyclic with slack $s = 4$. The arrows show the only differential d^2 which can be non-zero. Here $H_i(\mathcal{M}(F))$ is abbreviated in $H_i(\mathcal{M})$.

4.1 Barycentric Subdivisions of Simplicial Posets

We first need some definitions related to the notion of barycentric subdivision of a simplicial poset [42]. Again, these definitions, when specialized to simplicial complexes, are the standard definitions from combinatorial topology; see the textbook of Matoušek [35, Chapter 1.7].

If $(Y, \preceq^Y)$ is a partially ordered set, a *chain* of Y is a subset of Y where any two elements are comparable with respect to $\preceq^Y$. The *order complex* of $(Y, \preceq^Y)$ is the set of all chains of Y ordered by inclusion; it is a simplicial complex.

Now, let X be a simplicial poset; we denote by 0 its least element. The *barycentric subdivision* of X, denoted by $\mathrm{sd}(X)$, is the order complex of $X \setminus \{0\}$. It turns out that (the geometric realizations of) X and $\mathrm{sd}(X)$ are homeomorphic; see Figure 3(c).

More generally, let σ be a simplex of X. We denote by $\dot{D}_X(\sigma)$ the order complex of $(\sigma, \cdot]$ in X. Thus, $\dot{D}_X(\sigma)$ is a subcomplex of the barycentric subdivision $\mathrm{sd}(X)$ of X (see Figure 3(d)), and coincides with $\mathrm{sd}(X)$ if $\sigma = 0$.

Finally, given $S \subseteq V$ of vertices, we let $X[S]$ denote the sub-poset of X *induced* by S, that is, the set of simplices of X whose set of vertices is contained in S.

4.2 Proof of Theorem 4

Recall that the *Helly number* of a family of sets with empty intersection is the size of its largest subfamily F such that (i) the intersection of all elements of F is empty, and (ii) for any proper subfamily $G \subsetneq F$, the intersection of the elements of G is non-empty.

Let Γ denote a locally arc-wise connected topological space. We let d_Γ denote the smallest integer such that every open subset of Γ has trivial $\mathbb{Q}$-homology in dimension d_Γ and higher; in particular, when Γ is a d-dimensional manifold, we have $d_\Gamma = d$ if Γ is non-compact or non-orientable and $d_\Gamma = d + 1$ otherwise (see, e.g., Greenberg [26, p. 121]); for example, $d_{\mathbb{R}^d} = d$. In this section, we prove the following Helly-type theorem.

THEOREM 4. *Let F be a finite family of open subsets of Γ. If F is acyclic and any subfamily of F intersects in at most r connected components, then the Helly number of F is at most $r(d_\Gamma + 1)$.*

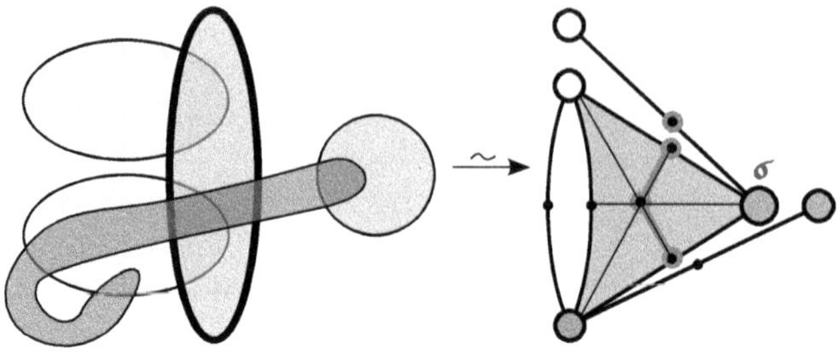

Figure 7. Continuation of Figure 4: On the left, the family F; on the right, the barycentric subdivision $\mathrm{sd}(\mathcal{M}(F))$ of the multinerve $\mathcal{M}(F)$. In this example, σ is a vertex of $\mathcal{M}(F)$ corresponding to one component C of an object in F. We see that $\dot{D}_{\mathcal{M}(F)}(\sigma)$ (in bold) is a subcomplex of $\mathrm{sd}(\mathcal{M}(F))$ that is the disjoint union of two homology cells. This is reflected in the fact that G_σ, the trace of the other objects of F inside C, is also the disjoint union of two homology cells.

We prove Theorem 4 in three steps. First, we associate to any simplicial poset X an index $J(X)$ that can be used to control Helly numbers. Specifically, given a simplicial poset X we let $J(X)$ be the smallest integer ℓ such that for every $j \geq \ell$, every $S \subseteq V$, and every simplex σ of $X[S]$, we have $\tilde{H}_j(\dot{D}_{X[S]}(\sigma)) = 0$.

LEMMA 5. *Any family F of sets has Helly number at most $J(\mathcal{N}(F)) + 1$.*

PROOF. Let $G \subseteq F$ be an inclusion-wise minimal subfamily with empty intersection. The nerve of G is the boundary of a $(|G| - 1)$-simplex and has therefore nontrivial homology in dimension $|G| - 2$. The nerve of G is the subcomplex of the nerve of F induced by G, that is $\mathcal{N}(G) = \mathcal{N}(F)[G]$, so we have that $\tilde{H}_{|G|-2}(\mathcal{N}(F)[G]) \neq 0$. Now, since $\dot{D}_{X[S]}(0)$ is the barycentric subdivision of $X[S]$, and therefore has the same homology groups as $X[S]$, it follows that $\tilde{H}_{|G|-2}(\dot{D}_{\mathcal{N}(F)[G]}(0)) \neq 0$ and $J(\mathcal{N}(F))$ is at least $|G| - 1$. The statement follows. □

As a side remark, we have only used the fact that for every $j \geq J(X)$ and every $S \subseteq V$, we have $\tilde{H}_j(X[S]) = 0$, which trivially follows from the definition of $J(X)$. The fact that J has a more subtle definition will be used later.

We next show that, under the assumptions of Theorem 4, the multinerve theorem yields an upper bound on $J(\mathcal{M}(F))$.

LEMMA 6. *If F is an acyclic family of open subsets of Γ then $J(\mathcal{M}(F)) \leq d_\Gamma$.*

PROOF. Let $G \subseteq F$ be a subfamily of F, and let σ be a simplex of $\mathcal{M}(F)[G] = \mathcal{M}(G)$. We need to prove that $\dot{D}_{\mathcal{M}(G)}(\sigma)$ has trivial reduced homology in dimension d_Γ and higher. The main idea is to show that $\dot{D}_{\mathcal{M}(G)}(\sigma)$ has the same homology as an open set in Γ. More precisely, let $\sigma = (C, A) \in \mathcal{M}(G)$. We define G_σ as the non-empty traces of the elements of $G \setminus A$ on C:

$$G_\sigma = \{U \cap C \mid U \in G \setminus A, U \cap C \neq \emptyset\}.$$

Figure 8. The projection of a disjoint union of 4 triangles (which has non-trivial homology only in dimension 0) onto the boundary of a simplex (which has non-trivial homology in dimension 2).

(Formally, G_σ is a multiset, as a given element may appear more than once.) We shall prove that $\dot{D}_{\mathcal{M}(G)}(\sigma)$ has the same homology as $\bigcup_{G_\sigma}$. See Figure 7.

Indeed, the map

$$\begin{cases} \mathcal{M}(G_\sigma) & \to & [\sigma, \cdot] \\ (C', A') & \mapsto & (C' \cap C, A \cup A') \end{cases}$$

is an isomorphism of posets. In particular, $[\sigma, \cdot]$ is a simplicial poset. Both posets have a least element, and removing them yields that $\mathcal{M}(G_\sigma) \setminus \{(\bigcup_G, \emptyset)\}$ and $(\sigma, \cdot]$ are isomorphic posets. Taking their order complexes, we get that $\dot{D}_{\mathcal{M}(G)}(\sigma)$ and $\mathrm{sd}(\mathcal{M}(G_\sigma))$ are isomorphic simplicial complexes.

Therefore, $\dot{D}_{\mathcal{M}(G)}(\sigma)$ has the same homology as $\mathcal{M}(G_\sigma)$. Since F is acyclic, the family G_σ is acyclic as well. Theorem 3 now ensures that the homology of $\mathcal{M}(G_\sigma)$ is the same as the homology of the union of the elements in G_σ. Since $\bigcup_{G_\sigma}$ is an open subset of Γ, it has homology zero in dimension d_Γ and higher. This concludes the proof. □

We now analyze what happens to the J-index when the simplicial poset $\mathcal{M}(F)$ is projected onto the simplicial complex $\mathcal{N}(F)$ by mapping each simplex of $\mathcal{M}(F)$ to its set of vertices:

$$\pi : \begin{cases} \mathcal{M}(F) & \to & \mathcal{N}(F) \\ (C, A) & \mapsto & A \end{cases}$$

While, already for simplicial complexes, a projection can create homology (see Figure 8), this can be controlled under certain assumptions. Let $\varphi : X \to Y$ be a monotone map between two simplicial posets. We say that φ is *dimension-preserving* if for any $\sigma \in X$ the dimension of $\varphi(\sigma)$ equals the dimension of σ; the *multiplicity* of φ is the maximum number of elements in the preimage of a simplex, that is, $\max_{\tau \in Y} |\varphi^{-1}(\tau)|$.

THEOREM 7. *If $\varphi : X \to Y$ is a surjective, dimension-preserving, monotone map of multiplicity r from a simplicial poset X to a simplicial complex Y, then $J(Y)+1 \leq r(J(X)+1)$.*

We only outline the strategy of the proof. In the special case where X is a simplicial complex, the statement was proven by Kalai and Meshulam [33, Theorem 1.3]. (Kalai and Meshulam use *Leray numbers* instead of J-indices, but for simplicial complexes the two notions can be seen to coincide, from a characterization of Leray numbers in terms of the homology of links [32, Proposition 3.1]. Incidentally, we do not know whether these two notions coincide for simplicial posets.) Their proof uses a spectral sequence of Goryunov and Mond [25] that, given a map $f : X \to Y$ between two topological spaces, computes the homology of Y in terms of the homology of X provided f is "nice" enough. The main technical difficulty in extending that proof to maps between simplicial posets is that it is not clear whether there is a generalization of the notion of links to simplicial posets that leads to a similar characterization. We introduced the index J precisely as a way around that technical difficulty, and the rest of the proof extends with minor technical modifications.

We can finally conclude the proof of the announced bound on Helly numbers.

PROOF OF THEOREM 4. Let F be a finite acyclic family of open subsets of Γ, and assume that any subfamily of F intersects in at most r connected components. Let $\mathcal{N}(F)$ and $\mathcal{M}(F)$ denote, respectively, the nerve and the multinerve of F. We consider the projection:

$$\pi : \begin{cases} \mathcal{M}(F) & \to & \mathcal{N}(F) \\ (C, A) & \mapsto & A. \end{cases}$$

Each simplex in the pre-image $\pi^{-1}(\sigma)$ of a simplex $\sigma \in \mathcal{N}(F)$ is of the form (C, σ) where C is a connected component of $\bigcap_\sigma$. The projection π is therefore at most r-to-one and we can apply Theorem 7 with $X = \mathcal{M}(F)$ and $Y = \mathcal{N}(F)$. We obtain that $J(\mathcal{N}(F)) \leq rJ(\mathcal{M}(F)) + r - 1$. With Lemma 6, this becomes $J(\mathcal{N}(F)) \leq r(d_\Gamma + 1) - 1$. With Lemma 5, we get that the Helly number of F is at most $J(\mathcal{N}(F)) + 1$, which concludes the proof. □

4.3 Extension of Theorem 4 and Relation to Previous Work

The conclusion of Theorem 4 does, in fact, hold even if the intersection of small subfamilies has more than r connected components and is acyclic with some slack. Specifically, we prove the following strengthening of Theorem 4:

THEOREM 8. *Let F be a finite family of open subsets of a locally arc-wise connected topological space Γ. If (i) F is acyclic with slack s and (ii) any subfamily of F of cardinality at least t intersects in at most r connected components, then the Helly number of F is at most $r(\max(d_\Gamma, s, t) + 1)$.*

Due to space limitations, we omit the proof. In both Theorems 4 and 8 the openness condition can be replaced by a compactness condition under an additional mild assumption (for instance that the elements in F are subcomplexes of a given triangulation).

Finally, we remark that Theorem 8 differs from Matoušek's topological theorem on two accounts. First, his proof only gives a loose bound on the Helly number (in fact, no explicit bound is given), whereas our approach gives sharp, explicit, bounds. Second, his theorem allows the connected components to have nontrivial *homotopy* in *high* dimension, whereas Theorem 8 lets them have nontrivial *homology* in *low* dimension.

5. APPLICATION TO HELLY NUMBERS IN GEOMETRIC TRANSVERSAL THEORY

Let $\mathcal{H} = \{A_1, \ldots, A_n\}$ be a family of pairwise disjoint convex sets in $\mathbb{R}^d$ and let $T(\mathcal{H})$ denote the set of *line transversal* to $\mathcal{H}$, that is of lines[5] intersecting every member in $\mathcal{H}$. An important problem in geometric transversal theory is to understand which geometric conditions on the A_i allow to

[5]All lines are considered non-oriented.

bound the Helly number of $\{T(A_1), \ldots, T(A_n)\}$ (we refer to the classical surveys [16, 19, 31, 44] for the history of that problem). In this section we show that Theorem 8 gives, in a single stroke, three bounds on Helly numbers on sets of lines:

(i) Santaló's theorem [39] asserts that when the A_i form a finite family of parallelotopes in $\mathbb{R}^d$ with edges parallel to the coordinate axis, the Helly number of $\{T(A_i)\}_i$ is at most $2^{d-1}(2d-1)$. The same bound follows from Theorem 8.

(ii) Tverberg's transversal theorem [43] states that when the A_i are disjoint translates of a compact convex subset $D \subset \mathbb{R}^2$ with non-empty interior, the Helly number of $\{T(A_i)\}_i$ is at most 5. A bound of 10 follows from Theorem 8.

(iii) Cheong et al. [13] showed that if the A_i are disjoint unit balls in $\mathbb{R}^d$, then the Helly number of $\{T(A_i)\}_i$ is at most $4d-1$. Theorem 8 implies a sharper bound of $4d-2$ for $d \geq 6$ (and weaker bounds for $2 \leq d \leq 5$).

That (some weaker form of) these three results can be traced back to the same principle is new and interesting since so far, only ad hoc proofs were known. Also note that the theorems of Tverberg and Cheong et al. were conjectured in the 1950's by Grünbaum and Danzer, respectively, but only proven in 1989 and 2006, respectively. Having a tool, such as Theorem 8, that allows to identify situations where the Helly numbers of sets of transversals are bounded is therefore interesting in its own right.

Due to space limitations, we give complete details for example (iii) and refer to the full version for examples (i) and (ii).

5.1 Ambient Space

The space of lines in $\mathbb{R}^d$ can be considered as a subspace of the space of lines in $\mathbb{RP}^d$, which is the Grassmannian $\mathbb{RG}_{2,d+1}$ of all 2-planes through the origin in $\mathbb{R}^{d+1}$; the space $\mathbb{RG}_{2,d+1}$ is a compact manifold of dimension $2d-2$ and can be seen as an algebraic sub-variety of some $\mathbb{RP}^m$ via Grassmann coordinates (also known as Plücker coordinates for $d=3$). We note that $d_{\mathbb{RG}_{2,d+1}} \leq 2d-1$ [26, p. 121]. However, in the applications below, we consider the set Γ of lines in $\mathbb{R}^d$, which is a non-compact submanifold of dimension $2d-2$ of $\mathbb{RG}_{2,d+1}$. It follows that $d_\Gamma \leq 2d-2$ (again by [26, p. 121]).

5.2 Number and Homology of Connected Components

For any $A_i \in \mathcal{H}$ the set $T_i = T(\{A_i\})$ has the homotopy type of $\mathbb{RP}^{d-1}$, and is therefore homologically trivial in dimension d and higher. Let $G \subseteq \mathcal{H}$ be a subset of size 2 or more. To discuss the number and homology of the connected components of $T(G)$ we first consider the map $p : \mathbb{RG}_{2,d+1} \to \mathbb{RP}^{d-1}$ associating each line to its direction. Since the restriction $p_{|T(\mathcal{H})}$ is a closed map with contractible fibers, it follows from the Vietoris-Begle mapping theorem [41, Theorem 15, Section 6.9] that $T(G)$ and $p(T(G))$ have the same homology. The *geometric permutation* on F induced by a line transversal to a family F of disjoint convex sets is the pair of orderings, one reverse of the other, induced on F by the two orientations of the line. Since G consists of disjoint balls, the connected components of $p(T(G))$ are convex[6] and in one-to-one correspondence with the geometric permutations of G [7]. It follows that the connected components of $T(G)$ have the same homology as convex sets, that is are homology cells. Altogether, we get that $\mathcal{H}$ is acyclic with slack d. Moreover, for $G \subseteq \mathcal{H}$ the number of connected components of $T(G)$ is the same as the number of geometric permutations of G, which is at most 3 in general and at most 2 when $n \geq 9$ [14] (independently of the dimension).

5.3 Wrapping Up

We can now apply Theorem 8 with $d_\Gamma = 2d-2$, $s = d+1$, $t=9$, and $r=2$, obtaining the upper bound of $2\max(2d-1, 10)$. For $d \geq 6$, this yields the upper bound of $4d-2$, but for $d \in \{2,3,4,5\}$ this bound is only 20. In the case $d=2$ (resp. $d=3$) it can be improved to 12 (resp. 15) by using $d_\Gamma = 2d-2$, $s=d+1$, $t=1$, and $r=3$. Note that it is conjectured that any family of 4 or more disjoint equal-radius balls in $\mathbb{R}^d$ has at most two geometric permutations. If this is true, then our bounds would improve to $4d-2$ for any $d \geq 3$.

When $\mathcal{H} = \{A_1, \ldots, A_n\}$ is a collection of disjoint unit balls in $\mathbb{R}^d$, the Helly number of $\{T(A_i)\}_i$ is at least $2d-1$ [12], so this number is known up to a factor of 2. If we allow the balls to have arbitrary radii then the maximum number of geometric permutations becomes $\Theta(n^{d-1})$ [40] and the Helly number of $\{T(A_i)\}_i$ is unbounded. If the radii are required to be in some fixed interval $[1, \rho]$, the upper bound on the number of geometric permutations reduces to $O(\rho^{\log \rho})$ [45] and Theorem 8 similarly implies that the Helly number of $\{T(A_i)\}_i$ is $O(d\rho^{\log \rho})$, where the constant in the $O()$ is independent of ρ, n and d.

6. REFERENCES

[1] N. Alon and G. Kalai. Bounding the piercing number. *Discrete & Computational Geometry*, 13:245–256, 1995.

[2] N. Amenta. Helly-type theorems and generalized linear programming. *Discrete & Computational Geometry*, 12:241–261, 1994.

[3] N. Amenta. A new proof of an interesting Helly-type theorem. *Discrete & Computational Geometry*, 15:423–427, 1996.

[4] D. Attali and H. Edelsbrunner. Inclusion-exclusion formulas from independent complexes. *Discrete & Computational Geometry*, 37:59–77, 2007.

[5] D. Attali, A. Lieutier, and D. Salinas. Efficient data structure for representing and simplifying simplicial complexes in high dimensions. In *Proceedings of the 27th Annual ACM Symposium on Computational Geometry*, SoCG '11, pages 501–509, 2011.

[6] D. Attali, A. Lieutier, and D. Salinas. Vietoris-Rips complexes also provide topologically correct reconstructions of sampled shapes. In *Proceedings of the 27th Annual ACM Symposium on Computational Geometry*, SoCG '11, pages 491–500, 2011.

[7] C. Borcea, X. Goaoc, and S. Petitjean. Line transversals to disjoint balls. *Discrete & Computational Geometry*, 1-3:158–173, 2008.

[8] K. Borsuk. On the imbedding of systems of compacta in simplicial complexes. *Fundamenta Mathematicae*, 35:217–234, 1948.

[6]Convexity in $\mathbb{RP}^{d-1}$ is relative to the metric induced through the identification $\mathbb{RP}^{d-1} = \mathbb{S}^{d-1}/\mathbb{Z}_2$.

[9] R. Bott and L. W. Tu. *Differential forms in algebraic topology*, volume 82 of *Graduate Texts in Mathematics*. Springer-Verlag, New York, 1982.

[10] G. E. Bredon. *Sheaf theory*, volume 170 of *Graduate Texts in Mathematics*. Springer-Verlag, New York, second edition, 1997.

[11] F. Chazal and S. Y. Oudot. Towards persistence-based reconstruction in Euclidean spaces. In *Proceedings of the 24th Annual Symposium on Computational Geometry*, SoCG '08, pages 232–241, 2008.

[12] O. Cheong, X. Goaoc, and A. Holmsen. Lower bounds to Helly numbers of line transversals to disjoint congruent balls. *Israel Journal of Mathematics*, 2012. To appear.

[13] O. Cheong, X. Goaoc, A. Holmsen, and S. Petitjean. Hadwiger and Helly-type theorems for disjoint unit spheres. *Discrete & Computational Geometry*, 1-3:194–212, 2008.

[14] O. Cheong, X. Goaoc, and H.-S. Na. Geometric permutations of disjoint unit spheres. *Computational Geometry: Theory & Applications*, 30:253–270, 2005.

[15] É. Colin de Verdière, G. Ginot, and X. Goaoc. Helly numbers of acyclic families. arXiv:1101.6006, 2011.

[16] L. Danzer, B. Grünbaum, and V. Klee. Helly's theorem and its relatives. In V. Klee, editor, *Convexity*, Proceedings of Symposia in Pure Mathematics, pages 101–180. AMS, 1963.

[17] V. de Silva and G. Carlsson. Topological estimation using witness complexes. *IEEE Symposium on Point-Based Graphics*, pages 157–166, 2004.

[18] H. Debrunner. Helly type theorems derived from basic singular homology. *American Mathematical Monthly*, 77:375–380, 1970.

[19] J. Eckhoff. Helly, Radon and Caratheodory type theorems. In J. E. Goodman and J. O'Rourke, editors, *Handbook of Convex Geometry*, pages 389–448. North Holland, 1993.

[20] J. Eckhoff and K.-P. Nischke. Morris's pigeonhole principle and the Helly theorem for unions of convex sets. *Bulletin of the London Mathematical Society*, 41:577–588, 2009.

[21] H. Edelsbrunner and N. R. Shah. Triangulating topological spaces. *International Journal of Computational Geometry and Applications*, 7(4):365–378, 1997.

[22] B. Farb. Group actions and Helly's theorem. *Advances in Mathematics*, 222:1574–1588, 2009.

[23] R. Godement. *Topologie algébrique et théorie des faisceaux*. Hermann, Paris, 1973. Troisième édition revue et corrigée, Publications de l'Institut de Mathématique de l'Université de Strasbourg, XIII, Actualités Scientifiques et Industrielles, No. 1252.

[24] P. G. Goerss and J. F. Jardine. *Simplicial homotopy theory*, volume 174 of *Progress in Mathematics*. Birkhäuser Verlag, Basel, 1999.

[25] V. Goryunov and D. Mond. Vanishing cohomology of singularities of mappings. *Compositio Mathematica*, 89(1):45–80, 1993.

[26] M. J. Greenberg. *Lectures on Algebraic Topology*. Benjamin, Reading, MA, 1967.

[27] B. Grünbaum and T. Motzkin. On components in some families of sets. *Proceedings of the American Mathematical Society*, 12:607–613, 1961.

[28] A. Hatcher. *Algebraic topology*. Cambridge University Press, 2002. Available at `http://www.math.cornell.edu/~hatcher/`.

[29] E. Helly. Über Mengen konvexer Körper mit gemeinschaftlichen Punkten. *Jahresbericht der Deutschen Mathematiker-Vereinigung*, 32:175–176, 1923.

[30] E. Helly. Über Systeme von abgeschlossenen Mengen mit gemeinschaftlichen Punkten. *Monatshefte für Mathematik und Physik*, 37:281–302, 1930.

[31] A. Holmsen. Recent progress on line transversals to families of translated ovals. In J. E. Goodman, J. Pach, and R. Pollack, editors, *Computational Geometry - Twenty Years Later*, pages 283–298. AMS, 2008.

[32] G. Kalai and R. Meshulam. Intersections of Leray complexes and regularity of monomial ideals. *Journal of Combinatorial Theory, Series A*, 113:1586–1592, 2006.

[33] G. Kalai and R. Meshulam. Leray numbers of projections and a topological Helly-type theorem. *Journal of Topology*, 1:551–556, 2008.

[34] M. Kashiwara and P. Schapira. *Sheaves on manifolds*, volume 292 of *Grundlehren der Mathematischen Wissenschaften [Fundamental Principles of Mathematical Sciences]*. Springer-Verlag, Berlin, 1990.

[35] J. Matoušek. *Using the Borsuk-Ulam Theorem*. Springer-Verlag, 2003.

[36] J. Matoušek. A Helly-type theorem for unions of convex sets. *Discrete & Computational Geometry*, 18:1–12, 1997.

[37] J. P. May. *Simplicial objects in algebraic topology*. Chicago Lectures in Mathematics. University of Chicago Press, Chicago, IL, 1992. Reprint of the 1967 original.

[38] H. Morris. *Two pigeon hole principles and unions of convexly disjoint sets*. Phd thesis, California Institute of Technology, 1973.

[39] L. Santaló. Un theorema sobre conjuntos de paralelepipedos de aristas paralelas. *Publ. Inst. Mat. Univ. Nat. Litoral*, 2:49–60, 1940.

[40] S. Smorodinsky, J. S. B. Mitchell, and M. Sharir. Sharp bounds on geometric permutations for pairwise disjoint balls in $\mathbb{R}^d$. *Discrete & Computational Geometry*, 23:247–259, 2000.

[41] E. H. Spanier. *Algebraic topology*. McGraw-Hill Book Co., New York, 1966.

[42] R. Stanley. f-vectors and h-vectors of simplicial posets. *J. Pure and Applied Algebra*, 71:319–331, 1991.

[43] H. Tverberg. Proof of Grünbaum's conjecture on common transversals for translates. *Discrete & Computational Geometry*, 4:191–203, 1989.

[44] R. Wenger. Helly-type theorems and geometric transversals. In J. E. Goodman and J. O'Rourke, editors, *Handbook of Discrete & Computational Geometry*, chapter 4, pages 73–96. CRC Press LLC, Boca Raton, FL, 2nd edition, 2004.

[45] Y. Zhou and S. Suri. Geometric permutations of balls with bounded size disparity. *Computational Geometry: Theory & Applications*, 26:3–20, 2003.

Duals of Orphan-Free Anisotropic Voronoi Diagrams are Embedded Meshes

Guillermo D. Canas
Massachusetts Institute of Technology
Cambridge, MA
guilledc@mit.edu

Steven J. Gortler
Harvard University
Cambridge, MA
sjg@seas.harvard.edu

ABSTRACT

Given an anisotropic Voronoi diagram, we address the fundamental question of when its dual is embedded. We show that, by requiring only that the primal be orphan-free (have connected Voronoi regions), its dual is always guaranteed to be an embedded triangulation. Further, the primal diagram and its dual have properties that parallel those of ordinary Voronoi diagrams: the primal's vertices, edges, and faces are connected, and the dual triangulation has a simple, closed boundary. Additionally, if the underlying metric has bounded anisotropy (ratio of eigenvalues), the dual is guaranteed to triangulate the convex hull of the sites. These results apply to the duals of anisotropic Voronoi diagrams of *any* set of sites, so long as their Voronoi diagram is orphan-free. By combining this general result with existing conditions for obtaining orphan-free anisotropic Voronoi diagrams, a simple and natural condition for a set of sites to form an embedded anisotropic Delaunay triangulation follows.

Categories and Subject Descriptors

I.3.5 [**Computational Geometry and Object Modeling**]: Geometric algorithms, languages, and systems.

General Terms

Algorithms, Theory.

Keywords

Anisotropic Voronoi diagrams, Delaunay triangulations, Orphan-free.

1. INTRODUCTION

Voronoi diagrams and their dual Delaunay triangulations are fundamental constructions with numerous associated guarantees, and extensive application in practice [6, 8]. At their heart is the use of a distance between points, which in the original version is taken to be Euclidean. This suggests that, by considering generalizations of the Euclidean distance, it may be possible to obtain variants which can be well-suited to a wider range of applications.

Attempts in this direction have been met with some success. Power diagrams [9] generalize Euclidean distance by associating a bias-term to each site. The duals of these diagrams are guaranteed to be embedded triangulations, in any number of dimensions. Although this is a strict generalization of Euclidean distance, it is a somewhat limited one. The effect of the bias term is to locally enlarge or shrink the region associated to each site, loosely-speaking "equally in every direction". It allows some freedom in choosing local scale, with no preference for specific directions.

Another way to generalize Voronoi diagrams is to endow Euclidean space with a continuously varying Riemannian metric, and use its associated (geodesic) distance in the definition of the Voronoi diagram. This generalization carries a significant amount of freedom, allowing for local scale and "directionality" to be freely specified at each point.

Despite being of potentially great interest, this latter approach has faced several obstacles. The dual of such an anisotropic Voronoi diagram is, in general, *not* an embedded triangulation, even if the primal is orphan-free (the Voronoi regions are connected), and may produce element inversions and edge crossings. An additional obstacle is that the geodesic distance is extremely expensive to compute in practice, rendering traditional methods for constructing Voronoi diagrams impractical.

Two notable attempts to resolve these problems have been proposed. Independently, Labelle and Shewchuk [13], and Du and Wang [7] propose two efficient approximations of geodesic distance from a site to any point in the domain. Although their associated Voronoi diagrams are, in general, no longer orphan-free, Labelle and Shewchuk show that a set of sites exists with an orphan-free diagram, whose dual is embedded, in two dimensions. They accomplish this by proposing an iterative site-insertion algorithm that, for any given metric, constructs one such set of sites. Some recent work making use of the above definitions are [2, 1].

In this paper, we show that if a set of sites produces an orphan-free anisotropic Voronoi diagram, using the definition of [7], then its dual is always an embedded triangulation (or a polygonal mesh with convex faces in general), in two dimensions (Thm. 6.2). This effectively states that, regardless of the sites' positions, if the primal is well-behaved, then the dual is as well. Further, in a way that parallels the ordinary Delaunay case, the dual has no degenerate elements (Lem. 4.9), its elements (vertices, edges, faces) are unique (Cor. 6.3), and, under mild assumptions on the metric, is guaranteed to triangulate the convex hull of the sites (Thm. 4.7). Note that, while [13] prove a property of the output of an algorithm, the results in this paper are fundamental properties of the diagrams themselves (independent of how the orphan-freedom of the primal was obtained).

In practice, we may combine our results with those of [4] to conclude that duals of anisotropic Voronoi diagrams of appropriate ϵ-nets are always embedded triangulations (Cor. 6.4). This may be particularly useful in (asymptotically-optimal) function approximation applications, where we are often interested in constructing the anisotropic Delaunay triangulations of appropriate ϵ-nets [5, 12], and was the initial motivation for the current work. Finally, we note that, as discussed in Sec. 7, in practice, algorithms for constructing anisotropic diagrams of the type discussed here may have a more numerical flavor (i.e. front-propagation, fast marching methods), in contrast with the more combinatorial nature of other Voronoi diagrams.

2. SETUP

Given a finite set $V \subset \mathbb{R}^2$ of sites on the plane, a Voronoi diagram decomposes Euclidean space into regions, each distinguished by the site its points are closest to. The notion of closeness is defined as follows: consider a continuous metric (in coordinates: $Q : \mathbb{R}^2 \to \mathbb{R}^{2\times 2}$, symmetric, positive definite) over two-dimensional Euclidean space and, following [7], define the (asymmetric) "distance" between a site ($v \in V$) and point ($p \in \mathbb{R}^2$) as

$$D(v,p) = \left[(p-v)^t Q_p (p-v)\right]^{1/2}$$

The Voronoi region of site v is the set of points no further from v than from any other site:

$$R(v) = \{p \in \mathbb{R}^2 : D(v,p) \le D(w,p), \forall w \in V\}$$

Note that, because V is a set, and not a multiset, there are no coincident sites.

This definition of anisotropic Voronoi diagram results in diagrams that take the usual shape, but whose regions may have curved boundaries, and tend to be elongated along certain directions, depending on the metric Q.

Following the notation of [13], we say that a diagram is *orphan-free* if its regions $R(v)$ are connected, for all $v \in V$ (equivalently: each *cell*, or connected component of a Voronoi region, contains its generating site).

An important distinction in this construction is that the "interfaces" in-between Voronoi regions do not, in general, have null measure (and points equidistant to three or more sites are not always isolated), unlike geodesic-distance diagrams [14] (see [5] Lemma 5.2), and those of [13].

Note that, given one such degenerate diagram, any interior point p of a Voronoi edge $R(v) \cap R(w)$ of non-zero measure has an open neighborhood $N_p \subset R(v) \cap R(w)$. By adding to Q an arbitrarily small perturbation supported on N_p, an orphan is created. That is, with respect to the metric, any degenerate diagram is arbitrarily close to a diagram that has orphans, and therefore the non-degeneracy requirement is only slightly more restrictive than the orphan-freedom one.

We begin with three simple lemmas, proved in Appendix A, which provide an introduction to some of the techniques used in the sequel. The first is a simple extension of Lemma 2.1 of [7], and the second follows directly from the continuity of Q and the fact that every site is strictly closer to itself than to all other sites.

LEMMA 2.1. *Given two sites $v, w \in V$, $v \neq w$, the only point equidistant to v, w in their supporting line is the midpoint $(v+w)/2$.*

LEMMA 2.2. *Every site is an interior point of its corresponding Voronoi region.*

LEMMA 2.3. *Every Voronoi region of an orphan-free anisotropic Voronoi diagram in $\mathbb{R}^2$ is simply connected.*

In the sequel, unless otherwise stated, proofs are omitted in the interest of conciseness, but they can be found in a longer version of this paper in [3].

Primal diagram. The Voronoi diagram is a collection of Voronoi regions $R(v)$, one for every site in V, as well as a structure induced by the sets of points closest and equidistant to two or more sites. Let $\tilde{P}$ to be the primal Voronoi diagram embedded on the sphere $\mathbb{S}^2$ by stereographically projecting the plane onto the punctured sphere and completing it with a "point at infinity" p_∞. $\tilde{P}$ includes not just regions, but also embedded edges $\tilde{E}_{vw}$, which are *connected* sets of points closest and equidistant to two sites v, w, as well as vertices (points closest and equidistant to three or more sites). Since we are assuming that sets of equidistant points to two sites have null measure, and that points equidistant to three or more sites are isolated, we can consider the abstract mesh $P = (V_p, E_p, F_p)$ with structure derived from $\tilde{P}$, where V_p, E_p, F_p are the vertices, edges, and faces of P.

A simple induction argument reveals that every edge in E_p connects two vertices in V_p. To see this, start with an orphan-free diagram with two sites. The only edge of $\tilde{E}$ is the unbounded set of points equidistant to the sites, passing through their midpoint (Lem. 2.1), and with endpoints at p_∞. The insertion of an additional site can only split existing edges of $\tilde{E}$ at points where two edges cross (a vertex by definition), and thus the original edge is split into pieces whose endpoints are now either p_∞ or an edge crossing (a vertex).

Clearly, $\tilde{P}$ is a planar embedded diagram since two edges cannot meet except at vertices: any point p where edge $D(u,p) = D(v,p)$ crosses edge $D(v,p) = D(w,p)$ is equidistant to three sites, u, v, w, and thus a vertex. Therefore P is planar.

Dual of an orphan-free Voronoi diagram. The dual of the primal mesh structure P is another abstract mesh $G = (V, E, F)$, with one vertex per site, and edges connecting adjacent regions of $\tilde{P}$. Its face structure is derived from that of P by duality and, in particular, (V, E) is planar, since (V_p, E_p) is planar. Using the embedded primal $\tilde{P}$, we can define an embedding $\tilde{G}$ of G on the plane, with vertices coinciding with sites (and curved edges). Although $\tilde{P}$ was defined on the sphere, we embed $\tilde{G}$ on the plane instead, and therefore we exclude (from both G and $\tilde{G}$) the unbounded face dual to the point at infinity p_∞.

Since (V, E) is planar, G derives the following properties from the primal:

(i) Every edge in E is incident to two dual faces, except for *boundary edges* (whose corresponding primal edge in $\tilde{P}$

connects unbounded regions), which are only incident to one dual face.

(ii) To every dual face in F, connecting vertices $v_1, \ldots, v_m$, corresponds a primal vertex $c \in R(v_1) \cap \cdots \cap R(v_m)$ of $\tilde{P}$, which is equidistant to the corresponding sites $v_1, \ldots, v_m$.

The property (ii) requires that the Voronoi diagram be orphan-free, as noted in the remark after Thm. 2.7.

Notice that either $\tilde{G}$ or $\tilde{P}$ could be multigraphs (e.g. $\tilde{P}$ would be a multigraph if an edge $\tilde{E}_p$ of $\tilde{P}$ has multiple connected components). However, the following lemma ensures that both are simple graphs.

LEMMA 2.4. *$\tilde{G}$ and $\tilde{P}$ are simple (have no multi-edges or self-loops).*

By its definition, an edge of $\tilde{P}$ is a connected set of points equidistant and closest to two sites v, w. Since $\tilde{P}$ has no multi-edges, this implies the following:

COROLLARY 2.5. *The set of points closest and equidistant to two sites of an orphan-free diagram is connected.*

Property (ii) above can be restated in a more useful way that parallels an equivalent property of duals of ordinary Voronoi diagrams.

DEFINITION 2.6 (**Empty circum-ellipse property**). *A face of G incident to vertices $v_1, \ldots, v_m$ satisfies the empty circum-ellipse (ECE) property if there is* some *ellipse that circumscribes the sites corresponding to $v_1, \ldots, v_m$, and contains no site in its interior.*

Note that *any* empty circum-ellipse serves as witness to the ECE property. We show that the ellipse centered at a Voronoi vertex c, with axis given by Q_c, is always an empty circum-ellipse of its dual polygon.

THEOREM 2.7. *Every face of G satisfies the empty circum-ellipse property.*

PROOF. The proof is constructive. To every face in G connecting vertices with corresponding sites $v_1, \ldots, v_m$, by property (ii), corresponds a Voronoi vertex $c \in R(v_1) \cap \cdots \cap R(v_m)$ of $\tilde{P}$ that is equidistant to $v_1, \ldots, v_m$. If $\mu = D(v_1, c) = \cdots = D(v_m, c)$, then the open ellipse

$$\theta = \{p \in \mathbb{R}^2 : D(p, c) < \mu\} = \{p \in \mathbb{R}^2 : (p-c)^t Q_c (p-c) < \mu^2\}$$

circumscribes the $v_1, \ldots, v_m$ and does not contain any site (otherwise, c would be closer to that site than to $v_1, \ldots, v_m$, and therefore not in $R(v_1) \cap \cdots \cap R(v_m)$, a contradiction). □

Although this may not at first seem apparent, the above property uses the fact that the primal diagram is orphan free. The critical fact in the proof of Thm. 2.7 is not that there is a Voronoi vertex c equidistant to $v_1, \ldots, v_m$, but rather that it is $c \in R(v_1) \cap \cdots \cap R(v_m)$. Consider Figs. 1(a) and 1(b), in which sites v_i, v_j, v_k form a face of G. In 1(a), the diagram is orphan-free, and the fact that $c_{ijk} \in R(v_i) \cap R(v_j) \cap R(v_k)$ implies Thm. 2.7. Figure 1(b) shows part of a Voronoi diagram that has an orphan cell (corresponding to site w) "covering" c_{ijk}. If the dual is constructed such that the face v_i, v_j, v_k is not in the dual then a "hole" in the triangulation may occur. If the face v_i, v_j, v_k is in the dual, then,

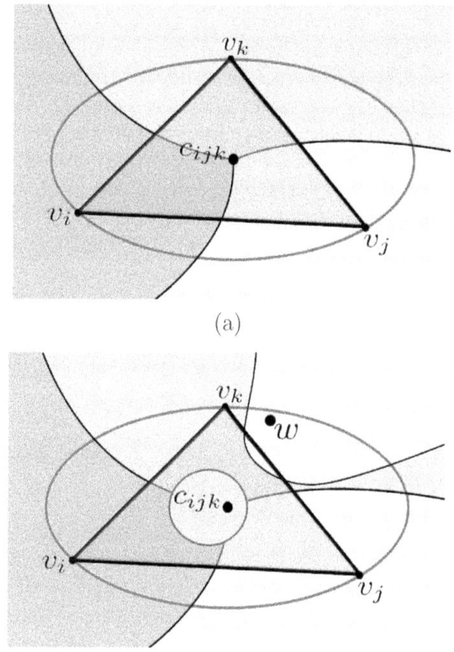

(a)

(b)

Figure 1: In an orphan-free diagram (a), every face satisfies the empty-circum ellipse (ECE) property. If the diagram is not orphan-free, then some faces may not satisfy the ECE property.

despite the fact that c_{ijk} is equidistant to v_i, v_j, v_k, this c_{ijk} is no longer in $R(v_i) \cap R(v_j) \cap R(v_k)$, but in the Voronoi region of w. Since c_{ijk} is closer to w than to v_i, v_j, v_k, any open ellipse centered at c_{ijk}, and having v_i, v_j, v_k on its boundary, contains w. Therefore, in Fig. 1(b), the ECE property does not hold for the dual face of c_{ijk}. Ensuring that Thm. 2.7 holds is one of the main reasons for requiring orphan-freedom.

3. SUMMARY OF RESULTS AND OUTLINE

The two main results in this paper assume that we are given an orphan-free anisotropic Voronoi diagram (in two dimensions) and prove that: 1) *the dual is an embedded triangulation* (Thm. 6.2), and 2) *if the metric has bounded anisotropy (ratio of eigenvalues), the dual triangulates the convex hull of the sites* (Thm. 6.1).

We prove a number of additional results. Possibly the most important shows that the elements of an orphan-free diagram (vertices, edges, faces), are connected sets, and therefore, in some sense, unique (Cor. 6.3). This is important, since this is a natural well-behave-ness condition on the primal, which, together with the above results, indicates that orphan-freedom is sufficient to guarantee well-behaveness of both primal and dual.

A smaller result of interest is: the dual faces satisfy an *empty circum-ellipse* property (Thm. 2.7), which parallels the empty-circumcircle property of ordinary Voronoi diagrams, and could have further practical implications from the ones described here.

The order of the proofs is, however, different from the one just stated. In effect, we begin by proving the most restrictive case: that the dual of an orphan-free diagram with metric of bounded anisotropy is embedded (Sec. 5), and that its boundary is the boundary of the convex hull of the sites (Sec. 4). We then relax the bounded anisotropy condition (Sec. 6), and show that, in the general case, we loose the convex hull property, but the dual remains embedded. Finally, once these results are established, we prove that vertices of the primal are unique (uniqueness of edges was proven in Sec. 2, and uniqueness of primal faces is the same as orphan-freedom).

We assume in the remainder of the paper that there are more than two sites, not all of which are colinear, and relegate the (considerably simpler) colinear case to Appendix B.

4. DUAL OF ORPHAN-FREE DIAGRAM (PART I: BOUNDARY)

In this section, we assume that the metric has bounded anisotropy, and conclude that the boundary of the dual of an orphan-free diagram is the same as the boundary of the convex hull of the sites (and in particular is simple and closed). If $R_p \Lambda_p R_p^t$ is the eigendecomposition of Q_p at $p \in \mathbb{R}^2$, with $\Lambda_p = diag[\lambda_2(p), \lambda_1(p)]$, $\lambda_2(p) \geq \lambda_1(p) > 0$, and R_p orthonormal, then we assume that there is some bound γ on the anisotropy of Q, such that: $1 \leq \lambda_2(p)/\lambda_1(p) < \gamma^2$ for $p \in \mathbb{R}^2$. Note that this condition may commonly hold in practice, if the metric is sampled on a compact domain (and possibly extended to the plane by reusing sampled values only).

We begin by defining $\bar{G}$ to be the straight-edge drawing of $\tilde{G}$ with vertices at the sites. For the moment, we assume that every Voronoi vertex is equidistant to no more than three sites, and therefore that all faces in $\bar{G}$ are triangles (we extend the results to the general case in Appendix G). We associate to $\bar{G}$ a mapping (from the vertices, edges, and faces of $\bar{G}$ to $\mathbb{R}^2$) for which, because all faces of $\bar{G}$ are triangles, it is well defined whether a point in $\mathbb{R}^2$ belongs to any given face of $\bar{G}$, a fact that will be used in the proofs of Sec. 5. This mapping will be shown in Sec. 5 to be an embedding. In the sequel, it is assumed that $\bar{G}$ encompasses both the mesh structure, and the mapping.

The boundary vertices of G are those whose corresponding primal regions in $\tilde{P}$ are unbounded, while boundary edges of G connect boundary vertices. Note that this is a *topological* property of G, rather than a geometric one (boundary elements of G may, in principle, not lie in the boundary of the convex hull of the sites). For convenience, we call $B \subseteq E$ the set of boundary edges of G.

The convex hull $\mathcal{CH}(V)$ of V is the minimal (w.r.t. set containment) convex set that contains V. For convenience, we name $W = \{w_i \in V : i = 1, \ldots, m\}$ the sites that are part of the boundary of the convex hull $\mathcal{CH}(V)$, and order them in clock-wise order around $\mathcal{CH}(V)$. The boundary $\mathcal{B}$ of the convex hull is a simple circular chain $\mathcal{B} = \{(w_i, w_{i\oplus 1}) : i = 1, \ldots, m\}$. We prove that it is $B = \mathcal{B}$ (loosely speaking: the topological boundary of G, and the geometric boundary of its straight-edge embedding $\bar{G}$, are the same), which implies that $\bar{G}$ covers the convex hull of the sites, and its boundary edges form a simple, closed polygonal chain.

LEMMA 4.1. *To every boundary edge (v, w) of G corresponds a segment in the boundary of $\mathcal{CH}(V)$.* $[B \subseteq \mathcal{B}]$

We now turn to the converse claim: that to every segment $(w_i, w_j) \in \mathcal{B}$ corresponds a boundary edge $(w_i, w_j) \in B$ in G. Since B is the set of boundary edges of G, whose primal edges in $\tilde{P}$ are unbounded, the claim is equivalent to proving that, to every segment (w_i, w_j) in the boundary of the convex hull corresponds an edge of the dual G, whose primal edge $\tilde{E}_{w_i, w_j}$ is unbounded.

The proof proceeds as follows. First, assume w.l.o.g. that the origin is in the interior of $\mathcal{CH}(V)$. Let $C(\sigma) = \{x \in \mathbb{R}^2 : \|x\| = \sigma\}$ be a sufficiently large origin-centered circle. We define two functions:

$$\pi : C(\sigma) \to \partial\mathcal{CH}(V), \quad \pi(p) = \underset{r \in \mathcal{CH}(V)}{\textbf{argmin}}\, D(r, p)$$
$$\nu : \partial\mathcal{CH}(V) \to C(\sigma), \quad \nu(r) = \sigma \cdot r / \|r\|$$

π simply projects every point in C to its closest in $\mathcal{CH}(V)$ (with respect to the distance D), and ν projects a point back to C.

LEMMA 4.2. $\pi(p) = \underset{x \in \mathcal{CH}(V)}{\operatorname{argmin}} D(x, p)$ *is a continuous function in* $\mathbb{R}^2$.

PROOF. We first prove by contradiction that $\pi(p)$ is unique. Let M_p be the unique [4], symmetric positive definite square-root matrix such that $M_p^t M_p = Q_p$. Consider distinct $q_1, q_2 \in \mathcal{CH}(V)$ closest to p. Since $q_1 \neq q_2$, by the convexity of the Euclidean norm, and the positive-definiteness of M_p, it is

$$\begin{aligned} D((q_1 + q_2)/2, p) &= \|M_p\left((q_1 + q_2)/2 - p\right)\| \\ &< \|M_p\left(q_1 - p\right)\| = \|M_p\left(q_2 - p\right)\| \end{aligned}$$

Thus p is closer to $(q_1 + q_2)/2$ than to q_1, q_2, a contradiction. Therefore the closest point $\pi(p)$ to p is unique.

If π were not continuous at p, then there is $\epsilon > 0$ such that for all $\delta > 0$ there is a point $\bar{p} = \bar{p}(\delta)$ such that $\|p - \bar{p}\| < \delta$ and $\|\pi(p) - \pi(\bar{p})\| \geq \epsilon$. Consider the sequence $\{\kappa_i = \pi(\bar{p}(1/i)) : i \in \mathbb{N}\}$ of points in $\mathcal{CH}(V)$. Because $\mathcal{CH}(V)$ is compact, $\{\kappa_i\}$ has a subsequence that converges to some $\kappa \in \mathcal{CH}(V)$. By continuity of D (which follows from the continuity of Q), it is

$$\begin{aligned} D(\pi(p), p) &= \min_{x \in \mathcal{CH}(V)} D(x, p) = \lim_{i \to \infty} \min_{x \in \mathcal{CH}(V)} D(x, \bar{p}_i) \\ &= \lim_{i \to \infty} D(\kappa_i, \bar{p}_i) = D(\kappa, p) \end{aligned}$$

and therefore the closest point in $\mathcal{CH}(V)$ to p is not unique, a contradiction. □

Clearly, ν is continuous in $\partial\mathcal{CH}(V)$. Note that, because $\mathcal{CH}(V)$ contains the origin, then, as shown in Fig. 2, ν projects every point $\pi(p) \in (w_i, w_j)$ on a segment of $\partial\mathcal{CH}(V)$, *outwards* from the convex hull (and on the empty side of (w_i, w_j)); that is, so that $\nu(\pi(p)) \in H_{ij}^+ \cap C$ ($\nu(\pi(p))$ is in the empty half-space of (w_i, w_j)).

Conveniently, Lem. 4.3 shows that, if $\pi(p) = (w_i + w_j)/2$, then $p \in \tilde{E}_{w_i, w_j}$, and so the claim now reduces to showing that for each segment (w_i, w_j) of $\partial\mathcal{CH}(V)$, and for *every* sufficiently large circle C, there is $p \in C$ with $\pi(p) = (w_i + w_j)/2$. Since this implies that $\tilde{E}_{w_i, w_j}$ is unbounded, it means that the corresponding edge (w_i, w_j) is in B.

The proof is by contradiction. Lem. 4.5 uses Brouwer's fixed point theorem to show that, for every segment (w_i, w_j) of $\mathcal{B}$, if there were no $p \in C$ with $\pi(p) = (w_i + w_j)/2$, then the function $\nu \circ \pi : C \to C$ must have a point $q \in C$ such that

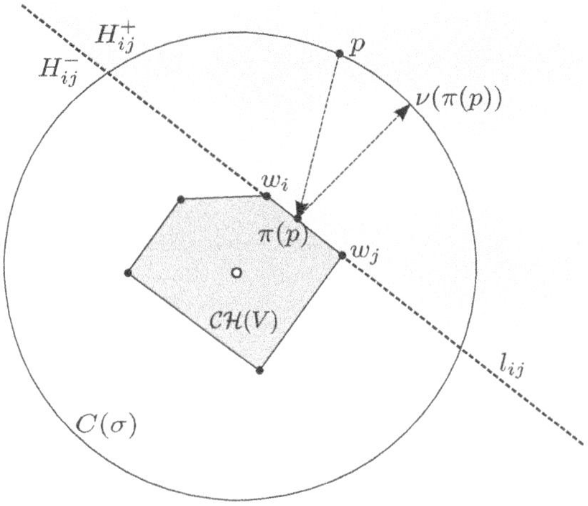

Figure 2: The construction for the proof of Thm. 4.7.

$\nu(\pi(q)) = -q$, or equivalently such that q is "behind" the segment $(w_i, w_j) \in \partial\mathcal{CH}(V)$ to which it is closest ($q \in H_{ij}^-$). On the other hand, Lem. 4.4 shows that, for all sufficiently large circles C, no point $q \in C$ can be closest to some segment $(w_i, w_j) \in \partial\mathcal{CH}(V)$ that it is *behind* of, creating a contradiction.

As mentioned above, we are interested in identifying points in $C(\sigma)$ that are also in some primal edge $\tilde{E}_{w_i,w_j}$. The following lemma will be used to show that, for sufficiently large σ, such a point p can be characterized by satisfying $\pi(p) = (w_i + w_j)/2$.

LEMMA 4.3. *If $\pi(p) = (w_i+w_j)/2$, with $p \in \mathbb{R}^2$, $w_i, w_j \in W$, then $D(w_i, p) = D(w_j, p)$.*

The following technical lemma is the key in constructing a contradiction by showing that, for sufficiently large circles $C(\sigma)$, no point $q \in C(\sigma)$ can be closest to some segment $(w_i, w_j) \in \partial\mathcal{CH}(V)$ that it is behind of ($q \notin H_{ij}^-$); where, as before, the open half space H_{ij}^- is chosen to be the only of the two half spaces on either side of the supporting line l_{ij} of (w_i, w_j) such that $H_{ij}^- \cap V \neq \phi$.

LEMMA 4.4. *There is ρ such that, for any segment $(w_i, w_j) \in \mathcal{B}$, with supporting line l_{ij}, every $p \in H_{ij}^-$ with $\|p\| > \rho$ whose closest point in l_{ij} is $m_p \in \overline{w_i w_j}$ is closer to a site in $V \setminus \{w_i, w_j\}$ than to l_{ij}.*

LEMMA 4.5. *Every continuous function $F : \mathbb{S}^n \to \mathbb{S}^n$ that is not onto has a fixed point.*

LEMMA 4.6. *To every segment in the boundary of $\mathcal{CH}(V)$ corresponds a boundary edge of G. $[B \supseteq \mathcal{B}]$*

Therefore,

THEOREM 4.7. *If the metric Q has bounded ratio of eigenvalues, then the boundary edges of $\bar{G}$ are the same as the boundary edges of the convex hull of V. $[B = \mathcal{B}]$*

COROLLARY 4.8. *If the metric Q has bounded ratio of eigenvalues, then the boundary edges of $\bar{G}$ form a simple, closed, convex polygonal chain.*

The ECE property is sufficient to show that no face in $\bar{G}$ is degenerate, a fact that will be used in the following section.

LEMMA 4.9. *$\bar{G}$ has no degenerate (null area) elements.*

In Thm. 5.9, we show that, even if Voronoi vertices are incident to more than three Voronoi regions (equidistant to more than three sites), every face in $\bar{G}$ is a (strictly) convex polygon, and therefore can be trivially triangulated (say in a fan arrangement). The resulting triangulation has no degenerate elements, and all its triangles satisfy the ECE condition (using the same witness ellipsoid as the convex polygon from which they are triangulated).

5. DUAL OF ORPHAN-FREE DIAGRAM (PART II: INTERIOR)

In this section, we assume that the (topological) boundary of $\bar{G}$ is simple and closed, and prove that $\bar{G}$ must be embedded. The main argument in the proof uses Theorems 2.7 and 4.7, Cor. 4.8, as well as the theory of discrete one-forms on graphs, to show that there are no "edge foldovers" in $\bar{G}$ (edges whose two incident faces are on the same side of its supporting line), and use this to conclude that $\bar{G}$ is embedded (Thm. 5.9). As in Sec. 4, we assume that not all sites are colinear (the simpler colinear case was addressed in Appendix B). We distinguish between the sites $W \subseteq V$ that lie on the boundary of the convex hull, and the remaining, or *interior sites* $(V \setminus W)$.

The following definition, from [11], assumes that, for each edge (v_i, v_j) in G, we distinguish the two opposing half-edges (v_i, v_j) and (v_j, v_i).

DEFINITION 5.1 (GORTLER ET AL. [11]). *A non-vanishing (discrete) one-form ξ is an assignment of a real value $\xi_{ij} \neq 0$ to each half edge (v_i, v_j) in G, such that $\xi_{ji} = -\xi_{ij}$.*

Since $\bar{G}$ has the same structure as G, we can construct a non-vanishing one-form over $\bar{G}$ as follows. Given some unit direction vector $n \in \mathbb{S}^1$ (in coordinates $n = [n_1, n_2]^t$), we assign a real value $z(v) = n^t v$ to each vertex v in $\bar{G}$, and define $\xi_{ij} = z(v_i) - z(v_j)$, which clearly satisfies $\xi_{ji} = -\xi_{ij}$. The one-form, denoted by ξ^n, is non-vanishing if, for all edges $(v_i, v_j) \in E$, it is $\xi_{ij} = n^t(v_i - v_j) \neq 0$. That is, if the direction n is not orthogonal to any edge. The set of edges has cardinality $|E| \leq |V|(|V|-1)/2$, and in particular it is finite. Therefore *almost all* directions $n \in \mathbb{S}^1$ generate a non-vanishing one-form ξ^n.

Since $G = (V, E, F)$ is an planar graph with a well-defined face structure, there is, for each face $f \in F$, a cyclically ordered set ∂f of half-edges round the face. Likewise, for each vertex $v \in V$, the set δv of cyclically ordered (oriented) half-edges emanating from each vertex is well-defined.

DEFINITION 5.2 (GORTLER ET AL. [11]). *Given non-vanishing one-form ξ^n corresponding to $n \in \mathbb{S}^1$, the index of vertex v with respect to ξ^n is $\mathbf{ind}_{\xi^n}(v) = 1 - \mathbf{sc}_{\xi^n}(v)/2$, where $\mathbf{sc}_{\xi^n}(v)$ is the number of sign changes of ξ^n as one visits the half-edges of δv in order.*
The index of face f is $\mathbf{ind}_{\xi^n}(f) = 1 - \mathbf{sc}_{\xi^n}(f)/2$ where $\mathbf{sc}_{\xi^n}(f)$ is the number of sign changes of ξ^n as one visits the half-edges of ∂f in order.

Note that, by definition, it is always $\mathbf{ind}_{\xi^n}(v) \leq 1$. A discrete analog of the Poincaré-Hopf index theorem relates the two indices above:

THEOREM 5.3 (GORTLER ET AL. [11]). *For any non-vanishing one-form ξ^n, it is*

$$\sum_{v \in V} \mathbf{ind}_{\xi^n}(v) + \sum_{f \in F} \mathbf{ind}_{\xi^n}(f) = 2$$

Note that this follows from Theorem 3.5 of [11] because the unbounded, outside face, which is not in G, is assumed in this section to be closed and simple, and therefore would have null index. Note that the machinery from [11] to deal with degenerate cases isn't needed here because vertices, by definition, cannot coincide (V is not a multiset).

The one-forms defined above satisfy the following property.

LEMMA 5.4. *Given a non-vanishing ξ^n, the sum of indices of interior vertices $(V \setminus W)$ of $\bar{G}$ is non-negative.*

The next two lemmas relate the presence of edge foldovers and the ECE property of Definition 2.6 to the indices of vertices in $\bar{G}$.

LEMMA 5.5. *If $\bar{G}$ has an edge foldover, then there is a non-vanishing one-form ξ^n such that $\mathbf{ind}_{\xi^n}(v) < 0$ for some interior vertex $v \in V \setminus W$.*

LEMMA 5.6. *Given $n \in \mathbb{S}^1$ and non-vanishing one-form ξ^n, if $\bar{G}$ has an interior vertex $v \in V \backslash W$ with index $\mathbf{ind}_{\xi^n}(v) = 1$, then there is a face f of G that does not satisfy the empty circum-ellipse property.*

The above provides the necessary tools to prove the following key lemma.

LEMMA 5.7. *$\bar{G}$ has no edge foldovers.*

Finally, the absence of edge foldovers, together with a simple and closed boundary, is sufficient to show that $\bar{G}$ is embedded.

LEMMA 5.8. *If its (topological) boundary is simple and closed, then the straight-line dual $\bar{G}$ of an orphan-free diagram, with vertices incident to at most three sites, is an embedded triangulation.*

As shown in Appendix C, even if there are non-generic Voronoi vertices that are incident to more than three sites, the dual is composed of faces each of which is convex and satisfies the ECE condition (Def. 2.6). Each convex face can be triangulated (e.g. in a fan arrangement) in such a way that individual triangles satisfy the ECE condition with the same witness ellipse as the face from which they are derived. This leads to the following:

THEOREM 5.9. *If its (topological) boundary is simple and closed, then the straight-line dual $\bar{G}$ of an orphan-free diagram is an embedded polygonal mesh with convex faces.*

6. FINAL RESULTS

We can combine Corollary 4.8 and Theorem 5.9 into

THEOREM 6.1. *If the metric Q has bounded ratio of eigenvalues, then the dual of an orphan-free Voronoi diagram with respect to Q is an embedded polygonal mesh with convex faces, and covers the convex hull of the sites.*

This result can be generalized, dropping the bounded anisotropy condition on Q, but at the cost of losing the convex hull property, as shown in Appendix D:

THEOREM 6.2. *The dual of an orphan-free Voronoi diagram is an embedded polygonal mesh with convex faces.*

From the above results, Corollary 2.5, and the definition of the dual, it follows (see Appendix D) that

COROLLARY 6.3. *An orphan-free anisotropic Voronoi diagram is composed of unique (connected) vertices, edges, and faces.*

Finally, we note that Thm. 5.9 can be combined with existing conditions for orphan-freedom [4], resulting in a simple and natural condition for a set of sites to induce an embedded polygonal mesh as the dual of their anisotropic Voronoi diagram:

COROLLARY 6.4. *If V is an asymmetric ϵ-net w.r.t. D, Q a continuous metric with metric variation σ, and $\epsilon\sigma \leq 0.09868$, then the dual of the anisotropic Voronoi diagram of V is an embedded polygonal mesh with convex faces.*

where an asymmetric ϵ-net is simply a weaker form of ϵ-net defined on non-symmetric functions D, which can be computed with the iterative algorithm of [10], and the metric variation σ is a Lipschitz-type condition on Q [4]. The above condition is known to be conservative, and there may be simpler conditions to achieve orphan-freedom. As a practical observation, Du and Wang [7] report orphans to be a rare occurrence in their experiments.

7. PROOF-OF-CONCEPT IMPLEMENTATION

Though not aiming for an efficient implementation, we implemented a simple proof-of-concept that constructs anisotropic Voronoi diagrams like the ones considered in this paper, and their duals (Fig. 3). A closed-form metric, which has bounded ratio of eigenvalues, is discretized on a fine regular grid, and linearly interpolated inside grid elements, resulting in a continuous metric. The sites are generated randomly (Figs. 3(a) and 3(b)), or using a combination of random, and equispaced points forming an asymmetric ϵ-net (remaining figures).

The primal diagram was obtained using front propagation from the sites outwards, until fronts meet at Voronoi edges. The runtime is proportional to the grid size, since every grid-vertex is visited exactly six times (equal to their valence). The implementation does not guarantee the correctness of the diagram unless it *is* orphan-free, and serves to verify the claims of the paper since well-behave-ness of the dual is predicated on that of the primal.

The two main claims of the paper are clearly illustrated in these examples. In all examples, the dual covers the convex hull of the vertices (Thm. 4.7), is a single cover, embedded with straight edges without edge crossings (Thm. 5.9), and has no degenerate faces (Lem. 4.9).

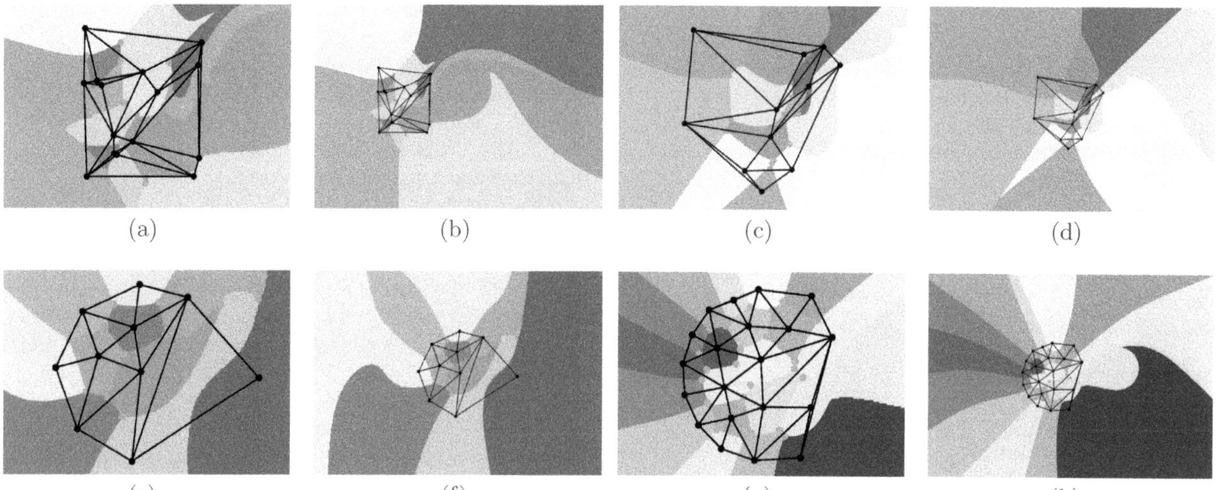

Figure 3: Anisotropic Voronoi diagrams, and their duals generated by our proof-of-concept implementation. Voronoi vertices are marked in red, while dual vertices (sites) and edges are drawn in black.

By focusing on the primal diagrams (second and fourth column), further claims in the paper become apparent, namely that Voronoi regions are simply connected (Lem. 2.3), Voronoi vertices are unique (Cor. 6.3), Voronoi edges are connected (Cor. 2.5), unbounded regions correspond to boundary dual vertices, and unbounded edges of the Voronoi diagram correspond to boundary dual edges.

8. CONCLUSION AND OPEN QUESTIONS

We studied the properties of duals of orphan-free anisotropic Voronoi diagrams, for the purposes of constructing triangulations on the plane. The main result (Theorems 5.9 and 4.7, Cor. 6.4) is that the dual, with straight edges and having the sites as vertices, is embedded and covers the convex hull of the sites, mirroring similar results for ordinary Voronoi diagrams and their duals.

A few, somewhat less important properties are proven, including the fact that every primal region is simply connected, that elements of the primal are unique (Cor. 6.3, but perhaps most significantly that every face in the dual satisfies an *empty circum-ellipse* property that has a direct parallel in the empty circum-circle property of ordinary diagrams, and is the basis for proving that it is embedded with straight edges.

Perhaps the most important outstanding question may be whether these ideas extend to higher dimensions. The results in Secs. 4 and 5, except for Lem. 2.3, can be trivially extended to n dimensions. Sec. 4 has been written only for the two-dimensional case, but a similar construction, and the same argument would work in higher dimensions (Lem. 4.5 being a hint of this). It is the argument in Sec. 5 that becomes problematic. While the ECE property is shown to be sufficient to prevent foldovers in the triangulation, it is not sufficient in higher dimensions. In particular, fixing the boundary to be simple and convex, there are simple arrangements of tetrahedra in $\mathbb{R}^3$ that contain face foldovers but do not break the ECE property. We plan to study these question next.

9. REFERENCES

[1] J.-D. Boissonnat, C. Wormser, and M. Yvinec. Curved Voronoi diagrams. In *Effective Computational Geometry for Curves and Surfaces*, Mathematics + Visualization, pages 67–116. Springer, 2007.

[2] J.-D. Boissonnat, C. Wormser, and M. Yvinec. Locally uniform anisotropic meshing. In *Proceedings of the twenty-fourth annual symposium on Computational geometry*, SCG '08, pages 270–277, New York, NY, USA, 2008. ACM.

[3] G. D. Cañas and S. J. Gortler. Duals of orphan-free anisotropic Voronoi diagrams are triangulations. *arXiv preprint 1102.3673v1, cs.CG*, 2011.

[4] G. D. Cañas and S. J. Gortler. Orphan-free anisotropic Voronoi diagrams. *Discrete & Computational Geometry*, 46(3):526–541, 2011.

[5] K. L. Clarkson. Building triangulations using epsilon-nets. In *STOC 2006: Proceedings of the Thirty-eighth Annual SIGACT Symposium*, 2006.

[6] B. N. Delaunay. Sur la sphère vide. *Bulletin of Academy of Sciences of the USSR*, (6):793–800, 1934.

[7] Q. Du and D. Wang. Anisotropic centroidal Voronoi tessellations and their applications. *SIAM Journal of Scientific Computing*, 26(3):737–761, 2005.

[8] H. Edelsbrunner. Triangulations and meshes in computational geometry. *Acta Numerica*, pages 133–213, 2000.

[9] H. E. F. Aurenhammer. An optimal algorithm for constructing the weighted Voronoi diagram in the plane. *Pattern Recognition 17 (1984), 251–257.*

[10] T. F. Gonzalez. Clustering to minimize the maximum intercluster distance. *Theoretical Computer Science*, 38:293–306, 1985.

[11] S. J. Gortler, C. Gotsman, and D. Thurston. Discrete one-forms on meshes and applications to 3d mesh parameterization. *Computer Aided Geometric Design*, 23:83–112, 2006.

[12] P. M. Gruber. Optimum quantization and its

applications. *Advances in Mathematics*, 186(2):456 – 497, 2004.

[13] F. Labelle and J. R. Shewchuk. Anisotropic Voronoi diagrams and guaranteed-quality anisotropic mesh generation. In *SCG '03: Proceedings of the Nineteenth Annual Symposium on Computational Geometry*, pages 191–200, New York, NY, USA, 2003. ACM.

[14] G. Leibon and D. Letscher. Delaunay triangulations and Voronoi diagrams for riemannian manifolds. In *SCG '00: Proceedings of the Sixteenth Annual Symposium on Computational Geometry*, pages 341–349, New York, NY, USA, 2000. ACM.

APPENDIX

A. PRELIMINARY PROOFS

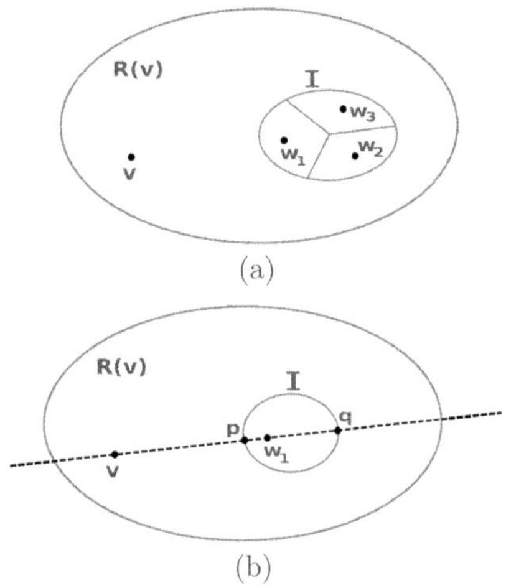

Figure 4: Diagram for the proof of Lem. 2.3.

Lemma 2.1 *Given two sites $v, w \in V$, $v \neq w$, the only point equidistant to v, w in their supporting line is the midpoint $(v+w)/2$.*

PROOF. If p is in the supporting line of v, w, then $p = (1-\lambda)v + \lambda w$, $\lambda \in \mathbb{R}$. p is equidistant to v, w (i.e. $D(v,p) = D(w,p)$) if and only if $\lambda^2(v-w)^t Q_p(v-w) = (1-\lambda)^2(v-w)^t Q_p(v-w)$.

Since Q is positive definite and $v \neq w$, it is $(v-w)^t Q_p(v-w) > 0$, and so the above equality holds iff $\lambda = 1/2$ (p is the midpoint). □

Lemma 2.3 *Every Voronoi region of an orphan-free anisotropic Voronoi diagram in $\mathbb{R}^2$ is simply connected.*

PROOF. A multiply connected region $R(v) \subset \mathbb{R}^2$ is path connected, and is such that there is a map $f : \mathbb{S}^1 \to R(v)$ that cannot be continuously contracted to a point. We can assume f injective (simple), since a non-injective f can always be broken up into injective pieces, at least one of which must be such that it cannot be continuously contracted to a point (or else f would be). By the Jordan curve theorem, f encloses a bounded set B. Since f cannot be continuously contracted to a point, there must be $I \subseteq B$ with $I \subset \mathbb{R}^2 \setminus R(v)$. Since I is a subset of B, and B is bounded, I is bounded.

I is part of the Voronoi diagram, so it must be composed of one or more Voronoi cells. Because the diagram is orphan-free, each cell in the Voronoi diagram contains its generating site: $I = \cup_i R(w_i), i = 1, \ldots, m$ (Fig 4(a)), for some number $m > 0$.

We can now consider what the diagram would look like if we remove, one by one, all of the sites w_i in I, until only one (w_1) is left. The new set of sites is $V' = V \setminus \{w_i : i = 2, \ldots, m\}$. From the definition of Voronoi diagram it is clear that the region corresponding to site w_1 in the new diagram is $R'(w_1) \subseteq I$ (since no point is *strictly* closer to w_1 than to $\{w_i : i = 1, \ldots, m\}$), and $R'(w_1) \neq \phi$ (since, by Lem. 2.2, it is $w_1 \in \textbf{int } R'(w_1)$).

In particular, since $R'(w_1) \neq \phi$, and $R'(w_1)$ is connected, $R'(v)$ is *still* multiply-connected, with $R'(w_1)$ playing the role of I in the new diagram (Fig. 4(b)). We show that this cannot be the case, resulting in a contradiction.

Since I is bounded, then $R'(w_1) \subseteq I$ must be bounded as well. And since w_1 is an interior point of $R'(w_1)$, and v is an interior point of $R'(v)$ (and therefore an interior point of the complement of $R'(w_1)$), then the line passing through v and w_1 must intersect $R'(v)$ at least at two distinct points p and q.

Since the boundary between $R'(v)$ and $R'(w_1)$ is composed of equidistant points to v, w_1, then both p, q must be equidistant to v, w_1. By Lem. 2.1, only one point on the line connecting v, w_1 can be equidistant to them, a contradiction.

□

B. COLINEAR SITES

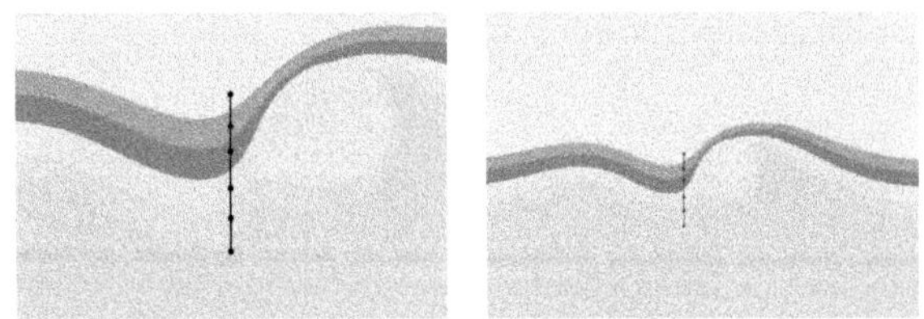

Figure 5: If all sites are colinear, the dual is always a polygonal chain.

If all the sites are colinear then the structure of the Voronoi diagram $\tilde{P}$ is greatly simplified, and always has the form shown in Fig. 5.In particular, $\tilde{P}$ has no vertices aside from p_∞ since vertices are equidistant to three or more sites, and no point $p \in \mathbb{R}^2$ can be equidistant to three colinear sites (since points equidistant to p form an ellipse, and a line intersects an ellipse at most twice).

Consider the set of sites $V = \{w_l : l = 1, \ldots, m\}$ ordered linearly along their supporting line. We shown that the graph dual to P, having V as vertices, has edges $\{(w_l, w_{l+1}) : l = 1, \ldots, m-1\}$.

For every pair of sites w_l, w_{l+1}, because they are consecutive, and by Lem. 2.1, the midpoint $m_{l,l+1} = (w_l + w_{l+1})/2$ is closest and equidistant to w_l, w_{l+1}, and therefore the edge (w_l, w_{l+1}) is in the dual.

We finally show that every dual edge (w_i, w_j) is of the form (w_l, w_{l+1}). Assume otherwise, and therefore that, since w_i, w_j are not consecutive, there is some w_k between them. Because there is a dual edge (w_i, w_j), the corresponding primal edge $\tilde{E}_{w_i,w_j}$ in $\tilde{P}$ is not empty, and therefore there is some point $p_{ij} \in \tilde{E}_{w_i,w_j}$ that is closest and equidistant to w_i, w_j. By the convexity of $D(\cdot, p_{ij})$, and the fact that

$w_k \in \overline{w_i w_j}$, it is $D(w_k, p_{ij}) < D(w_i, p_{ij}) = D(w_j, p_{ij})$, a contradiction. Therefore (w_i, w_j) are consecutive sites.

Since we have characterized the set of edges and vertices, and there are no (interior) faces in the dual, this completely determines the dual when all sites are colinear.

Perhaps interestingly, this means that, in the colinear case, the structure of the dual does not depend on the metric.

C. GENERAL CASE

So far we have assumed that Voronoi vertices are incident (equidistant) to no more than three sites. In general, however, Voronoi vertices may be incident to three or more sites, and the straight-edge dual $\bar{G}$ will be a polygonization, instead of a triangulation (simplicial complex). We show here that, even in this, more general case, the polygonization $\bar{G}$ is embedded, and can be easily triangulated into an embedded simplicial complex. The argument is quite simple and is given in summary.

Theorem 5.9 *If its (topological) boundary is simple and closed, then the straight-line dual $\bar{G}$ of an orphan-free diagram is an embedded polygonal mesh with convex faces.*

PROOF. First assume that only Voronoi vertex c is incident to more than three sites $\{v_i : i = 1 \ldots m\}$, $m > 3$, ordered in, say, clockwise order around c (which is possible since they are equidistant to c and therefore lie at the boundary of the ellipse $\theta_c(v_1)$). Note that, since the $\{v_i : i = 1 \ldots m\}$ are equidistant to c, then any polygon (or triangle) connecting any of the $\{v_i\}$ satisfies the empty circum-ellipse property (with empty circum-ellipse $\theta_c(v_1)$). The Voronoi regions $\{R_i : i = 1, \ldots, m\}$, corresponding to those sites, are the only ones incident to c. We can order these regions in clockwise order around c: $\{R_{i_k} : k \in K\}$ (for some index sequence K with $k \in K$ satisfying $k \in \{1, \ldots, m\}$), and therefore the polygon Π dual to c will connect vertices $\{v_{i_k} : k \in K\}$, in that order. Now choose any v_j in Π and triangulate Π in a fan arrangement centered at v_j. As pointed out above, both Π and every triangle in the triangulation of Π satisfies the empty circum-ellipse property. Therefore the new, triangulated $\bar{G}$ (which replaces Π by its triangulation), satisfies all the conditions needed for Theorems 5.9 and 4.7 (as well as Lem. 4.9) to hold, and in particular, it is embedded.

The claim now easily follows: any closed polygon with vertices $\{v_{i_k} : k \in K\}$, such that all of its fan-arranged triangulations centered at every one of its vertices is embedded, must be convex (given a polygon with non-convex vertex w, triangulate it in a fan centered at a vertex incident to w: this triangulation is not embedded). Lastly, since vertices $\{v_i : i = 1 \ldots m\}$ are equidistant from c, the only polygon connecting them that is also convex is the one where vertices are arranged in clockwise (counter-clockwise) order around c.

The argument has been made for a single Voronoi vertex incident to more than three sites, but the same argument applies to diagrams in which there are more such vertices (the key being that the triangulations of each dual polygon are independent of one another). □

D. BOUNDED ANISOTROPY

We show here that we may drop the bounded anisotropy condition from Thm. 6.1, at the expense of loosing the property of covering the convex hull of the sites; thus concluding that the dual of an orphan-free anisotropic diagram is an embedded polygonal mesh with convex faces. The proof proceeds by reducing this case to that of Thm. 6.1. Because of its simplicity, it is given here in summary.

Assume given a continuous metric Q defined over $\mathbb{R}^2$, and a set V of sites forming an orphan-free anisotropic Voronoi diagram V_Q. For now, assume that the set of Voronoi vertices is finite, and therefore bounded. Since there is $\rho > 0$ such that all Voronoi vertices are inside the origin-centered ball $B(0;\rho)$ of radius ρ, we can construct a new metric $Q'(p) = Q(p) \cdot (1 - \lambda(p)) + I \cdot \lambda(p)$, where

$$\lambda(p) = \begin{cases} 0, & \text{if } \|p\| < \rho \\ \|p\| - \rho, & \text{if } \rho \le \|p\| \le 2\rho \\ 1, & \text{if } \|p\| > 2\rho \end{cases}$$

Clearly, Q' is continuous, since both Q and λ are. Since it is $Q' = I$ outside of $B(0;2\rho)$, and $B(0;2\rho)$ is compact, then Q' has bounded anisotropy (ratio of eigenvalues). We may then apply Thm. 6.1 to conclude that the dual $\bar{G}_{Q'}$ of $V_{Q'}$ is an embedded polygonal mesh with convex faces. Since $B(0;\rho)$ includes all Voronoi vertices of V_Q, and it is $Q' = Q$ inside $B(0;\rho)$, then all Voronoi vertices in V_Q are also in $V_{Q'}$. Therefore, by duality, all faces of $\bar{G}_Q$ are also in $\bar{G}_{Q'}$. Finally, since $\bar{G}_{Q'}$ is embedded with convex faces, and removing faces from a polygonal mesh preserves both properties, then $\bar{G}_Q$ is also embedded with convex faces.

The only assumption we have made is that the set of Voronoi vertices of V_Q is finite, which can be justified as follows.

First, we show that Voronoi vertices cannot be isolated, and are always incident to some Voronoi edge. Consider a Voronoi vertex c_{ijk} closest and equidistant to sites v_i, v_j, v_k (the reasoning for vertices equidistant to more sites is analogous). Since Q, and therefore D, is continuous, there is a (possibly small) closed ball $B(c_{ijk}, \epsilon)$ centered at c_{ijk} where all points are *strictly* closer to v_i, v_j, v_k than to any other site, and therefore in $B(c_{ijk}, \epsilon)$, the Voronoi diagram of V is the same as that of $\{v_i, v_j, v_k\}$. We consider the Voronoi diagrams $\{v_i, v_j\}_Q$ and $\{v_j, v_k\}_Q$ of $\{v_i, v_j\}$ and $\{v_j, v_k\}$ (restricted to $B(c_{ijk}, \epsilon)$), respectively. By Corollary 2.5, the only Voronoi edge of $\{v_i, v_j\}_Q$ is connected, and likewise for $\{v_j, v_k\}_Q$. Since c_{ijk} is in their intersection, then it must be incident to both, and therefore it is not isolated.

By Corollary 2.5, the edges (connected sets of points equidistant to two given sites) of V_Q are unique (thus finite since V is finite), and by the induction argument of Sec. 2, each edge connects two Voronoi vertices. If there are n_e Voronoi edges then, since Voronoi vertices are always incident to some Voronoi edge, and each edge connects two vertices, there cannot be more than $2n_e$ Voronoi vertices. In particular, the number of Voronoi vertices is finite. This concludes the proof.

Finally, note that, from the way we have defined the dual $\bar{G}_Q$, and the fact that it is embedded, we can conclude that Voronoi vertices are unique (otherwise, multiple Voronoi vertices would result in coincident dual polygons, which would contradict the fact that $\bar{G}_Q$ is embedded). This, together with Corollary 2.5, and the orphan-freedom assumption means that, as stated in Corollary 6.3, orphan-freedom is sufficient to ensure that all the elements (vertices, edges, faces) of the primal diagram are unique.

Stability of Delaunay-type Structures for Manifolds

[Extended Abstract]

Jean-Daniel Boissonnat
INRIA, Geometrica
Sophia-Antipolis, France
Jean-Daniel.Boissonnat@inria.fr

Ramsay Dyer
INRIA, Geometrica
2004 route des Lucioles
06902 Sophia-Antipolis
Ramsay.Dyer@inria.fr

Arijit Ghosh
INRIA, Geometrica
2004 route des Lucioles
Sophia-Antipolis, France
arijitiitkgpster@gmail.com

ABSTRACT

We introduce a parametrized notion of genericity for Delaunay triangulations which, in particular, implies that the Delaunay simplices of δ-generic point sets are thick. Equipped with this notion, we study the stability of Delaunay triangulations under perturbations of the metric and of the vertex positions. We then show that, for any sufficiently regular submanifold of Euclidean space, and appropriate ϵ and δ, any sample set which meets a localized δ-generic ϵ-dense sampling criteria yields a manifold intrinsic Delaunay complex which is equal to the restricted Delaunay complex.

Categories and Subject Descriptors

I.5.1 [**Pattern Recognition**]: Geometric Models

General Terms

Theory

Keywords

Manifold reconstruction, intrinsic Delaunay complex, restricted Delaunay complex, sampling criteria

1. INTRODUCTION

One of the central properties of Delaunay complexes, which was demonstrated when they were introduced [Del34], is that under very mild assumptions they triangulate Euclidean space. This paper addresses issues that arise when the Delaunay paradigm is employed for triangulating non-Euclidean manifolds whose dimension may exceed two.

For a submanifold of Euclidean space, the restricted Delaunay complex [ES97], which is defined by the ambient metric restricted to the submanifold, was employed by Cheng et al. [CDR05] as the basis for a triangulation. However, it was found that sampling density alone was insufficient to ensure a triangulation, and manipulations of the complex were employed.

In an earlier work, Leibon and Letscher [LL00] announced sampling density conditions which would ensure that the Delaunay complex defined by the intrinsic metric of the manifold was a triangulation. In fact, the stated result is incorrect [BDG12b]: sampling density alone is insufficient to guarantee an intrinsic Delaunay triangulation. Topological defects can arise when the vertices lie too close to a degenerate or "quasi-cospherical" configuration.

We address this problem with the introduction of a parameterized notion of genericity for Delaunay complexes. Our interest in the intrinsic Delaunay complex stems from its close relationship with other Delaunay-like structures that have been proposed in the context of non-homogeneous metrics. For example, anisotropic Voronoi diagrams [LS03] and anisotropic Delaunay triangulations emerge as natural structures when we want to mesh a domain of $\mathbb{R}^m$ while respecting a given metric tensor field.

This paper builds over preliminary results on anisotropic Delaunay meshes [BWY11] and manifold reconstruction using the tangential Delaunay complex [BG11]. The central idea in both cases is to define Euclidean Delaunay triangulations locally and to glue these local triangulations together by removing inconsistencies between them. We view the inconsistencies as arising from instability in the Delaunay triangulations, and provide explicit bounds on the stability with respect to the genericity parameter.

The stability of Delaunay triangulations has not previously been studied in this way. Related work can be found in the context of kinetic data structures [AGG$^+$10] or in the context of robust computation [BS04], and in particular, the concept of protection we introduce in Section 3 is embodied in the guarded insphere predicate which has been employed in a 2D Delaunay triangulation algorithm [FKMS05].

After presenting background material in Section 2, we introduce the concept of δ-generic point sets for Euclidean Delaunay triangulations in Section 3. We show that Delaunay simplices of δ-generic point sets are thick; they satisfy a quality bound. Then in Section 4 we quantify how δ leads to robustness in the Delaunay triangulation when either the points or the metric are perturbed. The primary challenge is bounding the displacement of simplex circumcentres. Finally, we use these results in Section 5 to demonstrate conditions under which the intrinsic Delaunay complex and the restricted Delaunay complex coincide and are manifold. In this extended abstract, we have omitted proofs for some of the lemmas. In all cases, the missing proofs are routine exercises, and they can be found in the full version of the paper [BDG12b].

SCG'12, June 17–20, 2012, Chapel Hill, North Carolina, USA.

2. BACKGROUND

Within the context of the standard m-dimensional Euclidean space $\mathbb{R}^m$, when distances are determined by the standard norm, $\|\cdot\|$, we use the following conventions. The distance between a point p and a set $X \subset \mathbb{R}^m$, is the infimum of the distances between p and the points of X, and is denoted $d_{\mathbb{R}^m}(p, X)$. We refer to the distance between two points a and b as $\|b - a\|$ or $d_{\mathbb{R}^m}(a, b)$ as convenient. A ball $B_{\mathbb{R}^m}(c; r) = \{x \mid \|x - c\| < r\}$ is open, and $\overline{B}_{\mathbb{R}^m}(c; r)$ is its topological closure. Generally, we denote the topological closure of a set X by $\overline{X}$, the interior by $\operatorname{int} X$, and the boundary by ∂X. The convex hull is denoted $\operatorname{ch}(X)$, and the affine hull is $\operatorname{aff}(X)$.

We will make use of other metrics besides the Euclidean one. A generic metric is denoted d, and the associated open and closed balls are $B(c; r)$, and $\overline{B}(c; r)$. If a specific metric is intended, it will be indicated by a subscript, for example in Section 5 we introduce $d_{\mathcal{M}}$, the intrinsic metric on a manifold $\mathcal{M}$, which has associated balls $B_{\mathcal{M}}(c; r)$.

If A is a $k \times j$ matrix, we denote its ith singular value by $s_i(A)$. We use the operator norm $\|A\| = s_1(A)$, and employ the following standard observation:

Lemma 1. *If $\eta > 0$ is an upper bound on the norms of the columns of A, then $\|A\| \leq \sqrt{j}\eta$.*

We will also be interested in obtaining a lower bound on the smallest singular value, for which the following observation is useful:

Lemma 2. *If A is a $k \times j$ matrix of rank $j \leq k$, then the* pseudo inverse *$A^\dagger = (A^\mathsf{T} A)^{-1} A^\mathsf{T}$ is the unique left inverse of A whose kernel is the orthogonal complement of the column space of A. Furthermore, $s_i(A^\dagger) = s_{j-i+1}(A)^{-1}$.*

If U and V are vector subspaces of $\mathbb{R}^m$, with $\dim U \leq \dim V$, the *angle* between them is defined by

$$\cos \angle(U, V) = \inf_{u \in U} \sup_{v \in V} \frac{u^\mathsf{T} v}{\|u\| \, \|v\|}.$$

This is the largest principal angle between U and V. The angle between affine subspaces K and H is defined as the angle between the corresponding parallel vector subspaces.

2.1 Sampling parameters and perturbations

The structures of interest will be built from a finite set $\mathsf{P} \subset \mathbb{R}^m$, which we consider to be a set of *sample points*. If $D \subset \mathbb{R}^m$ is a bounded set, then P is an *ϵ-sample set* for D if $d_{\mathbb{R}^m}(x, \mathsf{P}) < \epsilon$ for all $x \in \overline{D}$. We say that ϵ is a *sampling radius* for D satisfied by P. If no domain D is specified, we say P is an ϵ-sample set if $d(x, \mathsf{P} \cup \partial\operatorname{ch}(\mathsf{P})) < \epsilon$ for all $x \in \operatorname{ch}(\mathsf{P})$. The set P is *λ-sparse* if $d_{\mathbb{R}^m}(p, q) > \lambda$ for all $p, q \in \mathsf{P}$. We usually assume that the sparsity of a ϵ-sample set is proportional to ϵ, thus: $\lambda = \mu_0 \epsilon$.

We consider a perturbation of the points $\mathsf{P} \subset \mathbb{R}^m$ given by a function $\zeta : \mathsf{P} \to \mathbb{R}^m$. If ζ is such that $d_{\mathbb{R}^m}(p, \zeta(p)) \leq \rho$, we say that ζ is a *ρ-perturbation*. As a notational convenience, we frequently define $\tilde{\mathsf{P}} = \zeta(\mathsf{P})$, and let $\tilde{p}$ represent $\zeta(p) \in \tilde{\mathsf{P}}$. We will only be considering ρ-perturbations where ρ is less than half the sparsity of P, so $\zeta : \mathsf{P} \to \tilde{\mathsf{P}}$ is a bijection.

Points in P which are not on the boundary of $\operatorname{ch}(\mathsf{P})$ are *interior points* of P.

Lemma 3. *Suppose P is an ϵ-sample set, and $\zeta : \mathsf{P} \to \tilde{\mathsf{P}}$ is a ρ-perturbation with $2\rho \leq \epsilon$. If point $p \in \mathsf{P}$ satisfies $d_{\mathbb{R}^m}(p, \partial\operatorname{ch}(\mathsf{P})) \geq 3\epsilon$, then $\tilde{p} = \zeta(p)$ is an interior point of $\tilde{\mathsf{P}}$.*

Proof. Let $S = \partial B$ be the bounding sphere for $B = B_{\mathbb{R}^m}(\tilde{p}; 3\epsilon/2)$. Then $d_{\mathbb{R}^m}(p, S) \geq \epsilon$ and for any $x \in S$, $d_{\mathbb{R}^m}(x, \partial\operatorname{ch}(\mathsf{P})) \geq \epsilon$. Thus the sampling assumption ensures that for every point $x \in S$, there is a point $q \in \mathsf{P}$ with $p \neq q$ and $d_{\mathbb{R}^m}(x, q) < \epsilon$. It follows that $d_{\mathbb{R}^m}(x, \zeta(q)) < 3\epsilon/2$, and thus that $\tilde{p}$ is not the closest point in $\tilde{\mathsf{P}}$ for any point on S.

Thus $\tilde{p}$ cannot belong to $\partial\operatorname{ch}(\tilde{\mathsf{P}})$. Indeed if $\tilde{p} \in \partial\operatorname{ch}(\tilde{\mathsf{P}})$, then take a unit vector v in an outward direction orthogonal to a closed half-space supporting $\operatorname{ch}(\tilde{\mathsf{P}})$ at $\tilde{p}$. The ray from $\tilde{p}$ defined by v must intersect S at some point y, and $\tilde{p}$ would be the closest point in $\tilde{\mathsf{P}}$ to y, a contradiction. $\square$

2.2 Simplices

Given a set of $j + 1$ points $\{p_0, \ldots, p_j\} \subset \mathsf{P} \subset \mathbb{R}^m$, a (geometric) *$j$-simplex* $\sigma = [p_0, \ldots, p_j]$ is defined by the convex hull: $\sigma = \operatorname{ch}(\{p_0, \ldots, p_j\})$. The points p_i are the *vertices* of σ. Any subset $\{p_{i_0}, \ldots, p_{i_k}\}$ of $\{p_0, \ldots, p_j\}$ defines a k-simplex τ which we call a *face* of σ. We write $\tau \leq \sigma$ if τ is a face of σ, and $\tau < \sigma$ if τ is a *proper face* of σ, i.e., if the vertices of τ are a proper subset of the vertices of σ.

The *boundary* of σ, is the union of its proper faces: $\partial\sigma = \bigcup_{\tau < \sigma} \tau$. In general this is distinct from the topological boundary defined above, but we denote it with the same symbol. The *interior* of σ is $\operatorname{int}\sigma = \sigma \setminus \partial\sigma$. Again this is generally different from the topological interior. Other geometric properties of σ include its diameter (its longest edge), $\Delta(\sigma)$, and its shortest edge, $L(\sigma)$.

For any vertex $p \in \sigma$, the *face oppposite* p is the face determined by the other vertices of σ, and is denoted σ_p. If τ is a j-simplex, and p is not a vertex of τ, we may construct a $(j+1)$-simplex $\sigma = p * \tau$, called the *join* of p and τ. It is the simplex defined by p and the vertices of τ, i.e., $\tau = \sigma_p$.

Our definition of a simplex has made an important departure from standard convention: we do not demand that the vertices of a simplex be affinely independent. A j-simplex σ is a *degenerate simplex* if $\dim \operatorname{aff}(\sigma) < j$. If we wish to emphasise that a simplex is a j-simplex, we write j as a superscript: σ^j; but this always refers to the *combinatorial* dimension of the simplex.

If σ is non-degenerate, then it has a *circumcentre*, $C(\sigma)$, which is the centre of the smallest circumscribing ball for σ. The radius of this ball is the *circumradius* of σ, denoted $R(\sigma)$. In the context of the Euclidean Delaunay complexes we will work with, the degenerate simplices we may encounter also have these properties. The ratio of the circumradius to the shortest edge is denoted $\Phi(\sigma) = R(\sigma)/L(\sigma)$. We will make use of the affine space $N(\sigma)$ composed of the centres of the balls that circumscribe σ. This space is orthogonal to $\operatorname{aff}(\sigma)$ and intersects it at the circumcentre of σ. Its dimension is $m - \dim \operatorname{aff}(\sigma)$.

The *altitude* of p in σ is $D(p, \sigma) = d_{\mathbb{R}^m}(p, \operatorname{aff}(\sigma_p))$. A poorly-shaped simplex can be characterized by the existence of a relatively small altitude. The *thickness* of a j-simplex σ is the dimensionless quantity given by $\Upsilon(\sigma) = 1$ if $j = 0$, and $\Upsilon(\sigma) = \min_{p \in \sigma} \frac{D(p,\sigma)}{j\Delta(\sigma)}$ otherwise. We say that σ is Υ_0-thick, if $\Upsilon(\sigma) \geq \Upsilon_0$. If σ is Υ_0-thick, then so are all of its faces. Indeed if $\tau \leq \sigma$, then the smallest altitude in τ cannot be smaller than that of σ, and also $\Delta(\tau) \leq \Delta(\sigma)$.

Other parameters such as the volume [Whi57], or the radius of the largest contained ball centred at the barycentre [Mun68], can be used to characterize simplex quality. We find a direct bound on the altitudes to be more convenient, in part due to the following consequence of Lemma 2:

LEMMA 4 (THICKNESS AND SINGULAR VALUE). *Let $\sigma = [p_0, \ldots, p_j]$ be a non-degenerate j-simplex in $\mathbb{R}^m$, with $j > 0$, and let P be the $m \times j$ matrix whose i^{th} column is $p_i - p_0$. Then $s_j(P) \geq \sqrt{j}\Upsilon(\sigma)\Delta(\sigma)$.*

PROOF. Let w_i^T be the ith row of $P^\dagger$. Then w_i belongs to the column space of P, and it is orthogonal to all $(p_{i'} - p_0)$ for $i' \neq i$. Let $u_i = w_i / \|w_i\|$. It follows from the definition that $u_i^\mathsf{T}(p_i - p_0) = D(p_i, \sigma)$. Thus $w_i = D(p_i, \sigma)^{-1} u_i$. Since $s_i(A^\mathsf{T}) = s_i(A)$ for any matrix A, we have $s_1(P^\dagger) \leq \sqrt{j} \max_{1 \leq i \leq j} D(p_i, \sigma)^{-1}$, by Lemma 1. Thus Lemma 2 yields $s_j(P) \geq \frac{1}{\sqrt{j}} \min_{1 \leq i \leq j} D(p_i, \sigma) = \sqrt{j}\Upsilon(\sigma)\Delta(\sigma)$. □

The proof of Lemma 4 shows that the pseudoinverse of P has a natural geometric interpretation in terms of the altitudes of σ, and thus the altitudes provide a convenient lower bound on $s_j(P)$. By Lemma 1, $s_1(P) \leq \sqrt{j}\Delta(\sigma)$, and thus $\Upsilon(\sigma) \leq \frac{s_j(P)}{s_1(P)}$. In other words, $\Upsilon(\sigma)^{-1}$ provides a convenient upper bound on the *condition number* of P.

Whitney [Whi57, p. 127] proved that the affine hull of a thick simplex makes a small angle with any hyperplane which lies near all the vertices of the simplex. Employing Lemma 4 in the proof of Whitney's Lemma yields a simpler proof [BDGO11, Lemma 2.6] and a sharper result:

LEMMA 5 (WHITNEY ANGLE BOUND). *Suppose σ is a j-simplex whose vertices all lie within a distance η from a hyperplane, $H \subset \mathbb{R}^m$. Then $\sin\angle(\text{aff}(\sigma), H) \leq \frac{2\eta}{\Upsilon(\sigma)\Delta(\sigma)}$.*

2.3 Complexes

The Delaunay complexes we study are abstract simplicial complexes, but their simplices carry a canonical geometry induced from the inclusion map $\iota : \mathsf{P} \hookrightarrow \mathbb{R}^m$. (We assume ι is injective on P, and so do not distinguish between P and $\iota(\mathsf{P})$.) To each abstract simplex $\sigma \in K$, we have an associated geometric simplex $\text{ch}(\iota(\sigma))$, and normally when we write $\sigma \in K$, we are referring to this geometric object. Occasionally, when it is convenient to emphasise a distinction, we will write $\iota(\sigma)$ instead of σ. Thus we view such a K as a set of simplices in $\mathbb{R}^m$, and we refer to it as a *complex*, but it is not generally a (geometric) simplicial complex.

The *carrier* of an abstract complex K is the underlying topological space $|K|$, associated with a geometric realization of K. For our complexes, the inclusion map ι induces a continous map $\iota : |K| \to \mathbb{R}^m$, defined by barycentric interpolation on each simplex. If this map is injective, we say that K is *embedded*. In this case ι also defines a geometric realization of K, and we may identify the carrier of K with the image of ι.

A subset $K' \subset K$ is a *subcomplex* of K if it is also a complex. The *star* of a subcomplex $K' \subseteq K$ is the subcomplex generated by the simplices incident to K'. I.e., it is all the simplices that share a face with a simplex of K', plus all the faces of such simplices. This is a departure from a common usage of this same term in the topology literature. The star of K' is denoted $\text{star}(K')$ when there is no risk of ambiguity, otherwise we also specify the parent complex, as in $\text{star}(K'; K)$.

A *triangulation* of $\mathsf{P} \subset \mathbb{R}^m$ is an embedded complex K with vertices P such that $|K| = \text{ch}(\mathsf{P})$.

DEFINITION 6 (TRIANGULATION AT A POINT). *A complex K is a* triangulation at $p \in \mathbb{R}^m$ *if:*

- *p is a vertex of K.*
- *$\text{star}(p)$ is embedded.*
- *p lies in $\text{int}\,|\text{star}(p)|$.*
- *For all $\tau \in K$, and $\sigma \in \text{star}(p)$, if $\text{int}\,\tau \cap \sigma \neq \emptyset$, then $\tau \in \text{star}(p)$.*

A complex K is a *j-manifold complex* if the star of every vertex is isomorphic to the star of a triangulation of $\mathbb{R}^j$.

If σ is a simplex with vertices in P, then any map $\zeta : \mathsf{P} \to \tilde{\mathsf{P}} \subset \mathbb{R}^m$ defines a simplex $\zeta(\sigma)$ whose verticies in $\tilde{\mathsf{P}}$ are the images of vertices of σ. If K is a complex on P, and $\tilde{K}$ is a complex on $\tilde{\mathsf{P}}$, then ζ induces a *simplicial map* $K \to \tilde{K}$ if $\zeta(\sigma) \in \tilde{K}$ for every $\sigma \in K$. We denote this map by the same symbol, ζ. We are interested in the case when ζ is an *isomorphism*, which means it establishes a bijection between K and $\tilde{K}$. We then say that K and $\tilde{K}$ are *isomorphic*, and write $K \cong \tilde{K}$, or $K \overset{\zeta}{\cong} \tilde{K}$ if we wish to emphasise that the correspondence is given by ζ. We observe [BDG12b] the following local version of a standard result:

LEMMA 7. *Suppose K is a complex with vertices $\mathsf{P} \subset \mathbb{R}^m$, and $\tilde{K}$ a complex with vertices $\tilde{\mathsf{P}} \subset \mathbb{R}^m$. Suppose also that K is a triangulation at $p \in \mathsf{P}$, and that $\zeta : \mathsf{P} \to \tilde{\mathsf{P}}$ induces an injective simplicial map $\text{star}(p) \to \text{star}(\zeta(p))$. If $\tilde{K}$ is a triangulation at $\zeta(p)$, then $\zeta(\text{star}(p)) = \text{star}(\zeta(p))$.*

3. PARAMETERIZED GENERICITY

In this section we examine the Delaunay complex of $\mathsf{P} \subset \mathbb{R}^m$, taking the view that poorly-shaped simplices arise from almost degenerate configurations of points. We introduce the concept of a protected Delaunay ball, which leads to a parameterized definition of genericity. We then show that a lower bound on the protection of the maximal simplices yields a lower bound on their thickness.

3.1 The Delaunay complex

An *empty ball* is one that contains no point from P.

DEFINITION 8 (DELAUNAY COMPLEX). *A* Delaunay ball *is a maximal empty ball. Specifically, $B = B_{\mathbb{R}^m}(x; r)$ is a Delaunay ball if any empty ball centred at x is contained in B. A simplex σ is a* Delaunay simplex, *if there exists some Delaunay ball B such that the vertices of σ belong to $\partial B \cap \mathsf{P}$. The* Delaunay complex *is the set of Delaunay simplices, and is denoted* $\text{Del}(\mathsf{P})$.

The Delaunay complex has the combinatorial structure of an abstract simplicial complex, but $\text{Del}(\mathsf{P})$ is embedded only when P satisfies appropriate genericity requirements, as discussed in Section 3.2. Otherwise, $\text{Del}(\mathsf{P})$ contains degenerate simplices. We make here some observations that are not dependent on assumptions of genericity.

The union of the Delaunay simplices is $\text{ch}(\mathsf{P})$. A simplex $\sigma \in \text{Del}(\mathsf{P})$ is a *boundary simplex* if all its vertices lie on $\partial\text{ch}(P)$. If none of the vertices of σ lie on $\partial\text{ch}(P)$, then it is an *interior simplex*. We observe [BDG12b]:

LEMMA 9 (MAXIMAL SIMPLICES). *If $\text{aff}(\mathsf{P}) = \mathbb{R}^m$, then every Delaunay j-simplex, σ, is a face of a Delaunay simplex σ' with $\dim \text{aff}(\sigma') = m$. In particular, if $j \leq m$, then σ is a face of a Delaunay m-simplex. If σ is not a boundary simplex, and $\dim \text{aff}(\sigma) < m$, then there are at least two Delaunay $(j+1)$-simplices that have σ as a face.*

Lemma 9 gives rise to the following observation, which plays an important role in Section 3.3, where we argue that protecting the Delaunay m-simplices yields a thickness bound on the simplices.

LEMMA 10 (SEPARATION). *If $\tau \in \mathrm{Del}(\mathsf{P})$ is a j-simplex that is not a boundary simplex, and $q \in \mathsf{P} \setminus \tau$, then there is a Delaunay m-simplex σ^m which has τ as a face, but does not include q.*

PROOF. Assume $j < m$, for otherwise there is nothing to prove. If $\sigma = q * \tau$ is not Delaunay, the assertion follows from the first part of Lemma 9. Assume σ is Delaunay and let $\tilde{\sigma}^m$ be a Delaunay m-simplex that has σ as a face. Thus $\tilde{\sigma}^m = q * \sigma^{m-1}$ for some Delaunay $(m-1)$-simplex, σ^{m-1}. Since $\tau \leq \sigma^{m-1}$ does not belong to the boundary of $\mathrm{ch}(\mathsf{P})$, neither does σ^{m-1}, so by the second part of Lemma 9, there is another Delaunay m-simplex σ^m that has σ^{m-1} (and therefore τ) as a face. Since σ^m is distinct from $\tilde{\sigma}^m$, it does not have q as a vertex. □

3.1.1 The Delaunay complex in other metrics

We will also consider the Delaunay complex defined with respect to a metric d on $\mathbb{R}^m$ which differs from the Euclidean one. Specifically, if $\mathsf{P} \subset U \subset \mathbb{R}^m$ and $d : U \times U \to \mathbb{R}$ is a metric, then we define the Delaunay complex $\mathrm{Del}_d(\mathsf{P})$ with respect to the metric d.

The definitions are exactly analogous to the Euclidean case: A Delaunay ball is a maximal empty ball $B(x; r)$ in the metric d. The resulting Delaunay complex $\mathrm{Del}_d(\mathsf{P})$ consists of all the simplices which are circumscribed by a Delaunay ball with respect to the metric d. The simplices of $\mathrm{Del}_d(\mathsf{P})$ are, possibly degenerate, geometric simplices in $\mathbb{R}^m$. As for $\mathrm{Del}(\mathsf{P})$, $\mathrm{Del}_d(\mathsf{P})$ has the combinatorial structure of an abstract simplicial complex, but unlike $\mathrm{Del}(\mathsf{P})$, $\mathrm{Del}_d(\mathsf{P})$ may fail to be embedded even when there are no degenerate simplices.

3.1.2 Deep interior simplices

In order to focus on the local properties of Delaunay triangulations, we wish to identify a subcomplex of the interior simplices of $\mathrm{Del}(\mathsf{P})$ consisting of those simplices whose neighbour simplices are also all interior simplices with small circumradius. An interior simplex near the boundary of $\mathrm{ch}(\mathsf{P})$ does not necessarily have its circumradius constrained by the sampling radius. However, we have the following:

LEMMA 11. *If P is an ϵ-sample set, and $\sigma \in \mathrm{Del}(\mathsf{P})$ has a vertex p such that $d_{\mathbb{R}^m}(p, \partial\mathrm{ch}(\mathsf{P})) \geq 2\epsilon$, then $R(\sigma) < \epsilon$ and σ is an interior simplex.*

PROOF. Let $B_{\mathbb{R}^m}(c; r)$ be a Delaunay ball for σ. We will show $r < \epsilon$. Suppose to the contrary. Let x be the point on $[c, p]$ such that $d_{\mathbb{R}^m}(p, x) = \epsilon$. Then p is the closest point in P to x, and so the sampling criteria imply that $d_{\mathbb{R}^m}(x, \partial\mathrm{ch}(\mathsf{P})) < \epsilon$. But then $d_{\mathbb{R}^m}(p, \partial\mathrm{ch}(\mathsf{P})) \leq d_{\mathbb{R}^m}(p, x) + d_{\mathbb{R}^m}(x, \partial\mathrm{ch}(\mathsf{P})) < 2\epsilon$, contradicting the hypothesis on p.

Thus $r < \epsilon$, and it follows that σ is an interior simplex because if $q \in \sigma$, then $d_{\mathbb{R}^m}(p, q) \leq 2r < d_{\mathbb{R}^m}(p, \partial\mathrm{ch}(\mathsf{P}))$. □

DEFINITION 12. *Suppose $\mathsf{P} \subset \mathbb{R}^m$ is an ϵ-sample set. The subset $\mathsf{P}_I \subset \mathsf{P}$ consisting of all $p \in \mathsf{P}$ with $d_{\mathbb{R}^m}(p, \partial\mathrm{ch}(\mathsf{P})) \geq 4\epsilon$ is the set of* safe interior points. *The* deep interior simplices *are the simplices in* $\mathrm{star}(P_I; \mathrm{Del}(\mathsf{P}))$.

By Lemma 11, all the neighbour simplices of deep interior simplices will be interior simplices with small circumradius.

3.2 Protection

A Delaunay simplex σ is *$\tilde{\delta}$-protected* if it has a Delaunay ball B such that $d_{\mathbb{R}^m}(q, \partial B) > \tilde{\delta}$ for all $q \in \mathsf{P} \setminus \sigma$. We say that B is a $\tilde{\delta}$-protected Delaunay ball for σ. If $\tau < \sigma$, then B is also a Delaunay ball for τ, but it cannot be a $\tilde{\delta}$-protected Delaunay ball for τ. We say that σ is *protected* to mean that it is $\tilde{\delta}$-protected for some unspecified $\tilde{\delta} > 0$.

DEFINITION 13 (δ-GENERIC). *A finite set of points $\mathsf{P} \subset \mathbb{R}^m$ is δ-generic if all the Delaunay m-simplices are δ-protected. The set P is simply* generic *if it is δ-generic for some unspecified $\delta > 0$.*

In his seminal work, Delaunay [Del34] demonstrated that if there is no empty ball with $m + 2$ points from P on its boundary, then $\mathrm{Del}(\mathsf{P})$ is realized as a simplicial complex in $\mathbb{R}^m$. In other words it is a triangulation, the *Delaunay triangulation*. If P is generic according to Definition 13, then Delaunay's criterion will be met. This is obvious if there are no degenerate m-simplices, and Definition 13 ensures that a degenerate m-simplex cannot exist in $\mathrm{Del}(\mathsf{P})$: If σ^m is degenerate, then by Lemma 9, there is a simplex σ with $\mathrm{aff}(\sigma) = \mathbb{R}^m$, and $\sigma^m < \sigma$. An affinely independent set of $m + 1$ vertices from σ defines a non degenerate m-simplex $\tilde{\sigma}^m$, and since its unique circumball is also a Delaunay ball for σ, it cannot be protected, a contradiction.

In particular, if P is generic if and only if there are no Delaunay simplices with dimension higher than m. We can say more. There are no degenerate Delaunay simplices. This can be inferred directly from Delaunay's result [Del34], but is also easily established directly. In Section 3.3 we will quantify this observation.

Delaunay avoided boundary complications by assuming a periodic point set, but we will be working in a local setting where we cannot assume that our conditions apply globally. This is not a problem because Delaunay's argument applies locally [BDG12b]:

LEMMA 14 (LOCAL DELAUNAY). *If $p \in \mathsf{P}$ is an interior point, and the Delaunay m-simplices incident to p are protected, then* $\mathrm{Del}(\mathsf{P})$ *is a triangulation at p.*

For technical reasons it is inconvenient to demand that *all* the Delaunay m-simplices be δ-protected. For our purposes it will be sufficient to have protection on the deep interior m-simplices and their neighbouring m-simplices.

DEFINITION 15 (δ-GENERIC FOR P_I). *The set $\mathsf{P} \subset \mathbb{R}^m$ is* δ-generic for P_I *if all the m-simplices in* $\mathrm{star}(\mathrm{star}(\mathsf{P}_I))$ *are δ-protected.*

If the following we will make assertions about deep interior simplices based on the assumption that P is δ-generic for P_I. These assertions will be equally valid if P_I is taken to be some subset of the safe interior points, provided "deep interior simplices" is restricted accordingly.

3.3 Thickness from protection

Our goal here is to demonstrate that the deep interior simplices on a δ-generic point set are Υ_0-thick. If $\delta = \nu_0\epsilon$, for some constant ν_0, then we obtain a constant Υ_0 which depends only on ν_0. The key observation is that together with Lemma 10, protection imposes constraints on all the Delaunay simplices; they cannot be too close to being degenerate. In the particular case that $j = 0$, Lemma 10 immediately implies that any Delaunay edge has length greater than δ:

LEMMA 16 (SPARSITY FROM PROTECTION). *If* P *is* δ-*generic for* P_I, *then* P_I *is* δ-*sparse.*

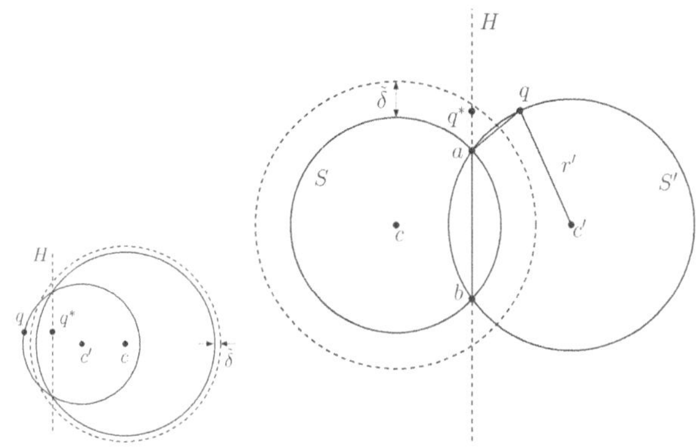

(a) opposite sides (b) q and c' on same side of H

Figure 1: Diagram for Lemma 17. (a) When H separates q and c' then $d_{\mathbb{R}^d}(q, q^*) > \tilde{\delta}$. (b) Otherwise, a lower bound on the distance between q and its projection q^* on H is obtained by an upper bound on the angle $\angle qab$.

LEMMA 17. *Let* P *be* λ-*sparse with sampling radius* ϵ *and* $\lambda \leq \epsilon$. *Suppose* B *is a* $\tilde{\delta}$-*protected Delaunay ball for* $\tilde{\sigma}$, *and* τ *is a deep interior simplex that is a (not necessarily proper) face of* $\tilde{\sigma}$. *If* $\sigma = q * \tau$, *is a Delaunay simplex that is not a face of* $\tilde{\sigma}$, *and* $H = \operatorname{aff}(\partial B \cap \partial B')$, *where* B' *is a Delaunay ball for* σ, *then* $d_{\mathbb{R}^m}(q, H) > \frac{\sqrt{3}\tilde{\delta}}{4\epsilon}(\lambda + \tilde{\delta})$. *In particular, if* P *is* δ-*generic for* P_I, *then* $D(q, \sigma) > \frac{\sqrt{3}\delta^2}{2\epsilon}$.

PROOF. Let $B = B_{\mathbb{R}^m}(c; r)$ be the $\tilde{\delta}$-protected Delaunay ball for $\tilde{\sigma}$, and $B' = B_{\mathbb{R}^m}(c'; r')$ the Delaunay ball for σ, so $r' < \epsilon$. Our geometry will be performed in the plane, Q, defined by c, c', and q. This plane is orthogonal to the $(m-1)$-flat H, and it follows that the distance $d_{\mathbb{R}^m}(q, H)$ is realized by a segment in the plane Q: the projection, q^*, of q onto H lies in Q, and $d_{\mathbb{R}^m}(q, H) = d_{\mathbb{R}^m}(q, q^*)$.

If H separates q from c', then ∂B separates q from q^*, and $d_{\mathbb{R}^m}(q, q^*) > d_{\mathbb{R}^m}(q, \partial B) > \tilde{\delta}$, since B is $\tilde{\delta}$-protected (Figure 1(a)). The lemma then follows since λ and $\tilde{\delta}$ are both less than ϵ. Thus assume that q and c' lie on the same side of H, as shown in Figure 1(b). Let $S = Q \cap \partial B$, and $S' = Q \cap \partial B'$, and let a and b be the points of intersection $S \cap S'$. Thus $H \cap Q$ is the line through a and b.

We will bound $d_{\mathbb{R}^m}(q, q^*)$ by finding an upper bound on the angle $\gamma = \angle qab$. This is the same as the standard calculation for upper-bounding the angles in a triangle with bounded circumradius to shortest edge ratio. Without loss of generality, we may assume that $\gamma \geq \angle qba$, and we will assume that $\gamma \geq \pi/2$ since otherwise $d_{\mathbb{R}^m}(q, q*) > \tilde{\delta}$ and the lemma is again trivially satisified.

Since $d_{\mathbb{R}^m}(a, q) > \tilde{\delta}$, we have $d_{\mathbb{R}^m}(q, q^*) = d_{\mathbb{R}^m}(a, q)\sin\gamma > \tilde{\delta}\sin\gamma$. Also note that $d(a, b) \geq 2R(\tau) \geq L(\tau) \geq \lambda$. Let $\alpha = \angle qac'$. Then $\cos\alpha = \frac{d_{\mathbb{R}^m}(a,q)}{2r'} \geq \frac{\tilde{\delta}}{2\epsilon}$. Similarly, with $\beta = \angle c'ab$, we have $\cos\beta \geq \frac{\lambda}{2\epsilon}$. Thus since $\gamma = \alpha + \beta \geq \pi/2$, we have

$$\begin{aligned} d_{\mathbb{R}^m}(q, q^*) &> \tilde{\delta}\sin\left(\arccos\frac{\tilde{\delta}}{2\epsilon} + \arccos\frac{\lambda}{2\epsilon}\right) \\ &\geq \tilde{\delta}\left(\frac{\lambda}{2\epsilon}\sin\left(\arccos\frac{\tilde{\delta}}{2\epsilon}\right) + \frac{\tilde{\delta}}{2\epsilon}\sin\left(\arccos\frac{\lambda}{2\epsilon}\right)\right) \\ &\geq \frac{\sqrt{3}\tilde{\delta}}{4\epsilon}(\lambda + \tilde{\delta}), \end{aligned}$$

where the last inequality follows from $\lambda \leq \epsilon$ and $\tilde{\delta} \leq \epsilon$.

Since $\operatorname{aff}(\tau) \subset H$, it follows that $D(q, \sigma) \geq d_{\mathbb{R}^m}(q, H)$, and if P is δ-generic for P_I, then $\lambda \geq \delta$, and Lemma 10 ensures that there is a δ-protected $\tilde{\sigma}$ that contains τ but not q. $\square$

We thus obtain a bound on the thickness of the deep interior simplices when P is δ-generic. Since Lemma 17 yields a lower bound of $\frac{\sqrt{3}\delta^2}{2\epsilon}$ on the altitudes of the deep interior simplices, and since $\Delta(\sigma) \leq 2\epsilon$, we have that $\Upsilon(\sigma) \geq \frac{\sqrt{3}\delta^2}{4\epsilon^2}$ for all deep interior σ. If $\delta = \nu_0\epsilon$, we obtain a constant thickness bound.

THEOREM 18 (THICKNESS FROM PROTECTION). *If* $\mathsf{P} \subset \mathbb{R}^m$ *is* δ-*generic for* P_I *with* $\delta = \nu_0\epsilon$, *where* ϵ *is a sampling radius for* P, *then the deep interior simplices are* Υ_0-*thick, with*

$$\Upsilon_0 = \frac{\sqrt{3}\nu_0^2}{4}.$$

4. DELAUNAY STABILITY

We find upper bounds on the magnitude of a perturbation for which a protected Delaunay ball remains a Delaunay ball. We consider both perturbations of the sample points in Euclidean space, and perturbations of the metric itself. The primary technical challenge is bounding the effect of a perturbation on the circumcentre of an m-simplex. We then find the relationship between the perturbation parameter ρ and the protection parameter δ which ensures that a δ-protected Delaunay simplex will remain a Delaunay simplex.

4.1 Perturbations and circumcentres

As expected, a bound on the displacement of the circumcentre requires a bound on the thickness of the simplex.

4.1.1 *Almost circumcentres*

If σ is thick, a point whose distances to the vertices of σ are all almost the same, will lie close to $N(\sigma)$.

LEMMA 19. *If* $\sigma = [p_0, ..., p_k] \subset \mathbb{R}^m$ *is a non-degenerate* k-*simplex, and* $x \in \mathbb{R}^m$ *is such that*

$$\left|\|p_i - x\|^2 - \|p_j - x\|^2\right| \leq \tilde{\rho}^2 \quad \textit{for all } i, j \in [0, ..., k], \quad (1)$$

then there is a $c \in N(\sigma)$ *such that* $\|c - x\| \leq \eta$, *where*

$$\eta = \frac{\tilde{\rho}^2}{2\Upsilon(\sigma)\Delta(\sigma)}.$$

In particular, if σ *is an* m-*simplex then* $x \in \overline{B}_{\mathbb{R}^m}(C(\sigma); \eta)$.

If the inequalities in Equations (1) are made strict, then the conclusions may also be stated with strict inequalities.

PROOF. First suppose $k = m$. The circumcentre of σ is given by the linear equations $\|C(\sigma) - p_i\|^2 = \|C(\sigma) - p_0\|^2$, or

$$(p_i - p_0)^{\mathsf{T}} C(\sigma) = \frac{1}{2}(\|p_i\|^2 - \|p_0\|^2).$$

Letting b be the vector whose i^{th} component is defined by the right hand side of the equation, and letting P be the $m \times m$ matrix, whose i^{th} column is $(p_i - p_0)$, we write the equations in matrix form:

$$P^{\mathsf{T}} C(\sigma) = b. \tag{2}$$

Without loss of generality, assume p_0 is the vertex that minimizes the distance to x. Then, defining x_a to be the vector whose i^{th} component is $\frac{1}{2}(\|p_i - x\|^2 - \|p_0 - x\|^2)$, we have $\|p_i - x\|^2 = \|p_0 - x\|^2 + 2(x_a)_i$, and we find

$$P^{\mathsf{T}} x = b - x_a. \tag{3}$$

From Equations (2) and (3) we have

$$\|C(\sigma) - x\| = \left\|P^{-T} x_a\right\| \leq \|P^{-1}\| \, \|x_a\| .$$

From Equation (1), it follows that $\|x_a\| \leq \frac{\sqrt{m}\tilde{\rho}^2}{2}$, and from Lemmas 2 and 4 we have $\|P^{-1}\| \leq (\sqrt{m}\Upsilon(\sigma)\Delta(\sigma))^{-1}$, and thus the result holds for full dimensional simplices.

If σ is a k-simplex with $k \leq m$, then we consider $\hat{x}$, the orthogonal projection of x into $\operatorname{aff}(\sigma)$. We observe that $\hat{x}$ also must satisfy Equation (1), and we conclude from the above argument that $\|C(\sigma) - \hat{x}\| \leq \eta$. Then letting $c = C(\sigma) + (x - \hat{x})$ we have that $c \in N(\sigma)$ and $\|c - x\| \leq \eta$. □

With a bound on the distance from x to the vertices of σ, we obtain a bound without the squared distances:

LEMMA 20. *If $\sigma = [p_0, \ldots, p_k] \subset \mathbb{R}^m$ is non-degenerate, and x satisfies $\|p_i - x\| < \tilde{\epsilon}$ for all $i, j \leq k$, and*

$$\big| \|p_i - x\| - \|p_j - x\| \big| \leq \tilde{\rho} \qquad \text{for all } i, j \leq k, \tag{4}$$

then there exists a $c \in N(\sigma)$ such that $\|x - c\| < \eta$, where

$$\eta = \frac{\tilde{\epsilon}\tilde{\rho}}{\Upsilon(\sigma)\Delta(\sigma)}.$$

In particular, if σ is an m-simplex, then $x \in B_{\mathbb{R}^m}(C(\sigma); \eta)$.

PROOF. Let $R = \max_i \|p_i - x\|$ and $r = \min_i \|p_i - x\|$. Then $R^2 - r^2 = (R + r)(R - r) < 2\tilde{\epsilon}(R - r) \leq 2\tilde{\epsilon}\tilde{\rho}$, and the result then follows from Lemma 19. □

It will be convenient to have a name for points x which satisfy Equation (4): we call such a point a *$\tilde{\rho}$-centre* for σ.

4.1.2 Circumcentres and metric perturbations

We will show here that for an Υ_0-thick m-simplex σ in $\mathbb{R}^m$, and a metric d that is close to $d_{\mathbb{R}^m}$, there will be a metric circumcentre c near $C(\sigma)$. We require the metric d to be *topologically equivalent* to $d_{\mathbb{R}^m}$: both metrics define the same topology. Henceforth, whenever we refer to a *perturbation of the Euclidean metric*, a topologically equivalent metric will always be assumed. We will demonstrate the following:

LEMMA 21 (CIRCUMCENTRES: METRIC PERTURBATION). *Let $U \subset \mathbb{R}^m$, and let $d : U \times U \to \mathbb{R}$ be a metric topologically equivalent to $d_{\mathbb{R}^m}$ and such that for any $x, y \in U$ with $d_{\mathbb{R}^m}(x, y) < 2\epsilon$, we have $|d(x, y) - d_{\mathbb{R}^m}(x, y)| \leq \rho$, with $\rho \leq \frac{\Upsilon_0 \mu_0 \epsilon}{4\sqrt{m}}$. If $\sigma = [p_0, \ldots, p_m] \subset U$ is an Υ_0-thick m-simplex with $R(\sigma) < \epsilon$, and $L(\sigma) \geq \mu_0\epsilon$, and such that $d_{\mathbb{R}^m}(p_i, \partial U) \geq 2\epsilon$, then there is a point*

$$c \in B = B_{\mathbb{R}^m}(C(\sigma); \frac{4\sqrt{m}\rho}{\Upsilon_0\mu_0})$$

such that $d(c, p_i) = d(c, p_j)$ for all $p_i, p_j \in \sigma$.

In order to prove Lemma 21, we will use a particular case of Lemma 20:

LEMMA 22. *Suppose σ is an Υ_0-thick m-simplex such that $L(\sigma) \geq \mu_0\epsilon$. If x is a $\tilde{\rho}$-centre for σ with $d_{\mathbb{R}^m}(x, p) < 2\epsilon$ for all $p \in \sigma$, then $x \in B_{\mathbb{R}^m}(C(\sigma); \eta)$, where $\eta = \frac{2\tilde{\rho}}{\Upsilon_0\mu_0}$.*

Let $B = B_{\mathbb{R}^m}(C(\sigma); \eta)$ be the open ball which contains the $\tilde{\rho}$-centres for σ. We wish to show that if $\tilde{\rho} = 2\sqrt{m}\rho$, then a circumcentre for σ with respect to d will also lie in B. Note that $\overline{B} \subset U$.

Consider the function $f_e : \overline{B} \to \mathbb{R}^m$ given by

$$f_e(x) = \\ (d_{\mathbb{R}^m}(x, p_1) - d_{\mathbb{R}^m}(x, p_0), \ldots, d_{\mathbb{R}^m}(x, p_m) - d_{\mathbb{R}^m}(x, p_0))^{\mathsf{T}}. \tag{5}$$

Lemma 22 implies that $\|f_e(x)\| > \tilde{\rho}$, for all $x \in \partial B$. Also, f_e maps the circumcentre of σ, and only the circumcentre, to the origin: $f_e^{-1}(0) = \{C(\sigma)\}$.

We construct a similar function for the metric d,

$$f(x) = (d(x, p_1) - d(x, p_0), \ldots, d(x, p_m) - d(x, p_0))^{\mathsf{T}}, \tag{6}$$

and use a topological argument to show that there must be a $c \in f^{-1}(0) \subset B$. We first show that there is a homotopy between f and f_e such that the image of $\partial\overline{B}$ never touches the origin. Then we observe that 0 is a regular value for f_e, and thus, since $f_e^{-1}(0) = \{C(\sigma)\}$, consideration of the *degree* of maps demonstrates that $f^{-1}(0) \neq \emptyset$.

LEMMA 23. *If $\rho = \frac{\tilde{\rho}}{2\sqrt{m}}$, there is a homotopy $F : \overline{B} \times [0, 1] \to \mathbb{R}^m$ with $F(x, t) \neq 0$ for all $x \in \partial\overline{B}$ and $t \in [0, 1]$.*

PROOF. We define the homotopy $F : \overline{B} \times [0, 1] \to \mathbb{R}^m$ by $F(x, t) = (1 - t)f_e(x) + tf(x)$. By the bounds on ρ and $R(\sigma)$, for every $x \in \overline{B}$, and $p \in \sigma$, we have $d_{\mathbb{R}^m}(x, p) \leq \frac{2\tilde{\rho}}{\Upsilon_0\mu_0} + R(\sigma) < 2\epsilon$. Thus Lemma 22 yields $\|f_e(x)\| > \tilde{\rho}$ for all $x \in \partial\overline{B}$. Also, from the hypothesis on d, we have $\|f_e(x) - f(x)\| \leq \tilde{\rho}$, for all $x \in \partial\overline{B}$. It follows that $F(x, t) \neq 0$ when $x \in \partial B$. □

By a direct calculation [BDG12b] we find:

LEMMA 24. *The origin is a regular value for the function f_e defined in Equation (5).*

Lemma 21 now follows from a consideration of the degree of the mappings f and f_e relative to zero. The *degree* of a smooth map $g : \overline{B} \to \mathbb{R}^m$ at a regular point $p \in g(B)$ is defined by

$$\deg(g, p, B) = \sum_{x \in g^{-1}(p)} \operatorname{sign} \det J(g)_x ,$$

where $J(g)_x$ is the Jacobian matrix of g at x. The exposition by Dancer [Dan00] is a good reference for the degree of maps from manifolds with boundary. It is shown that the definition of $\deg(g, p, B)$ extends to continuous maps g that are not necessarily differentiable. If $h : \overline{B} \to \mathbb{R}^m$ is homotopic to g by a homotopy $H : \overline{B} \times [0, 1] \to \mathbb{R}^m$ such that $H(x, t) \neq p$ for all $t \in [0, 1]$, and $x \in \partial B$, then $\deg(g, p, B) = \deg(h, p, B)$.

Since $f_e^{-1}(0) = \{C(\sigma)\}$, it follows from Lemma 24 that $\deg(f_e, 0, B) = \pm 1$. Then Lemma 23 implies $\deg(f, 0, B) = \deg(f_e, 0, B)$, and since this is nonzero, it must be that $f^{-1}(0) \neq \emptyset$. The demonstration of Lemma 21 is complete.

4.1.3 Circumcentres and point perturbations

The exact same argument as was used to demonstrate Lemma 21 can be used to show that an m-simplex $\tilde{\sigma} = [\tilde{p}_0, \ldots, \tilde{p}_m]$ whose vertices lie close to a thick m-simplex σ, will also have a circumcentre that lies close to $C(\sigma)$. We replace the function f defined in Equation (6) by the function

$$\tilde{f}(x) = (d_{\mathbb{R}^m}(x,\tilde{p}_1) - d_{\mathbb{R}^m}(x,\tilde{p}_0), \ldots, d_{\mathbb{R}^m}(x,\tilde{p}_m) - d_{\mathbb{R}^m}(x,\tilde{p}_0))^{\mathsf{T}},$$

and the same argument goes through. We obtain:

LEMMA 25 (CIRCUMCENTRES: POINT PERTURBATION). *Suppose that $\sigma = [p_0, \ldots, p_m]$ is an Υ_0-thick m-simplex with $R(\sigma) < \epsilon$ and $L(\sigma) \geq \mu_0\epsilon$. Suppose also that $\tilde{\sigma} = [\tilde{p}_0, \ldots, \tilde{p}_m]$ is such that $\|\tilde{p}_i - p_i\| \leq \rho$ for all $i \in [0, \ldots, m]$. If $\rho \leq \frac{\Upsilon_0\mu_0\epsilon}{4\sqrt{m}}$, then $d_{\mathbb{R}^m}(C(\tilde{\sigma}), C(\sigma)) < \frac{4\sqrt{m}\rho}{\Upsilon_0\mu_0}$.*

We will also make use of a similar result [BDGO11, Lemma 3.7], which depends on the square of the thickness bound, but applies also to simplices that are not full dimensional:

LEMMA 26. *Suppose $B = B_{\mathbb{R}^N}(c; r)$, with $r < \epsilon$, is a circumscribing ball for a j-simplex $\sigma = [p_0, \ldots, p_j]$. Suppose also that $\tilde{\sigma} = [\tilde{p}_0, \ldots, \tilde{p}_j]$ is such that $\tilde{p}_0 = p_0$ and $\|\tilde{p}_i - p_i\| \leq \rho$ for all $i \in [1, \ldots, j]$. If $\rho \leq \frac{\Upsilon(\sigma)^2\Delta(\sigma)}{5}$, then there is a circumscribing ball $\tilde{B} = B_{\mathbb{R}^d}(\tilde{c}; \tilde{r})$ for $\tilde{\sigma}$ with $|\tilde{r} - r| \leq \|\tilde{c} - c\| < \left(\frac{12\epsilon}{\Upsilon(\sigma)^2\Delta(\sigma)}\right)\rho$.*

4.2 Perturbations and protection

Suppose $\zeta : \mathsf{P} \to \tilde{\mathsf{P}}$ is a ρ-perturbation. If σ is a δ-protected m-simplex in $\mathrm{Del}(\mathsf{P})$, then we want an upper bound on ρ that will ensure that $\tilde{\sigma} = \zeta(\sigma)$ is protected in $\mathrm{Del}(\tilde{\mathsf{P}})$.

LEMMA 27 (PROTECTION AND POINT PERTURBATION). *Suppose that $\mathsf{P} \subset \mathbb{R}^m$ and $\sigma \in \mathrm{Del}(\mathsf{P})$ is a δ-protected, Υ_0-thick m-simplex with $R(\sigma) < \epsilon$ and $L(\sigma) \geq \mu_0\epsilon$. If $\zeta : \mathsf{P} \to \tilde{\mathsf{P}}$ is a ρ-perturbation with $\rho \leq \frac{\Upsilon_0\mu_0}{10\sqrt{m}}\delta$, then $\zeta(\sigma) = \tilde{\sigma} \in \mathrm{Del}(\tilde{\mathsf{P}})$ and has a $(\delta - \frac{10\sqrt{m}}{\Upsilon_0\mu_0}\rho)$-protected Delaunay ball.*

PROOF. Let $B = B_{\mathbb{R}^m}(c; r)$ be the δ-protected Delaunay ball for $\sigma \in \mathrm{Del}(\mathsf{P})$, and let $\tilde{B} = B_{\mathbb{R}^m}(\tilde{c}; \tilde{r})$ be the circumball for the corresponding perturbed simplex $\tilde{\sigma}$. We wish to establish a bound on ρ that will ensure that $\tilde{B}$ is protected with respect to $\tilde{\mathsf{P}}$.

Let $q \in \mathsf{P}$ be a point not in σ. We need to ensure that the corresponding $\tilde{q}$ lies outside the closure of $\tilde{B}$, i.e., that $d_{\mathbb{R}^m}(\tilde{q}, \tilde{c}) > \tilde{r}$.

Since $\delta \leq \epsilon$, the hypothesis of Lemma 25 is satisfied by ρ, and we have $d_{\mathbb{R}^m}(\tilde{c}, c) < \eta\rho$, where $\eta = \frac{4\sqrt{m}}{\Upsilon_0\mu_0}$. Thus for $p \in \sigma$ and corresponding $\tilde{p} \in \tilde{\sigma}$ we have

$$\begin{aligned}\tilde{r} &\leq d_{\mathbb{R}^m}(c,p) + d_{\mathbb{R}^m}(p,\tilde{p}) + d_{\mathbb{R}^m}(c,\tilde{c}) \\ &< r + (\eta + 1)\rho.\end{aligned}$$

Also

$$\begin{aligned}d_{\mathbb{R}^m}(\tilde{q},\tilde{c}) &\geq d_{\mathbb{R}^m}(q,c) - d_{\mathbb{R}^m}(\tilde{c},c) - d_{\mathbb{R}^m}(\tilde{q},q) \\ &> r + \delta - \rho(\eta + 1).\end{aligned}$$

Therefore $\tilde{q}$ will be outside of the closure of $\tilde{B}$ provided $r + \delta - \rho(\eta+1) \geq r + (1+\eta)\rho$, i.e., when $\delta \geq 2(\eta+1)\rho$. The result follows from the definition of η and the observation that μ_0 and Υ_0 are each no larger than one. □

A similar argument [BDG12b] yields a bound on the metric perturbation that will ensure the Delaunay balls for the m-simplices remain protected.

LEMMA 28 (PROTECTION AND METRIC PERTURBATION). *Suppose $U \subset \mathbb{R}^m$ contains $\mathrm{ch}(\mathsf{P})$ and $d : U \times U \to \mathbb{R}$ is a metric such that $|d_{\mathbb{R}^m}(x,y) - d(x,y)| \leq \rho$ for all $x, y \in U$. Suppose also that $\sigma \in \mathrm{Del}(\mathsf{P})$ is a δ-protected, Υ_0-thick m-simplex with $R(\sigma) < \epsilon$ and $\Delta(\sigma) \geq \mu_0\epsilon$. If $\rho \leq \frac{\Upsilon_0\mu_0}{12\sqrt{m}}\delta$, and $d_{\mathbb{R}^m}(p, \partial U) \geq 2\epsilon$ for every vertex $p \in \sigma$, then σ also belongs to $\mathrm{Del}_d(\mathsf{P})$, and has a $(\delta - \frac{12\sqrt{m}}{\Upsilon_0\mu_0}\rho)$-protected Delaunay ball in the metric d.*

4.3 Perturbations and Delaunay stability

The results of Section 4.2 translate into stability results for Delaunay triangulations. In the case of point perturbations in Euclidean space, the connectivity of the Delaunay triangulation cannot change as long as the simplices corresponding to the initial m-simplices remain protected. This is a direct consequence of Delaunay's original result [Del34], but we explicitly lay out the argument.

In the case of metric perturbation, we can no longer take for granted that the Delaunay complex cannot change its connectivity if the m-simplices remain protected. This is because we are no longer guaranteed that the Delaunay complex will be a triangulation. Using the consequences of the point-perturbation result, we establish bounds that ensure that the Delaunay complex in the perturbed metric will be the same as the original Delaunay triangulation.

4.3.1 Point perturbations

Lemma 27 establishes bounds on a ρ-perturbation $\zeta : \mathsf{P} \to \tilde{\mathsf{P}}$ which will guarantee that $\zeta(\mathrm{star}(\mathsf{P}_I)) \subseteq \mathrm{Del}(\tilde{\mathsf{P}})$. Since Lemma 27 also guarantees that, if ρ is small enough, the vertices of $\mathrm{star}(\zeta(\mathsf{P}_I); \mathrm{Del}(\tilde{\mathsf{P}}))$ will be generic, we know from Delaunay's Lemma 14 that $\mathrm{star}(\zeta(\mathsf{P}_I); \mathrm{Del}(\tilde{\mathsf{P}}))$ will be a triangulation at each $\tilde{p} \in \zeta(\mathsf{P}_I)$, and thus $\mathrm{star}(\zeta(\mathsf{P}_I)) = \zeta(\mathrm{star}(\mathsf{P}_I))$ follows from Lemma 7. Explicitly, we have:

LEMMA 29. *Suppose $\mathsf{P} \subset \mathbb{R}^m$ is a generic sample set, and $\zeta : \mathsf{P} \to \tilde{\mathsf{P}}$ is a perturbation such that $\zeta(\mathrm{star}(\mathsf{P}_I; \mathrm{Del}(\mathsf{P}))) \subseteq \mathrm{star}(\zeta(\mathsf{P}_I); \mathrm{Del}(\tilde{\mathsf{P}}))$, and every $\tilde{\sigma} \in \zeta(\mathrm{star}(\mathsf{P}_I))$ is protected in $\mathrm{Del}(\tilde{\mathsf{P}})$. Then $\zeta(\mathrm{star}(\mathsf{P}_I)) = \mathrm{star}(\zeta(\mathsf{P}_I))$.*

Thus, considering Theorem 18, Lemma 16, and Lemma 27, we have the following stability theorem for Delaunay triangulations of δ-protected points:

THEOREM 30 (STABILITY UNDER POINT PERTURBATION). *Suppose $\mathsf{P} \subset \mathbb{R}^m$ is δ-generic for P_I, with sampling radius ϵ and $\delta = \nu_0\epsilon$. If $\zeta : \mathsf{P} \to \tilde{\mathsf{P}}$ is a ρ-perturbation, with*

$$\rho \leq \frac{\nu_0^3}{24\sqrt{m}}\delta = \frac{\nu_0^4}{24\sqrt{m}}\epsilon,$$

then

$$\mathrm{star}(\mathsf{P}_I; \mathrm{Del}(\mathsf{P})) \overset{\zeta}{\cong} \mathrm{star}(\zeta(\mathsf{P}_I); \mathrm{Del}(\tilde{\mathsf{P}})).$$

The ρ-relaxed Delaunay complex for P was defined by de Silva [dS08] by the criterion that $\sigma \in \mathrm{Del}^{\rho}(\mathsf{P})$ if and only if there is a ball $B = B_{\mathbb{R}^m}(c; r)$ such that $\sigma \in \overline{B}$, and $d_{\mathbb{R}^m}(c, q) \geq r - \rho$ for all $q \in \mathsf{P}$. Thus the simplices in $\mathrm{Del}^{\rho}(\mathsf{P})$ all have "almost empty", balls centred on a ρ-centre for σ. We have the following consequence of Theorem 30:

COROLLARY 31 (STABILITY UNDER RELAXATION). *Let* $\mathsf{P} \subset \mathbb{R}^m$ *be* δ*-generic for* P_I*, with sampling radius* ϵ *and* $\delta = \nu_0\epsilon$*. If* $\rho \leq \frac{\nu_0^3}{24\sqrt{m}}\delta = \frac{\nu_0^4}{24\sqrt{m}}\epsilon$*, then* $\mathrm{star}(\mathsf{P}_I; \mathrm{Del}^\rho(\mathsf{P})) = \mathrm{star}(\mathsf{P}_I; \mathrm{Del}(\mathsf{P}))$.

PROOF. Suppose that $\sigma \in \mathrm{star}(\mathsf{P}_I; \mathrm{Del}^\rho(\mathsf{P}))$. Then there is a ball B for σ such that any point $q \in B$ is within a distance ρ from ∂B. Project all such points radially out to ∂B. Then we have a ρ-perturbation $\zeta : \mathsf{P} \to \tilde{\mathsf{P}}$, and σ has become $\tilde{\sigma} \in \mathrm{star}(\zeta(\mathsf{P}_I); \mathrm{Del}(\tilde{\mathsf{P}}))$. By Theorem 30, $\mathrm{star}(\zeta(\mathsf{P}_I); \mathrm{Del}(\tilde{\mathsf{P}})) \overset{\zeta}{\cong} \mathrm{star}(\mathsf{P}_I; \mathrm{Del}(\mathsf{P}))$, and therefore $\sigma \in \mathrm{star}(\mathsf{P}_I; \mathrm{Del}(\mathsf{P}))$. □

4.3.2 Metric perturbation

For a perturbation of the metric, we can exploit the stability results obtained for perturbations of points in the Euclidean metric to ensure that no simplices can appear in $\mathrm{star}(\mathsf{P}_I; \mathrm{Del}_d(\mathsf{P}))$ that do not already exist in $\mathrm{star}(\mathsf{P}_I; \mathrm{Del}(\mathsf{P}))$.

LEMMA 32. *Suppose* P *is* δ*-generic for* P_I*, with sampling radius* ϵ *and* $\delta = \nu_0\epsilon$*. Suppose also that* $\mathrm{ch}(\mathsf{P}) \subseteq U$ *and* $d : U \times U \to \mathbb{R}$ *is such that* $|d(x,y) - d_{\mathbb{R}^m}(x,y)| \leq \rho$ *for all* $x, y \in U$*. If* $\rho \leq \frac{\nu_0^3}{48\sqrt{m}}\delta$*, then* $\mathrm{star}(\mathsf{P}_I; \mathrm{Del}_d(\mathsf{P})) \subseteq \mathrm{star}(\mathsf{P}_I; \mathrm{Del}(\mathsf{P}))$.

PROOF. Let $B(c; r)$ be a Delaunay ball for simplex $\sigma \in \mathrm{star}(\mathsf{P}_I; \mathrm{Del}_d(\mathsf{P}))$. Then $d(c,p) \leq d(c,q)$ for all $p \in \sigma$, and $q \in \mathsf{P}$. By the hypothesis on d, this implies that $d_{\mathbb{R}^m}(c,p) \leq d_{\mathbb{R}^m}(c,q) + 2\rho$ for all $p \in \sigma$ and $q \in \mathsf{P}$, and therefore $\sigma \in \mathrm{Del}^{2\rho}(\mathsf{P})$. The result now follows from Corollary 31. □

The perturbation bounds required by Lemma 32, also satisfy the requirements of Lemma 28, when Theorem 18 and Lemma 16 are taken into account. This gives us the reverse inclusion, and thus we can quantify the stability under metric perturbation for Delaunay triangulations on δ-generic point sets:

THEOREM 33 (STABILITY UNDER METRIC PERTURBATION). *Suppose* P *is* δ*-generic for* P_I*, with sampling radius* ϵ *and* $\delta = \nu_0\epsilon$*. Suppose also that* $\mathrm{ch}(\mathsf{P}) \subseteq U$*, and* $d : U \times U \to \mathbb{R}$ *is such that* $|d(x,y) - d_{\mathbb{R}^m}(x,y)| \leq \rho$ *for all* $x, y \in U$*. If*

$$\rho \leq \frac{\nu_0^3}{48\sqrt{m}}\delta = \frac{\nu_0^4}{48\sqrt{m}}\epsilon,$$

then

$$\mathrm{star}(\mathsf{P}_I; \mathrm{Del}_d(\mathsf{P})) = \mathrm{star}(\mathsf{P}_I; \mathrm{Del}(\mathsf{P})).$$

5. EQUATING DELAUNAY STRUCTURES

We apply the results of the previous sections to the task of triangulating a compact m-manifold, $\mathcal{M}$ embedded in $\mathbb{R}^N$. We wish to build a Delaunay complex on a finite set $\mathcal{P} \subset \mathcal{M}$.

The *restricted Delaunay complex* is the Delaunay complex $\mathrm{Del}_{\mathbb{R}^N|_\mathcal{M}}(\mathcal{P})$ obtained when distances on the manifold are measured with the metric $d_{\mathbb{R}^N|_\mathcal{M}}$. This is the Euclidean metric of the ambient space, restricted to the manifold. In other words, $d_{\mathbb{R}^N|_\mathcal{M}}(x,y) = d_{\mathbb{R}^N}(x,y)$. We use this notation to avoid ambiguities in conjunction with the local Euclidean metrics discussed below. The Delaunay complex $\mathrm{Del}_{\mathbb{R}^N|_\mathcal{M}}(\mathcal{P})$ is a substructure of $\mathrm{Del}_{\mathbb{R}^N}(\mathcal{P})$.

Alternatively, distances on the manifold may be measured with $d_\mathcal{M}$, the *intrinsic metric* of the manifold. This metric defines the distance between x and y as the infimum of the lengths of the paths on $\mathcal{M}$ which connect x and y. This metric is also induced from $d_{\mathbb{R}^N}$. The *intrinsic Delaunay complex* is the Delaunay structure $\mathrm{Del}_\mathcal{M}(\mathcal{P})$ associated with this metric.

Although neither of these metrics are Euclidean, the idea is that locally, in a small neighbourhood of any point, these metrics may be well approximated by $d_{\mathbb{R}^m}$. Then, if the sampling satisfies appropriate δ-generic and ϵ-dense criteria in these local Euclidean metrics, the global Delaunay complex in the metric of the manifold will coincide locally with a Euclidean Delaunay triangulation, and we can thus guarantee a manifold complex.

5.1 Local Euclidean metrics

A *local parameterization* at a point $p \in \mathcal{M}$, is a pair (U, ψ_p), where $U \subset \mathbb{R}^m$ is an open neighbourhood of the origin, and $\psi_p : U \to \psi_p(U) = W \subset \mathcal{M}$ is a homeomorphism onto its image, and maps the origin to p. We will use ψ_p to pull back the metric of the manifold to U, and to simplify the notation we will write $d_\mathcal{M}(x,y)$ for $x, y \in U$, where it is to be understood that this means $d_\mathcal{M}(\psi_p(x), \psi_p(y))$, and likewise for $d_{\mathbb{R}^N|_\mathcal{M}}(x,y)$. Indeed, once W and U have been coupled together by a homeomorphism, we can transfer the metrics between them and the distinction becomes only one of perspective. We refer to the standard metric $d_{\mathbb{R}^m}$ on U as a *local Euclidean metric* for p on W. Clearly this metric depends upon the choice of ψ_p; there are different ways to impose a Euclidean metric on W. In this work the only local parameterization we consider is defined by the projection map π_p described in Section 5.3.

We wish to generate a sample set $\mathcal{P} \subset \mathcal{M}$ that will allow us to exploit the results of Section 4. We will require that for each $p \in \mathcal{P}$ there be a local parameterization (U, ψ_p) such that $B = B_{\mathbb{R}^m}(0; 5\epsilon) \subset U$, and $\mathsf{P} = \psi_p^{-1}(\mathcal{P} \cap W)$ is an ϵ-sample set for B in the Euclidean metric. We identify p with the origin, and note that $p \in \mathsf{P}_I$, since $d_{\mathbb{R}^m}(p, \partial\mathrm{ch}(\mathsf{P})) \geq 4\epsilon$. We will demand that P be δ-generic for $\{p\}$, with $\delta = \nu_0\epsilon$.

In order to exploit Theorem 33 we need to find a bound on the metric distortion induced by ψ_p, and this will depend on the sampling radius ϵ, through the requirement that $B \subset U$.

5.2 Background results for manifolds

We will make use of some established nomenclature and properties related to a smooth compact m-dimensional submanifold of $\mathbb{R}^N$.

For $p \in \mathcal{M}$, $T_p\mathcal{M}$ denotes the tangent space at p, which we may identify with an m-flat in the ambient space. The normal space, $N_p\mathcal{M}$, is the orthogonal complement of $T_p\mathcal{M}$ in $T_p\mathbb{R}^N$, and we likewise treat it as the affine space of dimension $m - k$ orthogonal to $T_p\mathcal{M} \subset \mathbb{R}^N$.

A ball $B = B_{\mathbb{R}^N}(c; r)$ is a *medial ball* at p if $B \cap \mathcal{M} = \emptyset$, it is tangent to $\mathcal{M}$ at p, and it is maximal in the sense that any ball which is centred on the line through p and c, and contains B, either coincides with B or intersects $\mathcal{M}$. The *local reach* at p is the infimum of the radii of the medial balls at p, and the *reach* of $\mathcal{M}$, denoted $\mathrm{rch}(\mathcal{M})$, is the infimum of the local reach over all points of $\mathcal{M}$. In order to approximate the geometry and topology with a simplical complex, manifolds with small reach require a higher sampling density than those with a larger reach. As is typical, an upper bound on our sampling radius will be proportional to $\mathrm{rch}(\mathcal{M})$. We require $\mathcal{M}$ to have positive reach.

Giesen and Wagner [GW04, Lemma 6] estimate how the tangent space locally deviates from the manifold:

LEMMA 34 (DISTANCE TO TANGENT SPACE). *If $x, y \in \mathcal{M} \subset \mathbb{R}^N$ and $d_{\mathbb{R}^N}(x,y) \leq r < rch(\mathcal{M})$, then $d_{\mathbb{R}^N}(y, T_x\mathcal{M}) \leq \frac{r^2}{2rch(\mathcal{M})}$, and thus $\sin\alpha \leq \frac{r}{2rch(\mathcal{M})}$, where α is the angle between $[x,y]$ and $T_x\mathcal{M}$.*

A convenient bound on the angle between nearby tangent spaces [BDGO11, Prop. A.3]:

LEMMA 35 (TANGENT SPACE VARIATION). *Let $x, y \in \mathcal{M}$ be such that $d_{\mathbb{R}^N}(x,y) \leq \frac{rch(\mathcal{M})}{10}$. Then, $\sin\angle(T_x\mathcal{M}, T_y\mathcal{M}) \leq \frac{5d_{\mathbb{R}^N}(x,y)}{rch(\mathcal{M})}$.*

Niyogi et al [NSW08, Prop 6.3] demonstrate a bound on the geodesic distance between nearby points, with respect to the ambient distance. We will use a modified statement [BDG12b] of this result:

LEMMA 36 (GEODESIC DISTANCE BOUND). *Let $x, y \in \mathcal{M}$ be such that $d_{\mathbb{R}^N}(x,y) \leq \frac{rch(\mathcal{M})}{2}$. Then*

$$d_{\mathcal{M}}(x,y) \leq d_{\mathbb{R}^N}(x,y)\left(1 + \frac{2d_{\mathbb{R}^N}(x,y)}{rch(\mathcal{M})}\right).$$

5.3 Sampling criteria for manifolds

We will confine ourselves to smooth compact manifolds without boundaries. A local parameterization at $p \in \mathcal{M}$ will be constructed with the aid of the orthogonal projection $\pi_p : \mathbb{R}^N \to T_p\mathcal{M}$, restricted to $\mathcal{M}$. Niyogi et al. [NSW08, Lemma 5.4] demonstrated that, when $r < \frac{\mathrm{rch}(\mathcal{M})}{2}$, π_p is a diffeomorphism from $W = B_{\mathbb{R}^N|_{\mathcal{M}}}(p;r)$ onto its image $U \subset T_p\mathcal{M}$. We will identify $T_p\mathcal{M}$ with $\mathbb{R}^m$, and define

$$\psi_p = \pi_p|_W^{-1} : U \xrightarrow{\cong} W. \tag{7}$$

Using ψ_p to pull back the metrics $d_{\mathcal{M}}$ and $d_{\mathbb{R}^N|_{\mathcal{M}}}$ to $\mathbb{R}^m$, we can view them as perturbations of $d_{\mathbb{R}^m}$. The magnitude of the perturbation is governed by the radius of the ball used to define W.

LEMMA 37. *Using the local parameterization (U, ψ_p), with $\mathbb{R}^m \supset U \overset{\psi_p}{\cong} W \subset \mathcal{M}$, if $W \subset B_{\mathbb{R}^N|_{\mathcal{M}}}(p;r)$, with $r \leq \frac{rch(\mathcal{M})}{10}$, then for all $x, y \in U$, $|d_{\mathcal{M}}(x,y) - d_{\mathbb{R}^m}(x,y)| \leq \frac{10r^2}{rch(\mathcal{M})}$.*

PROOF. Let $u, v \in W \subset B_{\mathbb{R}^N|_{\mathcal{M}}}(p;r)$, and let θ be the angle between the line segments $[u,v]$ and $[\pi_p(u), \pi_p(v)]$, θ_1 the angle between $[u,v]$ and $T_u\mathcal{M}$, and θ_2 the angle between the $T_p\mathcal{M}$ and $T_u\mathcal{M}$. Thus

$$\theta \leq \theta_1 + \theta_2 \tag{8}$$

and $d_{\mathbb{R}^m}(\pi_p(u), \pi_p(v)) = d_{\mathbb{R}^N}(u,v)\cos\theta$. Defining $\eta = \frac{r}{\mathrm{rch}(\mathcal{M})}$, Lemma 36 yields $d_{\mathcal{M}}(u,v) \leq d_{\mathbb{R}^N}(u,v)\,(1+4\eta)$, and so

$$d_{\mathbb{R}^m}(\pi_p(u), \pi_p(v)) \geq \frac{d_{\mathcal{M}}(u,v)\,\cos\theta}{1+4\eta}.$$

Using Lemma 34, we find $\theta_1 \leq \sin\theta_1 \leq \eta$, and Lemma 35, yields $\theta_2 \leq \sin\theta_2 \leq 5\eta$. Therefore from Equation (8), $\theta \leq 6\eta < \frac{\pi}{2}$, and thus $\cos\theta \geq 1 - \theta \geq 1 - 6\eta$. We get

$$\begin{aligned} d_{\mathbb{R}^m}(\pi_p(u), \pi_p(v)) &\geq d_{\mathcal{M}}(u,v)\left(\frac{1-6\eta}{1+4\eta}\right) \\ &\geq d_{\mathcal{M}}(u,v)(1-6\eta)(1-4\eta) \\ &\geq d_{\mathcal{M}}(u,v)(1-10\eta) \\ &\geq d_{\mathcal{M}}(u,v) - \frac{10r^2}{\mathrm{rch}(\mathcal{M})}. \end{aligned}$$

The result follows since $d_{\mathcal{M}}(u,v) \geq d_{\mathbb{R}^m}(\pi_p(u), \pi_p(v))$. □

We call a neighbourhood W of p *admissible* if it satisfies the hypothesis of Lemma 37.

We need to define the local parameterization on a domain that is large enough to contain an ϵ-sampled ball around p that will ensure that p is a safe interior point. Thus we translate Lemma 37 into a sampling density requirement:

LEMMA 38. *If $1 < a \leq 400$ and $\epsilon \leq \frac{rch(\mathcal{M})}{10a}$, and $U = B_{\mathbb{R}^m}(p;(a-1)\epsilon)$, then $\psi_p(U) = W \subseteq B_{\mathbb{R}^N|_{\mathcal{M}}}(p;a\epsilon)$, and*

$$|d_{\mathbb{R}^N|_{\mathcal{M}}}(x,y) - d_{\mathbb{R}^m}(x,y)| \leq |d_{\mathcal{M}}(x,y) - d_{\mathbb{R}^m}(x,y)| \leq \frac{10a^2\epsilon^2}{rch(\mathcal{M})}.$$

PROOF. Since $a\epsilon \leq \frac{\mathrm{rch}(\mathcal{M})}{10}$, the perturbation bound follows from Lemma 37, and the fact that

$$d_{\mathbb{R}^m}(x,y) \leq d_{\mathbb{R}^N|_{\mathcal{M}}}(x,y) \leq d_{\mathcal{M}}(x,y),$$

for any $x, y \in U = \pi_p(W)$. It remains to show that $B_{\mathbb{R}^m}(p;(a-1)\epsilon) \subset \pi_p(W)$. Using Lemma 34, we have that $B_{\mathbb{R}^m}(p;r) \subseteq \pi_p(W)$ if

$$\begin{aligned} r^2 \leq a^2\epsilon^2 - \left(\frac{a^2\epsilon^2}{2\mathrm{rch}(\mathcal{M})}\right)^2 &= a^2\epsilon^2\left(1 - \left(\frac{a\epsilon}{2\mathrm{rch}(\mathcal{M})}\right)^2\right) \\ &\leq a^2\epsilon^2\left(1 - \left(\frac{1}{20}\right)^2\right). \end{aligned}$$

Thus we require $r \leq \sqrt{\frac{399}{400}}a\epsilon$, which is satisfied by $r = (a-1)\epsilon$ if $a \leq 400$. □

With Lemma 38 we can establish the sampling requirements that ensure that we locally meet the criteria needed to apply Theorem 33.

THEOREM 39 (EQUATING STRUCTURES). *Suppose $\mathcal{P} \subset \mathcal{M}$ is such that for every $p \in \mathcal{P}$, the set $\mathsf{P} = B_{\mathbb{R}^N|_{\mathcal{M}}}(p;7\epsilon) \cap \mathcal{P}$ is an ϵ-sample set for $B_{\mathbb{R}^m}(p;5\epsilon)$ in the local Euclidean metric on $W = B_{\mathbb{R}^N|_{\mathcal{M}}}(p;7\epsilon)$, and P is δ-generic for $\{p\}$, with $\delta = \nu_0\epsilon$. If*

$$\epsilon \leq \frac{\nu_0^4\, rch(\mathcal{M})}{23520\sqrt{m}}, \tag{9}$$

then $\mathrm{Del}_{\mathbb{R}^N|_{\mathcal{M}}}(\mathcal{P}) = \mathrm{Del}_{\mathcal{M}}(\mathcal{P})$ and they are manifold complexes.

PROOF. As discussed in Section 5.1, if P is an ϵ-sample set for $B_{\mathbb{R}^m}(p;5\epsilon)$, then $p \in \mathsf{P}_I$. Theorem 33 guarantees that $\mathrm{star}(p; \mathrm{Del}(\mathsf{P})) = \mathrm{star}(p; \mathrm{Del}_{\mathbb{R}^N|_{\mathcal{M}}}(\mathsf{P})) = \mathrm{star}(p; \mathrm{Del}_{\mathcal{M}}(\mathsf{P}))$ if $\rho \leq \frac{\nu_0^4\epsilon}{48\sqrt{m}}$, and, by Lemma 38, this will be true with $a = 7$ if $\frac{490\epsilon}{\mathrm{rch}(\mathcal{M})} \leq \frac{\nu_0^4}{48\sqrt{m}}$, and we obtain the required bound on ϵ. Thus the star of every vertex in $\mathrm{Del}_{\mathcal{M}}(\mathcal{P})$ is equal to the star of that point in the local Euclidean metric, and likewise for

$\mathrm{Del}_{\mathbb{R}^N|_{\mathcal{M}}}(\mathcal{P})$. The claim follows since $\sigma \in \mathrm{Del}_{\mathcal{M}}(\mathcal{P})$ if and only if it is in the local Euclidean Delaunay triangulation of every one of its vertices, and likewise for the simplices in $\mathrm{Del}_{\mathbb{R}^N|_{\mathcal{M}}}(\mathcal{P})$. □

The awkwardness of the stated sampling conditions in Theorem 39 can be alleviated somewhat by recognizing that if d is a metric on U such that $d(x, y) > d_{\mathbb{R}^m}(x, y)$ for all $x, y \in U$, then if P is an ϵ-sample with respect to d, then it is also an ϵ-sample with respect to $d_{\mathbb{R}^m}$. Thus if $\mathcal{P} \subset \mathcal{M}$ is an ϵ-sample with respect to the intrinsic metric $d_{\mathcal{M}}$, with a sampling radius satisfying Equation (9), then it will satisfy the density requirements in the local Euclidean metrics.

The condition of δ-genericity is not so easily transferred to the metric of the manifold. Our results in Section 4 depend on the m-simplices of Del(P) being δ-protected, and exploit the fact that Del(P) is then necessarily a triangulation. It is not clear that similar assertions can be made if protection is assumed only on the m-simplices of $\mathrm{Del}_d(\mathsf{P})$.

6. CONCLUSIONS

We have quantified the close relationship between the genericity of a point set, the quality of the simplices in the Delaunay complex, and its stability under perturbation.

We have produced extrinsic sampling conditions which will guarantee that the intrinsic Delaunay complex is a manifold and coincides with the restricted Delaunay complex. In a companion paper [BDG12a], we have adapted an algorithm for the tangential Delaunay complex so as to meet these sampling conditions, guaranteeing that the tangential Delaunay complex also coincides with the intrinsic Delaunay complex. In this way we are able to exploit existing structural results [BG11] and demonstrate that the intrinsic Delaunay complex is a triangulation.

We have relied on an embedding of $\mathcal{M}$ in $\mathbb{R}^N$. In future work we aim to develop intrinsic sampling conditions. Another possible direction is to develop sampling conditions for manifolds that are not smooth, since our stability results impose no such constraints on the metric.

Acknowledgements

The authors would like to thank David Cohen-Steiner for suggesting the proof technique employed in Lemma 21.

7. REFERENCES

[AGG+10] P. K. Agarwal, J. Gao, L. Guibas, H. Kaplan, V. Koltun, N. Rubin, and M. Sharir. Kinetic stable Delaunay graphs. In *SoCG*, pages 127–136, 2010.

[BDG12a] J.-D. Boissonnat, R. Dyer, and A. Ghosh. Construction of stable Delaunay structures for manifolds. Technical Report CGL-TR-??, Computational Geometric Learning, 2012. http://cgl.uni-jena.de/Publications/WebHome.

[BDG12b] J.-D. Boissonnat, R. Dyer, and A. Ghosh. Stability of Delaunay structures for manifolds. Technical Report CGL-TR-??, Computational Geometric Learning, 2012. http://cgl.uni-jena.de/Publications/WebHome.

[BDGO11] J.-D. Boissonnat, R. Dyer, A. Ghosh, and S. Oudot. Equating the witness and restricted Delaunay complexes. Technical Report CGL-TR-24, Computational Geometric Learning, 2011. http://cgl.uni-jena.de/Publications/WebHome.

[BG11] J.-D. Boissonnat and A. Ghosh. Manifold reconstruction using tangential Delaunay complexes. Technical Report N° 7142 v.3, INRIA, 2011.

[BS04] D. Bandyopadhyay and J. Snoeyink. Almost-Delaunay simplices: nearest neighbor relations for imprecise points. In *SODA*, pages 410–419, 2004.

[BWY11] J.-D. Boissonnat, C. Wormser, and M. Yvinec. Anisotropic Delaunay mesh generation. Research Report RR-7712, INRIA, 2011.

[CDR05] S.-W. Cheng, T. K. Dey, and E. A. Ramos. Manifold reconstruction from point samples. In *SODA*, pages 1018–1027, 2005.

[Dan00] E. N. Dancer. Degree theory on convex sets and applications to bifurcation. In *Calculus of Variations and Partial Differential Equations*, pages 185–225. Springer-Verlag, 2000.

[Del34] B. Delaunay. Sur la sphère vide. *Izv. Akad. Nauk SSSR, Otdelenie Matematicheskii i Estestvennyka Nauk*, 7:793–800, 1934.

[dS08] V. de Silva. A weak characterisation of the Delaunay triangulation. *Geometriae Dedicata*, 135:39–64, 2008.

[ES97] H. Edelsbrunner and N. R. Shah. Triangulating topological spaces. *Int. J. Comput. Geometry Appl.*, 7(4):365–378, 1997.

[FKMS05] S. Funke, C. Klein, K. Mehlhorn, and S. Schmitt. Controlled perturbation for Delaunay triangulations. In *SODA*, pages 1047–1056, 2005.

[GW04] J. Giesen and U. Wagner. Shape dimension and intrinsic metric from samples of manifolds. *Discrete & Comp. Geom.*, 32:245–267, 2004.

[LL00] G. Leibon and D. Letscher. Delaunay triangulations and Voronoi diagrams for Riemannian manifolds. In *SoCG*, pages 341–349, 2000.

[LS03] F. Labelle and J. R. Shewchuk. Anisotropic Voronoi diagrams and guaranteed-quality anisotropic mesh generation. In *SoCG*, pages 191–200, 2003.

[Mun68] J. R. Munkres. *Elementary differential topology*. Princton University press, second edition, 1968.

[NSW08] P. Niyogi, S. Smale, and S. Weinberger. Finding the homology of submanifolds with high confidence from random samples. *Discrete & Comp. Geom.*, 39(1-3):419–441, 2008.

[Whi57] H. Whitney. *Geometric Integration Theory*. Princeton University Press, 1957.

Linear-Size Approximations to the Vietoris-Rips Filtration

Donald R. Sheehy
Inria Saclay
Geometrica Group
4 rue Jacques Monod
91893 ORSAY, France
don.r.sheehy@gmail.com

ABSTRACT

The Vietoris-Rips filtration is a versatile tool in topological data analysis. It is a sequence of simplicial complexes built on a metric space to add topological structure to an otherwise disconnected set of points. It is widely used because it encodes useful information about the topology of the underlying metric space. This information is often extracted from its so-called persistence diagram. Unfortunately, this filtration is often too large to construct in full. We show how to construct an $O(n)$-size filtered simplicial complex on an n-point metric space such that its persistence diagram is a good approximation to that of the Vietoris-Rips filtration. This new filtration can be constructed in $O(n \log n)$ time. The constant factors in both the size and the running time depend only on the doubling dimension of the metric space and the desired tightness of the approximation. For the first time, this makes it computationally tractable to approximate the persistence diagram of the Vietoris-Rips filtration across all scales for large data sets.

We describe two different sparse filtrations. The first is a zigzag filtration that removes points as the scale increases. The second is a (non-zigzag) filtration that yields the same persistence diagram. Both methods are based on a hierarchical net-tree and yield the same guarantees.

Categories and Subject Descriptors

F.2.2 [**Analysis of Algorithms and Problem Complexity**]: Nonnumerical Algorithms and Problems—*Computations on discrete structures, Geometrical problems and computations*

General Terms

Algorithms, Theory

Keywords

Vietoris-Rips Filtration, Persistent Homology, Metric Spanners, Doubling Dimension

1. INTRODUCTION

There is an extensive literature on the problem of computing sparse approximations to metric spaces (see the book [29] and references therein). There is also a growing literature on topological data analysis and its efforts to extract topological information from metric data (see the survey [3] and references therein). One might expect that topological data analysis would be a major user of metric approximation algorithms, especially given that topological data analysis often considers simplicial complexes that grow exponentially in the number of input points. Unfortunately, this is not the case. The benefits of a sparser representation are sorely needed, but it is not obvious how an approximation to the metric will affect the underlying topology. The goal of this paper is to bring together these two research areas and to show how to build sparse metric approximations that come with topological guarantees.

The target for approximation is the Vietoris-Rips complex, which has a simplex for every subset of input points with diameter at most some parameter α. The collection of Vietoris-Rips complexes at all scales yields the Vietoris-Rips filtration. The persistence algorithm takes this filtration and produces a persistence diagram representing the changes in topology corresponding to changes in scale [33]. The Vietoris-Rips filtration has become a standard tool in topological data analysis because it encodes relevant and useful information about the topology of the underlying of the underlying metric space [8]. It also extends easily to high dimensional data, general metric spaces, or even non-metric distance functions.

Unfortunately, the Vietoris-Rips filtration has a major drawback: It's huge! Even the k-skeleton (the simplices up to dimension k) has size $O(n^{k+1})$ for n points.

This paper proposes an alternative filtration called the sparse Vietoris-Rips filtration, which has size $O(n)$ and can be computed in $O(n \log n)$ time. Moreover, the persistence diagram of this new filtration is provably close to that of the Vietoris-Rips filtration. The constants depend only on the doubling dimension of the metric (defined below) and a user-defined parameter ε governing the tightness of the approximation. For the k-skeleton, the constants are bounded by $\left(\frac{1}{\varepsilon}\right)^{O(kd)}$.

The main tool we use to construct the sparse filtration is the net-tree of Har-Peled and Mendel [25]. Net-trees are closely related to hierarchical metric spanners [22, 23] and their construction is analogous to data structures used for nearest neighbor search in metric spaces [16, 12, 13].

Outline. After reviewing some related work and definitions in Sections 2 and 3, we explain how to perturb the input metric using weighted distances in Section 4. This perturbation is used in the definition of a sparse zigzag filtration in Section 5, i.e. one in which simplices are both added and removed as the scale increases. The full definition of the net-trees is given in Section 6. Using the properties of the net-tree and the perturbed distances, we prove in Section 7 that removing points from the filtration does not change the topology. This implies that the zigzag filtration does not actually zigzag at the homology level (Subsection 8.1). The zigzag filtration can then be converted into an ordinary (i.e. non-zigzag) filtration that also approximates the Vietoris-Rips filtration (Subsection 8.2). The theoretical guarantees are proven in Section 9. Subsection 9.1 proves that the resulting persistence diagrams are good approximations to the persistence diagram of the full Vietoris-Rips filtration. The size complexity of the sparse filtration is shown to be $O(n)$ in Subsection 9.2. Finally, in Section 10, we outline the $O(n \log n)$-time construction, which turns out to be quite easy once you have a net-tree.

2. RELATED WORK

The theory of persistent homology [21, 33] gives an algorithm for computing the persistent topological features of a complex that grows over time. It has been applied successfully to many problem domains, including image analysis[6], biology[30, 11], and sensor networks[19, 18]. See also the survey by Carlsson for background on the topological view of data [3]. It is also possible to consider the complexes that alternate between growing and shrinking in what is known as zigzag persistence [31, 4, 5, 27].

Due to the rapid blowup in the size of the Vietoris-Rips filtration, some attempts have been made to build approximations. Some notable examples include witness complexes [2, 24, 17] as well as the mesh-based methods of Hudson et al. in Euclidean spaces [26].

The work most similiar to the current paper is by Chazal and Oudot [10]. In that paper, they looked at a sequence of persistence diagrams on denser and denser subsamples. However, they were not able to combine these diagrams into a single diagram with a provable guarantee. Moreover, they were not able to prove general guarantees on the size of the filtration except under very strict assumptions on the data.

Recently, Zomorodian [32] and Attali et al. [1] have presented new methods for simplifying Vietoris-Rips complexes. These methods depend only on the combinatorial structure. However, they have not yielded results in simplifying filtrations, only static complexes. In this paper, we exploit the geometry to get topologically equivalent sparsification of an entire filtration.

3. BACKGROUND

Doubling metrics. For a point $p \in P$ and a set $S \subseteq P$, we will write $\mathbf{d}(p, S)$ to denote the minimum distance from p to S, i.e. $\mathbf{d}(p, S) = \min_{q \in S} \mathbf{d}(p, q)$. In a metric space $\mathcal{M} = (P, \mathbf{d})$, a **metric ball** centered at $p \in P$ with radius $r \in \mathbb{R}$ is the set $\mathbf{ball}(p, r) = \{q \in P : \mathbf{d}(p, q) \leq r\}$.

DEFINITION 1. *The* ***doubling constant*** λ *of a metric space* $\mathcal{M} = (P, \mathbf{d})$ *is the minimum number of metric balls of radius* r *required to cover any ball of radius* $2r$*. The* ***doubling dimension*** *is* $\mathbf{dim} = \lceil \lg \lambda \rceil$*. A metric space whose doubling dimension is bounded by a constant is called a* ***doubling metric****.*

The **spread** Δ of a metric space $\mathcal{M} = (P, \mathbf{d})$ is the ratio of the largest to smallest interpoint distances. A metric with doubling dimension d and spread Δ has at most $\Delta^{O(d)}$ points.

Simplicial Complexes. A **simplicial complex** X is a collection of **vertices** denoted $V(X)$ and a collection of subsets of $V(X)$ called **simplices** that is closed under the subset operation, i.e. $\sigma \subset \psi$ and $\psi \in X$ together imply $\sigma \in X$. The **dimension** of a simplex σ is $|\sigma| - 1$, where $|\cdot|$ denotes cardinality. Note that this definition is combinatorial rather than geometric. These abstract simplicial complexes are not necessarily embedded in a geometric space.

Homology. In this paper we will use simplicial homology over a field (see Munkres [28] for an accessible introduction to algebraic topology). Thus, given a space X, the homology groups $\mathrm{H}_i(X)$ are vector spaces for each i. Let $\mathrm{H}_*(X)$ denote the collection of these homology groups for all i.

The star subscript denotes the homomorphism of homology groups induced by a map between spaces, i.e. $f : X \to Y$ induces $f_\star : \mathrm{H}_*(X) \to \mathrm{H}_*(Y)$. We recall the functorial properties of the Homology operator, $\mathrm{H}_*(\cdot)$. In particular, $(f \circ g)_\star = f_\star \circ g_\star$ and $\mathrm{id}_{X\star} = \mathrm{id}_{\mathrm{H}_*(X)}$, where id indicates the identity map.

Persistence Modules and Diagrams. A **filtration** is a nested sequence of topological spaces: $X_1 \subseteq X_2 \subseteq \cdots \subseteq X_n$. If the spaces are simplicial complexes (as with all the filtrations in this paper), then it is called a **filtered simplicial complex**.

A **persistence module** is a sequence of Homology groups connected by homomorphisms:

$$\mathrm{H}_*(X_1) \to \mathrm{H}_*(X_2) \to \cdots \to \mathrm{H}_*(X_n).$$

The homology functor turns a filtration with inclusion maps $X_1 \hookrightarrow X_2 \hookrightarrow \cdots$ into a persistence module, but as we will see, this is not the only way to get one.

One can also consider **zigzag filtrations**, which allow the inclusions to go in both directions: $X_1 \subseteq X_2 \supseteq X_3 \subseteq \cdots$. The resulting module is called a **zigzag module**.

$$\mathrm{H}_*(X_1) \to \mathrm{H}_*(X_2) \leftarrow \mathrm{H}_*(X_3) \to \cdots.$$

The **persistence diagram** of a persistence module is a multiset of points in $(\mathbb{R} \cup \{\infty\})^2$. Each point of the diagram represents a topological feature. The x and y coordinates of the points are the *birth* and *death* times of the feature and correspond to the indices in the persistence module where that feature appears and disappears. Points far from the diagonal persisted for a long time, while those "non-persistent" points near the diagonal may be considered *topological noise*. By convention, the persistence diagram also contains every point (x, x) of the diagonal with infinite multiplicity.

Given a filtration $\mathcal{F}$, we let $\mathrm{D}\mathcal{F}$ denote the persistence diagram of the persistence module generated by $\mathcal{F}$. The persistence algorithm computes a persistence diagram from $\mathcal{F}$ [33]. It is also known how to compute a persistence diagram when $\mathcal{F}$ is a zigzag filtration [4, 27].

Approximating Persistence Diagrams. Given two filtrations $\mathcal{F}$ and $\mathcal{G}$, we say that the persistence diagram $\mathrm{D}\mathcal{F}$ is a **c-approximation to the diagram** $\mathrm{D}\mathcal{G}$ if there is a bijection $\phi : \mathrm{D}\mathcal{F} \to \mathrm{D}\mathcal{G}$ such that for each $p \in \mathrm{D}\mathcal{F}$, the birth times of p and $\phi(p)$ differ by at most a factor of c and the death times also differ by at most a factor of c. The reader familiar with stability results for persistent homology [15, 7] will recognize this as bounding the ℓ_∞-bottleneck distance between the persistence diagrams after reparameterizing the filtrations on a log-scale.

We will make use of two standard results on persistence diagrams. The first gives a sufficient condition for two persistence modules to yield identical persistence diagrams.

THEOREM 3.1. *[Persistence Equivalence Theorem [20, page 159]] Consider two sequences of vector spaces connected by homomorphisms $\phi_i : U_i \to V_i$:*

$$\begin{array}{ccccccccc} V_0 & \longrightarrow & V_1 & \longrightarrow & \cdots & \longrightarrow & V_{n-1} & \longrightarrow & V_n \\ \uparrow & & \uparrow & & & & \uparrow & & \uparrow \\ U_0 & \longrightarrow & U_1 & \longrightarrow & \cdots & \longrightarrow & U_{n-1} & \longrightarrow & U_n \end{array}$$

If the vertical maps are isomorphisms and all squares commute then the persistence diagrams defined by the U_i is the same as that defined by the V_i.

We prove approximation guarantees for persistence diagrams using the following lemma, which is a direct corollary of the Strong Stability Theorem of Chazal et al. [7] rephrased in the language of approximate persistence diagrams.

LEMMA 3.2. *[Persistence Approximation Lemma] For any two filtrations $\mathcal{A} = \{A_\alpha\}_{\alpha \ge 0}$ and $\mathcal{B} = \{B_\alpha\}_{\alpha \ge 0}$, if $A_{\alpha/c} \subseteq B_\alpha \subseteq A_{c\alpha}$ for all $\alpha \ge 0$ then the persistence diagram $\mathrm{D}\mathcal{A}$ is a c-approximation to the persistence diagram of $\mathrm{D}\mathcal{B}$.*

Contiguous Simplicial Maps. Contiguity gives a discrete version of homotopy theory for simplicial complexes.

DEFINITION 2. *Let X and Y be simplicial complexes. A **simplicial map** $f : X \to Y$ is a function that maps vertices of X to vertices of Y and $f(\sigma) = \bigcup_{v \in \sigma} f(v)$ is a simplex of Y for all $\sigma \in X$.*

A simplicial map is determined by its behavior on the vertex set. Consequently, we will abuse notation slightly and identify maps between vertex sets and maps between simplices. When it is relevant and non-obvious, we will always prove that the resulting map between simplicial complexes is simplicial.

DEFINITION 3. *Two simplicial maps $f, g : X \to Y$ are **contiguous** if $f(\sigma) \cup g(\sigma) \in Y$ for all $\sigma \in X$.*

DEFINITION 4. *For any pair of topological spaces $X \subseteq Y$, a map $f : Y \to X$ is a **retraction** if $f(x) = x$ for all $x \in X$. Equivalently, $f \circ i = \mathrm{id}_X$ where $i : X \hookrightarrow Y$ is the inclusion map.*

The theory of contiguity is a simplicial analogue of homotopy theory. If two simplicial maps are contiguous then they induce identical homomorphisms at the homology level [28, §12]. The following lemma gives a homology analogue of a deformation retraction.

LEMMA 3.3. *Let X and Y be simplicial complexes such that $X \subseteq Y$ and let $i : X \hookrightarrow Y$ be the canonical inclusion map. If there exists a simplicial retraction $\pi : Y \to X$ such that $i \circ \pi$ and id_Y are contiguous, then i induces an isomorphism $i_\star : \mathrm{H}_*(X) \to \mathrm{H}_*(Y)$ between the corresponding homology groups.*

PROOF. Since $i \circ \pi$ and id_Y are contiguous, the induced homomorphisms $(i \circ \pi)_\star : \mathrm{H}_*(Y) \to \mathrm{H}_*(Y)$ and $\mathrm{id}_{Y\star} : \mathrm{H}_*(Y) \to \mathrm{H}_*(Y)$ are identical [28, §12]. Since $\mathrm{id}_{Y\star} = (i \circ \pi)_\star = i_\star \circ \pi_\star$ is an isomorphism, it follows that $i_\star$ is surjective.

Since π is a retraction, $\pi \circ i = \mathrm{id}_X$ and thus $(\pi \circ i)_\star : \mathrm{H}_*(X) \to \mathrm{H}_*(X)$ and $\mathrm{id}_{X\star} : \mathrm{H}_*(X) \to \mathrm{H}_*(X)$ are identical. Since $\mathrm{id}_{X\star} = (\pi \circ i)_\star = \pi_\star \circ i_\star$ is an isomorphism, it follows that $i_\star$ is injective.

Thus, $i_\star$ is an isomorphism because it is both injective and surjective. □

4. THE RELAXED VIETORIS-RIPS FILTRATION

In this section, we relax the input metric so that it is no longer a metric, but it will still be provably close to the input. The new distance adds a small weight to each point that grows with α. The intuition behind this process is illustrated in Figure 1.

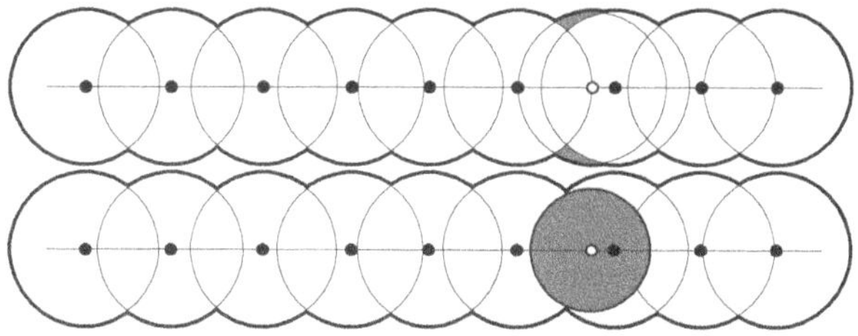

Figure 1: Top: Some points on a line. The white point contributes little to the union of α-balls. Bottom: Using the relaxed distance, the new α-ball is completely contained in the union of the other balls. Later, this property is the key that allows us to remove this point without changing the topology.

Throughout, we assume the user-defined parameter $\varepsilon \le \frac{1}{2}$ is fixed. Each point p is assigned a **deletion time** $t_p \in \mathbb{R}_{\ge 0}$. The specific choice of t_p will come from the net-tree construction in Section 6. For now, we will assume the deletion times are given, assuming only that they are nonnegative. The weight $w_i(\alpha)$ of point p at scale α is defined as

$$w_p(\alpha) = \begin{cases} 0 & \text{if } \alpha \le (1-2\varepsilon)t_p \\ \frac{1}{2}(\alpha - (1-2\varepsilon)t_p) & \text{if } (1-2\varepsilon)t_p < \alpha < t_p \\ \varepsilon\alpha & \text{if } t_p \le \alpha \end{cases}$$

The **relaxed distance** at scale α is defined as

$$\hat{\mathbf{d}}_\alpha(p,q) = \mathbf{d}(p,q) + w_p(\alpha) + w_q(\alpha).$$

For any pair $p, q \in P$, the relaxed distance $\hat{\mathbf{d}}_\alpha(p,q)$ is monotonically non-decreasing in α. In particular, $\hat{\mathbf{d}}_\alpha \ge \hat{\mathbf{d}}_0 = \mathbf{d}$ for all $\alpha \ge 0$. Although distances can grow as α grows, this growth is sufficiently slow to allow the following lemma which will be useful later.

LEMMA 4.1. *If* $\hat{\mathbf{d}}_\alpha(p,q) \le \alpha \le \beta$ *then* $\hat{\mathbf{d}}_\beta(p,q) \le \beta$.

PROOF. The weight of a point is $\frac{1}{2}$-Lipschitz in α, so $w_p(\beta) \le w_p(\alpha) + \frac{1}{2}|\beta - \alpha|$, and similarly, $w_q(\beta) \le w_q(\alpha) + \frac{1}{2}|\beta - \alpha|$. So,

$$\begin{aligned}\hat{\mathbf{d}}_\beta(p,q) &= \mathbf{d}(p,q) + w_p(\beta) + w_q(\beta)\\ &\le \mathbf{d}(p,q) + w_p(\alpha) + w_q(\alpha) + (\beta - \alpha)\\ &= \hat{\mathbf{d}}_\alpha(p,q) + \beta - \alpha\\ &\le \beta\end{aligned}$$

□

Given a set P, a distance function $\mathbf{d}' : P \times P \to \mathbb{R}$, and a scale parameter $\alpha \in \mathbb{R}$, we can construct a **Vietoris-Rips complex**

$$\mathrm{VR}(P, \mathbf{d}', \alpha) = \{\sigma \subset P : \mathbf{d}'(p,q) \le \alpha \text{ for all } p,q \in \sigma\}.$$

The Vietoris-Rips complex associated with the input metric space $(P, \mathbf{d})$ is $\mathcal{R}_\alpha = \mathrm{VR}(P, \mathbf{d}, \alpha)$. The **relaxed Vietoris-Rips complex** is $\hat{\mathcal{R}}_\alpha = \mathrm{VR}(P, \hat{\mathbf{d}}_\alpha, \alpha)$.

By considering the family of Vietoris-Rips complexes for all values of $\alpha \ge 0$, we get the **Vietoris-Rips filtration**, $\mathcal{R} = \{\mathcal{R}_\alpha\}_{\alpha \ge 0}$. Similarly, we may define the **relaxed Vietoris-Rips filtration**, $\hat{\mathcal{R}} = \{\hat{\mathcal{R}}_\alpha\}_{\alpha \ge 0}$. Lemma 4.1 implies that $\hat{\mathcal{R}}$ is indeed a filtration. The filtrations $\mathcal{R}$ and $\hat{\mathcal{R}}$ are very similar. The following lemma makes this similarity precise via a multiplicative interleaving.

LEMMA 4.2. *For all* $\alpha \ge 0$, $\mathcal{R}_{\frac{\alpha}{1+\varepsilon_0}} \subseteq \hat{\mathcal{R}}_\alpha \subseteq \mathcal{R}_\alpha$, *where* $\varepsilon_0 = \frac{2\varepsilon}{1-2\varepsilon}$.

PROOF. To prove inclusions between Vietoris-Rips complexes, it suffices to prove inclusion of the edge sets. For the first inclusion, we must prove that for any pair p,q, if $\mathbf{d}(p,q) \le \frac{\alpha}{1+\varepsilon_0}$ then $\hat{\mathbf{d}}_\alpha(p,q) \le \alpha$. Fix any such pair p,q. By definition, $w_p(\alpha) \le \varepsilon\alpha$ and $w_q(\alpha) \le \varepsilon\alpha$. So,

$$\hat{\mathbf{d}}_\alpha(p,q) = \mathbf{d}(p,q) + w_p(\alpha) + w_q(\alpha) \le \frac{\alpha}{1+\varepsilon_0} + 2\varepsilon\alpha = \alpha.$$

For the second inclusion, $\hat{\mathbf{d}}_\alpha \ge \mathbf{d}$. So, if $\hat{\mathbf{d}}_\alpha(p,q) \le \alpha$ then $\mathbf{d}(p,q) \le \alpha$ as well. Thus any edge of $\hat{\mathcal{R}}_\alpha$ is also an edge of $\mathcal{R}_\alpha$. □

5. THE SPARSE ZIGZAG VIETORIS-RIPS FILTRATION

We will construct a sparse subcomplex of the relaxed Vietoris-Rips complex $\hat{\mathcal{R}}_\alpha$ that is guaranteed to have linear size for any α. In fact, we will get a zigzag filtration that only has a linear total number of simplices, yet its persistence diagram is identical to that of the relaxed Vietoris-Rips filtration.

We define the **open net** $\mathcal{N}_\alpha$ at scale α to be the subset of P with deletion time greater than α, i.e.

$$\mathcal{N}_\alpha = \{p \in P : t_p > \alpha\}.$$

Similarly, the **closed net** at scale α is

$$\overline{\mathcal{N}}_\alpha = \{p \in P : t_p \ge \alpha\}.$$

The **sparse zigzag Vietoris-Rips complex** $\mathcal{Q}_\alpha$ at scale α is just the subcomplex of $\hat{\mathcal{R}}_\alpha$ induced on the vertices of $\mathcal{N}_\alpha$. Formally,

$$\mathcal{Q}_\alpha = \{\sigma \in \hat{\mathcal{R}}_\alpha : \sigma \subseteq \mathcal{N}_\alpha\} = \mathrm{VR}(\mathcal{N}_\alpha, \hat{\mathbf{d}}_\alpha, \alpha).$$

We also define a closed version of the sparse zigzag Vietoris-Rips complex:

$$\overline{\mathcal{Q}}_\alpha = \mathrm{VR}(\overline{\mathcal{N}}_\alpha, \hat{\mathbf{d}}_\alpha, \alpha).$$

Note that if $\alpha \ne t_p$ for all $p \in P$ then $\mathcal{N}_\alpha = \overline{\mathcal{N}}_\alpha$ and $\mathcal{Q}_\alpha = \overline{\mathcal{Q}}_\alpha$.

The complexes $\mathcal{R}_\alpha$, $\hat{\mathcal{R}}_\alpha$, $\mathcal{Q}_\alpha$, and $\overline{\mathcal{Q}}_\alpha$ are well-defined for all $\alpha \ge 0$, however, they only changes at discrete scales. Let $A = \{a_i\}_{i \in \mathbb{N}}$ be an ordered, discrete set of nonnegative real numbers such that $t_p \in A$ for all $p \in P$ and $\alpha \in A$ for any pair p,q such that $\alpha = \hat{\mathbf{d}}_\alpha(p,q)$. That is, A contains every scale at which a combinatorial changes happens, either a point deletion or an edge insertion. This implies that $\mathcal{N}_{a_i} = \overline{\mathcal{N}}_{a_{i+1}}$ and thus, using Lemma 4.1, that $\mathcal{Q}_{a_i} \subseteq \overline{\mathcal{Q}}_{a_{i+1}}$.

The sparse Vietoris-Rips complexes can be arranged into a zigzag filtration $\mathcal{Q}$ as follows.

$$\overline{\mathcal{Q}}_{a_1} \hookleftarrow \mathcal{Q}_{a_1} \hookrightarrow \overline{\mathcal{Q}}_{a_2} \hookleftarrow \mathcal{Q}_{a_2} \hookrightarrow \cdots$$

We will return to $\mathcal{Q}$ later as it has some interesting properties. However, at this point, it is underspecified as we have not yet shown how to compute the deletion times for the vertices. The next section will fill this gap.

6. HIERARCHICAL NET-TREES

The following treatment of net-trees is adapted from the paper by Har-Peled and Mendel [25].

DEFINITION 5. *A **net-tree** of a metric* $\mathcal{M} = (P, \mathbf{d})$ *is a rooted tree* T *with vertex* $v \in T$ *having a **representative point*** $\mathrm{rep}(v) \in P$. *There are* $n = |P|$ *leaves, each represented by a different point of* P. *Each non-root vertex* $v \in T$ *has a unique **parent*** $\mathrm{par}(v)$. *The set of vertices with the same parent* v *are called the **children** of* v, *denoted* $\mathrm{child}(v)$. *If* $\mathrm{child}(v)$ *is nonempty then for some* $u \in \mathrm{child}(v)$, $\mathrm{rep}(u) = \mathrm{rep}(v)$. *The set* $P_v \subseteq P$ *denotes the points represented by the leaves of the subtree rooted at* v. *Each vertex* $v \in T$ *has an associated radius* $\mathrm{rad}(v)$ *satisfying the following two conditions.*

1. ***Covering Condition:*** $P_v \subset \mathbf{ball}(\mathrm{rep}(v), \mathrm{rad}(v))$, *and*

2. ***Packing Condition:*** *if* v *is not the root, then*

$$P \cap \mathbf{ball}(\mathrm{rep}(v), K_p \mathrm{rad}(\mathrm{par}(v))) \subseteq P_v,$$

where K_p *(the "p" is for "packing") is a constant independent of* $\mathcal{M}$ *and* n.

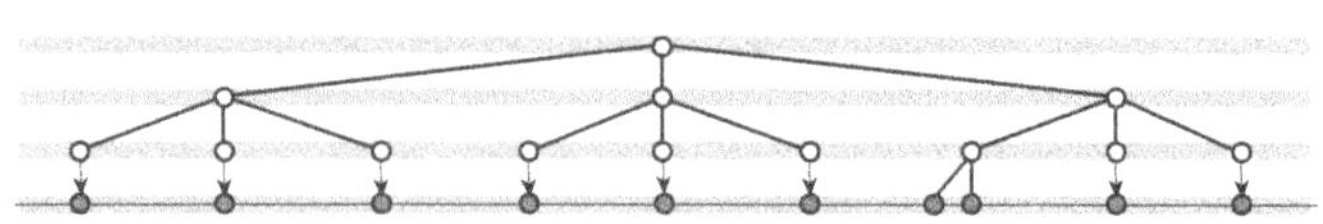

Figure 2: A net-tree is built over the set points from Figure 1. Each level of the tree represents a sparse approximation to the original point set at a different scale.

The radii of the net tree nodes are always some constant times larger than the radius of their children. Simple packing arguments guarantee that no node of the tree has more than $\lambda^{O(1)}$ children, where lambda is the doubling constant of the metric. The whole tree can be constructed in $O(n \log n)$ randomized time or $O(n \log \Delta)$ deterministically [25]. Moreover, it is important to note the construction does not require that we know the doubling dimension in advance.

Given a net-tree T for $\mathcal{M} = (P, \mathbf{d})$ and a point $p \in P$, let v_p denote the least common ancestor of all vertices in T represented by p. For each $p \in P$ the deletion time t_p is defined as

$$t_p = \frac{1}{\varepsilon(1-2\varepsilon)} \mathrm{rad}(\mathrm{par}(v_p)).$$

This is just the radius of the parent of v_p with a small scaling factor included for technical reasons. When the scale α reaches t_p, we remove point p from the (zigzag) filtration.

For a fixed scale $\alpha \in \mathbb{R}$, the set $\mathcal{N}_\alpha$ is a subset of points of P induced by the net-tree. The sets $\mathcal{N}_\alpha$ are the **nets** of the net-tree. For any α it satisfies a packing condition and a covering condition as defined in the following lemma.

LEMMA 6.1. *Let $\mathcal{M} = (P, \mathbf{d})$ be a metric space and let T be a net-tree for $\mathcal{M}$. For all $\alpha \geq$, the net $\mathcal{N}_\alpha$ induced by T at scale α satisfies the following two properties.*

1. ***Covering Condition:*** *For all $p \in P$, $\mathbf{d}(p, \mathcal{N}_\alpha) \leq \varepsilon(1-2\varepsilon)\alpha$.*
2. ***Packing Condition:*** *For all distinct $p, q \in \mathcal{N}_\alpha$, $\mathbf{d}(p,q) \geq K_p \varepsilon(1-2\varepsilon)\alpha$*

PROOF. First, we prove that the covering condition holds. Fix any $p \in P$. The statement is trivial if $p \in \mathcal{N}_\alpha$ so we may assume that $t_p < \alpha$.

Let v be the lowest ancestor of p in T such that $\mathrm{rep}(v) \in \mathcal{N}_\alpha$. Let u be the ancestor of p among the children of v. If $q = \mathrm{rep}(u)$ then $t_q = \frac{\mathrm{rad}(v)}{\varepsilon(1-2\varepsilon)}$. By our choice of v, $q \notin \mathcal{N}_\alpha$ and thus $t_q \leq \alpha$. It follows that $\mathrm{rad}(v) \leq \varepsilon(1-2\varepsilon)\alpha$. Thus, $\mathbf{d}(p, \mathcal{N}_\alpha) \leq \mathbf{d}(p, \mathrm{rep}(v)) \leq \mathrm{rad}(v) < \varepsilon(1-2\varepsilon)\alpha$.

We now prove that the packing condition holds. Let p, q be any two distinct points of $\mathcal{N}_\alpha$. Without loss of generality, assume $t_p \leq t_q$. Thus, $q \notin P_{v_p}$, where (as before) $v_p = \mathrm{lca}\{u \in T : \mathrm{rep}(u) = p\}$. Since $p \in \mathcal{N}_\alpha$, $\alpha \leq t_p = \frac{\mathrm{rad}(\mathrm{par}(v_p))}{\varepsilon(1-2\varepsilon)}$. Therefore, using the packing condition on the net-tree T, $\mathbf{d}(p,q) \geq K_p \mathrm{rad}(\mathrm{par}(v_p)) \geq K_p \varepsilon(1-2\varepsilon)\alpha$. □

A subset that satisfies this type of packing and covering conditions is sometimes referred to as a metric space net (not to be confused with a range space net) or, more accurately, as a Delone set [14].

7. TOPOLOGY-PRESERVING SPARSIFICATION

In this Section, we make the intuition of Figure 1 concrete by showing that deleting a vertex p (and its incident simplices) from the relaxed Vietoris-Rips complex $\hat{\mathcal{R}}_{t_p}$ does not change the topology.

For any $\alpha \geq 0$, we define the "projection" of P onto $\mathcal{N}_\alpha$ as

$$\pi_\alpha(p) = \begin{cases} p & \text{if } p \in \mathcal{N}_\alpha \\ \operatorname*{argmin}_{q \in \mathcal{N}_\alpha} \hat{\mathbf{d}}_\alpha(p,q) & \text{otherwise.} \end{cases} \quad (1)$$

The following lemma shows that the distance from a point to its projection is bounded by the difference in the weights of the point and its projection.

LEMMA 7.1. *For all $p \in P$, $\mathbf{d}(p, \pi_\alpha(p)) \leq w_p(\alpha) - w_{\pi_\alpha(p)}(\alpha)$.*

PROOF. Fix any $p \in P$. We first prove that if $\mathbf{d}(p,q) \leq w_p(\alpha) - w_q(\alpha)$ for some $q \in \mathcal{N}_\alpha$, then it holds for $q = \pi_\alpha(p)$. If we have such a q, then the definitions of $\hat{\mathbf{d}}_\alpha$ and π_α imply the following.

$$\begin{aligned} \mathbf{d}(p, \pi_\alpha(p)) &= \hat{\mathbf{d}}_\alpha(p, \pi_\alpha(p)) - w_p(\alpha) - w_{\pi_\alpha(p)}(\alpha) \\ &\leq \hat{\mathbf{d}}_\alpha(p,q) - w_p(\alpha) - w_{\pi_\alpha(p)}(\alpha) \\ &\leq \mathbf{d}(p,q) + w_q(\alpha) - w_{\pi_\alpha(p)}(\alpha) \\ &\leq w_p(\alpha) - w_{\pi_\alpha(p)}(\alpha). \end{aligned}$$

So, it will suffice to find a $q \in \mathcal{N}_\alpha$ such that $\mathbf{d}(p,q) \leq w_p(\alpha) - w_q(\alpha)$. If $p \in \mathcal{N}_\alpha$ then this is trivial. So we may assume $p \notin \mathcal{N}_\alpha$ and therefore $t_p \leq \alpha$ and $w_p(\alpha) = \varepsilon\alpha$.

Let $u \in T$ be the ancestor of p such that $\mathrm{rad}(u) < \frac{\varepsilon\alpha}{1-\varepsilon}$ and $\mathrm{rad}(\mathrm{par}(u)) \geq \frac{\varepsilon\alpha}{1-\varepsilon}$. Let $q = \mathrm{rep}(u)$. Since

$$t_q \geq \frac{1}{\varepsilon(1-2\varepsilon)} \mathrm{rad}(\mathrm{par}(u)) \geq \frac{\alpha}{1-2\varepsilon},$$

it follows that $w_q(\alpha) = 0$ and that $q \in \mathcal{N}_\alpha$. Finally, since $p \in P_u$, $\mathbf{d}(p,q) \leq \mathrm{rad}(u) \leq \varepsilon\alpha = w_p(\alpha) - w_q(\alpha)$. □

By bounding the distance between points and their projections, we can now show that distances in the projection do not grow.

LEMMA 7.2. *For all $p, q \in P$ and all $\alpha \geq 0$, $\hat{\mathbf{d}}_\alpha(\pi_\alpha(p), q) \leq \hat{\mathbf{d}}_\alpha(p,q)$.*

PROOF. The bound follows from the definition of $\hat{\mathbf{d}}_\alpha$, the triangle inequality, and Lemma 7.1. □

LEMMA 7.3. *Let $\alpha \geq 0$ be a fixed constant. Let X be a set of points such that $\mathcal{N}_\alpha \subseteq X \subseteq P$ and let $K = \mathrm{VR}(X, \hat{\mathbf{d}}_\alpha, \alpha)$. The inclusion map $i : \mathcal{Q}_\alpha \hookrightarrow K$ induces an isomorphism at the homology level.*

PROOF. The map $K \to \mathcal{Q}_\alpha$ induced by π_α is a retraction because $\mathcal{Q}_\alpha \subseteq K$ and π_α is a retraction onto $\mathcal{N}_\alpha$, the vertex set of $\mathcal{Q}_\alpha$. By Lemma 3.3, it will suffice to prove that π_α is simplicial and that $i \circ \pi_\alpha$ is contiguous to the identity map on K. Since $\mathcal{Q}_\alpha$ and K are Vietoris-Rips complexes, it will suffice to prove these facts for the edges:

1. π_α is simplicial: for all $p, q \in X$, if $\hat{\mathbf{d}}_\alpha(p,q) \leq \alpha$ then $\hat{\mathbf{d}}_\alpha(\pi_\alpha(p), \pi_\alpha(q)) \leq \alpha$, and
2. $i \circ \pi_\alpha$ and id_K are contiguous: for all $p, q \in X$, if $\hat{\mathbf{d}}_\alpha(p,q) \leq \alpha$ then all six edges of the tetrahedron $\{p, q, \pi_\alpha(p), \pi_\alpha(q)\}$ are in K.

The first statement follows from two successive applications of Lemmas 7.2. The second statement follows from Lemma 7.1 for the edges $\{p, \pi_\alpha(p)\}$ and $\{q, \pi_\alpha(q)\}$ and from Lemma 7.2 for the other edges. □

COROLLARY 7.4. *For all $\alpha \in A$, the inclusions $f : \mathcal{Q}_\alpha \hookrightarrow \overline{\mathcal{Q}}_\alpha$, $g : \mathcal{Q}_\alpha \hookrightarrow \hat{\mathcal{R}}_\alpha$, and $h : \overline{\mathcal{Q}}_\alpha \hookrightarrow \hat{\mathcal{R}}_\alpha$ induce isomorphisms at the homology level.*

PROOF. The inclusions f and g induce isomorphisms by applying Lemma 7.3 with $X = \overline{\mathcal{N}}_\alpha$ and $X = P$ respectively. Composing the inclusions, we get that $g = h \circ f$. Thus, at the homology level, we get $h_\star = g_\star \circ f_\star^{-1}$ is also an isomorphism. □

8. STRAIGHTENING OUT THE ZIGZAGS

In this section, we show two different ways in which zigzag persistence may be avoided. First, in Subsection 8.1, we show that the sparse zigzag filtration $\mathcal{Q}$ does not zigzag at the homology level. Then, in Subsection 8.2, we show how to modify the zigzag filtration so it does not zigzag as a filtration either.

The advantage of the non-zigzagging filtration is that it allows one to use the standard persistence algorithm, but it has larger size in the intermediate complexes. As we will see in Subsection 9.2, the total size is still linear.

8.1 Reversing Homology Isomorphisms

The backwards arrows in the zigzag filtration $\mathcal{Q}$ all induce isomorphisms. At the homology level, these isomorphisms can be replaced by their inverses to give a persistence module that does not zigzag. That is, the zigzag module

$$\cdots \to \mathrm{H}_*(\overline{\mathcal{Q}}_\alpha) \overset{\cong}{\leftarrow} \mathrm{H}_*(\mathcal{Q}_\alpha) \to \mathrm{H}_*(\overline{\mathcal{Q}}_\beta) \overset{\cong}{\leftarrow} \mathrm{H}_*(\mathcal{Q}_\beta) \to \cdots$$

can be transformed into

$$\cdots \to \mathrm{H}_*(\overline{\mathcal{Q}}_\alpha) \overset{\cong}{\to} \mathrm{H}_*(\mathcal{Q}_\alpha) \to \mathrm{H}_*(\overline{\mathcal{Q}}_\beta) \overset{\cong}{\to} \mathrm{H}_*(\mathcal{Q}_\beta) \to \cdots.$$

The latter module implies the existence of another that only uses the closed sparse Vietoris-Rips complexes:

$$\cdots \to \mathrm{H}_*(\overline{\mathcal{Q}}_\alpha) \to \mathrm{H}_*(\overline{\mathcal{Q}}_\beta) \to \cdots.$$

Note that this module does away with the duplication of indices needed for the zigzag. In these various transformations, we have only reversed or concatenated isomorphisms, thus we have not changed the rank of any induced map $\mathrm{H}_*(\overline{\mathcal{Q}}_\alpha) \to \mathrm{H}_*(\overline{\mathcal{Q}}_\beta)$. As a result the persistence diagram $\mathrm{D}\mathcal{Q}$ is unchanged.

This is novel in that we construct a zigzag filtration and we apply the zigzag persistence algorithm, but we are really computing the diagram of a persistence module that does not zigzag. The zigzagging can then be interpreted as sparsifying the complex without changing the topology.

8.2 A Sparse Filtration without the Zigzag

The preceding subsection showed that the sparse zigzag Vietoris-Rips filtration does not zigzag as a persistence module. This hints that it is possible to construct a filtration that does not zigzag with the same persistence diagram. Indeed, this is possible using the filtration

$$\mathcal{S} = \{\mathcal{S}_{a_k}\}_{a_k \in A}, \quad \text{where} \quad \mathcal{S}_{a_k} = \bigcup_{i=1}^{k} \overline{\mathcal{Q}}_{a_i}.$$

We first prove that $\mathrm{H}_*(\overline{\mathcal{Q}}_{a_k})$ and $\mathrm{H}_*(\mathcal{S}_{a_k})$ are isomorphic.

LEMMA 8.1. *For all $a_k \in A$, the inclusion $h : \overline{\mathcal{Q}}_{a_k} \hookrightarrow \mathcal{S}_{a_k}$ induces a homology isomorphism.*

PROOF. We define some intermediate complexes that interpolate between $\overline{\mathcal{Q}}_{a_k}$ and $\mathcal{S}_{a_k}$.

$$T_{i,k} = \bigcup_{j=i}^{k} \overline{\mathcal{Q}}_{a_j}.$$

In particular, we have that $\overline{\mathcal{Q}}_{a_k} = T_{k,k}$ and $\mathcal{S}_{a_k} = T_{1,k}$. The map h can be expressed as $h = h_1 \circ \cdots \circ h_{k-1}$, where $h_i : T_{i+1,k} \hookrightarrow T_{i,k}$ is an inclusion. It will suffice to prove that each h_i induces an isomorphism at the homology level for each $i = 1 \ldots k-1$. By Lemma 3.3, it will suffice to show that the projection $\pi_{a_i} : T_{i,k} \to T_{i+1,k}$ is a simplicial retraction and $h_i \circ \pi_{a_i}$ and $\mathrm{id}_{T_{i,k}}$ are contiguous.

Let $\sigma \in T_{i,k}$ be any simplex. So, $\sigma \in \overline{\mathcal{Q}}_{a_j}$ for some integer j such that $i \leq j \leq k$.

First, we prove that π_{a_i} is a retraction. If $\sigma \in T_{i+1,k}$ then $j \geq i+1$. So, $\sigma \subseteq \overline{\mathcal{N}}_{a_j} \subseteq \mathcal{N}_{a_i}$ and thus $\pi_{a_i}(\sigma) = \sigma$ because π_{a_i} is a retraction onto $\mathcal{N}_{a_i}$ by definition when viewed as a function on the vertex sets.

Second, we show that π_{a_i} is a simplicial map from $T_{i,k}$ to $T_{i+1,k}$. Since it is a retraction, it only remains to show that $\pi_{a_i}(\sigma) \in T_{i,k}$ when $\sigma \in T_{i,k} \setminus T_{i+1,k}$, i.e. when $j = i$. In this case, $\pi_{a_i}(\sigma) \in \mathcal{Q}_{a_i}$ because $\pi_{a_i} : \overline{\mathcal{Q}}_{a_i} \to \mathcal{Q}_{a_i}$ is simplicial (as shown in the proof of Lemma 7.3). Since $\mathcal{Q}_{a_i} \subseteq \overline{\mathcal{Q}}_{a_{i+1}} \subseteq T_{i,k}$, it follows that $\pi_{a_i}(\sigma) \in T_{i,k}$ as desired.

Last, we prove contiguity. We need to prove that $\sigma \cup \pi_{a_i}(\sigma) \in T_{i,k}$. If $j > i$, then $\sigma \cup \pi_{a_i}(\sigma) = \sigma \in T_{i,k}$ as desired. If $i = j$, then $\sigma \cup \pi_{a_i}(\sigma) \in \overline{\mathcal{Q}}_{a_i}$ as shown in the proof of Lemma 7.3. Since $\overline{\mathcal{Q}}_{a_i} \subseteq T_{i,k}$, it follows that $\sigma \cup \pi_{a_i}(\sigma) \in T_{i,k}$ as desired. □

THEOREM 8.2. *The persistence diagrams of $\mathcal{Q}$ and $\mathcal{S}$ are identical.*

PROOF. For any $a_i, a_{i+1} \in A$, we get the following commutative diagram where all maps are induced by inclusions.

$$\begin{array}{ccccc}
\mathrm{H}_*(\overline{\mathcal{Q}}_{a_i}) & \overset{\cong}{\longleftarrow} & \mathrm{H}_*(\mathcal{Q}_{a_i}) & \longrightarrow & \mathrm{H}_*(\overline{\mathcal{Q}}_{a_{i+1}}) \\
& \searrow^{\cong} & \swarrow & & \swarrow^{\cong} \\
& \mathrm{H}_*(\mathcal{S}_{a_i}) & \longrightarrow & \mathrm{H}_*(\mathcal{S}_{a_{i+1}}) &
\end{array}$$

Lemma 8.1 and Corollary 7.4 show that the indicated maps are isomorphisms. As in Section 8.1, we reverse the isomorphism $\mathrm{H}_*(\mathcal{Q}_{a_i}) \to \mathrm{H}_*(\overline{\mathcal{Q}}_{a_i})$ to get the following diagrams, which also commutes.

$$\begin{array}{ccc}
\mathrm{H}_*(\overline{\mathcal{Q}}_{a_i}) & \longrightarrow & \mathrm{H}_*(\overline{\mathcal{Q}}_{a_{i+1}}) \\
\cong \downarrow & & \downarrow \cong \\
\mathrm{H}_*(\mathcal{S}_{a_i}) & \longrightarrow & \mathrm{H}_*(\mathcal{S}_{a_{i+1}})
\end{array}$$

Therefore, by the Persistence Equivalence Theorem, $\mathrm{D}\mathcal{Q} = \mathrm{D}\mathcal{S}$. □

9. THEORETICAL GUARANTEES

There are two main theoretical guarantees regarding the sparse Vietoris-Rips filtrations. First, in Subsection 9.1, we show that the resulting persistence diagrams are good approximations to the true Vietoris-Rips filtration. Second, in Subsection 9.2, we show that the filtrations have linear size.

9.1 The Approximation Guarantee

In this subsection we prove that the persistence diagram of the sparse Vietoris-Rips filtration is a multiplicative $(1+\varepsilon)$-approximation to the persistence diagram of the standard Vietoris-Rips filtration. The approach has two parts. First, we show that the relaxed filtration is a multiplicative $(1+\varepsilon)$-approximation to the classical Vietoris-Rips filtration. Second, we show that the sparse and relaxed Vietoris-Rips filtrations have the same persistence diagrams, i.e. that $\mathrm{D}\mathcal{Q} = \mathrm{D}\mathcal{R}$. By passing through the filtration $\mathcal{R}$, we obviate the need to develop new stability results for zigzag persistence.

THEOREM 9.1. *For any metric space $\mathcal{M} = (P, \mathbf{d})$, the persistence diagrams of the corresponding sparse Vietoris-Rips filtrations $\mathcal{Q} = \mathcal{Q}(\mathcal{M})$ and $\mathcal{S} = \mathcal{S}(\mathcal{M})$ both yield $(1+\varepsilon)$-approximations to the persistence diagram of the Vietoris-Rips filtration $\mathcal{R} = \mathcal{R}(\mathcal{M})$.*

PROOF. By Lemma 4.2, we have a multiplicative $(1+\varepsilon)$-interleaving between $\mathcal{R}$ and $\hat{\mathcal{R}}$. Thus, the Persistence Approximation Lemma implies that $\mathrm{D}\hat{\mathcal{R}}$ is a $(1+\varepsilon)$-approximation to $\mathrm{D}\mathcal{R}$.

We have shown in Theorem 8.2 that $\mathrm{D}\mathcal{Q} = \mathrm{D}\mathcal{S}$, so it will suffice to prove that $\mathrm{D}\mathcal{Q} = \mathrm{D}\hat{\mathcal{R}}$. The rest of the proof follows the same pattern as in Theorem 8.2. For any $a_i, a_{i+1} \in A$, we get the following commutative diagram induced by inclusion maps.

$$\begin{array}{ccccc}
\mathrm{H}_*(\overline{\mathcal{Q}}_{a_i}) & \xleftarrow{\cong} & \mathrm{H}_*(\mathcal{Q}_{a_i}) & \longrightarrow & \mathrm{H}_*(\overline{\mathcal{Q}}_{a_{i+1}}) \\
& \searrow^{\cong} \quad \swarrow^{\cong} & & \swarrow^{\cong} & \\
& \mathrm{H}_*(\hat{\mathcal{R}}_{a_i}) & \longrightarrow & \mathrm{H}_*(\hat{\mathcal{R}}_{a_{i+1}}) &
\end{array}$$

Corollary 7.4 implies that many of these inclusions induce isomorphisms at the homology level (as indicated in the diagram). As a consequence, the following diagram also commutes and the vertical maps are isomorphisms.

$$\begin{array}{ccc}
\mathrm{H}_*(\overline{\mathcal{Q}}_{a_i}) & \longrightarrow & \mathrm{H}_*(\overline{\mathcal{Q}}_{a_{i+1}}) \\
\cong \downarrow & & \downarrow \cong \\
\mathrm{H}_*(\hat{\mathcal{R}}_{a_i}) & \longrightarrow & \mathrm{H}_*(\hat{\mathcal{R}}_{a_{i+1}})
\end{array}$$

So, the Persistence Equivalence Theorem implies that $\mathrm{D}\mathcal{Q} = \mathrm{D}\hat{\mathcal{R}}$ as desired. $\square$

9.2 The Linear Complexity of the Sparse Filtration

In this subsection, we prove that the total number of simplices in the sparse Rips filtration is only linear in the number of input points. We start by showing that the graph of all edges appearing in the filtration has only a linear number of edges.

For a point $p \in P$, let $E(p)$ be the set of edges from p to a point whose removal time is at least as large as that of p:

$$E(p) = \{q \in P : t_p \leq t_q \text{ and } (p, q) \in \overline{\mathcal{Q}}_{t_p}\}.$$

To compute the filtrations $\mathcal{Q}$ and $\mathcal{S}$, it suffices to compute $E(p)$ for each $p \in P$. In fact $\mathcal{S}_\infty$ is just the clique complex on the graph of all edge (p, q) such that $q \in E(p)$.

LEMMA 9.2. *Given a set of n points in a metric space $\mathcal{M} = (P, \mathbf{d})$ with doubling dimension d, the cardinality $|E(p)|$ is at most $\frac{1}{\varepsilon}^{O(d)}$ for each $p \in P$.*

PROOF. Let $\Delta(E(p))$ denote the spread of $E(p)$. Since $E(p)$ is a finite metric with doubling dimension at most d, the number of points is at most $\Delta(E(p))^{O(d)}$. So, it will suffice to prove that for all $p \in P$, $\Delta(E(p)) = O(\frac{1}{\varepsilon})$.

The definition of $E(p)$ implies that $E(p) \subseteq \mathcal{N}_{t_p}$ and so by Lemma 6.1, the nearest pair in $E(p)$ are at least $K_p\varepsilon(1 - 2\varepsilon)t_p$ apart. For $q \in E(p)$, since $(p, q) \in \mathcal{Q}_{t_p}$, $\mathbf{d}(p, q) \leq \hat{\mathbf{d}}_{t_p}(p, q) \leq t_p$. It follows that the farthest pair in $E(p)$ are at most $2t_p$ apart. So, we get that $\Delta(E(p)) \leq \frac{2t_p}{K_p\varepsilon(1-2\varepsilon)t_p} = O(\frac{1}{\varepsilon})$ as desired. $\square$

We see that the size of the graph in the filtration is governed by three variables: the doubling dimension, d; the packing constant of the net-tree, K_p; and the desired tightness of the approximation, ε. The preceding Lemma easily implies the following bound on the higher order simplices.

THEOREM 9.3. *Given a set of n points in a metric space $\mathcal{M} = (P, \mathbf{d})$ with doubling dimension d, the total number of k-simplices in the sparse Vietoris-Rips filtrations $\mathcal{Q}$ and $\mathcal{S}$ is at most $\left(\frac{1}{\varepsilon}\right)^{O(kd)} n$.*

10. AN ALGORITHM TO CONSTRUCT THE SPARSE FILTRATION

The net-tree defines the deletion times of the input points and thus determines the perturbed metric. It also gives the necessary data structure to efficiently find the neighbors of a point in the perturbed metric in order to compute the filtration. In fact, this is exactly the kind of search that the net-tree makes easy. Then we find all cliques, which takes linear time because each is subset of $E(p)$ for some $p \in P$ and each $E(p)$ has constant size.

As explained in the Har-Peled and Mendel paper, it is often useful to augment the net-tree with "cross" edges connecting nodes at the same level in the tree that are represented by geometrically close points. The set of *relatives* of a node $u \in T$ is defined as

$$\begin{aligned}\mathrm{Rel}(u) = \{v \in T : &\mathrm{rad}(v) \leq \mathrm{rad}(u) < \mathrm{rad}(\mathrm{par}(v)) \text{ and} \\ &\mathbf{d}(\mathrm{rep}(u), \mathrm{rep}(v)) \leq C\mathrm{rad}(u)\},\end{aligned}$$

where C is a constant bigger than 3.* The size of $\mathrm{Rel}(u)$ is a constant using the same packing arguments as in Lemma 9.2.

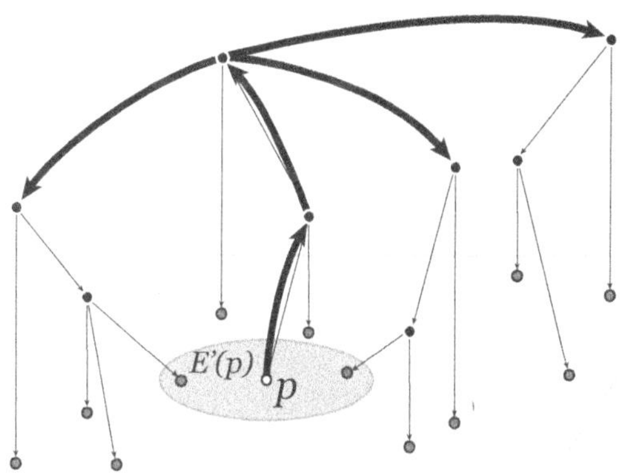

Figure 3: To find the nearby points at or above the level of v_p in the net-tree only requires traveling up a constant number of levels and then searching the relative trees.

This makes it easy to do a range search to find the points of $E(p)$. In fact, we will find the slightly larger set $E'(p) = \mathcal{N}_{t_p} \cap \mathbf{ball}(p, t_p)$. The search starts by finding u the highest ancestor of v_p whose radius is at most some fixed constant times t_p. Since the radius increases by a constant factor on each level, this is only a constant number of levels. Then

*The precise value of C depends on some constants chosen in the construction of the net-tree and can be extracted from the HM paper. For our purposes, we only need the fact that it is bigger than 3.

the subtrees rooted at each $v \in \text{Rel}(u)$ are searched down to the level of v_p. Thus, we search a constant number of trees of constant degree down a constant number of levels. The resulting search finds all of the points of $E(p)$ in constant time.

Since the work is only constant time per point, the only superlinear work is in the computation of the net-tree. As noted before, this requires only $O(n \log n)$ time.

11. CONCLUSIONS AND DIRECTIONS FOR FUTURE WORK

We have presented an efficient method for approximating the persistent homology of the Vietoris-Rips filtration. Computing these approximate persistence diagrams at all scales has the potential to make persistence-based methods on metric spaces tractable for much larger inputs.

Adapting the proofs given in this paper to the Čech filtration is a simple exercise. Moreover, it may be possible to apply a similar sparsification to complexes filtered by alternative distance-like functions like the distance to a measure introduced by Chazal et al. [9].

Another direction for future work is to identify a more general class of hierarchical structures that may be used in such a construction. The net-tree used in this paper is just one example chosen primarily because it can be computed efficiently.

The analytic technique used in this paper may find more uses in the future. We effectively bounded the difference between the persistence diagrams of a filtration and a zigzag filtration by embedding the zigzag filtration in a topologically equivalent filtration that does not zigzag at the homology level. It may be that other zigzag filtrations can be analyzed in this way.

Acknowledgements

I would like to thank Steve Oudot and Frédéric Chazal for their helpful comments. I would also like to acknowledge the keen insight of David Cohen-Steiner who suggested that I look for a non-zigzagging filtration. This insight became the filtration $\mathcal{S}$ of Section 8.2. This work was partially supported by GIGA grant ANR-09-BLAN-0331-01 and the European project CG-Learning No. 255827.

12. REFERENCES

[1] D. Attali, A. Lieutier, and D. Salinas. Efficient data structure for representing and simplifying simplicial complexes in high dimension. In *Proceedings of the 27th ACM Symposium on Computational Geometry*, 2011.

[2] J.-D. Boissonat, L. J. Guibas, and S. Y. Oudot. Manifold reconstruction in arbitrary dimensions using witness complexes. In *Proceedings of the 23rd ACM Symposium on Computational Geometry*, 2007.

[3] G. Carlsson. Topology and data. *Bull. Amer. Math. Soc.*, 46:255–308, 2009.

[4] G. Carlsson and V. de Silva. Zigzag persistence. *Foundations of Computational Mathematics*, 10(4):367–405, 2010.

[5] G. Carlsson, V. de Silva, and D. Morozov. Zigzag persistent homology and real-valued functions. In *Proceedings of the 25th ACM Symposium on Computational Geometry*, pages 247–256, 2009.

[6] G. Carlsson, T. Ishkhanov, V. de Silva, and A. Zomorodian. On the local behavior of spaces of natural images. *International Journal of Computer Vision*, 76(1):1–12, 2008.

[7] F. Chazal, D. Cohen-Steiner, M. Glisse, L. J. Guibas, and S. Y. Oudot. Proximity of persistence modules and their diagrams. In *Proceedings of the 25th ACM Symposium on Computational Geometry*, pages 237–246, 2009.

[8] F. Chazal, D. Cohen-Steiner, L. J. Guibas, F. Mémoli, and S. Y. Oudot. Gromov-hausdorff stable signatures for shapes using persistence. In *Eurographics Symposium on Geometry Processing*, 2009.

[9] F. Chazal, D. Cohen-Steiner, and Q. Mérigot. Geometric inference for probability measures. *Foundations of Computational Mathematics*, 11:733–751, 2011.

[10] F. Chazal and S. Y. Oudot. Towards persistence-based reconstruction in euclidean spaces. In *Proceedings of the 24th ACM Symposium on Computational Geometry*, pages 232–241, 2008.

[11] M. K. Chung, P. Bubenik, and P. T. Kim. Persistence diagrams of cortical surface data. In *Information Processing in Medical Imaging*, volume 5636, pages 386–397, 2009.

[12] K. L. Clarkson. Nearest neighbor queries in metric spaces. *Discrete & Computational Geometry*, 22(1), 1999.

[13] K. L. Clarkson. Nearest neighbor searching in metric spaces: Experimental results for 'sb(s)'. Preliminary version presented at ALENEX99, 2003.

[14] K. L. Clarkson. Building triangulations using epsilon-nets. In *STOC: ACM Symposium on Theory of Computing*, pages 326–335, 2006.

[15] D. Cohen-Steiner, H. Edelsbrunner, and J. Harer. Stability of persistence diagrams. In *Proceedings of the 21st ACM Symposium on Computational Geometry*, pages 263–271, 2005.

[16] R. Cole and L.-A. Gottlieb. Searching dynamic point sets in spaces with bounded doubling dimension. In *Proceedings of the 38th Annual ACM Symposium on Theory of Computing*, pages 574–583, 2006.

[17] V. de Silva and G. Carlsson. Topological estimation using witness complexes. In *Proc. Sympos. Point-Based Graphics*, pages 157–166, 2004.

[18] V. de Silva and R. Ghrist. Coverage in sensor networks via persistent homology. *Algorithmic & Geometric Topology*, 7:339–358, 2007.

[19] V. de Silva and R. Ghrist. Homological sensor networks. *Notices Amer. Math. Soc.*, 54(1):10–17, 2007.

[20] H. Edelsbrunner and J. L. Harer. *Computational Topology: An Introduction.* Amer. Math. Soc., 2009.

[21] H. Edelsbrunner, D. Letscher, and A. Zomorodian. Topological persistence and simplification. *Discrete & Computational Geometry*, 4(28):511–533, 2002.

[22] J. Gao, L. J. Guibas, and A. Nguyen. Deformable spanners and applications. In *Proceedings of the 20th ACM Symposium on Computational Geometry*, 2004.

[23] L.-A. Gottlieb and L. Roditty. An optimal dynamic spanner for doubling metric spaces. In *Proceedings of*

the 16th annual European symposium on Algorithms, pages 468–489, 2008.

[24] L. J. Guibas and S. Y. Oudot. Reconstruction using witness complexes. In *Proceedings 18th ACM-SIAM Symposium: Discrete Algorithms*, pages 1076–1085, 2007.

[25] S. Har-Peled and M. Mendel. Fast construction of nets in low dimensional metrics, and their applications. *SIAM Journal on Computing*, 35(5):1148–1184, 2006.

[26] B. Hudson, G. L. Miller, S. Y. Oudot, and D. R. Sheehy. Topological inference via meshing. In *Proceedings of the 26th ACM Symposium on Computational Geometry*, pages 277–286, 2010.

[27] N. Milosavljevic, D. Morozov, and P. Skraba. Zigzag persistent homology in matrix multiplication time. In *Proceedings of the 27th ACM Symposium on Computational Geometry*, 2011.

[28] J. R. Munkres. *Elements of Algebraic Topology*. Addison-Wesley, 1984.

[29] G. Narasimhan and M. H. M. Smid. *Geometric Spanner Networks*. Cambridge University Press, 2007.

[30] G. Singh, F. Mémoli, T. Ishkhanov, G. Sapiro, G. Carlsson, and D. L. Ringach. Topological analysis of population activity in visual cortex. *Journal of Vision*, 8(8):1–18, 2008.

[31] A. Tausz and G. Carlsson. Applications of zigzag persistence to topological data analysis. arxiv:1108.3545, Aug 2011.

[32] A. Zomorodian. The tidy set. In *Proceedings of the 26th ACM Symposium on Computational Geometry*, pages 257–266, 2010.

[33] A. Zomorodian and G. Carlsson. Computing persistent homology. *Discrete & Computational Geometry*, 33(2):249–274, 2005.

Alexander Duality for Functions: the Persistent Behavior of Land and Water and Shore

Herbert Edelsbrunner*
IST Austria
Klosterneuburg, Austria
edels@ist.ac.at

Michael Kerber
IST Austria
Klosterneuburg, Austria
mkerber@ist.ac.at

ABSTRACT

This note contributes to the point calculus of persistent homology by extending Alexander duality from spaces to real-valued functions. Given a perfect Morse function $f : \mathbb{S}^{n+1} \to [0, 1]$ and a decomposition $\mathbb{S}^{n+1} = \mathbb{U} \cup \mathbb{V}$ into two $(n + 1)$-manifolds with common boundary $\mathbb{M}$, we prove elementary relationships between the persistence diagrams of f restricted to $\mathbb{U}$, to $\mathbb{V}$, and to $\mathbb{M}$.

Categories and Subject Descriptors

F.2.2 [**Analysis of Algorithms and Problem Complexity**]: Nonnumerical Algorithms and Problems—*Geometrical problems and computations, Computations on discrete structures*

Keywords

Algebraic topology, homology, Alexander duality, Mayer-Vietoris sequences, persistent homology, point calculus.

1. INTRODUCTION

Persistent homology is a recent extension of the classical theory of homology; see e.g. [6]. Given a real-valued function on a topological space, it measures the importance of a homology class by monitoring when the class appears and when it disappears in the increasing sequence of sublevel sets. A technical requirement is that the function be *tame*, which means it has only finitely many homological critical values, and each sublevel set has finite rank homology groups. Pairing up the births and deaths, and drawing each pair of values as a *dot* (a point in the plane), we get a multiset which we refer to as the *persistence diagram* of the function. It is a combinatorial summary of the homological information contained in the sequence of sublevel sets. If we substitute reduced for standard homology groups, we get a slightly modified *reduced persistence diagram.*

As between homology groups, we can observe relationships between persistence diagrams. A prime example is Poincaré duality, which says that the p-th and the $(n - p)$-th homology groups of an orientable n-manifold are isomorphic. More precisely, this is true if homology is defined for field coefficients, which is what we assume throughout this paper. The extension to functions says that the diagram is symmetric with respect to reflection across the vertical axis; see [3]. Here, we change the homological dimension of a dot from p to $n - p$ whenever we reflect it across the axis. This paper contributes new relationships by extending Alexander duality from spaces to functions. To state our results, we assume a perfect Morse function, $f : \mathbb{S}^{n+1} \to [0, 1]$, which for the sphere has no critical points other than a minimum and a maximum, and a decomposition of the $(n + 1)$-dimensional sphere into two $(n + 1)$-manifolds with boundary, $\mathbb{S}^{n+1} = \mathbb{U} \cup \mathbb{V}$, whose intersection is the common n-manifold boundary $\mathbb{M} = \mathbb{U} \cap \mathbb{V}$. Our first result says that the reduced persistence diagrams of f restricted to $\mathbb{U}$ and to $\mathbb{V}$ are reflections of each other. We call this the Land and Water Theorem. Our second result relates land with shore. Ignoring some modifications, it says that the persistence diagram of f restricted to $\mathbb{M}$ is the disjoint union of the diagram of f restricted to $\mathbb{U}$ and of its reflection. The modifications become unnecessary if we assume that the minimum and maximum of f both belong to a common component of $\mathbb{V}$. We call this the Euclidean Shore Theorem.

In the example that justifies the title of this paper, and the names of our theorems, we let $\mathbb{U}$ be the planet Earth, not including the water and the air. To a coarse approximation, $\mathbb{U}$ is homeomorphic to a 3-ball, sitting inside the Universe, which we model as a 3-sphere, $\mathbb{S}$. The function we consider is the negative gravitational potential of the Earth, which is defined on the entire Universe. The sea is then a sublevel set of this function restricted to $\mathbb{V}$, which is the closure of $\mathbb{S} - \mathbb{U}$. With these definitions, our results relate the persistence diagram of the gravitational potential restricted to the Earth with the shape of the sea as its water level rises. Also, the Euclidean Shore Theorem applies, unmodified, expanding the relationship to include the sea floor, which is swept out by the shoreline as the water level rises.

Besides developing the mathematical theory of persistent homology, there are pragmatic reasons for our interest in the extension of Alexander duality to functions. Persistence has fast algorithms, so that the bulk of the work is often

*... and Departments of Computer Science and of Mathematics, Duke University, Durham, North Carolina; Delaunay Lab of Discrete and Computational Geometry, Yaroslavl' State University, Russia; Geomagic, Research Triangle Park, North Carolina.

SCG'12, June 17–20, 2012, Chapel Hill, North Carolina, USA.

in the construction of the space and the function for which we compute persistence. A point in case is the analysis of the biological process of cell segregation started in [7]. Modeling the process as a subset of space-time, the function of interest is time which, after compactifying space-time to $\mathbb{S}^4$, has no critical points other than a minimum and a maximum. The subset $\mathbb{U}$ of $\mathbb{S}^4$ is a union of cells times time, whose boundary is a 3-manifold. We can represent $\mathbb{U}$ by a 1-parameter family of alpha complexes, whose disjoint union has the same homotopy type; see e.g. [6, Chapter III]. However, the boundary of that disjoint union is not necessarily a 3-manifold. Using our Euclidean Shore Theorem, we can compute the persistence diagram of the function on the 3-manifold without ever constructing the 3-manifold.

Outline. Sections 2 and 3 introduce background on homology and persistent homology. Sections 4 and 5 present our two results. Section 6 concludes the paper.

2. HOMOLOGY

Starting with a brief introduction of classical homology groups, we present the relevant background on Alexander duality and Mayer-Vietoris sequences. More comprehensive discussions of these topics can be found in textbooks of algebraic topology, such as [8, 10].

Background. The p-dimensional homology of a topological space, $\mathbb{X}$, is a mathematical language to define, count, and reason about the p-dimensional connectivity of $\mathbb{X}$. There are different but essentially equivalent theories depending on the choices one makes in the representation of the space, the selection of cycles, and the meaning of addition. For our purpose, the most elementary of these theories will suffice: a simplicial complex, K, that triangulates $\mathbb{X}$, formal sums of p-simplices with zero boundary as p-cycles, and adding with coefficients in a field, F. Most algorithms on homology assume this model, in particular the ones developed within persistent homology. In this model, we call a formal sum of p-simplices a *p-chain*, a *p-cycle* if its boundary is empty, and a *p-boundary* if it is the boundary of a $(p+1)$-chain. The p-boundaries form a subgroup of the p-cycles, which form a subgroup of the p-chains: $\mathsf{B}_p \subseteq \mathsf{Z}_p \subseteq \mathsf{C}_p$. The *p-th homology group* is the quotient of the p-cycles over the p-boundaries: $\mathsf{H}_p = \mathsf{Z}_p/\mathsf{B}_p$. Its elements are sets of p-cycles that differ from each other by p-boundaries. More fully, we denote the p-th homology group by $\mathsf{H}_p(K,\mathsf{F})$, or by $\mathsf{H}_p(\mathbb{X},\mathsf{F})$ to emphasize that the group is independent of the simplicial complex we choose to triangulate the space. However, we will fix an arbitrary field F and drop it from the notation. For field coefficients, $\mathsf{H}_p(\mathbb{X})$ is necessarily a vector space, which is fully described by its rank, $\beta_p(\mathbb{X}) = \operatorname{rank} \mathsf{H}_p(\mathbb{X})$ such that $\mathsf{H}_p(\mathbb{X}) \simeq \mathsf{F}^{\beta_p(\mathbb{X})}$, called the *p-th Betti number* of $\mathbb{X}$. Finally, we will often drop the dimension from the notation and write $\mathsf{H}(\mathbb{X}) = \bigoplus_{p\in\mathbb{Z}} \mathsf{H}_p(\mathbb{X})$ for the direct sum.

Besides standard homology, we will frequently use *reduced homology groups*, $\tilde{\mathsf{H}}_p(\mathbb{X})$, which are isomorphic to the non-reduced groups except possibly for dimensions $p = 0, -1$. To explain the difference, we note that $\beta_0(\mathbb{X})$ counts the components of $\mathbb{X}$. In contrast, $\tilde{\beta}_0(\mathbb{X}) = \operatorname{rank} \tilde{\mathsf{H}}_0(\mathbb{X})$ counts the gaps between components or, equivalently, the edges that are needed to merge the components into one. Hence, $\tilde{\beta}_0(\mathbb{X}) = \beta_0(\mathbb{X}) - 1$, except when $\mathbb{X}$ is empty, in which case $\tilde{\beta}_0(\mathbb{X}) = \beta_0(\mathbb{X}) = 0$. To distinguish this case from a single component, we have $\tilde{\beta}_{-1}(\mathbb{X})$ equal to 1 if $\mathbb{X} = \emptyset$, and equal to 0, otherwise. Furthermore, we use relative homology, which is defined for pairs of spaces, $\mathbb{Y} \subseteq \mathbb{X}$. Taking a pair relaxes the requirement of a chain to be called a cycle, namely whenever its boundary is contained in $\mathbb{Y}$, which includes the case when the boundary is empty. We write $\mathsf{H}_p(\mathbb{X},\mathbb{Y})$ for the *p-th relative homology group* of the pair, and $\beta_p(\mathbb{X},\mathbb{Y}) = \operatorname{rank} \mathsf{H}_p(\mathbb{X},\mathbb{Y})$ for the *p-th relative Betti number*. As before, we will suppress the dimension from the notation by introducing $\mathsf{H}(\mathbb{X},\mathbb{Y}) = \bigoplus_{p\in\mathbb{Z}} \mathsf{H}_p(\mathbb{X},\mathbb{Y})$.

As examples, consider the $(n+1)$-dimensional sphere, $\mathbb{S} = \mathbb{S}^{n+1}$, and a closed hemisphere, $\mathbb{H} \subseteq \mathbb{S}$, which is a ball of dimension $n+1$. In standard homology, we have $\beta_0(\mathbb{S}) = \beta_0(\mathbb{H}) = \beta_{n+1}(\mathbb{S}) = 1$ while all other Betti numbers are zero. In reduced homology, we have $\tilde{\beta}_{n+1}(\mathbb{S}) = 1$ while all other reduced Betti numbers are zero. In particular, $\tilde{\beta}_p(\mathbb{H}) = 0$ for all p. In relative homology, we have $\beta_{n+1}(\mathbb{S},\mathbb{H}) = 1$ while all other relative Betti numbers are zero. In particular, $\beta_0(\mathbb{S},\mathbb{H}) = 0$, which may be confusing at first but makes sense because every point on the sphere can be connected by a path to a point in the hemisphere and is thus a 0-boundary.

Alexander duality. Recall that a *perfect Morse function* is one whose number of critical points equals the sum of Betti numbers of the space. For a sphere, this number is 2: a minimum and a maximum. Throughout the remainder of this section, we assume a perfect Morse function $f : \mathbb{S} \to [0,1]$, whose values at the minimum and the maximum are 0 and 1. We also assume two $(n+1)$-manifolds with boundary, $\mathbb{U}$ and $\mathbb{V}$, whose union is $\mathbb{S}$, and whose intersection is the common boundary, $\mathbb{M}$. For technical reasons, we require that $\mathbb{U}$, $\mathbb{V}$, and $\mathbb{M}$ are embedded in the sphere in a way that is compatible with taking the simplicial homology and with applying the Mayer-Vietoris sequence and Alexander duality. Specifically, we assume that $\mathbb{U}$ and $\mathbb{V}$ are triangulable closed cofibrations. In other words, there is a deformation retraction from an open neighborhood of $\mathbb{U}$ to $\mathbb{U}$, and similarly for $\mathbb{V}$. Furthermore, we assume that the restriction of f to $\mathbb{M}$ is tame, which implies that its restrictions to $\mathbb{U}$ and $\mathbb{V}$ are also tame. For each $t \in \mathbb{R}$, we write $\mathbb{S}_t = f^{-1}[0,t]$ for the *sublevel set* of f, and $\mathbb{U}_t = \mathbb{U} \cap \mathbb{S}_t$, $\mathbb{V}_t = \mathbb{V} \cap \mathbb{S}_t$, $\mathbb{M}_t = \mathbb{M} \cap \mathbb{S}_t$ for the sublevel sets of the restrictions of f to $\mathbb{U}$, $\mathbb{V}$, $\mathbb{M}$. Similarly, we write $\mathbb{S}^t = f^{-1}[t,1]$ for the *superlevel set*, and $\mathbb{U}^t = \mathbb{U} \cap \mathbb{S}^t$, $\mathbb{V}^t = \mathbb{V} \cap \mathbb{S}^t$, $\mathbb{M}^t = \mathbb{M} \cap \mathbb{S}^t$ for the superlevel sets of the restrictions. Since we will apply Mayer-Vietoris and Alexander duality to sub- and superlevel sets, we make the same technical assumptions as formulated above for $\mathbb{U}$, $\mathbb{V}$, and $\mathbb{M}$. For later reference, we call a pair of $(n+1)$-manifolds with boundary that satisfy those conditions a *complementary decomposition* of $\mathbb{S}$, noting that the two $(n+1)$-manifolds are indeed complementary except for the shared boundary.

The basic version of Alexander duality is a statement about a complementary decomposition of the sphere; see [10, page 424]. More specifically, it states that $\mathsf{H}_q(\mathbb{U})$ and $\mathsf{H}_p(\mathbb{V})$ are isomorphic, where $q = n-p$, except for $p = 0$ and $q = 0$ when the 0-dimensional group has an extra generator. This implies

$$\beta_0(\mathbb{V}) = \beta_n(\mathbb{U}) + 1, \quad (1)$$

$$\beta_p(\mathbb{V}) = \beta_q(\mathbb{U}), \quad (2)$$

$$\beta_n(\mathbb{V}) = \beta_0(\mathbb{U}) - 1, \quad (3)$$

for $1 \leq p \leq n-1$. We will also need the version that

deals with a complementary decomposition of the $(n+1)$-dimensional ball, $\mathbb{B}^{n+1}$; see [10, page 426]. Let $t \in (0,1)$ denote a regular value of $f|_{\mathbb{M}}$. Note first that $\mathbb{S}_t$ is homeomorphic to $\mathbb{B}^{n+1}$. By excision, the homology groups of $\mathbb{U}_t$ relative to $\mathbb{U}_t \cap f^{-1}(t)$ are isomorphic to those of $\mathbb{U}$ relative to $\mathbb{U}^t$. Alexander duality states that $\mathsf{H}_q(\mathbb{U}, \mathbb{U}^t)$ and $\mathsf{H}_p(\mathbb{V}_t)$ are isomorphic, where $q+p=n$, as before, except for $p=0$, when $\mathbb{V}_t$ has an extra component. This implies

$$\beta_0(\mathbb{V}_t) = \beta_n(\mathbb{U}, \mathbb{U}^t) + 1, \quad (4)$$
$$\beta_p(\mathbb{V}_t) = \beta_q(\mathbb{U}, \mathbb{U}^t), \quad (5)$$

for $1 \le p \le n$.

Mayer-Vietoris. We can connect the homology groups of $\mathbb{U}$ and $\mathbb{V}$ with those of $\mathbb{M}$ and $\mathbb{S}$ using the Mayer-Vietoris sequence of the decomposition. This sequence is *exact*, meaning the image of every map equals the kernel of the next map. Counting the trivial homology groups, the sequence is infinitely long, with three terms per dimension:

$$\ldots \to \mathsf{H}_{p+1}(\mathbb{S}) \to \mathsf{H}_p(\mathbb{M}) \to \mathsf{H}_p(\mathbb{U}) \oplus \mathsf{H}_p(\mathbb{V}) \to \mathsf{H}_p(\mathbb{S}) \to \ldots,$$

where the maps between homology groups of the same dimension are induced by the inclusions. The only non-trivial homology groups of $\mathbb{S}$ are in dimensions 0 and $n+1$, with ranks $\beta_0(\mathbb{S}) = \beta_{n+1}(\mathbb{S}) = 1$. It follows that for $1 \le p \le n-1$, the groups defined by $\mathbb{M}$, $\mathbb{U}$, $\mathbb{V}$ are surrounded by trivial groups. This implies that the groups of $\mathbb{M}$ and the direct sums of the groups of $\mathbb{U}$ and $\mathbb{V}$ are isomorphic. For $p=0$ and $p=n$, the non-trivial homology groups of $\mathbb{S}$ prevent this isomorphism, and we get

$$\beta_0(\mathbb{M}) = \beta_0(\mathbb{U}) + \beta_0(\mathbb{V}) - 1, \quad (6)$$
$$\beta_p(\mathbb{M}) = \beta_p(\mathbb{U}) + \beta_p(\mathbb{V}), \quad (7)$$
$$\beta_n(\mathbb{M}) = \beta_n(\mathbb{U}) + \beta_n(\mathbb{V}) + 1, \quad (8)$$

for $1 \le p \le n-1$. We also have a Mayer-Vietoris sequence for the spaces $\mathbb{S}_t$, $\mathbb{U}_t$, $\mathbb{V}_t$ and $\mathbb{M}_t$:

$$\ldots \to \mathsf{H}_{p+1}(\mathbb{S}_t) \to \mathsf{H}_p(\mathbb{M}_t) \to \mathsf{H}_p(\mathbb{U}_t) \oplus \mathsf{H}_p(\mathbb{V}_t) \to \mathsf{H}_p(\mathbb{S}_t) \to \ldots$$

For $0 \le t < 1$, $\mathbb{S}_t$ is a point or a closed $(n+1)$-dimensional ball, and its only non-trivial homology group is of dimension 0, with rank $\beta_0(\mathbb{S}_t) = 1$. It follows that for $1 \le p \le n$, the groups defined by $\mathbb{M}_t$, $\mathbb{U}_t$, $\mathbb{V}_t$ are surrounded by trivial groups, which implies that the groups of $\mathbb{M}_t$ and the direct sums of the groups of $\mathbb{U}_t$ and $\mathbb{V}_t$ are isomorphic. We get

$$\beta_0(\mathbb{M}_t) = \beta_0(\mathbb{U}_t) + \beta_0(\mathbb{V}_t) - 1 \quad (9)$$
$$\beta_p(\mathbb{M}_t) = \beta_p(\mathbb{U}_t) + \beta_p(\mathbb{V}_t), \quad (10)$$

for $1 \le p \le n$. We may also consider the relative homology groups, again connected by a Mayer-Vietoris sequence:

$$\ldots \to \mathsf{H}_{p+1}(\mathbb{S}, \mathbb{S}^t) \to \mathsf{H}_p(\mathbb{M}, \mathbb{M}^t)$$
$$\to \mathsf{H}_p(\mathbb{U}, \mathbb{U}^t) \oplus \mathsf{H}_p(\mathbb{V}, \mathbb{V}^t) \to \mathsf{H}_p(\mathbb{S}, \mathbb{S}^t) \to \ldots$$

For $0 < t \le 1$, the only non-trivial homology group of the pair $(\mathbb{S}, \mathbb{S}^t)$ is in dimension $n+1$, which implies

$$\beta_p(\mathbb{M}, \mathbb{M}^t) = \beta_p(\mathbb{U}, \mathbb{U}^t) + \beta_p(\mathbb{V}, \mathbb{V}^t), \quad (11)$$
$$\beta_n(\mathbb{M}, \mathbb{M}^t) = \beta_n(\mathbb{U}, \mathbb{U}^t) + \beta_n(\mathbb{V}, \mathbb{V}^t) + 1, \quad (12)$$

for $0 \le p \le n-1$.

Separating manifold. Combining Alexander duality with Mayer-Vietoris, we get relations between the homology of the sublevel sets of $\mathbb{M}$ and $\mathbb{U}$. More specifically, we get

$$\beta_p(\mathbb{M}) = \beta_p(\mathbb{U}) + \beta_q(\mathbb{U}), \quad (13)$$
$$\beta_p(\mathbb{M}_t) = \beta_p(\mathbb{U}_t) + \beta_q(\mathbb{U}, \mathbb{U}^t), \quad (14)$$
$$\beta_p(\mathbb{M}, \mathbb{M}^t) = \beta_p(\mathbb{U}, \mathbb{U}^t) + \beta_q(\mathbb{U}_t), \quad (15)$$

for all $0 \le p \le n$. Here, we combine (6), (7), (8) with (1), (2), (3) to get (13). Similarly, we combine (9), (10) with (4), (5) to get (14). Finally, we exploit the symmetry between $\mathbb{U}$ and $\mathbb{V}$ and combine (11), (12) with (4), (5) to get (15). Note that (14) and (15) imply $\beta_p(\mathbb{M}_t) = \beta_q(\mathbb{M}, \mathbb{M}^t)$, which is a consequence of Lefschetz duality; see [10].

An example. We illustrate the above relationships with an example. Let $\mathbb{M}$ be a 2-dimensional torus in $\mathbb{S}^3$. Accordingly, $\mathbb{U}$ and $\mathbb{V}$ are the two solid tori that decompose the 3-sphere and intersect in $\mathbb{M}$. This is sketched in Figure 1, where we assume that $\mathbb{U}$ is the part of space surrounded by $\mathbb{M}$, while $\mathbb{V}$ is the space surrounding $\mathbb{M}$. The only non-zero Betti numbers of the solid torus are $\beta_0(\mathbb{U}) = \beta_1(\mathbb{U}) = 1$. We get

$$\beta_0(\mathbb{M}) = \beta_0(\mathbb{U}) + \beta_2(\mathbb{U}) = 1,$$
$$\beta_1(\mathbb{M}) = \beta_1(\mathbb{U}) + \beta_1(\mathbb{U}) = 2,$$
$$\beta_2(\mathbb{M}) = \beta_2(\mathbb{U}) + \beta_0(\mathbb{U}) = 1$$

from (13). These are the correct Betti numbers of the 2-dimensional torus. Next, choose t so that the sublevel set of $\mathbb{U}$ is half the solid torus. Its only non-zero Betti number is

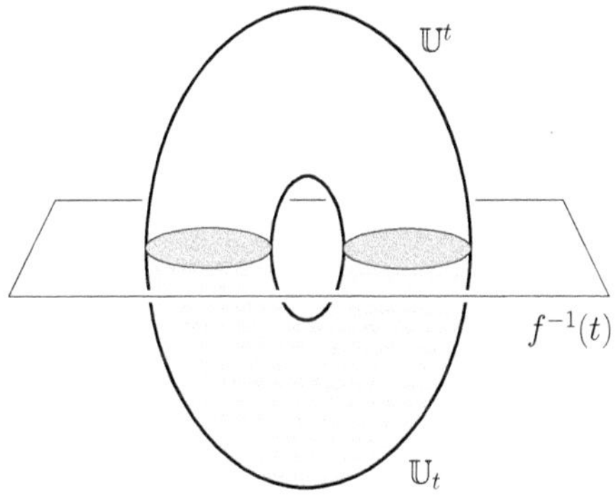

Figure 1: The level set defined by t splits the solid torus into two halves.

$\beta_0(\mathbb{U}_t) = 1$, and the only non-zero relative Betti number of the pair is $\beta_1(\mathbb{U}, \mathbb{U}^t) = 1$. We thus get

$$\beta_0(\mathbb{M}_t) = \beta_0(\mathbb{U}_t) + \beta_2(\mathbb{U}, \mathbb{U}^t) = 1,$$
$$\beta_1(\mathbb{M}_t) = \beta_1(\mathbb{U}_t) + \beta_1(\mathbb{U}, \mathbb{U}^t) = 1,$$
$$\beta_2(\mathbb{M}_t) = \beta_2(\mathbb{U}_t) + \beta_0(\mathbb{U}, \mathbb{U}^t) = 0$$

from (14). These are the correct Betti numbers of the sublevel set of $\mathbb{M}$. Finally, we get

$$\beta_0(\mathbb{M}, \mathbb{M}^t) = \beta_2(\mathbb{U}_t) + \beta_0(\mathbb{U}, \mathbb{U}^t) = 0,$$
$$\beta_1(\mathbb{M}, \mathbb{M}^t) = \beta_1(\mathbb{U}_t) + \beta_1(\mathbb{U}, \mathbb{U}^t) = 1,$$
$$\beta_2(\mathbb{M}, \mathbb{M}^t) = \beta_0(\mathbb{U}_t) + \beta_2(\mathbb{U}, \mathbb{U}^t) = 1$$

from (15). These are the correct relative Betti numbers of the pair $(\mathbb{M}, \mathbb{M}^t)$. Note again that the vector of Betti numbers of $\mathbb{M}_t$ is the reverse of that of $(\mathbb{M}, \mathbb{M}_t)$.

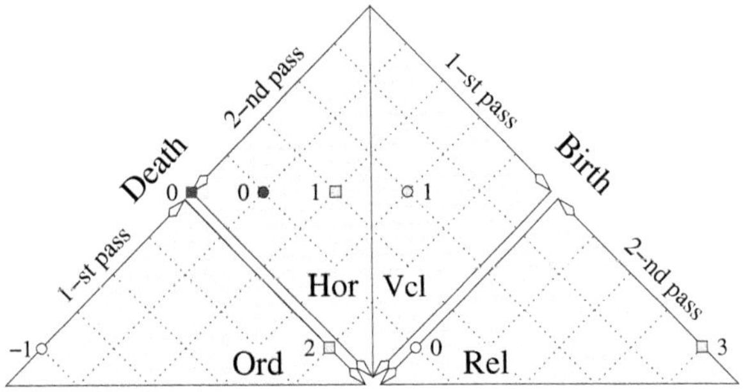

Figure 2: The persistence diagram of the function restricted to the solid torus (dots drawn as circles), and to the complementary solid torus (dots drawn as squares), as sketched on the left. The white dots belong to the reduced diagram, the black dots belong to the standard diagram, and the shaded dots belong to both.

3. PERSISTENCE

Starting with a brief introduction of persistent homology, we review its combinatorial expression as a point calculus; see [6] and [1, 2] for more comprehensive discussions.

Background. We take the step from homology to persistent homology by replacing a space with the sequence of (closed and open) sublevel sets of a function on the space. To explain this, let $\mathbb{X}$ be compact and $g : \mathbb{X} \to [0, 1]$ tame. As before, we write $\mathbb{X}_t = g^{-1}[0, t]$ for the sublevel set and $\mathbb{X}^t = g^{-1}[t, 1]$ for the superlevel set defined by t. Since g is tame, it has finitely many homological critical values, s_0 to s_m. We assume w.l.o.g. that $s_0 = 0$ and $s_m = 1$, else, we add them to the list. We interleave the s_i with homological regular values t_i, such that $s_0 < t_0 < s_1 < \ldots < t_{m-1} < s_m$. Taking the homology of closed and open sublevel sets at the regular values, we get the *persistent module*

$$\mathsf{X}_0 \to \ldots \to \mathsf{X}_{m-1} \to \mathsf{X}_m \to \mathsf{X}_{m+1} \to \ldots \to \mathsf{X}_{2m},$$

where X_i is $\mathsf{H}(\mathbb{X}_{t_i})$, if $0 \leq i < m$, $\mathsf{H}(\mathbb{X})$, if $i = m$, and $\mathsf{H}(\mathbb{X}, \mathbb{X}^{t_{2m-i}})$, if $m < i \leq 2m$. For notational convenience, we add the trivial groups, $\mathsf{X}_{-1} = 0$ and $\mathsf{X}_{2m+1} = 0$, at the beginning and end. The maps connecting the homology groups are induced by the inclusions $\mathbb{X}_s \subseteq \mathbb{X}_t$ and $\mathbb{X}^t \subseteq \mathbb{X}^s$, whenever $s \leq t$. The maps compose and we write $\mathsf{g}_{i,j} : \mathsf{X}_i \to \mathsf{X}_j$. Reading the module from left to right, we see homology classes appear and disappear. To understand these events, we say a class $\alpha \in \mathsf{X}_i$ is *born* at X_i if it does not belong to the image of $\mathsf{g}_{i-1,i}$. If furthermore $\mathsf{g}_{i,j}(\alpha)$ belongs to the image of $\mathsf{g}_{i-1,j}$ but $\mathsf{g}_{i,j-1}(\alpha)$ does not belong to the image of $\mathsf{g}_{i-1,j-1}$, then we say α *dies entering* X_j. Since the module starts and ends with trivial groups, every homology class has well-defined birth and death values. Every event is associated with the immediately preceding homological critical value, namely with s_i if the event is at X_i, and with s_{m-i} if the event is at X_{m+i+1}, for $0 \leq i \leq m-1$. Events at X_m are associated with $s_m = 1$. We represent a class by a dot in the plane whose two coordinates mark its birth and its death. More precisely, the coordinates signal the increase and decrease of Betti numbers, and the dot represents an entire coset of classes that are born and die with α. All dots have coordinates in $[0, 1]$, by construction.

Note that we use each s_i twice, once during the first pass in which the sublevel set grows from $\emptyset$ to $\mathbb{X}$, and then again during the second pass in which $\mathbb{X}$ minus the superlevel set shrinks back from $\mathbb{X}$ to $\emptyset$. When we collect the dots, we separate the passes by drawing each coordinate axis twice. The result is a multiset of dots, which we refer to as the *persistence diagram* of the function, and denote as $\mathsf{Dgm}(g)$, or $\mathsf{Dgm}_p(g)$ if we restrict ourselves to the dots representing p-dimensional classes; see Figure 2, where we label each dot with the dimension of the classes it represents. We further distinguish four regions within the diagram: the *ordinary*, the *horizontal*, the *vertical*, and the *relative subdiagrams*, denoted as $\mathsf{Ord}(g)$, $\mathsf{Hor}(g)$, $\mathsf{Vcl}(g)$, and $\mathsf{Rel}(g)$. For example, a dot belongs to the ordinary subdiagram if its birth and death both happen during the first pass; see again Figure 2.

Reduced persistence diagrams. It is sometimes convenient to use reduced instead of standard homology groups in the filtration. There are small differences caused by the (-1)-dimensional class, which exists if the space is empty. As a first step, we introduce reduced versions of the relative homology groups, which are isomorphic to the standard relative groups, for all p. To define them, let ω be a new (dummy) vertex, write $\omega \cdot \mathbb{X}^t$ for the cone of ω over $\mathbb{X}^t$, and let $\mathbb{X}^t_\omega = \mathbb{X} \cup \omega \cdot \mathbb{X}^t$ be the result of gluing the cone along $\mathbb{X}^t$ to $\mathbb{X}$. Then the *reduced relative homology group* of the pair is $\tilde{\mathsf{H}}_p(\mathbb{X}, \mathbb{X}^t) = \tilde{\mathsf{H}}_p(\mathbb{X}^t_\omega)$. For example, if $\mathbb{X}^t = \emptyset$, then ω forms a separate component, so that the reduced 0-th Betti number is equal to the number of components of $\mathbb{X}$, just as $\beta_0(\mathbb{X}, \mathbb{X}^t)$. The *reduced persistence diagram* is now defined as before, but for the filtration of reduced homology groups. To describe this, we suppress the dimension and write $\tilde{\mathsf{X}}_i$ for $\tilde{\mathsf{H}}(\mathbb{X}_{t_i})$, if $0 \leq i < m$, for $\tilde{\mathsf{H}}(\mathbb{X})$, if $i = m$, and for $\tilde{\mathsf{H}}(\mathbb{X}^{t_{2m-i}}_\omega)$, if $m < i \leq 2m$. The resulting sequence of reduced homology groups, from $\tilde{\mathsf{X}}_0$ to $\tilde{\mathsf{X}}_{2m}$, is connected from left to right by maps induced by inclusion. Finally, we define $\tilde{\mathsf{D}}\mathsf{gm}(g)$ by matching up the births and the deaths and by drawing each pair as a dot in the plane. Similar to before, we write $\tilde{\mathsf{D}}\mathsf{gm}_p(g)$ when we restrict ourselves to homology classes of dimension p.

For dimension $p \geq 1$, the reduced diagrams are the same as the standard ones, simply because the groups are the same. More formally, the persistence diagrams are the same because the two persistence modules form a commutative diagram with vertical isomorphisms:

$$\begin{array}{ccccccccccc} 0 & \to & \mathsf{X}^p_0 & \to & \mathsf{X}^p_1 & \to & \ldots & \to & \mathsf{X}^p_{2m} & \to & 0 \\ \downarrow & & \downarrow & & \downarrow & & & & \downarrow & & \downarrow \\ 0 & \to & \tilde{\mathsf{X}}^p_0 & \to & \tilde{\mathsf{X}}^p_1 & \to & \ldots & \to & \tilde{\mathsf{X}}^p_{2m} & \to & 0; \end{array}$$

see the Persistence Equivalence Theorem [6, page 159]. This is not true for $p = -1$, where the standard diagram is empty, while the reduced diagram contains a single dot marking the transition from an empty to a non-empty space. To describe the difference for $p = 0$, we call a dot (u, w) in the 0-th standard persistence diagram *extreme* if no other dot has a first coordinate smaller than u and a second coordinate larger than w. Here, we assume for simplicity that no two dots in the 0-th diagram share the same first coordinate or the same second coordinate.

We note that only dots in the horizontal subdiagram can be extreme. The non-extreme dots also belong to the reduced diagram of g, while the extreme dots exchange their coordinates to form new dots in the reduced diagram; see Figure 2. The way the coordinates are exchanged will be important later, so we describe the details now.

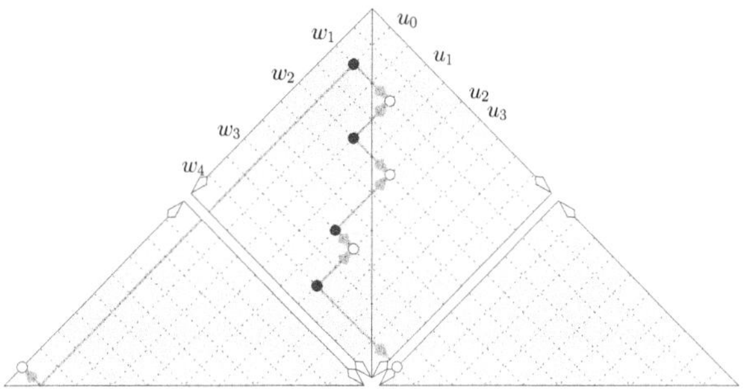

Figure 3: The cascade combines the coordinates of the four black dots with 0 and 1 to form the five white dots.

Cascades. We let $\ell+1$ be the number of extreme dots and, for a reason that will become clear shortly, denote their coordinates with different indices as (u_k, w_{k+1}), for $0 \leq k \leq \ell$, ordering them such that $u_0 < u_1 < \ldots < u_\ell$. By definition of extreme, the corresponding second coordinates satisfy $w_1 < w_2 < \ldots < w_{\ell+1}$. Observe that u_0 is the globally minimum value and $w_{\ell+1}$ is the globally maximum value. Let us now construct the reduced diagram, focusing on dimension 0. We start the upward pass with the birth of the (-1)-dimensional homology class at 0, which dies at u_0, so we have $(0, u_0)$ in $\tilde{\mathsf{D}}\mathsf{gm}_{-1}(g)$. The minima of all other connected components of $\mathbb{X}$ mark the births of 0-dimensional classes. They can be interpreted as gaps between components. We start the downward pass with the birth of a 0-dimensional class at 1, which dies at $w_{\ell+1}$, so we have $(1, w_{\ell+1})$ in $\tilde{\mathsf{D}}\mathsf{gm}_0(g)$. The maxima of all other connected components mark the deaths of 0-dimensional classes. If w is the second coordinate of a non-extreme dot, then it marks the death of the gap that opened up when we passed the corresponding first coordinate, u, during the upward pass. Hence, (u, w) belongs to $\tilde{\mathsf{D}}\mathsf{gm}_0(g)$, and we note that this dot also belongs to the standard diagram. However, if $w = w_k$ is the second coordinate of a extreme dot, then it marks the death of the gap that has opened up when we passed the minimum of the next component with extreme dot during the upward pass. This minimum value is u_k, so (u_k, w_k) belongs to $\tilde{\mathsf{D}}\mathsf{gm}_0(g)$, for $1 \leq k \leq \ell$.

In summary, we see that contiguous extreme dots exchange one pair of coordinates, using 0 on the left and 1 on the right to complete the process. More precisely, the reduced diagram is the same as the standard one, except for substituting the dots (17) for the dots (16):

$$(u_0, w_1), (u_1, w_2), \ldots, (u_{\ell-1}, w_\ell), (u_\ell, w_{\ell+1}), \quad (16)$$

$$(0, u_0), (u_1, w_1), (u_2, w_2), \ldots, (u_\ell, w_\ell), (1, w_{\ell+1}). \quad (17)$$

For later use, we call this substitution a *cascade* and write $\tilde{\mathsf{D}}\mathsf{gm}(g) = \mathsf{Dgm}(g)^C$; see Figure 3. Note that all dots belong to the horizontal and vertical subdiagrams of dimension 0, except for $(1, w_{\ell+1})$, which belongs to the relative subdiagram of dimension 0, and $(0, u_0)$, which belongs to the ordinary subdiagram of dimension -1. We will often have $u_0 = 0$ and $w_{\ell+1} = 1$, in which case the first and the last dots in (17) lie on the baseline and can be ignored.

Point calculus. As described in [1, 2], the persistence diagram can be harvested for a wealth of homological information. This includes the ranks of the homology groups of the

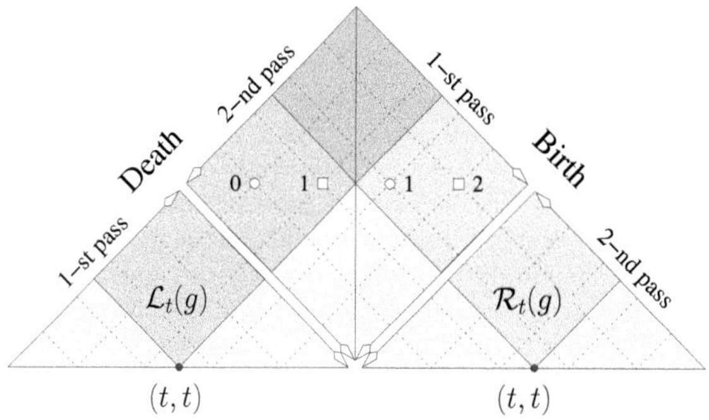

Figure 4: The persistence diagram of the function $g = f|_{\mathbb{M}}$, as defined in Figure 1; the circle dots also belong to the diagram of $f|_{\mathbb{U}}$, but the square dots do not. The rectangle on the left, $\mathcal{L}_t(g)$, represents the homology of $\mathbb{M}_t$, while the rectangle on the right, $\mathcal{R}_t(g)$, represents the homology of $(\mathbb{M}, \mathbb{M}^t)$; compare with the Betti numbers computed for the example at the end of Section 2.

sets $\mathbb{X}_t$ and of the pairs $(\mathbb{X}, \mathbb{X}^t)$. To explain this, we write $\mathcal{L}_t^p(g)$ for the multiset of dots in the rectangle with lower corner (t,t) on the base edge of $\mathsf{Ord}_p(g)$. Similarly, we write $\mathcal{R}_t^p(g)$ for the multiset of dots if the lower corner, (t,t), lies on the base edge of $\mathsf{Rel}_p(g)$; see Figure 4. Often, we drop the dimension from the notation and write $\mathcal{L}_t(g)$ for the disjoint union of the multisets $\mathcal{L}_t^p(g)$, over all p, and similar for $\mathcal{R}_t(g)$. We can read the Betti numbers of $\mathbb{X}_t$ by counting the dots in $\mathcal{L}_t(g)$, and we can read the relative Betti numbers of $(\mathbb{X}, \mathbb{X}^t)$ by counting the dots in $\mathcal{R}_t(g)$. Writing $\tilde{\mathcal{L}}_t^p(g)$ and $\tilde{\mathcal{R}}_t^p(g)$ for the corresponding multisets in the reduced diagrams, we note that the same relations hold between the reduced Betti numbers and the reduced persistence diagrams.

Betti numbers and persistence diagrams. To motivate the first result of this paper, we rewrite (1) to (5) for reduced Betti numbers:

$$\tilde{\beta}_p(\mathbb{V}_t) = \beta_q(\mathbb{U}, \mathbb{U}^t), \quad (18)$$

for all $0 \leq p \leq n$ and all regular values t. Observe also that $\mathbb{V}_t = \emptyset$ iff $\mathbb{U} - \mathbb{U}^t = \mathbb{S} - \mathbb{S}^t$, which implies that (18) also holds for $p = -1$. In words, for every dimension p and every regular value t, the rectangles $\tilde{\mathcal{L}}_t^p(f|_{\mathbb{V}})$ and $\tilde{\mathcal{R}}_t^q(f|_{\mathbb{U}})$ have equally many dots. This 1-parameter family of relations is satisfied if the reduced diagrams of $f|_{\mathbb{V}}$ and $f|_{\mathbb{U}}$ are reflections of each other. This is our first result, stated as the Land and Water Theorem in Section 4. We see an illustration in Figure 2, which shows the reduced and non-reduced diagrams of the function f in Figure 1 restricted to the solid torus, $\mathbb{U}$, and to the complementary solid torus, $\mathbb{V}$. Removing the black dots, we are left with the two reduced diagrams, which are indeed reflections of each other.

To motivate our second result, we consider the relations (13) to (15). They say that for every regular value t, the rectangle $\mathcal{L}_t^p(f|_{\mathbb{M}})$ has the same number of dots as $\mathcal{R}_t^q(f|_{\mathbb{U}})$ and $\mathcal{L}_t^p(f|_{\mathbb{U}})$ together. Similarly, $\mathcal{R}_t^p(f|_{\mathbb{M}})$ has the same number of dots as $\mathcal{L}_t^q(f|_{\mathbb{U}})$ and $\mathcal{R}_t^p(f|_{\mathbb{U}})$ together. This 1-parameter family of relations is satisfied if the persistence diagram of $f|_{\mathbb{M}}$ is the disjoint union of the diagram of $f|_{\mathbb{U}}$ and of its reflection. This is our second result, stated as the Euclidean Shore Theorem in Section 5. It is illustrated in Figure 4, which shows the persistence diagram of the function $f|_{\mathbb{M}}$ in Figure 1. Comparing Figures 2 and 4, we see that the two circle dots in Figure 4 also belong to the diagram of $f|_{\mathbb{U}}$, whereas the two square dots are reflections of the circles.

While (1) to (5) and (13) to (15) motivate our two results, these relations are not sufficient to prove them. Indeed, (13) to (15) hold in general, but the Euclidean Shore Theorem requires that the minimum and maximum of the perfect Morse function belong to a common component of $\mathbb{V}$. If this condition is violated, then we can have coordinate exchanges among the dots that contradict the Euclidean Shore Theorem without affecting the relations (13) to (15). An example of this phenomenon can be seen in Figure 5, which will be discussed in Section 5.

4. LAND AND WATER

In this section, we present our first result, which extends Alexander duality from spaces to functions.

Compatible pairings. A *pairing* between two finite-dimensional vector spaces X and Y over the field F is a bilinear map $\langle , \rangle : \mathsf{X} \times \mathsf{Y} \to \mathsf{F}$. The pairing is *non-degenerate* if every $\xi \in \mathsf{X}$ has at least one $\eta \in \mathsf{Y}$ with $\langle \xi, \eta \rangle = 1$, and every $\eta \in \mathsf{Y}$ has at least one $\xi \in \mathsf{X}$ with $\langle \xi, \eta \rangle = 1$. A non-degenerate pairing implies an isomorphism between the vector spaces, by choosing bases of both spaces and reducing the induced matrix to diagonal form.

Now consider two filtrations, each consisting of $m+1$ finite-dimensional vector spaces over F, and non-degenerate pairings connecting the filtrations contravariantly:

$$\begin{array}{ccccccc} \mathsf{X}_0 & \to & \mathsf{X}_1 & \to & \ldots & \to & \mathsf{X}_m \\ \times & & \times & & & & \times \\ \mathsf{Y}_m & \leftarrow & \mathsf{Y}_{m-1} & \leftarrow & \ldots & \leftarrow & \mathsf{Y}_0. \end{array}$$

Write $\mathsf{g}_i^j : \mathsf{X}_i \to \mathsf{X}_j$ and $\mathsf{h}_i^j : \mathsf{Y}_i \to \mathsf{Y}_j$ for the maps upstairs and downstairs. We call the pairings *compatible* with these maps if $\langle \xi, \eta \rangle = \langle \xi', \eta' \rangle$ for every $\xi' \in \mathsf{X}_i$ and $\eta \in \mathsf{Y}_{m-j}$, where $\xi = \mathsf{g}_i^j(\xi')$ and $\eta' = \mathsf{h}_{m-j}^{m-i}(\eta)$. The reason for considering compatible pairings is the following result.

CONTRAVARIANT PE THEOREM. *Let X_0 to X_m and Y_0 to Y_m be two filtrations contravariantly connected by non-degenerate and compatible pairings. Then there is a bijection between the groups upstairs and downstairs such that a class is born at X_i and dies entering X_j iff the corresponding class is born at Y_{m-j+1} and dies entering Y_{m-i+1}.*

PROOF. Let X_i^* be the dual vector space of X_i, that is, the space of homomorphisms from X_i to F, and similarly let Y_{m-j}^* be the dual vector space of Y_{m-j}. We consider the linear maps

$$\phi : \mathsf{X}_i \to \mathsf{Y}_{m-j}^* \quad \text{and} \quad \psi : \mathsf{Y}_{m-j} \to \mathsf{X}_i^*$$

defined by mapping x to $\phi(x)$, which sends $y \in \mathsf{Y}_{m-j}$ to $\langle g_i^j(x), y \rangle$, and by mapping y to $\psi(y)$, which sends $x \in \mathsf{X}_i$ to $\langle x, h_{m-j}^{m-i}(y) \rangle$. Since the pairings are compatible, we have

$$\phi(x)(y) = \langle g_i^j(x), y \rangle = \langle x, h_{m-j}^{m-i}(y) \rangle = \psi(y)(x).$$

Therefore, ϕ and ψ are adjoint, which means that the associated matrices are transposes of each other. It follows that ϕ and ψ have the same rank. Note that $\phi = \phi' \circ g_i^j$, where $\phi' : \mathsf{X}_j \to \mathsf{Y}_{m-j}^*$ is defined by mapping x' to $\phi'(x')$, which sends $y \in \mathsf{Y}_{m-j}$ to $\langle x', y \rangle$. Since the pairing is non-degenerate, ϕ' is an isomorphism. Hence, $\operatorname{rank} g_i^j = \operatorname{rank} \phi$, and by a similar argument, $\operatorname{rank} h_{m-j}^{m-i} = \operatorname{rank} \psi$. This now implies

$$\operatorname{rank} g_i^j \;=\; \operatorname{rank} h_{m-j}^{m-i}. \tag{19}$$

These ranks are the persistent Betti numbers, which characterize the births and deaths in the two persistence modules. Specifically, the number of classes born at X_i and dying entering X_j is

$$\operatorname{rank} g_i^{j-1} - \operatorname{rank} g_{i-1}^{j-1} - \operatorname{rank} g_i^j + \operatorname{rank} g_{i-1}^j.$$

Similarly, the number of classes born at Y_{m-j+1} and dying entering Y_{m-i+1} is

$$\operatorname{rank} h_{m-j+1}^{m-i} - \operatorname{rank} h_{m-j}^{m-i} - \operatorname{rank} h_{m-j+1}^{m-i+1} + \operatorname{rank} h_{m-j}^{m-i+1}.$$

But these two sums are equal by (19). $\square$

Alexander duality for functions. Let $f : \mathbb{S} \to [0,1]$ be a perfect Morse function on the $(n+1)$-dimensional sphere, and $\mathbb{S} = \mathbb{U} \cup \mathbb{V}$ a complementary decomposition into two $(n+1)$-manifolds with shared boundary $\mathbb{M} = \mathbb{U} \cap \mathbb{V}$. Assuming $f|_{\mathbb{M}}$ is tame, we have only finitely many homological critical values, including 0 and 1. Letting these critical values be s_0 to s_m, we interleave them with homological regular values to get $s_0 < t_0 < s_1 < \ldots < t_{m-1} < s_m$. We are interested in the filtrations defined by the restrictions of f to $\mathbb{U}$ and to $\mathbb{V}$. To describe them, we write $\tilde{\mathsf{U}}_i^p$ for the p-th reduced homology group of $\mathbb{U}_{t_i}$, if $0 \leq i < m$, of $\mathbb{U}$, if $i = m$, and of $(\mathbb{U}, \mathbb{U}^{t_{2m-i}})$, if $m < i \leq 2m$. In the same way, we define $\tilde{\mathsf{V}}_i^q$ for $0 \leq i \leq 2m$ with respect to $\mathbb{V}$. Using Alexander duality to form isomorphisms, we get two contravariantly connected persistence modules:

$$\begin{array}{ccccccccc} \tilde{\mathsf{U}}_0^p & \to & \ldots & \to & \tilde{\mathsf{U}}_m^p & \to & \ldots & \to & \tilde{\mathsf{U}}_{2m}^p \\ \downarrow & & & & \downarrow & & & & \downarrow \\ \tilde{\mathsf{V}}_{2m}^q & \leftarrow & \ldots & \leftarrow & \tilde{\mathsf{V}}_m^q & \leftarrow & \ldots & \leftarrow & \tilde{\mathsf{V}}_0^q, \end{array}$$

where $p + q = n$. At the end of this section, we will prove that there exists a sequences of compatible pairings between $\tilde{\mathsf{U}}_i^p$ and $\tilde{\mathsf{V}}_{2m-i}^q$, for $0 \leq i \leq 2m$. Using these pairings, we get our first result, which we formulate using the superscript 'T' for the operation that reflects a dot across the vertical axis of a persistence diagram and, at the same time, changes its dimension from p to $q = n - p$.

LAND AND WATER THEOREM. *Let $\mathbb{U}$ and $\mathbb{V}$ be a complementary decomposition of $\mathbb{S} = \mathbb{S}^{n+1}$, and let $f : \mathbb{S} \to [0,1]$ be a perfect Morse function whose restriction to the n-manifold $\mathbb{M} = \mathbb{U} \cap \mathbb{V}$ is tame. Then $\tilde{\mathsf{D}}\mathsf{gm}(f|_{\mathbb{V}}) = \tilde{\mathsf{D}}\mathsf{gm}(f|_{\mathbb{U}})^T$.*

PROOF. We apply the Contravariant PE Theorem to the persistence modules $\tilde{\mathsf{U}}_i^p$ and $\tilde{\mathsf{V}}_i^q$: If a class α is born at $\tilde{\mathsf{U}}_i^p$ and dies entering $\tilde{\mathsf{U}}_j^p$ then its corresponding class is born at $\tilde{\mathsf{V}}_{2m-j+1}^q$ and dies entering $\tilde{\mathsf{V}}_{2m-i+1}^q$. If $1 \leq i < j \leq m$, then the class α is represented by (s_i, s_j) in the p-th ordinary subdiagram of $f|_{\mathbb{U}}$. The corresponding class is represented by (s_j, s_i) in the q-th relative subdiagram of $f|_{\mathbb{V}}$. The reflection maps the first and second coordinate-axes of the ordinary subdiagram to the second and first coordinate-axes of the relative subdiagram. It follows that it maps (s_i, s_j) in the former to (s_j, s_i) in the latter subdiagram. Other cases are similar, and we conclude that $\tilde{\mathsf{D}}\mathsf{gm}_p(f|_{\mathbb{U}})$ and $\tilde{\mathsf{D}}\mathsf{gm}_q(f|_{\mathbb{V}})$ are reflections of each other. Writing this more succinctly gives the claimed relationship between the reduced persistence diagrams of f restricted to the two complementary subsets of the sphere. $\square$

Alexander pairing. We fill the gap in the proof of the Land and Water Theorem by establishing compatible pairings between the groups upstairs and downstairs. For the sake of simplicity, we restrict ourselves to the case $\mathsf{F} = \mathbb{Z}_2$. While the argument for general fields is similar, it requires oriented simplices, which is a technical formalism we prefer to avoid. We begin with the pairing implicit in Lefschetz duality for manifolds with possibly non-empty boundary:

$$\langle , \rangle_L : \tilde{\mathsf{H}}_p(\mathbb{U}_t) \times \tilde{\mathsf{H}}_{q+1}(\mathbb{U}_t, \partial\mathbb{U}_t) \to \mathbb{Z}_2,$$

which is defined by mapping two classes to the parity of the number of intersections between representing cycles. We get such a pairing for every regular value t, and these pairings are compatible with the horizontal maps induced by inclusion of the sublevel sets; see [3]. Next, we reduce the dimension of the second factor using a mapping, φ, which we compose from four simpler mappings:

$$\begin{aligned} \tilde{\mathsf{H}}_{q+1}(\mathbb{U}_t, \partial\mathbb{U}_t) &\overset{\varphi_1}{\to} \tilde{\mathsf{H}}_{q+1}(\mathbb{S}, \mathbb{S}^t \cup \mathbb{V}_t) \\ &\overset{\varphi_2}{\to} \tilde{\mathsf{H}}_q(\mathbb{S}^t \cup \mathbb{V}_t) \\ &\overset{\varphi_3}{\to} \tilde{\mathsf{H}}_q(\mathbb{S}^t \cup \mathbb{V}_t, \mathbb{S}^t) \\ &\overset{\varphi_4}{\to} \tilde{\mathsf{H}}_q(\mathbb{V}, \mathbb{V}^t). \end{aligned}$$

The first mapping, φ_1, is an isomorphism defined by excision. Next, φ_2, is the connecting homomorphism of the exact sequence of the pair $(\mathbb{S}, \mathbb{S}^t \cup \mathbb{V}_t)$. Since the q-th reduced homology of $\mathbb{S}$ is trivial, for all $q \neq n+1$, φ_2 is an isomorphism for all $0 \leq q < n$. It is a surjection for $q = n$. The third mapping, φ_3, is induced by inclusion. It occurs in the exact sequence of the pair $(\mathbb{S}^t \cup \mathbb{V}_t, \mathbb{S}^t)$. For $0 < t \leq 1$, the reduced homology groups of $\mathbb{S}^t$ are all trivial, implying that φ_3 is an isomorphism. For $t = 0$, φ_3 is the trivial isomorphism. Finally, φ_4 is again an isomorphism defined by excision. In summary, we get $\varphi : \tilde{\mathsf{H}}_{q+1}(\mathbb{U}_t, \partial\mathbb{U}_t) \to \tilde{\mathsf{H}}_q(\mathbb{V}, \mathbb{V}^t)$, which is an isomorphism for $0 \leq q < n$ and a surjection for $q = n$. More specifically, each class $\beta \in \tilde{\mathsf{H}}_n(\mathbb{V}, \mathbb{V}^t)$ has two preimages. Indeed, an n-cycle representing the corresponding class in $\tilde{\mathsf{H}}_n(\mathbb{S}^t \cup \mathbb{V}_t)$ partitions the components of $\mathbb{U}_t$ into two subsets, and each subset generates an $(n+1)$-dimensional relative homology class that maps to β.

Note that the roles of $\mathbb{U}$ and $\mathbb{V}$ can be interchanged to get a mapping from $\tilde{\mathsf{H}}_{q+1}(\mathbb{V}_t, \partial\mathbb{V}_t)$ to $\tilde{\mathsf{H}}_q(\mathbb{U}, \mathbb{U}^t)$. With this, we are ready to construct the *Alexander pairing*:

$$\langle , \rangle_A : \tilde{\mathsf{H}}_p(\mathbb{U}_t) \times \tilde{\mathsf{H}}_q(\mathbb{V}, \mathbb{V}^t) \to \mathbb{Z}_2,$$

defined by $\langle \alpha, \beta \rangle_A = \langle \alpha, \beta' \rangle_L$, where β' is a preimage of β under φ. This fixes the pairing of $\tilde{\mathsf{U}}^p_i$ and $\tilde{\mathsf{V}}^q_{2m-i}$ for $0 \leq i < m$. Similarly, we define $\langle , \rangle_A$ for $\tilde{\mathsf{H}}_p(\mathbb{U}, \mathbb{U}^t) \times \tilde{\mathsf{H}}_q(\mathbb{V}_t)$ which fixes the pairing for $m < i \leq 2m$. The pairing is clearly well-defined for $0 < p < n$, where φ is an isomorphism. The remaining two cases are symmetric, and we consider $p = 0$ and $\beta \in \tilde{\mathsf{H}}_n(\mathbb{V}, \mathbb{V}^t)$. As noted before, β has two preimages, β' and β'', which are generated by complementary subsets of $\mathbb{U}_t$. Every $\alpha \in \tilde{\mathsf{H}}_0(\mathbb{U}_t)$ is represented by an even number of points, which the complementary subsets partition into two even or two odd subsets. In either case, we have $\langle \alpha, \beta' \rangle_L = \langle \alpha, \beta'' \rangle_L$, so the pairing is well-defined in all cases. Moreover, the compatibility of the Alexander pairings within the left and right halves follows from the compatibility of the Lefschetz pairings.

It remains to define the paring in the middle of the sequence, for $\tilde{\mathsf{U}}^p_m = \tilde{\mathsf{H}}_p(\mathbb{U})$ and $\tilde{\mathsf{V}}^q_m = \tilde{\mathsf{H}}_q(\mathbb{V})$. To this end, let α be a reduced p-cycle in $\mathbb{U}$, β a reduced q-cycle in $\mathbb{V}$, and β' a $(q+1)$-chain whose boundary is β. We define $\langle \alpha, \beta \rangle_A$ by counting the intersections between α and β'. Note that β' is the preimage of β under the previously defined mapping, $\varphi : \tilde{\mathsf{H}}_{q+1}(\mathbb{U}_t, \partial\mathbb{U}_t) \to \tilde{\mathsf{H}}_q(\mathbb{V}, \mathbb{V}^t)$, for any regular value t. Therefore, the pairing in the middle is compatible with the pairings in left half, again exploiting the compatibility of the Lefschetz pairing. Alternatively, we can define the pairing by taking a $(p+1)$-chain α' with boundary α and counting its intersections with β. In this case, α' is the preimage of α under the map $\varphi : \tilde{\mathsf{H}}_{p+1}(\mathbb{V}_t, \partial\mathbb{V}_t) \to \tilde{\mathsf{H}}_p(\mathbb{U}, \mathbb{U}^t)$, and compatibility with the pairings in the right half follows. Indeed, both definitions are equivalent, as already observed by Lefschetz [9]. We give a simple proof for the case $\mathsf{F} = \mathbb{Z}_2$:

BRIDGE LEMMA. *Let α and β be non-intersecting reduced cycles on the $(n+1)$-sphere whose dimensions add up to n. Let α' and β' be chains whose boundaries are α and β, respectively. Then $\langle \alpha', \beta \rangle_L = \langle \alpha, \beta' \rangle_L$.*

PROOF. We can assume that α and β do not intersect. The intersection of α' and β' is a 1-chain. Its boundary consists of an even number of points, and is the disjoint union of intersections of α' with β and of α with β'. This implies that both types of intersections occur with the same parity. □

5. SHORE

This section presents our second result, which extends the combination of Alexander duality and Mayer-Vietoris sequences from spaces to functions.

Mayer-Vietoris sequence of filtrations. Assuming that $f|_{\mathbb{M}}$ is tame, we write $s_0 < t_0 < s_1 < \ldots < t_{m-1} < s_m$ for the interleaved sequence of homological critical and regular values, as before. The main tool in this section is the diagram obtained by connecting the filtrations of f and its restrictions with Mayer-Vietoris sequences. We describe this using shorthand notation for the groups. Consistent with earlier notation, we write S^p_i for the p-th homology group of $\mathbb{S}_{t_i}$, for $0 \leq i < m$, of $\mathbb{S}$, for $i = m$, and of $(\mathbb{S}, \mathbb{S}^{t_{2m-i}})$, for $m < i \leq 2m$. Similarly, we write M^p_i for the p-th homology group of $\mathbb{M}_{t_i}$, of $\mathbb{M}$, and of $(\mathbb{M}, \mathbb{M}^{t_{2m-i}})$. Finally, we write D^p_i for the direct sum of the p-th homology groups of $\mathbb{U}_{t_i}$ and $\mathbb{V}_{t_i}$, of $\mathbb{U}$ and $\mathbb{V}$, and of $(\mathbb{U}, \mathbb{U}^{t_{2m-i}})$ and $(\mathbb{V}, \mathbb{V}^{t_{2m-i}})$. Drawing the filtrations from left to right and the Mayer-Vietoris sequences from top to bottom, we get

$$\begin{array}{ccccccccc} \downarrow & & & & \downarrow & & & & \downarrow \\ \mathsf{S}^{p+1}_0 & \to & \ldots & \to & \mathsf{S}^{p+1}_m & \to & \ldots & \to & \mathsf{S}^{p+1}_{2m} \\ \downarrow & & & & \downarrow & & & & \downarrow \\ \mathsf{M}^p_0 & \to & \ldots & \to & \mathsf{M}^p_m & \to & \ldots & \to & \mathsf{M}^p_{2m} \\ \downarrow & & & & \downarrow & & & & \downarrow \\ \mathsf{D}^p_0 & \to & \ldots & \to & \mathsf{D}^p_m & \to & \ldots & \to & \mathsf{D}^p_{2m} \\ \downarrow & & & & \downarrow & & & & \downarrow \\ \mathsf{S}^p_0 & \to & \ldots & \to & \mathsf{S}^p_m & \to & \ldots & \to & \mathsf{S}^p_{2m} \\ \downarrow & & & & \downarrow & & & & \downarrow \end{array}$$

All squares commute, which is particularly easy to see for the squares that connect groups of the same dimension, whose maps are all induced by inclusion. For $1 \leq p \leq n$, all groups S^p_i are trivial. By exactness of the Mayer-Vietoris sequences, this implies that the maps $\mathsf{m}^p_i : \mathsf{M}^p_i \to \mathsf{D}^p_i$ are isomorphisms, for $1 \leq p \leq n-1$. The persistence diagram of $f|_{\mathbb{M}}$ is therefore the disjoint union of the persistence diagrams of $f|_{\mathbb{U}}$ and $f|_{\mathbb{V}}$. More generally, we claim:

General Shore Theorem. *Let n be a positive integer, let $\mathbb{U}$ and $\mathbb{V}$ be a complementary decomposition of $\mathbb{S} = \mathbb{S}^{n+1}$, and let $f : \mathbb{S} \to [0,1]$ be a perfect Morse function whose restriction to the n-manifold $\mathbb{M} = \mathbb{U} \cap \mathbb{V}$ is tame. Then*

$$\mathsf{Dgm}_0(f|_{\mathbb{M}}) = [\mathsf{Dgm}_0(f|_{\mathbb{U}}) \sqcup \mathsf{Dgm}_0(f|_{\mathbb{V}})]^C, \quad (20)$$

$$\mathsf{Dgm}_p(f|_{\mathbb{M}}) = \mathsf{Dgm}_p(f|_{\mathbb{U}}) \sqcup \mathsf{Dgm}_p(f|_{\mathbb{V}}), \quad (21)$$

$$\mathsf{Dgm}_n(f|_{\mathbb{M}}) = [\mathsf{Dgm}_0(f|_{\mathbb{U}}) \sqcup \mathsf{Dgm}_0(f|_{\mathbb{V}})]^{CT}, \quad (22)$$

for $1 \leq p \leq n-1$, where C stands for applying the cascade and T stands for reflecting the diagram.

We note that the assumption of n being positive is necessary since the formulas do not hold for $n = 0$. Concerning the proof of the theorem, note that (21) is clear, and that (22) follows from (20) using the duality of persistence diagrams from [3]. We will need to study the impact of the non-trivial groups S_i^0 to prove (20).

Latitudinal manifolds. Call the minimum of f the *south-pole* and the maximum the *north-pole* of the sphere. Let $\mathbb{M}'$ be a component of $\mathbb{M}$, and note that it is an n-manifold that decomposes $\mathbb{S}$ into two complementary subsets. If it separates the two poles, we refer to $\mathbb{M}'$ as a *latitudinal n-manifold.* Assuming neither pole lies on $\mathbb{M}$, we order the latitudinal n-manifolds from south to north as $\mathbb{M}_1, \mathbb{M}_2, \ldots, \mathbb{M}_\ell$. Letting u_k and w_k be the minimum and maximum values of f restricted to $\mathbb{M}_k$, we get $u_1 < u_2 < \ldots < u_\ell$ as well as $w_1 < w_2 < \ldots < w_\ell$.

For each component $\mathbb{M}'$ of $\mathbb{M}$, there are *neighboring components* $\mathbb{U}'$ of $\mathbb{U}$ and $\mathbb{V}'$ of $\mathbb{V}$ defined such that $\mathbb{U}' \cap \mathbb{V}' = \mathbb{M}'$. A single component of $\mathbb{U}$ or $\mathbb{V}$ can be neighbor to an arbitrary number of n-manifolds, but not to more than two latitudinal n-manifolds. Specifically, there are components $\mathbb{S}_0, \mathbb{S}_1, \ldots, \mathbb{S}_\ell$ of $\mathbb{U}$ and $\mathbb{V}$ such that $\mathbb{M}_k = \mathbb{S}_{k-1} \cap \mathbb{S}_k$ for each $1 \leq k \leq \ell$. We refer to the $\mathbb{S}_k$ as *latitudinal components.* For example, in Figure 5, we have $\ell = 2$ latitudinal 1-manifolds and $\ell + 1 = 3$ latitudinal components. Setting $u_0 = 0$ and $w_{\ell+1} = 1$, the minimum and maximum values of f restricted to $\mathbb{S}_k$ are u_k and w_{k+1}, for $0 \leq k \leq \ell$. Assuming $\mathbb{S}_k$ belongs to $\mathbb{U}$, it gives rise to zero or more dots in the ordinary subdiagram of $\mathsf{Dgm}_0(f|_{\mathbb{U}})$ and to exactly one dot, (u_k, w_{k+1}), in the horizontal subdiagram. We say the dot in the horizontal subdiagram *represents* the 0-dimensional homology class defined by $\mathbb{S}_k$. Note that its coordinates are indexed consistently with the notation used in the introduction of the cascade. There is indeed a connection, namely the dots in the 0-th horizontal subdiagram representing latitudinal components are the extreme ones in the multiset:

Latitudinal Component Lemma. *The latitudinal components of $\mathbb{U}$ and $\mathbb{V}$ correspond bijectively to the extreme dots in the disjoint union of $\mathsf{Dgm}_0(f|_{\mathbb{U}})$ and $\mathsf{Dgm}_0(f|_{\mathbb{V}})$.*

We omit the proof, which is not difficult. The statement includes the case in which there is no latitudinal n-manifold so that $\mathbb{S}_0$ is the only latitudinal component. It contains both poles and is therefore represented by $(0,1)$, which is the only extreme dot in the disjoint union of 0-th diagrams.

Proof of (20). We compare $\mathsf{Dgm}_0(f|_{\mathbb{M}})$ with the disjoint union of $\mathsf{Dgm}_0(f|_{\mathbb{U}})$ and $\mathsf{Dgm}_0(f|_{\mathbb{V}})$, noting that their dots all belong to the ordinary and horizontal subdiagrams. Consider first a dot (a,b) in the ordinary subdiagram $\mathsf{Ord}_0(f|_{\mathbb{M}})$. It represents a component in the sublevel set that is born at a minimum $x \in \mathbb{M}$, with $f(x) = a$, and that dies at a saddle point $y \in \mathbb{M}$, with $f(y) = b$. Let $\mathbb{M}' \subseteq \mathbb{M}$ be the connected n-manifold that contains x and y, and recall that $\mathbb{M}' = \mathbb{U}' \cap \mathbb{V}'$. Assume that x is a local minimum of $f|_{\mathbb{U}}$. In the sequence of sublevel sets of $f|_{\mathbb{U}}$, we see the birth of a component at $f(x) = a$ and its death at $f(y) = b$. It follows that (a,b) is also a dot in $\mathsf{Ord}_0(f|_{\mathbb{U}})$ and therefore of the disjoint union of the diagrams of $f|_{\mathbb{U}}$ and $f|_{\mathbb{V}}$. The argument can be reversed, which implies that the 0-th ordinary subdiagrams are the same.

Consider second a dot (a,b) in the horizontal subdiagram $\mathsf{Hor}_0(f|_{\mathbb{M}})$. It represents a connected n-manifold $\mathbb{M}' \subseteq \mathbb{M}$, which splits $\mathbb{S}$ into two subsets. If $\mathbb{M}'$ is non-latitudinal, then one subset contains both poles and the other contains neither. The latter subset contains a (non-latitudinal) neighboring component, which is represented by (a,b) in $\mathsf{Hor}_0(f|_{\mathbb{U}})$ or in $\mathsf{Hor}_0(f|_{\mathbb{V}})$. Again, the argument can be reversed. If on the other hand, $\mathbb{M}'$ is latitudinal, then $(a,b) = (u_i, w_i)$, for some i, where we write $u_1, u_2, \ldots, u_\ell$ for the minimum values and $w_1, w_2, \ldots, w_\ell$ for the maximum values of the latitudinal n-manifolds, as before. More generally, we get the dots (u_i, w_i) in $\mathsf{Hor}_0(f|_{\mathbb{M}})$, for $1 \leq i \leq \ell$. Adding $u_0 = 0$ and $w_{\ell+1} = 1$, we get the dots (u_i, w_{i+1}) in $\mathsf{Hor}_0(f|_{\mathbb{U}}) \sqcup \mathsf{Hor}_0(f|_{\mathbb{V}})$, for $1 \leq i \leq \ell$. After adding $(0,0)$ and $(1,1)$, which are both irrelevant, we get precisely the dots specified in (16) and in (17). This implies that the two diagrams are related to each other by a cascade, which completes the proof of (20).

The General Shore Theorem can also be proven with algebraic arguments: Di Fabio and Landi [4, 5] consider decompositions $\mathbb{X} = \mathbb{A} \cup \mathbb{B}$, for a general topological space $\mathbb{X}$, and relate the ordinary, relative, and extended (horizontal plus vertical) persistent Betti numbers of a function restricted to these sets with each other. These relationships depend on the ranks of maps between certain absolute and relative homology groups; see Theorems 3.1, 3.3, and 3.4 in [5] for details. Applied to the case of a sphere, these maps are trivial except for the extended case in dimensions 0 and n. Investigating this case more carefully leads to the same result as (20).

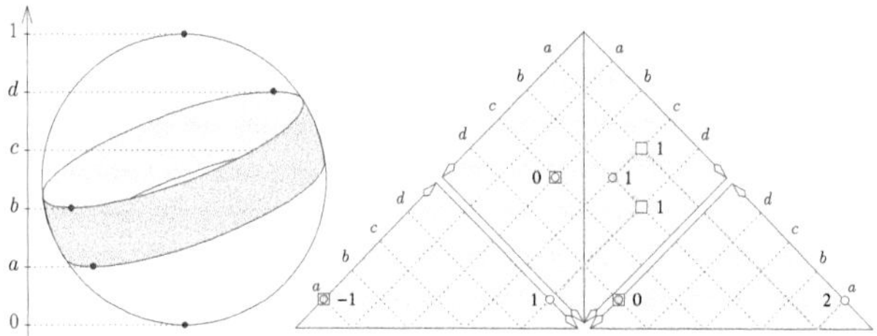

Figure 5: Left: the height function on the 2-sphere decomposed into an annulus and a pair of disks. Right: the reduced persistence diagram of f restricted to the annulus (shaded circular dots), to the pair of disks (white circular dots), and to the two circles (squares).

Euclidean case. Considering that (13) to (15) give elegant relations between the Betti numbers of the shore and the land, it is perhaps surprising that we need cascades to formulate a similar result for persistence diagrams. Indeed, (13) to (15) suggest that the persistence diagrams of $f|_{\mathbb{M}}$ be the disjoint union of the persistence diagram of $f|_{\mathbb{U}}$ and of its reflection. The example in Figure 5 shows that this simple relation does not hold in general. Indeed, we have the dots (a,c), (b,d), (c,a), (d,b) in the standard diagram of $f|_{\mathbb{M}}$, while the standard diagram of $f|_{\mathbb{U}}$ contains (a,d) and (c,b). This clearly violates the suggested relation. However, there is a natural setting in which the relation does hold. To state this result, we let $\mathbb{A}$ be a compact $(n+1)$-manifold with boundary in $\mathbb{R}^{n+1}$. To be compatible with the rest of this paper, we assume $\mathbb{A}$ is triangulable and there exists an open neighborhood and a deformation retraction of that neighborhood to $\mathbb{A}$.

EUCLIDEAN SHORE THEOREM. *Let n be a positive integer, let $\mathbb{A} \subset \mathbb{R}^{n+1}$ be a compact $(n+1)$-manifold with boundary satisfying the technical conditions as described, and suppose that $e : \mathbb{R}^{n+1} \to \mathbb{R}$ has no homological critical values and its restriction to $\partial\mathbb{A}$ is tame. Then*

$$\mathsf{Dgm}(e|_{\partial\mathbb{A}}) = \mathsf{Dgm}(e|_{\mathbb{A}}) \sqcup \mathsf{Dgm}(e|_{\mathbb{A}})^T.$$

PROOF. We can extend e to a perfect Morse function $f : \mathbb{S}^{n+1} \to \mathbb{R}$ and $\mathbb{A}$ to a subset $\mathbb{U}$ of $\mathbb{S} = \mathbb{S}^{n+1}$ such that the persistence diagrams of e restricted to $\mathbb{A}$ and to $\partial\mathbb{A}$ are the same as those of f restricted to $\mathbb{U}$ and $\mathbb{M} = \partial\mathbb{U}$. It thus suffices to show that the persistence diagram of $f|_{\mathbb{M}}$ is the disjoint union of the diagram of $f|_{\mathbb{U}}$ and of its reflection. For $0 < p < n$, this follows from (21), from

$$\tilde{\mathsf{D}}\mathsf{gm}_p(f|_{\mathbb{V}}) = \tilde{\mathsf{D}}\mathsf{gm}_{n-p}(f|_{\mathbb{U}})^T,$$

as stated in the Land and Water Theorem, and the fact that the reduced diagrams are equal to the non-reduced ones. For $p = 0$, we start with (20) and note that there is only one extreme dot in $\mathsf{Dgm}_0(f|_{\mathbb{U}}) \sqcup \mathsf{Dgm}_0(f|_{\mathbb{V}})$, namely the one representing the component of $\mathbb{V}$ that contains both the minimum and the maximum of f. It follows that the cascade leaves $\mathsf{Dgm}_0(f|_{\mathbb{U}})$ unchanged while it turns $\mathsf{Dgm}_0(f|_{\mathbb{V}})$ into $\tilde{\mathsf{D}}\mathsf{gm}_0(f|_{\mathbb{V}})$. Using the Land and Water Theorem again, we obtain:

$$\begin{aligned}\mathsf{Dgm}_0(f|_{\mathbb{M}}) &= [\mathsf{Dgm}_0(f|_{\mathbb{U}}) \sqcup \mathsf{Dgm}_0(f|_{\mathbb{V}})]^C \\ &= \mathsf{Dgm}_0(f|_{\mathbb{U}}) \sqcup \tilde{\mathsf{D}}\mathsf{gm}_0(f|_{\mathbb{V}}) \\ &= \mathsf{Dgm}_0(f|_{\mathbb{U}}) \sqcup \tilde{\mathsf{D}}\mathsf{gm}_n(f|_{\mathbb{U}})^T \\ &= \mathsf{Dgm}_0(f|_{\mathbb{U}}) \sqcup \mathsf{Dgm}_n(f|_{\mathbb{U}})^T.\end{aligned}$$

For $p = n$, we exploit Poincaré duality for manifolds, which implies $\mathsf{Dgm}_n(f|_{\mathbb{M}}) = \mathsf{Dgm}_0(f|_{\mathbb{M}})^T$; see [3]. □

The 4-dimensional version of the theorem, for $n = 3$, brings us back full circle to the motivation for this work, namely the computation of the persistence diagram of the space-time shape formed by a moving collection of biological cells [7]. Modeling space-time as $\mathbb{R}^4$, the shape is compact, and we consider the time function restricted to that shape. This data satisfies the assumptions of the Euclidean Shore Theorem, so we can infer the persistence diagram of the function on the boundary from the diagram of the function on the solid 4-dimensional shape.

6. DISCUSSION

The main contributions of this paper are two extensions of Alexander duality from spaces to functions. The first extension is direct and relates the persistence diagrams of a perfect Morse function restricted to two $(n+1)$-manifolds with shared boundary, forming a complementary decomposition of the $(n+1)$-dimensional sphere with each other. The second extension relates the persistence diagram of the function restricted to the n-manifold boundary with the diagram of the function restricted to one $(n+1)$-manifold. A key tool in its proof is the persistence module of Mayer-Vietoris sequences (or the Mayer-Vietoris sequence of persistence modules). This suggests the study of more general modules of exact sequences. Besides the hope to develop a general purpose device that pervades persistent homology in the same way exact sequences pervade homology, we motivate the study with a few concrete questions.

- Can our results be extended to the more general setting of a decomposition $\mathbb{X} = \mathbb{A} \cup \mathbb{B}$ of a topological space $\mathbb{X}$, as considered by Di Fabio and Landi [4, 5].
- As pointed out to us by Yuriy Mileyko, an alternative approach to generalizing our results for functions can be based on the generalization of Alexander duality given in [8, Proposition 3.46]. Combining this statement of duality with the Mayer-Vietoris sequence has the potential to generalize our two Shore Theorems.
- Recall that our Euclidean Shore Theorem relates the persistence diagram of a function f on an $(n+1)$-manifold with boundary $\mathbb{A} \subseteq \mathbb{R}^{n+1}$ with that of its restriction to the boundary. Here, $\mathbb{A}$ is assumed to be compact. As suggested to us by Amit Patel, we can consider other inclusions $\mathbb{M} \subseteq \mathbb{A}$ and ask how much information about the diagram of f restricted to $\mathbb{M}$ we can infer from the diagram of f restricted to $\mathbb{A}$. Similarly, how much can we say about $\mathbb{M}$ by comparing the persistence diagrams of f restricted to the two spaces?

Acknowledgments

The authors thank an anonymous referee for suggesting the simplified proof of the Contravariant PE Theorem given in this paper. They also thank Frederick Cohen, Yuriy Mileyko and Amit Patel for helpful discussions. This research is partially supported by the National Science Foundation (NSF) under grant DBI-0820624, the European Science Foundation under the Research Networking Programme, and the Russian Government Project 11.G34.31.0053.

7. REFERENCES

[1] P. BENDICH, S. CABELLO AND H. EDELSBRUNNER. A point calculus for interlevel set homology. *Pattern Rec. Lett.*, to appear.

[2] G. CARLSSON, V. DE SILVA AND D. MOROZOV. Zigzag persistent homology and real-valued functions. *In* "Proc. 25th Ann. Sympos. Comput. Geom., 2009", 247–256.

[3] D. COHEN-STEINER, H. EDELSBRUNNER AND J. HARER. Extending persistence using Poincaré and Lefschetz duality. *Found. Comput. Math.* **9** (2009), 79–103.

[4] B. DI FABIO AND C. LANDI. A Mayer-Vietoris formula for persistent homology with an application to shape recognition in the presence of occlusions. *Found. Comput. Math.* **11** (2011), 499–527.

[5] B. DI FABIO AND C. LANDI. Persistent homology and partial similarity of shapes. *Pattern Recognition Letters*, to appear.

[6] H. Edelsbrunner and J. L. Harer. *Computational Topology. An Introduction.* Amer. Math. Soc., Providence, Rhode Island, 2010.

[7] H. Edelsbrunner, C.-P. Heisenberg, M. Kerber and G. Krens. The medusa of spatial sorting. Manuscript, IST Austria, Klosterneuburg, Austria, 2012.

[8] A. Hatcher. *Algebraic Topology.* Cambridge Univ. Press, Cambridge, England, 2002.

[9] S. Lefschetz. Intersections and transformations of complexes and manifolds. *Amer. Math. Soc.* **28** (1926), 1–49.

[10] J. R. Munkres. *Elements of Algebraic Topology.* Perseus, Cambridge, Massachusetts, 1984.

Certified Computation of planar Morse-Smale Complexes

Amit Chattopadhyay
Applied Mechanics and Mathematics (MEMA)
Université Catholique de Louvain
B-1348, Louvain-la-Neuve, Belgium
amit.chattopadhyay@uclouvain.be

Gert Vegter
Johann Bernoulli Institute of Mathematics and Computer Science
University of Groningen
P.O. Box 407,
9700 AK Groningen, The Netherlands
G.Vegter@rug.nl

Chee K. Yap
Courant Institute of Mathematical Sciences
New York University
New York, NY-10012, USA
yap@cs.nyu.edu

ABSTRACT

The Morse-Smale complex is an important tool for global topological analysis in various problems of computational geometry and topology. Algorithms for Morse-Smale complexes have been presented in case of piecewise linear manifolds [9]. However, previous research in this field does not provide certified methods in the case of smooth functions. In the current paper we use interval arithmetic to compute a topologically correct approximation of Morse-Smale complex of smooth functions of two variables. The algorithm can also compute geometrically close Morse-Smale complex.

Categories and Subject Descriptors

G.1 [**Mathematics of Computing**]: Numerical Analysis; G.1.0 [**Numerical Analysis**]: Interval Arithmetic; G.1.2 [**Numerical Analysis**]: Approximation

General Terms

Algorithms

Keywords

Morse-Smale Complex, Certified Computing, Interval Arithmetic

1. INTRODUCTION

Geometrical shapes occurring in the real world are often extremely complex. To analyze them, one associates a sufficiently smooth scalar field with the shape, e.g., a density function or a function interpolating gray values. Using this function, topological and geometrical information about the shape may be extracted, e.g., by computing its *Morse-Smale complex*. The cells of this complex are maximal connected sets consisting of orthogonal trajectories of the contour lines—curves of steepest ascent—with the same critical point of the function as origin and the same critical point as destination. The leftmost plots in Figures 1(a) and 1(b) illustrate the level sets of such a density function h, and the rightmost pictures the Morse-Smale complex of h as computed by the algorithm in this paper. This complex reveals the global topology of the shape. Recently, the Morse-Smale complex has been successfully applied in different areas like molecular shape analysis, image analysis, data and vector field simplification, visualization and detection of voids and clusters in galaxy distributions [6, 11].

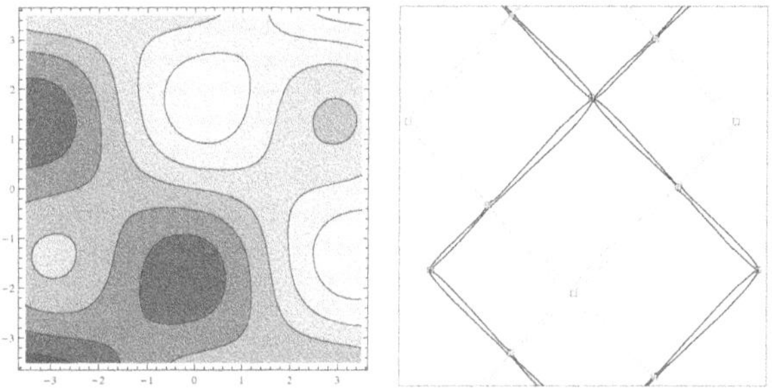

(a) Contour plot (left) and Morse-Smale complex (right) of $h(x, y) = \cos x\ \sin y + 0.2\,(x + y)$ inside box $[-3.5, 3.5] \times [-3.5, 3.5]$. CPU-time: 11 seconds.

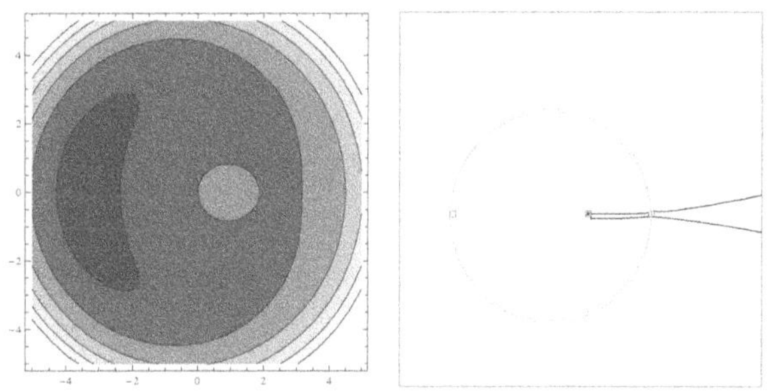

(b) Contour plot (left) and Morse-Smale Complex (right) of $h(x, y) = 10x - \frac{13}{2}(x^2 + y^2) + \frac{1}{3}(x^2 + y^2)^2$ inside box $[-5, 5] \times [-5, 5]$. CPU-time: 0.5 seconds.

Figure 1: Contour plots of Morse-Smale functions, and their Morse-Smale complexes.

1.1 Problem statement

A Morse function $h : \mathbb{R}^2 \to \mathbb{R}$ is a real-valued function with *non-degenerate critical points* (i.e., critical points with non-singular Hessian matrix). Non-degenerate critical points are *isolated*, and are either maxima, or minima, or

SCG'12, June 17–20, 2012, Chapel Hill, North Carolina, USA.

saddle points. They correspond to singular points of the *gradient vector field* ∇h of h, of type sink, source or saddle, respectively. Regular integral curves of the gradient vector field ∇h are orthogonal trajectories of the regular level curves of h. We are interested in the configuration of integral curves of the gradient vector field. An *unstable (stable) separatrix* of a saddle point is the set of all regular points whose forward (backward) integral curve emerges from the saddle point. Appendix A contains a more precise definition. A non-degenerate saddle point has two stable and two unstable separatrices. A *Morse-Smale function* is a Morse function whose stable and unstable separatrices are disjoint. In particular, the unstable separatrices flow into a sink, and separate the unstable regions of two sources. Similarly, the stable separatrices emerge from a source, and separate the stable regions of two sinks. The corresponding gradient vector field is called a *Morse-Smale system* (MS-system). The *Morse-Smale complex* (MS-complex for short) is a complex consisting of all singularities, separatrices and the open regions forming their complement, of the *MS-system*. In other words, a cell of the MS-complex is a maximal connected subset of the domain of h consisting of points whose integral curves have the same origin and destination. See also [8, 9, 20] and Appendix A. The MS-complex describes the global structure of a Morse-Smale function.

Existing algorithms for MS-complexes [8, 9] compute the complex of a piecewise linear function on a piecewise linear manifold, or, in other words, of a discrete gradient-like vector field. When h is an analytic function, we cannot use these algorithms without first creating a piecewise linear approximation $\widetilde{h}$. However, the MS-complex of $\widetilde{h}$ is not guaranteed to be combinatorially equivalent to the MS-complex of the smooth vector field. The topological correctness depends on how close the approximation $\widetilde{h}$ is to h. Here "topological correctness" of the computed MS-complex $\widetilde{M}$ means that there is a homeomorphism f of the domain that induces a homeomorphism of each cell $\widetilde{c} \in \widetilde{M}$ to a cell $c \in M$ where M is the true MS-complex, and, moreover, this induced map $\widetilde{c} \mapsto c$ is an isomorphism of $\widetilde{M}$ and M. An isomorphism of two MS-complexes preserves the types of cells and their incidence relations. We can also require f to be an *ε-homeomorphism* for some specified $\varepsilon > 0$, i.e., the distance of a point and its f-image does not exceed ε. As far as we know this problem has never been rigorously studied. Therefore, the main problem of this paper is *to compute a piecewise-linear complex that is ε-homeomorphic to the MS-complex of a smooth Morse-Smale function h.* In short, we seek an exact computation in the sense of the Exact Geometric Computation (EGC) paradigm [15]. Note that it is unclear whether many fundamental problems from analysis are exactly computable in the EGC sense. In particular, the current state-of-the-art in EGC does not (yet) provide a good approach for coping with degenerate situations, and, in fact, this paradigm needs to be extended to incorporate degeneracies. Therefore, we have to *assume* that the gradients we start out with are Morse-Smale systems. However, *generic* gradients are Morse-Smale systems [20], so the presence of degenerate singularities and of saddle-saddle connections is exceptional. Note that in restricted contexts, like the class of polynomial functions, absence of degenerate critical points (the first, and local, Morse-Smale condition) can be detected. However, even (most) polynomial gradient systems cannot be integrated explicitly, so absence of saddle-saddle connections (the second, and global, Morse-Smale condition) cannot be detected with current approaches. Detecting such connections even in a restricted context remains a challenging open problem.

1.2 Overview

We present a solution for the main problem which produces output such as illustrated in Figures 1(a)-(b). In particular, our algorithm produces:

- (arbitrarily small) isolated *certified boxes* each containing a unique saddle, source or sink;
- *certified initial and terminal intervals* (on the boundary of saddleboxes), each of which is guaranteed to contain a unique point corresponding to a stable or unstable separatrix;
- disjoint *certified funnels (strips)* around each separatrix, each of which contains exactly one separatrix, and as close to the separatrix as desired.

The construction of the MS-complex of a gradient system ∇h consists of two main steps: constructing disjoint certified boxes for its singular points, and constructing disjoint certified strips (funnels) enclosing its separatrices. Singular points of the gradient system are computed by solving the system of equations $h_x(x,y) = 0, h_y(x,y) = 0$. This is a special instance of the more general problem of solving a *generic system* of two equations $f(x,y) = 0$, $g(x,y) = 0$. Here, generic means that the Jacobi matrix at any solution is non-singular (geometrically it means the two curves $f = 0$ and $g = 0$ intersect transversally). Section 3 presents a method to compute disjoint isolating boxes for all solutions of such generic systems of two equations in two unknowns. In Section 4 these boxes are refined further. Saddle-boxes are augmented with four disjoint intervals in their boundary, one for each intersection of the boundary with the stable and unstable separatrices of the enclosed saddle point. We also show that these intervals can be made arbitrarily small, which is crucial in the second stage of the algorithm. Sink- and source-boxes are refined by computing boxes—not necessarily axis-aligned—around the sink or source on the boundaries of which the gradient system is transversal (pointing into the sink-box and out of the source-box). This implies that all integral curves reaching (emerging from) such a refined sink-box (source-box) lie inside this box beyond (before) the points of intersection.

Having constructed isolating boxes for the critical points of the MS-function h, we isolate the two unstable (stable) separatrices of each saddle point by tracing them in forward (backward) direction to the sinks (sources) that are their destination (origin). Reliable techniques for this tracing step are not readily available. In the simplest case where h is a polynomial (with rational coefficients), isolating the critical points can be done using known algebraic algorithms. But even in this restricted context integral curves are non-algebraic, so computing isolating sets for separatrices is a problem. To do so, we turn the Euler method for tracing integral curves into a certified algorithm, applying quantitative error bounds on the Euler method from [14] to construct isolating strips for all separatrices. Section 5 describes the second stage of the algorithm, in which isolating strips (funnels) for the stable and unstable separatrices are constructed. The boundary curves of funnels enclosing an unstable separatrix are polylines with initial point on a saddle box and terminal point on a sink box. The gradient vector field is transversally pointing inward at each point of these polylines. The initial points of the polylines are connected by the unstable interval through which the separatrix leaves its saddle box, and, hence, enters the funnel. The terminal points of these polylines lie on the boundary of the same sink-box. See also Figure 6. Given this direction of the gradient system on the boundary of the funnel, the unstable separatrix enters the sink-box and tends to the enclosed sink, which is its destination. Although the width of the funnel may grow exponentially in the distance from the saddle-box, this growth is controlled in a known way. We exploit the computable (although very conservative) upper bound on this growth rate to obtain funnels that isolate separatrices from each other, and, hence, form a good approximation of the Morse-Smale complex together with the source- and sink-boxes. These upper bounds are also used to prove that the algorithm, which may need several subdivision steps, terminates. We have implemented this algorithm, using Interval Arithmetic. Section 7 presents sample output of our algorithm.

REMARK. In this paper, the domain $\mathbb{D}$ of h is a finite union of axis-aligned dyadic boxes. Furthermore, we (have to) assume that all stable and unstable separatrices of the saddle points are transversal to the boundary. Computationally this means that any sufficiently close approximation of these separatrices is transversal to the boundary as well.

1.3 Related Work

Milnor [18] provides a basic set-up for Morse theory. The survey paper [5], focusing on geometrical-topological properties of real functions, gives an excellent overview of recent works on MS-complexes. Originally, Morse theory was developed for smooth functions on smooth manifolds. Banchoff [4] introduced the equivalent definition of critical points on polyhedral surfaces. Many of the recent developments on MS-complexes are based on this definition. A completely different discrete version of Morse theory is provided by Forman [10].

Different methods for computation. In the literature there are two different method for computing the Morse-Smale complexes: (a) boundary based approaches and (b) region based approaches. Boundary based methods compute boundaries of the cells of the MS-complex, i.e., the integral curves connecting a saddle to a source, or a saddle to a sink [24, 3, 9]. On the other hand, watershed algorithms for image segmentation are considered as region based approaches [17]. Edelsbrunner et.al [9] computes the Morse-Smale complex of piecewise linear manifolds using a paradigm called *Simulation of Differentiability*. In higher dimensions they give an algorithm for computing Morse Smale complexes of piecewise linear 3-manifolds [8].

Morse-Smale complexes have also been applied in shape analysis and data simplification. Computing MS-complexes is strongly related to vector field visualization [12]. In a similar context, designing vector fields on surfaces has been studied for many graphics applications [26]. Cazals et.al. [6] applied discrete Morse theory to molecular shape analysis.

This paper contributes to the emerging area of Exact Numerical Algorithms for geometric problems [25]. Recent algorithms of this genre (e.g., [21, 16]) are numerical subdivision algorithms based on interval function evaluation and sign evaluation.

2. PRELIMINARIES

Our computational model[1] has two simple foundations: (1) BigFloat packages and (2) interval arithmetic (IA) [19]. Together, these are able to deliver efficient and guaranteed results in the implementation of our algorithms. A BigFloat package is a software implementation of exact ring $(+,-,\times)$ operations, division by 2, and exact comparisons, over the set $\mathbb{F} = \{m2^n : m,n \in \mathbb{Z}\}$ of *dyadic numbers*. In practice, we can use IEEE machine arithmetic as a filter for BigFloat computation to speed up many computation. *Range functions* form a basic tool of IA: given any function $F : \mathbb{R}^m \to \mathbb{R}^n$, a *range function* $\Box F$ for F computes for each m- dimensional interval I (i.e., an m-box) an n-dimensional interval $\Box F(I)$, such that $F(I) \subset \Box F(I)$. A range function is said to be *convergent* if the diameter of the output interval converges to 0 when the diameter of the input interval shrinks to 0. Convergent range functions exist for the basic operators and functions, so all range functions are assumed to be convergent. Moreover, we assume that the sign of functions can be evaluated exactly at dyadic numbers. All our boxes are *dyadic boxes*, meaning that their corners have dyadic coordinates.

If $I = I_x \times I_y$ is a 2D-interval (box) in $\mathbb{R}^2$, the *interval Jacobian determinant* $\Box\frac{\partial(f,g)}{\partial(x,y)}$ is the 2×2 interval determinant given by

$$\Box\frac{\partial(f,g)}{\partial(x,y)}(I) = \begin{vmatrix} \Box\dfrac{\partial f}{\partial x}(I) & \Box\dfrac{\partial f}{\partial y}(I) \\ \Box\dfrac{\partial g}{\partial x}(I) & \Box\dfrac{\partial g}{\partial y}(I) \end{vmatrix}$$

PROPOSITION 1. ***(Interval Implicit Function Theorem, Snyder [22, 23])*** *Let $F : \mathbb{R}\times\mathbb{R} \to \mathbb{R}$ be a C^1-map with components f and g. If $I \subset \mathbb{R}\times\mathbb{R}$ is a box for which $0 \notin \Box\frac{\partial(f,g)}{\partial(x,y)}(I)$, then the system $f(x,y) = 0$, $g(x,y) = 0$ has at most one solution in I.*

3. ISOLATING BOXES FOR SINGULARITIES OF GRADIENT FIELDS

As a first step towards the construction of the Morse-Smale complex of h we construct disjoint isolating boxes for the singular points of ∇h. To this end, we first show how to compute isolating boxes for the solutions of a generic system of two equations in two unknowns, which are confined to a bounded domain in the plane, which is a finite union of dyadic boxes. Applying this general method to the case in which the two equations are defined by the components of the gradient vector field ∇h we obtain isolating boxes for the singularities of this gradient field.

3.1 Certified solutions of systems of equations

We consider a system of equations

$$f(x,y) = 0, \quad g(x,y) = 0, \tag{1}$$

where f and g are C^1-functions defined on a bounded axis-parallel box $\mathbb{D} \subset \mathbb{R}^2$ with *dyadic vertices*. Furthermore, we

[1]See Appendix A for additional details.

assume that the system has only non-degenerate solutions, i.e., the Jacobian determinant

$$\frac{\partial(f,g)}{\partial(x,y)} := \frac{\partial f}{\partial x}\frac{\partial g}{\partial y} - \frac{\partial f}{\partial y}\frac{\partial g}{\partial x}$$

is non-zero at a solution (x_0, y_0). Geometrically this means that the curves Z_f and Z_g, given by $f(x,y) = 0$ and $g(x,y) = 0$, respectively, are regular near a point of intersection (x_0, y_0) and the intersection is transversal. Since $\mathbb{D}$ is compact, the system has finitely many solutions in $\mathbb{D}$.

Our goal is to construct a collection of axis-aligned boxes $\mathbb{B}_1, \ldots, \mathbb{B}_m$ and $\mathbb{B}'_1, \ldots, \mathbb{B}'_m$ such that (i) box $\mathbb{B}_i$ is concentric with and strictly contained in $\mathbb{B}'_i$, (ii) the boxes $\mathbb{B}'_i$ are disjoint, (iii) each solution of (1) is contained in one of the boxes $\mathbb{B}_i$, and (iv) each box $\mathbb{B}'_i$ contains exactly one solution (contained inside the enclosed box $\mathbb{B}_i$). The box pair $(\mathbb{B}_i, \mathbb{B}'_i)$ is *certified*: $\mathbb{B}_i$ contains a solution, and $\mathbb{B}'_i$ provides positive clearance to other solutions.

In fact, the sequence of boxes will satisfy the following stronger conditions:

1. The curves Z_f and Z_g each intersect the boundary of $\mathbb{B}'_i$ transversally in two points.
2. There are disjoint intervals $\mathbb{I}_i^0(f)$, $\mathbb{I}_i^0(g)$, $\mathbb{I}_i^1(f)$ and $\mathbb{I}_i^1(g)$ in the boundary of $\mathbb{B}'_i$ (in this order), where the first and the third interval each contain one point of intersection of Z_f and $\partial\mathbb{B}'_i$, and the second and fourth interval each contain one point of intersection of Z_g and $\partial\mathbb{B}'_i$.
3. The (interval) Jacobian determinant of f and g does not vanish on $\mathbb{B}'_i$, i.e., $0 \notin \square\frac{\partial(f,g)}{\partial(x,y)}(\mathbb{B}'_i)$.

3.2 Construction of certified box pairs

We first subdivide the domain $\mathbb{D}$ into equal-sized boxes $\mathbb{I}$ (called grid-boxes), until all boxes satisfy certain conditions to be introduced now. For a (square, axis-aligned) box $\mathbb{I}$ let $N_\varrho(\mathbb{I})$ be the box obtained by multiplying box $\mathbb{I}$ from its center by a factor of $1+\varrho$, where $\frac{1}{2} \le \varrho \le 1$. We also denote $N_1(\mathbb{I})$ by $N(\mathbb{I})$; this is the box formed by the union of $\mathbb{I}$ and its eight neighbor grid-boxes. The algorithm subdivides $\mathbb{D}$ until all grid-boxes $\mathbb{I}$ satisfy $C_0(\mathbb{I}) \vee \big(C_1(\mathbb{I}) \wedge C_2(\mathbb{I})\big)$, where the clauses $C_i(\mathbb{I})$, $i = 0, 1, 2$, are the following predicates:

$$C_0(\mathbb{I}): \quad 0 \notin \square f(\mathbb{I}) \vee 0 \notin \square g(\mathbb{I})$$

$$C_1(\mathbb{I}): \quad 0 \notin \square\frac{\partial(f,g)}{\partial(x,y)}(N(\mathbb{I}))$$

$$C_2(\mathbb{I}): \quad C_2(\mathbb{I}, f) \wedge C_2(\mathbb{I}, g),$$

where

$$C_2(\mathbb{I}, f) = \langle\square\frac{\nabla f}{||\nabla f||}(N(\mathbb{I})), \square\frac{\nabla f}{||\nabla f||}(N(\mathbb{I}))\rangle \ge \cos\frac{\pi}{30}.$$

If $C_0(\mathbb{I})$ holds, box $\mathbb{I}$ does not contain a solution, so it is discarded. The second predicate guarantees that $N(\mathbb{I})$ contains at most one solution. This is a consequence of Theorem 1. Condition $C_2(\mathbb{I})$ is a *small angle variation condition*, guaranteeing that the variation of the unit normals of the curves $Z_f \cap N(\mathbb{I})$ and $Z_g \cap N(\mathbb{I})$ do not vary too much, so these curves are regular, and even 'nearly linear' (the unit normal of Z_f is the normalized gradient of f).

REMARK 2. Condition $C_1(\mathbb{I})$ implies that there is a computable positive lower bound $\alpha(\mathbb{I})$ on the angle between $\nabla f(p)$ and $\nabla g(q)$ where p, q range over the box $\mathbb{I}$. More precisely, to compute $\alpha(\mathbb{I})$, we first compute a lower bound L on the quantity $\frac{\partial(f,g)}{\partial(x,y)}\frac{1}{\|\nabla f\|\cdot\|\nabla g\|}$. This L may be obtained by an interval evaluation of this quantity at $N(\mathbb{I})$; note that $L > 0$ iff condition $C_1(\mathbb{I})$ holds. We define $\alpha(\mathbb{I})$ as $\arcsin(L)$. □

Our algorithm will construct disjoint certified boxes surrounding a box $\mathbb{I}$. Since the surrounding boxes $N(\mathbb{I})$ and $N(\mathbb{J})$ of disjoint boxes $\mathbb{I}$ and $\mathbb{J}$ may intersect, we consider smaller surrounding boxes for a grid box $\mathbb{I}$. If $N(\mathbb{J}) \cap \mathbb{I} = \emptyset$, then $N_{1/2}(\mathbb{I})$ and $N_{1/2}(\mathbb{J})$ have disjoint interiors. This is a key observation with regard to the correctness of our algorithm.

LEMMA 3. *Let $\mathbb{I}$ be a box such that conditions $\neg C_0(\mathbb{I})$, $C_1(\mathbb{I})$ and $C_2(\mathbb{I})$ hold. Let d be the length of its edges, and let $\frac{1}{2} \le \varrho \le 1$.*
1. If Z_f intersects $\mathbb{I}$, it intersects the boundary of $N_\varrho(\mathbb{I})$ transversally at exactly two points. At a point of intersection of Z_f and an edge e of $\partial N_\varrho(\mathbb{I})$ the angle between Z_f and e is at least $\frac{1}{15}\pi$.
2. If $\mathbb{I}$ contains a point of intersection of Z_f and Z_g, then the points of intersection of Z_f and $\partial N_\varrho(\mathbb{I})$ are at distance at least $2\varrho d \tan\frac{1}{2}\alpha(\mathbb{I})$ from the points of intersection of Z_g and $\partial N_\varrho(\mathbb{I})$. On $\partial N_\varrho(\mathbb{I})$, the points of intersection with Z_f and with Z_g are alternating.

If f and g are linear functions, the previous result is easy to check. The proof of the general case uses the small angle variation of ∇f and ∇g. Note that this Lemma already establishes the existence and isolation of a solution of (1) in case we can certify that Z_f and Z_g intersect $\mathbb{I}$.

3.3 Towards an algorithm

For each grid-box $\mathbb{I}$, the algorithm calls one of the following:

- DISCARD($\mathbb{I}$), if it decides that $\mathbb{I}$ does not contain a solution. It marks box $\mathbb{I}$ as processed.
- REPORTSOLUTION($\mathbb{I}$). It returns the certified pair of boxes $(N_{1/2}(\mathbb{I}), N(\mathbb{I}))$, and marks all boxes contained in $N(\mathbb{I})$ as processed.

In the latter case a solution is found inside $N_{1/2}(\mathbb{I})$, but, as will become clear later, it may not be contained in the smaller box $\mathbb{I}$. In view of $C_1(\mathbb{I})$ none of the grid-boxes in $N(\mathbb{I})$ contain a solution different from the one reported, so they are marked as being processed.

Decisions are based on evaluation of the signs of f and g at the vertices of the grid-boxes (or at certain dyadic points on edges of grid-boxes). An edge of a box is called *bichromatic* (*monochromatic*) for f if the signs of the value of f at its vertices are opposite (equal).

Algorithm, case 1: $\mathbb{I}$ has a bichromatic edge for f and a bichromatic edge for g. Then Z_f and Z_g intersect $\mathbb{I}$, and, according to Lemma 3, part 1, both curves intersect the boundary of $N_{1/2}(\mathbb{I})$ transversally in exactly two points. For each of the two points in $\partial N_{1/2}(\mathbb{I})$ the algorithm computes an isolating interval—called an f-interval—on $\partial N_{1/2}(\mathbb{I})$ of length $\frac{1}{2}d\tan\frac{1}{2}\alpha(\mathbb{I})$. The two g-intervals are computed similarly. If the f- and g-intervals are not interleaving, there is no solution of (1) in box $\mathbb{I}$—even though there may be a solution in $N_{1/2}(\mathbb{I})$—and DISCARD($\mathbb{I}$) is called.

This follows from Lemma 3, part 2. If the intervals are interleaving, then there is a point of intersection inside $N_{1/2}(\mathbb{I})$, so the algorithm calls REPORTSOLUTION($\mathbb{I}$).

Algorithm, case 2: $\mathbb{I}$ contains no bichromatic edge for f (g), and at least one bichromatic edge for g (f, respectively). We only consider the case in which all edges of $\mathbb{I}$ are monochromatic for f. Then the algorithm also evaluates the sign of f at the vertices of the box $N_{1/2}(\mathbb{I})$. If $N_{1/2}(\mathbb{I})$ has no *disjoint* bichromatic edges (as in the fourth and fifth configuration of Figure 2), the isocurve Z_f does not intersect $\mathbb{I}$, so the algorithm calls DISCARD($\mathbb{I}$). To deal

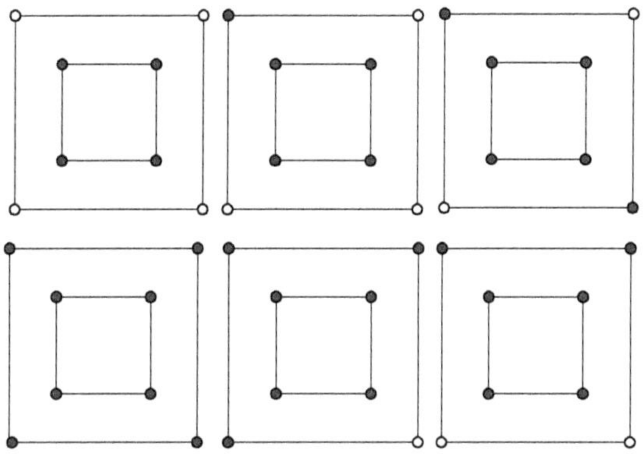

Figure 2: Sign patterns of the box $N_{1/2}(\mathbb{I})$ enclosing the grid-box $\mathbb{I}$ with monochromatic edges for f. The three top configurations are ruled out by the small normal variation condition $C_2(\mathbb{I})$. The fourth, fifth and sixth configuration are all possible, but only in the sixth situation Z_f may intersect the inner box.

with the remaining case, in which $N_{1/2}(\mathbb{I})$ has two *disjoint* bichromatic edges (as in the sixth configuration in Figure 2) we need to evaluate the sign of f at certain dyadic points of these bichromatic edges.

To make this more precise, let e be the edge of $\mathbb{I}$ closest to the (monochromatic) edge of $N_{1/2}(\mathbb{I})$ at whose vertices the sign of f is the opposite of the sign of f at the vertices of $\mathbb{I}$ (this is the bottom edge of the outer box in the sixth configuration of Figure 2). Let l and r be the points of intersection of the line through e and the edges of the surrounding box $N_\varrho(\mathbb{I})$, perpendicular to e. See also Figure 3. The intervals on the boundary of this surrounding box with length $\frac{2}{\sqrt{3}}(1+\varrho)d$, centered at l and r, respectively, are denoted by $L_\varrho(e)$ and $R_\varrho(e)$, where d is the length of the edges of $\mathbb{I}$, and $\frac{1}{2} \leq \varrho \leq 1$.

LEMMA 4. *Let $\mathbb{I}$ be a box such that $\neg C_0(\mathbb{I})$, $C_1(\mathbb{I})$ and $C_2(\mathbb{I})$ hold, and let e be one of its edges. Let $\frac{1}{2} \leq \varrho \leq 1$.*
1. If Z_f intersects an edge e of the boundary of $\mathbb{I}$ in at least two points, then it transversally intersects $\partial N_\varrho(\mathbb{I})$ in exactly two points, one in each of the intervals $L_\varrho(e)$ and $R_\varrho(e)$.
2. If Z_f intersects $\partial N_\varrho(\mathbb{I})$ in the intervals $L_\varrho(e)$ and $R_\varrho(e)$, then these intersections are transversal, and Z_f intersects $\partial N_\varrho(\mathbb{I})$ at exactly two points, one in each of these intervals.

Algorithm, case 2 (cont'd). By evaluating the signs of f at the (dyadic) endpoints of the interval $L_\varrho(e)$ and $R_\varrho(e)$ the algorithm decides whether they contain a point of intersection with Z_f. If at least one of these intervals

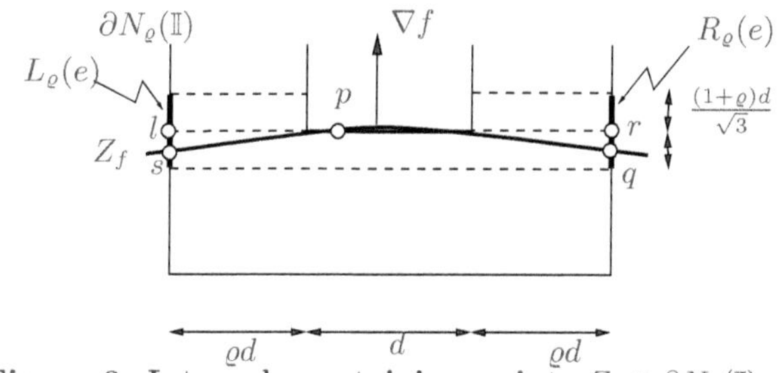

Figure 3: Intervals containing points $Z_f \cap \partial N_\varrho(\mathbb{I})$.

is disjoint from Z_f, then DISCARD($\mathbb{I}$) is called. Otherwise, the algorithm computes isolating f- and g-intervals of length $\frac{1}{2}d\tan\frac{1}{2}\alpha(I)$. As in case 1, the algorithm calls REPORTSOLUTION ($\mathbb{I}$) if these intervals are interleaving, and DISCARD($\mathbb{I}$) otherwise.

Algorithm, case 3: all edges of $\mathbb{I}$ are monochromatic for both f and g. Again, let e be the (unique) edge of $\mathbb{I}$ closest to the edge of $N_{1/2}(\mathbb{I})$ which is monochromatic for f, at whose vertices the sign of f is the opposite of the sign of f at the vertices of $\mathbb{I}$. Edge e' of $\mathbb{I}$ is defined similarly for g.
Case 3.1: $e = e'$. In this case Z_f or Z_g does not intersect $\mathbb{I}$. Indeed, if Z_f intersects $\mathbb{I}$, it intersects e in at least two points, so there is a point $p \in Z_f$ at which $\nabla f(p)$ is perpendicular to e. Condition $C_1(\mathbb{I})$ guarantees that ∇g is nowhere parallel to $\nabla f(p)$, so Z_g does not intersect e, and, hence, does not intersect $\mathbb{I}$. Therefore, DISCARD($\mathbb{I}$) is called.
Case 3.2: $e \neq e'$. If Z_f does not intersect $L_\varrho(e)$ or $R_\varrho(e)$, or if Z_g does not intersect $L_\varrho(e')$ or $R_\varrho(e')$, then, as in case 2, the algorithm calls DISCARD($\mathbb{I}$). Otherwise, $L_\varrho(e)$ or $R_\varrho(e)$ are isolating f-intervals which are disjoint from the isolating g-intervals $L_\varrho(e')$ or $R_\varrho(e')$. If e and e' are perpendicular, then these f- and g-intervals are interleaving, and, hence, REPORTSOLUTION($\mathbb{I}$) is called. Otherwise, there is no solution in $\mathbb{I}$, so DISCARD($\mathbb{I}$) is called.

3.4 Refinement: disjoint surrounding boxes

We would like distinct isolating boxes $\mathbb{I}, \mathbb{J}$ to have disjoint surrounding boxes $N(\mathbb{I}), N(\mathbb{J})$. There is a simple way to ensure this: we just use the predicate $C_1(N(\mathbb{I}))$ instead of $C_1(\mathbb{I})$ in the above subdivision process. Then, if the interior of $N(\mathbb{I}) \cap N(\mathbb{J})$ is non-empty, we can discard any one of $\mathbb{I}$ or $\mathbb{J}$.

4. COMPUTING SOURCE-, SADDLE- AND SINK-BOXES

Using the previous method we construct disjoint isolating boxes $\mathbb{B}'_1, \ldots, \mathbb{B}'_m$ for the singularities of ∇h, which are refined further to facilitate the construction of the MS-complex.

Each saddlebox is augmented by computing four disjoint intervals in its boundary, one for each intersection of a stable or unstable separatrix with the box boundary. See the leftmost picture in Figure 4. These *isolating separatrix intervals* can be made *arbitrarily small*. These intervals are constructed as the intersection of disjoint stable and unstable wedges along the boundary of which the gradient vector field is pointing outward and inward, respectively. Therefore, these wedges enclose the stable and unstable separatrices, respectively. In the full paper we show how the width of these wedges can be made arbitrarily small, provided the saddlebox satisfies some additional (computable) criteria.

Inside each source- and sink-box we construct a box on

the boundary of which the ∇h is pointing outward or inward, respectively. These boxes are contained in the source- and sinkboxes constructed in the previous section, but are not necessarily axes-aligned. They are aligned with approximations of the (orthogonal) eigenvectors of the Hessian of h at the center of the source- or sinkbox, however. See the rightmost picture in Figure 4.

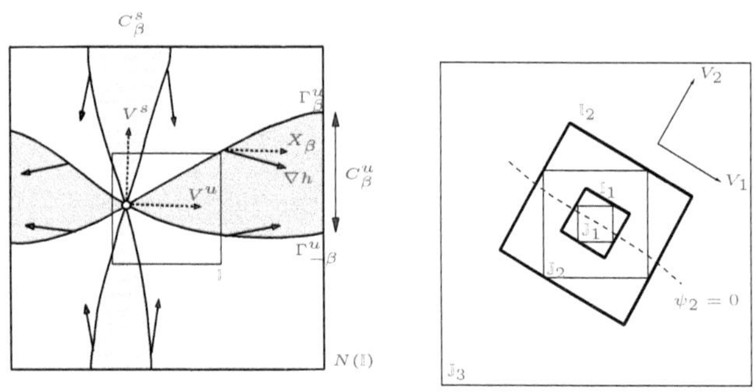

Figure 4: Left: The unstable wedge C^u_β enclosing the unstable separatrix, and the stable wedge C^s_β enclosing the stable separatrix. The gradient vector field ∇h, represented by solid arrows, is transversally pointing inward along the boundary of the unstable wedge, and outward along the boundary of the stable wedge. At points of the unstable wedge boundary the vector field X_β is (anti)parallel to the unstable eigenvector V^u of the Hessian of h at the center of the box, so ∇h makes an angle $-\beta$ with V^u or $-V^u$ there. Right: Sink-/source-boxes, nearly aligned with the eigenvectors of the Hessian matrix of h at the center of the box, at the boundaries of which ∇h is transversal (pointing inward/outward).

5. ISOLATING FUNNELS AROUND SEPARATRICES

We continue by constructing certified *funnels* around separatrices inside the region $\mathbb{D}^* = \mathbb{D} \backslash \cup_{i=1}^m \mathbb{B}'_i$, where $\mathbb{B}'_1, \ldots, \mathbb{B}'_m$ are the boxes isolating the singular points of ∇h. If the forward orbits of the endpoints of an unstable separatrix interval have the same sink of ∇h as destination, these forward orbits bound a region around the unstable separatrix leaving the saddle box via this unstable segment. This region is called a *funnel* for the separatrix (this terminology is borrowed from [14]). We now describe how to construct a certified funnel for an unstable separatrix. For each isolating unstable separatrix interval $\mathbb{J}$ on the boundary of a saddle box we construct two polylines $L_{-\vartheta}(\mathbb{J})$ and $L_\vartheta(\mathbb{J})$ as follows. The initial points of these polylines are the endpoints of $\mathbb{J}$, q_- and q_+, where q_- comes before q_+ in the counterclockwise orientation of the boundary of the saddle box. The polyline $L_\vartheta(\mathbb{J})$ is uniquely defined by requiring that its vertices $q_+ = p_0, p_1, \ldots, p_n$ lie on grid edges, with the property that

(1) The line segment $p_{i-1}p_i$, $0 < i \leq n$, lies in a (closed) grid box of $\mathbb{D}$, and has direction $X_\vartheta(p_{i-1})$, and

(2) p_n, the last vertex, lies on the boundary of $\mathbb{D}$.

Here X_ϑ is the vector field obtained by rotating the vector field $X = \nabla h$ over an angle ϑ, i.e.,

$$X_\vartheta = \begin{pmatrix} \cos\vartheta & -\sin\vartheta \\ \sin\vartheta & \cos\vartheta \end{pmatrix} \begin{pmatrix} h_x \\ h_y \end{pmatrix}.$$

See Figure 5 (Right). The polyline $L_{-\vartheta}(\mathbb{J})$ is defined similarly. The polylines $L_{\pm\vartheta}(\mathbb{J})$ are called *fences* of the (unique) unstable separatrix of ∇h intersecting $\mathbb{J}$.

It is not hard to see that that a grid box contains at most two *consecutive* edges of each of these polylines, but it is not obvious a priori that each box cannot contain more than two edges of each polyline in total. It follows from the next result that the intersection of a grid box with any of these polylines is connected, and, hence, that these polylines are finite.

5.1 Small angle variation

We subdivide the region $\mathbb{D}^*$ into square boxes over which the angle variation of ∇h is at most ϑ, where ϑ is to be determined later. If $X = (f, g)$ is a vector field on $\mathbb{R}^2$, then the angle variation over a regular curve Γ is given by [2, Section 36.7]: $\int_\Gamma \frac{g\,df - f\,dg}{f^2+g^2}$. If $X = \nabla h$, this angle variation is equal to

$$\int_\Gamma \frac{(h_x h_{xy} - h_y h_{xx})\,dx + (h_x h_{yy} - h_y h_{xy})\,dy}{h_x^2 + h_y^2}.$$

The next result gives an upper bound for the size of grid-boxes such that the angle variation of ∇h over any box in $\mathbb{D}^*$ is less than ϑ. It also provides a construction of a 'box-wise' funnel around integral curves of ∇h.

LEMMA 5. *Let C_0 be a constant such that*

$$\max_{\mathbb{D}^*}\left(\left| \frac{h_x h_{xy} - h_y h_{xx}}{h_x^2 + h_y^2} \right|, \left| \frac{h_x h_{yy} - h_y h_{xy}}{h_x^2 + h_y^2} \right| \right) \leq C_0,$$

and let $0 < \vartheta < \frac{\pi}{2}$, and let the width w of grid-boxes in $\mathbb{D}^*$ satisfy $w \leq \dfrac{\vartheta}{2C_0\sqrt{2}}$. Then

1. The angle variation of ∇h on any gridbox is less than ϑ.
2. Let p be a point on an edge of a gridbox, and let q_ϑ ($q_{-\vartheta}$) be the point on the boundary of the gridbox into which $X_\vartheta(p)$ ($X_{-\vartheta}(p)$) is pointing, such that the line segment pq_ϑ ($pq_{-\vartheta}$) has direction $X_\vartheta(p)$ ($X_{-\vartheta}(p)$). See Figure 5. Then ∇h is pointing rightward along the (directed) line segment pq_ϑ and leftward along $pq_{-\vartheta}$.
3. The function h is strictly increasing on the line segments pq_ϑ and $pq_{-\vartheta}$.

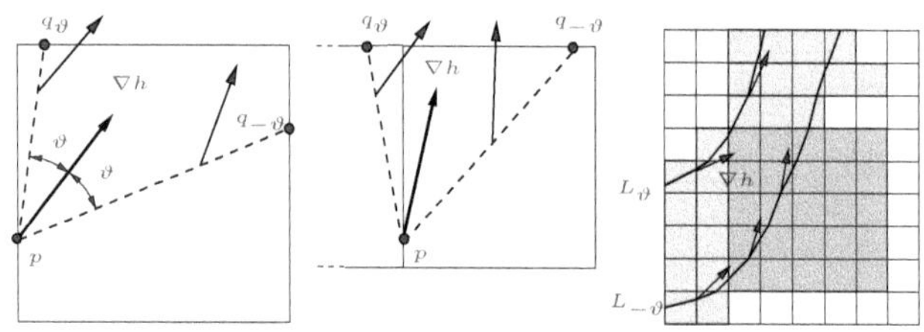

Figure 5: Left: The orientation of ∇h with respect to $X_\vartheta(p)$ does not change over a grid box. Right: A funnel formed by two polylines covered by two vertical ε-strips, one ε-box and three horizontal ε-strips. Here ε is six times the width of a grid box. $L_{+\vartheta}$ intersects the ε-box in a single edge, which is quasivertical but not quasihorizontal. All edges of $L_{-\vartheta}$ inside the ε-box are quasivertical as well (and some of them are also quasihorizontal). ∇h points inward, and the enclosed separatrix (not shown) runs from left to right.

The third property will be crucial for constructing fences around separatrices that cross each grid-box at most once.

5.2 Fencing in the separatrices

For each isolating unstable separatrix interval $\mathbb{J}$ on the boundary of a saddle box we construct two polylines (called *fences*) $L_{-\vartheta}(\mathbb{J})$ and $L_{\vartheta}(\mathbb{J})$ as follows. The initial points of these polylines are the endpoints q_- and q_+ of $\mathbb{J}$, where q_- comes before q_+ in the counterclockwise orientation of the boundary of the saddle box. The polyline $L_{\vartheta}(\mathbb{J})$ is uniquely defined by requiring that its vertices $q_+ = p_0, p_1, \ldots, p_n$ lie on grid edges, with the property that (i) the line segment $p_{i-1}p_i$, $0 < i \leq n$, lies in a (closed) grid box of $\mathbb{D}^*$, and has direction $X_{\vartheta}(p_{i-1})$, and (ii) p_n, the last vertex, lies on the boundary of $\mathbb{D}^*$. The polyline $L_{-\vartheta}(\mathbb{J})$ is defined similarly with respect to q_- and $X_{-\vartheta}$. If the endpoints of the fences $L_{\vartheta}(\mathbb{J})$ and $L_{-\vartheta}(\mathbb{J})$ lie on the same connected component of the boundary of $\mathbb{D}^*$, then these fences split $\mathbb{D}^*$ into two connected regions. See Figure 6. In this case, the region

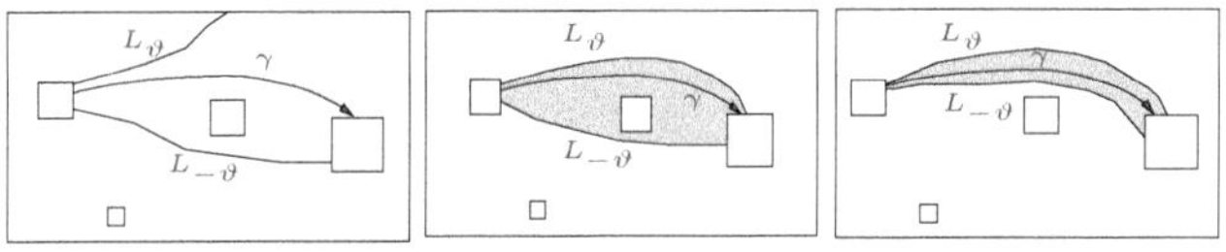

Figure 6: Fences around a separatrix γ. If the fences end in the same connected component of the boundary of $\mathbb{D}^*$, *and* enclose a simply connected region in $\mathbb{D}^*$, then this region isolates the separatrix from the source-, sink- and saddle-boxes (rightmost picture).

containing the separatrix interval $\mathbb{J}$ in its boundary is called the *funnel* of $\mathbb{J}$ (with angle ϑ) denoted by $F_{\vartheta}(\mathbb{J})$. Its boundary consists of $\mathbb{J}$, the two fences $L_{\vartheta}(\mathbb{J})$ and $L_{-\vartheta}(\mathbb{J})$, and a curve $\mathbb{J}^*$ on the boundary of $\partial\mathbb{D}$ connecting the endpoints of these fences. The fences $L_{\pm\vartheta}(\mathbb{J})$ start from the endpoints of $\mathbb{J}$. New line segments are added until a fence hits the boundary of $\mathbb{D}^*$, or until the width of the enclosed funnel becomes larger than a positive number ε (to be specified later). To define the width of a funnel more precisely, we distinguish quasihorizontal and quasivertical parts of a funnel, and show that the width of a funnel does not increase substantially at transitions between these quasihorizontal and quasivertical parts. A nonzero vector $v = (v_1, v_2)$ is called *quasihorizontal* if $|v_2| \leq 2|v_1|$, and *quasivertical* if $|v_1| \leq 2|v_2|$. Note that each nonzero vector is quasihorizontal, quasivertical, or both. A *horizontal ε-strip* is the union of a sequence of boxes where successive boxes share a vertical edge, such that the horizontal edge of the rectangle thus obtained has length at most ε. A *vertical ε-strip* is defined similarly. An *ε-box* is a square box with edge length at most ε which is the union of a number of boxes.

The construction of a funnel is illustrated in Figure 5 (Right). If both initial segments of $L_{\pm\vartheta}$ are quasihorizontal (quasivertical), the width of the funnel (under construction) is measured in the vertical (horizontal) direction. As long as the width is less than ε, the funnel crosses the next vertical (horizontal) ε-strip by extending $L_{\pm\vartheta}$ with one or two line segments. As soon as one of the new segments of L_{ϑ} or $L_{-\vartheta}$ ceases to be quasihorizontal (quasivertical), the funnel is extended through a square box of minimal width. Upon exit of this square box the edges of L_{ϑ} are both quasivertical (quasihorizontal). See Figure 5 (Right). This observation, together with a standard result from Dynamical Systems Theory [14, Theorem 4.4.1] on the growth of the deviation of the Euler approximation from the integral curve it approximates, yields the following (rather conservative) upper bound on the *growth of the funnel width* of a separatrix having M transitions between quasihorizontal and quasivertical parts (which are bounded by singularity-boxes or ε-boxes).

LEMMA 6. *Let T be the (computable) edge length of a bounding square of the domain $\mathbb{D}$ of h, and let M be the total number of quasihorizontal and quasivertical parts of the polylines bounding a separatrix funnel. There are computable constants[2] ϑ_0, c_1, c_2 and C, which **only depend on** $\mathbb{D}^*$ **and** h, such that the width of the funnel does not exceed $(c_1\vartheta + c_2 w)\, e^{CMT}$, provided $\vartheta \leq \vartheta_0$. In particular, this width is at most ε if $c_1\vartheta + c_2 w \leq \varepsilon\, e^{-CMT}$.*

6. CONSTRUCTING THE MS-COMPLEX

6.1 The algorithm

The construction of the MS-complex of h is a rather straightforward application of the preceding results. It uses a parameter M, the (a priori unknown) number of transitions (at ε-boxes) between quasihorizontal and quasivertical parts of a funnel. Let T be the edge length of a bounding square of the domain $\mathbb{D}$ of h. Then the algorithm performs the following steps.

Step 1. Construct certified isolating boxes $\mathbb{B}'_1, \ldots, \mathbb{B}'_m$ for the singularities of ∇h (cf Section 3).

Step 2. Let $\mathbb{D}^*$ be the closure of $\mathbb{D} \setminus (B'_1 \cup \cdots \cup \mathbb{B}'_m)$. Compute the constants ϑ_0, c_1, c_2 and C, which depend only on h and $\mathbb{D}^*$. Set ε to the minimum of the width of the source-, sink- and saddle-boxes.

Step 3. Let ϑ and w be such that $w \leq \frac{\vartheta}{2C_0\sqrt{2}}$, $\vartheta \leq \min(\frac{\pi}{40}, \vartheta_0)$, and $c_1\vartheta + c_2 w \leq \varepsilon\, e^{-CMT}$ (cf Lemma 6). Subdivide $\mathbb{D}^*$ until all gridboxes have maximum width w. For each saddle box, compute four separatrix intervals on its boundary, of width at most w.

Step 4. For each stable and unstable separatrix interval do the following. Start the computation of a funnel for a separatrix by setting M to a small number M_0 (say 4). Compute the fences $L_{-\vartheta}$ and L_{ϑ}, keeping track of the width of the enclosed funnel under construction and of the number m of transitions between quasihorizontal and quasivertical parts of this funnel.

If *the width of the funnel exceeds ε or the number of transitions m exceeds M*, then abort the computation of the current funnel, discard all funnels constructed so far, set M to twice its current value and goto Step 3.

If *the funnel intersects an already constructed funnel, or a source- or sinkbox on which it does not terminate* (i.e., if only one of its fences intersects this box, as in the leftmost part of Figure 6, or it encloses such a box, as in the middle part), then set ε to half its current value, discard all funnels constructed so far, and goto Step 3.

If *the funnel intersects a saddlebox* $\mathbb{B}'_i$, then decrease the size of $\mathbb{B}'_i$ by a factor of two via *subdivision*, discard all funnels constructed so far, set ε to half its current value, and goto Step 2. (Note that $\mathbb{D}^*$ gets larger, so the constants in Step 2 have to be recomputed.)

Otherwise, the fences end on the same component of the boundary of $\partial\mathbb{D}^*$, and the enclosed funnel is simply connected, and does not intersect any of the funnels constructed so far. See the rightmost part of Figure 6. Add the funnel

[2] See Appendix B for a precise definition of these constants, and for a sketch of the proof of Lemma 6.

to the output, and reset M to M_0 (and repeat until all separatrices have been processed).

6.2 Termination

Since the gradient field ∇h is a $2D$ Morse-Smale system, its separatrices are disjoint. Their intersections with $\mathbb{D}^*$ are compact, and have positive distance (although this distance is not known a priori). In the main loop of the algorithm, the maximal funnel width ε is bisected if funnels intersect, and saddleboxes intersected by the funnel are subdivided, so after a finite number of iterations of the main loop its value is less than half the minimum distance between any pair of distinct separatrices, and funnels stay clear from saddleboxes (apart from the one containing the origin or destination of the enclosed separatrix).

Separatrices that intersect $\partial\mathbb{D}$ do so transversally, cf. the remark at the end of Section 1.2. Therefore, after a finite number of subdivision steps, both fences around such separatrices will intersect $\partial\mathbb{D}$ transversally. Hence, eventually all funnels become disjoint, at which point the algorithm terminates after returning a topologically correct MS-complex for ∇h.

7. IMPLEMENTATION AND EXPERIMENTAL RESULTS

The algorithm has been implemented using the `Boost` library [1] for IA. All experiments have been performed on a 3GHz Intel Pentium 4 machine under Linux with 1 GB RAM using the `g++` compiler, version 3.3.5. Figures 1(a)-(b) and 7(a)-(b) depict the output of our algorithm, for several Morse-Smale functions. In our implementation the parameter ϑ, used in the construction of separatrix-funnels, is $\frac{\pi}{30}$, which is larger than the theoretical bound given by Lemma 6. The algorithm halves this angle several times, depending on the input, until the funnels are simply connected, mutually disjoint, and connect a saddle-box to a source-box (for stable separatrices) or sink-box (for unstable separatrices), in which case a topologically correct MS-complex has been computed. Each of the funnels with deep black boundaries contains an unstable separatrix, whereas a funnel with light black boundaries contains a stable separatrix. The CPU-time for computing a MS-system increases with the number of critical points and the complexity of the vector field, as indicated in the captions of the figures.

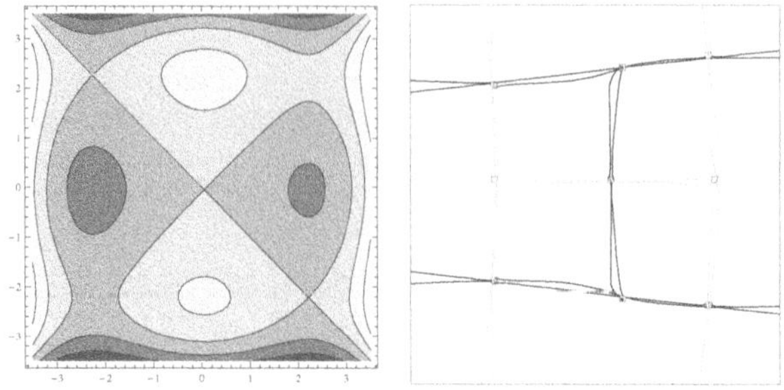

(a) Contour plot (left) and MS-Complex (right) of $h(x,y) = -10\,x^2 + x^4 + 10\,y^2 - y^4 + x + xy^2$, on the box $[-4, 3.5] \times [-4, 3.5]$. CPU-time: 11 seconds.

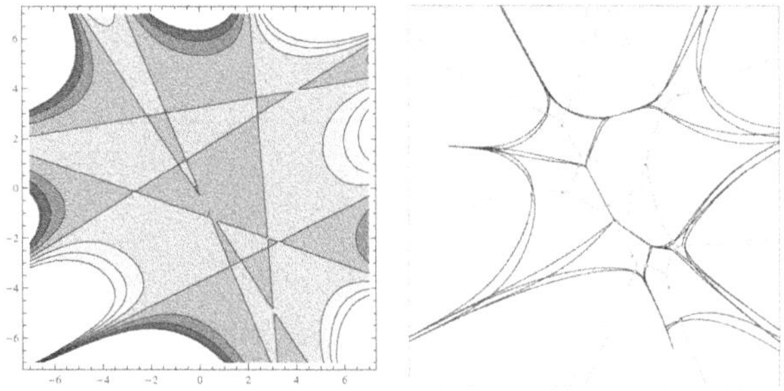

(b) Contour plot (left) and MS-complex (right) of a product of seven linear functions, on the box $[-7, 7] \times [-7, 7]$. CPU-time: 11.5 minutes.

Figure 7: Contour plots of MS-functions and their Morse-Smale complexes.

8. REFERENCES

[1] `Boost` interval arithmetic library. `http://www.boost.org`.

[2] V.I. Arnol'd. *Ordinary Differential Equations.* Universitext. Springer-Verlag, New York, Heidelberg, Berlin, 2006.

[3] C.L. Bajaj and D.R. Schikore. Topology preserving data simplification with error bounds. *Comput. Graph.*, 22(1):3–12, 1998.

[4] T. F. Banchoff. Critical points and curvature for embedded polyhedral surfaces. *Amer. Math. Month.*, 77:475–485, 1970.

[5] S. Biasotti, L. De Floriani, B. Falcidieno, P. Frosini, D. Giorgi, C. Landi, L. Papaleo, and M. Spagnuolo. Describing shapes by geometrical-topological properties of real functions. *ACM Computing Surveys*, 40(4):12.1–12:87, 2008.

[6] F. Cazals, F. Chazal, and T. Lewiner. Molecular shape analysis based upon the Morse-Smale complex and the Connolly function. In *In SCG 2003: Proceedings of the 19th Annual Symposium on Computational Geometry*, pages 351–360, ACM Press, New York, NY, 351-360, 2003.

[7] A. Chattopadhyay. *Certified Geometric Computation. Radial Basis Function based Isosurfaces and Morse-Smale Complexes.* Ph.D. thesis, University of Groningen, The Netherlands, 2011. `www.math.rug.nl/~gert/documents/amit.pdf`.

[8] H. Edelsbrunner, J. Harer, V. Natarajan, and V. Pascucci. Morse-Smale complexes for piecewise linear 3-manifolds. In *Proc. 19th Ann. Sympos. Comput. Geom.*, pages 361–370, 2003.

[9] H. Edelsbrunner, J. Harer, and A. Zomorodian. Hierarchical Morse-Smale complexes for piecewise linear 2-manifolds. *Discrete Comput. Geom*, 30:87–107, 2003.

[10] R. Forman. Morse theory for cell complexes. *Adv. Math.*, 134:90–145, 1998.

[11] A. Gyulassy, P. Bremer, B. Hamann, and V. Pascucci. A practical approach to Morse-Smale complex computation. *IEEE Transactions on Visualization and Computer*, 14:1619–1626., 2008.

[12] J. L. Helman and L. Hesselink. Visualizing vector field topology in fluid flows. *IEEE Computer Graphics and Applications*, 11(3):36–46, 1991.

[13] M. W. Hirsch and S. Smale. *Differential Equations, Dynamical Systems, and Linear Algebra.* Academic Press, 1974.

[14] J.H. Hubbard and B.H. West. *Differential Equations. A Dynamical Systems Approach. Part I*, volume 5 of *Texts in Applied Mathematics*. Springer Verlag, New York, Heidelberg, Berlin, 1991.

[15] C. Li, S. Pion, and C. Yap. Recent progress in Exact Geometric Computation. *J. of Logic and Algebraic Programming*, 64(1):85–111, 2004. Special issue on "Practical Development of Exact Real Number Computation".

[16] L. Lin and C.K. Yap. Adaptive isotopic approximation of nonsingular curves: the parameterizability and nonlocal isotopy approach. *Discrete and Computational Geometry*, 45(4):760–795, 2011.

[17] F. Meyer. Topographic distance and watershed lines. *Signal Process.*, 38:113–125, 1994.

[18] J. Milnor. *Morse Theory*. Princeton University Press, 1968.

[19] R.E. Moore. *Interval Analysis*. Prentice-Hall., 1996.

[20] J. Palis and W. de Melo. *Geometric Theory of Dynamical Systems: An Introduction*. Springer-Verlag, 1982.

[21] S. Plantinga and G. Vegter. Isotopic meshing of implicit surfaces. *The Visual Computer*, 23:45–58, 2007.

[22] J.M. Snyder. *Generative Modeling for Cimputer Graphics and CAD. Symbolic Shape Design Using Interval Analysis*. Academic Press Professional, Inc., San Diego, CA, USA, 1992.

[23] J.M. Snyder. Interval analysis for computer graphics. *SIGGRAPH Comput. Graph.*, 26(2):121–130, 1992.

[24] S. Takahashi, T. Ikeda, Y. Shinagawa, and I. Fujishiro. Algorithms for extracting correct critical points and constructing topological graphs from discrete geographic elevation data. *Comput. Graph. For.*, 14(3):181–192, 1995.

[25] C.K. Yap. In praise of numerical computation. In S. Albers, H. Alt, and S. Näher, editors, *Efficient Algorithms*, volume 5760 of *Lecture Notes in Computer Science*, pages 308–407. Springer-Verlag, 2009. Essays Dedicated to Kurt Mehlhorn on the Occasion of His 60th Birthday.

[26] E. Zhang, K. Mischaikow, and G. Turk. Vector field design on surfaces. *ACM Transactions on Graphics*, 25(4):1294–1326, 2006.

APPENDIX

A. MATHEMATICAL BACKGROUND

Morse functions. A function $h : \mathcal{D} \subset \mathbb{R}^2 \to \mathbb{R}$ is called a *Morse function* if all its critical points are non-degenerate. The *Morse lemma* [18] states that near a non-degenerate critical point a it is possible to choose local co-ordinates x, y in which h is expressed as $h(x,y) = h(a) \pm x^2 \pm y^2$. Existence of these local co-ordinates implies that non-degenerate critical points are isolated. The number of minus signs is called the *index* $i_h(a)$ of h at a. Thus a two variable Morse function has three types of non-degenerate critical points: minima (index 0), saddles (index 1) and maxima (index 2).

Integral curves. An integral curve $\mathbf{x} : I \subset \mathbb{R} \to \mathcal{D}$ passing through a point p_0 on $\mathcal{D}$ is a unique maximal curve satisfying: $\dot{\mathbf{x}}(t) = \nabla h(\mathbf{x}(t)), \quad \mathbf{x}(0) = p_0$, for all t in the *interval* I.

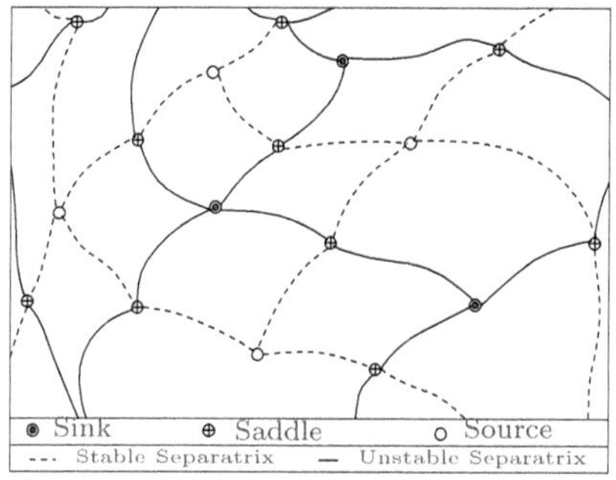

Figure 8: Morse-Smale complex

Integral curves corresponding to the gradient vector field of a smooth function $h : \mathcal{D} \to \mathbb{R}$ have the following properties:

1. Two integral curves are either disjoint or same.
2. The integral curves cover all the points of $\mathcal{D}$.
3. The integral curves of the gradient vector field of h form a partition of $\mathcal{D}$.
4. The integral curve $\mathbf{x}(t)$ through a critical point p_0 of h is the constant curve $\mathbf{x}(t) = p_0$.
5. The integral curve $\mathbf{x}(t)$ through a regular point p of h is injective, and if $\lim_{t\to\infty} \mathbf{x}(t)$ or $\lim_{t\to-\infty} x(t)$ exists, it is a critical point of h. This implies integral curves corresponding to gradient vector field are never closed curves.
6. The function h is strictly increasing along the integral curve of a regular point of h.
7. Integral curves are perpendicular to the regular level sets of h.

Stable and unstable manifolds. Consider the integral curve $x(t)$ passing through a point p. If the limit $\lim_{t\to\infty} \mathbf{x}(t)$ exists, it is called the ω-limit of p and is denoted by $\omega(p)$. Similarly, $\lim_{t\to-\infty} \mathbf{x}(t)$ is called the α-limit of p and is denoted by $\alpha(p)$ – again provided this limit exists. The stable manifold of a singular point p is the set $W^s(p) = \{q \in \mathcal{D} \mid \omega(q) = p\}$. Similarly, the unstable manifold of a singular point p is the set $W^u(p) = \{q \in \mathcal{D} \mid \alpha(q) = p\}$. Here we note that both $W^s(p)$ and $W^u(p)$ contain the singular point p itself [13].

Now, the stable and unstable manifolds of a saddle point are 1-dimensional manifolds. A stable manifold of a saddle point consists of two integral curves converging to the saddle point. Each of these integral curves (not including the saddle point) are called the stable separatrices of the saddle point. Similarly, an unstable manifold of a saddle point consists of two integral curves diverging from the saddle point and each of these integral curves (not including the saddle point) are called the unstable separatrices of the saddle point.

The Morse-Smale complex. A Morse function on $\mathcal{D}$ is called a Morse-Smale (MS) function if its stable and unstable separatrices are disjoint. In particular, a Morse-Smale function on a two-dimensional domain has no integral curve connecting two saddle points, since in that case a stable separatrix of one of the saddle points would coincide with an unstable separatrix of the other saddle point. The MS-complex associated with a MS-function h on $\mathcal{D}$ is the subdivision of $\mathcal{D}$ formed by the connected components of the intersections $W^s(p) \cap W^u(q)$, where p, q range over all singular points of h. If $\mathcal{D} = \mathbb{R}^2$, then, according to the Quadrangle Lemma [9], each region of the MS-complex is a quadrangle with vertices of index $0, 1, 2, 1$, in this order around the region.

Stability of equilibrium points. We note that a gradient vector field of a MS-function $h : \mathcal{D} \to \mathbb{R}$ can have three kinds of equilibria or *singular points*, namely, sinks (corresponding to maxima of h), saddles (saddles of h) and sources (corresponding to minima of h). These singular points can be distinguished based on the local behavior of the integral curves around those points. Locally, a sink has a neighborhood, which is a stable 2-manifold. Similarly, locally a source has a neighborhood, which is an unstable 2-manifold. Locally, a saddle has a stable 1-manifold and an unstable 1-manifold crossing each other at the saddle point. A sink is called a stable equilibrium point, where as a source or a saddle is called unstable equilibrium point. We note that, a source corresponding to a MS-function h is a sink corresponding to the function $-h$.

B. THE CONSTANTS IN LEMMA 6

A gridbox is called *quasihorizontal* (*quasivertical*) if it contains a point at which ∇h is quasihorizontal (quasivertical). Again, a gridbox may be both quasihorizontal and quasivertical.

An integral curve of ∇h in a quasihorizontal gridbox $[x_0, x_1] \times [y_0, y_1]$ is the graph of a function $x \mapsto y(x)$, where $y(x)$ is a solution of the ODE $y'(x) = F(x, y(x))$, with $y(x_0) = y_0$, where $F(x, y) = \dfrac{h_y(x, y)}{h_x(x, y)}$. Here x ranges over the full interval $[x_0, x_1]$ if $y_0 \le y(x) \le y_1$. Otherwise, the range of x is restricted to a suitable maximal subinterval $[\xi_0, \xi_1]$, such that $(\xi_0, y(\xi_0))$ and $(\xi_1, y(\xi_1))$ are points on the boundary of the gridbox. Similarly, a trajectory of X_ϑ in a quasihorizontal gridbox $[x_0, x_1] \times [y_0, y_1]$ is the graph of a function $x \mapsto y(x)$, where $y(x)$ is a solution of the ODE $\dfrac{dy}{dx} = F_\vartheta(x, y)$, with $F_\vartheta = \dfrac{h_x \sin\vartheta + h_y \cos\vartheta}{h_x \cos\vartheta - h_y \sin\vartheta}$.

The next result provides an upper bound for the growth of the funnel width along a quasihorizontal part of its bounding polylines. We assume that the funnel runs from left to right, so its initial points are on the line with smallest x-coordinate. If the funnel runs from right to left, a similar result is obtained.

LEMMA 7. *Let $y_{\vartheta,w}, y_{-\vartheta,w} : [a, b] \to [c, d]$ be the piecewise linear functions the graphs of which are quasihorizontal parts of the polylines L_ϑ and $L_{-\vartheta}$ for a grid with edge length w, respectively. Let Δ be an upper bound for the distance of the initial points of these polylines, i.e., $|y_{\vartheta,w}(a) - y_{-\vartheta,w}(a)| \le \Delta$. The width of the fence bounded by L_ϑ and $L_{-\vartheta}$ is bounded:*

$$|y_{\vartheta,w}(x) - y_{-\vartheta,w}(x)| \le \Delta\, e^{C_2(x-a)} + (C_0 w + C_1 \sin\vartheta)\,\frac{e^{C_2(x-a)} - 1}{C_2},$$

where C_0, C_1 and C_2 are computable constants determined by F and its first derivatives on the union of all quasihorizontal gridboxes.

A similar result holds for quasivertical trajectories.

Transitions: bounded increase of funnel width. Transition from a quasihorizontal to a quasivertical, or from a quasivertical to a quasihorizontal part of the funnel takes place at an ε-box. If the width of the funnel at the 'entry' of the box is less than the width w of a grid box, then the width may increase, but it will not be greater than $2w$ at the exit. This is made more precise by the following result.

LEMMA 8. *Let $\mathbb{J}$ be a ε-box, where, moreover, the initial points p and q of $L_\vartheta \cap \mathbb{J}$ and $L_{-\vartheta} \cap \mathbb{J}$, respectively, are on the boundary of the gridboxes containing the vertices of edge e of $\mathbb{J}$. If the distance between p and q is at least w, then the distance between the terminal points $\overline{p}$ and $\overline{q}$ of $L_\vartheta \cap \mathbb{J}$ and $L_{-\vartheta} \cap \mathbb{J}$, respectively, is less than the distance of p and q. If the distance between p and q is less than w, then the distance of $\overline{p}$ and $\overline{q}$ is at most $2w$.*

PROOF. Assume that the first edge of L_ϑ is quasivertical, but not quasihorizontal. Edge e of $\mathbb{J}$ is then vertical. Assume that this polyline consists of a single edge, namely the line segment $p\overline{p}$. Let β_+ be the angle between $p\overline{p}$ and edge e, then

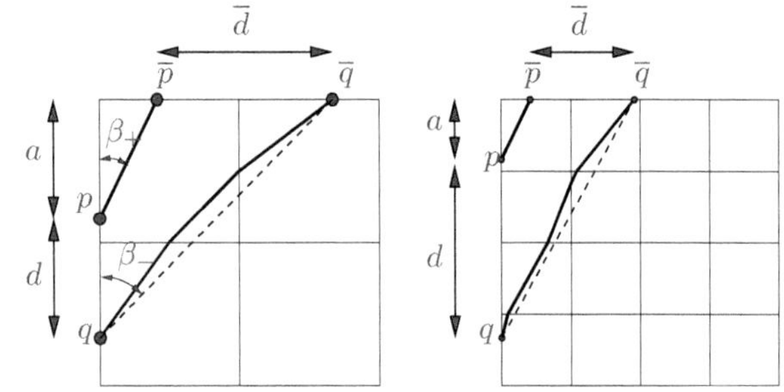

Figure 9: The distance between the two polylines upon entry and exit of a box. Left: If the distance d between the initial points p and q of the polylines is less than the edge-length w of a grid box, then the distance $\overline{d}$ between the terminal points $\overline{p}$ and $\overline{q}$ is less than $2w$. Right: Otherwise, the distance $\overline{d}$ between the terminal points is less than d.

$\arctan\frac{1}{2} \le \beta_+ \le \vartheta + \arctan\frac{1}{2}$. Let β_- be the angle between the line segment $q\overline{q}$ and edge e, then β_- is in between the smallest and largest slope of any edge of $L_{-\vartheta}$. Since the angle variation of X over $\mathbb{J}$ is less than $\frac{\pi}{20}$, the angle β_- is greater than $\beta_+ - \frac{\pi}{20}$. Let a be the distance of p to the nearest vertex of e, then $a \le w$. If $d \ge w$, the distance $\overline{d}$ between $\overline{p}$ and $\overline{q}$ satisfies

$$\begin{aligned}\overline{d} &= (d + a)\tan\beta_- - a\tan\beta_+ \\ &\le d\tan(\beta_+ + \tfrac{\pi}{20}) + a\,(\tan(\beta_+ + \tfrac{\pi}{20}) - \tan\beta_+) \\ &< \tfrac{3}{4}d + \tfrac{1}{4}a \le d,\end{aligned}$$

since $a \le w \le d$. Here we used $\tan\beta_+ \le \frac{1}{2}$ to get

$$\tan(\beta_+ + \tfrac{\pi}{20}) = \frac{\tan\beta_+ + \tan\frac{\pi}{20}}{1 - \tan\beta_+ \tan\frac{\pi}{20}} \le \frac{\frac{1}{2} + \tan\frac{\pi}{20}}{1 - \frac{1}{2}\tan\frac{\pi}{20}} \le \frac{3}{4}.$$

Since $\arctan\frac{1}{2} - \frac{\pi}{40} \le \arctan\frac{1}{2} - \vartheta \le \beta_+ \le \arctan\frac{1}{2}$, a short computation shows that $\tan(\beta_+ + \frac{1}{20}\pi) - \tan\beta_+ < \frac{1}{4}$.

If $d < w$, then q lies in the same gridbox as p, or in a gridbox adjacent to it. Then it is easy to see that $\overline{p}$ lies in the same grid box as p, and $\overline{q}$ also lies in this box, or in a box adjacent to it. Therefore, $\overline{d} \le 2w$ in this case. If $L_{-\vartheta}$ consists of a single edge, then the argument is similar. $\square$

Lemmas 7 and 8 provide the necessary tools for the proof of Lemma 6. We refer to the full version [7] for a proof.

A Deterministic $O(m \log m)$ Time Algorithm for the Reeb Graph*

Salman Parsa
Dept of Computer Science
Duke University, Durham
North Carolina, USA
salparsa@cs.duke.edu

ABSTRACT

We present a deterministic algorithm to compute the Reeb graph of a PL real-valued function on a simplicial complex in $O(m \log m)$ time, where m is the size of the 2-skeleton. The problem reduces to dynamic graph connectivity. We obtain the running time by using offline graph connectivity which assumes that the sequence of operations is known in advance. The algorithm is implemented and experimental results are given. In addition, we reduce the offline graph connectivity problem to computing the Reeb graph.

Categories and Subject Descriptors

F.2.2 [**Analysis of Algorithms and Problem Complexity**]: Nonnumerical Algorithms and Problems—*Geometrical problems and computations, Computations on discrete structures*

Keywords

Algorithms, Reeb graph, PL topology, graph connectivity

1. INTRODUCTION

Let f be a continuous function from a topological space to the real line. Roughly speaking, the Reeb graph of f is achieved by considering connected components of preimages (or level-sets) $f^{-1}(r)$ as points, for $r \in \mathbb{R}$. When the domain contains no nontrivial loops, such as $\mathbb{R}^d$, Reeb graph is the same as the contour tree. For example, for the altitude function on a terrain, contours (preimage components) are drawn on the plane, and their evolution traces out the contour tree. On the other hand, if the domain is a torus, contours should be drawn on a torus and its no longer true that their evolution traces a tree, it traces the Reeb graph.

Reeb graph of f provides a compact description of the domain, seen through f. Reeb graphs have found applications in a wide range of areas such as shape matching and retrieval [4, 16, 31], shape segmentation and simplification [23, 32], animation [3, 18], high dimensional data visualization and analysis [14, 19] and robotics [22]. We refer to [5, 6] for a survey of the Reeb graph and some applications. In this paper we consider the efficient computation of the Reeb graph of a piecewise linear function on an arbitrary simplicial complex. This setting is general enough to approximate functions and spaces usually considered in practice.

Our contributions are the following:

- We give the first deterministic algorithm that runs in worst-case time $O(m \log m)$, where m is the size of the 2-skeleton. This is optimal, if the number of edges of the complex is in the same order as the number of vertices.

- We implemented a simple version of the algorithm including a trick, making its running time output sensitive. This implementation gives running times superior to those of existing algorithms. Given the simplicity and possible choices for the implementation, it might turn into the algorithm of choice for various applications.

- We use an offline variant of the dynamic graph connectivity and show that it is equivalent to Reeb graph construction. This might be of independent interest.

*This work was partially done when the author was visiting IST Austria, Klosterneuburg, Austria.

SCG'12, June 17–20, 2012, Chapel Hill, North Carolina, USA.

2. RELATED WORK

Carr et al. [7] gave an efficient algorithm for computing the contour tree on a simplical complex domain in time $O(n \log n + m\alpha(m))$, where n is the number of vertices and m is the number of edges of the input complex. The first algorithm to compute the Reeb graph was given by Shinagawa and Kunii [24] that works for the triangulation of a 2-manifold and runs in time $\Theta(n^2)$. This algorithm sweeps the vertices in increasing order of function values and maintains the preimage. For the case of 2-manifolds, Cole-McLaughlin et al. [8] improved the running time to $O(n \log n)$. They use circular lists to maintain the preimage circles.

The Reeb graph of a d-dimensional simplicial complex for $d \geq 2$, depends only on the 2-skeleton, whose size we denote by m. One can maintain the preimage components as graphs. The problem reduces to that of dynamic graph connectivity on a graph of size m. Then, for an arbitrary simplicial complex, the sweep algorithm runs in time m times the bound for an operation in the dynamic graph connectivity data structure. Doraiswamy and Natarajan [10] were the first to use this reduction to compute the Reeb graph, see section 4 for details.

Holm et al. [17] give a deterministic algorithm for dynamic graph connectivity with the worst-cast time $O(\log^2 m)$ per operation. As used in [10], this connectivity algorithm resulted in the best deterministic algorithm for the Reeb graph for a general simplicial complex before this paper. Moreover, Thorup [29] presents an algorithm with $O(\log m(\log\log m)^3)$ expected amortized running time per operation for the dynamic graph connectivity. For computing the Reeb graph on a 3-manifold, Doraiswamy and Natarajan [10] give an algorithm that runs in expected time $O(m\log m + m\log g(\log\log g)^3)$, where g is the maximum genus over all preimages. This algorithm maintains a tree/co-tree partition of the graph, and uses Thorup's randomized graph connectivity.

Tierny et al. [30] present an algorithm that works on 3-manifolds with boundary embedded in $\mathbb{R}^3$. Their algorithm runs in time $O(m\log m + hm)$, where h is the number of independent loops in the Reeb graph. This algorithm is not general but is very efficient. A streaming algorithm for computing the Reeb graph of an arbitrary simplicial complex is presented in [20] with $\Theta(nm)$ running time. Harvey et al. [15] presented a randomized algorithm with the expected running time $O(m\log m)$. The algorithm works by collapsing triangles adjacent to a vertex. In the end, the complex collapses to a representation of the Reeb graph. The evolution of the Reeb graph, as the function varies over time, is studied in [11]. In [9], the authors study approximation of the Reeb graph and its persistence. Higher dimensional analogs of Reeb graphs are called Reeb spaces. They are more difficult to compute, however. These spaces are studied in [12].

We also mention that there has been extensive research on the dynamic graph connectivity and related problems. Both from upper bound and from lower bound point of view. In addition to the above, Patrascu and Demaine [21] proved an $\Omega(\log m)$ lower bound in the cell-probe model. Eppstein [13] uses a linear time minimum spanning tree algorithm to solve the offline minimum spanning tree problem in $O(\log m)$ time per operation. This is the only reference to offline graph connectivity that we know of.

3. BACKGROUND

Simplices and simplicial complexes. A *d-simplex* is the convex hull of $d+1$ affinely independent points $V = \{v_0, \ldots, v_d\}$ in some Euclidean space, e.g. $\mathbb{R}^d$, where $d \geq 0$. The set $V = V(\sigma)$ is called the *vertex* set of the simplex. Let σ be a d_1-simplex and δ a d_2-simplex. If $V(\sigma) \subset V(\delta)$, we say σ is a d_1*-face* of δ and denote it by $\sigma \leq \delta$. We also call a 0-simplex, a 1-simplex and a 2-simplex a *vertex*, an *edge* and a *triangle*, respectively. If K is a finite set of simplices, all in the same Euclidean space, then K is a *simplicial complex* provided i) $\delta \in K$ and $\sigma < \delta \Rightarrow \sigma \in K$, ii) $\sigma_1, \sigma_2 \in K \Rightarrow \sigma_1 \cap \sigma_2 < \sigma_1, \sigma_2$ if $\sigma_1 \cap \sigma_2$ is non-empty. We define K_0, K_1 and K_2 to be the set of vertices, edges, and triangles, respectively, of the simplicial complex K. Moreover, set $n_0 = \#K_0$, $n_1 = \#K_1$ and $n_2 = \#K_2$, where $\#$ denotes number of elements of the set. By $|K|$ we mean the underlying space of K, i.e. $|K| = \bigcup_{\sigma\in K} \sigma$ with the topology inherited from the ambient Euclidean space. For convenience, if no ambiguity is caused, we also write K instead of $|K|$. The *dimension* of a simplicial complex is the highest dimension of its simplices. The *k-skeleton* of the simplicial complex, denoted K^k, is the set of its simplices of dimension at most k. We denote by m the size of the 2-skeleton, i.e. $m = n_0 + n_1 + n_2$.

If $x \in |K|$, then there is a unique simplex with smallest dimension that contains x, say σ'. By definition of a simplex x can be written as a convex combination of the vertices of σ'. Setting the coefficients of other vertices in K_0 to zero, we can write x as a convex combination of points in K_0 in a unique way: $x = \sum_{i=1}^{n_0} b_i v_i$ where $K_0 = \{v_1, v_2, \ldots, v_{n_0}\}$ and $\sum_i b_i = 1$. and $b_i \geq 0$ for all i. The b_i are called the *barycentric coordinates* of $x \in |K|$.

Unless otherwise stated, graphs in this paper will be abstract over a fixed, labeled, set of nodes. To express the difference with a complex, we call the vertices of a graph *nodes* and its edges *arcs*.

Reeb graph. Let $f : K_0 \to \mathbb{R}$ be a function. We say that f is *generic* if it is injective. In this paper, we always assume the function on the set of vertices is generic. One can extend f to all of $|K|$ by setting $f(x) = \sum_i b_i f(v_i)$, where b_i are the barycentric coordinates of x. Then, the extended function f, is called a *piecewise linear* function from $|K|$ to $\mathbb{R}$. Now, fix $r \in \mathbb{R}$ and consider the preimage of r: $f^{-1}(r) = \{x \in |K|, f(x) = r\}$. If $\sigma \in K$ is a d-simplex, then $f^{-1}(r) \cap \sigma$ is the intersection of a $(d-1)$-plane with σ. It is not hard to see that $f^{-1}(r) \cap K^2$ is also a simplicial complex, namely, $f^{-1}(r) \cap K^2 = \{\sigma \cap f^{-1}(r) : \sigma \in K^2\}$. Therefore, every vertex of $f^{-1}(r) \cap K^2$, $r \notin f(K_0)$, corresponds to one edge of K and every edge of it comes from one triangle of K.

We are interested in the connected components of $f^{-1}(r)$, and their behavior as r varies. The *Reeb graph* of a function $f : |K| \to \mathbb{R}$ is the topological graph obtained by contracting every connected component of $f^{-1}(r)$ to a point, for every $r \in \mathbb{R}$. So it is a quotient space of $|K|$ with the quotient topology. Intuitively, the points of the Reeb graph are connected together as the preimage components were connected together. In other words, $f^{-1}(r)$ reduces to a finite set of points and as r varies, these points trace out arcs of a graph which meet at points where corresponding connected components meet. One can see that when r changes continuously, without passing any $f(v_i)$ value, the connected components of the preimages $f^{-1}(r)$ remain unchanged. Given a simplicial complex and a generic piecewise linear function, we are concerned with finding an efficient algorithm that computes its Reeb graph. Note that if the complex K is d-dimensional, one can easily embed the Reeb graph in $|K| \subset \mathbb{R}^k$, where k is as in the definition of the complex. However, there is no natural embedding of the Reeb graph on the plane (page) for example. This is true, even if the Reeb graph is planar. In the following we look at the Reeb graph as an abstract graph. We also note that, the Reeb graph is in fact a multigraph.

If we sweep the Reeb graph in increasing function value, a node is a point where a component (arc) is created, merged with others, split or destroyed. Since these events can happen only at preimage of some $f(v)$ and the preimage only includes one vertex v, vertices can be used to identify Reeb graph nodes. Consider the component of $f^{-1}([f(v)])$ containing v. If the contraction of this component in the Reeb graph is not a node, we call v *Reeb-regular*. In other words, Reeb-regular vertices are exactly those that when the preimage changes from $f^{-1}(f(v) - \epsilon)$ to $f^{-1}(f(v) + \epsilon)$ no arcs of

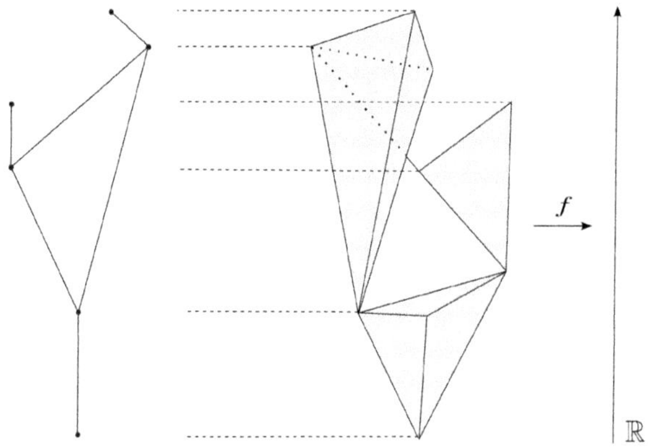

Figure 1: a drawing of the Reeb graph for the height function on a simplicial complex

the Reeb graph are created, get destroyed, merge or split. If a vertex is not Reeb-regular, we call it *Reeb-critical*. Therefore, a Reeb-critical vertex identifies a node of the Reeb graph. Moreover, a *Reeb-critical value* is the function value of a Reeb-critical vertex. Other values are *Reeb-regular*.

4. ALGORITHM

We describe the algorithm in two parts. First we show how to reduce the problem to that of maintaining connected components of a graph through insertion and deletion of arcs, then we explain how the latter problem can be solved in our setting within optimal time bounds. The Reeb graph of a d-dimensional simplicial complex depends only on its 2-skeleton. In the following, we assume that the input is therefore the 2-skeleton of the original complex, then $f^{-1}(r)$ is a 1-dimensional simplicial complex.

The input to our algorithm is described as a list of vertices, edges and triangles. Edges and triangles are defined by indexing their vertices. Also, for every vertex we need to know edges and triangles incident on it, which we compute in time linear in number of edges and triangles.

4.1 The Reduction

The outline of the algorithm is as follows. To obtain the Reeb graph of f, we need to know its nodes and its arcs. Since the components of the preimage are created and/or destroyed only at Reeb-critical values, we need to find the Reeb-critical vertices of K. Sort the v_i so that $f(v_1) < f(v_2) < \cdots < f(v_{n_0})$. We find the critical vertices by sweeping f from $-\infty$ to $+\infty$ and for each value $f(v_i)$, we look how the preimage components change in number, when we pass that value. If the vertex is recognized as Reeb-critical, we add it to the Reeb graph as a node and connect it to other nodes appropriately.

Before going into more detail, we introduce some terms. We say an edge of K *starts* at its vertex of lower function value and *ends* at the one with higher function value. Similarly, a triangle starts at its vertex of lowest function value and ends at its vertex of highest function value. The vertex with the middle function value of a triangle we call its *middle* vertex. We denote by v_1v_2 the edge connecting vertices v_1 and v_2. In this notation, we always first write the vertex with smaller function value.

The preimage $f^{-1}(r)$ can be abstracted into a graph G_r, which we call the *preimage graph* at value r. Nodes of G_r are edges of K intersecting the preimage and arcs of G_r are triangles of K contributing to the preimage, so arcs indeed connect nodes to each other. G_r changes if and only if we pass a function value $f(v_i)$. The graph G_r is $f^{-1}(r)$, when viewed as an abstract graph. In the following, we use a data structure to maintain the connected components of the preimage graph which we call *DynTrees*. Its description is the topic of next section. The pseudo-code for the sweep algorithm follows:

Algorithm 1 Sweep Algorithm

```
set DynTrees to be an empty graph
for i = 1 to n_0 do
  Lc = LowerComps(v_i)
  UpdatePreimage()
  Uc = UpperComps(v_i)
  UpdateReebGraph(Lc,Uc)
end
```

In Algorithm 1, we use four subroutines. The *LowerComps*(v_i) subroutine considers edges ending at the vertex v_i, one by one, and finds their corresponding components in the current preimage graph (which is $G_{f(v_i)-\epsilon}$). At the end, *LowerComps*(v_i) provides us with a set of nodes of the preimage graph, each representing a component. A pseudo-code for this procedure is given in Algorithm 2.

The *UpdatePreimage*() subroutine updates the preimage graph from that of immediately before $f(v_i)$ to that of immediately after $f(v_i)$. Triangles and edges ending at v_i are removed from the graph and edges and triangles starting at v_i are inserted into the preimage graph. Moreover, for every triangle of K that has v_i as a middle vertex, we delete the arc of the preimage graph that will no longer be in the graph and insert the new arc; see Figure 2. A pseudo-code for this subroutine is given in Algorithm 3. This code assumes that edges not intersecting the preimage are also in the preimage graph as isolated nodes so there is no need to add and remove isolated nodes. The subroutine *UpperComps*(v_i) is symmetric to *LowerComps*(v_i).

Algorithm 2 LowerComps(v)

```
Lc = empty list
for all edges e ending at v do
  c = DynTrees.find(e)
  if c is not marked then
    Lc.add(c)
    mark c as listed
  end
end
```

The *UpdateReebGraph(Lc, Uc)* subroutine, updates the Reeb graph. Note that all components in Lc will be merged into one component at v_i and this one, will split into the components in Uc. The vertex v_i is Reeb-regular, if and only if Lc and Rc both have exactly one element, therefore it is easy to decide if the vertex corresponds to a node of the Reeb graph. For such a vertex, we create a new node in the Reeb graph, ν_{v_i}, and associate it to nodes (representing components) in Uc. Intuitively, those components were first generated at v_i. We make an arc between corresponding Reeb graph nodes of components in Lc to the node ν_{v_i}. A pseudo-code for this subroutine is given in Algorithm 4.

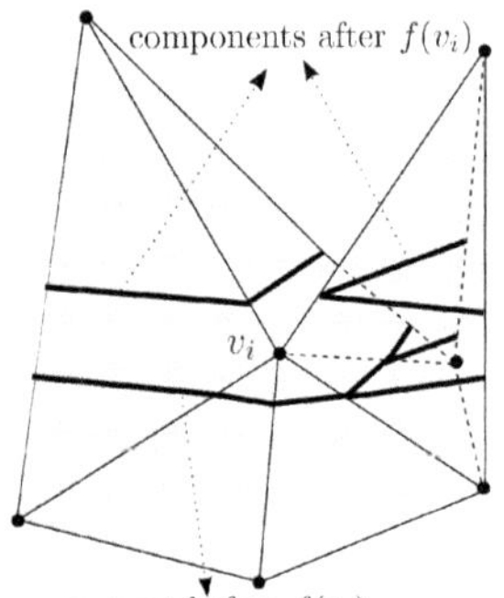

Figure 2: processing of a Reeb-critical (split) vertex in a 2-d simplicial complex

Algorithm 3 UpdatePreimage(v)

```
for all triangles t = {v1, v2, v3} incident on v while f(v1) < f(v2) < f(v3) do
  if v = v3 then DynTrees.delete((v1 v)(v2 v)) end
  if v = v2 then
    DynTrees.delete((v1 v3)(v1 v))
    DynTrees.insert((v v3)(v1 v3))
  end
  if v = v1 then DynTrees.insert((v v2)(v v3)) end
end
```

The total time spent in *UpdateReebGraph(Lc, Uc)* is linear in the size of the Reeb graph. The *DynTrees* data structure supports three types of operations: finding component of a node, inserting an arc, and deleting an arc from the preimage graph. Assuming n is the number of nodes in this graph, we write $U(n)$ for the time needed for any of these operations. Every edge of K is considered once in *LowerComps*(v_i) and once in *UpperComps*(v_i). For each, there is one *find* operation for finding the node representing the component of the edge in preimage graph, therefore the total running time of the two subroutines is $O(n_1 U(n_1))$ in the worst case. Moreover, every triangle gives rise to two arc insertions and two arc deletions in the preimage graph, so the total running time of *UpdateGraph*() is $O(n_2 U(n_1))$. Summing all of these together, the algorithm runs in time $O(mU(m))$ where m is the size of the 2-skeleton.

4.2 Graph Connectivity

We complete the description of the algorithm by explaining how to implement the three operations *find*, *delete* and *insert* on the preimage graph, required by the sweep algorithm. The *DynTrees* data structure keeps track of the connected components of the graph, when arcs are inserted and deleted over a fixed node set. This is called the *dynamic graph connectivity* problem.

A dynamic graph connectivity algorithm usually works by maintaining a rooted spanning forest of the graph. The root is used to identify the component, so a *find* query will be just finding the root of the tree containing the node. This is the approach we will also take, but we exploit the fact that the operations requested by the sweep algorithm can be predicted to choose our spanning trees. In order to do so, we assign weights to arcs of the preimage graph. The *weight* of an arc is the time that the arc is going to be deleted. In

Algorithm 4 UpdateReebGraph(Lc,Uc)

```
if #Lc = #Uc = 1 then
  return
else
  create a new node ν in Reeb graph
  assign the node to all c ∈ Uc
  create an arc between ν and ν_c, for all c ∈ Lc
end
```

other words, if the arc $(v_1 v_2)(v_3 v_4)$ corresponds to a triangle (so the v_i are not distinct), then the weight of the arc is the smallest function value of endpoints, i.e. $\min f(v_2), f(v_4)$. The weight of an arc is computed in constant time when the arc is inserted, and assigned to the arc.

The main idea is now to maintain the maximum spanning forest of the preimage graph. It has the important property that, the arc that is going to be deleted is not in this forest, unless it is absolutely necessary. To maintain the forest, we use a dynamic tree data structure. These data structures can keep a forest of node-disjoint trees and support various operations on those trees. Edges can have weights and information about the weights on a tree or a path can be obtained. The operations that we require are as follows:

- *parent*(x): return the parent of node x, or **null** if x is the root.
- *root*(x): return the root of the tree containing node x.
- *link*(x_1,x_2,w): link distinct trees containing the two nodes x_1 and x_2 by adding the arc $x_1 x_2$. Assigning the weight w to this arc.
- *cut*(x_1,x_2): remove the arc between x_1 and x_2, splitting the tree in two.
- *minWeight*(x): return a node with minimum weight arc to its parent on the path from x to the root of its tree, or **null** if x is the root.
- *evert*(x): make the node x the root of its tree.

All of the above operations are supported by existing dynamic tree data structures that allow path operations, for example, ST-trees or Link-Cut trees [25,26], top trees [2,27] and RC-trees [1]. See [28] for an experimental comparison of these data structures. Different implementations of ST-trees and top trees support all of the above operations in worst-case or amortized time $O(\log n)$, where n is the number of nodes in the forest [28]. RC-trees achieve the same bound in expected running time.

Using the above, we implement our three tree operations as follows:

- *find*(x):

```
return root(x)
```

- *insert*($x_1 x_2$):

```
ω = weight of arc x1x2
if root(x1) = root(x2) then
  evert(x1)
  x' = minWeight(x2)
  ω' = x'.weight
  if ω' < ω then
    cut(x', parent(x'))
    link(x1,x2, ω)
  end
else
```

```
    link(x1,x2,ω)
  end
```

• *delete*(x_1,x_2):

```
  if (x1,x2) is a tree edge then
    cut(x1,x2)
  end
```

The Reeb graph node associated with a root transfers as the root node changes in *evert*. Reeb graph nodes are assigned to the roots of new trees generated, after each *cut* or *link* operation. The overall cost of keeping track of Reeb graph nodes is then constant number of *find* calls per every edge. So, we can maintain the Reeb graph data in $O(n_1 U(n_1))$ time. Considering the above, we have $U(n) = O(\log n)$ using this offline graph connectivity algorithm.

Correctness. At the beginning of the algorithm, the preimage graph has no arcs. We should show that the above operations result in a maximum spanning forest, if they are applied to one such forest. The *insert*(a) operation, breaks the cycle that is formed by adding the arc a between two nodes of a tree. It does that by removing the arc of minimum weight, say b, on this cycle, if that weight is less than the weight of a. Therefore the overall weight of the tree increases. If a tree with higher weight existed, then it should have a as an arc, by removing a and inserting a missing edge of the cycle (with weight at least that of b), we get a new tree with higher weight for the original graph before insertion, which is a contradiction.

In *delete*(a), note that every other arc in the whole graph has weight at least as large as weight of a, since the weights are deletion times. Every arc that exists, either is deleted during the current call to *UpdatePreimage*() or has a higher weight. If all of the weights are higher than that of a, then, the deletion indeed splits the maximum spanning tree in two. Otherwise, any arc reconnecting the resulting trees is also deleted before update process finishes and before any *find* queries, therefore there is no harm in not connecting back the split trees. Essentially what we are doing is keeping an online maximum spanning forest while deletions are guaranteed to cut trees, when they happen for edges of the trees. We have the following theorem:

THEOREM 1. *Reeb graph of a piecewise linear function on a simplicial complex can be constructed in $O(m \log m)$ time in the worst case, where m is the size of the 2-skeleton.*

5. IMPLEMENTATION

We use "lazy insertion" to make the implementation faster and output-sensitive. Roughly speaking, our goal is not to insert arcs that die (i.e. get deleted) before any Reeb-critical value is met. For this, we need to decide whether a vertex is Reeb-regular or not, before inserting arcs. In general we cannot do this. However, for many vertices we can.

Simple Reeb-regular vertices. We recall some concepts before going into details. The *star* of a simplex $\sigma \in K$ is the collection of simplices that have σ as a face. This might not be a simplicial complex, however, if we add the missing faces of the simplices in the star, we obtain a subcomplex of K which is called *closed star* of σ. Here we are only concerned with stars of vertices. With $f : |K| \to \mathbb{R}$ as above, the *lower star* of a vertex $v \in K$ contains all the simplices in the star for which $f(v)$ is the maximum value among its vertices. Again, the *closed lower star* is obtained by adding missing simplicies. *Upper star* and *closed upper star* are defined symmetrically.

If the upper star and the lower star of a vertex both have just one component, then the vertex is Reeb-regular. We use this, to decide quickly if a vertex is Reeb-regular of this kind, which we call *simple* Reeb-regular. Note that, if this is not the case, the vertex still can be Reeb-regular. However, this latter kind of Reeb-regular vertices are those that the first homology of the preimage changes when passing their function values, i.e. a nontrivial loop is generated and/or gets destroyed in the preimage. They tend to be smaller in number compared to simple Reeb-regular vertices in practice.

In our implementation, we first check if the vertex is simple Reeb-regular, if so, we do not insert the corresponding arcs into the preimage, rather, we merely keep them for insertion later in an *insertion list*. For such a vertex, the arcs that should be removed, will be removed from the insertion list and the current (incomplete) preimage spanning trees. If a vertex is found to be not simple Reeb-regular, we build the preimage trees completely from the edges survived in the insertion list and continue the algorithm as usual.

We remark that building the preimage from insertion list involves also keeping track of the corresponding Reeb graph nodes of the spanning tree roots.

Performance comparison. We did a preliminary implementation of the algorithm using the simple "linear" tree data structure: every node contains a pointer to its parent and the weight of the arc to its parent if it is not a root. In the worst case, tree operations will need $O(n)$ time over a graph of n nodes, however, as the running times below demonstrate, it is promising.

We compare our implementation with that of [15] which we call RANDREEB. Asymptotically, this algorithm takes $O(m \log m)$ time in expectation, and has actual running times superior to earlier ones, see [15]. The exception is the surgery method of [30]. This approach cannot handle arbitrary 3-manifolds or simplicial complexes, however. The running times are shown in Table **??**. The input data sets are almost the same as those of Harvey et al. [15] and were kindly provided to us by the authors. We ran the experiments on a 64-bit computer with Dual-Core 3.00 GHz CPU and 8 GB's of memory running a Linux operating system.

With the exception of the *Camel* and *Simulation*, all data sets are manifolds. For further information on the data sets we refer to [15,30] and the aim@shape database. As can be seen from the table, the algorithm shows an almost linear performance. We note that this linearity is mostly the result of lazy insertion trick. The IV (important vertex) column shows the fraction of Reeb-critical vertices and non-simple Reeb regular vertices of all the vertices. they cause the non-linearity of the algorithm, which is barely noticeable, since the fraction is small. As long as this fraction remains low, the algorithm works specially well, even with trivial implementation of the tree data structure. With a full fledged dynamic tree data structure, we expect the running times to improve substantially, at least for large data sets, as building the preimage at the value of a IV involves a large number of tree operations. The source code is available upon request.

Data Set	m	IV	RandReeb	Ours
Camel	110,785	0.03	0.32	0.99
Simulation	190,165	0.6	1.64	1.39
Fighter	245,300	0.49	6.70	1.95
Blunt	762,683	0.05	13.29	5.12
Post	2,086,950	0.003	17.32	13.45
Buckyball	4,322,620	0.05	69.11	36.63
Plasma	4,530,561	0.02	135.79	42.35
Earthquake	7,085,157	0.05	177.68	71.41

Table 1: Comparison of running times. The running times are in seconds. The size of the 2-skeleton, m, is the total number of vertices, edges, and triangles of the input complex. IV is the fraction of Reeb-critical and non-simple Reeb-regular vertices of all the vertices.

6. MORE ON GRAPH CONNECTIVITY

In this section we first show how the ideas used to solve the offline graph connectivity problem can be used in more general settings. Later, we will give a linear time reduction from offline graph connectivity to the Reeb graph computation. Let $C = c_1c_2 \ldots c_m$ be a sequence of three types of dynamic graph connectivity operations over a fixed node set of size n. The parameters of operations are fixed and we think of them as indexing the nodes of the graph, thus integers in $\{1, \ldots, n\}$. The problem of answering the queries in a given sequence like C is what we call the *offline* graph connectivity problem. C consists of some arc insertions and deletions mixed with queries. We say i is the time when operation c_i happens. We can determine the deletion time of any arc in constant amortized time as follows. We make indexes for arcs inserted and deleted using the two integers indexing its endpoint nodes, and then sort them using a linear time algorithm for sorting integers. Then we can find operations that index the same arc. This takes time linear in the number of edges inserted. Having find the deletion times in constant time, we can solve this problem in $O(\log n)$ time per operation, as above.

Other than in this offline setting, i.e. in an application where the sequence of operations is not known beforehand, we still might be able to compute the deletion time of every arc at the time of its insertion, say with worst case time $d(n)$. If $d(n)$ is less than the worst-case time for a fully dynamic data structure (which is $\Omega(\log n)$, even in the amortized and expected running times), then we may apply the technique above and get an algorithm that runs in time $O(\log n + d(n))$ for all of the operations.

Reduction to the Reeb graph. Here we show a linear time reduction from the offline graph connectivity problem, as defined above, to the Reeb graph construction. This implies that the two problems are equivalent in terms of computational complexity. Given a sequence C of m graph connectivity operations, we construct a simplicial complex whose Reeb graph contains the answers to the queries and these can be read off in constant time. For this, we use the *augmented* Reeb graph. It is the same as the Reeb graph except every vertex of the input complex has a corresponding node in it. Reeb-regular vertices are degree-two nodes with one incoming and one outgoing arc. It is clear that with some modifications, our algorithm can compute the augmented Reeb graph within the same time bound. Moreover, every Reeb-regular vertex can contain an identifier of the preimage component (Reeb graph arc) that it belongs to.

We build the complex and the function at the same time. For simplicity, we assume that at the end of the sequence C, all arcs are deleted. For every node of the graph that will have an incident arc we create a sufficiently long vertical edge, say of length $m + 1$. So, at the beginning we have a collection of edges all of the same height. We consider each c_i in turn. If c_i is an insertion of an arc $a = xy$ with deletion time d_a, then we connect two edges x and y as in Figure 3 (a), where the vertices are assigned the heights as in the figure. In Figure 3, i' and d_a' are slightly perturbed values. If c_i is a query we add a regular vertex as in Figure 3 (b) with height i. If c_i is a deletion, we do nothing. The function on this complex is the height function. It is easy to verify that when the augmented Reeb graph of this complex is computed, the answer to the query is the component identifier kept with the node corresponding to the regular vertex.

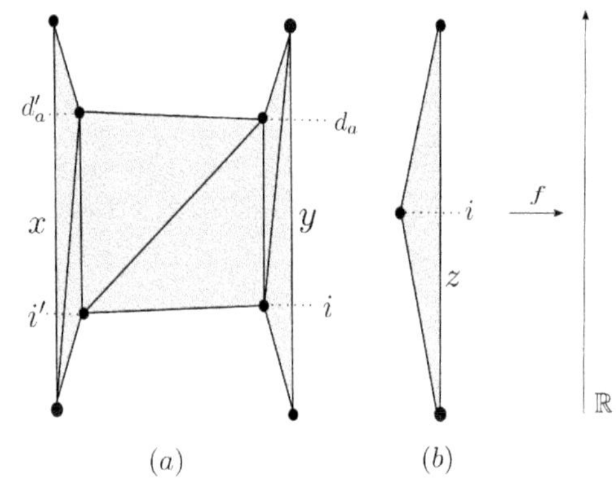

Figure 3: making a complex

7. ACKNOWLEDGMENT

The author thanks Herbert Edelsbrunner for support and valuable suggestions on the paper.

8. REFERENCES

[1] U. A. Acar, G. E. Blelloch, R. Harper, J. L. Vittes, and S. L. M. Woo. Dynamizing static algorithms, with applications to dynamic trees and history independence. In *Proceedings of the Fifteenth Annual ACM-SIAM Symposium on Discrete algorithms*, SODA '04, pages 531–540, Philadelphia, PA, USA, 2004. Society for Industrial and Applied Mathematics.

[2] S. Alstrup, J. Holm, K. D. Lichtenberg, and M. Thorup. Maintaining information in fully dynamic trees with top trees. *ACM Trans. Algorithms*, 1:243–264, October 2005.

[3] G. Aujay, F. Hétroy, F. Lazarus, and C. Depraz. Harmonic skeleton for realistic character animation. In *Proceedings of the 2007 ACM SIGGRAPH/Eurographics symposium on Computer animation*, SCA '07, pages 151–160, Aire-la-Ville, Switzerland, Switzerland, 2007. Eurographics Association.

[4] D. Bespalov, W. C. Regli, and A. Shokoufandeh. Reeb graph based shape retrieval for cad. *ASME Conference Proceedings*, 2003(36991):229–238, 2003.

[5] S. Biasotti, L. De Floriani, B. Falcidieno, P. Frosini, D. Giorgi, C. Landi, L. Papaleo, and M. Spagnuolo. Describing shapes by geometrical-topological properties of real functions. *ACM Comput. Surv.*, 40(4):12:1–12:87, Oct. 2008.

[6] S. Biasotti, D. Giorgi, M. Spagnuolo, and B. Falcidieno. Reeb graphs for shape analysis and applications. *Theoretical Computer Science*, 392(1-3):5 – 22, 2008. Computational Algebraic Geometry and Applications.

[7] H. Carr, J. Snoeyink, and U. Axen. Computing contour trees in all dimensions. *Computational Geometry*, 24(2):75 – 94, 2003.

[8] K. Cole-McLaughlin, H. Edelsbrunner, J. Harer, V. Natarajan, and V. Pascucci. Loops in reeb graphs of 2-manifolds. In *Proceedings of the Nineteenth Annual Symposium on Computational Geometry*, SCG '03, pages 344–350, New York, NY, USA, 2003. ACM.

[9] T. K. Dey and Y. Wang. Reeb graphs: approximation and persistence. In *Proceedings of the 27th Annual ACM Symposium on Computational Geometry*, SoCG '11, pages 226–235, New York, NY, USA, 2011. ACM.

[10] H. Doraiswamy and V. Natarajan. Efficient algorithms for computing reeb graphs. *Computational Geometry*, 42(6-7):606 – 616, 2009.

[11] H. Edelsbrunner, J. Harer, A. Mascarenhas, V. Pascucci, and J. Snoeyink. Time-varying reeb graphs for continuous space-time data. *Computational Geometry*, 41(3):149 – 166, 2008.

[12] H. Edelsbrunner, J. Harer, and A. K. Patel. Reeb spaces of piecewise linear mappings. In *Proceedings of the Twenty-fourth Annual Symposium on Computational Geometry*, SCG '08, pages 242–250, New York, NY, USA, 2008. ACM.

[13] D. Eppstein. Offline algorithms for dynamic minimum spanning tree problems. *J. Algorithms*, 17(2):237–250, Sept. 1994.

[14] I. Fujishiro, Y. Takeshima, T. Azuma, and S. Takahashi. Volume data mining using 3d field topology analysis. *IEEE Comput. Graph. Appl.*, 20:46–51, September 2000.

[15] W. Harvey, Y. Wang, and R. Wenger. A randomized $O(m \log m)$ time algorithm for computing reeb graphs of arbitrary simplicial complexes. In *Proceedings of the 2010 Annual Symposium on Computational Geometry*, SoCG '10, pages 267–276, New York, NY, USA, 2010. ACM.

[16] M. Hilaga, Y. Shinagawa, T. Kohmura, and T. L. Kunii. Topology matching for fully automatic similarity estimation of 3d shapes. In *Proceedings of the 28th Annual Conference on Computer Graphics and Interactive Techniques*, SIGGRAPH '01, pages 203–212, New York, NY, USA, 2001. ACM.

[17] J. Holm, K. de Lichtenberg, and M. Thorup. Poly-logarithmic deterministic fully-dynamic algorithms for connectivity, minimum spanning tree, 2-edge, and biconnectivity. *J. ACM*, 48:723–760, July 2001.

[18] P. Kanongchaiyos and Y. Shinagawa. Articulated reeb graphs for interactive skeleton animation. *Modeling Modeling Multimedia Information and System*, pages 451–467, 2000.

[19] M. Natali, S. Biasotti, G. Patane, and B. Falcidieno. Graph-based representations of point clouds. *Graphical Models*, 73(5):151 – 164, 2011.

[20] V. Pascucci, G. Scorzelli, P.-T. Bremer, and A. Mascarenhas. Robust on-line computation of reeb graphs: simplicity and speed. *ACM Trans. Graph.*, 26, July 2007.

[21] M. Patrascu and E. D. Demaine. Logarithmic lower bounds in the cell-probe model. *SIAM J. Comput.*, 35:932–963, April 2006.

[22] I. Rekleitis, V. Lee-Shue, and H. Choset. Limited communication, multi-robot team based coverage. *IEEE International Conference on Robotics and Automation 2004 Proceedings ICRA 04 2004*, 4(April):3462–3468, 2004.

[23] Y. Shi, R. Lai, S. Krishna, N. Sicotte, I. Dinov, and A. W. Toga. Anisotropic laplace-beltrami eigenmaps: Bridging reeb graphs and skeletons. *Computer Vision and Pattern Recognition Workshop*, 0:1–7, 2008.

[24] Y. Shinagawa and T. L. Kunii. Constructing a reeb graph automatically from cross sections. *IEEE Computer Graphics and Applications*, 11:44–51, 1991.

[25] D. D. Sleator and R. E. Tarjan. A data structure for dynamic trees. *Journal of Computer and System Sciences*, 26(3):362 – 391, 1983.

[26] D. D. Sleator and R. E. Tarjan. Self-adjusting binary search trees. *J. ACM*, 32:652–686, July 1985.

[27] R. E. Tarjan and R. F. Werneck. Self-adjusting top trees. In *Proceedings of the Sixteenth Annual ACM-SIAM Symposium on Discrete Algorithms*, SODA '05, pages 813–822, Philadelphia, PA, USA, 2005. Society for Industrial and Applied Mathematics.

[28] R. E. Tarjan and R. F. Werneck. Dynamic trees in practice. *J. Exp. Algorithmics*, 14:5:4.5–5:4.23, January 2010.

[29] M. Thorup. Near-optimal fully-dynamic graph connectivity. In *Proceedings of the Thirty-second Annual ACM Symposium on Theory of Computing*, STOC '00, pages 343–350, New York, NY, USA, 2000. ACM.

[30] J. Tierny, A. Gyulassy, E. Simon, and V. Pascucci. Loop surgery for volumetric meshes: Reeb graphs reduced to contour trees. *IEEE Transactions on Visualization and Computer Graphics*, 15:1177–1184, 2009.

[31] T. Tung and F. Schmitt. The augmented multiresolution reeb graph approach for content-based retrieval of 3d shapes. *International Journal of Shape Modeling*, 11(1):91–120, 2005.

[32] Z. Wood, H. Hoppe, M. Desbrun, and P. Schröder. Removing excess topology from isosurfaces. *ACM Trans. Graph.*, 23:190–208, April 2004.

Self-improving Algorithms for Coordinate-wise Maxima

[Extended Abstract]

Kenneth L. Clarkson
IBM Almaden Research Center
San Jose, USA
klclarks@us.ibm.com

Wolfgang Mulzer
Institut für Informatik
Freie Universität Berlin
Berlin, Germany
mulzer@inf.fu-berlin.de

C. Seshadhri
Sandia National Laboratories
Livermore, USA
scomand@sandia.gov

ABSTRACT

Computing the coordinate-wise maxima of a planar point set is a classic and well-studied problem in computational geometry. We give an algorithm for this problem in the *self-improving setting*. We have n (unknown) independent distributions $\mathcal{D}_1, \mathcal{D}_2, \ldots, \mathcal{D}_n$ of planar points. An input pointset $(p_1, p_2, \ldots, p_n)$ is generated by taking an independent sample p_i from each $\mathcal{D}_i$, so the input distribution $\mathcal{D}$ is the product $\prod_i \mathcal{D}_i$. A self-improving algorithm repeatedly gets input sets from the distribution $\mathcal{D}$ (which is *a priori* unknown) and tries to optimize its running time for $\mathcal{D}$. Our algorithm uses the first few inputs to learn salient features of the distribution, and then becomes an optimal algorithm for distribution $\mathcal{D}$. Let $\text{OPT}_\mathcal{D}$ denote the expected depth of an *optimal* linear comparison tree computing the maxima for distribution $\mathcal{D}$. Our algorithm eventually has an expected running time of $O(\text{OPT}_\mathcal{D} + n)$, even though it did not know $\mathcal{D}$ to begin with.

Our result requires new tools to understand linear comparison trees for computing maxima. We show how to convert general linear comparison trees to very restricted versions, which can then be related to the running time of our algorithm. An interesting feature of our algorithm is an interleaved search, where the algorithm tries to determine the likeliest point to be maximal with minimal computation. This allows the running time to be truly optimal for the distribution $\mathcal{D}$.

Categories and Subject Descriptors

F.2.2 [**Analysis of Algorithms and Problem Complexity**]: Nonnumerical Algorithms and Problems—*Geometrical problems and computations*

General Terms

Algorithms, Theory

Keywords

Coordinate-wise maxima; Self-improving algorithms

1. INTRODUCTION

Given a set P of n points in the plane, the *maxima* problem is to find those points $p \in P$ for which no other point in P has a larger x-coordinate and a larger y-coordinate. More formally, for $p \in \mathbb{R}^2$, let $x(p)$ and $y(p)$ denote the x and y coordinates of p. Then p' *dominates* p if and only if $x(p') \geq x(p)$, $y(p') \geq y(p)$, and one of these inequalities is strict. The desired points are those in P that are not dominated by any other points in P. The set of maxima is also known as a *skyline* in the database literature [BKS01] and as a *Pareto frontier*.

This algorithmic problem has been studied since at least 1975 [KLP75], when Kung et al. described an algorithm with an $O(n \log n)$ worst-case time and gave an $\Omega(n \log n)$ lower bound. Results since then include average-case running times of $n + \tilde{O}(n^{6/7})$ point-wise comparisons [Gol94]; output-sensitive algorithms needing $O(n \log h)$ time when there are h maxima [KS86]; and algorithms operating in external-memory models [GTVV93]. A major problem with worst-case analysis is that it may not reflect the behavior of real-world inputs. Worst-case algorithms are tailor-made for extreme inputs, none of which may occur (with reasonable frequency) in practice. Average-case analysis tries to address this problem by assuming some fixed distribution on inputs; for maxima, the property of coordinate-wise independence covers a broad range of inputs, and allows a clean analysis [Buc89], but is unrealistic even so. The right distribution to analyze remains a point of investigation. Nonetheless, the assumption of randomly distributed inputs is very natural and one worthy of further research.

The self-improving model. Ailon et al. introduced the self-improving model to address this issue [ACCL06]. In this model, there is some fixed but unknown input distribution $\mathcal{D}$ that generates independent inputs, that is, whole input sets P. The algorithm initially undergoes a *learning phase*, where it processes inputs with a worst-case guarantee but tries to learn information about $\mathcal{D}$. The aim of the algorithm is to become optimal *for the distribution* $\mathcal{D}$. After seeing some (hopefully small) number of inputs, the algorithm shifts into the *limiting phase*. Now, the algorithm is tuned for $\mathcal{D}$ and the expected running time is (ideally) optimal for $\mathcal{D}$. A self-improving algorithm can be thought of as an algorithm that attains the optimal average-case running time for all, or at least a large class of, distributions $\mathcal{D}$.

Following earlier self-improving algorithms, we assume the input has a product distribution. An input is a set of n points $P = (p_1, p_2, \ldots, p_n)$ in the plane. Each p_i is generated independently from a distribution $\mathcal{D}_i$, so the probability distribution of P is the product $\prod_i \mathcal{D}_i$. The $\mathcal{D}_i$s themselves are arbitrary, and the only assumption made is their independence. There are lower bounds showing that some restriction on $\mathcal{D}$ is necessary for a reasonable self-improving algorithm, as we explain later.

The first self-improving algorithm was for sorting; this was extended to Delaunay triangulations, with these results eventually merged [CS08, ACC+11]. A self-improving algorithm for planar convex hulls was given by Clarkson et al. [CMS10], however their analysis was recently discovered to be flawed.

1.1 Main result

Our main result is a self-improving algorithm for planar coordinate-wise maxima over product distributions. We need some basic definitions before stating our main theorem. We explain what it means for a maxima algorithm to be optimal for a distribution $\mathcal{D}$. This in turn requires a notion of *certificates* for maxima, which allow the correctness of the output to be verified in $O(n)$ time. Any procedure for computing maxima must provide some "reason" to deem an input point p non-maximal. The simplest certificate would be to provide an input point dominating p. Most current algorithms implicitly give exactly such certificates [KLP75, Gol94, KS86].

DEFINITION 1.1. *A* certificate γ *has: (i) the sequence of the indices of the maximal points, sorted from left to right; (ii) for each non-maximal point, a* per-point certificate *of non-maximality, which is simply the index of an input point that dominates it. We say that a certificate γ is* valid *for an input P if γ satisfies these conditions for P.*

The model of computation that we use to define optimality is a linear computation tree that generates query lines using the input points. In particular, our model includes the usual CCW-test that forms the basis for many geometric algorithms.

Let ℓ be a directed line. We use ℓ^+ to denote the open halfplane to the left of ℓ and ℓ^- to denote the open halfplane to the right of ℓ.

DEFINITION 1.2. *A* linear comparison tree *$\mathcal{T}$ is a binary tree such that each node v of $\mathcal{T}$ is labeled with a query of the form "$p \in \ell_v^+$?". Here p denotes an input point and ℓ_v denotes a directed line. The line ℓ_v can be obtained in three ways: (i) it can be a line independent of the input (but dependent on the node v); (ii) it can be a line with a slope independent of the input (but dependent on v) passing through a given input point; (iii) it can be a line through an input point and through a point q independent of the input (but dependent on v); (iv) it can be the line defined by two distinct input points. A linear comparison tree is* restricted *if it only makes queries of type (i).*

A linear comparison tree $\mathcal{T}$ computes the maxima *for P if each leaf corresponds to a certificate. This means that each leaf v of $\mathcal{T}$ is labeled with a certificate γ that is valid for every possible input P that reaches v.*

Let $\mathcal{T}$ be a linear comparison tree and v be a node of $\mathcal{T}$. Note that v corresponds to a region $\mathcal{R}_v \subseteq \mathbb{R}^{2n}$ such that an evaluation of $\mathcal{T}$ on input P reaches v if and only if $P \in \mathcal{R}_v$. If $\mathcal{T}$ is *restricted*, then $\mathcal{R}_v$ is the Cartesian product of a sequence $(R_1, R_2, \ldots, R_n)$ of polygonal regions. The *depth* of v, denoted by d_v, is the length of the path from the root of $\mathcal{T}$ to v. Given $\mathcal{T}$, there exists exactly one leaf $v(P)$ that is reached by the evaluation of $\mathcal{T}$ on input P. The *expected depth* of $\mathcal{T}$ over $\mathcal{D}$, $d_{\mathcal{D}}(\mathcal{T})$, is defined as $\mathbf{E}_{P\sim\mathcal{D}}[d_{v(P)}]$. Consider some comparison based algorithm A that is modeled by such a tree $\mathcal{T}$. The expected depth of $\mathcal{T}$ is a lower bound on the number of comparisons performed by A.

Let $\mathbf{T}$ be the set of trees that compute the maxima of n points. We define $\mathrm{OPT}_{\mathcal{D}} = \inf_{\mathcal{T}\in\mathbf{T}} d_{\mathcal{D}}(\mathcal{T})$. This is a lower bound on the expected time taken by *any* linear comparison tree to compute the maxima of inputs distributed according to $\mathcal{D}$. We would like our algorithm to have a running time comparable to $\mathrm{OPT}_{\mathcal{D}}$.

THEOREM 1.3. *Let $\varepsilon > 0$ be a fixed constant and $\mathcal{D}_1$, $\mathcal{D}_2$, $\ldots, \mathcal{D}_n$ be independent planar point distributions. The input distribution is $\mathcal{D} = \prod_i \mathcal{D}_i$. There is a self-improving algorithm to compute the coordinate-wise maxima whose expected time in the limiting phase is $O(\varepsilon^{-1}(n + OPT_{\mathcal{D}}))$. The learning phase lasts for $O(n^{\varepsilon})$ inputs and the space requirement is $O(n^{1+\varepsilon})$.*

There are lower bounds in [ACC+11] (for sorters) implying that a self-improving maxima algorithm that works for all distributions requires exponential storage, and that the time-space tradeoff (wrt ε) in the above theorem is optimal.

Challenges. One might think that since self-improving sorters are known, an algorithm for maxima should follow directly. But this reduction is only valid for $O(n \log n)$ algorithms. Consider Figure 1(i). The distributions $\mathcal{D}_1$, $\mathcal{D}_2$, $\ldots, \mathcal{D}_{n/2}$ generate the fixed points shown. The remaining distributions generate a random point from a line below L. Observe that an algorithm that wishes to sort the x-coordinates requires $\Omega(n \log n)$ time. On the other hand, there is a simple comparison tree that determines the maxima in $O(n)$ time. For all p_j where $j > n/2$, the tree simply checks if $p_{n/2}$ dominates p_j. After that, it performs a linear scan and outputs a certificate.

We stress that even though the points are independent, the collection of maxima exhibits strong dependencies. In Figure 1(ii), suppose a distribution $\mathcal{D}_i$ generates either p_h or p_ℓ; if p_ℓ is chosen, we must consider the dominance relations among the remaining points, while if p_h is chosen, no such evaluation is required. The optimal search tree for a distribution $\mathcal{D}$ must exploit this complex dependency.

Indeed, arguing about optimality is one of the key contributions of this work. Previous self-improving algorithms employed information-theoretic optimality arguments. These are extremely difficult to analyze for settings like maxima, where some points are more important to process that others, as in Figure 1. (The main error in the self-improving convex hull paper [CMS10] was an incorrect consideration of dependencies.) We focus on a somewhat weaker notion of optimality—linear comparison trees—that nonetheless covers most (if not all) important algorithms for maxima.

In Section 3, we describe how to convert linear comparison trees into restricted forms that use much more structured (and simpler) queries. Restricted trees are much more amenable to analysis. In some sense, a restricted tree decouples the individual input points and makes the maxima com-

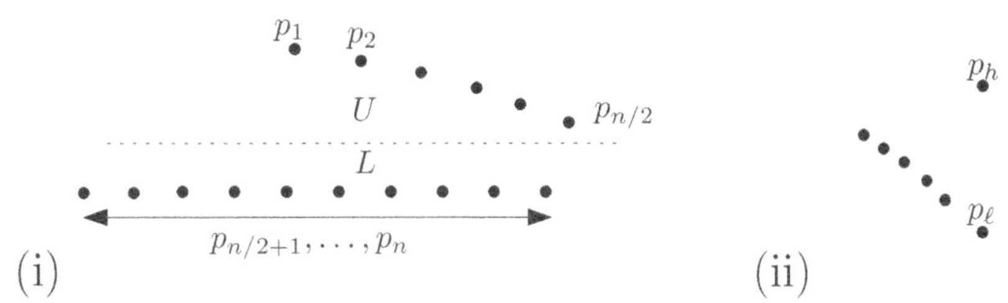

Figure 1: Examples of difficult distributions

putation amenable to separate $\mathcal{D}_i$-optimal searches. A leaf of a restricted tree is associated with a sequence of polygons $(R_1, R_2, \ldots, R_n)$ such that the leaf is visited if and only if every $p_i \in R_i$, and conditioned on that event, the p_i remain independent. This independence is extremely important for the analysis. We design an algorithm whose behavior can be related to the restricted tree. Intuitively, if the algorithm spends many comparisons involving a single point, then we can argue that the optimal restricted tree must also do the same. We give more details about the algorithm in Section 2.

1.2 Previous work

Afshani et al. [ABC09] introduced a model of *instance-optimality* applying to algorithmic problems including planar convex hulls and maxima. (However, their model is different from, and in a sense weaker than, the prior notion of instance-optimality introduced by Fagin et al. [FLN01].) All previous (e.g., output sensitive and instance optimal) algorithms require expected $\Omega(n \log n)$ time for the distribution given in Figure 1, though an optimal self-improving algorithm only requires $O(n)$ expected time. (This was also discussed in [CMS10] with a similar example.),

We also mention the paradigm of preprocessing regions in order to compute certain geometric structures faster (see, e.g., [BLMM11, EM11, HM08, LS10, vKLM10]). Here, we are given a set $\mathcal{R}$ of planar regions, and we would like to preprocess $\mathcal{R}$ in order to quickly find the (Delaunay) triangulation (or convex hull) for any point set which contains exactly one point from each region in $\mathcal{R}$. This setting is adversarial, but if we only consider point sets where a point is randomly drawn from each region, it can be regarded as a special case of our setting. In this view, these results give us bounds on the running time a self-improving algorithm can achieve if $\mathcal{D}$ draws its points from disjoint planar regions.

1.3 Preliminaries and notation

Before we begin, let us define some basic concepts and agree on a few notational conventions. We use c for a sufficiently large constant, and we write $\log x$ to denote the logarithm of x in base 2. All the probability distributions are assumed to be continuous. (It is not necessary to do this, but it makes many calculations a lot simpler.)

Given a polygonal region $R \subseteq \mathbb{R}^2$ and a probability distribution $\mathcal{D}$ on the plane, we call ℓ a *halving line* for R (with respect to $\mathcal{D}$) if

$$\Pr_{p \sim \mathcal{D}}[p \in \ell^+ \cap R] = \Pr_{p \sim \mathcal{D}}[p \in \ell^- \cap R].$$

Note that if $\Pr_{p \sim \mathcal{D}}[p \in R] = 0$, every line is a halving line for R. If not, a halving line exactly halves the conditional probability for p being in each of the corresponding halfplanes, conditioned on p lying inside R.

Define a *vertical slab structure* $\mathbf{S}$ as a sequence of vertical lines partitioning the plane into vertical regions, called *leaf slabs*. (We will consider the latter to be the open regions between the vertical lines. Since we assume that our distributions are continuous, we abuse notation and consider the leaf slabs to partition the plane.) More generally, a *slab* is the region between any two vertical lines of the $\mathbf{S}$. The *size* of the slab structure is the number of leaf slabs it contains. We denote it by $|\mathbf{S}|$. Furthermore, for any slab S, the probability that $p_i \sim \mathcal{D}_i$ is in S is denoted by $q(i, S)$.

A *search tree* T over $\mathbf{S}$ is a comparison tree that locates a point within leaf slabs of $\mathbf{S}$. Each internal node compares the x-coordinate of the point with a vertical line of $\mathbf{S}$, and moves left or right accordingly. We associate each internal node v with a slab S_v (any point in S_v will encounter v along its search).

1.4 Tools from self-improving algorithms

We introduce some tools that were developed in previous self-improving results. The ideas are by and large old, but our presentation in this form is new. We feel that the following statements (especially Lemma 1.6) are of independent interest.

We define the notion of *restricted searches*, introduced in [CMS10]. This notion is central to our final optimality proof. (The lemma and formulation as given here are new.) Let $\mathbf{U}$ be an ordered set and $\mathcal{F}$ be a distribution over $\mathbf{U}$. For any element $j \in \mathbf{U}$, q_j is the probability of j according to $\mathcal{F}$. For any interval S of $\mathbf{U}$, the total probability of S is q_S.

We let T denote a search tree over $\mathbf{U}$. It will be convenient to think of T as (at most) ternary, where each node has at most 2 children that are internal nodes. In our application of the lemma, $\mathbf{U}$ will just be the set of leaf slabs of a slab structure $\mathbf{S}$. We now introduce some definitions regarding restricted searches and search trees.

Definition 1.4. *Consider a distribution $\mathcal{F}$ and an interval S of U. An* S-restricted distribution *is given by the probabilities (for element $r \in U$) $q'_r / \sum_{j \in U} q'_j$, where the sequence $\{q'_j | j \in U\}$ has the following property. For each $j \in S$, $0 \le q'_j \le q_j$. For every other j, $q'_j = 0$.*

Suppose $j \in S$. An S-restricted search *is a search for j in T that terminates once j is located in any interval contained in S.*

For any sequence of numbers $\{q'_j | j \in U\}$ and $S \subseteq U$, we use q'_S to denote $\sum_{j \in S} q'_j$.

DEFINITION 1.5. *Let $\mu \in (0,1)$ be a parameter. A search tree T over $\mathbf{U}$ is μ-reducing if: for any internal node S and for any non-leaf child S' of S, $q_{S'} \leq \mu q_S$.*

A search tree T is c-optimal for restricted searches over $\mathcal{F}$ if: for all S and S-restricted distributions $\mathcal{F}_S$, the expected time of an S-restricted search over $\mathcal{F}_S$ is at most $c(-\log q'_S + 1)$. (The probabilities q' are as given in Definition 1.4.)

We give the main lemma about restricted searches. A tree that is optimal for searches over $\mathcal{F}$ also works for restricted distributions. The proof is given in the full version of the paper.

LEMMA 1.6. *Suppose T is a μ-reducing search tree for $\mathcal{F}$. Then T is $O(1/\log(1/\mu))$-optimal for restricted searches over $\mathcal{F}$.*

We list theorems about data structures that are built in the learning phase. Similar structures were first constructed in [ACC+11], and the following can be proved using their ideas. The data structures involve construction of slab structures and specialized search trees for each distribution $\mathcal{D}_i$. It is also important that these trees can be represented in small space, to satisfy the requirements of Theorem 1.3. The following lemmas give us the details of the data structures required. Because this is not a major contribution of this paper, we relegate the details to §5.

LEMMA 1.7. *We can construct a slab structure $\mathbf{S}$ with $O(n)$ leaf slabs such that, with probability $1 - n^{-3}$ over the construction of $\mathbf{S}$, the following holds. For a leaf slab λ of $\mathbf{S}$, let X_λ denote the number of points in a random input P that fall into λ. For every leaf slab λ of $\mathbf{S}$, we have $\mathbf{E}[X_\lambda^2] = O(1)$. The construction takes $O(\log n)$ rounds and $O(n \log^2 n)$ time.*

LEMMA 1.8. *Let $\varepsilon > 0$ be a fixed parameter. In $O(n^\varepsilon)$ rounds and $O(n^{1+\varepsilon})$ time, we can construct search trees T_1, T_2, ..., T_n over $\mathbf{S}$ such that the following holds. (i) the trees can be represented in $O(n^{1+\varepsilon})$ total space; (ii) with probability $1 - n^{-3}$ over the construction of the T_is, every T_i is $O(1/\varepsilon)$-optimal for restricted searches over $\mathcal{D}_i$.*

2. OUTLINE

We start by providing a very informal overview of the algorithm. Then, we shall explain how the optimality is shown.

If the points of P are sorted by x-coordinate, the maxima of P can be found easily by a right-to-left sweep over P: we maintain the largest y-coordinate Y of the points traversed so far; when a point p is visited in the traversal, if $y(p) < Y$, then p is non-maximal, and the point p_j with $Y = y(p_j)$ gives a per-point certificate for p's non-maximality. If $y(p) \geq Y$, then p is maximal, and can be put at the beginning of the certificate list of maxima of P.

This suggests the following approach to a self-improving algorithm for maxima: sort P with a self-improving sorter and then use the traversal. The self-improving sorter of [ACC+11] works by locating each point of P within the slab structure $\mathbf{S}$ of Lemma 1.7 using the trees T_i of Lemma 1.8.

While this approach does use $\mathbf{S}$ and the T_i's, it is not optimal for maxima, because the time spent finding the exact sorted order of non-maximal points may be wasted: in some sense, we are learning much more information about the input P than necessary. To deduce the list of maxima, we do not need the sorted order of *all* points of P: it suffices to know the sorted order of just the maxima! An optimal algorithm would probably locate the maximal points in $\mathbf{S}$ and would not bother locating "extremely non-maximal" points. This is, in some sense, the difficulty that output-sensitive algorithms face.

As a thought experiment, let us suppose that the maximal points of P are known to us, but not in sorted order. We search only for these in $\mathbf{S}$ and determine the sorted list of maximal points. We can argue that the optimal algorithm must also (in essence) perform such a search. We also need to find per-point certificates for the non-maximal points. We use the slab structure $\mathbf{S}$ and the search trees, but now we shall be very conservative in our searches. Consider the search for a point p_i. At any intermediate stage of the search, p_i is placed in a slab S. This rough knowledge of p_i's location may already suffice to certify its non-maximality: let m denote the leftmost maximal point to the right of S (since the sorted list of maxima is known, this information can be easily deduced). We check if m dominates p_i. If so, we have a per-point certificate for p_i and we promptly terminate the search for p_i. Otherwise, we continue the search by a single step and repeat. We expect that many searches will not proceed too long, achieving a better position to compete with the optimal algorithm.

Non-maximal points that are dominated by many maximal points will usually have a very short search. Points that are "nearly" maximal will require a much longer search. So this approach should derive just the "right" amount of information to determine the maxima output. But wait! Didn't we assume that the maximal points were known? Wasn't this crucial in cutting down the search time? This is too much of an assumption, and because the maxima are highly dependent on each other, it is not clear how to determine which points are maximal before performing searches.

The final algorithm overcomes this difficulty by interleaving the searches for sorting the points with confirmation of the maximality of some points, in a rough right-to-left order that is a more elaborate version of the traversal scheme given above for sorted points. The searches for all points p_i (in their respective trees T_i) are performed "together", and their order is carefully chosen. At any intermediate stage, each point p_i is located in some slab S_i, represented by some node of its search tree. We choose a specific point and advance its search by one step. This order is very important, and is the basis of our optimality. The algorithm is described in detail and analyzed in §4.

Arguing about optimality. A major challenge of self-improving algorithms is the strong requirement of optimality for the distribution $\mathcal{D}$. We focus on the model of linear comparison trees, and let $\mathcal{T}$ be an optimal tree for distribution $\mathcal{D}$. (There may be distributions where such an exact $\mathcal{T}$ does not exist, but we can always find one that is near optimal.) One of our key insights is that when $\mathcal{D}$ is a product distribution, then we can convert $\mathcal{T}$ to $\mathcal{T}'$, a restricted comparison tree whose expected depth is only a constant factor worse. In other words, there exists a near optimal restricted comparison tree that computes the maxima.

In such a tree, a leaf is labeled with a sequence of regions $\mathcal{R} = (R_1, R_2, \ldots, R_n)$. Any input $P = (p_1, p_2, \ldots, p_n)$ such that $p_i \in R_i$ for all i, will lead to this leaf. Since the distributions are independent, we can argue that the probability that an input leads to this leaf is $\prod_i \Pr_{p_i \sim \mathcal{D}_i}[p_i \in R_i]$. Furthermore, the depth of this leaf can be shown to be $-\sum_i \log \Pr[p_i \in R_i]$. This gives us a concrete bound that we can exploit.

It now remains to show that if we start with a random input from $\mathcal{R}$, the expected running time is bounded by the sum given above. We will argue that for such an input, as soon as the search for p_i locates it inside R_i, the search will terminate. This leads to the optimal running time.

3. THE COMPUTATIONAL MODEL AND LOWER BOUNDS

3.1 Reducing to restricted comparison trees

We prove that when P is generated probabilistically, it suffices to focus on restricted comparison trees. To show this, we provide a sequence of transformations, starting from the more general comparison tree, that results in a restricted linear comparison tree of comparable expected depth. The main lemma of this section is the following.

LEMMA 3.1. *Let $\mathcal{T}$ a finite linear comparison tree and $\mathcal{D}$ be a product distribution over points. Then there exists a restricted comparison tree $\mathcal{T}'$ with expected depth $d_{\mathcal{D}}(\mathcal{T}') = O(d_{\mathcal{D}}(\mathcal{T}))$, as $d_{\mathcal{D}}(\mathcal{T}) \to \infty$.*

We will describe a transformation from $\mathcal{T}$ into a restricted comparison tree with similar depth. The first step is to show how to represent a single comparison by a restricted linear comparison tree, provided that P is drawn from a product distribution. The final transformation basically replaces each node of $\mathcal{T}$ by the subtree given by the next claim. For convenience, we will drop the subscript of $\mathcal{D}$ from $d_{\mathcal{D}}$, since we only focus on a fixed distribution.

CLAIM 3.2. *Consider a comparison C as described in Definition 1.2, where the comparisons are listed in increasing order of simplicity. Let $\mathcal{D}'$ be a product distribution for P such that each p_i is drawn from a polygonal region R_i. Then either C is the simplest, type (i) comparison, or there exists a restricted linear comparison tree $\mathcal{T}_C'$ that resolves the comparison C such that the expected depth of $\mathcal{T}_C'$ (over the distribution $\mathcal{D}'$) is $O(1)$, and all comparisons used in $\mathcal{T}_C'$ are simpler than C.*

PROOF. v **is of type (ii).** This means that v needs to determine whether an input point p_i lies to the left of the directed line ℓ through another input point p_j with a fixed slope a. We replace this comparison with a binary search. Let R_j be the region in $\mathcal{D}'$ corresponding to p_j. Take a halving line ℓ_1 for R_j with slope a. Then perform two comparisons to determine on which side of ℓ_1 the inputs p_i and p_j lie. If p_i and p_j lie on different sides of ℓ_1, we declare success and resolve the original comparison accordingly. Otherwise, we replace R_j with the appropriate new region and repeat the process until we can declare success. Note that in each attempt the success probability is at least $1/4$. The resulting restricted tree $\mathcal{T}_C'$ can be infinite. Nonetheless, the probability that an evaluation of $\mathcal{T}_C'$ leads to a node of depth k is at most $2^{-\Omega(k)}$, so the expected depth is $O(1)$.

v **is of type (iii).** Here the node v needs to determine whether an input point p_i lies to the left of the directed line ℓ through another input point p_j and a fixed point q.

We partition the plane by a constant-sized family of cones, each with apex q, such that for each cone V in the family, the probability that line $\overline{qp_j}$ meets V (other than at q) is at most $1/2$. Such a family could be constructed by a sweeping a line around q, or by taking a sufficiently large, but constant-sized, sample from the distribution of p_j, and bounding the cones by all lines through q and each point of the sample. Such a construction has a non-zero probability of success, and therefore the described family of cones exists.

We build a restricted tree that locates a point in the corresponding cone. For each cone V, we can recursively build such a family of cones (inside V), and build a tree for this structure as well. Repeating for each cone, this leads to an infinite restricted tree $\mathcal{T}_C'$. We search for both p_i and p_j in $\mathcal{T}_C'$. When we locate p_i and p_j in two different cones of the same family, then comparison between p_i and $\overline{qp_j}$ is resolved and the search terminates. The probability that they lie in the same cones of a given family is at most $1/2$, so the probability that the evaluation leads to k steps is at most $2^{-\Omega(k)}$.

v **is of type (iv).** Here the node v needs to determine whether an input point p_i lies to the left of the directed line ℓ through input points p_j and p_k.

We partition the plane by a constant-sized family of triangles and cones, such that for each region V in the family, the probability that the line through p_j and p_k meets V is at most $1/2$. Such a family could be constructed by taking a sufficiently large random sample of pairs p_j and p_k and triangulating the arrangement of the lines through each pair. Such a construction has a non-zero probability of success, and therefore such a family exists. (Other than the source of the random lines used in the construction, this scheme goes back at least to [Cla87]; a tighter version, called a *cutting*, could also be used [Cha93].)

When computing C, suppose p_i is in region V of the family. If the line $\overline{p_jp_k}$ does not meet V, then the comparison outcome is known immediately. This occurs with probability at least $1/2$. Moreover, determining the region containing p_i can be done with a constant number of comparisons of type (i), and determining if $\overline{p_jp_k}$ meets V can be done with a constant number of comparisons of type (iii); for the latter, suppose V is a triangle. If $p_j \in V$, then $\overline{p_jp_k}$ meets V. Otherwise, suppose p_k is above all the lines through p_j and each vertex of V; then $\overline{p_jp_k}$ does not meet V. Also, if p_k is below all the lines through p_j and each vertex, then $\overline{p_jp_k}$ does not meet V. Otherwise, $\overline{p_jp_k}$ meets V. So a constant number of type (i) and type (iii) queries suffice.

By recursively building a tree for each region V of the family, comparisons of type (iv) can be done via a tree whose nodes use comparisons of type (i) and (iii) only. Since the probability of resolving the comparison is at least $1/2$ with each family of regions that is visited, the expected number of nodes visited is constant. □

PROOF OF LEMMA 3.1. We transform $\mathcal{T}$ into a tree $\mathcal{T}'$ that has no comparisons of type (iv), by using the construction of Claim 3.2 where nodes of type (iv) are replaced by a tree. We then transform $\mathcal{T}'$ into a tree $\mathcal{T}''$ that has no comparisons of type (iii) or (iv), and finally transform $\mathcal{T}'''$ into a restricted tree. Each such transformation is done in the same general way, using one case of Claim 3.2, so we focus on the first one.

We incrementally transform $\mathcal{T}$ into the tree $\mathcal{T}'$. In each such step, we have a partial restricted comparison tree $\mathcal{T}''$ that will eventually become $\mathcal{T}'$. Furthermore, during the process each node of $\mathcal{T}$ is in one of three different states. It is either *finished*, *fringe*, or *untouched*. Finally, we have a function S that assigns to each finished and to each fringe node of $\mathcal{T}$ a subset $S(v)$ of nodes in $\mathcal{T}''$.

The initial situation is as follows: all nodes of $\mathcal{T}$ are untouched except for the root which is fringe. Furthermore, the partial tree $\mathcal{T}''$ consists of a single root node r and the function S assigns the root of $\mathcal{T}$ to the set $\{r\}$.

Now our transformation proceeds as follows. We pick a fringe node v in $\mathcal{T}$, and mark v as finished. For each child v' of v, if v' is an internal node of $\mathcal{T}$, we mark it as fringe. Otherwise, we mark v' as finished. Next, we apply Claim 3.2 to each node $w \in S(v)$. Note that this is a valid application of the claim, since w is a node of $\mathcal{T}''$, a restricted tree. Hence $\mathcal{R}_w$ is a product set, and the distribution $\mathcal{D}$ restricted to $\mathcal{R}_w$ is a product distribution. Hence, replace each node $w \in S(v)$ in $\mathcal{T}''$ by the subtree given by Claim 3.2. Now $S(v)$ contains the roots of these subtrees. Each leaf of each such subtree corresponds to an outcome of the comparison in v. (Potentially, the subtrees are countably infinite, but the expected number of steps to reach a leaf is constant.) For each child v' of v, we define $S(v')$ as the set of all such leaves that correspond to the same outcome of the comparison as v'. We continue this process until there are no fringe nodes left. By construction, the resulting tree $\mathcal{T}'$ is restricted.

It remains to argue that $d_{\mathcal{T}'} = O(d_{\mathcal{T}})$. Let v be a node of $\mathcal{T}$. We define two random variables X_v and Y_v. The variable X_v is the indicator random variable for the event that the node v is traversed for a random input $P \sim \mathcal{D}$. The variable Y_v denotes the number of nodes traversed in $\mathcal{T}'$ that correspond to v (i.e., the number of nodes needed to simulate the comparison at v, if it occurs). We have $d_{\mathcal{T}} = \sum_{v\in\mathcal{T}} \mathbf{E}[X_v]$, because if the leaf corresponding to an input $P \sim \mathcal{D}$ has depth d, exactly d nodes are traversed to reach it. We also have $d_{\mathcal{T}'} = \sum_{v\in\mathcal{T}} \mathbf{E}[Y_v]$, since each node in $\mathcal{T}'$ corresponds to exactly one node v in $\mathcal{T}$. Claim 3.3 below shows that $\mathbf{E}[Y_v] = O(\mathbf{E}[X_v])$, which completes the proof. □

CLAIM 3.3. $\mathbf{E}[Y_v] \le c\mathbf{E}[X_v]$

PROOF. Note that $\mathbf{E}[X_v] = \Pr[X_v = 1] = \Pr[P \in \mathcal{R}_v]$. Since the sets $\mathcal{R}_w$, $w \in S(v)$, partition $\mathcal{R}_v$, we can write $\mathbf{E}[Y_v]$ as

$$\mathbf{E}[Y_v \mid X_v = 0]\Pr[X_v = 0] + \sum_{w\in S(v)} \mathbf{E}[Y_v \mid P \in \mathcal{R}_w]\Pr[P \in \mathcal{R}_w].$$

Since $Y_v = 0$ if $P \notin \mathcal{R}_v$, we have $\mathbf{E}[Y_v \mid X_v = 0] = 0$ and also $\Pr[P \in \mathcal{R}_v] = \sum_{w\in S(v)} \Pr[P \in \mathcal{R}_w]$. Furthermore, by Claim 3.2, we have $\mathbf{E}[Y_v \mid P \in \mathcal{R}_w] \le c$. The claim follows. □

3.2 Entropy-sensitive comparison trees

Since every linear comparison tree can be made restricted, we can incorporate the entropy of $\mathcal{D}$ into the lower bound. For this we define entropy-sensitive trees, which are useful because the depth of a node v is related to the probability of the corresponding region $\mathcal{R}_v$.

DEFINITION 3.4. *We call a restricted linear comparison tree* entropy-sensitive *if each comparison "$p_i \in \ell^+$?" is such that ℓ is a halving line for the current region R_i.*

LEMMA 3.5. *Let v be a node in an entropy-sensitive comparison tree, and let $\mathcal{R}_v = R_1 \times R_2 \times \cdots \times R_n$. Then $d_v = -\sum_{i=1}^{n} \log \Pr[R_i]$.*

PROOF. We use induction on the depth of v. For the root r we have $d_r = 0$. Now, let v' be the parent of v. Since $\mathcal{T}$ is entropy-sensitive, we reach v after performing a comparison with a halving line in v'. This halves the measure of exactly one region in $\mathcal{R}_v$, so the sum increases by one. □

As in Lemma 3.1, we can make every restricted linear comparison tree entropy-sensitive without affecting its expected depth too much.

LEMMA 3.6. *Let $\mathcal{T}$ a restricted linear comparison tree. Then there exists an entropy-sensitive comparison tree $\mathcal{T}'$ with expected depth $d_{\mathcal{T}'} = O(d_{\mathcal{T}})$.*

PROOF. The proof extends the proof of Lemma 3.1, via an extension to Claim 3.2. We can regard a comparison against a fixed halving line as simpler than an comparison against an arbitrary fixed line. Our extension of Claim 3.2 is the claim that any type (i) node can be replaced by a tree with constant expected depth, as follows. A comparison $p_i \in \ell^+$ can be replaced by a sequence of comparisons to halving lines. Similar to the reduction for type (ii) comparisons in Claim 3.2, this is done by binary search. That is, let ℓ_1 be a halving line for R_i parallel to ℓ. We compare p_i with ℓ. If this resolves the original comparison, we declare success. Otherwise, we repeat the process with the halving line for the new region R_i'. In each step, the probability of success is at least $1/2$. The resulting comparison tree has constant expected depth; we now apply the construction of Lemma 3.1 to argue that for a restricted tree $\mathcal{T}$ there is an entropy-sensitive version $\mathcal{T}'$ whose expected depth is larger by at most a constant factor. □

Recall that $\text{OPT}_{\mathcal{D}}$ is the expected depth of an optimal linear comparison tree that computes the maxima for $P \sim \mathcal{D}$. We now describe how to characterize $\text{OPT}_{\mathcal{D}}$ in terms of entropy-sensitive comparison trees. We first state a simple property that follows directly from the definition of certificates and the properties of restricted comparison trees.

PROPOSITION 3.7. *Consider a leaf v of a restricted linear comparison tree $\mathcal{T}$ computing the maxima. Let R_i be the region associated with non-maximal point $p_i \in P$ in $\mathcal{R}_v$. There exists some region R_j associated with an extremal point p_j such that every point in R_j dominates every point in R_i.*

We now enhance the notion of a certificate (Definition 1.1) to make it more useful for our algorithm's analysis. For technical reasons, we want points to be "well-separated" according to the slab structure $\mathbf{S}$. By Prop. 3.7, every non-maximal point is associated with a dominating region.

DEFINITION 3.8. *Let $\mathbf{S}$ be a slab structure. A certificate for an input P is called* $\mathbf{S}$-labeled *if the following holds. Every maximal point is labeled with the leaf slab of $\mathbf{S}$ containing it. Every non-maximal point is either placed in the containing leaf slab, or is separated from a dominating region by a slab boundary.*

We naturally extend this to trees that compute the **S**-labeled maxima.

Definition 3.9. *A linear comparison tree $\mathcal{T}$ computes the **S**-labeled maxima of P if each leaf v of $\mathcal{T}$ is labeled with a **S**-labeled certificate that is valid for every possible input $P \in \mathcal{R}_v$.*

Lemma 3.10. *There exists an entropy-sensitive comparison tree $\mathcal{T}$ computing the **S**-labeled maxima whose expected depth over $\mathcal{D}$ is $O(n + \mathrm{OPT}_{\mathcal{D}})$.*

Proof. Start with an optimal linear comparison tree $\mathcal{T}'$ that computes the maxima. At every leaf, we have a list M with the maximal points in sorted order. We merge M with the list of slab boundaries of **S** to label each maximal point with the leaf slab of **S** containing it. We now deal with the non-maximal points. Let R_i be the region associated with a non-maximal point p_i, and R_j be the dominating region. Let λ be the leaf slab containing R_j. Note that the x-projection of R_i cannot extended to the right of λ. If there is no slab boundary separating R_i from R_j, then R_i must intersect λ. With one more comparison, we can place p_i inside λ or strictly to the left of it. All in all, with $O(n)$ more comparisons than $\mathcal{T}'$, we have a tree $\mathcal{T}''$ that computes the **S**-labeled maxima. Hence, the expected depth is $\mathrm{OPT}_{\mathcal{D}} + O(n)$. Now we apply Lemmas 3.1 and 3.6 to $\mathcal{T}''$ to get an entropy-sensitive comparison tree $\mathcal{T}$ computing the **S**-labeled maxima with expected depth $O(n + \mathrm{OPT}_{\mathcal{D}})$. □

4. THE ALGORITHM

In the learning phase, the algorithm constructs a slab structure **S** and search trees T_i, as given in Lemmas 1.7 and 1.8. Henceforth, we assume that we have these data structures, and will describe the algorithm in the limiting (or stationary) phase. Our algorithm proceeds by searching progressively each point p_i in its tree T_i. However, we need to choose the order of the searches carefully.

At any stage of the algorithm, each point p_i is placed in some slab S_i. The algorithm maintains a set A of *active points*. An inactive point is either proven to be non-maximal, or it has been placed in a leaf slab. The active points are stored in a data structure $L(A)$. This structure is similar to a heap and supports the operations *delete*, *decrease-key*, and *find-max*. The key associated with an active point p_i is the right boundary of the slab S_i (represented as an element of $[|\mathbf{S}|]$).

We list the variables that the algorithm maintains. The algorithm is initialized with $A = P$, and each S_i is the largest slab in **S**. Hence, all points have key $|\mathbf{S}|$, and we insert all these keys into $L(A)$.

- $A, L(A)$: the list A of active points stored in data structure $L(A)$.
- $\widehat{\lambda}, B$: Let m be the largest key among the active points. Then $\widehat{\lambda}$ is the leaf slab whose right boundary is m and B is a set of points located in $\widehat{\lambda}$. Initially B is empty and m is $|S|$, corresponding to the $+\infty$ boundary of the rightmost, infinite, slab.
- $M, \hat{p}$: M is a sorted (partial) list of currently discovered maximal points and $\hat{p}$ is the leftmost among those. Initially M is empty and $\hat{p}$ is a "null" point that dominates no input point.

The algorithm involves a main procedure **Search**, and an auxiliary procedure **Update**. The procedure **Search** chooses a point and proceeds its search by a single step in the appropriate tree. Occasionally, it will invoke **Update** to change the global variables. The algorithm repeatedly calls **Search** until $L(A)$ is empty. After that, we perform a final call to **Update** in order to process any points that might still remain in B.

Search. Let p_i be obtained by performing a *find-max* in $L(A)$. If the maximum key m in $L(A)$ is less than the right boundary of $\widehat{\lambda}$, we invoke **Update**. If p_i is dominated by $\hat{p}$, we delete p_i from $L(A)$. If not, we advance the search of p_i in T_i by a single step, if possible. This updates the slab S_i. If the right boundary of S_i has decreased, we perform the appropriate *decrease-key* operation on $L(A)$. (Otherwise, we do nothing.)

Suppose the point p_i reaches a leaf slab λ. If $\lambda = \widehat{\lambda}$, we remove p_i from $L(A)$ and insert it in B (in time $O(|B|)$). Otherwise, we leave p_i in $L(A)$.

Update. We sort all the points in B and update the list of current maxima. As Claim 4.1 will show, we have the sorted list of maxima to the right of $\widehat{\lambda}$. Hence, we can append to this list in $O(|B|)$ time. We reset $B = \emptyset$, set $\widehat{\lambda}$ to the leaf slab to the left of m, and return.

We prove some preliminary claims. We state an important invariant maintained by the algorithm, and then give a construction for the data structure $L(A)$.

Claim 4.1. *At any time in the algorithm, the maxima of all points to the right of $\widehat{\lambda}$ have been determined in sorted order.*

Proof. The proof is by backward induction on m, the right boundary of $\widehat{\lambda}$. When $m = |S|$, then this is trivially true. Let us assume it is true for a given value of m, and trace the algorithm's behavior until the maximum key becomes smaller than m (which is done in **Update**). When **Search** processes a point p with a key of m then either (i) the key value decreases; (ii) p is dominated by $\hat{p}$; or (iii) p is eventually placed in $\widehat{\lambda}$ (whose right boundary is m). In all cases, when the maximum key decreases below m, all points in $\widehat{\lambda}$ are either proven to be non-maximal or are in B. By the induction hypothesis, we already have a sorted list of maxima to the right of m. The procedure **Update** will sort the points in B and all maximal points to the right of $m - 1$ will be determined. □

Claim 4.2. *Suppose there are x* find-max *operations and y* decrease-key *operations. We can implement the data structure $L(A)$ such that the total time for the operations is $O(n + x + y)$. The storage requirement is $O(n)$.*

Proof. We represent $L(A)$ as an array of lists. For every $k \in [|\mathbf{S}|]$, we keep a list of points whose key values are k. We maintain m, the current maximum key. The total storage is $O(n)$. A *find-max* can trivially be done in $O(1)$ time, and an *insert* is done by adding the element to the appropriate list. A *delete* is done by deleting the element from the list (supposing appropriate pointers are available). We now have to update the maximum. If the list at m is non-empty, no action is required. If it is empty, we check sequentially whether the list at $m - 1, m - 2, \ldots$ is empty. This will eventually lead to the maximum. To do a *decrease-key*, we *delete*, *insert*, and then update the maximum.

Note that since all key updates are *decrease-keys*, the maximum can only decrease. Hence, the total overhead for scanning for a new maximum is $O(n)$. $\square$

4.1 Running time analysis

The aim of this section is to prove the following lemma.

LEMMA 4.3. *The algorithm runs in* $O(n + \mathrm{OPT}_{\mathcal{D}})$ *time.*

We can easily bound the running time of all calls to **Update**.

CLAIM 4.4. *The expected time for all calls to* **Update** *is* $O(n)$.

PROOF. The total time taken for all calls to **Update** is at most the time taken to sort points within leaf slabs. By Lemma 1.7, this takes expected time

$$\mathbf{E}\Big[\sum_{\lambda \in \mathbf{S}} X_\lambda^2\Big] = \sum_{\lambda \in \mathbf{S}} \mathbf{E}[X_\lambda^2] = \sum_{\lambda \in \mathbf{S}} O(1) = O(n).$$

$\square$

The important claim is the following, since it allows us to relate the time spent by **Search** to the entropy-sensitive comparison trees. Lemma 4.3 follows directly from this.

CLAIM 4.5. *Let* $\mathcal{T}$ *be an entropy-sensitive comparison tree computing* **S***-labeled maxima. Consider a leaf* v *labeled with the regions* $\mathcal{R}_v = (R_1, R_2, \ldots, R_n)$*, and let* d_v *denote the depth of* v*. Conditioned on* $P \in \mathcal{R}_v$*, the expected running time of* **Search** *is* $O(n + d_v)$.

PROOF. For each R_i, let S_i be the smallest slab of **S** that completely contains R_i. We will show that the algorithm performs at most an S_i-restricted search for input $P \in \mathcal{R}_v$. If p_i is maximal, then R_i is contained in a leaf slab (this is because the output is **S**-labeled). Hence S_i is a leaf slab and an S_i-restricted search for a maximal p_i is just a complete search.

Now consider a non-maximal p_i. By the properties of **S**-labeled maxima, the associated region R_i is either inside a leaf slab or is separated by a slab boundary from the dominating region R_j. In the former case, an S_i-restricted search is a complete search. In the latter case, we argue that an S_i-restricted search suffices to process p_i. This follows from Claim 4.1: by the time an S_i-restricted search finishes, all maxima to the right of S_i have been determined. In particular, we have found p_j, and thus $\hat{p}$ dominates p_i. Hence, the search for p_i will proceed no further.

The expected search time taken conditioned on $P \in \mathcal{R}_v$ is the sum (over i) of the conditional expected S_i-restricted search times. Let $\mathcal{E}_i$ denote the event that $p_i \in R_i$, and $\mathcal{E}$ be the event that $P \in \mathcal{R}_v$. We have $\mathcal{E} = \bigwedge_i \mathcal{E}_i$. By the independence of the distributions and linearity of expectation

$$\begin{aligned}
&\mathbf{E}_{\mathcal{E}}[\text{search time}] \\
&= \sum_{i=1}^{n} \mathbf{E}_{\mathcal{E}}[S_i\text{-restricted search time for } p_i] \\
&= \sum_{i=1}^{n} \mathbf{E}_{\mathcal{E}_i}[S_i\text{-restricted search time for } p_i].
\end{aligned}$$

By Lemma 1.6, the time for an S_i-restricted search conditioned on $p_i \in R_i$ is $O(-\log \Pr[p_i \in R_i] + 1)$. By Lemma 3.5, $d_v = \sum_i -\log \Pr[p_i \in R_i]$, completing the proof. $\square$

We can now prove the main lemma.

PROOF OF LEMMA 4.3. By Lemma 3.10, there exists an entropy-sensitive comparison tree $\mathcal{T}$ that computes the **S**-labeled maxima with expected depth $O(\mathrm{OPT} + n)$. According to Claim 4.5, the expected running time of **Search** is $O(\mathrm{OPT} + n)$. Claim 4.4 tells us the expected time for **Update** is $O(n)$, and we add these bounds to complete the proof. $\square$

5. DATA STRUCTURES OBTAINED DURING THE LEARNING PHASE

Learning the vertical slab structure **S** is very similar to to learning the V-list in Ailon et al. [ACC[+]11, Lemma 3.2]. We repeat the construction and proof for convenience: take the union of the first $k = \log n$ inputs $P_1, P_2, \ldots, P_k$, and sort those points by x-coordinates. This gives a list $x_0, x_1, \ldots, x_{nk-1}$. Take the n values $x_0, x_k, x_{2k}, \ldots, x_{(n-1)k}$. They define the boundaries for **S**. We recall a useful and well-known fact [ACC[+]11, Claim 3.3].

CLAIM 5.1. *Let* $Z = \sum_i Z_i$ *be a sum of nonnegative random variables such that* $Z_i = O(1)$ *for all* i, $\mathbf{E}[Z] = O(1)$*, and for all* i, j, $\mathbf{E}[Z_i Z_j] = \mathbf{E}[Z_i]\mathbf{E}[Z_j]$*. Then* $\mathbf{E}[Z^2] = O(1)$.

Now let λ be a leaf slab in **S**. Recall that we denote by X_λ the number of points of a random input P that end up in λ. Using Claim 5.1, we quickly obtain the following lemma.

LEMMA 5.2. *With probability* $1 - n^{-3}$ *over the construction of* **S***, we have* $\mathbf{E}[X_\lambda^2] = O(1)$ *for all leaf slabs* $\lambda \in$ **S**.

PROOF. Consider two values x_i, x_j from the original list. Note that all the other $kn - 2$ values are independent of these two points. For every $r \notin \{i, j\}$, let $Y_t^{(r)}$ be the indicator random variable for $x_r \in t := [x_i, x_j)$. Let $Y_t = \sum_r Y_t^{(r)}$. Since the $Y_t^{(r)}$'s are independent, by Chernoff's bound [AS00], for any $\beta \in (0, 1]$,

$$\Pr[Y_t \le (1 - \beta)\mathbf{E}[Y_t]] \le \exp(-\beta^2 \mathbf{E}[Y_t]/2).$$

With probability at least $1 - n^{-5}$, if $\mathbf{E}[Y_t] > 12 \log n$, then $Y_t > \log n$. By applying the same argument for any pair x_i, x_j and taking a union bound over all pairs, with probability at least $1 - n^{-3}$ the following holds: for any pair t, if $Y_t \le \log n$, then $\mathbf{E}[Y_t] \le 12 \log n$.

For any leaf slab $\lambda = [x_{ak}, x_{(a+1)k}]$, we have $Y_\lambda \le \log n$. Let $X_\lambda^{(i)}$ be the indicator random variable for the event that $x_i \sim \mathcal{D}_i$ lies in λ, so that $X_\lambda = \sum_i X_\lambda^{(i)}$. Since $\mathbf{E}[Y_\lambda] \ge (\log n - 2)\mathbf{E}[X_\lambda]$, we get $\mathbf{E}[X_\lambda] = O(1)$. By independence of the $\mathcal{D}_i$'s, for all i, j, $\mathbf{E}[X_\lambda^{(i)} X_\lambda^{(j)}] = \mathbf{E}[X_\lambda^{(i)}]\mathbf{E}[X_\lambda^{(j)}]$, so $\mathbf{E}[X_\lambda^2] = O(1)$, by Claim 5.1. $\square$

Lemma 1.7 follows immediately from Lemma 5.2 and the fact that sorting the k inputs $P_1, P_2, \ldots, P_k$ takes $O(n \log^2 n)$ time. After the leaf slabs have been determined, the search trees T_i can be found using essentially the same techniques as before [ACC[+]11, Section 3.2]. The main idea is to use $n^{\varepsilon} \log n$ rounds to find the first $\varepsilon \log n$ levels of T_i, and to use a balanced search tree for searches that need to proceed to a deeper level. This only costs a factor of ε^{-1}. We restate Lemma 1.8 for convenience.

LEMMA 5.3. *Let $\varepsilon > 0$ be a fixed parameter. In $O(n^{\varepsilon})$ rounds and $O(n^{1+\varepsilon})$ time, we can construct search trees T_1, T_2, ..., T_n over $\mathbf{S}$ such that the following holds. (i) the trees can be totally represented in $O(n^{1+\varepsilon})$ space; (ii) probability $1-n^{-3}$ over the construction of the T_is: every T_i is $O(1/\varepsilon)$-optimal for restricted searches over $\mathcal{D}_i$.*

PROOF. Let $\delta > 0$ be some sufficiently small constant and c be sufficiently large . For $k = c\delta^{-2}n^{\varepsilon}\log n$ rounds and each p_i, we record the leaf slab of $\mathbf{S}$ that contains it. We break the proof into smaller claims.

CLAIM 5.4. *Using k inputs, we can compute estimates $\hat{q}(i,S)$ for each index i and slab S. The following guarantee holds (for all i and S) with probability $> 1 - 1/n^3$ over the choice of the k inputs. If at least $5\log n$ instances of p_i fell in S, then $\hat{q}(i,S) \in [(1-\delta)q(i,S),(1+\delta)q(i,S)]$*[1].

PROOF. For a slab S, let $N(S)$ be the number of times p_i was in S, and let $\hat{q}(i,S) = N(S)/k$ be the empirical probability for this event ($\hat{q}(i,S)$ is an estimate of $q(i,S)$). Fix a slab S. If $q(i,S) \leq 1/2n^{\varepsilon}$, then by a Chernoff bound we get $\Pr[N(S) \geq 5\log n \geq 10kq(i,S)] \leq 2^{-5\log n} = n^{-5}$. Furthermore, if $q(i,S) \geq 1/2n^{\varepsilon}$, then $q(i,S)k \geq (c/2\delta^2)\log n$ and $\Pr[N(S) \leq (1-\delta)q(i,S)k] \leq \exp(-q(i,S)\delta^2 k/4) \leq n^{-5}$ as well as $\Pr[N(S) \geq (1+\delta)q(i,S)k] \leq \exp(-\delta^2 q(i,S)k/4) \leq n^{-5}$. Thus, by taking a union bound, we get that with probability at least $1-n^{-3}$ for any slab S, if $N(S) \geq 5\log n$, then $q(i,S) \geq n^{-\varepsilon}/2$ and hence $\hat{q}(i,S) \in [(1-\delta)q(i,S),(1+\delta)q(i,S)]$. □

We will henceforth assume that this claims holds for all i and S. Based on the values $\hat{q}(i,S)$, we construct the search trees. The tree T_i is constructed recursively. We will first create a partial search tree, where some searches may end in non-leaf slabs (or, in other words, leaves of the tree may not be leaf slabs). The root is the just the largest slab. Given a slab S, we describe how the create the sub-tree of T_i rooted at S. If $N(S) < 5\log n$, then we make S a leaf. Otherwise, we pick a leaf slab λ such that for the slab S_l consisting of all leaf slabs (strictly) to the left of λ and the slab S_r consisting of all leaf slabs (strictly) to the right of λ we have $\hat{q}(i,S_l) \leq (2/3)\hat{q}(i,S)$ and $\hat{q}(i,S_r) \leq (2/3)\hat{q}(i,S)$. We make λ a leaf child of S. Then we recursively create trees for S_l and S_r and attach them as children to S. For any internal node of the tree S, we have $q(i,S) \geq n^{\varepsilon}/2$, and hence the depth is at most $O(\varepsilon\log n)$. Furthermore, this partial tree is β-reducing (for some constant β). The partial tree T_i is extended to a complete tree in a simple way. From each T_i-leaf that is not a leaf slab, we perform a basic binary search for the leaf slab. This yields a tree T_i of depth at most $(1+O(\varepsilon))\log n$. Note that we only need to store the partial T_i tree, and hence the total space is $O(n^{1+\varepsilon})$.

Let us construct, as a thought experiment, a related tree T'_i. Start with the partial T_i. For every leaf that is not a leaf slab, extend it downward using the true probabilities $q(i,S)$. In other words, let us construct the subtree rooted at a new node S in the following manner. We pick a leaf slab λ such that $q(i,S_l) \leq (2/3)q(i,S)$ and $q(i,S_r) \leq (2/3)q(i,S)$ (where S_l and S_r are as defined above). This ensures that T'_i is β-reducing. By Lemma 1.6, T'_i is $O(1)$-optimal for restricted searches over $\mathcal{D}_i$ (we absorb the β into $O(1)$ for convenience).

[1]We remind the reader that this the probability that $p_i \in S$.

CLAIM 5.5. *The tree T_i is $O(1/\varepsilon)$-optimal for restricted searches.*

PROOF. Fix a slab S and an S-restricted distribution $\mathcal{D}_S$. Let $q'(i,\lambda)$ (for each leaf slab λ) be the series of values defining $\mathcal{D}_S$. Note that $q'(i,S) \leq q(i,S)$. Suppose $q'(i,S) \leq n^{-\varepsilon/2}$. Then $-\log q'(i,S) \geq \varepsilon(\log n)/2$. Since any search in T_i takes at most $(1+O(\varepsilon))\log n$ steps, the search time is at most $O(\varepsilon^{-1}(-\log q'(i,S)+1))$.

Suppose $q'(i,S) > n^{-\varepsilon/2}$. Consider a single search for some p_i. We will classify this search based on the leaf of the partial tree that is encountered. By the construction of T_i, any leaf S' is either a leaf slab or has the property that $q(i,S') \leq n^{-\varepsilon}/2$. The search is of *Type 1* if the leaf of the partial tree actually represents a leaf slab (and hence the search terminates). The search is of *Type 2* (resp. *Type 3*) if the leaf of the partial tree is a slab S is an internal node of T_i and the depth is at least (resp. less than) $\varepsilon(\log n)/3$.

When the search is of Type 1, it is identical in both T_i and T'_i. When the search is of Type 2, it takes at $\varepsilon(\log n)/3$ in T'_i and at most (trivially) $(1+O(\varepsilon))(\log n)$ in T_i. The total number of leaves (that are not leaf slabs) of the partial tree at depth less than $\varepsilon(\log n)/3$ is at most $n^{\varepsilon/3}$. The total probability mass of $\mathcal{D}_i$ inside such leaves is at most $n^{\varepsilon/3} \times n^{-\varepsilon}/2 < n^{-2\varepsilon/3}$. Since $q'(i,S) > n^{-\varepsilon/2}$, in the restricted distribution $\mathcal{D}_S$, the probability of a Type 3 search is at most $n^{-\varepsilon/6}$.

Choose a random $p \sim \mathcal{D}_S$. Let $\mathcal{E}$ denote the event that a Type 3 search occurs. Furthermore, let X_p denote the depth of the search in T_i and X'_p denote the depth in T'_i. When $\mathcal{E}$ does not occur, we have argued that $X_p \leq O(X'_p/\varepsilon)$. Also, $\Pr(\mathcal{E}) \leq n^{-\varepsilon/6}$. The expected search time is just $\mathbf{E}[X_p]$. By Bayes' rule,

$$\begin{aligned}
\mathbf{E}[X_p] &= \Pr(\overline{\mathcal{E}})\mathbf{E}_{\overline{\mathcal{E}}}[X_p] + \Pr(\mathcal{E})\mathbf{E}_{\mathcal{E}}[X_p] \\
&\leq O(\varepsilon^{-1}\mathbf{E}_{\overline{\mathcal{E}}}[X'_p]) + n^{-\varepsilon/6}(1+O(\varepsilon))\log n \\
\mathbf{E}[X'_p] &= \Pr(\overline{\mathcal{E}})\mathbf{E}_{\overline{\mathcal{E}}}[X'_p] + \Pr(\mathcal{E})\mathbf{E}_{\mathcal{E}}[X_p] \\
\implies &\quad \mathbf{E}_{\overline{\mathcal{E}}}[X'_p] \leq \mathbf{E}[X'_p]/\Pr(\overline{\mathcal{E}}) \leq 2\mathbf{E}[X'_p]
\end{aligned}$$

Combining, the expected search time is $O(\varepsilon^{-1}(\mathbf{E}[X'_p]+1))$. Since T'_i is $O(1)$-optimal for restricted searches, T_i is $O(\varepsilon^{-1})$-optimal. □

6. ACKNOWLEDGMENTS

C. Seshadhri was funded by the Early-Career LDRD program at Sandia National Laboratories. Sandia National Laboratories is a multi-program laboratory managed and operated by Sandia Corporation, a wholly owned subsidiary of Lockheed Martin Corporation, for the U.S. Department of Energy's National Nuclear Security Administration under contract DE-AC04-94AL85000.

We would like to thank Eden Chlamtáč for suggesting a simple proof for Claim 3.3.

7. REFERENCES

[ABC09] Peyman Afshani, Jérémy Barbay, and Timothy M. Chan. Instance-optimal geometric algorithms. In *Proc. 50th Annu. IEEE Sympos. Found. Comput. Sci. (FOCS)*, pages 129–138, 2009.

[ACC+11] Nir Ailon, Bernard Chazelle, Kenneth L. Clarkson, Ding Liu, Wolfgang Mulzer, and C. Seshadhri. Self-improving algorithms. *SIAM Journal on Computing*, 40(2):350–375, 2011.

[ACCL06] Nir Ailon, Bernard Chazelle, Seshadhri Comandur, and Ding Liu. Self-improving algorithms. In *Proc. 17th Annu. ACM-SIAM Sympos. Discrete Algorithms (SODA)*, pages 261–270, 2006.

[AS00] Noga Alon and Joel H. Spencer. *The probabilistic method*. Wiley-Interscience, New York, second edition, 2000.

[BKS01] S. Borzsony, D. Kossmann, and K. Stocker. The skyline operator. In *Data Engineering, 2001. Proceedings. 17th International Conference on*, pages 421–430. IEEE, 2001.

[BLMM11] Kevin Buchin, Maarten Löffler, Pat Morin, and Wolfgang Mulzer. Preprocessing imprecise points for Delaunay triangulations: Simplified and extended. *Algorithmica*, 61(3):674–693, 2011.

[Buc89] C. Buchta. On the average number of maxima in a set of vectors. *Inform. Process. Lett.*, 33(2):63–66, 1989.

[Cha93] Bernard Chazelle. Cutting hyperplanes for divide-and-conquer. *Discrete Comput. Geom.*, 9:145–158, 1993. 10.1007/BF02189314.

[Cla87] Kenneth L. Clarkson. New applications of random sampling to computational geometry. *Discrete Comput. Geom.*, 2:195–222, 1987.

[CMS10] K. Clarkson, W. Mulzer, and C. Seshadhri. Self-improving algorithms for convex hulls. In *Proc. 21st Annu. ACM-SIAM Sympos. Discrete Algorithms (SODA)*, 2010.

[CS08] Kenneth L. Clarkson and C. Seshadhri. Self-improving algorithms for Delaunay triangulations. In *Proc. 24th Annu. ACM Sympos. Comput. Geom. (SoCG)*, pages 226–232, 2008.

[EM11] Esther Ezra and Wolfgang Mulzer. Convex hull of imprecise points in $o(nlogn)$ time after preprocessing. In *Proc. 27th Annu. ACM Sympos. Comput. Geom. (SoCG)*, pages 11–20, 2011.

[FLN01] Ronald Fagin, Amnon Lotem, and Moni Naor. Optimal aggregation algorithms for middleware. In *Proceedings of the twentieth ACM SIGMOD-SIGACT-SIGART symposium on Principles of database systems*, PODS '01, pages 102–113, New York, NY, USA, 2001. ACM.

[Gol94] M. J. Golin. A provably fast linear-expected-time maxima-finding algorithm. *Algorithmica*, 11:501–524, 1994. 10.1007/BF01189991.

[GTVV93] M. T. Goodrich, Jyh-Jong Tsay, D. E. Vengroff, and J. S. Vitter. External-memory computational geometry. In *Proceedings of the 1993 IEEE 34th Annual Foundations of Computer Science*, pages 714–723, Washington, DC, USA, 1993. IEEE Computer Society.

[HM08] Martin Held and Joseph S. B. Mitchell. Triangulating input-constrained planar point sets. *Inform. Process. Lett.*, 109(1):54–56, 2008.

[KLP75] H. T. Kung, F. Luccio, and F. P. Preparata. On finding the maxima of a set of vectors. *J. ACM*, 22:469–476, October 1975.

[vKLM10] Marc J. van Kreveld, Maarten Löffler, and Joseph S. B. Mitchell. Preprocessing imprecise points and splitting triangulations. *SIAM Journal on Computing*, 39(7):2990–3000, 2010.

[KS86] David G. Kirkpatrick and Raimund Seidel. The ultimate planar convex hull algorithm? *SIAM Journal on Computing*, 15(1):287–299, 1986.

[LS10] Maarten Löffler and Jack Snoeyink. Delaunay triangulation of imprecise points in linear time after preprocessing. *Comput. Geom. Theory Appl.*, 43(3):234–242, 2010.

Finding the Maximal Empty Disk Containing a Query Point

Haim Kaplan
School of Computer Science
Tel Aviv University
Tel Aviv, Israel
haimk@tau.ac.il

Micha Sharir
Scholl of Computer Science
Tel Aviv University
Tel Aviv, Israel, and
Courant Institute of Mathematical Sciences
New York University, New York, NY 10012, USA
michas@tau.ac.il

ABSTRACT

Let P be a set of n points in the plane. We present an efficient algorithm for preprocessing P, so that, for a given query point q, we can quickly report the largest disk that contains q but its interior is disjoint from P. The storage required by the data structure is $O(n \log n)$, the preprocessing cost is $O(n \log^2 n)$, and a query takes $O(\log^2 n)$ time.

1. INTRODUCTION

We consider the problem of preprocessing a given set P of n points in the plane into a data structure such that for any given query point q we can efficiently find the largest disk that contains q but its interior does not contain any point of P. This problem models scenarios where one wants to efficiently estimate the size of the "free space" (free of points of P, that is) surrounding a query point and locate the center of this free space. Concrete examples of such a scenario are: (i) We want to transmit a message over the largest possible area (which is a disk if propagation proceeds uniformly in all directions) such that we are sure that q receives the message but none of the eavesdroppers at the points of P hears it. (ii) We want to position a sprinkler to water the largest possible disk that includes our favorite plant at q but does not include any of the locations in P, which are facilities that we do not want to get wet.

Note that it suffices to consider query points q that lie inside the convex hull $CH(P)$ of P because otherwise there are arbitrarily large disks containing q and disjoint from P.

Our result.

We present an efficient algorithm that requires $O(n \log n)$ storage, $O(n \log^2 n)$ preprocessing time, and can answer a query in $O(\log^2 n)$ time.

Background.

To the best of our knowledge this problem was first introduced by Augustine *et al.* [1], who design a data structure that requires $O(n^2)$ space, $O(n^2 \log n)$ preprocessing time, and can answer a query in $O(\log n)$ time.

A much easier problem, which has been studied much earlier, is that of finding the largest empty disk for a set P of n points in the plane (where, to make sense of the problem, the center of the disk has to lie, say, inside the convex hull of P—recall the remark made earlier concerning points outside the hull). This problem arises in several areas such as data mining, database management, visualization, and VLSI design, and it has a very simple solution. Indeed, the largest empty disk has to be centered at a vertex of the Voronoi diagram of P, or at a crossing of an edge of the diagram with the convex hull boundary, and therefore can be found in $O(n \log n)$ time.

We note that Augustine *et al.* [1] also consider a similar problem, where the goal is to report tha largest-area axis-parallel rectangle that contains the query point, is contained in some fixed bounding box B, and its interior is disjoint from P. Their solution for this variant also requires nearly quadratic storage and preprocessing; in a companion paper [4] (see also [3]), we present a more efficient solution, which requires only nearly linear storage and preprocessing, for that latter problem.

2. PRELIMINARIES

Let P be a set of n points in the plane. We call a disk whose interior is disjoint from P a P-*empty disk*. We wish to preprocess P into a data structure so that, given a query point q, we can efficiently find the largest P-empty disk containing q. The following easy lemma completes the observation made earlier, that it suffices to consider only query points q that lie inside the convex hull $CH(P)$ of P.

LEMMA 2.1. *There are arbitrarily large P-empty disks containing q if and only if q does not lie in the interior of the convex hull $CH(P)$ of P.*

PROOF. If q does not lie in the interior of $CH(P)$ then there is a line ℓ through q that defines halfplane whose interior is P-empty; clearly there are arbitrarily large P-empty disks in this halfplane tangent to ℓ at q.

For the converse statement, let B be some fixed disk containing $P \cup \{q\}$. Clearly if there are arbitrarily large P-empty disks containing q then there are arbitrarily large P-empty disks containing q on their boundary. For any such disk D, $D \cap B$ contains the intersection of B with a wedge W whose apex is q and whose bounding rays go through the points in $\partial B \cap \partial D$. The wedge W is P-empty ($W \cap B$ is P-empty since it is contained in D and $W \setminus B$ is empty since there are

SCG'12, June 17–20,2012, Chapel Hill, Noth Carolina, USA.

no points of P outside B), and its opening angle tends to π as D grows. In the limit, W becomes a halfplane containing q and openly disjoint from P. Hence q does not lie in the interior of $CH(P)$. □

Let $\mathrm{Vor}(P)$ denote the Voronoi diagram of P.

For a point q in the interior of $CH(P)$, let $D_{\max}(q)$ denote an arbitrary largest P-empty disk containing q, and let $c_{\max}(q)$ denote its center. (In general this disk might not be unique. For example we may have two congruent Delaunay disks, that is, maximal P-empty disks centered at vertices of $\mathrm{Vor}(P)$, both containing q, and larger than any other P-empty disk containing q.)

The analysis proceeds through the following stages.

LEMMA 2.2. *Let q be in the interior of $CH(P)$. Then $c_{\max}(q)$ lies on a Voronoi edge or at a Voronoi vertex.*

PROOF. If D is a P-empty disk that contains q and is centered at some point c not on the Voronoi diagram then D touches at most one point of P. If it does not touch any point of P, we can get a larger P-empty disk containing q by expanding D about c until it touches some (unique) point $a \in P$. If D already touches a point a of P then we can get a larger P-empty disk containing q by moving the center of D along the ray from a through c, away from a, keeping it touching a, until it touches another point b of P. This has to occur, for otherwise we get arbitrarily large P-empty disks containing q, contrary to Lemma 2.1. It follows that a disk which is not centered at a Voronoi edge or vertex is not a maximal P-empty disk containing q. □

Suppose that $c_{\max}(q)$ lies in the relative interior of a Voronoi edge e_{ab} defined by two sites $a, b \in P$. There is at least one direction along e_{ab} so that we can increase the size of the disk by moving its center in that direction away from $c_{\max}(q)$ while keeping it touching a and b and P-empty. Since $D_{\max}(q)$ is the largest P-empty disk containing q, we conclude that in this case q must lie on the boundary of $D_{\max}(q)$. Moreover, this boundary is split into two arcs by a and b, and q has to lie on the *smaller* arc; see Figure 1. Note that in this case we have $\angle aqb > \pi/2$. (In particular, ab cannot be a diameter of $D_{\max}(q)$.)

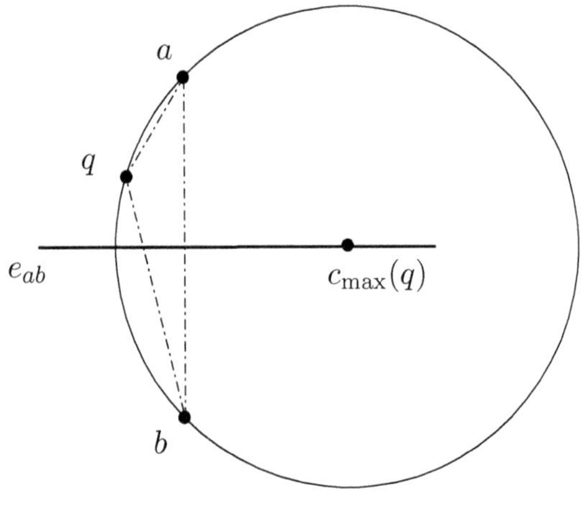

Figure 1: The case where $c_{\max}(q)$ lies in the relative interior a Voronoi edge.

Continuing the analysis of this scenario, we note that in this case $c_{\max}(q)$ is equally nearest to a, b, and q. That is, $c_{\max}(q)$ is a vertex of the Voronoi cell $V(q)$ of q in the augmented Voronoi diagram $\mathrm{Vor}(P \cup \{q\})$. Then a and b are two consecutive neighbors of q; that is, the Voronoi edges e_{aq}, e_{bq} are consecutive along $\partial V(q)$. The angle between these edges, within $V(q)$, is $\pi - \angle aqb$, and is therefore smaller than $\pi/2$. See Figure 2. Since the sum of the exterior angles of a convex polygon is 2π, it follows that $V(q)$ has at most three vertices that can be the location of $c_{\max}(q)$. We summarize the observations made so far in the following lemma.

LEMMA 2.3. *If $c_{\max}(q)$ lies in the relative interior of a Voronoi edge e_{ab} then q lies on the smaller arc connecting a and b along $\partial D_{\max}(q)$. In this case $c_{\max}(q)$ is a vertex of the Voronoi cell $V(q)$ of q in the Voronoi diagram $\mathrm{Vor}(P \cup \{q\})$. Furthermore, $c_{\max}(q)$ is one of the at most three vertices of $V(q)$ with an acute interior angle.*

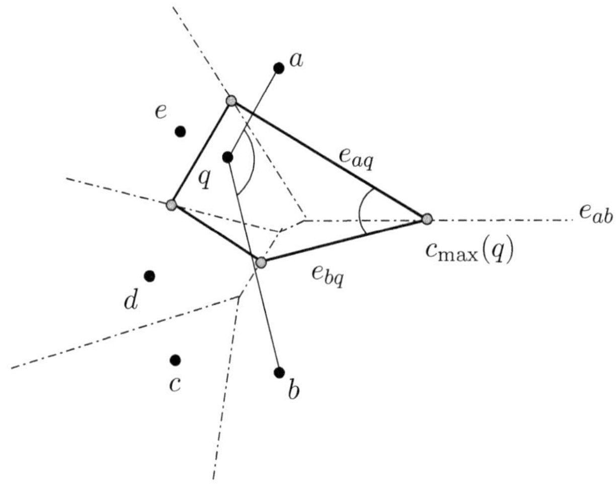

Figure 2: The cell $V(q)$ in $\mathrm{Vor}(P \cup \{q\})$, superimposed on $\mathrm{Vor}(P)$. The angle between e_{aq} and e_{bq} is smaller than $\pi/2$.

The preceding analysis therefore suggests the following high-level approach: Identify, without constructing $V(q)$ explicitly (whose complexity can be as high as $\Theta(n)$), the at most three vertices of this cell which might be the location of $c_{\max}(q)$, compute, in $O(1)$ time, the radii of the corresponding largest P-empty disks centered at these vertices, and output the center with the largest radius. This is however only one part of the solution, because we also have to consider Voronoi vertices of $\mathrm{Vor}(P)$ as possible locations for $c_{\max}(q)$.

2.1 A quick-and-dirty interim solution

Before continuing towards an efficient solution, we can already exploit the analysis given so far, to obtain the following quick-and-dirty solution. Consider a query point q, the cell $V(q)$ of q in $\mathrm{Vor}(P \cup \{q\})$, and its at most three special vertices having acute interior angles, which are the candidate placements for $c_{\max}(q)$ (other than Voronoi vertices of $\mathrm{Vor}(P)$). If we move q, $V(q)$ will vary continuously, and its combinatorial structure (namely, the sequence of its edges and the identity of the three acute vertices) will remain fixed until q reaches some critical placement in which one of these attributes changes. It is easy to verify that the sequence of edges of $V(q)$ can change only when one of its vertices is a vertex of the non-augmented diagram $\mathrm{Vor}(P)$; that is, when q lies on the boundary of the Delaunay disk which is centered at that vertex (and circumscribes its dual Delaunay triangle). The "acuteness" of a vertex of $V(q)$, lying on an original Voronoi edge e_{ab}, can change only when q lies on the boundary of the diametral disk determined by a and b, but only if this disk is P-empty. (Note that the diametral disk is P-empty if and only if ab is an edge of the *Gabriel graph* of P.)

So a quick-and-dirty solution is to draw the $O(n)$ disks of the above two types (Delaunay disks and diametral disks determined by edges of the Gabriel graph), form their arrangement $\mathcal{A}$, and store with each face f of $\mathcal{A}$ the list $L(f)$ of the at most three acute vertices of $V(q)$, for any $q \in f$ (by the preceding argument, the list is the same for all $q \in f$). We also store with each f the largest Delaunay disk, $D(f)$, centered at a vertex of $\mathrm{Vor}(P)$, which contains f. We can compute the lists $L(f)$ and the disks $D(f)$ incrementally, observing that it is straightforward to update $L(f)$ in $O(1)$ time as we cross from one face f to a neighboring face f'. Updating $D(f)$ can be done in $O(\log n)$ time, by dynamically maintaining the list of all Delaunay disks containing the present face f, sorted in the decreasing order of their radii. We then preprocess $\mathcal{A}$ for efficient point location, and obtain the desired data structure. It requires $O(n^2)$ storage, $O(n^2 \log n)$ preprocessing, and a query takes $O(\log n)$ time: Locate the query point q in $\mathcal{A}$, retrieve the list $L(f)$ of the face f containing q, and search each of its (at most three) elements for a possible location of $c_{\max}(q)$. We return the largest among the maximal P-empty disks centered at the vertices defined by $L(f)$ and the disk $D(f)$. Note that the elements of $L(f)$ are only stored symbolically, because the actual vertices of $V(q)$ vary continuously as q moves inside f; still, the combinatorial identification of each of these elements allows us to test it in $O(1)$ time for any concrete query point q.

This straightforward approach matches the performance of the algorithm in [1].

3. A MORE EFFICIENT DATA STRUCTURE

3.1 Efficient processing of Delaunay disks

It is relatively easy to improve the algorithm that finds the largest Delaunay disk containing q, using the approach given in [1]. That is, we sort the Voronoi vertices v of $\mathrm{Vor}(P)$ by the radius r_v of the Delaunay disk centered at v, and store them, in this inorder, at the leaves of a balanced binary tree T. For each node w of T, let Q_w denote the subset of vertices stored at the leaves of the subtree rooted at w, and let R_w denote the set of the corresponding radii r_v. Construct the union U_w of all the Delaunay disks centered at the points of Q_w, and store it at w. U_w has $O(|Q_w|)$ edges and vertices [5] and can be constructed in $O(|Q_w| \log |Q_w|)$ time (e.g., by lifting the disks to planes in three dimensions and by constructing the upper envelope of these planes), for a total of $O(n \log n)$ storage and $O(n \log^2 n)$ preprocessing time. We process each U_w for fast (logarithmic time) point location.

Now, given a query point q, we search through T. At each node w that we visit, we take its right child w_r and check whether $q \in U_{w_r}$. If so, then q is contained in at least one Delaunay disk whose radius belongs to R_{w_r}, and we continue the search at w_r. If $q \notin U_{w_r}$, we continue the search at the left child w_l of w. When the search terminates at some leaf, the disk stored at that leaf is the largest Delaunay disk containing q. The search takes a total of $O(\log^2 n)$ time.

3.2 Disks centered at Voronoi edges

We next consider the other situation, where $c_{\max}(q)$ lies in the relative interior of some Voronoi edge. Let e_{ab} be a Voronoi edge, determined by two points $a, b \in P$ and delimited by the Voronoi vertices v_{abc}, v_{abd}, for two additional points $c, d \in P$, so that v_{abc} (resp., v_{abd}) is equally nearest to a, b, c (resp., to a, b, d). Let D_{ab} denote the diametral disk on ab, and let D_{abc} (resp., D_{abd}) denote the Delaunay disk centered at v_{abc} (resp., v_{abd}), so it circumscribes Δabc (resp., Δabd).

LEMMA 3.1. *Any point q for which $c_{\max}(q)$ lies in the relative interior of e_{ab} must lie in*

$$K_{ab} = int\,(D_{ab} \setminus (D_{abc} \cap D_{abd}))\ .$$

PROOF. Indeed, we have already argued that such a point q must satisfy $\angle aqb > \pi/2$, which is equivalent to q belonging to the interior of D_{ab}. In addition, the center of the disk circumscribing Δaqb lies in the relative interior of e_{ab} if and only if q lies in the interior of $(D_{abc} \cup D_{abd}) \setminus (D_{abc} \cap D_{abd})$, as is easily verified. The conjunction of these two conditions is $q \in K_{ab}$. □

The case where ab is an edge of the *Gabriel graph* is shown in Figure 3. In this case ab crosses e_{ab} and D_{ab} is P-empty. We refer to such an edge in short as a *Gabriel edge*. In this case K_{ab} is the union of the interiors of the two shaded lunes.

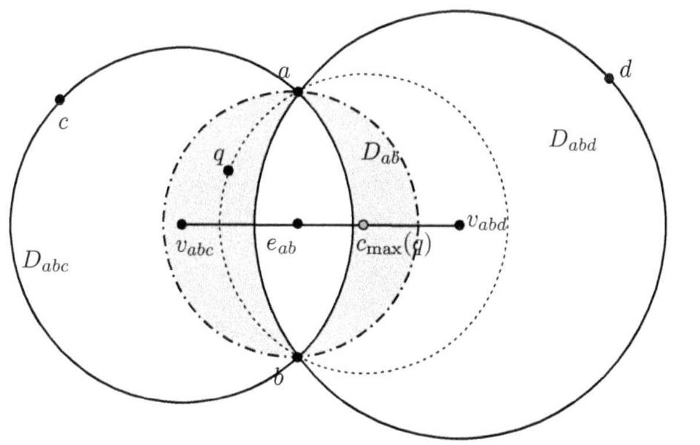

Figure 3: q has to lie in the interior of one of the shaded lunes if $c_{\max}(q)$ lies in the relative interior of the Voronoi edge e_{ab}; the case where ab is an edge of the Gabriel graph. The corresponding disk $D_{\max}(q)$ is shown dotted.

The case where the Delaunay edge ab does not belong to the Gabriel graph (a *non-Gabriel edge*) is shown in Figure 4. Here K_{ab} is the single shaded lune.

Note that for each lune, in both cases of Gabriel edges and non-Gabriel edges, the area enclosed by the "outer" (longer) arc of the lune and the edge ab is P-empty.

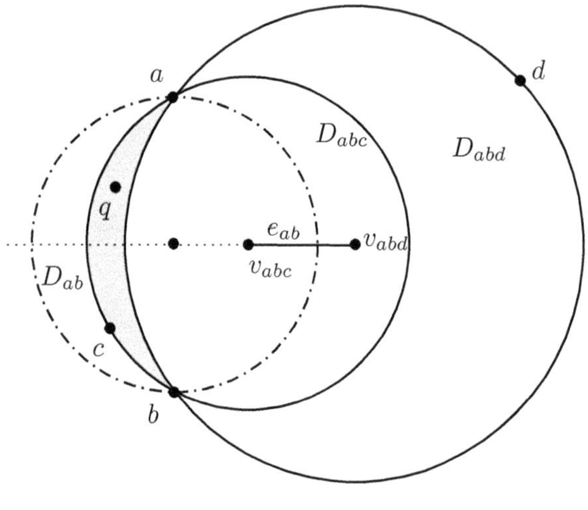

Figure 4: q has to lie in the interior of the shaded lune if $c_{\max}(q)$ lies in the relative interior of the Voronoi edge e_{ab}; the case where ab is not an edge of the Gabriel graph.

We construct the regions K_{ab}, for all Voronoi edges e_{ab} of Vor(P), and denote the collection of all these $O(n)$ lunes and double lunes as $\mathcal{K}$. A crucial property is that no point q can be contained in more than three regions of $\mathcal{K}$. Indeed, a point q belongs to K_{ab} if and only if the disk circumscribing Δaqb is P-empty, its center lie in the relative interior of the Voronoi edge e_{ab}, and $\angle aqb > \pi/2$. The claim then follows from Lemma 2.3.

Ideally, we could construct the arrangement $\mathcal{A}(\mathcal{K})$, preprocess it for fast point location, and store with each face f of $\mathcal{A}(\mathcal{K})$ the at most three Delaunay edges ab whose regions K_{ab} (fully) contain f. Then, given a query point q, we could locate q in the arrangement, retrieve the at most three corresponding Delaunay edges, and test, in constant time, each of the respective Voronoi edges for possible location of $c_{\max}(q)$. The problem with this approach is that the complexity of $\mathcal{A}(\mathcal{K})$ might be quadratic, as depicted in Figure 5. (Note that the arrangement $\mathcal{A}$ constructed in Section 2.1 is in fact a refinement of $\mathcal{A}(\mathcal{K})$.)

Figure 5: Illustrating a scenario in which the arrangement of the lunes K_{ab} has quadratic complexity.

We overcome this problem by first decomposing $\mathcal{K}$ and its lunes into a constant number of subcollections, each consisting of portions of the lunes, so that, within each subproblem that we create, the regions are *pairwise openly disjoint*, and so the planar map that they form has linear complexity. Overall, we obtain a structure that uses only linear storage, and the query time, which is dominated by the point location in these maps, is only $O(\log n)$. (The overall query time, though, is $O(\log^2 n)$, because of the search for the largest Delaunay disk containing q; see Section 3.1.)

The intersection pattern of the lunes.

The following simple lemma provides the main technical tool for decomposing the lunes.

LEMMA 3.2. *Let K_{ab} and K_{cd} be two distinct lunes (the four points a, b, c, d need not all be distinct) with a nonempty intersection. Then, for each point $q \in K_{ab} \cap K_{cd}$, there exists a line ℓ through q that (weakly) separates a, b from c, d. That is, one closed halfplane bounded by ℓ contains a and b, and the other closed halfplane contains c and d.*

PROOF. Refer to Figure 6. Let D_{abq} (resp., D_{cdq}) denote the disk circumscribing the triangle Δabq (resp., Δcdq). By the properties of the lunes, as analyzed in Lemma 3.1, both disks are P-empty. The circle C_{abq} bounding D_{abq} and the circle C_{cdq} bounding D_{cdq} meet at q and therefore either (i) they also intersect at a second point q', or (ii) they are tangent to each other at q. Let ℓ be the line connecting q and q' in case (i) or the common tangent line at q in case (ii). As is easily checked, ℓ separates the arcs $\alpha = C_{abq} \setminus D_{cdq}$ and $\gamma = C_{cdq} \setminus D_{abq}$. Moreover, since D_{abq} and D_{cdq} are both P-empty, a and b must lie in the closure of α and c and d must lie in the closure of γ, thereby completing the proof of the lemma. □

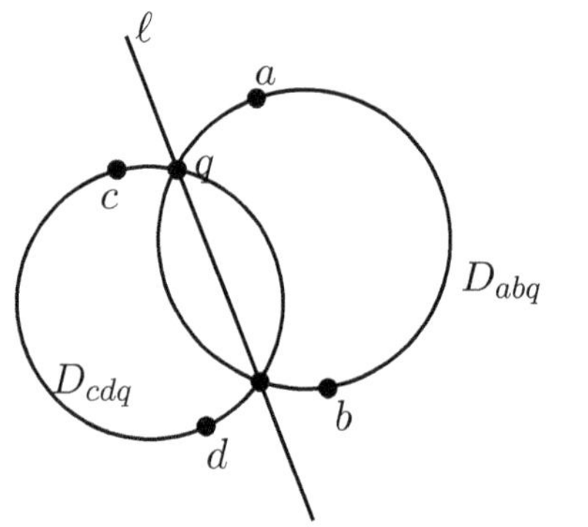

Figure 6: Illustrating the proof of Lemma 3.2.

3.3 Partitioning the lunes

Having these observations at hand, we now proceed to eliminate all intersections between the lunes by an appropriate problem decomposition, as promised above. Rotate the coordinate frame so that no two points have the same x- or y-coordinate. We first partition $\mathcal{K}$ into two subcollections $\mathcal{K}^+$ and $\mathcal{K}^-$, where $\mathcal{K}^+$ (resp., $\mathcal{K}^-$) consists of all the lunes which lie above (resp., below) the lines containing their "bases" (i.e., the corresponding Delaunay edges). We describe the processing of $\mathcal{K}^+$; processing $\mathcal{K}^-$ is done in a fully symmetric manner. We refer to the lunes in $\mathcal{K}^+$ as *upper lunes*.

We next distinguish between lunes in $\mathcal{K}^+$ whose bases have positive slope and those whose bases have negative slope, and treat each subcollection separately. We will only consider the first subcollection, since the other one is handled in a fully symmetric manner, and, for simplicity of notation, we will continue to denote it as $\mathcal{K}^+$.

Let K_{ab} be one of these upper lunes, with, say, a lying to the left of b (and with ab having positive slope). We note that some portion of K_{ab} extends to the left of the vertical line through a and that K_{ab} lies fully to the left of the vertical line through b (this latter property holds because K_{ab} is always contained in the diametral disk of ab).

We form from $\mathcal{K}^+$ two new collections. We take each upper lune K_{ab} in $\mathcal{K}^+$, with a to the left of (and below) b, and split it into two open regions by drawing a vertical line through a. (The union of the regions is K_{ab} without its intersection segment with the vertical line through a.) One region, denoted K^T_{ab}, lies vertically above ab, and the other, denoted K^L_{ab}, lies fully to the left of a. We denote by $\mathcal{K}^T$ (resp., $\mathcal{K}^L$) the collection of all the top portions K^T_{ab} (resp., left portions K^L_{ab}) of the lunes of $\mathcal{K}^+$. See Figure 7 for an illustration.

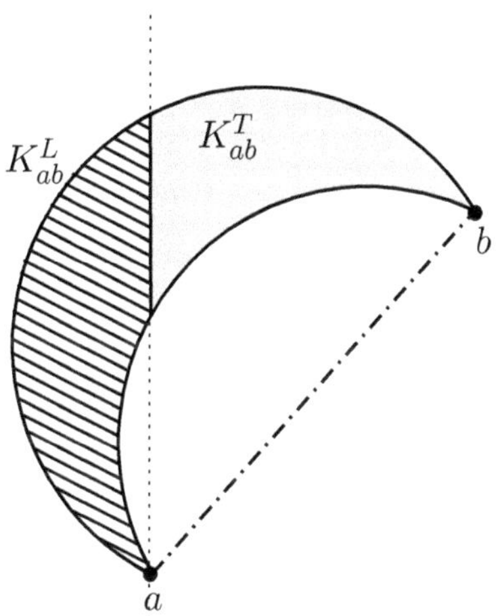

Figure 7: Partitioning an upper lune (with a base *ab* of positive slope) into a top portion and a left portion.

LEMMA 3.3. *The regions of $\mathcal{K}^T$ are pairwise disjoint.*

PROOF. Suppose to the contrary that there exist two distinct top regions K^T_{ab} and K^T_{cd} in $\mathcal{K}^T$, with a to the left of b and c to the left of d, such that K^T_{ab} and K^T_{cd} have a nonempty intersection, and let q be a point in both regions. By Lemma 3.2 there exists a line ℓ through q that weakly separates a and b from c and d. This however is impossible because q lies vertically above both segments ab and cd. See Figure 8. $\square$

LEMMA 3.4. *The regions of $\mathcal{K}^L$ are pairwise disjoint.*

PROOF. Suppose to the contrary that there exist two distinct left regions K^L_{ab} and K^L_{cd} in $\mathcal{K}^L$, with a to the left

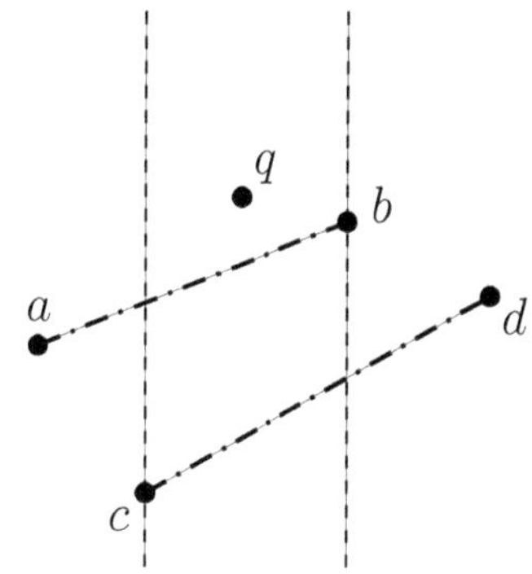

Figure 8: Illustrating the proof of Lemma 3.3.

of b and c to the left of d, such that K^L_{ab} and K^L_{cd} have a nonempty intersection, and let $q \in K^L_{ab} \cap K^L_{cd}$. By Lemma 3.2 there exists a line ℓ through q that weakly separates a and b from c and d. By construction, the two segments ab and cd lie to the right of the vertical line λ through q; one of them, say, ab lies in the upper-right quadrant Q^+ formed by ℓ and λ, and the other, cd, lies in the lower-right quadrant Q^-. See Figure 9. The angle at q of at least one of these quadrants is at most $\pi/2$; without loss of generality assume that this is the case for Q^-. Since q lies in K^L_{cd}, which is fully contained in the interior of the diametral disk of cd, it follows that $\angle cqd < \pi/2$, contradicting the properties of the lunes as studied at the beginning of this section. This contradiction completes the proof of the lemma. $\square$

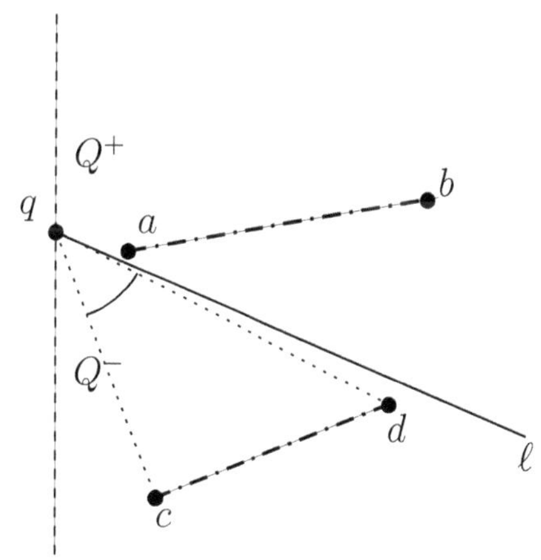

Figure 9: Illustrating the proof of Lemma 3.4.

3.4 The data structure

We thus obtain the following algorithm. Partition $\mathcal{K}$ into the four subcollections of upper lunes with positive-slope bases, upper lunes with negative-slope bases, lower lunes with positive-slope bases, and lower lunes with negative-slope bases. Each of these subcollections is processed in a similar manner, and we describe only the processing of the first subcollection, which, as above, we denote as $\mathcal{K}^+$.

We decompose each lune K in $\mathcal{K}^+$ into its top subregion K^T and its left subregion K^L, by the recipe given above, and let $\mathcal{K}^T$ (resp., $\mathcal{K}^L$) denote the collection of the resulting top (resp., left) subregions. By Lemmas 3.3 and 3.4, each collection consists of pairwise disjoint regions, and so forms a planar map of linear complexity (so that each subregion in the collection is a single separate face of the map). We process each of the two resulting maps for fast (logarithmic time) point location queries. Altogether we obtain eight such maps.

Now, given a query point q, we locate it in each of these eight maps, and retrieve the at most three lunes that contain q. Each such lune yields a candidate maximal empty disk containing q and centered at the respective Voronoi edge. (If q happens to lie on a vertical edge of a map, which bounds some region K_{ab}^T or K_{ab}^L, then the Voronoi edge e_{ab} is one of the three candidate edges.) This part of the data structure requires $O(n)$ space, take $O(n \log n)$ time to construct and querying it takes $O(\log n)$ time. We then find the largest empty Delaunay disk (centered at a Voronoi vertex) containing q, as described in Section 3.1, and output the largest of the at the four candidate disks. The overall complexity is dominated by the costs of handling Delaunay disks.

We thus obtain the main result of this paper.

THEOREM 3.5. *One can preprocess a set P of n points in the plane in $O(n \log^2 n)$ time into a data structure of size $O(n \log n)$, so that, given a query point q (in the interior of the convex hull of P), we can compute, in $O(\log^2 n)$ time, the largest P-empty disk containing q.*

Remarks: (1) In view of the applications discussed in the introduction, a natural interesting generalization of the problem studied in this paper is to handle queries, each consisting of a collection Q of k points, where, say, k is some constant, and the goal is to find the largest disk containing Q (or, rather, its convex hull $CH(Q)$) whose interior is disjoint from P. (Note that when $|Q| > 1$ the problem does not always have a solution.)

To answer such a query with a set Q we observe that the largest P-empty disk D containing Q must contain on its boundary either three points of P, in which case it is a Delaunay disk, or two points of P and one point of Q. Indeed, as is easy to see, in all other cases there is a larger P-empty disk containing Q. Finding the largest Delaunay disk containing a set of two or more points is harder than the case of a single query point since we cannot use the binary search procedure in the arrangements of unions of disks described in Section 3.1.

We can determine whether there exists a P-empty disk containing Q in polylogarithmic time with an appropriate preprocessing of P, as follows. We lift the points of P and the points of Q to the paraboloid $z = x^2 + y^2$ in $\mathbb{R}^3$. (For simplicity, we refer to the lifted sets of points also as P and Q, respectively.) In this representation a P-empty disk containing Q corresponds to a hyperplane separating the convex hull $CH(P)$ of P and the convex hull $CH(Q)$ of Q. Such a hyperplane exists if and only if $CH(Q)$ is disjoint from $CH(P)$. It is not difficult to show, using the technique of Dobkin and Kirkpatrick [2] that one can preprocess P in $O(n \log n)$ time, into a linear-size data structure, such that, given a query set Q, we can determine whether $CH(Q)$ and $CH(P)$ are disjoint, in $O(k \log n)$ time.

Briefly and informally, we note that since neither of $CH(Q)$, and $CH(P)$ fully contains the other hull, they intersect if and only if either an edge of $CH(Q)$ intersects a face of $CH(P)$ or (an edge of) $CH(P)$ intersects a face of $CH(Q)$. Intersection of edges of $CH(Q)$ with $CH(P)$ can be detected in $O(k \log n)$ time using the technique of [2]. Once we have determined that no edge of $CH(Q)$ meets $CH(P)$, we test, for each face f of $CH(Q)$, whether its supporting plane π_f meets $CH(P)$. (This can also be easily done in a total of $O(k \log n)$ time.) If no such intersection is found, f is disjoint from $CH(P)$. Otherwise we obtain a witness point $w \in \pi_f \cap CH(P)$, and we test whether $w \in f$ (and then f intersects $CH(P)$) or not (and then there is no intersection).

We also note that finding the largest P-empty disk whose boundary contains two points of P and one of Q is rather easy to do. We simply query the structure presented at Sections 3.2–3.4 with each point of Q separately, collect the at most $3k$ candidate disks, filter out those that do not fully contain Q, and return the longest among the surviving ones. With an appropriate implementation this takes a total of $O(k \log n + k \log k)$ time.

We leave the problem of finding the largest P-empty Delaunay disk containing Q as an open problem for further research.

(2) Another open problem is to reduce the query time to $O(\log n)$ without increasing significantly the storage and preprocessing costs. As already noted, the bottleneck in the query cost is at the stage that finds the largest Delaunay disk that contains q (see Section 3.1).

4. ACKNOWLEDGMENTS

Work by Haim Kaplan was partially supported by Grant 2006/204 from the U.S.–Israel Binational Science Foundation, Grant 822/10 from the Israel Science Fund, and the Israeli Centers of Research Excellence (I-CORE) program (Center No. 4/11).

Work by Micha Sharir was partially supported by NSF Grant CCR-08-30272, by Grant 2006/194 from the U.S.–Israel Binational Science Foundation, by Grant 338/09 from the Israel Science Fund, by the Hermann Minkowski–MINERVA Center for Geometry at Tel Aviv University, and by the Israeli Centers of Research Excellence (I-CORE) program (Center No. 4/11).

5. REFERENCES

[1] J. Augustine, S. Das, A. Maheshwari, S. C. Nandy, S. Roy, and S. Sarvattomananda, Recognizing the largest empty circle and axis-parallel rectangle in a desired location, in `arXiv.org:1004.0558`, 2010.

[2] D. P. Dobkin and D. G. Kirkpatrick, Fast detection of polyhedral intersection, *Theoret. Comput. Sci.* 27 (1983), 241–253.

[3] H. Kaplan, S. Mozes, Y. Nussbaum and M. Sharir, Submatrix maximum queries in Monge matrices and Monge partial matrices, and their applications, *Proc. 23rd ACM-SIAM Annu. Sympos. Discrete Algorithms*, 2012, 338-355.

[4] H. Kaplan and M. Sharir, Finding the maximal empty rectangle containing a query point, in `arXiv.org:1106.3628`, 2011.

[5] K. Kedem, R. Livne, J. Pach and M. Sharir, On the union of Jordan regions and collision-free translational motion amidst polygonal obstacles, *Discrete Comput. Geom.* 1 (1986), 59–71.

Conflict-Free Coloring of Points with Respect to Rectangles and Approximation Algorithms for Discrete Independent Set

Timothy M. Chan*
Cheriton School of Computer Science
University of Waterloo
tmchan@uwaterloo.ca

ABSTRACT

In the conflict-free coloring problem, for a given range space, we want to bound the minimum value $F(n)$ such that every set P of n points can be colored with $F(n)$ colors with the property that every nonempty range contains a unique color. We prove a new upper bound $O(n^{0.368})$ with respect to orthogonal ranges in two dimensions (i.e., axis-parallel rectangles), which is the first improvement over the previous bound $O(n^{0.382})$ by Ajwani, Elbassioni, Govindarajan, and Ray [SPAA'07]. This result leads to an $O(n^{1-0.632/2^{d-2}})$ upper bound with respect to orthogonal ranges (boxes) in dimension d, and also an $O(n^{1-0.632/(2^{d-3}-0.368)})$ upper bound with respect to dominance ranges (orthants) in dimension $d \geq 4$.

We also observe that combinatorial results on conflict-free coloring can be applied to the analysis of approximation algorithms for *discrete* versions of *geometric independent set* problems. Here, given a set P of (weighted) points and a set S of ranges, we want to select the largest(-weight) subset $Q \subset P$ with the property that every range of S contains at most one point of Q. We obtain, for example, a randomized $O(n^{0.368})$-approximation algorithm for this problem with respect to orthogonal ranges in the plane.

Categories and Subject Descriptors

F.2.2 [**Analysis of Algorithms and Problem Complexity**]: Nonnumerical Algorithms and Problems—*geometrical problems and computations*; G.2.1 [**Discrete Mathematics**]: Combinatorics

Keywords

Conflict-free coloring, independent set, range spaces, combinatorial geometry, approximation algorithms

*Work supported by an NSERC grant.

1. INTRODUCTION

The main topic of this paper concerns three inter-related problems in combinatorial geometry:

Definition 1. (CF-coloring of elements w.r.t. ranges) Let $\mathcal{S}$ be a range space (or set system) [22, 23]. A *conflict-free (or CF-)coloring* of a set P of elements is a coloring χ of P such that for every range s in $\mathcal{S}$ with $P \cap s \neq \emptyset$, the multiset $\{\chi(p) \mid p \in P \cap s\}$ contains a unique color. Let CF-COLOR$(n, \mathcal{S})$ be the minimum number f such that every set of n elements in $\mathcal{S}$ has a CF-coloring with f colors.

Definition 2. (DG_k-independent set of elements w.r.t. ranges) Given a set P of elements in a range space $\mathcal{S}$ and a value k, define the *"order-k" Delaunay graph* $\mathrm{DG}_k(P, \mathcal{S})$ to be the graph where the vertices are the elements of P, and pq is an edge iff there exists a range s in $\mathcal{S}$ such that $p, q \in s$ and $|P \cap s| \leq k$. A *DG_k-independent set* of P refers to an independent set of the graph $\mathrm{DG}_k(P, \mathcal{S})$. Let DG-INDEP$_k(n, \mathcal{S})$ be the maximum number g such that every set of n elements in $\mathcal{S}$ has a DG_k-independent set of size g.

Definition 3. (DG_k-coloring of elements w.r.t. ranges) A *DG_k-coloring* of a set P of elements refers to a vertex coloring of the graph $\mathrm{DG}_k(P, \mathcal{S})$. Let DG-COLOR$_k(n, \mathcal{S})$ be the minimum number f such that every set of n elements has a DG_k-coloring with f colors.

Of the above three, CF-coloring has garnered the most fame (hence, we use the term in the title); see [25] for a recent survey. Frequency assignment in radio networks is often mentioned as one possible application; another application will be described in Section 1.3.

The standard Delaunay graph corresponds to the $k = 2$ case; for simplicity, we drop the subscript in the above definitions when $k = 2$, e.g., DG-INDEP$(n, \mathcal{S})$ = DG-INDEP$_2(n, \mathcal{S})$. In the case of orthogonal ranges, i.e., boxes,[1] the concept is further simplified because for every two points p, q, there is a unique minimal enclosing box, which we denote by $B(p, q)$. We get similar definitions for CF-coloring/DG_k-independent set/DG_k-coloring of *ranges w.r.t. elements* when we dualize the range space.

As explained below, the problems of bounding CF-COLOR$(n, \mathcal{S})$, DG-INDEP$(n, \mathcal{S})$, and DG-COLOR$(n, \mathcal{S})$ are known to be equivalent up to logarithmic factors:

[1] All rectangles and boxes are axis-parallel in this paper.

PROPOSITION 1.1. *Let α, c be constants. Assume the following monotonicity property (which holds for all range spaces considered in this paper): for every set P of elements and every range s in $\mathcal{S}$ with $|P \cap s| > 2$, there exists a range s' in $\mathcal{S}$ with $P \cap s' \subset P \cap s$ and $|P \cap s'| = 2$. Then*[2]

(i) $\text{DG-INDEP}_k(n, \mathcal{S}) = \widetilde{\Omega}(n^\alpha / k^c)$ *iff* $\text{DG-COLOR}_k(n, \mathcal{S}) = \widetilde{O}(n^{1-\alpha} k^c)$;

(ii) $\text{DG-INDEP}(n, \mathcal{S}) = \widetilde{\Omega}(n^\alpha)$ *iff* $\text{CF-COLOR}(n, \mathcal{S}) = \widetilde{O}(n^{1-\alpha})$.

Proposition 1.1(i) is easy to see, due to connection between independent set and coloring in graphs (since our graph family DG_k is hereditary). Proposition 1.1(ii) follows from a simple algorithm which appeared in Even et al.'s original paper [16] on CF-coloring and was re-explained in Har-Peled and Smorodinsky's paper [19].

1.1 Main Result on Points w.r.t. Rectangles in Two Dimensions

Many results on CF-coloring are known in the literature [25]. For example, there is an $O(\log n)$ upper bound in the dual for disks w.r.t. points in two dimensions (or more generally, for planar objects with linear union complexity w.r.t. points), and polylogarithmic upper bounds for two-dimensional rectangles and three-dimensional boxes w.r.t. points [16, 19]. Perhaps the most perplexing open problem in the area—the simplest case for which no polylogarithmic upper bound is currently known and yet a sublinear bound is possible—is CF-coloring of *points w.r.t. rectangles* in the plane. Let $\mathcal{B}^d$ be the range space where the elements are all points in $\mathbb{R}^d$ and the ranges are all boxes in $\mathbb{R}^d$. We provide a brief history of the problem of bounding $\text{CF-COLOR}(n, \mathcal{B}^2)$:

There are at least two different proofs that $\text{CF-COLOR}(n, \mathcal{B}^2) = O(\sqrt{n})$, i.e., $\text{DG-INDEP}(n, \mathcal{B}^2) = \Omega(\sqrt{n})$. One proof [19] uses the Erdős–Szekeres theorem to find a chain of $\Omega(\sqrt{n})$ points which is monotone in x and y simultaneously; then an independent set can be obtained by taking, say, all the even-indexed points in the chain. Another proof is divided into two cases: every graph with maximum degree at most $\sqrt{n}$ has independence number $\Omega(\sqrt{n})$; on the other hand, if some vertex in the Delaunay graph has degree exceeding $\sqrt{n}$, then its neighborhood would contain a monotone chain of size $\Omega(\sqrt{n})$, from which an independent set of size $\Omega(\sqrt{n})$ can again be obtained. This second proof can be refined to yield a slight improvement $\text{CF-COLOR}(n, \mathcal{B}^2) = O(\sqrt{n/\log n})$ by using better bounds on the independence number of graphs with sparse neighborhoods [2], as several researchers have independently observed [19, 24].

The square-root barrier was finally broken in a more substantial way by Ajwani et al. [1], by building on ideas from a previous paper of Elbassioni and Mustafa [13]. Ajwani et al. showed that $\text{CF-COLOR}(n, \mathcal{B}^2) = O(n^{2-\phi+\varepsilon}) = O(n^{0.382})$, where ϕ is the golden ratio $(1+\sqrt{5})/2$ and ε is an arbitrarily small constant. Consequently, $\text{DG-INDEP}(n, \mathcal{B}^2) = \Omega(n^{\phi-1-\varepsilon}) = \Omega(n^{0.618})$. In the opposite direction, Chen et al. [10] proved that $\text{DG-INDEP}(n, \mathcal{B}^2) = O(n \log^2 \log n / \log n)$.

[2]The $\widetilde{O}$ or $\widetilde{\Omega}$ notation hides $\log^{O(1)} n$ and $\log^{O(1)} m$ factors in this paper.

In Section 2, we refine Ajwani et al.'s proof approach [1] to obtain the following result:

THEOREM 1.2. $\text{DG-INDEP}(n, \mathcal{B}^2) = \widetilde{\Omega}(n^\alpha)$ *for some* $\alpha > 0.632$.

This implies that $\text{CF-COLOR}(n, \mathcal{B}^2) = \widetilde{O}(n^{1-\alpha}) = O(n^{0.368})$. Although the improvement is not huge, it represents the first progress in several years, and deepens the mystery surrounding the true value of the exponent. Our value of 0.632 or 0.368 originates from an interesting recurrence relation, and is a result of a computer-assisted search (alas, not a nice number like the golden ratio).

1.2 Implications for Larger k and Higher Dimensions

We next examine consequences of our result, extending Theorem 1.2 to larger $k > 2$ as well as to boxes in higher dimensions. We also consider the case of dominance ranges, i.e., orthants. Let $\mathcal{O}^d$ be the range space where the elements are points in $\mathbb{R}^d$ and the ranges are orthants of the form $(-\infty, x_1] \times \cdots \times (-\infty, x_d]$. In Section 3, we prove the following:

THEOREM 1.3. *Suppose that* $\text{DG-INDEP}(n, \mathcal{B}^2) = \widetilde{\Omega}(n^\alpha)$. *Then*

(i) $\text{DG-INDEP}_k(n, \mathcal{B}^2) = \widetilde{\Omega}([n/k]^\alpha)$, *and more generally,* $\text{DG-INDEP}_k(n, \mathcal{B}^d) = \widetilde{\Omega}([n/k]^{\alpha/2^{d-2}})$ *for any fixed* $d \geq 2$;

(ii) $\text{DG-INDEP}_k(n, \mathcal{O}^d) = \widetilde{\Omega}([n/k]^{\alpha/(2^{d-3}-1+\alpha)})$ *for any fixed* $d \geq 4$.

In the two-dimensional case, the extension to larger k in Theorem 1.3(i) is not supposed to be trivial and exploits ideas specific to the range space $\mathcal{B}^2$. In higher dimenisons, Har-Peled and Smorodinksy [19] previously stated a weaker bound $\text{CF-COLOR}(n, \mathcal{B}^d) = O(n^{1-1/2^{d-1}})$, i.e., $\text{DG-INDEP}(n, \mathcal{B}^d) = \Omega(n^{1/2^{d-1}})$. The proof of Theorem 1.3(ii) requires further ideas, and in particular, the consideration of an intermediate case of points w.r.t. 3-dimensional 4-sided boxes. The bounds deteriorate quickly as the dimension increases, but the bounds for small d are still interesting, e.g., we have $\text{DG-INDEP}(n, \mathcal{O}^4) = \widetilde{\Omega}(n^{\alpha/(1+\alpha)}) = \Omega(n^{0.387})$.

The dual $(\mathcal{B}^d)^*$ of the range space $\mathcal{B}^d$ may be viewed as a special case of $\mathcal{O}^{2d}$ by mapping a box $[a_1, b_1] \times \cdots \times [a_d, b_d]$ to the point $(a_1, -b_1, \ldots, a_d, -b_d) \in \mathbb{R}^{2d}$. Thus, as a corollary of Theorem 1.3(ii), $\text{DG-INDEP}_k(n, (\mathcal{B}^d)^*) = \widetilde{\Omega}([n/k]^{\alpha/(2^{2d-3}-1+\alpha)})$.

1.3 Application to Discrete Independent Set

We also point out an application to approximation algorithms for the following optimization problem, which is actually the original motivation of this paper:

PROBLEM 1.4. *(Discrete independent set of elements w.r.t. ranges)* Given a set P of n (weighted) elements and a set S of m ranges in a range space $\mathcal{S}$, find the largest(-weight) subset $Q \subseteq P$ such that each range $s \in S$ contains at most one element of Q.

We obtain a similar optimization problem of discrete independent set of *ranges w.r.t. elements* when we dualize the range space: namely, given a set P of m elements and a set S of n (weighted) ranges in a range space $\mathcal{S}$, find the largest(-weight) subset $Q \subseteq S$ such that each element $p \in P$ is contained in at most one range of Q. When P is taken as the entire universe, this dual problem is equivalent to finding maximum independent sets in intersection graphs—this "continuous" version of geometric independent set has been studied extensively (e.g., see [6, 7, 9, 18]). The unweighted discrete independent set problem can be viewed as the LP-dual of geometric *set cover* and geometric *hitting set*, both of which have also received considerable attention (e.g., see [5, 8, 12]).

In contrast, the discrete version of geometric independent set has not been studied as often. We are aware of only two papers: Chan and Har-Peled's work [9] has a short section showing that their $O(1)$-approximation algorithm for independent set of pseudo-disks can be extended to discrete independent set of pseudo-disks w.r.t. points. Ene, Har-Peled, and Raichel's work [14] contains multiple results on a problem that generalizes discrete independent set (namely, replacing the "at most one element/range" with "at most the capacity of the element/range" in the constraints). For example, they gave polylogarithmic-approximation algorithms for discrete independent set of two-dimensional rectangles and three-dimensional boxes w.r.t. points.

We observe the following connection:

THEOREM 1.5. *Consider the following linear program (LP) for Problem 1.4:*

$$\begin{array}{lll} \text{maximize} & \sum_{p\in P} w_p x_p & \\ \text{subject to} & \sum_{p\in s} x_p \le 1 & \forall s \in S \\ & 0 \le x_p \le 1 & \forall p \in P, \end{array}$$

where $w_p > 0$ denotes the weight of p. Let opt *be the optimal value of the LP and* OPT *be the optimal value of the original problem. Suppose* DG-INDEP$_k(n, \mathcal{S}) = \widetilde{\Omega}(n^\alpha / k^{O(1)})$. *Then* $\mathrm{OPT} \ge \widetilde{\Omega}(\mathrm{opt}^\alpha) \ge \widetilde{\Omega}(\mathrm{OPT}^\alpha)$ *in the unweighted case, and* $\mathrm{OPT} \ge \widetilde{\Omega}(\mathrm{opt}/n^{1-\alpha}) \ge \widetilde{\Omega}(\mathrm{OPT}/n^{1-\alpha})$ *in the weighted case.*

The proof, given in Section 4, uses random sampling as in [9] and is straightforward (especially since we are not too concerned with optimizing logarithmic factors here, unlike in [9]). For this LP, there are unweighted instances where OPT $\le$ DG-INDEP$(n, \mathcal{S})$ but opt $\ge n/2$: just let S consist of all (combinatorially different) ranges in $\mathcal{S}$ that contain exactly two points in the set P that realizes DG-INDEP$(n, \mathcal{S})$, and set all x_p's to 1/2. Thus, the above theorem is in some sense tight (ignoring logarithmic factors).

From the bound on DG-INDEP$_k(n, \mathcal{B}^2)$ implied by Theorems 1.2 and 1.3(i), we immediately obtain an $O(n^{0.368})$-approximation algorithm for weighted discrete independent set of points w.r.t. rectangles in two dimensions, or an $\widetilde{O}(\mathrm{OPT}^{0.368})$-approximation algorithm in the unweighted case; to the author's knowledge, no previous approximation algorithms were given for this particular problem. From the bound on DG-INDEP$_k(n, (\mathcal{B}^d)^*)$ implied by Theorems 1.2 and 1.3(ii), we obtain the first sublinear approximation algorithm for discrete independent set of boxes w.r.t. points in dimensions $d \ge 4$; Ene et al. [14] previously stated the $d = 4$ case as an open problem. Since the proofs of all the theorems are constructive (using randomization in some places), it can be verified that the resulting approximation algorithms run in Las Vegas polynomial time.

We remark that our linking of LP-based approximation algorithms to combinatorial geometry problems continues an existing trend. For example, bounding the integrality gap of the standard LP for unweighted geometric set cover and hitting set problems is equivalent (up to constant factors) to the *ε-net* problem [17, 21]. For (continuous or discrete) independent set of planar objects w.r.t. points, Chan and Har-Peled [9] have shown previously that an upper bound on the integrality gap can be obtained from the *union complexity* (but unfortunately boxes may have high union complexity even in two dimensions).

2. POINTS W.R.T. RECTANGLES IN TWO DIMENSIONS (PROOF OF THEOREM 1.2)

In this section, we prove that DG-COLOR$(n, \mathcal{B}^2) = \widetilde{O}(n^{1-\alpha})$ for some $\alpha > 0.632$, thereby showing that DG-INDEP$(n, \mathcal{B}^2) = \widetilde{\Omega}(n^\alpha)$. Our proof heavily builds on Ajwani et al.'s grid-based approach [1], but our description will be kept self-contained. We will solve a generalization of the problem, introducing two additional parameters v and h:

PROBLEM 2.1. Given a set P of $\le n$ points in the plane and a set V of $< v$ vertical lines and a set H of $< h$ horizontal lines, define a *"generalized Delaunay graph"* $\mathrm{DG}(P, V, H, \mathcal{B}^2)$ where the vertices are the points in P, and pq is an edge iff $B(p, q) \cap P = \{p, q\}$, and *$p$ and q are separated by both a vertical line in V and a horizontal line in H.* We want to find a coloring of the graph $\mathrm{DG}(P, V, H, \mathcal{B}^2)$, which in this section will be referred simply as a *coloring of P relative to V and H.*

Let $F(n, v, h)$ be the minimum number of colors needed to solve this problem. It will be more convenient to work with the function $G(n, v, h) = n/F(n, v, h)$.

Initially, we can set V to consist of $n - 1$ vertical lines separating all the points and H to consist of $n-1$ horizontal lines separating all the points; then a coloring of P relative to V and H would be a DG-coloring of P. Thus, our goal is to bound $G(n, n, n)$.

For "base cases" of our recursive algorithm, we can start with a weak bound $F(n, v, h) = O(\sqrt{n})$ by one of the known methods [19]. Another upper bound is $F(n, v, h) \le v$, which follows from the trivial method of assigning the same color to all points between two consecutive vertical lines of V. Rewriting in terms of the function G, we have

$$G(n, v, h) \;\ge\; \Omega(\sqrt{n}) \qquad (1)$$

$$G(n, v, h) \;\ge\; n/v. \qquad (2)$$

Symmetry gives the following additional relationship:

$$G(n, v, h) \;=\; G(n, h, v). \qquad (3)$$

It turns out a further generalization involving a fourth parameter r_0 will be useful for our recursive algorithm:

Definition 4. We say that (x_1, y_1) *(++)-dominates* (x_2, y_2) if $x_1 > x_2$ and $y_1 > y_2$. Similarly, (x_1, y_1) *(+−)-dominates* (x_2, y_2) if $x_1 > x_2$ and $y_1 < y_2$. For any sign

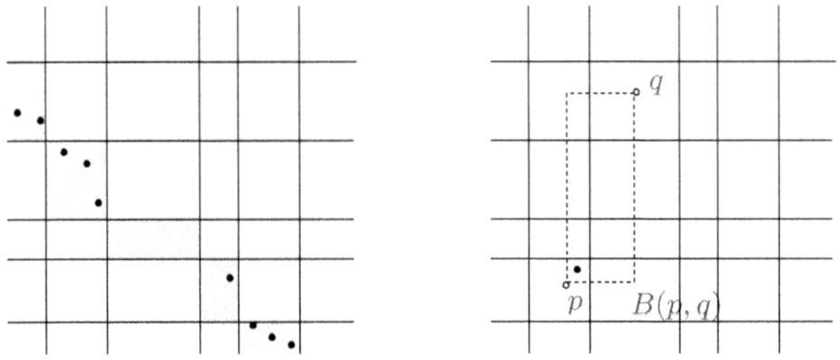

Figure 1: (Left) The (++)-maximal points in the cells of one diagonal of a grid $\mathcal{G}_0$. (Right) Two unmarked points p and q lying in different columns and different rows of a grid $\mathcal{G}$.

sequence σ, we say that a point p is *σ-maximal* in a set P if $p \in P$ and p is σ-dominated by no other point in P.

Given a grid $\mathcal{G}_0$ formed by a set of vertical and horizontal lines, we say that a point set P *respects* $\mathcal{G}_0$ if for every grid cell τ in $\mathcal{G}_0$, every point in $P \cap \tau$ is (++)-maximal or (+−)-maximal in $P \cap \tau$.

Let $F_{r_0}(n, v, h)$ be the minimum number of colors needed to solve Problem 2.1, under the knowledge that the point set P respects an $r_0 \times r_0$ grid $\mathcal{G}_0$. Let $G_{r_0}(n, v, h) = n/F_{r_0}(n, v, h)$. (Of course, every point set P respects a sufficiently fine, $n \times n$ grid $\mathcal{G}_0$, so $G(n, v, h) = G_n(n, v, h)$.)

For base case, we have the upper bound $F_{r_0}(n, v, h) = O(r_0)$: This follows by observing that the (++)-maxima (resp. (+−)-maxima) in each grid cell in $\mathcal{G}_0$ form a monotone chain, i.e., a downward (resp. upward) "staircase". Since the grid cells of $\mathcal{G}_0$ can be partitioned into $O(r_0)$ northwest-to-southeast or southwest-to-northeast diagonals (see Figure 1(left) for one such diagonal), the whole point set P can be decomposed into $O(r_0)$ monotone chains. In each chain, we can easily color the points alternately with two colors. The claim follows. In terms of the function G,

$$G_{r_0}(n, v, h) \ \geq\ \Omega(n/r_0). \qquad (4)$$

2.1 The Main Algorithm

We now present our main recursive algorithm to solve Problem 2.1. Let $r \in [1, \min\{v, h\}]$ be a parameter. The algorithm consists of the five simple steps:

1. Form a grid $\mathcal{G}$ such that each column contains $\leq n/r$ points of P and $< v/r$ vertical lines of V, and each row contains $\leq n/r$ points of P and $< h/r$ horizontal lines of H. At most $O(r)$ vertical and horizontal grid lines suffice.

2. For each grid cell τ of $\mathcal{G}$, mark the (++)-maxima and (+−)-maxima of $P \cap \tau$. Find a coloring of all the marked points in P. Let C_0 be the palette used.

3. For each column ξ of $\mathcal{G}$, recursively find a coloring of the unmarked points in $P \cap \xi$ *relative to the vertical lines of V inside ξ and the horizontal grid lines of $\mathcal{G}$*. Use a common palette C in all the columns. Make sure that each coloring is "distributed fairly"—see details below.

4. For each row η of $\mathcal{G}$ and each $i \in C$, recursively find a coloring of the unmarked points in $P \cap \eta$ that have received color i in Step 3, *relative to all the vertical lines of V and the horizontal lines of H inside η*. Use a common palette C' in all the columns.

5. An unmarked point that has received color i in Step 3 and color j in Step 4 is given a final color (i, j) in the palette $C \times C'$ (assumed to be disjoint from C_0).

Correctness. We now verify that the result is a valid coloring of $\mathrm{DG}(P, V, H, \mathcal{B}^2)$. Consider two points $p, q \in P$ such that $B(p, q) \cap P = \{p, q\}$ and p and q are separated by a vertical line of V and a horizontal line of H. If both p and q are marked, they must have different colors by Step 2. If one is marked and the other is unmarked, they obviously have different colors. From now on, we may assume that p and q are both unmarked. Three cases may arise:

- p and q lie in different columns and different rows of $\mathcal{G}$. If p is to the left of and below q as in Figure 1(right) (all other cases are similar), then $B(p, q)$ must contain a (++)-maximal point inside the grid cell containing p; this point must be different from p and q since p is unmarked—a contradiction.

- p and q lie in the same column but in different rows of $\mathcal{G}$. Then p and q are separated by a horizontal grid line of $\mathcal{G}$. They must receive different colors in Step 3, and thus different final colors.

- p and q lie in the same row of $\mathcal{G}$. If they receive the same color in Step 3, they will receive different colors in Step 4. Thus, they must have different final colors.

Details. Step 2 can be done using a palette C_0 of $F_{cr}(n, v, h)$ colors for some constant c, since the marked points respect the $cr \times cr$ grid $\mathcal{G}$.

Step 3 can be done using a palette C of $f = F(n/r, v/r, cr)$ colors. We ensure that the coloring of the unmarked points in $P \cap \xi$ for each column ξ is *distributed fairly*, i.e., satisfies the following properties: (a) the maximum size of the color classes is $O(|P \cap \xi|/f)$, and (b) each point has probability $O(1/f)$ of receiving any fixed color.

To ensure (a), when a color class contains more than $2|P \cap \xi|/f$ points, we can just subdivide it into color subclasses of size $\Theta(|P \cap \xi|/f)$. The number of colors remains $O(f)$. To ensure (b), we can just pick a random permutation π_ξ of the color palette and rename color i as color $\pi_\xi(i)$.

Let P_i denote the subset of all unmarked points that have received color i after Step 3. Consider a row η and a color i. We analyze the random variable $|P_i \cap \eta|$. Since each point has probability $O(1/f)$ of receiving color i by (b), $|P_i \cap \eta|$ has expected value $\mu = O(n/(rf))$. Furthermore, $|P_i \cap \eta| = \sum_\xi |P_i \cap \xi \cap \eta|$, where the sum is over all columns ξ. The terms in the sum are independent random variables, since the permutations π_ξ's are independent. Because $|P_i \cap \xi| \leq O(|P \cap \xi|/f) = O(n/(rf))$ by (a), each term in the sum is upper-bounded by $M = O(n/(rf))$. By the Chernoff–Hoeffding inequality [20] for sums of bounded random variables (after rescaling the variables and the mean by a factor of $1/M$),

$$\Pr[|P_i \cap \eta| > (1+\delta)\mu] \ \leq\ \left[\frac{e^\delta}{(1+\delta)^{1+\delta}}\right]^{\mu/M}$$

for any $\delta > 0$ (possibly large). Choosing $\delta = \Theta((1 + M/\mu)\log n)$, we have $|P_i \cap \eta| \leq O((n/(rf))\log n)$ with high probability (e.g., probability $1 - o(1/n^2)$). So, by repeating for an expected $O(1)$ number of trials, we can find a choice of random permutations that ensure $|P_i \cap \eta| \le \widetilde{O}(n/(rf))$ for all rows η and all colors i at the same time.

Step 4 can therefore be done using a palette C' of $F(c'n/(rf), v, h/r)$ colors for some $c' = \widetilde{O}(1)$.

The recurrence. The final palette $C_0 \cup (C \times C')$ has size $F_{cr}(n, v, h) + O(f)\, F(c'n/(rf), v, h/r)$, where $f = F(n/r, v/r, cr)$. We conclude that for any $r \in [1, \min\{v, h\}]$,

$$F(n,v,h) \le F_{cr}(n,v,h) + O(F(n/r,v/r,cr))\, F\left(\frac{c'n/r}{F(n/r,v/r,cr)}, v, h/r\right).$$

When expressed in terms of the function $G(n, v, h) = n/F(n, v, h)$, this peculiar recurrence is simplified somewhat. Make the following definition: $x \oplus y = z$ iff $1/x + 1/y = 1/z$. Then for any $r \in [1, \min\{v, h\}]$,

$$G(n,v,h) \ge G_{cr}(n,v,h) \oplus \widetilde{\Omega}(r)\, G\left(c'G\left(\frac{n}{r}, \frac{v}{r}, cr\right), v, \frac{h}{r}\right). \quad (5)$$

Note that one can think of the operator $\oplus$ as a shorthand for min, since they have the same behavior asymptotically when the number of terms is small: more precisely, $(1/k)\min\{x_1, \ldots, x_k\} \le x_1 \oplus \cdots \oplus x_k \le \min\{x_1, \ldots, x_k\}$ for all $x_1, \ldots, x_k > 0$.

2.2 First Analysis of the Recurrence

We now analyze the recurrence (5) (in conjunction with (1)–(4)). First, we ignore the effects of the parameters v and h. Assume that a bound $G(n, v, h) = \widetilde{\Omega}(n^\alpha)$ is known. Then (5) together with (4) implies that

$$G(n,v,h) \ge \widetilde{\Omega}\left(n/r \oplus r[n/r]^{\alpha^2}\right).$$

By setting $r = n^{(1-\alpha^2)/(2-\alpha^2)}$ to balance the two terms, we obtain a new bound $G(n, v, h) = \widetilde{\Omega}(n^{1/(2-\alpha^2)})$.

We can apply this process a constant number of times, starting with the base case (1). It is straightforward to verify that the sequence $\alpha_0 = 1/2$, $\alpha_{i+1} = 1/(2 - \alpha_i^2)$ converges to a root of the equation $\alpha = 1/(2 - \alpha^2)$, namely, $\phi - 1 = (\sqrt{5} - 1)/2 > 0.618$. For any arbitrarily small constant $\varepsilon > 0$, we thus get $G(n, n, n) = \widetilde{\Omega}(n^{\phi-1-\varepsilon}) = \Omega(n^{0.618})$ and $F(n, n, n) = O(n^{0.382})$, recovering the previous result by Ajwani et al. [1].

Although our explanation might be somewhat cleaner, our algorithm is identical to Ajwani et al.'s when V and H are absent. Our main innovation is in considering colorings *relative to the horizontal grid lines* during Step 3. Because Ajwani et al. did not do this, the case when unmarked points p and q lie in both the same column and the same row is "handled" twice—our idea eliminates this redundancy.

It may not be immediately obvious that taking into account v and h can indeed lead to an improvement, however. One needs more than one level of recursion to see an effect. The following is a shortest chain of applications of (5) that yield a new result. In each instance, we use (4) for the first term $G_r(n, r, h)$ and the bound $G(n, v, h) = \Omega(n^{0.618})$ for the outer G in the second term of the right-hand side of (5):

$$\begin{aligned}
&G(n,n,n) \\
&\quad \ge \widetilde{\Omega}\left(n^{0.622} \oplus n^{0.378}[G(n^{0.622}, n^{0.622}, cn^{0.378})]^{0.618}\right) \\
&G(n^{0.622}, n^{0.622}, cn^{0.378}) = G(n^{0.622}, cn^{0.378}, n^{0.622}) \\
&\quad \ge \widetilde{\Omega}\left(n^{0.395} \oplus n^{0.227}[G(n^{0.395}, cn^{0.151}, cn^{0.227})]^{0.618}\right) \\
&G(n^{0.395}, cn^{0.151}, cn^{0.227}) \\
&\quad \ge \widetilde{\Omega}\left(n^{0.272} \oplus n^{0.123}[G(n^{0.272}, cn^{0.028}, cn^{0.123})]^{0.618}\right) \\
&G(n^{0.272}, cn^{0.028}, cn^{0.123}) \\
&\quad \ge \Omega(n^{0.272-0.028}) = \Omega(n^{0.244}) \text{ by (2).}
\end{aligned}$$

Substituting in reverse, we conclude that $G(n, n, n) = \widetilde{\Omega}(n^{0.622})$, i.e., $F(n, n, n) = \widetilde{\Omega}(n^{0.378})$.

The recurrence appears hard to solve exactly, but clearly further improvements should be possible with more iterations. Later in Section 2.4, we will describe a way to make the setting of r and the verification process more systematic. However, we will not explore the best bound achievable by (5) at this point, since a different version of the algorithm and recurrence turns out to be more effective...

2.3 Variants of the Algorithm

For example, a similar algorithm gives the following alternative recurrence for any $r \in [1, \min\{v, h\}]$:

$$G(n,v,h) \ge G_{cr}(n,v,h) \oplus \widetilde{\Omega}(r)\, G\left(c'G\left(\frac{n}{r}, \frac{v}{r}, h\right), cr, \frac{h}{r}\right). \quad (6)$$

This is obtained by modifying the algorithm so that the colorings in Step 3 are now *relative to the vertical lines of* V *inside* ξ *and all the horizontal lines of* H and the colorings in Step 4 are now *relative to the vertical grid line of* $\mathcal{G}$ *and the horizontal lines of* H *inside* η. In the correctness proof, Step 3 takes care of the case when p and q lie in the same column, and Step 4 takes care of the case when p and q lie in the same row but in different columns.

Better still, we can prove the following for any $r \in [1, \min\{r_0, v, h\}]$:

$$G_{r_0}(n,v,h) \ge G_{cr}(n,v,h) \oplus \widetilde{\Omega}(r \cdot (r/r_0))\, G\left(c'G\left(\frac{n}{r}, \frac{v}{r}, cr\right), cr, \frac{h}{r}\right). \quad (7)$$

Here, we need several modifications to the algorithm, knowing that the given point set P respects an $r_0 \times r_0$ grid $\mathcal{G}_0$: First, in step 1, by adding $O(r)$ vertical and horizontal grid lines to $\mathcal{G}$, we can ensure that each column of $\mathcal{G}$ contains $< r_0/r$ vertical grid lines of $\mathcal{G}_0$, and each row of $\mathcal{G}$ contains $< r_0/r$ horizontal grid lines of $\mathcal{G}_0$. As before, the colorings in Step 3 are *relative to the vertical lines of* V *inside* ξ *and the horizontal grid lines of* $\mathcal{G}$, but the colorings in step 4 are now *relative to the vertical grid line of* $\mathcal{G}$ *and the horizontal lines of* H *inside* η. We perform an extra step: inside each grid cell τ of $\mathcal{G}$, we compute a coloring of the unmarked points in $P \cap \tau$ using a common palette C''; since $P \cap \tau$ respects an $\lfloor r_0/r \rfloor \times \lfloor r_0/r \rfloor$ grid, C'' requires $O(r_0/r)$ colors by (4) (which accounts for the extra r/r_0 factor in (7)). In Step 5, an unmarked point that has received color i in Step 3, color j in Step 4, color ℓ in the extra step is given a final color (i, j, ℓ) in the palette $C \times C' \times C''$.

In the correctness proof, Step 3 takes care of the case when p and q lie in the same column but in different rows of $\mathcal{G}$, and Step 4 takes care of the case when p and q lie in the same row but in different columns of $\mathcal{G}$. The extra step takes of the case when p and q lie in both the same row and the same column of $\mathcal{G}$. This establishes (7).

Using (7) to bound $G(n,v,h)$ requires some preparation, since we need to keep the ratio r_0/r small in each application. We first start with the following: for $r^* = \min\{v,h\}$,

$$\begin{aligned} G(n,v,h) &\geq G_{cr^*}(n,v,h) \oplus \\ &\quad \widetilde{\Omega}(r^*)\, G\left(c'G\left(\frac{n}{r^*},\frac{v}{r^*},cr^*\right),cr^*,\frac{h}{r^*}\right). \end{aligned}$$

If $h \leq v$, then $r^* = h$ and the above inequality follows from (5) since by (2), $G(x,v,1) = x = G(x,cr^*,1)$. If $v \leq h$, then $r^* = v$ and the above inequality follows from (6) since by (2), $G(x,1,h) = x = G(x,1,cr^*)$.

Next, by applying (7) with $r_0 = cr^*/2^i$ and $r = r^*/2^{i+1}$ repeatedly for logarithmically many iterations and terminating with the base case (4) with $r = r_*$, we obtain

$$\begin{aligned} G(n,v,h) &\geq \Omega(n/r_*) \oplus \\ &\quad \min_{r_* \leq r \leq \min\{h,v\}} \widetilde{\Omega}(r)\, G\left(c'G\left(\frac{n}{r},\frac{v}{r},cr\right),cr,\frac{h}{r}\right) \end{aligned} \tag{8}$$

for any $r_* \in [1,\min\{v,h\}]$. (Remember that $\oplus$ behaves like min, since the number of terms is $O(\log n) = \widetilde{O}(1)$.)

We will see that the min in the above formula tends to be attained by the term for $r = r_*$, and if so, (8) indeed subsumes both (5) and (6) simultaneously in some sense. From now on, we work with (8) in place of (5)–(7).

2.4 Better Analysis of the Alternative Recurrence

We now obtain bounds on $G(n,v,h)$ using the recurrence (8) together with (1)–(3). Specifically, we show how to derive a new bound from two known bounds.

Assume that $G(n,v,h) = \widetilde{\Omega}(n^{\alpha_1}v^{\beta_1}h^{\gamma_1})$ and $G(n,v,h) = \widetilde{\Omega}(n^{\alpha_2}v^{\beta_2}h^{\gamma_2})$, with $\alpha_1,\alpha_2 \in [0,1]$ and $\beta_1,\beta_2,\gamma_1,\gamma_2 \leq 0$. Then by (8),

$$\begin{aligned} G(n,v,h) &\geq \widetilde{\Omega}\Big(n/r_* \oplus \\ &\quad \min_{r_* \leq r \leq \min\{v,h\}} r\left[(n/r)^{\alpha_1}(v/r)^{\beta_1}r^{\gamma_1}\right]^{\alpha_2} r^{\beta_2}[h/r]^{\gamma_2}\Big) \\ &\geq \widetilde{\Omega}\Big(n/r_* \oplus \\ &\quad \min_{r_* \leq r \leq \min\{v,h\}} n^{\alpha_1\alpha_2}v^{\beta_1\alpha_2}h^{\gamma_2}r^{1-(\alpha_1+\beta_1-\gamma_1)\alpha_2+\beta_2-\gamma_2}\Big). \end{aligned}$$

Assuming $1-(\alpha_1+\beta_1-\gamma_1)\alpha_2+\beta_2-\gamma_2 \geq 0$, we see that the minimum above is attained when $r = r_*$. By setting $r_* = [n^{\alpha_1\alpha_2-1}v^{\beta_1\alpha_2}h^{\gamma_2}]^{-1/(2-(\alpha_1+\beta_1-\gamma_1)\alpha_2+\beta_2-\gamma_2)}$ to balance the two terms,[3] we conclude that $G(n,v,h) \geq \widetilde{\Omega}(n^{\alpha}v^{\beta}h^{\gamma})$, where

$$\begin{aligned} &(\alpha,\beta,\gamma) = \mathbf{g}((\alpha_1,\beta_1,\gamma_1),(\alpha_2,\beta_2,\gamma_2)) \\ &= (1,0,0) + \tfrac{1}{2-(\alpha_1+\beta_1-\gamma_1)\alpha_2+\beta_2-\gamma_2}(\alpha_1\alpha_2-1,\ \beta_1\alpha_2,\ \gamma_2). \end{aligned} \tag{9}$$

If $1-(\alpha_1+\beta_1-\gamma_1)\alpha_2+\beta_2-\gamma_2 < 0$, we make $\mathbf{g}$ undefined.

We can apply the above process in conjunction with (3) a constant number of times, in any order, starting with the base case (2) and the known bound $G(n,v,h) = \Omega(n^{0.618})$. In other words, letting

$$\mathbf{s}(\alpha,\beta,\gamma) = (\alpha,\gamma,\beta),\ \ \mathbf{u} = (1,-1,0),\ \ \mathbf{v}_0 = (0.618,0,0), \tag{10}$$

we have $G(n,v,h) = \widetilde{\Omega}(n^{\alpha}v^{\beta}h^{\gamma})$ for any vector (α,β,γ) that can be generated by a constant number of applications of the mappings $\mathbf{g}$ and $\mathbf{s}$ to the initial vectors $\mathbf{u}$ and $\mathbf{v}_0$.

Numerical calculations. For example, the specific chain of inequality used in Section 2.2 is analogous to taking the expression

$$\mathbf{g}(\mathbf{s}(\mathbf{g}(\mathbf{g}(\mathbf{u},\mathbf{v}_0),\mathbf{v}_0)),\mathbf{v}_0),$$

which indeed evaluates to a vector exceeding $(0.622,0,0)$. We can immediately do better by iterating:

$$\mathbf{v}_{i+1} = \mathbf{g}(\mathbf{s}(\mathbf{g}(\mathbf{g}(\mathbf{u},\mathbf{v}_i),\mathbf{v}_i)),\mathbf{v}_i)$$

gives a vector $\geq (0.624,0,0)$ for a sufficiently large constant i. (Of course, at this stage, the evaluations are best done on a computer than a hand calculator.) Modifying the expression to

$$\mathbf{v}_{i+1} = \mathbf{g}(\mathbf{s}(\mathbf{g}(\mathbf{g}(\mathbf{u},\mathbf{s}(\mathbf{g}(\mathbf{u},\mathbf{s}(\mathbf{u})))),\mathbf{v}_i)),\mathbf{v}_i)$$

gives $\geq (0.627,0,0)$. Expanding and further tweaking the expression, we find that letting $\mathbf{w}_0 = \mathbf{s}(\mathbf{u})$, $\mathbf{w}_{i+1} = \mathbf{g}(\mathbf{u},\mathbf{w}_i)$, and

$$\begin{aligned} \mathbf{v}_{i+1} &= \mathbf{g}(\mathbf{s}(\mathbf{g}(\mathbf{g}(\mathbf{w}_5,\mathbf{s}(\mathbf{g}(\mathbf{w}_3, \\ &\quad \mathbf{s}(\mathbf{g}(\mathbf{w}_6,\mathbf{s}(\mathbf{g}(\mathbf{w}_2,\mathbf{w}_0))))))),\mathbf{v}_i)),\mathbf{v}_i) \end{aligned}$$

gives $\geq (0.629,0,0)$. With more ad hoc experimentation, we discover that

$$\begin{aligned} \mathbf{v}_{i+1} &= \mathbf{g}(\mathbf{s}(\mathbf{g}(\mathbf{g}(\mathbf{g}(\mathbf{w}_7,\mathbf{g}(\mathbf{w}_7,\mathbf{w}_2)),\mathbf{s}(\mathbf{g}(\mathbf{g}(\mathbf{w}_3,\mathbf{g}(\mathbf{w}_2,\mathbf{w}_0)), \\ &\quad \mathbf{s}(\mathbf{g}(\mathbf{w}_6,\mathbf{s}(\mathbf{g}(\mathbf{w}_2,\mathbf{w}_0))))))),\mathbf{v}_i)),\mathbf{v}_i) \end{aligned}$$

gives $\geq (0.630,0,0)$. Continuing still further and letting $\mathbf{z}_0 = \mathbf{s}(\mathbf{w}_1)$, $\mathbf{z}_{i+1} = \mathbf{g}(\mathbf{u},\mathbf{z}_i)$, and

$$\begin{aligned} \mathbf{v}_{i+1} &= \mathbf{g}(\mathbf{s}(\mathbf{g}(\mathbf{g}(\mathbf{g}(\mathbf{g}(\mathbf{u},\mathbf{g}(\mathbf{u},\mathbf{g}(\mathbf{w}_3,\mathbf{g}(\mathbf{w}_2,\mathbf{s}(\mathbf{z}_1))))), \\ &\quad \mathbf{g}(\mathbf{g}(\mathbf{u},\mathbf{g}(\mathbf{w}_3,\mathbf{g}(\mathbf{w}_2,\mathbf{s}(\mathbf{z}_1)))),\mathbf{s}(\mathbf{g}(\mathbf{g}(\mathbf{z}_2,\mathbf{s}(\mathbf{g}(\mathbf{w}_2, \\ &\quad \mathbf{z}_0))),\mathbf{s}(\mathbf{g}(\mathbf{w}_3,\mathbf{g}(\mathbf{w}_2,\mathbf{s}(\mathbf{z}_1)))))))),\mathbf{s}(\mathbf{g}(\mathbf{g}(\mathbf{g}(\mathbf{z}_3, \\ &\quad \mathbf{g}(\mathbf{z}_2,\mathbf{s}(\mathbf{g}(\mathbf{u},\mathbf{s}(\mathbf{z}_1))))),\mathbf{g}(\mathbf{g}(\mathbf{z}_2,\mathbf{s}(\mathbf{g}(\mathbf{z}_2,\mathbf{s}(\mathbf{z}_1)))), \\ &\quad \mathbf{s}(\mathbf{g}(\mathbf{w}_3,\mathbf{g}(\mathbf{z}_2,\mathbf{s}(\mathbf{z}_1)))))),\mathbf{s}(\mathbf{g}(\mathbf{g}(\mathbf{w}_4,\mathbf{g}(\mathbf{w}_3,\mathbf{g}(\mathbf{z}_2, \\ &\quad \mathbf{s}(\mathbf{z}_1)))),\mathbf{s}(\mathbf{g}(\mathbf{g}(\mathbf{z}_2,\mathbf{s}(\mathbf{g}(\mathbf{w}_2,\mathbf{s}(\mathbf{z}_1)))),\mathbf{s}(\mathbf{g}(\mathbf{w}_3, \\ &\quad \mathbf{g}(\mathbf{z}_2,\mathbf{s}(\mathbf{z}_1)))))))))))),\mathbf{v}_i)),\mathbf{v}_i) \end{aligned}$$

give $\geq (0.631,0,0)$.

This last expression is not generated by hand (as the reader might guess), but automatically by a C program which employs a simple greedy expansion strategy. Roughly speaking, in each round, the program evaluates the arguments (n,v,h) at all nodes of the current expression tree; then at each leaf, it decides whether to apply the $\mathbf{s}$ map first before applying $\mathbf{g}$, based on whether $v > h$, to produce the next level of the tree.

The same program finds an expression that gives a vector exceeding $(0.632,0,0)$, although the expression is just a bit more complicated to write down—the curious reader can see the whole thing in Appendix A. (We have an even longer expression giving $(0.6322,0,0)$, but decide not to waste more paper.) We conclude that $G(n,n,n) = \widetilde{\Omega}(n^{\alpha})$ and $F(n,n,n) = \widetilde{O}(n^{1-\alpha})$ for some $\alpha > 0.632$. This proves Theorem 1.2.

[3]It is easy to check that $r_* \geq 1$. The bound remains true for $r_* > \min\{v,h\}$, since (2) gives $G(n,v,h) \geq \Omega(n/\min\{v,h\}) \geq \Omega(n/r_*)$.

Remarks. Although there is no guarantee the exponent obtained by our greedy heuristics should converge to the optimal value for our recurrence, new ideas are probably required to get a more substantial improvement.

In fact, our recurrence cannot yield a bound for α greater than $2/3 \approx 0.667$. For example, the function $G(n, v, h) = \max\{n^{2/3}, n/\sqrt{v}, n/\sqrt{h}\}$ (which has $G(n, n, n) = n^{2/3}$) obeys (8) and (1)–(3), as one can verify by hand with a tedious case analysis. One could try to find worse examples by using another computer search, experimenting with more complicated functions of the form $G(n, v, h) = \max_{(\alpha,\beta,\gamma)\in S} n^\alpha v^\beta h^\gamma$ for some finite sets S.

3. IMPLICATIONS

In this section, we consider the case of boxes in higher dimensions. Given $d \le j \le 2d$, let $\mathcal{B}^{d,j}$ be the range space where the elements are all points in $\mathbb{R}^d$ and the ranges are all j-sided boxes of the form $[x_1, x_1'] \times [x_{j-d}, x_{j-d}'] \times (-\infty, x_{j-d+1}] \times \cdots \times (-\infty, x_d]$. For example, $\mathcal{B}^{d,2d} = \mathcal{B}^d$ and $\mathcal{B}^{d,d} = \mathcal{O}^d$.

3.1 Points w.r.t. 3-Dimensional 4-Sided Boxes

For the case of points w.r.t. 3-dimensional orthants in $\mathcal{O}^3$ (3-sided boxes), which is equivalent to orthants in $\mathcal{O}^3$ w.r.t. points by self-duality, good bounds can be obtained because the union complexity is low, which generally implies that the order-k Delaunay graph is sparse. Aloupis et al. [4] observed this fact specifically for the case of pseudo-disks w.r.t. points; they used a sampling-based proof. Chan and Har-Peled [9] gave a different sampling-based proof in the style of "shallow cuttings" which can be easily adapted to the 3-dimensional orthant case (see the full paper):

PROPOSITION 3.1.

(i) *The graph $DG_k(P, \mathcal{O}^3)$ has $O(nk)$ edges.*

(ii) DG-INDEP$_k(P, \mathcal{O}^3) = \Omega(n/k)$.

We are thus led to the next nontrivial case of 3-dimensional 4-sided boxes in $\mathcal{B}^{3,4}$; this case turns out to be useful later in Section 3.3.

First the concept of dominance and maximality in Definition 4 can be extended to three dimensions in the obvious way: for example, (x_1, y_1, z_1) $(+--)$-dominates (x_2, y_2, z_2) if $x_1 > x_2$ and $y_1 < y_2$ and $z_1 < z_2$. The *σ-orthant* of a point q is the region of all points that σ-dominate q.

We reduce the case of 3-dimensional 4-sided boxes to the 2-dimensional rectangular case in a series of three stages:

LEMMA 3.2. *Suppose that* DG-INDEP$(n, \mathcal{B}^2) = \widetilde{\Omega}(n^\alpha)$.

(i) *Let P be an n-point set in $\mathbb{R}^3$ such that every point in P is both $(---)$-maximal and $(+--)$-maximal. Then we can find an independent set of $DG(P, \mathcal{B}^{3,4})$ of size $\widetilde{\Omega}(n^\alpha)$.*

(ii) *Let P be an n-point set in $\mathbb{R}^3$ such that every point in P is $(---)$-maximal. Then we can find an independent set of $DG(P, \mathcal{B}^{3,4})$ of size $\widetilde{\Omega}(n^\alpha)$.*

(iii) DG-INDEP$(n, \mathcal{B}^{3,4}) = \widetilde{\Omega}(n^\alpha)$.

PROOF.

(i) Consider the yz-projection $\widehat{P}$ of P. All points in $\widehat{P}$ are $(--)$-maximal, i.e., they form a descending staircase. Thus, for any value z, the range $\mathbb{R} \times (-\infty, z]$ contains the same points of $\widehat{P}$ as the range $[y, \infty) \times \mathbb{R}$ for some y. Consequently, the 3-dimensional 4-sided box $[x, x'] \times (-\infty, y'] \times (-\infty, z]$ contains the same points of P as the range $[x, x'] \times [y, y'] \times \mathbb{R}$. We can find a DG-independent set of P w.r.t. $\mathcal{B}^{3,4}$ by finding a DG-independent set of the xy-projection of P w.r.t. $\mathcal{B}^2$.

(ii) We describe a simple divide-and-conquer algorithm to color DG$(P, \mathcal{B}^{3,4})$ with $\widetilde{O}(n^{1-\alpha})$ colors. Use a vertical line to divide P into a left point set P_L and a right point set P_R, each of size $n/2$. Mark the $(+--)$-maximal points of P_L. Color the marked points by applying (i), which requires a palette of $\widetilde{O}(n^{1-\alpha})$ colors. Recursively color the unmarked points in P_L and recursively color the points in P_R, using a common new palette. The number of colors satisfies the recurrence $G(n) = G(n/2) + \widetilde{O}(n^{1-\alpha})$, which gives $\widetilde{O}(n^{1-\alpha})$ as desired.

To prove correctness of this coloring, consider a 4-sided range s in $\mathcal{B}^{3,4}$ that contains exactly two points p, q of P. If both p and q are marked, they must have different colors. If one is marked and the other is unmarked, they obviously have different colors. So, assume both are unmarked. If $p, q \in P_L$ or $p, q \in P_R$, they would have different colors by induction. If $p \in P_L$ and $q \in P_R$, then s must contain a $(+--)$-maximal point of P_L; this point must be different from p and q since p is unmarked—a contradiction.

(iii) We describe a similar simple divide-and-conquer algorithm to color DG$(P, \mathcal{B}^{3,4})$ with $O(n^{1-\alpha})$ colors. Use a vertical line to divide P into a left point set P_L and a right point set P_R, each of size $n/2$. Mark the $(---)$-maximal points of P_R. Color the marked points by applying (ii), which requires a palette of $\widetilde{O}(n^{1-\alpha})$ colors. Recursively color the points in P_L and recursively color the unmarked points in P_R, using a common new palette. The number of colors satisfies the recurrence $G'(n) = G'(n/2) + \widetilde{O}(n^{1-\alpha})$, which gives $\widetilde{O}(n^{1-\alpha})$ again. Correctness follows by the same argument as in (ii).

□

Lemma 3.2 addresses only the $k = 2$ case. We next show that the $k > 2$ case can be directly reduced to the $k = 2$ case for 3-dimensional 4-sided boxes, which implies the same for 2-dimensional rectangles as well. The key is the following proposition whose proof can be found in the full version of the paper (a similar proof was given by Ene, Har-Peled, and Raichel [14] in the special case of 2-dimensional rectangles).

PROPOSITION 3.3. *Let P be any n-point set in $\mathbb{R}^3$ and k be a given value. Then we can find a subset Q with $\Omega(n/(k \log n))$ points such that every range $s \in \mathcal{B}^{3,4}$ with $|P \cap s| \le k$ contains at most 2 points of Q.*

COROLLARY 3.4. *Suppose that* DG-INDEP$(n, \mathcal{B}^2) = \widetilde{\Omega}(n^\alpha)$. *Then*

(i) DG-INDEP$_k(n, \mathcal{B}^{3,4}) = \widetilde{\Omega}([n/k]^\alpha)$;

(ii) DG-INDEP$_k(n, \mathcal{B}^2) = \widetilde{\Omega}([n/k]^\alpha)$.

PROOF. To prove (i), we first apply Proposition 3.3 to obtain a subset Q of the given n-point set and then apply Lemma 3.2(iii) to output a DG-independent set of Q of size $\widetilde{\Omega}(|Q|^\alpha) = \widetilde{\Omega}([n/(k \log n)]^\alpha)$.

(ii) is a special case of (i), since after the transformation $(a, b) \mapsto (a, b, -b)$, a 2-dimensional rectangle $[x', x] \times [y', y]$ corresponds to a 3-dimensional 4-sided box $[x', x] \times (-\infty, y] \times (-\infty, -y']$. □

Remark. Note that since 3-dimensional 4-sided boxes are at least as expressive as 2-dimensional rectangles, as the above transformation illustrates, the bound in Lemma 3.2(iii) is essentially tight.

3.2 Points w.r.t. Boxes in Higher Dimensions (Proof of Theorem 1.3(i))

The case of higher-dimensional boxes can be handled by repeated application of the Erdős–Szekeres theorem, as shown by Har-Peled and Smorodinsky [19]. We notice that the bound can be slightly improved using our new result for 2-dimensional rectangles as the base case:

LEMMA 3.5. *If* DG-INDEP$_k(n, \mathcal{B}^{d-1}) = \widetilde{\Omega}(n^\beta/k^c)$, *then* DG-INDEP$_k(n, \mathcal{B}^d) = \widetilde{\Omega}(n^{\beta/2}/k^c)$.

PROOF. Let P be a sequence of n points in $\mathbb{R}^d$ sorted according to the first coordinate. By the Erdős–Szekeres theorem [15], there exists a subsequence Q of $\Omega(\sqrt{n})$ points that is monotone increasing or decreasing in the last coordinate.

For any x_d and x'_d, the range $\mathbb{R}^{d-1} \times [x_d, x'_d]$ contains the same points of Q as the range $[x_1, x'_1] \times \mathbb{R}^{d-1}$ for some x_1 and x'_1. Thus, the d-dimensional box $[x''_1, x'''_1] \times [x_2, x'_2] \times \cdots \times [x_d, x'_d]$ contains the same points of Q as the box $[\max\{x_1, x''_1\}, \min\{x'_1, x'''_1\}] \times [x_2, x'_2] \times \cdots \times [x_{d-1}, x'_{d-1}] \times \mathbb{R}$. We can find a DG$_k$-independent set of Q (and thus P) w.r.t. $\mathcal{B}^d$ by finding a DG$_k$-independent set w.r.t. $\mathcal{B}^{d-1}$ of the projection of Q along the last coordinate. This independent set has size $\widetilde{\Omega}(|Q|^\beta/k^c) = \widetilde{\Omega}(n^{\beta/2}/k^c)$. □

By repeated applications of Lemma 3.5 with Corollary 3.4(ii) as the base case, we obtain DG-INDEP$_k(n, \mathcal{B}^d) = \widetilde{\Omega}(n^{\alpha/2^{d-2}}/k)$. This proves Theorem 1.3(i) except for a weaker exponent in the k factor, which can be fixed by the following general reduction:

PROPOSITION 3.6. *For a range space $\mathcal{S}$ with a constant VC dimension,* DG-INDEP$_k(n, \mathcal{S}) = \widetilde{\Omega}(n^\beta/k^{O(1)})$ *implies* DG-INDEP$_k(n, \mathcal{S}) = \widetilde{\Omega}([n/k]^\beta)$.

PROOF. The claim follows from a standard random sampling argument, or as a special case of Theorem 1.5. For the latter, given a set P of n elements, just let S consist of all combinatorial different ranges $s \in \mathcal{S}$ and $|P \cap s| \le k$. By the bounded VC-dimension assumption (Sauer's lemma) [23], S has polynomial size. In the LP for the unweighted setting, OPT equals the size of the maximum independent set of DG$_k(P, \mathcal{S})$, and opt $\ge n/k$ since we can set all x_p's to $1/k$. Theorem 1.5 implies that OPT $\ge \widetilde{\Omega}([n/k]^\beta)$. □

Remark. It would be nice if Ajwani et al.'s grid-based method could be directly extended to handle higher-dimensional boxes to yield better bounds, but such an extension does not appear obvious.

3.3 Points w.r.t. Orthants in Higher Dimensions (Proof of Theorem 1.3(ii))

For higher-dimensional orthants, better results can be obtained by applying the Erdős–Szekeres theorem in a slightly more interesting way:

LEMMA 3.7. *If* DG-INDEP$_k(n, \mathcal{B}^{d-1,d}) = \widetilde{\Omega}(n^\beta/k^c)$ *and* DG-INDEP$_k(n, \mathcal{O}^{d-1}) = \widetilde{\Omega}(n^\gamma/k^c)$, *then*

(i) DG-INDEP$_k(n, \mathcal{B}^{d,d+1}) = \widetilde{\Omega}(n^{\beta/2}/k^c)$, *and*

(ii) DG-INDEP$_k(n, \mathcal{O}^d) = \widetilde{\Omega}(n^{\beta \oplus \gamma}/k^c)$ *where* $\beta \oplus \gamma = 1/(1/\beta + 1/\gamma)$.

PROOF.

(i) The proof of this part is almost identical to the proof of Lemma 3.5. We just observe that the d-dimensional $(d+1)$-sided box $[x''_1, x'''_1] \times (-\infty, x'_2] \times \cdots \times (-\infty, x'_d]$ contains the same points of Q as the box $[\max\{x_1, x''_1\}, \min\{x'_1, x'''_1\}] \times (-\infty, x'_2] \times \cdots \times (-\infty, x'_{d-1}] \times \mathbb{R}$ for some x_1 and x'_1. The latter box is in $\mathcal{B}^{d-1,d}$ after projection along the last coordinate.

(ii) The proof of this part is also similar but requires the use of a parameter r. Let P be a sequence of n points in $\mathbb{R}^d$ sorted in increasing order according to the first coordinate. By the Erdős–Szekeres theorem, there exists either (a) a subsequence Q of $\Omega(r)$ points that is monotone increasing in the last coordinate, or (b) a subsequence Q of $\Omega(n/r)$ points that is monotone decreasing in the last coordinate.

In Case (a), for any value x_d, the range $\mathbb{R}^{d-1} \times (-\infty, x_d]$ contains the same points of Q as the range $(-\infty, x_1] \times \mathbb{R}^{d-1}$ for some x_1. Thus, the d-dimensional orthant $(-\infty, x'_1] \times (-\infty, x_2] \times \cdots \times (-\infty, x_d]$ contains the same points of Q as the range $(-\infty, \min\{x_1, x'_1\}] \times (-\infty, x_2] \times \cdots \times (-\infty, x_{d-1}] \times \mathbb{R}$. We can find a DG$_k$-independent set of Q (and thus P) w.r.t. $\mathcal{O}^d$ by finding a DG$_k$-independent set w.r.t. $\mathcal{O}^{d-1}$ of the projection of Q along the last coordinate.

In Case (a), for any value x_d, the range $\mathbb{R}^{d-1} \times (-\infty, x_d]$ contains the same points of Q as the range $[x_1, \infty) \times \mathbb{R}^{d-1}$ for some x_1. Thus, the d-dimensional orthant $(-\infty, x'_1] \times (-\infty, x_2] \times \cdots \times (-\infty, x_d]$ contains the same points of Q as the range $[x_1, x'_1] \times (-\infty, x_2] \times \cdots \times (-\infty, x_{d-1}] \times \mathbb{R}$. We can find a DG$_k$-independent set of Q (and thus P) w.r.t. $\mathcal{O}^d$ by finding a DG$_k$-independent set w.r.t. $\mathcal{B}^{d-1,d}$ of the projection of Q along the last coordinate.

In any case, we can find an independent set of DG$_k(P, \mathcal{O}^d)$ of size

$$\widetilde{\Omega}(\min\{r^\gamma,\ (n/r)^\beta\}/k^c).$$

By setting $r = n^{\beta/(\beta+\gamma)}$ to balance the two terms, we obtain a bound of $\widetilde{\Omega}(n^{\beta \oplus \gamma}/k^c)$.

□

By repeated applications of Lemma 3.7(i) with Corollary 3.4(i) as the base case, we have DG-INDEP$_k(n, \mathcal{B}^{d-1,d}) = \widetilde{\Omega}(n^{\alpha/2^{d-4}}/k)$.

By repeated applications of Lemma 3.7(ii) with Proposition 3.1(ii) as the base case, we then have DG-INDEP$_k(n, \mathcal{O}^d) = \widetilde{\Omega}(n^{\gamma_d}/k)$ with $\gamma_3 = 1$ and $\gamma_d = \alpha/2^{d-4} \oplus \gamma_{d-1}$, i.e.,

$$\begin{aligned}\gamma_d &= \alpha/2^{d-4} \oplus \cdots \oplus \alpha/2 \oplus \alpha \oplus 1 \\ &= \frac{1}{2^{d-4}/\alpha + \cdots + 2/\alpha + 1/\alpha + 1} = \frac{1}{(2^{d-3}-1)/\alpha + 1}.\end{aligned}$$

The exponents in the k factors can be fixed again by Proposition 3.6. This proves Theorem 1.3(ii).

4. DISCRETE INDEPENDENT SET (PROOF OF THEOREM 1.5)

In this last section, we present an approximation algorithm for discrete independent set by rounding the LP from Theorem 1.5. Let $\{x_p \mid p \in P\}$ be the solution to the LP. Draw a random sample R of the elements where element p is chosen independently with probability x_p. For a fixed range s, consider the random variable $|R \cap s|$, which has expected value $\mu_s = \sum_{p \in s} x_p \leq 1$ and is a sum of independent 0-1 random variables. By the Chernoff bound,

$$\Pr[|R \cap s| > (1+\delta)\mu_s] \;\leq\; \left[\frac{e^\delta}{(1+\delta)^{1+\delta}}\right]^{\mu_s}$$

for any $\delta > 0$ (possibly large). Choosing[4] $\delta = \Theta((\log m)/\mu_s)$, we have $|R \cap s| = O(\log m)$ with high probability (e.g., probability $1 - o(1/m)$).

Next consider the total weight $w(R)$ of R, which has expected value $\mu = \sum_{p \in s} w_p x_p = \text{opt}$ and variance $\sigma^2 = \sum_{p \in s} w_p^2 (x_p - x_p^2) \leq W \cdot \text{opt}$, where W denotes the maximum weight in P. We may assume that $W \leq \text{opt}/c$ for any fixed constant c, for otherwise we can just return the singleton with weight W and be done. So, $\sigma \leq \text{opt}/\sqrt{c}$. By Chebyshev's inequality,

$$\Pr[|w(R) - \mu| > t\sigma] \;\leq\; 1/t^2$$

for any $t > 0$. Choosing constants c and t appropriately, we have $|w(R)| \geq \Omega(\text{opt})$ with probability $\Omega(1)$. So, by repeating for an expected $O(1)$ number of trials, we can find a subset R satisfying both (a) $|R \cap s| \leq O(\log m)$ for all $s \in S$ at the same time, and (b) $w(R) \geq \Omega(\text{opt})$.

In the unweighted case, we return a $\text{DG}_{c' \log m}$-independent set Q of R of size $\widetilde{\Omega}(|R|^\alpha / \log^{O(1)} m)$, which is $\widetilde{\Omega}(\text{opt}^\alpha)$ by (b). We know that Q is a feasible solution for a sufficiently large constant c' by (a) and the definition of $\text{DG}_{c' \log m}$-independence.

In the weighted case, we compute a $\text{DG}_{c' \log m}$-coloring of R with $\widetilde{O}(|R|^{1-\alpha} \log^{O(1)} m) = \widetilde{O}(n^\alpha)$ colors, and return a color class Q of the largest weight. Then $w(Q) \geq \widetilde{\Omega}(w(R)/n^{1-\alpha}) \geq \widetilde{\Omega}(\text{opt}/n^{1-\alpha})$ by (b). Again, Q is a feasible solution by (a). This completes the proof of Theorem 1.5.

Remark. It is possible to make small improvements in the logarithmic factors by experimenting with a different choice of probability values, e.g., selecting element p with probability $\min\{bx_p, 1\}$ for some parameter $b \geq 1$.

Acknowledgement. I thank Sariel Har-Peled for asking about discrete independent set of higher-dimensional boxes w.r.t. points, which started this work. Thanks also to Vinayak Pathak for discussions two years ago on conflict-free coloring of points w.r.t. rectangles, and Shakhar Smorodinsky for introducing me to the problem many more years ago.

[4]Actually $\delta = \Theta((\log m/\log\log m)/\mu_s)$ works too, as noted in [9].

5. REFERENCES

[1] D. Ajwani, K. M. Elbassioni, S. Govindarajan, and S. Ray. Conflict-free coloring for rectangle ranges using $O(n^{.382})$ colors. In *19th ACM Sympos. Parallelism in Algorithms and Architectures*, pages 181–187, 2007. *Discrete Comput. Geom.*, to appear.

[2] N. Alon, M. Krivelevich, and B. Sudakov. Coloring graphs with sparse neighborhoods. *J. Combin. Theory Ser. B*, 77:73–82, 1999.

[3] N. Alon and J. H. Spencer. *The Probabilistic Method.* Wiley-Interscience, 2nd ed., 2000.

[4] G. Aloupis, J. Cardinal, S. Collette, S. Langerman, and S. Smorodinsky. Coloring geometric range spaces. *Discrete Comput. Geom.*, 41:348–362, 2009.

[5] H. Brönnimann and M. T. Goodrich. Almost optimal set covers in finite VC-dimension. *Discrete Comput. Geom.*, 14:463–479, 1995.

[6] P. Chalermsook and J. Chuzhoy. Maximum independent set of rectangles. In *Proc. 20th ACM–SIAM Sympos. Discrete Algorithms*, pages 892–901, 2009.

[7] T. M. Chan. Polynomial-time approximation schemes for packing and piercing fat objects. *J. Algorithms*, 46(2):178–189, 2003.

[8] T. M. Chan, E. Grant, J. Könemann, and M. Sharpe. Weighted capacitated, priority, and geometric set cover via improved quasi-uniform sampling. In *Proc. 23rd ACM–SIAM Sympos. Discrete Algorithms*, pages 1576–1585, 2012.

[9] T. M. Chan and S. Har-Peled. Approximation algorithms for maximum independent set of pseudo-disks. *Discrete Comput. Geom.*, to appear. (Preliminary version in *Proc. 25th ACM Sympos. Comput. Geom.*, pages 333–340, 2009.)

[10] X. Chen, J. Pach, M. Szegedy, and G. Tardos. Delaunay graphs of point sets in the plane with respect to axis-parallel rectangles. *Random Struct. Algorithms*, 34:11–23, 2009.

[11] K. L. Clarkson and P. W. Shor. Applications of random sampling in computational geometry, II. *Discrete Comput. Geom.*, 4:387–421, 1989.

[12] K. L. Clarkson and K. R. Varadarajan. Improved approximation algorithms for geometric set cover. *Discrete Comput. Geom.*, 37:43–58, 2007.

[13] K. Elbassioni and N. H. Mustafa. Conflict-free colorings of rectangular ranges. In *Proc. 23rd Sympos. Theoret. Aspects Comput. Sci.*, Lect. Notes Comput. Sci., vol. 3884, Springer-Verlag, pages 254–263, 2006.

[14] A. Ene, S. Har-Peled, and B. Raichel. Geometric packing under non-uniform constraints. To appear in *Proc. 28th ACM Sympos. Comput. Geom.*, 2012.

[15] P. Erdős and G. Szekeres. A combinatorial problem in geometry. *Compositio Mathematica*, 2:463–470, 1935.

[16] G. Even, Z. Lotker, D. Ron, and S. Smorodinsky. Conflict-free colorings of simple geometric regions with applications to frequency assignment in cellular networks. *SIAM J. Comput.*, 33:94–136, 2003.

[17] G. Even, D. Rawitz, and S. Shahar. Hitting sets when the VC-dimension is small. *Inform. Process. Lett.*, 95:358–362, 2005.

[18] J. Fox and J. Pach. Computing the independence number of intersection graphs. In *Proc. 22nd ACM–SIAM Sympos. Discrete Algorithms*, pages 1161–1165, 2011.

[19] S. Har-Peled and S. Smorodinsky. Conflict-free coloring of points and simple regions in the plane. *Discrete Comput. Geom.*, 34:47–70, 2005.

[20] W. Hoeffding. Probability inequalities for sums of bounded random variables. *J. American Statistical Assoc.*, 58:13–30, 1963.

[21] P. M. Long. Using the pseudo-dimension to analyze approximation algorithms for integer programming. In *Proc. 7th Algorithms Data Struct. Sympos.*, Lect. Notes Comput. Sci., vol. 2125, Springer-Verlag, pages 26–37, 2001.

[22] J. Matoušek. *Lectures on Discrete Geometry*. Springer-Verlag, 2002.

[23] J. Pach and P. K. Agarwal. *Combinatorial Geometry*. Wiley-Interscience, New York, 1995.

[24] J. Pach and G. Tóth. Conflict free colorings. In *Discrete and Computational Geometry: The Goodman–Pollack Festschrift*, Springer-Verlag, Berlin, pages 665–671, 2003.

[25] S. Smorodinsky. Conflict-free coloring and its applications. http://arxiv.org/abs/1005.3616, 2010.

APPENDIX

A. VERIFICATION OF 0.632 IN MAPLE

The following script in Maple outputs a vector $\geq (0.6320002, 0, 0)$:

```
g := proc(a1,b1,c1,a2,b2,c2)
        if (1-(a1+b1-c1)*a2+b2-c2 < 0) then
          printf("undefined!\n")
        end if;
      (1,0,0) + (a1*a2-1, b1*a2, c2) / (2-(a1+b1-c1)*a2+b2-c2)
    end proc;
s := proc(a,b,c) a,c,b end proc;
u := (1,-1,0); v := (0.618,0,0);
w0 := s(u); w1 := g(u,w0); w2 := g(u,w1); w3 := g(u,w2);
w4 := g(u,w3); w5 := g(u,w4);
z0 := s(w1); z1 := g(u,z0); z2 := g(u,z1); z3 := g(u,z2);
z4 := g(u,z3); z5 := g(u,z4);
for i from 1 to 11 do v := \
g(s(g(g(g(g(g(z5,g(z5,g(z4,g(g(u,g(u,g(u,s(g(w1,w0))))),s(g(g(w\
2,s(z1)),s(z2))))))),g(g(z5,g(z4,g(g(u,g(u,g(w3,s(g(w1,w0))))),\
g(g(u,g(w3,s(g(w1,s(w2))))),s(g(g(w2,g(w1,s(w2))),s(g(u,s(g(w1,\
w0))))))))))),g(g(g(u,g(u,g(u,g(w3,s(g(w1,w0)))))),g(g(u,g(u,g(w\
3,s(g(w1,s(w2)))))),g(g(w3,g(w3,s(g(z1,s(w2))))),s(g(g(w2,g(z1,\
s(w2))),s(g(u,s(g(w1,s(w2))))))))))),g(g(g(u,g(w3,g(w3,s(g(z1,s(\
w2)))))),g(g(w4,g(z3,g(z2,s(g(w2,s(z1)))))),s(g(g(w3,g(w2,s(z1)\
)),s(g(u,((s(g(w1,w0)))))))))),s(g(g(g(w3,g(w2,g(z1,s(w2)))),g(\
g(z2,g(z1,s(w2))),s(g(w3,s(g(z1,s(w2))))))),s(g(g(u,g(w3,s(g(z1\
,s(w2))))),s(g(g(w2,g(z1,s(w2))),s(g(u,s(g(w1,s(w2))))))))))))\
)),g(g(g(z5,g(g(u,g(u,g(u,g(w3,s(g(w1,w0)))))),g(g(u,g(u,g(w3,s\
(g(w1,s(w2)))))),g(g(w4,g(w3,g(z2,s(z1)))),s(g(g(w2,g(z1,s(w2))\
),s(g(u,s(g(w1,s(w2))))))))))),g(g(g(u,g(w4,g(w4,g(w3,s(g(w1,s(\
w2))))))),g(g(w4,g(w4,g(w3,g(w2,s(z1))))),g(g(w4,g(z3,g(z2,s(g(\
u,s(z1)))))),s(g(g(z2,s(g(w2,s(z1)))),s(g(w3,g(z2,s(z1)))))))))\
,g(g(g(w4,g(w4,g(w3,g(z2,s(z1))))),g(g(w4,g(z3,g(z2,s(g(w2,s(z1\
)))))),s(g(g(z2,s(g(z2,s(z1)))),s(g(z3,g(z2,s(g(u,s(z1)))))))))\
),s(g(g(g(z3,g(z2,s(g(w2,s(z1))))),g(g(z2,s(g(z2,s(z1)))),s(g(w\
3,g(z2,s(z1)))))),s(g(g(w4,g(z3,g(z2,s(g(u,s(z1)))))),s(g(g(z2,\
s(g(w2,s(z1)))),s(g(w3,g(z2,s(z1)))))))))))),s(g(g(g(g(g(u,g(u\
,g(w2,z0))),g(g(w3,g(w2,s(z1))),g(g(w2,s(g(w2,z0))),s(g(w3,g(w2\
,s(z1))))))),g(g(g(w3,g(w2,s(g(w2,z0)))),g(g(z2,g(w2,s(g(u,((w1\
)))))),s(g(u,((s(g(w1,w0)))))))),s(g(g(u,g(u,((g(w2,z0))))),s(g\
(g(w2,s(g(w2,z0))),s(g(w3,g(w2,s(z1))))))))),s(g(g(g(u,g(u,g(w\
3,g(w2,s(z1))))),g(g(w4,g(w3,g(z2,s(z1)))),s(g(g(z2,s(g(w2,z0))\
),s(g(w3,g(w2,s(z1)))))))),s(g(g(g(w3,g(w2,s(z1))),g(g(w2,s(g(w\
2,z0))),s(g(w3,g(w2,s(z1)))))),s(g(g(u,g(w3,g(w2,s(z1)))),s(g(g\
(w2,s(z1)),s(g(u,g(w2,z0)))))))))))),s(g(g(g(u,g(u,g(u,g(u,s(g(\
w1,w0)))))),g(g(u,g(u,g(u,s(g(w1,s(w2)))))),g(g(w4,g(w3,g(z2,s(\
g(w2,z0))))),s(g(g(w3,g(w2,s(z1))),s(g(u,((z0))))))))),g(g(g(w4\
,g(w4,g(w3,g(z2,s(z1))))),g(g(w4,g(z3,g(z2,s(g(u,s(z1)))))),s(g\
(g(z2,s(g(w2,s(z1)))),s(g(w3,g(z2,s(z1)))))))),s(g(g(g(w3,g(z2,\
s(g(w2,z0)))),g(g(z2,s(g(w2,s(z1)))),s(g(w3,g(z2,s(z1)))))),s(g\
(g(w4,g(w3,g(z2,s(z1)))),s(g(g(w2,s(g(w2,z0))),s(g(w3,g(w2,s(z1\
)))))))))))))))))),s(g(g(g(g(g(z3,g(g(u,g(u,g(w2,z0))),g(g(w3,g(\
w2,s(z1))),s(g(z2,s(g(w2,z0))))))),g(g(g(w3,g(w3,g(w2,s(z1)))),\
g(g(w3,g(w2,g(z1,s(w2)))),s(g(g(u,g(w2,z0)),s(g(w2,s(z1))))))),\
g(g(g(w3,g(w2,g(z1,s(w2)))),g(g(z2,g(z1,s(w2))),s(g(w3,s(g(z1,s\
(w2))))))),s(g(w3,((s(g(g(w2,z0),s(z2))))))))),g(g(g(g(w3,g(w3\
,g(w2,s(z1)))),g(g(w3,g(z2,s(g(w2,z0)))),g(g(z2,s(g(w2,s(z1))))\
,s(g(w3,g(z2,s(z1))))))),g(g(g(z3,g(z2,s(g(w2,s(z1))))),g(g(z2,\
g(z1,s(w2))),s(g(u,((((z1)))))))),s(g(g(u,g(w3,((g(w2,s(z1))))\
),s(g(g(z2,s(g(w2,s(z1)))),s(g(w3,g(z2,s(z1)))))))))),s(g(g(g(w\
4,g(w4,g(w3,g(z2,s(z1))))),g(g(w4,g(w3,g(w2,s(z1)))),s(g(g(z2,s\
(g(w2,s(z1)))),s(g(w3,g(z2,s(z1)))))))),s(g(g(g(w3,g(z2,s(g(w2,\
z0)))),g(g(z2,s(g(w2,s(z1)))),s(g(w3,g(z2,s(z1)))))),s(g(g(u,g(\
w3,g(w2,s(z1)))),s(g(g(z2,s(g(w2,z0))),s(g(w3,g(w2,s(z1))))))))\
)))))),g(g(g(g(g(w3,g(w3,g(z2,s(z1)))),g(g(z3,g(z2,s(g(w2,s(z1)\
)))),g(g(z2,s(g(w2,s(z1)))),s(g(w3,g(z2,s(z1))))))),g(g(g(z3,g(\
z2,s(g(w2,s(z1))))),g(g(z2,g(w2,s(g(u,((w1)))))),s(g(w3,((s(g(z\
1,s(w2))))))))),s(g(g(w3,g(w3,((g(w2,s(z1)))))),s(g(g(z2,s(g(z2\
,s(z1)))),s(g(w3,g(z2,s(z1)))))))))),s(g(g(g(w4,g(w4,g(w3,g(z2,\
s(z1))))),g(g(w4,g(z3,g(z2,s(g(w1,s(z1)))))),s(g(g(z2,s(g(z2,s(\
z1)))),s(g(w3,g(z2,s(z1)))))))),s(g(g(g(z3,g(z2,s(g(w2,s(z1))))\
),g(g(z2,s(g(w2,s(z1)))),s(g(w3,g(z2,s(z1)))))),s(g(g(w4,g(w3,g\
(z2,s(z1)))),s(g(g(z2,s(g(w2,s(z1)))),s(g(w3,g(w2,s(z1))))))))\
)))),s(g(g(g(u,g(w4,g(w4,g(w3,s(g(z1,s(w2))))))),g(g(w4,g(w3,g(\
w3,s(g(z1,s(w2)))))),g(g(w4,g(w3,s(g(z1,s(z2))))),s(g(g(z2,g(z1\
,s(w2))),s(g(w3,s(g(z1,s(w2)))))))))),g(g(g(w4,g(w4,g(w3,s(g(z1\
,s(w2)))))),g(g(w4,g(z3,g(z2,s(g(w2,s(z1)))))),s(g(g(z2,g(z1,s(\
w2))),s(g(w3,((g(w2,s(z1))))))))))),s(g(g(g(z3,g(z2,s(g(w2,s(z1)\
)))),g(g(z2,g(w2,s(g(u,((w1)))))),s(g(w3,((s(g(z1,s(w2))))))))\
,s(g(g(w3,g(w3,((g(w2,s(z1)))))),s(g(g(z2,s(g(z2,s(z1)))),s(g(w\
3,g(z2,s(z1))))))))))))))),s(g(g(g(g(u,g(u,g(u,g(w3,g(w3,s(z1))\
)))),g(g(u,g(w4,g(w4,g(w3,s(g(w1,s(w2))))))),g(g(w4,g(w4,g(w3,s\
(g(z1,s(w2)))))),g(g(w4,g(w3,g(z2,s(z1)))),s(g(g(z2,g(z1,s(w2))\
),s(g(w3,s(g(z1,s(w2))))))))))),g(g(g(u,g(w4,g(w3,g(w3,s(g(g(u,\
((w0))),s(w2))))))),g(g(w4,g(w4,g(w3,g(w2,s(z1))))),g(g(w4,g(z3\
,g(z2,s(g(w2,s(z1)))))),s(g(g(z2,s(g(z2,s(z1)))),s(g(w3,g(z2,s(\
z1))))))))),g(g(g(z4,g(w4,g(w3,g(z2,s(g(u,s(z1))))))),g(g(w4,g(\
z3,g(z2,s(g(w2,s(z1)))))),s(g(g(g(u,g(u,s(z1))),s(g(z2,s(g(u,s(\
z1)))))),s(g(z3,g(z2,s(g(w2,s(z1))))))))),s(g(g(g(z3,g(z2,s(g(\
z2,s(z1))))),g(g(z2,s(g(z2,s(z1)))),s(g(w3,g(z2,s(z1)))))),s(g(\
g(w4,g(w3,g(w2,s(z1)))),s(g(g(z2,s(g(z2,s(z1)))),s(g(w3,g(z2,s(\
z1))))))))))))))),s(g(g(g(g(g(w3,g(w3,g(w2,s(z1)))),g(g(w3,g(z2,s\
(g(w2,z0)))),g(g(z2,s(g(w2,s(z1)))),s(g(w3,g(z2,s(z1))))))),g(g\
(g(w3,g(z2,s(g(w2,s(z1))))),g(g(z2,g(w2,s(g(u,((w1)))))),s(g(u,\
((s(g(w1,s(w2))))))))),s(g(g(w3,g(w3,((g(w2,s(z1)))))),s(g(g(z2\
,s(g(w2,s(z1)))),s(g(w3,g(z2,s(z1)))))))))),s(g(g(g(w4,g(w4,g(w\
3,g(z2,s(z1))))),g(g(w4,g(w3,g(w2,s(z1)))),s(g(g(z2,s(g(w2,s(z1\
)))),s(g(w3,g(z2,s(z1)))))))),s(g(g(g(w3,g(z2,s(g(w2,z0)))),g(g\
(z2,s(g(w2,s(z1)))),s(g(w3,g(z2,s(z1)))))),s(g(g(w4,g(w3,g(z2,s\
(z1)))),s(g(g(w2,s(g(w2,z0))),s(g(w3,g(w2,s(z1))))))))))))),s(g\
(g(g(u,g(u,g(w3,g(w3,s(g(g(u,((u))),s(w2))))))),g(g(u,g(w3,g(w3\
,s(g(z1,s(w2)))))),g(g(w4,g(z3,g(z2,s(g(w2,s(z1)))))),s(g(g(w3,\
g(w2,s(z1))),s(g(u,((s(g(w1,w0))))))))))),g(g(g(w4,g(w4,g(w3,g(\
z2,s(z1))))),g(g(w4,g(z3,g(z2,s(g(w2,s(z1)))))),s(g(g(z2,s(g(z2\
,s(z1)))),s(g(w3,g(z2,s(z1)))))))),s(g(g(g(z3,g(z2,s(g(w2,s(z1)\
)))),g(g(z2,s(g(z2,s(z1)))),s(g(w3,g(z2,s(z1)))))),s(g(g(w4,g(z\
3,g(z2,s(g(u,s(z1)))))),s(g(g(z2,s(g(w2,s(z1)))),s(g(w3,g(z2,s(\
z1))))))))))))))))))))),v)),v);
end do;
```

Approximating Tverberg Points in Linear Time for Any Fixed Dimension

Wolfgang Mulzer
Institut für Informatik, Freie Universität Berlin
Takustr. 9
14195 Berlin, Germany
mulzer@inf.fu-berlin.de
http://page.mi.fu-berlin.de/mulzer

Daniel Werner*
Institut für Informatik, Freie Universität Berlin
Takustr. 9
14195 Berlin, Germany
daniel.werner@fu-berlin.de
http://page.mi.fu-berlin.de/dawerner

ABSTRACT

Let $P \subseteq \mathbb{R}^d$ be a d-dimensional n-point set. A *Tverberg* partition of P is a partition of P into r sets $P_1, \dots, P_r$ such that the convex hulls $\text{conv}(P_1), \dots, \text{conv}(P_r)$ have non-empty intersection. A point in $\bigcap_{i=1}^{r} \text{conv}(P_i)$ is called a *Tverberg point* of depth r for P. A classic result by Tverberg implies that there always exists a Tverberg partition of size $\lceil n/(d+1) \rceil$, but it is not known how to find such a partition in polynomial time. Therefore, approximate solutions are of interest.

We describe a deterministic algorithm that finds a Tverberg partition of size $\lceil n/4(d+1)^3 \rceil$ in time $d^{O(\log d)}n$. This means that for every fixed dimension we can compute an approximate Tverberg point (and hence also an approximate *centerpoint*) in *linear* time. Our algorithm is obtained by combining a novel lifting approach with a recent result by Miller and Sheehy [10].

Categories and Subject Descriptors

F.2.2 [**Analysis of Algorithms and Problem Complexity**]: Nonnumerical Algorithms and Problems—*Geometrical problems and computations*

General Terms

Algorithms, Theory

Keywords

discrete geometry, Tverberg's theorem, centerpoint, approximation, high dimension

*This research was funded by Deutsche Forschungsgemeinschaft within the Research Training Group (Graduiertenkolleg) "Methods for Discrete Structures".

SCG'12, June 17–20, 2012, Chapel Hill, North Carolina, USA.

1. INTRODUCTION

Let $P \subseteq \mathbb{R}^d$ be a d-dimensional point set with n points. In many applications (such as statistical analysis or finding sparse geometric separators in meshes) we would like to have a way to generalize the one-dimensional notion of a median to the high-dimensional point set P. A very natural means to accomplish this are *centerpoints*; a point $c \in \mathbb{R}^d$ is a centerpoint for P if every halfspace that contains c meets P in at least $n/(d+1)$ points. Equivalently, for every direction $\vec{v}$, the projection of c onto $\vec{v}$ splits the projection of P onto $\vec{v}$ into pieces with no more than $dn/(d+1)$ points [3]. A classic result in discrete geometry, the centerpoint theorem, shows that there exists a centerpoint for every point set [5, 12].

However, if we actually would like to compute a centerpoint for a given point set, the situation becomes more involved. Helly's theorem implies that the set of all centerpoints is given by the intersection of $O(n^d)$ halfspaces [6], so we can find a centerpoint in $O(n^d)$ time through linear programming. Chan [1] shows how to improve this running time to $O(n^{d-1})$ steps in expectation. He actually solves the harder problem of finding a point with maximum *Tukey depth*. The Tukey depth of a point c' is defined as the minimum number of points in P that are met by any halfspace containing c'. A centerpoint has Tukey depth at least $n/(d+1)$. If the dimension is not fixed, a result by Teng shows that it is co-NP-hard to check whether a given point is a centerpoint [13]. In two dimensions, a center point can be found in linear time [7].

Since a running time of $O(n^{d-1})$ is not feasible for large d, it makes sense to look for faster approximate solutions. A classic approach uses ε-approximations [2]: in order to obtain a point of Tukey depth $n(1/(d+1) - \varepsilon)$ take a random sample $A \subseteq P$ of size $O((d/\varepsilon^2)\log(d/\varepsilon))$ and compute a centerpoint for A, using the linear-programming method. This gives the desired approximation with constant probability, and the resulting running time after the sampling step is constant. What more could we possibly wish for? For one, the algorithm is Monte-Carlo: with a certain probability, the reported point fails to be a centerpoint, and we know of no fast algorithm to check its validity. This problem can be solved by constructing the ε-approximation deterministically [2], at the expense of a more complicated algorithm. Nonetheless, in either case the resulting running time, though constant, still grows exponentially with d, an undesirable feature for large dimensions.

This situation motivated Clarkson et al. [3] to look for more efficient randomized algorithms for approximate centerpoints. They give a simple probabilistic algorithm that computes a point of Tukey depth $O(n/(d+1)^2)$ in time $O(d^2(d\log n + \log(1/\delta))^{\log(d+2)})$, where δ is the error probability. They also describe a more sophisticated algorithm that finds such a point in time polynomial in n, d, and $\log(1/\delta)$. Both algorithms are based on a repeated algorithmic application of Radon's theorem (see below). Unfortunately, there remains a probability of δ that the result is not correct, and we do not know how to detect a failure efficiently.

More than ten years later, Miller and Sheehy [10] launched a new attack at the problem. Their goal is to develop a deterministic algorithm for approximating centerpoints whose running time is subexponential in the dimension. For this, they use a different proof of the centerpoint theorem that is based on a result by Tverberg: any d-dimensional n-point set can be partitioned into $r = \lceil n/(d+1) \rceil$ sets $P_1, \dots, P_r$ such that the convex hulls $\text{conv}(P_1), \dots, \text{conv}(P_r)$ intersect. Such a partition is called a *Tverberg* partition of P. Any point in $\bigcap_{i=1}^r \text{conv}(P_i)$ must be a center point.

More generally, for any partition $P_1, \dots, P_{r'}$ of P into r' sets and any point c that is contained in the convex hull of each of the sets, we say that c has *Tverberg depth* r' with respect to P. Consequently, c is called an *approximate Tverberg point* (of depth r'). See Figure 1.

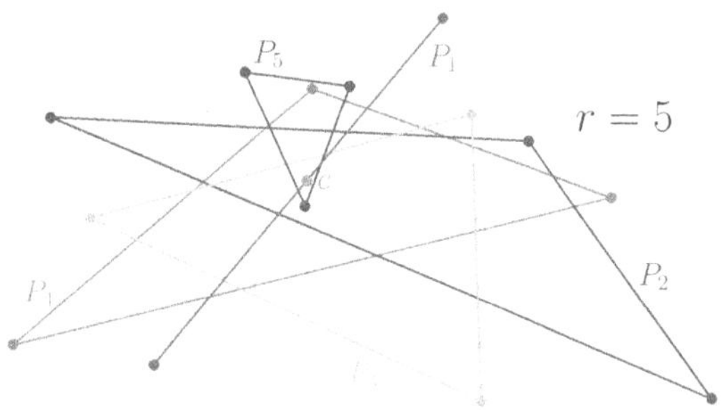

Figure 1: c is a point of Tverberg depth $r = 5$.

Miller and Sheehy describe how to find disjoint subsets $Q_1, \dots, Q_{r'}$, $r' = \lceil n/2(d+1)^2 \rceil$, of P and a point $c \in \mathbb{R}^d$, such that each Q_i contains $d+1$ points and such that $c \in \text{conv}(Q_i)$. Hence, c constitutes an approximate centerpoint for P, and the Q_i provide a certificate for this fact. The algorithm is deterministic and runs in time $n^{O(\log d)}$. At the same time, it is the first algorithm that also finds an approximate *Tverberg partition* of P. The running time is subexponential in d, but it is still the case that n depends exponentially on $\log d$.

In this paper, we show that the running time for finding approximate Tverberg partitions (and hence approximate centerpoints) can be improved. In particular, we show how to find a Tverberg partition for a set of points P that contains $\lceil n/4(d+1)^3 \rceil$ sets in deterministic time $d^{O(\log d)} n$. This is linear in n for any fixed dimension, and the dependence on d is only quasipolynomial.

1.1 Some discrete geometry

We begin by recalling some basic facts and definitions from discrete geometry [9]. A classic fact about convexity is Radon's theorem.

THEOREM 1.1 (RADON'S THEOREM). *For any $P \subseteq \mathbb{R}^d$ with $d+2$ points there exists a partition (P_1, P_2) of P such that $conv(P_1) \cap conv(P_2) \neq \emptyset$.*

As mentioned above, Tverberg [14] generalized this theorem for larger point sets.

THEOREM 1.2 (TVERBERG'S THEOREM). *Any set $P \subseteq \mathbb{R}^d$ with $n = (r-1)(d+1)+1$ points can be partitioned into r sets $P_1, \dots, P_r$ such that $\bigcap_{i=1}^r conv(P_i) \neq \emptyset$.*

Let P be a set of n points in $\mathbb{R}^d$. We say that $x \in \mathbb{R}^d$ has *Tverberg depth* r (with respect to P) if there is a partition of P into sets $P_1, \dots, P_r$ such that $x \in \bigcap_{i=1}^r \text{conv}(P_i)$. Tverberg's theorem thus states that, for any set P in $\mathbb{R}^d$, there is a point of Tverberg depth at least $\lfloor (n-1)/(d+1) + 1 \rfloor = \lceil n/(d+1) \rceil$. Note that every point with Tverberg depth r also has Tukey depth r. Thus, from now on we will use the term *depth* as a shorthand for Tverberg depth. As remarked above, Tverberg's theorem immediately implies the famous centerpoint theorem (see [9]):

THEOREM 1.3 (CENTERPOINT THEOREM). *For any set P of n points in $\mathbb{R}^d$ there is a point c such that all half-spaces containing c contain at least $\lceil n/(d+1) \rceil$ points from P.*

Finally, another classic theorem will be useful for us.

THEOREM 1.4 (CARATHÉODORY'S THEOREM). *Suppose that P is a set of n points in $\mathbb{R}^d$ and $x \in conv(P)$. Then there is a set of $d+1$ points $P' \subseteq P$ such that $x \in conv(P')$.*

This means that, in order to describe a Tverberg partition of depth r, we only need $r(d+1)$ points from P. This observation is also used by Miller and Sheehy [10]. They also observe that replacing $d+2$ by $d+1$ points can be done in $O(d^3)$ time by using Gaussian elimination. We denote the process of replacing larger sets by sets of size $d+1$ as *pruning*.

1.2 Our contribution

We now describe our results in more detail. In Section 2, we present a simple lifting argument which leads to an easy Tverberg approximation algorithm.

THEOREM 1.5. *Let P be a set of n points in $\mathbb{R}^d$ in general position. One can compute a Tverberg point of depth $\lceil n/2^d \rceil$ for P and the corresponding partition in time $d^{O(1)}n$.*

While this does not yet give a good approximation ratio (though constant for any fixed d), it is a very natural approach to the problem: it computes a higher dimensional Tverberg point via successive median partitions — just as a Tverberg point is a higher dimensional generalization of the 1-dimensional median.

By collecting several low-depth points and afterwards applying the brute-force algorithm on small point sets, we get an even higher depth in linear time for any *fixed* dimension:

THEOREM 1.6. *Let P be a set of n points in $\mathbb{R}^d$. Then one can compute a Tverberg point of depth $\lceil n/2(d+1)^2 \rceil$ and a corresponding partition in time $f(2^{d+1}) + d^{O(1)}n$, where $f(m)$ is the time for computing a Tverberg point of depth $\lceil m/(d+1) \rceil$ for m points brute force.*

Finally, by combining our approach with that of Miller and Sheehy, we improve our algorithm to yield an algorithm whose running time is quasipolynomial in d:

THEOREM 1.7. *Let P be a set of n points in $\mathbb{R}^d$. Then one can compute a Tverberg point of depth $\lceil n/4(d+1)^3 \rceil$ and a corresponding pruned partition in time $d^{O(\log d)}n$.*

In Section 4, we compare these results to the Miller-Sheehy algorithm and its extensions.

2. A SIMPLE FIXED PARAMETER ALGORITHM

First, we present a simple algorithm that runs in linear time for any fixed dimension and computes a point of depth $\lceil n/2^d \rceil$. For this, we show how to compute a Tverberg point by recursion on the dimension. As a byproduct, we obtain a quick proof of a weaker version of Tverberg's theorem.

2.1 The lifting argument and a simple algorithm

Let P be a d-dimensional point set. A natural way to compute a Tverberg point for P is to first project P to some lower-dimensional space, then to recursively compute a good Tverberg point for this projection, and use this point to find a solution in the higher-dimensional space. Surprisingly, we are not aware of any argument along these lines having appeared in the literature so far.

In what follows, we will describe how to *lift* a lower-dimensional Tverberg point into some higher dimension. Unfortunately, this process will come at the cost of a decreased depth for the lifted Tverberg point. For clarity of presentation, we first explain the lifting lemma in its simplest form. In Section 3.1, we then state the lemma in its full generality.

LEMMA 2.1. *Let P be a set of n points in $\mathbb{R}^d$, and let h be a hyperplane in $\mathbb{R}^d$. Let $c' \in h$ be a Tverberg point of depth r for the projection of P onto h, and suppose we know a corresponding Tverberg partition. Then we can find a Tverberg point $c \in \mathbb{R}^d$ of depth $\lceil r/2 \rceil$ for P and a corresponding Tverberg partition in time $O(n)$.*

PROOF. For every point $p \in P$, let $\mathrm{pr}(p)$ denote the projection of p onto h, and for every $Q \subseteq P$, let $\mathrm{pr}(Q)$ be the projections of all the points in Q. Let $P_1, \dots, P_r \subseteq P$ such that $\mathrm{pr}(P_1), \dots, \mathrm{pr}(P_r)$ is a Tverberg partition for $\mathrm{pr}(P)$ with Tverberg point c'. Let ℓ be the line orthogonal to h that passes through c'.

Since our assumption implies $c' \in \mathrm{conv}(\mathrm{pr}(P_i))$ for $i = 1, \dots, r$, it follows that ℓ intersects each $\mathrm{conv}(P_i)$ at some point x_i. Let $\widehat{Q}_i = \{x_{i_1}, x_{i_2}\}$, $i = 1, \dots, \lceil r/2 \rceil$, be a Tverberg partition of $x_1, \dots, x_r$. (If r is odd, one of the sets contains only one point, the median.) Since the points x_i lie on the line ℓ, such a Tverberg partition exists and can be computed in time $O(r)$ by finding the median c, i.e., the element of rank $\lceil r/2 \rceil$, according to the order along ℓ [4].

We claim that c is a Tverberg point for P of depth $\lceil r/2 \rceil$. Indeed, we have

$$c \in \mathrm{conv}(\widehat{Q}_i) = \mathrm{conv}(\{x_{i_1}, x_{i_2}\}) \subset \mathrm{conv}(P_{i_1} \cup P_{i_2}),$$

for $1 \le i \le \lceil r/2 \rceil$. Thus, if we set $Q_i := P_{i_1} \cup P_{i_2}$, then $Q_1, \dots, Q_{\lceil r/2 \rceil}$ is a Tverberg partition for the point c. The total time to find c and the Q_i is $O(n)$, as claimed. See Figure 2 for a two-dimensional illustration of the lifting argument. □

The proof of Theorem 1.5 is now a direct consequence of Lemma 2.1.

PROOF OF THEOREM 1.5. If $d = 1$, we can immediately find a Tverberg point and a corresponding partition by finding the median c of P [4] and pairing each point to the left of the median with exactly one point to the right of the median.

If $d > 1$, we project the points onto the hyperplane defined by $x_d = 0$. This results in an n-point set $P' \subseteq \mathbb{R}^{d-1}$. We recursively find a Tverberg point and a Tverberg partition of depth $\lceil n/2^{d-1} \rceil$ for P', and then apply Lemma 2.1: in each step, the depth it at most halved. Thus, we end of with a point of depth at least $\lceil n/2^d \rceil$. We apply the pruning step described in the introduction to the resulting partition in each step, in order to ensure that the size of the resulting sets is $d+1$.

Then the running time for each level of the recursion is $d^{O(1)}n$, and there are d levels, which implies the result. □

In particular, we obtain a weak version of Tverberg's theorem with a very elementary proof.

COROLLARY 2.2 (WEAK TVERBERG THEOREM). *Let P be a set of n points in $\mathbb{R}^d$. Then P can be partitioned into $\lceil n/2^d \rceil$ sets $P_1, \dots, P_{\lceil n/2^d \rceil}$ such that*

$$\bigcap_{i=1}^{\lceil n/2^d \rceil} conv(P_i) \neq \emptyset.$$

□

2.2 Improving the approximation factor

In order to improve the approximation factor, we will use an easy lemma to bootstrap the Tverberg depth.

LEMMA 2.3. *Suppose for any m-point set $Q \subseteq \mathbb{R}^d$ we can compute a point of Tverberg depth $\lceil m/\rho \rceil$ and a corresponding Tverberg partition in time $q(m,d)$. Let $P \subseteq \mathbb{R}^d$ with $|P| = n$, and let $c \in [2, n/\rho]$ be a constant. Then we can find $\alpha := \lceil \frac{n(1-1/c)}{\delta(d+1)} \rceil$ many Tverberg points of depth $\delta := \lceil n/c\rho \rceil$ for P and their corresponding partitions. These partitions have the additional property that each point of P appears in at most one partition. This takes total time*

$$O\Big(\frac{(c-1)\rho}{d+1}(p(n,d) + q(n,d))\Big),$$

where $p(n,d)$ is the time for the pruning phase.

PROOF. Let $P_1 := P$. We take an arbitrary subset $P'_1 \subseteq P_1$ with $\lceil n/c \rceil$ points and find a Tverberg point c_1 of depth $\lceil n/c\rho \rceil$ and a corresponding Tverberg partition $\mathcal{P}'_1$ for P'_1. Then we prune $\mathcal{P}'_1$ to get a Tverberg partition $\mathcal{P}_1$. Note that this takes time $p(n,d) + q(n,d)$ and that $Q_1 := \bigcup_{Z \in \mathcal{P}_1} Z$ contains at most $\lceil n/c\rho \rceil (d+1)$ points. Set $P_2 := P_1 \setminus Q_1$ and continue.

Each partition $\mathcal{P}_i$ partitions a set $Q_i \subseteq P$ such that the Q_i are pairwise disjoint. We can repeat this process until

$$n - i\delta(d+1) < \frac{n}{c},$$

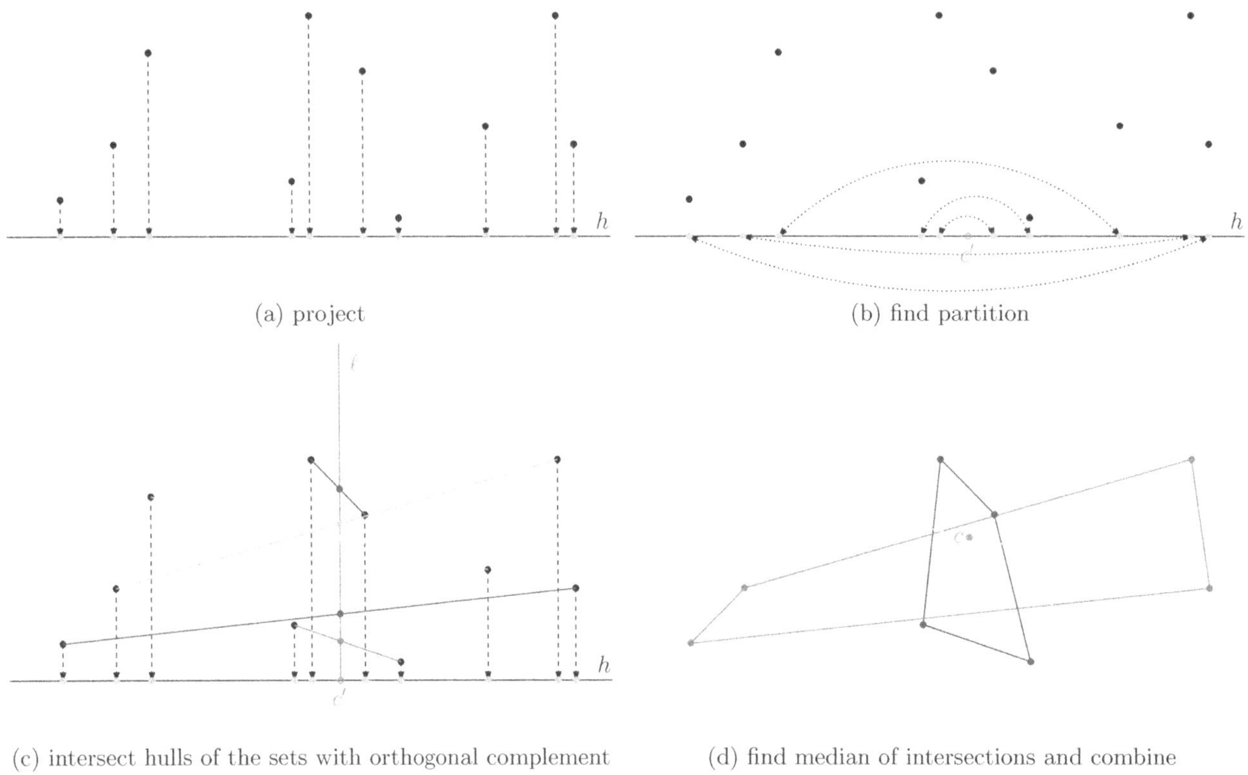

(a) project

(b) find partition

(c) intersect hulls of the sets with orthogonal complement

(d) find median of intersections and combine

Figure 2: Illustrating the lifting lemma in the plane: we project the point set P to the line h and find a Tverberg partition and a Tverberg point c' for the projection. Then, we construct the line ℓ through c' that is perpendicular to h, and we take the intersection with the lifted convex hulls of the Tverberg partition. We then find the median c and the corresponding partition for the intersections along ℓ. Finally, we group the points according to this partition.

which so solves to

$$\alpha \geq i > \left\lceil \frac{n(1-1/c)}{\lceil n/c\rho \rceil (d+1)} \right\rceil.$$

Thus, we obtain α many sets of points $c_1, \dots, c_\alpha$ with corresponding Tverberg partitions $\mathcal{P}_1, \dots, \mathcal{P}_\alpha$, each of depth at least $\lceil n/c\rho \rceil$, as desired. □

For example, by Theorem 1.5 we can find a point of depth $\lceil n/2^d \rceil$ and a corresponding pruned partition in time $d^{O(1)}n$. Thus, by applying Lemma 2.3 with $c = 2$, $\rho = 2^d$, we can also find $\lceil n/(2\lceil n/2^{d+1} \rceil (d+1)) \rceil \approx 2^d/(d+1)$ points of depth $\lceil n/2^{d+1} \rceil$ in linear time.

In order to make use of Lemma 2.3, we will also need a lemma that states that sometimes by *combining* Tverberg points we can multiply their depth. This generalizes a similar lemma by Miller and Sheehy [10, Lemma 4.1].

LEMMA 2.4. *Let P be a set of n points in $\mathbb{R}^d$, and let $P = \biguplus_{i=1}^{k} P_i$ be a partition of P. Furthermore, suppose that for each P_i we have a Tverberg point $c_i \in \mathbb{R}^d$ of depth r, together with a corresponding pruned Tverberg partition $\mathcal{P}_i$. Let $C := \{c_i \mid 1 \leq i \leq k\}$ and c be a point of depth r' for C, with corresponding pruned Tverberg partition $\mathcal{C}$. Then c is a point of depth rr' for P. Furthermore, we can find a corresponding pruned Tverberg partition in time $d^{O(1)}n$.*

PROOF. For $i = 1, \dots, k$, write $\mathcal{P}_i = \{Q_{i1}, \dots, Q_{ir}\}$, and write $\mathcal{C} = \{D_1, \dots, D_{r'}\}$. For $a = 1, \dots, r'$, $b = 1, \dots, r$, we define sets Z_{ab} as

$$Z_{ab} := \bigcup_{c_i \in D_a} Q_{ib}.$$

We claim that the set $\mathcal{Z} := \{Z_{ab} \mid a = 1, \dots, r'; b = 1, \dots, r\}$ is a Tverberg partition of depth rr' for P with Tverberg point c. Clearly, by definition $\mathcal{Z}$ is a partition with the appropriate number of elements. It only remains to check that $c \in \operatorname{conv}(Z_{ab})$ for each Z_{ab}. We have

$$c \in \operatorname{conv}(D_a) \subseteq \operatorname{conv}\Big(\bigcup_{c_i \in D_a} Q_{ib} \Big) = \operatorname{conv}(Z_{ab}),$$

for $a = 1 \dots r', b = 1 \dots r$. Since certainly $|\mathcal{Z}| \leq n$, and since each Z_{ab} can be pruned in $d^{O(1)}$ time, the lemma follows. □

Combining Lemmas 2.3 and 2.4, we can now prove Theorem 1.6.

PROOF OF THEOREM 1.6. If $n \leq 2^{d+1}$, we use the brute-force algorithm. This takes $f(2^{d+1})$ time.

Otherwise, we apply Lemma 2.3 (as in the remark following that lemma) to obtain a set C of

$$|C| = \left\lceil \frac{n}{2\lceil n/2^{d+1}\rceil(d+1)} \right\rceil$$

Tverberg points for P of depth $\lceil n/2^{d+1}\rceil$ with corresponding pruned partitions in linear time. We then use the brute-force algorithm to get a Tverberg point for C with depth $\lceil |C|/(d+1)\rceil$ with a corresponding partition, in time $f(|C|)$. Finally, we apply Lemma 2.4 to obtain a Tverberg point and corresponding partition in time $d^{O(1)}n$. The total running time is $f(2^d) + d^{O(1)}n$, and the resulting depth is

$$\lceil |C|/(d+1)\rceil \cdot \lceil n/2^{d+1}\rceil \geq \left\lceil \frac{n}{2^{d+1}} \frac{2^d}{(d+1)^2} \right\rceil = \left\lceil \frac{n}{2(d+1)^2} \right\rceil,$$

as desired. □

Alternatively, instead of the brute-force algorithm, we can use the algorithm by Miller and Sheehy to find a point of depth $\lceil n/4(d+1)^3\rceil$ in time $2^{O(d\log d)} + d^{O(1)}n$.

3. AN IMPROVED ALGORITHM

The algorithm in the previous section runs in linear time for any fixed dimension, but the constants are huge. Thus, finally we show how to improve our approach through an improved recursion and obtain an algorithm with running time $d^{O(\log d)}n$ while losing a depth factor of $1/2(d+1)$. First, however, we describe the more general version of Lemma 2.1 we promised above.

3.1 A more general version of the lifting argument

We present a more general version of the lifting argument in Lemma 2.1. For this we need some more notation. Let $P \subseteq \mathbb{R}^d$. Recall that a k-dimensional *flat* $F \subseteq \mathbb{R}^d$ (often abbreviated as *k-flat*) is defined as a k-dimensional affine subspace of $\mathbb{R}^d$ (or, equivalently, as the affine hull of $k+1$ affinely independent points in $\mathbb{R}^d$). A k-dimensional flat $F \subseteq \mathbb{R}^d$ is called a *Tverberg k-flat* of depth r for P, if there is a partition of P into sets $P_1, \ldots, P_r$ such that $\operatorname{conv}(P_i) \cap F \neq \emptyset$ for all $i = 1, \ldots, r$.

LEMMA 3.1. *Let P be a set of n points in $\mathbb{R}^d$, and let $h \subseteq \mathbb{R}^d$ be a k-flat. Suppose we have a Tverberg point $c \in h$ of depth r for $\operatorname{pr}(P)$, as well as a corresponding Tverberg partition. Let $h_c^\perp$ be the $(d-k)$ flat orthogonal to h that passes through c. Then $h_c^\perp$ is a Tverberg $(d-k)$-flat for P of depth r, with the same Tverberg partition.*

PROOF. Let $\operatorname{pr}(P_1), \ldots, \operatorname{pr}(P_r)$ be the Tverberg partition for the projection $\operatorname{pr}(P)$. It suffices to show that $\operatorname{conv}(P_i)$ intersects $h_c^\perp$ for $i = 1, \ldots, r$. Indeed, for $P_i = \{p_{i1}, \ldots, p_{il_i}\}$ let $c = \sum_{j=1}^{l_i} \lambda_j \operatorname{pr}(p_{ij})$ be a convex combination that witnesses $c \in \operatorname{conv}(\operatorname{pr}(P_i))$. We now write each $p_{ij} = \operatorname{pr}(p_{ij}) + \operatorname{pr}^\perp(p_{ij})$, where $\operatorname{pr}^\perp(\cdot)$ denotes the projection onto the orthogonal complement $h^\perp$ of h. Then,

$$\sum_{j=1}^{l_i} \lambda_j p_{ij} = \sum_{j=1}^{l_i} \lambda_j \operatorname{pr}(p_{ij}) + \sum_{j=1}^{l_i} \lambda_j \operatorname{pr}^\perp(p_{ij}) \in c + h^\perp = h_c^\perp,$$

as claimed. □

First of all, this shows how a good algorithm for any fixed dimension improves the general case:

COROLLARY 3.2. *Let $\delta \geq 1$ be a fixed integer. Suppose we have an algorithm $\mathcal{A}$ with the following property: for every point set $Q \subseteq \mathbb{R}^\delta$, the algorithm $\mathcal{A}$ constructs a Tverberg point of depth $\lceil |Q|/\rho\rceil$ for Q as well as a corresponding pruned Tverberg partition in time $f(|Q|)$.*

Then, for any n-point set $P \subseteq \mathbb{R}^d$ and for any $d \geq \delta$, we can find a Tverberg point of depth $n/\rho^{\lceil d/\delta\rceil}$ and a corresponding pruned partition in time $\lceil d/\delta\rceil f(n) + d^{O(1)}n$.

PROOF. We use induction on $k := \lceil d/\delta\rceil$ to show that such an algorithm exists with running time $k(f(n) + d^{O(1)}n)$. If $k = 1$, we can just use algorithm $\mathcal{A}$ and there is nothing to show.

Now suppose $k > 1$. Let $h \subseteq \mathbb{R}^d$ be a δ-flat in $\mathbb{R}^d$, and let $\operatorname{pr}(P)$ be the projection of P onto h. We use algorithm $\mathcal{A}$ to find a Tverberg point c of depth $\lceil n/\rho\rceil$ for $\operatorname{pr}(P)$ as well as a corresponding pruned partition $\operatorname{pr}(P_1), \ldots, \operatorname{pr}(P_{\lceil n/\rho\rceil})$. This takes time $f(n)$. By Lemma 3.1, the $(d-\delta)$-flat $h_c^\perp$ is a Tverberg flat of depth $\lceil n/\rho\rceil$ for P, with corresponding Tverberg partition $P_1, \ldots, P_{\lceil n/\rho\rceil}$. For each i, we can find a point q_i in $\operatorname{conv}(P_i) \cap h_c^\perp$ in time $d^{O(1)}$.

Now consider the point set $Q = \{q_1, \ldots, q_{\lceil n/\rho\rceil}\} \subseteq h_c^\perp$. The set Q is a $(d-\delta)$-dimensional point set. Since we have $\lceil (d-\delta)/\delta\rceil = k-1$, by induction we can find a Tverberg point c' for Q of depth $|Q|/\rho^{\lceil d/\delta\rceil - 1} = n/\rho^{\lceil d/\delta\rceil}$ and a corresponding pruned Tverberg partition $\mathcal{Q}$ in total time $(k-1)(f(n) + d^{O(1)}n)$. Now, c' is a Tverberg point of depth $n/\rho^{\lceil d/\delta\rceil}$ for P, and a corresponding Tverberg partition is obtained by replacing each point q_i in the partition $\mathcal{Q}$ by the corresponding subset P_i. The resulting partition can be pruned in time $d^{O(1)}n$.

Thus, the total running time is

$$(k-1)(f(n) + d^{O(1)}n) + f(n) + d^{O(1)}n = k(f(n) + d^{O(1)}n),$$

and since $k = O(d)$, the claim follows. □

3.2 An improved algorithm

We will now show how to combine the above techniques for an algorithm with a better running time. The idea of the new algorithm is as follows: using Corollary 3.2, we reduce solving a d-dimensional instance to solving two instances of dimension $d/2$. This can be done recursively. Unfortunately, applying Corollary 3.2 reduces the depth of the partition. To fix this, we apply Lemmas 2.3, 2.4 and the Miller-Sheehy algorithm to increase the depth again. For clarity of presentation, we omit the roundings, which can be taken care of easily.

PROOF OF THEOREM 1.7. We prove the theorem by induction on d. As stated before, for $d = 1$ the claim is immediate, as in this case the problem reduces to a median computation.

Thus, suppose that $d > 1$. By induction, for any $(d/2)$-dimensional point set $Q \subseteq \mathbb{R}^{d/2}$ there exists an algorithm that returns a Tverberg point of depth $|Q|/4(d/2+1)^3$ and a corresponding pruned Tverberg partition in time $d^{\alpha\log(d/2)}n$, for some sufficiently large constant $\alpha > 0$.

Thus, by Corollary 3.2 (with $\delta = d/2$), there exists an algorithm that can compute a Tverberg point for P of depth $n/16(d/2+1)^6$ and a corresponding Tverberg partition in total time $2d^{\alpha\log(d/2)} + d^{O(1)}n$.

Now we apply Lemma 2.3. The lemma shows that we can compute $16(d/2+1)^6/(d+1)$ many points with depth

$n/32(d/2+1)^6$ and corresponding (disjoint) pruned partitions in time $d^{\alpha \log(d/2)+O(1)}n$.

Let C be the set of these Tverberg points. Applying the Miller-Sheehy algorithm, we can find a Tverberg point for C of depth $|C|/2(d+1)^2$ and a corresponding pruned Tverberg partition in time $|C|^{O(\log d)}$. Now, Lemma 2.4 shows that in additional $d^{O(1)}n$ time, we obtain a Tverberg point and a corresponding Tverberg partition for P of size

$$\frac{n}{2 \cdot 16(d/2+1)^6} \frac{16(d/2+1)^6}{2(d+1)^2(d+1)} = \frac{n}{4(d+1)^3},$$

as desired.

It remains to analyze the running time. Adding the various terms, we end up with a time bound of

$$T(n,d) = d^{\alpha \log(d/2)+O(1)}n + |C|^{O(\log d)} + d^{O(1)}n.$$

Since $|C| = d^{O(1)}$, we get

$$\begin{aligned} T(n,d) &\leq d^{\alpha \log(d/2)+O(1)}n + d^{O(\log d)}n \\ &\leq d^{\alpha \log d - \alpha/2}n + d^{\beta \log d}n, \end{aligned}$$

for α large enough and some $\beta > 0$, independent of d. Hence, it follows that for large enough α we have

$$T(n,d) \leq d^{\alpha \log d}n = d^{O(\log d)}n,$$

as claimed. This completes the proof. □

4. COMPARISON TO MILLER-SHEEHY

In the table below, we compare our algorithm in more detail to the Miller-Sheehy algorithm and its extensions. They give a generalization of their approach that shows that by computing higher order Tverberg points of depth r by brute-force, the running time can be improved for small d. This comes with the loss of factor r in the output. No exact values are given, but as far as we can tell, one can achieve a polynomial $O(f(d)n^2)$ running time for fixed d by setting the parameter $r = (d+1)$, while losing a factor of $(d+1)$ in the approximation. Further, even though it is not explicitly mentioned in the paper, we think that it is possible to also bootstrap their own algorithm (for a better running time in terms of d, while losing another factor of $(d+1)$ in the output). Table 4 shows a rough comparison of the different approaches. Again, f denotes the running time of the brute force algorithm.

Algorithm	Running time	Depth
Theorem 1.5	$O(dn)$	$n/2^d$
Miller-Sheehy	$n^{O(\log d)}$	$n/2(d+1)^2$
Theorem 1.6	$O\left(f(2^d) + d^{O(1)}n\right)$	$n/2(d+1)^2$
Miller-Sheehy generalized ($r = d+1$)	$O\left(f(d)n^2\right)$	$\approx n/2(d+1)^3$
Theorem 1.6 with M.-S.	$O\left(2^{O(d \log d)} + n\right)$	$n/4(d+1)^3$
Miller-Sheehy bootstrapped	$d^{O(\log d)}n^3$	$\approx n/2(d+1)^4$
Theorem 1.7	$d^{O(\log d)}n$	$n/4(d+1)^3$

5. CONCLUSION AND OUTLOOK

We have presented a very simple algorithm for finding an approximate Tverberg point, which runs in linear time for any fixed dimension. Using more sophisticated methods and combining our methods with known results, we managed to improve the running time to $d^{O(\log d)}n$, while getting within a factor of $1/4(d+1)^2$ of the guaranteed optimum.

Unfortunately, the resulting running time is still quasi-polynomial in d, and we still do not know whether there exists a polynomial algorithm (in n and d) for finding an approximate Tverberg point. However, we are hopeful that our techniques constitute a further step in this direction and that such an algorithm will eventually be discovered—maybe even by a more clever combination of our algorithm with that of Miller and Sheehy.

A promising approach, as also pointed out by one of the reviewers, would be to have a look at Sarkaria's proof of Tverberg's theorem using Colorful Carathéodory's theorem (see Matoušek [9, Chapter 8]). If one can compute a colorful simplex in truly polynomial time in any dimension, one can also compute a Tverberg point efficiently.

As mentioned before, the problem of deciding whether a given point has at least a certain depth is NP-complete. It is possible to strengthen this result to show that in $\mathbb{R}^{d+1}$, the problem is d-SUM hard, using the approach by Knauer et al. [8]. However, this does not tell us anything about the actual problem of computing a point of depth $n/(d+1)$. As such a point is guaranteed to exist, it is not clear how to prove the problem to be hard using the "standard" NP-hardness reduction. Rather, we think that a hardness proof along the lines of complexity classes such as PPAD (see Papadimitriou [11]) should be pursued.

Acknowledgments. We would like to thank Nabil Mustafa for suggesting the problem to us. We also thank him and Don Sheehy for helpful discussions.

We would further like to thank the anonymous referees for their helpful and detailed comments.

6. REFERENCES

[1] T. M. Chan. An optimal randomized algorithm for maximum Tukey depth. In *Proc. 15th Annu. ACM-SIAM Sympos. Discrete Algorithms (SODA)*, pages 430–436, 2004.

[2] B. Chazelle. *The discrepancy method: randomness and complexity*. Cambridge University Press, Cambridge, 2000.

[3] K. L. Clarkson, D. Eppstein, G. L. Miller, C. Sturtivant, and S.-H. Teng. Approximating center points with iterative Radon points. *Internat. J. Comput. Geom. Appl.*, 6(3):357–377, 1996.

[4] T. H. Cormen, C. E. Leiserson, R. L. Rivest, and C. Stein. *Introduction to algorithms*. MIT Press, Cambridge, MA, third edition, 2009.

[5] L. Danzer, B. Grünbaum, and V. Klee. Helly's theorem and its relatives. In *Proc. Sympos. Pure Math., Vol. VII*, pages 101–180. Amer. Math. Soc., Providence, R.I., 1963.

[6] H. Edelsbrunner. *Algorithms in combinatorial geometry*. Springer-Verlag, Berlin, 1987.

[7] S. Jadhav and A. Mukhopadhyay. Computing a centerpoint of a finite planar set of points in linear time. *Discrete Comput. Geom.*, 12(3):291–312, 1994.

[8] C. Knauer, H. R. Tiwary, and D. Werner. On the computational complexity of Ham-Sandwich cuts, Helly sets, and related problems. In *28th International Symposium on Theoretical Aspects of Computer Science (STACS 2011)*, volume 9, pages 649–660. Schloss Dagstuhl–Leibniz-Zentrum fuer Informatik, 2011.

[9] J. Matoušek. *Lectures on Discrete Geometry*. Springer, 2002.

[10] G. L. Miller and D. R. Sheehy. Approximate centerpoints with proofs. *Comput. Geom. Theory Appl.*, 43(8):647–654, 2010.

[11] C. H. Papadimitriou. On the complexity of the parity argument and other inefficient proofs of existence. *Journal of Computer and System Sciences*, 48(3):498 – 532, 1994.

[12] R. Rado. A theorem on general measure. *J. London Math. Soc.*, 21:291–300, 1946.

[13] S.-H. Teng. *Points, spheres, and separators: a unified geometric approach to graph partitioning*. PhD thesis, School of Computer Science, Carnegie Mellon University, 1992.

[14] H. Tverberg. A generalization of Radon's theorem. *J. London Math. Soc.*, 41:123–128, 1966.

Tessellation Analysis of the Cosmic Web

Rien van de Weygaert
Kapteyn Astronomical Institute
P.O. Box 800, 9700AV Groningen, the Netherlands
weygaert@astro.rug.nl

ABSTRACT

The large scale distribution of matter and galaxies features a complex network of interconnected filamentary galaxy associations. This network, which has become known as the *Cosmic Web*, contains structures from a few megaparsecs up to tens and even hundreds of Megaparsecs of size[1]. Galaxies and mass exist in a wispy weblike spatial arrangement consisting of dense compact clusters, elongated filaments, and sheetlike walls, amidst large near-empty void regions. An important additional aspect of this mass distribution is that it is marked by substructure over a wide range of scales and densities. The vast Megaparsec cosmic web is one of the most striking examples of complex geometric patterns found in nature, and certainly the largest in terms of sheer size.

The overwhelming complexity of both the individual structures as well as their connectivity, the lack of structural symmetries, the intrinsic multiscale nature and the wide range of densities that one finds in the cosmic matter distribution has prevented the use of simple and straightforward instruments. In this lecture, I describe the considerable advances that have been made over the past decade towards unravelling the structure of the Cosmic Web, enabled by a range of tools and concepts from computational geometry and computational topology. This will include our own work, in which Voronoi and Delaunay tessellations figure prominently through their high sensitivity to density and local shape of the local galaxy distribution, or particle distribution in the case of computer simulations of cosmic structure formation. It has led to the development of the Delaunay Tessellation Field Estimator (DTFE) formalism, which forms the basis of a range of techniques to identify different aspects of the Cosmic Web [1]. Examples are the Watershed Void Finder to trace voids, the Nexus multiscale morphology formalism and the Morse-based SpineWeb formalism to find walls, filaments and clusters. Recently, we used alpha shapes to study the multiscale topology of the Cosmic Web, in terms of Betti numbers and persistence diagrams. I will also review a number of other astronomical applications of tesssellations, motivated by their quickly proliferaating use in astrophysics and cosmology.

[1]The main measure of length in astronomy is the parsec. Technically a parsec is the distance at which we would see the distance Earth-Sun at an angle of 1 arcsec. It is equal to 3.262 lightyears $= 3.086 \times 10^{13}$km. Cosmological distances are substantially larger, so that a Megaparsec ($= 10^6\ pc$) is the regular unit of distance. Usually this goes along with h, the cosmic expansion rate (Hubble parameter) H in units of 100 km/s/Mpc ($h \approx 0.71$).

SCG'12, June 17–20, 2012, Chapel Hill, North Carolina, USA.
ACM 978-1-4503-1299-8/12/06.

Categories and Subject Descriptors

J.2 [**Computer Applications**]: Physical Sciences and Engineering—*Astronomy,Physics*; I.3.5 [**Computer Graphics**]: Computational Geometry and Object Modeling—*physically based modeling*

Keywords

Cosmology, Large Scale Structure of the Universe, Computational Geometry, Computational Topology, Tessellations

1. SHORT BIOGRAPHY

Rien van de Weygaert studied astronomy and physics at the University of Leiden, where he obtained his PhD cum laude in 1991 on the thesis "Voids and the Geometry of Large Scale Structure". Subsequently, he worked as an NSERC fellow at the Canadian Institute of Astrophysics (CITA) in Toronto, Canada and as a research fellow at the Max Planck Institut für Astrophysik in Garching, Germany before taking up a Royal Academy fellowship at the Kapteyn Astronomical Institute at the University of Groningen. Since 2004 he is professor of cosmological structure formation at the Kapteyn Astronomical Insitute.

His research interests concern cosmology, the large scale structure of the Universe and the formation of structure in the Universe. In addition, he has a keen interest in computational geometry and topology and pattern recognition. Within these research subjects he has had a particular interest in the formation and dynamics of the Cosmic Web and the structure and evolution of cosmic voids. Recently he started to wander into the history of astronomy, motivated by world's oldest astronomical computer, the complex bronze geared Antikythera mechanism from ancient Greek times.

2. REFERENCES

[1] R. van de Weygaert and W. Schaap. The Cosmic Web: Geometric Analysis. In V. J. Martínez, E. Saar, E. Martínez-González, & M.-J. Pons-Bordería, editor, *Data Analysis in Cosmology*, volume 665 of *Lecture Notes in Physics, Berlin Springer Verlag*, pages 291–413, 2009.

Stabbing Horizontal Segments with Vertical Rays*

Yufei Tao[1,2]

[1]Department of Computer Science and Engineering
Chinese University of Hong Kong
New Territories, Hong Kong
taoyf@cse.cuhk.edu.hk

[2]Division of Web Science and Technology
Korea Korea Advanced Institute of Science and Technology
Daejeon, Republic of Korea
taoyf@kaist.ac.kr

ABSTRACT

We consider maintaining a dynamic set $\mathcal{S}$ of N horizontal segments in $\mathbb{R}^2$ such that, given a vertical ray Q in $\mathbb{R}^2$, the segments in $\mathcal{S}$ intersecting Q can be reported efficiently. In the external memory model, we give a structure that consumes $O(N/B)$ space, answers a query in $O(\log_B N + K/B)$ time (where K is the number of reported segments), and can be updated in $O(\log_B N)$ amortized time per insertion and deletion. With B set to a constant, the structure also works in internal memory, consuming space $O(N)$, answering a query in $O(\log N + K)$ time, and supporting an update in $O(\log N)$ amortized time.

Categories and Subject Descriptors

F.2.2 [**Analysis of algorithms and problem complexity**]: Nonnumerical Algorithms and Problems—*computations on discrete structures*

General Terms

Theory

Keywords

Ray stabbing, segment intersection, data structure, dynamic

1. INTRODUCTION

In the *orthogonal ray stabbing problem* (henceforth, the ray stabbing problem), we want to store a set $\mathcal{S}$ of N horizontal segments in $\mathbb{R}^2$ such that, given a vertical ray $Q = x \times (-\infty, y]$, all the segments in $\mathcal{S}$ intersecting Q can be reported efficiently. See Figure 1 for an example. We consider the problem in a fully dynamic setting, where segments can be inserted and deleted in $\mathcal{S}$. Applications of this problem have been described in databases [1, 10, 14], GIS [6], networking [12], and so on. Throughout the paper, every logarithm is assumed to be at least 1. In other words, $\log_x y$ should be understood as $\max\{1, \log_x y\}$.

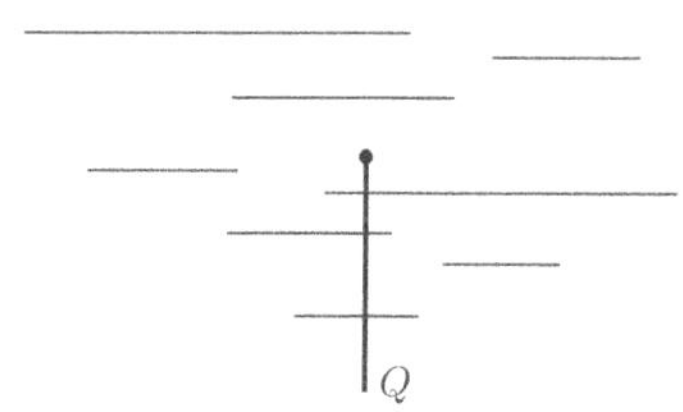

Figure 1: A ray stabbing query that reports 3 segments

Previous results. The ray stabbing problem is a special instance of the *segment intersection problem*, where a query reports the segments of $\mathcal{S}$ intersecting a vertical segment (instead of a ray). In internal memory, when $\mathcal{S}$ is static, the segment intersection problem can be solved with a persistent binary tree that uses $O(N)$ space and answers a query in $O(\log N + K)$ time, where K is the number of reported segments. For dynamic data, Cheng and Janardan [9] gave a linear-size structure that solves a query in $O(\log^2 N + K)$ time, and supports an insertion and a deletion in $O(\log N)$ time. Mortensen [13] proposed a structure that has space complexity $O(N \log N / \log\log N)$, $O(\log N + K)$ query time and $O(\log N)$ update time. Blelloch [7] presented a linear-size structure that supports an update in $O(\log N)$ amortized time and a query in $O(\log N + K \log N/ \log\log N)$ time.

In the *external memory* (EM) model [2], a machine has a disk formatted into *blocks* of B words, and memory of M words satisfying $M \geq 2B$. Time complexity is measured as the number of I/Os performed, whereas space complexity is measured as the number of disk blocks occupied. *Linear* complexity on an input of size N is interpreted as $O(1 + N/B)$. In this model, the static version of the ray stabbing problem can be settled with the persistent B-tree [5] that consumes linear space and solves a query in $O(\log_B N + K/B)$ I/Os. For the dynamic version, we are not aware of any published result. Perhaps as a folklore, using standard techniques [3, 4, 5], one can obtain a linear-size structure that has query time $O(\log_B^2 N + K/B)$ and supports an update in $O(\log_B N)$ amortized I/Os.

Our results. We present a structure for solving the ray stabbing problem:

THEOREM 1. *For the ray stabbing problem, there is a structure in external memory that uses linear space, answers a query in*

*This work was supported in part by (i) projects GRF 4169/09, 4166/10, 4165/11 from HKRGC, and (ii) the WCU (World Class University) program under the National Research Foundation of Korea, and funded by the Ministry of Education, Science and Technology of Korea (Project No: R31-30007).

$O(\log_B N + K/B)$ *I/Os, and can be updated in* $O(\log_B N)$ *amortized I/Os per insertion and deletion.*

The theorem also holds in internal memory by setting B to an appropriate constant.

Technique overview. Our solution is based on the *external interval tree* [4]. To improve over standard techniques, however, we adopt a different approach to handle the so-called "left segments". Conventionally, a query reports the qualifying left segments of a node when the node is reached, which necessitates $\Omega(\log_B^2 N)$ query time. To break this barrier, we take a delayed approach which does not report a left segment until its left endpoint has "departed" from the query, namely, the endpoint is no longer on the search path. With a filtering-search argument, we can charge most of the cost incurred at a node of the search path on the reported segments, which is the key to achieving logarithmic query time.

Central to our techniques is a new notion called *bottom set*. Concerning the interval tree (or its external version), intuitively, a bottom set includes some low segments (by y-coordinates) among those that are stored at an internal node. Each node can have a large number of bottom sets, each of which is defined with respect to a distinct descendant of the node. A side message delivered by this paper is that bottom sets may equip us with an extra weapon in solving problems (e.g., ray stabbing) for which the interval tree appears to be the right structure. Because of this, the mechanism presented in this paper for maintaining bottom sets may be of independent interest (even though most challenges in developing the mechanism are technical), since it handles an update in logarithmic amortized time.

2. A STATIC STRUCTURE

This section describes a linear-size, static, structure that answers a query in $O(\log_B N + K/B)$ I/Os. Although the same performance can also be achieved by a persistent B-tree, our structure manages segments in an alternative manner. We will make the structure fully dynamic in the later sections, but its static version illustrates some central ideas behind our techniques without the distraction from update handling.

2.1 Structure

Given horizontal segments s_1 and s_2, we say that s_1 is *lower* if it has a smaller y-coordinate than s_2; otherwise, s_1 is *higher*. Also, we will refer to the x-coordinate of a segment's left endpoint simply as its *left x-coordinate*. As before, let $\mathcal{S}$ be the input set of data segments, and $N = |\mathcal{S}|$.

Base tree. In our structure, the base tree is a B-tree $\mathcal{T}$ indexing the x-coordinates of the segments in $\mathcal{S}$. All the x-coordinates are stored at the leaf level. A leaf node contains $\Theta(b)$ coordinates where $b = B\log_B^3 N$, while an internal node has $\Theta(f)$ child nodes where $f = \max\{B^{1/3}, (\log_B N)^{1/3}\}$. A leaf node is at *level* 0; and in general, the parent of a level-l node is at level $l+1$.

Naturally, a total order exists on the leaf nodes. The *slab* of a leaf node z is an interval $[x, x')$ where x and x' are the smallest x-coordinates stored in z and the leaf node succeeding z, respectively ($x' = \infty$ if no leaf succeeds z). We denote the slab as σ_z. The slab σ_u of an internal node u, in general, is the union of the slabs of its child nodes. In $\mathbb{R}^2$, a slab σ corresponds to a vertical strip, enclosing all the points (of the data space) with x-coordinates in σ. Henceforth, we will interpret a slab as a strip.

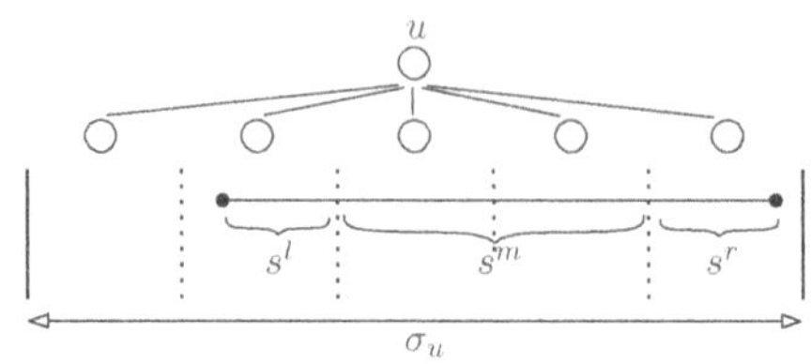

Figure 2: A segment s in the stabbing set of node u

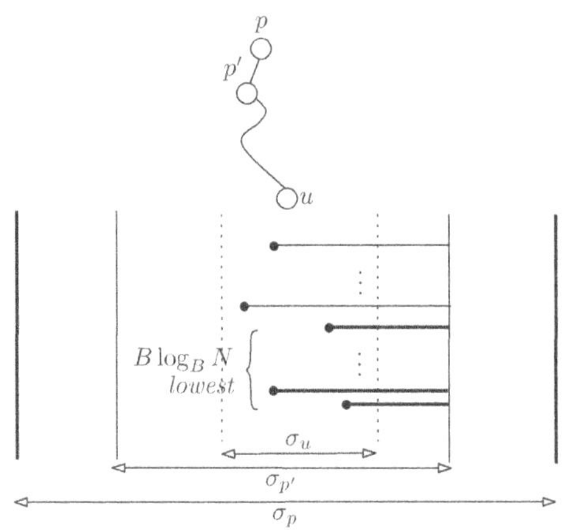

Figure 3: Illustration of $bottom_u(p)$: The segments shown are the left segments at p whose left endpoints are in σ_u. The $B\log_B N$ lowest ones constitute $bottom_u(p)$. All these segments are grounded on the right boundary of $\sigma_{p'}$, where p' is a child node of p and a proper ascenstor of u.

As in the external interval tree [4], each segment $s \in \mathcal{S}$ is assigned to the lowest node u whose slab covers s. The set of segments assigned to u is the *stabbing set* of u, denoted as $stab(u)$.

Now, consider u as an internal node. We refer to σ_v as a *child slab* if v is a child node of u. As in the external interval tree, every segment $s \in stab(u)$ defines a *left segment* s^l, a *right segment* s^r, and perhaps a *middle segment* s^m. Specifically, s^l (s^r) is the part of s in the only child slab of u covering the left (right) endpoint of s. On the other hand, s^m is the remainder of s after trimming s^l and s^r. See Figure 2. We say that s^l is a *left segment at* u, and adopt similar terms for s^r and s^m. Let $\mathcal{L}_u, \mathcal{R}_u$ and $\mathcal{M}_u$ be the sets of left, right and middle segments at u, respectively. $\mathcal{L}_u, \mathcal{R}_u$ and $\mathcal{M}_u$ are managed by secondary structures. We will focus on $\mathcal{L}_u$ in this section. Symmetric structures exist for $\mathcal{R}_u$, whereas the management of $\mathcal{M}_u$ is left to Section 3.

Bottom set. Let u be a leaf/internal node in $\mathcal{T}$, and $parent(u)$ be its parent. Consider a proper ancestor p of $parent(u)$. Clearly, σ_u is within σ_p. Some left segments at p may have their left endpoints covered by σ_u. Among them, we are interested in the few *lowest* ones. Let $bottom_u(p)$ be $B\log_B N$ lowest segments in $\mathcal{L}_p$ whose left endpoints are in σ_u. In case less than $B\log_B N$ such segments exist, $bottom_u(p)$ includes all of them. Refer to $bottom_u(p)$ as the *bottom set of p at u*. See Figure 3. Note that the definition of $bottom_u(p)$ requires that the levels of p and u differ by at least 2.

Now, consider u as an internal node with child nodes $v_1, ..., v_f$. For each $i \in [1, f]$, define:

$$\pi_u(i) = \bigcup_{\forall \text{ proper ancestor } p \text{ of } u} bottom_{v_i}(p). \quad (1)$$

Let $\psi_u(i)$ be the set of B lowest segments in $\pi_u(i)$. In case $|\pi_u(i)| < B$, $\psi_u(i)$ includes all of them.

Secondary structures. An internal node u with children $v_1, ..., v_f$ is associated with:

- for each v_i, a B-tree $\Pi_u(i)$ on the y-coordinates of the segments in $\pi_u(i)$. We call $\Pi_u(i)$ a *bottom structure*.
- (only if the level of u is at least 2) for each v_i, a *priority search tree* [3] $F_u(i)$ on the left endpoints in $\mathcal{L}_u$ that appear in σ_{v_i}. $F_u(i)$ answers *lower-open* 3-sided range queries efficiently[1]. Such a query reports all the endpoints in a rectangle of the form $[x_1, x_2] \times (-\infty, y_2]$, i.e., the lower edge of the rectangle is grounded on the bottom of $\mathbb{R}^2$. We call $F_u(i)$ a *flank structure*.
- a priority search tree P_u on the left endpoints in $\psi_u(1) \cup ... \cup \psi_u(f)$. P_u is also for answering lower-open 3-sided range queries. We call P_u a *pilot structure*.

Consider a leaf node z of $\mathcal{T}$. If a segment $s \in \mathcal{S}$ has its left x-coordinate in z, we store s at z. Note that z stores at most b segments.

Space. $\mathcal{T}$ itself obviously occupies $O(N/B)$ space. It has height $h = O(\log_f N)$. A bottom structure indexes at most $hB \log_B N$ segments in $O(h \log_B N)$ blocks. An internal node has f bottom structures using $O(fh \log_B N)$ space in total. The node's pilot structure stores at most fB points in $O(f)$ blocks. Since there are $O(N/(fb))$ internal nodes, all the bottom and pilot structures require $O(\frac{N}{fb} fh \log_B N) = O((N/b)\log_B^2 N) = o(N/B)$ space in total. A flank structure consumes linear space. The left endpoint of a data segment is stored in at most one flank structure. Hence, all the flank structures use $O(N/B)$ space. Finally, each segment is stored constant times at the leaf level. Overall, our structure occupies linear space.

Remark. The idea of storing information about a node's stabbing set in its subtree is due to Kaplan, Molad and Tarjan [12]. The way that segments are managed by bottom, flank, and pilot structures is developed in this paper. So are the algorithms for querying and updating them, as will be presented later.

2.2 Query

Given a query $Q = x \times (-\infty, y]$, we visit a root-to-leaf path of $\mathcal{T}$, formed by the nodes whose slabs contain Q. Refer to the path as the *query path*. We will report segments only from the secondary structures of the nodes on this path. For a data segment, only one of its left, right and middle segments can intersect Q. Hence, we concentrate on reporting the left, right and middle segments, respectively, without worrying that a data segment may be reported twice. Next, we discuss first left segments.

[1]A priority search tree on N 2d points consumes $O(N/B)$ space. Given a 3-sided axis-parallel rectangle, it can be used to report all the points in the rectangle in $O(\log_B N + K/B)$ I/Os, where K is the number of such points. It supports an update in $O(\log_B N)$ time per insertion and deletion.

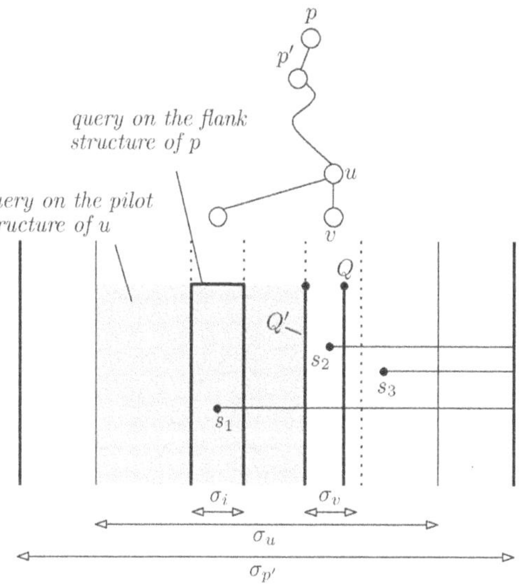

Figure 4: Query processing at an internal node u: s_1, s_2 and s_3 are left segments at p. They are in the spanning, pending, and disposable groups of p, respectively.

Internal node. We examine the nodes of the query path in top-down order. Suppose that we are currently at an internal node u, which has f child slabs $\sigma_1, ..., \sigma_f$. Let v be the child node of u on the query path. Hence, Q is contained in σ_v (which is one of $\sigma_1, ...\sigma_f$).

Consider any proper ancestor p of u. Let S be the set of segments in $\mathcal{L}_p$ whose left endpoints are in σ_u. Based on left endpoints, S can be divided into three disjoint groups as follows. The *spanning group* includes the segments of S whose left endpoints are to the left of σ_v. Every such segment must span σ_v, and hence the group's name. The *pending group*, on the other hand, includes the segments whose left endpoints are in σ_v. In the final *disposable group*, all the segments are to the right of σ_v, and thus can be safely ignored. Figure 4 illustrates this with s_1, s_2 and s_3.

When the processing finishes at u, we ensure an invariant that *all* the qualifying segments in the spanning group of p have been reported. There is no such guarantee for the pending group. Nevertheless, as we descend further along the query path, the qualifying segments in the pending group either will appear in a spanning group later, or are stored in the path's leaf node. Our invariant implies that we will find those segments in the former situation, whereas in the latter we will also find them by inspecting the leaf node. No left segment can be reported twice, noticing that the spanning group of p is completely different if u changes. Specifically, as we move from u to its child node on the search path, the pending group at u will unfold into the spanning, pending and disposable groups at its child node.

Next, we explain how to report segments from the spanning groups of the proper ancestors of u. Obtain a ray Q' by snapping Q onto the left boundary of σ_v. Namely, the resulting ray Q' is $x' \times (-\infty, y]$, where x' is the x-coordinate of the left boundary of σ_v. No segment in the pending group of any proper ancestor of u can intersect Q'. On the other hand, each qualifying segment from a spanning group must intersect Q'; its left endpoint must lie in rectangle $(-\infty, x'] \times (-\infty, y]$. See the gray area in Figure 4.

We search the pilot structure of u to retrieve all the left endpoints covered by $(-\infty, x'] \times (-\infty, y]$. Recall that every such endpoint belongs to a left segment in $\psi_u(i)$ for some $i \in [1, f]$. The segments whose endpoints are retrieved definitely satisfy the query, and hence, are reported. If less than B segments from $\psi_u(i)$ are found, no other segment in $\pi_u(i)$ can intersect Q' because $\psi_u(i)$ collects the lowest segments of $\pi_u(i)$. In this case, $\pi_u(i)$ is pruned from further consideration.

If, on the other hand, B segments from $\psi_u(i)$ have been reported, we have earned ourselves a *free I/O*, namely, an extra I/O that we can charge on those B segments without worrying that the output cost can become super-linear – the standard idea of *filtering search* [8]. We spend this I/O jumping to the first leaf node of the bottom structure $\Pi_u(i)$. From there, scan the segments of $\pi_u(i)$ in ascending order of y-coordinates. Keep reporting the segments[2] until encountering the first one that does not intersect Q', i.e., the first segment whose y-coordinate is greater than y.

Every segment in $\pi_u(i)$ is a left segment at some proper ancestor p of u. If less than $B \log_B N$ segments from $\mathcal{L}_p$ have been reported from $\pi_u(i)$, we have found all the segments from $\mathcal{L}_p$ whose left endpoints are in σ_i, because the lowest $B \log_B N$ of those segments are in $\pi_u(i)$. Otherwise, $B \log_B N$ segments from $\mathcal{L}_p$ have been reported. In this case, we have earned ourselves $O(\log_B N)$ free I/Os, which are utilized to launch a 3-sided range query to report the other segments of $\mathcal{L}_p$ with left endpoints in σ_i, as explained next.

Recall that p has f flank structures, each corresponding to a different child node of p. Let F_p be the flank structure for the child node of p on the query path. We formulate a 3-sided rectangle r that has the same x-range as σ_i, and has the same y-range as Q'. See Figure 4. A segment in $\mathcal{L}(p)$ intersects Q' and has its left endpoint in σ_i, if and only if r covers its left endpoint. The cost of answering the 3-sided range query on F_p is $O(\log_B N)$ plus the linear output time.

Leaf node. At the leaf node z of the query path, we simply report all the qualifying segments stored in z.

A sentinel trick. As mentioned before, in searching an internal node u, we need to know whether B segments from a $\psi_u(i)$ have been reported in the pilot structure, for each $i \in [1, f]$. Since f may be greater than $2B$ (the smallest amount of memory that may be available), we cannot simply keep a counter for each i in memory. A similar situation occurs when we are scanning through $\pi_u(i)$ for some i – we need to know whether $B \log_B N$ segments from the $\mathcal{L}_p$ of a proper ancestor p of u have been found. Again, as there can be $O(h) = O(\log_f N)$ different p, we cannot keep a memory-resident counter for each p if h is far greater than $2B$. Maintaining the counters in external memory would lead to cost penalty we cannot afford.

We resolve this issue by marking some segments as *sentinels* such that enough segments (from a certain set) have been reported if and only if a sentinel is reported. Specifically, in the pilot structure, the sentinel from $\psi_u(i)$ is the B-th lowest segment in $\psi_u(i)$; if $\psi_u(i)$ has less than B segments, no sentinel is marked. Likewise, in a $\pi_u(i)$, the sentinel from an $\mathcal{L}_p$ is the $(B \log_B N)$-th lowest segment in $\mathcal{L}_p$, or undefined if there are less than $B \log_B N$ segments in $\mathcal{L}_p$.

Cost. The query path has $O(\log_f N)$ nodes. At each internal node u of the path, searching its pilot structure requires $O(\log_B f)$ I/Os (recall that the pilot structure indexes at most fB points). The other cost at u can be charged on the output cost, as discussed earlier. Hence, overall, only $O(h \log_B f) = O(\log_B N)$ I/Os has not been charged on the output. At the leaf node of the query path, we spend $O(b/B)$ I/Os reading its contents. Therefore, the total query cost is $O(b/B + \log_B N)$ plus the linear output time. A symmetric algorithm can be used to retrieve the qualifying right segments. We have not discussed the reporting of middle segments. As will be shown later, this can be done in $O(\log_B N)$ I/Os plus the linear output time, by augmenting our current structure with extra information occupying $O(N/B)$ space.

2.3 Bootstrapping

We have obtained a linear-space structure whose query complexity is $O(b/B + \log_B N) = O(\log_B^3 N)$, plus the linear output time. The term b/B arises from finding the qualifying segments in a leaf node z. Currently, no specialized structure exists on the segments in z. Now, let us bootstrap by indexing those segments with the structure described above. As z stores at most b segments, those satisfying a query can be reported in $O(\log_B^3 b) = O((\log_B \log_B N)^3)$ I/Os plus the linear output time. Therefore, the overall query cost is improved to $O((\log_B \log_B N)^3 + \log_B N + K/B) = O(\log_B N + K/B)$. The space consumption of the bootstrapped structure remains linear.

3. THE MIDDLE-INDEXING PROBLEM

As before, let $\mathcal{S}$ be a set of N horizontal segments. Let L be a set of vertical lines $\{\ell_1, ..., \ell_f\}$ where $f = \Theta(\max\{B^{1/3}, (\log_B N)^{1/3}\})$. ℓ_i is to the left of ℓ_j for any i, j with $1 \le i < j \le f$. For each $i \in [1, f-1]$, refer to the vertical strip between ℓ_i and ℓ_{i+1} as a *slab*. A *multi-slab* is the union of a non-empty set of *consecutive* slabs. There are $f-1$ slabs, and hence, $f(f-1)/2$ multi-slabs (note that a slab is also a multi-slab).

All segments of $\mathcal{S}$ lie between ℓ_1 and ℓ_f. A segment $s \in \mathcal{S}$ is *long* if it spans at least one multi-slab, i.e., its x-range covers that of the multi-slab. Otherwise, s is *short*. Every short segment is required to intersect one of $\ell_2, ..., \ell_{f-1}$. A long segment s defines a middle segment, which is the part of s in the widest multi-slab it spans. Denote by $\mathcal{M}$ the set of middle segments thus defined. The goal is to store $\mathcal{S}$ in a structure so that, given a vertical ray Q, all the segments in $\mathcal{M}$ intersecting Q can be reported efficiently (i.e., we are interested in retrieving only middle segments). We refer to the above problem as the *middle-indexing problem*.

LEMMA 1. *In the middle-indexing problem, there is a structure of $O(f + N/B)$ space such that a query can be answered in $O(\log_B f + K/B)$ I/Os. The structure can be updated in $O(\log_B N)$ amortized I/Os per insertion and deletion.*

We have encapsulated some standard details into lemmas (such as the above one), whose proofs can be found in the appendix.

Completing the static structure. Refer to the structure of Lemma 1 as a *middle structure*. In the static structure of Section 2, each internal node u in the base tree has a stabbing set $stab(u)$. We use a middle structure to manage the segments of $stab(u)$, with L being the set of boundary lines of the child slabs of u. In answering

[2]Of course, skip the first B segments in $\pi_u(i)$ because they have been reported from $\psi_u(i)$. Henceforth, we will not discuss trivial duplicate removal like this.

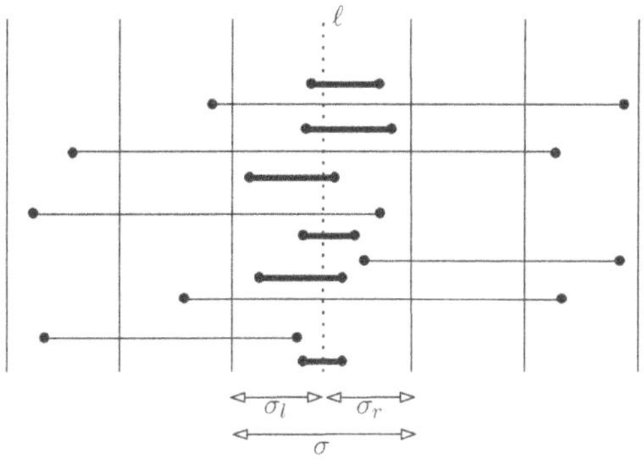

Figure 5: A refine operation: The solid vertical lines constitute the current L. S_{new} consists of the thick segments. The other segments are in the current $\mathcal{S}$.

a query Q, we report the qualifying middle segments by searching the middle structures of the internal nodes along the query path. As there are $O(\log_f N)$ such nodes, by Lemma 1, the total cost is $O(\log_f N \cdot \log_B f) = O(\log_B N)$ plus the linear output time.

Each middle structure uses $O(f)$ extra blocks, on top of space linear to the number of segments indexed. There are $o(N/B)$ extra blocks for all middle structures because the base tree has $O(N/(bf))$ internal nodes. Hence, the overall space consumption is still linear.

Refinement. Next, we describe a *refine* operation which will be useful later. In this operation, a slab σ is divided into two slabs σ_l and σ_r by a vertical line ℓ, where σ_l is on the left of ℓ. Accordingly, ℓ is added to L. Furthermore, ℓ is associated with a set S_{new} of horizontal segments, all of which are inside σ. See Figure 5. S_{new} is to be incorporated into the dataset $\mathcal{S}$. S_{new} is given in two copies where its segments are sorted by x- and y-coordinates, respectively. The objective is to update the middle structure to index the resulting $\mathcal{S}$ and L.

Let α be the total number of segments in the final $\mathcal{S}$ having an endpoint in σ. In other words, α equals the sum of $|S_{new}|$ and the number of such segments in the original $\mathcal{S}$. The lemma below explains the cost of refinement:

LEMMA 2. *A refine operation can be performed in* $O(f^2 \log_2 f + \alpha/B)$ *I/Os.*

Split. Let ℓ be a line in L. A *split* operation cuts S and L into S_1, S_2 and L_1, L_2, respectively. Specifically, S_1 (S_2) includes all the segments in S entirely on the left (right) of ℓ, whereas L_1 (L_2) includes all the lines of L on the left (right) of ℓ. Let S_{cross} be the segments of S intersecting ℓ. See Figure 6. The operation has two goals. First, output S_{cross} in two copies, with its segments sorted by x- and y-coordinates, respectively. Second, create middle structures for the two middle-indexing problems defined on (S_1, L_1) and (S_2, L_2), respectively. As for the efficiency of the operation, we have:

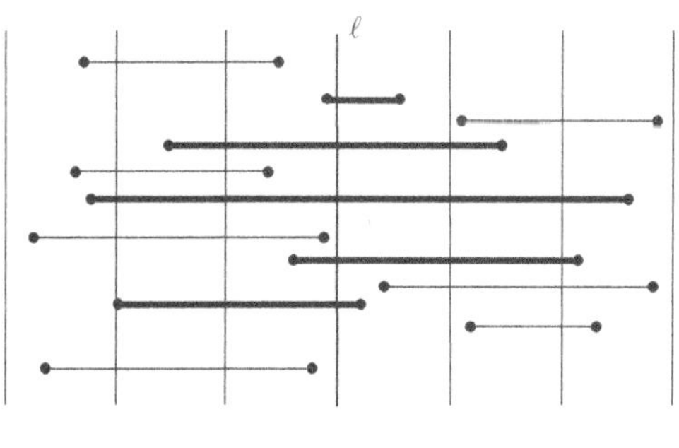

Figure 6: A split operation: S_{cross} consists of the thick segments.

LEMMA 3. *A split operation can be performed in* $O((N/B)\log_2 f + f^2 \log_2 f)$ *I/Os.*

4. A DYNAMIC STRUCTURE

In this section, we make our structure in Section 2 dynamic. The main challenge is to efficiently maintain bottom sets. In general, changes to bottom sets may occur for two reasons. First, segments are moved from a stabbing set to another during tree rebalancing. Second, a newly inserted segment enters a bottom set, or an existing segment there is deleted. We will describe techniques to handle these situations.

4.1 Structure

We now implement the base tree $\mathcal{T}$ as a *weight-balanced B-tree* [4], where a leaf node stores $\Theta(b)$ x-coordinates and an internal node has $\Theta(f)$ child nodes, with b and f defined as in Section 2. All the secondary structures in Sections 2 and 3 are inherited by the nodes of $\mathcal{T}$. Next, we describe the additional structures required on left segments. Symmetric structures exist for right segments, whereas the middle structure has been completely presented in the previous section. Let $h = O(\log_f N)$ be the height of $\mathcal{T}$.

Let u be an internal/leaf node of $\mathcal{T}$, and p a proper ancestor of $parent(u)$. Recall that $bottom_u(p)$, the bottom set of p at u, is the $B \log_B N$ lowest left segments at p whose left endpoints are in σ_u. We maintain a B-tree indexing the segments of $bottom_u(p)$ by their y-coordinates. $\mathcal{T}$ has $O(N/b)$ nodes, each of which has $O(h)$ proper ancestors. Hence, the B-trees on all the bottom sets occupy $O(\frac{N}{b} h \log_B N) = o(N/B)$ space in total.

Now, consider u instead as an internal node with child nodes $v_1, ..., v_f$. Define:

$$\gamma_u(p) = \bigcup_{i=1}^{f} bottom_{v_i}(p). \qquad (2)$$

See Figure 7. Notice that, if $bottom_u(p)$ is defined (i.e., the levels of p and u differ by at least 2), the segments of $bottom_u(p)$ are in fact the $B \log_B N$ lowest in $\gamma_u(p)$.

We associate u with a (slightly augmented) B-tree $\Gamma_u(p)$, which indexes the y-coordinates of the segments in $\gamma_u(p)$. Given an integer k, $\Gamma_u(p)$ allows us to find efficiently the k-th lowest segment in $\gamma_u(p)$. Refer to $\Gamma_u(p)$ as a *rank B-tree*. As $\gamma_u(p)$ has at most $fB \log_B N$ segments, a rank B-tree occupies $O(f \log_B N)$ blocks. There are $O(N/(bf))$ internal nodes, each of which has $h = O(\log_f N)$ proper ancestors. Hence, the total space of all rank B-trees is $O(\frac{Nh}{bf} f \log_B N) = o(N/B)$.

Let z be a leaf node in $\mathcal{T}$, and p a proper ancestor of z. Define:

$\phi_z(p)$ = the set of left segments at p whose left endpoints are in σ_z.

If $bottom_z(p)$ is defined, the $B \log_B N$ lowest segments of $\phi_z(p)$ are exactly the segments in $bottom_z(p)$. We maintain a B-tree

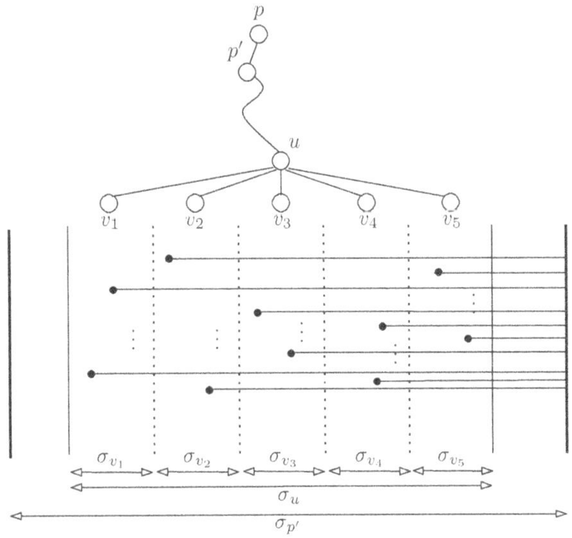

Figure 7: Illustration of $\gamma_u(p)$: For each $bottom_{v_i}(p)$, only the highest and lowest segments are shown, whereas the other segments are omitted. All these segments together constitute $\gamma_u(p)$.

on the y-coordinates of the segments in $\phi_z(p)$. The B-trees on the $\phi_z(p)$ of all p together occupy $O(h + b/B) = O(b/B)$ space, because only b endpoints are in σ_z and z has $O(h)$ proper ancestors. Thus, all these B-trees require $O(N/B)$ blocks in total. Finally, there is a B-tree on the y-coordinates of the segments stored in z – recall that every such segment has at least one x-coordinate in z. The space of our overall structure still remains linear.

The query algorithm remains the same as in the static structure. The cost is $O(b/B + \log_B N + K/B)$, where the b/B term stems from reading the segments stored in the leaf node of the query path. We will eliminate the term by bootstrapping later.

4.2 Updates

In the sequel, whenever we insert or delete a segment in some $bottom_u(p)$, the B-tree on $bottom_u(p)$ is updated accordingly. The same convention applies to $\phi_z(p)$, and the segment set stored in a leaf node. Recall that our structure uses some sentinels to facilitate query processing. We, however, will not elaborate specifically on the maintenance of sentinels because the relevant details are straightforward extensions to the update procedures presented below.

Insertion. To insert a segment s, first update the base tree $\mathcal{T}$ by inserting the x-coordinates of s. This may trigger one or more node splits in $\mathcal{T}$. Deferring node split handling to Section 4.3, we now proceed assuming that all the splits have been taken care of. Let p be the node whose stabbing set $stab(p)$ contains s. If p is a leaf, we finish by storing s there. Next, we consider that p is an internal node.

Insert s into the middle structure of p in $O(\log_B N)$ I/Os (Lemma 1). Now, consider the left segment s^l of s (a similar algorithm applies to its right segment). Let φ be the path from p to the leaf node of $\mathcal{T}$ that stores the left x-coordinate of s^l. Recall that p has a flank structure for each child. In $O(\log_B N)$ I/Os, we insert the left endpoint of s^l in the flank structure of p for its child on φ.

Let u be an internal node that is a proper descendant of p on φ. Let v be the child of u on φ. Recall that u has a bottom structure for v – denote the structure as $\Pi_u(i)$ for some $i \in [1, f]$. Whether $\Pi_u(i)$ needs to be updated depends on whether s^l enters $bottom_v(p)$. This is also true for the rank B-tree $\Gamma_u(p)$. There are two situations where s^l should be added to $bottom_v(p)$:

- first, $bottom_v(p)$ has less than $B \log_B N$ segments;
- second, s^l is lower than the highest segment, say s', in $bottom_v(p)$.

Using the B-tree on $bottom_v(p)$, which (if any) situation occurs can be decided in one I/O: simply read the last leaf node of the B-tree. This also allows us to retrieve the segment s'.

If neither of the aforementioned situation happens, there is no more processing on v. Otherwise, insert s^l in $bottom_v(p)$, and in case $bottom_v(p)$ has more than $B \log_B N$ segments, also remove s' from $bottom_v(p)$. In any case, the cost is $O(\log_B \log_B N)$. Accordingly, s^l and s' (if applicable) are inserted in and deleted from $\Pi_u(i)$, respectively. As $\Pi_u(i)$ indexes $O(hB \log_B N)$ segments, this can be done in $O(\log_B(hB \log_B N)) = O(\log_B \log_B N)$ I/Os. Likewise, insert s^l and delete s' (again, if applicable) in the rank B-tree $\Gamma_u(p)$, which takes $O(\log_B(fB \log_B N)) = O(\log_B f)$ I/Os.

Changes to $\Pi_u(i)$ may necessitate updating the pilot structure P_u, because only the lowest B segments in $\Pi_u(i)$ should appear in P_u. Those B segments can be obtained in one I/O from $\Pi_u(i)$. At most one insertion and one deletion are then need on P_u. Their cost is $O(\log_B f)$ as P_u stores at most fB segments.

In summary, the processing at u incurs $O(\log_B \log_B N + \log_B f) = O(\log_B f)$ I/Os. We need to do the same to $O(h)$ nodes on φ, the cost of which adds up to $O(h \log_B f) = O(\log_B N)$. Finally, at the leaf z of φ, insert s^l to $\phi_z(p)$ and the segment set stored in z with $O(\log_B b) = o(\log_B N)$ I/Os.

Deletion. To delete a segment s, we first find the node p such that $s \in stab(p)$. If p is a leaf, finish by deleting s from the segments stored at p. Now, consider that p is an internal node. Remove s from the middle structure of p in $O(\log_B N)$ I/Os. Next we explain the rest of processing on only the left segment s^l. Let φ be the path from p to the leaf node of $\mathcal{T}$ storing the left x-coordinate of s^l. Delete in $O(\log_B N)$ I/Os the left endpoint of s^l from the relevant flank structure at p.

Let u be an internal node that is a proper descendant of p on φ. Let v be the child of u on φ. In one I/O, we check whether s^l is in $bottom_v(p)$. If so, delete s^l from $bottom_v(p)$, the rank B-tree $\Gamma_u(p)$, and the bottom structure of u for v. Update the pilot structure P_u accordingly. The cost is $O(\log_B f)$, as should have been clear from the earlier analysis on insertion. Currently, u is *dirty* because $bottom_v(p)$ may have one less segment than required. At the leaf z of φ, we remove s^l from $\phi_z(p)$ and the segment set stored in z with $o(\log_B N)$ I/Os.

Next, we fix the dirty nodes in bottom-up order. Let u be the lowest dirty node, and v its child on φ. If v is a leaf node, we check whether $\phi_v(p)$ has at least $B \log_B N$ segments. If not, $bottom_v(p)$ is already accurate. Otherwise, retrieve the $(B \log_B N)$-th lowest segment s' from $\phi_v(p)$ using $O(\log_B N)$ I/Os, by simply reading the entire $\phi_v(p)$. Insert s' into $bottom_v(p)$, $\Gamma_u(p)$, the bottom structure of u for v, and update P_u in totally $O(\log_B f)$ I/Os. At this time, $bottom_v(p)$ has been brought up-to-date, and so have the secondary structures of u.

On the other hand, if v is an internal node, we check whether $\gamma_v(p)$ (see (2)) has at least $B \log_B N$ segments. If not, $bottom_v(p)$ is accurate. Otherwise, retrieve the $(B \log_B N)$-th lowest segment s' from $\gamma_v(p)$. Using the rank B-tree $\Gamma_v(p)$, this can be done in $O(\log_B f)$ I/Os. Then, insert s' into the relevant structures (same as the case where v is a leaf) in $O(\log_B f)$ I/Os.

In summary, to delete a segment s, we spend $O(\log_B f)$ I/Os per internal level, and $O(\log_B N)$ I/Os at the leaf level. Hence, the total cost is $O(\log_B N)$. Note that we keep the x-coordinates of the endpoints of s in the base tree $\mathcal{T}$. This does not affect the correctness of our query algorithm since it reports segments only from secondary structures, which are always properly maintained. On the other hand, it brings the benefit of not having to deal with node underflows. With global rebuilding, we ensure that the height of $\mathcal{T}$ should remain $O(\log_f N)$.

4.3 Handling node splits

We have explained that each insertion and deletion can be performed in $O(\log_B N)$ I/Os, excluding the cost of handling node splits in the base tree $\mathcal{T}$. This subsection completes the description of our update procedure by elaborating the split algorithms. Our discussion concentrates on the secondary structures for managing left and middle segments.

We start by giving some basic facts. Recall that, a lower-open 3-sided range query specifies a rectangle of the form $[x_1, x_2] \times (-\infty, y]$. The flank structures of our index are external priority search trees for answering such queries efficiently. The lemma below explains how fast a flank structure can be constructed.

LEMMA 4. *An external priority search tree for answering lower-open 3-sided range queries on N points, can be built in $O(N/B)$ I/Os, provided that all the points have been sorted by x-coordinates.*

The next lemma indicates that, once the bottom sets at the child nodes of a node u are ready, we can build several secondary structures of u efficiently. Recall that $h = O(\log_f N)$ is the height of $\mathcal{T}$.

LEMMA 5. *Let u be an internal node with child nodes $v_1, ..., v_f$. Suppose that the segments of $bottom_{v_i}(p)$ have been sorted by y-coordinates, for each v_i ($1 \le i \le f$) and each proper ancestor p of u. In $O(hf \log_B N \cdot \log_2 f)$ I/Os, we can build the following secondary structures of u: the bottom structure $\Pi_u(i)$ of each i, the pilot structure P_u, the B-tree on $bottom_u(p)$ of each relevant p, and the rank B-tree $\Gamma_u(p)$ of each p.*

Leaf split. Next, we discuss the split of a leaf node z. Let u be the parent of z, and z_1, z_2 be the leaf nodes that z splits into. Let ℓ be the vertical line that divides σ_{z_1} and σ_{z_2}.

The stabbing set $stab(z)$ can be divided into three groups. The first (second) group includes those segments on the left (right) of ℓ, while the last group, denoted as S_{cross}, includes those intersecting ℓ. The first (second) group becomes $stab(z_1)$ ($stab(z_2)$), whereas S_{cross} will be merged into $stab(u)$.

Let S_z be the set of (at most b) segments stored at z, and similarly, S_{z_1} (S_{z_2}) the set of segments to be stored at z_1 (z_2) after the split. Recall that $stab(z)$ is a subset of S_z (because a segment in S_z has at least an endpoint in σ_z, whereas σ_z covers both endpoints of a segment in $stab(z)$). Each segment in S_z should be added to S_{z_1} if its left endpoint is on the left of ℓ, or to S_{z_2} otherwise. Utilizing the B-tree on S_z, in $O(b/B)$ I/Os, we can generate S_{z_1}, S_{z_2} and S_{cross} such that the segments of each set are sorted by y-coordinates. Then, the B-trees on S_{z_1} and S_{z_2} can be built in $O(b/B)$ I/Os.

Let p be an ancestor of u (note: p can be u). After the split, a segment $s \in \phi_z(p)$ appears in $\phi_{z_1}(p)$ if its left endpoint is to the left of ℓ; otherwise, s appears in $\phi_{z_2}(p)$. Moreover, if $p = u$, $\phi_{z_1}(p)$ also includes those segments of S_{cross} whose left endpoints are to the left of ℓ. Using the B-tree on $\phi_z(p)$ (and S_{cross}, if $p = u$), we create the B-trees on $\phi_{z_1}(p)$ and $\phi_{z_2}(p)$ in $O(1 + |\phi_z(p)|/B)$ I/Os (and plus $O(b/B)$ if $p = u$). As there are $O(h)$ different p, and the sum of $|\phi_z(p)|$ for all p is at most b, the B-trees on the $\phi_{z_1}(p)$ and $\phi_{z_2}(p)$ of all p can be built in $O(h + b/B)$ I/Os. If p is at level 2 or above, $bottom_{z_1}(p)$ is simply a prefix of $\phi_{z_1}(p)$. Hence, the B-trees on the $bottom_{z_1}(p)$ of all p can also be built in $O(h + b/B)$ I/Os. The same is true for $bottom_{z_2}(p)$.

At this point, the secondary structures of z_1 and z_2 are ready, so we proceed to fix those of u. Create an additional copy of S_{cross} where the segments are sorted by x-coordinates. This can be achieved in $O(\frac{b}{B} \log_2 b)$ I/Os. Perform a refine operation (Section 3) on the middle structure of u, with ℓ and S_{cross} as the inputs. By Lemma 2, the cost is $O(f^2 \log_2 f + b/B)$, noticing that the value of α in the lemma is at most b. The other secondary structures of u mentioned in Lemma 5 can be built in $O(hf \log_B N \cdot \log_2 f)$ I/Os.

In summary, a leaf split can be performed in $O(\frac{b}{B} \log_2 b + hf \log_B N \cdot \log_2 f)$ I/Os.

Internal split. Consider that an internal node u is split into u_1 and u_2 along a vertical line ℓ that divides σ_{u_1} and σ_{u_2}. Define $w(u)$ as the number of x-coordinates stored at the leaf nodes in the subtree of u, and $w(u_1), w(u_2)$ similarly. We start with a split operation on the middle structure of u with input ℓ. This operation creates the middle structures of u_1 and u_2, and generates a segment set S_{cross} in two copies: one sorted by y-coordinates, and the other by x-coordinates. S_{cross} includes all the segments from $stab(u)$ that intersect ℓ, and will be added to the stabbing set of $parent(u)$. By Lemma 3, the operation requires $O(\frac{w(u)}{B} \log_2 f + f^2 \log_2 f)$ I/Os, noticing that the middle structure of u indexes at most $w(u)$ segments.

If u is at level 2 or above, we need to build the flank structures of u_1 and u_2. Let v be a child node of u that becomes a child node of u_1 after the split (the case of u_2 is analogous). Let $F_{u_1}(i)$ be the flank structure of u_1 for v, and F_{old} the flank structure of u for v before the split. $F_{u_1}(i)$ should store those endpoints in F_{old} that do not belong to the segments of S_{cross}. As the endpoints indexed by F_{old} are sorted by x-coordinates, in linear I/Os, we can generate the list of endpoints to be stored in $F_{u_1}(i)$, preserving their ordering. After this, $F_{u_1}(i)$ can be built in linear I/Os (Lemma 4). Hence, the flank structures of u_1 and u_2 can be constructed using $O(f + w(u)/B)$ I/Os in total, because they together manage at most $w(u)$ endpoints.

At $parent(u)$, we update its middle structure with a refine operation using $O(w(u)/B + f^2 \log_2 f)$ (Lemma 2), feeding ℓ and the two copies of S_{cross} obtained earlier. Next, we discuss how to build the flank structures of $parent(u)$ for u_1 and u_2 – denote them, respectively, as $F_{parent(u)}(i)$ and $F_{parent(u)}(i+1)$ for some $i \in [1, f-1]$. Also, let F'_{old} be the flank structure of $parent(u)$ for u before the split. $F_u(i+1)$ should store the endpoints in F'_{old} that are on the right of ℓ. $F_u(i)$ should store (i) the other endpoints of F'_{old}, and (ii) the left endpoints of the segments in S_{cross}. In F'_{old}, the at most $w(u)$ endpoints are sorted in ascending order of x-coordinates. Given that a copy of S_{cross} is sorted in the same way, in $O(w(u)/B)$ I/Os, we generate the lists of endpoints managed by $F_u(i)$ and $F_u(i+1)$ respectively, where each list is sorted by x-coordinates. Then, $F_u(i)$ and $F_u(i+1)$ are created in $O(w(u)/B)$ I/Os with Lemma 4.

Recall that u may have bottom sets at its proper descendants. After the splits, those bottom sets no longer exist, but instead we should build the bottom sets of u_1 and u_2. Furthermore, as new segments have been added to the stabbing set of $parent(u)$, some bottom sets of $parent(u)$ may contain errors. Also affected are the secondary structures dependent on the aforementioned bottom sets. Next, we explain a *bottom overhaul* to fix all these issues, by processing the descendants of u in bottom-up manner.

Let z be a leaf descendant of u that is now a descendant of u_1 (the case of u_2 is similar). Create the B-tree of $\phi_z(u_1)$ by scanning the B-tree on $\phi_z(u)$ in $O(b/B)$ I/Os (ignoring segments intersecting ℓ). Regarding $\phi_z(parent(u))$, the segments in S_{cross} whose left endpoints are in σ_z should now be added to $\phi_z(parent(u))$. As those segment must be in $\phi_z(u)$ before the split, in $O(b/B)$ I/Os, we can extract them into a list sorted by y-coordinates, and then, merge the list into the B-tree of $\phi_z(parent(u))$. After this, $bottom_z(parent(u))$ and if applicable, also $bottom_z(u_1)$ and $bottom_z(u_2)$, can be built in $O(b/B)$ I/Os. Note that the set of segments stored at z is not affected; hence, its B-tree before the split can be used directly. As u has $O(w(u)/b)$ leaf descendants, processing all of them takes $O(\frac{w(u)}{b}\frac{b}{B}) = O(w(u)/B)$ I/Os in total.

For every proper internal descendant of u, we rebuild its secondary structures mentioned in Lemma 5 in $O(hf \log_B N \cdot \log_2 f)$ time. Since there $O(w(u)/(bf))$ such internal nodes, the total cost is

$$O\left(\frac{w(u)}{bf} hf \log_B N \cdot \log_2 f\right) = O\left(\frac{w(u)}{B \log_B N} \log_2 f\right) = o(w(u)).$$

We complete the bottom overhaul by applying Lemma 5 to u_1, u_2 and p, necessitating another $O(hf \log_B N \cdot \log_2 f)$ I/Os.

In summary, an internal split takes $O(\frac{w(u)}{B} \log_2 f + w(u) + f^2 \log_2 f + hf \log_B N \cdot \log_2 f)$ time, which is $O(\frac{w(u)}{B} \log_2 f + w(u))$ given the fact $w(u) = \Omega(bf)$.

Amortized cost. The WBB-tree guarantees that when a node u is split, $\Omega(w(u))$ insertions must have been performed in its subtree since the creation of u. Hence, if u is a leaf, we charge the split cost over those $\Omega(b)$ insertions, so that each insertion accounts for

$$O\left(\frac{\log_2 b}{B} + \frac{hf \log_B N \cdot \log_2 f}{b}\right) = O\left(\frac{\log_2 b}{B} + \frac{f \cdot \log_2 f}{B \log_B N}\right) = o(\log_2 \log_B N)$$

I/Os. If u is an internal node, after amortization, each insertion bears $O(1 + (1/B) \log_2 f)$ I/Os incurred from splitting u. As an insertion bears such cost for $O(h)$ nodes, its amortized cost increases by

$$O\left(h + \frac{h \log_2 f}{B}\right) = O\left(h + \frac{\log_f N \cdot \log_2 f}{B}\right) = O\left(h + \frac{\log_2 N}{B}\right)$$

which is $O(\log_B N)$.

4.4 Bootstrapping

Currently, our ray-stabbing structure handles an update in $O(\log_B N)$ I/Os amortized, but its query time is $O(b/B + \log_B N + K/B) = O(\log_B^3 N + K/B)$. Similar to the bootstrapping in Section 2, we use the above structure to index the (at most b) segments stored in a leaf node, thus improving the query cost to $O(\log_B N + K/B)$. In a leaf split, this structure can be rebuilt in $O(b \log_B b) = O(b \log_B \log_B N)$ time by doing at most b insertions. Note that the structure does not need to be touched in a bottom overhaul (which does not affect the segments stored in a leaf, as mentioned earlier). Hence, each update (on the bootstrapped structure) is charged only additional $O(\log_B \log_B N)$ I/Os. This completes the proof of Theorem 1.

5. REFERENCES

[1] P. K. Agarwal, S.-W. Cheng, Y. Tao, and K. Yi. Indexing uncertain data. In *Proceedings of ACM Symposium on Principles of Database Systems (PODS)*, pages 137–146, 2009.

[2] A. Aggarwal and J. S. Vitter. The input/output complexity of sorting and related problems. *Communications of the ACM (CACM)*, 31(9):1116–1127, 1988.

[3] L. Arge, V. Samoladas, and J. S. Vitter. On two-dimensional indexability and optimal range search indexing. In *Proceedings of ACM Symposium on Principles of Database Systems (PODS)*, pages 346–357, 1999.

[4] L. Arge and J. S. Vitter. Optimal external memory interval management. *SIAM Journal of Computing*, 32(6):1488–1508, 2003.

[5] B. Becker, S. Gschwind, T. Ohler, B. Seeger, and P. Widmayer. An asymptotically optimal multiversion B-tree. *The VLDB Journal*, 5(4):264–275, 1996.

[6] E. Bertino, B. Catania, and B. Shidlovsky. Towards optimal indexing for segment databases. In *Proceedings of Extending Database Technology (EDBT)*, pages 39–53, 1998.

[7] G. E. Blelloch. Space-efficient dynamic orthogonal point location, segment intersection, and range reporting. In *Proceedings of the Annual ACM-SIAM Symposium on Discrete Algorithms (SODA)*, pages 894–903, 2008.

[8] B. Chazelle. Filtering search: A new approach to query-answering. *SIAM Journal of Computing*, 15(3):703–724, 1986.

[9] S.-W. Cheng and R. Janardan. Efficient dynamic algorithms for some geometric intersection problems. *Information Processing Letters (IPL)*, 36(5):251–258, 1990.

[10] J. Enderle, N. Schneider, and T. Seidl. Efficiently processing queries on interval-and-value tuples in relational databases. In *Proceedings of Very Large Data Bases (VLDB)*, pages 385–396, 2005.

[11] M. T. Goodrich, J.-J. Tsay, D. E. Vengroff, and J. S. Vitter. External-memory computational geometry. In *Proceedings of Annual IEEE Symposium on Foundations of Computer Science (FOCS)*, pages 714–723, 1993.

[12] H. Kaplan, E. Molad, and R. E. Tarjan. Dynamic rectangular intersection with priorities. In *Proceedings of ACM Symposium on Theory of Computing (STOC)*, pages 639–648, 2003.

[13] C. W. Mortensen. Fully-dynamic two dimensional orthogonal range and line segment intersection reporting in logarithmic time. In *Proceedings of the Annual ACM-SIAM Symposium on Discrete Algorithms (SODA)*, pages 618–627, 2003.

[14] B. Salzberg and V. J. Tsotras. Comparison of access methods for time-evolving data. *ACM Computing Surveys*, 31(2):158–221, 1999.

Appendix

Proof of Lemma 1. We assign each long segment $s \in \mathcal{S}$ to the widest multi-slab it spans. A multi-slab is *large* if it has at least B segments (assigned to it), and is *small* otherwise. Let S_{low} be the union of the B lowest segments in each multi-slab. All the segments of a small multi-slab are in S_{low}. We index the $O(f^2 B)$ segments in S_{low} with a persistent B-tree [5] so that the segments intersecting a vertical ray can be reported efficiently. For each large multi-slab X, create a B-tree on the y-coordinates of its segments. Call the B-tree a *multi-slab B-tree.*

To answer a query with a ray Q, first search the persistent B-tree to report all the segments intersecting Q. For each large multi-slab X, if B segments have been reported (as can be detected using the sentinel trick in Section 2.2, namely, by marking some segments in the persistent tree as sentinels), deploy the B-tree of X to report the other qualifying segments of X in ascending order of y-coordinates. The query cost is $O(\log_B f)$ plus the linear output cost, where the logarithmic term comes from querying the persistent B-tree.

To support updates, we maintain a B-tree $\mathcal{T}_{low}$ that indexes the segments in $\mathcal{S}_{low}$ by the ids of their origin multi-slabs. Given any multi-slab X, $\mathcal{T}_{low}$ allows us to retrieve the lowest B segments of X in $O(\log_B f^2) = o(\log_B N)$ I/Os. Furthermore, additional structures are needed in order to facilitate some operations required later in Lemmas 2 and 3. For each ℓ_i $(2 \le i \le f-1)$, we create a B-tree on the y-coordinates of the short segments intersecting ℓ_i. Call the B-tree a *line B-tree*. Furthermore, for each slab σ, create a B-tree on the x-coordinates of all (long and short) the segments whose endpoints are covered by σ. Call this B-tree a *slab B-tree*.

All structures use linear space, except that a line/slab B-tree occupies at least one block in any case. As each segment exists in constant structures, the space consumption is $O(f + N/B)$.

Next, we describe updates. First note that the persistent B-tree can be constructed in $O(f^2 \log_2 f)$ I/Os with the algorithm in [11]. Using a block to buffer updates, we apply global rebuilding to update the structure in $O(\frac{f^2}{B} \log_2 f) = o(\log_B N)$ I/Os per insertion and deletion.

Inserting/deleting a short segment s only requires updating a line B-tree and a slab B-tree. Consider inserting a long segment s to $\mathcal{S}$. First, update the relevant slab and multi-slab B-trees. Assume that s belongs to multi-slab X. Using $\mathcal{T}_{low}$, check in $o(\log_B N)$ I/O whether s is among the B lowest segments in X. If no, the insertion is complete. Otherwise, insert s in S_{low} (updating the relevant structures), and if S_{low} now has $B+1$ segments from X, delete the highest segment of X in S_{low}. Accordingly, the sentinel segment of X can be maintained, if necessary, with an insertion and perhaps also a deletion in the persistent B-tree.

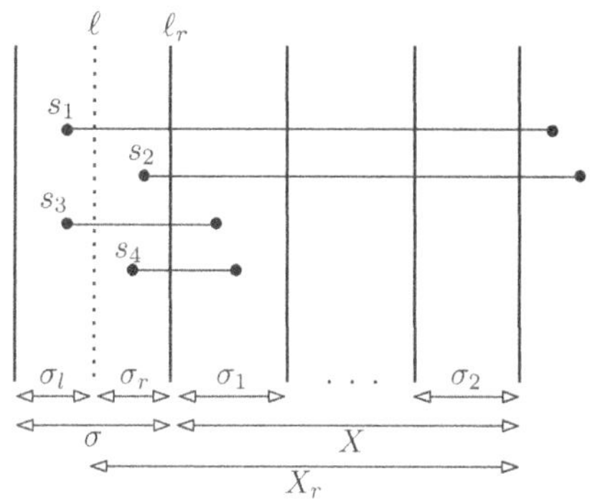

Figure 8: Proof of Lemma 2

Deleting a long segment from $\mathcal{S}$ can be handled in a reverse manner with the same overhead.

Proof of Lemma 2. We will utilize the structure described in the proof of Lemma 1.

Let σ_1 be the slab next to σ on the right. Let X be a multi-slab that starts from σ_1 and ends at some slab σ_2 (it is possible that $\sigma_2 = \sigma_1$). Denote by X_r the multi-slab after refinement that starts from σ_r and ends at σ_2. See Figure 8. Consider a segment s_{long} of X. After refinement, s_{long} belongs to either X_r or X, depending on whether the left endpoint of s_{long} is in σ_l or σ_r. For example, in Figure 8, s_1 (s_2) is a long segment that will appear in X_r (X). In $O(\log_B f)$ I/Os, we can obtain a list of segments in X sorted by y-coordinates: if X is large, such a list is already availabe from its multi-slab B-tree; otherwise, we retrieve all the (less than B) segments of X from $\mathcal{T}_{low}$. Then, we can generate two lists of segments that should belong to X_r and X after refinement, respectively. In both lists, segments are sorted by y-coordinates. The cost of the generation is linear to the number of segments in X before refinement.

Let ℓ_r be the right boundary of σ. Consider a short segment s_{short} intersecting ℓ_r before refinement. After refinement, s_{short} may or may not become long. It will, if its left endpoint is in σ_l, in which case it will belong to multi-slab σ_r. Otherwise, it will not. For example, in Figure 8, s_3 (s_4) is a short segment that will become long (remain short). Hence, from the line B-tree of ℓ_r, we can generate a list of segments that have become long or remain short, respectively. In both lists, segments are sorted by y-coordinates. The cost is linear to the number of short segments intersecting ℓ_r before refinement.

So far we have obtained a sorted list for each new/affected multi-slab on the right of ℓ, and a sorted list of the short segments intersecting ℓ_r after refinement. Symmetrically, we can do the same with respect to each new/affected multi-slab on the left of ℓ, and the left boundary ℓ_l of σ. The total cost is $O(f \log_B f + \alpha/B)$. After this, all the corresponding multi-slab/line B-trees can be constructed in $O(\alpha/B + f)$ I/Os. $\mathcal{T}_{low}$ and the persistent B-tree can be rebuilt in $O(f^2 \log_2 f)$ I/Os.

Finally, we rebuild the line B-tree of ℓ, and the slab B-trees of σ_l and σ_r in $O(\alpha/B)$ I/Os. In total, a refine operation takes $O(f \log_B f + f^2 \log_2 f + \alpha/B) = O(f^2 \log_2 f + \alpha/B)$ I/Os.

Proof of Lemma 3. Again, our discussion is based on the structure described in the proof of Lemma 1.

Refer to the middle structure for the problem defined on (S_1, L_1) as the *first middle structure*, and to the one on (S_2, L_2) as the *second middle structure*. As mentioned in the proof of Lemma 2, in $O(\log_B f)$ I/Os, we can obtain a list of all the segments in a multi-slab, sorted by y-coordinates. Hence, the lists of all f^2 multi-slabs can be generated in $O(f^2 \log_B f)$ I/Os, after which they can be merged on y-coordinates in $O((N/B) \log_2 f)$ I/Os. By scanning this merged list and the line B-tree of ℓ, in $O(N/B)$ I/Os, we create S_{cross} with its segments sorted by y-coordinates. Another copy of S_{cross} where its segments are sorted by x-coordinates can be obtained easily from the slab B-trees in $O(f + N/B)$ I/Os.

If a multi-slab does not intersect ℓ, its B-tree can be re-used directly in either the first or second middle structure. Likewise, every line B-tree (except, obviously, the one on ℓ) can also be re-used. $\mathcal{T}_{low}$ and the persistent B-trees of the two new middle structures can be constructed in $O(f^2 \log_2 f)$ I/Os from the multi-slab lists obtained earlier. Finally, the slab B-trees of the two middle structures are built in $O(f + N/B)$ I/Os from the original slab B-trees.

Proof of Lemma 4. First, build the base tree, namely, a B-tree on the x-coordinates of all the data points. Each leaf node stores between $B/4$ and B coordinates, while an internal node node has between $B/4$ and B child nodes. If a data point's x-coordinate is in a leaf, we store the point in the leaf as well. Assume that the leaf nodes are at level 0, and that the parent of a level-l node is at level $l + 1$.

As discussed in [3], each internal node u is associated with at most B^2 *prior points* defined as follows. Let $v_1, ..., v_B$ be the child nodes of u. If u is the root of the base tree, for each child v_i, u stores the B highest points from the subtree of v_i. Otherwise, for each v_i, u stores the B highest points from the subtree of v_i that are not a prior point at any proper ancestor of u. In any case, denote the set of those B points as $prior_u(i)$. Let $prior_u = prior_u(1) \cup ... \cup prior_u(B)$. The (at most) B^2 points in $prior_u$ are indexed by a structure called a B^2-*structure*. Consuming $O(B)$ space, the B^2-structure answers a lower-open 3-sided range query in $O(1)$ I/Os plus the linear output time. If the B^2 points indexed have been sorted by x-coordinates, the structure can be built in $O(B)$ I/Os [3], and supports an update in $O(1)$ I/Os per insertion and deletion. In addition, create a B-tree on the y-coordinates of the points in $prior_u$.

We build the B^2-structures of internal nodes in bottom-up order, i.e., finishing with all the nodes at a level before attending to the parent level. Let u be a node at level 1, with child nodes $v_1, ..., v_B$ (which are leaves). For each $i \in [1, B]$, obtain the B points stored in v_i in one I/O. These points form $prior_u(i)$. After this, the B^2-structure of u is constructed in $O(B)$ I/Os. The B-tree on $prior_u$ be easily built in $O(B)$ I/Os.

Now, consider u instead as a node at level $l \geq 2$, with $v_1, ..., v_B$ as its child nodes. For each i, with the B-tree on $prior(v_i)$, we obtain the B highest points from $prior(v_i)$ in one I/O; and they constitute $prior_u(i)$. The B^2-structure of u is then created in $O(B)$ I/Os. Some additional processing is necessary at v_i. Suppose that a point o from $prior(v_i)$ has been included in $prior_u(i)$. Accordingly, o is deleted from the B^2-structures at v_i with $O(1)$ I/Os. We also need to *promote* a point, say o', by inserting it in $prior(v_i)$. Specifically, o' is the highest of the prior points at the child nodes of v_i. Recursively, the promotion necessitates similar promotions along a path from v_i to a leaf. As shown in [3], all these promotions can be performed with a *bubble-up* procedure in $O(l)$ I/Os. In total, at most B^2 bubble-up procedures are performed due to u, entailing $O(lB^2)$ I/Os.

In summary, the secondary structures of each node at level 1 can be built in $O(B)$ I/Os, whereas those of a node at level $l \geq 2$ in $O(lB^2)$ I/Os. As there are at most $N/(B/4)^{l+1}$ nodes at level $l \geq 1$, the total cost of creating all the secondary structures is

$$O\left(\frac{N}{B^2} \cdot B + \sum_{l=2}^{h} \frac{N}{(B/4)^{l+1}} lB^2\right) = O(N/B)$$

where h is the level of the base tree's root.

Proof of Lemma 5. For each p, we first merge the segments of $bottom_{v_1}(p)$, ..., $bottom_{v_f}(p)$ into a list where the segments are sorted by y-coordinates. As each $bottom_{v_f}(p)$ has at most $B \log_B N$ segments, the merge can be done in $O(\frac{1}{B} f B \log_B N \cdot \log_2 f) = O(f \log_B N \cdot \log_2 f)$ I/Os. Then, $bottom_u(p)$ can be acquired by reading the first $B \log_B N$ segments of the merged list, after which the B-tree on $bottom_u(p)$ can be created in $O(\log_B N)$ I/Os. Hence, the B-trees of $bottom_u(p)$ for all p can be constructed with cost $O(hf \log_B N \cdot \log_2 f)$.

To build the bottom structure $\Pi_u(i)$, first generate $\pi_u(i)$ – as defined in (1) – by merging the $bottom_{v_i}(p)$ of all p. As u has at most h proper ancestors, the merge takes $O(\frac{1}{B} hB \log_B N \cdot \log_2 h) = O(h \log_B N \cdot \log_2 h)$ time. Then, $\Pi_u(i)$ can be created in $O(h \log_B N)$ I/Os. Hence, all the bottom structures can be constructed with cost $O(hf \log_B N \cdot \log_2 h)$. To build the pilot structure P_u, collect the lowest B segments from each $\pi_u(i)$, and sort the at most fB segments collected by x-coordinates. This requires $O((fB/B) \log_2 f) = O(f \log_2 f)$ time. P_u can then be constructed in $O(f)$ I/Os using Lemma 4.

To build the rank B-tree $\Gamma_u(p)$, first obtain $\gamma_u(p)$ by merging the $bottom_{v_i}(p)$ of all $i \in [1, f]$ in $O(f \log_B N \cdot \log_2 f)$ I/Os, and then create $\Gamma_u(p)$ from $\gamma_u(p)$ in $O(f \log_B N)$ I/Os. Hence, all the rank B-trees can be constructed with cost $O(hf \log_B N \cdot \log_2 f)$.

We have built all the secondary structures stated in the lemma using $O(hf \log_B N \cdot (\log_2 f + \log_2 h))$ I/Os. Note that $\log_2 h = O(\log_2 \log_B N)$. If $\log_B N \leq B$, then $\log_2 h = O(\log_2 B) = O(\log_2 f)$ because $f = B^{1/3}$. Otherwise, $f = (\log_B N)^{1/3}$ and $\log_2 h$ is clearly still $O(\log_2 f)$. The lemma thus follows.

Higher-dimensional Orthogonal Range Reporting and Rectangle Stabbing in the Pointer Machine Model

Peyman Afshani*
peyman@cs.au.dk

Lars Arge
large@cs.au.dk

Kasper Green Larsen†
larsen@cs.au.dk

MADALGO‡
Department of Computer Science
Aarhus University
Aarhus, Denmark

ABSTRACT

In this paper, we consider two fundamental problems in the pointer machine model of computation, namely orthogonal range reporting and rectangle stabbing. Orthogonal range reporting is the problem of storing a set of n points in d-dimensional space in a data structure, such that the t points in an axis-aligned query rectangle can be reported efficiently. Rectangle stabbing is the "dual" problem where a set of n axis-aligned rectangles should be stored in a data structure, such that the t rectangles that contain a query point can be reported efficiently. Very recently an optimal $O(\log n + t)$ query time pointer machine data structure was developed for the three-dimensional version of the orthogonal range reporting problem. However, in four dimensions the best known query bound of $O(\log^2 n/\log\log n + t)$ has not been improved for decades.

We describe an orthogonal range reporting data structure that is the first structure to achieve significantly less than $O(\log^2 n + t)$ query time in four dimensions. More precisely, we develop a structure that uses $O(n(\log n/\log\log n)^d)$ space and can answer d-dimensional orthogonal range reporting queries (for $d \geq 4$) in $O(\log n(\log n/\log\log n)^{d-4+1/(d-2)}+t)$ time. Ignoring $\log\log n$ factors, this speeds up the best previous query time by a $\log^{1-1/(d-2)} n$ factor. For the rectangle stabbing problem, we show that any data structure that uses nh space must use $\Omega(\log n(\log n/\log h)^{d-2} + t)$ time to answer a query. This improves the previous results by a $\log h$ factor, and is the first lower bound that is optimal for a large range of h, namely for $h \geq \log^{d-2+\varepsilon} n$ where $\varepsilon > 0$ is an arbitrarily small constant. By a simple geometric transformation, our result also implies an improved query lower bound for orthogonal range reporting.

Categories and Subject Descriptors

F.2.2 [**Range Searching**]: Orthogonal Range Reporting

General Terms

Theory

Keywords

Orthogonal range reporting, dominance reporting, lower bounds, pointer machine

1. INTRODUCTION

In this paper we study two fundamental range searching problems, namely rectangle stabbing and orthogonal range reporting. In the orthogonal range reporting problem the goal is to store a set of n points in d-dimensional space in a data structure such that the t points contained in[1] an axis-aligned query rectangle[2] can be reported efficiently. Rectangle stabbing (also called "point enclosure") is the "dual" problem where the goal is to store a set of n rectangles in d-dimensional space such that the t rectangles containing a query point can be reported efficiently.

Orthogonal range reporting is a central problem in several fields, including spatial databases and computational geometry, and it has been studied extensively [5, 6]. Rectangle stabbing is also a classical problem in computational geometry [15]. We study both problems in the pointer machine model of computation [20]. In two dimensions, orthogonal range reporting was completely characterized more than two decades ago [11, 12]. Very recently, a complete characterization was also achieved in three dimensions [4]. However, in higher dimensions the exact complexity of the problem remain open. Unlike orthogonal range reporting, only the

*This research was done while the author was a postdoctoral researcher at Dalhousie University, and was supported by Natural Sciences and Engineering Research Council of Canada (NSERC) through the Postdoctoral Fellowships Program (PDF).

†Is supported in part by MADALGO and in part by a Google Fellowship in Search and Information Retrieval

‡Center for Massive Data Algorithmics, a center of the Danish National Research Foundation.

[1] A point on the boundary of a rectangle is not assumed to be contained in the rectangle.

[2] In this paper, by rectangle we actually mean a hyperrectangle, the Cartesian product of d intervals.

SCG'12, June 17–20, 2012, Chapel Hill, North Carolina, USA.

two-dimensional case has been characterized for rectangle stabbing.

In this paper we present two main results. For orthogonal range reporting, we improve on the best known query time in dimensions four and above. Curiously, the improvement grows with dimension. For rectangle stabbing, we obtain the first tight query time lower bound. By known techniques, this gives an improved query time lower bound for orthogonal range reporting as well.

1.1 Previous Results

Below we review previous results on orthogonal range reporting and rectangle stabbing in the pointer machine. We only review the results most related to ours, that is, results for static structures that answer queries in poly-logarithmic time and with near-linear space usage. We refer the reader to the surveys [5, 6] for further results. For the best data structures in the RAM model see [7, 9].

1.2 Orthogonal Range Reporting

The complexity of the two-dimensional version of the orthogonal range reporting problem was completely settled more than two decades ago. In [11], Chazelle provided a structure that can answer queries in $O(\log n + t)$ time using $O(n \log n / \log\log n)$ space. This is optimal, since Chazelle later proved that any structure for d-dimensional orthogonal range reporting that answers queries in $O(\log^{O(1)} n + t)$ time must use $\Omega(n(\log n/\log\log n)^{d-1})$ space [12]. For the special three-sided query case, where only three of the query coordinates are finite (e.g., $(a, b) \times (-\infty, c)$), McCreight [19] presented the priority search tree that answers queries in optimal $O(\log n + t)$ time and linear space.

Until recently the complexity of orthogonal range reporting remained unresolved in three dimensions. For three-dimensional dominance queries, that is, where at most one of the query coordinates in each dimension is finite, Afshani [2] was the first to present an optimal structure that uses linear space and answers queries in $O(\log n + t)$ time. Using a standard reduction, this provides a three-dimensional orthogonal range reporting structure with optimal $O(\log n + t)$ query time, but with suboptimal $O(n \log^3 n)$ space. This matched two previous results by Chazelle and Guibas [14], and Bozanis et al. [8]. A structure using optimal $O(n(\log n/\log\log n)^2)$ space is also known, but it answers queries in $O(\log^{2+\varepsilon} n + t)$ time, where $\varepsilon > 0$ is an arbitrarily small constant [12]. Recently, Afshani et al. [3, 4] presented an optimal structure for the general problem in three dimensions, namely a structure that uses $O(n(\log n/\log\log n)^2)$ space and answers queries in $O(\log n + t)$ time. They also provided optimal structures for queries with four or five finite coordinates ($O(\log n + t)$ query time and $O(n \log n/\log\log n)$ space), thus completely closing the problem in three dimensions.

For higher dimensions ($d \geq 4$), the best structure answers queries in $O(\log n(\log n/\log\log n)^{d-3} + t)$ time and uses optimal $O(n(\log n/\log\log n)^{d-1})$ space [4]. For dominance queries, the same query bound can be obtained using $O(n(\log n/\log\log n)^{d-3})$ space; in this case, the space bound is known to be optimal only in four dimensions, and the query time is not known to be optimal in any dimension beyond three.

For lower bounds, Afshani et al. [4] proved that any structure for d-dimensional dominance reporting that uses nh space must have $\Omega((\log n/\log h)^{\lfloor d/2 \rfloor - 1} + t)$ query time. Thus setting $h = \log^{O(1)} n$, the result shows that the query time must be $\Omega((\log n/\log\log n)^2 + t)$ for $d = 6$, while for $d = 2, 3$, $O(\log n + t)$ query time is possible. Determining the exact dimension where $O(\log n + t)$ query time is not possible using near-linear space remains an intriguing open problem.

1.3 Rectangle Stabbing

One-dimensional rectangle stabbing (the classical *interval stabbing* problem) can be solved with a variety of techniques [15, 16]. The best results use linear space and answers queries in $O(\log n + t)$ time. Note that it is possible to reduce the problem to two-dimensional dominance reporting by simply mapping an input interval $[a, b]$ to the point (a, b) and a query point x to the dominance query $(-\infty; x) \times (x; -\infty)$.

In two dimensions, an optimal data structure that uses linear space and answers queries in $O(\log n + t)$ time was developed by Chazelle [11]. Using range trees, the structure can be generalized to higher dimensions, by paying a $\log n$ factor per dimension in space and query time, which gives a data structure that uses $O(n \log^{d-2} n)$ space and can answer queries in $O(\log^{d-1} n + t)$ time; it is also possible to obtain $O(\log n(\log n/\log\log n)^{d-2} + t)$ query time using $O(n \log^{d-2+\varepsilon} n)$ space, for any constant $\varepsilon > 0$, using range trees of $\log^\varepsilon n$ fan out. Note that d-dimensional dominance reporting is also a special case of d-dimensional rectangle stabbing, and that a d-dimensional stabbing query can be reduced to a $2d$-dimensional dominance query using a similar reduction outlined for the one dimensional case.

For lower bounds, Afshani et al. [4] proved the first non trivial query lower bound, showing that with nh space, rectangle stabbing queries require $\Omega((\log n/\log h)^{d-1} + t)$ time. Combined with the above reduction, this gives the aforementioned query lower bound for orthogonal range reporting.

1.4 Our Results

Our main upper bound result is a structure that uses $O(n(\log n/\log\log n)^d)$ space and answers d-dimensional range reporting queries in $O(\log n(\log n/\log\log n)^{d-4+1/(d-2)} + t)$ time. Ignoring $\log\log n$ factors, this is a $\log^{1-1/(d-2)} n$ factor improvement in the query time over the fastest previous data structure. Additionally, for the special case of four-dimensional dominance, we provide a data structure that answers queries in $O(\log n\sqrt{\log n/\log\log n} + t)$ time using optimal $O(n \log n/\log\log n)$ space.

From a technical point of view, we obtain our improved data structures using a clever combination of techniques: shallow cuttings and rectangle stabbing. Our key idea is that shallow cuttings can be used to reduce dominance reporting to a rectangle stabbing problem with sublinear input size, which in turn can be exploited to give us our query speed up. At a high level, this approach is inspired by recent results of Chan et al. [10] for offline four-dimensional dominance reporting in the word-RAM model.

In terms of lower bounds, we prove that with nh space, answering d-dimensional rectangle stabbing queries requires $\Omega(\log n(\log n/\log h)^{d-2} + t)$ time. This lower bound is tight for any $h = \log^{d-2+\Omega(1)} n$. To obtain the lower bound, we use a novel geometric argument, rather than the combinatorial framework that was pioneered by Chazelle [12] and which was used in all the previous lower bounds. In fact, our new technique has already led to an improved lower bound for simplex range reporting [1], a long standing open problem. We thus suspect that our new technique might have even further implications.

By the reduction from d-dimensional rectangle stabbing to $2d$-dimensional dominance reporting, we also obtain an improved lower bound of $\Omega(\log n(\log n/\log h)^{\lfloor d/2\rfloor-2}+t)$ for answering d-dimensional dominance reporting queries using nh space.

We describe our orthogonal range reporting data structures in Section 2 and our rectangle stabbing lower bound in Section 3.

2. THE DATA STRUCTURES

In this section, we describe our new orthogonal range reporting data structures. We start with a brief preliminaries section to introduce some of the basic tools we make use of. We then describe a simplified version of our four-dimensional dominance reporting data structure, which answers queries in $O(\log^{3/2} n+t)$ time using $O(n\log n)$ space. We believe it carries most of our important ideas and it is significantly easier to understand. In Subsection 2.3, we present our best dominance reporting structure in higher dimensions. Our final structure appears in Subsection 2.4 and it answers general d-dimensional orthogonal range reporting queries.

2.1 Preliminaries

We now introduce some convenient notations for talking about special cases of orthogonal range reporting.

Restricted Queries.

We adopt the terminology defined in [3]: We use $Q(d,k)$ to refer to the special case of d-dimensional orthogonal range reporting, in which the query rectangles have finite ranges in k of the d dimensions, that is, they are unbounded in $d-k$ dimensions. The $Q(2,1)$ and $Q(d,0)$ problems are the 3-sided planar range reporting and d-dimensional dominance reporting problems, respectively.

3-D Dominance.

For two points p and q in d dimensions, we say p dominates q if every coordinate of p is greater than that of q. Thus, the dominance reporting problem is the problem of outputting the subset of the input that is dominated by a query point. In 3-D, dominance reporting can be solved optimally using an important combinatorial structure known as shallow cuttings [2].

Consider a set S of points in three dimensions. A *shallow cutting for the h-level of S*, or an *h-shallow cutting* for short, is a set $\mathcal{C}$ of $O(|S|/h)$ points such that any point q that dominates at most h points of S is dominated by a point p in $\mathcal{C}$; furthermore, every point of $\mathcal{C}$ dominates $O(h)$ points of S. The existence of such shallow cuttings was proved by Afshani [2], and more general shallow cuttings have been used extensively in the computational geometry literature (see e.g. [18]). To be useful in data structures, for a given point q, we also need a method that finds the point $p\in\mathcal{C}$ that dominates q. This is done using the following lemma.

LEMMA 1 ((MAKRIS AND TSAKALIDIS [17])). *Let S be a set of points in 3-D. It is possible to construct a subdivision $\mathcal{A}_S$ of the plane into $O(|S|)$ rectangles such that: for any query point q in 3-D, if one projects q onto the plane (i.e., the first two coordinates of q) and finds the rectangle in $\mathcal{A}_S$ that contains the projection, then one can find a point in S that dominates q or conclude that no such point exists. The point location query can be answered in $O(\log|S|)$ time using $O(|S|)$ space.*

Rectangle Stabbing.

As discussed, a subproblem that we encounter is rectangle stabbing. It turns out that we need a fast solution for rectangle stabbing when given a budget of nh space. The previous results only focus on the case when h is polylogarithmic but for our purposes, we need to go far beyond polylogarithmic space. Using range trees with fan out h, we can prove the following lemma.

LEMMA 2. *Given n rectangles in d dimensions, rectangle stabbing queries can be answered in $O(\log n\cdot(\log n/\log h)^{d-2}+t)$ time using $O(nh\log^{d-2} n)$ space, in which $h\geq 2$ is any parameter.*

PROOF. Let Q be the input set of rectangles and let m be the total number of corners in the input. We divide the d-dimensional space into h regions, using $h-1$ hyperplanes perpendicular to the d-th dimension, such that each region contains roughly m/h rectangle corners. This creates $h-2$ slabs, the regions sandwiched between two consecutive hyperplanes. We say an input rectangle spans a slab b if it crosses b but it does not have any corners inside b. Let $Q_s(b)$ be the subset of Q that spans b and let $Q_c(b)$ be the subset that crosses but does not span b. Observe that we can use a $(d-1)$-dimensional rectangle stabbing data structure on $Q_s(b)$ and output those rectangles in $Q_s(b)$ that contain the query point; to find the output among $Q_c(b)$ we can simply recurse.

Thus, we build a $(d-1)$-dimensional rectangle stabbing data structure on $Q_s(b)$ for each slab and then recurse on $Q_c(b)$. We use Chazelle's data structure as a base case for $d=2$ [11]. Let $S_k(m)$ denote the space complexity of the algorithm on a k-dimensional input with m corners. We have,

$$S_d(m)\leq hS_d(m/h)+hS_{d-1}(m)$$

which solves to $S_d(m)=O(n\log_h^{d-2})$ using with $S_2(m)=O(m)$ as the base case. If we denote the query time by $Q_d(m)+O(t)$ then the recursion for the query time is

$$Q_d(m)\leq Q_d(m/h)+\log h+Q_{d-1}(m)$$

which solves to $O_d(m)=O(\log n\cdot(\log_h n)^{d-2}n)$.

□

2.2 Simple 4-D Dominance

In this section, we present our simple solution for 4-D dominance that achieves $O(\log^{3/2} n+t)$ query time and uses $O(n\log n)$ space. We obtain the data structure by refining on the standard range tree solution which we describe below.

We use a range tree on the fourth dimension, which is a complete binary tree, T, with the input points $p_1,\dots,p_n$ stored in sorted order of their fourth coordinate in the leaves. Associate every node v in T with the points $p_i,\dots,p_j$ stored in the leaves of the subtree rooted at v. We use $S_v=\{p_i,\dots,p_j\}$ to denote the set of points associated to v. We also associate an interval I_v to v as follows. If $j=n$, then $I_v=(p_{i-1}^{(4)};\infty)$, otherwise, $I_v=(p_{i-1}^{(4)};p_j^{(4)}]$, in which $p_i^{(4)}$ denotes the fourth coordinate of a point p_i and $p_0^{(4)}=-\infty$.

Consider a node $u \in T$ and its left child v. We define a $Q(3,0)$ *query on node* u as a $Q(3,0)$ query on the projection of the points of S_v onto the first three dimensions. It turns out that this is the subproblem that we need to solve.

LEMMA 3. *A $Q(4,0)$ query $q = (x_1, \dots, x_4)$ can be answered using a $Q(3,0)$ query on $O(\log n)$ nodes of T that lie on a root to leaf path of T. Furthermore, for every node u on the path we have $x_4 \in I_u$, and $x_4 \notin I_v$ for nodes v not on the path.*

PROOF. Let $(-\infty; a)$ be the range of q in the last dimension. Start at the root r of T; let v_1 and v_2 be its left and the right child respectively. Clearly we have $x_4 \in I_r$. We have two cases. (i) a is contained in I_{v_1}: in this case, $(-\infty; a)$ does not intersect I_{v_2} so the right subtree does not contain any output points. We simply recurse on the left child. (ii) a is contained in I_{v_2}: in this case, all the points associated to v_1 have their fourth coordinate smaller than a. Thus, we can consider the projection of S_{v_1} and q onto the first three dimensions and find the portion of the output that lies in S_{v_1} using a $Q(3,0)$ query, i.e., a $Q(3,0)$ query on node r. Next, we recurse on v_2. □

One obvious solution to answer the $Q(3,0)$ queries on nodes of T is to use the data structure of Afshani [2]. But this results in $O(\log^2 n + t)$ query time for the $Q(4,0)$ problem. Until now, this has been the only way to solve 4-D dominance (with a possible increase in fanout of the range tree to $\log^\varepsilon n$). We now show that using shallow cuttings, we can do better. We have encapsulated the description of our data structure in two parts: "outputting" data structures and "finder" data structures. The outputting data structures use finder data structures as black boxes and do the actual reporting of the output points while the finder data structures find a small number of crucial elements in the data structure.

The outputting data structures.
Consider a node $u \in T$. Define P_u as the projection of S_v into the first three dimensions in which v is the left child of u. With this notation, a $Q(3,0)$ query on u is equivalent to a dominance query on P_u. Now for every $u \in T$, we build an h-shallow cutting $\mathcal{C}_u$ for P_u in which h is a parameter to be determined later; the only restriction that we place is that $h > \log^2 n$. For every point $p \in \mathcal{C}_u$, we implement an optimal dominance reporting structure [2] on the subset, $P_u(p) \subseteq P_u$ that is dominated by p. Finally, we store the $O(\log^2 n / \log\log n + t)$ query time and $O(n \log n / \log\log n)$ space $Q(4,0)$ data structure of Afshani et al. [4] on the entire input set.

Generating the output of the query.
Let $q = (x_1, \dots, x_4)$ be a $Q(4,0)$ query and define $q' = (x_1, x_2, x_3)$. By Lemma 3, we can answer q by querying q' on $O(\log n)$ nodes, $u_1, \dots, u_{O(\log n)}$, of T. Assume we have a finder structure that for each u_i can find a point $p_i \in \mathcal{C}_{u_i}$ that dominates q' or conclude that no such point exists. If for some i, no point in $\mathcal{C}_{u_i}$ dominates q', then it follows that the output size is at least $h > \log^2 n$; in this case, we simply query the structure of Afshani et al. [4] to report our output. This takes $O(t)$ time. Otherwise, to find points in P_{u_i} that are dominated by q', we query the dominance structure that stores $P_{u_i}(p_i)$; this takes $O(\log |P_{u_i}(p_i)| + t') = O(\log h + t')$ time in which t' is the output size of q' on u_i. Over all the $O(\log n)$ nodes, this takes $O(\log n \cdot \log h + t)$ time.

The finder data structures.
Consider a node $u \in T$ and the h-shallow cutting $\mathcal{C}_u$ for P_u. Using Lemma 1, we decide if a point in $\mathcal{C}_u$ dominates q' and if so find it using a point location query on $\mathcal{A}_{\mathcal{C}_u}$ (remember $\mathcal{A}_{\mathcal{C}_u}$ is the planar subdivision given for $\mathcal{C}_u$ by Lemma 1). Unfortunately, this naive approach only yields $O(\log^2 n)$ query time. Our key idea is to perform all these point location queries *simultaneously*. We lift each rectangle $[x_1; x_2] \times [y_1; y_2]$ in $\mathcal{A}_{\mathcal{C}_u}$ to 3-D by forming the 3-D rectangle $[x_1; x_2] \times [y_1; y_2] \times I_u$. This creates $O(|P_u|/h)$ rectangles for a given node u and thus $O(n \log n / h)$ rectangles in total. We collect all the rectangles (over all the nodes in T) and store them in a 3-D rectangle stabbing data structure given by Lemma 2, with the space usage set to $O(n \log n)$.

The finder query.
Let $q = (x_1, \dots, x_4)$ be the query and define $q' = (x_1, x_2, x_3)$ and $q'' = (x_1, x_2, x_4)$. We claim that to obtain the result of a point location query for (x_1, x_2) on all the sets $\mathcal{A}_{\mathcal{C}_u}$, where we are to perform a $Q(3,0)$ query on u, it suffices to query the stabbing data structure with q'': By Lemma 3, we have $x_4 \in I_u$ for nodes u on the path where we are to query the $Q(3,0)$ structures and $x_4 \notin I_v$ for nodes v not on the path. By Lemma 1, we need to find the rectangle r that contains the point (x_1, x_2) which is equivalent to finding the 3-D rectangle $r \times I_u$ that contains q''. Thus, the point location query for u can be answered by looking at the result of the stabbing query.

Query time analysis.
First, observe that for two nodes u_1 and u_2 at the same depth of T, the intervals I_{u_1} and I_{u_2} are disjoint. Thus, the output size of the stabbing query is $O(\log n)$. Since the ratio between the input size and space usage is $\Omega(h)$, by Lemma 2, the stabbing query takes $O(\log n \cdot \log_h n + \log n) = O(\log n \cdot \log_h n)$ time. Thus, the total query time, including the time to generate the output is $O(\log n \log_h n + \log n \log h + t)$. We pick $h = 2^{\sqrt{\log n}}$ and obtain the following.

THEOREM 1. *The 4-D dominance problem can be solved with $O(n \log n)$ space and $O(\log^{3/2} n + t)$ query time.*

2.3 Higher Dimensions

In this section we show how to extend the simple 4-D dominance result to d-dimensional dominance reporting. Before we begin, we note that similar to the 4-D case, the most interesting case of the problem is when the output size is small. For example, if we realize that the output size is larger than say $\log^{d-1} n$, we can switch to the best previous result and the query time will be optimal.

As with the 4-D case, we start with range trees. However, to get the best performance, we use range trees with large fan outs. Let $\alpha = \log^\varepsilon n$ in which ε is a sufficiently small constant to be determined later. We build a range tree T_d on the d-th dimension with fan out α (i.e., all nodes except leaves have α children). For a node $v_d \in T_d$, the set S_{v_d} and the interval I_{v_d} is defined as in our simple 4-D dominance solution, using the value of the d-th coordinate of the points. We also store a pointer from v_d to a range tree T_{d-1,v_d} with fan out α, built on S_{v_d} and on the $(d-1)$-th dimension. We continue this recursive construction until we reach dimension three. No range tree is built on the first three dimensions. At the end, we have a nested hierarchy of range trees, $T_{i,v_d,v_{d-1},\ldots,v_{i+1}}$, for $4 \le i \le d-1$, in which v_{i+1} is a node in the range tree $T_{i+1,v_d,v_{d-1},\ldots,v_{i+2}}$. A node $v_i \in T_{i,v_d,v_{d-1},\ldots,v_{i+1}}$ is associated with a point set S_{v_i} and consists of all the points whose j-th coordinate, $i \le j \le d$, is inside the interval I_{v_j}. To every node v_4 in a range tree $T_{4,v_d,\ldots,v_5}$, we associate a subproblem $Q_{3,v_d,\ldots,v_4}$ on S_{v_4}. Note that S_{v_4} is the set of input points that lie in the region $R(Q) = (-\infty;\infty) \times (-\infty;\infty) \times (-\infty;\infty) \times I_4 \times \cdots \times I_{v_d}$. Before describing this subproblem, we need to examine how the configuration of the range trees is traversed at query time.

Query traversal.

Let $q = (x_1, \ldots, x_d)$ be the point representing a $Q(d,0)$ query. We define the *query traversal* to be the set of subproblems that are reached using the following procedure. Start from the root r of the range tree T_d. Let v be the child of r such that $x_d \in I_v$. We follow the link to the range tree $T_{d-1,v}$, traverse it and then recurse on v; the recursion stops at the subproblems. To be more precise, assume a node $v_i \in T_{i,v_d,v_{d-1},\ldots,v_{i+1}}$ is reached in this traversal. We follow two pointers for v_i: (i) If $i = 4$, then $Q_{3,v_d,\ldots,v_4}$ is added to the query traversal; otherwise, we follow the pointer to the root of the range tree $T_{i-1,v_d,v_{d-1},\ldots,v_i}$ and traverse it and (ii) we find a child u of v_i such that $x_i \in I_u$ and then recurse on u.

The subproblem.

Consider a $Q := Q_{3,v_d,\ldots,v_4}$ subproblem that is in the query traversal for the query q. Observe that the crucial property of the query traversal is that if Q is in the query traversal of q, then $q \in R(Q)$. The children of v_i decompose I_{v_i} into α smaller intervals which in geometric terms corresponds to cutting the i-th dimension of $R(Q)$ into α smaller parts. Based on this division, for every point $p = (p_1, \cdots, p_d) \in R(Q)$, we associate a *child-coordinate point* $f_Q(p)$, $f_Q(p) \subset [\alpha]^{d-3}$, to p in which the i-th coordinate is the index of the child of v_i whose interval contains p_i. Now, the subproblem Q is defined as the problem of outputting all the points $p \in S_{v_4}$ such that the point (x_1, x_2, x_3) dominates the projection of p into the first three dimensions and every coordinate of $f_Q(q)$ is larger than that of $f_Q(p)$ (i.e., $f_Q(q)$ dominates $f_Q(p)$).

Before describing our data structures, we need an equivalent of Lemma 3.

LEMMA 4. *A $Q(d,0)$ query q can be solved by solving all the subproblems that are reached during the query traversal of q. For each such subproblem Q, we have $q \in R(Q)$. Furthermore, the number of subproblems Q such that $q \in R(Q)$ is $O(\log_\alpha^{d-3} n)$.*

PROOF. Consider a point $p = (p_1, \ldots, p_d) \in S$ that is dominated by q. We claim that there exists a subproblem Q in the query traversal of q that outputs p. To see this, we begin from the root r of T_d and essentially follow the procedure that builds the query traversal. Assume we have reached a node $v_i \in T_{i,v_d,v_{d-1},\ldots,v_{i+1}}$. In contrast to the procedure that builds the query traversal, we follow a single pointer at each step: (i) if there is a child u of v_i such that p_i and $x_i \in I_u$ then we follow the pointer to u (ii) otherwise, if $i > 4$, we follow the pointer to $T_{i-1,v_d,v_{d-1},\ldots,v_i}$ but if $i = 4$, then $Q := Q_{3,v_d,\ldots,v_4}$ is the desired subproblem. It is easily checked that $q \in R(Q)$ and that every coordinate of $f_Q(q)$ is greater than that of $f_Q(p)$.

It remains to prove that the number of subproblems Q, s.t. $q \in R(Q)$ is $O(\log_\alpha^{d-3})$. In a range tree with fanout α, there are only $O(\log_\alpha n)$ nodes v such that I_v contains a given point. Since we have build $d-3$ levels of range trees, our claim can be proved by a simple inductive argument. □

Now we are ready to describe our data structures.

The outputting data structures.

Consider a $Q := Q_{3,v_d,\ldots,v_4}$ subproblem. Let P be the projection of S_{v_4} onto the first three dimensions. We partition P into $\beta := \alpha^{d-3}$ subsets as follows. For every element $\gamma \in [\alpha]^{d-3}$, we place all the points $p \in P$ with $f_Q(p) = \gamma$ in the set P_γ. We also define the set $\hat{P}_\gamma$ to be the union of all the sets $P_{\gamma'}$, $\gamma' \in [\alpha]^{d-3}$, in which γ dominates γ'. We build three categories of shallow cuttings: First, for every P_γ, we build a $\log^d n$-shallow cutting $\mathcal{C}_{d,\gamma}$ and then store the points dominated by every point $p \in \mathcal{C}_{d,\gamma}$ in an optimal 3-D dominance reporting structure. Second, for every $\hat{P}_\gamma$, we build an h-shallow cutting $\hat{\mathcal{C}}_{h,\gamma}$; the value of h will be determined later and the only restriction that we impose is that $h > \log^{2d} n$. Finally, for every point $p \in \hat{\mathcal{C}}_{h,\gamma}$, we build a $\log^{d-1} n$-shallow cutting $\hat{\mathcal{C}}_{d-1,\gamma,p}$, as well as the corresponding point location data structure of Lemma 1, on the subset of $\hat{P}_\gamma$ dominated by p. Consider a point $\gamma' \in [\alpha]^{d-3}$ that is dominated by γ. We know $P_{\gamma'} \subset \hat{P}_\gamma$ and thus every point $\hat{p} \in \hat{\mathcal{C}}_{d-1,\gamma,p}$ contains $O(\log^{d-1} n)$ points in $P_{\gamma'}$ which implies there exists a point $p \in \mathcal{C}_{d,\gamma'}$ such that p dominates $\hat{p}$. We place a pointer from $\hat{p}$ to p. Since the sets P_γ partition P, the first category of shallow cuttings take up linear space. The number of cells in the second category is $O(\alpha^{d-3}|S_{v_4}|/h)$ which is sublinear by our assumptions on h and α. Finally, the number of points in $\hat{\mathcal{C}}_{h,\gamma}$ is $O(|S_{v_4}|/h)$ and for each we store a shallow cutting that has $O(h/\log^{d-1} n)$ size and for each point in the second shallow cutting we store $O(\alpha^{d-3})$ additional pointers; thus, the total space is $O(|S_{v_4}|)$ for the subproblem Q. Over all the subproblems this sums up to $O(n \log_\alpha^{d-3} n)$.

Answering output queries.

Let $q = (x_1, \ldots, x_d)$ be a $Q(d,0)$ query and let $q' = (x_1, x_2, x_3)$. By Lemma 4, we need to answer q on $O(\log_\alpha^{d-3} n)$ subproblems. Consider one such subproblem $Q := Q_{3,v_d,\ldots,v_4}$. If we let $\gamma = f_Q(q)$ the subproblem translates to outputting the subset of $\hat{P}_\gamma$ that is dominated by q'. If no point in $\hat{\mathcal{C}}_{h,\gamma}$ dominates q', then the output size is larger than $h > \log^{d-1} n$, which means we can switch to the previous best result on dominance reporting. Assume we have a finder data structures that can find a point $p \in \hat{\mathcal{C}}_{h,\gamma}$ that dominates q' (or can tell us no such point exists). Using the point location data structure implemented for $\hat{\mathcal{C}}_{d-1,\gamma,p}$ we can check if there exists a point $p' \in \hat{\mathcal{C}}_{d-1,\gamma,p}$ that dominates q'. If no such point exists, then the output size is larger than $\log^{d-1} n$ and we are done. Thus, assume we locate such a point p'. Note that finding p' takes $O(\log|\hat{\mathcal{C}}_{d-1,\gamma,p}|) = O(\log h)$ time by Lemma 1. Next, for every $\gamma' \in [\alpha]^{d-3}$, that is dominated by γ, we follow the pointer from p' to the corresponding point $p'' \in \mathcal{C}_{d,\gamma'}$ that dominates p'. As p'' also dominates

q', the query can answered by querying the 3-D dominance data structure built for p''. The query time for Q becomes $O(\log h + \alpha^{d-3} \log\log n)$ which results in the overall query time of $O((\log h + \alpha^{d-3} \log\log n) \log_\alpha^{d-3} n)$.

The finder data structures.
Consider a subproblem $Q := Q_{3,v_d,\ldots,v_4}$ and the shallow cutting $\hat{C}_{h,\gamma}$ for some $\gamma \in [\alpha]^{d-3}$. We build the corresponding orthogonal planar subdivision $\hat{\mathcal{A}}_{h,\gamma}$ and lift a rectangle $r \in \hat{\mathcal{A}}_{h,\gamma}$ to a $d-1$-dimensional rectangle $L_Q(r) = r \times I_{v_4} \times \cdots \times I_{v_d}$. The total number of rectangles created from all the subproblems is $O(n \log_\alpha^{d-3} n/h)$ and we place them all in a $(d-1)$-dimensional rectangle stabbing structure given by Lemma 2 in which the space usage is set to $O(n)$.

The finder query.
Let $q = (x_1, \ldots, x_d)$ be a query point and define $q' = (x_1, x_2, x_4, \ldots, x_d)$. We claim the finder queries for all the subproblems can be answered by looking at the result of q' on the rectangle stabbing data structure. Consider a subproblem $Q := Q_{3,v_d,\ldots,v_4}$ that is reached during the query traversal of q. By Lemma 4, $q \in R(Q)$. By Lemma 1, to answer the finder query for Q, it suffices to do a point location query on $\hat{\mathcal{A}}_{h,\gamma}$ using (x_1, x_2) as the query point. Observe that if a rectangle $r \in \hat{\mathcal{A}}_{h,\gamma}$ contains (x_1, x_2), then $L_Q(r)$ contains q'. To bound the running time, notice that the number of different values of γ is α^{d-3}, and the output size of the stabbing query is at most $O(\alpha^{d-3} \log_\alpha^{d-3} n)$ by Lemma 4. The input size of the stabbing data structure is $O(n \log_\alpha^{d-3} n/h)$ while its space usage is set to n. The ratio of the space usage to the input size is thus $\Omega(h/\log_\alpha^{d-3} n) = \Omega(\sqrt{h})$. Since this is a $(d-1)$-dimensional stabbing problem, by Lemma 2, the query time is $O((\log_h n)^{d-3} \log n + \alpha^{d-3} \log_\alpha^{d-3} n)$.

Optimizing the running time.
We get that the total query time is $O((\log_h n)^{d-3} \log n + \alpha^{d-3} \log_\alpha^{d-3} n + (\log h + \alpha^{d-3} \log\log n) \log_\alpha^{d-3} n)$. By setting $\alpha = \log^\varepsilon n$ for a small enough ε, and picking h such that $\log h = \log^{1/(d-2)} n (\log\log n)^{(d-3)/(d-2)}$ we get the following:

THEOREM 2. *The d dimensional dominance reporting problem can be solved using* $O(n(\log n/\log\log n)^{d-3})$ *space and with query time of* $O(\log n(\log n/\log\log n)^{d-4+1/(d-2)} + t)$.

In the next subsection, we show that with some modifications, our data structure can in fact answer $Q(d, d-3)$ queries which in turn is used to obtain our general d-dimensional orthogonal range reporting data structure.

2.4 General orthogonal range reporting queries

The main result of this subsection is the following theorem.

THEOREM 3. *The* $Q(d, d-3)$ *problem can be solved using* $O(n(\log n/\log\log n)^{d-3})$ *space and with query time of* $O(\log n(\log n/\log\log n)^{d-4+1/(d-2)} + t)$.

The data structure that we build is very similar to one used for Theorem 2. We begin by building the exact same set of range trees. The definition of the subproblem $Q_{3,v_d,\ldots,v_4}$ will be different. However, as with the dominance structure, it is more convenient to define the query traversal first.

Query traversal.
Without loss of generality, we can assume the $Q(d, d-3)$ query interested is $q = (-\infty; y_1) \times (-\infty; y_2) \times (-\infty; y_3) \times (x_4; y_4) \times \cdots \times (x_d; y_d)$. We now define the query traversal. As before, the traversal starts at the root of T_d. Assume a node $v_i \in T_{i+1,v_d,v_{d-1},\ldots,v_{i+1}}, i \geq 4$, is reached in this traversal. We maintain the invariant that at least one of x_i or y_i is contained in I_{v_i}. Let $u_1, \cdots, u_\alpha$ be the children of v_i. We consider four cases:

(i) $x_i \in I_{v_i}, y_i \in I_{v_i}$ and x_i and y_i are in I_{u_j} for some j: In this case, we follow the pointer to u_j.

(ii) $x_i \in I_{v_i}, y_i \in I_{v_i}$ but $x_i \in I_{u_j}$ and $y_i \in u_\ell$ for $j < \ell$: In this case, we follow three pointers, first to u_j, second to u_ℓ, and third to the root of the range tree $T_{i,v_d,\ldots,v_i}$ if $i > 4$ but if $i = 4$, we add $Q_{3,v_d,\ldots,v_4}$ to the query traversal.

(iii) $x_i \in I_{v_i}, y_i \notin I_{v_i}$: In this case, x_i will be contained in I_{u_j} for some j. We follow two pointers, one to u_j and another to $T_{i,v_d,\ldots,v_i}$ if $i > 4$ but if $i = 4$, we add $Q_{3,v_d,\ldots,v_4}$ to the query traversal.

(iv) $x_i \notin I_{v_i}, y_i \in I_{v_i}$: Similar to the previous case, for an index j, I_{u_j} will contains y_i. We follow two pointers, one to u_j and another to $T_{i,v_d,\ldots,v_i}$ if $i > 4$ but if $i = 4$, we add $Q_{3,v_d,\ldots,v_4}$ to the query traversal.

The subproblem.
Consider a $Q := Q_{3,v_d,\ldots,v_4}$ subproblem that is in the query traversal for the query q. As before, the region $R(Q)$ is defined as $R(Q) = (-\infty, \infty) \times (-\infty, \infty) \times (-\infty, \infty) \times I_4 \times \cdots \times I_{v_d}$ and it contains all the points in S_{v_4}. The crucial property here is that $R(Q)$ contains at least one corner of q. For a point $p \in R(Q)$, the child-coordinate point $f_Q(p)$ is defined in the exact same way. The query in the subproblem Q is defined using a $(d-3)$-dimensional rectangle and is denoted by $\tau_Q(q)$, in which the coordinates are either integers from $[\alpha]$, $-\infty$, or ∞. Let $\tau_Q(q) = (a_4, b_4) \times \cdots \times (a_d, b_d)$. For an index i, $4 \leq i \leq d$, the values a_i and b_i are defined below, depending on which condition of the query traversal was true during the query traversal at node v_i. Note that for v_i only cases, (ii), (iii), and (iv) could be valid as case (i) does not result in a traversal of a lower dimensional range tree. Consider the notation used in the case analysis of the query traversal.

- If case (ii) was true at v_i, then $a_i = j$ and $b_i = \ell$.
- If case (iii) was true at v_i, then $a_i = j$ and $b_i = \infty$.
- If case (iv) was true at v_i, then $a_i = -\infty$ and $b_i = j$.

With this definition, the subproblem Q is defined as outputting all the points $p \in S_{v_4}$ such that the point (y_1, y_2, y_3) dominates the projection of p into the first three dimensions and $f_Q(p)$ is contained in the rectangle $\tau_Q(q)$.

Before describing our data structures, we need an equivalent of Lemma 4.

LEMMA 5. *A $Q(d, d-3)$ query q can be answered by solving all the subproblems that are reached during the query traversal of q. For each such subproblem Q, $R(Q)$ contains at least one corner of q. Furthermore, the number of subproblems Q that contains a corner of q is $O(\log_\alpha^{d-3} n)$.*

PROOF. Let $q = (-\infty; y_1)\times(-\infty; y_2)\times(-\infty; y_3)\times(x_4; y_4)\times \cdots \times (x_d; y_d)$ be the query rectangle. Consider a point $p = (p_1, \dots, p_d) \in S$ that is contained in q. We claim that there exists a subproblem Q in the query traversal of q that outputs p. To find Q, we begin from the root r of T_d and follow the procedure that builds the query traversal, and trace the footsteps of the query traversal, except that we follow exactly one pointer at each case. Assume we have reached a node $v_i \in T_{i,v_d,v_{d-1},\dots,v_{i+1}}$. We maintain the invariant that at least one of x_i or y_i is contained in I_{v_i}.

We review the four cases that were considered for the query traversal.

(i) $x_i \in I_{v_i}, y_i \in I_{v_i}$ and x_i and y_i are in I_{u_j} for some j: In this case, we have $p_i \in I_{u_j}$ and we follow the pointer to u_j.

(ii) $x_i \in I_{v_i}, y_i \in I_{v_i}$ but $x_i \in I_{u_j}$ and $y_i \in u_\ell$ for $j < \ell$: we follow one the three pointers followed at the query traversal. If $p_i \in I_{u_j}$ we follow the pointer to u_j, if $p_i \in I_{u_\ell}$ we follow the pointer to u_ℓ, but otherwise, we follow the pointer to the root of the range tree $T_{i,v_d,\dots,v_i}$ if $i > 4$ but if $i = 4$, then $Q_{3,v_d,\dots,v_4}$ is the desired subproblem.

(iii) $x_i \in I_{v_i}, y_i \notin I_{v_i}$: In this case, x_i will be contained in I_{u_j} for some j. If $p_i \in I_{u_j}$, then we follow the pointer to u_j, otherwise, we follow the pointer to $T_{i,v_d,\dots,v_i}$ if $i > 4$ but if $i = 4$, then $Q_{3,v_d,\dots,v_4}$ is the desired subproblem.

(iv) $x_i \notin I_{v_i}, y_i \in I_{v_i}$: Similar to the previous case, for an index j, I_{u_j} will contains y_i. If $p_i \in I_{u_j}$, then we follow the pointer to u_j, otherwise, we follow the pointer to $T_{i,v_d,\dots,v_i}$ if $i > 4$ but if $i = 4$, then $Q_{3,v_d,\dots,v_4}$ is the desired subproblem.

It is straightforward to verify that when a subproblem Q is reached, $R(Q)$ contains a corner of q and the rectangle $\tau_Q(q)$ contains the point $f_Q(p)$. Thus, p will be outputted by solving Q.

Rectangle q has a constant number of corners and thus the number of subproblems Q such that $R(Q)$ contains a corner of q is $O(\log_\alpha^{d-3})$. □

We now describe our data structures.

The outputting data structures.

Consider a $Q := Q_{3,v_d,\dots,v_4}$ subproblem. Let P be the projection of S_{v_4} onto the first three dimensions. Like the $Q(d, 0)$ case, we partition P into α^{d-3} subsets. For every element $\gamma \in [\alpha]^{d-3}$, we place all the points $p \in P$ with $f_Q(p) = \gamma$ in the set P_γ. Observe that the number of possible rectangles $\tau_Q(q)$ that can be queried on this subproblem is less than α^{2d}. For every such $\tau = \tau_Q(q)$, define the set $\hat{P}_\tau$ to be the union of all the sets P_γ in which γ is in the interior of τ. With these definitions, the rest of our outputting data structures is almost identical to the $Q(d, 0)$ case: we build three categories of shallow cuttings: First, for every P_γ, we build a $\log^d n$-shallow cutting $\mathcal{C}_{d,\gamma}$ and then store the points dominated by every point $p \in \mathcal{C}_{d,\gamma}$ in an optimal 3-D dominance reporting structure. Second, for every $\hat{P}_\tau$, we build an h-shallow cutting $\hat{\mathcal{C}}_{h,\tau}$; the value of h will be determined later and the only restriction that we impose is that $h > \log^{2d} n$. Finally, for every point $p \in \hat{\mathcal{C}}_{h,\tau}$, we build a $\log^{d-1} n$-shallow cutting $\hat{\mathcal{C}}_{d-1,\tau,p}$, as well as the corresponding point location data structure, on the subset of $\hat{P}_\tau$ dominated by p. As before, every point $\hat{p} \in \hat{\mathcal{C}}_{d-1,\tau,p}$ is dominated by a point $p \in \mathcal{C}_{d,\gamma'}$ and we place a pointer from $\hat{p}$ to p. It is easy to see that the total space consumption is linear for Q and thus $O(n \log_\alpha^{d-3} n)$ over all the subproblems.

Answering output queries.

Let $q = (-\infty; y_1) \times (-\infty; y_2) \times (-\infty; y_3) \times (x_4; y_4) \times \cdots \times (x_d; y_d)$ be the $Q(d, d-3)$ query and let $q' = (y_1, y_2, y_3)$. By Lemma 5, we need to answer q on $O(\log_\alpha^{d-3} n)$ subproblems. Consider one such subproblem $Q = Q_{3,v_d,\dots,v_4}$. If we let $\tau = \tau_Q(q)$, then the subproblem translates to outputting the subset of $\hat{P}_\tau$ that is dominated by q'. If no point in $\hat{\mathcal{C}}_{h,\tau}$ dominates q', then the output size is larger than $h > \log^{d-1} n$, and we are done by switching to the best previous orthogonal reporting data structure. Assume we have a finder data structures that can find a point $p \in \hat{\mathcal{C}}_{h,\tau}$ that dominates q' (or tell us no such point exists). Using the point location data structure implemented for $\hat{\mathcal{C}}_{d-1,\tau,p}$, in $O(\log h)$ time we can check if there exists a point $p' \in \hat{\mathcal{C}}_{d-1,\tau,p}$ that dominates q'. If no such point exists, then the output size is larger than $\log^{d-1} n$ and we are done. Thus, assume we locate such a point p'. Next, for every $\gamma' \in [\alpha]^{d-3}$, s.t., γ' is contained in τ, we follow the pointer from p' to the corresponding point $p'' \in \mathcal{C}_{d,\gamma'}$ that dominates p'. As p'' also dominates q', the query can answered by querying the 3-D dominance data structure built for p''. The query time for Q becomes $O(\log h + \alpha^{d-3} \log\log n)$ which results in the overall query time of $O((\log h + \alpha^{d-3} \log\log n) \log_\alpha^{d-3} n)$.

The finder data structures.

Consider a subproblem $Q := Q_{3,v_d,\dots,v_4}$ and the shallow cutting $\hat{\mathcal{C}}_{h,\tau}$ for all the $O(\alpha^{2d})$ choices of τ. We build the corresponding orthogonal planar subdivision $\hat{\mathcal{A}}_{h,\tau}$ and lift a rectangle $r \in \hat{\mathcal{A}}_{h,\gamma}$ to a $d-1$-dimensional rectangle $L_Q(r) = r \times I_{v_4} \times \cdots \times I_{v_d}$. The total number of rectangles created from all the subproblems is $O(n \log_\alpha^{d-3} n/h)$ and we place them all in a $(d-1)$-dimensional rectangle stabbing structure given by Lemma 2 in which the space usage is set to $O(n \log^{d-3} n)$.

The finder query.

Let $q = (-\infty; y_1) \times (-\infty; y_2) \times (-\infty; y_3) \times (x_4; y_4) \times \cdots \times (x_d; y_d)$ be the $Q(d, d-3)$ query, and let $q' = (y_1, y_2, y_3)$. We query the stabbing data structure using every corner of q, after removing the third coordinate. We claim the results of these queries is sufficient to answer all the finder queries.

Consider a subproblem Q that is in the query traversal of q. By Lemma 5, $R(Q)$ contains at least one corner $\delta = (y_1, y_2, y_3, \delta_4, \cdots, \delta_d)$ of q in which δ_i, for $4 \le i \le d$, is either x_i or y_i. As discussed, the stabbing data structure will be queried with point $\delta' = (y_1, y_2, \delta_4, \cdots, \delta_d)$. Let $\tau = \tau_Q(q)$ and consider the shallow cutting $\hat{\mathcal{C}}_{h,\tau}$ and the corresponding planar subdivision $\hat{\mathcal{A}}_{h,\tau}$. By Lemma 1, it suffices to find a rectangle $r \in \mathcal{A}_{h,\tau}$ that contains the point (y_1, y_2). Observe that since $\delta \in R(Q)$, we have $\delta_i \in I_{v_i}, 4 \le i \le d$. Thus, if r contains (y_1, y_2), then it follows that $L_Q(r)$ contains the point δ'.

As before, the output size of the stabbing query is at most $O(\log^{d-3} n)$ by Lemma 5 and by Lemma 2, the query time is $O(\log n (\log_h n)^{d-3} + \log^{d-3} n)$. Using the same value of h as before, we get the following theorem.

THEOREM 4. *$Q(d, d-3)$ queries can be answered in of $O(\log n(\log n/\log\log n)^{d-4+1/(d-2)} + t)$ time using a data structure that consumes $O(n(\log n/\log\log n)^{d-3})$ space.*

Using an standard application of range trees with fan out $\alpha = \log^{\varepsilon} n$ we get the following theorem.

THEOREM 5. *There exists an orthogonal range reporting data structure that uses $O(n(\log n/\log\log n)^d)$ space and can answer queries in $O(\log n(\log n/\log\log n)^{d-4+1/(d-2)} + t)$ time.*

3. A TIGHT LOWER BOUND FOR RECTANGLE STABBING

In this section we prove our lower bound for the rectangle stabbing problem. Our model of computation is identical to one used by Chazelle [12] but for sake of completeness, we include a brief description.

Consider an input set S of n elements. A data structure is a collection of (memory) cells and its space complexity is the number of cells used. Each cell can store an input element or an auxiliary data type. Furthermore, each cell can have two pointers to two other cells. Such a data structure can be represented by a graph G with $V(G)$ being the set of cells used in the data structure. A pointer from a cell u to a cell v is represented with a directed edge from u to v. The crucial restriction is that a cell can only be accessed through pointers and thus random accesses are disallowed. (It is assumed that we always begin by a pointer to a root cell.) A query q is answered by exploring (a subgraph of) G. In this model, to output an element $p \in S$, the query algorithm must visit a cell that stores p. We do not impose any restriction on how the algorithm navigates G to reach such a node. We measure the query time by the number of vertices of G visited by the query algorithm.

Consider a data structure D that answers rectangle stabbing queries in d dimensions in the above model. Let nh be the number of cells used by D. Furthermore, assume D answers every query in $f(n) + Ct$ time in which t is the output size, C is a constant, and $f(n)$ is the search time of the query. Our goal is to prove that $f(n) = \Omega(\log n \cdot \log_h^{d-2} n)$.

We build an input set, Q, of rectangles that is in fact a simplification of the ones used before [4, 12, 13]. Consider a d-dimensional cube $\mathcal{R}$ with side length m, $m > \sqrt{n}$, in which m is a parameter to be determined later. Let $r = c_r h^3$, in which c_r is a large enough constant, and let $\text{Log}_r m = \lfloor \log_r m \rfloor - 1$. For every choice of $d-1$ indices, $1 \le i_1, \cdots, i_{d-1} \le \text{Log}_r m$, we divide the j-th dimension of $\mathcal{R}$ into r^{i_j} equal length pieces for $1 \le j \le d-1$, and the d-th dimension of $\mathcal{R}$ into $\lfloor m^{d-1}/r^{i_1+\cdots+i_{d-1}} \rfloor$ equal length pieces. The number of such choices is $k = \text{Log}_r^{d-1} m$. With each such choice, we create between $m/2$ and m rectangles and thus $\Theta(mk)$ rectangles are created in total. We pick m such that this number equals n, thus $n = \Theta(mk)$. Also, note that the volume of each rectangle is between m^{d-1} and $2m^{d-1}$. The crucial property of this input set is the following.

OBSERVATION 1. *The volume of the intersection of any two rectangles in $\mathcal{R}$ is at most $2m^{d-1}/r$.*

Unlike the previous attempts, our query set is very simple: a single query q chosen uniformly at random inside $\mathcal{R}$. Note that the output size of q is k. The rest of this section is devoted to prove that with positive probability answering q needs $\Omega(k \log r) = \Omega(\log_r^{d-1} m \log r) = \Omega(\log n \cdot \log_h^{d-2} n)$ time which proves our claim.

For a vertex $u \in G$, let $\text{In}(u)$ denote the set of vertices in G that have a directed path of size at most $\log h$ to u. Similarly, define $\text{Out}(u)$ to be the set vertices in G that can be reached from u using a path of size at most $\log h$. For a rectangle b, let V_b be the set of vertices of G that store b. We define $\text{In}(b) = \sum_{v \in V_b} \text{In}(v)$ and $\text{Out}(b) = \sum_{v \in V_b} \text{Out}(v)$. We say b is *popular* if $\text{In}(b) > ch^2$.

LEMMA 6. *There are at most n/c popular rectangles.*

PROOF. As each vertex in G has two out edges, for every $u \in G$, the size of $\text{Out}(u)$ is at most $2^{\log h} = h$. Also, if $v \in \text{Out}(u)$, then $u \in \text{In}(v)$ and vice versa. Thus,

$$\sum_{b \in Q} |\text{In}(b)| = \sum_{b \in Q} |\text{Out}(b)| = \sum_{v \in G} |\text{In}(v)| = \sum_{u \in G} |\text{Out}(u)| \le nh^2.$$

This implies the number of rectangles b with $\text{In}(b) > ch^2$ is at most n/c. □

LEMMA 7. *With probability at least $3/5$, q is enclosed by at most $k/4$ rectangles that are popular, if c is to be a chosen sufficiently large constant.*

PROOF. By Lemma 6, the number of popular rectangles is at most n/c. As each rectangle has volume at most $2m^{d-1}$, the total volume of popular rectangles is at most $4m^{d-1}n/c$. Let A be the region of $\mathcal{R}$ that contains all the points covered by more than $k/4$ popular rectangles. We have,

$$\text{Vol}(A)k/4 \le 4m^{d-1}n/c = \Theta(m^d k/c)$$

which implies $\text{Vol}(A) = O(m^d/c) < 2m^d/5$ if c is large enough. Thus, with probability at least $3/5$, q will not be picked in A. □

Let S' be the subset of rectangles that not popular and let $n' = |S'|$. We say two rectangles $b_1, b_2 \in S'$ are *close*, if there exists a vertex $u \in G$ such that from u we can reach a cell that stores b_1 and a cell that stores b_2 with directed paths of size at most $\log h$ each.

LEMMA 8. *A rectangle $b \in S'$ can be close to at most ch^3 other rectangles in S'.*

PROOF. If a rectangle b is close to a rectangle b', then there exists a vertex $u \in \text{In}(b)$ such that b' is stored in a vertex in $\text{Out}(u)$. For every $u \in G$ we have $\text{Out}(u) \le h$, and $|\text{In}(b)| \le ch^2$. Thus, b can be close to ch^3 different rectangles. □

Consider a rectangle $b \in S'$. Let B be the subset of rectangles that are close to b. We call $\cup_{b' \in B}(b \cap b')$ the *close region* of b and denote it with C_b.

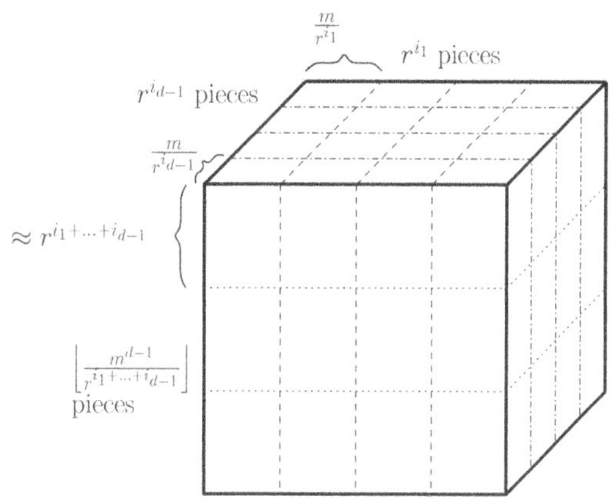

Figure 1: An illustration of the d-dimensional cube $\mathcal{R}$ and its division into smaller rectangles for $d = 3$.

LEMMA 9. *With probability at least 1/5, answering q needs $\Omega(\log n \cdot \log_h^{d-2} n)$ time.*

PROOF. Consider a rectangle $b \in S'$ and set B defined above. By Lemma 8, $|B| \le ch^3$. By Observation 1, for every $b' \in B$, $\mathrm{Vol}(b \cap b') \le 2m^{d-1}/r$ and thus

$$\mathrm{Vol}(C_b) \le \sum_{b' \in B} \mathrm{Vol}(b \cap b') \le 2ch^3 m^{d-1}/r = 2cm^{d-1}/c_r.$$

As there are at most n rectangles in S', the sum of the volumes of the close regions of all the rectangles in S' is at most $n \cdot 2cm^{d-1}/c_r = \Theta(ckm^d/c_r)$. Consider the region R of all the points p such that p is inside the close region of at least $k/4$ rectangles of S'. We have,

$$\mathrm{Vol}(R)k/4 \le \sum_{b \in S'} \mathrm{Vol}(C_b) \le \Theta(ckm^d/c_r)$$

which means $\mathrm{Vol}(R) = O(cm^d/c_r)$. If we choose c_r large enough, this volume is less than $2m^d/5$, which means with probability at least 3/5, the selected query will not be inside R. Combined with Lemma 7, this gives the following: with probability at least 1/5, q will be inside at least $3k/4$ rectangles that are in S' (by Lemma 7) and it will also be inside the close region of at most $k/4$ rectangles. Let q be the query when this happens.

Consider the subgraph H_q of G that is explored by the query procedure while answering q. Assume q needs to output rectangles $b_1, \cdots, b_{k'}$ from S'. We have $k' \ge 3k/4$. Let v_i be a node in H_q that stores b_i, $1 \le i \le k'$. Also, let W_i be the set of nodes in H_q that are reachable by traversing up to $\log h$ edges from v_i, where the direction of edged have been reversed. If two sets W_i and W_j, $1 \le i < j \le k'$, share a common vertex, it means that q is inside the close region of both b_i and b_j. However, we know that q is inside the close region of at most $k/4$ rectangles. This means that there are at least $k/2$ sets W_i that do not share any vertices. Thus, the size of H_q is at least $k/2 \cdot \log h = \Omega(\log n \cdot \log_h^{d-2} n)$. □

Remarks.

Our technique is quite different from the previous lower bounds results in the points machine model [4, 12]. The previous results rely on a combinatorial framework that deals with the subsets of the input returned by the queries. This framework is non-geometric and could even be applied to abstract reporting problems. Our argument on the other hand is inherently geometric and simpler but it cannot be applied to non-geometric problems.

4. CONCLUSIONS

We believe our most surprising result is the $O(n \log n / \log\log n)$ space structure for 4-D dominance reporting that answers queries in $O(\log n \sqrt{\log n / \log\log n} + t)$ time. The existence of such a structure raises the obvious open question of whether this query time can be reduced to $O(\log n + t)$. Unfortunately, our rectangle stabbing lower bound shows that improving our query time might be difficult.

Our idea of using geometry and volume based arguments to prove lower bounds is quite novel, and as mentioned, it has already led to improved simplex range reporting lower bounds [1]. Finding further applications of our technique is still open.

5. REFERENCES

[1] P. Afshani. Improved pointer machine and I/O lower bounds for simplex range reporting and related problems. submitted to SoCG'12.

[2] P. Afshani. On dominance reporting in 3D. In *ESA'08: Proc. of the 16th conference on Annual European Symposium*, pages 41–51, 2008.

[3] P. Afshani, L. Arge, and K. D. Larsen. Orthogonal range reporting in three and higher dimensions. In *FOCS'09: Proc. of the 50th Annual Symposium on Foundations of Computer Science*, pages 149–158, 2009.

[4] P. Afshani, L. Arge, and K. D. Larsen. Orthogonal range reporting: query lower bounds, optimal structures in 3-d, and higher-dimensional improvements. In *SCG'10: Proc. of the 26th Annual Symposium on Computational Geometry*, pages 240–246, 2010.

[5] P. K. Agarwal. Range searching. In J. E. Goodman and J. O'Rourke, editors, *CRC Handbook of Discrete and Computational Geometry*. CRC Press, Inc., 2004.

[6] P. K. Agarwal and J. Erickson. Geometric range searching and its relatives. In B. Chazelle, J. E. Goodman, and R. Pollack, editors, *Advances in*

Discrete and Computational Geometry. AMS Press, 1999.

[7] S. Alstrup, G. S. Brodal, and T. Rauhe. New data structures for orthogonal range searching. In *FOCS '00: Proc. of the 41st Annual Symposium on Foundations of Computer Science*, pages 198–207. IEEE Computer Society, 2000.

[8] P. Bozanis, N. Kitsios, C. Makris, and A. Tsakalidis. New results on intersection query problems. *The Computer Journal*, 40:22–29, 1997.

[9] T. M. Chan, K. G. Larsen, and M. Pătraşcu. Orthogonal range searching on the RAM, revisited. In *SCG '11: Proc. of the 27th Annual Symposium on Computational Geometry*, SoCG '11, pages 1–10. ACM, 2011.

[10] T. M. Chan, K. G. Larsen, and M. Pătraşcu. Orthogonal range searching on the RAM, revisited. In *SCG'11: Proc. of the 27th Annual Symposium on Computational Geometry*, pages 1–10, 2011.

[11] B. Chazelle. Filtering search: a new approach to query answering. *SIAM Journal on Computing*, 15(3):703–724, 1986.

[12] B. Chazelle. Lower bounds for orthogonal range searching: I. the reporting case. *Journal of the ACM*, 37(2):200–212, 1990.

[13] B. Chazelle. Lower bounds for off-line range searching. In *STOC '95: Proc. of the 27th annual ACM symposium on Theory of computing*, pages 733–740, 1995.

[14] B. Chazelle and L. J. Guibas. Fractional cascading: II. Applications. *Algorithmica*, 1:163–191, 1986.

[15] M. de Berg, O. Cheong, M. van Kreveld, and M. Overmars. *Computational Geometry: Algorithms and Applications*. Springer-Verlag, 2008.

[16] E. Hanson. The interval skip list: A data structure for finding all intervals that overlap a point. In *Algorithms and Data Structures*, volume 519 of *Lecture Notes in Computer Science*, pages 153–164. Springer Berlin / Heidelberg, 1991.

[17] C. Makris and A. Tsakalidis. Algorithms for three-dimensional dominance searching in linear space. *Inf. Process. Lett.*, 66(6):277–283, 1998.

[18] J. Matoušek. Reporting points in halfspaces. *Computational Geometry Theory and Applications*, 2(3):169–186, 1992.

[19] E. M. McCreight. Priority search trees. *SIAM Journal on Computing*, 14(2):257–276, 1985.

[20] R. E. Tarjan. A class of algorithms that require nonlinear time to maintain disjoint sets. *Journal of Computer and System Sciences*, 18:110–127, 1979.

A Theorem of Bárány Revisited and Extended

Nabil H. Mustafa
Department of Computer Science
Université Paris-Est, ESIEE Paris, 93162
Noisy-Le-Grand Cedex, France
mustafan@esiee.fr

Saurabh Ray
Max-Planck-Institut für Informatik
66123 Saarbrücken, Germany
saurabh@mpi-inf.mpg.de

ABSTRACT

The colorful Carathéodory theorem [Bár82] states that given $d+1$ sets of points in $\mathbb{R}^d$, the convex hull of each containing the origin, there exists a simplex (called a 'rainbow simplex') with at most one point from each point set, which also contains the origin. Equivalently, either there is a hyperplane separating one of these $d+1$ sets of points from the origin, or there exists a rainbow simplex containing the origin. One of our results is the following extension of the colorful Carathéodory theorem: given $\lfloor d/2 \rfloor + 1$ sets of points in $\mathbb{R}^d$, and a convex object C, then either one set can be separated from C by a *constant* (depending only on d) number of hyperplanes, or there is a $\lfloor d/2 \rfloor$-dimensional rainbow simplex intersecting C.

Categories and Subject Descriptors

G.2 [**Combinatorics**]: Discrete Geometry

General Terms

Algorithms, Theory

Keywords

Discrete Geometry, Weak ϵ-nets, Caratheodory's theorem, Colorful Caratheodory theorem, Hadwiger-Debrunner (p, q)-theorem

1. INTRODUCTION

The goal of this paper is to study the behavior of *low-dimensional* simplices with respect to convex sets in $\mathbb{R}^d$. We examine a number of classical theorems in discrete geometry – Radon's theorem [Rad47], Carathéodory's theorem [Mat02], colorful Carathéodory theorem [Bár82] – and prove extensions that demonstrate the phenomenon of low-dimensional intersections.

Three classical theorems.

One of the starting theorems in discrete geometry is the following result. For a set $P \subset \mathbb{R}^d$, let $\text{conv}(P)$ denote the convex hull of P.

THEOREM 1 (RADON'S THEOREM). *Given any set P of $d+2$ points in $\mathbb{R}^d$, one can partition P into two sets P_1 and P_2 such that $\text{conv}(P_1) \cap \text{conv}(P_2) \neq \emptyset$.*

Note here that one of the two sets P_1 and P_2 can be large, e.g., P_1 can consist of $d+1$ points. So only the trivial bound $|P_1|, |P_2| \leq d+1$ holds. Therefore one cannot get a better upper bound on the dimension of the simplices $\text{conv}(P_1)$ or $\text{conv}(P_2)$.

We say a point p can be *separated* from a convex set C if there exists a hyperplane h with C and p in the interior of the two different halfspaces defined by h.

THEOREM 2 (CARATHÉODORY'S THEOREM). *If a convex set C intersects the convex hull of some point set P, then it also intersects a simplex spanned by P. Equivalently, either P can be separated from C with one hyperplane, or C intersects the convex hull of some $(d+1)$ points of P.*

Here we have stated the theorem in a slightly more general form; usually it is stated where C is just a point.

A beautiful extension of Carathéodory's theorem was discovered by Imre Bárány [Bár82]:

THEOREM 3 (COLORFUL CARATHÉODORY THEOREM). *Given $d+1$ sets of points $P_1, \ldots, P_{d+1}$ in $\mathbb{R}^d$ and a convex set C such that $C \cap \text{conv}(P_i) \neq \emptyset$ for all $i = 1, \ldots, d+1$, there exists a set Q with $C \cap \text{conv}(Q) \neq \emptyset$ and where $|Q \cap P_i| = 1$ for all i. Such a Q is called a 'rainbow set'. Equivalently, either some P_i can be separated from C with one hyperplane, or C intersects the convex hull of a rainbow set of $d+1$ points.*

This theorem is also commonly stated for the case where C is a point, but the above slight generalization follows immediately from Bárány's proof technique [Bár82]. Also, Carathéodory's theorem follows by applying the colorful Carathéodory theorem to $d+1$ copies of the same pointset.

Our results.

The starting point of our work is the following well-known generalization of the Erdös-Szekeres theorem:

THEOREM 4 (GENERALIZED ERDÖS-SZEKERES THEOREM). *Given positive integers d, k, n such that $\lceil d/2 \rceil + 1 \leq k \leq d$, there exists an integer $n_0 = ES_d(n, k)$ such that any set of n_0 points in $\mathbb{R}^d$ contains a subset P of size n with the following property: the simplex spanned by every $(d+1) - k$ points of P lies on the boundary of $\text{conv}(P)$. This statement is optimal, in the sense that this is not true for $k < \lceil d/2 \rceil + 1$ for arbitrarily large pointsets.*

The case $k = d$ simply corresponds to the Erdös-Szekeres theorem (that any large-enough set contains a lot of points in convex position). Of course the 'large-enough' size for the above theorem increases with decreasing k; but if one pays that price, one can get more properties. For example, for $d = 4, k = 3$, any large-enough set of points in $\mathbb{R}^4$ contains a large subset Q where every edge spanned by points of Q lies on $\texttt{conv}(Q)$.

We now observe that this immediately carries over to an at-first nonobvious extension of Radon's theorem: if one is willing to increase the number of points, then a better upper-bound can be achieved on the sizes of the Radon partition:

THEOREM 1.1. *Given an integer $\lfloor d/2 \rfloor + 1 \leq k \leq d$, any set P of $ES_d(d+2, k)$ points in $\mathbb{R}^d$ contains two sets P_1, P_2 such that $\texttt{conv}(P_1) \cap \texttt{conv}(P_2) \neq \emptyset$ and additionally, $|P_1|, |P_2| \leq k$. Furthermore, this is optimal in the sense that the statement does not hold for $k \leq \lfloor d/2 \rfloor$.*

PROOF. Apply Theorem 4 to P to get a set of $d+2$ points P'. Apply Radon's theorem to P' to get a partition $P_1, P_2 \subset P'$ whose convex hulls intersect. Now note that if $|P_1| > k$, then $|P_2| \leq (d+1) - k$. But then $\texttt{conv}(P_2)$ lies on the convex hull of P', and so cannot intersect $\texttt{conv}(P_1)$, a contradiction.

Optimality is obvious as $|P| \geq d+2$ for such a partition to exist (for P in general position), and so one set has to have at least $\lfloor d/2 \rfloor + 1$ points. □

Our first result is to show that a similar extension is possible for Carathéodory's theorem (Section 4):

THEOREM 1.2. *Given a set P of n points in $\mathbb{R}^d$ and a convex object C, either P can be separated from C by $O(d^4 \log d)$ hyperplanes (i.e., each $p \in P$ is separated from C by one of the hyperplanes), or C intersects the convex hull of some $\lfloor d/2 \rfloor + 1$-sized subset of P.*

We show the above result by relating this problem to another well-known problem; in fact we prove that the bounds for these two problems are within a factor of d of each other, a result of independent interest.

Unfortunately the above approach does not work for proving an extension of the colorful Carathéodory theorem, for which we give a proof using a different technique (Section 5):

THEOREM 1.3. *For any d, there exists a constant N_d such that given $k = \lfloor d/2 \rfloor + 1$ sets of points $P_1, \ldots, P_k$ in $\mathbb{R}^d$ and a convex object C, either one of the sets P_i can be separated from C by N_d hyperplanes, or there is a rainbow set of size k whose convex hull intersects C.*

Remark 1: Unlike the small polynomial bound in the extension of Carathéodory's theorem, the constant N_d is exponential in d. We leave improving N_d as an open problem.

Remark 2: The case where there are $d+1$ sets, and a set can be separated by one hyperplane is exactly the colorful Carathéodory theorem.

Remark 3: Note also that, as before, Theorem 1.3 implies the corresponding extension for Carathéodory's theorem, although with much worse quantitative bound than that given in Theorem 1.2.

Remark 4: Theorem 1.3 implies that given C and a set of red and blue points in $\mathbb{R}^3$, either the red set or the blue set can be separated from C by a constant N_3 number of planes. Or there is a red-blue edge intersecting C. In fact, for this case, a direct elementary proof with better constant exists, as follows. First, elementary considerations show that given a pointset and a convex set C, either there is a triangle Δ spanned by the points so that each edge of Δ intersects C or all the point can be separated from C using twelve hyperplanes. Now suppose that we have some red and blue points and a convex set C. Then applying the above result to each set of points, we conclude that either one of the sets can be separated from C with twelve planes or there is a red triangle and a blue triangle each of whose edges intersect C. For each vertex of these triangles consider the region on the boundary of C that it can see (considering C to be opaque). Since each red (resp. blue) edge intersects C, no two of the red (resp. blue) regions intersect, i.e., no three of the six regions intersect at a common point. Since the regions are pseudodisks, their intersection graph is planar. As $K_{3,3}$ is not planar, there is a red region and a blue region which do not intersect. This implies that the red-blue edge defined by the points corresponding to these regions intersects C.

We conclude with some open problems and future directions of research in Section 6.

2. MOTIVATION & RELATED PROBLEMS

Besides building on some basic theorems of discrete geometry, there are a few other reasons why statements of the type considered in this paper are useful.

Weak ϵ-nets.

The weak ϵ-net problem asks, given a set P of n points in $\mathbb{R}^d$, for the existence of a small-sized set $Q \subseteq \mathbb{R}^d$ such that any convex set containing at least ϵn points of P contains a point of Q. The current best upper-bound on the size of Q required is $\tilde{O}(1/\epsilon^d)$, while the best lower-bound is $\Omega(1/\epsilon \log^{d-1} 1/\epsilon)$. It is not clear what upper-bound to expect. But if one believes that "the truth is probably much smaller, maybe around $O(1/\epsilon^{\lfloor d/2 \rfloor})$" [Mat99], then consider the following natural approach from [MR08].

Pick a strong ϵ-net Q (with respect to the intersection of some constant k halfspaces). Now suppose a convex object C containing ϵn points of P is not hit by this strong ϵ-net. Then it must be that no set of k halfspaces separate C from points of Q; otherwise the intersection of these halfspaces contains C and so contains at least ϵn points without containing any point of Q, a contradiction. Intuitively, this means that the points of Q are 'close' to the boundary of C. It seems likely that then C must contain a large fraction of the volume (w.r.t. some measure) of some low-dimensional simplex spanned by points of Q (in which case we are done by picking weak nets of constant size within each low dimensional simplex). Unfortunately even the existence of an intersection of some low-dimensional simplex with C is not clear. The best one can hope for is a large intersection with a $\lfloor d/2 \rfloor$-dimensional simplex (better than this is not possible – to see this consider P to be the vertex set of a cyclic polytope and let C to be a slightly shrunk copy of the polytope). This is somewhat similar to the inductive approach in [ABFK92, CEG[+]93]. Hence the above discussion provides one reason to study the separation interaction of low-dimensional simplices spanned by some set P with a convex set C.

High-depth edge.

The following is a well-known problem open for more than 20 years: given a set P of n points in the plane, show that there exist two points of P such that any disk containing them contains at least $n/4$ points of P. The current-best bound is $n/4.73$ [EHSS89]. We now show a close relation to the problems we study in this paper.

Given P, map P to P' via a standard lifting to the paraboloid in

$\mathbb{R}^3$. So each disk in $\mathbb{R}^2$ corresponds to a downward-facing halfspace in $\mathbb{R}^3$. Now let C be the centerpoint region of P'. By definition, each plane supporting a facet of C has less than $n/4$ points on the side not containing C. Now our goal is to find two points, say p_1, p_2 of P' such that any downward-facing halfspace containing p_1p_2 must contain at least one point of C – and so contain at least $n/4$ points. For contradiction, assume that for every pair of points, the "slab" swept by projecting the segment p_1p_2 onto the xy-plane avoids C. Then one would prove the conjecture by showing that there exist 4 planes supporting facets of C that separate all the points of P' from C. This would yield a contradiction, as each such halfspace contains less than $n/4$ points.

Therefore, the goal is to show that either the vertical slab for a pair intersects C, or all the points of P' can be separated by 4 supporting hyperplanes. Then the easier question of separating P' from C with *any* separating hyperplanes if no *edge* intersects is exactly the extension of Carathéodory's theorem for $\mathbb{R}^3$.

Gallai-type problems for Pseudodisks.

Danzer proved that given any set of caps of a sphere in three dimensions such that every pair of them intersect, it is possible to pierce all of them using four points. While at first sight this appears unrelated to the problems we study, it is intimately related. Consider a sphere S and a set of points in $\mathbb{R}^3$ outside it. For each point p, the set of points $x \in S$ such that a tangent plane to the sphere at x separates S from p is a cap of S. Two points p and q can be separated from S using a single plane iff their corresponding regions intersect. Danzer's theorem therefore implies that either the set of points can be separated from S using four planes or there is an edge spanned by the points that intersects S. If we replace this sphere in $\mathbb{R}^3$ with an arbitrary convex set, then caps are replaced by regions that are pseudodisks. Danzer's result is believed to hold for pseudodisks as well but is currently known only with worse bounds. Therefore such separation problems can be seen as generalized Gallai-type problems.

3. PRELIMINARIES

In this section, we describe some basic results in discrete geometry that we will need.

Hadwiger-Debrunner (*p,q*)-theorem [AK92]. Given a set S of convex sets in $\mathbb{R}^d$ such that out of every $p \geq d+1$ sets, there is a point common to $q \geq d+1$ of them, then S has a hitting set of finite size and the minimum size of such a set is denoted by $HD_d(p,q)$ (so this is independent of $|S|$).

ϵ-nets w.r.t. halfspaces in $\mathbb{R}^d$ [HW87]. Given a set P of n points in $\mathbb{R}^d$ and a parameter $\epsilon > 0$, a set $Q \subseteq P$ is an ϵ-net w.r.t. halfspaces if any halfspace containing at least ϵn points of P contains a point of Q. A famous theorem of Haussler-Welzl [HW87] shows that ϵ-nets of size $O(d/\epsilon \log d/\epsilon)$ exist, independent of n. This bound was later improved in [KPW92] to a near-optimal bound of $(1+o(1))(\frac{d}{\epsilon}\log(1/\epsilon))$.

Centerpoint depth [Mat02]. Given any set P of n points in $\mathbb{R}^d$, the *Tukey depth* of a point $q \in \mathbb{R}^d$ is the minimum number of points of P contained in any halfspace containing q. It is known that there always exists a point of Tukey depth at least $n/(d+1)$.

In general, the set of points of Tukey depth at least βn form a convex region called the β-deep region of P. The β-deep region is non-empty for any $\beta \leq 1/(d+1)$. It is the intersection of all halfspaces containing more than $(1-\beta)n$ points of P. Each facet of this region is supported by a hyperplane that passes through d points of P. We will need the following fact:

Fact 1 [PA95]: If P is a set of n points and h is a hyperplane defining a facet of the β-deep region C of P, then the halfspace defined by h that does not intersect the interior of C contains less than βn points of P.

4. PROOF OF EXTENDED CARATHÉODORY'S THEOREM

We first show that this problem is related to another problem involving low-dimensional simplices.

Let $f(d)$ be the smallest positive number such that for any set P of points in $\mathbb{R}^d$, there exists a $\lfloor d/2 \rfloor + 1$-sized subset $P' \subseteq P$ such that any halfspace containing P' contains at least $|P|/f(d)$ points of P.

Let $g(d)$ be the smallest positive number such that given any set P of points in $\mathbb{R}^d$ and a convex set C, if P cannot be separated from C using at most $g(d)$ hyperplanes, then C must intersect the convex hull of some $\lfloor d/2 \rfloor + 1$ size subset of P.

We now show that $g(d)$ and $f(d)$ are related within a factor of d.

THEOREM 4.1. $g(d) \leq d \cdot f(d) \log f(d)$. *In other words, given a set P of points in $\mathbb{R}^d$ and a convex set C such that P cannot be separated from C by $df(d)\log f(d)$ hyperplanes, then C must intersect the convex hull of some $\lfloor d/2 \rfloor + 1$ points of P.*

PROOF. Assume that no convex hull of any $\lfloor d/2 \rfloor + 1$ points of P intersects C. Then we show that P can be separated from C using $df(d)\log f(d)$ hyperplanes.

CLAIM 4.2. *Let P be a weighted set of points in $\mathbb{R}^d$, with weight of the point $p_i \in P$ to be w_i. Assume all w_i's are rationals, and let $W = \sum w_i$. If the convex hull of no $\lfloor d/2 \rfloor + 1$ points of P intersects C, then there exists a hyperplane separating points of total weight at least $\frac{W}{f(d)}$ from C.*

PROOF. As each w_i is a rational, assume $w_i = \hat{w}_i/D$, where $\hat{w}_i$ and D are integers. Let Q be the pointset gotten by replacing each point p_i with $\hat{w}_i$ copies of p_i. Crucially, if the convex hull of no $\lfloor d/2 \rfloor + 1$ subset of P intersects C, then the convex hull of no $\lfloor d/2 \rfloor + 1$ subset of Q can intersect C. Take the $\lfloor d/2 \rfloor + 1$-sized subset Q' of Q such that any halfspace containing Q' contains at least $|Q|/f(d)$ points of Q. As the convex hull of Q' does not intersect C, there is a halfspace h which does not intersect C and contains Q'. Let P' be the set of points of P contained in h. Then h contains exactly $\sum_{p_i \in P'} \hat{w}_i$ points of Q, which by definition of Q' must be at least $|Q|/f(d)$. Then the sum of weights of points of P contained in h is bounded by

$$\sum_{p_i \in P'} w_i = \frac{\sum_{p_i \in P'} \hat{w}_i}{D} \geq \frac{|Q|/f(d)}{D} = \frac{(\sum \hat{w}_i)/f(d)}{D}$$

$$= \frac{\sum w_i D}{Df(d)} = \frac{W}{f(d)}$$

□

Discretize the set of all combinatorially distinct hyperplanes separating some subset of P from C to get a set $\mathcal{H} = \{h_1, \ldots, h_m\}$ of $O(|P|^d)$ hyperplanes. Now consider assigning weights $w(h_i)$ to each halfspace such that the total weight $\sum w(h_i)$ is minimized, and the sum of weights of halfspaces containing any point of P is at least 1. Let $W(\mathcal{H})$ denote the minimum value.

Similarly, assign weights $w(p_i)$ to each point of P such that the total weight $\sum w(p_i)$ is maximized, and the sum of weights of points contained in any halfspace $h \in \mathcal{H}$ is at most 1. Let $W(P)$ denote the maximum value. Then the above two problems are dual to each other (as linear programs), and so by the Strong Duality Theorem, $W(\mathcal{H}) = W(P)$.

Now note that $W(P) \leq f(d)$: by Claim 4.2, there exists a halfspace in $\mathcal{H}$ of weight at least $W(P)/f(d)$, which by the definition of $W(P)$ is at most 1.

Therefore there exists an assignment of weights to halfspaces in $\mathcal{H}$ such that $W(\mathcal{H}) \leq f(d)$, and each point is contained in halfspaces of total weight at least 1. Now using the ϵ-net theorem for halfspaces [KPW92], with $\epsilon = 1/W(\mathcal{H})$, there exists a set of

$$d/\epsilon \log 1/\epsilon = dW(\mathcal{H}) \log W(\mathcal{H}) = df(d) \log f(d)$$

halfspaces of $\mathcal{H}$ containing all points of P. As all halfspaces in $\mathcal{H}$ were separating halfspaces, we are done.

Remark 1: The above technique is similar to the one used in the proof of Hadwiger-Debrunner (p, q) theorem [AK92], with some crucial differences. In their use, they get an exponential bound, which we are able to avoid due to three reasons: ϵ-nets for halfspaces have a near-linear bound, avoiding double-counting arguments that they use, and finally, the weighted version (Claim 4.2) gives exactly the same quantitative bound as the unweighted version.

Proof of Theorem 1.2: the paper [SSW08] proves that $f(d) \leq O(d^3)$. And the proof is complete by using Theorem 4.1.

Similarly we now show that a bound on $g(d)$ gives an upper-bound on $f(d)$:

THEOREM 4.3. $f(d) \leq d \cdot g(d)$. *In other words, given a set P of points in $\mathbb{R}^d$, there always exists a subset P' of size $\lfloor d/2 \rfloor + 1$ such that any halfspace containing P' contains at least $|P|/dg(d)$ points of P.*

PROOF. Consider the β-deep region C of P; by the Centerpoint theorem, for $\beta \leq 1/(d+1)$, such a region always exists. Now we claim that for $\beta = 1/dg(d)$, there exists a $\lfloor d/2 \rfloor + 1$-sized subset P' whose convex hull intersects C. Then any halfspace containing P' contains at least one point of C, and so contains at least $|P|/dg(d)$ points by the definition of the centerpoint region.

Otherwise, for contradiction assume that the convex hull of no $\lfloor d/2 \rfloor + 1$-sized subset intersects C. Then by definition of $g(d)$, P can be separated from C using $g(d)$ hyperplanes, say the set $\mathcal{H}$.

Now any halfspace not intersecting C contains less than $d \cdot \beta|P|$ points: each halfspace supporting a facet of C contains less than $\beta|P|$ points, and any other halfspace not intersecting C is contained in the union of at most d halfspaces supported by facets of C.

Therefore each halfspace of $\mathcal{H}$ contains less than $d \cdot \beta|P|$ points. And so the union of halfspaces in $\mathcal{H}$ contains less than $g(d) \cdot d \cdot \beta|P|$ points of P, a contradiction for $\beta = 1/dg(d)$. ☐

5. PROOF OF EXTENDED COLORFUL CARATHÉODORY THEOREM

The goal of this section is to prove Theorem 1.3. We use a slightly different language for convenience: instead of saying that "a point set P can be separated from a convex body C using k hyperplanes", we say that "there exists a polyhedron $\mathcal{Q}$ with k facets such that $C \subseteq Q$ and $Q \cap P = \emptyset$". In such a case we also say that *$\mathcal{Q}$ separates P from C*. We re-state Theorem 1.3 in this language.

THEOREM 5. *For any positive d and $l > \lfloor d/2 \rfloor$, there exists a constant $N_{d,l}$ s.t. the following is true. Given any compact convex body C and l finite sets of points $P_1, \ldots P_l$ in $\mathbb{R}^d$, at least one of the following holds:*

1. *There exists a polyhedron $\mathcal{Q}$ with at most $N_{d,l}$ facets such that for some i, $\mathcal{Q}$ separates P_i from C.*
2. *There exists a rainbow subset $P' \subseteq \cup_{i=1}^{d} P_i$ whose convex hull intersects the interior of C.*

The approach of the previous section does not work for proving the extension of the colorful Carathéodory's theorem. The method gave *one* low-dimensional simplex intersecting C, or the existence of a hyperplane separating many points. Unfortunately with multiple sets, one needs to find many such intersecting simplices for each set P_i. Therefore one is forced to use a more Ramsey-theoretic technique, and this causes the constant to become exponential in d.

We next present some preliminary definitions and claims required to prove the above theorem.

Call a convex body C *fine* if it is compact and its boundary ∂C is smooth and has positive curvature everywhere. Let C be a fine convex body and let P be a finite set of points in $\mathbb{R}^d$. We say that a point p can *see* a point y if the relative interior of the segment py does not intersect C. For any $p \in P$, let U_p be the set of points in ∂C that p can see.

Let h_y be the tangent plane to C at the point $y \in \partial C$ and let h_y^+ be the closed halfspace defined by it that contains C. Observe that any point $p \in P$ sees a point $y \in \partial C$ iff $p \notin \text{int}(h_y^+)$, where $\text{int}(S)$ denotes the interior of the set S.

LEMMA 5.1. *For any positive numbers d and $t \geq d$, there exists a constant $H_{t,d}$ such that given any fine convex body C and a finite set of points P such that $P \cap C = \emptyset$, at least one of the following hold:*

- *There exists a set $X \subset \partial C$ of size at most $H_{t,d}$ such that each point in P is seen by some $x \in X$.*
- *There is a subset $P' \subseteq P$ of size at least t such that no $y \in \partial C$ sees more than $d-1$ points in P'.*

PROOF. Fix any point $\nu \in \partial C$ as a reference point. Let h_ν be the tangent plane to C at ν. Let h be the unique tangent hyperplane parallel to h_ν such that C is contained in the strip between h_ν and h. Let π be the continuous bijective map that maps any $y \in \partial C \setminus \{\nu\}$ to $l(\nu, y) \cap h$, where $l(\nu, y)$ denotes the line through ν and y.

For any $p \in P$ that does not see ν (i.e., $\nu \notin U_p$), let $V_p = \pi(U_p) = \{\pi(y) : y \in U_p\}$ and let $\mathcal{V} = \{V_p : p \in P,\ p \text{ does not see } \nu\}$. Clearly, for each $p \in P$ that does not see ν, V_p lies on the plane h, is convex and has dimension $d-1$. Note that if two points $p, q \in P$ see a point $y \in \partial C$ then $y \in U_p \cap U_q$. So $\pi(y) \in V_p \cap V_q$.

Suppose that the second part of the theorem does not hold; i.e., in every subset of P of size t, there are at least d points which can be seen by a single point $y \in \partial C$. Equivalently, any subset of $\mathcal{V}$ of size t has at least d sets which have a common intersection. By the Hadwiger-Debrunner theorem [AK92], there exists a constant $HD_{d-1}(t, d)$ of points in h that hit all the sets in $\mathcal{V}$. Let X' be the set of these points. Let $X = \pi^{-1}(X') \cup \{\nu\}$. Each point in P is seen by at least one point in X (if V_p is hit by the point $y' \in h$, then $p \in P$ is seen by $\pi^{-1}(y') \in \partial C$). The theorem is therefore proved by setting $H_{t,d} = HD_{d-1}(t, d) + 1$. ☐

Now we can finish the proof of the main theorem of this section:

PROOF. Let $P = \cup_{i=1}^{l} P_i$. We set $N_{d,l} = H_{t,d}$ for some t to be fixed later.

If $P \cap C \neq \emptyset$ then the second part of the theorem is trivially satisfied. We therefore assume that $P \cap C = \emptyset$. Without loss of generality we also assume that C is fine since we can always find a fine convex body C' that contains C and does not intersect P and furthermore for each point $y' \in C'$, there is a point $y \in C$ such that the Euclidean distance between y and y' is smaller than any prescribed $\delta > 0$. Proving the theorem for such C's also proves it for arbitrary closed convex bodies.

For each i, apply Lemma 5.1 to C and P_i with the parameter t. This gives us either a set X_i of at most $H_{t,d}$ points in ∂C such that each point in P_i is seen by at least one of these or we get a set $Q_i \subseteq P_i$ of t points so that no d of them is seen by the same point in ∂C. If the first possibility happens for some j, then $\bigcap_{x \in X_j} h_x^+$ gives us the polyhedron $\mathcal{Q}$ with at most $H_{t,d}$ facets and where $\mathcal{Q}$ contains C while P_j lies outside $\mathcal{Q}$. This satisfies the first part of the Theorem and we're done.

We therefore assume the second possibility for each i; namely, each P_i has a subset Q_i of t points such that no d points of Q_i are seen by the same point of ∂C. Equivalently, the convex hull of any d points of Q_i intersects C.

Let $Q = \cup_{i=1}^{l} Q_i$. Consider any rainbow set $R \subseteq Q$ with one point from each Q_i. There are t^l such sets. If the convex hull of R intersects C, then the second part of the theorem is satisfied, and we're done. Assume for contradiction that this is not the case. Then for each rainbow set R, there exists a hyperplane h separating R from C. The closed halfspace h^- bounded by h and not intersecting C contains at most $d-1$ points from any particular Q_i due to the fact that any d-sized subset of Q_i intersects C. Therefore $|h^- \cap Q| \leq (d-1)l$, and hence is a $\leq k$-set of Q with $k = l(d-1)$.

If no rainbow set intersects C, then we get such a $\leq k$-set for each rainbow set R of size l. As there are t^l such rainbow sets, we get $t^l \leq k$-sets. However each such $\leq k$-set can be overcounted at most $\binom{k}{l} = \binom{l(d-1)}{l}$ times. This implies that there are at least $L(t) = t^l / \binom{l(d-1)}{l}$ distinct $\leq k$-sets. On the other hand, it is known that the number of $\leq k$-sets of a set of n points in $\mathbb{R}^d$ is at most $O(n^{\lfloor d/2 \rfloor}(k+1)^{\lceil d/2 \rceil})$ [Mat02]. This gives an upper bound of $U(t) = O((tl)^{\lfloor d/2 \rfloor}((d-1)l+1)^{\lceil d/2 \rceil})$ on the number of $\leq k$-sets. Since $l > \lfloor d/2 \rfloor$, for some large enough t depending only on l and d, $L(t) > U(t)$. Thus we get a contradiction implying that one of the rainbow sets must intersect C. □

6. CONCLUSIONS

We think that the phenomenon studied in this paper is much more widely applicable. In fact, statements of this type might shed light on interactions between several classical theorems (see below). Besides improving quantitative bounds presented in this paper – for both extended Carathéodory and the extended colorful Carathéodory theorems – we end with a number of questions for which we think the answer is affirmative:

1. **Extension of Kirchberger's theorem.** Given a set P of n red points and n blue points in $\mathbb{R}^d$ either there exist a constant number of hyperplanes $\mathcal{H}$ such that every red-blue pair is separated by a plane in $\mathcal{H}$, or a $\lfloor d/2 \rfloor$-dimensional red simplex intersects a $\lfloor d/2 \rfloor$-dimensional blue simplex.

2. **Extension of Tverberg's theorem.** Given a set P of n points in $\mathbb{R}^d$, there exists a Tverberg partition on a large subset of P where two sets have size $\lfloor d/2 \rfloor + 1$. It is not too hard to see that this implies a version of the extension of Carathéodory's theorem.

3. **Extension of Colored Tverberg's theorem.** Given $(d+2)$ sets $P_1, \ldots, P_{d+2}$, there exists a colored Tverberg partition on a large subset where two sets have size $\lfloor d/2 \rfloor + 1$. Again, this version implies the extension of the colorful Carathéodory theorem.

7. ACKNOWLEDGEMENTS

We thank an anonymous reviewer for meticulous and very insightful comments. The first author would also like to thank Janos Pach for his support and encouragement. Part of this work was done by the first author while visiting the wonderful algorithms group at the Max-Planck Institute for Informatik in Saarbrucken.

References

[ABFK92] Noga Alon, Imre Bárány, Zoltán Füredi, and Daniel J. Kleitman. Point selections and weak ε-nets for convex hulls. *Combinatorics, Probability & Computing*, 1:189–200, 1992.

[AK92] N. Alon and D. Kleitman. Piercing convex sets and the Hadwiger Debrunner (p; q)-problem. *Adv. Math.*, 96(1):103–112, 1992.

[Bár82] Imre Bárány. A generalization of Carathéodory's theorem. *Discrete Mathematics*, 40(2-3):141–152, 1982.

[CEG+93] Bernard Chazelle, Herbert Edelsbrunner, Michelangelo Grigni, Leonidas Guibas, Micha Sharir, and Emo Welzl. Improved bounds on weak ε-nets for convex sets. In *Proceedings of the twenty-fifth annual ACM Symposium on Theory of Computing*, pages 495–504, 1993.

[EHSS89] H. Edelsbrunner, N. Hasan, R. Seidel, and X.J. Shen. Circles through two points that always enclose many points. *Geom. Dedicata*, 32:1–12, 1989.

[HW87] D. Haussler and E. Welzl. Epsilon-nets and simplex range queries. *Discrete Comput. Geom.*, 2:127–151, 1987.

[KPW92] J. Komos, J. Pach, and G. Woeginger. Almost tight bounds for epsilon nets. *Discrete & Computational Geometry*, pages 163–173, 1992.

[Mat99] J. Matousek. Piercing and selection theorems in convexity, 1999.

[Mat02] Jirí Matousek. *Lectures in Discrete Geometry*. Springer-Verlag, New York, NY, 2002.

[MR08] N. Mustafa and S. Ray. Weak ϵ-nets have a basis of size $O(1/\epsilon \log 1/\epsilon)$. *Comp. Geom: Theory and Appl.*, 40(1):84–91, 2008.

[PA95] Janos Pach and Pankaj K. Agarwal. *Combinatorial Geometry*. John Wiley & Sons, New York, NY, 1995.

[Rad47] R. Rado. A theorem on general measure. *J. London. Math. Soc.*, 21:291–300, 1947.

[SSW08] Shakhar Smorodinsky, Marek Sulovský, and Uli Wagner. On center regions and balls containing many points. In *Proceedings of the 14th annual International Conference on Computing and Combinatorics*, COCOON '08, pages 363–373, 2008.

Improved Pointer Machine and I/O Lower Bounds for Simplex Range Reporting and Related Problems

Peyman Afshani*

MADALGO†, Department of Computer Science
Aarhus University
Aarhus, Denmark
peyman@cs.au.dk

ABSTRACT

We investigate one of the fundamental areas in computational geometry: lower bounds for range reporting problems in the pointer machine and the external memory models. We develop new techniques that lead to new and improved lower bounds for simplex range reporting as well as some other geometric problems.

Simplex range reporting is the problem of storing n points in a data structure such that the k points that lie inside a query simplex can be outputted efficiently. This is one of the fundamental and extensively studied problems in computational geometry. Currently, the best data structures for the problem achieve $Q(n)+O(k)$ query time using $\tilde{O}((n/Q(n))^d)$ space in which the $\tilde{O}(.)$ notation either hides a polylogarithmic or an n^ε factor for any constant $\varepsilon > 0$, (depending on the data structure and $Q(n)$). The best lower bound on this problem is due to Chazelle and Rosenberg who proved a space lower bound of $\Omega(n^{d-\varepsilon-d\gamma})$ for pointer machine data structures that can answer queries in $O(n^\gamma + k)$ time.

For data structures with $Q(n) + O(k)$ query time, we improve the space lower bound to $\Omega((n/Q(n))^d/2^{O(\sqrt{\log Q(n)})})$. Not only this reduces the overhead from polynomial to sub-polynomial, it also offers a smooth trade-off curve. For instance, for polylogarithmic values of $Q(n)$, our lower bound is within a $o(\log n)$ factor of the conjectured trade-off curve.

By a simple geometric transformation, we also improve the best lower bounds for the halfspace range reporting problem. Furthermore, we also study the external memory model and offer a new framework for proving lower bound in this model. For the first time we show that answering simplex range reporting queries with $Q(n) + k/B$ I/Os requires $\Omega(B(n/(BQ(n)))^d/2^{O(\sqrt{\log Q(n)})})$ space in which B is the block size.

*This research was done while the author was a post-doctoral research at Dalhousie University, supported by Natural Sciences and Engineering Research Council of Canada (NSERC) through the Postdoctoral Fellowships Program (PDF).

†Center for Massive Data Algorithmics, a center of the Danish National Research Foundation.

SCG'12, June 17–20, 2012, Chapel Hill, North Carolina, USA.

Categories and Subject Descriptors

E.1 [**Pointer Machine Data Structures**]: Lower Bounds; I.3.5 [**Range Searching**]:

Keywords

Simplex range reporting, lower bounds, the I/O model

1. INTRODUCTION

Range reporting is one of the most fundamental areas in computational geometry. In a range reporting problem the goal is to preprocess an input set of n points such that given a query range r, the points inside r can be found efficiently. Three types of queries are studied much more than others: simplices, halfspaces, and axis-aligned boxes and they lead to simplex range reporting, halfspace range reporting, and orthogonal range reporting, respectively.

The progress on range reporting has been significant and research in this area has resulted in development of many classical tools and techniques that have been proved useful in many other areas of computational geometry as well. However, there are still many important open problems remaining, even for classical problems such as simplex range reporting. Questions of lower bound are among them. In this paper, we provide improved lower bounds for simplex range reporting in two fundamental models of computation, the pointer machine model and the I/O model.

1.1 Background

Many of the techniques that are used to solve simplex range reporting are in fact developed for the general problem of simplex range searching. In this problem, the input points are assigned weights from a semi-group and for a query simplex r, the output is the sum of the semi-group weights of the points inside r. The model of computation assumes that the sum of any number of semi-group elements fits into one register and thus the output has constant size. For more details, refer to surveys on range searching [5, 3].

This is different from simplex range reporting in which the output can have non-constant size. The greatest distinction, however, lies in the way data structures are built. A data structure for a range searching problem in the semi-group model stores a number of precomputed sums and this takes space equal to the number of sums; the query is answered by selecting some of the stored precomputed sums that cover the query range and then summing them. In a range reporting data structure storing precomputed answers would take too much space but on the other hand, the existence of an output term allows the data structure to spend more time on queries that produce large outputs. This idea, known as the "filtering search" [8], is fundamental when designing efficient range reporting data structures. Fortunately, many of the existing techniques for building data structures (such as partition trees) can be easily adapted to apply to both problems. Thus, very few papers that deal with building data structures distinguish these two variants. Unfortunately, from a lower bound point of view, the two problems are fundamentally different and there is no general technique that can be applied to both problems.

Previous results.

Much of the focus on simplex range searching is devoted to building linear-size or near linear-size data structures. Such results were published as early as 1982 [22] and they offer $O(n^\gamma + k)$ query time, (the constant γ depends on the data structure) while using linear space. After a number of results that lowered the exponent γ in two and three dimensions, significant breakthroughs came via works of researchers such as Welzl, Chazelle, Sharir, and Matoušek [21, 15, 20] culminating in Matoušek's work [19] that presented a data structure with linear space and $O(n^{1-1/d} + k)$ query time (also an $O(n^{1-1/d})$ query time for semi-group version). Very recently, Timothy Chan proved a new partition theorem that simplifies Matoušek's approach and offers other advantages [7]. In the external memory model and using linear space, it is possible to answer simplex range reporting queries with $O((n/B)^{1-1/d+\varepsilon} + k/B)$ I/Os in which ε is any positive constant and B is the block size [4].

At the other side of the spectrum, data structures with polylogarithmic query time were obtained through *cuttings* [12, 16, 11]. Chazelle et al. [15] showed that using $O(n^{d+\varepsilon})$ space it is possible to achieve query time of $O(\log n + k)$. Matoušek showed query time of $O(\log^{d+1} n + k)$ is possible using $O(n^d)$ space [19]. By combining these with linear space data structures, it is possible to strike a trade-off between query time and space usage: one can build a data structure that uses $m + O(n)$ space, for any parameter m, that can answer queries in $O((n/m^{1/d}) \log^{d+1}(m/n) + k)$ time [19].

All of the previous simplex range reporting results can be implemented in the pointer machine model of computation and currently there are no better bounds in the RAM model and it seems random accesses do not help to reduce the complexity of this problem.

While there are a number of lower bounds for range searching problems (in semi-group or even group models), we are aware of only three papers that are relevant to simplex range reporting. The first paper by Erickson [17] deals with hyperplane emptiness queries. As such queries can be answered using simplex range reporting queries, his lower bounds also apply to simplex range reporting problem. His results are proved in a restricted model of computation known as the *partition graph model* (intuitively, it models data structures that work by partitioning the space). Let $S(n), Q(n)$, and $P(n)$ be the space complexity, query time, and the preprocessing time of a data structure that can answer hyperplane emptiness queries, respectively. He proved that $P(n)Q(n)^{\frac{(d+2)(d-1)}{2}} = \Omega(n^d), P(n)Q(n)^{\frac{2}{d-1}} = \Omega(n^{\frac{d+2}{d}})$, and $S(n)Q(n)^d = \Omega((n/\log n)^{d-\frac{(d-1)}{(d+1)}})$. Unfortunately, the exponent of none of these bounds seems to be optimal except in special cases (i.e., when $d = 2$, or $P(n) = O(n \log^{O(1)} n)$, or $Q(n) = \log^{O(1)} n$)). However, these results are significant because proving lower bounds for emptiness queries is extremely difficult and these are the only known non-trivial lower bounds for such queries. The second paper is due to Chazelle and Liu [13] in which they proved that given a planar subdivision with n vertices, computing all the k edges intersected by a query line in $Q(n) + O(k)$ time requires $\Omega(n^2/Q(n)^2)$ space. While they do not mention simplex range reporting in the paper, the particular input instance they build for this problem does also give the same lower bound for simplex range reporting queries in the plane[1]. Unfortunately, the technique does not seem to generalize to higher dimensions.

The third paper by Chazelle and Rosenberg [14] is the most relevant one. It operates in a general pointer machine model. It builds on Chazelle's earlier works [9, 10] and offers a space lower bound of $\Omega(n^{d-d\gamma-\varepsilon})$ assuming the data structure can answer queries in $O(n^\gamma + k)$ time in which ε is any positive constant. This lower bound is a polynomial factor away from the (conjectured optimal) trade-off curve.

Our results.

Simplex range reporting.

We show any pointer machine data structure that can answer simplex range reporting queries in $Q(n) + O(k)$ time needs to consume $\Omega((n/Q(n))^d / 2^{O(\sqrt{\log Q(n)})})$ space. This improves Chazelle and Rosenberg's lower bound and achieves a sub-polynomial overhead. Furthermore, it offers a smooth trade-off curve. For instance, for polylogarithmic values of $Q(n)$, we are only a sub-logarithmic factor away from the the trade-off curve $S(n) = \Omega((n/Q(n))^d)$!

Slab stabbing.

Our lower bound is first obtained for a different problem we call the *slab stabbing problem*: preprocess a set of slabs (the region between two parallel hyperplanes) such that given a query point, the slabs containing the query point can be reported efficiently.

Halfspace range reporting.

Using a known geometric transformation [17], our slab stabbing lower bound offers the same space/query trade-off (i.e., $S(n) = \Omega((n/Q(n))^d / 2^{O(\sqrt{\log n})})$) for halfspace range reporting in $d(d+3)/2$ dimensions. This reduction from slab stabbing is the only known technique that gives lower bounds for halfspace range reporting, making our lower bound the best known lower bound for this problem as well.

External memory.

Our main result here is a new framework for proving range reporting lower bounds in the external memory model. Our lower bound for simplex range reporting in the ex-

[1]It is possible the authors are unaware of this important consequence of their results!

ternal memory model follows easily from this framework: answering simplex range reporting queries in the external memory model using $Q(n) + O(k/B)$ I/Os requires $S(n) = \Omega(B(n/(BQ(n)))^d/2^{O(\sqrt{\log Q(N)})}$ space in which B is the block size. This is the first known external memory lower bound for simplex range reporting.

2. A SIMPLE GEOMETRIC LOWER BOUND FRAMEWORK

In this section, we describe our simple framework that is tailored for geometric problems. First, we describe our model of computation.

We operate in a powerful variant of the pointer machine model which also has been employed by the previous lower bound papers [10]. Consider an abstract reporting problem in which given an input set S containing of n elements, each query is to output a subset of S. A data structure that solves such an abstract reporting problem is represented by a directed graph G with maximum out-degree of two. The set of vertices of G, $V(G)$, represents the set of memory cells used in the data structure. Each memory cell stores an input element as well as two pointers to two other memory cells. Any other information can be stored and accessed for free by the data structure. There are two main restrictions: First, it is assumed that to output an element $p \in S$, the query algorithm must visit a vertex $v \in V(G')$ that stores p. Second a cell can only be accessed through navigating pointers and thus random accesses are disallowed. This implies that at the query time a connected subgraph G' of G is explored. The query time is obviously $\Omega(|V(G')|)$. At the query time, we call such a vertex v an *output vertex*. We are only interested in query times of the form of $Q(n) + O(k)$ in which k is the output size and $Q(n)$ is a search overhead that depends on the input size. Thus we must have $|V(G')| \leq Q(n) + O(k)$.

We assume S is set of n geometric regions inside a d-dimensional square Q of volume one. The queries are points inside Q and the output of a query q is the set of regions that contain q. We call such a problem a *geometric stabbing problem*. The main result of this section is the following.

THEOREM 1. *Assume we have a data structure for a geometric stabbing problem that uses at most $S(n)$ space and answers queries in at most $Q(n) + O(k)$ time in which n is the input size and k is the output size. Assume for this problem we can construct an input set S of n regions such that (i) every point of Q is contained in exactly t regions in which t is a parameter greater than $Q(n)$ and (ii) the volume of the intersection of every α regions is at most v, for two parameter $\alpha < t$ and v. Then, we must have $S(n) = \Omega(tv^{-1}/2^{O(\alpha)}) = \Omega(Q(n)v^{-1}/2^{O(\alpha)})$.*

To prove the above theorem, we consider the graph G that represents the memory layout of the data structure. We say a subgraph H of G is a *hub (of size α)* if H is a tree with at most α vertices. Before giving the intuition behind our proof, we bound the maximum possible number of hubs.

LEMMA 1. *The number of hubs is $O(|V(G)|2^{2\alpha})$.*

PROOF. Consider a vertex v. We count the maximum number of hubs with v as their root. Let u_1 and u_2 be the two nodes that v points to. If H is a hub of size i rooted at v, and H_1 and H_2 are subtrees of H rooted on u_1 and u_2, then the size of H_2 is $i - j - 1$ in which j is the size of H_1. Let $f(i)$ denote the maximum number of hubs of size i that can share a common root. We have

$$f(i) = \sum_{0 \leq j < i} f(j)f(i-j-1)$$

and $f(0) = 1$. The solution to this recursive formula is known as a Catalan number and it is not too hard to show that $f(i) \leq 2^{2i}$ (e.g., it counts number of expressions with i matching parentheses which is less than the number of binary strings of size $2i$). Summing this for $i = 1$ to α gives the maximum number of hubs with a common root, which is $O(2^{2\alpha})$. Thus, the total number of hubs is $O(|V(G)|2^{2\alpha})$. □

Now consider the subgraph G' explored at the query time. As we are operating in the pointer machine model, the query algorithm starts from a root vertex of G and begins exploring G; the final set of vertices explored forms $V(G')$. For every $v \in V(G')$, we mark the in-edge of v that is used to visit v for the first time during the query time. Note that each vertex (except the root) has a single marked in-edge. Thus, the marked edges form a spanning tree T of G'. We call a subtree H of T an *α-heavy hub* if size of H is $O(\alpha)$ and it contains α output vertices. Note that the definition of an α-heavy hub only makes sense with respect to the subgraph explored at the query time.

Our general attack plan is the following. Intuitively, a hub represents a set of vertices that can be explored together with little cost. If the output size is large enough, then query time is forced to be $O(k)$. Thus, intuitively, the query algorithm needs to produce many output hubs (this is captured by the next lemma). However, an output hub that contains α output regions implies that the query point is inside all of the α regions which subsequently means that the query is inside the geometric region formed by the α regions. If this region is small, then it follows that the output hub will be useful only for a small number of queries which in turn means the data structure must contain a lot of hubs. Since the number of hubs is proportional to the size of the data structure (by the previous lemma), it follows at the data structure must be large.

LEMMA 2. *Consider the subgraph G' explored at the query time. If G' contains $\Omega(|G'|)$ output vertices, then there exists a set $\mathcal{H}$ that contains $\Omega(|G'|/\alpha)$ α-heavy hubs.*

PROOF. Let T be the spanning tree of G' that is formed by the marked edges and let k be the number of output vertices in T. We find $\mathcal{H}$ by playing the following game. We initialize the game by placing one pebble on each output vertex of T and assign a size value to every node of T which is initially set to one. Thus, in total k pebbles are placed on T and total size of all the vertices is $|V(T)| = O(k)$.

At each step of the game we operate on a leaf v of T. Let u be the parent of v. In a typical step, we remove all the pebbles on v and place them on u and then add the size of v to u. The exception is when there are α pebbles on v. In this case, we do not perform the typical step and instead look at the size of v: if it is greater than $c\alpha$, for a large enough constant c, then we delete v and throw away all the existing pebbles on v and we call these pebbles *wasted.* Otherwise, we have found a hub of size at most $c\alpha$ rooted on v that contains α output nodes. This is an α-heavy hub and thus we add it to $\mathcal{H}$ and then remove v and its pebbles from the game. At the end, we will have a set of at most $\alpha - 1$ pebbles left at the root of T and all the other pebbles will either be wasted or placed in a hub in $\mathcal{H}$. As wasting α pebbles corresponds to deleting $c\alpha$ vertices of T, it follows that at most $k/2$ pebbles will be wasted if c is chosen large enough. The rest of the pebbles, $k/2-\alpha+1$, will be in hubs. This means $\mathcal{H}$ contains at least $(k/2 - \alpha + 1)/\alpha$ hubs. □

Now it is useful to switch to the geometric view of the problem. Remember that in a geometric stabbing problem, the input is a set of geometric regions inside Q and the queries are points of Q. The output of the query is the subset of regions that contain the query point. With this in mind, consider an output hub $H \in \mathcal{H}$ and let $v_1, \dots, v_\alpha$ be the output nodes in H and let r_i be the region stored by v_i. We must have that the query point is inside r_i for $1 \le i \le \alpha$. We call the region formed by intersecting r_i for $1 \le i \le \alpha$ the *geometric region of* H. If the volume of the geometric region of H is small, then H can only be useful for a small set of queries, meaning, the data structure must contain many hubs to cover all the queries.

PROOF OF THEOREM 1. By Lemma 1, the total number of hubs of size $O(\alpha)$ is $O(|V(G)|2^{O(\alpha)})$. Consider a hub H of size $O(\alpha)$. There are $\binom{O(\alpha)}{\alpha} = O(2^{O(\alpha)})$ possible ways for H to be an α-heavy hub at the query time, over all the possible queries. However, each α-heavy hub contains α output vertices that store α regions $r_1, \dots, r_\alpha$ which means it will be used only by queries that are contained in the intersection of $r_1, \dots, r_\alpha$, a geometric region of volume at most v.

Let G' be the subgraph explored at the query time, for a given query. Remember that each point of Q is covered by t regions and since the query time is $Q(n) + O(t)$, it follows that $|V(G')| \le Q(n)+O(t) = O(t)$. But $V(G')$ must also contain t output nodes and thus $|V(G')| = \Theta(t)$. By Lemma 2, each query point must be inside the geometric region of $\Omega(|V(G')|/\alpha) = \Omega(t/\alpha)$ α-heavy hubs. As every single point of Q can be used as a query, it follows that the geometric regions of all the possible α-heavy hubs in G have to cover Q $\Omega(t/\alpha)$ times. Thus, we must have

$$2^{O(\alpha)} v O\left(|V(G)|2^{O(\alpha)}\right) \ge \Omega\left(\frac{t}{\alpha}\right)$$

and since $S(n) \ge |V(G)|$

$$S(n) = \Omega\left(\frac{t}{v2^{O(\alpha)}}\right).$$

□

Remark I.

For technical readers, here we compare our framework with the one offered by Chazelle and Rosenbern [14]. First, in their proof, regions are queries rather than input. Second, their framework is purely combinatoric: it requires a set of m query regions such that (i) each query outputs at least $Q(n)$ elements and (ii) the output of every α queries have at most a constant number, C, of elements in common. If q_i is the output of the i-th query, then they offer a space lower bound of $\Omega((\sum|q_i|)/\alpha)$. Unfortunately, to guarantee condition (ii) for simplex range reporting, they have to sacrifice a factor n^ε at the final space bound (ε depends on C). In our theorem, we still require (i) but replace (ii) with the requirement that (ii') the intersection of every α elements has small volume v. In fact, Chazelle and Rosenberg obtain their condition (ii) by first building a set of query regions that satisfies (ii') and then they place their input points uniformly at random. As every α regions have intersection volume of at most v, they will have at most vn points in common on average; unfortunately, the framework strongly requires a worst case bound rather than an average case bound. Because of this, n cannot be set as high as $1/v$. The biggest value of n for one can satisfy (ii) in the worst case is when $nv = 1/n^\varepsilon$, for a constant ε (in other words, when every α regions have $1/n^\varepsilon$ points in common, on average). Our method does not have this drawback but instead our theorem has an extra $1/2^{O(\alpha)}$ term compared to the $1/\alpha$ term of Chazelle and Rosenberg's theorem. Unfortunately, reducing this extra factor seems very difficult.

Remark II.

The idea of studying geometric stabbing problems and using volume-based arguments comes from a very recent work of Afshani et al. [2] where an optimal lower bound for rectangle stabbing in d-dimensions was proved. However, there are two major differences: first, their proof technique is insufficient for us (e.g., only considers the case $\alpha = 2$) and second, instead of having $Q(n) \le t$, they build a specific input instance and prove that $Q(n) = \Omega(t \log(S(n)/n))$.

Remark III.

It is easy to see that our volume-based argument can be generalized to any reasonable measure. Although we cannot immediately find any algorithmic consequences for such generalizations.

3. LOWER BOUNDS

In this section, we use our framework to prove our lower bounds.

3.1 Slab Enclosure

Preliminaries.

Let O be the origin. In this article, we use $\mathbb{R}^d$ to denote the d-dimensional Euclidean space. For d real values $A_1, \dots, A_d$, a hyperplane is set of points $(X_1, \dots, X_d) \in \mathbb{R}^d$ that satisfy the equation $X_d = A_d + \sum_{1\le i\le d-1} A_i X_i$ and its dual is the point $(A_1, \dots, A_d)$. For a real value, τ, we define a *slab with thickness* τ as the region of space between two parallel hyperplanes with distance τ from each other. Consider a slab whose lower boundary is hyperplane h_1 with equation $X_d = A_d + \sum_{1\le i\le d-1} A_i X_i$ and its upper boundary is hyperplane h_2 with equation $X_d = A_d + A + \sum_{1\le i\le d-1} A_i X_i$. We call A the *vertical distance* between h_1 and h_2 or the vertical distance of the slab. The dual of this slab is a vertical line segment of length A (note that A is larger than τ, the distance between h_1 and h_2).

We define the *slab stabbing problem* as follows: given a set S of n slabs, store them in a data structure such that the slabs containing a given query point q can be found efficiently. Thus, the dual of the slab stabbing problem is the following: given a set $\overline{S}$ of vertical line segments, store them in a data structure such that the line segments intersecting a query hyperplane $\overline{q}$ can be found efficiently.

To prove a lower bound for the slab stabbing problem, we use Theorem 1 and thus we need an input set of slabs such that the volume of the intersection of a few of them is small. In [9], Chazelle showed the existence of a set of slabs such that the intersection of any $\log n$ of them has a very small volume. Later with Rosenberg, they used this construction to prove their simple range reporting lower bound [15]. Unfortunately for us, setting $\alpha = \log n$ in Theorem 1 destroys any chance of obtaining a reasonable lower bound (although with some easy tweaks, it is possible to set $\alpha = \varepsilon \log n$ and reclaim Chazelle and Rosenberg's result). In essence, we need a more sensitive lemma.

All of our input slabs have the same thickness τ for a parameter τ to be determined later. This means each slab is determined by a single point: for a given point $p \in \mathbb{R}^d$, we define the slab S_p as $\left\{q \in \mathbb{R}^d : |< q, p > -\|p\|^2| \le \tau\|p\|\right\}$. Let Q be the d-dimensional cube $[1,2]^d$ and let Q^{top} be its top face, i.e., $Q^{\text{top}} = [1,2]^{d-1} \times [2]$. Let $\mathcal{P}$ be a set of t points placed on Q^{top}. For every $p \in \mathcal{P}$, we take the slab S_p and tile Q using S_p (Figure 1(left and center)). Let $\mathcal{S}_p$ denote the set of slabs obtained by tiling Q using S_p. Our final set of slabs $\mathcal{S}$ is defined as $\mathcal{S} = \bigcup_{p \in \mathcal{P}} \mathcal{S}_p$.

The following two lemmas follow closely from the previous work [9, 14] although nowhere they have been explicitly listed as lemma. Because of this, and also since we use a different set of parameters, we provide short proofs. The first lemma computes the volume of a parallelotope formed by the intersection of d slabs using elementary techniques.

LEMMA 3. *[14] Consider d points $p_1, \dots, p_d \in \mathcal{P}$ and let $V = \bigcap_{1 \le i \le d} S_{p_i}$. We have $\text{Vol}(V) = O\left(\tau^d / v_p\right)$ in which v_p is the $(d-1)$-dimensional volume of convex hull of $p_1, \dots, p_d$.*

PROOF. It is easy to see that V is in indeed a parallelotope. First, S_{p_i} is the region between two parallel hyperplanes: $h_i = \left\{x \in \mathbb{R}^d :< x, p_i > -\|p_i\|^2 = -\tau\|p_i\|\right\}$ describes the lower boundary and the upper boundary is described by $h'_i = \left\{x \in \mathbb{R}^d :< x, p_i > -\|p_i\|^2 = \tau\|p_i\|\right\}$. As there are d such slabs (Figure 1(right)) and no two slabs contain parallel hyperplanes, the intersection becomes a parallelotope. The lowest vertex of V, v, satisfies

$$< v, p_i >= \|p_i\|^2 - \tau\|p_i\|, \quad \text{for } 1 \le i \le d.$$

For every $1 \le i \le d$, there exists another vertex, v_i, adjacent to v. v_i satisfies

$$\begin{cases} < v_i, p_j >= \|p_j\|^2 - \tau\|p_j\| & \text{for } 1 \le j \le n, j \ne i \\ < v_i, p_i >= \|p_i\|^2 + \tau\|p_i\|. \end{cases}$$

Note that $\text{Vol}(V) = \det(v_1 - v, v_2 - v, \dots, v_d - v)$. Observe that $< v_i - v, p_j >= 0$ for $i \ne j$ and $< v_i - v, p_i >= 2\tau\|p_i\|$.

Let A_v and A_p be the matrices in which the i-th row is $v_i - v$, and p_i, respectively. Let A be the diagonal matrix in which $A(i,i) = 2\tau\|p_i\|$. With this notation $\text{Vol}(V) = \det(A_v)$. It is easily verified that $A_v A_p^T = A$. Note that $\det(A) = (2\tau)^d \prod_{1 \le i \le d} \|p_i\|$. Furthermore, $\det(A_p)/d!$ is the volume the simplex formed by points $p_1, \dots, p_d$ and the origin; since the points $p_1, \dots, p_d$ lie on a hyperplane that has distance two from origin, it follows that $\det(A_p)/d! = 2v_p/d$. Thus,

$$\text{Vol}(V) = \frac{(2\tau)^d \prod_{1 \le i \le d} \|p_i\|}{2(d-1)! v_p} = O\left(\frac{\tau^d}{v_p}\right).$$

□

LEMMA 4. *[14] Consider α points $p_1, \dots, p_\alpha \in \mathcal{P}$ and let $V = \bigcap_{1 \le i \le \alpha} S_{p_i}$. We have $\text{Vol}(V) = O(\tau^d \alpha^{d-1}/v)$ in which v is the $(d-1)$-dimensional volume of convex hull of $p_1, \dots, p_\alpha$.*

PROOF. Since $p_1, \dots, p_\alpha$ are inside a $(d-1)$-dimensional space, we can triangulate the convex hull of $p_1, \dots, p_\alpha$ into $O(\alpha^{d-1})$ simplices. We pick the simplex σ with the largest volume and let v_σ be its $(d-1)$-dimensional volume. Thus, $v_\sigma = \Omega(v/\alpha^{d-1})$. V is a subset of the intersection of slabs corresponding to the vertices of σ and thus from the previous lemma we know $\text{Vol}(V) = O(\tau^d / v_\sigma) = O(\tau^d \alpha^{d-1}/v)$. □

The above lemmas imply that a set of slabs have small intersection if and only if the convex hull of the points that give rise to those slabs is large. The next lemma is the main tool that we borrow from [9].

LEMMA 5. *[9] Consider the unit square in d dimensions in which d is a constant. For any real value $0 < \rho < 1$, there exists a set $\mathcal{C}_\rho$ of $1/\rho^{O(1)}$ convex shapes inside the unit square such that each shape in $\mathcal{C}_\rho$ has volume $O(\rho)$ and for any convex shape A of volume at least ρ, there exists a convex shape in $\mathcal{C}_\rho$ that completely contains A.*

We are now ready to describe the set $\mathcal{P}$ which defines our set of slabs: $\mathcal{P}$ is chosen by placing t points uniformly at random inside Q^{top}, for a parameter t, and the thickness τ of the hyperplanes is chosen to be t/n. With this choice of parameters, it is easy to verify that for each point $p \in \mathcal{P}$, we can tile Q using $\Theta(n/t)$ copies of slab S_p. Remember that $\mathcal{S}$ is defined as $\mathcal{S} = \bigcup_{p \in \mathcal{P}} \mathcal{S}_p$. Thus, $|\mathcal{S}| = \Theta(n)$. Clearly, each point of Q is covered by exactly t slabs. From this point forward, for a region r, we use $\text{Vol}(r)$ to denote the d-dimensional volume of $r \cap Q$. It is easy to see that for each $r \in \mathcal{S}_p$, $\text{Vol}(r) = O(t/n)$.

We apply Lemma 5 to Q^{top} (since Q^{top} is a translation of the unit square in the $(d-1)$-dimensional space) with parameter $\rho = (tx)^{-1}$ in which x is a parameter to be adjusted later. For a convex region $C \in \mathcal{C}_\rho$, we bound the probability of the event, E_C, that more than α points are placed inside C, for another parameter α. This probability is

$$\Pr[E_C] \le \binom{t}{\alpha} \text{Vol}(C)^\alpha = O((t\rho)^\alpha)$$

and thus

$$\Pr[\bigvee_{C \in \mathcal{C}_\rho} E_C] \le O\left(\frac{(t\rho)^\alpha}{\rho^{O(1)}}\right).$$

If we set $\alpha = \Theta((\log t)/\log x)$ (with a large enough constant hidden in the Θ notation), the above probability will be less than one and thus there exists an input $\mathcal{P}$ such that the convex hull of any α points in $\mathcal{P}$ has $(d-1)$-dimensional volume greater than $\rho = (tx)^{-1}$. By Lemma 4, it follows that for any α slabs $S_1, \dots, S_\alpha \in \mathcal{S}$, we have

$$\begin{aligned} \text{Vol}(S_1 \cap \dots \cap S_\alpha) &\le \tau^d \alpha^{d-1} t x = \left(\frac{t}{n}\right)^d \alpha^{d-1} t x \\ &\le \left(\frac{t}{n}\right)^d \alpha^{d-1} t 2^{\Theta(\alpha^{-1} \log t)}. \end{aligned}$$

LEMMA 6. *Given parameters t and α, there exists a set of $\Theta(n)$ slabs $\mathcal{S}$ in d dimensions such that (i) the thickness*

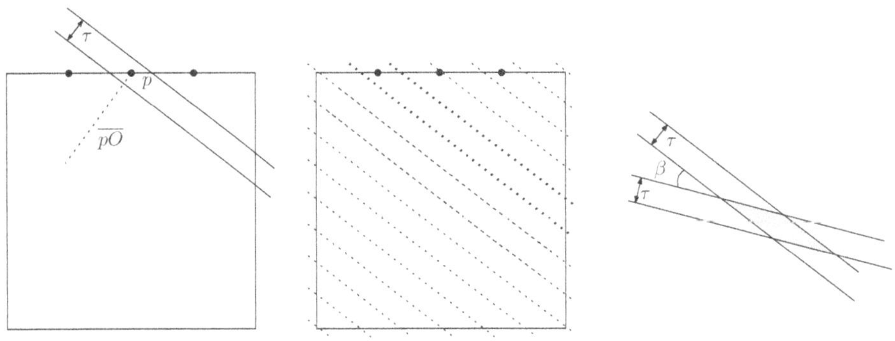

Figure 1: (left) The slab $S_p = \{q \in \mathbb{R}^d : |< q, p > -\|p\|^2| \le \tau\|p\|\}$ is depicted. The thickness (the distance between two defining hyperplanes) is τ and both hyperplanes are perpendicular to the line segment that connects p to the origin. (center) Tiling Q using copies of S_p. (right) The intersection of d slabs forms a parallelotope.

of each slab in $\mathcal{S}$ is t/n (ii) every point in Q is contained in t slabs and (iii) for every α slabs $S_1, \dots, S_\alpha \in \mathcal{S}$ the volume of $S_1 \cap \dots \cap S_\alpha$ is $O(t^{d+1}\alpha^{d-1}2^{O(\alpha^{-1}\log t)}/n^d)$.

To prove our lower bound, we simply need to use Theorem 1 and adjust the parameters.

THEOREM 2. *Answering d-dimensional slab stabbing queries in $Q(n) + O(k)$ time requires $\Omega\left((n/Q(n))^d 2^{-O(\sqrt{\log Q(n)})}\right)$ space.*

PROOF. We pick $t = Q(n)$ in Lemma 6. Combined by Theorem 1, we obtain that

$$S(n) = \Omega\left(Q(n)\frac{n^d}{Q(n)^{d+1}\alpha^{d-1}2^{O(\alpha^{-1}\log t)}}\frac{1}{2^{O(\alpha)}}\right).$$

By setting $\alpha = \sqrt{\log Q(n)}$ we get

$$S(n) = \Omega\left(\left(\frac{n}{Q(n)}\right)^d 2^{-O(\sqrt{\log Q(n)})}\right).$$

□

3.2 Simplex Range Reporting Lower Bound

In this section we prove our lower bound for simplex range reporting. We use our result on slab stabbing problem. Unfortunately, we are unable to give a general reduction from simplex range reporting to slab stabbing problem. However, we can do it for some special cases which include the instances that we used to obtain our slab stabbing lower bound.

LEMMA 7. *Consider an input I for the slab stabbing problem. Let d_{min} and d_{max} be the minimum and the maximum vertical distance of slabs in I. If $d_{max}/d_{min} = O(1)$, then the slab stabbing problem can be solved using a data structure for simplex range reporting with no asymptotical increase in query or space complexities.*

PROOF. For an integer i, $0 \le i \le \log(d_{\max}/d_{\min})$, let I_i be the subset of I that contains slabs with vertical distance of at least $2^i d_{\min}$ and less than $2^{i+1}d_{\min}$. Let H_i^{low} and H_i^{high} be the set of hyperplanes that form the lower boundary and the upper boundary of the slabs in I_i respectively. We build simplex range reporting data structures one for each set of points dual to H_i^{low} and H_i^{high} and for each $0 \le i \le \log(d_{\max}/d_{\min})$.

Let q be the query point. Consider an index $i, 0 \le i \le \log(d_{\max}/d_{\min})$. Let q^{high} (resp. q^{low}) be the point obtained from q by adding (resp. subtracting) $2^i d_{\min}$ to the d-th coordinate of q. Using simplex range reporting data structures, we find all the hyperplanes in H_i^{low} that pass between q and q_{low} (in primal space this translates to finding all the points between two query hyperplanes). Similarly, we find all the hyperplanes in H_i^{high} that pass between q and q^{high}. We output the slabs corresponding to the hyperplanes found. Since $\log d_{\max}/d_{\min} = O(1)$, we only impose a constant factor overhead on the query time and space complexity of the simplex range reporting data structure used.

To see the correctness, consider a slab $S \in I_i$ and let A be the vertical distance of S. As $A < 2^{i+1}d_{\min}$, if S contains q, then either the lower boundary of S will pass above q^{low} or its upper boundary will pass below q^{high} and thus it will be outputted. However, since $A \ge 2^i d_{\min}$, if S does not contain q, then the lower boundary of S will not pass between q and q^{low} and the upper boundary of S will not pass between q and q^{high} as well. Thus, S will not be outputted if it does not contain q. □

Thus, we simply need to bound the vertical distance of the slabs that we built in the previous section.

THEOREM 3. *Any structure that answers d-dimensional simplex range reporting queries in $Q(n) + O(k)$ time must use $\Omega\left((n/Q(n))^d 2^{-O(\sqrt{\log Q(n)})}\right)$ space.*

PROOF. Consider a slab S_p with its lower boundary being $h = \{x \in \mathbb{R}^d :< x, p > -\|p\|^2 = -\tau\|p\|\}$ and its upper boundary being $h' = \{x \in \mathbb{R}^d :< x, p > -\|p\|^2 = \tau\|p\|\}$. Let x and $x + v$ be two points on lower and upper boundary respectively in which v is a vertical vector, i.e., $v = (0, 0, \dots, 0, A)$ where A is the vertical distance of S_p. We have

$$< x, p > -\|p\|^2 = -\tau\|p\| \quad \text{and} \quad < x+v, p > -\|p\|^2 = -\tau\|p\|.$$

As p is chosen on Q^{top}, its d-th coordinate is two. Thus,

$$2A =< v, p >= 2\tau\|p\| = \Theta(\tau).$$

As all the slabs we constructed in the previous section had the same thickness, the theorem follows from Lemma 7 and Theorem 2. □

3.2.1 Halfspace Range Reporting Lower Bounds

Currently the best known lower bounds for halfspace range reporting problem are obtained through reductions from the slab stabbing problem (or its dual) via a simple geometric transformation: Consider a slab S with hyperplanes h_1 and h_2 being its lower and upper boundaries respectively. Let $X_d = A_d + \sum_{1 \le i \le d-1} A_i X_i$ and $X_d = A_d + A + \sum_{1 \le i \le d-1} A_i X_i$ be equations describing h_1 and h_2 respectively. S contains points $q = (q_1, \dots, q_d)$ such that $q_d \ge A_d + \sum_{1 \le i \le d-1} A_i q_i$ and $q_d \le A_d + A + \sum_{1 \le i \le d-1} A_i q_i$ and thus $|q_d - A_d - A/2 - \sum_{1 \le i \le d-1} A_i q_i| \le A/2$. Thus, $(q_d - A_d - A/2 - \sum_{1 \le i \le d-1} A_i q_i)^2 \le A^2/4$. By introducing some new variables, one for each q_i^2, $1 \le i \le d$, and one for each $q_i q_j$, $1 \le i < j \le d$, this can be turned into a linear inequality in $d(d+3)/2$ variables, a halfspace in $(d(d+3)/2)$-dimensional space. Thus, slab enclosure can be solved with a $(d(d+3)/2)$-dimensional halfspace range reporting structure.

THEOREM 4. *Any data structure that answers $(d(d+3)/2)$-dimensional halfspace range reporting queries in $Q(n)+O(k)$ time requires $\Omega\left((n/Q(n))^d 2^{-O(\sqrt{\log Q(n)})}\right)$ space.*

4. I/O MODEL LOWER BOUNDS

The I/O model is a widely used model of computation that deals with massive data. In this model, data sits on a disk of conceptually infinite size which is divided into blocks and each block can store up to B records. All computation must be performed in a main memory of size M. By performing an I/O, it is possible to transfer one block between the main memory and the disk and the goal is to minimize number of I/Os performed by the algorithm.

Our lower bound is proved in the stronger indexibility model [18] which only considers the I/O sequences needed to report the output elements: Let S be a set of n elements. We assume each block can store up to B elements and to output an element r a block containing r must be loaded into the main memory. All the other informations can be stored and accessed for free.

Previously, only one lower bound framework was used for reporting problems. The heart of this framework is the Redundancy Theorem [18] and its refinement [6] that ties the space overhead of the data structure to a combinatorial structure of the query set. Here we use our new techniques to provide another lower bound framework.

THEOREM 5. *Assume that there exists an input S of n regions for a geometric stabbing problem such that (i) every point of Q is contained in exactly t regions and (ii) the volume of the intersection of every α regions is at most v in which $\alpha > t$ and t are two parameters. Then, there exists an input S' of $N = \Theta(nB/\alpha)$ regions such that any external memory data structure that answers geometric stabbing queries on S' using $Q(N) + k/B$ I/Os, for $Q(N) = O(t/\alpha)$, consumes $B\Omega(tv^{-1}/2^{O(\alpha)}) = B\Omega(Q(N)v^{-1}/2^{O(\alpha)})$ space.*

Remark.

It is crucial that we build another input instance using S. An easy generalization of Theorem 1 that only uses S yields a space lower bound that is a factor B smaller.

The rest of this subsection is devoted to the proof of the above theorem.

S' is made by duplicating every region in S β times. We set $\beta = \Theta(B/\alpha)$ but its exact value will be determined later. For every $r \in S$, we create β identical copies of r and place them at the same location as r. S' is the set of all these copies. We denote the size of S' by N and thus $N = \beta n$. We say two regions $r, r' \in S'$ are *different* if they are not copies of the same region in S. Now consider a data structure that stores S' using $S(N)$ space and can answer geometric stabbing queries using $Q(N) + O(k/B)$ I/Os.

We call a tuple of α elements a *hub (of size α)* if they are all different and there exists a block that stores all of them. Since the data structure contains $S(N)/B$ blocks, it follows that the total number of hubs is at most $\binom{B}{\alpha} S(N)/B$.

Now consider a particular query. Note that every point of Q is contained in $t\beta$ regions in S' and thus the output size is $t\beta$. A hub whose all elements are outputted at the query time is called an *output hub (of size α)*. The query algorithm reads $Q(N) + O(k/B)$ blocks and since $k = \beta t$ and $Q(N) = O(t/\alpha)$, it follows that $Q(N) + O(k/B) = O(k/B)$. It is easy to show that there are at least $k/(2B)$ blocks that each outputs at least k_b elements and $k_b = \Theta(B)$; we call these blocks *heavy*. We pick $\beta = k_b/\alpha = \Theta(B/\alpha)$ (for simplicity assume β is an integer). We now lower bound the number of output hubs.

LEMMA 8. *Every heavy block* B *has at least β^α output hubs.*

PROOF. Let A be a set of k_b regions that are outputted from B. Let x be the maximum number of pairwise different regions in A. We can decompose A into x equivalent classes $A_1, \dots, A_x$ in which the regions in each A_i are copies of the same region. Note that since $|A| = k_b = \beta\alpha$ we have $x \ge \alpha$. Let X be the number of output hubs of size α and let $X_{i,0}$ be the number of output hubs of size i that have no element from A_1 or A_2. Each output hub has either zero or one element from A_1 or A_2 and thus we have

$$X = X_{\alpha,0} + (|A_1| + |A_2|)X_{\alpha-1,0} + |A_1||A_2|X_{\alpha-2,0}.$$

If we keep $|A_1| + |A_2|$ fixed, then X is minimized when the difference between $|A_1|$ and $|A_2|$ is maximized. Thus, the minimum of X is achieved when $x = \alpha$ and $|A_i| = \beta$ for $1 \le i \le \alpha$. And in that case we have $X = \beta^\alpha$. □

For a hub H to be used as an output hub, the query must be inside the geometric region formed by the intersection of all the α regions that comprise H. Since the regions that form a hub are different, the intersection has volume at most v. Thus, the total volume of the geometric regions of all the hubs in the data structure is $\binom{B}{\alpha} vS(n)/B$. However, as we saw, to answer any query we must access $k/(2B)$ heavy blocks and each heavy block has β^α output hubs. Thus, the total number of output hubs for any query is at least $\beta^\alpha k/(2B)$. Each hub can only produce one output hub. Since every point of Q can be used as a query, the geometric regions of the hubs must cover Q at least $\beta^\alpha k/(2B)$ times. We must have

$$\begin{aligned} \frac{\binom{B}{\alpha} vS(n)}{B} &\ge \beta^\alpha \frac{k}{2B} = \beta^\alpha \frac{t\beta}{2B} = \beta^\alpha \frac{\Omega(Q(N)\alpha)\beta}{2B} \\ &= \Omega(\beta^\alpha Q(N)) = O\left(\frac{(B/\alpha)^\alpha Q(N)}{2^{O(\alpha)}}\right) \end{aligned}$$

and thus

$$\frac{S(N)}{B} = \Omega\left(\frac{(B/\alpha)^{\alpha}Q(N)}{\binom{B}{\alpha}v2^{O(\alpha)}}\right) = \Omega\left(\frac{Q(N)}{v2^{O(\alpha)}}\right).$$

Using Lemma 6 and setting $\alpha = \sqrt{\log Q(N)}$, we obtain the following theorem.

THEOREM 6. *In the I/O model, any structure that answers d-dimensional simplex range reporting or d-dimensional slab stabbing queries, or $(d(d+3)/2)$-dimensional halfspace range reporting queries with $Q(N)+O(k/B)$ I/Os must use* $\Omega\left(BN^d(BQ(N))^{-d}2^{-O(\sqrt{\log Q(N)})}\right)$ *space.*

5. CONCLUSIONS

In this paper, we presented a number of improved lower bounds for simplex range reporting, slab stabbing, and halfspace range reporting in the pointer machine. We also presented new lower bounds for the problems in the I/O model of computation. On the way to obtain our lower bounds, we built new frameworks for proving lower bounds in both models of computation.

An interesting aspect of our frameworks is that building a single input instance is enough to prove good lower bounds in both models of computation. This is unlike the previous attempts where obtaining I/O lower bounds always required tweaking the internal memory constructions in non-trivial ways. For instance, for orthogonal range reporting, the pointer machine lower bounds were obtained as early as 1990 [10] but the equivalent I/O lower bounds were obtained almost two decades later [1].

At the end, we must mention that there are still very interesting questions that are left open. Recently, an important open problem that is directly related to our work came from Timothy Chan [7] who asked whether logarithmic factor increases are necessary in multi-level partition trees. For instance, we can now ask whether it is possible to prove a space lower bound of $\Omega((n/Q(n))^d \log n)$ for data structures that answer simplex-simplex containment queries (preprocess a set of simplices such that the simplices contained in a query simplex can be outputted efficiently) in $O(Q(n)+k)$ time for polylogarithmic values of $Q(n)$.

6. REFERENCES

[1] P. Afshani, L. Arge, and K. D. Larsen. Orthogonal range reporting in three and higher dimensions. In *FOCS '09: Proc. of the 50th Annual Symposium on Foundations of Computer Science*, pages 149–158, 2009.

[2] P. Afshani, L. Arge, and K. G. Larsen. Higher-dimensional orthogonal range reporting and rectangle stabbing in the pointer machine model. submitted to SoCG'12.

[3] P. K. Agarwal. Range searching. In J. E. Goodman and J. O'Rourke, editors, *CRC Handbook of Discrete and Computational Geometry*. CRC Press, Inc., 2004.

[4] P. K. Agarwal, L. Arge, J. Erickson, P. G. Franciosa, and J. S. Vitter. Efficient searching with linear constraints. *J. Comput. Syst. Sci.*, 61:194–216, 2000.

[5] P. K. Agarwal and J. Erickson. Geometric range searching and its relatives. In B. Chazelle, J. E. Goodman, and R. Pollack, editors, *Advances in Discrete and Computational Geometry*. AMS Press, 1999.

[6] L. Arge, V. Samoladas, and K. Yi. Optimal external-memory planar point enclosure. In *ESA'04: Proc. of the 12th conference on Annual European Symposium*, pages 40–52, 2004.

[7] T. M. Chan. Optimal partition trees. In *SCG '10: Proc. of the 26th Annual Symposium on Computational Geometry*, pages 1–10. ACM, 2010.

[8] B. Chazelle. Filtering search: a new approach to query answering. *SIAM Journal on Computing*, 15(3):703–724, 1986.

[9] B. Chazelle. Lower bounds on the complexity of polytope range searching. *Journal of the American Mathematical Society*, 2:637–666, 1989.

[10] B. Chazelle. Lower bounds for orthogonal range searching: I. the reporting case. *Journal of the ACM*, 37(2):200–212, 1990.

[11] B. Chazelle. Cutting hyperplanes for divide-and-conquer. *Discrete and Computational Geometry*, 9(2):145–158, 1993.

[12] B. Chazelle and J. Friedman. A deterministic view of random sampling and its use in geometry. *Combinatorica*, 10:229–249, 1990.

[13] B. Chazelle and D. Liu. Lower bounds for intersection searching and fractional cascading in higher dimension. *Journal of Computer and System Sciences*, 68(2):269 – 284, 2004.

[14] B. Chazelle and B. Rosenberg. Simplex range reporting on a pointer machine. *Computational Geometry*, 5(5):237 – 247, 1996.

[15] B. Chazelle, M. Sharir, and E. Welzl. Quasi-optimal upper bounds for simplex range searching and new zone theorems. *Algorithmica*, 8:407–429, December 1992.

[16] K. Clarkson. New applications of random sampling in computational geometry. *Discrete and Computational Geometry*, 2:195–222, 1987.

[17] J. Erickson. Space-time tradeoffs for emptiness queries. *SIAM Journal on Computing*, 29(6):1968–1996, 2000.

[18] J. M. Hellerstein, E. Koutsoupias, D. P. Miranker, C. H. Papadimitriou, and V. Samoladas. On a model of indexability and its bounds for range queries. *Journal of the ACM*, 49(1):35–55, 2002.

[19] J. Matoušek. Range searching with efficient hierarchical cuttings. *Discrete and Computational Geometry*, 10(2):157–182, 1993.

[20] J. Matoušek. Efficient partition trees. *Discrete and Computational Geometry*, 8(3):315–334, 1992.

[21] E. Welzl. On spanning trees with low crossing numbers. In B. Monien and T. Ottmann, editors, *Data structures and efficient algorithms*, volume 594 of *Lecture Notes in Computer Science*, pages 233–249. Springer Berlin / Heidelberg, 1992.

[22] D. E. Willard. Polygon retrieval. *SIAM Journal on Computing*, 11:149–165, 1982.

Periodic Body-and-bar Frameworks

Ciprian Borcea
Department of Mathematics,
Rider University
Lawrenceville, NJ 08648
borcea@rider.edu

Ileana Streinu
Department of Computer
Science, Smith College
Northampton, MA 01063, USA
istreinu@smith.edu

Shin-ichi Tanigawa
Research Institute for
Mathematical Sciences
Kyoto University, Japan
tanigawa@kurims.kyoto-u.ac.jp

ABSTRACT

Flexibility studies of macromolecules modeled as mechanical frameworks rely on computationally expensive, yet numerically imprecise simulations. Much faster approaches for degree-of-freedom counting and rigid component calculations are known for finite structures characterized by theorems of Maxwell-Laman type, but such results are exceedingly rare and difficult to obtain. The situation is even more complex for infinite, periodic structures such as those appearing in the study of crystalline materials. Here, an adequate rigidity theoretical formulation has been proposed only recently, opening the way to a combinatorial treatment.

Abstractions of crystalline materials known as periodic body-and-bar frameworks are made of rigid bodies connected by fixed-length bars and subject to the action of a group of translations. In this paper, we give a Maxwell-Laman characterization for generic minimally rigid periodic body-and-bar frameworks in terms of their quotient graphs. As a consequence we obtain efficient polynomial time algorithms for their recognition based on matroid partition and pebble games.

Categories and Subject Descriptors

I.3.5 [**Computational Geometry and Object Modeling**]: [Geometric algorithms, languages, and systems]

General Terms

Theory, Algorithms

Keywords

Rigidity, periodicity, crystalline matter

1. INTRODUCTION

In this paper, we study d-dimensional periodic frameworks made of rigid bodies connected with rigid bars. Fig. 1 gives an example. We prove a combinatorial characterization for the quotient graphs of *generic* minimally rigid frameworks, in terms of matroid unions of graphs satisfying Maxwell-sparsity conditions. As a consequence, we obtain an efficient, polynomial time algorithm for their recognition, based on matroid partition and pebble games.

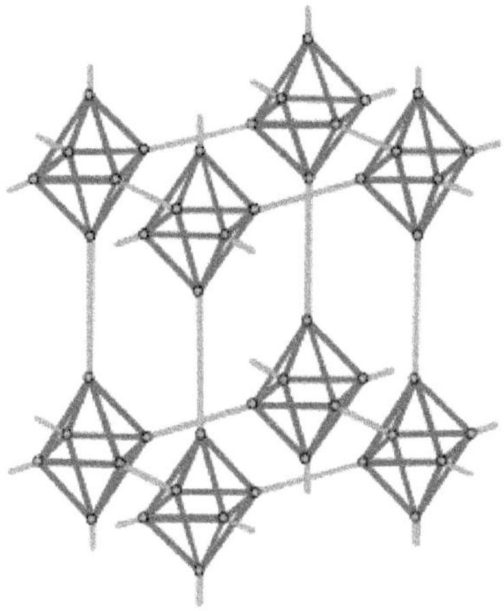

Figure 1: Fragment of a periodic body-and-bar framework associated to Prussian Blue crystalline materials [12]. Bodies are shown as octahedra. The framework is flexible and its geometric deformations are determined by a system of quadratic equations (1) described below. A bar-and-joint model and the deformation theory of [1] may be also be used for this purpose.

Crystalline materials. *Periodic structures* are naturally associated with *crystallographic studies* [23]. The central role played by a *periodicity lattice* and corresponding fundamental domains was recognized even before X-ray diffraction opened up experimental possibilities. Crystals known as *framework materials* [26, 9, 27] have distinctive substructures which can be modeled and investigated as articulated systems made of rigid pieces and joints described abstractly by an *infinite periodic graph*. Similar models appear in structural engineering in connection with infinite trusses, foams or cellular materials [6, 13, 8, 16]. While a fair number of traditional areas of mathematics are related or dedicated to lattice structures and periodicity, from crystallographic groups and quadratic forms to the geometry of numbers and sphere packings [15, 38, 4], mathematical studies of periodic frameworks are of relatively recent date [20, 5, 1].

Displacive phase transitions. Crystallographers have long observed that under variations of temperature or pressure, materials may undergo structural changes while still retaining a distinctively regular, crystalline arrangement. Understanding these phase transitions remains a challenging problem [7]. The role of geometry is

Figure 2: Left: A 2D periodic bar-and-joint framework containing smaller rigid components. Right: the same framework, viewed as a body-and-bar framework; the rigid components form the bodies, and the remaining bars connect distinct bodies.

particularly relevant in *displacive* phase transitions which may be interpreted as continuous deformations of a given framework [9]. A rigorous mathematical understanding of rigidity and flexibility properties of periodic frameworks may assist, clarify and complement approaches based on computational physics and materials science [18, 1].

Computational challenges. Flexibility studies of macromolecules modeled as mechanical (finite or periodic) frameworks [12, 18] are often conducted with computationally expensive yet numerically imprecise simulations. For finite structures characterized by theorems of Maxwell-Laman type (described below), much faster approaches based on degree-of-freedom counting and rigid component calculations are known. But such theorems are exceedingly rare and difficult to obtain. For periodic structures, one would expect, intuitively, that the number and distribution of connections in a "unit cell" would dictate the flexibility properties of the framework, but the occurrence of certain 'paradoxes' [18] related to choices of periodicity has stymied the progress. Making precise such an intuition with an appropriate mathematical definition has emerged only very recently [1], opening the way to a combinatorial treatment and efficient algorithms [2] for rigidity and flexibility analysis of periodic *bar-and-joint* structures (made from rigid bars connected through rotatable joints).

The approach adopted in this paper is rooted in the theory developed in these recent papers [1, 2], where we gave a Maxwell-Laman theorem for *generic* periodic bar-and-joint rigidity, thus reducing both the long-standing finite case and several other periodic situations to genericity refinement conjectures. It is therefore important to understand in this context the intrinsic theoretical difficulties we were facing in proving the results of this paper: the rest of the introduction gives this perspective.

Modeling crystalline frameworks. Because of interatomic bond length and angle constraints, a crystalline material such as the one illustrated in Fig. 1 can be abstracted as a periodic framework made from rigid bars connected at vertices, to which, in principle, the theory developed in [1, 2] applies. Often, substructures making individual *rigid bodies* are identified from the outset, such as the tetrahedra in zeolites, the octahedra in Fig. 1 or, in a 2D example, the gray bodies from Fig. 2(b). In this paper we focus on those made of rigid bodies connected by rigid bars, as illustrated in Fig. 1 and 2(b).

The practical difficulty in applying the results of [2] to these frameworks comes from the *generic* nature of the characterization theorem of [2]. Indeed, the placement of the "bars" in actual crystal models, with several bars sharing endpoints and connecting pairs of atoms with specified regularity, is not guaranteed, *a priori*, to be generic. This is a very subtle distinction for which the reader may need further guidance. Hence we discuss now briefly the challenges inherent in addressing questions of generic rigidity, at the same time presenting an overview of the techniques we use.

Maxwell-Laman sparsity conditions. The theory of finite frameworks has a distinguished tradition going back almost 150 years to Maxwell [25], where a *sparsity condition* was shown to be necessary for minimal rigidity of bar-and-joint frameworks: in dimension d, for any subset of $d \leq n' \leq |V|$ vertices, the graph should span at most $dn' - \binom{d+1}{2}$ edges, with equality for the whole set of $n = |V|$ vertices. Its sufficiency for generic frameworks in dimension two was proven over 100 years later (Laman's theorem [21]), and is known to fail in higher dimensions. The problem of completing the combinatorial characterization for bar-and-joint frameworks in arbitrary dimensions remains the most conspicuous open question in rigidity theory. However, some restricted classes of finite frameworks have been shown to have similar Maxwell-Laman counting characterizations: body-and-bar and body-and-hinge frameworks [17, 39, 35, 34, 33], panel-and-hinge frameworks [19] and $(d-2)$-dimensional plate-and-bar frameworks [36, 35, 33].

Generic structures. All known theorems giving combinatorial characterizations of framework rigidity and flexibility proceed by associating a *rigidity matrix* M to the framework. A set of algebraic constraints (e.g. fixed distances between some pairs of points) are applied on a collection of rigid objects (e.g. a finite set of points), with the goal of obtaining a rigid structure, i.e. one in which all the pairwise distances are determined. Linearizing on the tangent space at a generic point of the variety defined by these algebraic constraints leads to a matrix, called the *rigidity matrix*, whose rank we seek to determine. The maximal number of independent constraints that can be imposed is also the minimum number that would make the structure rigid. Together with additional constraints that eliminate the trivial rigid motions, they lead to a square rigidity matrix. Maximal rank, attained when the determinant $det(M)$ of the rigidity matrix is non-zero, indicates that the framework is *minimally rigid*: it is rigid and the removal of any constraint will make it flexible. The condition of maximal rank is therefore an algebraic condition in the free variables of the system. A framework (with a square rigidity matrix) is *generic* if its actual geometric realization falls outside the set of zeros of the algebraic equation $det(M) = 0$. This extends naturally to frameworks whose rigidity matrix is not square, by expressing the rank condition in terms of corresponding minors of M.

Theorems of Maxwell-Laman type for minimally rigid structures characterize classes of graphs underlying frameworks whose rigidity matrices have maximum rank, generically, for certain structural models as described above. A random geometric realization of a graph belonging to such a class is, with probability 1, rigid. But the measure-zero set when it is not rigid is in general non-empty, and *may* contain important cases that appear in practice. Refining the genericity conditions to prove that they include such practical occurrences is in general a very difficult problem. An important success of this nature is the recent proof of the *Molecular Conjecture* [19]. Our contribution here, as a refinement of genericity conditions, is of a similar flavor.

Maxwell-Laman sparsity for periodic frameworks. The connection pattern of a periodic body-and-bar system determines an infinite multi-graph $G = (V, E)$, with vertices corresponding to bodies and edges corresponding to bars between pairs of connected bodies.

See Fig. 2 for an example in 2D and Fig. 1 for one in 3D. Periodicity, or more precisely d-periodicity (where d represents the dimension of the ambient space in which the graph is realized geometrically) requires a free Abelian automorphism subgroup $\Gamma \subset Aut(G)$ of rank d. We work under the assumption that the quotient graph G/Γ has a finite number n of vertex orbits and a finite number m of edge orbits.

The problem of characterizing periodic frameworks is dramatically different from the finite case: the generic periodic bar-and-joint frameworks have been characterized in all dimensions [2] by a Maxwell-sparsity condition *on the quotient graph*. The finite bar-and-joint frameworks, as well as the periodic body-and-bar, body-and-hinge, body-and-pin, etc. appear as special cases where *additional algebraic dependencies* are present. As indicated, such cases are not guaranteed to be generic *a priori*, and even getting a necessary sparsity condition (something that was trivial in the finite case) is not easy.

Matroidal sparsity conditions. An additional difficulty in the quest for generic rigidity characterizations, relevant from an algorithmic point of view, arises from the fact that not all types of sparsity conditions are *matroidal*. For instance, Maxwell sparsity for finite 3D bar-and-joint frameworks (the "3n-6" condition) is not. For the periodic body-and-bar structures studied in this paper, the conditions *are matroidal*. In this case, efficient recognition algorithms exist. Relevant for our setting are matroid partition [10] and pebble game algorithms [22].

We can now state our contributions in more technical terms.

Contribution and novelty. Motivated by foundational questions in theoretical crystallography, we present a complete mathematical and algorithmic solution of the *periodic body-and-bar problem* by giving a Maxwell-Laman characterization. We formally define the periodicity model, identify the necessary sparsity conditions, prove their sufficiency and, finally, give efficient algorithms for deciding rigidity, based on matroid partition and pebble games.

As already seen in the bar-and-joint case [1, 2], the mutual relations between the finite and the periodic settings are intricate. We emphasize here that our approach considers the periodicity group Γ as part of the initial data (G,Γ) but allows the variation of its representation as a lattice of translations of full rank. Moreover, the problem of minimal rigidity may be formulated at several levels. The most basic level looks only at the structure of the quotient graph G/Γ and provides characterizations 'up to a generic lifting of edges' from G/Γ to a covering periodic graph. The main results presented below are formulated at this level. For $n = 1$, that is, in the case of a single body orbit under Γ, an answer is given at all levels and offers a first measure of the contrast between solving in terms of G/Γ or fully in terms of the given periodic graph (G,Γ). We discuss aspects implicated at this higher level after solving the problem at the basic level.

Related work. Among other recent contributions to the subject, we mention [24] which addresses the two-dimensional case of periodic bar-and-joint frameworks and [32, 29] where certain classes with 'added symmetry' are investigated. The final part of the thesis [31] includes some considerations about body-and-bar frameworks on "the fixed torus". This setup corresponds to a fixed lattice of periods. A necessary condition for minimal rigidity is presented and conjectured to be also sufficient. It will be seen below that this is a restricted version of the minimal rigidity problem in terms of (G,Γ).

Overview. We describe our setting in Section 2, where we also derive the rigidity matrix. We indicate different levels for the problem of minimal rigidity in Section 3 and solve it at all levels for the base case $n = 1$. Our main result characterizing minimal rigidity in terms of the structure of the quotient graph G/Γ is proven in Section 4. In the full version of the paper [3], we present alternative, more technical proofs for the sufficiency part of our central Theorem 2. In Section 5, we generalize our main theorem to mixed periodic plate-and-bar frameworks. We conclude with the algorithmic consequences.

2. PERIODIC FRAMEWORKS

In this section we introduce our formal definition of periodic rigidity for periodic body-and-bar frameworks. For standard rigidity theoretic definitions such as infinitesimal rigidity and rigidity matrix, see [1, 39].

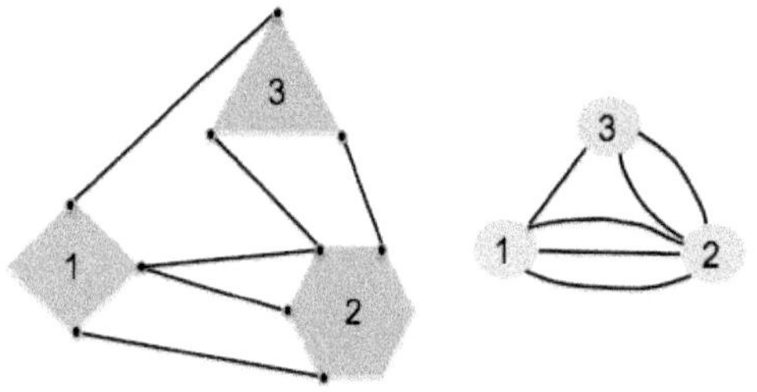

Figure 3: A finite, minimally rigid body-and-bar framework in 2D, with 3 bodies and 6 bars, and its associated multi-graph.

We remind the reader of a classical concept from finite rigidity theory: a finite **body-and-bar** d-framework is a finite collection of rigid bodies in R^d connected through bars. The endpoints of a bar lie on two distinct bodies and act as rotatable joints, i.e. they permit rigid motions of the incident objects (bodies or bars) relative to each other, while contact is maintained at the joints. Several bars may be placed between the same pair of bodies. See Fig. 3 and Fig. 4. The combinatorial (incidence) structure of such a framework is captured by a multi-graph $G = (V,E)$, whose vertices V correspond to bodies and edges to bars. Body-and-bar frameworks have been studied in several foundational papers (e.g. [34, 35]) and have practical application in studies of molecular flexibility [37, 11]. We proceed now to our new definitions.

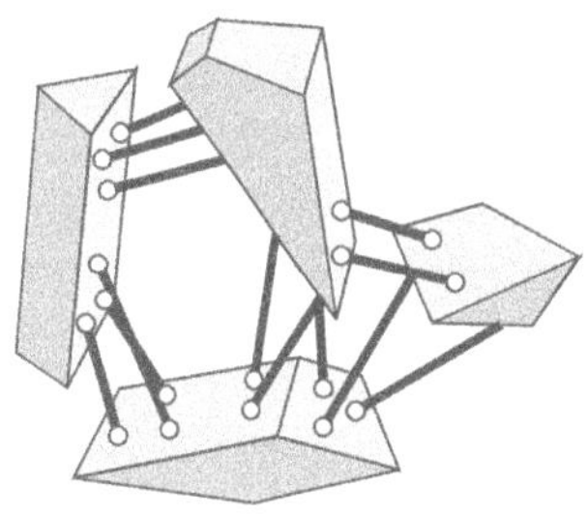

Figure 4: A finite body-and-bar framework in 3D.

A d**-periodic body-and-bar graph** is a pair (G,Γ) of an infinite multi-graph $G = (V,E)$ and a group Γ acting on it. The group Γ, called the *periodicity group* of G, is a rank d free Abelian subgroup of the automorphism group $Aut(G)$ of G, which acts without fixed points and has a finite number of vertex and edge orbits. The elements $\gamma \in \Gamma$ may be called *periods* of G. See Figure 6 for an example. Since both $n = card(V/\Gamma)$ and $m = card(E/\Gamma)$ are finite, this setting induces a finite quotient multi-graph $G/\Gamma = (V/\Gamma, E/\Gamma)$.

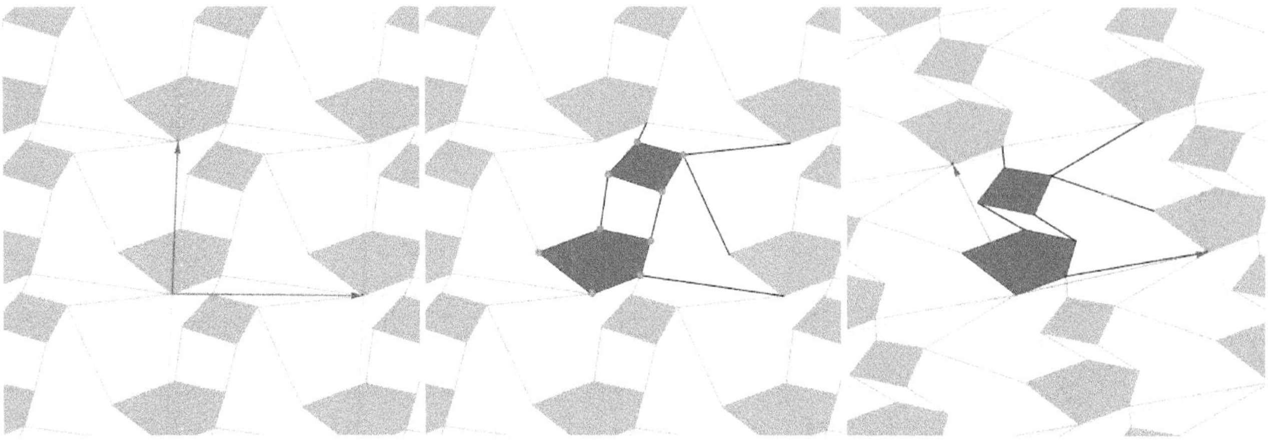

Figure 5: **Left: fragment of a 2-periodic body-and-bar framework ($d = 2$), with the generators (which are not bars) of the periodicity lattice marked by arrows and the orbit of one bar endpoint visualized as a "lattice". Center: representatives of the $n = 2$ equivalence classes of bodies and $m = 6$ classes of bars are shown in darker shades. The endpoints of the bars, marked in red, are not all distinct in this case. Right: the framework is *periodically flexible*, as illustrated by this deformation, affecting both the framework and of the periodicity lattice.**

Isometries. We denote by $E(d)$ the Euclidean group in dimension d, that is, the isometry group of the Euclidean space R^d. The connected component of the identity, denoted by $SE(d)$, is made of all the orientation preserving isometries and is referred to as the special Euclidean group or the group of rigid motions in R^d. The subgroup of translations is denoted by $\mathcal{T}(R^d) \subset SE(d)$. A transformation in $E(d)$, respectively $SE(d)$, can be represented by a pair $T = (p, M)$, where $p \in R^d$ denotes a translation and $M \in O(d)$, respectively $M \in SO(d)$, is an orthogonal transformation, resp. special orthogonal one. M is an orthogonal matrix with $|det(M)| = 1$, resp. $det(M) = 1$. A Euclidean transformation $T : R^d \to R^d$ operates by the formula $T(x) = Mx + p$. In this notation translations correspond to pairs (p, I_d), where I_d is the $d \times d$ identity matrix.

To define a body-and-bar framework, we will specify a placement in R^d of the bodies. We assume that all our bodies are oriented as the ambient space R^d. Then the placement of a body in R^d is described by a pair $T = (p, M) \in SE(d)$. In other words, we conceive of the body as described by a (positively oriented) Cartesian frame with origin at p and basis vectors corresponding to the columns of the orthogonal matrix $M \in SO(d)$.

Our next goal is to define a **periodic body-and-bar framework**. A 2D example is given in Fig. 5.

A **presentation** in R^d of a d-periodic body-and-bar graph (G, Γ) is given by an assignment of Cartesian frames to the vertices $\tau : V \to SE(d)$, alongside with an injective representation of the periodicity group Γ, $\pi : \Gamma \to \mathcal{T}(R^d)$. We refer to $\pi(\Gamma)$ as the *lattice of periods* and require that it has rank d. Furthermore, for each edge $e = (i, j) \in E$ two (arbitrary but fixed) endpoints are given, indicating where the bar is attached on the bodies corresponding to vertices i and j. Let us denote these endpoints with $q^i = q^i(e)$, $q^j = q^j(e)$. The coordinates $q^i \in R^d$ are given with respect to the Cartesian frame marking the body corresponding to i, so that for a frame $\tau(i) = (p_i, M_i)$, this end of the bar is at the point of R^d with coordinates $\tilde{q}^i = M_i q^i + p_i$.

All this data must respect the conditions required by periodicity, i.e. when we act on a vertex i by a period $\gamma \in \Gamma$, we have $\tau(\gamma i) = \pi(\gamma)\tau(i)$ and $q^i(e) = q^{\gamma i}(\gamma e)$ etc. The main definition can now be given:

A d-**periodic body-and-bar framework** is a connected d-periodic graph (G, Γ) together with a presentation (τ, π, q) where all bars have non-zero lengths.

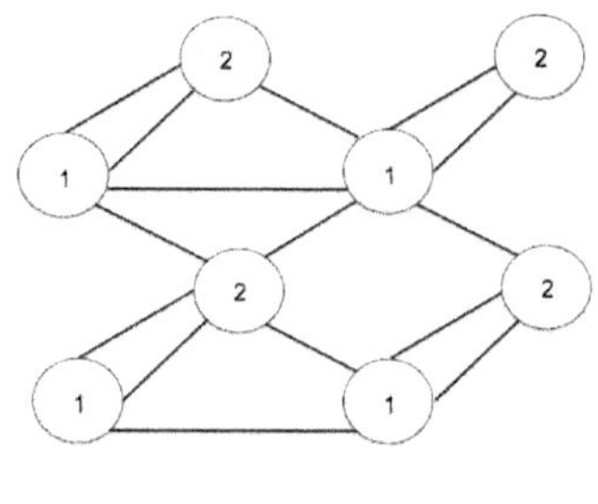

Figure 6: **A small fragment of an abstract periodic graph. Only the edges between depicted vertices are drawn. Equivalent vertices under the action of the group of periods have the same label. A realization is shown in Fig. 5 and the quotient multi-graph in Fig. 7.**

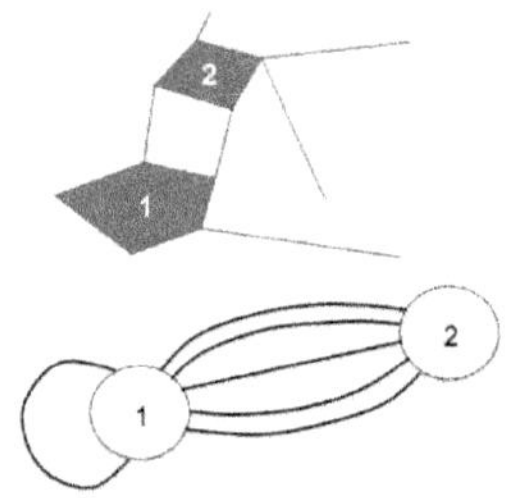

Figure 7: **Top: representatives of orbits for bodies and bars for the periodic framework from Fig. 5. Bottom: the quotient multi-graph.**

Realizations, configurations, rigid and flexible frameworks. Presentations which use the same attachment points q (relative to the corresponding bodies) and induce the same system ℓ of bar lengths form the *realization space* for the data (G, Γ, q, ℓ).

Realizations that differ by an isometry of R^d will be considered as the same *configuration*, and the *configuration space* of $(G, \Gamma, \tau, \pi, q)$ is defined as the *quotient space* of all realizations by the group $E(d)$ of all isometries of R^d. The deformation space of a periodic framework $(G, \Gamma, \tau, \pi, q)$ is the connected component of the corresponding configuration. When the deformation space consists of a single point, the framework is called *rigid*, otherwise it is *flexible*. An example of a flexible periodic framework is illustrated in Fig. 5, with its quotient graph shown in Fig. 7.

As in [1], this deformation theory has the following useful characteristics: **(i)** The periodicity group Γ is part of the structure and the deformations of a framework are those preserving this specified periodicity. **(ii)** The lattice of translations $\pi(\Gamma)$ representing Γ is allowed to vary as the framework deforms. **(iii)** The realization space is the solution space of a finite algebraic system of quadratic equations. **(iv)** There is an infinitesimal deformation theory, obtained by differentiating the equations. This leads to the concept of *infinitesimally rigid framework*, characterized by a *rigidity matrix* of maximum rank.

In particular, standard arguments from algebraic geometry (inverse function theorem) can now be used to prove that infinitesimal rigidity implies rigidity.

Minimal rigidity. Following a heuristic which goes back to Maxwell [25], we do a quick calculation of the minimal number of parameters needed to specify a rigid framework. This gives us an indication of the number of bars we would expect in a minimally rigid graph. A vertex orbit requires $\binom{d+1}{2}$ parameters to be specified, and an additional d^2 parameters specify the periodicity lattice. The total number of variables in the algebraic system is thus $n\binom{d+1}{2}+d^2$. Each edge orbit leads to a constraint (equation), and generically it eliminates one degree of freedom. Since we cannot eliminate the trivial isometries, we expect to have at most $n\binom{d+1}{2}+d^2-\binom{d+1}{2}=(n-1)\binom{d+1}{2}+d^2$ independent equations. By definition:

A **minimally rigid d-periodic body-and-bar graph** (G,Γ) is a d-periodic graph with $m=(n-1)\binom{d+1}{2}+d^2$ which has an infinitesimally rigid presentation.

For $n=2$ bodies in dimension 2 this number is 7. Since the quotient graph from Fig. 7 of the periodic framework from Fig. 5 has $6<7$ bars, we expect the periodic framework to be flexible with one degree of freedom (1dof).

We now have all the ingredients to start developing the rigidity theory of periodic body-and-bar frameworks. The first step is to define and analyze the rigidity matrix associated to a generic framework.

The rigidity matrix. We express the rigidity matrix in terms of the following coordinates: we choose a basis $\lambda_1,\dots,\lambda_d$ for the lattice of periods $\pi(\Gamma)$ and denote by Λ the $d\times d$ matrix with columns λ_i; then we choose representatives for the n equivalence classes of vertices modulo Γ and consider the corresponding bodies as marked by (p_i,M_i), $i=1,\dots,n$. We then choose an orientation of the edge set in the quotient graph G/Γ and select representatives for the equivalence classes of edges modulo Γ by the rule that all edge representatives originate at the already chosen vertex representatives. We remind the reader that the endpoints q^u, q^v of an edge representative are fixed parameters, but their choice is only subject to $\tilde{q}^u\neq\tilde{q}^v$. In order to keep the notation as simple as possible, we do not pursue the details related to marking distinctions between several bars possibly joining the same pair of bodies.

Since, by definition of representatives, one end (the origin) of any edge representative is on a body representative, the other end is on the corresponding representative translated by a period λ_e for some $\lambda_e\in\pi(\Gamma)$. Using Λ, we can define a column vector $C(e)$ with integer entries c_i, $i=1,\dots,d$ by $\lambda_e=\Lambda C(e)$.

Then, a bar constraint corresponding to an edge e between bodies i and j, with endpoints $q^i(e)$, resp. $q^j(e)$, on these two bodies, takes the form

$$||\Lambda C(e)+M_jq^j(e)-M_iq^i(e)+p_j-p_i||^2=\ell_e^2 \qquad (1)$$

with e running over all $m=|E/\Gamma|$ edge representatives.

We will eliminate the equivalence under rigid motions by assuming that the first representative body is fixed, with $(p_1,M_1)=(0,I_d)$. We note here that for an orthogonal matrix M, an 'infinitesimal variation' $\dot{M}$ takes the form $\dot{M}=MA$, with A skew-symmetric. Thus, if we differentiate the system at $(p_2,M_2),\dots,(p_n,M_n),\Lambda$, we obtain a linear system in infinitesimal variations $(\dot{p}_i,A_i)$, $i=2,\dots,n$ and $\dot{\Lambda}$ with rows

$$\langle\dot{\Lambda}c+M_jA_jq^j-M_iA_iq^i+\dot{p}_j-\dot{p}_i,\Lambda c+M_jq^j-M_iq^i+p_j-p_i\rangle=0 \qquad (2)$$

where we have dropped the notational dependency on e and used indices i,j for the chosen body-representatives, with $M_1=I_d$, $p_1=0$ and $A_1=0$. The **rigidity matrix** is the matrix of this linear system (for some specific ordering of the unknowns).

Normalized coordinates. Given any d-periodic body-and-bar framework, the manner of marking chosen representatives in each vertex orbit by Cartesian frames (p_i,M_i) is also a matter of choice and we may adopt the *normalization* which has all $p_i=0$ and $M_i=I_d$. The rigidity matrix is then the matrix of the linear system in $\dot{p}_i$, $A_i=-A_i^t$ and $\dot{\Lambda}$:

$$\langle\dot{\Lambda}c+A_jq^j-A_iq^i+\dot{p}_j-\dot{p}_i,\Lambda c+q^j-q^i\rangle=0 \qquad (3)$$

where the edge vectors $\Lambda c+q^j-q^i$ run over a complete set of edge representatives.

3. MINIMAL RIGIDITY

As already observed, up to Euclidean rigid motions, a periodic body-and-bar framework is described by $\binom{d+1}{2}(n-1)+d^2$ parameters. A bar equivalence class imposes a single constraint on these parameters, hence rigidity requires at least $\binom{d+1}{2}(n-1)+d^2$ constraints. A d-periodic body-and-bar graph (G,Γ), with $n=|E/\Gamma|$ and $m=|E/\Gamma|=\binom{d+1}{2}(n-1)+d^2$, is *minimally rigid* if it has a placement as a d-periodic body-and-bar framework in R^d with a rigidity matrix of maximal rank m. In this section we characterize these graphs in terms of a sparsity condition on their quotient graphs.

Liftings and marked quotient graphs. The problem of characterizing the structure of minimally rigid periodic graphs can be formulated at *three distinct levels*, illustrated in Fig. 8. The most discerning and demanding level concerns the periodic graph (G,Γ). Two other less discerning levels focus only on the quotient multi-graph G/Γ, with or without some additional *lifting* information retained.

The information lost upon passing from the periodic framework (G,Γ) to the quotient graph G/Γ is the *lifting* of the edges to specific vertex representatives of the same orbit. This information can be retained in the following form. We fix a basis of Γ, that is, an isomorphism $\Gamma\approx Z^d$. Then we choose vertex representatives and an orientation of the quotient graph. By using edge representatives which originate at the chosen vertex representatives, we have a well defined *lifting function* $C:E/\Gamma\to Z^d$ which allows retrieval of (G,Γ) from G/Γ. When several bars connect the same bodies, we can retain this information by joining the vertex representatives with a single edge, marked with a number indicating the edge multiplicity, as in Fig. 8(b). This 'multi-graph with multiplicities' is denoted by $\mathcal{G}/\Gamma$. Combining the edge multiplicities with the lifting function yields a labeled multi-graph with multiplicities, such as illustrated in Fig. 8(c). This contains all the information needed to reconstruct (G,Γ).

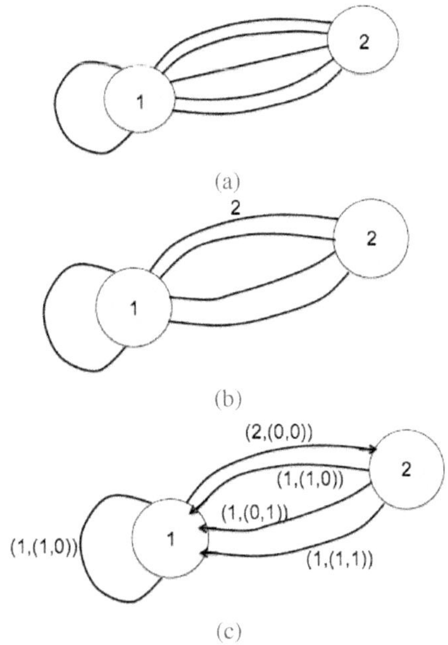

Figure 8: (a) The basic quotient graph G/Γ. (b) The quotient graph with multiplicities $\mathcal{G}/\Gamma$. (c) The labeled and oriented quotient multi-graph with multiplicities, equivalent with (G,Γ).

Remark: An *invariant* way for describing the lifting data can be given in terms of the first *homology group* (with integer coefficients) $H_1(G/\Gamma)$, where G/Γ is now seen as the topological space corresponding to the one-dimensional CW-complex defined by the quotient graph [20]. With G considered in the same topological perspective, we have a (connected) Abelian covering $G \to G/\Gamma$ with covering group Γ, and this is equivalent to a *surjective group homomorphism* $H_1(G/\Gamma) \to \Gamma$.

In the rest of this paper, we give a complete characterization of (G,Γ) for the case of a single body $n = 1$, and a characterization in terms of the quotient graph G/Γ for arbitrary n.

The $n = 1$ case. The characterization of minimal rigidity in terms of G/Γ is trivial for $n = 1$, since the only condition about the quotient graph is that of having d^2 loops. This is enough for finding generic liftings which have infinitesimally rigid framework presentations. We address here the more refined questions related to (G,Γ) and $\mathcal{G}/\Gamma$, assuming $m = d^2$. The quotient multi-graph with multiplicities $\mathcal{G}/\Gamma$ has a single vertex and a number of loops with multiplicities k_i, $i = 1,\dots,\ell$ with $\sum_{i=1}^{\ell} k_i = d^2$. Minimal rigidity in this setting is equivalent with the condition that all multiplicities k_i be at most d. This will follow from the solution of the sharper problem about the structure of (G,Γ).

Recalling that $n = 1, m = d^2$ and that we eliminate the equivalence under rigid motions by fixing $(p_1, M_1) = (0, I_d)$, we obtain from (2) a rigidity matrix with d^2 rows of the form $C(e) \otimes [\Lambda C(e) + (q^j(e) - q^i(e))]$, where the tensor notation stands for a listing of all the d^2 products of the d components of the two indicated vectors. Since the endpoints of bars can be chosen at will, the edge vectors $\Lambda C(e) + (q^j(e) - q^i(e))$ are arbitrary (non-zero) vectors, say $h = h(e)$. With simplified notation c_i for the coordinates in $C(e)$, such a row can be presented as:

$$[\, h_1(c_1\ c_2 \dots c_d),\ h_2(c_1\ c_2 \dots c_d),\ \dots,\ h_d(c_1\ c_2 \dots c_d)\,]$$

In this form we recognize a union operation on d copies of a linear matroid.

THEOREM 1. *Let (G,Γ) be a d-periodic body-and-bar graph with $n = |V/\Gamma| = 1$ and $m = |E/\Gamma| = d^2$. Then (G,Γ) has infinitesimally rigid framework presentations in R^d if and only if, for any set of equivalence classes of edges $F \subset E/\Gamma$, we have:*

$$|F| \le d \cdot dim(spanC(F)) \tag{4}$$

where $C(F) = \{C(e) : e \in F\}$.

Proof: The necessity of (4) follows directly from the form, described above, of the row constraints. The sufficiency follows from results in matroid theory, which will be used again in arguments for subsequent statements. Specifically, $d \cdot dim(spanC(F))$ is an increasing submodular function with values in the non-negative integers [28]. By [30], the associated matroid is the union of d copies of the matroid associated to the function $dim(spanC(F))$. Hence a basis of the former is a disjoint union of bases in the latter which yield the maximal rank of the rigidity matrix. □

This proof also implies our earlier claim that a quotient graph $\mathcal{G}/\Gamma$ comes from a minimally rigid (G,Γ) if and only if all multiplicities k_i, $i = 1,\dots,\ell$ are at most d. Indeed, any choice of lifting function $C : E/\Gamma \to Z^d$, with image consisting of classes of at most d vectors (corresponding to the given multiplicities) and with representatives in general position, will satisfy the criterion of Theorem 1.

4. THE QUOTIENT GRAPH

In this section we present the main theoretical result. We assume that $m = \binom{d+1}{2}(n-1) + d^2$ and characterize the quotient graphs G/Γ which are obtained from minimally rigid d-periodic body-and-bar graphs (G,Γ). In other words, we characterize minimal rigidity 'up to a generic lifting of the edge representatives'.

Graph sparsity. We say that a multi-graph $G = (V,E)$ (possibly with loops) is (a,b)-sparse, or has *$an - b$ sparsity type*, if any subset of $n' \le |V|$ vertices spans at most $an' - b$ edges. When equality holds we have a *tight* (sub)graph. See [22] for a comprehensive treatment of this kind of graph sparsity.

When $0 \le b < 2a$, the family of (a,b)-sparse graphs on a fixed number of vertices n form a matroid. When $b = a$, this matroid is the union of b graphic (tree) matroids. When $b = 0$, it is the union of a bi-cycle matroids, whose independent sets are the $(1,0)$-sparse graphs (also known in the literature as map-graphs, pseudo-forests or bi-cycles). We also obtain a matroid when $b < 0$: as shown in Corollary 4 below, it is the union of the $(a,0)$-sparsity matroid and the uniform matroid on b elements, or, equivalently, it is any graph which has $an + b$ edges and contains a $(a,0)$-sparse subgraph with n vertices.

We characterize now the quotient graphs of periodic minimally rigid graphs in terms of matroid unions of graph sparsity matroids.

THEOREM 2. *A quotient graph G/Γ with n vertices and $m = \binom{d+1}{2}(n-1) + d^2$ edges is the quotient graph of an infinitesimally rigid periodic body-and-bar framework in R^d if and only if it satisfies one and hence both of the following equivalent conditions:*

(i) it decomposes into a disjoint union of two graphs on the full set of n vertices, one with $dn-d$ edges and sparsity type, the other with $\binom{d}{2}n+\binom{d+1}{2}$ edges and sparsity type;
(ii) it contains the disjoint union of two graphs on the full set of n vertices, one with $dn-d$ edges and sparsity type, the other with $\binom{d}{2}n$ edges and sparsity type.

As shown in [2], graphs with $dn-d$ edges on n vertices and with $dn-d$ sparsity type are minimally rigid graphs for the problem of rigidly connecting with bars a set of n translated bodies (i.e. parallel Cartesian frames). Thus, the presence of a $dn-d$ sparse subgraph is related to the fact that vertex-orbits (when 'frozen solid' together into a body) provide a system of n translated bodies which has to become rigid.

Proof of sufficiency: We begin with some combinatorial background about sparsity of type $an+b$, with a and b non-negative integers. Again, in this discussion, graphs may have multiple edges and loops.

Loop breaking. When a graph has loops, we shall describe as *loop breaking* the operation of replacing a loop by an edge connecting the loop vertex with some other vertex of the graph. It is immediate that loop breaking preserves $an+b$ sparsity. The following lemma is the relevant converse.

LEMMA 3. *Let G be a graph with $an+b$ edges on n vertices, with a and b non-negative integers. Suppose G is $an+b$ sparse, that is, any subset of n' vertices has at most $an'+b$ edges between them. Then, there is a graph $\tilde{G}$, made of $an+b$ loops on n vertices, which is $an+b$ sparse and yields G after an adequate sequence of loop breaking.*

Proof: We claim that we can always replace an edge of G between vertices u and v by a loop at one of these two vertices so that $an+b$ sparsity is preserved. Suppose this were not the case. Then there are subsets of vertices U and V, with u in U but not in V and v in V but not in U with $a|U|+b$ edges in U and $a|V|+b$ edges in V. Since the union $U\cup V$ has at most $a|U\cup V|+b$ edges, the intersection $U\cap V$ must be non-void and tight. Even so, the union $U\cup V$ has already $a|U\cup V|+b$ edges without counting the edge we started with. This contradiction proves the claim and the lemma follows by iteration. Note that the resulting graph $\tilde{G}$ must have at least a loops at each vertex. □

COROLLARY 4. *With the notations of the above lemma, G is "$an+b$"-sparse if and only if it contains a spanning subgraph of sparsity type "an", with a number of edges $m=an$ on the n vertices.*

We show now that a *loop breaking* operation may be performed on minimal infinitesimally rigid frameworks. Let us assume that such a framework has a loop at vertex j in the quotient graph G/Γ. Using normalized coordinates as in (3), the loop corresponds to an equation of the form:

$$\langle \dot{\Lambda}c + A_j(q^j - q'^j), \Lambda c + q^j - q'^j\rangle = 0 \tag{5}$$

which involves only the difference $q^j - q'^j$. This allows the assumption $q'^j=0$ for the origin of the bar. We are going to replace this loop at vertex j of G/Γ by an edge from vertex i to vertex j. The corresponding bar has one endpoint at the origin of the body representative indexed by i and the other endpoint at kq^j on the body representative indexed by j translated by period Λkc, for some integer k. This gives an equation of the form:

$$\langle \dot{\Lambda}kc + A_jkq^j + \dot{p}_j - \dot{p}_i, \Lambda kc + kq^j\rangle = 0 \tag{6}$$

which, divided by k^2, takes the form:

$$\langle \dot{\Lambda}c + A_jq^j + \frac{1}{k}(\dot{p}_j - \dot{p}_i), \Lambda c + q^j\rangle = 0. \tag{7}$$

With all other rows unchanged, the rigidity matrices for the two frameworks involve only the exchange of row (5), where $q'_j=0$, with row (7) and for a sufficiently large integer k, the rank won't be affected. This shows that loop breaking operations can be performed on quotients of minimally rigid graphs with preservation of minimal rigidity.

The sufficiency proof follows from these results. Indeed, if G/Γ has edges separated into a $dn-d$ tight sparse graph and a $\binom{d}{2}n+\binom{d+1}{2}$ tight sparse graph, we can produce minimally rigid d-periodic graphs with that quotient by a two-step procedure. First, we use the $dn-d$ sparse subgraph, but replace the remainder graph with a graph made only of loops, as guaranteed by Lemma 3. Infinitesimally rigid frameworks corresponding to this quotient structure can be constructed in a natural way, as indicated immediately below. Then, as a second step, we break the appropriate loops and obtain the initial graph while maintaining minimal rigidity.

When the $\binom{d}{2}n+\binom{d+1}{2}$ tight sparse subgraph is made entirely of loops, natural liftings of G/Γ to minimally rigid (G,Γ) are constructed as follows. First, we use $\binom{d+1}{2}$ loops to make rigid the lattice of periods, as in [1, 2]. Note that bars placed as periods can be interpreted as loops at any vertex of the quotient graph. We distribute them so that exactly $\binom{d}{2}$ loops remain for bar assignment at each vertex class. Now in each vertex orbit we place the available $\binom{d}{2}$ classes of bars so that the whole body-and-bar orbit becomes a single rigid body. Thus, with the $\binom{d}{2}n+\binom{d+1}{2}$ loop subgraph we have turned the system into n parallel bodies and now the $dn-d$ sparse subgraph is precisely what we need to turn this system into a single rigid body, as in [2]. □

Proof of necessity: We consider the $m\times u$ rigidity matrix of an infinitesimally rigid periodic body-and-bar framework in R^d. The number of rows $m=\binom{d+1}{2}(n-1)+d^2$ is the number of edge orbits and the number of columns $u=\binom{d+1}{2}n+d^2=m+\binom{d+1}{2}$ is the number of *'unknowns'* in the linear constraint system for infinitesimal variations. By assumption, the rows are linearly independent and the $\binom{d+1}{2}$-dimensional kernel of our rigidity matrix consists precisely of the trivial infinitesimal rigid motions.

When we consider some subset of rows, denoted say by F, we have $|F|=rk(F)=u-dim(Ker(F))$. We denote by n_F the number of vertex orbits incident to the given edge orbits and by ω_F the number of connected components of F as a set of edges in the quotient graph. We observe that:

$$dim(Ker(F)) \geq \binom{d+1}{2} + (n-n_F)\binom{d+1}{2} + d(\omega_F - 1) \tag{8}$$

where the numbers on the right hand side count the following independent infinitesimal deformations respecting the constraint system F: trivial infinitesimal rigid motions, arbitrary infinitesimal deformations for the body representatives for non-incident vertices and relative infinitesimal translations of the connected components in F. Thus, we have a necessary *edge sparsity* condition:

$$|F| = u - dim(Ker(F)) \leq d(n_F - \omega_F) + \binom{d}{2}n_F + \binom{d+1}{2} \tag{9}$$

Written in this form, the right hand side becomes very suggestive from the point of view of *matroid theory* [28]. Indeed, if we consider a ground set of sufficiently many edges and loops on n vertices (so that any graph with m edges can be found as a subgraph), then the function defined by the right hand side is an increasing submodular function with values in the non-negative integers and determines a matroid which is, by a theorem of Pym and Perfect [30], the union of d copies of the *graphic matroid* determined by $f_1(F) = n_F - \omega_F$, then $\binom{d}{2}$ copies of the *bi-cycle matroid* determined by $f_2(F) = n_F$ and the *uniform matroid* determined by $f_3(F) = \binom{d+1}{2}$.

A basis in a union of d graphic matroids is a union of d spanning trees and this is precisely what $dn - d$ tight sparse graphs are, while $\binom{d}{2}n + \binom{d+1}{2}$ tight sparse graphs give the bases of the remaining union. Thus, the m edges corresponding to our maximal rank rigidity matrix have the claimed decomposition. □

This completes the proof of our main Theorem 2.

5. MIXED PLATE-AND-BAR

The result proven above for minimally rigid d-periodic body-and-bar frameworks is obviously related to the characterization obtained in [2] for minimally rigid d-periodic bar-and-joint graphs. Actually, these two types of periodic framework structures may be seen as the extreme cases $k = d$, respectively $k = 0$, of the family obtained by articulating with bars k-dimensional *plates* in R^d, with $0 \le k \le d$. A plate is understood in this context as a rigid object marked by a k-frame in R^d. An example is illustrated in Fig. 9 through rod-and-bar frameworks in 3D, in the finite (non-periodic) case.

Remark: The distinction between articulations made of $(d-1)$-dimensional plates, also called *panels*, and those made of bodies, which are d-dimensional plates, is that bars have to be attached at points of the panel. Otherwise, any $(d-1)$-frame has a unique extension to a d-frame with standard orientation.

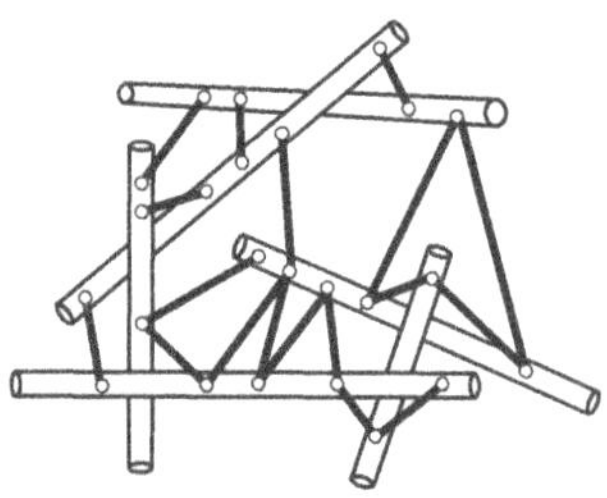

Figure 9: A finite rod-and-bar framework where each plate is one-dimensional.

We observe that, since k-frames in R^d can be parametrized by $SE(d)/SO(d-k)$, the periodic setting with variable period lattice will require for minimal rigidity the relation:

$$m = [d(k+1) - \binom{k+1}{2}]n + d^2 - \binom{d+1}{2} =$$

$$(dn-d) + (dk - \binom{k+1}{2})n + \binom{d+1}{2}.$$

The extended version of the main theorem is that a quotient graph comes from an minimally rigid d-periodic plate-and-bar framework (with all plates of dimension k) if and only if it contains the disjoint union of two graphs on the full set of n vertices, one with $dn - d$ edges and sparsity type, the other with $(dk - \binom{k+1}{2})n$ edges and sparsity type. Note that, as in Theorem 2 (ii), there is no condition on the remaining $\binom{d+1}{2}$ edges of G/Γ.

We may consider a more general setting by *mixing* various types of plates. The associated d-periodic graph (G,Γ) will have vertices labeled according to the dimension of the plate they represent. This leads to a quotient graph G/Γ with labels, or weights $k_v \in \{0,1,\ldots,d\}$ for each $v \in V/\Gamma$. With vertex orbits labeled from 1 to n, we write k_i, $i = 1,...,n$ for the corresponding (dimensional) weights. Then, the same type of reasoning will establish the following general characterization of minimal rigidity:

THEOREM 5. *A vertex weighted quotient graph G/Γ with*

$$m = d(n-1) + \sum_{i=1}^{n}[dk_i - \binom{k_i+1}{2}] + \binom{d+1}{2}$$

is the quotient graph of an infinitesimally rigid mixed plate-and-bar periodic framework in R^d if and only if it satisfies the sparsity condition:

$$|F| \le d(n_F - \omega_F) + \sum_{v\in V_F}[dk_v - \binom{k_v+1}{2}] + \binom{d+1}{2} \quad (10)$$

for every nonempty $F \subseteq E/\Gamma$, where V_F is the set of vertices incident to F.

Remark: The edge sparsity formula (10) replaces formula (9) of the 'pure' top case $k_v = d$. Its role and combinatorial unfolding can be treated similarly. Let $k'_v = [dk_v - \binom{k_v+1}{2}]$. The function defined by the right hand side of (10) determines a matroid which is the union of d copies of the graphic matroid, the uniform matroid with rank $\binom{d+1}{2}$, and the matroid determined by $f_4(F) = \sum_{v\in V_F} k'_v$.

The *'loop breaking scenario'* used in Section 3 remains valid for the mixed case and shows in particular that any basis of the matroid determined by f_4 is obtained from the graph with k'_v loops at each vertex of V/Γ by loop breaking.

In this sense, *'archetypal'* infinitesimally rigid frameworks are obtained by fixing the periodicity lattice through $\binom{d+1}{2}$ bars which are periods, then the k'_v loops at the vertex orbit of v are used to eliminate the relative motion of the k_v-frame with respect to the lattice and turn the whole orbit of v into a rigid body. Finally, the n Γ-orbits are a sysytem of translated bodies which is rendered rigid by using the remaining d spanning trees. An alternative, more detailed argument for the proof of Theorem 5 is the following.

Proof: The necessity follows from the same argument as Theorem 2 by counting the number of independent motions. We show that the sufficiency also follows from the loop breaking argument given in Theorem 2.

We first provide the corresponding combinatorial part. Let $k'_v = [dk_v - \binom{k_v+1}{2}]$ for simplicity. By a theorem of Pym and Perfect on increasing submodular functions again, the function defined by the right hand side of (10) determines the matroid, which is the union of d copies of the graphic matroid, the uniform matroid with rank $\binom{d+1}{2}$, and the matroid determined by $f_4(F) = \sum_{v\in V_F} k'_v$ which is also an increasing submodular function with values in non-negative integers.

To further decompose the last matroid, let $V_\ell = \{v \in V : k_v \ge \ell\}$ for $\ell = 0,\ldots,d$ and define $g_\ell(F) = |V_\ell \cap V_F|$. Observe $f_4 = \sum_{\ell=1}^{d-1}(d-\ell)g_\ell$, and hence the matroid determined by f_4 can be further decomposed into $\binom{d}{2}$ matroids. This gives us a decomposition of $\sum_{v\in V} k'_v$ edges into $\binom{d}{2}$ subsets $F_{k\ell}$ for $1 \le \ell \le d-1$ and $1 \le k \le d-\ell$ such that $F_{k\ell}$ is a pseudo-forest spanning only vertices of V_ℓ (i.e., $F_{k\ell}$ is an edge set on V_ℓ with $|F_{k\ell}| = |V_\ell|$ such that each connected component contains exactly one cycle). Each $F_{k\ell}$ can be oriented so that each vertex in V_ℓ has in-degree exactly one, and this orientation implies the following lemma, which supplies the converse direction of loop breaking operations.

LEMMA 6. *Let G be a graph with $\sum_{v\in V} k'_v$ edges on n vertices, with k'_v non-negative integers. Suppose $|F| \le \sum_{v\in V_F} k'_v$ for $F \subseteq E$. Then, there is a graph $\tilde{G}$, made of k'_v loops for each vertex $v \in V$, which yields G after an adequate sequence of loop breakings.*

We thus convert G/Γ to a graph which consists of $dn-d$ tight sparse subgraph, $\binom{d+1}{2}$ edges, and k'_v loops for each $v \in V$. As before $\binom{d+1}{2}$ is further converted to $\binom{d+1}{2}$ loops, and we have the base graph. $\binom{d+1}{2}$ loops are used to make rigid the lattice of periods. In each vertex orbit v, we place k'_v classes of bars, corresponding to k'_v loops attached to v, so that the whole k_v-dimensional plate-and-bar orbit becomes a single body. Thus, using the $\sum_{v\in V} k'_v + \binom{d+1}{2}$ loops we have turned the system into n parallel bodies, and the remaining $dn-d$ sparse subgraph makes this system into a single rigid body [2].

The proof is complete by performing a sequence of loop breaking operations we have shown in Theorem 2. From above construction, we may assume that each k_v-dimensional plate in the realization is the one determined by the first k_v column vectors of I_d. Then the rigidity matrix is written in the same form:

$$\langle \dot{\Lambda}c + A_j q^j - A_i q^i + \dot{p}_j - \dot{p}_i, \Lambda c + q^j - q^i \rangle = 0 \qquad (11)$$

over all edge representatives, where A_i is a skew-symmetric matrix with k'_v non-zero entries. Thus we can apply the exactly same argument as the proof of Theorem 2 to show that a loop breaking operation preserves infinitesimal rigidity. □

6. ALGORITHMS

We conclude with the algorithmic consequences of Theorem 2. Since our Maxwell-Laman characterization decomposes the matroids associated with minimal rigidity of periodic frameworks into smaller pieces of well-known (graphic and bi-cycle) matroids, we can simply apply a matroid partitioning algorithm (e.g. [10]) for checking whether a given quotient graph G/Γ is realizable as a minimally rigid periodic framework.

However, we can give a better algorithm using condition (ii) of Theorem 2. It follows from a theorem of [14] that a graph is the disjoint union of a $dn-d$ tight subgraph and a $\binom{d}{2}n$ tight subgraph if and only if it is a $\binom{d+1}{2}n-d$ tight graph. This type of sparsity is *matroidal* (i.e., the family of tight subgraphs are the bases of a matroid on the edges of the complete graph), allowing for the use of the pebble game algorithms of [22]. Thus condition (ii) can be checked in $O(n^2)$ time. The same algorithm computes the degrees of freedom of a generic realization of a given quotient graph, where m may not be exactly $\binom{d+1}{2}(n-1)+d^2$. This is done by first computing a *maximum* $\binom{d+1}{2}n-d$-sparse subgraph and then adding at most $\binom{d+1}{2}$ unused edges. The number of edges of the resulting graph gives the maximum size of independent equations, its difference from $\binom{d+1}{2}(n-1)+d^2$ being the degree of freedom.

7. CONCLUSION

Motivated by computational studies of crystalline materials, we introduced periodic body-and-bar frameworks. We gave a combinatorial characterization of those which are generically minimally rigid, obtaining the first polynomial time algorithms for recognizing them, and for computing flexibility parameters (degrees of freedom) of those which are not rigid.

Particularly relevant for molecular models are the *body-hinge* frameworks, which allow rotational motions of rigid bodies connected through rotatable hinges, as in Fig. 10. It is well known that the formal rigidity analysis of *finite* body-hinge structures can be reduced to body-and-bar frameworks, by replacing a hinge with 5 bars. This reduction, when applied to the periodic case, would have to require that the endpoints of these five bars be placed on the same body, and not lifted to some other periodic copy of it. Finding a suitable generic characterization for periodic body-and-hinge frameworks requires further investigation.

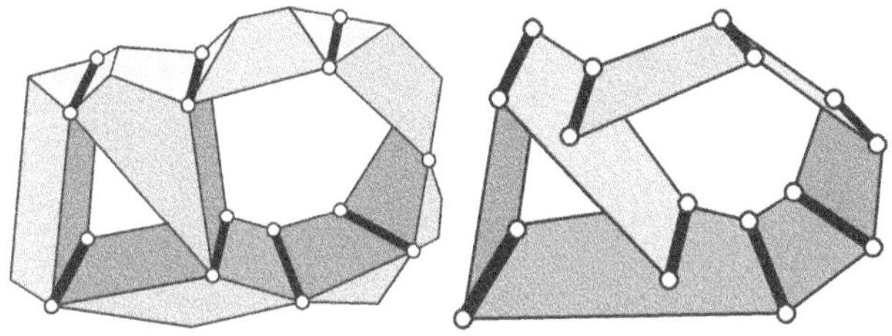

Figure 10: Finite body-and-hinge and panel-and-hinge frameworks.

We conclude with a short list of open questions: (a) characterize body-and-hinge and (b) body-and-pin periodic frameworks, and (c) verify whether the *Molecular conjecture* [19] holds as well in the periodic case. Besides intrinsic theoretical interest, such results would have immediate applications, as these structures appear in modeling families of crystalline materials.

Acknowledgments

Research of the first and second authors is supported by a DARPA "23 Mathematical Challenges" grant, under the topic "Algorithmic Origami and Biology". The second author is also funded by NSF grants CCF-1016988 and DMS-0714934. The third author is supported by Grant-in-Aid for JSPS Research Fellowships for Young Scientists and Kyoto University Start-Up Grant for Young Researchers.

8. REFERENCES

[1] C. S. Borcea and I. Streinu. Periodic frameworks and flexibility. *Proceedings of the Royal Society A 8*, 466(2121):2633–2649, September 2010.

[2] C. S. Borcea and I. Streinu. Minimally rigid periodic graphs. *Bulletin of the London Mathematical Society*, 2011.

[3] C. S. Borcea, I. Streinu, and S. Tanigawa. Periodic body-and-bar frameworks. arXiv:1110.4660, 2011.

[4] J. H. Conway and N. J. A. Sloane. *Sphere packings, lattices and groups*, volume 290 of *Grundlehren der Mathematischen Wissenschaften*. Springer-Verlag, New York, 3rd edition, 1999.

[5] O. Delgado-Friedrichs. Equilibrium placement of periodic graphs and convexity of plane tilings. *Discrete and Computational Geometry*, 33:67–81, 2005.

[6] V. S. Deshpande, M. F. Ashby, and N. A. Fleck. Foam topology: bending versus stretching dominated architectures. *Acta Materialia*, 49(6):1035–1040, 2001.

[7] G. Dolino. The α-inc-β transitions in quartz: a century of research on displacive phase transitions. *Phase Transitions*, 21:59–72, 1990.
[8] A. Donev and S. Torquato. Energy-efficient actuation in infinite lattice structures. *Journal of the Mechanics and Physics of Solids*, 51:1459–1475, 2003.
[9] M. T. Dove. Theory of displacive phase transitions in minerals. *American Mineralogist*, 82:213–244, 1997.
[10] J. Edmonds. Minimum partition of a matroid into independent sets. *J. Res. Nat. Bur. Standards Sect. B*, 69B:67–72, 1965.
[11] N. Fox, F. Jagodzinski, Y. Li, and I. Streinu. KINARI-Web: A server for protein rigidity analysis. *Nucleic Acids Research*, 39(Web Server Issue), 2011.
[12] A. L. Goodwin. Rigid unit modes and intrinsic flexibility of linearly bridged framework structures. *Physical Review B*, 74:132302–132312, 2006.
[13] S. Guest and J. W. Hutchinson. On the determinacy of repetitive structures. *Journal of the Mechanics and Physics of Solids*, 51:383–391, 2003.
[14] R. Haas, A. Lee, I. Streinu, and L. Theran. Characterizing sparse graphs by map decompositions. *Journal of Combinatorial Mathematics and Combinatorial Computing*, 62:3–11, 2007.
[15] D. Hilbert and S. Cohn-Vossen. *Geometry and the Imagination*. Chelsea Publishing Company, New York, 1962.
[16] R. G. Hutchinson and N. A. Fleck. The structural performance of the periodic truss. *Journal of the Mechanics and Physics of Solids*, 54:765–782, 2006.
[17] B. Jackson and T. Jordán. The generic rank of body-bar-and-hinge frameworks. *European Journal of Combinatorics*, 31(2):574–588, 2009.
[18] V. Kapko, M. M. J. Treacy, M. F. Thorpe, and S. Guest. On the collapse of locally isostatic networks. *Proceedings of the Royal Society A*, 465:3517–3530, 2009.
[19] N. Katoh and S. Tanigawa. A proof of the molecular conjecture. *Discrete and Computational Geometry*, 45(4):647–700, 2011. Prelim. version in Proc. Symp. Comp. Geometry (SoCG), 2008.
[20] M. Kotani and T. Sunada. Standard realizations of crystal lattices via harmonic maps. *Transactions of the American Mathematical Society*, 353:1–20, 2001.
[21] G. Laman. On graphs and rigidity of plane skeletal structures. *Journal of Engineering Mathematics*, 4:331–340, 1970.
[22] A. Lee and I. Streinu. Pebble game algorithms and sparse graphs. *Discrete Mathematics*, 308(8):1425–1437, April 2008.
[23] J. Lima-de Faria, editor. *Historical Atlas of Crystallography*. Kluwer, IUC, 1990.
[24] J. Malestein and L. Theran. Generic combinatorial rigidity of periodic frameworks. arxiv.1008.1837, 2010.
[25] J. C. Maxwell. On the calculation of the equilibrium and stiffness of frames. *Philosophical Magazine*, 27:294–299, 1864.
[26] H. D. Megaw. *Crystal structures: a working approach*. W.B.Saunders Co., Philadelphia, 1973.
[27] M. O'Keeffe, M. Eddaoudi, H. Li, T. Reineke, and O. M. Yaghi. Frameworks for extended solids: geometrical design principles. *Journal of Solid State Chemistry*, 152:3–20, 2000.
[28] J. G. Oxley. *Matroid theory*. The Clarendon Press, Oxford University Press, New York, 1992.
[29] S. C. Power. Crystal frameworks, symmetry and affinely periodic flexes. arXiv:1103.1914, 2011.
[30] J. S. Pym and H. Perfect. Submodular functions and independence structures. *Journal of Mathematical Analysis and Applications*, 30:1–31, 1970.
[31] E. Ross. *Geometric and combinatorial rigidity of periodic frameworks as graphs on the torus*. PhD thesis, York University, Ontario, Canada, Toronto, May 2011.
[32] E. Ross, B. Schulze, and W. Whiteley. Finite motions from periodic frameworks with added symmetry. *International Journal Solids and Structures*, 48:1711–1729, 2011.
[33] S. Tanigawa. Generic rigidity matroids with Dilworth truncations. arXiv:1010.5699, 2010.
[34] T.-S. Tay. Rigidity of multigraphs I: linking rigid bodies in n-space. *Journal of Combinatorial Theory, Series B*, 26:95–112, 1984.
[35] T.-S. Tay. Linking $(n-2)$-dimensional panels in n-space II: $(n-2,2)$-frameworks and body and hinge structures. *Graphs and Combinatorics*, 5:245–273, 1989.
[36] T.-S. Tay. Linking $(n-2)$-dimensional panels in n-space I:$(k-1,k)$-graphs and $(k-1,k)$-frames. *Graphs and Combinatorics*, 7(3):289–304, 1991.
[37] M. F. Thorpe, M. V. Chubynsky, B. M. Hespenheide, S. Menor, D. J. Jacobs, L. A. Kuhn, M. I. Zavodszky, M. Lei, A. J. Rader, and W. Whiteley. *Flexibility in Biomolecules*, chapter 6, pages 97–112. Current Topics in Physics. Imperial College Press, London, 2005. R.A. Barrio and K.K. Kaski, eds.
[38] H. Weyl. *Symmetry*. Princeton University Press, 1952.
[39] W. Whiteley. The union of matroids and the rigidity of frameworks. *SIAM Journal Discrete Mathematics*, 1(2):237–255, May 1988.

On Sub-Determinants and the Diameter of Polyhedra

Nicolas Bonifas
LIX, École Polytechnique
Palaiseau (France)
and IBM, Gentilly (France)
bonifas@lix.polytechnique.fr

Marco Di Summa
Dipartimento di Matematica
Università di Padova (Italy)
disumma@math.unipd.it

Friedrich Eisenbrand
Technische Universität Berlin
(Germany)
eisenbrand@tu-berlin.de

Nicolai Hähnle
Technische Universität Berlin
(Germany)
haehnle@math.tu-berlin.de

Martin Niemeier
Technische Universität Berlin
(Germany)
martin.niemeier@tu-berlin.de

ABSTRACT

We derive a new upper bound on the diameter of the graph of a polyhedron $P = \{x \in \mathbb{R}^n : Ax \leqslant b\}$, where $A \in \mathbb{Z}^{m\times n}$. The bound is polynomial in n and the largest absolute value of a sub-determinant of A, denoted by Δ. More precisely, we show that the diameter of P is bounded by $O\left(\Delta^2 n^4 \log n\Delta\right)$. If P is bounded, then we show that the diameter of P is at most $O\left(\Delta^2 n^{3.5} \log n\Delta\right)$.

For the special case in which A is a totally unimodular matrix, the bounds are $O\left(n^4 \log n\right)$ and $O\left(n^{3.5} \log n\right)$ respectively. This improves over the previous best bound of $O(m^{16} n^3 (\log mn)^3)$ due to Dyer and Frieze [5].

Categories and Subject Descriptors

G.2 [**Mathematics of Computing**]: Discrete Mathematics

Keywords

Diameter of polyhedra, polyhedral graph, totally unimodular matrices, isoperimetric inequality

1. INTRODUCTION

One of the fundamental open problems in optimization and discrete geometry is the question whether the diameter of a polyhedron can be bounded by a polynomial in the dimension and the number of its defining inequalities. The problem is readily explained. A *polyhedron* is a set of the form $P = \{x \in \mathbb{R}^n : Ax \leqslant b\}$, where $A \in \mathbb{R}^{m\times n}$ is a matrix and $b \in \mathbb{R}^m$ is an m-dimensional vector.[1] A *vertex* of P is a point x^* such that there exist n linearly independent rows of A whose corresponding inequalities of $Ax \leqslant b$ are satisfied by x^* with equality. Two different vertices x^* and y^* are *neighbors* if there exist $n-1$ linearly independent rows of A whose corresponding inequalities of $Ax \leqslant b$ are satisfied with equality both by x^* and y^*. In this way, we obtain the undirected *polyhedral graph* with edges being pairs of neighboring vertices of P. This graph is connected. The *diameter* of P is the smallest natural number that bounds the length of a shortest path between any pair of vertices in this graph, where the length of a path is defined as the number of edges in the path. The question is now as follows:

> Can the diameter of a polyhedron $P = \{x \in \mathbb{R}^n : Ax \leqslant b\}$ be bounded by a polynomial in m and n?

Despite a lot of research effort during the last 50 years, the gap between lower and upper bounds on the diameter remains huge. While, when the dimension n is fixed, the diameter can be bounded by a linear function of m [14, 2], for the general case the best upper bound, due to Kalai and Kleitman [11], is $O(m^{1+\log n})$. The best lower bound is of the form $(1+\varepsilon)\cdot m$ for some $\varepsilon > 0$ in fixed and sufficiently large dimension n. This is due to a celebrated recent result of Santos [19] who disproved the until then longstanding *Hirsch conjecture*. The Hirsch conjecture stated that the diameter of a bounded polyhedron[2] is at most $m-n$. Interestingly, this huge gap (polynomial versus quasi-polynomial) is also not closed in a very simple combinatorial abstraction of polyhedral graphs [6]. However, it was shown by Vershynin [20] that every polyhedron can be perturbed by a small random amount so that the expected diameter of the resulting polyhedron is bounded by a polynomial in m and n. See Kim and Santos [12] for a recent survey.

In light of the importance and apparent difficulty of the open question above, many researchers have shown that it can be answered in an affirmative way in some special cases. Naddef [17] proved that the Hirsch conjecture holds true for 0/1-polytopes. Orlin [18] provided a quadratic upper bound for flow-polytopes. Brightwell et al. [3] showed that the diameter of the transportation polytope is linear in m and n,

[1] Throughout this paper we assume that the matrix A has full column-rank, i.e., the columns of A are linearly independent. If this is not the case, then P has no vertices.

[2] A counterexample to the same conjecture for unbounded polyhedra was found in 1967 by Klee and Walkup [13].

SCG'12, June 17–20, 2012, Chapel Hill, North Carolina, USA.

and a similar result holds for the dual of a transportation polytope [1] and the axial 3-way transportation polytope [4].

The results on flow polytopes and classical transportation polytopes concern polyhedra defined by *totally unimodular matrices*, i.e., integer matrices whose sub-determinants are $0, \pm 1$. For such polyhedra Dyer and Frieze [5] had previously shown that the diameter is bounded by a polynomial in n and m. Their bound is $O(m^{16}n^3(\log mn)^3)$. Their result is also algorithmic: they show that there exists a randomized simplex-algorithm that solves linear programs defined by totally unimodular matrices in polynomial time.

Our main result is a generalization and considerable improvement of the diameter bound of Dyer and Frieze. We show that the diameter of a polyhedron $P = \{x \in \mathbb{R}^n : Ax \leqslant b\}$, with $A \in \mathbb{Z}^{m\times n}$ is bounded by $O\left(\Delta^2 n^4 \log n\Delta\right)$. Here, Δ denotes the largest absolute value of a *sub-determinant* of A. If P is bounded, i.e., a *polytope*, then we can show that the diameter of P is at most $O\left(\Delta^2 n^{3.5} \log n\Delta\right)$. To compare our bound with the one of Dyer and Frieze one has to set Δ above to one and obtains $O\left(n^4 \log n\right)$ and $O\left(n^{3.5}\log n\right)$ respectively. Notice that our bound is independent of m, i.e., the number of rows of A.

The proof method

Let u and v be two vertices of P. We estimate the maximum number of iterations of two breadth-first-search explorations of the polyhedral graph, one initiated at u, the other initiated at v, until a common vertex is discovered. The diameter of P is at most twice this number of iterations. The main idea in the analysis is to reason about the normal cones of vertices of P and to exploit a certain volume expansion property.

We can assume that $P = \{x \in \mathbb{R}^n : Ax \leqslant b\}$ is *non-degenerate*, i.e., each vertex has exactly n tight inequalities. This can be achieved by slightly perturbing the right-hand side vector b: in this way the diameter can only grow. We denote the polyhedral graph of P by $G_P = (V, E)$. Let $v \in V$ now be a vertex of P. The *normal cone* C_v of v is the set of all vectors $c \in \mathbb{R}^n$ such that v is an optimal solution of the linear program $\max\{c^T x : x \in \mathbb{R}^n,\ Ax \leqslant b\}$. The normal cone C_v of a vertex of v is a full-dimensional simplicial polyhedral cone. Two vertices v and v' are adjacent if and only if C_v and $C_{v'}$ share a facet. No two distinct normal cones share an interior point. Furthermore, if P is a polytope, then the union of the normal cones of vertices of P is the complete space $\mathbb{R}^n$.

We now define the *volume* of a set $U \subseteq V$ of vertices as the volume of the union of the normal cones of U intersected with the *unit ball* $B_n = \{x \in \mathbb{R}^n : \|x\|_2 \leqslant 1\}$, i.e.,

$$\operatorname{vol}(U) := \operatorname{vol}\left(\bigcup_{v\in U} C_v \cap B_n\right).$$

Consider an iteration of breadth-first-search. Let $I \subseteq V$ be the set of vertices that have been discovered so far. Breadth-first-search will next discover the neighborhood of I, which we denote by $\mathcal{N}(I)$. Our diameter bound for polytopes is a consequence of the following volume-expansion property. We provide a proof of this lemma in the next section.

Lemma 1. *Let $P = \{x \in \mathbb{R}^n : Ax \leqslant b\}$ be a polytope where all sub-determinants of $A \in \mathbb{Z}^{m\times n}$ are bounded by Δ in absolute value and let $I \subseteq V$ be a set of vertices of G_P with $\operatorname{vol}(I) \leqslant (1/2)\cdot \operatorname{vol}(B_n)$. Then the volume of the neighborhood of I is at least*

$$\operatorname{vol}(\mathcal{N}(I)) \geqslant \sqrt{\frac{2}{\pi}}\frac{1}{\Delta^2 n^{2.5}}\cdot \operatorname{vol}(I).$$

With this lemma at hand, one can easily derive our bound on the diameter of polytopes.

Theorem 2. *Let $P = \{x \in \mathbb{R}^n : Ax \leqslant b\}$ be a polytope where all subdeterminants of $A \in \mathbb{Z}^{m\times n}$ are bounded by Δ in absolute value. The diameter of P is bounded by $O\left(\Delta^2 n^{3.5}\log n\Delta\right)$.*

Proof. We estimate the maximum number of iterations of breadth-first-search until the total volume of the discovered vertices exceeds $(1/2)\cdot\operatorname{vol}(B_n)$. This is an upper bound on the aforementioned maximum number of iterations of two breadth-first-search explorations until a common vertex is discovered.

Suppose we start at vertex v and let I_j be the vertices that have been discovered during the first j iterations. We have $I_o = \{v\}$. If $j \geqslant 1$ and $\operatorname{vol}(I_{j-1}) \leqslant (1/2)\cdot\operatorname{vol}(B_n)$ we have by Lemma 1

$$\begin{aligned}\operatorname{vol}(I_j) &\geqslant \left(1+\sqrt{\frac{2}{\pi}}\frac{1}{\Delta^2 n^{2.5}}\right)\operatorname{vol}(I_{j-1})\\ &\geqslant \left(1+\sqrt{\frac{2}{\pi}}\frac{1}{\Delta^2 n^{2.5}}\right)^j \operatorname{vol}(I_0).\end{aligned}$$

The condition $\operatorname{vol}(I_j) \leqslant (1/2)\cdot\operatorname{vol}(B_n)$ implies

$$\left(1+\frac{1}{\sqrt{\frac{\pi}{2}}\Delta^2 n^{2.5}}\right)^j \operatorname{vol}(I_0) \leqslant 2^n.$$

This is equivalent to

$$j\cdot\ln\left(1+\frac{1}{\sqrt{\frac{\pi}{2}}\Delta^2 n^{2.5}}\right) \leqslant \ln(2^n/\operatorname{vol}(I_0)).$$

For $0 \leqslant x \leqslant 1$ one has $\ln(1+x) \geqslant x/2$ and thus the inequality above implies

$$j \leqslant \sqrt{2\pi}\Delta^2 n^{2.5}\cdot\ln(2^n/\operatorname{vol}(I_0)). \tag{1}$$

To finish the proof we need a lower bound on $\operatorname{vol}(I_0)$, i.e., the n-dimensional volume of the set $C_v \cap B_n$. The normal cone C_v contains the full-dimensional simplex spanned by 0 and the n row-vectors $a_{i_1},\dots,a_{i_n}$ of A that correspond to the inequalities of $Ax \leqslant b$ that are tight at v. Since A is integral, the volume of this simplex is at least $1/n!$. Furthermore, if this simplex is scaled by $1/\max\{\|a_{i_k}\| : k = 1,\dots,n\}$, then it is contained in the unit ball. Since each component of A is itself a sub-determinant, one has $\max\{\|a_{i_k}\| : k = 1,\dots,n\} \leqslant \sqrt{n}\Delta$ and thus $\operatorname{vol}(I_0) \geqslant 1/(n!\cdot n^{n/2}\Delta^n)$. It follows that (1) implies $j = O\left(\Delta^2 n^{3.5}\log n\Delta\right)$. □

Remarks. The result of Dyer and Frieze [5] is also based on analyzing expansion properties via isoperimetric inequalities. It is our choice of normal cones as the natural geometric representation, and the fact that we only ask for volume expansion instead of expansion of the graph itself, that allows us to get a better bound. Expansion properties of the graphs of general classes of polytopes have also been studied elsewhere in the literature, e.g. [10, 9].

Organization of the paper

The next section is devoted to a proof of the volume-expansion property, i.e., Lemma 1. The main tool that is used here is a classical *isoperimetric inequality* that states that among measurable subsets of a sphere with fixed volume, spherical caps have the smallest circumference. Section 3 deals with unbounded polyhedra. Compared to the case of polytopes, the problem that arises here is the fact that the union of the normal cones is not the complete space $\mathbb{R}^n$. To tackle this case, we rely on an isoperimetric inequality of Lovász and Simonovits [15]. Finally, we discuss how our bound can be further generalized. In fact, not all sub-determinants of A need to be at most Δ but merely the entries of A and the $(n-1)$-dimensional sub-determinants have to be bounded by Δ, which yields a slightly stronger result.

2. VOLUME EXPANSION

This section is devoted to a proof of Lemma 1. Throughout this section, we assume that $P = \{x \in \mathbb{R}^n : Ax \leqslant b\}$ is a polytope. We begin with some useful notation. A (not necessarily convex) *cone* is a subset of $\mathbb{R}^n$ that is closed under the multiplication with non-negative scalars. The intersection of a cone with the unit ball B_n is called a *spherical cone*. Recall that C_v denotes the normal cone of the vertex v of P. We denote the spherical cone $C_v \cap B_n$ by S_v and, for a subset $U \subseteq V$, the spherical cone $\bigcup_{v\in U} S_v$ by S_U. Our goal is to show that the following inequality holds for each $I \subseteq V$ with $\operatorname{vol}(S_I) \leqslant \frac{1}{2}\operatorname{vol}(B_n)$:

$$\operatorname{vol}(S_{\mathcal{N}(I)}) \geqslant \sqrt{\frac{2}{\pi}}\frac{1}{\Delta^2 n^{2.5}} \cdot \operatorname{vol}(S_I). \tag{2}$$

Recall that two vertices are adjacent in G_P if and only if their normal cones have a common facet. This means that the neighbors of I are those vertices u for which S_u has a facet which is part of the surface of the spherical cone S_I. In an iteration of breadth-first-search we thus augment the set of discovered vertices I by those vertices u that can "dock" on S_I via a common facet. We call the $(n-1)$-dimensional volume of the surface of a spherical cone S that is not on the sphere, the *dockable surface* $D(S)$, see Figure 1.

The *base* of S is the intersection of S with the unit sphere. We denote the area of the base by $B(S)$. By *area* we mean the $(n-1)$-dimensional measure of some surface. Furthermore, $L(S)$ denotes the length of the relative boundary of the base of S. We use the term *length* to denote the measure of an $(n-2)$-dimensional volume, see Figure 1.

Given any spherical cone S in the unit ball, the following well-known relations follow from basic integration:

$$\operatorname{vol}(S) = \frac{B(S)}{n}, \quad D(S) = \frac{L(S)}{n-1}. \tag{3}$$

To obtain the volume expansion relation (2) we need to bound the dockable surface of a spherical cone from below by its volume and, for a simplicial spherical cone, we need an upper bound on the dockable surface by its volume. More precisely, we show that for every simplicial spherical cone S_v one has

$$\frac{D(S_v)}{\operatorname{vol}(S_v)} \leqslant \Delta^2 n^3 \tag{4}$$

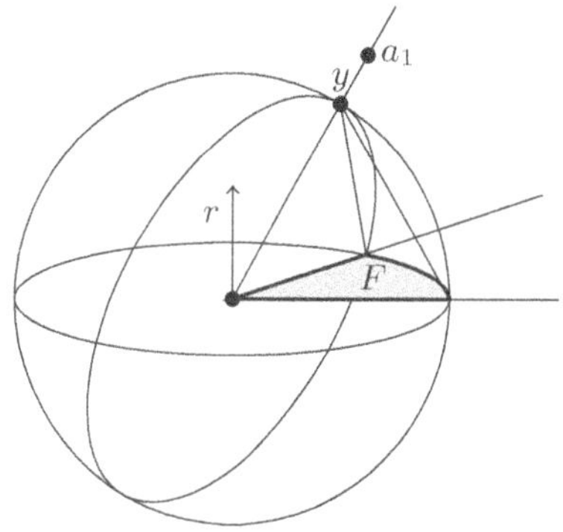

Figure 2: Proof of Lemma 3.

and for any spherical cone one has

$$\frac{D(S)}{\operatorname{vol}(S)} \geqslant \sqrt{\frac{2n}{\pi}}. \tag{5}$$

Once inequalities (4) and (5) are derived, the bound (2) can be obtained as follows. All of the dockable surface of S_I must be "consumed" by the neighbors of I. Using (5) one has thus

$$\sum_{v\in\mathcal{N}(I)} D(S_v) \geqslant D(S_I) \geqslant \sqrt{\frac{2n}{\pi}} \cdot \operatorname{vol}(S_I). \tag{6}$$

On the other hand, (4) implies

$$\sum_{v\in\mathcal{N}(I)} D(S_v) \leqslant \Delta^2 n^3 \cdot \sum_{v\in\mathcal{N}(I)} \operatorname{vol}(S_v) = \Delta^2 n^3 \cdot \operatorname{vol}(S_{\mathcal{N}(I)}). \tag{7}$$

These last two inequalities imply inequality (2). The remainder of this section is devoted to proving (4) and (5).

2.1 Area to volume ratio of a spherical simplicial cone

We will first derive inequality (4).

Lemma 3. *Let v be a vertex of P. One has*

$$\frac{D(S_v)}{\operatorname{vol}(S_v)} \leqslant \Delta^2 n^3.$$

Proof. Let F be a facet of a spherical cone S_v. Let y be the vertex of S_v not contained in F. Let Q denote the convex hull of F and y (see Figure 2). We have $Q \subseteq S_v$ because S_v is convex. Moreover, if h_F is the Euclidean distance of y from the hyperplane containing F, then

$$\operatorname{vol}(S_v) \geqslant \operatorname{vol}(Q) = \frac{\operatorname{area}(F) \cdot h_F}{n}.$$

Summing over the facets of S_v, we find

$$\frac{D(S_v)}{\operatorname{vol}(S_v)} = \sum_{\text{facet } F} \frac{\operatorname{area}(F)}{\operatorname{vol}(S_v)} \leqslant n \cdot \sum_{\text{facet } F} \frac{1}{h_F}. \tag{8}$$

It remains to provide a lower bound on h_F. Let $a_1, \ldots, a_n$ be the row-vectors of A defining the extreme rays of the normal cone of v, and let A_v be the non-singular matrix whose rows are $a_1, \ldots, a_n$. Furthermore, suppose that the vertex y lies on the ray generated by a_1. Let H be the hyperplane generated by $a_2, \ldots, a_n$. The distance $d(y, H)$ of y to H is

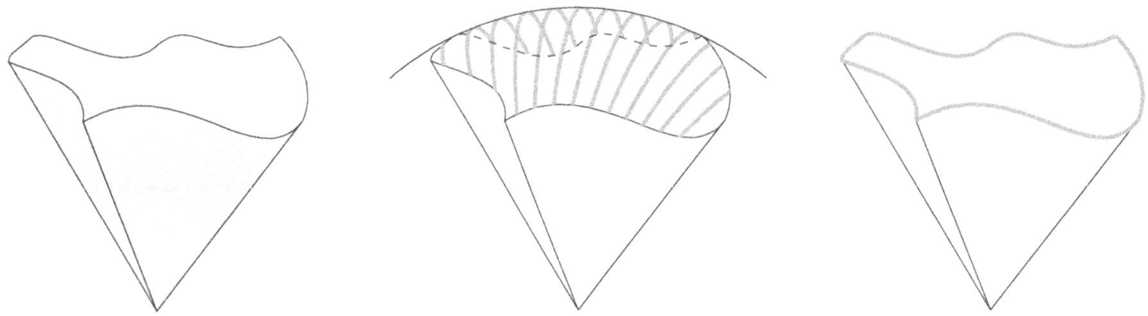

(a) Dockable surface of S.

(b) Base of S.

(c) Relative boundary of the base of S.

Figure 1: Illustration of $D(S)$, $B(S)$ and $L(S)$.

equal to $d(a_1, H)/\|a_1\|$. Let $b_1, \ldots, b_n$ be the columns of the adjugate of A_v. The column-vector b_1 is integral and each component of b_1 is bounded by Δ. Furthermore b_1 is orthogonal to each of $a_2, \ldots, a_n$. Thus $d(a_1, H)$ is the length of the projection of a_1 to b_1, which is $|\langle a_1, b_1\rangle|/\|b_1\| \geqslant 1/(\sqrt{n} \cdot \Delta)$, since a_1 and b_1 are integral. Thus

$$h_F = d(y, H) \geqslant \frac{1}{n\Delta^2}.$$

Plugging this into (8) completes the proof. □

2.2 An isoperimetric inequality for spherical cones

We now derive the lower bound (5) on the area to volume ratio for a general spherical cone. To do that, we assume that the spherical cone has the least favorable shape for the area to volume ratio and derive the inequality for cones of this shape. Here one uses classical isoperimetric inequalities. The basic isoperimetric inequality states that the measurable subset of $\mathbb{R}^n$ with a prescribed volume and minimal area is the ball of this volume. In this paper, we need Lévy's isoperimetric inequality, see e.g. [8, Theorem 2.1], which can be seen as an analogous result for spheres: it states that a measurable subset of the sphere of prescribed area and minimal boundary is a spherical cap.

A spherical cone S is a *cone of revolution* if there exist a vector v and an angle $0 < \theta \leqslant \pi/2$ such that S is the set of vectors in the unit ball that form an angle of at most θ with v:

$$S = \left\{ x \in B_n : \frac{v^T x}{\|v\|\|x\|} \geqslant \cos\theta \right\}.$$

Note that a spherical cone is a cone of revolution if and only if its base is a spherical cap. We also observe that two spherical cones of revolution, defined by two different vectors but by the same angle, are always congruent. Therefore, in the following we will only specify the angle of a cone of revolution.

LEMMA 4. *The spherical cone of given volume with minimum lateral surface is a cone of revolution.*

PROOF. By the first equation of (3), every spherical cone of volume V intersects the unit sphere in a surface of area nV. Furthermore, by the second equation of (3), the length of the boundary of this surface is proportional to the area of the lateral surface of the cone. Then the problem of finding the spherical cone of volume V with the minimum lateral surface can be rephrased as follows: Find a surface of area nV on the unit sphere having the boundary of minimum length. By Lévy's isoperimetric inequality for spheres, the optimal shape for such a surface is a spherical cap. As observed above, this corresponds to a cone of revolution. □

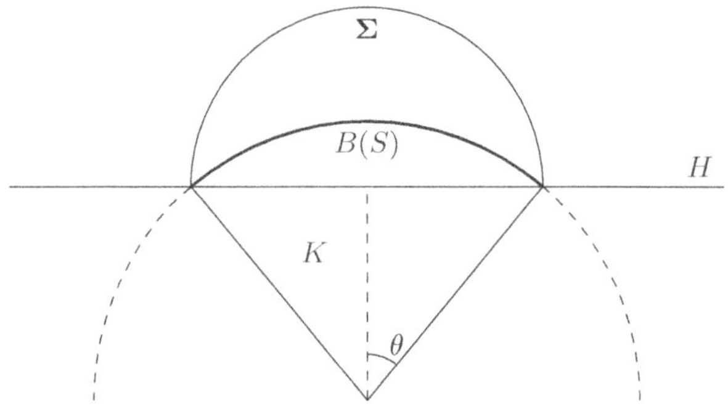

Figure 3: Proof of Lemma 5.

LEMMA 5. *Let S be a spherical cone of revolution of angle $0 < \theta \leqslant \pi/2$. Then*

$$\frac{D(S)}{\mathrm{vol}(S)} \geqslant \sqrt{\frac{2n}{\pi}}.$$

PROOF. Using (3), we have to show that

$$\frac{L(S)}{B(S)} \geqslant \sqrt{\frac{2}{\pi}} \frac{n-1}{\sqrt{n}}. \tag{9}$$

This is done in two steps. We first prove that this ratio is minimal for S being the half-ball, i.e., $\theta = \pi/2$. Then we show that $\frac{L(S)}{B(S)} \geqslant \sqrt{\frac{2}{\pi}} \frac{n-1}{\sqrt{n}}$ holds for the half-ball.

Let H be the hyperplane containing the boundary of the base of S. Then H divides S into two parts: a truncated cone K and the convex hull of a spherical cap. The radius r of the base of K is bounded by one.

Consider now the half-ball that contains $B(S)$ and that has $H \cap B_n$ as its flat-surface, see Figure 3, and let Σ denote the area of the corresponding half-sphere. One has $B(S) \leqslant \Sigma$ and thus

$$\frac{L(S)}{B(S)} \geqslant \frac{L(S)}{\Sigma}.$$

Now Σ and $L(S)$ are the surface of an $(n-1)$-dimensional half-sphere of radius r and the length of its boundary respectively. If we scale this half-sphere by a factor of $1/r$, we obtain the unit half-ball and its length respectively. Since scaling by a factor of $1/r$ increases areas by a factor of $1/r^{n-1}$ and lengths by a factor of $1/r^{n-2}$, we have that $\frac{L(S)}{\Sigma}$ is at least the length of the unit-half-ball divided by the area of the base of the half-ball.

Suppose now that S is the half-unit-ball. We show that the inequality $L(S)/B(S) \geqslant \sqrt{\frac{2}{\pi}\frac{n-1}{\sqrt{n}}}$ holds. The base of S is a half unit sphere and $L(S)$ is the length of the boundary of a unit ball of dimension $n-1$. Thus

$$B(S) = \frac{n}{2}\frac{\pi^{n/2}}{\Gamma\left(\frac{n}{2}+1\right)}, \quad L(S) = \frac{(n-1)\pi^{(n-1)/2}}{\Gamma\left(\frac{n-1}{2}+1\right)},$$

where Γ is the well-known Gamma function. Using the fact that $\Gamma(x+1/2)/\Gamma(x) \geqslant \sqrt{x-\frac{1}{4}}$ for all $x > \frac{1}{4}$ (see, e.g., [16]), one easily verifies that

$$\Gamma\left(\frac{n}{2}+1\right) \geqslant \sqrt{\frac{n}{2}} \cdot \Gamma\left(\frac{n-1}{2}+1\right).$$

It follows that

$$\frac{L(S)}{B(S)} = \frac{2}{\sqrt{\pi}}\frac{n-1}{n}\frac{\Gamma\left(\frac{n}{2}+1\right)}{\Gamma\left(\frac{n-1}{2}+1\right)} \geqslant \sqrt{\frac{2}{\pi}} \cdot \frac{n-1}{\sqrt{n}}.$$

□

Finally we are now ready to consider the case of an arbitrary spherical cone.

LEMMA 6. *Let S be a (not necessarily convex) spherical cone with* $\text{vol}(S) \leqslant \frac{1}{2}\text{vol}(B_n)$. *Then*

$$\frac{D(S)}{\text{vol}(S)} \geqslant \sqrt{\frac{2n}{\pi}}.$$

PROOF. Let S^* be a spherical cone of revolution with the same volume as S. By Lemma 4, $D(S) \geqslant D(S^*)$. Now, using Lemma 5 one has

$$\frac{D(S)}{\text{vol}(S)} \geqslant \frac{D(S^*)}{\text{vol}(S^*)} \geqslant \sqrt{\frac{2n}{\pi}}.$$

□

This was the final step in the proof of Lemma 1 and thus we have also proved Theorem 2, our main result on polytopes. The next section is devoted to unbounded polyhedra.

3. THE CASE OF AN UNBOUNDED POLYHEDRON

If the polyhedron P is unbounded, then the union of the normal cones of all vertices of P forms a proper subset K' of $\mathbb{R}^n$: namely, K' is the set of objective functions c for which the linear program $\max\{c^T x : x \in P\}$ has finite optimum. Similarly, the set $K' \cap B_n$ is a proper subset of B_n. Then, given the union of the spherical cones that have already been discovered by the breadth-first-search (we denote this set by S), we should redefine the dockable surface of S as that part of the lateral surface of S that is shared by some neighboring cones. In other words, we should exclude the part lying on the boundary of $K' \cap B_n$. However, this implies that the result of Lemma 6 does not hold in this context.

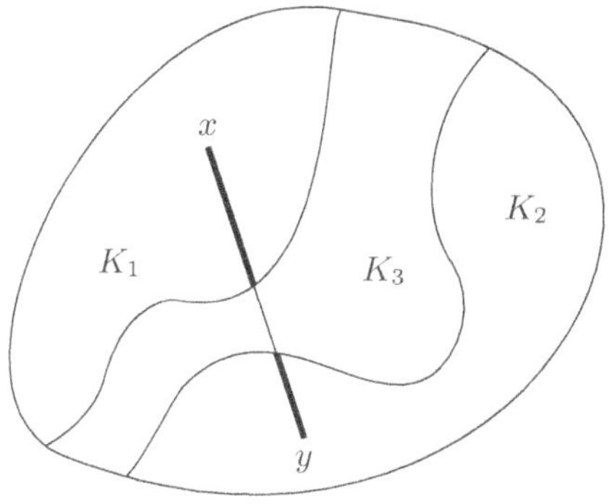

Figure 4: Illustration of the Lovász-Simonovits inequality.

To overcome this difficulty, we make use of the Lovász-Simonovits inequality, which we now recall. Below we use notation $d(X,Y)$ to indicate the Euclidean distance between two subsets $X, Y \subseteq \mathbb{R}^n$, i.e., $d(X,Y) = \inf\{\|x-y\| : x \in X, y \in Y\}$. Also, $[x,y]$ denotes the segment connecting two points $x, y \in \mathbb{R}^n$ (see Figure 4).

THEOREM 7. *[15] Let $K \subseteq \mathbb{R}^n$ be a convex compact set, $0 < \varepsilon < 1$ and (K_1, K_2, K_3) be a partition of K into three measurable sets such that*

$$\forall x, y \in K, \quad d([x,y] \cap K_1, [x,y] \cap K_2) \geqslant \varepsilon \cdot \|x-y\|. \quad (10)$$

Then

$$\text{vol}(K_3) \geqslant \frac{2\varepsilon}{1-\varepsilon} \min\left(\text{vol}(K_1), \text{vol}(K_2)\right).$$

We now illustrate how the above result can be used in our context. Let $K = K' \cap B_n$ and observe that K is a convex and compact set. Let $S \subseteq K$ be the union of the spherical cones that have already been discovered by the breadth-first-search. We define the dockable surface of S as that part of the lateral surface of S that is disjoint from the boundary of K. We denote by $D'(S)$ the area of the dockable surface of S. We can prove the following analogue of Lemma 6:

LEMMA 8. *If* $\text{vol}(S) \leqslant \frac{1}{2}\text{vol}(K)$, *then* $D'(S) \geqslant \text{vol}(S)$.

PROOF. Let F denote the dockable surface of S (thus $D'(S)$ is the area of F). For every $\varepsilon > 0$ we define

$$K_{3,\varepsilon} = (F + \varepsilon B_n) \cap K,$$
$$K_{1,\varepsilon} = S \setminus K_{3,\varepsilon},$$
$$K_{2,\varepsilon} = K \setminus (K_{1,\varepsilon} \cup K_{3,\varepsilon}),$$

where $X + Y$ denotes the Minkowski sum of two subsets $X, Y \in \mathbb{R}^n$, i.e., $X + Y = \{x + y : x \in X, y \in Y\}$. Clearly $(K_{1,\varepsilon}, K_{2,\varepsilon}, K_{3,\varepsilon})$ is a partition of K into three measurable sets. Furthermore, condition (10) is satisfied. Thus Theorem 7 implies that

$$\frac{\text{vol}(K_{3,\varepsilon})}{2\varepsilon} \geqslant \frac{1}{1-\varepsilon} \min\left(\text{vol}(K_{1,\varepsilon}), \text{vol}(K_{2,\varepsilon})\right).$$

We observe that

$$\begin{aligned}\mathrm{vol}(K_{2,\varepsilon}) &\geqslant \mathrm{vol}(K \setminus S) - \mathrm{vol}(K_{3,\varepsilon})\\ &\geqslant \mathrm{vol}(S) - \mathrm{vol}(K_{3,\varepsilon})\\ &\geqslant \mathrm{vol}(K_{1,\varepsilon}) - \mathrm{vol}(K_{3,\varepsilon}).\end{aligned}$$

Combining those two inequalities, we find

$$\frac{\mathrm{vol}(F+\varepsilon B_n)}{2\varepsilon} \geqslant \frac{\mathrm{vol}(K_{3,\varepsilon})}{2\varepsilon} \geqslant \frac{1}{1-\varepsilon}(\mathrm{vol}(K_{1,\varepsilon}) - \mathrm{vol}(K_{3,\varepsilon})). \tag{11}$$

By a well-known result in geometry (see, e.g., [7],) as ε tends to 0 the left-hand side of (11) tends to the area of F, which is precisely the dockable surface $D'(S)$. Moreover, as ε tends to 0, $\mathrm{vol}(K_{3,\varepsilon})$ tends to 0 and $\mathrm{vol}(K_{1,\varepsilon})$ tends to $\mathrm{vol}(S)$. We conclude that $D'(S) \geqslant \mathrm{vol}(S)$. □

Following the same approach as that used for the case of a polytope, one can show the following result for polyhedra.

THEOREM 9. *Let $P = \{x \in \mathbb{R}^n : Ax \leqslant b\}$ be a polyhedron, where all sub-determinants of $A \in \mathbb{Z}^{m\times n}$ are bounded by Δ in absolute value. Then the diameter of P is bounded by $O\left(\Delta^2 n^4 \log n\Delta\right)$. In particular, if A is totally unimodular, then the diameter of P is bounded by $O(n^4 \log n)$.*

Remark

For simplicity, we have assumed that a bound Δ was given for the absolute value of all sub-determinants of A. However, our proof only uses the fact the the sub-determinants of size 1 (i.e., the entries of the matrix) and $n-1$ are bounded. Calling Δ_1 (resp. Δ_{n-1}) the bound on the absolute value of the entries of A (resp. on the sub-determinants of A of size $n-1$), one easily verifies that all the results discussed above remain essentially unchanged, except that the statement of Lemma 3 becomes

$$\frac{D(S_v)}{\mathrm{vol}(S_v)} \leqslant \Delta_1\Delta_{n-1}n^3$$

and the lower bound on $\mathrm{vol}(I_0)$ becomes

$$\mathrm{vol}(I_0) \geqslant \frac{1}{n! n^{n/2}\Delta_1^n}.$$

This implies the following strengthened result:

THEOREM 10. *Let $P = \{x \in \mathbb{R}^n : Ax \leqslant b\}$ be a polyhedron, where the entries of A (respectively the sub-determinants of A of size $n-1$) are bounded in absolute value by Δ_1 (respectively Δ_{n-1}). Then the diameter of P is bounded by $O\left(\Delta_1\Delta_{n-1}n^4 \log n\Delta_1\right)$. Moreover, if P is a polytope, its diameter is bounded by $O\left(\Delta_1\Delta_{n-1}n^{3.5} \log n\Delta_1\right)$.*

4. ACKNOWLEDGEMENTS

This work was carried out while all authors were at EPFL (École Polytechnique Fédérale de Lausanne), Switzerland. The authors acknowledge support from the DFG Focus Program 1307 within the project "Algorithm Engineering for Real-time Scheduling and Routing" and from the Swiss National Science Foundation within the project "Set-partitioning integer programs and integrality gaps".

5. REFERENCES

[1] M. L. Balinski. The Hirsch conjecture for dual transportation polyhedra. *Math. Oper. Res.*, 9(4):629–633, 1984.

[2] D. Barnette. An upper bound for the diameter of a polytope. *Discrete Math.*, 10:9–13, 1974.

[3] G. Brightwell, J. van den Heuvel, and L. Stougie. A linear bound on the diameter of the transportation polytope. *Combinatorica*, 26(2):133–139, 2006.

[4] J. A. De Loera, E. D. Kim, S. Onn, and F. Santos. Graphs of transportation polytopes. *J. Combin. Theory Ser. A*, 116(8):1306–1325, 2009.

[5] M. Dyer and A. Frieze. Random walks, totally unimodular matrices, and a randomised dual simplex algorithm. *Math. Program.*, 64(1, Ser. A):1–16, 1994.

[6] F. Eisenbrand, N. Hähnle, A. Razborov, and T. Rothvoß. Diameter of polyhedra: Limits of abstraction. *Math. Oper. Res.*, 35(4):786–794, 2010.

[7] H. Federer. *Geometric Measure Theory*. Springer, 1969.

[8] T. Figiel, J. Lindenstrauss, and V. Milman. The dimension of almost spherical sections of convex bodies. *Acta Math.*, 139(1):53–94, 1977.

[9] V. Kaibel. On the expansion of graphs of 0/1-polytopes. In *The sharpest cut*, MPS/SIAM Ser. Optim., pages 199–216. SIAM, Philadelphia, PA, 2004.

[10] G. Kalai. The diameter of graphs of convex polytopes and f-vector theory. In *Applied geometry and discrete mathematics*, volume 4 of *DIMACS Ser. Discrete Math. Theoret. Comput. Sci.*, pages 387–411. Amer. Math. Soc., Providence, RI, 1991.

[11] G. Kalai and D. J. Kleitman. A quasi-polynomial bound for the diameter of graphs of polyhedra. *Bull. Amer. Math. Soc. (N.S.)*, 26(2):315–316, 1992.

[12] E. D. Kim and F. Santos. An update on the Hirsch conjecture. *Jahresber. Dtsch. Math.-Ver.*, 112(2):73–98, 2010.

[13] V. Klee and D. W. Walkup. The d-step conjecture for polyhedra of dimension $d < 6$. *Acta Math.*, 133:53–78, 1967.

[14] D. G. Larman. Paths of polytopes. *Proc. London Math. Soc. (3)*, 20:161–178, 1970.

[15] L. Lovász and M. Simonovits. Random walks in a convex body and an improved volume algorithm. *Rand. Struct. Algor.*, 4(4):359–412, 1993.

[16] M. Merkle. Logarithmic convexity and inequalities for the gamma function. *J. Math. Anal. Appl.*, 203(2):369–380, 1996.

[17] D. Naddef. The Hirsch conjecture is true for (0,1)-polytopes. *Math. Program.*, 45:109–110, 1989.

[18] J. B. Orlin. A polynomial time primal network simplex algorithm for minimum cost flows. *Math. Program.*, 78(2, Ser. B):109–129, 1997.

[19] F. Santos. A counterexample to the Hirsch conjecture. 2010. arXiv:1006.2814v1 [math.CO]. To appear in *Ann. Math.*

[20] R. Vershynin. Beyond Hirsch conjecture: walks on random polytopes and smoothed complexity of the simplex method. *SIAM J. Comput.*, 39(2):646–678, 2009.

Optimal Area-Sensitive Bounds for Polytope Approximation

Sunil Arya[*]
Department of Computer Science and Engineering
The Hong Kong University of Science and Technology
Clear Water Bay, Kowloon
Hong Kong
arya@cse.ust.hk

Guilherme D. da Fonseca[†]
Dept. de Informática Aplicada
Universidade Federal do Estado do Rio de Janeiro (UniRio)
Rio de Janeiro, Brazil
fonseca@uniriotec.br

David M. Mount[‡]
Department of Computer Science and Institute for Advanced Computer Studies
University of Maryland
College Park, Maryland 20742
mount@cs.umd.edu

ABSTRACT

Approximating convex bodies is a fundamental question in geometry and has applications to a wide variety of optimization problems. Given a convex body K in $\mathbb{R}^d$ for fixed d, the objective is to minimize the number of vertices or facets of an approximating polytope for a given Hausdorff error ε. The best known uniform bound, due to Dudley (1974), shows that $O((\mathrm{diam}(K)/\varepsilon)^{(d-1)/2})$ facets suffice. While this bound is optimal in the case of a Euclidean ball, it is far from optimal for skinny convex bodies.

We show that, under the assumption that the width of the body in any direction is at least ε, it is possible to approximate a convex body using $O(\sqrt{\mathrm{area}(K)}/\varepsilon^{(d-1)/2})$ facets, where $\mathrm{area}(K)$ is the surface area of the body. This bound is never worse than the previous bound and may be significantly better for skinny bodies. This bound is provably optimal in the worst case and improves upon our earlier result (which appeared in SODA 2012).

Our improved bound arises from a novel approach to sampling points on the boundary of a convex body in order to stab all (dual) caps of a given width. This approach involves the application of an elegant concept from the theory of convex bodies, called *Macbeath regions*. While Macbeath regions are defined in terms of volume considerations, we show that by applying them to both the original body and its dual, and then combining this with known bounds on the Mahler volume, it is possible to achieve the desired width-based sampling.

[*]This author's work was supported by the Research Grants Council, Hong Kong, China under project number 610108.

[†]This author's work was supported by CNPq and FAPERJ grants.

[‡]This author's work was supported in part by the National Science Foundation under grant CCF-1117259 and the Office of Naval Research under grant N00014-08-1-1015.

SCG'12, June 17–20, 2012, Chapel Hill, North Carolina, USA.

Categories and Subject Descriptors

F.2.2 [**Analysis of Algorithms and Problem Complexity**]: Nonnumerical Algorithms and Problems—*Geometrical problems and computations*

General Terms

Algorithms, Theory

Keywords

Convexity, polytopes, approximation, Macbeath regions

1. INTRODUCTION

Approximating convex bodies by polytopes is a fundamental problem, which has been extensively studied in the literature. (See Bronstein [13] for a recent survey.) At issue is the minimum number of vertices (alternatively, the minimum number of facets) needed in an approximating polytope for a given error $\varepsilon > 0$. Consider a convex body K in Euclidean d-dimensional space. A polytope P is said to *ε-approximate* K if the Hausdorff distance [13] between K and P is at most ε. Throughout, we will restrict attention to the Hausdorff metric, and we assume that the dimension d is a constant.

Our interest is in establishing bounds on the combinatorial complexity of approximating general convex bodies. Approximation bounds are of two common types. In both cases, it is shown that there exists $\varepsilon_0 > 0$ such that the bounds hold for all $\varepsilon \leq \varepsilon_0$. In the first type, which we call *nonuniform bounds*, the value of ε_0 depends on K (for example, on K's maximum curvature). Such bounds are often stated as holding "in the limit" as ε approaches zero, or equivalently as the combinatorial complexity of the approximating polytope approaches infinity. Examples include bounds by Gruber [22], Clarkson [16], and others [9, 26, 28, 29].

In the second type, which we call *uniform bounds*, the value of ε_0 is independent of K. For example, these include the results of Dudley [19] and Bronshteyn and Ivanov [12]. These bounds hold without any smoothness assumptions. Dudley showed that, for $\varepsilon \leq 1$, any convex body K can be ε-approximated by a polytope P with $O((\mathrm{diam}(K)/\varepsilon)^{(d-1)/2})$

facets. Bronshteyn and Ivanov showed the same bound holds for the number of vertices. Constants hidden in the O-notation depend only on d. These results have many applications, for example, in the construction of coresets [1].

The approximation bounds of both Dudley and Bronshteyn and Ivanov are tight up to constant factors (specifically when K is a Euclidean ball). These bounds may be significantly suboptimal if K is skinny, however. In an earlier paper [3], we presented an upper bound that is based not on diameter, but on surface area. In particular, let $\text{area}(K)$ denote the $(d-1)$-dimensional Hausdorff measure of ∂K. We showed that, under the assumption that the width of the body in any direction is at least ε, there exists an ε-approximating polytope whose number of facets is $O(t \log t)$, where $t = \sqrt{\text{area}(K)}/\varepsilon^{(d-1)/2}$. For a given diameter, the surface area of a convex body is maximized for a Euclidean ball, implying that $\text{area}(K) = O(\text{diam}(K)^{d-1})$. Thus, this bound is tight in the worst case up to the logarithmic factor. The additional log factor is disconcerting since it implies that the bound is suboptimal even for the simple case of a Euclidean ball. In this paper we show that the logarithmic factor can be eliminated. In particular, we prove the following result, which is worst-case optimal, up to constant factors.

THEOREM 1.1. *Consider real d-space, $\mathbb{R}^d$. There exists a positive ε_0 and constant c_d such that for any convex body $K \subset \mathbb{R}^d$ and any ε, $0 < \varepsilon \leq \varepsilon_0$, if the width of K in any direction is at least ε, then there exists an ε-approximating polytope P whose number of facets is at most*

$$c_d \sqrt{\text{area}(K)}/\varepsilon^{(d-1)/2}.$$

Note that the width assumption seems to be a technical necessity. For example, consider a $(d-2)$-dimensional unit ball B embedded within $\mathbb{R}^d$, and let B' denote its Minkowski sum with a d-dimensional ball of radius $\delta \ll \varepsilon$. By the optimality of Dudley's bound for Euclidean balls, $\Omega(1/\varepsilon^{(d-3)/2})$ facets are needed to approximate B and hence to approximate B'. But, the surface area of B' can be made arbitrarily small as a function of δ.

The width assumption is not a fundamental impediment, however. If the body is of width less than ε in some direction, then by projecting the body onto a hyperplane orthogonal to this direction, it is possible to reduce the problem to a convex approximation problem in one lower dimension. This can be repeated until the body's width is sufficiently large in all remaining dimensions, and the stated bound can be applied in this lower dimensional subspace.

1.1 Overview of Methods

It is well known (see, e.g., [11, 12, 15]) that computing a Hausdorff approximation of a convex body K by a polytope can be reduced to the problem of computing an economical set cover or an economical hitting set in which the set systems involve appropriately defined surface patches on K's boundary. Depending on the nature of the approximation (e.g., whether minimizing the number of vertices or facets and whether an inner or outer approximation is desired) the surface patches of interest are either *caps* or *dual caps*. For our purposes, a cap is the portion of K's boundary that lies within a halfspace and a dual cap is the portion of K's boundary that is visible from an external point. (Formal definitions are given in Section 2.4.) For both caps and dual caps, we define a hyperplanar surface, which we call a *base*, whose area is less than or equal to the associate cap or dual cap, respectively. Of particular interest are caps and dual caps whose defining width is ε. Depending on the formulation, the approximation problem reduces to computing a small set of points on the boundary of K such that every cap of width ε contains one of these points or every dual cap of width ε contains one of these points.

The source of slackness in the bound of [3] arises from a sampling method that is based on a relatively heavy-handed tool, namely ε-nets for halfspace ranges. In light of recent lower bounds on the size of ε-nets for halfspace ranges [27], it is clear that the elimination of the log factor requires a sampling process that is specially tailored to caps or dual caps. The principal contribution of this paper is such a sampling method.

Our new approach makes use of a classical structure from the theory of convexity, called *Macbeath regions*. Intuitively, for any convex body K and a volume parameter v, there exists a collection of pairwise disjoint bodies, each of volume $\Omega(v)$, such that for every halfspace H where the cap $K \cap H$ has volume v, one of these bodies will be completely contained within this cap. (The formal statement is given in Section 2.4.) Macbeath regions have found numerous uses in the theory of convex sets and the geometry of numbers (see Bárány [7]). To date, the application of Macbeath regions in the field of computational geometry has been quite limited. For example, they have been used as a technical device in proving lower bounds for range searching (see, e.g., [4,5,10]).

Because their definition is based on volume, not width, the use of Macbeath regions in the context of uniform bounds for convex approximation has been limited to volume-based notions of distance, such as the Nikodym metric (which is based on the volume of the symmetric difference) [6,8]. The difficulty in adapting Macbeath regions to width-based sampling is that caps of a given volume may have widely varying widths. Our approach to dealing with this is through the application of a two-pronged strategy, which combines Macbeath-based sampling in both the original body and its dual.

This strategy relies on a combination of two well known dual transformations, the polar dual (which is based on distances to the origin) and the functional dual (which is the dual transform most widely used in computational geometry and is based on vertical distances). The problem with either form of dual is that distances are not generally preserved, and this makes it difficult to relate approximations in the primal and dual settings. The principal feature of the functional dual transform is that it does preserve vertical distances between points and hyperplanes.

To exploit this, we decompose the approximation problem into a constant number of subproblems, where each involves approximating just a portion of the body in which the surface normals have similar directions. After a suitable rotation, within each subproblem, the distance from an external point to the boundary can be approximated by its vertical distance. In Lemma 4.1 in Section 4, we establish an intriguing relationship between the base of a dual cap in the original body K and the base of the corresponding cap in the functional dual set K^*. In particular, we show that the base of a dual cap in K is essentially the polar of the base of the corresponding cap in K^*. This polar relationship is an essential ingredient in our construction, because it allows us to apply the classical concept of the *Mahler volume*, to show that small dual caps in K correspond to large caps in K^*. Our two-pronged sampling works because dual caps that are too small to be sampled in K will be sampled as caps in K^*.

2. PRELIMINARIES

Let K denote a convex body in $\mathbb{R}^d$, that is, a compact convex subset with nonempty interior. Let ∂K denote its boundary. Throughout, we assume that the dimension d is a constant. We say that a convex body is *smooth*, if at each point $q \in \partial K$, there exists a ball of positive radius that lies entirely within K and contains q on its boundary.

If K is smooth, there is a unique supporting hyperplane at every point on its boundary. Since we do not assume smoothness, we augment our representation of boundary points. Given any convex body K, we define an *augmented point* on ∂K to consist of a pair (q, u), where q is a point on ∂K and u is an outward-directed unit vector that is orthogonal to some supporting hyperplane passing through q. We call such a vector a *surface normal*. To keep our notation simple, we will usually refer to an augmented point simply as q, but it is understood throughout that every augmented point is associated with a unique surface normal and hence a unique supporting hyperplane, which we will denote by $h(q)$. Define $H^+(q)$ to be the closed halfspace bounded by $h(q)$ that contains K, and define $H^-(q)$ to be the other closed halfspace bounded by $h(q)$.

Given a convex body K in $\mathbb{R}^d$, let $\text{vol}(K)$ denote its d-dimensional Hausdorff measure. Given a $(d-1)$-dimensional manifold Ψ in $\mathbb{R}^d$, for example, a surface patch on a convex body, let $\text{area}(\Psi)$ denote its $(d-1)$-dimensional Hausdorff measure. We use $\text{area}(K)$ as a convenient shorthand for $\text{area}(\partial K)$.

For $\varepsilon > 0$, we say that a polytope P is an *ε-approximation* of K if the Hausdorff distance between K and P is at most ε. Observe that simultaneously scaling K and ε by any positive factor, does not alter the ratio $\sqrt{\text{area}(K)}/\varepsilon^{(d-1)/2}$. Therefore, for the sake of proving Theorem 1.1, we may assume that K has been uniformly scaled to lie within the hypercube $[-(1-2\varepsilon), (1-2\varepsilon)]^d$. This means that the Minkowski sum of K with a ball of radius 2ε (a shape that will be useful to us later) lies entirely within $[-1, 1]^d$.

To avoid specifying many real-valued constants that arise in our analysis, we will often hide them using asymptotic notation. For positive real x, we use the notation $O(x)$ (resp., $\Omega(x)$) to mean a quantity whose value is at most (resp., at least) cx for an appropriately chosen constant c. We use $\Theta(x)$ to denote a quantity that lies within the interval $[cx, c'x]$ for appropriate constants c and c'. These constants will generally be functions of d, but not of ε.

2.1 Nonuniform Area-Based Bounds

Before presenting our analysis, we note that a nonuniform bound very similar to ours can be derived from Gruber's result [22]. (We thank Quentin Merigót for pointing this out.) Gruber shows that if K is a strictly convex body and ∂K is twice differentiable, then as ε approaches zero, the number of bounding halfspaces needed to achieve an ε-approximation of K is

$$O\left(\left(\frac{1}{\varepsilon}\right)^{(d-1)/2} \int_{\partial K} \kappa(x)^{1/2} dx\right), \qquad (1)$$

where $\kappa(x)$ denotes the Gaussian curvature of K at x, and dx is a differential surface element. (Böröczky showed that the requirement that K be "strictly" convex can be eliminated [9].) Because square root is concave and $\int_{\partial K} dx = \text{area}(K)$, we may apply Jensen's inequality to obtain

$$\frac{1}{\text{area}(K)} \int_{\partial K} \kappa(x)^{1/2} dx \;\leq\; \left(\frac{1}{\text{area}(K)} \int_{\partial K} \kappa(x) dx\right)^{1/2}.$$

By the Gauss-Bonnet theorem [18], the total Gaussian curvature of K is bounded, from which we conclude that the number of approximating halfspaces is

$$O\left(\frac{\sqrt{\text{area}(K)}}{\varepsilon^{(d-1)/2}}\right),$$

which matches the bound of Theorem 1.1.

We hasten to add that this approach cannot be used to produce a uniform bound, however. To show this, we will present a two-dimensional counterexample (but the result can be readily extended to any constant dimension). Consider a fixed value of ε, and let $0 < \delta \leq \varepsilon$. Let $m = \lfloor 1/\sqrt{\delta} \rfloor$, and define K_δ to be the Minkowski sum of a regular m-gon inscribed in a unit circle and a Euclidean ball of radius δ. Observe that K_δ consists of m flat edges, each of length $\Theta(\sqrt{\varepsilon})$, connected by m circular arcs, each of radius δ and subtending an angle of $2\pi/m$. Since $\delta \leq \varepsilon$, it is straightforward to show that any convex polygon that ε-approximates K_δ requires $\Omega(1/\sqrt{\varepsilon})$ sides. (This follows from the same argument that shows that Dudley's bound is tight for circles.) We also assert that

$$\int_{\partial K_\delta} \kappa(x)^{1/2} dx \;=\; \Theta\left(\sqrt{\delta}\right).$$

To see this, observe that the flat sides of K_δ contribute zero to the integral. Each circular arc has curvature $1/\delta$, and so altogether they contribute a total of $2\pi\delta/\sqrt{\delta} = 2\pi\sqrt{\delta}$ to the integral.

Thus, Equation (1) provides a bound[1] on the number of sides of an ε-approximating polygon of $O(\sqrt{\delta/\varepsilon})$, which is clearly incorrect given our hypothesis that $\delta \leq \varepsilon$. Indeed, this hypothesis is exactly the sort of assumption that the nonuniform analysis forbids. In contrast, since Theorem 1.1 is a uniform bound, it can be applied to this type of example.

2.2 Support Sets

Our analysis makes use of a representation of the body K as the intersection of $2d$ unbounded convex sets, based on the orientations of surface normals. Let I denote the index set $\{1, \ldots, d\}$, and let $I^\pm$ denote the $(2d)$-element index set consisting of $\{\pm 1, \ldots, \pm d\}$. For $i \in I^\pm$, if $i > 0$, let x_i denote the unit vector associated with the ith coordinate axis, and if $i < 0$, let x_i be its negation. Let X_i denote the hyperplane passing through the origin and orthogonal to x_i. (For the sake of illustration, we will think of x_i as

[1]Note that we cannot apply Gruber's or Böröczky's theorems directly to K_δ, since its boundary is not twice differentiable. In particular, the second derivative is discontinuous at the joints where each edge meets a circular arc. We can easily fix this by creating a sufficiently small gap at each joint and introducing a smooth polynomial spline of constant degree to fill the gap. Although the resulting body is not strictly convex, Böröczky showed that this assumption is not necessary for the bound to hold.

being directed vertically upwards and X_i as horizontal.) The *vertical projection* of a point onto X_i means setting its ith coordinate to zero.

For a vector $\vec{u} = (u_1, \ldots, u_d) \in \mathbb{R}^d$, let $\|\vec{u}\|_\infty = \max_i |u_i|$. For $i \in I^\pm$ we say that i is a *signed principal axis* of $\vec{u}$ if $\text{sign}(i) \cdot u_i = \|\vec{u}\|_\infty$. (For example, the vector $(-2, 1, 2, -2, 0)$ has the signed principal axes -1, 3, and -4.) For each $i \in I^\pm$, let Ψ_i denote the subset of augmented points $q \in \partial K$ such that i is a signed principal axis of the inward-directed surface normal at q. These subsets subdivide K's boundary into $2d$ (relatively closed) surface patches. For each $i \in I^\pm$, let $S_i = \bigcap_{q \in \Psi_i} H^+(q)$ (see Figure 1(a)). We call this the ith *support set* of K. Each S_i is an unbounded convex set that properly contains K. Clearly, $\bigcap_{i \in I^\pm} S_i = K$.

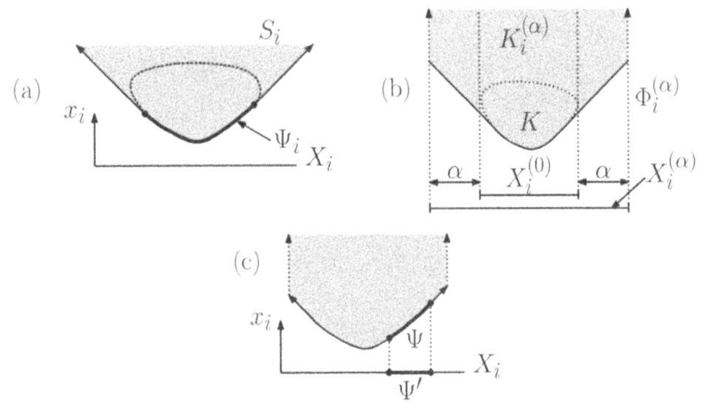

Figure 1: Support sets and projections.

It will be useful to further restrict each support set S_i so that its vertical projection covers only a bounded region of X_i that is somewhat larger than the vertical projection of K. For a given $\alpha \geq 0$, take the Minkowski sum of K with a ball of radius α and project this set vertically onto X_i. Let $X_i^{(\alpha)}$ denote the resulting $(d-1)$-dimensional convex body on X_i. Let $\Phi_i^{(\alpha)}$ denote the infinite cylinder whose central axis is aligned with x_i and whose horizontal cross section is $X_i^{(\alpha)}$ (see Figure 1(b)). Define $K_i^{(\alpha)}$ to be $S_i \cap \Phi_i^{(\alpha)}$, which we call the *$\alpha$-restricted support set*. As we shall see below, the body $K_i^{(2\varepsilon)}$, will play an important role in our analysis. Indeed, to simplify notation, henceforth we let K_i denote $K_i^{(2\varepsilon)}$. By our initial scaling of K, it follows that the vertical projection of K_i lies within the hypercube $[-1, 1]^{d-1}$.

A closed convex set is *U-shaped* (with respect to a given vertical direction) if for every point on the set's boundary, the vertical ray directed upwards from this point lies entirely within the set. When dealing with U-shaped sets, such as K_i, we will be most interested in the lower hull, not the vertical sides. For this reason, given a U-shaped set U, we define ∂U to consist of the augmented points of the boundary of U such that the associated surface normal for each point is not horizontal. In fact, we will impose the stronger restriction that the absolute tangent of the angle between the surface normal and the vertical axis is at most a constant (whose exact value is implicit in the radius of the ball B of Lemma 4.2 below). Define area(U) to be the surface area of ∂U.

Given a parameter $\sigma > 0$, a U-shaped set U is *σ-steep* if for every augmented point $q \in \partial U$, the absolute tangent of the angle between the vertical axis and the surface normal at q does not exceed σ. The halfspace above a horizontal hyperplane is 0-steep, and K_i is easily seen to be $O(1)$-steep. The following are straightforward consequences of the above definitions.

LEMMA 2.1.

(i) If the width of K in any direction is at least ε, then for any $i \in I^+$, area$(K_i) = O($area$(K))$.

(ii) Consider a surface patch Ψ of an $O(1)$-steep U-shaped set, and let Ψ' denote its vertical projection. Then area$(\Psi') = \Theta($area$(\Psi))$ *(see Figure 1(c)).*

2.3 Dual Transforms

Our results are based on two commonly used dual transforms in geometry. Such transforms map points to hyperplanes and vice versa, while preserving point-hyperplane incidences. The first transform is a generalization of the standard polar transform, and the second is a dual transform frequently used in computational geometry, which we call the functional dual.

Given a vector $v \in \mathbb{R}^d$ other than the origin and $r > 0$, define the *generalized polar transform* of v, denoted polar$_r(v)$, to be the halfspace containing the origin, whose bounding hyperplane is orthogonal to v and is at distance $r^2/\|v\|$ from the origin (see Figure 2(a)). More formally, polar$_r(v) = \{u \in \mathbb{R}^d : u_1 v_1 + \cdots + u_d v_d \leq r^2\}$. Given a convex body K that contains the origin in its interior, define polar$_r(K) = \bigcap_{v \in K}$ polar$_r(v)$ (see Figure 2(b)).

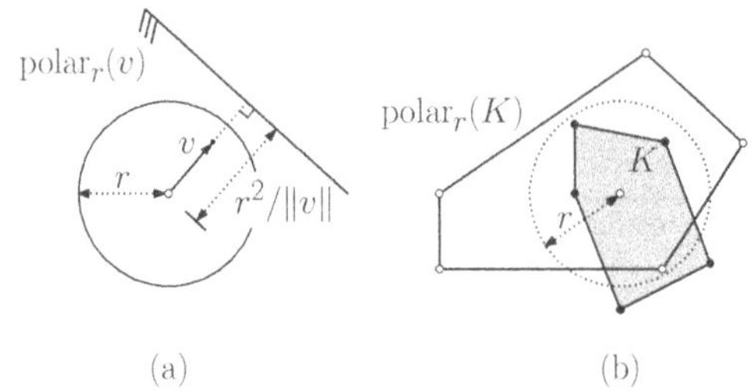

Figure 2: The generalized polar transform.

The standard geometric polar transform [20] arises as a special case when $r = 1$. In particular, polar$_r(K)$ is a scaled copy of polar$_1(K)$ by a factor of r^2.

Later, we will make use of an important result from the theory of convex sets, which states that, given a convex body K, the product vol$(K) \cdot$ vol(polar$_1(K)$), which is called K's *Mahler volume*, is bounded below by a constant (see, e.g., [24]).

Next, let us define the *functional dual transform* [14]. Let $x_1, \ldots, x_{d-1}$ denote the first $d-1$ coordinates of a point, and let y denote the d-th coordinate. Any nonvertical hyperplane $y = \sum_{j=1}^{d-1} a_j x_j - a_d$ can be represented by a d-tuple $(a_1, \ldots, a_d)$. Given a point $p = (p_1, \ldots, p_d) \in \mathbb{R}^d$, define its functional dual, denoted p^*, to be the non-vertical hyperplane:

$$p^* : y = \sum_{j=1}^{d-1} p_j x_j - p_d.$$

(The term "functional" comes from the fact that y is expressed as a function of the x_j's.) The dual of a nonvertical hyperplane is defined so that $p^{**} = p$. When applying this transform in the context of support sets, such as K_i, the ith coordinate axis will take on the role of the dth (vertical) axis in the above definitions.

It is easily verified that the functional dual transform negates vertical distances, in the sense that the signed vertical distance from a point p to a hyperplane h is equal to the negation of the signed vertical distance from p^* to h^* [17]. (The fact that point-hyperplane incidences are preserved is a direct consequence.) Given any U-shaped convex set U, define its dual U^* to be the intersection of the upper halfspaces of q^* for all $q \in \partial U$. It is easily verified that the dual is also U-shaped. Further, if U's vertical projection lies within a ball of constant radius centered at the origin, then U^* is $O(1)$-steep.

Let U be a U-shaped convex body and let U^* be its dual. There exists a natural correspondence between the augmented points on ∂U and augmented points on ∂U^*. Given an augmented point (q, u) on ∂U, let h be the supporting plane at q that is orthogonal to u. Let $p = h^*$. Clearly, p lies on ∂U^*, and the dual hyperplane q^* is a supporting hyperplane at p. Letting v denote the outward-directed unit vector orthogonal to q^*, we define the augmented point on ∂U^* *corresponding* to (q, u) to be (p, v). By the involutory nature of the dual, the augmented point on ∂U corresponding to (p, v) is (q, u).

2.4 Caps, Dual Caps, and Macbeath Regions

Bronshteyn and Ivanov [12] and Dudley [19] demonstrated the relevance of caps and dual-caps (defined below) to convex approximation. Let U be a U-shaped convex set, and let $\varepsilon > 0$. For any augmented point $q \in \partial U$, recall that $h(q)$ is the supporting hyperplane at this point, and let $h(q) + \varepsilon$ denote its translate vertically upwards by ε. The intersection of ∂U and the lower halfspace of $h(q) + \varepsilon$ is called the *ε-cap* induced by q, which we denote by $C_q(U)$ (see Figure 3(a)). (In contrast to standard usage, where a cap consists of the subset of the body lying in the halfspace, for us a cap is a subset of the nonvertical boundary.) The intersection of $h(q) + \varepsilon$ with U is called the *base* of the cap, and is denoted by $\Gamma_q(U)$. When U is clear from context, we simply write C_q and Γ_q.

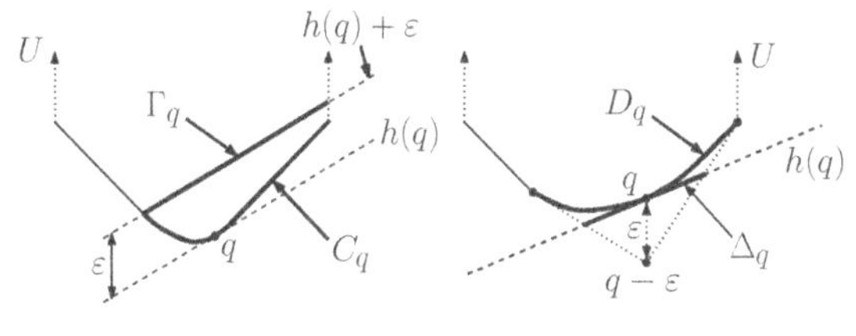

(a) (b)

Figure 3: Caps and dual caps.

For any augmented point $q \in \partial U$, let $q - \varepsilon$ be the vertical translate of q downwards by distance ε. Define the *ε-dual cap*, denoted $D_q(U)$, to be the portion of ∂U that is visible from $q - \varepsilon$ (see Figure 3(b)). The intersection of the bounding halfspaces of U that contain $q - \varepsilon$ defines an infinite cone. The intersection of this cone with $h(q)$ is called the dual cap's *base*, and is denoted $\Delta_q(U)$. Again, when U is clear, we simply write D_q and Δ_q.

Given an augmented point $q \in \partial U$, consider cap $C_q(U)$ and dual cap $D_q(U)$. We say that an augmented point $p \in \partial U$ *stabs* $C_q(U)$ if $p \in C_q(U)$, and we say that p *stabs* $D_q(U)$ if the hyperplane $h(p)$ passing through p separates U from the point $q - \varepsilon$. The importance of ε-dual caps to approximation is established in the following lemma. We say that an ε-dual cap of K_i is *useful*, if its inducing point q lies on $\partial K_i^{(\varepsilon)}$. Recall that $K_i = K_i^{(2\varepsilon)}$, so the inducing point is at horizontal distance at least ε from the vertical portion of K_i.

LEMMA 2.2. *Let K be a convex body in $\mathbb{R}^d$. For any $\varepsilon > 0$ and $i \in I^{\pm}$, let Q_i be a set of augmented points on ∂K_i that stab all useful ε-dual caps of K_i. Let $P_i = \bigcap_{q \in Q_i} H^+(q)$, and $P = \bigcap_{i \in I^{\pm}} P_i$. Then P is an ε-approximation to K. The number of facets in P is at most $\sum_{i \in I^{\pm}} |Q_i|$.*

PROOF. Observe that P is the intersection of supporting halfspaces for K, and hence $K \subseteq P$. Thus, every point of K is within distance 0 of P. To show that every point of P is within distance ε of K, we will prove the contrapositive. In particular, we will show that any point $p \in \mathbb{R}^d$ whose distance from ∂K exceeds ε cannot be in P.

Let q be the closest point of ∂K to p. Let p' be a point along the segment $\overline{pq}$ whose distance from q is exactly ε (see Figure 4). Let i be any signed principal axis for the inward surface normal for K at q. Clearly, the horizontal component of the distance between p' and q is at most ε.

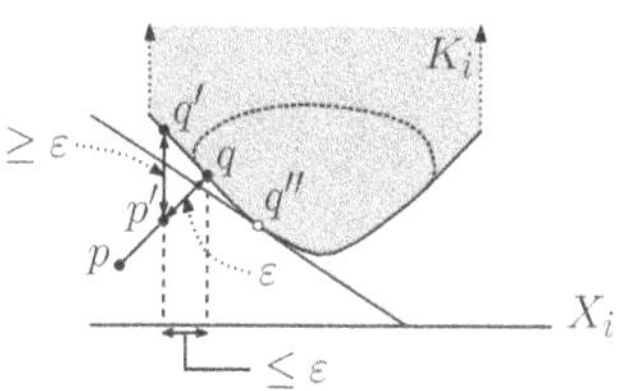

Figure 4: Proof of Lemma 2.2.

The vertical projection of q lies within $X_i^{(0)}$. Therefore the vertical projection of p' lies within $X_i^{(\varepsilon)}$, which is subset of $X_i^{(2\varepsilon)}$, above which K_i lies. This implies that that there exists a point $q' \in \partial K_i^{(\varepsilon)} \subset \partial K_i$ that is vertically above p' (see Figure 4). Augment q' by associating it with any valid surface normal for which i is a signed principal axis. It follows that any ε-dual cap induced by q' is useful. By local minimality considerations, it is easy to see that q is the closest point to p' on ∂K_i. Therefore, the distance from p' to q' is at least ε. By hypothesis, the ε-dual cap induced by q' is stabbed by the supporting hyperplane of some augmented point $q'' \in Q_i$. This implies that p' lies outside this bounding hyperplane, and hence is external to P_i. It follows directly that p is also external to P_i, and hence it is external to P. This completes the proof. $\square$

In order to prove the results of the next two sections, we will make use of the following result. It demonstrates the existence of a collection of convex bodies, such that all caps of sufficiently large volume contain at least one such body. These bodies are closely related to the concept of *Macbeath regions* (also called *M-regions*). This concept was introduced by Macbeath [25], and its relevance to cap coverings was explored by Ewald, Larman, and Rogers [21]. Applications of Macbeath regions have appeared in numerous works, include Bárány and Larman [8], Bárány [7], and Brönnimann et al. [10]. The following lemma follows directly from these earlier works.

Lemma 2.3. *Given a convex body $K \subset \mathbb{R}^d$ and a parameter $0 < v \leq \text{vol}(K)$, there exist two collections of convex bodies, $\mathcal{M}$ and $\mathcal{M}'$, such that the bodies of $\mathcal{M}$ are contained within K and are pairwise disjoint. Each $M \in \mathcal{M}$ is associated with a corresponding body in $\mathcal{M}'$, denoted M', such that $M \subseteq M'$. M' is called M's* expanded body. *These sets satisfy the following:*

(i) *For all $M \in \mathcal{M}$, $\text{vol}(M)$ and $\text{vol}(M')$ are both $\Theta(v)$.*
(ii) *For any halfspace H, if $\text{vol}(K \cap H) = v$, then there exists $M \in \mathcal{M}$ such that $M \subseteq K \cap H \subseteq M'$, where M' is M's expanded body.*

3. STABBING DUAL CAPS IN THE PRIMAL

The purpose of this section is to establish bounds on the size of stabbing sets for ε-dual caps. Our approach will involve the use of the Macbeath region machinery from Lemma 2.3. We begin with the following utility lemma, which establishes a relationship between the areas of the bases of the cap and dual cap induced by the same point. Recall that, given an augmented point q on the boundary of a U-shaped set, D_q denotes the ε-dual cap induced by q, and Δ_q is its base. Also, C_q denotes the ε-cap induced by q, and Γ_q denotes its base.

Lemma 3.1. *Let U be any $O(1)$-steep U-shaped convex set in $\mathbb{R}^d$, and let $\varepsilon > 0$. Consider any augmented point $q \in \partial U$ such that the vertical projection of C_q is bounded. Let h be the supporting hyperplane at q, and let $h + \varepsilon$ denote the hyperplane containing Γ_q. Let H be the lower halfspace bounded by $h + \varepsilon$. Then $U \cap H$ contains a (generalized) cone whose base is a vertical translate of Δ_q onto $h + \varepsilon$, and whose apex is at vertical distance ε from the base. Further, $\text{vol}(U \cap H) = \Theta(\varepsilon \cdot \text{area}(\Delta_q))$.*

Proof. Recall that $q - \varepsilon$ denotes the point at distance ε vertically below q, which forms the apex of D_q (see Figure 5).

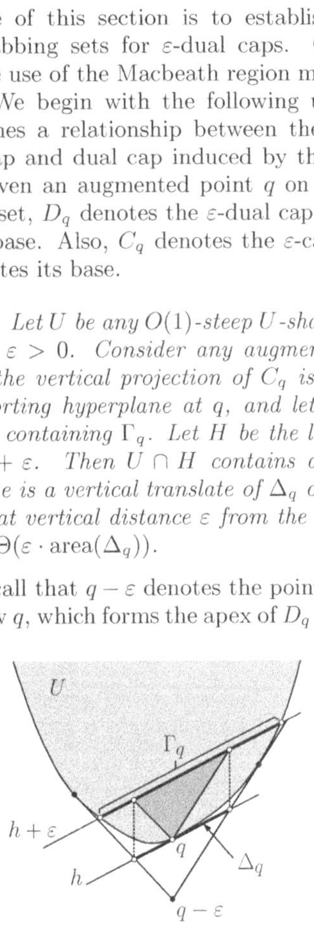

Figure 5: Proof of Lemma 3.1.

Consider the cone T bounded by the supporting hyperplanes of the dual cap and h. The apex of this cone is $q - \varepsilon$, and its base is Δ_q. Let T' denote the vertical translate of T upwards by distance ε. It is easy to see that T' lies entirely within $U \cap H$, which establishes the first claim.

Next, consider the cone T'' bounded by the same supporting hyperplanes as T, but whose base is $h + \varepsilon$. Because T'' is bounded by the supporting hyperplanes for U, it follows that Γ_q is contained within the base of this cone. Clearly, T'' is just a factor-2 scaling of T. Therefore, their volumes are related to one another by a constant factor. Since U is $O(1)$-steep, it follows that $\text{vol}(T') = \text{vol}(T) = \Theta(\varepsilon \cdot \text{area}(\Delta_q))$. It is easy to see that $T' \subseteq U \cap H \subseteq T''$, and therefore $\text{vol}(U \cap H) = \Theta(\varepsilon \cdot \text{area}(\Delta_q))$, which establishes the second claim. □

Our next lemma is the main result of this section. It states that there exists a small set of points that stab all sufficiently large ε-dual caps of K_i. The notion of "large" is based on the area of the dual cap's base and a threshold parameter t.

Lemma 3.2. *Let K be a convex body in $\mathbb{R}^d$, and consider any $\varepsilon > 0$ and $i \in I^{\pm}$. Given any $t > 0$, there exists a set of $O(\text{area}(K)/t)$ augmented points $Q_i \subset \partial K_i$ such that, for any useful ε-dual cap D_q induced by a point $q \in \partial K_i^{(\varepsilon)}$, if $\text{area}(\Delta_q) \geq t$, there exists a point in Q_i that stabs D_q.*

Proof. Apply Lemma 2.3 to K_i with $v = c \cdot \varepsilon \cdot t$, for a suitable constant c, and let $\mathcal{M}$ and $\mathcal{M}'$ denote the resulting collections of bodies. (Note that K_i is not bounded, but for our purposes it suffices to bound it crudely, say by intersection with the lower halfspace of a horizontal hyperplane that is high enough to contain every point at vertical distance ε above ∂K_i.)

For each $M \in \mathcal{M}$, if M does not lie entirely within vertical distance ε of ∂K_i, then discard M from further consideration. (Such a body cannot lie within any useful ε-cap, and hence will be of no value to us.) Otherwise, let $M' \in \mathcal{M}'$ denote M's expanded body as described in Lemma 2.3(ii).

Let N be a net [2] chosen so that, for a suitable constant c', any ellipsoid contained within M' of volume at least $c'v$ contains at least one point of the net. By Lemma 2.3(i), $\text{vol}(M') = \Theta(v)$, so any ellipsoid of volume $\Omega(v)$ covers a constant fraction of the volume of M'. Since the range space of ellipsoids is known to have constant VC-dimension, it follows that the size of the resulting net is $O(1)$. For each point of the net that lies within K_i, project it vertically downward onto ∂K_i, and augment it by associating it with any valid surface normal whose signed principal axis is i. Add the resulting augmented point to Q_i. Repeating this process for all $M \in \mathcal{M}$ yields the desired set Q_i.

To establish the correctness of this construction, consider any augmented point $q \in \partial K_i$ whose ε-dual cap (D_q) has base (Δ_q) of area at least t. Let h denote the supporting hyperplane at q, and let H denote the lower halfspace bounded by $h + \varepsilon$, which contains the base of q's cap, C_q. Clearly, the vertical projection of C_q is bounded, and so by Lemma 3.1, $\text{vol}(K_i \cap H) = \Omega(\varepsilon \cdot \text{area}(\Delta_q)) = \Omega(\varepsilon t)$. By choosing c suitably, we can ensure that $\text{vol}(K_i \cap H) \geq v$. In order to apply Lemma 2.3, observe that $\text{vol}(K_i \cap H)$ is a continuous monotonic function of ε, which decreases to 0 when $\varepsilon = 0$. Therefore, if $\text{vol}(K_i \cap H) > v$, there exists a value $\varepsilon' < \varepsilon$ such that $\text{vol}(K_i \cap H) = v$. The resulting ε'-dual cap and ε'-cap are subsets of the original dual cap and cap, respectively, and therefore there is no loss of generality in using these reduced objects throughout the rest of the analysis in place of the originals.

By Lemma 2.3(ii), there exists $M \in \mathcal{M}$ such that $M \subseteq K_i \cap H \subseteq M'$. (Note that M could not have been discarded in the construction process.) By Lemma 3.1, $K_i \cap H$ contains a cone T whose base is a translate of Δ_q and whose vertical height is ε. This cone is a convex body, and thus by John's Theorem [23], it contains an ellipsoid E such that

$$\text{vol}(E) = \Omega(\text{vol}(T)) = \Omega(\varepsilon \cdot \text{area}(\Delta_q)) = \Omega(v).$$

Therefore, by choosing the constant c' in the net construction appropriately, there exists a point of the net that lies within T, and hence the vertical projection of this point will be included in Q_i. It is easy to see that the horizontal extents of a dual cap's base lie entirely within the horizontal extents of the dual cap itself. Therefore, this point of Q_i stabs the dual cap, as desired.

Finally, we bound the number of points in Q_i. To do this, consider the set of points of K_i that lie within vertical distance ε of its boundary. The volume of this region is clearly $O(\varepsilon \cdot \text{area}(K_i))$, which by Lemma 2.1(i) is $O(\varepsilon \cdot \text{area}(K))$. By Lemma 2.3(i), each body M is of volume $\Omega(v) = \Omega(\varepsilon t)$. These bodies are pairwise disjoint, and (after discarding) each of them lies within vertical distance ε of ∂K_i. Therefore, by a simple packing argument, the number of such bodies is

$$O\left(\frac{\varepsilon \cdot \text{area}(K_i)}{v}\right) = O\left(\frac{\varepsilon \cdot \text{area}(K)}{\varepsilon \cdot t}\right) = O\left(\frac{\text{area}(K)}{t}\right),$$

as desired. □

4. STABBING CAPS IN THE DUAL

In this section we will consider the problem of stabbing caps of the (functional) dual of the convex body (recall Section 2.3). Consider a U-shaped set U and let U^* be its dual. The following lemma establishes a polar relationship between the bases of the ε-dual caps of U and the bases of the ε-caps in the dual U^*.

Consider any augmented point $q \in \partial U$ and let p be the corresponding augmented point on ∂U^* (recall the definition from the end of Section 2.3). The lemma shows that the base of q's dual cap and the base of p's cap, if viewed as convex bodies in $\mathbb{R}^{d-1}$, are general polar duals of each other. Given a convex body K and a point p, let $K - p$ denote the translate of K so that p coincides with the origin.

Lemma 4.1. *Let q be an augmented point of ∂U, and let p be the corresponding augmented point on ∂U^*. Let q' and p' be the respective vertical projections of q and p. Let Δ_q be the base of q's ε-dual cap in U, and let Γ_p be the base of p's ε-cap in U^*. Let Δ'_q and Γ'_p denote the respective vertical projections of these bases. Then, $\Delta'_q - q' = \text{polar}_r(\Gamma'_p - p')$, for $r = \sqrt{\varepsilon}$.*

Proof. We begin by showing that Δ_q consists of the set of points $v \in h(q)$ such that, for all $u \in \Gamma_p$,

$$\sum_{j=1}^{d-1}(u_j - p_j)(v_j - q_j) \le \varepsilon.$$

By the incidence-preserving property of the dual transform, the supporting hyperplane passing through p is q^*. From basic properties of the dual transform, the hyperplane containing Γ_p is the vertical translate of q^* upwards by a distance of ε, which is just $(q - \varepsilon)^*$ (see Figure 6).

The base Δ_q of the ε-dual cap induced by q is the intersection of the supporting hyperplane $h(q)$ with $\text{conv}(\{q - \varepsilon\} \cup U)$. Another way to express this set is to define an infinite cone C formed by the intersection all the upper halfspaces of the hyperplanes that pass through $q - \varepsilon$ such that U lies within each upper halfspace. Clearly, $\Delta_q = C \cap h(q) = C \cap p^*$. By the incidence-preserving property of the dual transform, these hyperplanes are the duals of points u lying on $(q - \varepsilon)^*$. By the order-reversing property of the

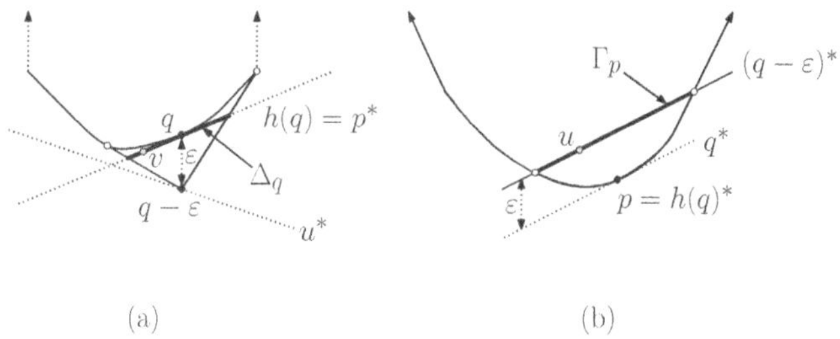

Figure 6: Proof of Lemma 4.1.

dual, each such point u lies within U^* (since it lies above all the supporting hyperplanes of U^*). But, the set of points $u \in (q - \varepsilon)^* \cap U^*$ is easily seen to be Γ_p (see the point u in Figure 6(b) and hyperplane u^* in Figure 6(a)). Letting $(u^*)^+$ denote the upper halfspace of u's dual, we have $\Delta_q = \left(\bigcap_{u \in \Gamma_p} (u^*)^+\right) \cap p^*$.

The equations of the hyperplanes q^* and $(q - \varepsilon)^*$ are

$$q^* \quad : \quad y = \sum_{j=1}^{d-1} q_j x_j - q_d$$

$$(q - \varepsilon)^* \quad : \quad y = \sum_{j=1}^{d-1} q_j x_j - q_d + \varepsilon.$$

Since p lies on q^*, we have $p_d = \sum_{j=1}^{d-1} q_j p_j - q_d$. Any $u \in \Gamma_p$ lies on $(q - \varepsilon)^*$, and so we have $u_d = \sum_{j=1}^{d-1} q_j u_j - q_d + \varepsilon$.

From the remarks made earlier, a point v is in Δ_q if and only if: (1) it lies on p^*, and (2) it lies in the upper halfspace $(u^*)^+$ for each $u \in \Gamma_p$. From condition (1) we have

$$\begin{aligned} v_d &= \sum_{j=1}^{d-1} p_j v_j - p_d = \sum_{j=1}^{d-1} p_j v_j - \left(\sum_{j=1}^{d-1} q_j p_j - q_d\right) \\ &= \sum_{j=1}^{d-1} p_j (v_j - q_j) + q_d. \end{aligned}$$

From condition (2) we have

$$\begin{aligned} v_d &\ge \sum_{j=1}^{d-1} u_j v_j - u_d = \sum_{j=1}^{d-1} u_j v_j - \left(\sum_{j=1}^{d-1} q_j u_j - q_d + \varepsilon\right) \\ &= \sum_{j=1}^{d-1} u_j (v_j - q_j) + q_d - \varepsilon. \end{aligned}$$

Combining these observations, we have $v \in \Delta_q$ if and only if $v \in p^*$ and for all $u \in \Gamma_p$,

$$\sum_{j=1}^{d-1} p_j (v_j - q_j) + q_d \ge \sum_{j=1}^{d-1} u_j (v_j - q_j) + q_d - \varepsilon,$$

or equivalently, $\sum_{j=1}^{d-1}(u_j - p_j)(v_j - q_j) \le \varepsilon$, as desired.

Recall that Δ'_q and Γ'_p are the respective vertical projections of Δ_q and Γ_p. Recall that p' and q' are the respective vertical projections of p and q. Then, we have shown that $v \in \Delta'_q$ if and only if, for all $u \in \Gamma'_p$, $\sum_{j=1}^{d-1}(u_j - p_j)(v_j - q_j) \le \varepsilon$. If we translate Δ'_q and Γ'_p so that q' and p' each coincide with the origin, then these bodies are generalized polars of each other, for $r = \sqrt{\varepsilon}$. That is, $\Delta'_q - q' = \text{polar}_r(\Gamma'_p - p')$. □

We will apply the above result to K_i to convert the problem of stabbing ε-dual caps to the problem of stabbing ε-caps in K_i^*. Although we have restricted K_i so that the nonvertical elements of its boundary are bounded, the same cannot be said for its dual K_i^*, which (due to the vertical sides of K_i) has a vertical projection that covers the entire horizontal coordinate hyperplane X_i. Therefore, it would be meaningless to apply the Macbeath region approach that was used in the previous section. What rescues us is the observation that we are only interested in ε-caps of K_i^* that arise from the application of the previous lemma to useful ε-dual caps of K_i.

Recall that each augmented point $q \in \partial K_i^{(\varepsilon)}$ induces a useful ε-dual cap. We say that a cap of K_i^* is *useful* if its inducing augmented point corresponds to an augmented point of $\partial K_i^{(\varepsilon)}$. The next lemma shows that each such cap is of constant horizontal extents.

LEMMA 4.2. *Let C_p be any useful ε-cap induced by an augmented point $p \in \partial K_i^*$. Then, the vertical projection of C_p lies entirely within a ball B of radius $O(1)$ centered at the origin.*

PROOF. Recall that Γ_p denotes the base of p's cap in K_i^*, and let Γ'_p be its vertical projection. Let q be p's corresponding point. Since p is useful, we have $q \in \partial K_i^{(\varepsilon)}$. Also, recall that Δ_q denotes the base of q's ε-dual cap, and let Δ'_q be its vertical projection. By Lemma 4.1, $\Delta'_q - q' = \text{polar}_r(\Gamma'_p - p')$, for $r = \sqrt{\varepsilon}$, which by the involutory nature of the polar transform implies that $\Gamma'_p - p' = \text{polar}_r(\Delta'_q - q')$.

We assert that there exists a positive constant c such that Δ'_q contains a Euclidean ball centered at q' of radius $c\varepsilon$ (see Figure 7(a)). Assuming this assertion for now, observe that, by basic properties of the generalized polar transform, $\text{polar}_r(\Delta'_q - q')$ is contained within a ball of radius $r^2/c\varepsilon = 1/c$ centered at the origin. Since $\Gamma'_p - p' = \text{polar}_r(\Delta'_q - q')$, it follows that Γ'_p is contained within a ball of radius $1/c$ centered at the p' (see Figure 7(b)). Since K_i is $O(1)$-steep, the coordinates of p' (which are all slopes) are of constant absolute values. Therefore, p' is within constant distance of the origin, and hence so are all the points of Γ'_p.

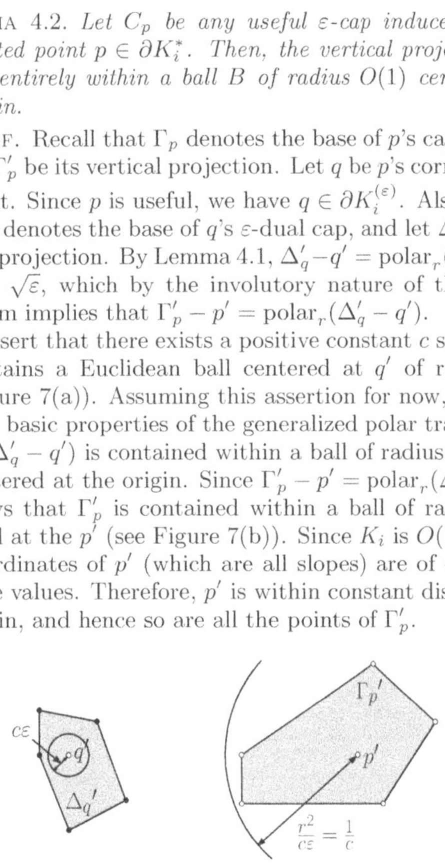

Figure 7: Proof of Lemma 4.2.

To complete the proof, it suffices to establish the above assertion. We begin by showing that the cone with apex $q - \varepsilon$ that defines D_q contains a cone $\Psi(q)$ whose apex is at $q - \varepsilon$ and whose central angle is $\Omega(1)$ (see Figure 8(a)). To see this, consider any augmented point $u \in \partial K_i$ that lies on the boundary of Δ_q. This means that, the supporting hyperplane at u passes through $q - \varepsilon$. It suffices to show that the acute angle between the ray from $q - \varepsilon$ to u and the vertical axis is $\Omega(1)$.

If u is a point in the relative interior of K_i (such as u' in Figure 8(a)), then, by the slope constraints of the associated

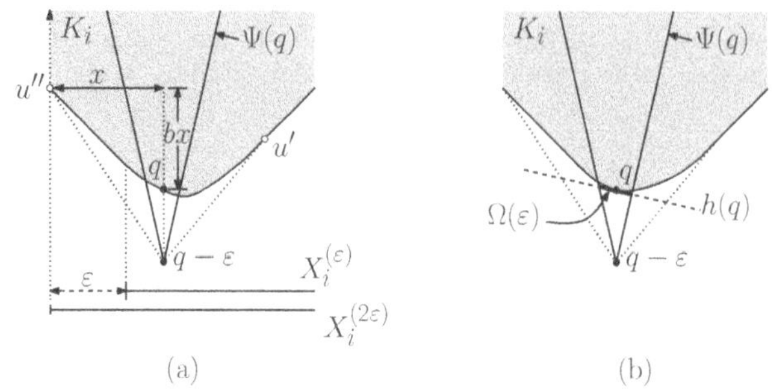

Figure 8: Proof of Lemma 4.2 (continued).

support set, the angle of the ray from $q - \varepsilon$ to u and the vertical axis is $\Omega(1)$. On the other hand, suppose that u is a point on the relative boundary of K_i (such as u'' in Figure 8(a)). Since C_p is useful, it follows that D_q is useful, which implies that $q \in \partial K_i^{(\varepsilon)}$.

Recall that K_i is the intersection of S_i with the infinite cylinder $\Phi_i^{(2\varepsilon)}$ whose cross section is $X_i^{(2\varepsilon)}$. By definition, $X_i^{(2\varepsilon)}$ contains a region of distance at least ε around the vertical projection of any $q \in \partial K_i^{(\varepsilon)}$. If we let x denote the horizontal component of the distance from q to u, we have $x \geq \varepsilon$. By the constraint on the slopes of K_i's support set, since both q and u lie on ∂K_i, the vertical component of the distance between q and u is at most bx, for some positive constant b. Since $q - \varepsilon$ lies directly below q by a distance of ε, the horizontal component of the distance between $q - \varepsilon$ and u is x, and the vertical component of the distance is at most $bx + \varepsilon \leq (b+1)x$. Therefore, the tangent of angle between the ray from $q - \varepsilon$ to u and the vertical axis is at least $x/(b+1)x = 1/(b+1) = \Omega(1)$. Therefore, for a suitably chosen (constant) central angle, the cone $\Psi(q)$ lies within the cone defining D_q.

Next, consider the supporting hyperplane $h(q)$ that passes through the augmented point q. Because $q \in \partial K_i^{(\varepsilon)}$, this hyperplane satisfies the slope constraints of the associated support set. Because q is located at distance ε above $\Psi(q)$'s apex and $h(q)$ is of constant slope, it follows that $h(q) \cap \Psi(q)$ is an ellipse whose smallest radius is $\Omega(\varepsilon)$ (see Figure 8(b)). Therefore, the vertical projection of this ellipse contains a ball of radius $\Omega(\varepsilon)$. Because $\Psi(q)$ lies within the cone defining D_q, it follows that this ball lies entirely within the vertical projection Δ'_q. This establishes the assertion, and completes the proof. □

The following lemma establishes a cap-based variant of the sampling process of Lemma 3.2. The proof is a straightforward adaptation of the proof of Lemma 3.2 to the case of caps. Caps are actually easier to deal with because the Macbeath-region machinery can be applied directly.

LEMMA 4.3. *Let K be a convex body in $\mathbb{R}^d$, and consider any $\varepsilon > 0$ and $i \in I^{\pm}$. Given any $t > 0$, there exists a set of $O(1/t)$ augmented points $P_i \subset \partial K_i^*$ such that, for any useful ε-cap C_p induced by an augmented point $p \in \partial K_i^*$, if $\text{area}(\Gamma_p) \geq t$, there exists a point in P_i that stabs C_p.*

PROOF. Apply Lemma 2.3 to K_i^* with $v = c \cdot \varepsilon \cdot t$, for a suitable constant c, and let $\mathcal{M}$ denote the resulting collection of disjoint bodies of volume $\Omega(v)$. (Note that K_i^* is not bounded, but it suffices to bound it crudely, say by a ball of suitably large radius.) Let B denote the $(d-1)$-dimensional ball of radius $O(1)$ that lies on X_i, as described in Lemma 4.2.

As in the proof of Lemma 3.2, for each $M \in \mathcal{M}$, if M does not lie entirely within vertical distance ε of ∂K_i^* or if its vertical projection does not lie entirely within B, then discard M from further consideration. For each remaining body M, select an arbitrary point from it, project this point vertically downward onto ∂K_i^*. Augment the point by associating it with any valid surface normal. Add the resulting augmented point to P_i. Repeating this process for all $M \in \mathcal{M}$ yields the desired set P_i.

To establish the correctness of this construction, consider any useful ε-cap C_p induced by an augmented point $p \in \partial K_i^*$ whose base Γ_p is of area at least t. Let h denote the supporting hyperplane at p, and let H denote the lower halfspace bounded by $h + \varepsilon$, which contains Γ_p. Recall that K_i^* is U-shaped and $O(1)$ steep. By Lemma 4.2 the vertical projection of C_p is bounded, and thus, by Lemma 3.1, $\text{vol}(K_i^* \cap H) = \Omega(\varepsilon \cdot \text{area}(\Gamma_p)) = \Omega(\varepsilon t)$. By choosing c suitably, we can ensure that $\text{vol}(K_i^* \cap H) \geq v$. As observed in Lemma 3.2, it is possible to reduce the volume of this region by translating the halfspace downwards until the volume equals v. By Lemma 2.3(ii), there exists $M \in \mathcal{M}$ such that M is contained within the reduced region, and so $M \subseteq K_i^* \cap H$. (Note that M could not have been discarded in the construction process because, by Lemma 4.2, any useful ε-cap is contained within ball B.) It is easy to see that the point of P_i associated with M stabs C_p, as desired.

In order to bound the number of points in P_i, consider the set of points of K_i^* that lie within vertical distance ε of its boundary and whose vertical projection lies in B. The volume of this region is clearly $O(\varepsilon)$. By Lemma 2.3(i), each body M is of volume $\Omega(v) = \Omega(\varepsilon t)$. These bodies are pairwise disjoint, and (after discarding) each of them lies within the considered region. Therefore, by a simple packing argument, the number of such bodies is $O(\varepsilon/v) = O(1/t)$, as desired. □

5. PUTTING THE PIECES TOGETHER

Lemma 4.3 provides a set of augmented points $P_i \subset \partial K_i^*$ that stab all useful caps of K_i^* for which $\text{area}(\Gamma_p)$ exceeds a given parameter t. Let's consider how to exploit this for stabbing dual caps in K_i. For each $p' \in P_i$, augment it by associating it with any valid surface normal, and let q' be the corresponding augmented point on ∂K_i. By basic properties of the dual transform, p' stabs a useful ε-cap $C_p(K_i^*)$ if and only if q' stabs the corresponding useful ε-dual cap $D_q(K_i)$. We can therefore map P_i to a set $Q_i' \subset \partial K_i$ in order to stab all the useful ε-dual caps of K_i that correspond to the ε-caps of K_i^* that are stabbed by P_i.

What area properties do these dual caps satisfy? Let $D_q(K_i)$ be any useful ε-dual cap and $C_p(K_i^*)$ be the corresponding useful cap. Let Δ_q' and Γ_p' denote the respective projections of the bases of $D_q(K_i)$ and $C_p(K_i^*)$. We claim that if $\text{area}(\Delta_q') < c\varepsilon^{d-1}/t$, where c is a suitable constant, then D_q will be stabbed by some point in Q_i'. We will show that $\text{area}(\Gamma_p') > t$, which will imply this claim since $\text{area}(\Gamma_p)$ is clearly more than $\text{area}(\Gamma_p')$.

By Lemma 4.1, $\Delta_q' = \text{polar}_r(\Gamma_p')$, for $r = \sqrt{\varepsilon}$. Recall that $\text{polar}_r(K)$ is a scaled copy of $\text{polar}_1(K)$ by $r^2 = \varepsilon$. Since these are $(d-1)$-dimensional sets, we have

$$\text{area}(\Delta_q') = \text{area}(\varepsilon \cdot \text{polar}_1(\Gamma_p')) = \varepsilon^{d-1} \cdot \text{area}(\text{polar}_1(\Gamma_p')).$$

Thus, in order to bound $\text{area}(\Gamma_p')$ in terms of $\text{area}(\Delta_q')$, we need to establish a relationship between $\text{area}(\Gamma_p')$ and $\text{area}(\text{polar}_1(\Gamma_p'))$. To do this, we make use of results on the Mahler volume. Recall that the Mahler volume of a convex body K is $\text{vol}(K) \cdot \text{vol}(\text{polar}_1(K))$. It is known that the Mahler volume is bounded below by a constant [24]. Applying this to the $(d-1)$-dimensional set Γ_p', we have $\text{area}(\Gamma_p') = \Omega(1/\text{area}(\text{polar}_1(\Gamma_p')))$. Therefore, $\text{area}(\Gamma_p') = \Omega(\varepsilon^{d-1}/\text{area}(\Delta_q'))$. Since $\text{area}(\Delta_q') < c\varepsilon^{d-1}/t$, for a suitable c, it follows that $\text{area}(\Gamma_p') > t$. Thus, adjusting for the constant factors, we have established the following analog to Lemma 3.2, but for small dual caps.

Lemma 5.1. *Let K be a convex body in $\mathbb{R}^d$, and consider any $\varepsilon > 0$ and $i \in I^{\pm}$. Given any $t > 0$, there exists a set of $O(1/t)$ augmented points $Q_i' \subset \partial K_i$ such that, for any useful ε-dual cap D_q induced by an augmented point $q \in \partial K_i^{(\varepsilon)}$, if $\text{area}(\Delta_q) \leq \varepsilon^{d-1}/t$, there exists a point in Q_i' that stabs D_q.*

We are now ready to provide the proof of Theorem 1.1. Recall that $K_i = K_i^{(2\varepsilon)}$. By Lemma 2.2, it suffices to stab all the useful ε-dual caps of each of the K_i's, for $i \in I^{\pm}$. Fix any $i \in I^{\pm}$, and let $t = \sqrt{\text{area}(K)} \cdot \varepsilon^{(d-1)/2}$. For any augmented point $q \in \partial K_i^{(\varepsilon)}$, we say that the associated ε-dual cap is *large* if $\text{area}(\Delta_q) \geq t$, and otherwise it is *small*. We use two different strategies for stabbing the two types of dual caps.

For large dual caps, we apply Lemma 3.2 with the value of t defined above. This yields a set $Q_i \subset \partial K_i$ of size $O(\text{area}(K)/t) = O(\sqrt{\text{area}(K)}/\varepsilon^{(d-1)/2})$, such that every useful large ε-dual cap is stabbed by one of these points.

In order to handle small dual caps, we apply Lemma 5.1 with the value of t set to $t' = c'\varepsilon^{(d-1)/2}/\sqrt{\text{area}(K)}$, for a suitably chosen constant c'. This yields a set $Q_i' \subset \partial K_i$ of size $O(1/t') = O(\sqrt{\text{area}(K)}/\varepsilon^{(d-1)/2})$, such that every useful ε-dual cap whose base is of area at most $\varepsilon^{d-1}/t' = O(\sqrt{\text{area}(K)} \cdot \varepsilon^{(d-1)/2})$ is stabbed. By choosing c' suitably, every useful small ε-dual cap is stabbed.

In summary, $Q_i \cup Q_i'$ is a set of size $O(\sqrt{\text{area}(K)}/\varepsilon^{(d-1)/2})$ that stabs all useful ε-dual caps. By repeating this for all $i \in I^{\pm}$ and taking the union of all these sets, by Lemma 2.2, the resulting set provides the desired ε-approximation to K. This completes the proof of Theorem 1.1.

6. REFERENCES

[1] P. K. Agarwal, S. Har-Peled, and K. R. Varadarajan. Geometric approximation via coresets. In J. E. Goodman, J. Pach, and E. Welzl, editors, *Combinatorial and Computational Geometry* MSRI Publications, 2005.

[2] N. Alon and J. H. Spencer. *The probabilistic method.* Wiley, 2nd edition, 2008.

[3] S. Arya, G. D. da Fonseca, and D. M. Mount. Polytope approximation and the Mahler volume. In *Proc. 23rd Annu. ACM-SIAM Sympos. Discrete Algorithms*, pages 29–42, 2012.

[4] S. Arya, T. Malamatos, and D. M. Mount. The effect of corners on the complexity of approximate range searching. *Discrete Comput. Geom.*, 41:398–443, 2009.

[5] S. Arya, D. M. Mount, and J. Xia. Tight lower bounds for halfspace range searching. *Discrete Comput. Geom.*, 2012. (doi: 10.1007/s00454-012-9412-x).

[6] I. Bárány. Intrinsic volumes and f-vectors of random polytopes. *Math. Ann.*, 285:671–699, 1989.

[7] I. Bárány. Random polytopes, convex bodies, and approximation. In W. Weil, editor, *Stochastic Geometry*, volume 1892 of *Lecture Notes in Mathematics*, pages 77–118. Springer, 2007.

[8] I. Bárány and D. G. Larman. Convex bodies, economic cap coverings, random polytopes. *Mathematika*, 35:274–291, 1988.

[9] K. Böröczky, Jr. Approximation of general smooth convex bodies. *Advances in Mathematics*, 153:325–341, 2000.

[10] H. Brönnimann, B. Chazelle, and J. Pach. How hard is halfspace range searching. *Discrete Comput. Geom.*, 10:143–155, 1993.

[11] H. Brönnimann and M. T. Goodrich. Almost optimal set covers in finite VC-dimension,. *Discrete Comput. Geom.*, 14:463–479, 1995.

[12] E. M. Bronshteyn and L. D. Ivanov. The approximation of convex sets by polyhedra. *Siberian Math. J.*, 16:852–853, 1976.

[13] E. M. Bronstein. Approximation of convex sets by polytopes. *Journal of Mathematical Sciences*, 153(6):727–762, 2008.

[14] B. Chazelle, L. J. Guibas, and D. T. Lee. The power of geometric duality. *BIT*, 25:76–90, 1985.

[15] K. L. Clarkson. Algorithms for polytope covering and approximation. In *Proc. 3th Workshop Algorithms Data Struct.*, pages 246–252, 1993.

[16] K. L. Clarkson. Building triangulations using epsilon-nets. In *Proc. 38th Annu. ACM Sympos. Theory Comput.*, pages 326–335, 2006.

[17] M. de Berg, O. Cheong, M. van Kreveld, and M. Overmars. *Computational Geometry: Algorithms and Applications*. Springer, 3rd edition, 2010.

[18] M. P. do Carmo. *Differential Geometry of Curves and Surfaces*. Prentice Hall, 1976.

[19] R. M. Dudley. Metric entropy of some classes of sets with differentiable boundaries. *Approx. Theory*, 10(3):227–236, 1974.

[20] H. G. Eggleston. *Convexity*. Cambridge University Press, 1958.

[21] G. Ewald, D. G. Larman, and C. A. Rogers. The directions of the line segments and of the r-dimensional balls on the boundary of a convex body in Euclidean space. *Mathematika*, 17:1–20, 1970.

[22] P. M. Gruber. Asymptotic estimates for best and stepwise approximation of convex bodies I. *Forum Math.*, 5:521–537, 1993.

[23] F. John. Extremum problems with inequalities as subsidiary conditions. In *Studies and Essays Presented to R. Courant on his 60th Birthday*, pages 187–204. Interscience Publishers, Inc., New York, 1948.

[24] G. Kuperberg. From the Mahler conjecture to Gauss linking integrals. *Geometric And Functional Analysis*, 18:870–892, 2008.

[25] A. M. Macbeath. A theorem on non-homogeneous lattices. *Ann. of Math.*, 56:269–293, 1952.

[26] D. E. McClure and R. A. Vitalie. Polygonal approximation of plane convex bodies. *J. Math. Anal. Appl.*, 51:326–358, 1975.

[27] J. Pach and G. Tardos. Tight lower bounds for the size of epsilon-nets. In *Proc. 27th Annu. Sympos. Comput. Geom.*, pages 458–463, 2011.

[28] R. Schneider. Polyhedral approximation of smooth convex bodies. *J. Math. Anal. Appl.*, 128:470–474, 1987.

[29] L. F. Toth. Approximation by polygons and polyhedra. *Bull. Amer. Math. Soc.*, 54:431–438, 1948.

Object-Image Correspondence for Curves under Central and Parallel Projections

Joseph M. Burdis *
Department of Mathematics
North Carolina State University
Raleigh, NC 27659
joe.burdis@gmail.com

Irina A. Kogan †
Department of Mathematics
North Carolina State University
Raleigh, NC 27659
iakogan@ncsu.edu

ABSTRACT

We present a novel algorithm for deciding whether a given planar curve is an image of a given spatial curve, obtained by a central or a parallel projection with unknown parameters. A straightforward approach to this problem consists of setting up a system of conditions on the projection parameters and then checking whether or not this system has a solution. The computational advantage of the algorithm presented here, in comparison to algorithms based on the straightforward approach, lies in a significant reduction of a number of real parameters that need to be eliminated in order to establish existence or non-existence of a projection that maps a given spatial curve to a given planar curve. Our algorithm is based on projection criteria that reduce the projection problem to a certain modification of the equivalence problem of planar curves under affine and projective transformations. The latter problem is then solved using a separating set of rational differential invariants. A similar approach can be used to decide whether a given finite list of points on a plane is an image of a given finite list of points in $\mathbb{R}^3$. The motivation comes from the problem of establishing a correspondence between an object and an image, taken by a camera with unknown position and parameters.

Categories and Subject Descriptors

I.1.4 [**Symbolic and Algebraic Manipulations**]: Applications

General Terms

Algorithms, Theory

*Supported in part by NSF grants CCF-0728801 and CCF-0347506.

†Supported in part by NSF grant CCF-0728801 and NSA grant H98230-11-1-0129.

SCG'12, June 17–20, 2012, Chapel Hill, North Carolina, USA.

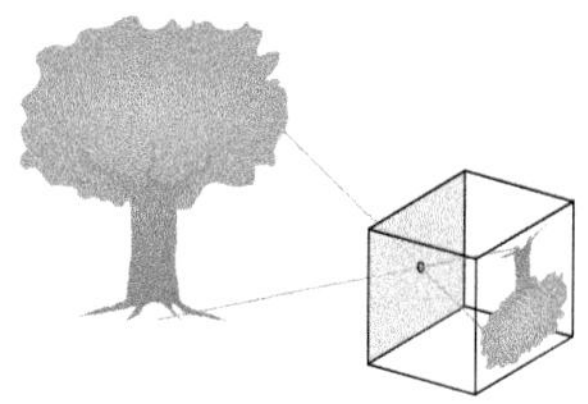

Figure 1: A pinhole camera [24].

Keywords

Central and parallel projections; finite and affine cameras; curves; projective and affine transformations; separating differential invariants; signatures; machine vision.

1. INTRODUCTION

A central projection from $\mathbb{R}^3$ to $\mathbb{R}^2$ models a simple pinhole camera pictured in Figure 1. A generic central projection is described by a linear fractional transformation (1):

$$x = \frac{p_{11}\,z_1 + p_{12}\,z_2 + p_{13}\,z_3 + p_{14}}{p_{31}\,z_1 + p_{32}\,z_2 + p_{33}\,z_3 + p_{34}},$$
$$y = \frac{p_{21}\,z_1 + p_{22}\,z_2 + p_{23}\,z_3 + p_{24}}{p_{31}\,z_1 + p_{32}\,z_2 + p_{33}\,z_3 + p_{34}}, \qquad (1)$$

where (z_1, z_2, z_3) denote coordinates in $\mathbb{R}^3$, (x, y) denote coordinates in $\mathbb{R}^2$ and p_{ij}, $i = 1\dots3$, $j = 1\dots4$, are real parameters of the projection, such that the left 3×3 submatrix of 3×4 matrix $P = (p_{ij})$ has a non-zero determinant. Parameters represent the freedom to choose the center of the projection, the position of the image plane and (in general non-orthogonal) coordinate system on the image plane.[1] In the case when the distance between a camera and an object is significantly greater than the object depth, a parallel projection provides a good camera model. A parallel projection has 8 parameters and can be described by a 3×4 matrix of rank 3, whose last row is $(0, 0, 0, 1)$. We review various camera models and related geometry in Section 2 (see also [17]). In most general terms, the projection problem considered here is formulated as follows:

[1]It is clear from (1) that multiplication of P by a non-zero constant does not change the projection map. Therefore, we can identify P with a point of the projective space $\mathbb{PR}^{11}$, rather than a point in $\mathbb{R}^{12}$. However, since we do not know which of the parameters are non-zero, in computations we have to keep all 12 parameters.

PROBLEM 1. *Given a subset $\mathcal{Z}$ of $\mathbb{R}^3$ and a subset $\mathcal{X}$ of $\mathbb{R}^2$, does there exist a projection $P\colon \mathbb{R}^3 \to \mathbb{R}^2$ such that $\mathcal{X} = P(\mathcal{Z})$?*

A straightforward approach to this problem consists of setting up a system of conditions on the projection parameters and then checking whether or not this system has a solution. In the case when $\mathcal{Z}$ and $\mathcal{X}$ are finite lists of points, a solution based on the straightforward approach can be found in [17]. For curves and surfaces under central projections, this approach is taken in [12]. However, internal parameters of the camera are considered to be known in that paper and, therefore, there are only 6 camera parameters in that study vs. 12 considered here. The method presented in [12] also uses an additional assumption that a planar curve $\mathcal{X} \subset \mathbb{R}^2$ has at least two points, whose tangent lines coincide. An alternative approach to the problem in the case when $\mathcal{Z}$ and $\mathcal{X}$ are *finite lists of points under parallel projections* was presented in [2, 1]. In these articles, the authors establish polynomial relationships that have to be satisfied by coordinates of the points in the sets $\mathcal{Z}$ and $\mathcal{X}$ in order for a projection to exists. Our approach to the projection problem for *curves* is closer in spirit to [2, 1], as we also establish necessary and sufficient conditions that curves $\mathcal{Z}$ and $\mathcal{X}$ must satisfy in order for a projection to exist. However, unlike [2, 1], we exploit the relationship between the projection problem and equivalence problem under group-actions. We will show below that, in comparison with the straightforward approach, our solution leads to a significant reduction of the number of parameters that have to be eliminated in order to solve Problem 1 for curves.

In this paper, we assume that $\mathcal{Z}$ and $\mathcal{X}$ are *rational algebraic curves*, i. e. $\mathcal{Z} = \{\Gamma(s)|s \in \mathbb{R}\}$ and $\mathcal{X} = \{\gamma(t)|t \in \mathbb{R}\}$, where $\Gamma\colon \mathbb{R} \to \mathbb{R}^3$ and $\gamma\colon \mathbb{R} \to \mathbb{R}^2$ are rational maps, and that $\mathcal{Z}$ is not a straight line (and, therefore, its image under any projection is a one-dimensional constructible set). We also relax the projection condition to $\mathcal{X} = \overline{P(\mathcal{Z})}$, *where bar throughout the paper denotes the algebraic closure of a set.* The case of non-rational algebraic curves in more technical and is discussed in [6]. Then Problem 1, for central projections, can be reformulated as the following real quantifier elimination problem:

Reformulation 1. (STRAIGHTFORWARD APPROACH) Given two rational maps $\Gamma\colon \mathbb{R} \to \mathbb{R}^3$ and $\gamma\colon \mathbb{R} \to \mathbb{R}^2$, determine the truth of the statement:

$$\begin{aligned} &\exists P \in \mathbb{R}^{3\times 4} \quad \det(p_{ij})_{i=1\ldots 3}^{j=1\ldots 3} \neq 0 \\ &\forall s \text{ in the domain of } \Gamma(s) \quad \exists t \in \mathbb{R} \quad \gamma(t) = P(\Gamma(s)). \end{aligned}$$

Real quantifier elimination problems are algorithmically solvable [22]. A survey of subsequent developments in this area can be found, for instance, in [18] and [9]. High computational complexity of these problems makes a reduction in the number of parameters to be desirable.

The projection criteria developed in this paper reduces the projection problem to the problem of deciding whether the given planar curve $\mathcal{X}$ is equivalent to a curve in a certain family of *planar* curves under an action of the projective group in the case of central projections, and under the action of the affine group in the case of parallel projections. The family of curves depends on 3 parameters in the case of central projections, and on 2 parameters in the case of parallel projections.

These group-equivalence problems can be solved by an adaptation of differential signature construction developed in [7] to solve *local* equivalence problems for smooth curves. In Section 4, we give an algebraic formulation of signature construction and show that it leads to a solution of *global* equivalence problems for algebraic curves. For this purpose we introduce a notion of *differentially separating set of invariants*. Following this method for the case of central projections, when $\mathcal{Z}$ and $\mathcal{X}$ are rational algebraic curves, we define two rational signature maps $\mathcal{S}_{\mathcal{X}}\colon \mathbb{R} \to \mathbb{R}^2$ and $\mathcal{S}_{\mathcal{Z}}\colon \mathbb{R}^4 \to \mathbb{R}^2$. Construction of these signature maps requires only differentiation and arithmetic operations and is computationally trivial. Problem 1 becomes equivalent to

Reformulation 2. (SIGNATURE APPROACH) Given two rational maps $S_{\mathcal{X}}\colon \mathbb{R} \to \mathbb{R}^2$ and $S_{\mathcal{Z}}\colon \mathbb{R}^4 \to \mathbb{R}^2$, determine the truth of the statement:

$$\begin{aligned} &\exists c \in \mathbb{R}^3 \quad \forall s \text{ in the domain of } S_{\mathcal{Z}}(c,s) \\ &\exists t \in \mathbb{R} \quad S_{\mathcal{Z}}(c,s) = S_{\mathcal{X}}(t). \end{aligned}$$

We note that Reformulation 1 and Reformulation 2 have similar structure, but the former requires elimination of 14 parameters, while the latter requires elimination of only 5 parameters.[2] The case of parallel projection is treated in the similar manner and leads to the reduction of the number of real parameters that need to be eliminated from 10 to 4.

Our method can be easily adapted to solve projection problems for curves in $\mathbb{R}^n$ projected to a hyperplane. The problem and the solution remains valid if we replace $\mathbb{R}$ with $\mathbb{C}$ and, in fact, as with most problems in algebraic geometry, the implementation of the algorithms is easier over the complex numbers.. The existence of a complex projection provides a necessary but not a sufficient condition for the existence of a real projection. In Section 6, we discuss how this method leads to an alternative solution to the projection problem for finite lists of points (see [5] for more details).

Although the relation between projections and group actions is known, our literature search did not yield algorithms that exploit this relationship to solve the projection problem for curves in the generic setting of cameras with unknown internal and external parameters. The goal of the paper is to introduce such algorithms. Although the development of efficient implementations of these algorithms and their complexity study lie outside of the scope of this paper, we made a preliminary implementations in MAPLE of projection algorithms over *complex numbers*. We implemented both an algorithm based on signature construction presented here and an algorithm based on the straightforward approach and timed their performance. The code and the experiments are posted on the internet [25].

In order to become practically useful in real-life applications, the algorithmic solution presented here, would have to be adapted to curves given by a finite sampling of points. Some directions of such adaptation are indicated in Section 7 of the paper.

2. PROJECTIONS AND CAMERAS

We embed $\mathbb{R}^n$ into projective space $\mathbb{PR}^n$ and use homogeneous coordinates on $\mathbb{PR}^n$ to express the map (1) by matrix multiplication.

[2]We include parameters t and s into the count.

NOTATION 2. Square brackets *around matrices (and, in particular, vectors) will be used to denote an equivalence class with respect to multiplication of a matrix by a nonzero scalar. Multiplication of equivalence classes of matrices A and B of appropriate sizes is well-defined by* $[A]\,[B] := [A\,B]$.

With this notation, a point $(x, y) \in \mathbb{R}^2$ corresponds to a point $[x, y, 1] = [\lambda x, \lambda y, \lambda] \in \mathbb{PR}^2$ for all $\lambda \neq 0$, and a point $(z_1, z_2, z_3) \in \mathbb{R}^3$ corresponds to $[z_1, z_2, z_3, 1] \in \mathbb{PR}^3$. We will refer to the points in $\mathbb{PR}^n$ whose last homogeneous coordinate is zero as *points at infinity*. In homogeneous coordinates projection (1) is a map $[P]\colon \mathbb{PR}^3 \to \mathbb{PR}^2$ given by

$$[x,\, y,\, 1]^{\mathrm{T}} = [P]\,[z_1,\, z_2,\, z_3,\, 1]^{\mathrm{T}}, \tag{2}$$

where P is 3×4 matrix of rank 3 and superscript T denotes transposition. Matrix P has a 1-dimensional kernel. Therefore, there exists a point $[z_1^0, z_2^0, z_3^0, z_4^0] \in \mathbb{PR}^3$ whose image under the projection is undefined (recall that $[0, 0, 0]$ is not a point in $\mathbb{PR}^2$). Geometrically, this point is the center of the projection.

In computer science literature (e.g. [17]), a camera is called *finite* if its center is not at infinity. A finite camera is modeled by a matrix P, whose left 3×3 submatrix is non-singular. Geometrically, finite cameras correspond to central projections from $\mathbb{R}^3$ to a plane. On the contrary, an *infinite* camera has its center at an infinite point of $\mathbb{PR}^3$. An infinite camera is modeled by a matrix P whose left 3×3 submatrix is singular. An infinite camera is called *affine* if the preimage of the line at infinity in $\mathbb{PR}^2$ is the plane at infinity in $\mathbb{PR}^3$. An affine camera is modeled by a matrix P whose last row is $(0, 0, 0, 1)$. In this case map (1) becomes $x = p_{11}\, z_1 + p_{12}\, z_2 + p_{13}\, z_3 + p_{14}\,,\quad y = p_{21}\, z_1 + p_{22}\, z_2 + p_{23}\, z_3 + p_{24}$. Geometrically, affine cameras correspond to *parallel projections* from $\mathbb{R}^3$ to a plane.[3] Eight degrees of freedom reflect a choice of the direction of a projection, a position of the image plane and a choice of linear system of coordinates on the image plane. An image plane may be assumed to be perpendicular to the direction of the projection, since other choices are absorbed in the freedom to choose, in general, non-orthogonal coordinate system on the image plane.

DEFINITION 3. *A set of equivalence classes* $[P]$, *where* $P = (p_{ij})_{j=1\ldots4}^{i=1\ldots3}$ *is a* 3×4 *matrix whose left* 3×3 *submatrix is non-singular, is called* the set of central projections *and is denoted* $\mathcal{CP}$.

A set of equivalence classes $[P]$, *where* $P = (p_{ij})_{j=1\ldots4}^{i=1\ldots3}$ *has rank 3 and its last row is* $(0, 0, 0, \lambda)$, $\lambda \neq 0$, *is called* the set of parallel projections *and is denoted* $\mathcal{PP}$.

Equation (1) determines a *central projection* when $[P] \in \mathcal{CP}$ and it determines a parallel projection when $[P] \in \mathcal{PP}$. Sets $\mathcal{CP}$ and $\mathcal{PP}$ are disjoint. Projections that are not included in these two classes correspond to infinite, non-affine cameras. These are not frequently used in computer vision and are not considered in this paper.

3. PROJECTION CRITERIA FOR CURVES

Recall that for every algebraic curve $\mathcal{X} \subset \mathbb{R}^n$ there exists a unique projective algebraic curve $[\mathcal{X}] \subset \mathbb{PR}^n$ such that $[\mathcal{X}]$ is the smallest projective variety containing $\mathcal{X}$ (see [14]).

[3]Parallel projections are also called *generalized weak perspective projections* [2, 1].

DEFINITION 4. *We say that a curve* $\mathcal{Z} \subset \mathbb{R}^3$ projects to $\mathcal{X} \subset \mathbb{R}^2$ *if there exists a* 3×4 *matrix* P *of rank 3 such that* $[\mathcal{X}] = \{[P][\mathbf{z}] \mid \mathbf{z} \in \mathcal{Z},\, P\mathbf{z} \neq 0\}$. *We then write* $\mathcal{X} = P(\mathcal{Z})$ *or* $[\mathcal{X}] = [P][\mathcal{Z}]$.

DEFINITION 5. *The* projective group $\mathcal{PGL}(n{+}1)$ *is a quotient of the general linear group* $\mathcal{GL}(n+1)$, *consisting of* $(n+1) \times (n+1)$ *non-singular matrices, by a 1-dimensional abelian subgroup* λI, *where* $\lambda \neq 0 \in \mathbb{R}$ *and* I *is the identity matrix. Elements of* $\mathcal{PGL}(n+1)$ *are equivalence classes* $[B] = [\lambda B]$, *where* $\lambda \neq 0$ *and* $B \in \mathcal{GL}(n+1)$.

The affine *group* $\mathcal{A}(n)$ *is a subgroup of* $\mathcal{PGL}(n+1)$ *whose elements* $[B]$ *have a representative* $B \in \mathcal{GL}(n+1)$ *with the last row equal to* $(0, \ldots, 0, 1)$.

The special affine *group* $\mathcal{SA}(n)$ *is a subgroup of* $\mathcal{A}(n)$ *whose elements* $[B]$ *have a representative* $B \in \mathcal{GL}(n+1)$ *with determinant* 1 *and the last row equal to* $(0, \ldots, 0, 1)$.

In homogeneous coordinates, the standard action of the projective group $\mathcal{PGL}(n{+}1)$ on $\mathbb{PR}^n$ is defined by multiplication:

$$[z_1, \ldots, z_n, z_0]^{\mathrm{T}} \to [B]\,[z_1, \ldots, z_n, z_0]^{\mathrm{T}}. \tag{3}$$

The action (3) induces linear-fractional action of $\mathcal{PGL}(n{+}1)$ on $\mathbb{R}^n$. The restriction of (3) to $\mathcal{A}(n)$ induces an action on $\mathbb{R}^n$ consisting of compositions of linear transformations and translations.

DEFINITION 6. *We say that two curves* $\mathcal{X}_1 \subset \mathbb{R}^n$ *and* $\mathcal{X}_2 \subset \mathbb{R}^n$ *are* $\mathcal{PGL}(n+1)$*-equivalent if there exists* $[A] \in \mathcal{PGL}(n{+}1)$, *such that* $[\mathcal{X}_2] = \{[A][\mathbf{p}] \mid [\mathbf{p}] \in [\mathcal{X}_1]\}$. *We then write* $\mathcal{X}_2 = A(\mathcal{X}_1)$ *or* $[\mathcal{X}_2] = [A][\mathcal{X}_1]$. *If* $[A] \in G$, *where* G *is a subgroup of* $\mathcal{PGL}(n+1)$, *we say that* $\mathcal{X}_1$ *and* $\mathcal{X}_2$ *are* G*-equivalent.*

Before stating the projection criteria, we make the following simple, but important observations.

PROPOSITION 7. *(i) If* $\mathcal{Z} \subset \mathbb{R}^3$ *projects to* $\mathcal{X} \subset \mathbb{R}^2$ *by a* parallel projection, *then any curve that is* $\mathcal{A}(3)$*-equivalent to* $\mathcal{Z}$ *projects to any curve that is* $\mathcal{A}(2)$*-equivalent to* $\mathcal{X}$ *by a* parallel projection. *In other words, parallel projections are defined on affine equivalence classes of curves.*

(ii) If $\mathcal{Z} \subset \mathbb{R}^3$ *projects to* $\mathcal{X} \subset \mathbb{R}^2$ *by a* central projection *then any curve in* $\mathbb{R}^3$ *that is* $\mathcal{A}(3)$*-equivalent to* $\mathcal{Z}$ *projects to any curve on* $\mathbb{R}^2$ *that is* $\mathcal{PGL}(3)$*-equivalent to* $\mathcal{X}$ *by a* central projection.

PROOF. (i) Assume that there exists a parallel projection $[P] \in \mathcal{PP}$ such that $[\mathcal{X}] = [P][\mathcal{Z}]$. Then for all $(A, B) \in \mathcal{A}(2) \times \mathcal{A}(3)$ we have $[A]\,[\mathcal{X}] = [A]\,[P]\,[B^{-1}]\,[B]\,[\mathcal{Z}]$. Since $[A]\,[P]\,[B^{-1}] \in \mathcal{PP}$, curve $B(\mathcal{Z})$ projects to $A(\mathcal{X})$. (ii) is proved similarly. □

REMARK 8. *It is known that if* $\mathcal{X}_1$ *and* $\mathcal{X}_2$ *are images of a curve* $\mathcal{Z}$ *under two central projections* with the same center, *then* $\mathcal{X}_1$ *and* $\mathcal{X}_2$ *are* $\mathcal{PGL}(3)$*-equivalent, but if the centers of the projections are not the same this is no longer true (see Example 33). Similarly, images of* $\mathcal{Z}$ *under different parallel projections may not be* $\mathcal{A}(2)$*-equivalent (see Example 35).*

THEOREM 9. *(*CENTRAL PROJECTION CRITERION.*)*
A curve $\mathcal{Z} \subset \mathbb{R}^3$ *projects to a curve* $\mathcal{X} \subset \mathbb{R}^2$ *by a* central

projection *if and only if there exist* $c_1, c_2, c_3 \in \mathbb{R}$ *such that* $\mathcal{X}$ *is* $\mathcal{PGL}(3)$*-equivalent to a planar curve*

$$\tilde{\mathcal{Z}}_{c_1,c_2,c_3} = \overline{\left\{\left(\frac{z_1+c_1}{z_3+c_3},\quad \frac{z_2+c_2}{z_3+c_3}\right) \,\middle|\, (z_1,z_2,z_3)\in\mathcal{Z}\right\}} \quad (4)$$

PROOF. ($\Rightarrow$) Assume there exists a central projection $[P]$ such that $[\mathcal{X}] = [P]\,[\mathcal{Z}]$. Then P is a 3×4 matrix $P = (p_{ij})_{j=1\ldots4}^{i=1\ldots3}$ whose left 3×3 submatrix is non-singular. Therefore there exist c_1, c_2, $c_3 \in \mathbb{R}$ such that $p_{*4} = c_1\, p_{*1}+c_2\, p_{*2}+c_3\, p_{*3}$, where p_{*j} denotes the j-th column of the matrix P. We observe that

$$[A][P^0_{\mathcal{C}}][B] = [P]\,, \quad (5)$$

where $A := (p_{ij})_{j=1\ldots3}^{i=1\ldots3}$ is 3×3 submatrix of P,

$$P^0_{\mathcal{C}} := \begin{pmatrix} 1&0&0&0\\ 0&1&0&0\\ 0&0&1&0\end{pmatrix} \text{ and } B := \begin{pmatrix} 1&0&0&c_1\\ 0&1&0&c_2\\ 0&0&1&c_3\\ 0&0&0&1\end{pmatrix}. \quad (6)$$

Note that $[A]$ belongs to $\mathcal{PGL}(3)$. Since

$$[P^0_{\mathcal{C}}][B][z_1,z_2,z_3,\,1]^{\mathrm{T}} = [z_1+c_1, z_2+c_2, z_3+c_3]^{\mathrm{T}},$$

then $[\mathcal{X}] = [A][\tilde{\mathcal{Z}}_{c_1,c_2,c_3}]$, where $\tilde{\mathcal{Z}}_{c_1,c_2,c_3}$ is defined by (4).

($\Leftarrow$) To prove the converse direction we assume that there exists $[A] \in \mathcal{PGL}(3)$ and $c_1, c_2, c_3 \in \mathbb{R}$ such that $[\mathcal{X}] = [A][\tilde{\mathcal{Z}}_{c_1,c_2,c_3}]$, where $\tilde{\mathcal{Z}}_{c_1,c_2,c_3}$ is defined by (4). A direct computation shows that $\mathcal{Z}$ is projected to $\mathcal{X}$ by the central projection $[P] = [A]\,[P^0_{\mathcal{C}}]\,[B]$, where B and $[P^0_{\mathcal{C}}]$ are given by (6). □

THEOREM 10. (PARALLEL PROJECTION CRITERION.) *A curve* $\mathcal{Z}\subset\mathbb{R}^3$ *projects to a curve* $\mathcal{X}\subset\mathbb{R}^2$ *by a parallel projection if and only if there exist* $c_1, c_2 \in \mathbb{R}$ *and an ordered triplet* $(i,j,k) \in \{(1,2,3), (1,3,2), (2,3,1)\}$ *such that* $\mathcal{X}$ *is* $\mathcal{A}(2)$*-equivalent to*

$$\tilde{\mathcal{Z}}^{i,j,k}_{c_1,c_2} = \overline{\left\{(z_i + c_1\, z_k,\quad z_j + c_2\, z_k) \,\middle|\, (z_1,z_2,z_3)\in\mathcal{Z}\right\}} \quad (7)$$

PROOF. ($\Rightarrow$)Assume there exists a parallel projection $[P]$ such that $[\mathcal{X}] = [P]\,[\mathcal{Z}]$. Then $[P]$ can be represented by a matrix:

$$P = \begin{pmatrix} p_{11}&p_{12}&p_{13}&p_{14}\\ p_{21}&p_{22}&p_{23}&p_{24}\\ 0&0&0&1\end{pmatrix} \quad (8)$$

of rank 3. Therefore there exist $1\le i<j\le 3$ such that the rank of the submatrix $\begin{pmatrix} p_{1i}&p_{1j}\\ p_{2i}&p_{2j}\end{pmatrix}$ is 2. Then for $1\le k\le 3$, such that $k\neq i$ and $k\neq j$, there exist c_1, $c_2\in\mathbb{R}$, such that $\begin{pmatrix} p_{1k}\\ p_{2k}\end{pmatrix} = c_1\begin{pmatrix} p_{1i}\\ p_{2i}\end{pmatrix} + c_2\begin{pmatrix} p_{1j}\\ p_{2j}\end{pmatrix}$. We define $A := \begin{pmatrix} p_{1i}&p_{1j}&p_{14}\\ p_{2i}&p_{2j}&p_{24}\\ 0&0&1\end{pmatrix}$ and define B to be the matrix whose columns are vectors $b_{*i} := (1,0,0,0)^{\mathrm{T}}$, $b_{*j} := (0,1,0,0)^{\mathrm{T}}$, $b_{*k} := (c_1,c_2,1,0)^{\mathrm{T}}$, $b_{*4} = (0,0,0,1)^{\mathrm{T}}$. We observe that $[A][P^0_{\mathcal{P}}][B] = [P]$, where $P^0_{\mathcal{P}} := \begin{pmatrix} 1&0&0&0\\ 0&1&0&0\\ 0&0&0&1\end{pmatrix}$. Since $[P^0_{\mathcal{P}}][B][\mathcal{Z}] = [\tilde{\mathcal{Z}}^{i,j,k}_{c_1,c_2}]$, then $[\mathcal{X}] = [A][\tilde{\mathcal{Z}}^{i,j,k}_{c_1,c_2}]$. Observe that $[A]\in\mathcal{A}(2)$ and the direct statement is proved.

($\Leftarrow$) To prove the converse direction we assume that there exist $[A]\in\mathcal{A}(2)$, two real numbers c_1 and c_2, and a triplet of indices such that $(i,\,j,\,k) \in \{(1,2,3), (1,3,2), (2,3,1)\}$, such that $[\mathcal{X}] = [A][\tilde{\mathcal{Z}}^{i,j,k}_{c_1,c_2}]$, where a planar curve $\tilde{\mathcal{Z}}^{i,j,k}_{c_1,c_2}$ is given by (7). Let B be a matrix defined in the first part of the proof. A direct computation shows that $\mathcal{Z}$ is projected to $\mathcal{X}$ by the parallel projection $[P] = [A][P^0_{\mathcal{P}}][B]$. □

The families of curves $\tilde{\mathcal{Z}}^{i,j,k}_{c_1,c_2}$ given by (7) have a large overlap. The following corollary eliminates this redundancy and, therefore, is useful for practical computations. A proof can be found in [6].

COROLLARY 11. (REDUCED PARALLEL PROJECTION CRITERION.) *A curve* $\mathcal{Z}\subset\mathbb{R}^3$ *projects to* $\mathcal{X}\subset\mathbb{R}^2$ *by a parallel projection if and only if there exist* $a_1, a_2, b\in\mathbb{R}$ *such that the curve* $\mathcal{X}$ *is* $\mathcal{A}(2)$*-equivalent to one of the following* planar *curves:*

$$\begin{aligned}
\tilde{\mathcal{Z}}_{a_1,a_2} &= \overline{\left\{(z_1 + a_1\, z_3,\ z_2 + a_2\, z_3) \,\middle|\, (z_1,z_2,z_3)\in\mathcal{Z}\right\}},\\
\tilde{\mathcal{Z}}_{b} &= \overline{\left\{(z_1 + b\, z_2,\ z_3) \,\middle|\, (z_1,z_2,z_3)\in\mathcal{Z}\right\}}, \quad (9)\\
\tilde{\mathcal{Z}} &= \overline{\left\{(z_2,\ z_3) \,\middle|\, (z_1,z_2,z_3)\in\mathcal{Z}\right\}}.
\end{aligned}$$

4. GROUP-EQUIVALENCE PROBLEM

Theorems 9 and 10 reduce the projection problem to the problem of establishing group-action equivalence between a given curve and a curve from a certain family. A variety of methods exist to solve group-equivalence problem for curves. We base our algorithm on the differential signature construction described in [7]. Differential signature proposed there solves *local* equivalence problem for smooth curves. We adapt this construction to the algebraic setting and prove that differential signature gives a solution of *global* equivalence problem in the case of algebraic curves. For this purpose, we introduce a notion of a *differentially separating set of rational invariants.* We discuss the possibility of using some other methods for solving equivalence problem in Section 7.

4.1 Differential invariants for planar curves

A rational action of an algebraic group G on $\mathbb{R}^2$ can be prolonged to an action on the n-th jet space $J^n = \mathbb{R}^{n+2}$ with coordinates $(x, y, y^{(1)}, \ldots, y^{(n)})$ as follows.[4] For a fixed $g\in G$, let $(\bar{x},\bar{y}) = g\cdot(x,y)$. Then $\bar{x},\bar{y}$ are rational functions of (x,y) and

$$\begin{aligned}
g\cdot(x,y,y^{(1)},\ldots,y^{(n)}) &= (\bar{x},\bar{y},\bar{y}^{(1)},\ldots,\bar{y}^{(n)}), \text{ where}\\
\bar{y}^{(1)} &= \frac{\frac{d}{dx}\left[\bar{y}(x,y)\right]}{\frac{d}{dx}\left[\bar{x}(x,y)\right]} \text{ and for } k=1,\ldots,n-1\\
\bar{y}^{(k+1)} &= \frac{\frac{d}{dx}\left[\bar{y}^{(k)}(x,y,y^{(1)},\ldots,y^{(k)})\right]}{\frac{d}{dx}\left[\bar{x}(x,y)\right]}. \quad (10)
\end{aligned}$$

In (10), $\frac{d}{dx}$ is the total derivative, applied under assumption that y is function of x.[5] We note that a natural projection $\pi^n_k\colon J^n \to J^k$, $k<n$ is equivariant with respect to action (10).

[4]Here $y = y^{(0)}$ and $J^0 = \mathbb{R}^2$.

[5]We note the duality of our view of variables $y^{(k)}$. On one hand, they are viewed as independent coordinate functions on J^n. On the other hand, operator $\frac{d}{dx}$ is applied under

DEFINITION 12. *A function on J^n is called* a differential function. *The* order of a differential function *is the maximum value of k such that the function explicitly depends on the variable $y^{(k)}$.*

A differential function which is invariant under action (10) is called a differential invariant.

REMARK 13. *Due to equivariant property of the projection $\pi^n_k\colon J^n \to J^k$, $k < n$, a differential invariant of order k on J^k can be viewed as a differential invariant on J^n for all $n \geq k$.*

DEFINITION 14. *Let G act on $\mathbb{R}^N$. A set $\mathcal{I}$ of rational invariants is* separating *on a subset $W \subset \mathbb{R}^N$ if W is contained in the domain of definition of each $I \in \mathcal{I}$ and $\forall w_1, w_2 \in W$*

$$I(w_1) = I(w_2),\ \forall I \in \mathcal{I} \iff \exists g \in G \text{ such that } w_1 = g \cdot w_2.$$

DEFINITION 15. *Let r-dimensional algebraic group G act on $\mathbb{R}^2$. Let K and T be rational differential invariants of orders $r-1$ and r, respectively. The set $\mathcal{I} = \{K, T\}$ is called* differentially separating *if K separates orbits on a Zariski open subset $W^{r-1} \subset J^{r-1}$ and $\mathcal{I} = \{K, T\}$ separate orbits on a Zariski open subset of $W^r \subset J^r$.*

4.2 Jets of curves and signatures

In this section, we assume that $\mathcal{X} \subset \mathbb{R}^2$ is an irreducible algebraic curve, different from a vertical line. Let $F(x, y)$ be an irreducible polynomial, whose zero set equals to $\mathcal{X}$. Then the derivatives of y with respect to x are rational functions on $\mathcal{X}$, whose explicit formulas are obtained by implicit differentiation:

$$y^{(1)}_{\mathcal{X}} = -\frac{F_x}{F_y},$$
$$y^{(2)}_{\mathcal{X}} = \frac{-F_{xx}\,F_y^2 + 2\,F_{xy}\,F_x\,F_y - F_{yy}\,F_x^2}{F_y^3}, \dots$$

DEFINITION 16. *The n-th jet of a curve $\mathcal{X} \subset \mathbb{R}^2$ is a rational map $j^n_{\mathcal{X}}\colon \mathcal{X} \to J^n$, where for $\mathbf{p} \in \mathcal{X}$*

$$j^n_{\mathcal{X}}(\mathbf{p}) = \left(x(\mathbf{p}),\, y(\mathbf{p}),\, y^{(1)}_{\mathcal{X}}(\mathbf{p}), \dots, y^{(n)}_{\mathcal{X}}(\mathbf{p})\right).$$

From (10) it follows that $\quad j^n_{g\cdot\mathcal{X}}(g\cdot \mathbf{p}) = g \cdot [j^n_{\mathcal{X}}(\mathbf{p})] \quad$ (11)

DEFINITION 17. *A* restriction of a rational differential function *$\Phi\colon J^n \to \mathbb{R}$ to a curve $\mathcal{X}$ is a composition of Φ with the n-th jet of curve, i. e. $\Phi|_{\mathcal{X}} = \Phi \circ j^n_{\mathcal{X}}$. If defined, such composition produces a rational function $\mathcal{X} \to \mathbb{R}$.*

DEFINITION 18. *Let $\mathcal{I} = \{K, T\}$ be a differentially separating set of invariants for the G-action (see Definition 15). Then a point $\mathbf{p} \in \mathcal{X}$ is called $\mathcal{I}$-regular if: (1) $\mathbf{p}$ is a non-singular point of $\mathcal{X}$; (2) $j^{r-1}_{\mathcal{X}}(\mathbf{p}) \in W^{r-1}$ and $j^r_{\mathcal{X}}(\mathbf{p}) \in W^r$; (3) $\left.\frac{\partial K}{\partial y^{(r-1)}}\right|_{j^{r-1}_{\mathcal{X}}(\mathbf{p})} \neq 0$ and $\left.\frac{\partial T}{\partial y^{(r)}}\right|_{j^{r}_{\mathcal{X}}(\mathbf{p})} \neq 0$.*
An algebraic curve $\mathcal{X} \subset \mathbb{R}^2$ is called non-exceptional *with respect to $\mathcal{I}$ if all but a finite number of its points are $\mathcal{I}$-regular.*

REMARK 19. *If $\mathcal{X}$ is non-exceptional with respect to $\mathcal{I}$ then $K|_{\mathcal{X}}$ and $T|_{\mathcal{X}}$ are rational functions on $\mathcal{X}$. The $\mathcal{I}$-regular points of $\mathcal{X}$ are in the domain of the definition of these functions.*

---

assumption that y is a function of x and, therefore, $y^{(k)}$ is also viewed as the k-th derivative of y with respect to x. (The same duality of view appears in calculus of variations.)

DEFINITION 20. *Let $\mathcal{I} = \{K, T\}$ be a differentially separating set of invariants with respect to G-action and $\mathcal{X}$ be non-exceptional with respect to $\mathcal{I}$. The* signature $\mathcal{S}_{\mathcal{X}}$ *is the image of the rational map $S|_{\mathcal{X}}\colon \mathcal{X} \to \mathbb{R}^2$ defined by $S_{\mathcal{X}}(\mathbf{p}) = (K|_{\mathcal{X}}(\mathbf{p}),\, T|_{\mathcal{X}}(\mathbf{p}))$.*

THEOREM 21. (GROUP-EQUIVALENCE CRITERION.) *Assume that irreducible algebraic curves $\mathcal{X}_1$ and $\mathcal{X}_2$ are non-exceptional with respect to differentially separating invariants $\mathcal{I} = (K, T)$ under G-action. Then $\mathcal{X}_1$ and $\mathcal{X}_2$ are G-equivalent if and only if their signatures are equal:*

$$\mathcal{X}_1 \cong_G \mathcal{X}_2 \iff \mathcal{S}_{\mathcal{X}_1} = \mathcal{S}_{\mathcal{X}_2}.$$

We make the following remarks before proving the theorem.

REMARK 22. *Let $\mathcal{I} = \{K, T\}$ be a set of differentially separating invariants (see Definition 15). Let $\mathcal{X} \subset \mathbb{R}^2$ be a non $\mathcal{I}$-exceptional curve defined by an irreducible implicit equation $F(x, y) = 0$. Then $K|_{\mathcal{X}}(x, y) = \frac{k_1(x,y)}{k_2(x,y)}$, $T|_{\mathcal{X}} = \frac{t_1(x,y)}{t_2(x,y)}$, where k_1, k_2 and t_1, t_2 are pairs of polynomials with no non-constant common factors modulo F. Consider an ideal*

$$X := \langle F,\, k_2\,\varkappa - k_1,\, t_2\,\tau - t_1,\, k_2\,t_2\,\sigma - 1\rangle \subset \mathbb{R}[\varkappa, \tau, x, y, \sigma].$$

The algebraic closure $\overline{\mathcal{S}_{\mathcal{X}}}$ of the signature set $\mathcal{S}_{\mathcal{X}}$ is the variety of the radical of the elimination ideal $\hat{X} = X \cap \mathbb{R}[\varkappa, \tau]$.

REMARK 23. *We note that $\dim \overline{\mathcal{S}_{\mathcal{X}}} = 0$ if and only if $K_{\mathcal{X}}$ and $T_{\mathcal{X}}$ are constant functions on $\mathcal{X}$ and $\dim \overline{\mathcal{S}_{\mathcal{X}}} = 1$ otherwise. In the latter case, $\overline{\mathcal{S}_{\mathcal{X}}}$ is an algebraic planar curve with a single irreducible defining equation $\hat{S}_{\mathcal{X}}(\varkappa, \tau) = 0$. The equality of signatures for two curves, $\mathcal{S}_{\mathcal{X}_1} = \mathcal{S}_{\mathcal{X}_2}$, implies $\hat{S}_{\mathcal{X}_1}(\varkappa, \tau)$ is equal up to a constant multiple to $\hat{S}_{\mathcal{X}_2}(\varkappa, \tau)$. The converse is true over $\mathbb{C}$, but not over $\mathbb{R}$, because the latter is not an algebraically closed field (see [10]).*

PROOF OF THEOREM 21. Direction $\Longrightarrow$ follows immediately from the definition of invariants. Below we prove $\Longleftarrow$. We notice that there are two cases. Either $K|_{\mathcal{X}_1}$ and $K|_{\mathcal{X}_2}$ are constant maps on $\mathcal{X}_1$ and $\mathcal{X}_2$, respectively, and these maps take the same value. Otherwise both $K|_{\mathcal{X}_1}$ and $K|_{\mathcal{X}_2}$ are non-constant rational maps on $\mathcal{X}_1$ and $\mathcal{X}_2$, respectively.

Case 1: There exists $c \in \mathbb{R}$ such that $K|_{\mathcal{X}_1}(\mathbf{p}_1) = c$ and $K_{\mathcal{X}_2}(\mathbf{p}_2) = c$ for all $\mathbf{p}_1 \in \mathcal{X}_1$ and for all $\mathbf{p}_2 \in \mathcal{X}_2$. Since $\mathcal{X}_1$ and $\mathcal{X}_2$ are non-exceptional, we may fix $\mathcal{I}_G$-regular points $\mathbf{p}_1 = (x_1, y_1) \in \mathcal{X}_1$ and $\mathbf{p}_2 = (x_2, y_2) \in \mathcal{X}_2$. Then, due to separation property of the invariant K, $\exists g \in G$ such that $j^{r-1}_{\mathcal{X}_1}(\mathbf{p}_1) = g \cdot [j^{r-1}_{\mathcal{X}_2}(\mathbf{p}_2)]$. We consider a new algebraic curve $\mathcal{X}_3 = g \cdot \mathcal{X}_2$. Then due to (11), we have

$$j^{r-1}_{\mathcal{X}_1}(\mathbf{p}_1) = j^{r-1}_{\mathcal{X}_3}(\mathbf{p}_1) =: \mathbf{p}^{(r-1)}. \quad (12)$$

Since $\mathbf{p}_1$ is a $\mathcal{I}$-regular point of $\mathcal{X}_1$, it follows from (12) that it is also a $\mathcal{I}$-regular point of $\mathcal{X}_3$ and, in particular, is non-singular. Let $F_1(x, y) = 0$ and $F_3(x, y) = 0$ be implicit equations of $\mathcal{X}_1$ and $\mathcal{X}_3$, respectively. We may assume that $\frac{\partial F_1}{\partial y}(\mathbf{p}_1) \neq 0$ and $\frac{\partial F_3}{\partial y}(\mathbf{p}_1) \neq 0$ (otherwise, $\frac{\partial F_1}{\partial x}(\mathbf{p}_1) \neq 0$ and $\frac{\partial F_3}{\partial x}(\mathbf{p}_1) \neq 0$ and we may use a similar argument). Then, there exist functions $f_1(x)$ and $f_3(x)$, analytic on an interval $I \ni x_1$, such that $F_1(x, f_1(x)) = 0$ and $F_3(x, f_3(x)) = 0$ for $x \in I_1$.

Functions $y = f_1(x)$ and $y = f_3(x)$ are local analytic solutions of differential equation

$$K(x, y, y^{(1)}, \dots, y^{(r-1)}) = c \quad (13)$$

with the same initial condition $f_1^{(k)}(x_1) = f_3^{(k)}(x_1)$, $k = 0, \dots, r-1$ prescribed by (12). From the $\mathcal{I}$-regularity of $\mathbf{p}_1$, we have that $\left.\frac{\partial K}{\partial y^{(r-1)}}\right|_{\mathbf{p}^{(r-1)}} \neq 0$ and so (13) can be solved for $y^{(r-1)}$:

$$y^{(r-1)} = H(x, y, y^{(1)}, \dots, y^{(r-2)}), \quad (14)$$

where function H is smooth in a neighborhood $\mathbf{p}^{(r-1)} \in J^{r-1}$. From the uniqueness theorem for the solutions of ODEs, it follows that $f_1(x) = f_3(x)$ on an interval $I \ni x_1$. Since $\mathcal{X}_1$ and $\mathcal{X}_3$ are irreducible algebraic curves it follows that $\mathcal{X}_1 = \mathcal{X}_3$. Therefore, $\mathcal{X}_1 = g \cdot \mathcal{X}_2$.

Case 2: $K|_{\mathcal{X}_1}$ and $K|_{\mathcal{X}_2}$ are non-constant rational maps. Then $\mathcal{S}_{\mathcal{X}_1} = \mathcal{S}_{\mathcal{X}_2}$ is a one-dimensional set that we will denote $\mathcal{S}$. Let $\hat{S}(\varkappa, \tau) = 0$ be the implicit equation for $\mathcal{S}$ (see Remark 22). We know that $\frac{\partial \hat{S}}{\partial \tau}(\varkappa, \tau) \neq 0$ for all but finite number of values $(\varkappa, \tau)$, because, otherwise, $K|_{\mathcal{X}_1}$ and $K|_{\mathcal{X}_2}$ are constant maps. Therefore, since the curves are non-exceptional, there exists $\mathcal{I}$-regular points $\mathbf{p}_1 = (x_1, y_1) \in \mathcal{X}_1$ and $\mathbf{p}_2 = (x_2, y_2) \in \mathcal{X}_2$ such that

$$K|_{\mathcal{X}_1}(\mathbf{p}_1) = K|_{\mathcal{X}_2}(\mathbf{p}_2) =: \varkappa_0, \; T|_{\mathcal{X}_1}(\mathbf{p}_1) = T|_{\mathcal{X}_2}(\mathbf{p}_2) =: \tau_0$$
$$\text{and } \frac{\partial \hat{S}}{\partial \tau}(\varkappa_0, \tau_0) \neq 0. \quad (15)$$

Due to separation property of the set $\mathcal{I}_G = \{K, T\}$, $\exists g \in G$ such that $j^r_{\mathcal{X}_1}(\mathbf{p}_1) = g \cdot [j^r_{\mathcal{X}_2}(\mathbf{p}_2)]$. We consider a new algebraic curve $\mathcal{X}_3 = g \cdot \mathcal{X}_2$. Then due to (11), we have

$$j^r_{\mathcal{X}_1}(\mathbf{p}_1) = j^r_{\mathcal{X}_3}(\mathbf{p}_1) =: \mathbf{p}^{(r)}. \quad (16)$$

From (15), (16) and $\mathcal{I}$-regularity of the point $\mathbf{p}_1 \in \mathcal{X}_1$ it follows that

$$K(\mathbf{p}^{(r)}) = \varkappa_0, \quad T(\mathbf{p}^{(r)}) = \tau_0 \text{ and } \left.\frac{\partial T}{\partial y^{(r)}}\right|_{\mathbf{p}^{(r)}} \neq 0 \quad (17)$$

Since $\mathbf{p}_1$ is a $\mathcal{I}$-regular point of $\mathcal{X}_1$, it follows from (16) that it is also a $\mathcal{I}$-regular point of $\mathcal{X}_3$ and, in particular, is non-singular. Let $F_1(x, y) = 0$ and $F_3(x, y) = 0$ be implicit equations of $\mathcal{X}_1$ and $\mathcal{X}_3$, respectively. We may assume that $\frac{\partial F_1}{\partial y} \neq 0$ and $\frac{\partial F_3}{\partial y} \neq 0$ (otherwise, $\frac{\partial F_1}{\partial x} \neq 0$ and $\frac{\partial F_3}{\partial x} \neq 0$ and we may use a similar argument). Then, there exist functions $f_1(x)$ and $f_3(x)$, analytic on an interval $I \ni x_1$, such that $F_1(x, f_1(x)) = 0$ and $F_3(x, f_3(x)) = 0$ for $x \in I_1$.

Then functions $y = f_1(x)$ and $y = f_3(x)$ are local analytic solutions of differential equation

$$\hat{S}\left(K(x, y, y^{(1)}, \dots, y^{(r-1)}), T(x, y, y^{(1)}, \dots, y^{(r)})\right) = 0 \quad (18)$$

with the same initial condition $f_1^{(k)}(x_1) = f_3^{(k)}(x_1)$, $k = 0, \dots, r$, dictated by (16).

Since $\frac{\partial \hat{S}}{\partial \tau}(\varkappa_0, \tau_0) \neq 0$ and $\left.\frac{\partial T}{\partial y^{(r)}}\right|_{\mathbf{p}^{(r)}} \neq 0$ (see (15) and (17)), equation (18) can be solved for $y^{(r)}$:

$$y^{(r)} = H(x, y, y^{(1)}, \dots, y^{(r-1)}), \quad (19)$$

where function H is smooth in a neighborhood $\mathbf{p}^{(r)} \in J^r$. From the uniqueness theorem for the solutions of ODE it follows that $f_1(x) = f_3(x)$ on an interval $I \ni x_1$. Since $\mathcal{X}_1$ and $\mathcal{X}_3$ are irreducible algebraic curves it follows that $\mathcal{X}_1 = \mathcal{X}_3$. Therefore, $\mathcal{X}_1 = g \cdot \mathcal{X}_2$. □

4.3 Separating sets of invariants for affine and projective actions

In this section, we construct a differentially separating set of rational invariants for affine and projective actions. We will build them from classical invariants from differential geometry [8, 3]. We start with Euclidean curvature $\kappa = \frac{y^{(2)}}{(1+[y^{(1)}]^2)^{3/2}}$ which is, up to a sign[6], a differential invariant of the lowest order. Higher order Euclidean differential invariants are obtained by differentiating the curvature with respect to the Euclidean arclength $ds = \sqrt{1 + [y^{(1)}]^2}\, dx$, i. e. $\kappa_s = \frac{d\kappa}{ds} = \frac{1}{\sqrt{1+[y^{(1)}]^2}} \frac{d\kappa}{dx}$, $\kappa_{ss} = \frac{d\kappa_s}{ds}, \dots$.

Affine and projective curvatures and infinitesimal arclengths are well known, and can be expressed in terms of Euclidean invariants [11, 19]. In particular, $\mathcal{SA}$-curvature μ and infinitesimal $\mathcal{SA}$-arclength $d\alpha$ are expressed in terms of their Euclidean counterparts as follows:

$$\mu = \frac{3\,\kappa\,(\kappa_{ss} + 3\,\kappa^3) - 5\,\kappa_s^2}{9\,\kappa^{8/3}}, \quad d\alpha = \kappa^{1/3} ds. \quad (20)$$

By considering effects of scalings and reflections on $\mathcal{SA}(2)$-invariants, we obtain two lowest order $\mathcal{A}(2)$-invariants:

$$K_{\mathcal{A}} = \frac{(\mu_\alpha)^2}{\mu^3}, \quad T_{\mathcal{A}} = \frac{\mu_{\alpha\alpha}}{3\,\mu^2}. \quad (21)$$

They are of order 5 and 6, respectively, and are *rational functions* in jet variables.

$\mathcal{PGL}(3)$-curvature η and infinitesimal arclength $d\rho$ are expressed in terms of their $\mathcal{SA}$-counterparts:

$$\eta = \frac{6\mu_{\alpha\alpha\alpha}\mu_\alpha - 7\,\mu_{\alpha\alpha}^2 - 9\mu_\alpha^2\,\mu}{6\mu_\alpha^{8/3}}, \quad d\rho = \mu_\alpha^{1/3} d\alpha. \quad (22)$$

The two lowest order *rational* $\mathcal{PGL}(3)$-invariants are of differential order 7 and 8, respectively:

$$K_{\mathcal{P}} = \eta^3, \quad T_{\mathcal{P}} = \eta_\rho. \quad (23)$$

THEOREM 24. *According to Definition 15:*

(1) The set $\mathcal{I}_{\mathcal{A}} = \{K_{\mathcal{A}}, T_{\mathcal{A}}\}$ given by (21) is differentially separating for the $\mathcal{A}(2)$-action on $\mathbb{R}^2$.

(2) The set $\mathcal{I}_{\mathcal{PGL}} = \{K_{\mathcal{P}}, T_{\mathcal{P}}\}$ given by (23) is differentially separating for the $\mathcal{PGL}(3)$-action on $\mathbb{R}^2$.

PROPOSITION 25.

(1) $\mathcal{I}_{\mathcal{A}}$-exceptional algebraic curves are lines and parabolas.

(2) $\mathcal{I}_{\mathcal{PGL}}$-exceptional algebraic curves are lines and conics.

COROLLARY 26. *An $\mathcal{I}_{\mathcal{A}}$-exceptional algebraic curve is* not *$\mathcal{A}(2)$ equivalent to a non $\mathcal{I}_{\mathcal{A}}$-exceptional algebraic curve. An $\mathcal{I}_{\mathcal{PGL}}$-exceptional algebraic curve is* not *$\mathcal{PGL}(3)$ equivalent to a non $\mathcal{I}_{\mathcal{PGL}}$-exceptional algebraic curve.*

Theorem 24, in combination with Theorem 21, leads to a solution for the projective and the affine equivalence problems for curves. For rather technical proofs of these theorems we refer a reader to [6]. See also [15] for related results for smooth curves.

[6]The sign of κ changes when a curve is reflected, rotated by π radians or traced in the opposite direction. A rational function κ^2 is invariant under the full Euclidean group.

5. ALGORITHMS AND EXAMPLES

The algorithms for solving projection problems based on a combination of the projection criteria of Section 3 and the group equivalence criterion of Section 4.

5.1 Explicit formulas for invariants

Before stating the algorithms we write out explicit formulas for invariants in terms of jet coordinates. Let

$$\Delta_1 = 3\,y^{(4)}\,y^{(2)} - 5\,[y^{(3)}]^2, \quad (24)$$

$$\Delta_2 = 9\,y^{(5)}\,[y^{(2)}]^2 - 45\,y^{(4)}\,y^{(3)}\,y^{(2)} + 40\,[y^{(3)}]^3, \quad (25)$$

then a differentially separating set of rational $\mathcal{A}(2)$-invariants (21) is given by:

$$\begin{aligned} K_{\mathcal{A}} &= \frac{(\Delta_2)^2}{(\Delta_1)^3}; \qquad (26)\\ T_{\mathcal{A}} &= \frac{1}{(\Delta_1)^2}\,\big(9\,y^{(6)}\,[y^{(2)}]^3 - 63\,y^{(5)}\,y^{(3)}\,[y^{(2)}]^2\\ &- 45\,[y^{(4)}]^2\,[y^{(2)}]^2 + 255\,y^{(4)}\,[y^{(3)}]^2\,y^{(2)} - 160\,[y^{(3)}]^4\big), \end{aligned}$$

while a separating set of rational $\mathcal{PGL}(3)$-invariants (23) is given by:

$$\begin{aligned} K_{\mathcal{P}} &= \frac{729}{8\,(\Delta_2)^8}\Big(18\,y^{(7)}\,[y^{(2)}]^4\,\Delta_2 - 189\,[y^{(6)}]^2\,[y^{(2)}]^6 \qquad (27)\\ &+ 126\,y^{(6)}\,[y^{(2)}]^4\,(9\,y^{(5)}\,y^{(3)}\,y^{(2)} + 15\,[y^{(4)}]^2\,y^{(2)}\\ &- 25\,y^{(4)}\,[y^{(3)}]^2) - 189\,[y^{(5)}]^2\,[y^{(2)}]^4\,(4\,[y^{(3)}]^2\\ &+ 15\,y^{(2)}\,y^{(4)}) + 210\,y^{(5)}\,y^{(3)}\,[y^{(2)}]^2\,(63\,[y^{(4)}]^2\,[y^{(2)}]^2\\ &- 60\,y^{(4)}\,[y^{(3)}]^2\,y^{(2)} + 32\,[y^{(3)}]^4) - 525\,y^{(4)}y^{(2)}\\ &\ (9\,[y^{(4)}]^3\,[y^{(2)}]^3 + 15\,[y^{(4)}]^2\,[y^{(3)}]^2\,[y^{(2)}]^2\\ &- 60\,y^{(4)}\,[y^{(3)}]^4\,y^{(2)} + 64\,[y^{(3)}]^6) + 11200\,[y^{(3)}]^8\Big)^3 \end{aligned}$$

$$\begin{aligned} T_{\mathcal{P}} &= \frac{243[y^{(2)}]^4}{2(\Delta_2)^4}\Big(2y^{(8)}y^{(2)}\,(\Delta_2)^2 - 8y^{(7)}\,\Delta_2(9y^{(6)}[y^{(2)}]^3\\ &- 36y^{(5)}\,y^{(3)}[y^{(2)}]^2 - 45[y^{(4)}]^2[y^{(2)}]^2 + 120y^{(4)}\,[y^{(3)}]^2\\ &- 40\,[y^{(3)}]^4) + 504\,[y^{(6)}]^3\,[y^{(2)}]^5 - 504\,[y^{(6)}]^2\,[y^{(2)}]^3\\ &\ (9\,y^{(5)}\,y^{(3)}\,y^{(2)} + 15\,[y^{(4)}]^2\,y^{(2)} - 25\,y^{(4)}\,[y^{(3)}]^2)\\ &+ 28y^{(6)}(432[y^{(5)}]^2[y^{(3)}]^2[y^{(2)}]^3 + 243[y^{(5)}]^2y^{(4)}\,[y^{(2)}]^4\\ &- 1800\,y^{(5)}\,y^{(4)}\,[y^{(3)}]^3\,[y^{(2)}]^2 - 240y^{(5)}\,[y^{(3)}]^5\,y^{(2)}\\ &+ 540y^{(5)}\,[y^{(4)}]^2\,[y^{(3)}]\,[y^{(2)}]^3 + 6600\,[y^{(4)}]^2\,[y^{(3)}]^4\,y^{(2)}\\ &- 2000y^{(4)}\,[y^{(3)}]^6 - 5175\,[y^{(4)}]^3\,[y^{(3)}]^2\,[y^{(2)}]^2\\ &+ 1350\,[y^{(4)}]^4\,[y^{(2)}]^3) - 2835\,[y^{(5)}]^4\,[y^{(2)}]^4\\ &+ 252\,[y^{(5)}]^3y^{(3)}\,[y^{(2)}]^2\,(9y^{(4)}\,y^{(2)} - 136\,[y^{(3)}]^2)\\ &- 35840\,[y^{(5)}]^2\,[y^{(3)}]^6 - 630\,[y^{(5)}]^2\,[y^{(4)}]\,[y^{(2)}]\\ &\ (69[y^{(4)}]^2[y^{(2)}]^2 - 160[y^{(3)}]^4 - 153y^{(4)}[y^{(3)}]^2[y^{(2)}])\\ &+ 2100\,y^{(5)}\,[y^{(4)}]^2\,y^{(3)}\,(72\,[y^{(3)}]^4 + 63\,[y^{(4)}]^2\,[y^{(2)}]^2\\ &- 193y^{(4)}[y^{(3)}]^2y^{(2)}) - 7875[y^{(4)}]^4(8[y^{(4)}]^2[y^{(2)}]^2\\ &- 22y^{(4)}\,[y^{(3)}]^2\,[y^{(2)}] + 9\,[y^{(3)}]^4)\Big). \end{aligned}$$

We adapt Definition 17 to rational curves as follows. Let $\mathcal{X}$ is a rational curve parametrized by $\gamma(t) = (x(t), y(t))$, such that $x(t)$ is not a constant function.[7] Make a recursive definition of the following rational functions of t:

$$y^{(1)} = \frac{\dot{y}}{\dot{x}}\ ,\ \ldots,\ \ y^{(k)} = \frac{\dot{y^{(k-1)}}}{\dot{x}}, \quad (28)$$

where $\dot{}$ *denotes the derivative with respect to the parameter.*

Let Φ be a rational differential function. Then the restriction of $\Phi|_\gamma$ is computed by substituting (28) into Φ. If defined, $\Phi|_\gamma$ is a rational function of t. The following proposition follows from Proposition 25.

PROPOSITION 27. *Let $\mathcal{X}$ be a rational curve with a parameterization $\gamma(t) = (x(t), y(t))$, such that $x(t)$ is not a constant function. Then the restrictions $\Delta_1|_\gamma$ and $\Delta_2|_\gamma$ are rational functions of t.*

If $\Delta_1|_\gamma(t)$ is zero for more than a finite number of values t, then $\mathcal{X}$ is either a line or a parabola and $\Delta_1|_\gamma(t)$ is zero for all t. Otherwise, restrictions $K_{\mathcal{A}}|_\gamma$ and $T_{\mathcal{A}}|_\gamma$ are rational functions of t.

If $\Delta_2|_\gamma(t)$ is zero for more than a finite number of values t, then γ is either a line or an irreducible conic and then $\Delta_2|_\gamma(t)$ is zero for all t. Otherwise $K_{\mathcal{P}}|_\gamma$ and $T_{\mathcal{P}}|_\gamma$ are rational functions on t.

5.2 Central projections

The following algorithm is based on the central projection criterion stated in Theorem 9.

ALGORITHM 28. (CENTRAL PROJECTIONS.)

INPUT: Parameterizations $\Gamma = (z_1, z_2, z_3) \in \mathbb{Q}(s)^3$ and $\gamma = (x, y) \in \mathbb{Q}(t)^2$ of rational algebraic curves $\mathcal{Z} \subset \mathbb{R}^3$ and $\mathcal{X} \subset \mathbb{R}^2$, respectively, such that $\dot{\Gamma} \times \ddot{\Gamma} \neq 0$. [8]

OUTPUT: The truth of the statement:

$$\exists [P] \in \mathcal{CP},\ \text{such that } \mathcal{X} = P(\mathcal{Z}).$$

STEPS:

1. *if* $\begin{vmatrix} \dot{\gamma} \\ \ddot{\gamma} \end{vmatrix} \underset{\mathbb{Q}(t)}{=} 0$ *then*

   if $\begin{vmatrix} \dot{\Gamma} \\ \ddot{\Gamma} \\ \dddot{\Gamma} \end{vmatrix} \underset{\mathbb{Q}(s)}{=} 0$

   then return TRUE

   else return FALSE;

2. $\epsilon := \left(\frac{z_1+c_1}{z_3+c_3},\ \frac{z_2+c_2}{z_3+c_3}\right) \in \mathbb{Q}(c_1, c_2, c_3, s)^2$;

3. *if* $\Delta_2|_\gamma \underset{\mathbb{Q}(t)}{=} 0$ *then*

   if $\exists (c_1, c_2, c_3) \in \mathbb{R}^3$

   $$z_3 + c_3 \underset{\mathbb{R}(s)}{\neq} 0 \wedge \begin{vmatrix} \dot{\epsilon} \\ \ddot{\epsilon} \end{vmatrix} \underset{\mathbb{R}(s)}{\neq} 0 \wedge \Delta_2|_\epsilon \underset{\mathbb{R}(s)}{=} 0$$

   then return TRUE

   else return FALSE.

4. *return the truth of the statement:*

   $\exists\,(c_1, c_2, c_3) \in \mathbb{R}^3$

   $$z_3 + c_3 \underset{\mathbb{R}(s)}{\neq} 0 \wedge \begin{vmatrix} \dot{\epsilon} \\ \ddot{\epsilon} \end{vmatrix} \underset{\mathbb{R}(s)}{\neq} 0 \wedge \Delta_2|_\epsilon \underset{\mathbb{R}(s)}{\neq} 0 \quad (29)$$

   $$\wedge\, \forall s \in \mathbb{R}\quad \Delta_2|_\epsilon \underset{\mathbb{R}}{\neq} 0 \Rightarrow \exists t \in \mathbb{R}\quad K_{\mathcal{P}}|_\epsilon \underset{\mathbb{R}}{=} K_{\mathcal{P}}|_\gamma \wedge T_{\mathcal{P}}|_\epsilon \underset{\mathbb{R}}{=} T_{\mathcal{P}}|_\gamma.$$

[7]Equivalently, $\mathcal{X}$ is not a vertical line.

[8]Equivalently, $\mathcal{Z}$ is not a line.

Remark 29. *In Algorithm 28, we use restrictions of differential functions to a family of curves parametrized by $\epsilon(c, s) \in \mathbb{Q}(c, s)^2$, where $c = (c_1, c_2, c_3)$ determines a member of the family and s serves to parametrize a curve in the family. In this case, derivatives in (28) are taken with respect to s.*

Remark 30. *On the first step of Algorithm 28, we consider the case when $\mathcal{X}$ is a line. Then $\mathcal{Z}$ can be projected to $\mathcal{X}$ if and only if $\mathcal{Z}$ is coplanar. If $\mathcal{X}$ is not a line we define, on Step 2, a rational map ϵ that parametrizes a family of curves. On Step 3, we consider the case when $\mathcal{X}$ is an irreducible conic. Then $\mathcal{Z}$ can be projected to $\mathcal{X}$ if and only if $\exists c$ such that the algebraic closure $\tilde{Z}_c$ of the image of $\epsilon(c, s)$ is an irreducible conic. If $\mathcal{X}$ is neither a line nor a conic we proceed to Step 4, where we decide if there exists $c \in \mathbb{R}^3$ such that (1) $\epsilon(c, s)$ is neither a line nor a conic [9]; (2) the signatures of the algebraic curve parametrized by $\gamma(t)$ and the curve $\tilde{Z}_c$ are the same.*

Remark 31. *If at least one of $K_{\mathcal{P}}|_\gamma(t)$, $T_{\mathcal{P}}|_\gamma(t)$ is a non-constant function, Step 4 of the algorithm can be performed as follows. Compute the implicit equation*

$$\hat{S}_{\mathcal{X}}(\varkappa, \tau) = 0 \tag{30}$$

for the signature of $\mathcal{X}$ by eliminating t from $\varkappa = K_{\mathcal{P}}|_\gamma(t)$ and $\tau = T_{\mathcal{P}}|_\gamma(t)$. Then substitute $\varkappa = K_{\mathcal{P}}|_\epsilon(c, s)$ and $\tau = T_{\mathcal{P}}|_\epsilon(c, s)$ into (30). The left-hand side of (30) becomes a rational function, which we denote $H(c, s)$. Decide if there exists $c \in \mathbb{R}^3$ such that conditions (29) are satisfied, $K_{\mathcal{P}}|_\epsilon(c, s)$, and $T_{\mathcal{P}}|_\epsilon(c, s)$ are non-constant rational functions and the numerator of $H(c, s)$ is a zero polynomial in s. If we were solving the problem over $\mathbb{C}$, this would provide a sufficient condition for existence of a projection, but over $\mathbb{R}$ some additional steps have to be taken (see Remark 23). If both $K_{\mathcal{P}}|_\gamma(t)$ and $T_{\mathcal{P}}|_\gamma(t)$ are constant functions, then Step 4 can be preformed by deciding if there exists c, such that both $K_{\mathcal{P}}|_\epsilon(c, s)$ and $T_{\mathcal{P}}|_\epsilon(c, s)$ are constant functions with the same values as $K_{\mathcal{P}}|_\gamma(t)$, $T_{\mathcal{P}}|_\gamma(t)$, respectively. An implementation over $\mathbb{C}$ is posted on the internet [25].

Remark 32. *If the output is TRUE, then, in many cases, we can, in addition to establishing the existence of c_1, c_2, c_3 in Step 4 of the algorithm, find at least one of such triplets explicitly. We then know that $\mathcal{Z}$ can be projected to $\mathcal{X}$ by a projection centered at $(-c_1, -c_2, -c_3)$. We can also, in many cases, determine explicitly a transformation $[A] \in \mathcal{PGL}(3)$ that maps $\mathcal{X}$ to the algebraic closure $\tilde{Z}_c$ of the image of the map $\epsilon(c, s)$. We then know that $\mathcal{Z}$ can be projected to $\mathcal{X}$ by the projection $[P] = [A][P_C^0][B]$, where P_C^0 and B are defined by (6).*

Example 33. *We would like to decide if the spatial curve $\mathcal{Z}$ parametrized by*

$$\Gamma(s) = \left(s^3,\, s^2,\, s\right),\ s \in \mathbb{R}\ \textit{(Twisted Cubic)}$$

projects to any of the four given planar curves $\mathcal{X}_1$, $\mathcal{X}_2$, $\mathcal{X}_3$ and $\mathcal{X}_4$ parametrized, respectively, by:

$$\gamma_1(t) = \left(t^2,\, t\right), \quad \gamma_2(t) = \left(\frac{t^3}{t+1},\, \frac{t^2}{t+1}\right),$$

$$\gamma_3(t) = \left(\frac{t}{t^3+1},\, \frac{t^2}{t^3+1}\right), \quad \gamma_4(t) = \left(t,\, t^5\right).$$

Let $\epsilon(c_1, c_2, c_3, s) = \left(\frac{s^3+c_1}{s+c_3}, \frac{s^2+c_2}{s+c_3}\right)$. Curve $\mathcal{X}_1$ is a parabola and so is $\mathcal{PGL}(3)$-exceptional. It is known that all irreducible planar conics are $\mathcal{PGL}(3)$-equivalent and so, from Theorem 9, we know that $\mathcal{Z}$ can be projected on $\mathcal{X}_1$ if there exist $c \in \mathbb{R}^3$, such that $\epsilon(c, s)$ parametrizes a conic. This is obviously true for $c_1 = c_2 = c_3 = 0$. Indeed, on can check that $\mathcal{Z}$ can be projected to $\mathcal{X}_1$ by projection $x = \frac{z_1}{z_3}$, $y = \frac{z_2}{z_3}$.

The curve $\mathcal{X}_2$ is not $\mathcal{PGL}(3)$-exceptional. Its signature is parametrized by a constant map:

$$K_{\mathcal{P}}|_{\gamma_2}(t) = \frac{250047}{12800} \text{ and } T_{\mathcal{P}}|_{\gamma_2}(t) = 0, \quad \forall t \in \mathbb{R}.$$

Following Algorithm 28, we need to decide whether there exists $c \in \mathbb{R}^3$, such that $\epsilon(c, s)$ does not parametrize a line or a conic and

$$K_{\mathcal{P}}|_\epsilon(s) = \frac{250047}{12800} \text{ and } T_{\mathcal{P}}|_\epsilon(s) = 0,\ \forall s \in \mathbb{R}.$$

This is, indeed, true for $c_1 = c_2 = 0$ and $c_3 = 1$. We can check that $\mathcal{Z}$ can be projected to $\mathcal{X}_2$ by the a central projection $x = \frac{z_1}{z_3+1}$, $y = \frac{z_2}{z_3+1}$.

The signature of $\mathcal{X}_3$ is parameterized by a non-constant map:

$$K_{\mathcal{P}}|_{\gamma_3}(t) = -\frac{9261}{50}\,\frac{t^7 - t^4 + t}{(t^3-1)^8}, \quad T_{\mathcal{P}}|_{\gamma_3}(t) = -\frac{21}{10}\,\frac{(t^3+1)^4}{(t^3-1)^4}.$$

Evaluation of step 4 of Algorithm 28 yields TRUE. One can, in fact, check that for $c^* = (1, 0, 0)$, the map $\epsilon(c^*, s)$ does not parametrize a line or a conic and

$$K_{\mathcal{P}}|_\epsilon(c^*, s) = K_{\mathcal{P}}|_{\gamma_3}(s) \text{ and } T_{\mathcal{P}}|_\epsilon(c^*, s) = K_{\mathcal{P}}|_{\gamma_3}(s),\ \forall s \in \mathbb{R}.$$

We conclude that $\mathcal{Z}$ can be projected to $\mathcal{X}_3$. It is not difficult to determine a possible projection: $x = \frac{z_3}{z_1+1}$, $y = \frac{z_2}{z_1+1}$.

It is important to observe that although $\mathcal{Z}$ can be projected to each of the planar curves $\mathcal{X}_1$, $\mathcal{X}_2$, and $\mathcal{X}_3$, these planar curves are not *$\mathcal{PGL}(3)$-equivalent. This underscores an observation made in Remark 8.*

The signature of $\mathcal{X}_4$ is parametrized by a constant map:

$$K_{\mathcal{P}}|_{\gamma_4}(t) = \frac{1029}{128} \text{ and } T_{\mathcal{P}}|_{\gamma_4}(t) = 0, \quad \forall t.$$

Following Algorithm 28, we need to decide whether there exists $c \in \mathbb{R}^3$, such that $\epsilon(c, s)$ does not parametrize a line or a conic and

$$K_{\mathcal{P}}|_\epsilon(c, s) = \frac{1029}{128} \text{ and } T_{\mathcal{P}}|_\epsilon(c, s) = 0,\ \forall s \in \mathbb{R}.$$

Substitution of several values of s in the above equation yields a system of polynomial equations for $c_1, c_2, c_3 \in \mathbb{R}$ that has no solutions. We conclude that there is no central projection from $\mathcal{Z}$ to $\mathcal{X}_4$.

5.3 Parallel projections

The algorithm for parallel projections is based on the reduced parallel projection criterion stated in Corollary 11. This algorithm follows the same logic but has more steps than Algorithm 28, because we need to decide whether a

[9] For such values of c, we can prove that whether we first specialize a value of c in $\epsilon(c, s)$ and then compute rational functions $K_{\mathcal{P}}|_\epsilon(c, s)$ and $T_{\mathcal{P}}|_\epsilon(c, s)$ or we perform these operations in the reverse order, we obtain the same rational functions of s.

given planar curve is $\mathcal{A}(2)$-equivalent to a curve parametrized by $\alpha(s) = (z_2(s), z_3(s))$, or to a curve parametrized by $\beta(b,s) = (z_1(s) + b\, z_2(s),\, z_3(s))$ for some $b \in \mathbb{R}$, or to a curve parametrized by $\delta(a_1, a_2, s) = (z_1(s) + a_1 z_3(s), z_2 + a_2 z_3(s))$ for some $a = (a_1, a_2) \in \mathbb{R}^2$. Since the affine transformations are considered, the differential function Δ_2 is replaced with Δ_1 (see (24)) and projective invariants are replaced with affine invariants (see (26)). Due to its similarity to Algorithm 28, we refrain from writing out the steps of the parallel projection algorithm and content ourselves with presenting examples. The algorithm can be found in [6].

EXAMPLE 34. *To decide whether the spatial curve $\mathcal{Z}$ parametrized by*

$$\Gamma(s) = \left(s^4 + 1,\, s^2,\, s\right),\ s \in \mathbb{R}$$

can be projected to $\mathcal{X}$ parametrized by

$$\gamma(t) = \left(t\,,\, t^4 + t^2\right),\ t \in \mathbb{R}$$

by a parallel projection, we start by noticing that $\mathcal{X}$ is not an $\mathcal{A}(2)$-exceptional curve. Its signature is parametrized by a non-constant map:

$$T_{\mathcal{A}}|_\gamma(t) = \frac{100t^2(3 - 14t^2)^2}{(1 - 14t^2)^3},\ K_{\mathcal{A}}|_\gamma(t) = -5\frac{(140t^4 - 56t^2 + 1)}{(1 - 14t^2)^2}.$$

We first check whether $\mathcal{X}$ is $\mathcal{A}(2)$-equivalent to a curve parametrized by $\alpha(s) = (z_2(s), z_3(s)) = \left(s^2, s\right)$. The answer is no, since $\alpha(s)$ defines a parabola, which is an $\mathcal{A}(2)$-exceptional curve and $\mathcal{X}$ is not $\mathcal{A}(2)$-exceptional. We next check if there exists $b \in \mathbb{R}$ such that $\mathcal{X}$ is $\mathcal{A}(2)$-equivalent to a curve parametrized by $\beta_b(s) = \left(z_1(s) + b\, z_2(s),\, z_3(s)\right) = \left(s^4 + 1 + b\, s^2,\, s\right)$. We evaluate invariants (21):

$$T_{\mathcal{A}}|_\beta(b,s) = \frac{100\, s^2\,(3\,b - 14s^2)^2}{(b - 14s^2)^3},$$

$$K_{\mathcal{A}}|_\beta(b,s) = \frac{-5\,(140\, s^4 - 56\, b\, s^2 + b^2)}{(b - 14\, s^2)^2}.$$

Since $T_{\mathcal{A}}|_\beta(1,s) = K_{\mathcal{A}}|_\gamma(s)$ and $T_{\mathcal{A}}|_\beta(1,s) = T_{\mathcal{A}}|_\gamma(s), \forall s$ we conclude that a curve parametrized by $\beta(1,s)$ is $\mathcal{A}(2)$-equivalent to $\mathcal{X}$ and, therefore, $\mathcal{Z}$ projects to $\mathcal{X}$ by a parallel projection.

EXAMPLE 35. *We would like to decide if the spatial curve $\mathcal{Z}$ parametrized by*

$$\Gamma(s) = \left(s^3,\, s^2,\, s\right),\ s \in \mathbb{R}\ (\textit{Twisted Cubic})$$

projects to any of the three given planar curves $\mathcal{X}_1$, $\mathcal{X}_2$ and $\mathcal{X}_3$ parametrized, respectively, by:

$$\gamma_1(t) = \left(t^4 + t\,,\, t^2\right),$$

$$\gamma_2(t) = \left(t^3 - t\,,\, t^3 + t^2\right),$$

$$\gamma_3(t) = \left(\frac{t}{(1 + t^3)},\, \frac{t^2}{(1 + t^3)}\right)\ (\textit{Folium of Descartes}).$$

None of the given planar curves is $\mathcal{A}(2)$-exceptional and so none of them is $\mathcal{A}(2)$-equivalent to a parabola parametrized by $\alpha(s) = \left(s^2, s\right)$. We then consider a family of curves parametrized by $\beta(b,s) = \left(s^3 + b\, s^2,\, s\right)$ and establish that none of the curves in the family is $\mathcal{A}(2)$-equivalent to either of $\mathcal{X}$'s. We proceed, by considering a family of curves $\delta(a_1, a_2, s) = \left(s^3 + a_1\, s,\, s^2 + a_2\, s\right)$ and establish that a curve parametrized by $\delta(0, 1/2, s)$ is $\mathcal{A}(2)$-equivalent to $\mathcal{X}_1$ and a curve parametrized by $\delta(0,0,s)$ is $\mathcal{A}(2)$-equivalent to $\mathcal{X}_2$, but there are no real values of a_1 and a_2, such that a curve parametrized by $\delta(a_1, a_2, s)$ is $\mathcal{A}(2)$-equivalent to $\mathcal{X}_3$.

We conclude that there are parallel projections of $\mathcal{Z}$ to both $\mathcal{X}_1$ and $\mathcal{X}_2$, but not to $\mathcal{X}_3$. Note that $\mathcal{X}_1$ and $\mathcal{X}_2$ are not $\mathcal{A}(2)$-equivalent (their signatures have different implicit equations). This underscores an observation made in Remark 8.

6. PROJECTION PROBLEM FOR FINITE LISTS OF POINTS

In [2, 1], the authors present a solution to the problem of deciding whether or not there exists a parallel projection of a list $Z = (\mathbf{z}^1, \ldots, \mathbf{z}^m)$ of m points in $\mathbb{R}^3$ to a list $X = (\mathbf{x}^1, \ldots, \mathbf{x}^m)$ of m points in $\mathbb{R}^2$, without finding a projection explicitly. They identify the lists Z and X with the elements of certain Grassmanian spaces and use Plüker embedding of Grassmanians into projective spaces to explicitly define the algebraic variety that characterizes pairs of sets related by a parallel projection.

We indicate here how our approach leads to an alternative solution for the projection problem for lists of points. Details of this adaptation appear in the dissertation [5]. The projection criterion of Theorem 9 adapt to finite lists of points as follows:

THEOREM 36. (CENTRAL PROJECTION CRITERION FOR FINITE LISTS.) *A given list $Z = (\mathbf{z}^1, \ldots, \mathbf{z}^m)$ of m points in $\mathbb{R}^3$ with coordinates $\mathbf{z}^l = (z_1^l, z_2^l, z_3^l)$, $l = 1 \ldots m$, projects onto a given list $X = (\mathbf{x}^1, \ldots, \mathbf{x}^m)$ of m points in $\mathbb{R}^2$ with coordinates $\mathbf{x}^l = (x^l, y^l)$, $l = 1 \ldots m$, by a central projection if and only if there exist $c_1, c_2, c_3 \in \mathbb{R}$ and $[A] \in \mathcal{PGL}(3)$, such that*

$$[x^l, y^l, 1]^{\mathrm{T}} = [A][z_1^l + c_1,\, z_2^l + c_2,\, z_3^l + c_3]^{\mathrm{T}}\ \textit{for}\ l = 1 \ldots m.$$

The proof of Theorem 36 is a straightforward adaptation of the proof of Theorem 9. The parallel projection criteria for curves, given in Theorem 10 and Corollary 11, are adapted to the finite lists in an analogous way.

The central and the parallel projection problems for lists of m points is therefore reduced to a modification of the problems of equivalence of two lists of m points in $\mathbb{PR}^2$ under the action of $\mathcal{PGL}(3)$ and $\mathcal{A}(2)$ groups, respectively. A separating set of invariants for lists of m points in $\mathbb{PR}^2$ under $\mathcal{A}(2)$-action consists of ratios of certain areas and is listed, for instance, in Theorem 3.5 of [20]. Similarly, a separating set of invariants for lists of m ordered points in $\mathbb{PR}^2$ under $\mathcal{PGL}(3)$-action consists of cross-ratios of certain areas and is listed, for instance, in Theorem 3.10 in [20]. In the case of central projections we, therefore, obtain a system of polynomial equations on c_1, c_2 and c_3 that have solutions if and only if the given set Z projects to the given set X and analog of Algorithms 28 follows. The parallel projections are treated in a similar way.

A solution of the projection problem for lists of points does not provide an immediate solution to the discretization of the projection problem for curves. Indeed, let $Z = (\mathbf{z}^1, \ldots, \mathbf{z}^m)$ be a discrete sampling of a spatial curve $\mathcal{Z}$ and $X = (\mathbf{x}^1, \ldots, \mathbf{x}^m)$ be a discrete sampling of a planar curve $\mathcal{X}$. It might be impossible to project the list Z onto X, even when the curve $\mathcal{Z}$ can be projected to the curve $\mathcal{X}$. Some approaches to the discretization of the projection algorithms for curves are discussed in the next section.

7. FURTHER RESEARCH

The projection criteria developed in Section 3 reduce the problem of object-image correspondence for curves under a projection from $\mathbb{R}^3$ to $\mathbb{R}^2$ to a variation of the group-equivalence problem for curves in $\mathbb{R}^2$. We use differential invariants to address the group-equivalence problem. In practical applications, curves are often given by samples of points. In this case, invariant numerical approximations of differential invariants presented in [7, 4] may be used. Differential invariants and their approximations are highly sensitive to image perturbations and, therefore, are not practical in many situations. Other types of invariants, such as semi-differential (or joint) invariants [23, 20] and integral invariants [21, 16, 13] are less sensitive to image perturbations and may be employed to solve the group-equivalence problem.

One of the essential contributions of [2, 1] is the definition of an object/image distance between ordered sets of m points in $\mathbb{R}^3$ and $\mathbb{R}^2$, such that the distance is zero if and only if these sets are related by a projection. Since, in practice, we are given only an approximate position of points, a "good" object/image distance provides a tool for deciding whether a given set of points in $\mathbb{R}^2$ is a good approximation of a projection of a given set of points in $\mathbb{R}^3$. Defining such object/image distance in the case of curves is an important direction of further research.

Our algorithm establishes existence of a projection, but does not compute a possible projection map (see Remark 32). Algorithmic reconstruction of a projection is another interesting question to consider.

8. ACKNOWLEDGEMENTS

This project was inspired by a discussion with Peter Stiller of paper [1]. We are grateful to Hoon Hong, Peter Olver and Agnes Szanto for suggestions and support. We thank anonymous referees for their valuable comments.

9. REFERENCES

[1] G. Arnold and P. F. Stiller. Mathematical aspects of shape analysis for object recognition. In *proceedings of IS&T/SPIE Symposium*, 11 pp, San Jose, CA, 2007.

[2] G. Arnold, P. F. Stiller, and K. Sturtz. Object-image metrics for generalized weak perspective projection. In *Statistics and analysis of shapes*, Model. Simul. Sci. Eng. Technol., 253 – 279. Birkhäuser, Boston, 2006.

[3] W. Blaschke. *Vorlesungen über Differentialgeometrie und geometrische Grundlagen von Einsteins Relativitätstheorie, vol. II, "Affine Differentialgeometrie"*. Springer, Berlin, 1923.

[4] M. Boutin. Numerically invariant signature curves. *Int. J. Computer Vision*, 40: 235 – 248, 2000.

[5] J. M. Burdis. *Object-image correspondence under projections*. PhD thesis, NCSU, 2010.

[6] J. M. Burdis and I. A. Kogan. Object-image correspondence for curves under projections, 2012. `http://arxiv.org/abs/1202.1303`

[7] E. Calabi, P. J. Olver, C. Shakiban, A. Tannenbaum, and S. Haker. Differential and numerically invariant signature curves applied to object recognition. *Int. J. Computer Vision*, 26:107–135, 1998.

[8] É. Cartan. *Leçons sur la géométrie projective complexe*. Gauthier-Villars, Paris, 1950.

[9] B. F. Caviness and J. R. Johnson, editors. Texts and monographs in symbolic computation. 1998.

[10] D. Cox, J. Little, and D. O'Shea. *Ideals, Varieties, and Algorithms*. Undergraduate Text in Mathematics. Springer-Verlag, New-York, 1997.

[11] O. Faugeras. Cartan's moving frame method and its application to the geometry and evolution of curves in the Euclidean, affine and projective planes. *Application of Invariance in Computer Vision, J.L Mundy, A. Zisserman, D. Forsyth (eds.) Springer-Verlag Lecture Notes in Computer Science*, 825: 11 – 46, 1994.

[12] J. Feldmar, N. Ayache, and F. Betting. 3D-2D projective registration of free-form curves and surfaces. In *Proceedings of the Fifth International Conference on Computer Vision, ICCV, IEEE Computer Society*, 549 – 556, Washington, DC, 1995.

[13] S. Feng, I. A. Kogan, and H. Krim. Classification of curves in 2D and 3D via affine integral signature. *Acta Appl. Math.*, 109(3): 903–937, 2010.

[14] W. Fulton. *Algebraic curves*. Advanced Book Classics. Addison-Wesley Publishing Company, Redwood City, CA, 1989. Notes written with the collaboration of Richard Weiss, Reprint of 1969 original.

[15] H. W. Guggenheimer. *Differential Geometry*. McGraw-Hill, New York, 1963.

[16] C. Hann and M. Hickman. Projective curvature and integral invariants. *Acta Appl. Math.*, 74: 549 – 556, 2002.

[17] R. I. Hartley and A. Zisserman. *Multiple View Geometry in Computer Vision*. Cambridge University Press, second edition, 2004.

[18] H. Hong, editor. *Special issue on computational quantifier elimination*, volume 36 of *The Computer Journal*. 1993.

[19] I. A. Kogan. Two algorithms for a moving frame construction. *Canad. J. Math.*, 55: 266–291, 2003.

[20] P. J. Olver. Joint invariant signatures. *Found. Comp. Math*, 1: 3 – 67, 2001.

[21] J. Sato and R. Cipolla. Affine integral invariants for extracting symmetry axes. *Image and Vision Computing*, 15:627–635, 1997.

[22] A. Tarski. *A Decision Method for Elementary Algebra and Geometry*. Univ. of California Press, Berkeley, second edition, 1951.

[23] L. Van Gool, T. Moons, E. Pauwels, and A. Oosterlinck. Semi-differential invariants. *Geometric Invariance in Computer Vision, J.L. Mundy and A. Zisserman (eds), MIT Press*, pages 157–192, 1992.

[24] `http://en.wikipedia.org/wiki/File:Pinhole-camera.svg`

[25] `http://www.math.ncsu.edu/~iakogan/symbolic/projections.html`

Disjoint Edges in Complete Topological Graphs

Andrew Suk*
Massachusetts Institute of Technology
Cambridge MA, 02139
asuk@math.mit.edu

ABSTRACT

It is shown that every complete n-vertex simple topological graph has at least $\Omega(n^{1/3})$ pairwise disjoint edges, and these edges can be found in polynomial time. This proves a conjecture of Pach and Tóth, which appears as problem 5 from chapter 9.5 in *Research Problems in Discrete Geometry* by Brass, Moser, and Pach.

Categories and Subject Descriptors

G.2.1 [**Discrete Mathematics**]: Combinatorics

General Terms

Theory

Keywords

geometric graphs, topological graphs, independent set

1. INTRODUCTION

Given a collection of objects $\mathcal{C}$ in the plane, the *intersection graph* $G(\mathcal{C})$ has vertex set $\mathcal{C}$ and two objects are adjacent if and only if they have a nonempty intersection. The *independence number* of $G(\mathcal{C})$ is the size of the largest independent set, that is, the size of the largest subfamily of pairwise disjoint objects in $\mathcal{C}$. It is known that computing the independence number of an intersection graph is NP-hard [12], even for very simple objects such as rectangles and disks in the plane [3, 11]. However, due to its applications in VLSI design [13], map labeling [2], and elsewhere, a lot of research has been devoted to developing polynomial-time approximation schemes (PTAS) for computing the independence number of intersection graphs (see [9] for more references). In this paper, we study the independence number of the intersection graph of edges in a complete simple topological graph.

*Research funded by an NSF Postdoctoral Fellowship.

SCG'12, June 17–20, 2012, Chapel Hill, North Carolina, USA.

A *topological graph* is a graph drawn in the plane such that its vertices are represented by points and its edges are represented by non-self-intersecting arcs connecting the corresponding points. The arcs are allowed to intersect, but they may not intersect vertices except for their endpoints. Furthermore, no three edges have a common interior point and no two edges are tangent, i.e., if two edges share an interior point, then they must properly cross at that point in common. A topological graph is *simple* if every pair of its edges intersect at most once. Two edges of a topological graph *cross* if their interiors share a point, and are *disjoint* if they neither share a common vertex nor cross. If the edges are drawn as straight-line segments, then the graph is *geometric*.

Let F denote the graph obtained from the complete graph on 5 vertices by subdividing each edge with an extra vertex. It is easy to see that every intersection graph G of curves in the plane does not contain F as an induced subgraph [6]. By applying a theorem of Erdős and Hajnal [7], every complete n-vertex simple topological graph contains $e^{\Omega(\sqrt{\log n})}$ edges that are either pairwise disjoint or pairwise crossing. However, it was suspected [16] that this bound is far from optimal. Fox and Pach [8] showed that there exists a constant $\delta > 0$, such that every complete n-vertex simple topological graph contains $\Omega(n^{\delta})$ pairwise crossing edges. However, much weaker bounds were previously known for pairwise disjoint edges.

In 2003, Pach, Solymosi, and Tóth [16] showed that every complete n-vertex simple topological graph has at least $\Omega(\log^{1/6} n)$ pairwise disjoint edges. This lower bound was later improved by Pach and Tóth [17] to $\Omega(\log n / \log\log n)$. Recently, Fox and Sudakov [10] gave a modest improvement of $\Omega(\log^{1+\epsilon} n)$, where ϵ is a very small constant. We note that the previous two bounds hold for dense simple topological graphs. Pach and Tóth conjectured (see problem 5 of chapter 9.5 in [4]) that there exists a constant $\delta > 0$ such that every complete n-vertex simple topological graph has at least $\Omega(n^{\delta})$ pairwise disjoint edges. Our main result settles the conjecture in the affirmative.

THEOREM 1. *Every complete n-vertex simple topological graph contains $\Omega(n^{1/3})$ pairwise disjoint edges.*

Note that Theorem 1 does not remain true if the *simple* condition is dropped. Indeed, in [17], Pach and Tóth gave a construction of a complete n-vertex topological graph such that every pair of edges intersect exactly once or twice.

2. THE DUAL SHATTER FUNCTION

In this section, we will recall one of the most useful parameters measuring the complexity of a set system: the dual shatter function. All of the following concepts and results can be found in Chapter 5 of [14]. Let $(X, \mathcal{S})$ be a set system with ground set X, such that X is finite.

Definition 1. The *dual shatter function* of $(X, \mathcal{S})$ is a function, denoted by $\pi^*_{\mathcal{S}}$, whose value at m is defined as the maximum number of equivalence classes on X defined by an m-element subfamily $\mathcal{Y} \subset \mathcal{S}$, where two points $x, y \in X$ are equivalent with respect to $\mathcal{Y}$ if x belongs to the same sets of $\mathcal{Y}$ as y does. In other words, $\pi^*_{\mathcal{S}}(m)$ is the maximum number of nonempty cells in the Venn diagram of m sets of $\mathcal{S}$.

One of the main tools used to prove Theorem 1 is the following result of Chazelle and Welzl on matchings with low stabbing number. A similar approach was done by Pach in [15], who showed that every n-vertex complete geometric graph contains at least $\Omega(n^{1/2})$ pairwise parallel edges, where two edges are *parallel* if they are the opposite sides of a convex quadrilateral.

Given a set system $(X, \mathcal{S})$ and a graph $G = (X, E)$, we say that a set $S \in \mathcal{S}$ *stabs* edge $uv \in E(G)$ if $|S \cap \{u, v\}| = 1$. The *stabbing number of* G *with respect to the set* S is the number of edges of G stabbed by S, and the *stabbing number of* G is the maximum of stabbing numbers of G with respect to all sets of $\mathcal{S}$.

Lemma 2 ([5]). *Let n be an even integer, and $\mathcal{S}$ be a set system on an n-point set X with $\pi^*_{\mathcal{S}}(m) \leq Cm^d$ for all m, where C and d are constants. Then there exists a perfect matching M on X (i.e. a set of $n/2$ vertex disjoint edges) whose stabbing number is at most $C_1 n^{1-1/d}$, where C_1 is a constant that depends only on C and d.*

We will also need the following two lemmas on the dual shatter function of a set system.

Lemma 3. *Let $\mathcal{S}_1, \mathcal{S}_2$ be set systems on a set X and let $\mathcal{S} = \mathcal{S}_1 \cup \mathcal{S}_2$. Then $\pi^*_{\mathcal{S}}(m) \leq \pi^*_{\mathcal{S}_1}(m)\pi^*_{\mathcal{S}_2}(m)$.*

Proof. Suppose that $\pi^*_{\mathcal{S}}(m)$ is maximized for some m_1 sets in $\mathcal{S}_1$ and m_2 sets in $\mathcal{S}_2$ where $m_1 + m_2 = m$. Then the number of nonempty cells in the Venn diagram of these m sets is at most

$$\pi^*_{\mathcal{S}_1}(m_1)\pi^*_{\mathcal{S}_2}(m_2) \leq \pi^*_{\mathcal{S}_1}(m)\pi^*_{\mathcal{S}_2}(m).$$

□

Lemma 4. *Let $\Phi(X_1, X_2, ..., X_t)$ be a fixed set-theoretic expression (using the operations of union, intersection, and difference) with variables $X_1, ..., X_t$ standing for sets. Let $\mathcal{S}$ be a set system on a set X. Let $\mathcal{T}$ consist of all sets $\Phi(S_1, ..., S_t)$ for all possible choices $S_1, ..., S_t \in \mathcal{S}$. If $\pi^*_{\mathcal{S}}(m) \leq Cm^d$, then $\pi^*_{\mathcal{T}}(m) \leq Ct^d m^d$.*

Proof. Each set in $\mathcal{T}$ is a disjoint union of some cells of the Venn diagram of some t sets in $\mathcal{S}$. Therefore, given any m sets in $\mathcal{T}$, each cell in the Venn diagram of these m sets is a disjoint union of some cells in the Venn diagram of at most tm sets of $\mathcal{S}$. Hence

$$\pi^*_{\mathcal{T}}(m) \leq \pi^*_{\mathcal{S}}(tm) \leq Ct^d m^d.$$

□

3. PROOF OF THEOREM 1

Let $K = (V, E)$ be a complete simple topological graph with $n+1$ vertices. Without loss of generality, we can assume n is even. Notice that the edges of K divide the plane into several cells (regions), one of which is unbounded. We can assume that there is a vertex $v_0 \in V$ such that v_0 lies on the boundary of the unbounded cell. Indeed, otherwise we can project K onto a sphere, and then project K back to the plane such that v_0 lies on the boundary of the unbounded cell, and moreover two edges cross (are disjoint) in the new drawing if and only if they crossed (were disjoint) in the original drawing.

Given a subset $V' \subset V$, we denote by $K[V']$ the topological subgraph induced by the vertex set V'. Since we will be dealing with both combinatorial and topological graphs, the term *edge* will refer to an edge in a combinatorial graph and will be denoted by pairs of vertices v_iv_j. The term *topological edge* will refer to edges in the topological graph K, and will be denoted by $K[v_i, v_j]$ (i.e. the topological subgraph induced by vertices v_i and v_j).

Consider the topological edges emanating out from v_0, and label their endpoints $v_1, ..., v_n$ in counterclockwise order so that for $i < j$, the vertices (v_0, v_i, v_j) appear in counterclockwise order along the simple closed curve $K[v_0, v_i, v_j]$. This is possible since v_0 lies on the unbounded cell and these three edges do not cross. For all pairs i, j, we will call the topological subgraph $K[v_0, v_i, v_j]$ a *triangle* in K. Note that each "triangle" is incident to v_0, and is a simple closed curve since K is simple.

Set $X = \{v_1, ..., v_n\}$, and now we will define the set system $(\mathcal{S}_1, X)$ as follows. For each pair i, j that satisfies $1 \leq i < j \leq n$, we define the set $S_{i,j}$ to be the set of vertices in X that lie in the interior of triangle $K[v_0, v_i, v_j]$. See Figure 1(a) for a small example. Then set

$$\mathcal{S}_1 = \bigcup_{1 \leq i < j \leq n} \{S_{i,j}\}.$$

Lemma 5. *Let $(\mathcal{S}_1, X)$ be the set system defined as above. Then $\pi^*_{\mathcal{S}_1}(m) \leq 5m^2$.*

Proof. In order to obtain an upper bound on $\pi^*_{\mathcal{S}_1}(m)$, we simply need to bound the maximum number of regions for which m triangles in K partitions the plane into. We proceed by induction on m. The base case when $m = 1$ is trivial. Now assume that the statement holds up to $m - 1$. Notice that any set of $m - 1$ triangles consists of at most $3(m - 1)$ topological edges of K. Therefore, when we add the mth triangle T, each topological edge in T creates at most $3(m - 1)$ new crossing points (since K is simple). Hence triangle T creates at most $9(m - 1)$ new regions in the arrangement. By the induction hypothesis, we have at most

$$5(m - 1)^2 + 9(m - 1) \leq 5m^2$$

cells created by the m triangles. □

Now we define the set system $(\mathcal{S}_2, X)$ as follows. For each pair i, j that satisfies $1 \leq i < j \leq n$, we define the set $S'_{i,j} = \{v_k \in X : K[v_0, v_k]$ crosses $K[v_i, v_j]\}$. See Figure 1(b) for a small example. Then set

$$\mathcal{S}_2 = \bigcup_{1 \leq i < j \leq n} \{S'_{i,j}\}.$$

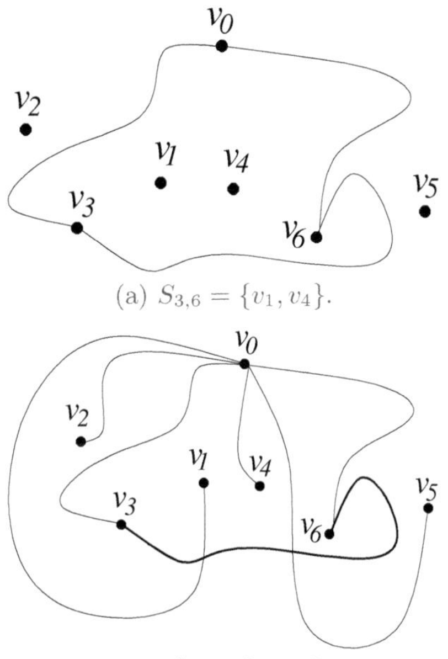

(a) $S_{3,6} = \{v_1, v_4\}$.

(b) $S'_{3,6} = \{v_1, v_5\}$.

Figure 1: Sets $S_{3,6} \in \mathcal{S}_1$ and $S'_{3,6} \in \mathcal{S}_2$.

LEMMA 6. *Let $\mathcal{S} = \mathcal{S}_1 \cup \mathcal{S}_2$ where $\mathcal{S}_1, \mathcal{S}_2$ are defined as above. Then $\pi^*_{\mathcal{S}}(m) \leq 120m^3$.*

PROOF. Let i, j be fixed such that $i < j$. Notice that if k is an integer that satisfies $k < i$ or $k > j$, then $v_k \in S'_{i,j} \in \mathcal{S}_2$ if and only if $v_k \in S_{i,j} \in \mathcal{S}_1$. Moreover, if k is an integer that satisfies $i < k < j$, then $v_k \in S'_{i,j} \in \mathcal{S}_2$ if and only if $v_k \notin S_{i,j} \in \mathcal{S}_1$. Therefore, in what follows, we will apply Lemmas 3 and 4 to $\mathcal{S}_1$ in order to obtain the set system $\mathcal{S} = \mathcal{S}_1 \cup \mathcal{S}_2$. This will ensure that the dual shatter function $\pi^*_{\mathcal{S}}$ is *well behaved.*

Let $(\mathcal{I}, X)$ be a set system such that

$$\mathcal{I} = \bigcup_{1 \leq i < j \leq n} \{I_{ij}\},$$

where $I_{ij} = \{v_k : i < k < j\}$. It is easy to see that $\pi^*_{\mathcal{I}}(m) \leq 2m + 1 \leq 3m$, since this is equivalent to determining the number of cells created by m intervals on the real line. Let $\mathcal{T}_1$ be the set system defined as $\mathcal{T}_1 = \mathcal{S}_1 \cup \mathcal{I} \cup \{\emptyset\}$. By Lemmas 3 and 5, we have $\pi^*_{\mathcal{T}_1}(m) \leq 15m^3$. Notice that each set $S'_{ij} \in \mathcal{S}_2$ can be expressed as

$$S'_{ij} = (S_{ij} \setminus I_{ij}) \cup (I_{ij} \setminus S_{ij})$$

where $S_{ij} \in \mathcal{S}_1$ and $I_{ij} \in \mathcal{I}$ are defined above. Therefore, let $\Phi(X_1, X_2)$ be the set-theoretic expression defined as

$$\Phi(X_1, X_2) = (X_1 \setminus X_2) \cup (X_2 \setminus X_1),$$

and let $\mathcal{T}_2$ consist of all sets $\Phi(X_1, X_2)$ for all possible choices $X_1, X_2 \in \mathcal{T}_1$. Hence $\mathcal{S} = \mathcal{S}_1 \cup \mathcal{S}_2 \subset \mathcal{T}_2$. By Lemma 4, we have

$$\pi^*_{\mathcal{S}}(m) \leq \pi^*_{\mathcal{T}_2}(m) \leq \pi^*_{\mathcal{T}_1}(2m) \leq 15(2m)^3 = 120m^3.$$

□

By setting $\mathcal{S} = \mathcal{S}_1 \cup \mathcal{S}_2$, we know that $\pi^*_{\mathcal{S}}(m) \leq 120m^3$. Therefore we can apply Lemma 2 to $(\mathcal{S}, X)$ and find a perfect matching M on X whose stabbing number is at most $Cn^{2/3}$, where C is an absolute constant. In other words, each set in $S \in \mathcal{S}$ stabs at most $Cn^{2/3}$ members in M.

Now let $G = (V, E)$ be an $(n/2)$-vertex graph, where $V(G) = M$, and two vertices v_iv_j, v_kv_l in G (edges in M) are adjacent if and only if

1. $S_{i,j} \in \mathcal{S}_1$ stabs v_kv_l, or
2. $S'_{i,j} \in \mathcal{S}_2$ stabs v_kv_l, or
3. $S_{k,l} \in \mathcal{S}_1$ stabs v_iv_j, or
4. $S'_{k,l} \in \mathcal{S}_2$ stabs v_iv_j.

Since M has a stabbing number at most $Cn^{2/3}$, the two sets corresponding to each vertex in G stabs (in total) at most $2{\cdot}Cn^{2/3}$ other edges in M. A simple counting argument shows that G contains at most $Cn^{5/3}$ edges. Therefore we can use the following well known theorem of Turán to find a large independent set in M.

THEOREM 7. *Every graph with n vertices and e edges contains an independent set of size at least $n^2/(2e + n)$.*

In particular, the graph G defined above contains an independent set of size $\Omega(n^{1/3})$. Let $M' \subset M$ be the edges corresponding to this independent set of size $\Omega(n^{1/3})$. Now we claim that the corresponding topological edges in K are pairwise disjoint. Indeed, let $e_1, e_2 \in M'$ such that $e_1 = v_iv_j$ and $e_2 = v_kv_l$. Since the set $S_{i,j} \in \mathcal{S}_1$ does not stab v_kv_l, both vertices v_k, v_l must either lie inside or outside of triangle $K[v_0, v_i, v_j]$. See Figure 2.

Case 1. Suppose that v_k, v_l lie outside of triangle $K[v_0, v_i, v_j]$. For sake of contradiction, suppose that the topological edges $K[v_i, v_j]$ and $K[v_k, v_l]$ cross. By a simple parity argument, $K[v_k, v_l]$ must cross either $K[v_0, v_i]$ or $K[v_0, v_j]$, but not both. However, this implies that the set $S'_{k,l} \in \mathcal{S}_2$ stabs v_iv_j which is a contradiction. Therefore, $K[v_i, v_j]$ and $K[v_k, v_l]$ must be disjoint.

Case 2. If v_k, v_l lie inside of triangle $K[v_0, v_i, v_j]$, then by the same parity argument as above, $K[v_i, v_j]$ and $K[v_k, v_l]$ must be disjoint.

4. CONCLUDING REMARKS

1. **Remarks on algorithms.** Since the matching in Lemma 2 and the independent set in Theorem 7 can be found in polynomial time (see [5] and [1]), the proof above gives a polynomial time algorithm for finding $\Omega(n^{1/3})$ pairwise disjoint edges in a complete n-vertex simple topological graph.
2. We proved that every complete n-vertex simple topological graph has at least $\Omega(n^{1/3})$ pairwise disjoint edges. It would be interesting to see if a polynomial bound holds for dense simple topological graphs.
3. We proved that the dual shatter function $\pi^*_{\mathcal{F}}(m) = O(m^3)$. Recently Géza Tóth [18] gave a construction showing that this bound is tight.

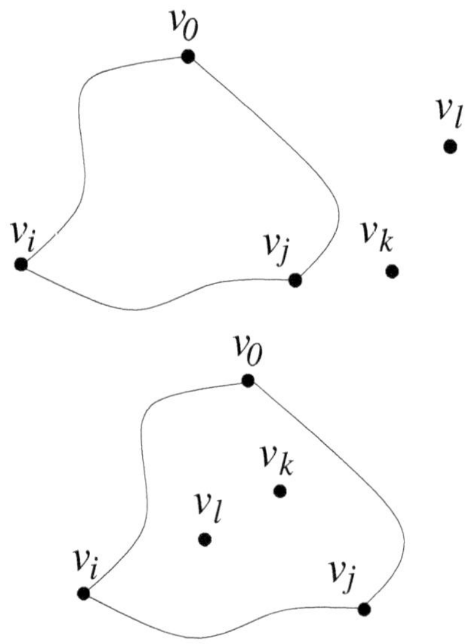

Figure 2: Case 1 and 2.

4. To our knowledge, there are no constructions of a complete n-vertex simple topological graph with sublinear number of pairwise disjoint edges. However, let us remark that Valtr [4] gave a construction of a complete n-vertex simple topological graph, such that every edge crosses at least $\Omega(n^{3/2})$ other edges.

5. REFERENCES

[1] N. Alon, J. Spencer, *The Probabilistic Method*, Wiley, 3rd Edition (2008).

[2] P. K. Agarwal, M. van Kreveld, S. Suri, Label placement by maximum independent set in rectangles, *Comput. Geom. Theory Appl.* **11** (1998), 209-218.

[3] T. Asano, H. Imai, Finding the connected components and a maximum clique of an intersection graph of rectangles in the plane, *J. Algorithms* **4** (1983), 310-323.

[4] P. Brass, W. Moser, J. Pach, *Research Problems in Discrete Geometry.* Berlin, Germany: Springer-Verlag, 2005.

[5] B. Chazelle and E. Welzl. Quasi-optimal range searching in spaces of finite VC-dimension. *Discrete Comput. Geom.* **4** (1989), 467-489.

[6] G. Ehrlich, S. Even, R.E. Tarjan, Intersection graphs of curves in the plane, *J. Combinatorial Theory, Ser. B* **21** (1979), 8-20.

[7] P. Erdős, A. Hajnal, Ramsey-type theorems, *Discrete Appl. Math.* **25** (1989), 37-52.

[8] J. Fox, J. Pach, Coloring K_k-free intersection graphs of geometric objects in the plane. *In Proceedings of the Twenty-Fourth Annual Symposium on Computational Geometry*, 2008, 346-354.

[9] J. Fox, J. Pach, Computing the Independence Number of Intersection Graphs, SODA 2011, 1161-1165.

[10] J. Fox, B. Sudakov, Density theorems for bipartite graphs and related Ramsey-type results. *Combinatorica* **29** (2009), 153-196.

[11] R. J. Fowler, M.S. Paterson, S. L. Tanimoto, Optimal packing and covering in the plane are NP-complete, *Inform. Process. Lett.* **12** (1981), 133-137.

[12] M. R. Garey, D. S. Johnson, *Computers and Intractability: A Guide to the Theory of NP-Completeness*, Freeman, New York, 1979.

[13] D. S. Hochbaum, W. Maass, Approximation schemes for covering and packing problems in image processing and VLSI, *J. ACM* **32** (1985), 130-136.

[14] J. Matousek, *Geometric Discrepancy: An Illustrated Guide.* Algorithms and combinatorics. 18. Berlin: Springer, 1999.

[15] J. Pach, Notes on geometric graph theory, in: Discrete and Computational Geometry (J.E. Goodman et al, eds.), DIMACS Series, *Amer. Math. Soc.* **6** (1991), 273-285.

[16] J. Pach, J. Solymosi, G. Tóth, Unavoidable Configurations in Complete Topological Graphs. *Discrete Comput. Geom.* **30** (2003), 311-320.

[17] J. Pach and G. Tóth, Disjoint edges in topological graphs, in: Combinatorial Geometry and Graph Theory (J. Akiyama et al., eds.), Lecture Notes in Computer Science 3330, Springer-Verlag, Berlin, 2005, 133-140.

[18] G. Tóth, Personal Communcation (2012).

Representing a Cubic Graph as the Intersection Graph of Axis-parallel Boxes in Three Dimensions

Abhijin Adiga *
Network Dynamics and Simulation Science Laboratory
Virginia Bioinformatics Institute, Virginia Tech
Blacksburg, VA 24061, USA
abhijin@vbi.vt.edu

L. Sunil Chandran
Department of Computer Science and Automation
Indian Institute of Science
Bangalore–560012, India
sunil@csa.iisc.ernet.in

ABSTRACT

We show that every graph of maximum degree 3 can be represented as the intersection graph of axis parallel boxes in three dimensions, that is, every vertex can be mapped to an axis parallel box such that two boxes intersect if and only if their corresponding vertices are adjacent. In fact, we construct a representation in which any two intersecting boxes just touch at their boundaries. Further, this construction can be realized in linear time.

Categories and Subject Descriptors

F.2.2 [**Nonnumerical Algorithms and Problems**]: Geometrical problems and computations

General Terms

Theory

Keywords

cubic graphs, intersection graphs, axis parallel boxes, boxicity

1. INTRODUCTION

We will be considering only simple, undirected and finite graphs. Let $\mathcal{F} = \{S_1, S_2, \ldots, S_n\}$ be a family of sets. An intersection graph associated with $\mathcal{F}$ has $\mathcal{F}$ as the vertex set and two vertices S_i and S_j are adjacent if and only if $i \neq j$ and $S_i \cap S_j \neq \emptyset$. It is interesting to study intersection graphs of sets with some restriction, for example, sets which correspond to geometric objects such as intervals, spheres, boxes, axis-parallel lines, etc. Many important graph classes arise out of such restrictions: interval graphs, circular arc graphs, unit-disk graphs and grid-intersection graphs, to name a few. In this paper, we are concerned with intersection graphs of 3-dimensional boxes. A *3-dimensional axis parallel box* (3-box in short) is a Cartesian product of 3 closed intervals on the real line. A graph is said to have a *3-box representation* if it can be represented as the intersection graph of 3-boxes.

In the literature there are several results on representing a planar graph as the intersection graph of various geometric objects. Among these, the most noted result is the circle packing theorem (also known as the Koebe-Andreev-Thurston theorem) from which it follows that planar graphs are exactly the intersection graphs of closed disks in the plane such that the intersections happen only at the boundaries. In [9], Thomassen gave a similar representation for planar graphs with 3-boxes. He showed that every planar graph has a *strict 3-box representation*, that is, intersections occur only in the boundaries of the boxes and two boxes which intersect have precisely a 2-box (a rectangle) in common. Very recently, Felsner and Francis [5] strengthened this result by showing that there exists a strict 3-box representation for a planar graph such that each box is an isothetic cube. In [7, 9], it was shown that every planar graph has a strict representation using at most two rectangles per vertex. Scheinerman and West [8] showed that every planar graph is an intersection graph of intervals such that each vertex is represented by at most three intervals on the real line.

We consider the question of whether a graph of maximum degree 3 has a 3-box representation. We note that there exist graphs with maximum degree greater than 3 which do not have a 3-box representation. For example, it is easy to show that a K_8 minus a perfect matching does not have a 3-box representation [6]. Considering the effort that has gone into discovering geometric representation theorems for planar graphs, it is surprising that no such results are known up to now in the case of cubic graphs. It may be because of the fact that intuitively cubic graphs are farther away from "geometry" compared to planar graphs. In this paper we present the first such theorem (as far as we know) for cubic graphs:

THEOREM 1. *Every graph of maximum degree 3 has a 3-box representation with the restriction that two boxes can intersect only at their boundaries.*

1.1 k-box representations and boxicity:

The concept of 3-box representation can be extended to

*This work was carried out when the first author was a research scholar in the Department of Computer Science and Automation, Indian Institute of Science, Bangalore–560012, India.

SCG'12, June 17–20, 2012, Chapel Hill, North Carolina, USA.

higher dimensions. A k-box is a Cartesian product of closed intervals $[a_1, b_1] \times [a_2, b_2] \times \cdots \times [a_k, b_k]$. A graph G has a *k-box representation* if it is the intersection graph of a family of k-boxes in the k-dimensional Euclidean space. The *boxicity* of G denoted by $\text{box}(G)$, is the minimum integer k such that G has a k-box representation. Clearly, Theorem 1 can be rephrased as: Every graph with maximum degree 3 has boxicity at most 3. The best known upper bound for the boxicity of cubic graphs is 10; it follows from the bound $\text{box}(G) \leq 2\left\lfloor \frac{\Delta^2}{2} \right\rfloor + 2$ by Esperet [4], where Δ is the maximum degree of the graph. In [3], it was conjectured that boxicity of a graph is $O(\Delta)$. However, this was disproved in [1] by showing the existence of graphs with boxicity $\Omega(\Delta \log \Delta)$. Theorem 1 implies that the conjecture is true for $\Delta = 3$. Our result also implies that any problem which is hard for cubic graphs is also hard for graphs with a 3-box representation. Some such problems are: crossing number, minimum vertex cover, Hamiltonian cycle, maximum independent set, minimum dominating set and maximum cut. Now we present an alternate characterization of k-box representation in terms of interval graphs. This is used more frequently than its geometric definition.

LEMMA 1. *A graph G has a k-box representation if and only if there exist k interval graphs $I_1, I_2, \ldots, I_k$ such that $V(I_i) = V(G)$, $i = 1, 2, \ldots, k$ and $E(I_1) \cap E(I_2) \cap \cdots \cap E(I_k) = E(G)$.*

Our proof of Theorem 1 uses Lemma 1; we construct 3 interval graphs such that the given cubic graph is the intersection of these interval graphs.

Tight examples:.

We observe that there exist graphs with maximum degree 3 (and hence cubic graphs) with boxicity strictly greater than 2. For example, let G be a non-planar cubic graph and G_s be the graph obtained by subdividing each edge once. Then, $\text{box}(G_s) > 2$. It is an easy exercise to prove this.

The following is easy to verify.

LEMMA 2. *If every cubic graph has a 3-box representation, then, every graph of maximum degree 3 also has a 3-box representation. The statement holds even when the intersections are restricted to the boundaries of the boxes.*

In view of Lemma 2, we note that it is enough to prove that a cubic graph has a 3-box representation. Therefore, in our proof of Theorem 1, we will assume that the graph is cubic.

1.2 Notation

Let G be a graph. For $U \subseteq V(G)$, $G[U]$ denotes the graph induced by U in G. $N(U, G)$ denotes the set

$$\{x \in V(G) \setminus U | \exists y \in U \text{ such that } (x, y) \in E(G)\}.$$

The length of a path is the number of edges in the path. We consider an isolated vertex as a path of length 0. Suppose G and H are graphs defined on the same vertex set. $G \cap H$ denotes the graph with $V(G \cap H) = V(G)$ $(= V(H))$ and $E(G \cap H) = E(G) \cap E(H)$. Consider a non-empty set X and let Π be an ordering of the elements of X. $\overline{\Pi}$ denotes the reverse of Π, that is, for any $x, y \in X$, $\overline{\Pi}(x) < \overline{\Pi}(y)$ if and only if $\Pi(x) > \Pi(y)$. Let A and B be disjoint subsets of X. The notation $\Pi(A) < \Pi(B)$ implies the following: $\forall a \in A$, $b \in B$, $\Pi(a) < \Pi(b)$.

1.3 An outline of the proof of Theorem 1:

We first partition the vertex set of G in such a way that each part induces a simple graph in G (Section 2). For example, a part may induce only a collection of isolated vertices, disjoint paths or cycles in G. In addition to this, the partitioning is such that the adjacency relation between the parts is restricted; there may be two parts such that there are no edges between them or there is exactly one edge per vertex from one part to the other. This structural decomposition of cubic graphs is central to our construction, but it could be of independent interest. The second step is the construction of 3 interval graphs I_1, I_2 and I_3 such that $G = I_1 \cap I_2 \cap I_3$ (Section 3). In each of these interval graphs, a vertex is assigned an interval based on which part it belongs to. The last step is to verify this construction. This involves proving three statements: (1) Each interval graph is a supergraph of G; (2) if two vertices are not adjacent in G, then, they are not adjacent in at least one of the interval graphs and finally, (3) if two vertices are adjacent, then, in at least one interval graph their intervals intersect only at one point. The last statement is sufficient to show that two boxes can intersect only at their boundaries. Due to lack of space we only mention the key observations so that the reader gets a feel for the construction. A more rigorous proof is available in the complete version of this paper [2].

2. STRUCTURAL PREREQUISITES

An induced cycle C is a **special cycle** if for all $x \in C$, $C \setminus \{x\}$ is not a subgraph of an induced cycle or path of size $\geq |C| + 1$. An induced path P is a **special path** if (1) it is maximal in the sense that it is not a subgraph of an induced cycle or a longer induced path, and (2) for any end point of P, say x, $P \setminus \{x\}$ is not a subgraph of an induced cycle of size $\geq |P|$ or an induced path of length $\geq |P| + 1$.

OBSERVATION 1. *Any connected graph with at least 3 vertices contains a special cycle or path. This is easy to see. At least one of the largest induced cycles or paths in a graph is special.*

2.1 Partitioning the vertex set

Let G be a cubic graph and let $V = V(G)$. We partition V in two stages. In Algorithm 1, we obtain the primary partition: $V = \mathcal{S} \uplus \mathcal{N}_1 \uplus \mathcal{A}_1$. Here, $\mathcal{S}$ induces a collection of cycles and paths which is obtained by iteratively extracting special cycles and paths from G. $\mathcal{N}_1$ is the neighborhood of $\mathcal{S}$ and $\mathcal{A}_1$ is the remaining set of vertices.

Algorithm 1:

input : Cubic graph G
output: $\mathcal{S}, \mathcal{N}_1, \mathcal{A}_1$ such that $V = \mathcal{S} \uplus \mathcal{N}_1 \uplus \mathcal{A}_1$.
Let $V' = V$ and $\mathcal{S} = \varnothing$;
while *there is a connected component in $G[V']$ with at least 3 vertices* **do**
  Let $T \subseteq V'$ be a set which induces a special cycle or path; `// which exists by Observation 1`
  $\mathcal{S} \longleftarrow \mathcal{S} \cup T$;
  $V' \longleftarrow V' \setminus \{T \cup N(T, G[V'])\}$;
end
$\mathcal{A}_1 = V'$;
$\mathcal{N}_1 = N(\mathcal{S}, G)$;

Note that every vertex in $\mathcal{S}$ has at least one neighbor in $\mathcal{N}_1$. Therefore, every vertex in $\mathcal{N}_1$ is adjacent to at most two vertices in $\mathcal{A}_1$. For any $u \in \mathcal{S}$ and $v \in \mathcal{A}_1$, u and v are not adjacent. From Observation 1, it follows that $\mathcal{A}_1$ induces a collection of isolated vertices and edges in G. We have an important observation to make:

OBSERVATION 2. *Every vertex in $\mathcal{N}_1$ is adjacent to at least one vertex in $\mathcal{C} \cup \mathcal{P}_i$. If this is not true, it implies that in $\mathcal{S}$, this vertex has neighbor(s) in $\mathcal{P}_e$ only. Under this condition, we can show that one of the cycles or paths extracted in Algorithm 1 is not special, a contradiction.*

In Algorithm 2, we define a finer partitioning: $\mathcal{N}_1 = \mathcal{R} \uplus \mathcal{N}$ and $\mathcal{A}_1 = \mathcal{B} \uplus \mathcal{A}$. $\mathcal{R}$ is obtained by iteratively extracting from $\mathcal{N}_1$, vertices which have two neighbors in $\mathcal{A}_1$. These neighbors along with their neighbors (if any) within $\mathcal{A}_1$ are moved to $\mathcal{B}$.

Algorithm 2:

input : Cubic graph G, $\mathcal{N}_1$, $\mathcal{A}_1$
output: $\mathcal{N}, \mathcal{A}, \mathcal{R}, \mathcal{B}$ such that $\mathcal{N}_1 = \mathcal{R} \uplus \mathcal{N}$ and $\mathcal{A}_1 = \mathcal{B} \uplus \mathcal{A}$.
Let $\mathcal{R} = \mathcal{B} = \varnothing$;
foreach $v \in \mathcal{N}_1$ **do**
  // Recall that v is adjacent to at least one vertex in $\mathcal{S}$ and therefore to at most 2 vertices in $\mathcal{A}_1$.
  if *v is adjacent to two vertices $\mathcal{A}_1 \setminus \mathcal{B}$,* **then**
    Let $X_1(v) = N(v, G[\mathcal{A}_1 \setminus \mathcal{B}])$; // the two neighbors of v in $\mathcal{A}_1 \setminus \mathcal{B}$
    Let $X_2(v) = N(X_1(v), G[\mathcal{A}_1 \setminus \mathcal{B}])$; // neighbors of neighbors of v in $\mathcal{A}_1 \setminus \mathcal{B}$
    $\mathcal{R} \longleftarrow \mathcal{R} \cup \{v\}$;
    $\mathcal{B} \longleftarrow \mathcal{B} \cup (X_1(v) \cup X_2(v))$;
  end
end
$\mathcal{N} = \mathcal{N}_1 \setminus \mathcal{R}$;
$\mathcal{A} = \mathcal{A}_1 \setminus \mathcal{B}$;

We observe the following from Algorithms 1 and 2: Every vertex in $\mathcal{N}$ is adjacent to at most one vertex in $\mathcal{A}$. Every vertex in $\mathcal{R}$ is adjacent to one vertex in $\mathcal{S}$ and two vertices in $\mathcal{B}$. This immediately implies that (a) $\mathcal{R}$ is an independent set and (b) for any $u \in \mathcal{R}$ and $v \in \mathcal{N} \cup \mathcal{A}$, u and v are not adjacent. Since $\mathcal{B} \subseteq \mathcal{A}_1$, for any $u \in \mathcal{B}$ and $v \in \mathcal{S}$, u and v are not adjacent. For any $u \in \mathcal{B}$ and $v \in \mathcal{A}$, u and v are not adjacent.

We partition $\mathcal{S}$ into $\mathcal{C}$, the set of vertices which induce special cycles and $\mathcal{P}$, the set of vertices which induce special paths. $\mathcal{P}$ is further partitioned into $\mathcal{P}_e$, the set of end points and $\mathcal{P}_i$, the set of interior points of all the paths. We define the **second end points** of a path of length at least 2 as the interior vertices of the path which are adjacent to at least one of its end points. The set of second end points of the paths in $\mathcal{P}$ is denoted by $\mathcal{P}_{2e}$ and $\mathcal{P}_{2i} = \mathcal{P}_i \setminus \mathcal{P}_{2e}$.

2.2 The graph induced by $\mathcal{S} \cup \mathcal{R} \cup \mathcal{B}$

For each $u \in \mathcal{R}$, let $\Gamma(u) = \{u\} \cup X_1(u) \cup X_2(u)$, where $X_1(\cdot)$ and $X_2(\cdot)$ are as defined in Algorithm 2.

OBSERVATION 3. *We have some observations regarding $X_1(\cdot)$ and $X_2(\cdot)$. We recall that $\mathcal{A}_1$ induces a collection of isolated vertices and edges. Let $v \in \mathcal{R}$: (1) $|X_1(v)| = 2$; (2) since $X_1(v) \cup X_2(v) \subseteq \mathcal{A}_1$, every vertex in $X_1(v) \cup X_2(v)$ has at most one neighbor in $\mathcal{A}_1$ and this neighbor is in $X_1(v) \cup X_2(v)$; (3) if the two vertices in $X_1(v)$ are adjacent, then, $X_2(v)$ is empty and (4) If $X_2(v)$ is not empty, then, every vertex in $X_2(v)$ is adjacent to exactly one vertex in $X_1(v)$ and $|X_2(v)| \leq 2$.*

From Algorithm 2 and Observation 3, one can show that $\Gamma(u)$ is a component in the graph induced by $\mathcal{R} \cup \mathcal{B}$ and it is isomorphic to one of the graphs illustrated in Figure 1. Recall that each vertex of $\mathcal{R}$ has a unique neighbor in $\mathcal{S}$. In fact, we can infer more:

OBSERVATION 4. *Let $u \in \mathcal{R}$. The unique vertex of $\mathcal{S}$ to which u is adjacent to belongs to $\mathcal{P}_{2i}$. The proof follows by contradiction. If u is adjacent to a vertex in $\mathcal{C} \cup \mathcal{P}_e \cup \mathcal{P}_{2e}$, then, it will imply that in Algorithm 1, a non-special cycle or path is extracted in some iteration.*

The above observation implies that (1) each vertex in $\mathcal{C} \cup \mathcal{P}_{2e}$ is adjacent to exactly one vertex in $\mathcal{N}$ (2) every vertex in $\mathcal{P}_e$ is adjacent to exactly two vertices in $\mathcal{N}$ and (3) every vertex in $\mathcal{P}_{2i}$ is adjacent to exactly one vertex in $\mathcal{R} \cup \mathcal{N} = \mathcal{N}_1$. From Figure 1, we note that every vertex in $\mathcal{B}$ is adjacent to either 1 or 2 vertices in $\mathcal{N}$. Let $\mathcal{B}_1$ be the set of vertices of $\mathcal{B}$ which have one neighbor in $\mathcal{N}$ and $\mathcal{B}_2$ be the set of vertices which have two neighbors in $\mathcal{N}$.

2.3 The graph induced by $\mathcal{B}_2 \cup \mathcal{P}_e \cup \mathcal{N}$

It can be easily verified that $\mathcal{B}_2 \cup \mathcal{P}_e$ is an independent set in G. Consider a vertex $v \in \mathcal{B}_2 \cup \mathcal{P}_e$. From the earlier discussions, it follows that v is adjacent to exactly two vertices in $\mathcal{N}$ and since it is not adjacent to any vertex in $\mathcal{B}_2 \cup \mathcal{P}_e$, its degree is 2 in $G[\mathcal{B}_2 \cup \mathcal{P}_e \cup \mathcal{N}]$. Since every vertex in $\mathcal{N}$ is adjacent to at least one vertex in $\mathcal{S}$, it follows that it is adjacent to at most two vertices in $\mathcal{B}_2 \cup \mathcal{P}_e \cup \mathcal{N}$. From these two observations, we can infer the following about the graph induced by $\mathcal{B}_2 \cup \mathcal{P}_e \cup \mathcal{N}$: (1) It is a collection of paths and cycles, (2) all the end points of paths (which also includes isolated vertices) belong to $\mathcal{N}$, and (3) a vertex in $\mathcal{N}$ is adjacent to a vertex in $\mathcal{A}$ only if it is an end point of a path. Based on these inferences, $\mathcal{N}$ is partitioned into $\mathcal{N}_e$, the set of end points of paths (which includes isolated vertices) and $\mathcal{N}_{int}$, the set of interior points of cycles and paths in $G[\mathcal{B}_2 \cup \mathcal{P}_e \cup \mathcal{N}]$.

Type 1 and Type 2 cycles.
We classify the cycles induced by $\mathcal{B}_2 \cup \mathcal{P}_e \cup \mathcal{N}_{int}$ in the following manner: Type 1 cycles whose vertices alternate between $\mathcal{N}$ and $\mathcal{B}_2 \cup \mathcal{P}_e$ and Type 2 cycles which are not Type 1. Since $\mathcal{B}_2 \cup \mathcal{P}_e$ induces an independent set, it is easy to infer that a Type 2 cycle has at least one pair of adjacent vertices which belong to $\mathcal{N}_{int}$. We recall that every vertex in $\mathcal{C} \cup \mathcal{P}_{2e}$ is adjacent to exactly one vertex in $\mathcal{N}$. The following lemma implies that any vertex in $\mathcal{N}$ which belongs to a Type 1 cycle cannot be adjacent to a vertex in $\mathcal{C} \cup \mathcal{P}_{2e}$. This is crucial to the construction of the first interval graph.

LEMMA 3. *If a vertex $v \in \mathcal{N}$ is adjacent to 2 vertices in $\mathcal{B}_2 \cup \mathcal{P}_e$, then its remaining neighbor belongs to $\mathcal{P}_{2i}$. In other words, v has no neighbor in $\mathcal{C} \cup \mathcal{P}_{2e}$.*

The proof is by contradiction; if v has a neighbor in $\mathcal{C} \cup \mathcal{P}_{2e}$, then, it will imply that in Algorithm 1, a non-special cycle

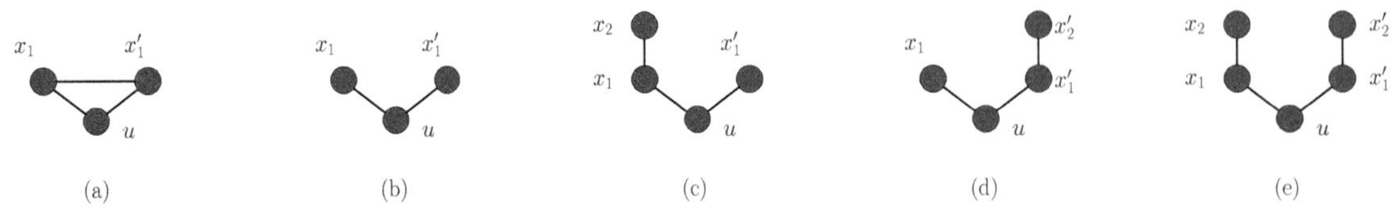

Figure 1: In the graph induced by $\mathcal{R} \cup \mathcal{B}$, each component $\Gamma(u)$, $u \in \mathcal{R}$ is isomorphic to one of the graphs illustrated in the figure. Here, u has exactly two neighbors in $\mathcal{B}$, x_1 and x_1'. These neighbors if not adjacent can each have at most one neighbor in $\mathcal{B}$ which are denoted by x_2 and x_2' respectively.

or path is extracted in some iteration. We have omitted it due to lack of space.

A summary.
The partitioning of the vertex set of G is illustrated in Figure 2. The relationship between the parts is summarized in Tables 1 and 2 so that the reader can easily recall the observations stated in this section.

1	$\mathcal{A}$	$\mathcal{C} \cup \mathcal{P}$
2	$\mathcal{A} \cup \mathcal{N}$	$\mathcal{R}$
3	$\mathcal{A}$	$\mathcal{B}$
4	$\mathcal{C}$	$\mathcal{P}$
5	$\mathcal{C} \cup \mathcal{P}_e \cup \mathcal{P}_{2e}$	$\mathcal{R}$
6	$\mathcal{C} \cup \mathcal{P}$	$\mathcal{B}$
7	$\mathcal{P}_e$	$\mathcal{P}_e$
8	$\mathcal{P}_e$	$\mathcal{P}_{2i}$
9	$\mathcal{N}_{int}$	$\mathcal{A}$
10	$\mathcal{R}$	$\mathcal{R}$

Table 1: Non-adjacency table: In every row, there is no edge between the set in the 1st column and the set in the 2nd column in G.

3. CONSTRUCTION OF A 3-BOX REPRESENTATION OF G

We define the three interval graphs I_1, I_2 and I_3. Let $n = |V|$, the number of vertices in G. While defining the intervals, we also make certain important structural observations of the concerned interval graph to help the reader visualize the construction and the subsequent verification

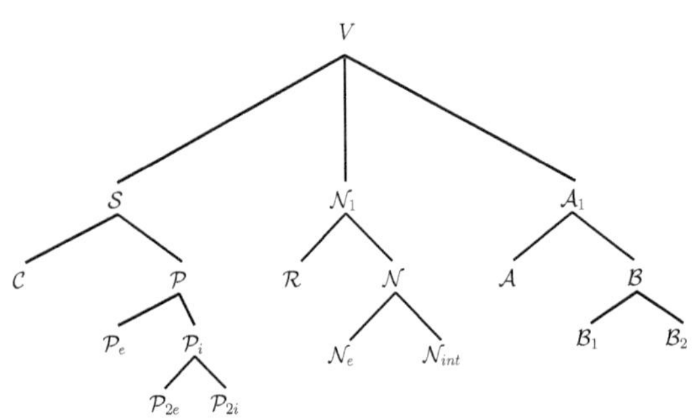

Figure 2: The partition of the vertex set V of the cubic graph.

1	$\mathcal{N}_e$	$\mathcal{A}$	at most one neighbor
2	$\mathcal{B}_1$	$\mathcal{N}$	exactly one neighbor
3	$\mathcal{C} \cup \mathcal{P}_{2e}$	$\mathcal{N}$	exactly one neighbor
4	$\mathcal{P}_{2i}$	$\mathcal{R} \cup \mathcal{N}$	exactly one neighbor
5	$\mathcal{R}$	$\mathcal{P}_{2i}$	exactly one neighbor

Table 2: Unique neighbor table: In every row, each vertex belonging to the set in the 1st column has either (i) at most one neighbor OR (ii) exactly one neighbor in the set given in the 2nd column.

process which has otherwise been omitted. For more details, we refer to [2].

3.1 Construction of I_1

Vertices of $\mathcal{A}$.
Note that $\mathcal{A} \subseteq \mathcal{A}_1$ and therefore, induces a collection of isolated vertices and edges in G. Let Π_A be an ordering of $\mathcal{A}$ which satisfies the condition that the two end points of every (isolated) edge are consecutively ordered. An isolated vertex u is given a point interval, $f(u, I_1) = [2n + \Pi_A(u), 2n + \Pi_A(u)]$. For the end points of an isolated edge (u, v), assuming without loss of generality that $\Pi_A(v) = \Pi_A(u) + 1$, the intervals assigned to u and v are $f(u, I_1) = [2n + \Pi_A(u), 2n + \Pi_A(u) + 0.5]$ and $f(v, I_1) = [2n + \Pi_A(v) - 0.5, 2n + \Pi_A(v)]$.

OBSERVATION 5. *From the construction it is easy to infer that $I_1[\mathcal{A}] = G[\mathcal{A}]$.*

OBSERVATION 6. *Let $x, y \in \mathcal{A}$ such that $\Pi_A(x) < \Pi_A(y)$. If they are adjacent in G, then, $r(x, I_1) = l(y, I_1)$. This implies that the boxes corresponding to x and y intersect only at their boundaries.*

Vertices of $\mathcal{B}_2 \cup \mathcal{P}_e \cup \mathcal{N}$.
We recall that each component induced by $G[\mathcal{B}_2 \cup \mathcal{P}_e \cup \mathcal{N}]$ is either a path or a cycle. We define an ordering on this set with the following structure: (1) The vertices are ordered componentwise in the sense that vertices of a component occur as a contiguous block; (2) within the component (which is either a path or a cycle), the vertices follow a natural order and (3) Type 1 cycles are ordered first followed by Type 2 cycles and finally by paths. A more precise definition is given below. The placement of intervals corresponding to the vertices of this set is determined by this order and it will help the reader to keep in mind that the construction for this set in I_1 is componentwise.

Definition 1. *Let Π_1 be an ordering of $\mathcal{B}_2 \cup \mathcal{P}_e \cup \mathcal{N}$ which satisfies the following properties. Let S be a component induced by $\mathcal{B}_2 \cup \mathcal{P}_e \cup \mathcal{N}$.*

1. *Suppose S is a path with at least two vertices. Then, for one of the natural orderings of the vertices of S, say $p_1 p_2 \dots p_t$, we have $\Pi_1(p_i) = \Pi_1(p_{i-1}) + 1$, $2 \le i \le t$.*
2. *If S is a cycle, then recall that it is either Type 1 or Type 2. Consider one of the natural orderings of the vertices of S, say $c_1 c_2 \dots c_t c_1$ such that if S is a Type 1 cycle, then $c_1 \in \mathcal{N}$ and if it is a Type 2 cycle, then $c_1, c_t \in \mathcal{N}$. Then, we have, $\Pi_1(c_i) = \Pi_1(c_{i-1}) + 1$, $2 \le i \le t$.*
3. *Let $\mathcal{B}_2 \cup \mathcal{P}_e \cup \mathcal{N} = \Lambda_{Type\ 1} \uplus \Lambda_{Type\ 2} \uplus \Lambda_{Paths}$ where, $\Lambda_{Type\ 1}$, $\Lambda_{Type\ 2}$ and Λ_{Paths} are the sets of vertices belonging to Type 1 cycles, Type 2 cycles and paths respectively. Then, we have $\Pi_1(\Lambda_{Type\ 1}) < \Pi_1(\Lambda_{Type\ 2}) < \Pi_1(\Lambda_{Paths})$.*

The interval assignments for the vertices of $\mathcal{B}_2 \cup \mathcal{P}_e \cup \mathcal{N}$ are as follows:

For a vertex in a Type 1 cycle:.
Let $S = c_1 c_2 \dots c_t c_1$ be a Type 1 cycle such that $\Pi_1(c_{i+1}) = \Pi_1(c_i) + 1$, $1 \le i < t$. The construction is such that $c_2, c_3, \dots, c_t$ induces a path while c_1 is adjacent to all the other vertices of the cycle.

$$f(c_1, I_1) = [\Pi_1(c_1), \Pi_1(c_t)]; \quad (1)$$
$$f(c_i, I_1) = [\Pi_1(c_i), \Pi_1(c_i) + 1], 1 < i < t; \quad (2)$$
$$f(c_t, I_1) = [\Pi_1(c_t), \Pi_1(c_t) + 0.5]. \quad (3)$$

Observation 7. *We have the following observations: (1) $I_1[S \backslash c_1] = G[S \backslash c_1]$ and (2) if two vertices belong to different Type 1 cycles, then, they are not adjacent in I_1.*

For a vertex in a Type 2 cycle.
Let $S = c_1 c_2 \dots c_t c_1$ be a Type 2 cycle such that $\Pi_1(c_{i+1}) = \Pi_1(c_i) + 1$, $1 \le i < t$. We recall from the definition of Π_1 that $c_1, c_t \in \mathcal{N}$. They are assigned intervals as follows:

$$f(c_1, I_1) = [n + \Pi_1(c_1), n + \Pi_1(c_t)], \quad (4)$$
$$f(c_t, I_1) = [n + \Pi_1(c_t), n + \Pi_1(c_t) + 0.5]. \quad (5)$$

The remaining vertices are assigned intervals as follows. For $1 < i < t$,

$$\text{if } c_i \in \mathcal{N},\ f(c_i, I_1) = [n + \Pi_1(c_i), n + \Pi_1(c_i) + 1], (6)$$
$$\text{if } c_i \in \mathcal{B}_2 \cup \mathcal{P}_e,\ f(c_i, I_1) = [n, n + \Pi_1(c_i) + 1]. \quad (7)$$

Note that for every vertex in the cycle the left end point is $\ge n$. We can visualize this construction as being obtained by using the following 3-step procedure. We first construct it exactly the way a Type 1 cycle is constructed, followed by shifting all the intervals to the right by n. Finally, we extend the intervals of vertices of $\mathcal{B}_2 \cup \mathcal{P}_e$ to the left to the common end point n.

For a vertex in a path.
Let $S = p_1 p_2 \dots p_t$ be a path such that $\Pi_1(c_{i+1}) = \Pi_1(c_i) + 1$, $1 \le i < t$. By Definition $p_1, p_t \in \mathcal{N}_e$. From Table 2 (row 1), they can be adjacent to at most one vertex in $\mathcal{A}$. Taking this into consideration, the interval assignments are as follows: Let $p \in \{p_1, p_t\} \subseteq \mathcal{N}_e$:

if p is not adjacent to any vertex in $\mathcal{A}$:
$$f(p, I_1) = [n + \Pi_1(p), 2n], \quad (8)$$
if p is adjacent to a vertex a in $\mathcal{A}$:
$$f(p, I_1) = [n + \Pi_1(p), l(a, I_1)]. (9)$$

If p is an interior point of the path, its interval assignment is as follows:

$$\text{if } p \in \mathcal{N}_{int},\ f(p, I_1) = [n + \Pi_1(p), n + \Pi_1(p) + 1], (10)$$
$$\text{if } p \in \mathcal{B}_2 \cup \mathcal{P}_e,\ f(p, I_1) = [n, n + \Pi_1(p) + 1]. \quad (11)$$

Observation 8. *We have some observations regarding the intervals assigned to vertices of $\mathcal{B}_2 \cup \mathcal{P}_e \cup \mathcal{N}$.*

1. *If z belongs to a Type 1 cycle, then, (a) $l(z, I_1) = \Pi_1(z)$ and (b) $1 \le l(z, I_1) < r(z, I_1) < n$.*
2. *If $z \in \mathcal{N}$ belongs to a Type 2 cycle or a path then, $l(z, I_1) = n + \Pi_1(z)$ and therefore, $n < l(z, I_1) < 2n - 1$.*
3. *If $z \in \mathcal{B}_2 \cup \mathcal{P}_e$ belongs to a Type 2 cycle or a path then, $l(z, I_1) = n$ and $r(z, I_1) = n + \Pi_1(z) + 1 < 2n$.*
4. *If $x, y \in \mathcal{N}$ such that $\Pi_1(x) < \Pi_1(y)$, then, $l(x, I_1) + 1 \le l(y, I_1)$.*

Observation 9. *If $v \in \mathcal{N}$ is adjacent to a vertex $x \in \mathcal{A}$, then, it is adjacent to x but not adjacent to any $y \in \mathcal{A}$ which satisfies $\Pi_A(y) > \Pi_A(x)$ in I_1. Moreover, $r(v, I_1) = l(x, I_1)$ implying that their boxes intersect only at their boundaries.*

Observation 10. *If x belongs to a Type 1 cycle and y belongs to a Type 2 cycle or a path, then, they are not adjacent in I_1.*

Vertices of $\mathcal{B}_1 \cup \mathcal{P}_{2e}$.
Let $v \in \mathcal{B}_1 \cup \mathcal{P}_{2e}$. From Table 2 (rows 2 and 3), we note that v has a unique neighbor in $\mathcal{N}$, say v'. Note that $f(v', I_1)$ is already defined.

$$\text{if } v \in \mathcal{B}_1, \text{ then, } f(v, I_1) = [0, l(v', I_1)], \quad (12)$$
$$\text{if } v \in \mathcal{P}_{2e}, \text{ then, } f(v, I_1) = [-1, l(v', I_1)]. \quad (13)$$

Observation 11. *Here, from Lemma 3 we observe that $r(v, I_1) = l(v', I_1) > n$ and therefore, every vertex in P_{2e} is adjacent to every vertex in P_e in I_1.*

Vertices of $\mathcal{R}$.
$\forall v \in \mathcal{R}, f(v, I_1) = [-1, n]$.

Vertices of $\mathcal{P}_{2i}$.
Suppose $v \in \mathcal{P}_{2i}$. Let v' be its unique neighbor in $\mathcal{R} \cup \mathcal{N}$ (see Table 2 row 4). Note that $f(v', I_1)$ is already defined.

$$\text{if } v' \in \mathcal{R}, \text{ then, } f(v, I_1) = [-1, -1], \quad (14)$$
$$\text{if } v' \in \mathcal{N}, \text{ then, } f(v, I_1) = [-1, l(v', I_1)]. \quad (15)$$

Observation 12. *If $v \in \mathcal{P}_{2i}$ and $x \in \mathcal{R} \cup \mathcal{N}$, then, $r(v, I_1) = l(x, I_1)$ and hence, their boxes touch only at their boundaries.*

Vertices of $\mathcal{C}$.

We recall that $\mathcal{C}$ induces a collection of cycles in G.

Definition 2. **Notation $\eta(\cdot)$ and special vertex:** *We recall from Table 2 (row 3) that every vertex $x \in \mathcal{C}$ has a unique neighbor in $\mathcal{N}$. We denote this neighbor by $\eta(x)$. We define a vertex $c \in C$ as the special vertex of C if $l(\eta(c), I_1) = \min_{c' \in C} l(\eta(c'), I_1)$. Note that $\eta(c)$ is already assigned an interval.*

Suppose C is a cycle in $\mathcal{C}$. Let $C = c_1c_2 \dots c_tc_1$ be a natural ordering of the vertices of C such that c_1 is the special vertex of C. The interval assignments are as follows:

$$f(c_1, I_1) = [l(\eta(c_1), I_1), l(\eta(c_1), I_1)],$$
$$f(c_i, I_1) = [l(\eta(c_1), I_1), l(\eta(c_i), I_1) + 0.5], i = 2, t, \quad (16)$$
$$f(c_i, I_1) = [l(\eta(c_1), I_1) + 0.5, l(\eta(c_i), I_1) + 0.5], \text{otherwise.} \quad (17)$$

Observation 13. *Let $C = c_1c_2 \dots c_tc_1$ be a cycle in $\mathcal{C}$ with c_1 being the special vertex. We observe the following: (1) c_1 is adjacent to only c_2 and c_t in I_1. Further, $r(c_1, I_1) = l(c_2, I_1) = l(c_t, I_1)$ and therefore, their boxes touch only at their boundaries; (2) $C \setminus \{c_1\}$ is a clique in I_1 and (3) for every vertex $c \in \mathcal{C}$, $l(\eta(c), I_1) \in f(c, I_1)$ and therefore, c is adjacent to $\eta(c)$ in I_1.*

We recall from Lemma 3 that any vertex $v \in \mathcal{C}$ is not adjacent to a vertex in $\mathcal{N}$ which belongs to a Type 1 cycle. Therefore, $l(v, I_1) > n$ and v is not adjacent to any vertex belonging to a Type 1 cycle in I_1. The following is easy to observe from the interval assignments.

Observation 14. *Consider $v \in \mathcal{B}_1 \cup \mathcal{P}_{2e} \cup \mathcal{P}_{2i} \cup \mathcal{C}$. If $v \in \mathcal{P}_{2i}$ and has no neighbor in $\mathcal{N}$, then it is not adjacent to any vertex in $\mathcal{N}$ in I_1. Otherwise, v has exactly one neighbor in $\mathcal{N}$, say x. For every $y \in \mathcal{N}$ such that $\Pi_1(y) > \Pi_1(x)$, v is not adjacent to y in I_1.*

Observation 15. *Let $x \in \mathcal{A}$ and $y \in V \setminus (\mathcal{A} \cup \mathcal{N})$. Then, $(x, y) \notin E(I_1)$. This follows from the interval assignments; $r(y, I_1) \leq 2n$ while $l(x, I_1) > 2n$.*

3.2 Construction of I_2

Vertices of $\mathcal{A}$.

We recall the interval assignment for $\mathcal{A}$ in I_1. Let $\overline{\Pi_A}$ be the reverse of Π_A. The interval assignments for vertices of $\mathcal{A}$ in I_2 is identical to that of I_1 with Π_A replaced by $\overline{\Pi_A}$.

Vertices of $\mathcal{N}$.

Let $v \in \mathcal{N}$. From Table 2 (row 1), v is adjacent to at most one vertex in $\mathcal{A}$.

$$\text{if } v \text{ is not adjacent to any vertex in } \mathcal{A}, f(v, I_2) = [0, n], \quad (18)$$
$$\text{if } v \text{ is adjacent to vertex } a \text{ in } \mathcal{A}, f(v, I_2) = [0, l(a, I_2)]. \quad (19)$$

Observation 16. *If $v \in \mathcal{N}$ is adjacent to a vertex $x \in \mathcal{A}$, then, it is adjacent to x but not adjacent to any $y \in \mathcal{A}$ which satisfies $\overline{\Pi_A}(y) > \overline{\Pi_A}(x)$. Combining with Observation 9, it follows that if $v \in \mathcal{N}$ and $y \in \mathcal{A}$, such that $(u, v) \notin E(G)$, then, $(u, v) \notin E(I_1 \cap I_2)$.*

Vertices of $\mathcal{C} \cup \mathcal{P}$.

We recall that $\mathcal{C} \cup \mathcal{P}$ induces a collection of cycles and paths in G.

Definition 3. *Π_2 is an ordering of $\mathcal{C} \cup \mathcal{P}$ such that the following properties are satisfied:*

1. *Let P be a path in $\mathcal{P}$. For a natural ordering of P, say $p_1p_2 \dots p_t$, we have $\Pi_2(p_i) = \Pi_2(p_{i-1}) + 1$, $2 \leq i \leq t$.*
2. *Suppose C is a cycle in $\mathcal{C}$. For a natural ordering of C, say $c_1c_2 \dots c_tc_1$, where c_1 is the special vertex (recall Definition 2), we have $\Pi_2(c_i) = \Pi_2(c_{i-1}) + 1$, $2 \leq i \leq t$.*

For the vertices of a path.

Suppose $P = p_1p_2 \dots p_t$ such that $\Pi_2(p_{i+1}) = \Pi_2(p_i) + 1$, $1 \leq i < t$.

$$\begin{aligned} f(p_i, I_2) &= [\Pi_2(p_i), \Pi_2(p_i) + 1], 1 \leq i < t, \\ f(p_t, I_2) &= [\Pi_2(p_t), \Pi_2(p_t) + 0.5]. \end{aligned} \quad (20)$$

For the vertices of a cycle.

Suppose $C = c_1c_2 \dots c_tc_1$ such that $\Pi_2(c_{i+1}) = \Pi_2(c_i) + 1$, $1 \leq i < t$.

$$\begin{aligned} f(c_1, I_2) &= [\Pi_2(c_1), \Pi_2(c_t)], \\ f(c_i, I_2) &= [\Pi_2(c_i), \Pi_2(c_i) + 1], 1 < i < t, \\ f(c_t, I_2) &= [\Pi_2(c_t), \Pi_2(c_t) + 0.5]. \end{aligned} \quad (21)$$

Observation 17. *From Observation 13 and the definitions of the intervals of (20) and (21), it follows that the graph induced by $\mathcal{C} \cup \mathcal{P}$ in G and $I_1 \cap I_2$ are identical, that is, $G[\mathcal{C} \cup \mathcal{P}] = (I_1 \cap I_2)[\mathcal{C} \cup \mathcal{P}]$.*

Observation 18. *Suppose $C = c_1c_2 \dots c_tc_1$ is a special cycle from $\mathcal{C}$ such that $\Pi_2(c_{i+1}) = \Pi_2(c_i) + 1$, $1 \leq i < t$. For $2 \leq i \leq t$, $l(c_{i+1}, I_2) = r(c_i, I_2)$ and in I_1, $r(c_1, I_1) = l(c_2, I_1) = l(c_t, I_1)$. Therefore, the boxes of adjacent vertices in the cycle intersect only at their boundaries. Suppose $P = p_1p_2 \dots p_t$ is a special path from $\mathcal{P}$ such that $\Pi_2(p_{i+1}) = \Pi_2(p_i) + 1$, $1 \leq i < t$. Then, for $1 < i \leq t$, $l(p_i, I_2) = r(p_{i-1}, I_2)$ and hence, the boxes of adjacent vertices intersect only at their boundaries.*

Vertices of $\mathcal{R} \cup \mathcal{B}$.

We recall that each component in $\mathcal{R} \cup \mathcal{B}$ is isomorphic to one of the graphs shown in Figure 1. Further, each component contains exactly one vertex from $\mathcal{R}$ and is uniquely identified by it; By the notation introduced in the figure, for every $u \in \mathcal{R}$, $\Gamma(u)$ denotes the component containing u in $G[\mathcal{R} \cup \mathcal{B}]$.

Definition 4. **$\beta(\cdot)$:** *In the graph induced by $\mathcal{R} \cup \mathcal{B}$, consider each component $\Gamma(u)$, $u \in \mathcal{R}$. From Table 2 (row 5), u is adjacent to a unique vertex in $\mathcal{P}_{2i}$. We denote this vertex by $\beta(u)$.*

Let us consider a component of $G[\mathcal{R} \cup \mathcal{B}]$, say $\Gamma(u)$, $u \in \mathcal{R}$. From (20), we note that $\beta(u)$ is assigned a unit interval in I_2. The interval assignments for the vertices of $\Gamma(u)$ is illustrated in Figure 3. Note that every vertex of $\Gamma(u)$ is assigned a strict sub-interval of $f(\beta(u), I_2)$ and none of these intervals contains any end point of $f(\beta(u), I_2)$.

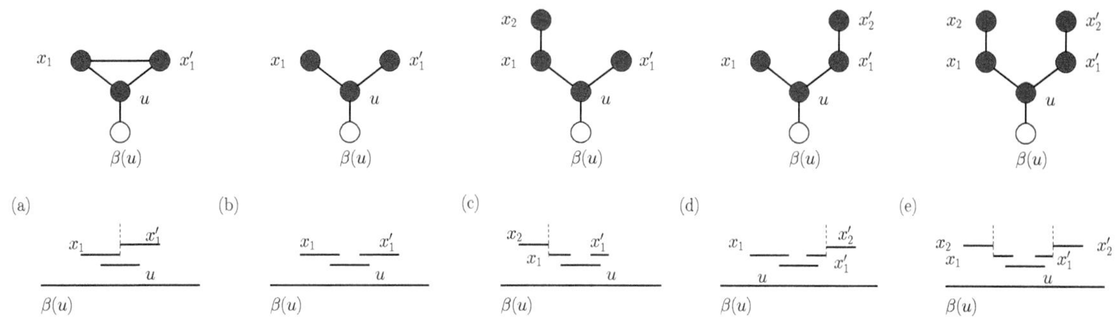

Figure 3: The interval assignments for each component $\Gamma(u)$, $u \in \mathcal{R}$ induced by $\mathcal{R} \cup \mathcal{B}$ in the interval graph I_2. The dotted vertical lines are used to indicate that the concerned intervals intersect exactly at their end points.

OBSERVATION 19. *Let $z \in \mathcal{R}$ and $\Gamma(z)$ be a component of $G[\mathcal{R} \cup \mathcal{B}]$. We have the following observations: (1) Every vertex in $\Gamma(z)$ is adjacent to only one vertex in $\mathcal{C} \cup \mathcal{P}$ and that is $\beta(z)$; (2) the graph induced by $\mathcal{R} \cup \mathcal{B}$ in I_2 and in G are identical, that is, $I_2[\mathcal{R} \cup \mathcal{B}] = G[\mathcal{R} \cup \mathcal{B}]$; (3) for any $x \in \mathcal{R} \cup \mathcal{B}$ and $y \in \mathcal{C} \cup \mathcal{P}_e \cup \mathcal{P}_{2e}$, $(x, y) \notin E(I_2)$ and (4) for any $x, y \in \mathcal{B}$ which are adjacent, either $l(x, I_2) = r(y, I_2)$ or $l(y, I_2) = r(x, I_2)$ implying that there boxes intersect only at their boundaries.*

3.3 Construction of I_3

Vertices of $\mathcal{B}_2 \cup \mathcal{P}_e \cup \mathcal{N}$.

We recall the notations developed for this set in I_1. The interval assignment in this set is very similar to that in I_1.

For a vertex in a Type 1 cycle.

Let $S = c_1c_2 \ldots c_tc_1$ be a Type 1 cycle such that $\Pi_1(c_{i+1}) = \Pi_1(c_i)+1$, $1 \le i < t$. We recall that $c_1 \in \mathcal{N}$ and $c_t \in \mathcal{B}_2 \cup \mathcal{P}_e$. The interval assignments are as follows: For $1 \le i < t$,

$$\text{if } c_i \in \mathcal{N}, \text{ then, } f(c_i, I_3) = [\Pi_1(c_i), \Pi_1(c_i) + 1], \quad (22)$$
$$\text{if } c_i \in \mathcal{B}_2, \text{ then, } f(c_i, I_3) = [\Pi_1(c_i), n], \quad (23)$$
$$\text{if } c_i \in \mathcal{P}_e, \text{ then, } f(c_i, I_3) = [\Pi_1(c_i), n + 1]. \quad (24)$$

The interval assignment for c_t is as follows:

$$\text{if } c_t \in \mathcal{B}_2, \text{ then, } f(c_t, I_3) = [\Pi_1(c_1) + 1, n], \quad (25)$$
$$\text{if } c_t \in \mathcal{P}_e, \text{ then, } f(c_t, I_3) = [\Pi_1(c_1) + 1, n + 1], \quad (26)$$

OBSERVATION 20. *Let $x \in \mathcal{N}$ and $y \in \mathcal{B}_2 \cup \mathcal{P}_e \cup \mathcal{N}$ be such that $x \in S_x$ and $y \in S_y$ where both S_x and S_y induce Type 1 cycles. If $(x, y) \notin E(G)$, then $(x, y) \notin E(I_1 \cap I_3)$. The proof is as follows: If S_x and S_y are distinct, then, by Observation 7(2), $(x, y) \notin E(I_1)$. Otherwise, by Observation 7(1) and the interval assignments in I_3, it follows that $(x, y) \notin E(I_1 \cap I_3)$.*

For a vertex in a Type 2 cycle.

Let $S = c_1c_2 \ldots c_tc_1$ be a Type 2 cycle such that $\Pi_1(c_{i+1}) = \Pi_1(c_i) + 1$, $1 \le i < t$. We recall from Definition 1 (Point 2) that $c_1, c_t \in \mathcal{N}$. They are assigned intervals as follows:

$$f(c_1, I_3) = [\Pi_1(c_1), \Pi_1(c_1) + 1], \quad (27)$$
$$f(c_t, I_3) = [\Pi_1(c_1) + 1, \Pi_1(c_t) + 0.5], \quad (28)$$

For c_i, $1 < i < t$,

$$\text{if } c_i \in \mathcal{N}, \text{ then, } f(c_i, I_3) = [\Pi_1(c_i), \Pi_1(c_i) + 1], \quad (29)$$
$$\text{if } c_i \in \mathcal{B}_2, \text{ then, } f(c_i, I_3) = [\Pi_1(c_i), n], \quad (30)$$
$$\text{if } c_i \in \mathcal{P}_e, \text{ then, } f(c_i, I_3) = [\Pi_1(c_i), n + 1]. \quad (31)$$

For a vertex in a path.

Let $S = p_1p_2 \ldots p_t$ be a path such that $\Pi_1(p_{i+1}) = \Pi_1(p_i) +1$, $1 \le i < t$. We recall that $p_1, p_t \in \mathcal{N}_e$. The interval assignment for p_t is as follows:

$$f(p_t, I_3) = [\Pi_1(p_t), \Pi_1(p_t) + 0.5]. \quad (32)$$

The interval assignment for p_i, $i < t$ is as follows:

$$\text{if } p_i \in \mathcal{N}, \text{ then, } f(p_i, I_3) = [\Pi_1(p_i), \Pi_1(p_i) + 1], \quad (33)$$
$$\text{if } p_i \in \mathcal{B}_2, \text{ then, } f(p_i, I_3) = [\Pi_1(p_i), n], \quad (34)$$
$$\text{if } p_i \in \mathcal{P}_e, \text{ then, } f(p_i, I_3) = [\Pi_1(p_i), n + 1]. \quad (35)$$

OBSERVATION 21. *Here are some observations regarding the intervals assigned to vertices of $\mathcal{B}_2 \cup \mathcal{P}_e \cup \mathcal{N}$.*

1. *If $z \in \mathcal{N}$, $\Pi_1(z) + 0.5 \le r(z, I_3) \le \Pi_1(z) + 1 < n$.*
2. *If $z \in \mathcal{B}_2 \cup \mathcal{P}_e$ belongs to a Type 2 cycle or a path, then, $l(z, I_3) = \Pi_1(z) < n - 1$.*
3. *If $z \in \mathcal{B}_2$, $r(x, I_3) = n$.*
4. *If $z \in \mathcal{P}_e$, $r(x, I_3) = n + 1$.*

OBSERVATION 22. *Let $x \in \mathcal{N}$ and $y \in \mathcal{B}_2 \cup \mathcal{P}_e \cup \mathcal{N}$ be such that $x \in S_x$ and $y \in S_y$ where S_x and S_y induce a Type 2 cycle or a path. If $(x, y) \notin E(G)$, then $(x, y) \notin E(I_1 \cap I_3)$.*

OBSERVATION 23. *If $x \in \mathcal{N}$ and $y \in \mathcal{B}_2 \cup \mathcal{P}_e \cup \mathcal{N}$ such that $(x, y) \notin E(G)$, then, $(x, y) \notin E(I_1 \cap I_3)$. This follows from Observations 10, 20 and 22.*

Vertices of $\mathcal{B}_1 \cup \mathcal{C} \cup \mathcal{P}_{2e}$.

Let $v \in \mathcal{B}_1 \cup \mathcal{C} \cup \mathcal{P}_{2e}$. By Table 2 (rows 2 and 3), it follows that v is adjacent to exactly one vertex in $\mathcal{N}$; let v' be this vertex.

$$\text{If } v \in \mathcal{C} \cup \mathcal{P}_{2e}, \text{ then, } f(v, I_3) = \left[r(v', I_3), n + 1\right], \quad (36)$$
$$\text{If } v \in \mathcal{B}_1, \text{ then, } f(v, I_3) = \left[r(v', I_3), n\right], \quad (37)$$

Note that the interval for v' is already defined.

Vertices of $\mathcal{P}_{2i}$.
Let $v \in \mathcal{P}_{2i}$. By Table 2 (row 4), v has a unique neighbor in $\mathcal{R} \cup \mathcal{N}$; let v' be this vertex.

$$\text{if } v' \in \mathcal{R}, \text{ then, } f(v, I_3) = [n+1, n+1], \quad (38)$$
$$\text{if } v' \in \mathcal{N}, \text{ then, } f(v, I_3) = \left[r(v', I_3), n+1\right]. \quad (39)$$

OBSERVATION 24. *Let $x \in \mathcal{P}_{2i}$ and $y \in \mathcal{B}$. $(x, y) \notin E(I_2 \cap I_3)$.*

Vertices of $\mathcal{R}$.
$\forall v \in \mathcal{R}, f(v, I_3) = [n, n+1]$.

OBSERVATION 25. *If $x \in \mathcal{R}$ and $y \in \mathcal{B}$ are adjacent in G, then, $l(x, I_3) = r(y, I_3) = n$ and therefore, their boxes intersect only at their boundaries.*

Vertices of $\mathcal{A}$.
$\forall v \in \mathcal{A}, f(v, I_3) = [1, n+1]$.

OBSERVATION 26. *Let $x \in \mathcal{N}$ and $y \in V \setminus \mathcal{A}$ such that $(x, y) \in E(G)$. Then, for some $I \in \{I_1, I_3\}$, either $l(x, I) = r(y, I)$ or $l(y, I) = r(x, I)$.*

This completes the construction of the 3 interval graphs. In the following section, we consider the computational complexity of constructing this representation.

4. ALGORITHMIC ASPECTS

Our construction of the 3-box representation can be realized in $O(n)$ time, where n is the number of vertices in the graph. Firstly, we note that the process can be split into three stages: (1) Partitioning the vertex set as illustrated in Figure 2, (2) ordering the vertices of $\mathcal{A}$, $\mathcal{B}_2 \cup \mathcal{P}_e \cup \mathcal{N}$ and $\mathcal{C} \cup \mathcal{P}$ and finally, (3) assigning intervals to all the vertices. In Stage (1) the first step is to extract special cycles and special paths as described in Algorithm 1. This is the only non-trivial part of the construction. We will show that a special cycle or path can be extracted from a graph with maximum degree 3 in time linear to the number of vertices in it. This will imply that Algorithm 1 can be implemented in $O(n)$ time. Algorithm 2 takes $O(n)$ time since in every iteration we need to only check if a vertex in $\mathcal{N}_1$ has two neighbors in $\mathcal{A}_1$ in that iteration and accordingly move or retain the vertex and its neighbors. It is easy to see that the finer partitioning of $\mathcal{P}$, $\mathcal{N}$ and $\mathcal{B}$ can be accomplished in linear time. Stage (2) involves ordering the vertices of sets $\mathcal{A}$, $\mathcal{B}_2 \cup \mathcal{P}_e \cup \mathcal{N}$ and $\mathcal{C} \cup \mathcal{P}$ component wise. Since each of these sets induces a graph of maximum degree 2, they can be ordered in linear time. Stage (3) only involves assignment of intervals to the vertices and can be achieved in linear time. Now we will analyze the complexity of Algorithm 1.

4.1 Extending a non-special induced cycle

Suppose C is a non-special induced cycle in G. Then, by the definition of special cycle, it follows that there exists a vertex $x \in C$, such that $C \setminus \{x\}$ belongs to a cycle or path of length $|C| + 1$. We call such a vertex a **removable vertex**. Extending a non-special induced cycle C corresponds to removing a removable vertex x and adding two new vertices a and b such that $(C \setminus \{x\}) \cup \{a, b\}$ is an induced cycle or a path. There are only two possible ways in which a non-special induced cycle can be extended and these are illustrated in Figure 4. The vertices which are added to or removed from C in an extension operation are called **participating vertices**. In the figure, x, a and b are the participating vertices.

To verify if x is a removable vertex, we need to only check if the vertices a and b exist. Similarly, given that x is a removable vertex, we need to only find a and b to extend C by removing x. Recalling that $\Delta(G) \leq 3$, it is easy to see that this can be done in constant time. Hence, we have the following lemma.

LEMMA 4. *Suppose C is a non-special induced cycle and $x \in C$. It takes constant time to verify whether x is a removable vertex or not. If x is a removable vertex, then, the extension of C by removing x can again be achieved in constant time.*

4.2 Extending a non-special induced path

Let P be a non-special induced path in G. By the definition of special path, it implies that either: (1) it is not maximal in the sense that it is part of an induced cycle or a longer induced path, or (2) for some end point of P, say x, $P \setminus \{x\}$ belongs to an induced cycle of size $\geq |P|$ or an induced path of length $\geq |P| + 1$. We call x a **removable end point** of P.

Extending a non-special path P corresponds to the following operations: (1) If P is not maximal, then, we add a new vertex y to P such that $P \cup \{y\}$ is an induced path or cycle. Clearly, y must be a neighbor of an end point of P such that it is not adjacent to any interior vertex of P and (2) if P is maximal, then, we remove a removable end point x and either

(a) add a single new vertex a such that $(P \setminus \{x\}) \cup \{a\}$ is an induced cycle (note that in this case a has to be adjacent to the neighbor of x in P) OR

(b) add two new vertices a and b such that $(P \setminus \{x\}) \cup \{a, b\}$ is an induced cycle or a path (in this case, a and b are adjacent and a is adjacent to the neighbor of x in P).

This is illustrated in Figure 5.

As in the case of extending a cycle, the vertices which are added to or removed from P are called **participating vertices**. In case 1, y is the participating vertex. In case 2(a), x and a are participating vertices and in case 2(b), x, a and b are participating vertices.

To check if P is special, we need to first check if P is maximal or not, that is, whether it is part of a larger induced cycle or a longer induced path. This can be done in constant time. If P is maximal, then, we need to check if there is a removable end point and then extend P by removing it. For this, as shown in Figure 5, we need to check if the vertices x, a and b exist. Since $\Delta(G) \leq 3$, it is easy to see that this can be done in constant time. It is also trivial to verify that the extension can be achieved in constant time. Hence, we have the following lemma.

LEMMA 5. *If P is an induced path, then, in constant time, it can be verified whether it is a special path or not. If not, then, in constant time it can be extended.*

4.3 An algorithm to find special cycle or path

We now give an iterative algorithm to obtain a special cycle or path. The outline of the algorithm is as follows: Let

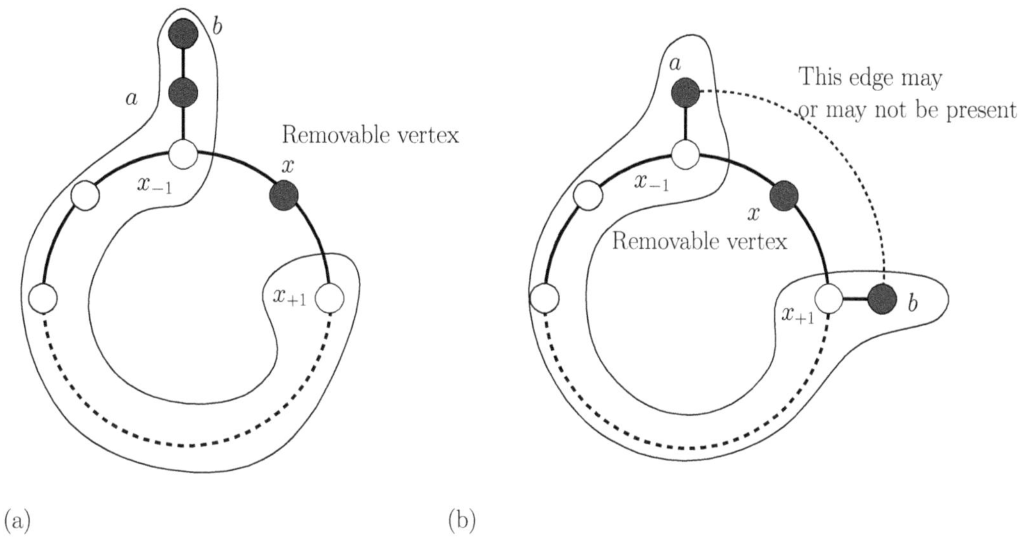

Figure 4: The possible ways in which a non-special induced cycle can be extended. The vertices marked black are the participating vertices.

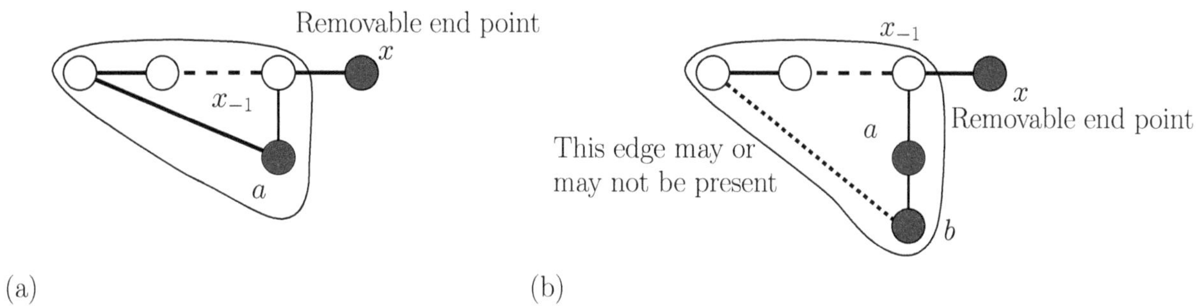

Figure 5: The ways in which a non-special induced path can be extended. The vertices marked black are the participating vertices.

S be the set which holds the vertices of the special cycle or path at the termination of the algorithm. We start with S containing an arbitrary vertex. In each iteration, we extend it as described in Sections 4.1 and 4.2. The algorithm terminates when S induces a special cycle or path. In Algorithm 3, we present an outline of this procedure.

4.4 Potential removable vertices

From Lemma 5, we note that in constant time we can recognize a non-special induced path and extend it. However, in view of Lemma 4, to recognize and extend a non-special induced cycle in constant time, we first need a strategy to find a removable vertex. For efficiently finding removable vertices, we maintain a list of potential removable vertices which is updated in each iteration.

DEFINITION 5. **Potential removable vertices:** *A vertex $x \in S$ is a potential removable vertex if it has two neighbors in S, say x_{-1} and x_{+1} and satisfies at least one of the following conditions: There are two vertices a and b such that*

1. *a is adjacent to only x_{-1} (or x_{+1}) in $S \setminus \{x\}$ and b is adjacent to a and not any vertex in $S \setminus \{x\}$. This corresponds to case (a) in Figure 4; OR*
2. *a is adjacent to x_{-1} in $S \setminus \{x\}$ and b is adjacent to only x_{+1} in $S \setminus \{x\}$. This corresponds to case (b) in Figure 4.*

From the definition of removable vertices at the beginning of Section 4.1, we infer the following:

OBSERVATION 27. *If S induces a cycle, then, all potential removable vertices in S will correspond to removable vertices. If there are no potential removable vertices, then, it implies that S is a special cycle.*

LEMMA 6. *In constant time, we can check if a particular vertex in S is a potential removable vertex.*

The proof is similar to that of Lemma 4.

In the algorithm, we maintain a set of all potential removable vertices, which we denote as R. From Observation 27, it follows that if S induces a cycle, we can decide whether it is special or not by just checking if R is empty or not. Therefore, given an induced cycle and the corresponding R,

```
Algorithm 3:
input : Graph G with maximum degree 3
output: Set S ⊆ V(G) which induces a special cycle or
        path in G
1  Let S = {x} where x is an arbitrary vertex;
2  Let R = ∅, the set of potential removable vertices of S;
3  Let specialFlag = 0; // If set to 1, it implies
       that S is a special cycle or path
4  while specialFlag = 0 do
5      if S induces a cycle then
6          if R = ∅ then
7              specialFlag = 1; // No removable
                   vertices and therefore, S is a
                   special cycle (see Observation 27)
8          else
9              Choose any vertex x from R;
10             Extend S by removing x as described in
               Section 4.1;
11         end
12     else if S induces a path then
13         if S is a special path then
14             specialFlag = 1;
15         else
16             Extend S as described in Section 4.2;
17         end
18     end
19     Update R as described in Section 4.4;
20 end
```

we can recognize in constant time whether it is a special cycle or not. Now, we show that after each extension of S, R can be updated in constant time.

LEMMA 7. *In each iteration of Algorithm 3, the set of potential removable vertices, R can be updated in constant time.*

PROOF. Let us consider a vertex $x \in R$ before extension of S. Let X denote the set containing x and the associated vertices x_{-1}, x_{+1}, a and b (see Definition 5). Since S is extended, there are participating vertices. We observe that x will remain as a potential removable vertex if no vertex in X and no neighbor of X is a participating vertex. This implies that if a vertex is at a distance 5 or more from any participating vertices, then, clearly its status as a potential removable vertex or not remains unchanged. Therefore, we need to only check all the vertices at a distance 4 or less from each participating vertex. The number of such vertices is a constant since $\Delta(G) = 3$ and by Lemma 6 verifying for each vertex takes only constant time. Hence proved. □

LEMMA 8. *If the special cycle or path extracted by Algorithm 3 is of size l, then, the total number of iterations required is at most $2l + 2$.*

PROOF. We will prove the lemma by showing that in Algorithm 3, for every two iterations (excluding the last two) the size of S increases by at least 1. If S induces a cycle at the beginning of the ith iteration, then from Section 4.1, it follows that S is extended to a cycle or path of size $|S| + 1$ at the end of the iteration. If S induces a path at the beginning of the ith iteration, either S is extended to a cycle or path of size $|S| + 1$ or to a cycle of size $|S|$ at the end of the iteration. In the latter case, assuming S is not a special cycle in the $(i + 1)$th iteration, it is extended to a cycle or path of size $|S| + 1$ at the end of the $(i + 1)$th iteration. Hence, proved. □

From Lemmas 4–7, it follows that Lines 10, 16 and 19 in Algorithm 3 require constant time. From Lemma 8, the number of iterations is bounded by $2l + 2$. Therefore, the algorithm takes $O(l)$ time to terminate. Since the total number of vertices in the set of special cycles and paths will be bounded above by n, the overall running time of Algorithm 1 is $O(n)$.

5. CONCLUSION

We showed that every graph of maximum degree 3 has a 3-box representation and therefore, its boxicity is at most 3. One interesting question which remains open is whether we can characterize cubic graphs which have a 2-box representation. Answering this will also determine if the boxicity of a cubic graph can be computed in polynomial time. Another question is whether the proof techniques employed in this paper can be used to improve the bounds on the boxicity of graphs with maximum degree 4.

6. ACKNOWLEDGMENTS

Abhijin Adiga has been partially supported by the following grants: DTRA R&D Grant HDTRA1-09-1-0017, DTRA CNIMS Contract HDTRA1-11-D-0016-000, DTRA Grant HDTRA1-11-1-0016, DOE Grant DE-SC0003957, NSF NetSE Grant CNS-1011769, NSF OCI-1032677.

7. REFERENCES

[1] A. Adiga, D. Bhowmick, L. S. Chandran, Boxicity and poset dimension., SIAM J. Discrete Math. 25 (4) (2011) 1687–1698.

[2] A. Adiga, L. S. Chandran, Representing a cubic graph as the intersection graph of axis-parallel boxes in three dimensions, archived version of this paper. URL `http://arxiv.org/abs/1108.5635`

[3] L. S. Chandran, M. C. Francis, N. Sivadasan, Boxicity and maximum degree, J. Comb. Theory Ser. B 98 (2) (2008) 443–445.

[4] L. Esperet, Boxicity of graphs with bounded degree, European J. Combin. 30 (5) (2009) 1277–1280.

[5] S. Felsner, M. C. Francis, Contact representations of planar graphs with cubes, in: Proc. 27th Annu. ACM Sympos. Comput. Geom. (SoCG 2011), 2011.

[6] F. S. Roberts, Recent Progresses in Combinatorics, chap. On the boxicity and cubicity of a graph, Academic Press, New York, 1969, pp. 301–310.

[7] E. R. Scheinerman, Intersection classes and multiple intersection parameters, Ph.D. thesis, Princeton University (1984).

[8] E. R. Scheinerman, D. B. West, The interval number of a planar graph: Three intervals suffice, J. Comb. Theory Ser. B 35 (1983) 224–239.

[9] C. Thomassen, Interval representations of planar graphs, J. Comb. Theory Ser. B 40 (1986) 9–20.

The 2-page Crossing Number of K_n

[Extended Abstract]

Bernardo M. Ábrego
California State University
Northridge, CA, USA
bernardo.abrego@csun.edu

Oswin Aichholzer
Graz University of Technology
Graz, Austria
oaich@ist.tugraz.at

Silvia Fernández-Merchant
California State University
Northridge, CA, USA
silvia.fernandez@csun.edu

Pedro Ramos
Universidad de Alcalá
Alcalá de Henares, Spain
pedro.ramos@uah.es

Gelasio Salazar
Universidad Autónoma de San Luis Potosí
San Luis Potosí, México
gsalazar@ifisica.uaslp.mx

ABSTRACT

Around 1958, Hill conjectured that the crossing number $\mathrm{cr}(K_n)$ of the complete graph K_n is

$$Z(n) := \frac{1}{4}\left\lfloor\frac{n}{2}\right\rfloor\left\lfloor\frac{n-1}{2}\right\rfloor\left\lfloor\frac{n-2}{2}\right\rfloor\left\lfloor\frac{n-3}{2}\right\rfloor$$

and provided drawings of K_n with exactly $Z(n)$ crossings. Towards the end of the century, substantially different drawings of K_n with $Z(n)$ crossings were found. These drawings are *2-page book drawings*, that is, drawings where all the vertices are on a line ℓ (the spine) and each edge is fully contained in one of the two half-planes (pages) defined by ℓ. The *2-page crossing number* of K_n, denoted by $\nu_2(K_n)$, is the minimum number of crossings determined by a 2-page book drawing of K_n. Since $\mathrm{cr}(K_n) \le \nu_2(K_n)$ and $\nu_2(K_n) \le Z(n)$, a natural step towards Hill's Conjecture is the weaker conjecture $\nu_2(K_n) = Z(n)$, that was popularized by Vrt'o. In this paper we develop a novel and innovative technique to investigate crossings in drawings of K_n, and use it to prove that $\nu_2(K_n) = Z(n)$. To this end, we extend the inherent geometric definition of k-edges for finite sets of points in the plane to topological drawings of K_n. We also introduce the concept of $\le\le k$-edges as a useful generalization of $\le k$-edges. Finally, we extend a powerful theorem that expresses the number of crossings in a rectilinear drawing of K_n in terms of its number of k-edges to the topological setting.

Categories and Subject Descriptors

G.2.2 [**Mathematics of Computing**]: Discrete Mathematics—*Graph Theory*

General Terms

Theory

Keywords

Crossing number, Topological drawing, Complete graph

1. INTRODUCTION

In a *drawing* of a graph G in the plane, each vertex is represented by a point and each edge $e = uv$ is represented by a simple curve, in such a way that the endpoints of this simple curve are the points representing u and v. A crossing in a drawing D of a graph G is an ordered pair $(x, \{e_1, e_2\})$ consisting of a non-vertex-point x on the plane and distinct edges e_1, e_2 of G whose representing curves both contain x. The *crossing number* $\mathrm{cr}(D)$ of D is the number of crossings in D, and the *crossing number* $\mathrm{cr}(G)$ of G is the minimum $\mathrm{cr}(D)$, taken over all drawings D of G.

A drawing is *good* if (i) no edge contains vertices different from its extreme points; (ii) no three edges meet at a point other than a vertex; (iii) two edges share at most one point (including vertices); and (iv) if two edges share a point and they are not both incident to the same vertex, then they cross at that point; that is, such a common point is a *crossing*, rather than tangential. It is well-known (and easy to prove) that every graph has a crossing-minimal drawing which is good. Thus, when our aim (as in this paper) is to estimate the crossing number of a graph, we may assume that all drawings under consideration are good.

Around 1958, Hill conjectured that

$$\mathrm{cr}(K_n) = Z(n) := \frac{1}{4}\left\lfloor\frac{n}{2}\right\rfloor\left\lfloor\frac{n-1}{2}\right\rfloor\left\lfloor\frac{n-2}{2}\right\rfloor\left\lfloor\frac{n-3}{2}\right\rfloor. \quad (1)$$

This conjecture appeared in print a few years later in papers by Guy [14] and Harary and Hill [15]. Hill described drawings of K_n with $Z(n)$ crossings, which were later corroborated by Blažek and Koman [4]. These drawings show that $\mathrm{cr}(K_n) \le Z(n)$. The best known general lower bound is $\lim_{n\to\infty} \mathrm{cr}(K_n)/Z(n) \ge 0.8594$, due to de Klerk et al. [10]. For more on the history of this problem we refer the reader to the excellent survey by Beineke and Wilson [2].

One of the major motivations for investigating crossing numbers is their application to VLSI design. With this motivation in mind, Chung, Leighton and Rosenberg [6] analyzed embeddings of graphs in books: the vertices lie on a line (the *spine*) and the edges lie on the *pages* of the book. Book embeddings of graphs have been extensively studied [3, 11]. Now if the book has k pages, and crossings among edges are allowed, the result is a *k-page book drawing.*

Here we concentrate on 2-page book drawings. The *2-page crossing number* $\nu_2(G)$ of a graph G is the minimum of $\mathrm{cr}(D)$ taken over all 2-page book drawings D of G. Alternative terminologies for the 2-page crossing number are *circular crossing number* [16] and *fixed linear crossing number* [7]. We may regard the pages as the closed half-planes defined by the spine, and so every 2-page book drawing can be realized as a plane drawing; it follows that $\mathrm{cr}(G) \leq \nu_2(G)$ for every graph G.

In 1968, Guy et al. [13] found 2-page book drawings of K_n with $Z(n)$ crossings, thus showing that $\nu_2(K_n) \leq Z(n)$ (see also the constructions by Damiani et al. [8], Harborth [16], and Shahrokhi et al. [19]). Once these constructions were known, the conjecture that $\nu_2(K_n) = Z(n)$ is implicit in the conjecture given by Equation (1) since $\mathrm{cr}(K_n) \leq \nu_2(K_n)$. However, the only explicit reference to this weaker conjecture is, as far as we know, from Vrt'o [20].

Buchheim and Zhang [5] reformulated the problem of finding $\nu_2(K_n)$ as a maximum cut problem on associated graphs, and then solved exactly this maximum cut problem for all $n \leq 13$, thus confirming Equation (1) for 2-page book drawings for all $n \leq 14$ (the case $n = 14$ follows from the case $n = 13$ by an elementary counting argument). Very recently, De Klerk and Pasechnik [9] used this max cut reformulation to find the exact value of $\nu_2(K_n)$ for all $n \leq 21$ and $n = 24$, and moreover, by using semidefinite programming techniques, to obtain the asymptotic bound $\lim_{n\to\infty} \mathrm{cr}(K_n)/Z(n) \geq 0.9245$. Both of these results [5, 9] were obtained with the aid of computers.

In this paper we prove that $\nu_2(K_n) = Z(n)$. The main technique for the proof is the extension of the concept of *k-edge* of a finite set of points to topological drawings of the complete graph. We do this in a way such that the identity proved by Ábrego and Fernández-Merchant [1] and Lovász et al. [18], that expresses the crossing number of a rectilinear drawing of K_n in terms of the k-edges of its set of vertices, is also valid in the topological setting.

We recall that a drawing D is *rectilinear* if the edges of D are straight line segments, and the *rectilinear crossing number* $\overline{\mathrm{cr}}(G)$ of a graph G is the minimum of $\mathrm{cr}(D)$ taken over all rectilinear drawings D of G. An edge pq of D is a *k-edge* if the line spanned by pq divides the remaining set of vertices into two subsets of cardinality k and $n-2-k$. Thus a k-edge is also an $(n-2-k)$-edge. Denote by $E_k(D)$ the number of k-edges of D. The following identity [1, 18] has been key to the recent developments on the rectilinear crossing number of K_n.

$$\overline{\mathrm{cr}}\,(D) = 3\binom{n}{4} - \sum_{k=0}^{\lfloor n/2 \rfloor -1} k\,(n-2-k)\,E_k\,(D). \qquad (2)$$

In Section 2 we generalize the concept of k-edge to arbitrary (that is, not necessarily rectilinear) drawings of K_n. This allows us to extend Equation (2) to (good) topological drawings of K_n. The key observation to extend the definition of k-edge to the new setting is to observe that, although half-planes are not well defined, we can use the orientation of the triangles defined by three points: the edge pq will be a k-edge of the topological drawing if the set of triangles adjacent to pq is divided, according to its orientation, into two subsets with cardinality k and $n-k-2$. In Section 3 we use this tool to show that $\nu_2(K_n) = Z(n)$. In order to do that, we need to introduce the new concept of $\leq\leq k$-edges, because for topological drawings the lower bound for $\leq k$-edges, $E_{\leq k}(D) \geq 3\binom{k+2}{2}$ does not hold. In Section 4 we analyze optimal 2-page drawings of K_n, and in Section 5 we present some open questions and directions for future research.

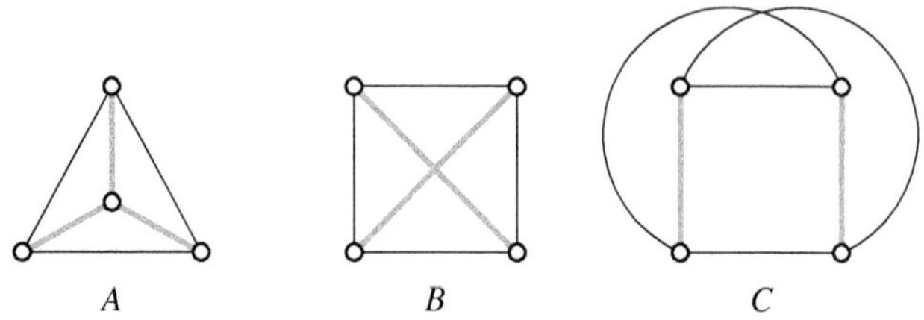

Figure 1: The three different types A, B, and C of nonisomorphic good drawings of K_4, with 3, 2, and 2 separations. The edge of each separation is shown bold.

2. CROSSINGS AND k-EDGES

In this section we generalize the concept of k-edges, which has so far only been used in the geometric setting of finite sets of points in the plane, to topological drawings of K_n. Let D be a good drawing of K_n, let $\overrightarrow{pq}$ be a directed edge of D and r a vertex of D other than p or q. We say that r *is on the left (respectively, right) side of* $\overrightarrow{pq}$ if the topological triangle pqr traced in that order (its vertices and edges correspond to those in D) is oriented counterclockwise (respectively, clockwise). Note that this is well defined as the three edges pq, qr, and rp in D do not self intersect and do not intersect each other, since D is good. We say that the edge pq is a *k-edge of D* if it has exactly k points of D on the same side (left or right), and thus $n-2-k$ points on the other side. Hence, as before, a k-edge is also an $(n-2-k)$-edge. Note that the direction of the edge pq is no longer relevant and every edge of D is a k-edge for some unique k such that $0 \leq k \leq \lfloor n/2 \rfloor - 1$. Let $E_k(D)$ be the number of k-edges of D.

THEOREM 1. *For any good drawing D of K_n in the plane the following identity holds,*

$$\mathrm{cr}\,(D) = 3\binom{n}{4} - \sum_{k=0}^{\lfloor n/2 \rfloor -1} k\,(n-2-k)\,E_k\,(D).$$

PROOF. In a good drawing of K_n, we say that an edge pq *separates* the vertices r and s if the orientations of the triangles pqr and pqs are opposite. In this case, we say that the set $\{pq, r, s\}$ is a *separation.* It is straightforward to check that any good drawing of K_4 is isomorphic to one of the three drawings shown in Figure 1.

We denote by T_A, T_B, and T_C the number of induced

subdrawings of D of type A, B, and C, respectively. Then

$$T_A + T_B + T_C = \binom{n}{4}, \tag{3}$$

and since the subdrawings of types B or C are in one-to-one correspondence with the crossings of D, it follows that

$$\mathrm{cr}(D) = T_B + T_C. \tag{4}$$

We count the number of separations in D in two different ways: First, each subdrawing of type A has 3 separations (the edge in each separation is bold in Figure 1), and each subdrawing of types B or C has 2 separations. This gives a total of $3T_A + 2T_B + 2T_C$ separations in D. Second, each k-edge belongs to exactly $k(n-2-k)$ separations. Summing over all k-edges for $0 \le k \le \lfloor n/2 \rfloor - 1$ gives a total of $\sum_{k=0}^{\lfloor n/2\rfloor-1} k(n-2-k)E_k(D)$ separations in D. Therefore

$$3T_A + 2T_B + 2T_C = \sum_{k=0}^{\lfloor n/2\rfloor-1} k\,(n-2-k)\,E_k(D). \tag{5}$$

Finally, subtracting Equation (5) from three times Equation (3) we get

$$T_B + T_C = 3\binom{n}{4} - \sum_{k=0}^{\lfloor n/2\rfloor-1} k\,(n-2-k)\,E_k(D),$$

and thus by Equation (4) we obtain the claimed result. □

For $0 \le k \le \lfloor n/2 \rfloor - 1$ and D a good drawing of K_n, we define the set of *$\le k$-edges* of D as all j-edges in D for $j = 0, \ldots, k$. The number of $\le k$-edges of D is denoted by

$$E_{\le k}(D) := \sum_{j=0}^{k} E_j(D).$$

Similarly, we denote the number of *$\le\le k$-edges* of D by

$$\begin{aligned} E_{\le\le k}(D) &:= \sum_{j=0}^{k} E_{\le j}(D) = \sum_{j=0}^{k}\sum_{i=0}^{j} E_i(D) \\ &= \sum_{i=0}^{k} (k+1-i)\,E_i(D). \end{aligned}$$

To avoid special cases we define $E_{\le\le -1}(D) = E_{\le\le -2}(D) = 0$.

The following result restates Theorem 1 in terms of the number of $\le\le k$-edges.

PROPOSITION 2. *Let D be a good drawing of K_n. Then*

$$\begin{aligned} \mathrm{cr}(D) = 2\sum_{k=0}^{\lfloor n/2\rfloor-2} E_{\le\le k}(D) - \frac{1}{2}\binom{n}{2}\left\lfloor \frac{n-2}{2} \right\rfloor \\ - \frac{1}{2}(1 + (-1)^n) E_{\le\le \lfloor n/2\rfloor - 2}(D). \end{aligned}$$

PROOF. First note that for $2 \le k \le \lfloor n/2 \rfloor - 1$ we have that $E_{\le\le k}(D) - E_{\le\le k-1}(D) = E_{\le k}(D)$ and $E_{\le k}(D) - E_{\le k-1}(D) = E_k(D)$. Thus

$$E_k(D) = E_{\le\le k}(D) - 2E_{\le\le k-1}(D) + E_{\le\le k-2}(D).$$

We rewrite the last term in Theorem 1.

$$\begin{aligned} &\sum_{k=0}^{\lfloor n/2\rfloor-1} k(n-2-k)E_k(D) \\ &= \sum_{k=2}^{\lfloor n/2\rfloor-1} k(n-2-k)[E_{\le\le k}(D) - 2E_{\le\le k-1}(D) + E_{\le\le k-2}(D)] \\ &= \sum_{k=0}^{\lfloor n/2\rfloor-3} [k(n-2-k) - 2(k+1)(n-3-k) \\ &\qquad + (k+2)(n-4-k)]\, E_{\le\le k}(D) \\ &\quad + \left(\left\lfloor\frac{n}{2}\right\rfloor - 1\right)\left(n-1-\left\lfloor\frac{n}{2}\right\rfloor\right) E_{\le\le\lfloor n/2\rfloor-1}(D) \\ &\quad + \left[-2\left(\left\lfloor\frac{n}{2}\right\rfloor - 1\right)\left(n-1-\left\lfloor\frac{n}{2}\right\rfloor\right)\right. \\ &\qquad \left. + \left(\left\lfloor\frac{n}{2}\right\rfloor - 2\right)\left(n-\left\lfloor\frac{n}{2}\right\rfloor\right)\right] E_{\le\le\lfloor n/2\rfloor-2}(D) \\ &= -2\sum_{k=0}^{\lfloor n/2\rfloor-3} E_{\le\le k}(D) \\ &\quad + \left(\left\lfloor\frac{n}{2}\right\rfloor - 1\right)\left(n-1-\left\lfloor\frac{n}{2}\right\rfloor\right) E_{\le\le\lfloor n/2\rfloor-1}(D) \\ &\quad + \left[-2\left(\left\lfloor\frac{n}{2}\right\rfloor - 1\right)\left(n-1-\left\lfloor\frac{n}{2}\right\rfloor\right)\right. \\ &\qquad \left. + \left(\left\lfloor\frac{n}{2}\right\rfloor - 2\right)\left(n-\left\lfloor\frac{n}{2}\right\rfloor\right)\right] E_{\le\le\lfloor n/2\rfloor-2}(D). \end{aligned}$$

Since $E_{\le\le\lfloor n/2\rfloor-1}(D) = E_{\le\le\lfloor n/2\rfloor-2}(D) + E_{\le\lfloor n/2\rfloor-1}(D) = E_{\le\le\lfloor n/2\rfloor-2}(D) + \binom{n}{2}$, it follows by Theorem 1 that

$$\begin{aligned} \mathrm{cr}(D) &= 3\binom{n}{4} - \sum_{k=0}^{\lfloor n/2\rfloor-1} k\,(n-2-k)\,E_k(D) \\ &= 3\binom{n}{4} + 2\sum_{k=0}^{\lfloor n/2\rfloor-3} E_{\le\le k}(D) \\ &\quad + \left(n+1-2\left\lfloor\frac{n}{2}\right\rfloor\right) E_{\le\le\lfloor n/2\rfloor-2}(D) \\ &\quad - \left(\left\lfloor\frac{n}{2}\right\rfloor - 1\right)\left(n-1-\left\lfloor\frac{n}{2}\right\rfloor\right)\binom{n}{2} \\ &= 2\sum_{k=0}^{\lfloor n/2\rfloor-3} E_{\le\le k}(D) - \frac{1}{2}\binom{n}{2}\left\lfloor\frac{n-2}{2}\right\rfloor \\ &\quad + \begin{cases} E_{\le\le\lfloor n/2\rfloor-2}(D) & \text{if } n \text{ is even}, \\ 2E_{\le\le\lfloor n/2\rfloor-2}(D) & \text{if } n \text{ is odd}, \end{cases} \end{aligned}$$

which is equivalent to the claimed result. □

3. THE 2-PAGE CROSSING NUMBER

Consider a 2-page book drawing D of K_n with horizontal spine and label the vertices $1, 2, \ldots, n$ from left to right. Color the edges above or on the spine blue and below the spine red, respectively. We construct an upper triangular matrix which corresponds to the coloring of these edges, see Figure 2. We call this the *2-page matrix* of D. Label the columns of the 2-page matrix with $2, \ldots, n$ from left to right and the rows with $1, 2, \ldots, n-1$ from top to bottom. For $i < j$ an entry ij (row-column) in the 2-page matrix is a point with the same color as the edge ij in the drawing D.

We start by proving some basic properties of the 2-page matrix.

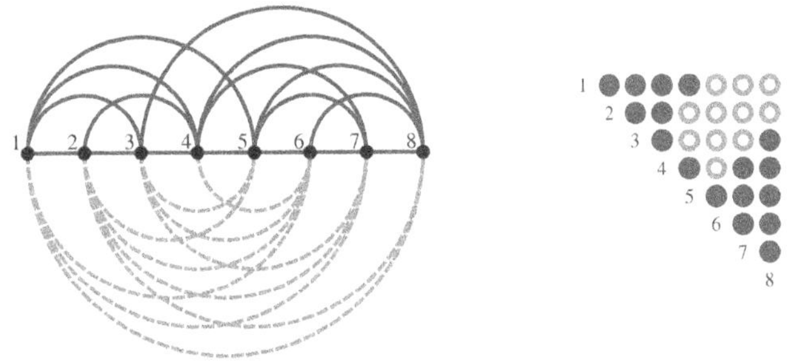

Figure 2: Two-colored diagram for a 2-page book drawing D of K_8 and the corresponding 2-page matrix. Solid dots and lines represent blue edges. Open dots and dashed lines represent red edges.

LEMMA 3. *Let D be a 2-page book drawing of K_n. For $1 \leq i < j \leq n$, let k be the sum of the number of points to the right plus the number of points above the entry ij in the 2-page matrix of D, which have the same color as ij. Then the edge ij is a k-edge. (It is possible to have $k > \lfloor n/2 \rfloor - 1$.)*

PROOF. Let $1 \leq i < j \leq n$ and assume that the edge ij is blue (red). We count the number of points l in D to the left (right) of ij. For $l \notin \{i, j\}$ the triangle ijl is oriented counter-clockwise (clockwise) if and only if either $l < i$ and the edge lj is blue (red), or $l > j$ and the edge il is blue (red). In the first case these edges correspond to blue (red) points above the entry ij, and in the second case to blue (red) points to the right of the entry ij, respectively. □

In view of Lemma 3 we say that the point in the cell ij of the 2-page matrix of D *represents* a k-edge if ij is a k-edge (or an $(n-2-k)$-edge) in D.

LEMMA 4. *For $k < n/2 - 1$ and for $1 \leq j \leq k+1$, in the 2-page matrix of a drawing D of K_n there are at least $2(k+2-j)$ points in row j representing $\leq k$-edges. Similarly, for $n-k \leq j \leq n$ there are at least $2(k+1-n+j)$ points in column j representing $\leq k$-edges.*

PROOF. For $1 \leq j \leq k+1$, in row j the rightmost $k+2-j$ points of each color represent $\leq k$-edges as they have at most $k+1-j$ points of their color to the right and at most $j-1$ on top. So if each color appears at least $k+2-j$ times in row j, we have guaranteed $2(k+2-j)$ $\leq k$-edges in row j. If one of the colors appears fewer than $k+2-j$ times, so that there are $k+2-j-e$ blue points in row j for some $1 \leq e \leq k+2-j$, then there are $n-j-(k+2-j-e) = n-2-k+e$ red points in this row. In this case we claim that also the leftmost e red points in this row represent $\leq k$-edges. In fact, for $1 \leq i \leq e$, the i-th red point (from the left) in row j, has exactly $n-2-k+e-i$ red points to the right and perhaps more red points on top. Since for $n \geq 2$ we have $n-2-k+e-i \geq n/2-k$, this i-th red point also represents a $\leq k$-edge. The equivalent result holds for the rightmost $k+1$ columns. □

LEMMA 5. *For $0 \leq j < n/2 - 1$, in the 2-page matrix of a drawing D of K_n there are two points in column n which correspond to j-edges in D. For n even there exists one such point in column n corresponding to an $(n/2-1)$-edge in D.*

PROOF. We follow the lines of the proof of Lemma 4. Consider the points in column n in order from top to bottom. By Lemma 3 the i-th vertex of a color corresponds to an $(i-1)$-edge. Thus, if there are at least $j+1$ vertices for each color we are done. Otherwise assume without loss of generality that there are $j+1-e$ blue points in column n for some $1 \leq e \leq j+1$. Then there are $n-1-(j+1-e) = n-j+e-2$ red points in this column. For $1 \leq i \leq \lfloor n/2 \rfloor$ the i-th red point corresponds to an $(i-1)$-edge, and for $\lfloor n/2 \rfloor + 1 \leq i \leq n-j+e-2$ the i-th red point corresponds to an $(i-1) = (n-i-1)$-edge. Thus we get two red points corresponding to j-edges for $i = j+1$ and $i = n-j-1$. Finally, observe that these two points are different for $j < n/2 - 1$. For n even we get only one such point for $j = n/2 - 1$. □

The next theorem gives a lower bound on the number of $\leq\leq k$-edges, which will play a central role in deriving our main result.

THEOREM 6. *Let $n \geq 3$. For every 2-page book drawing D of K_n and $0 \leq k < n/2 - 1$, we have*

$$E_{\leq\leq k}(D) \geq 3\binom{k+3}{3}.$$

PROOF. We proceed by induction on n. The induction base $n = 3$ holds trivially. For $n \geq 4$, consider a 2-page book drawing D of K_n with horizontal spine and label the vertices from left to right with $1, 2, \ldots, n$. Remove the point n and all incident edges to obtain a 2-page book drawing D' of K_{n-1}. To bound $E_{\leq\leq k}(D)$, recall that

$$E_{\leq\leq k}(D) = \sum_{j=0}^{k} (k+1-j) E_j(D). \tag{6}$$

All edges incident to n are in D but are not in D'. In fact, by Lemma 5, there are two j-edges adjacent to the vertex n for each $0 \leq j \leq k \leq \lfloor n/2 \rfloor - 2$. These edges contribute with $2\sum_{j=0}^{k}(k+1-j) = 2\binom{k+2}{2}$ to Equation (6). We next compare Equation (6) to

$$E_{\leq\leq k-1}(D') = \sum_{j=0}^{k-1} (k-j) E_j(D'). \tag{7}$$

Any edge contributing to Equation (7) also contributes to Equation (6), but possibly with a different value. A j-edge in D' is a j-edge or a $(j+1)$-edge in D. More precisely, if for $a < b$ the point ab in the 2-page matrix of D' represents a blue j-edge in D' (the equivalent argument holds if ab is red), with $0 \leq j \leq \lfloor (n-1)/2 \rfloor - 1$, then the sum of blue points above it or to its right is either j or $n-3-j$. If the edge an is blue, then the point ab has $j+1$ or $n-3-j+1$ blue points above it or to its right in the 2-page matrix of D. This means that the edge ab is a $(j+1)$- or an $n-2-(n-3-j+1) = j$-edge in D, respectively. If the edge an is red, then the point ab in the 2-page matrix of D has the same number of blue points above it or to its right as in the 2-page matrix of D'. So it is a j- or an $n-2-(n-3-j) = (j+1)$-edge in D, respectively. A j-edge in D' contributes to Equation (7) with $k-j$. A j-edge and a $(j+1)$-edge in D contribute to Equation (6) with $k+1-j$ and $k-j$, respectively. This is a gain of $+1$ or 0, respectively, towards $E_{\leq\leq k}(D)$ when compared to $E_{\leq\leq k-1}(D')$. Finally, a k-edge in both D and D' does not contribute to Equation (7) and contributes to

Equation (6) with +1. Therefore

$$E_{\leq\leq k}(D) = E_{\leq\leq k-1}(D') + 2\binom{k+2}{2} + \sum_{j=0}^{k} \#(j\text{-edges in } D' \text{ that are } j\text{-edges in } D).$$

By induction hypothesis, $E_{\leq\leq k-1}(D') \geq 3\binom{k+2}{3}$ and thus

$$\begin{aligned} E_{\leq\leq k}(D) &\geq 3\binom{k+2}{3} + 2\binom{k+2}{2} + \sum_{j=0}^{k} \#(j\text{-edges in } D' \text{ that are } j\text{-edges in } D) \\ &= 3\binom{k+3}{3} - \binom{k+2}{2} + \sum_{j=0}^{k} \#(j\text{-edges in } D' \text{ that are } j\text{-edges in } D). \end{aligned}$$

We finally prove that

$$\sum_{j=0}^{k} \#(j\text{-edges in } D' \text{ that are } j\text{-edges in } D) \geq \binom{k+2}{2}. \tag{8}$$

In fact, we prove that for each $1 \leq j \leq k+1$ there are at least $k+2-j$ points in row j of the 2-page matrix of D that represent i-edges in both D' and D for some $0 \leq i \leq k$. Suppose that the edge jn is blue (the equivalent argument holds when jn is red). Then any red point in row j with $i \leq k$ red points above it or to its right represents an i-edge in both D and D' for some $0 \leq i \leq k$; and any blue point in row j with $i \geq n-2-k$ blue points above it or to its right represents an i'-edge in both D and D' for some $0 \leq i' \leq k$. Thus, the first $k+2-j$ red points from the right in row j (if they exist) represent $\leq k$-edges, because they have at most $k+2-j-1$ red points to the right and at most $j-1$ red points above, and therefore they all contribute to the left-hand-side of Inequality (8). If there are fewer than $k+2-j$ red points in row j, say $k+2-j-e$ for some $1 \leq e \leq k+2-j$, then the first e blue points in row j from the left represent $\leq k$-edges, because they have at least $n-j-e \geq n-j-k-2+j = n-k-2$ blue points to their right. Hence there are at least $k+2-j-e$ red points and at least e blue points (for a total of at least $k+2-j$ points) that represent $\leq k$-edges in row j that contribute to the left-hand-side of Inequality (8). Summing over all $1 \leq j \leq k+1$, we get that

$$\sum_{j=0}^{k} \#(j\text{-edges in } D' \text{ that are } j\text{-edges in } D) \geq \sum_{j=1}^{k+1} (k+2-j) = \binom{k+2}{2}. \quad \square$$

We are now ready to prove our main result, namely that the 2-page crossing number of K_n is $Z(n)$.

THEOREM 7. *For every positive integer n, $\nu_2(K_n) = Z(n)$.*

PROOF. The cases $n = 1$ and $n = 2$ are trivial. Let $n \geq 3$. As we mentioned before, 2-page book drawings with $Z(n)$ crossings were constructed by Shahrokhi, et al. [19] showing $\nu_2(K_n) \leq Z(n)$. Let D be a 2-page book drawing of K_n. To prove the lower bound, we use Theorem 6 and Proposition 2.

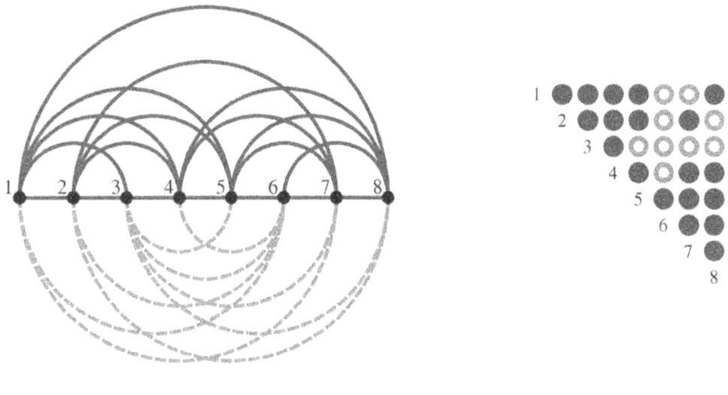

Figure 3: This 2-page book drawing of K_8 has four 0-edges, $(1,7)$, $(1,8)$, $(2,7)$, and $(2,8)$, and four 1-edges, $(1,5)$, $(1,6)$, $(3,8)$, and $(4,8)$.

$$\begin{aligned} \operatorname{cr}(D) &\geq 2\sum_{k=0}^{\lfloor n/2\rfloor-2} 3\binom{k+3}{3} - \frac{1}{2}\binom{n}{2}\left\lfloor\frac{n-2}{2}\right\rfloor - \frac{3}{2}\left(1+(-1)^n\right)\binom{\lfloor\frac{n}{2}\rfloor+1}{3} \\ &= 6\binom{\lfloor\frac{n}{2}\rfloor+2}{4} - \frac{1}{2}\binom{n}{2}\left\lfloor\frac{n-2}{2}\right\rfloor - \frac{3}{2}\left(1+(-1)^n\right)\binom{\lfloor\frac{n}{2}\rfloor+1}{3} \\ &= \begin{cases} \frac{1}{64}(n-1)^2(n-3)^2 & \text{if } n \text{ is odd,} \\ \frac{1}{64}n(n-2)^2(n-4) & \text{if } n \text{ is even,} \end{cases} \\ &= Z(n). \quad \square \end{aligned}$$

4. OPTIMAL CONFIGURATIONS

It was proved by Ábrego and Fernández-Merchant [1] and by Lovász et al. [18] that the inequality $E_{\leq k}(P) \geq 3\binom{k+2}{2}$ holds for every set P of n points in general position in the plane and for every k such that $0 \leq k \leq \lfloor n/2\rfloor - 2$. This inequality used with the rectilinear version of Theorem 1 gives $Z(n)$ as a lower bound for the rectilinear crossing number of K_n [1]. In contrast to the rectilinear case, the inequality $E_{\leq k}(D) \geq 3\binom{k+2}{2}$ does not hold in general for topological drawings D of K_n, not even for general 2-page book drawings as can be seen in Figure 3. This shows the relevance of introducing the number $E_{\leq\leq k}(D)$ and Theorem 6. However, the inequality $E_{\leq k}(D) \geq 3\binom{k+2}{2}$ does hold for crossing optimal drawings of K_n, where in fact the following stronger result is true.

PROPOSITION 8. *Let D be a 2-page book drawing of K_n and $I_n = \{k \in \mathbb{Z} : 0 \leq k \leq \lfloor n/2\rfloor - 2\}$. The following are equivalent: (i) $\operatorname{cr}(D) = Z(n)$, (ii) $E_k(D) = 3(k+1)$ for all $k \in I_n$, (iii) $E_{\leq k}(D) = 3\binom{k+2}{2}$ for all $k \in I_n$, and (iv) $E_{\leq\leq k}(D) = 3\binom{k+3}{3}$ for all $k \in I_n$.*

PROOF. In order to satisfy (i) equality must be achieved in the proof of Theorem 7, which is precisely (iv). Reciprocally, (iv) implies equality in the proof of Theorem 7, which proves the equivalency of (i) and (iv). The implications (ii) $\Rightarrow$ (iii) $\Rightarrow$ (iv) follow directly from the definitions of $E_{\leq k}(D)$ and $E_{\leq\leq k}(D)$, using the identity $\sum_{m=0}^{r}\binom{m}{s}=\binom{r+1}{s+1}$. It remains to show that (iv) implies (ii), which we do by applying induction on k. For the induction base note that $E_{\leq\leq 0}(D)=E_{\leq 0}(D)=E_0(D)=3$. For $1\leq k\leq\lfloor n/2\rfloor-2$, the identities $E_j(D)=3(j+1)$ for all $0\leq j\leq k-1$ and $E_{\leq\leq k}(D)=3\binom{k+3}{3}$ imply that

$$3\binom{k+3}{3}=E_{\leq\leq k}(D)=\sum_{j=0}^{k}(k+1-j)\,E_j\,(D)$$
$$=E_k(D)+3\sum_{j=0}^{k-1}(k+1-j)(j+1),$$

and thus

$$E_k(D)=3\binom{k+3}{3}-3\sum_{j=0}^{k-1}(k+1-j)(j+1)$$
$$=3\binom{k+3}{3}-\frac{1}{2}k\,(k+1)\,(k+5)=3\,(k+1)\,. \quad \square$$

An elementary counting argument (see for instance Erdős and Guy [12]) shows that if n is even and $\mathrm{cr}(K_{n-1})=Z(n-1)$, then $\mathrm{cr}(K_n)=Z(n)$. Moreover, under these assumptions any topological drawing of K_n with $Z(n)$ crossings satisfies that every induced subdrawing with $n-1$ points has $Z(n-1)$ crossings. Our technique allows a different proof of this fact for 2-page book drawings of K_n, but more importantly it gives for the first time a similar property when n is odd.

PROPOSITION 9. *Let D be a 2-page book drawing of K_n with $Z(n)$ crossings and D' an induced subdrawing of D with $n-1$ vertices. Then D' satisfies that $E_k(D')=3(k+1)$ for all k such that $0\leq k\leq\lfloor n/2\rfloor-3$. In other words, if D is crossing optimal, then D' is crossing optimal when n is even, and D' is* almost *crossing optimal when n is odd, in the sense that Proposition 8(ii) holds for every $k\in I_{n-1}$ except perhaps for $k=(n-5)/2$.*

PROOF. In order to achieve $\mathrm{cr}(D)=Z(n)$, also Proposition 8(iv) must hold. Thus, equality must be achieved throughout the proof of Theorem 6 for all $k\in I_n$. In particular, we must have $E_{\leq\leq k}(D')=3\binom{k+3}{3}$ for every

$$0\leq k\leq\lfloor n/2\rfloor-3=\begin{cases}\lfloor(n-1)/2\rfloor-2 \text{ if } n \text{ is even,}\\ \lfloor(n-1)/2\rfloor-3 \text{ if } n \text{ is odd.}\end{cases}$$

This means that D' satisfies Proposition 8(iv) and thus $\mathrm{cr}(D')=Z(n-1)$ if n is even. However, if n is odd, then D' satisfies Proposition 8(iv) for all $k\in I_{n-1}$ except perhaps for $k=\lfloor(n-1)/2\rfloor-2=(n-5)/2$. This is equivalent to satisfying Proposition 8(ii) for all k in the same range and thus being almost crossing optimal in the sense of the statement. $\square$

A more detailed analysis on the crossing optimal 2-page book drawings will be included in the journal version of this work.

5. CONCLUDING REMARKS

Our approach to determine k-edges in the topological setting is to define the orientation of three vertices by the orientation of the corresponding triangle in a good drawing of the complete graph. It is naturally to ask whether this defines an abstract order type. To this end, the setting would have to satisfy the axiomatic system described by Knuth [17]. But it is easy to construct an example which does not fulfill these axioms, that is, our setting does not constitute an abstract order type as described by Knuth [17]. It is an interesting question for further research how this new concept compares to the classic order type, both in terms of theory (realizability, etc.) and applications.

We believe that the developed techniques of generalized orientation, k-edge for topological drawings, and $\leq\leq k$-edges are of interest in their own. We will investigate their usefulness for related problems in future work. For example, they might also play a central role to approach the crossing number problem for general drawings of the complete and complete bipartite graphs.

6. ACKNOWLEDGMENTS

This work was initiated during the workshop Crossing Numbers Turn Useful, held at the Banff International Research Station (BIRS). The authors thank the BIRS authorities and staff for their support. O. Aichholzer is partially supported by the ESF EUROCORES programme EuroGIGA, CRP ComPoSe, under grant FWF [Austrian Fonds zur Förderung der Wissenschaftlichen Forschung] I648-N18. P. Ramos is partially supported by MEC grant MTM2011-22792 and by the ESF EUROCORES programme EuroGIGA, CRP ComPoSe, under grant EUI-EURC-2011-4306. G. Salazar is supported by CONACYT Grant 106432

7. REFERENCES

[1] B. M. Ábrego and S. Fernández-Merchant. A lower bound for the rectilinear crossing number. *Graphs and Combinatorics*, 21:293–300, 2005.

[2] L. Beineke and R. Wilson. The early history of the brick factory problem. *Math. Intelligencer*, 32:41–48, 2010.

[3] T. Bilski. Embeddings graphs in books: a survey. *Computers and Digital Techniques, IEEE Proceedings*, 139:134–138, 1992.

[4] J. Blažek and M. Koman. A minimal problem concerning complete plane graphs. In M. Fiedler, editor, *Theory of graphs and its applications*, pages 113–117. Czech. Acad. of Sci., 1964.

[5] C. Buchheim and L. Zheng. Fixed linear crossing minimization by reduction to the maximum cut problem. In *COCOON*, pages 507–516, 2006.

[6] F. R. K. Chung, F. T. Leighton, and A. L. Rosenberg. Embedding graphs in books: a layout problem with applications to vsli design. *SIAM J. Algebraic Disc. Math.*, 8:33–58, 1987.

[7] R. Cimikowski and B. Mumey. Approximating the fixed linear crossing number. *Disc. App. Math.*, 155:2202–2210, 2007.

[8] E. Damiani, O. D'Antona, and P. Salemi. An upper bound to the crossing number of the complete graph drawn on the pages of a book. *J. Combin. Inform. System Sci.*, 19:75–84, 1994.

[9] E. de Klerk and D. Pasechnik. Improved lower bounds for the 2-page crossing numbers of $k_{m,n}$ and k_n via semidefinite programming, 2011.

[10] E. de Klerk, D. V. Pasechnik, and A. Schrijver. Reduction of symmetric semidefinite programs using the regular *-representation. *Math. Program.*, 109:613–624, 2007.

[11] V. Dujmović and D. Wood. On linear layouts of graphs. *Discrete Mathematics and Theoretical Computer Science*, 6:339–358, 2004.

[12] P. Erdős and R. K. Guy. Crossing number problems. *Amer. Math. Monthly*, 80:52–57, 1973.

[13] R. Guy, T. Jenkyns, and J. Schaer. The toroidal crossing number of the complete graph. *J. Combinatorial Theory*, 4:376–390, 1968.

[14] R. K. Guy. A combinatorial problem. *Bull. Malayan Math. Soc.*, 7:68–72, 1960.

[15] F. Harary and A. Hill. On the number of crossings in a complete graph. *Proc. Edinburgh Math. Soc.*, 13:333–338, 1963.

[16] H. Harborth. Special numbers of crossings for complete graphs. *Disc. Math.*, 224:95–102, 2002.

[17] D. Knuth. *Axioms and hulls*, volume 606 of *Lecture notes in computer science*. Springer-Verlag, 1992.

[18] L. Lovász, K. Vesztergombi, U. Wagner, and E. Welzl. Convex quadrilaterals and k-sets. In J. Pach, editor, *Contemporary Mathematics Series, 342, AMS 2004*, volume 342, pages 139–148. American Mathematical Society, 2004.

[19] F. Shahrokhi, O. Sýkora, L. A. Székely, and I. Vrt'o. The book crossing number of a graph. *J. Graph Th.*, 21:413–424, 1996.

[20] I. Vrt'o. Two special crossing number problems for complete graphs. In *Abstracts of the 6th Slovenian International Conference on Graph Theory Bled'07*, page 60, 2007.

String Graphs and Incomparability Graphs

Jacob Fox*
Department of Mathematics MIT
77 Massachusetts Ave.
Cambridge, MA 02139
fox@math.mit.edu

János Pach†
Chair of Combin. Geometry EPFL
1015 Lausanne
Switzerland
pach@renyi.hu

ABSTRACT

Given a collection C of curves in the plane, its *string graph* is defined as the graph with vertex set C, in which two curves in C are adjacent if and only if they intersect. Given a partially ordered set $(P, <)$, its *incomparability graph* is the graph with vertex set P, in which two elements of P are adjacent if and only if they are incomparable.

It is known that every incomparability graph is a string graph. For "dense" string graphs, we establish a partial converse of this statement. We prove that for every $\varepsilon > 0$ there exists $\delta > 0$ with the property that if C is a collection of curves whose string graph has at least $\varepsilon|C|^2$ edges, then one can select a subcurve γ' of each $\gamma \in C$ such that the string graph of the collection $\{\gamma' : \gamma \in C\}$ has at least $\delta|C|^2$ edges and is an incomparability graph. We also discuss applications of this result to extremal problems for string graphs and edge intersection patterns in topological graphs.

Categories and Subject Descriptors

G.2.1 [**Discrete Mathematics**]: Combinatorics

General Terms

Theory

Keywords

String graphs, incomparability graphs, perfect graphs.

*Supported by an NSF Graduate Research Fellowship, a Princeton Centennial Fellowship, a Simons Fellowship, and NSF grant DMS-1069197.

†Supported by Hungarian Science Foundation EuroGIGA Grant OTKA NN 102029, by Swiss National Science Foundation Grant 200021-125287/1, and by NSF Grant CCF-08-30272.

1. INTRODUCTION

The *intersection graph* of a collection C of sets has vertex set C and two sets in C are adjacent if and only if they have nonempty intersection. A *curve* is a subset of the plane which is homeomorphic to the interval $[0, 1]$. A *string graph* is an intersection graph of a collection of curves. It is straightforward to show that the intersection graph of any collection of arcwise connected sets in the plane is a string graph.

String graphs have been intensively studied both for practical applications and theoretical interest. Benzer [4] was the first to introduce these graphs in 1959, while exploring the topology of genetic structures. In 1966, interested in electrical networks realizable by printed circuits, Sinden [41] considered the same constructs at Bell Labs. He showed that not every graph is a string graph but all planar graphs are. He also raised the question whether there exists any algorithm for recognizing string graphs.

In 1976, reporting on Sinden's work, Graham [20] introduced string graphs to the mathematics community. Later that year, Ehrlich, Even, and Tarjan [8] proved that computing the chromatic number of a string graph is NP-hard. A decade later, Kratochvíl, Goljan, and Kučera [28] wrote a tract devoted to string graphs. They showed that every string graph can be realized by a family of polygonal arcs with a finite number of intersections. Kratochvíl [27] proved that the recognition of string graphs is NP-hard. Kratochvíl and Matoušek [29] constructed string graphs on n vertices that require at least 2^{cn} intersection points in any realization, where c is a positive constant. They conjectured that every string graph on n vertices can be realized with at most 2^{cn^k} intersection points, for some constants c and k.

Sinden's question remained a challenging open problem for 35 years, until the conjecture of Kratochvíl and Matoušek was confirmed [36, 40], implying that the string graph problem is decidable. A short time later, Schaefer, Sedgewick, and Štefankovič [39] proved that recognizing string graphs is NP-complete. Despite these results, understanding the *structure* of string graphs has remained a wide open problem.

Given a partially ordered set (poset, for short) $(P, <)$, its *incomparability graph* is the graph with vertex set P, in which two elements are adjacent if and only if they are incomparable. Unlike string graphs, incomparability graphs are fairly well understood. In 1950, Dilworth [7] proved that every incomparability graph is a perfect graph, so the chromatic number of an incomparability graph is equal to its clique number (the analogous result for comparability graphs was earlier proved by Erdős and Szekeres [10]). In 1967, Gallai [16] gave a characterization of incomparability graphs in terms of minimal forbidden induced subgraphs. It is known that incomparability graphs can be recognized in polynomial time [17].

In 1983, Golumbic, Rotem, and Urrutia [18] and Lovász [33] proved that every incomparability graph is a string graph (see Proposition 2.1). There are many string graphs, such as odd cycles of length at least *five*, which are not incomparability graphs. In fact, the number of string graphs on n vertices is $2^{(3/4+o(1))\binom{n}{2}}$ [38], while the number of incomparability graphs on n vertices is only $2^{(1/2+o(1))\binom{n}{2}}$ [23]. Nevertheless, as the main result of this paper demonstrates, most string graphs contain huge subgraphs that are incomparability graphs.

THEOREM 1.1. *For every $\varepsilon > 0$ there exists $\delta > 0$ with the property that if C is a collection of curves whose string graph has at least $\varepsilon|C|^2$ edges, then one can select a subcurve γ' of each $\gamma \in C$ such that the string graph of the collection $\{\gamma' : \gamma \in C\}$ has at least $\delta|C|^2$ edges and is an incomparability graph.*

It follows from our proof that δ can be chosen to be a polynomial in ε, that is, we can choose $\delta = \varepsilon^c$ for an appropriate absolute constant c.

We say that a graph with n vertices is *dense* if, for some $\varepsilon > 0$, its number of edges is at least εn^2. Our theorem immediately implies that every dense string graph contains a dense spanning subgraph (with a different ε) which is an incomparability graph.

Can Theorem 1.1 be strengthened to say that every dense string graph contains a dense *induced* subgraph with a linear number of vertices that is an incomparability graph? The answer is no. Indeed, a construction of Kynčl [30] (improving on earlier constructions [31] and [22]) shows that there is a dense intersection graph of n segments in the plane whose largest clique or independent set is of size $O(n^{\frac{\log 8}{\log 169}})$. Since incomparability graphs are perfect graphs, then there are dense string graphs on n vertices whose largest induced subgraph which is an incomparability graph has $O(n^{2\frac{\log 8}{\log 169}})$ vertices, where $2\frac{\log 8}{\log 169} < .811$.

A *bi-clique* is a complete bipartite graph whose two parts are of equal size. It follows from a result of Kővári, Sós, and Turán [25] that every graph on n vertices or its complement contains a (not necessarily induced) subgraph which is a bi-clique with $\log n - \log\log n$ vertices in each of its parts.[1] Considering a random graph on n vertices, we obtain that this bound is tight apart from a constant factor. Fox [11] proved that every incomparability graph or its complement contains a bi-clique whose parts are of size $\frac{n}{4\log n}$, and that this bound is also tight up to a constant factor. This result was applied by Fox, Pach, and Cs. Tóth [14] to show that for every $\varepsilon > 0$ there exists $\delta > 0$ such that every incomparability graph with n vertices and εn^2 edges contains a bi-clique of size $\delta n/\log n$. Here δ can be taken to be a polynomial in ε. We have the following immediate corollary of Theorem 1.1 and this result.

[1]Throughout this paper, for the sake of simplicity, we systematically omit floor and ceiling signs whenever they are not crucial. All logarithms are base 2.

COROLLARY 1.2. *For every $\varepsilon > 0$, there exists $\delta > 0$ such that every string graph with n vertices and at least εn^2 edges contains a bi-clique with parts of size $\delta n/\log n$.*

In this corollary, δ again can be taken to be a polynomial in ε. In other words, every collection C of n curves in the plane with at least εn^2 intersecting pairs has two subcollections A and B each of size at least $\delta n/\log n$ such that every curve in A intersects every curve in B. By the construction in [11], the dependence on n in Corollary 1.2 is tight.

Pach and G. Tóth [37] conjectured that for every collection C of n curves in the plane, any pair of which intersect in at most k points, the intersection (string) graph of C or its complement contains a bi-clique of size $c_k n$, where $c_k > 0$ depends only on k. This conjecture was proved by Fox, Pach, and Cs. Tóth [13]. The main ingredient of the proof was a variant of Corollary 1.2 for intersection graphs of curves with a bounded number of intersection points per pair. A similar result for intersection graphs of convex sets was established in [14].

The importance of arrangements of curves and Theorem 1.1 in particular is exhibited in its applications to graph drawing problems. A *topological graph* is a graph drawn in the plane with vertices as points and edges as curves connecting corresponding endpoints. The well known Crossing Lemma discovered by Ajtai, Chvátal, Newborn, and Szemerédi [3] and independently by Leighton [32] says that every topological graph with n vertices and $m \geq 4n$ edges has $\Omega\left(m^3/n^2\right)$ pairs of crossing edges. By induction, this is equivalent to the statement that every topological graph with n vertices and $m \geq 3n$ edges has an edge that intersects $\Omega\left(m^2/n^2\right)$ other edges. An *ℓ-grid* in a topological graph is a pair of disjoint edge subsets E_1, E_2 such that every edge in E_1 crosses every edge in E_2. Is the following strengthening of the Crossing Lemma true: every topological graph with n vertices and $m \geq 3n$ edges contains an ℓ-grid with $\ell = \Omega\left(m^2/n^2\right)$? With Cs. Tóth in [15], we show that the answer is "yes" if we assume that every pair of curves in the topological graph intersect in at most a fixed constant number of points, but "no" in general. Indeed, we construct a drawing of the complete bipartite graph $K_{n,n}$, which does not contain an ℓ-grid with $\ell \geq cn^2/\log n$, where c is an absolute constant. This counterexample cannot be substantially improved: using Corollary 1.2, together with a result of Kolman and Matoušek [24] relating the bisection width and the pairwise crossing number of a graph, we prove in [15] that every topological graph with n vertices and $m \geq 3n$ edges contains an ℓ-grid with $\ell = \Omega\left(\frac{m^2/n^2}{\log^c m/n}\right)$. It was proved in [34] that for each positive integer ℓ there is a constant c_ℓ such that every topological graph with n vertices and at least $c_\ell n$ edges contains an ℓ-grid. Their proof gives that we may take $c_\ell = 16 \cdot 24^{4^\ell}\ell$, which is double-exponential in ℓ, while the above mentioned result shows that we may take $c_\ell = \sqrt{\ell}\log^c \ell$ for some absolute constant c, which is best possible up to the polylogarithmic factor. With Ackerman and Suk in [2], we show using Corollary 1.2 that every topological graph with n vertices and no ℓ-grid with *distinct vertices* has at most $c_\ell n \log^* n$ edges, where $c_\ell = \ell^{O(\log\log \ell)}$ and $\log^*$ is the iterated logarithm function.

It is a general question in geometric graph theory to investigate how much one can relax planarity while still ensuring that the graph is sparse? To formalize this question we need the following definition. A topological graph is k-

quasi-planar if no k edges pairwise intersect. A well known conjecture states that every k-quasi-planar topological graph on n vertices has at most $c_k n$ edges for some constant c_k depending only on k. This conjecture is only known for $k \leq 4$. The case $k = 2$ follows easily from Euler's polyhedral formula, the case $k = 3$ was proved by Pach, Radoičić, and G. Tóth [35], and the case $k = 4$ was proved by Ackerman [1]. The best known upper bound on the number of edges of a k-quasi-planar topological graph on n vertices is by Ackerman, who gave an upper bound of the form $c_k n(\log n)^{4k-16}$ for $k \geq 4$. In another paper [12], we again use Corollary 1.2 together with the result of Kolman and Matoušek [24] to obtain a new upper bound on the number of edges in a k-quasi-planar topological graph. We show that every k-quasi-planar topological graph on n vertices has at most $n(\log n)^{c \log k}$ edges, where c is an absolute constant. In particular for each $\varepsilon > 0$ there is $\delta > 0$ and n_0 such that every topological graph on $n \geq n_0$ vertices and at least $n^{1+\varepsilon}$ edges has $n^{\delta/\log\log n}$ pairwise crossing edges. This is a significant improvement on the previous bound of $\delta \log n/\log\log n$.

We remark that the proof of Theorem 1.1 can be made algorithmic. That is, there is an algorithm which runs in polynomial time such that, given a collection of n curves whose string graph has ϵn^2 edges, it finds a collection of subcurves whose string graph has at least δn^2 edges and is an incomparability graph. For this algorithm to work, the input includes the curves with a parametrization, the intersection points along with their location on the curves, and, at each intersection point, which side each curve intersects the other curve. Consequently, there are polynomial time algorithms for Corollary 1.2 and the other consequences of Theorem 1.1 mentioned above. We stress here that the input for these algorithms is not simply the string graph, but also the collection of curves as described above. The running time thus depends polynomially on the number of curves and intersection points. As discussed earlier, Kratochvíl and Matoušek [29] showed that for some string graphs on n vertices, the number of intersection points in any representation is exponentially large in n.

To make our paper self-contained, in the next section we present the (few lines long) proof of the fact discovered by Golumbic, Rotem, and Urrutia [18] and Lovász [33] that every incomparability graph is a string graph. In Section 3, we establish a simple lemma showing that every dense graph has a cubic number of triangles K_3 or a quartic number of induced claws $K_{1,3}$. In the proof of Theorem 1.1, we will distinguish between the case that the string graph of the collection of curves has few triangles and the case in which it has many triangles. These two proofs will be presented in Sections 5 and 7, respectively.

In Section 4, we introduce some important notions: we define "grounded", "double-grounded", "strongly double-grounded", and "split" families of curves. For instance, roughly speaking, a family C is called grounded if there is a special curve, a so-called "ground" curve γ, with the property that all elements of C have an endpoint on γ, but otherwise they are disjoint from it. (The other definitions will be similar.) We show that the intersection graphs of split families of curves are incomparability graphs (Lemma 4.2). This elementary fact plays an important role in our arguments. The proof of Theorem 1.1 in the case where the string graph has few triangles is presented in Section 5. In Section 6, we settle the case where the collection of curves is grounded. We wrap up the proof in the following section.

In most parts of this paper, we assume for simplicity that all families of curves we consider are in "general position", i.e., no point belongs to *three* curves of the family. In Section 8, we outline how to get rid of this assumption. For the clarity of the presentation, we do not make any serious attempt to optimize the absolute constants appearing in our statements and proofs.

2. INCOMPARABILITY GRAPHS AS STRING GRAPHS

We use the notation $[n] = \{1, \ldots, n\}$. A *linear extension* of a poset $(P, \prec)$ on n elements is a one-to-one map $\pi : P \to [n]$ such that if $y \prec z$, then $\pi(y) < \pi(z)$. The *intersection* of a set Π of one-to-one maps from a set P on n elements to $[n]$ is the poset $(P, \prec)$ such that $y \prec z$ if and only if $\pi(y) < \pi(z)$ for every $\pi \in \Pi$ and for all $y, z \in P$. It is straightforward to show that every poset is the intersection of its linear extensions. The *dimension* of a poset is the minimum number of linear extensions whose intersection is that poset. An old result of Hiraguchi [21] states that every poset on $n \geq 4$ elements has dimension at most $n/2$. See the book [42] by Trotter for more on the dimension theory for posets.

If $f_1, \ldots, f_n : [0, 1] \to \mathbb{R}$ are continuous functions, we can define a partial order $\prec$ on these functions by $f_i \prec f_j$ if $f_i(x) < f_j(x)$ for all $x \in [0, 1]$. The following proposition implies that every partially ordered set can be represented in this way. In particular, it implies that the a graph is an incomparability graph if and only if it is the intersection graph of a collection of curves given by continuous functions defined on $[0, 1]$.

PROPOSITION 2.1. *([18], [33]) For each partial order $\prec$ on $[n]$, there is a family of continuous functions $f_1, \ldots, f_n : [0, 1] \to \mathbb{R}$ such that $i \prec j$ if and only if $f_i(x) < f_j(x)$ for each $x \in [0, 1]$.*

PROOF. Let $([n], \prec)$ be a poset with dimension d, and let $\Pi = \{\pi_1, \ldots, \pi_d\}$ denote a set of d linear extensions whose intersection is the poset. Assign to each π_k a distinct point x_k of the interval $[0, 1]$, so that

$$0 = x_1 < x_2 < \ldots < x_d = 1.$$

For each i $(1 \leq i \leq n)$, define a continuous, piecewise linear function $f_i(x)$, as follows. For any k $(1 \leq k \leq d)$, set $f_i(x_k) = \pi_k(i)$, and let $f_i(x)$ change linearly over the interval $[x_k, x_{k+1}]$ for $k < m$.

Obviously, whenever $i \prec j$ for some $i \neq j$, we have that $\pi_k(i) < \pi_k(j)$ for every k, and hence $f_i(x) < f_j(x)$ for all $x \in [0, 1]$. On the other hand, if i and j are incomparable with respect to the ordering $\prec$, we find that there are indices k and k' $(1 \leq k \neq k' \leq m)$ such that $f_i(x_k) < f_j(x_k)$ and $f_i(x_{k'}) > f_j(x_{k'})$, therefore, by continuity, the curves of f_i and f_j must cross at least once in the interval $(x_k, x_{k'})$. This completes the proof. □

The proof of Proposition 2.1, together with Hiraguchi's theorem mentioned in the first paragraph of this section, implies that every incomparability graph with n vertices is the intersection graph of a collection of curves given by continuous functions defined on the interval $[0, 1]$ such that every pair intersect in at most $n/2 - 1$ points. Every bi-clique is the incomparability graph of a 2-dimensional poset and is the intersection graph of a collection of segments (intersecting in at most one point per pair).

On the other hand, according to a result of Kratochvíl and Matoušek [29], there are string graphs with n vertices that require an exponential number of intersection points in any of their realizations. Consequently, the "geometric complexity" of a string graph may be much larger than the complexity of the canonical substructures whose existence is guaranteed by Theorem 1.1 and Corollary 1.2.

3. TRIANGLES AND CLAWS

The *clique multiplicity* $k_s(G)$ is the number of cliques of size s in graph G. The *Ramsey multiplicity* $k_s(n) = \min\{k_s(G) + k_s(\bar{G}) : |G| = n\}$ is the minimum number of cliques or independent sets of size s over all graphs G on n vertices. The exact value of $k_3(n)$ was determined by Goodman [19]:

$$k_3(n) = \begin{cases} \frac{n(n-2)(n-4)}{24} & \text{if } n \text{ is even} \\ \frac{n(n-1)(n-5)}{24} & \text{if } n \equiv 1 \pmod 4 \\ \frac{(n+1)(n-3)(n-4)}{24} & \text{if } n \equiv 3 \pmod 4 \end{cases}$$

Note that $k_3(n)$ is asymptotic to $n^3/24$ and we will use the estimate $k_3(n) \geq n^3/32$ for $n \geq 24$. See the paper [6] by Conlon for more details on $k_s(n)$ with $s > 3$.

For any two subsets of vertices, $A, B \subseteq V(G)$, let $e(A, B)$ denote the number of edges of G with one endpoint in A and the other in B. A *claw* is a graph with *four* vertices and *three* edges, having a vertex, the *root*, which is adjacent to the remaining *three* vertices.

LEMMA 3.1. *Let G be a graph, and let A and B be (not necessarily disjoint) subsets of $V(G)$ with $e(A, B) \geq \varepsilon n^2$ and $\varepsilon \geq 48/n$. Then B contains at least $2^{-9}\varepsilon^3 n^3$ triangles or there are at least $2^{-9}\varepsilon^3 n^4$ induced claws whose root is in A and whose* three *other vertices are in B.*

PROOF. Delete all vertices from A that are adjacent to at most $\varepsilon n/2$ vertices in B, and let A' be the resulting subset of vertices. Since $e(A \setminus A', B) \leq (\varepsilon n/2)|A \setminus A'| \leq \varepsilon n^2/2$, then there are at least $\varepsilon n^2 - \varepsilon n^2/2 = \varepsilon n^2/2$ edges with one vertex in A' and the other vertex in B.

For $a \in A'$, let $N(a)$ denote the subset of vertices in B adjacent to a, so $|N(a)| \geq \varepsilon n/2 \geq 24$. By Goodman's theorem, $N(a)$ contains at least $|N(a)|^3/64$ triangles or at least $|N(a)|^3/64$ independent sets of size three. In the first case, $N(a)$ and hence B contains at least $|N(a)|^3/64 \geq (\varepsilon n/2)^3/64 = 2^{-9}\varepsilon^3 n^3$ triangles, and we are done. So we may suppose that there are at least $|N(a)|^3/64$ independent sets of size 3 in $N(a)$ for each $a \in A'$. Hence, the number of induced claws whose root is in A and whose other three vertices are in B is at least

$$\sum_{a \in A'} |N(a)|^3/64 \geq |A'| \left(\sum_{a \in A'} |N(a)|/|A'| \right)^3 /64$$

$$= \frac{e(A, B')^3}{64|A'|^2} \geq \frac{(\varepsilon n^2/2)^3}{64|A'|^2} \geq 2^{-9}\varepsilon^3 n^4,$$

where the first inequality is by Jensen's inequality for the convex function $f(x) = x^3$, $x \geq 0$, and the last inequality follows from $|A'| \leq n$. □

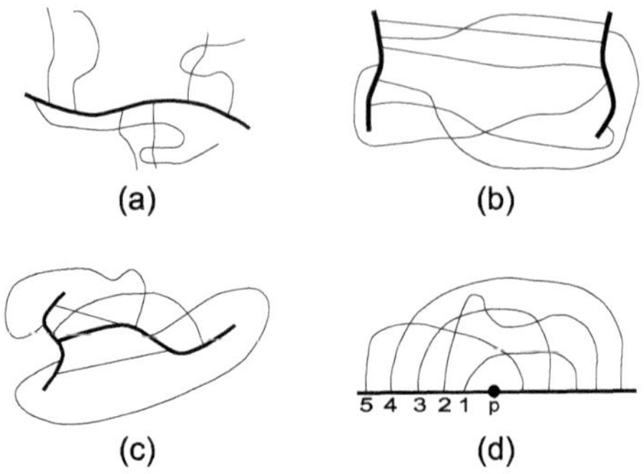

Figure 1: In (a), a grounded collection of curves. In (b), a double-grounded collection of curves with separator γ. In (c), a strongly double-grounded collection of curves. In (d), a split collection of curves with middle point p.

The complete bipartite graph $K_{a,b}$ on $a + b$ vertices has parts of size a and b with all edges between them. In particular, $K_{1,3}$ is the claw. The above lemma can be easily extended to show that for any fixed a, b, s, every dense graph on n vertices contains $\Omega(n^s)$ copies of the complete graph K_s or $\Omega(n^{a+b})$ induced copies of the complete bipartite graph $K_{a,b}$. One way to do this using the Ramsey multiplicity of the complete graph (see, e.g., [6]).

4. SPECIAL STRING GRAPHS THAT ARE INCOMPARABILITY GRAPHS

The purpose of this section is to develop notation and terminology necessary for the proof of Theorem 1.1 and to prove a simple lemma showing that the intersection graphs of rather special collections of curves are incomparability graphs.

Throughout this paper, unless it is stated otherwise, we always assume that the curves we consider are in *general position*, i.e., that no point belongs to three of them. After completing the proof of Theorem 1.1 for collections of curves in general position, in Section 8 we discuss how our arguments can be modified to deal with the degenerate cases.

Recall that a curve γ is a subset of the plane homeomorphic to the unit interval $[0, 1]$. That is, a curve γ is the image of a homeomorphism f from $[0, 1]$ to a subset of the plane $\mathbb{R}^2$. We associate the function f with the curve γ. In particular, each curve comes with an orientation from the *starting point* $f(0)$ to the *final point* $f(1)$. The points $f(0)$ and $f(1)$ are *endpoints of* γ. The other points of γ are *interior points of* γ. For distinct points $f(a), f(b)$ of γ, $f(a)$ *comes before* $f(b)$ *along* γ if $a < b$, and otherwise $f(a)$ *comes after* $f(b)$ *along* γ. A *subcurve* of a curve γ is the image of the function f restricted to a subinterval $[a, b]$ of $[0, 1]$. In particular, a subcurve is a curve if $a < b$, it consists of a single point if $a = b$, and it is the empty set if $a > b$.

A collection C of curves in the plane is said to be *grounded* if there is a curve α such that every member γ of C has precisely one endpoint on α and the rest of γ is disjoint from α (see Figure 1(a)). The curve α is called a *ground* for C. Since we consider curves in general position, so that $C \cup \{\alpha\}$ is in general position, in a grounded collection C, no point of the ground α can belong to *two* elements of C.

A collection C of curves is *double-grounded* if there are disjoint curves α, α' such that every member of C has one endpoint on α, the other endpoint on α', and the rest of the curve is disjoint from $\alpha \cup \alpha'$. A collection C of curves is *double-grounded with a separator* if there is a curve γ such that $C \cup \{\gamma\}$ is double-grounded and every curve in C is disjoint from γ (see Figure 1(b)). We then call γ a *separator* for C. A collection C of curves is *strongly double-grounded* if there is an ordered pair (α, α') of curves with no interior point in common such that one of the endpoints of α' lies on α, and every member of C has one endpoint on α, the other on α', and the rest of it is disjoint from $\alpha \cup \alpha'$ (see Figure 1(c)). Finally, we call a collection C of curves *split* if there is a curve α and a point p in the interior of α such that every member γ in C has one endpoint before p along α, the other endpoint after p along α, the interior of γ is disjoint from α, and the ends of curves on C all lie on the same side of α (see Figure 1(d)). In this case we say that α splits C and call p a middle point for C.

By tracing along the exterior of $\alpha \cup \alpha' \cup \gamma$ of a double grounded collection with grounds α, α' and separator γ, we see that every double-grounded collection of curves with a separator is split. Similarly, by tracing along the two grounds of a strongly double-grounded collection of curves, we see that every strongly double-grounded collection of curves is split. One can easily check the other cases of the following simple proposition.

PROPOSITION 4.1. *A collection of curves is double-grounded with a separator if and only if it is strongly double-grounded if and only if it is split.*

We are implicitly assuming that the collection of curves are in general position. The above proposition fails to hold if this assumption is not made, as there could be a pair of curves in the collection which share an endpoint on different sides of a ground, which would make it impossible for the collection of curves to be split. This technical issue is discussed in detail in Section 8.

For any pair of intersecting (oriented) curves (α, β), let $p(\alpha, \beta)$ denote the first point along α that belongs to β. (The existence of such a point follows from the fact that β is a closed set and α is homeomorphic to the unit interval.) Furthermore, let $\alpha(\beta)$ denote the subcurve of α with the same starting point as α and with final point $p(\alpha, \beta)$.

We finish this section with the following lemma followed by a useful remark.

LEMMA 4.2. *The intersection graph of every split collection S of curves is an incomparability graph.*

PROOF. Let α be a ground for S with middle point p. We can label the curves in S according to the order of their endpoints along α, as $\{\gamma_1, \ldots, \gamma_n\}$, starting at the middle point p and increasing label in the direction toward the starting point, so the curve γ_i has label i. See Figure 1(d). Define a binary relation $\prec$ on S, as follows. Let $\gamma_i \prec \gamma_j$ if and only if γ_i is disjoint from γ_j and $i < j$. For each curve γ_j, there is a closed Jordan curve β_j which consists of γ_j together with the subcurve of α whose endpoints are the endpoints of γ_j. By the Jordan curve theorem, if $i < j < k$, curve γ_i is disjoint from curve γ_j and curve γ_j is disjoint from curve γ_k, then the interior of curve γ_i lies in the interior of the Jordan region bounded by β_j and the interior of curve γ_k lies in the exterior of the Jordan region bounded by β_j, so curve i is disjoint from curve k. Thus, $\prec$ is a partial order and the intersection graph of S is an incomparability graph. □

By Proposition 4.1 and Lemma 4.2, the intersection graphs of double-grounded collections of curves with a separator and strongly double-grounded collections of curves are incomparability graphs.

The *join* of two graphs $G_1 = (V_1, E_1)$ and $G_2 = (V_2, E_2)$ is the graph union $G_1 \cup G_2$ together with all edges between V_1 and V_2. It is easy to check the following useful remark.

Remark: The join of two incomparability graphs is an incomparability graph. Similarly, the disjoint union of two incomparability graphs is an incomparability graph. In particular, an incomparability graph with added isolated vertices is also an incomparability graph, so a graph with just one edge is an incomparability graph.

In the proof of Theorem 1.1, we are allowed to take the empty subcurve for some of the curves in C, and as long as the the intersection graph of the nonempty subcurves is a dense incomparability graph, we are done, as the empty subcurves are isolated vertices in the intersection graph. So if C is a collection of curves in the plane whose intersection graph is sparse but has at least one edge, then we will take C' to be the collection of subcurves of the elements of C consisting of a pair of intersecting curves in C and the *empty* subcurve for each of the remaining elements. The intersection graph of C', consisting of a single edge and $|C|-2$ isolated vertices, is clearly an incomparability graph.

5. STRING GRAPHS WITH FEW TRIANGLES

The aim of this section is to prove Theorem 1.1 for every collection C of curves whose string graph has few triangles (see Theorem 5.4). This is the first half of the proof of Theorem 1.1. We start with three simple but useful lemmas whose proofs are omitted in this extended abstract.

LEMMA 5.1. *Let A, B be collections of curves such that there are q ordered quadruples $(\alpha, \beta_1, \beta_2, \beta_3) \in A \times B \times B \times B$ such that $\beta_1, \beta_2, \beta_3$ are pairwise disjoint and α intersects β_1, β_2, and β_3. Then there are disjoint curves $\beta, \beta' \in B$, a subcollection $B' \subset B$ such that every curve in B' is disjoint from β and β', and a double-grounded collection A' of subcurves of curves of A with grounds β, β' such that there are more than $\frac{q}{6|B|^2}$ intersecting pairs of curves in $A' \times B'$.*

LEMMA 5.2. *Let A be a double-grounded collection of at most n curves with grounds β, β', and let B be a collection of at most n curves disjoint from β, β' such that the number of intersecting pairs in $A \times B$ is at least εn^2 with $\varepsilon \geq 48/n$, and the number of pairwise intersecting triples in A is less than $2^{-9}\varepsilon^3 n^3$. Then we can find disjoint curves $\alpha, \alpha' \in A$ and a double-grounded collection B' of subcurves of curves in B with grounds α, α' such that the subcollection A' consisting of all curves in A that are disjoint from α and α' has the property that the the number of intersecting pairs of curves in $A' \times B'$ is at least $2^{-9}\varepsilon^3 n^4/|A|^2$.*

LEMMA 5.3. *Let $A \cup \{\alpha, \alpha'\}$ be a double-grounded collection of curves with grounds β, β', and let B be a double-grounded collection of curves with grounds α, α' such that every curve in A is disjoint from α and α', and every curve in B is disjoint from β and β'. Then $A \cup B$ is the disjoint* *union of two collections of split curves such that every curve in the first split collection is disjoint from every curve in the second split collection.*

THEOREM 5.4. *Let C be a collection of curves such that the intersection graph of C has at least εn^2 edges and fewer than $2^{-36}\varepsilon^9 n^3$ triangles. Then for each $\gamma \in C$, there is a subcurve γ' of γ such that the intersection graph of the collection $\{\gamma' : \gamma \in C\}$ is an incomparability graph with at least $2^{-45}\varepsilon^9 n^2$ edges.*

The proof follows by combining the Remark at the end of Section 4, Lemmas 3.1, 5.1, 5.2, and 5.3, Lemma 4.2.

6. GROUNDED COLLECTIONS OF CURVES

The aim of this section is to prove Theorem 1.1 for (nondegenerate) grounded collections of curves (see Theorem 6.6). This is an important step toward the proof of Theorem 1.1 in the general case, which will be presented in the next section. We first collect several useful lemmas for families of grounded curves.

In the previous section we proved Theorem 1.1 in the case when the string graph has few triangles. To prove Theorem 1.1 in the case when the string graph has many triangles, we have to classify the different ways how *three* curves can pairwise intersect. We start with a simple observation.

PROPOSITION 6.1. *Let $T = \{\alpha_1, \alpha_2, \alpha_3\}$ be a set of* three *pairwise intersecting curves. There is a permutation $\gamma_1, \gamma_2, \gamma_3$ of $\alpha_1, \alpha_2, \alpha_3$ such that $p(\gamma_3, \gamma_1)$ comes before $p(\gamma_3, \gamma_2)$ along γ_3 and $p(\gamma_1, \gamma_2)$ comes before $p(\gamma_1, \gamma_3)$ along γ_1. In particular, if γ is a subcurve of γ_3 with the same starting point as γ_3 and whose final point is at least $p(\gamma_3, \gamma_1)$ and before $p(\gamma_3, \gamma_2)$ along γ_3, then γ intersects γ_1 and is disjoint from γ_2, and γ_2 intersects $\gamma_1(\gamma)$. We call such a curve γ_3* nice *for the triple T.*

PROOF. We may suppose by symmetry that $p(\alpha_3, \alpha_1)$ comes before $p(\alpha_3, \alpha_2)$ along α_3.

Case 1: $p(\alpha_1, \alpha_2)$ comes before $p(\alpha_1, \alpha_3)$ along α_1. Then we may let $\gamma_i = \alpha_i$ for $i = 1, 2, 3$ and the lemma follows. Indeed, in this case $\gamma_1(\gamma_3)$ intersects γ_2, and if γ is a subcurve of γ_3, then $\gamma_1(\gamma_3)$ is a subcurve of $\gamma_1(\gamma)$ and so $\gamma_1(\gamma)$ intersects γ_2.

Thus, we may assume that we are in the following case.

Case 2: $p(\alpha_1, \alpha_3)$ comes before $p(\alpha_1, \alpha_2)$ along α_1. This case has two subcases.

Case 2a: Point $p(\alpha_2, \alpha_1)$ comes before $p(\alpha_2, \alpha_3)$ along α_2. Letting $\gamma_3 = \alpha_2$, $\gamma_2 = \alpha_3$, and $\gamma_1 = \alpha_1$, as in Case 1, the proposition follows.

Case 2b: Point $p(\alpha_2, \alpha_3)$ comes before $p(\alpha_2, \alpha_1)$ along α_2. Letting $\gamma_3 = \alpha_2$, $\gamma_2 = \alpha_1$, and $\gamma_1 = \alpha_3$, as in Case 1, the proposition follows. □

Proposition 6.1 is needed for the proof of the next lemma.

LEMMA 6.2. *If C is a collection of at most n curves with at least εn^3 triples of pairwise intersecting members, then there is a subcurve γ' of a curve $\gamma \in C$ with the same starting point as γ and disjoint subcollections A and B of C such that each curve in A intersects γ', no curve in B intersects γ', and the following holds. Letting A' denote the set of curves $\alpha(\gamma')$ with $\alpha \in A$, γ' is a ground for each curve in A' and there are more than $\frac{\varepsilon^2}{2}n^2$ pairs in $A' \times B$ that intersect.*

PROOF. By Lemma 6.1, in each triple of curves that pairwise intersect, there is a curve that is nice for that triple. By averaging, there is a curve $\gamma \in C$ that is nice for at least $\varepsilon n^3/|C| \geq \varepsilon n^2$ triples. A triple T of curves is *helpful* if $\gamma \in T$, the curves in T pairwise intersect, and γ is nice for T. So the number of helpful triples is at least εn^2. Let α_i denote the ith curve that intersects γ along γ, breaking ties arbitrarily. The number of quadruples $\{\alpha_i, \alpha_k, \alpha_j, \gamma\}$ with $i \leq k < j$ and $T = \{\alpha_i, \alpha_j, \gamma\}$ a helpful triple is the sum of $j - i$ over all helpful triples $T = \{\alpha_i, \alpha_j, \gamma\}$. Let $f(d)$ denote the number of helpful triples $T = \{\alpha_i, \alpha_j, \gamma\}$ with $d = j - i$. The number of helpful triples is $\sum_d f(d)$, which is at least εn^2. We also have $f(d) \leq n$ for each d. Hence, the number of quadruples $\{\alpha_i, \alpha_k, \alpha_j, \gamma\}$ with $i \leq k < j$ and $T = \{\alpha_i, \alpha_j, \gamma\}$ a helpful triple is $\sum_d df(d) \geq n\left(1 + 2 + \cdots + \frac{\varepsilon n^2}{n}\right) > \frac{\varepsilon^2 n^3}{2}$. By averaging, there is a value of k such that there are more than $\frac{\varepsilon^2 n^3/2}{n} = \frac{\varepsilon^2}{2}n^2$ helpful triples $T = \{\alpha_i, \alpha_j, \gamma\}$ such that $p(\gamma, \alpha_i)$ is at most $p(\gamma, \alpha_k)$ along γ and $p(\gamma, \alpha_k)$ comes before $p(\gamma, \alpha_j)$ along γ. Let γ' denote the subcurve of γ with the same starting point as γ and final point $p(\gamma, \alpha_k)$. Let A be those curves α_h that intersect γ with $h \leq k$. Let B be those curves α_h that intersect γ with $h > k$. By construction, every curve in A intersects γ' and no curve in B intersects γ'. Also, γ' is a ground for A', and more than $\frac{\varepsilon^2}{2}n^2$ pairs of curves in $A' \times B$ intersect. □

We apply the last lemma to a grounded collection of curves, in the second case of the next lemma.

LEMMA 6.3. *Let C be a grounded collection of at most n curves with ground γ, for which there are εn^2 pairs of curves in C that intersect with $\varepsilon \geq 1/n$. There is a subcurve α' of a curve $\alpha \in C$ with the same starting point on γ as α, a subcollection A of C, a strongly double-grounded collection A' of subcurves of A with grounds (γ, α'), and a subcollection $B \subset C$ of curves disjoint from α' such that there are at least $\frac{\varepsilon^4}{72}n^2$ pairs of curves in $A' \times B$ that intersect.*

PROOF. We may assume that the starting point of each curve in C is its endpoint on γ. Let X denote the set of ordered triples $(\gamma_1, \gamma_2, \gamma_3)$ of distinct curves in C with γ_3 intersecting γ_1 and γ_2 and $p(\gamma_3, \gamma_1)$ coming before $p(\gamma_3, \gamma_2)$ along γ_3. For each curve $\gamma_3 \in C$, let $C(\gamma_3)$ denote the collection of curves in C that intersect γ_3. So $\sum_{\gamma_3 \in C} |C(\gamma_3)| = 2\varepsilon n^2$. By convexity of the function $f(x) = x^2$, we have

$$\begin{aligned}
|X| &= \sum_{\gamma_3 \in C} \binom{|C(\gamma_3)|}{2} = \frac{1}{2}\sum_{\gamma_3 \in C} \left(|C(\gamma_3)|^2 - |C(\gamma_3)|\right) \\
&= -\varepsilon n^2 + \frac{1}{2}\sum_{\gamma_3 \in C} |C(\gamma_3)|^2 \\
&\geq -\varepsilon n^2 + \frac{|C|}{2}\left(\sum_{\gamma_3 \in C} |C(\gamma_3)|/|C|\right)^2 \\
&= -\varepsilon n^2 + \frac{1}{2|C|}(2\varepsilon n^2)^2 \geq \varepsilon^2 n^3.
\end{aligned}$$

Let X_1 denote the collection of ordered triples $(\gamma_1, \gamma_2, \gamma_3)$ in X for which γ_2 is disjoint from γ_1, and let $X_2 = X \setminus X_1$. The proof splits into two cases depending on which of the sets X_1 and X_2 is larger.

Case 1: $|X_1| \geq |X_2|$. In this case we have $|X_1| \geq |X|/2 \geq \frac{\varepsilon^2}{2}n^3$. By averaging, there is a curve α that is the second coordinate for at least $|X_1|/|C| \geq \frac{\varepsilon^2}{2}n^2$ triples in X_1. Let A denote the collection of those curves in C that intersect α, and let A' be the collection of curves $\kappa(\alpha)$ with $\kappa \in A$. So A' is strongly double grounded with grounds (γ, α). Let B be the subcollection of curves in C that are disjoint from α. Each triple $(\beta, \alpha, \kappa) \in X_1$ with second coordinate α satisfies $\kappa \in A$, $\beta \in B$, and $\kappa(\alpha)$ intersects β. Hence, there are at least $\frac{\varepsilon^2}{2}n^2$ pairs in $A' \times B$ that intersect.

Case 2: $|X_2| > |X_1|$. In this case we have $|X_2| > |X|/2 \geq \frac{\varepsilon^2}{2}n^3$. So there are at least $\frac{\varepsilon^2}{6}n^3$ unordered triples of pairwise intersecting curves in C. By Lemma 6.2, there is a subcurve α' of a curve $\alpha \in C$ with the same starting point as α (which is on γ) and there are disjoint subcollections A and B of C for which each curve in A intersects α' and no curve in B intersects α', and letting A' denote the set of curves $\kappa(\alpha')$ with $\kappa \in A$, the collection A' is strongly double-grounded with grounds (γ, α') and there are at least $(\frac{\varepsilon^2}{6})^2 n^2/2 = \frac{\varepsilon^4}{72}n^2$ pairs in $A' \times B$ that intersect. □

LEMMA 6.4. *Let A be a strongly double-grounded collection of at most n curves with grounds (γ, α). Let B be a grounded collection of at most n curves also with ground γ such that every curve in B is disjoint from α. If the number of pairs in A that intersect is at most $\frac{\varepsilon^2}{8}n^2$ and the number of pairs of curves in $A \times B$ that intersect is εn^2 with $\varepsilon \geq 2/n$, then there is a curve $\kappa \in A$ satisfying the following property. Letting A' denote the set of curves in A that are disjoint from κ, and letting B' denote the set of subcurves $\beta(\kappa)$ for which $\beta \in B$ intersects κ, B' is strongly double-grounded with grounds (γ, κ) and there are at least $\frac{\varepsilon^2}{8}n^2$ pairs of curves in $A' \times B'$ that intersect.*

PROOF. Let X denote the set of ordered triples $(\beta, \alpha_1, \alpha_2) \in B \times A \times A$ with α_1 and α_2 disjoint, β intersecting α_1 and α_2, and $p(\beta, \alpha_1)$ coming before $p(\beta, \alpha_2)$ along the curve β. For each curve $\beta \in B$, let $A(\beta)$ denote the collection of curves in A that intersect β. So $\sum_{\beta \in B} |A(\beta)| = \varepsilon n^2$. Since there are fewer than $\frac{\varepsilon^2}{8}n^2$ pairs of curves in A that intersect, then for a given β, there are at least $\binom{|A(\beta)|}{2} - \frac{\varepsilon^2}{8}n^2$ ordered triples $(\beta, \alpha_1, \alpha_2) \in B \times A \times A$ with α_1 and α_2 disjoint, β intersecting α_1 and α_2, and $p(\beta, \alpha_1)$ coming before $p(\beta, \alpha_2)$ along the curve β. By convexity of the function $f(x) = x^2$, we have

$$\sum_{\beta \in B} \binom{|A(\beta)|}{2} = -\frac{\varepsilon}{2}n^2 + \frac{1}{2}\sum_{\beta \in B} |A(\beta)|^2$$

$$\geq -\frac{\varepsilon}{2}n^2 + \frac{|B|}{2}\left(\sum_{\beta \in B} |A(\beta)|/|B|\right)^2$$

$$= -\frac{\varepsilon}{2}n^2 + \frac{1}{2|B|}\left(\varepsilon n^2\right)^2 \geq \frac{\varepsilon^2}{4}n^3.$$

Hence,

$$|X| \geq \sum_{\beta}\left(\binom{|A(\beta)|}{2} - \frac{\varepsilon^2}{8}n^2\right) \geq \frac{\varepsilon^2}{4}n^3 - |B|\frac{\varepsilon^2}{8}n^2 \geq \frac{\varepsilon^2}{8}n^3.$$

So there is a curve κ that is the second member in at least $\frac{\varepsilon^2}{8}n^3/|A| \geq \frac{\varepsilon^2}{8}n^2$ ordered triples in X. It follows that $A' \times B'$ has at least $\frac{\varepsilon^2}{8}n^2$ intersecting pairs. □

The next lemma is very similar to Lemma 5.3.

LEMMA 6.5. *Suppose $A \cup \{\kappa\}$ is a strongly double-grounded collection of curves with grounds (γ, α) and κ is disjoint from every curve in A. Suppose B is a strongly double-grounded collection of curves with grounds (γ, κ). Then $A \cup B$ is the disjoint union of two collections of split curves such that every curve in the first split collection is disjoint from every curve in the second split collection.*

PROOF. By the Jordan curve theorem, $\gamma \cup \alpha \cup \kappa$ partitions the plane into two regions, an inside region I and an outside region O. Each curve in $A \cup B$ either lies entirely in the closure $\bar{I}$ of I or the closure $\bar{O}$ of O. A curve β in $A \cup B$ that lies entirely in $\bar{I}$ is an *inside curve*, otherwise β is an *outside curve*. Since our curves are in general position, every outside curve is disjoint from every inside curve. By tracing along the outside, we see that the collection of outside curves is split. Similarly tracing along the inside, we see that the collection of inside curves is split. □

Putting together the last three lemmas, we obtain the main result of this section, according to which Theorem 1.1 is true for grounded collections.

THEOREM 6.6. *Suppose C is a grounded collection of at most n curves whose intersection graph has at least εn^2 edges. Then for each $\gamma \in C$, there is a subcurve γ' of γ such that the intersection graph of the collection $\{\gamma' : \gamma \in C\}$ is an incomparability graph with at least $2^{-16}\varepsilon^8 n^2$ edges.*

PROOF. By the remark at the end of Section 4, we may assume that $2^{-16}\varepsilon^8 n^2 > 1$ and hence $\varepsilon > 4n^{-1/4}$.

Let ψ be a ground for C. By Lemma 6.3, we can find a curve $\alpha \in C$ that has a subcurve α' with the same starting point on C as α, a subcollection A of C, a strongly double-grounded collection A' of subcurves of A with grounds (ψ, α'), and a subcollection $B \subset C$ consisting of curves disjoint from α' such that there are at least $\varepsilon_1 n^2$ intersecting pairs in $A' \times B$, where $\varepsilon_1 = \frac{\varepsilon^4}{72}$. Note that $\varepsilon_1 > \frac{(4n^{-1/4})^4}{72} = \frac{32}{9n} > \frac{2}{n}$ and $\varepsilon_1^2 \geq 2^{-13}\varepsilon^8$.

The intersection graph of A' is split by Proposition 4.1. Therefore, we may assume that the intersection graph of A' has fewer than $2^{-16}\varepsilon^8 n^2 < \frac{\varepsilon_1^2}{8}n^2$ edges, otherwise we are done. By Lemma 6.4, there is a curve $\kappa \in A'$ for which the following holds. Let A'' denote the family of all curves in A' that are disjoint from κ, and let B'' denote the family of subcurves $\beta(\kappa)$ with $\beta \in B'$ intersecting κ. The family B'' is strongly double-grounded with grounds (ψ, κ), and there are at least $\varepsilon_2 n^2$ pairs of curves in $A'' \times B''$ that intersect, where $\varepsilon_2 = \frac{\varepsilon_1^2}{8} = \frac{1}{8}\left(\frac{\varepsilon^4}{72}\right)^2 > 2^{-16}\varepsilon^8$. Lemma 6.5 implies that $A'' \cup B''$ is the disjoint union of two split collections such that every curve in the first split collection is disjoint from every curve in the second split collection. Since the intersection graph of every split collection of curves is an incomparability graph (Proposition 4.2), and the disjoint union of two incomparability graphs is an incomparability graph as remarked at the end of Section 4, then the intersection graph of $A'' \cup B''$ is an incomparability graph. This completes the proof. □

7. PROOF OF THEOREM 1.1 FOR NONDEGENERATE COLLECTIONS

In this section, after proving an auxiliary lemma, we prove Theorem 1.1 for nondegenerate collections of curves whose intersection graph has many triangles (Theorem 7.2). Recall that in Section 5 we proved Theorem 1.1 for nondegenerate collections of curves whose intersection graph have few triangles. At the end of this section, we put these two results together to establish a quantitative version of Theorem 1.1 for nondegenerate collections of curves (Theorem 7.3). In the next section, we discuss how the proof can be modified to handle degenerate collections of curves.

LEMMA 7.1. *Let n be a positive integer and $\varepsilon \geq \frac{4}{n}$. Let A be a grounded collection of at most n curves with ground γ. Let B be a collection of at most n curves that are disjoint from γ such that there are εn^2 pairs of intersecting curves in $A \times B$. Then the number of intersecting pairs of curves in A is at least $\frac{\varepsilon^2}{4}n^2$, or we can find a subcollection $A' \subset A$ and a collection B' of subcurves of curves of B such that $A' \cup B'$ is grounded and there are at least $\frac{\varepsilon^2}{8}n^2$ intersecting pairs of curves in $A' \times B'$.*

PROOF. For each curve $\beta \in B$, let $d(\beta)$ be the number of curves in A that intersect β, so that we have $\sum_{\beta \in B} d(\beta) = \varepsilon n^2$. Let β_i denote the ith curve in A that intersects β along β. The number of pairs of curves (β_i, β_j) in A that intersect β with $i < j$ is $\binom{d(\beta)}{2}$. Call a triple $(\beta_i, \beta_j, \beta) \in A \times A \times B$ with $i < j$ *great* if β_i is disjoint from β_j.

We may assume that the number of intersecting pairs of curves in A is less than $\frac{\varepsilon^2}{4}n^2$, so the number of great triples is at least

$$\sum_{\beta \in B} \left(\binom{d(\beta)}{2} - \frac{\varepsilon^2}{4}n^2 \right)$$

$$\begin{aligned} &= -|B|\frac{\varepsilon^2}{4}n^2 + \frac{1}{2}\sum_{\beta \in B}\left(d(\beta)^2 - d(\beta)\right) \\ &\geq -\frac{\varepsilon^2}{4}n^3 - \frac{\varepsilon}{2}n^2 + \frac{1}{2}|B|\Big(\sum_{\beta \in B} d(\beta)/|B|\Big)^2 \\ &= -\frac{\varepsilon^2}{4}n^3 - \frac{\varepsilon}{2}n^2 + \frac{1}{2|B|}(\varepsilon n^2)^2 \geq \frac{\varepsilon^2}{8}n^3. \end{aligned}$$

Here we used the convexity of the function $f(x) = x^2$.

Hence, there is a curve $\alpha \in A$ that is represented at least $\frac{\varepsilon^2}{8}n^3/|A| \geq \frac{\varepsilon^2}{8}n^2$ times as the second member of a great triple. Let B' denote the collection of subcurves $\beta(\alpha)$ with $\beta \in B$ intersecting α. Let A' denote the collection of curves in A that are disjoint from α. By construction, there are at least $\frac{\varepsilon^2}{8}n^2$ intersecting pairs of curves in $A' \times B'$. Tracing around $\gamma \cup \alpha$, we see that $A' \cup B'$ is a grounded collection of curves with at least $\frac{\varepsilon^2}{8}n^2$ intersecting pairs. □

THEOREM 7.2. *Let C be a collection of curves in the plane such that the intersection graph of C has at least εn^3 triangles. Then for each $\gamma \in C$, there is a subcurve γ' of γ such that the intersection graph of the collection $\{\gamma' : \gamma \in C\}$ is an incomparability graph with at least $2^{-56}\varepsilon^{32}n^2$ edges.*

PROOF. By the remark at the end of Section 4, we may assume that $2^{-56}\varepsilon^{32}n^2 > 1$ and hence $\varepsilon > 2^{7/4}n^{-1/16}$.

By Lemma 6.2, there is a subcurve κ' of a curve $\kappa \in C$ with the same starting point as κ and disjoint subcollections A and B of C such that each curve in A intersects κ', no curve in B intersects κ', and the following holds. Letting A' denote the set of curves $\alpha(\kappa')$ with $\alpha \in A$, the collection A' is grounded with ground κ' and there are at least $\varepsilon_1 n^2$ pairs in $A' \times B$ that intersect, where $\varepsilon_1 = \frac{\varepsilon^2}{2}$. Note that $\varepsilon_1 = \frac{\varepsilon^2}{2} > 4/n$.

By Lemma 7.1, there are $\frac{\varepsilon_1^2}{4}n^2$ intersecting pairs of curves in A', or there is a subcollection A'' of A' and a collection B'' of subcurves of curves of B' such that $A'' \cup B''$ is grounded and has at least $\frac{\varepsilon_1^2}{8}n^2$ intersecting pairs of curves. In either case, we have a grounded collection of subcurves with at least $\frac{\varepsilon_1^2}{8}n^2$ intersecting pairs of curves. By Theorem 6.6, for each curve γ in this grounded collection, there is a subcurve γ' of γ such that the intersection graph of the collection $\{\gamma' : \gamma \in C\}$ is an incomparability graph with at least

$$2^{-16}(\varepsilon_1^2/8)^8 n^2 = 2^{-40}\varepsilon_1^{16}n^2 = 2^{-40}(\varepsilon^2/2)^{16}n^2 = 2^{-56}\varepsilon^{32}n^2$$

edges. □

We can now establish Theorem 1.1 for nondegenerate collections of curves.

THEOREM 7.3. *Let C be a collection of curves such that the intersection graph of C has εn^2 edges. Then for each $\gamma \in C$, there is a subcurve γ' of γ such that the intersection graph of the collection $\{\gamma' : \gamma \in C\}$ is an incomparability graph with at least $2^{-1208}\varepsilon^{288}n^2$ edges.*

PROOF. The proof splits into two cases depending on the number of triangles in the intersection graph of C. If the intersection graph of C has fewer than $2^{-36}\varepsilon^9 n^3$ triangles, then we are done by Theorem 5.4. If the intersection graph of C has at least $2^{-36}\varepsilon^9 n^3$ triangles, then we are done by Theorem 7.2, noting that $2^{-56}\left(2^{-36}\varepsilon^9\right)^{32} = 2^{-1208}\varepsilon^{288}$. □

8. PROOF OF THEOREM 1.1 FOR DEGENERATE COLLECTIONS

So far we have made the assumption that the curves we consider are in general position, i.e., no point belongs to *three* or more curves. As we have indicated before, this assumption is not essential for the proof of Theorem 1.1, and it was made only for the clarity of the presentation. In this section, we discuss how the proof can be modified to handle collections of curves that are not necessarily in general position.

For any curve γ and any interior point p of γ, every sufficiently small neighborhood of p is partitioned into two regions by γ. If α is a curve that contains p as an endpoint and is otherwise disjoint from γ, α is either a *left curve* of γ at p or a *right curve* of γ at p, depending on which side of γ the curve α intersects. For a grounded collection C of curves with ground γ, an interior point p of γ is called *two-sided* with respect to C if there are $\alpha, \alpha' \in C$ such that α is a left curve of p and α' is a right curve of p.

We now discuss how to prove Theorem 1.1 in the case where the collection of curves is not necessarily in general position. Let C be a collection of n curves with εn^2 intersecting pairs. If there is a point p in the plane which belongs to at least $\delta_1 n$ members of C, where $\delta_1 = \varepsilon^c$ and c is a sufficiently large constant, then we may simply take the set consisting of the point p as the subcurve for each curve containing p and the empty set as the subcurve for each curve not containing p. In this special case, the intersection graph of these subcurves consists of a clique with a quadratic number of edges and the remaining vertices are isolated. It is easy to see that this intersection graph is an incomparability graph. Therefore, we may assume that no point p belongs to at least $\delta_1 n$ members of C. With this assumption, only one minor issue arises when trying to use the same argument as in the case where the curves were in general position. The problem is that Proposition 4.1, which says that a collection of curves is double-grounded with a separator if and only if it is strongly double-grounded if and only if it is split, does not hold in this case.

While the intersection graph of every split collection of curves is an incomparability graph, the intersection graph of a degenerate, strongly-double grounded collection of curves is not necessarily an incomparability graph. Indeed, it is not hard to give an example of such a collection of curves whose intersection graph is the cycle on *five* vertices, and which is not an incomparability graph. Proposition 4.1 is not true if we allow for there to be a point p of one of the two grounds which is two-sided with respect to the collection of curves which is either strongly-double grounded or double-grounded with a separator. If we try to repeat the argument as we did for non-degenerate collections, as we trace along the union of the two grounds of a strongly double grounded collection or a double-grounded collection with a separator to make a curve α which verifies the collection is split, the curve α will have to touch the two-sided point p twice, so that α is not a simple curve. The same problem arises in Lemmas 5.3 and 6.5, when claiming that there are no intersecting pairs of curves between the two split collections (the inside collection and the outside collection) of curves.

To handle these issues, we can use essentially the same proof of Theorem 1.1 as we did for collections of curves in general position, except that whenever we obtain a grounded collection of curves, we have to use the following lemma to find a grounded subcollection of curves with no two-sided points and we can still retain a constant fraction of the intersecting pairs.

LEMMA 8.1. *Let A be a grounded collection of at most n curves with ground α, and let B be a (possibly empty) collection of curves disjoint from α such that $A \times (A \cup B)$ has m intersecting unordered pairs of distinct curves and no point of α is contained in more than m/n elements of A. Then there exists a subcollection $A' \subset A$ such that no interior point of α is two-sided with respect to A' and at least $m/8$ unordered pairs of distinct curves in $A' \times (A' \cup B)$ intersect.*

PROOF. By convexity and by the fact that no point of α is contained in more than m/n elements of A, we obtain that the number of pairs of distinct curves in A that share an endpoint on α is at most $\binom{m/n}{2} n^2/m < m/2$. Thus, there are at least $m/2$ unordered pairs of distinct curves in $A \times (A \cup B)$ that intersect and do not share an endpoint on α. For each interior point p of α that is two-sided with respect to A, either we keep all left curves of p in A or all right curves of p in A, each with equal probability. All other curves we keep. The curves in A we keep make up A'. For each pair of distinct curves in A that do not share an endpoint on α, the probability that they are both kept in A' is at least $1/4$. Also, the probability that a pair of curves in $A \times B$ is in $A' \times B$ is at least $1/2$. Hence, by linearity of expectation, there is a choice of A' for which the number of unordered pairs of distinct curves in $A' \times (A' \cup B)$ that intersect is at least $\frac{1}{4} m/2 = m/8$. □

9. REFERENCES

[1] E. Ackerman, On the maximum number of edges in topological graphs with no four pairwise crossing edges, *Proc. 22nd ACM Symp. on Computational Geometry (SoCG)*, Sedona, AZ, June 2006, 259–263.

[2] E. Ackerman, J. Fox, J. Pach, and A. Suk, On grids in topological graphs, *Proc. 25nd ACM Symp. on Computational Geometry (SoCG)*, Aarhus, Denmark, June 2009, 403–412.

[3] M. Ajtai, V. Chvátal, M. Newborn, and E. Szemerédi, Crossing-free subgraphs, in: *Theory and Practice of Combinatorics,* vol. **60**, Mathematical Studies, North-Holland, Amsterdam, 1982, pp. 9–12.

[4] S. Benzer, On the topology of the genetic fine structure, *Proc. Nat. Acad. Sci.* **45** (1959), 1607–1620.

[5] D. Conlon, A new upper bound for diagonal Ramsey numbers, *Annals of Math.* **170** (2009), 941–960.

[6] D. Conlon, On the Ramsey multiplicity of complete graphs, *Combinatorica*, to appear.

[7] R. P. Dilworth, *Annals of Math.* **51**, (1950), 161–166.

[8] G. Ehrlich, S. Even, and R. E. Tarjan, Intersection graphs of curves in the plane, *Journal of Combinatorial Theory, Series B* **21** (1976), 8–20.

[9] P. Erdős, Some remarks on the theory of graphs, *Bull. Amer. Math. Soc.* **53** (1947), 292–294.

[10] P. Erdős and G. Szekeres, A combinatorial problem in geometry, *Compositio Math.* **2** (1935), 463–470.

[11] J. Fox, A bipartite analogue of Dilworth's theorem, *Order* **23** (2006), 197–209.

[12] J. Fox and J. Pach, Coloring K_k-free intersection graphs of geometric objects in the plane, *Proc. 24th ACM Sympos. on Computational Geometry*, College Park, Maryland, 2008, ACM Press, 346–354. Also in: *European J. Combinatorics*, to appear.

[13] J. Fox, J. Pach, and Cs. Tóth, Intersection patterns of curves, *J. London Math. Soc.* **83** (2011), 389–406.

[14] J. Fox, J. Pach, and Cs. Tóth, Turán-type results for partial orders and intersection graphs of convex sets, *Israel J. Math.* **178** (2010), 29–50.

[15] J. Fox, J. Pach, and Cs. Tóth, A bipartite strengthening of the Crossing Lemma, *J. Combin. Theory Ser. B* **100** (2010), 23–35.

[16] T. Gallai, Transitive Orientierbare Graphen, *Acta. Math. Hungar.* **18** (1967), 25–66.

[17] M. Golumbic, *Algorithmic Graph Theory and Perfect Graphs*, Academic Press, New York, 1980.

[18] M. Golumbic, D. Rotem, and J. Urrutia, Comparability graphs and intersection graphs, *Discrete Math.* **43** (1983), 37–46.

[19] A. Goodman, On sets of acquaintances and strangers at any party, *Amer. Math. Monthly* **66** (1959), 778-783.

[20] R. L. Graham, Problem, in: *Combinatorics, Vol. II* (A. Hajnal and V. T. Sós, eds.), North-Holland, Amsterdam, 1978, p. 1195.

[21] T. Hiraguchi, On the dimension of orders, *Sci. Rep. Kanazawa Univ.* Vol. 4, 1–20.

[22] G. Károlyi, J. Pach, and G. Tóth, Ramsey-type results for geometric graphs. I, *Discrete Comput. Geom.* **18** (1997), 247–255.

[23] D. J. Kleitman and B. L. Rothschild, Asymptotic enumeration of partial orders on a finite set, *Trans. Amer. Math. Soc.* **205** (1975), 205–220.

[24] P. Kolman and J. Matoušek, Crossing number, pair-crossing number, and expansion. *J. Combin. Theory Ser. B* **92** (2004), 99–113.

[25] T. Kővári, V. T. Sós, P. Turán, On a problem of K. Zarankiewicz, *Colloquium Math.* **3** (1954), 50–57.

[26] J. Kratochvíl, String graphs I: the number of critical nonstring graphs is infinite, *J. Combin. Theory Ser. B* **52** (1991), 53–66.

[27] J. Kratochvíl, String graphs II: recognizing string graphs is NP-hard, *J. Combin. Theory Ser. B* **52** (1991), 67–78.

[28] J. Kratochvíl, M. Goljan, and P. Kučera, *String graphs. Rozpravy Československé Akad. Věd Řada Mat. Přírod.* Věd **96** (1986), 96 pp.

[29] J. Kratochvíl and J. Matoušek, String graphs requiring exponential representations, *J. Combin. Theory Ser. B* **53** (1991), 1–4.

[30] J. Kynčl, Ramsey-type constructions for arrangements of segments, *Electron. Notes Discrete Math.*, **31** (2008), 265–269.

[31] D. Larman, J. Pach, J. Matoušek, and J. Töröcsik, A Ramsey-type result for convex sets, *Bull. London Math. Soc.* **26** (1994), 132–136.

[32] T. Leighton, New lower bound techniques for VLSI, *Math. Systems Theory* **17** (1984), 47–70.

[33] L. Lovász, Perfect graphs, in: *Selected Topics in Graph Theory, vol. 2*, Academic Press, London, 1983, 55–87.

[34] J. Pach, R. Pinchasi, M. Sharir, and G. Tóth, Topological graphs with no large grids, *Graphs and Combinatorics* **21** (2005), 355–364.

[35] J. Pach, R. Radoičić, and G. Tóth, Relaxing planarity for topological graphs, *Discrete and Computational Geometry* (J. Akiyama, M. Kano, eds.), Lecture Notes in Computer Science 2866 Springer-Verlag, Berlin, 2003, 221–232.

[36] J. Pach and G. Tóth, Recognizing string graphs is decidable, *Discrete Comput. Geom.* **28** (2002), 593–606.

[37] J. Pach and G. Tóth, Comment on Fox News, *Geombinatorics* **15** (2006), 150–154.

[38] J. Pach and G. Tóth, How many ways can one draw a graph? *Combinatorica* **26** (2006), 559–576.

[39] M. Schaefer, E. Sedgwick, and D. Štefankovič, Recognizing string graphs in NP. Special issue on STOC2002 (Montreal, QC). *J. Comput. System Sci.* **67** (2003), 365–380.

[40] M. Schaefer and D. Štefankovič, Decidability of string graphs, *J. Comput. System Sci.* **68** (2004), 319–334.

[41] F.W. Sinden, Topology of thin film RC-circuits, *Bell System Technological Journal* (1966), 1639-1662.

[42] W. T. Trotter, *Combinatorics and Partially Ordered Sets. Dimension Theory,* Johns Hopkins Series in the Mathematical Sciences, Johns Hopkins University Press, Baltimore, MD, 1992.

Configuration Space Visualization

Dror Atariah
Institut für Informatik
Freie Universität Berlin
dror.atariah@fu-berlin.de

Günter Rote
Institut für Informatik
Freie Universität Berlin
rote@inf.fu-berlin.de

ABSTRACT

Based on a simple parameterization of contact surfaces, we visualize both the workspace and the corresponding configuration space of a planar polygonal robot that moves amid planar polygonal obstacles.

Categories and Subject Descriptors

G.0 [**Mathematics of Computing**]: General

General Terms

Algorithms, Experimentation, Verification

Keywords

Configuration Space, Workspace, Motion

1. INTRODUCTION

Our video has two goals; the first, to visualize the configuration space of a planar polygonal robot moving amid polygonal obstacles, and its correspondence with the actual workspace. The second one is to introduce an explicit parameterization of contact surfaces in the configuration space. These two goals are, as a matter of fact, strongly related, as the actual visualization of the configuration space became feasible by applying the mentioned parameterization.

We consider a planar convex polygonal robot which is free to translate and rotate amid scattered convex polygonal obstacles. Given these three degrees of freedom, the configuration space is three-dimensional, and can be viewed either as $\mathbb{R}^2 \times \mathbb{R}$ or as $\mathbb{R}^2 \times S^1$. We use the latter representation, and briefly present the former one. We mention the notions of free and forbidden parts of the configuration space, contact surfaces and other related issues. Further background and rigorous discussions can be found in classical textbooks like [2].

The fundamental motion that we study is a rotation of the robot about one of its boundary points. This motion is represented as a helix in the configuration space, and thus can be easily parameterized. Based on the parameterization of the helices, we easily parameterize the contact object (i.e., surfaces patches or curves) corresponding to the various contact types.

Related Work.

A simple way to get an impression of the three-dimensional configuration space is by taking horizontal slices. This correspondes to keeping the rotation angle fixed and allowing the robot only to translate. The resulting two-dimensional slice of the configuration space is polygonal and can be obtained using Minkowski sums. The sequence of slices can be shown in discrete steps (see for example [1, Appendix F]) or animated on a computer. Our goal is more ambitious: we want to show the obstacles in the configuration space directly and exactly in high quality.

A similar video[1] is based on the dissertation of Trac [4]. His video deals with smooth objects, and therefore does not consider notions like the vertex-edge and edge-vertex contacts. In our video, we consider a simpler setting, but we aim for a deeper insight into the configuration space. Another, minor, difference is that we mainly consider the rotation parameter as the bounded interval $[0, 2\pi]$, whereas Trac shows the infinite periodic configuration space.

Finally, another visualization of the configuration space available is the *Snake scenario*[2] which is part of the Master's thesis of Salzman [3]. Here the video shows an instance of a configuration space with a narrow passages, but does not highlight the structure of the various patches of which the contact surfaces are built nor does it visualize any kind of motion.

2. HELIX IN CONFIGURATION SPACE

Given a convex polygonal robot, a boundary point a of the robot and some fixed point P in the workspace, we can consider the rotation of the robot about P such that a is fixed to P. In this case, the robot's *reference point* R_0 traverses, in the workspace, a circle of radius $\|R_0 - a\|$ around P. This circle lifts, in the configuration space, to a helix. From the parameterization of this helix, we can obtain the parameterization of a vertex-edge contact surface by letting P slide along an edge of an obstacle. Similarly, by choosing a varying boundary point a on one of the robot's edges and setting P to be an obstacle vertex, a parameterization of an edge-vertex contact surface can be found. Thus, the parameterization of these helices is the building block of the various contact surfaces.

SCG'12, June 17–20, 2012, Chapel Hill, North Carolina, USA.
ACM 978-1-4503-1299-8/12/06.

[1] `http://sites.google.com/site/justinscsstuff/configuration-space-visualization`

[2] `http://acg.cs.tau.ac.il/projects/mms/videos/configuration%20space.wmv`

3. OVERVIEW OF THE VIDEO

Along the video, the workspace is always on the right and the configuration space on the left of the frame. In the workspace, the robot is plotted in green, and the obstacles in red. The robot's reference point is marked as a black dot. In the configuration space, the robot is represented as an orange ball, which we call the *configuration point*; it carries a *fixed* coordinate frame, where the white arrows point in the (x, y) directions and the purple one points in the vertical (i.e., angular) direction. In the configuration space, blue helices and yellow straight lines correspond to *vertex-vertex* and *edge-edge* contacts respectively. Two-dimensional contacts, i.e., *vertex-edge* and *edge-vertex*, are plotted in shades of red and green respectively. In order to keep the scenes as clear and understandable as possible, the configuration space, in all but two scenes, is bounded to a gridded bounding box. This bounding box has no mathematical meaning.

3.1 Contents of the Video

Introduction.
In this scene the robot moves in a narrow passage between five obstacles. This scene is very involved and shows all elements of the video which are introduced and discussed in the following sections.

Fundamental Motions.
We first present a workspace *free* from obstacles, and we investigate some basic motions of the robot: *pure translation*, *pure rotation* and a simple combination of the two.

Helices in the Configuration Space.
Next, we consider a particular type of motion, namely rotation of the robot, in a workspace still free from obstacles, such that one of its boundary points is fixed to a point in the plane. In this case, the configuration point traverses a helix which is easy to parameterize. Since the contact surfaces are families of helices, we can explicitly construct the contact surfaces using their parameterization.

Contact Surfaces.
We then introduce an obstacle in the workspace and we study the various possible contact surfaces. In particular, we demonstrate

- a *Vertex-Edge contact* and the corresponding *cylindrical* surface patch;
- an *Edge-Vertex contact* and the corresponding *ruled* surface patch.
- Finally, we visualize all possible contacts of our triangular robot and an obstacle in detail (see Figure 1). We obtain an overview of this beautiful geometric object, and then we traverse a few particular contacts.

Visualizing a Narrow Passage.
In the final chapter of the video we come back to the first scene, namely, a *narrow passage*. Since this is a very complicated scene, we break it up and overlay one obstacle after the other. In this way, the viewer can have a better understanding of how such a passage looks like in the configuration space.

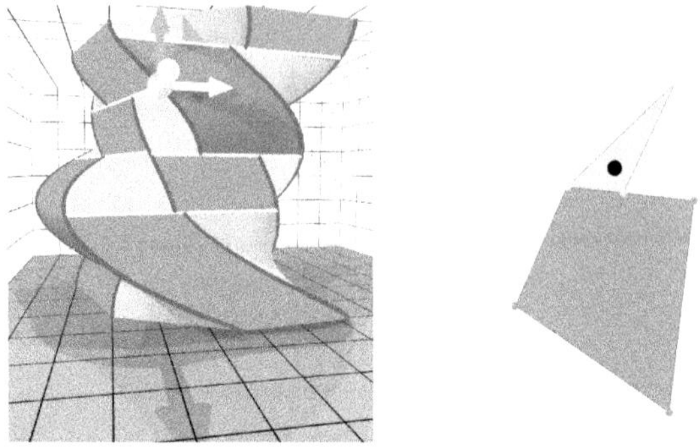

(a) An obstacle represented in the configuration space

(b) The corresponding configuration of the robot and the obstacle

Figure 1: The two components of a frame

4. PRODUCTION DETAILS

This video is the concatenation of a sequence of pairs of frames; each pair consists of an image of the workspace and a corresponding image of the configuration space. In order to describe a continuous motion in the workspace, it is sufficient to provide a continuous path in the configuration space. For each scene in the video, we defined such a path, and extracted, using `Mathematica`, a dense sample of it. For each point of a sample, we generated, using `Mathematica`, one frame of the workspace which included the robot in the corresponding pose and the obstacles (if they are present). In addition, from `Mathematica` we exported the three-dimensional models representing the obstacles into the `POV-Ray` format. Then, each corresponding image of the configuration space was rendered using `POV-Ray`.

Finally, the two images of a pair were joined using `ImageMagick` and the sequence of frames was concatenated into a video using `FFmpeg`.

5. ACKNOWLEDGMENT

This video was made within the project *Computational Geometric Learning*. The project CG Learning acknowledges the financial support of the Future and Emerging Technologies (FET) program within the Seventh Framework Program for Research of the European Commission, under FET-Open grant number 255827.

6. REFERENCES

[1] Howie Choset, Kevin M. Lynch, Seth Hutchinson, George A. Kantor, Wolfram Burgard, Lydia E. Kavraki, and Sebastian Thrun. *Principles of Robot Motion: Theory, Algorithms, and Implementations (Intelligent Robotics and Autonomous Agents series)*. The MIT Press, 5 2005.

[2] Jean-Claude Latombe. *Robot Motion Planning*. Kluwer Academic Publishers, 3rd edition, 1993.

[3] Oren Salzman. Motion planning via manifold samples. Master's thesis, Tel Aviv University, 2011.

[4] Steven Cy Trac. *Robust Explicit Construction Of 3D Configuration Spaces Using Controlled Linear Perturbation*. PhD thesis, University Of Miami, December 2008.

Surface Patches from Unorganized Space Curves

Fatemeh Abbasinejad
University of California-Davis
One Shields Avenue
Davis, CA, USA
fabbasinejad@ucdavis.edu

Pushkar Joshi
Motorola Mobility
807 11th Ave
Sunnyvale, CA, USA
pushkar@motorola.com

Nina Amenta
University of California-Davis
One Shields Avenue
Davis, CA, USA
amenta@cs.ucdavis.edu

ABSTRACT

Recent 3D sketch tools produce networks of three-space curves that suggest the contours of shapes. The shapes may be non-manifold, closed three-dimensional, open two-dimensional, or mixed. Our video demonstrates a system that automatically generates intuitively appealing, piecewise-smooth surface patches from such a curve network, and an intelligent user interface for modifying the automatically chosen surface patches. Both parts of the system use a linear algebra representation of the set of surface patches to track the topology.

Categories and Subject Descriptors

G.2.2 [**Graph Theory**]: Path and circuit problems; I.3.3 [**Picture/Image Generation**]: Line and curve generation; I.3.5 [**Computational Geometry and Object Modeling**]: Curve, surface, solid, and object representations

General Terms

Algorithms, Theory

Keywords

face identification, curve network, minimum cycle basis

1. INTRODUCTION

New intuitive interfaces for 3D curve drawing have made it easier to sketch curve models of 3D shapes, such as the rocket on the left of Figure 1. The outputs of these new design tools are networks of curves in $\mathbb{R}^3$. Such a curve network can be interpreted as a wireframe model of a piecewise-smooth surface, usually in more than one way. By "wireframe model" we mean a representation of a piecewise-smooth, not necessarily manifold surface by its set of patch edges alone. Constructing the 3D surface from the wireframe model is commonly referred to as "surfacing", "lofting", or sometimes "skinning." Surfacing is an important part of the design process; it disambiguates the often ambiguous curve network and produces clearer visualizations of the desired shape (see Figure 1). Of course, producing the surface intended by the user is an ill-posed problem, so the goal is to produce heuristics which give reasonable results.

In this video, we describe a heuristic, based on some simple homology computation, which seems to give good results, automating a significant portion of the surfacing process by selecting cycles from the curve network as surface patch boundaries. Our paper [1] gives a detailed explanation of our work and shows experiments on a set of 58 3D sketches produced by the ILoveSketch system [2].

SCG'12, June 17–20, 2012, Chapel Hill, North Carolina, USA.
ACM 978-1-4503-1299-8/12/06.

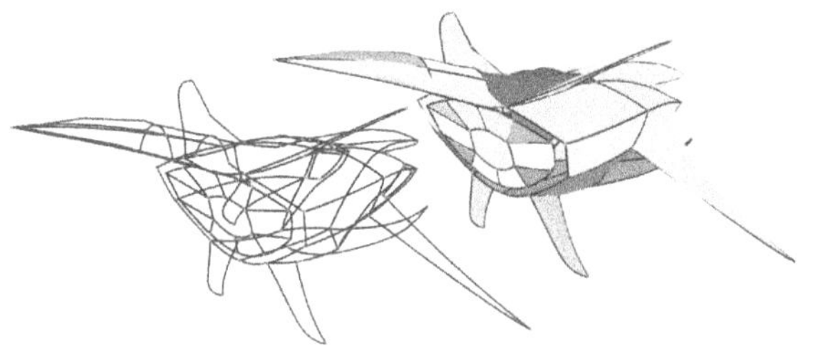

Figure 1: 3D curve input from "ILoveSketch" (left) for which our system automatically finds an acceptable, unambiguous patch representation (right)

System Overview The assumption that the input sketch represents the manifold boundary of a solid object would impose very strong topological requirements on the output, but this assumption cannot be made for 3D sketches; for instance, the rocket on the right in Figure 1 has two-dimensional fins and two connected components, forming a non-manifold surface. We instead begin with the weaker assumption that the surface does not bound *any* solids, and select a maximal set of patches that has that property. There are many such maximal sets, but as we shall see, we can find one that is optimal with respect to any heuristically chosen weighting function on the cycles, using a greedy algorithm.

This automatic phase is followed by a user-interaction phase, where we help the user refine the automatically selected patches. We suggest other low-weighted patches that could replace a deleted patch. We also suggest patches that might close off solids, using a different heuristic that scores not only cycles but also the solids that they create. In the video, we show some demos of the user interaction phase.

Both phases make use of the topological structure of the current set of patches. We demonstrate the system on networks of curves produced by "ILoveSketch", and another system called "JustDrawIt".

2. ALGORITHM

Any set of two-dimensional patches forms what is called a two-dimensional cell complex. If it separates three-space into two or more connected components, we treat all but the exterior as three-dimensional solids, producing a three-

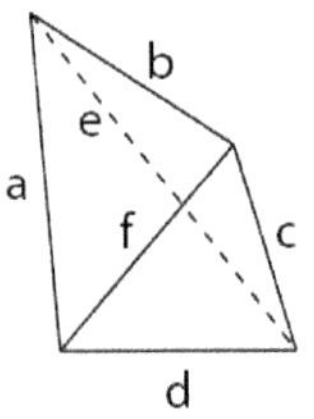

f1 = abf
f2 = aed
f3 = cdf
f4 = bec

	f1	f2	f3	f4
a	1	1	0	0
b	1	0	0	1
c	0	0	1	1
d	0	1	1	0
e	0	1	0	1
f	1	0	1	0

Figure 2: The matrix representing the four faces of a tetrahedron. We can verify that the fourth column of the matrix is the sum of the first three, using arithmetic modulo two. We also observe that geometrically the cycle bounding the fourth face is the sum of the first three, and that it closes off a solid (the interior of the tetrahedron). Any choice of three of these four columns forms a cycle basis (one of many possible cycle bases for this graph). The entire vector space of cycles defined by the wireframe model would consist of all possible sums of these four cycles; for example, the cycle a, b, c, d is the sum of a, b, f and f, c, d.

dimensional cell complex (possibly containing two-dimensional as well as three-dimensional parts). Our approach to selecting an initial set of patches is to construct the largest two-dimensional complex we can before closing off any solid three-dimensional regions. Such a set of patches is called a *cycle basis* for the set of cycles in the input graph (i.e. curve network). It is called a basis because the set of cycles - that is, the set of possible patches - forms a vector space. This vector space representation (described below) is the same as that used in topology when computing homology using the Smith Normal Form and in persistent homology computation.

Computationally, we can find the cycle basis using matrix arithmetic over the integers mod two, also known as the Galois Field of order two, $GF(2)$. The entire vector space could be represented by a single large matrix M with a row for each edge of the input graph, and a column for each possible cycle. There is a 1 in matrix location $m_{i,j}$ if edge i belongs to cycle j, and 0 otherwise. Adding two cycles c_j, c_k corresponds to adding the corresponding columns in $GF(2)$, so that $1 + 1 = 0$; notice that this produces the symmetric difference, which is another cycle c_l (i.e. $c_l = c_j + c_k$). The three cycles c_j, c_k and c_l are dependent, since any one is the sum of the other two. The rank r of M, over $GF(2)$, is the size of the largest *independent* set of columns. As usual, we can select many possible bases for the vector space M, each of which is a set of columns of size r. There is no requirement that our inputs be connected, and many are not. There may be, and often are, multiple curves connecting the same pair of vertices. There is no requirement that the output should be a manifold.

Our objective is to select an optimum cycle basis, where the optimum is defined as the minimum using a heuristic weighting criterion.

To produce the optimal basis, we use the standard *greedy matroid algorithm*, which can be applied to any vector space, with any weighting function on the elements. We write $W(c)$ to denote the weight of cycle c (or just W instead of $W(c)$ when clear). The weight of a cycle basis $\{c_1, \ldots, c_r\}$ is the sum of its cycle weights.

THEOREM 1. *For any weighting function W, the greedy matroid algorithm produces a minimum-weight basis. [4]*

Applied to minimum spanning tree computation, the greedy matroid algorithm is Kruskal's algorithm. Applied to cycle basis computation, the greedy matroid algorithm proceeds as follows. We begin with an empty set of cycles C, and at each step we greedily attempt to add the cycle c of minimum weight $W(c)$. If c is dependent on some subset of the cycles in C, we discard it. If, on the other hand, c is independent, we add it to C and continue.

The current set of cycles C is represented as a set of column vectors as described in the previous section, and to see if a new column vector is independent of the current set, we append it to the current matrix, and compute the rank. If the rank is increased by adding the new column, the new cycle is independent of the previous set of cycles. We use the FFLAS-FFPACK[3] package for linear algebra over finite fields to compute the rank of the $0 - 1$ matrix.

Weighting Function Our choice combines three heuristics into a single lexicographic function W.

Our first heuristic is that we prefer short cycles over long ones. The second heuristic is that non-separating cycles are preferred over separating cycles. A separating cycle is one for which removal of the cycle vertices and all of their adjacent edges (including the cycle edges) increases the number of connected components in the graph.

To combine these two heuristics, we define for each cycle the quantity

$$k = \text{number of edges} + (2 + \epsilon)s$$

where s is zero if the cycle is non-separating and one if it is separating. Cycles with the same value of k are ordered geometrically, by the volumes of their axis-aligned bounding boxes, from smallest to largest. Since the curves are given as ordered sets of sample points, the computation of bounding boxes is trivial. This choice favors nearly flat faces which are well-aligned with the coordinate planes.

Combined with the fact that we consider the cycles in order of edge length, this means that the cycle basis we find optimizes the following lexicographic objective function W: it finds a basis with minimum total summed value of k, and of the many bases with that total sum of k, it finds one that minimizes the sum of the volumes of the bounding boxes.

3. REFERENCES

[1] F. Abbasinejad, P. Joshi, and N. Amenta. Surface patches from unorganized space curves. *Computer Graphics Forum*, 30(5):1379–1387, 2011.

[2] S.-H. Bae, R. Balakrishnan, and K. Singh. ILoveSketch: as-natural-as-possible sketching system for creating 3d curve models. In *Proc. of the 21st annual ACM Symposium on User Interface Software and Technology*, UIST '08, pages 151–160, New York, NY, USA, 2008. ACM.

[3] P. G. Jean-Guillaume Dumas and C. Pernet. Ffpack: Finite field linear algebra package. In *ISSAC : International Symposium on Symbolic and Algebraic Computations*, 2004.

[4] E. Lawler. *Combinatorial Optimization: Networks and Matroids*. Holt, Reinhart and Winston, 1976.

Lang's Universal Molecule Algorithm

John C. Bowers
Department of Computer Science
University of Massachusetts
Amherst, MA 01003, USA.
jbowers@cs.umass.edu

Ileana Streinu
Department of Computer Science
Smith College
Northampton, MA 01063, USA.
istreinu@smith.edu, streinu@cs.umass.edu

ABSTRACT

We present a Java implementation of Lang's Universal Molecule algorithm, alongside with a visualization of its interconnected structures: the input metric tree and compatible convex polygon, whose 2D crease pattern and 3D uniaxial base are computed by the algorithm.

The Java applet, the video, as well as further references and accompanying materials are available on our web site `linkage.cs.umass.edu/origamiLang`. We also include a recent example, found with the help of this implementation, of a Universal Molecule crease pattern which, as a flat-faced origami, is completely rigid; in particular, its corresponding uniaxial base cannot be reached through continuous folding without bending of the paper.

Categories and Subject Descriptors

I.3.5 [**Computational Geometry and Object Modeling**]: [Geometric algorithms, languages, and systems]

General Terms

Theory, Algorithms

Keywords

computational origami, origami design, rigidity, Universal Molecule, TreeMaker

1. INTRODUCTION

In a beautiful paper from SoCG 1996 [4] (see also [5, 3]), Robert Lang introduced TreeMaker, an algorithm for designing origami crease patterns compatible with an input metric tree. His designs are distinguished by the existence of a flat-folded uniaxial base compatible with the input tree. At the core of TreeMaker lies the Universal Molecule algorithm, which is the focus of this video.

In origami design, line segments called *creases* are introduced on a sheet of paper to allow the paper to *fold* along them into a 3D state, called a *base*. The line segments form the *crease pattern*, which subdivides the paper into a set of polygonal *faces*. The TreeMaker algorithm creates creases that guarantee the existence of special 3D folded shapes, called *uniaxial bases*.

SCG'12, June 17–20, 2012, Chapel Hill, North Carolina, USA.
ACM 978-1-4503-1299-8/12/06.

A base in which each face is perpendicular to a common plane is called *projectable*. Lang's bases project to a tree called the *shadow tree*. A *uniaxial base* is a Lang projectable base in which the boundary of the paper folds exactly onto its shadow tree, and the rest of the base lies above the projection plane. Lang's TreeMaker algorithm solves the following version of the origami design problem: given a metric tree and a sheet of paper (typically, a square or rectangle), produce a crease pattern which has a realization as a uniaxial base whose shadow tree is the input tree. In a first step, TreeMaker subdivides the paper into regions, called molecules, which can be folded independently. The Universal Molecule algorithm then fills in the crease pattern of each molecule.

This video provides a visualization of Lang's Universal Molecule algorithm through a Java implementation, and is a companion to our recent papers [1, 2].

2. UNIVERSAL MOLECULE

The input to Lang's Universal Molecule algorithm is a topologically embedded metric tree and a special type of convex polygon associated to the tree, which we call a Lang polygon. The output is a crease pattern for the polygon for which there exists a folded realization as a uniaxial base with shadow tree equal to (a geometric embedding of) the input tree.

A Lang polygon is a special case of a doubling cycle for the input tree. A *doubling cycle* is a polygon created by walking around the topologically embedded input tree. Each arc traversed along the walk corresponds to an edge of the same length in the doubling cycle. The leaf nodes are encountered only once and thus have unique corresponding vertices in the polygon, which we call *corners*. An interior node of degree d is encountered d times and is each time marked on the boundary. We call these vertices *marker* vertices. If a doubling cycle is convex and for each pair of corners the distance between the corners is greater than or equal to the distance in the tree between the corresponding leaf nodes, then the doubling cycle is a *Lang polygon*. The interior angle of any marker vertex is π.

The algorithm works by a *parallel sweep* process in which the edges of the polygon are moved inwards in a parallel fashion at constant speed. The interior sweeping polygon is called a *contour* and the distance the edges have moved at any point in the sweep process is called the *height* of the contour. The edges of the crease pattern are created by tracing the paths of the vertices (both corner and marker) of the polygon.

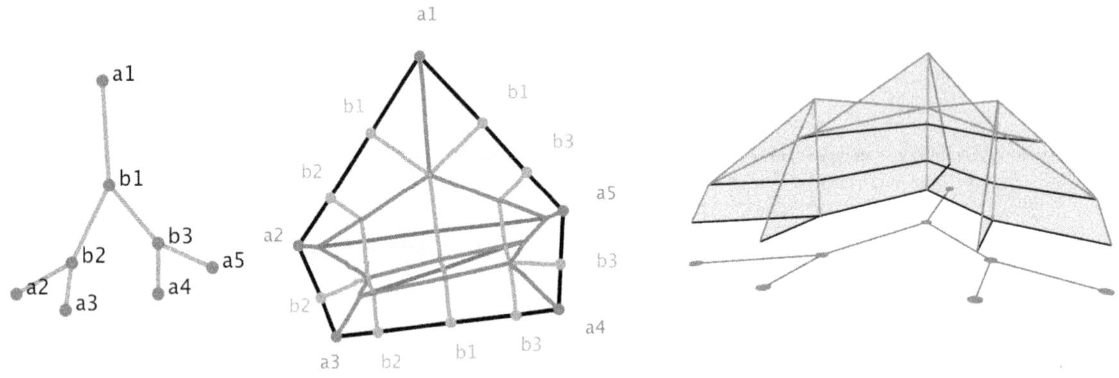

Figure 1: The applet illustrates the intercorrelation of the three structures that underlie the Universal Molecule algorithm.

As the contour shrinks inwards, an edge between two marker vertices maintains its length, while an edge between a marker and a corner shrinks. This corresponds to a shrinking of the tree by moving the leaves inward. At certain discrete heights one of two events occurs. A *contraction event* occurs when an edge between a corner and a marker vertex shrinks to a single point. All leaves incident to the internal node corresponding to the marker vertex shrink at once. This node is then a leaf node of the shrunken tree. A *branching event* occurs when the distance between two corner vertices becomes equal to the corresponding distance between the leaf nodes. If the sweeping process were to proceed past this point, the contour would cease to be a Lang polygon, since the distance between two leaf nodes would become less than the corresponding distance between the leaves. The contour is therefore split by adding an edge between the two corner nodes. Marker nodes are added so that the edge corresponds to the path in the tree between the two leaf nodes. This splits the contour into two contours, both of which are Lang polygons. The sub-tree for each of the contours is created by splitting the tree along the path between the two leaf nodes. The sweeping process then proceeds in parallel in each contour.

In 3D the sweeping contours correspond to a sweeping process of the shrinking tree upwards by the height of the contour. Conceptually, the sweeping process is a continuous process with discrete events. The algorithm, however, works by identifying the height of the next event and by recursing on the contour and sub-tree at that event. The base case is when the next event is a contraction of all edges to the same point.

Rigidity of Universal Molecules. Lang's Universal Molecule algorithm produces crease patterns for which uniaxial realizations exist. A question remains: can the uniaxial state be achieved by continuously folding the creased paper, in such a way that the faces are kept rigid and folds occur only along creases? In the video, we show an example found using our software implementation, which (as proven in [2]) answers this question in the negative. The example gives a totally inflexible crease pattern, which would not fold into its associated uniaxial base without further deformation (bending) of the paper.

3. SOFTWARE AND VIDEO

Our software is a Java Applet implementation in which the user defines the input tree and is then able to modify a doubling cycle for the tree to produce different Lang polygons. A slider allows the user to visualize the sweeping process. A 3D window shows the user a live visualization of the final folded base realizable with the computed crease-pattern.

After a brief overview of TreeMaker, the video discusses the intricacies of Lang's Universal Molecule algorithm, and shows a totally rigid example it may produce. The animations for the Universal Molecule segment were screen-captured from our implementation, with annotations added in post-processing with presentation software. The live demo of our implementation is available at our website `linkage.cs.umass.edu/origamiLang`.

4. ACKNOWLEDGEMENTS

Research funded by DARPA "23 Mathematical Challenges" (under "Algorithmic Origami and Biology") and NSF CCF-1016988 grants of the second author. The first author is supported by an NSF Graduate Fellowship.

5. REFERENCES

[1] J. C. Bowers and I. Streinu. Lang's universal molecule algorithm. Technical report, Dec 2011.

[2] J. C. Bowers and I. Streinu. Rigidity of Lang's origami universal molecules. Technical report, Feb 2012. (Submitted).

[3] R. J. Lang. *Origami design secrets: mathematical methods for an ancient art*, A. K. Peters, 2003.

[4] R. J. Lang. A computational algorithm for origami design. In *Proceedings of the 12th Annual ACM Symposium on Computational Geometry*, pages 98–105, 1996.

[5] R. J. Lang. Treemaker 4.0: A program for origami design, 1998. http://www.langorigami.com.

The Sticky Geometry of the Cosmic Web

Johan Hidding
Kapteyn Institute,
Groningen University
Groningen, the Netherlands
hidding@astro.rug.nl

Rien van de Weygaert
Kapteyn Institute,
Groningen University
Groningen, the Netherlands
weygaert@astro.rug.nl

Gert Vegter
Johan Bernoulli Institute
Groningen University
Groningen, the Netherlands
g.vegter@rug.nl

Bernard J.T. Jones
Kapteyn Institute,
Groningen University
Groningen, the Netherlands
bernard@astrag.demon.co.uk

Monique Teillaud
INRIA
Sophia Antipolis, France
monique.teillaud@inria.fr

ABSTRACT

In this video we highlight the application of Computational Geometry to our understanding of the formation and dynamics of the Cosmic Web. The emergence of this intricate and pervasive weblike structure of the Universe on Megaparsec[1] scales can be approximated by a well-known equation from fluid mechanics, the Burgers' equation. The solution to this equation can be obtained from a geometrical formalism. We have extended and improved this method by invoking weighted Delaunay and Voronoi tessellations. The duality between these tessellations finds a remarkable and profound reflection in the description of physical systems in Eulerian and Lagrangian terms.

The resulting Adhesion formalism provides deep insight into the dynamics and topology of the Cosmic Web. It uncovers a direct connection between the conditions in the very early Universe and the complex spatial patterns that emerged out of these under the influence of gravity.

Categories and Subject Descriptors

J.2 [**Physical Sciences and Engineering**]: Astronomy; F.2.2 [**Analysis of Algorithms and Problem Complexity**]: Nonnumerical Algorithms and Problems—*Geometrical problems and computation*

Keywords

Cosmology, Large Scale Structure of the Universe, Numerical Methods, Voronoi Diagram, Delaunay Triangulation

1. THE COSMIC WEB

The Cosmic Web is the fundamental spatial organization of matter on scales of a few up to a hundred Megaparsec. Galaxies, intergalactic gas and dark matter exist in a wispy weblike arrangement of dense compact clusters, elongated filaments, and sheetlike walls, amidst large near-empty void regions. The filaments are the transport channels along which matter and galaxies flow into massive high-density cluster located at the nodes of the web. The weblike network is shaped by the tidal force field accompanying the inhomogeneous matter distribution.

Structure in the Universe has risen out of tiny primordial (Gaussian) density and velocity perturbations by means of gravitational instability [8, 11]. The large-scale anisotropic force field induces anisotropic gravitational collapse, resulting in the emergence of elongated or flattened matter configurations. According to Zeldovich, the collapse of a primordial cloud (dark) matter passes through successive stages, first assuming a flattened sheetlike configuration as it collapses along its shortest axis. This is followed by a rapid evolution towards an elongated filament as the medium axis collapses and, if collapse continues along the longest axis, may ultimately produce a dense and compact cluster or halo [9]. The hierarchical setting of these processes, occurring simultaneously over a wide range of scales complicates the picture considerably.

2. ZELDOVICH & ADHESION MODELS

The simplest model that describes the emergence of structure and complex patterns in the Universe is the Zeldovich Approximation (ZA) [11, 9]. In essence, it describes a ballistic flow, driven by a constant (gravitational) potential. The resulting Eulerian position $x(t)$ at some cosmic epoch t is specified by the expression,

$$x(t) = q + D(t)\ u_0(q),$$

where q is the initial "Lagrangian" position of a particle, $D(t)$ the time-dependent structure growth factor and $u_0 = -\nabla_q \Phi_0$ its velocity. The nature of this approximation may be appreciated by the corresponding source-free equation of motion,

$$\frac{\partial u}{\partial D} + (u \cdot \nabla_x)u = 0.$$

The use of ZA is ubiquitous in cosmology. One major application is its key role in setting up initial conditions in cosmological N-body simulations. Of importance here is its nonlinear extension in terms of Adhesion Model.

[1]The main measure of length in astronomy is the parsec. Technically a parsec is the distance at which we would see the distance Earth-Sun at an angle of 1 arcsec on the sky. One Megaparsec is 3.26 million light-years; this is approximately the distance between our Milkyway and the nearest other galaxy: Andromeda.

SCG'12, June 17–20, 2012, Chapel Hill, North Carolina, USA.
ACM 978-1-4503-1299-8/12/06.

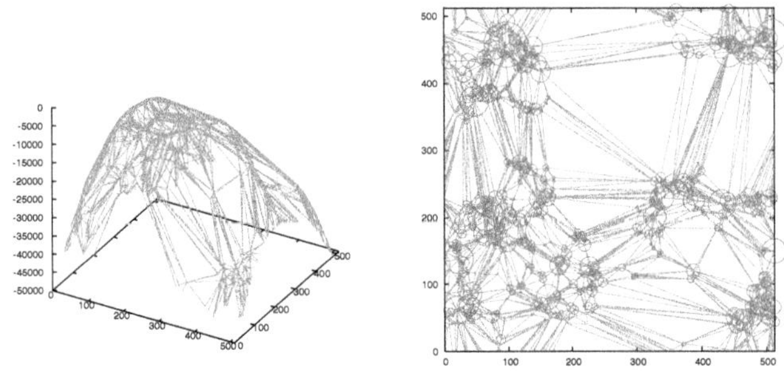

Figure 1: Convex hull of the modified potential and the resulting weighted Delaunay triangulation.

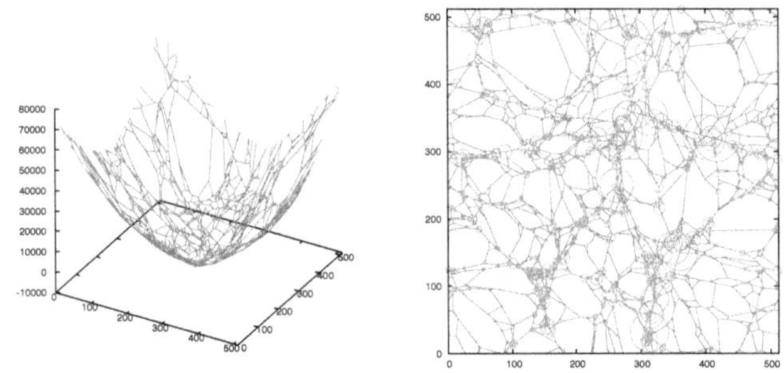

Figure 2: Convex conjugate of the modified potential and the resulting weighted Voronoi diagram.

The ZA breaks down as soon as self gravity of the forming structures becomes important. To 'simulate' the effects of self gravity, Gurbatov et al. [2, 3] included an artificial viscosity. This results in the Burgers' equation,

$$\frac{\partial u}{\partial D} + (u \cdot \nabla_x)u = \nu \nabla_x^2 u,$$

a well known PDE from fluid mechanics. This equation has an exact analytical solution [6],

$$u(x, D) = \frac{\int dq \, \frac{x-q}{D} \, \exp[G/(2\nu)]}{\int dq \exp[G/(2\nu)]},$$

where $G(x, q, D) = \Phi_0(q) - (x - q)^2/(2D)$. In the limit of $\nu \rightarrow 0$, the solution is

$$\Phi(x, D) = \max_q \left[\Phi_0(q) - \frac{(x-q)^2}{2D} \right].$$

This leads to a geometric interpretation of the Adhesion Model. The solution follows from the evaluation of the convex hull of the velocity potential modified by a quadratic term [10]. We found that the solution can also be found by computing the weighted Voronoi diagram [7] of a mesh weighted with the velocity potential. It is of special importance that this computation is done on a periodic space. This is essential to treat boundaries correctly.

3. ADHESION & TESSELLATIONS

The duality between the weighted Voronoi diagram and the weighted Delaunay triangulation has a deep connection with the concepts of Eulerian and Lagrangian coordinates in fluid mechanics. The Voronoi diagram represents volume. A vertex of the Delaunay triangulation represents a mass-element. Its dual Voronoi cell is the volume this mass-element occupies. As time moves on, the particle is expanding in a struggle for dominance over its neighbours, a struggle that is lost once the vertex becomes redundant (in the sense that it no longer contributes to the Delaunay triangulation). Physically the particle is then trapped in a larger structure. In general particles first stream into walls (objects of co-dimension 1), moving down into objects of increasing co-dimension. [4, 5].

4. THIS VIDEO

We use the Computational Geometry Algorithms Library [1] for our key geometric computations: weighted Delaunay triangulations in both 2D and 3D (at this point, the periodic triangulation package of CGAL only offers 3D periodic non-weighted Delaunay triangulations).

This video takes us from the initial conditions, through both the Zeldovich and Adhesion models, to a display of how the topology of the Cosmic Web depends directly on the starting point.

5. REFERENCES

[1] CGAL, *Computational Geometry Algorithms Library* http://www.cgal.org.

[2] S. N. Gurbatov and A. I. Saichev. Probability distribution and spectra of potential hydrodynamic turbulence. *Radiophysics and Quantum Electronics*, 27:303–313, Apr. 1984.

[3] S. N. Gurbatov, A. I. Saichev, and S. F. Shandarin. The large-scale structure of the universe in the frame of the model equation of non-linear diffusion. *MNRAS*, 236:385–402, Jan. 1989.

[4] J. Hidding. Adhesion: a sticky way of understanding large scale structure. Master's thesis, Groningen University, 2010.

[5] J. Hidding, R. van de Weygaert, G. Vegter, and B. Jones. The geometry of the adhesion generated web (to be subm.). *MNRAS*, 2012.

[6] E. Hopf. The partial differential equation $u_t + uu_x = \mu_{xx}$. *Communications on Pure and Applied Mathematics*, 3(3):201–230, 1950.

[7] A. Okabe, B. Boots, and K. Sugihara. *Spatial tessellations: concepts and applications of Voronoi diagrams.* John Wiley & Sons, Inc., New York, NY, USA, 2000.

[8] P. J. E. Peebles. *Large-Scale Structure of the Universe.* Princeton University Press, 1980.

[9] S. F. Shandarin and Y. B. Zeldovich. The large-scale structure of the universe: Turbulence, intermittency, structures in a self-gravitating medium. *Reviews of Modern Physics*, 61:185–220, Apr. 1989.

[10] M. Vergassola, B. Dubrulle, U. Frisch, and A. Noullez. Burgers' equation, Devil's staircases and the mass distribution for large-scale structures. *Astronomy & Astrophysics*, 289:325–356, Sept. 1994.

[11] Y. B. Zeldovich. Gravitational instability: An approximate theory for large density perturbations. *Astronomy & Astrophysics*, 5:84–89, Mar. 1970.

Author Index

www.ingramcontent.com/pod-product-compliance
Ingram Content Group UK Ltd.
Pitfield, Milton Keynes, MK11 3LW, UK
UKHW050952290926
15470UKWH00005B/1777

9 781450 312998